Contemporary Physicists

Gian-Carlo Rota
David Sharp
Editors

Birkhäuser
Boston · Basel · Stuttgart

Jeannette and T.D. Lee with their sons, James and Stephen, in October 1957, at the time of the announcement of the Nobel awards.

T.D. Lee

Selected Papers

Volume 2

Field Theory and Symmetry Principles

G. Feinberg
Editor

Birkhäuser
Boston · Basel · Stuttgart
1986

T.D. Lee
G. Feinberg *(Editor)*
Department of Physics
Columbia University
New York, NY 10027
USA

Library of Congress Cataloging in Publication Data
Lee, T.D., 1926–
T.D. Lee selected papers.
(Contemporary physicists)
Includes bibliographies.
1. Physics. 2. Nuclear physics. I. Feinberg,
Gerald, 1933– . II. Title. III. Series.
QC71.L4325 1986 530 86-6169

CIP-Kurztitelaufnahme der Deutschen Bibliothek
Lee, Tsung D.:
Selected papers / T. D. Lee. G. Feinberg, ed.—
Boston : Basel : Stuttgart : Birkhäuser
(Contemporary physicists)
ISBN 3-7643-3344-8 (Stuttgart . . .);
ISBN 0-8176-3344-8 (Boston . . .)
NE: Lee, Tsung D.: [Sammlung]
Vol. 2. Field theory and symmetry principles. —1986.
ISBN 3-7643-3342-1 (Stuttgart . . .)
ISBN 0-8176-3342-1 (Boston . . .)

Printed in USA.

ISBN 0-8176-3344-8 (complete set)	3-7643-3344-8 (complete set)
ISBN 0-8176-3341-3 (Vol. 1)	3-7643-3341-3 (Vol. 1)
ISBN 0-8176-3342-1 (Vol. 2)	3-7643-3342-1 (Vol. 2)
ISBN 0-8176-3343-X (Vol. 3)	3-7643-3343-X (Vol. 3)

Preface

In these volumes, the most significant of the collected papers of the Chinese–American theoretical physicist Tsung-Dao Lee are printed. A complete list of his published papers, in order of publication, appears in the Bibliography of T.D. Lee. The papers have been arranged into ten categories, in most cases according to the subject matter. At the beginning of each of the first eight categories of papers, there is a commentary on the content and significance of all of the papers in the category. The two short final categories do not have any commentaries.

The editor would like to thank Dr. Richard Friedberg for his assistance in the early stages of the editorial work on this project, as well as for writing commentaries on the papers of Categories III and IV. I would also like to thank Dr. Norman Christ for writing the commentary on the papers of Category VII. The assistance of Irene Tramm was invaluable in many aspects of preparing this collection, including locating copies of Lee's papers.

GERALD FEINBERG

List of Categories of T.D. Lee's Papers

Volume 1

I. Weak Interactions
II. Early Papers on Astrophysics and Hydrodynamics
III. Statistical Mechanics
IV. Polarons and Solitons

Volume 2

V. Quantum Field Theory
VI. Symmetry Principles

Volume 3

VII. Discrete Physics
VIII. Strong Interaction Models
IX. Historical Papers
X. Gravity (Continuum Theory)

Contents (Volume 2)*

Introduction (by G. Feinberg) ... xi
Bibliography of T.D. Lee ... xv

V. Quantum Field Theory

Commentary ... 1
[12] Intermediate Coupling Method for Meson-Nucleon Scattering (with R. Christian) ... 3
[13] Some Special Examples in Renormalizable Field Theory ... 11
[65] Spin ½ Wave Equation in De-Sitter Space (with F. Gursey) ... 17
[69] Degenerate Systems and Mass Singularities (with M. Nauenberg) ... 25
[100] Neutral Vector Mesons and the Hadronic Electromagnetic Current (with N.M. Kroll and B. Zumino) ... 39
[101] Algebra of Fields (with S. Weinberg and B. Zumino) ... 63
[103] Field-current Identities and the Algebra of Fields (with B. Zumino) ... 67
[117] Negative Metric and the Unitarity of the S-Matrix (with G.C. Wick) ... 83
[118] Unitarity in the Nθθ Sector of a Soluble Model with Indefinite Metric (with G.C. Wick) ... 119
[121] Finite Theory of Quantum Electrodynamics (with G.C. Wick) ... 129
[129] Questions of Lorentz Invariance in Field Theories with Indefinite Metric (with G.C. Wick) ... 145
[150] Vacuum Stability and Vacuum Excitation in a Spin-0 Field Theory (with G.C. Wick) ... 147
[173] Feynman Rules of QCD inside a Hadron ... 174
[174] Operator Ordering and Feynman Rules in Gauge Theories (with N.H. Christ) ... 193
[175] Is the Physical Vacuum a Medium? ... 213

VI. Symmetry Principles

Commentary ... 227
[14] Conservation of Heavy Particles and Generalized Gauge Transformations (with C.N. Yang) ... 230

*The numbers in brackets preceding the titles of the papers are from the Bibliography of T.D. Lee (pages xv–xxiv in this volume).

[16] Speculations on Heavy Mesons (with J. Orear) 231
[20] Charge Conjugation, a New Quantum Number G, and Selection Rules Concerning a Nucleon-antinucleon System (with C.N. Yang) 233
[21] Question of Parity Conservation in Weak Interactions (with C.N. Yang) 239
[21a] Errata to "Question of Parity Conservation in Weak Interactions (with C.N. Yang) 244
[24] Parity Nonconservation and a Two-component Theory of the Neutrino (with C.N. Yang) 245
[25] Remarks on Possible Noninvariance under Time Reversal and Charge Conjugation (with R. Oehme and C.N. Yang) 251
[27] Introductory Talk at the 1957 Rochester Conference, Session on Weak Interactions 257
[28] Possible Detection of Parity Nonconservation in Hyperon Decay (with J. Steinberger, G. Feinberg, P.K. Kabir and C.N. Yang) 269
[32] Elementary Particles and Weak Interactions (with C.N. Yang) 271
[33] Theoretical Implications of Parity Violations in Beta Interactions (see Category I, Volume 1) ....page viii
[35] Possible Determination of the Spin of Λ^0 from Its Large Decay Angular Asymmetry (with C.N. Yang) (see Category I, Volume 1)
[38] Weak Interactions and Nonconservation of Parity (see Category I, Volume 1)
[41] Decay Modes of a $(\theta + \overline{\theta})$ System (with M. Goldhaber and C.N. Yang) 337
[53] Elementary Particles (see Category I, Volume 1)
[54] Some Considerations on Global Symmetry (with C.N. Yang) 341
[73] CP Invariance and the 2π Decay Mode of the K_2^0 (with J. Bernstein and N. Cabibbo) 350
[75] Symmetry Properties of Leptons in the Zero-mass Limit 353
[77] Hypercharge Conservation, *CP* Invariance and the Possible Existence of a Zero-mass Zero-spin Field 365
[78] Analysis of *CP* Noninvariant Interactions and the K_1^0, K_2^0 System (with L. Wolfenstein) 373
[80] Possible *C, T* Noninvariance in the Electromagnetic Interaction (with J. Bernstein and G. Feinberg) 381
[85] An Elementary Discussion of Possible Noninvariance under *T, CP* and *CPT* in Hyperon Decays 391
[89] Weak Interactions and Questions of C, P, T Noninvariances 403
[91] Space Inversion, Time Reversal and Particle-antiparticle Conjugation 437
[93] Space Inversion, Time Reversal and Other Discrete Symmetries in Local Field Theories (with G.C. Wick) 445
[97] Other Aspects of Weak Interactions: High Energy Neutrino Physics and Questions of C, P, T Noninvariance 465
[110] *C, P, T* Symmetries 472
[114] Weak Interactions and CPT Asymmetries 487
[122] Symmetry Principles in Physics 502

CONTENTS (VOLUME 2)

[137] *CP* Nonconservation and Inequalities between $\mu^+\mu^-$ and 2γ Decay Rates of K_S^0 and K_L^0 (with N.H. Christ) 511
[146] A Theory of Spontaneous *T* Violation 516
[147] Spontaneous T Violation 531
[149] *CP* Nonconservation and Spontaneous Symmetry Breaking 556

Permissions 591

Introduction

Tsung-Dao Lee was born in Shanghai, China, on November 25, 1926. After early education in China he came to the United States in 1946, through the help of Ta-You Wu, to do graduate study in physics at the University of Chicago. In that period, the presence of Enrico Fermi at Chicago had attracted many talented physics students, including J. Anderson, O. Chamberlain, G. Chew, R. Garwin, M. Goldberger, J. Orear, M. Rosenbluth, J. Steinberger R. Sternheimer, S. Treiman, L. Wolfenstein and C.N. Yang. Lee became a doctoral student of Fermi, and while a student already published the important paper, with Rosenbluth and Yang ([1]), on what came to be known as the universal Fermi interaction and the intermediate boson. Fermi directed Lee's research towards topics in astrophysics, which eventually resulted in his thesis, on white dwarf stars ([2]), which demonstrated conclusively that these stars contain no hydrogen and must therefore be the end points of stellar evolution.

Following the completion of his thesis, Lee spent eight months working at Chicago's Yerkes Observatory, with S. Chandrasekhar, on problems of turbulence ([3] and [5]). After a brief stay at the University of California at Berkeley, he went to the Institute for Advanced Study, at Princeton, where he spent the next two years. At the Institute, Lee, in collaboration with Yang, wrote two important papers on statistical mechanics ([7] and [8]). The first of these provides a rigorous definition of the thermodynamic limit for various phases, and shows that the thermodynamic functions can differ from phase to phase. In the second, the "unit circle theorem" on the roots of the grand partition function was proven.

In 1953, Lee went to Columbia University as an Assistant Professor in Physics. Except for a three-year period in the early 1960s, he has remained on the Columbia faculty, becoming Professor of Physics in 1956, and University Professor in 1983. Lee's first work at Columbia was on the renormalizable field theory model known as the Lee model ([13]). His attention soon turned to particle physics, and the developing puzzle of *K* meson decays. Lee and various collaborators suggested several models to account for the phenomena, but the observations eventually impelled him, with Yang, to make a careful study of the evidence about various symmetries in particle reactions. The result, in [21], was a precise criterion for what experiments need to be done to test the discrete symmetries *C, P* and *T*. The point was made that existing experiments on weak interactions did not test them, and experiments to do so, on beta decay and other weak interactions, were proposed. The paper was soon followed by another paper ([24]), proposing a specific model of two-component neutrinos, which made definite predictions for the amount of parity non-conservation expected. At the same time, Lee and collaborators also examined *T* and *CP* violations, with particular reference to the neutral kaon system [25]. The experiments were soon carried out by Wu and collaborators, by Lederman, Garwin and collaborators, and by Telegdi and collaborators, with the result that parity and charge conjugation invariance were not satisfied by weak interactions. This confirmation of their suggestion earned Lee and Yang the 1957 Nobel Prize for Physics (see [38] for the text of Lee's Nobel Prize Lecture).

In the following year Lee, with several collaborators, extended the new ideas about weak interactions to several other processes ([25], [27], [36] and [41]). The cumulation of this work on symmetry principles was the basis of the dominant theme in particle physics through the 1960s.

In the period 1957–1960 Lee, Yang and others studied the statistical mechanics of a Bose system of hard spheres ([26], [42] and [43]), finding a way of eliminating divergences in the theory by selective summation of series. They also, in this period, produced a general formalism for studying many-body problems in statistical mechanics ([45]–[49]).

Beginning in 1960, it became clear that experiments could be done with high-energy neutrinos. Lee, with several collaborators, provided much of the theoretical impetus for these early experiments ([50], [56], [58] and [60]). These experiments have proven to be the major source of information about weak interactions through the present.

The successful description of low-energy weak interactions led many physicists to revive the idea suggested in [1] that these interactions result from the exchange of intermediate vector bosons with universal coupling. Beginning in 1960, Lee wrote a number of papers on the properties and implications of these then hypothetical particles ([58], [61], [62], [68], and [115]). Of these, paper [61] was particularly important because it indicated the need for careful consideration of the gauge invariance questions involved in the interaction of spin-one intermediate vector bosons.

In 1964, Lee, with M. Nauenberg, analyzed the divergences connected with particles of zero rest mass, and described a general method for dealing with these divergences, which plays an important role in contemporary work in QCD, with its massless, self-interacting gluons.

The discovery in 1964 of *CP* non-invariance focussed the attention of many theorists on this problem. Lee, with several collaborators, suggested a number of models for *CP* violation ([77] and [80]), which later proved not to be in agreement with experiment. Some years later, in the context of spontaneously broken symmetry, he suggested another model ([146], [147] and [149]) which remains one of the viable explanations for *CP* violation.

In the period 1969–1971, Lee, together with G.C. Wick, proposed a method of eliminating ultraviolet divergences in quantum field theory, through the introduction of an indefinite metric in Hilbert space ([117], [118], [121] and [129]). They concluded that this type of theory is not in disagreement with any existing experiments.

In 1974, Lee and Wick began a study of the hypothesis that in some circumstances, such as inside a heavy nucleus, the broken symmetry that sometimes occurs through the vacuum expectation value of a scalar field could be restored ([150], [152], [155], [156] and [170]). This would show up as a change in the mass of nucleons in the heavy nucleus, with a resultant increase in nuclear density. This suggestion may eventually be testable through the collisions of beams of relativistic heavy ions.

In the late 1970s, theoretical physicists became interested in solutions to classical field equations (such as magnetic monopoles) that are stabilized by boundary conditions. Lee and collaborators showed that in several field theory models, stable configurations could be produced by other mechanisms than boundary conditions ([150], [159], [162], [163] and

[165]). These ideas were applied to models of hadrons in [167], with R. Friedberg, where it is shown that it can reproduce the results of the "bag" model of hadrons.

In 1982, Lee's interest turned to the lattice approach to field theory. In three papers with N.H. Christ and R. Friedberg ([179]–[181]), the concept of a random lattice was introduced, and it was shown how field theory calculations could be performed on such a lattice. Applications of these ideas to actual problems in QCD are presently under investigation.

The discrete space–time approach used in lattice theories led Lee to make the suggestion not previously considered by other workers with lattices, that space and time were really discrete. The general consequences of this idea for classical and quantum mechanics were explored in [182] and [183]. The possible application of discrete space time to general relativity is under current investigation by Lee and collaborators.

Bibliography of T.D. Lee

Note: **Boldface** numbers in parentheses at the end of each entry denote the volume that the entry appears in.

1949

[1] Interaction of Mesons with Nucleons and Light Particles (with M. Rosenbluth and C.N. Yang). *Phys. Rev.* **75,** 905. **(I)**

1950

[2] Hydrogen Content and Engery-Productive Mechanism of White Dwarfs. *Ap. J.* **111,** 625. **(I)**
[3] Note on the Coefficient of Eddy Viscosity in Isotropic Turbulence. *Phys. Rev.* **77,** 842. **(I)**
*[4] On the Proton–Proton Reaction in White Dwarf Stars. *Ap. J.* **112,** 561.

1951

[5] Difference Between Turbulence in a Two-Dimensional Fluid and in a Three-Dimensional Fluid. *J. Ap. Phys.* **22,** 524. **(I)**

1952

*[6] On Some Statistical Properties of Hydrodynamical and Magneto-Hydrodynamical Fields. *Quart. Ap. Math.* **10,** 69.
[7] Statistical Theory of Equations of State and Phase Transitions. I. Theory of Condensation (with C.N. Yang). *Phys. Rev.* **87,** 404. **(I)**
[8] Statistical Theory of Equations of State and Phase Transitions. II. Lattice Gas and Ising Model (with C.N. Yang). *Phys. Rev.* **87,** 410. **(I)**
[9] Motion of Slow Electrons in Polar Crystals (with David Pines). *Phys. Rev.* **88,** 960. **(I)**

1953

[10] The Motion of Slow Electrons in Polar Crystals (with F. Low and D. Pines). *Phys. Rev.* **90,** 297. **(I)**
[11] Interaction of a Nonrelativistic Particle and a Scalar Field with Applications to Slow Electrons in Polar Crystal (with D. Pines). *Phys. Rev.* **92,** 883. **(I)**

1954

[12] Intermediate Coupling Theory of Meson-Nucleon Scattering (with R. Christian). *Phys. Rev.* **94,** 1760. **(II)**
[13] Some Special Examples in Renormalizable Field Theory. *Phys. Rev.* **95,** 1329. **(II)**

*Does not appear in these volumes.

1955

[14] Conservation of Heavy Particles and Generalized Gauge Transformations (with C.N. Yang). *Phys. Rev.* **98,** 1501. **(II)**

*[15] Absorption Experiments Involving Heavy Mesons. *Phys. Rev.* **99,** 337.

[16] Speculations on Heavy Mesons (with J. Orear). *Phys. Rev.* **100,** 932. **(II)**

*[17] Scattering of Pions by Nucleons in Intermediate Coupling (with M. Friedman and R. Christian). *Phys. Rev.* **100,** 1494.

1956

*[18] Model for Multiple Meson Production (with E.M. Henley). *Phys. Rev.* **101,** 1536.

*[19] Mass Degeneracy of the Heavy Mesons (with C.N. Yang). *Phys. Rev.* **102,** 290.

[20] Charge Conjugation, a New Quantum Number G, and Selection Rules Concerning a Nucleon-Antinucleon System (with C.N. Yang). *Nuovo Cimento* **3,** 749. **(II)**

[21] Question of Parity Conservation in Weak Interactions (with C.N. Yang). *Phys. Rev.* **104,** 254. **(II)**

[21a] Errata to "Question of Parity Conservation in Weak Interactions" (with C.N. Yang). *Phys. Rev.* **106,** 1371 (1957). **(II)**

*[22] Possible Interference Phenomena Between Parity Doublets (with C.N. Yang). *Phys. Rev.* **104,** 822.

1957

[23] Many-Body Problem in Quantum Mechanics and Quantum Statistical Mechanics (with C.N. Yang). *Phys. Rev.* **105,** 1119. **(I)**

[24] Parity Nonconservation and a Two-Component Theory of the Neutrino (with C.N. Yang). *Phys. Rev.* **105,** 1671. **(II)**

[25] Remarks on Possible Noninvariance Under Time Reversal and Charge Conjugation (with R. Oehme and C.N. Yang). *Phys. Rev.* **106,** 340. **(II)**

[26] Eigenvalues and Eigenfunctions of a Bose System of Hard Spheres and Its Low-Temperature Properties (with K. Huang and C.N. Yang). *Phys. Rev.* **106,** 1135. **(I)**

[27] Introductory Talk at the 1957 Rochester Conference, Session on Weak Interactions. Interscience Publishers, Section VII, pp. 1–12. **(II)**

[28] Possible Detection of Parity Nonconservation in Hyperon Decay (with J. Steinberger, G. Feinberg, P.K. Kabir and C.N. Yang). *Phys. Rev.* **106,** 1367. **(II)**

[29] Capture of μ^- Mesons by Protons (with K. Huang and C.N. Yang). *Phys. Rev.* **108,** 1340. **(I)**

[30] Possible Nonlocal Effects in μ Decay (with C.N. Yang). *Phys. Rev.* **108,** 1611. **(I)**

[31] General Partial Wave Analysis of the Decay of a Hyperon of Spin ½ (with C.N. Yang). *Phys. Rev.* **108,** 1645. **(I)**

[32] Elementary Particles and Weak Interactions (with C.N. Yang). BNL 443 (T-91). **(II)**

[33] Theoretical Implications of Parity Violations in Beta Interactions. *Proc. of Rehovoth Conference on Nuclear Structure,* North-Holland Publishing Co. (1958), p. 336. **(I, II)**

*[34] Theoretical Implications of Parity Violations in Beta Interactions. *Proc. of Padua–Venice Conference on Mesons and Newly Discovered Particles,* September.

1958

[35] Possible Determination of the Spin of Λ^0 from Its Large Decay Angular Asymmetry (with C.N. Yang). *Phys. Rev.* **109,** 1755. **(I, II)**

[36] Effect of the Hyperfine Splitting of a μ-Mesic Atom on Its Lifetime (with J. Bernstein, H. Primakoff and C.N. Yang). *Phys. Rev.* **111,** 313. **(I)**

*[37] Weak Interactions and Nonconservation of Parity (Nobel lecture). *Science* **127,** 569.

[38] Weak Interactions and Nonconservation of Parity, Le Prix Nobel Lecture, December 11, 1957. Stockholm. **(I, II)**

*[39] The Low Temperature Properties of a Bose System of Hard Spheres (with C.N. Yang). *Proc. of Midwest Conference on Theoretical Physics, St. Louis, March 14–15,* p. 149.

*[40] A Review of Weak Interactions. *Proc. of Midwest Conference on Theoretical Physics, St. Louis, March 14–15,* p. 167.

[41] Decay Modes of a $(\theta + \bar{\theta})$ System (with M. Goldhaber and C.N. Yang). *Phys. Rev.* **112,** 1796. **(II)**

[42] Low Temperature Behavior of a Dilute Bose System of Hard Spheres. I. Equilibrium Properties (with C.N. Yang). *Phys. Rev.* **112,** 1419. **(I)**

1959

[43] Low Temperature Behavior of a Dilute Bose System of Hard Spheres. II. Non-Equilibrium Properties (with C.N. Yang). *Phys. Rev.* **113,** 1406. **(I)**

[44] Possible Determination of the Helicity of Elementary Excitations in Liquid He II (with F. Mohling). *Phys. Rev. Letters* **2,** 284. **(I)**

[45] Many-Body Problem in Quantum Statistical Mechanics. I. General Formulation (with C.N. Yang). *Phys. Rev.* **113,** 1165. **(I)**

[46] Many-Body Problem in Quantum Statistical Mechanics. II. Virial Expansion for Hard-Sphere Gas (with C.N. Yang). *Phys. Rev.* **116,** 25. **(I)**

1960

[47] Many-Body Problem in Quantum Statistical Mechanics. III. Zero Temperature Limit for Dilute Hard Spheres (with C.N. Yang). *Phys. Rev.* **117,** 12. **(I)**

[48] Many-Body Problem in Quantum Statistical Mechanics. IV. Formulation in Terms of Average Occupation Number in Momentum Space (with C.N. Yang). *Phys. Rev.* **117,** 22. **(I)**

[49] Many-Body Problem in Quantum Statistical Mechanics. V. Degenerate Phase in Bose–Einstein Condensation (with C.N. Yang). *Phys. Rev.* **117,** 897. **(I)**

[50] Theoretical Discussions on Possible High Energy Neutrino Experiments (with C.N. Yang). *Phys. Rev. Letters* **4,** 307. **(I)**

[51] Implications of the Intermediate Boson Basis of the Weak Interactions: Existence of a Quartet of Intermediate Bosons and Their Dual Isotopic Spin Transformation Properties (with C.N. Yang). *Phys. Rev.* **119,** 1410. **(I)**

*[52] Intermediate Boson Hypothesis of Weak Interactions. 1960 Rochester Conference on High Energy Physics, p. 567.

[53] Elementary Particles. *Physics Today* **13,** 30. **(I, II)**

1961

[54] Some Considerations on Global Symmetry (with C.N. Yang). *Phys. Rev.* **122,** 1954. **(II)**

*[55] Some Theoretical Considerations on the Desirability of a 300 to 1000 Bev Proton Accelerator (with R. Serber, G.C. Wick and C.N. Yang). BNL 772 (T-290), p. 15.

[56] Production Cross Section of Intermediate Bosons by Neutrinos in the Coulomb Field of Protons and Iron (with P. Markstein and C.N. Yang). *Phys. Rev. Letters* **7,** 429. **(I)**

*[57] Theoretical Considerations on *W*-Production by "High"-Energy Neutrinos. CERN Conference on Theoretical Aspects of Very High Energy Phenomena, Session on Weak Interactions, June 1961, p. 44.

[58] High-Energy Neutrino Experiments. CERN 61-31: Particle Physics—CERN Seminars 1961, p. 65. **(I)**

*[59] Lepton Physics. Aix-en-Provence International Conference on Elementary Particles, September 1961, Vol. II, p. 15.

1962

[60] High Energy Neutrino Reactions Without Production of Intermediate Bosons (with C.N. Yang). *Phys. Rev.* **126,** 2239. **(I)**

[61] A Theory of Charged Vector Mesons Interacting with Electromagnetic Field (with C.N. Yang). *Phys. Rev.* **128,** 885. **(I)**

[62] Application of ξ-Limiting Process to Intermediate Bosons. *Phys. Rev.* **128,** 899. **(I)**

[62a] Errata to "Application of ξ-Limiting Processes to Intermediate Bosons." *Phys. Rev.* **173,** 1806 (1968). **(I)**

*[63] Speculations Concerning the $\Delta S = -\Delta Q$ Reactions and the Intermediate Boson Structures in Weak Interactions. *Phys. Rev. Letters* **9,** 319.

1963

*[64] Quantum Mechanical Many-Body Problem and the Low Temperature Properties of a Bose System of Hard Spheres (with K. Huang and C.N. Yang). *The Many-Body Problem,* Interscience Publishers, p. 165.

[65] Spin ½ Wave Equation in De-Sitter Space (with F. Gürsey). *Proc. of National Academy of Sciences* **49,** 179. **(II).**

*[66] Remarks Concerning the Recent High-Energy Neutrino Experiment (with G. Danby, J.M. Gaillard, K. Goulianos, L. Lederman, N. Mistry, M. Schwartz and J. Steinberger). *Phys. Rev. Letters* **10,** 260.

[67] Intensity of Upward Muon Flux Due to Cosmic-Ray Neutrinos Produced in the Atmosphere (with H. Robinson, M. Schwartz and R. Cool). *Phys. Rev.* **132,** 1297. **(I)**

[68] Electromagnetic Form Factor of the Neutrinos (with J. Bernstein). *Phys. Rev. Letters* **11,** 512. **(I)**

1964

[69] Degenerate Systems and Mass Singularities (with M. Nauenberg). *Phys. Rev.* **133,** B1549. **(II)**

[70] A Possible Method of Determining the Moment of Charge of the ν_e (with A. Sirlin). *Reviews of Modern Physics* **36,** 666. **(I)**

[71] Radiative Corrections to Electromagnetic Moments and Leptonic Decays of the Intermediate Boson *W. Phys. Rev. Letters* **12,** 569. **(I)**

[72] Implications of Approximate SU_3 Symmetry and Mass Formulas for the Mesons (with F. Gürsey and M. Nauenberg). *Phys. Rev.* **135,** B467. **(III)**

[73] CP Invariance and the 2π Decay Mode of the K_2^0 (with J. Bernstein and N. Cabibbo). *Physics Letters* **12,** 146. **(II)**

1965

*[74] Remarks on Triplet Models in SU_3 Symmetry. *Nuovo Cimento* **35,** 933.

[75] Symmetry Properties of Leptons in the Zero-Mass Limit. *Nuovo Cimento* **35,** 945. **(II)**

*[76] Comments on Elementary Particle Physics. *Nature of Matter, Purposes of High Energy Physics,* edited by L.C.L. Yuan, Brookhaven National Laboratory #888 (T-360), p. 50.

[77] Hypercharge Conservation, CP Invariance and the Possible Existence of a Zero-Mass Zero-Spin Field. *Phys. Rev.* **137,** B1621. **(II)**

[78] An Analysis of CP Noninvariant Interactions and the K_1^0, K_2^0 System (with L. Wolfenstein). *Phys. Rev.* **138,** B1490. **(II)**

*[79] Possible *C* Noninvariant Effects in the 3π Decay Modes of η^0 and ω^0. *Phys. Rev.* **139,** B1415.

[80] Possible *C, T* Noninvariance in the Electromagnetic Interaction (with J. Bernstein and G. Feinberg). *Phys. Rev.* **139,** B1650. **(II)**

*[81] Minimal Electromagnetic Interaction and *C, T* Noninvariance. *Phys. Rev.* **140,** B967.

*[82] Classification of All C Noninvariant Electromagnetic Interactions and the Possible Existence of a Charged, But $C = 1$, Particle. *Phys. Rev.* **140,** B959.

*[83] Neutrino Physics. *Informal Conference on Experimental Neutrino Physics,* CERN 1965, edited by C. Franzinetti, CERN #65-32, p. 275.

[84] Weak Interactions (with C.S. Wu). *Ann. Rev. Nuclear Science* **15,** 381. **(I)**

[84a] Corrections to "Weak Interactions" (with C.S. Wu). *Ann. Rev. Nuclear Science* **17,** 513 (1967). **(I)**

1966

[85] An Elementary Discussion of Possible Noninvariance Under *T, CP* and *CPT* in Hyperon Decays. *Preludes in Theoretical Physics* (in Honor of V.F. Weisskopf), edited by De-Shalit, Feshbach and Van Hove, North-Holland Publishing Co., p. 5. **(II)**

*[86] Theoretical Survey of High-Energy Neutrino Interactions. *Weak Interactions and High Energy Neutrino Physics* (Rendiconti della Scuola Internazionale di Fisica "E. Fermi"-XXXII Corso), Academic Press, p. 142.

*[87] Mass Formulae and SU_3 Symmetry. *Ibid,* p. 311.

*[88] Possible C Noninvariance in the Electromagnetic Interaction. *Particle Symmetries* (Vol. II of 1965 Brandeis University Summer Institute in Theoretical Physics), edited by M. Chretien and S. Deser, Gordon and Breach, p. 453.

[89] Weak Interactions and Questions of C, P, T Noninvariances. *Proc. Oxford International Conference on Elementary Particles, 19/25 September 1965,* Rutherford High Energy Laboratory, p. 225. **(II)**

*[90] Theoretical Remarks on CP Noninvariance. *Proc. of International Conference on Weak Interactions, October 25–27, 1965,* Argonne National Laboratory #7130, p. 431.

[91] Space Inversion, Time Reversal and Particle-Antiparticle Conjugation (based on a lecture delivered at the Fourth Annual Science Conference of the Belfer Graduate School of Science, Yeshiva University), *Physics Today* **19,** 23. **(II)**

*[92] Possible Tests of C_{st} and T_{st} Invariances in $l^{\pm} + N \to l^{\pm} + \Gamma$ and $A \to B + e^{+} + e^{-}$ (with N.H. Christ). *Phys. Rev.* **143,** 1310.

[93] Space Inversion, Time Reversal and Other Discrete Symmetries in Local Field Theories (with G.C. Wick). *Phys. Rev.* **148,** 1385. **(II)**

*[94] Reciprocity Relations in Photo-Pion Reactions (with N.H. Christ). *Phys. Rev.* **148,** 1520.

[95] Weak Interactions: Chapter 8—Decays of Charged K Mesons, and Chapter 9—Decays of Neutral K Mesons (with C.S. Wu). *Ann. Rev. Nuclear Science* **16,** 471. **(I)**

*[95a] Corrections to "Weak Interactions" (Chapters 8 and 9) (with C.S. Wu). *Ann. Rev. Nuclear Science* **17,** 513 (1967). **(I)**

1967

*[96] Space Inversion, Time Reversal and Charge Conjugation. *Cargese Lectures, 1966,* Gordon and Breach, p. 55.

[97] Other Aspects of Weak Interactions: High Energy Neutrino Physics and Questions of C, P, T Noninvariance. *Proc. of XIIIth International Conference on High Energy Physics, Berkeley, 1966,* Univ. California Press, p. 75. **(II)**

*[98] Time Reversal Symmetry. *Plenarvorträge,* Teil II (Physikertagung 1966 München), B.G. Teubner, p. 243.

*[99] Hadronic Electromagnetic Current and Neutral Vector Mesons. *Proc. of Fifth Eastern Theoretical Physics Conference, Brown University, 1966,* W.A. Benjamin, p. 101.

[100] Neutral Vector Mesons and the Hadronic Electromagnetic Current (with N.M. Kroll and Bruno Zumino). *Phys. Rev.* **157,** 1376. **(II)**

[101] Algebra of Fields (with S. Weinberg and Bruno Zumino). *Phys. Rev. Letters* **18,** 1029. **(II)**

*[102] High Energy Weak Interactions. *Proc. of Argonne High Energy Physics Meeting.*
[103] Field-Current Identities and the Algebra of Fields (with Bruno Zumino). *Phys. Rev.* **163,** 1667. **(II)**
*[104] CP Noninvariance, Electromagnetic Interaction and Field-Current Identities. *Proc. of International Symposium on Electron and Photon Interactions at High Energies, SLAC, 1967,* p. 389.

1968

*[105] Field Current Identities and Intermediate Bosons. *Phys. Rev.* **168,** 1714.
*[106] Hadronic Contributions to the Photon Propagator (with Bruno Zumino). *Old and New Problems in Elementary Particles* (a volume dedicated to G. Bernardini), edited by G. Puppi, Academic Press, p. 167.
*[107] Algebra of the Observed Current Operators. *Proc. of Second Hawaii Topical Conference in Particle Physics, Honolulu, 1967,* edited by S. Pakvasa and S.F. Tuan, Univ. Hawaii Press, p. 13.
*[108] CPT Symmetries. *Ibid.,* p. 67.
*[109] Algebra of the Observed Current Operators. *Fundamental Particle Physics* (1967 Tokyo Summer Lectures in Theoretical Physics), edited by G. Takeda and Y. Hara, W.A. Benjamin, Inc., p. 74.
[110] CPT Symmetries. *Ibid.,* p. 102. **(II)**
*[111] Field Current Identities. The Fermi Lectures, a series given at the Scuola Normale Superiore, Pisa, January 1968.
*[112] CPT Symmetries. *Ibid.*
*[113] Intermediate Bosons and the Electromagnetic Field. *Phys. Rev.* **171,** 1731.
[114] Weak Interactions and CPT Asymmetries. *Contemporary Physics,* Volume II, Trieste Symposium, 1968, International Atomic Energy Agency, p. 283. **(II)**

1969

[115] Analysis of Divergences in a Neutral Spin 1 Meson Theory with Parity Non-Conserving Interactions. *Nuovo Cimento* **59A,** 579. **(I)**
*[116] A Possible Way to Remove Divergence Difficulties in Physics. *Topical Conference on Weak Interactions, CERN, January 1969,* edited by J.S. Bell, CERN 69-7, p. 427.
[117] Negative Metric and the Unitarity of the S-Matrix (with G.C. Wick). *Nuclear Physics* **B9,** 209. **(II)**
[118] Unitarity in the Nθθ Sector of a Soluble Model with Indefinite Metric (with G.C. Wick). *Nuclear Physics* **B10,** 1. **(II)**
*[119] Current-Field Identities. *Theory and Phenomenology in Particle Physics,* Part A (1968 International School of Physics "Ettore Majorana"), edited by A. Zichichi, Academic Press, p. 262.

1970

*[120] A Relativistic Complex Pole Model with Indefinite Metric. *Quanta* (Essays in Theoretical Physics dedicated to Gregor Wentzel), Univ. Chicago Press, p. 260.
[121] Finite Theory of Quantum Electrodynamics (with G.C. Wick). *Phys. Rev.* **D 2,** 1033. **(II)**
[122] Symmetry Principles in Physics. *Evolution of Particle Physics* (a volume dedicated to Edoardo Amaldi), edited by M. Conversi, Academic Press, p. 176. **(II)**
*[123] On the Algebra of Fields. *Methods and Problems of Theoretical Physics* (in honour of R.E. Peierls), edited by J.E. Bowcock, North-Holland Publishing Co., p. 415; also "Concluding Lecture," p. 421.

[124] Remarks on the $|\Delta \vec{I}| = ½$ Rule in Nonleptonic Weak Decays. Brookhaven National Laboratory Report #50260 (T-590), printed by the National Technical Information Service, U.S. Department of Commerce. **(I)**

*[125] Symmetry Principles in Physics. Brookhaven National Laboratory Report #50261 (T-591), printed by the National Technical Information Service, U.S. Department of Commerce.

*[126] Possible Existence of Spin 0 $W^{\pm}$ and Some of Its Experimental Consequences. *Phys. Rev. Letters* **25,** 1144.

*[127] Recent Progress in Constructing Finite, Relativistic and Unitary Field Theories. *Comments on Nuclear and Particle Physics,* **IV** (6), 189.

1971

*[128] Scaling Properties in *e-p* Scattering. Lecture given at the National Accelerator Laboratory, January, Columbia University Report #NYO-1932(2)-189.

[129] Questions of Lorentz Invariance in Field Theories with Indefinite Metric (with G.C. Wick). *Phys. Rev.* **D3,** 1046. **(II)**

*[130] Speculative Determination of the Intermediate-Boson Mass. *Phys. Rev. Letters* **26,** 801.

*[131] History of Weak Interactions. Talk given at Columbia University, March, Report #NYO-1932(2)-185.

*[132] Scaling Properties in Inelastic Electron Scattering with a Fixed Final Multiplicity. *Annals of Physics* **66.**

*[133] A Finite Theory of Quantum Electrodynamics. Lectures given at the "Ettore Majorana" International School of Physics in Erice, Sicily during July 1970, in *Elementary Processes at High Energy*, edited by A. Zichichi, Academic Press.

*[134] Remarks on the $|\Delta \vec{I}| = ½$ Rule in Nonleptonic Weak Decays. *Ibid.*

*[135] Symmetry Principles in Physics. *Ibid.*

[136] History of Weak Interactions. *Ibid.* **(III)**

[137] *CP* Nonconservation and Inequalities between $\mu^+\mu^-$ and 2γ Decay Rates of K_S^0 and K_L^0 (with N.H. Christ). *Phys. Rev.* **D4,** 209. **(II)**

[138] High Energy Electromagnetic and Weak Interaction Processes, Brookhaven National Laboratory Report CRISP 71-57, Isabelle Project, August. **(I)**

1972

*[139] Scaling Properties in Inelastic Electron Scattering with a Fixed Final Multiplicity, in *Nuclear, Particle and Many-Body Physics,* Volume II, Academic Press, p. 857.

*[140] Some New Applications of Indefinite Metric to Quantum Electrodynamics and Weak Interactions, in *Elementary Particles* (Proc. of Amsterdam International Conference), edited by A.G. Tenner and M. Veltman, North-Holland Publishing Co., p. 399.

*[141] High Energy Physics and What Lies Beyond, a talk before the Joint Committee on Atomic Energy, Washington, D.C., March 1.

*[142] A New High-Energy Scale? *Physics Today* **25,** 23.

[143] Scaling Properties and the Bound-State Nature of the Physical Nucleon (with S.D. Drell). *Phys. Rev.* **D5,** 1738. **(III)**

*[143a] Scaling Properties in Weak and Electromagnetic Processes, in *Neutrino '72* (Proc. of Conference held at Balatonfüred, Hungary, 11–17 June), edited by A. Frenkel and G. Marx, OMKDK Technoinform, Budapest, Volume II, p. 1.

[144] A Simple Bound-State Model of the Physical Nucleon. *Phys. Rev.* **D6,** 1110. **(III)**

[145] Multiplicity Distribution in the Multiperipheral Model and a One-Dimensional Gas. *Phys. Rev.* **D6,** 3617. **(III)**

1973

[146] A Theory of Spontaneous *T* Violation. *Phys. Rev.* **D8,** 1226. **(II)**
[147] Spontaneous *T* Violation, a talk presented at the Topical Seminar on Weak Interactions, Trieste, Italy, June 29, 1973. **(II)**
*[148] Scaling Properties and the Bound-State Model of Physical Baryons, CERN Report 73-15.

1974

[149] *CP* Nonconservation and Spontaneous Symmetry Breaking. *Physics Reports* **9C,** No. 2. **(II)**
[150] Vacuum Stability and Vacuum Excitation in a Spin 0 Field Theory (with G.C. Wick). *Phys. Rev.* **D9,** 2291. **(II)**
*[151] Abnormal Nuclear States, *Proc. of XVII International Conference on High Energy Physics,* London, Science Research Council, III, 81.

1975

[152] Abnormal Nuclear States and Vacuum Excitations, an invited talk given at the Annual Bevatron Users' Meeting, Lawrence Berkeley Laboratory, 1974; *Revs. Mod. Phys.* **47,** 267. **(III)**
*[153] Abnormal Nuclear States and Vacuum Excitations, *Lepton and Hadron Structure,* ed. A. Zichichi, Academic Press, pp. 697–715.
*[154] Spontaneous *CP* Nonconservation, *Lepton and Hadron Structure,* ed. A. Zichichi, Academic Press, pp. 721–731.
[155] A Possible New Form of Matter at High Density, *Report of the Workshop on BeV/Nucleon Collisions of Heavy Ions—How and Why,* Bear Mountain, 1974 (BNL No. 50445), 1. **(III)**
[156] Interactions of a Dense Fermion Medium with a Scalar Meson Field (with M. Margulies). *Phys. Rev.* **D11,** 1591. **(III)**
[156a] Errata to "Interactions of a Dense Fermion Medium with a Scalar Meson Field" (with M. Margulies). *Phys. Rev.* **D12,** 4008 (1976). **(III)**
[157] Quantum Expansion of Soliton Solutions (with N.H. Christ). *Phys. Rev.* **D12,** 1606. **(I)**

1976

*[158] A Possible New Form of Matter at High Density, *Physics and Our World: Symposium in Honor of V.F. Weisskopf,* M.I.T. 1974, ed. K. Huang (AIP Conference Proc. No. 28), pp. 65–81.
[159] A Class of Scalar-Field Soliton Solutions in Three Space Dimensions (with R. Friedberg and A. Sirlin). *Phys. Rev.* **D13,** 2739. **(I)**
*[160] Unconfined, Heavy Fermion Model of Psions (with G. Feinberg). *Phys. Rev.* **D13,** 3071.
*[161] Examples of Four-Dimensional Soliton Solutions and Abnormal Nuclear States. *Phys. Reports* **23C,** 254.
[162] Gauge-Field Nontopological Solitons in Three Space Dimensions (I) (with R. Friedberg and A. Sirlin). *Nucl. Phys.* **B115,** 1. **(I)**
[163] Gauge-Field Nontopological Solitons in Three Space Dimensions (II) (with R. Friedberg and A. Sirlin). *Nucl. Phys.* **B115,** 32. **(I)**
[164] Nontopological Solitons, in *Proceedings of the Symposium on Frontier Problems in High Energy Physics* (in honour of G. Bernardini), edited by L. Foa and L.A. Radicati, Scuola Normale Superiore, Pisa, p. 47. **(I)**

1977

[165] Fermion-Field Nontopological Solitons I (with R. Friedberg). *Phys. Rev.* **D15,** 1694. **(I)**

*[166] Solitons and Hadrons, in *Problèmes Théorique Liés Aux Nouvelles Particules,* Centre National de la Recherche Scientifique, Montpellier, p. 413.

[167] Fermion-Field Nontopological Solitons II. Models for Hadrons (with R. Friedberg). *Phys. Rev.* **D16,** 1096. **(I)**

1978

[168] Nontopological Solitons and Applications to Hadrons. *Comments on Nuclear and Particle Physics* **7,** 165. **(I)**

*[169] QCD and the Soliton Model of Hadrons (with R. Friedberg). *Phys. Rev.* **D18,** 2623.

1979

[170] Abnormal Nuclear States (with G.C. Wick). *Mesons in Nuclei,* Vol. III, edited by M. Rho and D. Wilkinson, North-Holland Publishing Co., p. 869. **(III)**

*[171] Recent Results on Nontopological Soliton Solutions, *Science of Matter* (Festschrift in honor of T.Y. Wu), ed. S. Fujita, Gordon and Breach, p. 89.

*[172] Nontopological Solitons and Applications to Hadrons (Chalmers Symposium on Solitons, Göteborg), edited by H. Wilhelmsson, *Physica Scripta* **20,** 440.

[173] Feynman Rules of QCD inside a Hadron. *Phys. Rev.* **D19,** 1802. **(II)**

1980

[174] Operator Ordering and Feynman Rules in Gauge Theories (with N.H. Christ). *Phys. Rev.* **D22,** 939. **(II)**

[175] Is the Physical Vacuum a Medium?, in *A Festschrift for Maurice Goldhaber,* edited by G. Feinberg, A.W. Sunyar and J. Weneser, *Transactions of the NY Academy of Sciences,* Series II, **40,** 111. **(II)**

[176] QCD and the Bag Model of Hadrons, *Proceedings of the 1980 Guangzhou Conference on Theoretical Particle Physics,* edited by Ning Hu and Hung-yuan Tzu, Science Press and Van Nostrand Reinhold Co., p. 32. **(III)**

1981

*[177] Is the Vacuum a Physical Medium?, in *Statistical Mechanics of Quarks and Hadrons,* edited by H. Satz, North-Holland Publishing Co., 3.

*[178] *Particle Physics and Introduction to Field Theory,* Harwood Academic Publishers GmbH.

1982

[179] Random Lattice Field Theory: General Formulation (with N.H. Christ and R. Friedberg). *Nucl. Phys.* **B202,** 89. **(III)**

[180] Gauge Theory on a Random Lattice (with N.H. Christ and R. Friedberg). *Nucl. Phys.* **B210** [FS6], 310. **(III)**

[181] Weights of Links and Plaquettes in a Random Lattice (with N.H. Christ and R. Friedberg). *Nucl. Phys.* **B210** [FS6], 337. **(III)**

1983

[182] Can Time Be a Discrete Dynamical Variable? *Phys. Letters* **122B,** 217. **(III)**

[183] Discrete Quantum Mechanics (with R. Friedberg). *Nucl. Phys.* **B225** [FS9], 1. **(III)**

*[184] Time As a Dynamical Variable, in *Shelter Island II,* edited by R. Jackiw, N.N. Khuri, S. Weinberg and E. Witten, The MIT Press (1985).

[185] Discrete Mechanics, to be published in the *Proceedings of the International School of Subnuclear Physics,* Erice, Plenum Press. **(III)**

[186] Broken Parity. **(III)**

1984

[187] Relativistic Heavy Ion Collisions and Future Physics, in *Symmetries in Particle Physics,* Plenum Press, 93.

[188] Derivation of Regge's Action from Einstein's Theory of General Relativity (with R. Friedberg). *Nucl. Phys.* **B242,** 145. **(III)**

[189] Lattice Gravity Near the Continuum Limit (with G. Feinberg, R. Friedberg and H.C. Ren). *Nucl. Phys.* **B245,** 343. **(III)**

*[190] High Energy Theory, in *Pief-Fest,* edited by P. McDonough, Stanford Linear Accelerator Center, 1985, p. 664.

1985

*[191] Difference Equations As the Basis of Fundamental Physical Theories, to be published in the *Proceedings of the Symposium "Old and New Problems in Fundamental Physics."*

[192] Are Black Holes Black Bodies? *Nucl. Phys.* **B264,** (1986), 437. **(III)**

*[193] A Theorem on Black Hole Radiation, to be published in the *Proceedings of Meson 50.*

*[194] Difference Equations As the Basis of Fundamental Physical Theories, to be published in the *Proceedings of the APS Division of Particles and Fields,* Eugene, Oregon meeting.

[195] Physics in Terms of Difference Equations, in *The Lesson of Quantum Theory*, edited by J. de Boer, E. Dal, and O. Ulfbeck, North-Holland Physics Publishing (1986), p. 181.

V

Quantum Field Theory

Commentary

Paper [13] introduced the "Lee model," a soluble model in quantum field theory, involving the interaction of a neutral spin zero field with two states of a fermion. It is shown that renormalization can be carried out exactly in this model, using precise definitions of the renormalization constants, and that this exact renormalization procedure, when expanded in powers of a coupling constant, agrees with the usual perturbative renormalization scheme. This paper was very influential in subsequent work on field theory and renormalization.

Paper [65] discussed the spin ½ equation in DeSitter space and corrected a misconception implied in an earlier paper by Dirac.

Paper [69] describes a general procedure for eliminating mass singularities from observable quantities. It shows that by carrying out ensemble averages over degenerate sets of initial and final states, all such mass singularities are removed. Examples are given for the elimination of infrared divergences in QED. This method has later been applied to QCD, where the mass singularities are more severe, due to the gluon self-interactions and the zero quark masses.

Papers [100], [101], and [103] discuss various aspects of the hypothesis that the various isospin terms in the hadronic electromagnetic current are proportional to vector meson fields with the same isospin. In paper [100], it is shown that this hypothesis gives a gauge invariant explanation of the vector dominance model that had been previously used to explain some aspects of hadron electromagnetic properties. In paper [101], the same assumption is used to determine some current commutators that otherwise contain ambiguous terms, and these commutators are used to derive sum rules. In paper [103], it is shown that the hypothesis of field-current proportionality can only be maintained if the strong interactions of the vector meson fields have the form of a non-Abelian gauge theory.

Paper [117] discusses quantum field theory with an indefinite metric in Hilbert space. Such theories had first been considered by Dirac many years earlier. It is shown that if all real eigenvalues correspond to states of positive length, then the scattering matrix in the theory is unitary. The method is applied to the Lee model of paper [13], and it is shown that there are small, but probably unobservable, violations of causality due to the indefinite metric. The discussion of the unitarity of the S-matrix of the Lee model with indefinite metric is continued in paper [118]. In paper [121], this method is applied to the construction of a completely finite version of QED, in which the indefinite metric allows the propagators of the particles to fall off more rapidly than k^{-2} in momentum space. The paper discusses experimental implications of this model, and analyzes violations of causality in this theory. This discussion is continued in paper [129].

Paper [150] forms a bridge between the work on solitons described in Category VII, and work on properties of the vacuum. This work is continued in paper [173], where the SU(3) gauge theory known as QCD is analyzed under the assumption that the vacuum possesses a color dielectric constant that approaches zero in the limit of infinite quantization volume. This assumption may be regarded as a phenomenological consequence of the complex vacuum structure that QCD implies. It is shown that under this assumption, the Feynman rules for QCD are altered in a definite way, and that confinement of states with non-zero color emerges in the infinite volume limit. Paper [175] summarizes, in the context that the vacuum may be a physical medium, the ideas on the structure of the vacuum introduced in papers [150] and [173]. It is suggested that we may be able to alter the local properties of the vacuum, and so generate new physical phenomena. Finally, paper [176] discusses some of the properties of hadrons in the context of the ideas introduced in paper [173].

In quantizing non-Abelian gauge theories, the gauge invariance properties must be treated very carefully. In paper [174], it is shown that by paying careful attention to the question of operator ordering, Feynman rules can be derived for gauge theories in arbitrary covariant or non-covariant gauges. It is pointed out that for non-Abelian gauge theories, in non-covariant gauges such as the Coulomb gauge, additional terms arise in the Hamiltonian, corresponding to non-linear interactions of the gauge fields. These additional terms contribute to the matrix elements in higher orders of the perturbation expansion. The method used in the paper, a characteristic one in T.D. Lee's work on field theory, is to first examine an analogous situation in classical mechanics or ordinary quantum mechanics, find how to deal with it, and then generalize the solution to quantum field theory.

Reprinted from THE PHYSICAL REVIEW, Vol. 94, No. 6, 1760–1767, June 15, 1954
Printed in U. S. A.

Intermediate Coupling Method for Meson-Nucleon Scattering*

T. D. LEE, *Columbia University, New York, New York,*

AND

R. CHRISTIAN, *Los Alamos Scientific Laboratory, Los Alamos, New Mexico*

(Received March 1, 1954)

It is shown that for the meson-nucleon scattering processes an intermediate-coupling method can be applied which joins smoothly the results obtained from the weak- and strong-coupling limits. The method is illustrated by a detailed study of the charged scalar meson field with fixed nucleon.

I. INTRODUCTION

An intermediate-coupling approximation that is applicable to the bound states of a nucleon-meson system, i.e., a "physical" nucleon and its possible isobars, has been given by Tomonaga.[1] This method is limited to a nonrelativistic treatment of the nucleon but does afford some insight into the problem when the coupling constant is not sufficiently small or large that either the usual weak- or strong-coupling treatment is trustworthy. Furthermore, in the limit of very small or very large values of the coupling constant the result of the intermediate-coupling treatment agrees with that of the correct weak- and strong-coupling calculation.

This intermediate-coupling method is a simple Hartree-Fock approximation in which the total number of mesons is not limited, but all mesons of the same type are assumed to be only in a finite number of orbital states. These orbital state functions, together with the probability amplitude for finding any number of virtual mesons in the nucleon state, are determined by minimizing the total energy. Calculations so far have been limited to determining the self-energies of the stable state of a nucleon with charged scalar mesons neglecting recoil[1] and obtaining the interaction between a nucleon with neutral scalar mesons including recoil.[2]

The purpose of the present note is to extend Tomonaga's method to the scattering of a meson by a nucleon.[3] The charged scalar meson field with fixed nucleon is chosen to illustrate the essential features. The extension of this method to the pseudoscalar meson with pseudovector coupling will be discussed in a subsequent paper.

For the scattering system of a free meson plus a "physical" nucleon we first assume that we have a wave function in the Fock space representing the "physical" nucleon. This wave function is multiplied by a scattering function of the meson which represents an incident and scattered wave. This product function is then symmetrized with respect to all the mesons, and the best functional form of the scattering function is determined by the usual variational technique. As such, the approximation allows for a distortion of the meson wave but does not allow for a "polarization" of the "physical" nucleon.

In the weak-coupling limit the results for the cross section agree with those of the perturbation method. In the strong-coupling limit the results differ from the rigorous strong-coupling results by numerical factors close to unity (depending slightly upon the source size and incident meson energy).

For intermediate values of the coupling constant the procedure is to calculate (numerically) the Tomonaga wave function for the bound states from which one determines the scattering function and consequently the cross section by solving the variational problem. As the coupling constant varies from small to large values it is found that resonance features may occur. It is interesting to note that in the scattering calculation the effect of self-energy and renormalization of coupling constant appear in a very natural way so that they can be identified and eliminated in an unambiguous manner.

II. BOUND STATE

We consider a charged scalar meson field with the Hamiltonian

$$H=\int \omega_k(\alpha_k^\dagger\alpha_k+\beta_k^\dagger\beta_k)d^3k - g\int u(k)[\tau_+(\alpha_k+\beta_k^\dagger)+\tau_-(\alpha_k^\dagger+\beta_k)]d^3k, \quad (1)$$

where $\alpha_k^\dagger$, α_k, and $\beta_k^\dagger$, β_k are the creation and annihilation operators of the positive and negative mesons of momentum k, and

$$u(k)=(4\pi^2\omega_k)^{-\frac{1}{2}}\int U(r)e^{ik\cdot r}d^3r, \quad (2)$$

where $U(r)$ is the normalized source function. ω_k is the total energy of the meson, $\omega_k=(k^2+\mu^2)^{\frac{1}{2}}$.

In order to introduce our notation and have at hand some results already found by Tomonaga, we shall summarize his method for the bound state.

* Publication assisted by the Ernest Kempton Adams Fund.

[1] S. Tomonaga, Prog. Theoret. Phys. **2**, 6 (1947). F. Harlow and B. Jacobson, Phys. Rev. **93**, 333 (1954).

[2] T. D. Lee and D. Pines, Phys. Rev. **92**, 883 (1953).

[3] During the completion of this paper, a similar extension of the intermediate-coupling method to include scattering problem has been worked out by S. Tomonaga, Abstract Book, International Conference of Theoretical Physics, Kyoto and Tokyo, 1953 (unpublished).

TABLE I. Various quantities for the ground state of the nucleon, as functions of g.

g^2	E_0/μ	$\mathfrak{N}_+^0$	$\mathfrak{N}_-^0$	$\mathfrak{M}_+^0$	$\mathfrak{M}_-^0$	λ_+^0/μ	λ_-^0/μ
$\ll 1$	$-g^2\int \frac{u^2}{\omega}d^3k$	$g^2\int \frac{u^2}{\omega^2}d^3k$	$O(g^4)$	$g\left(\int \frac{u^2}{\omega^2}d^3k\right)^{\frac{1}{2}}$	$O(g^3)$	0	$\sim M/\mu$
2	-1.75	0.385	0.05	0.465	0.12	0.2	4
6	-5.45	0.77	0.33	0.555	0.32	0.2	1.8
$\gg 1$	$-\frac{g^2}{2}\int \frac{u^2}{\omega}d^3k$[a]	$\frac{g^2}{4}\int \frac{u^2}{\omega^2}d^3k$	$\frac{g^2}{4}\int \frac{u^2}{\omega^2}d^3k$	$\frac{g}{4}\left(\int \frac{u^2}{\omega^2}d^3k\right)^{\frac{1}{2}}$	$\frac{g}{4}\left(\int \frac{u^2}{\omega^2}d^3k\right)^{\frac{1}{2}}$	$\frac{1.22}{g^2}$[b]	$\frac{10.8}{g^2}$[b]

[a] See Eq. (3.38), reference 1, for the complete expression.
[b] See Eq. (3.37), reference 1, for the complete expression.

The values of the various integrals occurring in the table are

$$\int \frac{u^2(k)}{\omega_k}d^3k = 1.34; \quad \int \frac{u^2(k)}{\omega_k^2}d^3k = 0.410; \quad \int \frac{u^2(k)}{\omega_k^3}d^3k = 0.167.$$

A brief derivation of the form $\mathfrak{N}_+^0$ and $\mathfrak{M}_+^0$ is given in Appendix I.

We denote the probability amplitude for finding n positive and m negative mesons with wave numbers $k_1^+\cdots k_n^+$; $k_1^-\cdots k_m^-$ in the ground state of a single nucleon by

$$\psi_{n,m}^0 \equiv \langle k_1^+, \cdots k_n^+; k_1^-, \cdots k_m^- | \Psi^0 \rangle. \tag{3}$$

For definiteness we suppose Ψ^0 to be a proton state (the superscript 0 indicates a ground state); thus, the difference $n-m$ can only be 0 or 1, depending upon the charge of the "bare" nucleon.

Following Tomonaga we apply the Ritz variational principle,

$$\delta\langle \Psi^0 | H - E_0 | \Psi^0 \rangle = 0,$$

with trial functions of the form

$$\psi_{n,n}^0 = C_{n,n}^0 \prod_{i=1}^{n} f_+^0(k_i^+) \prod_{j=1}^{n} f_-^0(k_j^-),$$
$$\psi_{n,n-1}^0 = C_{n,n-1}^0 \prod_{i=1}^{n} f_+^0(k_i^+) \prod_{j=1}^{n-1} f_-^0(k_j^-), \tag{4}$$

where $f_+^0(k)$ and $f_-^0(k)$ are normalized but otherwise arbitrary functions to be determined, together with the numerical constants $C_{n,m}^0$, by the application of the Ritz principle. It is easy to carry out the variation with respect to $f_\pm^0$ and show that the best forms of f_+^0 and f_-^0 are

$$f_\pm^0(k) = N_\pm^0 u(k)/(\omega_k + \lambda_\pm^0), \tag{5}$$

where λ_+^0 and λ_-^0 are numerical constants depending on the coupling constant and $N_\pm^0$ are normalization constants. The result of carrying out the variation with respect to the constants $C_{n,m}^0$, λ_+^0, and λ_-^0 leads to a set of difference equations[4] in $C_{n,m}^0$, which can be evaluated analytically in the weak- and strong-coupling limits and give *identical* results for the binding energy with the rigorous treatments in these two limits.[5] For the intermediate value of coupling constant these equation s can only be solved numerically.

As shall be shown later, the determination of the quantities,

$$\mathfrak{N}_+^0 \equiv \sum_{n=1}^{\infty} \sum_{m=n-1}^{n} |C_{n,m}^0|^2 n,$$
$$\mathfrak{N}_-^0 \equiv \sum_{n=1}^{\infty} \sum_{m=n-1}^{n} |C_{n,m}^0|^2 m,$$
$$\mathfrak{M}_+^0 \equiv \sum_{n=0}^{\infty} C_{n+1,n}^0 C_{n,n}^0 (n+1)^{\frac{1}{2}},$$
$$\mathfrak{M}_-^0 \equiv \sum_{n=0}^{\infty} C_{n+1,n+1}^0 C_{n+1,n}^0 (n+1)^{\frac{1}{2}}, \tag{6}$$

and $\lambda_\pm^0$ will be necessary for the evaluation of the meson-nucleon scattering cross section. With these definitions the normalization $N_\pm^0$, may be written as

$$N_\pm^0 = g\mathfrak{M}_\pm^0/\mathfrak{N}_\pm^0. \tag{7}$$

The analytical forms of these quantities in the strong and weak-coupling limits, together with some of their numerical values for the intermediate region, are listed in Table I and shown graphically in Figs. 1

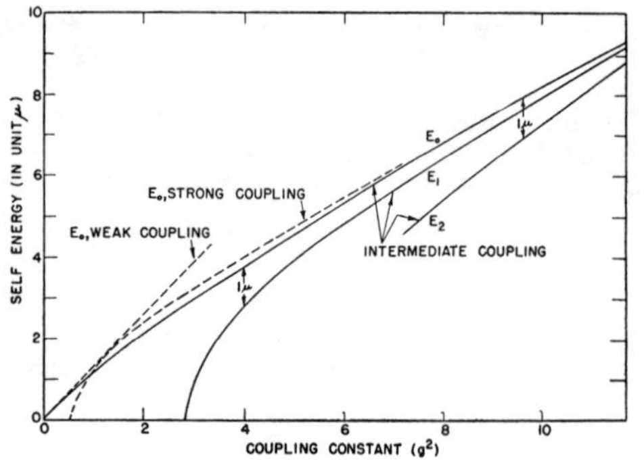

FIG. 1. The self-energy as calculated with the intermediate coupling approximation for the lowest state, E_0; 1st excited state with $m=\pm\frac{3}{2}$, E_1; and 2nd excited state with $m=\pm 5/2$, E_2.

[4] See reference 1, Eq. (3.15).

[5] The weak-coupling result is well known. The strong-coupling result is discussed by: G. Wentzel, Helv. Phys. Acta **13**, 269 (1940); **14**, 633 (1941); R. Serber and S. Dancoff, Phys. Rev. **63**, 143 (1943); S. Tomonaga, Prog. Theoret. Phys. **1**, 109 (1946).

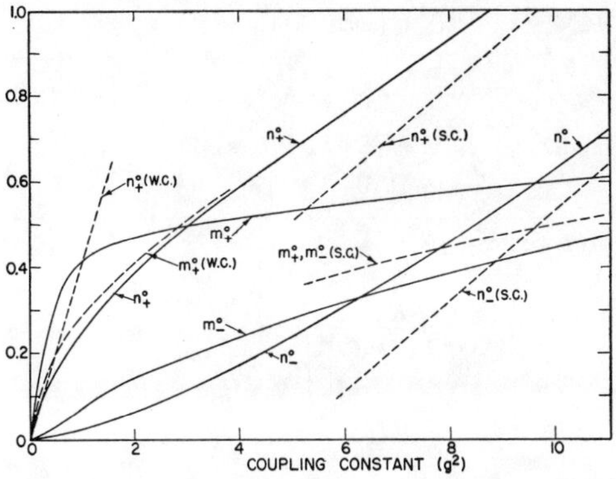

FIG. 2. Various quantities arising in meson-nucleon scattering as determined by the intermediate coupling approximation (solid lines).

and 2. For the numerical calculations the source function $U(r)=[M^2\exp(-M/r)]/4\pi r$, was used where $\mu/M=1/7$.

III. ELASTIC SCATTERING

To illustrate the extension of Tomonaga's method to meson-nucleon scattering, we shall consider first only elastic scattering and restrict ourselves to energies less than 2μ and to values of the coupling constant g such that no *stable* isobars exist. The method used can be easily extended to inelastic scattering and will be discussed in the next section.

The functional form of the state $\mathbf{\Psi}$, representing the scattering process $\pi^++p\rightarrow\pi^++p$, is assumed to be[6]:

$$\psi_{n+1,m}\equiv\langle k_1^+,\cdots k_n^+,k_{n+1}^+;k_1^-,\cdots k_m^-|\mathbf{\Psi}\rangle$$
$$=(n+1)^{-\frac{1}{2}}S_+[\psi_{n,m}{}^0\chi(k_{n+1}{}^+)],\tag{8}$$

where S_+ is the symmetrization operator with respect to positive meson, given by

$$S_+[\psi_{n,m}{}^0\chi(k_{n+1}{}^+)]$$
$$=\psi_{n,m}{}^0(k_1^+,\cdots k_n^+;k_1^-,\cdots k_m^-)\chi(k_{n+1}{}^+)$$
$$+\psi_{n,m}{}^0(k_1^+,\cdots k_{n-1}^+,k_{n+1}^+;k_1^-,\cdots k_m^-)\chi(k_n^+)$$
$$+\cdots$$
$$+\psi_{n,m}{}^0(k_2^+,\cdots k_{n+1}^+;k_1^-,\cdots k_m^-)\chi(k_1^+).$$

$\psi_{n,m}{}^0$ is the probability amplitude for the ground state of the nucleon given by (3), and $\chi(k)$ represents the wave function of the free meson. To determine the functional form of $\chi(k)$, we use the variational principle for scattering, which may be *formally* written[7] as

$$J\equiv\langle\mathbf{\Psi}|H-E_0-\omega_0|\mathbf{\Psi}\rangle,$$
$$\delta J/\delta\chi=0,\tag{9}$$

where ω_0 is the energy of the incident meson and E_0 is the self-energy of the proton.

For computational purposes it is convenient to write Eq. (8) in the equivalent form,

$$\mathbf{\Psi}=\int d^3k\chi(k)\alpha_k{}^\dagger\Psi^0.\tag{10}$$

J may then be easily calculated by commuting $\alpha_k{}^\dagger$ in Eq. (10) with the Hamiltonian, H, and noting that Ψ^0 is an eigenfunction of H with eigenvalues E_0. Thus

$$2J=\left\langle\Psi^0\left|\int d^3k'\chi^*(k')\alpha_{k'}\int d^3k(\omega_k-\omega_0)\chi(k)\alpha_k{}^\dagger\right|\Psi^0\right\rangle$$
$$-g\left\langle\Psi^0\left|\int d^3k'\chi^*(k')\alpha_{k'}\int d^3k\tau_+u(k)\chi(k)\right|\Psi^0\right\rangle$$
$$+\text{c.c.}\tag{11}$$

Upon varying J with respect to $\chi(k)$, we find the integral equation

$$(\omega_k-\omega_0)\chi(k)=\int K(k,k')\chi(k')d^3k',\tag{12}$$

where

$$2K(k,k')=g\langle\Psi^0|[\alpha_{k'}u(k)+\alpha_ku(k')]\tau_+|\Psi^0\rangle$$
$$-\langle\Psi^0|\alpha_k{}^\dagger(\omega_{k'}-\omega_0)\alpha_{k'}|\Psi^0\rangle$$
$$-\langle\Psi^0|\alpha_{k'}{}^\dagger(\omega_k-\omega_0)\alpha_k|\Psi^0\rangle.\tag{12a}$$

By using the Tomonaga approximate form for Ψ^0, Eq. (4), and the expressions for $\langle\alpha^\dagger\alpha\rangle_{\text{Av}}$ and $\langle\alpha\tau_+\rangle_{\text{Av}}$ given by Eqs. (A.3) and (6), Eq. (12a) becomes

$$2K(k,k')=g\mathfrak{M}_+{}^0[f_+{}^0(k)u(k')+u(k)f_+{}^0(k')]$$
$$-\mathfrak{N}_+{}^0[\omega_k+\omega_{k'}-2\omega_0]f_+{}^0(k)f_+{}^0(k').$$

Finally upon using Eqs. (5) and (7) we find,

$$K(k,k')=\mathfrak{N}_+{}^0(\omega_0+\lambda_+{}^0)f_+{}^0(k)f_+{}^0(k').\tag{13}$$

The solution for the scattering problem of Eq. (12) is

$$\chi(k)=\delta^3(k-k_0)+\frac{\mathfrak{N}_+{}^0(\omega_0+\lambda_+{}^0)f_+{}^0(k)f_+{}^0(k_0)}{(\omega_k-\omega_0-i\epsilon)\left[1-\mathfrak{N}_+{}^0(\omega_0+\lambda_+{}^0)\int d^3k'|f_+{}^0(k')|^2/(\omega_{k'}-\omega_0-i\epsilon)\right]},\tag{14}$$

[6] For π^-+p, an additional term corresponding to the formation of a neutron must be included. See Appendix II for a detailed discussion.

[7] A rigorous formulation of the variational principle is given in Appendix III.

corresponding to a differential cross section given by (see Appendix III):

$$\frac{d\sigma}{d\Omega}=\frac{[4\pi^2\omega_0 u^2(k_0)Rg^2]^2}{[4\pi^2\omega_0 k_0 u^2(k_0)Rg^2]^2+(\omega_0+\lambda_+{}^0)^2\left[1-Rg^2(\omega_0+\lambda_+{}^0)\mathrm{P}\int d^3k u^2/(\omega+\lambda_+{}^0)^2(\omega-\omega_0)\right]^2}, \tag{15}$$

where P indicates that the principal value of the integral is to be taken and $R=(\mathfrak{M}_+{}^0)^2/\mathfrak{N}_+{}^0$.

It may be emphasized that the self-energy E_0 has been explicitly eliminated in the kernel $K(k,k')$, Eq. (12), and that in (15), R is always multiplied by g^2 and plays the role of a charge renormalization factor. The result (15) is then completely convergent even for a delta function source.

Weak-Coupling Limit

For a sufficiently small coupling constant, $R=1$ and $\lambda_+{}^0=0$ so that with a delta function source the cross section becomes

$$d\sigma/d\Omega=g^4/\omega_0{}^2,$$

which agrees exactly with the usual weak-coupling calculation.

IV. INELASTIC SCATTERING

For sufficiently large value of the coupling constant a *stable* isobaric state of the nucleon with charge 3 is possible and the scattering will be inelastic; that is, the reaction

$$\text{(A)}\quad \pi^+ + p\rightarrow \pi^- + p^{+++}$$

competes with the scattering process

$$\text{(B)}\quad \pi^+ + p\rightarrow \pi^+ + p.$$

Therefore, instead of Eq. (9) we use the trial function[8]

$$\psi_{n+1,m}=(n+1)^{-\frac{1}{2}}\mathcal{S}_+[\psi_{n,m}{}^0\chi_+(k_{n+1}{}^+)] +m^{-\frac{1}{2}}\mathcal{S}_-[\psi_{n+1,m-1}{}^1\chi_-(k_m{}^-)], \tag{16}$$

where $\psi_{n+1,m-1}{}^1$ is the probability amplitude of the triply charged stable isobaric state p^{+++} for finding $n+1$ positive mesons and $m-1$ negative mesons. The procedure for the determination of the isobaric state wave function is identical with that used for the ground state. We shall in this paper use superscript 0 for quantities connected with the ground state and the superscript 1 for the corresponding quantities associated with the isobaric state p^{+++}. $\chi_-(k)$ represents the outgoing wave of a negative meson and $\chi_+(k)$ is defined as before.

Applying the variational principle (9) to the trial function (16) with independent variations of $\chi_+(k)$ and $\chi_-(k)$ we have

$$\begin{aligned}(\omega_k-\omega_0)\chi_+(k)&=\int K_{++}(k,k')\chi_+(k')d^3k' +\int K_{+-}(k,k')\chi_-(k')d^3k',\\ (\omega_k-\omega_1)\chi_-(k)&=\int K_{-+}(k,k')\chi_+(k')d^3k' +\int K_{--}(k,k')\chi_-(k')d^3k',\end{aligned} \tag{17}$$

where ω_0 and ω_1 are the total energy of the free positive meson and the outgoing negative meson, respectively. The kernels are

$$2K_{++}(k,k')=\mathfrak{M}_+{}^0 g[f_+{}^0(k)u(k')+u(k)f_+{}^0(k')] -\mathfrak{N}_+{}^0(\omega_k+\omega_k{}'-2\omega_0)f_+{}^0(k)f_+{}^0(k'),$$

$$\begin{aligned}2K_{+-}(k,k')&=2K_{-+}(k',k)\\ &=g\mathcal{O}f_+{}^1(k)u(k')+g\mathcal{P}u(k)f_-{}^0(k') -\mathcal{Q}(\omega_k+\omega_{k'}-\omega_0-\omega_1)f_+{}^1(k)f_-{}^0(k'),\end{aligned}$$

and

$$2K_{--}(k,k')=\mathfrak{M}_+{}^1 g[f_-{}^1(k)u(k')+u(k)f_-{}^1(k')] -\mathfrak{N}_+{}^1(\omega_k+\omega_{k'}-2\omega_1)f_-{}^1(k)f_-{}^1(k'),$$

where $\mathfrak{M}_+{}^0$, $\mathfrak{N}_+{}^0$ are defined by Eq. (6) and $\mathfrak{M}_+{}^1$, $\mathfrak{N}_+{}^1$ are the corresponding quantities for the state p^{+++}. The constants $\mathcal{O}$, $\mathcal{P}$, $\mathcal{Q}$, are

$$\begin{aligned}\mathcal{O}&\equiv\sum_{n=1}^{\infty}C_{n,n-1}{}^0C_{n+1,n-1}{}^1(n+1)^{\frac{1}{2}}\left[\int f_+{}^1(k)f_+{}^0(k)d^3k\right]^n \times\left[\int f_-{}^1(k)f_-{}^0(k)d^3k\right]^{n-1},\\ \mathcal{P}&\equiv\sum_{n=1}^{\infty}C_{n+1,n-1}{}^1C_{n+1,n}{}^0 n^{\frac{1}{2}}\left[\int f_+{}^0(k)f_+{}^1(k)d^3k\right]^{n+1} \times\left[\int f_-{}^0(k)f_-{}^1(k)d^3k\right]^{n-1},\end{aligned} \tag{18}$$

and

$$\mathcal{Q}\equiv\sum_{n=1}^{\infty}\sum_{m=n-1}^{n}C_{n,m}{}^0C_{n+1,m-1}{}^1[(n+1)m]^{\frac{1}{2}} \times\left[\int f_+{}^1(k)f_+{}^0(k)d^3k\right]^n\left[\int f_-{}^1(k)f_-{}^0(k)d^3k\right]^{m-1},$$

[8] For simplicity we have restricted ourselves to energy less than 2μ. At greater energy, multiple production of mesons is possible which must be taken into account by adding additional outgoing waves in the trial function.

where $C_{n,m}^0$ and $C_{n,m}^1$ are the total probability amplitude for finding n positive mesons and m negative mesons in the ground state and the triply charged isobaric state of the nucleon, respectively.

Upon noticing that the kernels of the integral equation (17) can be written as a sum of product functions each having only one variable, the solution for χ_+ and χ_- must, therefore, be of the form

$$\chi_+(k)=\delta^3(k-k_0)+\alpha_+^0 f_+^0(k)+\alpha_+^1 f_+^1(k) +[\beta_+^0 f_+^0(k)+\beta_+^1 f_+^1(k)+\gamma_+ u(k)] \times[\omega-\omega_0-i\epsilon]^{-1}, \quad (19)$$

and

$$\chi_-(k)=\alpha_-^0 f_-^0(k)+\alpha_-^1 f_-^1(k) +[\beta_-^0 f_-^0(k)+\beta_-^1 f_-^1(k)+\gamma_- u(k)][\omega-\omega_1-i\epsilon]^{-1},$$

where $\alpha_\pm^i$, $\beta_\pm^i$ ($i=0, 1$), and $\gamma_\pm$ are numerical constants which can be determined by substituting Eq. (19) back into the integral equation (17).

Strong-Coupling Limit

It is of interest to examine the solution when the coupling constant is very large. In this case the energy separation $(\omega_0-\omega_1)$ of the isobaric state with the ground state is of the order g^{-2} and is neglected. Furthermore, the distribution functions of mesons around the nucleon becomes identical for the isobaric and the ground state as $g\to\infty$. We can then *in this limit*[9] set

$$\mathfrak{N}_+^0=\mathfrak{N}_+^1=\mathfrak{Q},$$

$$\mathfrak{M}_+^0=\mathfrak{M}_+^1=\mathcal{O}=\mathcal{P},$$

and

$$f_+^1=f_+^0=f_-^1=f_-^0=(u(k)/\omega)\left(\int d^3k u^2/\omega^2\right)^{-\frac{1}{2}}.$$

Thus the kernels become

$$K_{++}=K_{+-}=K_{-+}=K_{--}=K(k,k'),$$

where K is identical with that in Eq. (12). By forming the linear combinations $\chi_+\pm\chi_-$ we readily find

$$(\omega-\omega_0)[\chi_+(k)-\chi_-(k)]=0, \quad (20)$$

and

$$(\omega-\omega_0)[\chi_+(k)+\chi_-(k)] =2\int K(k,k')[\chi_+(k')+\chi_-(k')]d^3k'.$$

The first equation shows that the difference between $\chi_+(k)$ and $\chi_-(k)$ can only be a $\delta^3(k-k_0)$ so that *in the strong-coupling limit the charge exchange and nonexchange cross sections are equal.* The second equation can now be solved by casting $\chi_+(k)+\chi_-(k)$ into the same form as $\chi(k)$ in Eq. (14). On using the limiting values for $\mathfrak{N}_\pm^0$ and $\mathfrak{M}_\pm^0$ as tabulated in Table I we find the differential cross section $d\sigma_{++}/d\Omega$ and $d\sigma_{+-}/d\Omega$ for the reaction (A) and (B) to be

$$d\sigma_{++}/d\Omega=d\sigma_{+-}/d\Omega =4\pi^4\frac{f^4(k_0)\omega_0^2}{m^{-2}+[4\pi^2 f^2(k_0)k_0\omega_0]^2}+O(\mu/M), \quad (21)$$

where M^{-1} is the size of the source function and

$$m^{-1}=\mathrm{P}\int[f^2(k)/(\omega-\omega_0)]d^3k, \quad (22)$$

where P indicates that the principal value of the integral is to be taken. For a δ-function source (i.e., $M\to\infty$) we get for the total cross section when $\omega_0=\mu$

$$\sigma_{++}=\sigma_{+-}=\pi\mu^{-2}(2^{-1}+\pi^{-1})^{-2}, \quad (23)$$

while the rigorous treatment in the strong-coupling limit[4] yields

$$\sigma_{++}=\sigma_{+-}=\pi\mu^{-2}. \quad (24)$$

Hence, unlike Tomonaga's treatments on the bound states where the variational method leads to exact agreement with both the rigorous calculations for weak- and strong-coupling limits, our calculations for meson scattering give exact agreement only in the weak-coupling limit while in the strong-coupling limit the rigorous treatment differs from our result by the factor[10]

$$(\tfrac{1}{2}+1/\pi)^2$$

at zero energy. At other energy this factor is

$$\left[\left(\frac{1}{2}+\frac{1}{\pi}\right)+\frac{\mu-\omega_0}{2\omega_0}+\frac{(\omega_0^2-\mu^2)^{\frac{1}{2}}}{\pi\omega_0}\ln\frac{(\omega_0+\mu)^{\frac{1}{2}}-(\omega_0-\mu)^{\frac{1}{2}}}{(\omega_0+\mu)^{\frac{1}{2}}+(\omega_0-\mu)^{\frac{1}{2}}}\right]^2 +\left[\frac{\omega_0^2-\mu^2}{\omega_0^2}\right],$$

which only varies from about 0.7 to 0.8 as ω_0 increases from μ to 2μ.

V. INTERMEDIATE-COUPLING RANGE

In this section we shall present the results of the numerical calculations in the intermediate-coupling range. Figure 3 shows $k^2(d\sigma/d\Omega)$ as a function of energy for π^++p and π^-+p scattering. It is of interest to notice that even as $g^2=10$ the deviation in cross section from the strong-coupling limit is still substantial. Thus, at least throughout the range $1<g^2<10$, both the weak-

[9] The constants $\lambda_\pm^i$ in the strong-coupling limit are of the order g^{-2}. Thus, their influence on each of the overlapping integrals in (18) is of order g^{-4} which can be neglected in the present calculation. On the other hand, in the calculation of isobaric separation it is necessary to include $\lambda_\pm^i$ since the isobaric separation energy is itself of the order g^{-2}.

[10] This discrepancy presumably is due to the absence of polarization effect in our approximation. The inclusion of such effect, though desirable, tends to make the calculations at intermediate coupling range much more complicated.

and strong-coupling results are in very poor agreement with the intermediate-coupling results. In Fig. 4, $k \cot\delta$ is plotted against energy and shows that the scattering amplitudes for $\pi^+ + p$ and $\pi^- + p$ are very similar to *that of a short-range attractive potential and a short-range repulsive potential, respectively*, while for $\pi^+ + p$, a 90° phase shift is always possible at a suitable value of the energy, provided $g^2 \geq 4.0$. Yet the phase shift for $\pi^- + p$ would always be smaller than 90° except may be at the region of extremely large g values. Figure 5 shows the manner in which the scattering amplitude for zero incident kinetic energy ($\omega_0 = \mu$) changes smoothly from weak-coupling value to the strong-coupling value. It may be noted that for zero-energy $\pi^+ + p$ scattering, a phase shift of 90° will occur at a value of $g^2 = 4.0$. From Fig. 1 one sees that, by using the Tomonaga's technique for bound state, also at $g^2 = 4.0$, the first excited state p^{++} has a zero binding energy. Thus, at least in this case, the method for the bound state agrees with that for the scattering state.

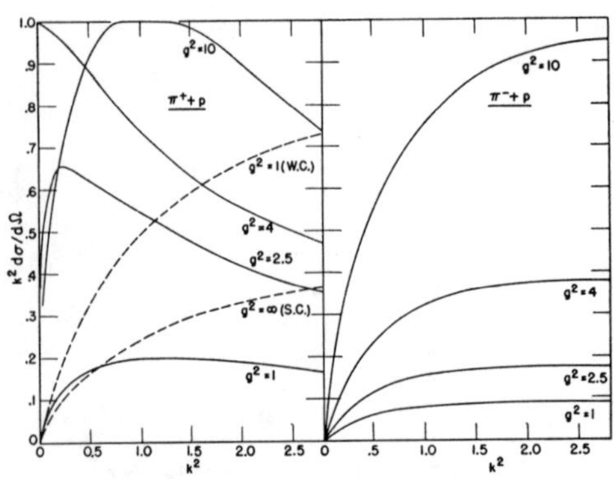

FIG. 3. $k^2 d\sigma/d\Omega$ *versus* k^2. Solid lines refer to intermediate coupling calculations.

While the detailed results are essentially of a numerical nature, some approximate expressions are found useful in understanding the over-all behavior of the cross sections. For $\pi^+ + p$ scattering it is found that with λ_+^0 in (15) taken equal to zero the phase shift δ may be written as[11]

$$\tan\delta = \frac{4\pi^2 (Rg^2) k_0 u^2(k_0)}{1-\Delta}, \tag{25}$$

$$\Delta = Rg^2 \omega_0 \mathrm{P} \int [u^2/\omega^2(\omega - \omega_0)] d^3k.$$

Since the integral is in general positive (if $\omega_0 < 2\mu$), it is possible to find one resonance at any energy less than 2μ for some value of g. On the other hand, since the integral is a decreasing function of ω_0, the phase shift will be less than $\pi/2$ if g is smaller than the value of g required to give a resonance at zero energy, $\omega_0 = \mu$.

[11] A similar form for the phase shifts is obtained by G. Chew for pseudoscalar meson theory, using the Dancoff-type wave functions [Phys. Rev. **89**, 591 (1952)].

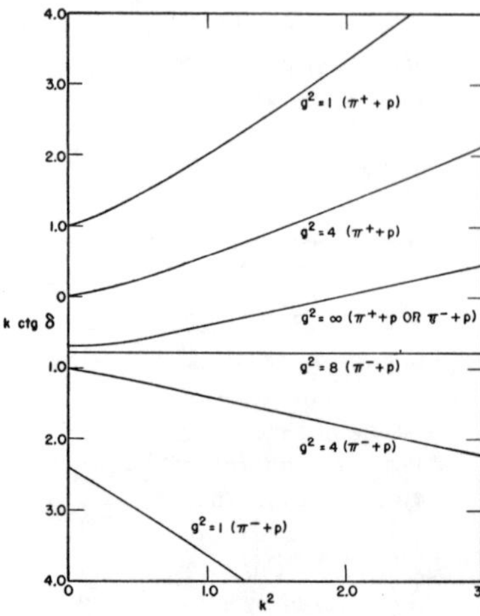

FIG. 4. $k \cot\delta$ *vs* k^2 for various g values.

Examination of the simplified form of the cross section indicates that essentially two distinct conditions must be satisfied to validate a perturbation calculation. These are (1) that the number of mesons about a nucleon be sufficiently small that the perturbation result agrees with the intermediate coupling result, namely

$$\mathfrak{N}_+^0 \cong g^2 \int [u^2(k)/\omega_k^2] d^3k < 1, \tag{26}$$

and (2) that the damping terms in the denominator be small compared with unity, i.e.,

$$\Delta < 1. \tag{27}$$

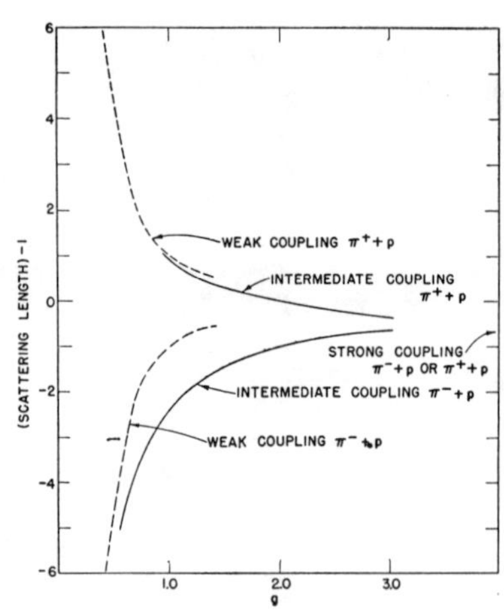

FIG. 5. (Scattering length)$^{-1}$ *vs* g for $\omega_0 = \mu$.

VI. CONCLUSIONS

It has been shown that for the scattering processes as well as the bound states the intermediate-coupling method may be applied, and that this treatment joins all results calculated for the weak- and strong-coupling limits smoothly. For the ground state it is found that $g^2 \lesssim 1$ for a 10 percent accuracy in the self-energy if the weak-coupling method is used and $g^2 \gtrsim 3$ if the strong-coupling method is used. More stringent limits are placed upon g^2 when the same accuracy is desired for more critical quantities such as average number of mesons, isobar separation, scattering, etc. For example, the calculated value of g^2 for a stable isobar of charge two is $g^2 \geq 4.0$ with the intermediate-coupling method and $g^2 \geq 5.8$ with the strong-coupling method. By referring to Figs. 3 and 5, it is seen that for scattering processes $g^2 < 1.0$ and $g^2 \gg 10$ for the applicability of the weak- and strong-coupling methods, respectively.

It has also been shown that $\pi^+ + p$ scattering is similar to scattering by a short-range attractive potential and that $\pi^- + p$ is similar to scattering by a repulsive potential.

The authors wish to thank Dr. C. N. Yang for some helpful discussions and express their gratitude to Professor Oppenheimer for the hospitality extended to them by the Institute for Advanced Study.

APPENDIX I

To derive the explicit forms of $\mathfrak{N}_{\pm}{}^0$ and $\mathfrak{M}_{\pm}{}^0$ we consider an alternative formulation of the Tomonaga method. In this formulation one approximates the annihilation and creation operators of the mesons by

$$\begin{aligned} \alpha_k &= f_+(k)\alpha; \quad \alpha_k{}^\dagger = f_+(k)\alpha^\dagger, \\ \beta_k &= f_-(k)\beta; \quad \beta_k{}^\dagger = f_-(k)\beta^\dagger, \end{aligned} \tag{A.1}$$

where α, β, and $\alpha^\dagger$, $\beta^\dagger$ are annihilation and creation operators independent of k. (In this section we omit the superscript indicating the state of the total charge as all equations are identical in form for any isobaric state.) The Hamiltonian (1) is then reduced to

$$\begin{aligned} H = \omega_+\alpha^\dagger\alpha + \omega_-\beta^\dagger\beta - \tau_+[g_+\alpha + g_-\beta^\dagger] \\ - \tau_-[g_+\alpha^\dagger + g_-\beta], \end{aligned} \tag{A.2}$$

where

$$\omega_\pm = \int \omega_k f_\pm{}^2(k) d^3k,$$

$$g_\pm = g \int u(k) f_\pm(k) d^3k.$$

It can then be shown that

$$\mathfrak{N}_+ = \sum |C_{n,m}|^2 n = \langle \alpha^\dagger \alpha \rangle_{\text{Av}},$$

and

$$\mathfrak{M}_+ = \sum C_{n+1,n} C_{n,n}(n+1)^{\frac{1}{2}} = \langle \alpha\tau_+ \rangle_{\text{Av}}, \tag{A.3}$$

where $\langle\ \rangle_{\text{Av}}$ means the diagonal matrix element. These averages can be obtained explicitly by converting α, β into real canonical conjugate variables, and the problem of determining the eigenstates of H [Eq. (A.2)] reduces to the problem of two coupled harmonic oscillators. In the strong-coupling limit, one may neglect higher orders in g^{-1} and set $f_+(k) = f_-(k)$. The results are listed in Table I.

APPENDIX II

The trial function Ψ representing the process,

$$\pi^- + p \to \pi^- + p,$$

differs slightly from that for π^+ scattering, Eq. (10), due to the possibility of finding a neutron with no mesons. The assumed form is now a combination of

$$\Psi = \int \chi_-(k) \beta_k{}^\dagger \Psi_P d^3k + c\Psi_N, \tag{A.4}$$

where Ψ_P and Ψ_N are the state vectors representing a "physical" proton and a "physical" neutron, respectively. c is a numerical constant. Upon varying

$$\langle \Psi | H - E_0 - \omega_0 | \Psi \rangle$$

with respect to $\chi_-(k)$ and c, one finds

$$c = -\mathfrak{R} \int f_+{}^0(k) \chi_-(k) d^3k, \tag{A.5}$$

and

$$(\omega - \omega_0)\chi_-(k) = \int K(k,k')\chi_-(k') d^3k', \tag{A.6}$$

where

$$\mathfrak{R} = \sum_{n=0}^{\infty} \sum_{m=n-1}^{n} C_{n,m}{}^0 C_{m+1,n}{}^0 (m+1)^{\frac{1}{2}} \times \left(\int f_+{}^0 f_-{}^0 d^3k \right)^{n+m}, \tag{A.7}$$

and

$$K(k,k') = \mathfrak{N}_-{}^0(\omega_0 + \lambda_-{}^0) f_-{}^0(k) f_-{}^0(k') - \mathfrak{R}^2 \omega_0 f_+{}^0(k) f_+{}^0(k'). \tag{A.8}$$

Equation (A.6) can then be solved by assuming a form for $\chi_-(k)$ similar to that of Eq. (14). In the weak-coupling limit it yields the rigorous scattering amplitude.

APPENDIX III

In the case of elastic scattering (i.e., $\omega < 2\mu$ and no stable isobar of the nucleon exists), the rigorous Schrödinger wave function for the scattering of a positive meson with incident wave number k_0 by a proton can be written as

$$\begin{aligned} \psi_{n,m} = n^{-\frac{1}{2}} \\ S_+\{\psi_{n-1,m}{}^0(k_1{}^+, \cdots, k_{i-1}{}^+, k_{i+1}{}^+, \cdots, k_n{}^+; k_1{}^-, \cdots, k_m{}^-) \\ \times [\delta^3(k_i - k_0) + B_{n,m}{}^i(\omega_k{}^+ - \omega_0 - i\epsilon)^{-1}]\}, \end{aligned} \tag{A.8}$$

where $B_{n,m}{}^i$ is a regular function of $k_1{}^+\cdots k_n{}^+$ and $k_1{}^-\cdots k_m{}^-$. The value of $B_{n,m}{}^i$ when $k_i{}^+=k_0$ represents the scattering amplitude of the ith meson; consequently

$$B_{n,m}{}^i|_{k_i{}^+=k_0}=B(k_0)$$

depends on k_0 only. These conditions are physically necessary for any elastic scattering problem and can be readily verified by a direct substitution into the Schrödinger equation. The differential cross section $d\sigma/d\Omega$ and the phase shift η are related to $B(k_0)$ by the relations:

$$\begin{aligned} d\sigma/d\Omega&=(4\pi^2)^2|B(k_0)|^2\omega_0{}^2,\\ B(k_0)&=(e^{2i\eta}-1)/8i\pi^2k_0\omega_0. \end{aligned} \tag{A.9}$$

Following Kohn's[12] variational treatment for scattering problem in momentum space, we find for the correct solution

$$\begin{aligned} J&=\langle\Psi|H-E_0-\omega|\Psi\rangle\\ &=B(k_0)-4i\pi^2k_0\omega_0|B(k_0)|^2\\ &=\sin 2\eta/8\pi^2k_0\omega_0, \end{aligned} \tag{A.10}$$

and the variation of J in the neighborhood of the correct Ψ is

$$\begin{aligned} \delta J&=\delta B(k_0)+\delta B^*(k_0)\\ &\quad+4i\pi^2k_0\omega_0[-B^*(k_0)\delta B(k_0)+B(k_0)\delta B^*(k_0)]\\ &=(4\pi^2k_0\omega_0)^{-1}(1+\cos 2\eta)\delta\eta. \end{aligned} \tag{A.11}$$

Equations (A.10) and (A.11) then give the variational calculation for η. It can be easily seen that if one assumes Ψ to be of the form (8) a formal variation with respect to χ [Eq. (9)] gives the same result as (A.10) and (A.11).

[12] W. Kohn, Phys. Rev. 84, 495 (1951).

Reprinted from THE PHYSICAL REVIEW, Vol. 95, No. 5, 1329–1334, September 1, 1954
Printed in U. S. A.

Some Special Examples in Renormalizable Field Theory

T. D. LEE
Department of Physics, Columbia University, New York, New York
(Received May 4, 1954)

Some special problems of interacting fields that contain removable divergences are treated in detail. Comparisons with the power series renormalization procedures are made. Examination of the closed forms of the solutions before and after renormalization shows that, in one special case solved, coupling-constant renormalization cannot be obtained by any limiting processes that involve only real values of the unrenormalized coupling constant.

I. INTRODUCTION

EVER since the overwhelming success of the applications of renormalization technique in quantum electrodynamics, the problem of understanding this renormalization procedure without the use of perturbation methods has been of great interest. Unfortunately, in all physically realistic cases the Hamiltonians are quite complicated in structure, and so far no solution other than the power series method has been found. Yet the renormalization methods as developed by Dyson,[1] Ward,[2] and others can certainly be applied to a large class of field-theoretical problems that involve removable divergencies. Thus some deeper insight as to the nature of the renormalization procedure may, perhaps, be obtained by examining problems that are similar to, but not as rich as, either the quantum electrodynamics or the relativistic pseudoscalar meson theory. With this motive we shall in this paper treat some simple problems of interacting fields that are both renormalizable and solvable.

We consider first a problem that involves the interactions between two neutral nonrelativistic fermion fields and one relativistic boson field. Although the Hamiltonian of this problem does involve infinities, the Schroedinger equation can still be solved directly. A close examination of the solutions indicates that all the divergent quantities can be removed by both a mass renormalization and a coupling-constant renormalization. The scattering amplitude is then calculated after renormalization and is indeed found to be finite (i.e., not zero). This problem can also be treated by the application of the customary power series renormalization procedure whereby identical conclusions concerning the nature of renormalization quantities are reached. Furthermore, even outside its radius of convergence, the formal sum of the power series can still be used to give the correct closed form. A rather unexpected and quite surprising feature is obtained by comparing the renormalized coupling constant with the unrenormalized coupling constant. In this particular case it can be shown that the result of the renormalization process *cannot* be obtained by any limiting process that involves only *real* values of the unrenormalized coupling constant. This difficulty may, however, be overcome by a modification of the present rules of quantum mechanics.

Next, the well-known soluble problem of neutral scalar mesons with fixed nucleons is studied. The closed forms of the nucleon propagation functions and vertex functions are listed in Appendix I.

In Appendix II, the charged scalar theory together with some interesting identities between the customary matrix elements and the renormalization quantities Z_1, Z_2 are discussed.

II. HAMILTONIAN

Let us consider the interaction between two neutral fermion fields, V and N, and a neutral scalar boson field, θ. The Hamiltonian for the free fields is

$$H_0=m_V\int\psi_V^\dagger\psi_V d\tau+m_N\int\psi_N^\dagger\psi_N d\tau +\tfrac{1}{2}\int[\pi^2+(\nabla\varphi)^2+\mu^2\varphi^2]d\tau, \quad (1)$$

where $\psi_V^\dagger$, ψ_V and $\psi_N^\dagger$, ψ_N obey the usual anticommutation relation and represent the field variables of V and N, φ and π are the field variable and conjugate momentum of the θ particles. The "observed" masses of V, N, and θ are denoted by m_V, m_N, and μ. Recoil effects of V and N are not included.

It is convenient to write

$$\varphi=A(\mathbf{r})+A^\dagger(\mathbf{r}),$$

where

$$A(\mathbf{r})=\sum(2\omega\Omega)^{-\frac{1}{2}}\alpha_k\exp(i\mathbf{k}\cdot\mathbf{r}),$$
$$A^\dagger(\mathbf{r})=\sum(2\omega\Omega)^{-\frac{1}{2}}\alpha_k^\dagger\exp(-i\mathbf{k}\cdot\mathbf{r}).$$

α_k and $\alpha_k^\dagger$ are the annihilation and creation operators of the θ particle with wave number $\mathbf{k}$. Ω is the volume of the system and ω is $(k^2+\mu^2)^{\frac{1}{2}}$. The interaction Hamiltonian H_1 that represents the reaction

$$V\rightleftarrows N+\theta \quad (2)$$

can be written as

$$H_1=g\int[\psi_V^\dagger(\mathbf{r})\psi_N(\mathbf{r})A(\mathbf{r})+\psi_N^\dagger(\mathbf{r})\psi_V(\mathbf{r})A^\dagger(\mathbf{r})]d\tau +\delta m_V\int\psi_V^\dagger\psi_V d\tau, \quad (3)$$

[1] F. J. Dyson, Phys. Rev. **75**, 1736 (1949).
[2] J. C. Ward, Proc. Phys. Soc. (London) **A64**, 54 (1951); Phys. Rev. **84**, 897 (1951).

where δm_V is used to cancel any change in mass of the V particle due to the reaction (2).

Upon examining the total Hamiltonian

$$H=H_0+H_1, \tag{4}$$

one sees that this system possesses two simple conservation laws:

$$\mathfrak{N}_V+\mathfrak{N}_N=\text{constant},$$

and

$$\mathfrak{N}_V+\mathfrak{N}_\theta=\text{constant}, \tag{5}$$

with $\mathfrak{N}_V$, $\mathfrak{N}_N$, $\mathfrak{N}_\theta$ the total number of V, N, and θ particles, respectively. Because of (5) the eigenfunctions of H contain only a finite number of particles and consequently can be solved directly.

III. GROUND STATES

In the following, we denote the state of a "bare" V particle and that of a "bare" N particle by $|V\rangle$ and $|N\rangle$ while the states of the corresponding "physical" particles are indicated by $|\mathbf{V}\rangle$ and $|\mathbf{N}\rangle$. Thus $|\mathbf{V}\rangle$ and $|\mathbf{N}\rangle$ are eigenstates of the Hamiltonian and satisfy

$$H|\mathbf{N}\rangle=m_N|\mathbf{N}\rangle,$$
$$H|\mathbf{V}\rangle=m_V|\mathbf{V}\rangle. \tag{6}$$

In order that V be a stable particle we assume $m_V-m_N<\mu$. Using (5), one see that

$$|\mathbf{N}\rangle=|N\rangle,$$

and

$$|\mathbf{V}\rangle=Z_2^{\frac{1}{2}}[|V\rangle+g\textstyle\sum_k f(k)\alpha_k^\dagger|N\rangle], \tag{7}$$

where $Z_2^{\frac{1}{2}}$ is a normalization constant and $f(k)$ is proportional to the probability amplitude for finding a θ particle with wave number k in a physical V state.

Applying the Hamiltonian (4) on $|\mathbf{V}\rangle$ and requiring that the eigenvalue should be the "observed" mass m_V, one finds

$$\delta m_V=-g^2\textstyle\sum(2\omega\Omega)^{-1}(m_V-m_N-\omega)^{-1}, \tag{8}$$

$$f(k)=(2\omega\Omega)^{-\frac{1}{2}}(m_V-m_N-\omega)^{-1}, \tag{9}$$

and

$$Z_2^{-1}=1+g^2\textstyle\sum(2\omega\Omega)^{-1}(m_V-m_N-\omega)^{-2}. \tag{10}$$

We remark that the divergent quantity δm_V serves as the renormalization in mass such that the eigenvalue m_V of the Hamiltonian is now finite. The divergent quantity Z_2^{-1} is related to the coupling-constant renormalization and will be studied in the next section.

IV. SCATTERING STATE

We consider first the scattering process

$$N+\theta\rightarrow N+\theta.$$

The eigenfunction $|\mathbf{N}+\boldsymbol{\theta}\rangle$ that represents this process can be written in terms of the physical one-particle states as

$$|\mathbf{N}+\boldsymbol{\theta}\rangle=\textstyle\sum_k\chi(k)\alpha_k^\dagger|\mathbf{N}\rangle+c|\mathbf{V}\rangle, \tag{11}$$

and furthermore it satisfies the Schroedinger equation

$$H|\mathbf{N}+\boldsymbol{\theta}\rangle=(m_N+\omega_0)|\mathbf{N}+\boldsymbol{\theta}\rangle. \tag{12}$$

On using the special form of H together with (6) and (7), the Schroedinger equation can be readily solved and one obtains

$$c=-\langle\mathbf{V}|\textstyle\sum\chi(k)\alpha_k^\dagger|\mathbf{N}\rangle, \tag{13}$$

and

$$(\omega-\omega_0)\chi(k)=g^2Z_2\int K(k,k')\chi(k')d^3k', \tag{14}$$

where

$$K(k,k')=(m_V-m_N-\omega_0)(4\omega\omega')^{-\frac{1}{2}}\times[8\pi^3(m_V-m_N-\omega)(m_V-m_N-\omega')]^{-1}.$$

One observes that although Z_2^{-1} involves a divergent sum, the cross section will not vanish if the renormalized coupling constant, given by

$$g_c^2=g^2Z_2, \tag{15}$$

is chosen to be finite. Equation (14) can then be easily solved, and the phase shift δ for the scattering process is

$$\tan\delta=\frac{g_c^2k_0}{4\pi(m_V-m_N-\omega_0)}\times\left[1-\frac{g_c^2}{16\pi^3}\mathcal{P}\int\frac{d^3k(m_V-m_N-\omega_0)}{\omega(\omega-\omega_0)(m_V-m_N-\omega)^2}\right]^{-1}, \tag{16}$$

where $\mathcal{P}$ indicates that in the integration the principal value is to be taken. Thus if g_c is finite the expression for the phase shift is indeed free from divergent quantities.

Next we consider the scattering process,

$$V+\theta\rightarrow V+\theta. \tag{17}$$

Again, the corresponding eigenstate that satisfies

$$H|\mathbf{V}+\boldsymbol{\theta}\rangle=(m_V+\omega_0)|\mathbf{V}+\boldsymbol{\theta}\rangle \tag{18}$$

can be written in terms of the physical one-particle states as

$$|\mathbf{V}+\boldsymbol{\theta}\rangle=\sum_k\psi(k)\alpha_k^\dagger|\mathbf{V}\rangle+\sum_{k_1,k_2}\phi(k_1,k_2)\alpha_{k_1}^\dagger\alpha_{k_2}^\dagger|\mathbf{N}\rangle. \tag{19}$$

The Schroedinger equation can also be solved and one finds

$$(\omega-\omega_0)h(\omega)\psi(k)=g^2Z_2\int K'(k,k')\psi(k')d^3k', \tag{20}$$

where

$$K'(k,k')=(16\pi^3)^{-1}(\omega\omega')^{-\frac{1}{2}}(\omega+\omega'-\omega_0-m_V+m_N)^{-1},$$

$$h(\omega)=1-g^2Z_2(16\pi^3)^{-1}\times\int\frac{d^3k'}{(\omega+\omega'-\omega_0-m_V+m_N)(m_V-m_N-\omega')^2\omega'},$$

and

$$(\omega_1+\omega_2-\omega_0-m_V+m_N)\phi(k_1,k_2)$$
$$=-\tfrac{1}{2}(gZ_2^{\frac{1}{2}})[(\omega_1-\omega_0)\psi(k_1)f(k_2)+(\omega_2-\omega_0)\psi(k_2)f(k_1)]. \quad (21)$$

$f(k)$ is given by Eq. (9). By counting the powers of ω and ω' in the kernel K' one sees that the solution $\psi(k)$ and $\phi(k_1,k_2)$ would be free from divergent integrals if the renormalized coupling constant $g_c=gZ_2^{\frac{1}{2}}$ is finite.

V. COMPARISON WITH POWER SERIES METHOD

The above problem can also be renormalized by the power series method. By constructing the S matrix in the interaction representation and commuting all the time ordered operators into S products[1,3] one finds the following rules for the Feynman diagram:

(i) Each V line gives a factor $(p_0-m_V+i\epsilon)^{-1}$, where p_0 is $-i$ times the fourth component of the momentum vector carried by the V particle. ϵ is a positive infinitesimal quantity.

(ii) Each N line gives a factor $(p_0-m_N+i\epsilon)^{-1}$.

(iii) Each θ line gives a factor $(-k_\mu^2-\mu^2+i\epsilon)^{-1}$ and each variable momentum k_μ gives an integration $\int d^4k/(2\pi)^4$.

(iv) Each vertex gives a factor g.

Let S_N', S_V', and Γ denote the sum of all diagrams that contribute to the propagation function of N, the propagation function of V, and the vertex function, respectively. From the nature of the interaction (2), these quantities can be calculated readily and they are

$$\Gamma=1,$$
$$(S_N')^{-1}=p_0-m_N, \quad (22)$$
$$(S_V')^{-1}=p_0-m_V-\delta m_V-\Sigma(p_0),$$

where

$$\Sigma(p_0)=g^2\sum_k(2\omega\Omega)^{-1}(p_0-m_N-\omega)^{-1}.$$

Thus, both Γ and S_N' need no renormalization. The renormalization of mass and propagation function of V are given by

$$(S_V')^{-1}=0 \quad \text{at} \quad p_0=m_V,$$

and

$$Z_2^{-1}=d[(S_V')^{-1}]/dp_0 \quad \text{at} \quad p_0=m_V. \quad (23)$$

Hence δm_V and S_V' can be written as

$$\delta m_V=-\Sigma(p_0=m_V)=-g^2\sum(2\omega\Omega)^{-1}(m_V-m_N-\omega)^{-1},$$

and

$$S_V'(p_0)=Z_2S_{Vc}(p_0), \quad (24)$$

where

$$Z_2^{-1}=1+g^2\sum(2\omega\Omega)^{-1}(m_V-m_N-\omega)^{-2},$$
$$S_{Vc}^{-1}(p_0)=(p_0-m_V)\left[1-g^2Z_2\sum\frac{p_0-m_V}{(2\omega\Omega)(p_0-m_N-\omega)(m_V-m_N-\omega)^2}\right]. \quad (25)$$

[3] G. C. Wick, Phys. Rev. 80, 268 (1950).

If, as in (15), g^2Z_2 is set to be g_c^2 and remains finite, then the renormalized propagation function $S_{Vc}(p_0)$ will be free from divergent quantities.[4]

For any other physical processes it can then be shown that one needs only to consider the irreducible diagrams and use S_{Vc}, $S_{Nc}=S_N'$, and g_c for the propagation lines of V, N, and the vertex, respectively. Although the number of irreducible diagrams may still be infinite, the contribution of each diagram will be free from divergent integrals. In this case, using (25), one can compute the scattering processes in powers of g_c and one obtains indeed the same results as discussed in the previous section.

Thus, identical conclusions concerning both the mass renormalization and coupling-constant renormalization are obtained either by the power series method or by directly solving the Schroedinger equation. A comparison between these two methods shows that Z_2, which in the power series method is defined as the residue of S_V' at its pole, is actually the probability[5] of finding a "bare" V particle in the state of a "physical" V particle as shown by (7). It may be of interest to notice that, for example, in (25) the radius of convergence for the power series expansion of $S_{Vc}(p_0)$ in g_c^2 depends on the variable p_0 and it is 0 if $p_0=\infty$. Yet, the power series can always be summed formally and still gives the correct result.

VI. DISCUSSION OF COUPLING CONSTANT RENORMALIZATION

The nature of the coupling constant renormalization may be investigated by expressing g^2 and Z_2 in terms of the renormalized coupling constant g_c. Using (10) and (15) one finds

$$g^2=g_c^2[1-g_c^2\sum(2\omega\Omega)^{-1}(m_V-m_N-\omega)^{-2}]^{-1}, \quad (26)$$

and

$$Z_2=1-g_c^2\sum(2\omega\Omega)^{-1}(m_V-m_N-\omega)^{-2}. \quad (27)$$

Hence, if g_c does not vanish and remains finite, the unrenormalized coupling constant becomes

$$g=i\infty^{-1},$$

while Z_2 which, being a probability, should be between 0 and 1 is actually

$$Z_2=-\infty.$$

This shows that the result of coupling-constant renormalization in this problem cannot be realized by any limiting process if g is restricted to be on the real axis. Instead, the expressions for cross sections in previous sections may be obtained by allowing in the original Hamiltonian an unrenormalized coupling constant

[4] It is of interest to notice that if g_c^2 is finite, then S_{Vc} has another pole besides $p_0=m_V$. This pole corresponds to another stable state for the V particle. Identical conclusions can be obtained either by studying the bound state solutions for the state $|\mathbf{N}+\boldsymbol{\theta}\rangle$ or by examining the energy of the outgoing θ particle in the scattering state $|\mathbf{V}+\boldsymbol{\theta}\rangle$.

[5] See Appendix II for a general discussion of this property.

which is pure imaginary and which approaches zero while the upper limit of the sum in momentum space approaches infinity as described by (26).

This raises immediately the question as to what changes in the present rules of quantum mechanics we have made by allowing the Hamiltonian to be non-Hermitian. As shown in the preceding sections, the eigenvalues of the Hamiltonian and the matrix elements of the collision matrix are only functions of the renormalized coupling constant g_c, which is always real. Thus, even if g is imaginary, the unitarity of the collision matrix and the reality of the energy spectrum can still be preserved. Consequently, it is possible to transform this non-Hermitian Hamiltonian into a Hermitian matrix by a similarity transformation. As is well known, a similarity transformation preserves all relations between matrices and vectors; the laws of quantum mechanics may still be applied to this non-Hermitian Hamiltonian provided some attention is given to the formal differences between a unitary transformation and a similarity transformation. In particular, the transformation between the bare particle states and the physical particle states is not unitary, which explains why Z_2 as shown by (27) is not confined between 0 and 1.

The author wishes to thank Professors R. Serber and N. Kroll for discussions.

APPENDIX I[6]

Another soluble problem is that of the neutral scalar meson with fixed nucleon. The Hamiltonian is H_0+H_1, where

$$H_0=\int m\psi_N{}^\dagger\psi_N d\tau+\tfrac{1}{2}\int[\pi^2+(\nabla\varphi)^2+\mu^2\varphi^2]d\tau,$$

and

$$H_1=g\int\psi_N{}^\dagger\psi_N\varphi d\tau+\delta m\int\psi_N{}^\dagger\psi_N d\tau,$$

where $\psi_N{}^\dagger$, ψ_N describe the nucleon field and π, φ the meson field, g the coupling constant, and m the physical mass of the nucleon. While the Schroedinger equation of the above Hamiltonian is well known to be solvable, it is of some interest to examine the nucleon propagation function S', the vertex function Γ, and the renormalization quantities δm and Z_1, Z_2.

The mass renormalization is

$$\delta m=-g^2\sum(2\omega^2\Omega)^{-1}, \tag{A1}$$

where Ω is the volume of the system. The nucleon propagation function can be expressed as[7]

$$\mathbf{S}'(x-x')=T\langle\text{vac}|\psi_N(x)\psi_N{}^\dagger(x')|\text{vac}\rangle, \tag{A2}$$

where ψ_N and $\psi_N{}^\dagger$ are operators in the Heisenberg representations and T is the T product as defined by Wick.[3] Using the known solutions in the Heisenberg representation, one finds

$$\mathbf{S}'(x-x')=Z_2\mathbf{S}_c(x-x'), \tag{A3}$$

where

$$Z_2=\exp[-g^2\sum(2\omega^3\Omega)^{-1}] \tag{A4}$$

and

$$\begin{aligned}\mathbf{S}_c(x-x')=&\delta^3(r-r')\exp[-im(t-t')]\\&\times\exp[g^2\sum(2\omega^3\Omega)^{-1}e^{-i\omega(t-t')}]\text{ if }t>t';\\=&0\text{ if }t<t'.\end{aligned} \tag{A5}$$

It is of interest to notice that Z_2 is again the probability of finding the bare nucleon in the state of a physical nucleon.[5] If $t\neq t'$, $\mathbf{S}_c(x-x')$ is always free from divergent quantities; furthermore it can always be expanded into convergent power series in g.

For convenience we shall in the following omit the trivial spatial dependence of $\mathbf{S}'(x)$ and denote its Fourier transform by $S'(p_0)$. The vertex function $\Gamma(p_0,p_0')$ is related to $S'(p_0)$ by[8]

$$\Gamma(p_0,p_0')(p_0-p_0')=[S'(p_0)]^{-1}-[S'(p_0')]^{-1}. \tag{A6}$$

Following Dyson's notation,[1] we write

$$\Gamma=Z_1^{-1}\Gamma_c. \tag{A7}$$

If we use (A3), Γ_c would be free from a divergent sum if

$$Z_1=Z_2. \tag{A8}$$

Thus, no coupling-constant renormalization is needed in this case and Z_2 is actually ∞^{-1} for any finite value of g.

APPENDIX II

There exist some interesting relations between the renormalization quantities Z_1, Z_2, etc., and the ordinary matrix elements. In this section we shall illustrate the proof of these relations by considering the problem of the charged scalar meson field with a fixed nucleon. For simplicity, we shall set the nucleon at the origin. The Hamiltonian is

$$H=(m+\delta m)\psi^\dagger\psi+\tfrac{1}{2}\int\sum_{i=1}^{2}[\pi_i^2+(\nabla\varphi_i)^2+\mu^2\varphi_i^2]d\tau+(g/\sqrt{2})\sum_{i=1}^{2}\psi^\dagger\tau_i\psi\varphi_i(0), \tag{A9}$$

where

$$\psi=\begin{pmatrix}\psi_P(t)\\\psi_N(t)\end{pmatrix}$$

describes the nucleon field and τ_1, τ_2 are the Pauli matrices.

By applying the power-series renormalization method, one can prove that this problem can be renormalized by both a mass renormalization and a coupling-constant

[6] Some of the results obtained in the appendices have been obtained by other authors: M. Gell-Mann and F. Low (private communication); S. Edwards and R. Peierls (to be published).

[7] M. Gell-Mann and F. Low, Phys. Rev. 84, 350 (1951).

[8] This can be obtained by generalizing the arguments used in proving Ward's identity. The author wishes to thank Dr. C. N. Yang for informing him of this generalization.

renormalization. The nucleon propagation function $S'(p_0)$ and the vertex function $\Gamma(p_0,p_0')$ can be written as

$$S'(p_0)=Z_2S_c(p_0), \tag{A10}$$

$$\Gamma(p_0,p_0')=Z_1^{-1}\Gamma_c(p_0,p_0'). \tag{A11}$$

The quantities Z_2 and Z_1 are determined by the conditions

$$(p_0-m)S_c(p_0)=1 \quad \text{when} \quad p_0=m, \tag{A12}$$

and

$$\Gamma_c(m,m)=1, \tag{A13}$$

where m is the physical mass of the nucleon. The renormalized coupling constant g_c is connected to the unrenormalized coupling constant g by

$$gZ_2Z_1^{-1}=g_c. \tag{A14}$$

S_c and Γ_c are then finite functions of g_c and p_0.

Let $|\mathbf{P}\rangle$ (or $|\mathbf{N}\rangle$) and $|P\rangle$ (or $|N\rangle$) represent one physical proton (neutron) state and one bare proton (neutron) state, respectively. Z_2 and Z_1 are related to these state vectors through the following theorems.

Theorem 1.

$$Z_2=|\langle P|\mathbf{P}\rangle|^2=|\langle N|\mathbf{N}\rangle|^2, \tag{A15}$$

Theorem 2.

$$Z_2Z_1^{-1}=\langle\mathbf{N}|\tau_-|\mathbf{P}\rangle, \tag{A16}$$

where τ_- is the operator $\frac{1}{2}(\tau_1-i\tau_2)$ which transforms a bare proton into a bare neutron.

(i) Proof of Theorem 1. The propagation function of a proton (or neutron) in time can be written as[7]

$$\mathbf{S}'(t_1-t_2)=T\langle\text{vac}|\boldsymbol{\psi}_P(t_1)\boldsymbol{\psi}_P^\dagger(t_2)|\text{vac}\rangle$$

in the Heisenberg representation. For convenience one chooses $t=0$ as the time when the Heisenberg representation coincides with the interaction representation; i.e.,

$$\boldsymbol{\psi}_P(t=0)=\psi_P(t=0), \tag{A17}$$

where $\boldsymbol{\psi}_P$ is in the Heisenberg representation and ψ_P in the interation representation.

$\mathbf{S}'(t_1-t_2)$ can be rewritten as

$$\mathbf{S}'(t_1-t_2)=\sum_n\langle\text{vac}|\boldsymbol{\psi}_P(0)|n\rangle\langle n|\boldsymbol{\psi}_P^\dagger(0)|\text{vac}\rangle \times\exp[-iE_n(t_1-t_2)] \text{ if } t_1>t_2, \tag{A18}$$

where $|n\rangle$ is eigenstate of the total Hamiltonian with eigenvalue E_n ($E_{\text{vac}}=0$). Thus,[9]

$$\underset{\substack{t_2\to-\infty\\ t_1\to+\infty}}{\mathfrak{L}} \exp[im(t_1-t_2)]\mathbf{S}'(t_1-t_2)=|\langle\text{vac}|\psi_P(0)|\mathbf{P}\rangle|^2.$$

By using (A12) and noticing that $\mathbf{S}'(t)$ is the Fourier transform of $S'(p_0)$, one has

$$Z_2=|\langle\mathbf{P}|P\rangle|^2.$$

(ii) To prove Theorem 2, we consider the quantity A_k defined by

$$A_k=T\langle\text{vac}|\boldsymbol{\psi}_N(t_1)\boldsymbol{\alpha}_k(t_3)\boldsymbol{\psi}_P^\dagger(t_2)|\text{vac}\rangle, \tag{A19}$$

where all operators are in the Heisenberg representation. We denote by $\alpha_k(t)$ the annihilation operator of a positive meson in the interaction representation and by $\boldsymbol{\alpha}_k(t)$ the corresponding operator in the Heisenberg representation. Similarly to (A17), we have

$$\boldsymbol{\alpha}_k(t=0)=\alpha_k(t=0). \tag{A17'}$$

By making similar arguments to those used by Gell-Mann and Low[7] in proving (A2), one can prove that

$$A_k=g\int\mathbf{S}'(t_1-t)\boldsymbol{\Gamma}(t-t'',t'-t'') \times\mathbf{S}'(t'-t_2)f(t''-t_3)dtdt'dt'', \tag{A20}$$

where $\boldsymbol{\Gamma}$, $\mathbf{S}'$ are Fourier transforms of Γ and S'. The function $f(t''-t_3)$ represents the retarded part of the propagation of a positive meson with wave number k from time t'' to time t_3. It can be written as

$$f(t''-t_3)=\frac{1}{2\pi}\int\frac{\exp[ip_0(t''-t_3)]}{(p_0-\omega+i\epsilon)(2\omega\Omega)^{\frac{1}{2}}}dp_0,$$

where Ω is the volume of the system. The factor $(2\omega\Omega)^{\frac{1}{2}}$ arises because we use α_k in the definition for A_k. Equation (A20) thus represents the totality of all diagrams in which a nucleon propagates from t_2 to t' and then a vertex part from t' to t with an emission of a positive meson at t''. The positive meson, then, propagates from t'' to t_3 and the nucleon from t to t_1.

As in the previous proof, we wish to examine the part of A_k that oscillates like $\exp[-im(t_1-t_2)]$ as $t_1\to+\infty$ and $t_2\to-\infty$. This can be achieved by two different ways. On the one hand, we can express the definition of A_k (A19) in terms of the eigenstates $|n\rangle$ of the total Hamiltonian. A_k can then be written as

$$A_k=\sum_{n,m}\langle\text{vac}|\boldsymbol{\psi}_N(0)|n\rangle\langle n|\boldsymbol{\alpha}_k(0)|m\rangle\langle m|\boldsymbol{\psi}_P^\dagger(0)|\text{vac}\rangle \times\exp[-iE_n(t_1-t_3)]\exp[-iE_m(t_3-t_2)]; \text{ if } t_1>t_3>t_2.$$

Using (A17) and (A15), we have

$$\underset{\substack{t_1\to+\infty\\ t_2\to-\infty}}{\mathfrak{L}} \exp[im(t_1-t_2)]A_k=Z_2\langle\mathbf{N}|\alpha_k(0)|\mathbf{P}\rangle. \tag{A21}$$

On the other hand, by (A18) and the definitions of $\mathbf{S}'$ and $\boldsymbol{\Gamma}$, one has

$$\underset{t_1\to+\infty}{\mathfrak{L}} \exp(imt_1)\mathbf{S}'(t_1-t)=Z_2\exp(imt),$$

$$\underset{t_2\to-\infty}{\mathfrak{L}} \exp(-imt_2)\mathbf{S}'(t'-t_2)=Z_2\exp(-imt'),$$

---

[9] $\mathfrak{L}_{t\to\pm\infty}$ represents an operator of the type $\lim_{t\to\pm\infty} t^{-1}\int_t^{2t}dt$ such that all oscillating terms can be taken to be zero.

and

$$\Gamma(m,m)=\int \Gamma(t,t')\exp[im(t-t')]dtdt'.$$

By using (A20), the same limit becomes

$$\mathop{\mathcal{L}}_{\substack{t_1\to+\infty\\ t_2\to-\infty}} \exp[im(t_1-t_2)]A_k=gZ_2{}^2\Gamma(m,m)(-\omega)^{-1}(2\omega\Omega)^{-\frac{1}{2}}, \tag{A22}$$

which, together with (A21), gives

$$Z_2\Gamma(m,m)=g^{-1}(-\omega)(2\omega\Omega)^{\frac{1}{2}}\langle\mathbf{N}|\alpha_k(0)|\mathbf{P}\rangle. \tag{A23}$$

The left-hand side of (A23) is the ratio of the renormalization constants, Z_2/Z_1. The k dependence of the right-hand side can be eliminated by using the identity:

$$\langle\mathbf{N}|[\alpha_k,H]|\mathbf{P}\rangle=0.$$

By commuting α_k with the total Hamiltonian, we have

$$\langle\mathbf{N}|\alpha_k|\mathbf{P}\rangle=-g\omega^{-1}(2\omega\Omega)^{-\frac{1}{2}}\langle\mathbf{N}|\tau_-|\mathbf{P}\rangle. \tag{A24}$$

By comparing (A24) with (A23), we can express the coupling constant renormalization (A14) as

$$g_c/g=Z_2/Z_1=\langle\mathbf{N}|\tau_-|\mathbf{P}\rangle.$$

Thus g_c/g, if real, must be less than unity.

These proofs can be obviously generalized to other renormalizable field theoretical problems. In the charged scalar theory these identities can be applied to calculate formally the values of Z_1 and Z_2 by using both the weak-coupling and strong-coupling solutions.[10] They are:

(i) weak-coupling solution:

$$Z_2=1-g_c{}^2\sum(2\omega^3\Omega)^{-1}+\cdots,$$
$$Z_2/Z_1=1-g_c{}^2\sum(2\omega^3\Omega)^{-1}+\cdots;$$

(ii) strong-coupling solution:

$$Z_2=\tfrac{1}{2}\exp[-g_c{}^2\sum(2\omega^3\Omega)^{-1}]+\cdots,\quad Z_2/Z_1=\tfrac{1}{2}.$$

[10] G. Wentzel, Helv. Phys. Acta **13**, 269 (1940); **14**, 633 (1941); R. Serber and S. Dancoff, Phys. Rev. **63**, 143 (1943); S. Tomonaga, Progr. Theoret. Phys. (Japan) **1**, 109 (1946).

Reprinted from the Proceedings of the NATIONAL ACADEMY OF SCIENCES
Vol. 49, No. 2, pp. 179–186. February, 1963.

SPIN 1/2 WAVE EQUATION IN DE-SITTER SPACE*

BY F. GÜRSEY AND T. D. LEE†

MIDDLE EAST TECHNICAL UNIVERSITY, ANKARA, TURKEY, AND COLUMBIA UNIVERSITY, NEW YORK

Communicated by Robert Serber, November 16, 1962

1. *Introduction.*—The de-Sitter space can be represented by a 4-dimensional pseudo-sphere embedded in a 5-dimensional space. Let $\xi_\mu(\mu = 1, 2, \ldots, 5)$ be the five real coordinates of the 5-dimensional space. The de-Sitter space is described by the surface

$$\xi^\mu \xi_\mu = \mathrm{R}^2, \tag{1}$$

where

$$\xi^\mu = \eta^{\mu\nu}\xi_\nu. \tag{2}$$

$\eta^{\mu\nu}$ is the (μ, ν)th matrix element of the diagonal matrix

$$\eta = \begin{pmatrix} 1 & & & & \\ & 1 & & & \\ & & 1 & & \\ & & & -1 & \\ & & & & 1 \end{pmatrix}, \tag{3}$$

and the real constant R is the radius of the pseudo-sphere. The totality of all homogeneous linear transformations among ξ^μ that leave the surface (1) invariant forms a ten-parameter group. In the limit $\mathrm{R} \to \infty$, the de-Sitter space becomes simply the usual flat 4-dimensional space, and this ten-parameter group reduces to that of the inhomogeneous Lorentz transformation.

Much of our present knowledge of particle physics are simple consequences of invariance requirements under certain group transformations such as Lorentz transformations, isotopic rotations, etc. It seems, therefore, useful to examine the physical consequences by demanding that the field equations of elementary particles be invariant under the ten-parameter transformations in de-Sitter space.

This problem has been studied by Dirac.[1] His equation for a spin $^1/_2$ particle in the de-Sitter space is given by[2]

$$(^1/_2)\gamma^\mu\gamma^\nu L_{\mu\nu}\chi = \mathrm{R}\lambda\chi, \tag{4}$$

where χ is the wave function,

$$L_{\mu\nu} = -i\left(\xi_\mu \frac{\partial}{\partial\xi^\nu} - \xi_\nu \frac{\partial}{\partial\xi_\mu}\right) \tag{5}$$

and γ^1, γ^2, γ^3, $i\gamma^4$, γ^5 are five *constant* (4×4) Hermitian matrices that satisfy

$$\gamma^\mu\gamma^\nu + \gamma^\nu\gamma^\mu = 2\eta^{\mu\nu}. \tag{6}$$

In the limit $\mathrm{R} \rightarrow \infty$, equation (4) becomes the usual Dirac equation in a flat space and λ the mass of this spin $^1/_2$ particle. However, Dirac found that for a finite R the eigenvalue λ must be a complex number with a small imaginary part:

$$\mathrm{Im}(\lambda) = -(2/\mathrm{R}). \tag{7}$$

The physical meaning of this nonzero imaginary part does not appear to be completely clear. The fact that $|\lambda| \geqq (2/\mathrm{R})$ makes, e.g., the equation of a neutrino in the de-Sitter space seem (at least at first sight) somewhat ambiguous, since any equation with $\lambda = 0(\mathrm{R}^{-1})$ would, in the limit of infinite R, approach that of a particle with zero mass.

In this paper, the same problem is investigated from a different approach, by starting from an invariant Lagrangian. The resulting equation differs in form from that obtained by Dirac. With this new form, certain aspects of the physical consequences of invariance in the de-Sitter space become more apparent. For example, the generalization of a two-component theory of the neutrino seems to appear in a natural way. The relation between this new form and equation (4) is discussed.

2. *The Lagrangian.*—We start from the Lagrangian

$$\mathcal{L} = i\int_D\left[\frac{1}{2}\left(\bar{\psi}\gamma^\mu \frac{\partial\psi}{\partial\xi^\mu} - \frac{\partial\bar{\psi}}{\partial\xi^\mu}\gamma^\mu\psi\right) + m\bar{\psi}\psi\right]\prod_1^5 d\xi^\nu, \tag{8}$$

where ψ is a (4×1) column matrix describing the spin $^1/_2$ particle, $\bar{\psi}$ is related to its hermitian conjugate $\psi\dagger$ by

$$\bar{\psi} = \psi\dagger\gamma^4, \tag{9}$$

and the mass of the particle is given by the real number m. In (8), the integration extends over the region D which is bounded by $\rho = \mathrm{R}$

and $$\rho = \mathrm{R} + \epsilon, \tag{10}$$

where $$\rho^2 = \xi^\mu\xi_\mu \tag{11}$$

and ϵ is a positive infinitesimal quantity. As $\epsilon \rightarrow 0$, D becomes simply the de-Sitter space.

From the definition (8), it is clear that the Hermitian operator $\mathfrak{L}$ is invariant under the ten-parameter transformations in the de-Sitter space. The equation for a spin $^1/_2$ particle is now given by

$$\delta \mathfrak{L} = 0 \tag{12}$$

for arbitrary small variation $\delta\psi$ of the wave function provided that

$$\psi \text{ is independent of } \rho. \tag{13}$$

3. *The Wave Equation.*—Without loss of generality, we introduce the curvilinear coordinates x^μ, which are related to the ξ^μ by

$$\xi^\mu = (x^5/\mathrm{R}) f^\mu(x^1, x^2, x^3, x^4), \tag{14}$$

where the five f^μ satisfy the condition[2]

$$f^\mu f_\mu = \mathrm{R}^2 \tag{15}$$

but are otherwise completely arbitrary functions of the four variables x^1, x^2, x^3, and x^4. From (15), it follows that

$$x^5 = \rho. \tag{16}$$

The line element in the five-dimensional space is given by[2]

$$(ds)^2 = d\xi^\mu d\xi_\mu = g_{ij} dx^i dx^j + (d\rho)^2, \tag{17}$$

where

$$g_{ij} = \left(\frac{\partial \xi^\mu}{\partial x^i}\right)\left(\frac{\partial \xi_\mu}{\partial x^j}\right) \tag{18}$$

and the part $g_{ij}dx^i dx^j$ determines its component in the de-Sitter space. It is useful to introduce the five matrices

$$\beta^\mu \equiv \left(\frac{\partial x^\mu}{\partial \xi^\nu}\right)\gamma^\nu. \tag{19}$$

By using (6), (14), and (16), we find

$$\beta^5 = (\xi_\mu \gamma^\mu)/\rho, \tag{20}$$

$$\beta^5\beta^i + \beta^i\beta^5 = 0 \tag{21}$$

and

$$\beta^i\beta^j + \beta^j\beta^i = g^{ij}, \tag{22}$$

where

$$g^{ij} = (\partial x^i/\partial \xi^\mu)(\partial x^j/\partial \xi_\mu). \tag{23}$$

In terms of x^i and β^i, the Lagrangian (8) becomes

$$\mathfrak{L} = i\epsilon \int \left[\frac{1}{2}\left(\bar{\psi}\beta^i \frac{\partial \psi}{\partial x^i} - \frac{\partial \bar{\psi}}{\partial x^i}\beta^i\psi\right) + m\bar{\psi}\psi\right](-g)^{1/2} \prod_1^4 dx^i, \tag{24}$$

where

$$g = [\text{determinant } |g_{ij}|\,] \times \left(\frac{\mathrm{R}}{\rho}\right)^8 \tag{25}$$

and ψ is a spinor function depending only on x^i. The variational derivative $(\delta\mathcal{L}/\delta\psi)$ can now be easily carried out. By using the identity (proved in Appendix 1)

$$\frac{\partial}{\partial x^i}\left[(-g)^{1/2}\frac{\partial x^i}{\partial \xi^\mu}\right] = -\frac{4}{\rho^2}\,\xi_\mu\,(-g)^{1/2}, \tag{26}$$

we find the wave equation for a spin $^1/_2$ particle in the de-Sitter space to be

$$\left(\beta^i\frac{\partial}{\partial x^i} - \frac{2}{\rho}\beta^5\right)\psi + m\psi = 0 \tag{27}$$

evaluated at $\rho = \mathrm{R}$.

The conjugate equation of (27) is given by

$$\left(\frac{\partial\bar{\psi}}{\partial x^i}\beta^i - \frac{2}{\rho}\bar{\psi}\beta^5\right) - m\bar{\psi} = 0. \tag{28}$$

4. *Discussion.*—(1) The invariance property of (27) under the ten-parameter transformations in the de-Sitter space can be easily inferred from the structure of $\mathcal{L}$. It can also be verified by establishing the following commutation relations:

$$\left[J_{u\nu},\,\beta^i\frac{\partial}{\partial x^i}\right] = 0 \tag{29}$$

and

$$[J_{u\nu},\,\beta^5] = 0, \tag{30}$$

where the ten operators,

$$J_{u\nu} \equiv L_{v\nu} - \frac{1}{2}i\gamma_u\gamma_\nu, \tag{31}$$

are the generators of the group of transformations in the de-Sitter space.

(2) By using (26), (27), and (28), it can be readily seen that the current conservation law is given by

$$\frac{\partial}{\partial x^i}\left[(-g)^{1/2}\bar{\psi}\beta^i\psi\right] = 0. \tag{32}$$

(3) For particles of zero mass, we have $m = 0$ and

$$\left(\beta^i\frac{\partial}{\partial x^i} - \frac{2}{\mathrm{R}}\beta^5\right)\psi = 0. \tag{33}$$

Utilizing the anti-commutation relation

$$\beta^5\left(\beta^i\frac{\partial}{\partial x^i} - \frac{2}{\rho}\beta^5\right) + \left(\beta^i\frac{\partial}{\partial x^i} - \frac{2}{\rho}\beta^5\right)\beta^5 = 0, \tag{34}$$

the four-component field ψ can be decomposed into two uncoupled two-component fields $(1 + \beta^5)\psi$ and $(1 - \beta^5)\psi$, each of which can be used as the equation for a two-component neutrino in the de-Sitter space.

(4) To discuss the connection between the de-Sitter space and the usual flat 4-space, it is necessary only to study the behavior of the wave equation within a

small region in the de-Sitter space. Since all points are equivalent, we may choose this region to be the neighborhood of

$$(\xi^1, \xi^2, \ldots, \xi^5) = (0, 0, 0, 0, R). \tag{35}$$

By setting in (14),

$$\rho = R,$$
$$\xi^i = x^i \tag{36}$$

and neglecting $0(x^i/R)$, we find

$$\beta^\mu = \gamma^\mu. \tag{37}$$

Therefore, all the above equations reduce to the usual ones in the flat space.

It will be shown in the subsequent section that the effects of de-Sitter space actually show up only if terms of the order of R^{-2} [not $0(R^{-1})$] are kept.

5. *Local Coordinate System (Vierbein).*—To understand more fully the behavior of the wave equation (27), it is convenient to introduce the local coordinate system. Without specifying the particular choice of our curvilinear coordinates x^i, we can choose at every point in the de-Sitter space a local system which is determined by four base vectors $e_a{}^i$ ($a = 1, 2, 3, 4$), where[2]

$$e_a{}^i e_b{}^j g_{ij} = \eta_{ab} \tag{38}$$

and η_{ab} is the (a, b)th matrix element of the matrix η given by (3). The metric of the local system is, therefore, simply given by η_{ab}.

It is useful to define

$$\alpha^i \equiv e_a{}^i \gamma^a. \tag{39}$$

By using (6), (22), and (23), we find that the matrices α^i satisfy the same anti-commutation relations as that between the β^i; i. e.

$$\alpha^i\alpha^j + \alpha^j\alpha^i = 2g^{ij}. \tag{40}$$

Therefore, there exists a transformation matrix S which relates the α^i with the β^i:

$$\alpha^i = S\beta^i S^{-1}. \tag{41}$$

It is shown in Appendix 2 that under the S transformation the wave equation (27) becomes

$$\alpha^i\left(\frac{\partial}{\partial x^i} + \Gamma_i\right)\phi + m\phi = 0, \tag{42}$$

where

$$\phi = S\psi, \tag{43}$$

$$\alpha^i\Gamma_j - \Gamma_j\alpha^i = \left(\frac{\partial \alpha^i}{\partial x^j}\right) + \Gamma^i{}_{jk}\alpha^k, \tag{44}$$

and $\Gamma^i{}_{jk}$ is the usual Christoffel symbol given by

$$\Gamma^i{}_{jk} = \frac{1}{2} g^{ia}\left(\frac{\partial g_{ja}}{\partial x^k} + \frac{\partial g_{ak}}{\partial x^j} - \frac{\partial g_{jk}}{\partial x^a}\right). \tag{45}$$

The above equation (42) is exactly the same as that given by the method[3, 4] of covariant derivative. It must be emphasized that by using covariant derivatives, the spin $^1/_2$ equation in a space of arbitrary metric can be easily written down. However, in such a derivation there is no requirement that the resulting equation should satisfy any invariance condition. Under a transformation of the coordinate system, the method of covariant derivative merely gives a compact way to transform the equation to the new system. Thus, as a by-product, we also establish the interesting result that, by applying the usual method of covariant derivative to the de-Sitter space, the resulting equation *automatically* satisfies the invariance requirement under the ten-parameter group of transformations.

It is interesting to notice that the matrix β^5 is related to the constant matrix γ^5 by

$$S\beta^5 S^{-1} = \gamma^5. \tag{46}$$

If the mass $m = 0$, then both $(1 + \gamma^5)\phi$ and $(1 - \gamma^5)\phi$ are solutions of (42). In the local coordinate system, the reduction from four to two components proceeds exactly in the same way as that in the flat space.

A particularly convenient and symmetrical coordinate system is given by

$$\xi^i = (x^5/\mathrm{R})(1 + s)^{-1}x^i$$

and

$$\xi^5 = x^5(1 + s)^{-1}(1 - s), \tag{47}$$

where

$$s = (4\mathrm{R}^2)^{-1}x^i x^j \eta_{ij}. \tag{48}$$

In the de-Sitter space,

$$x^5 = \mathrm{R}.$$

Therefore, we have

$$g_{ij} = (1 + s)^{-2}\eta_{ij} \tag{49}$$

and

$$\Gamma^i_{\ jk} = [2(1 + s)\mathrm{R}^2]^{-1}[x^i\eta_{jk} - \delta_j^{\ i}x_k - \delta_k^{\ i}x_j], \tag{50}$$

where $\delta_j^{\ i} = 1$ if $j = i$ and 0 otherwise. In this coordinate system, the wave equation (42) can be written in a particularly simple form in terms of the constant matrices γ^i:

$$(1 + s)\gamma^i \frac{\partial \phi}{\partial x^i} - \frac{3}{4}\mathrm{R}^{-2}(x^i\gamma^j\eta_{ij})\phi + m\phi = 0. \tag{51}$$

It is clear that neglecting terms proportional to R^{-2}, (51) is identical with the usual Dirac equation in the flat space.

6. *Dirac's Form.*—The remaining task is to establish the relation between our basic equation (27) and the equation obtained by Dirac. By using (13) and the identities

$$\beta^5(\gamma^\mu\gamma^\nu L_{\mu\nu}) = 2\rho\gamma^\mu \frac{\partial}{\partial \xi^\mu} - 2\beta^5\rho\frac{\partial}{\partial \rho} \tag{52}$$

and

$$(\gamma^\mu\gamma^\nu L_{\mu\nu})\beta^5 + \beta^5(\gamma^\mu\gamma^\nu L_{\mu\nu}) = 8\beta^5, \tag{53}$$

it can be readily shown that the solution ψ of (27) is related to the solution χ of

$$\frac{1}{2}(\gamma^\mu\gamma^\nu L_{uv})\chi = \mathrm{R}\lambda\chi \tag{54}$$

by
$$\chi = (1 \pm i\beta^5)\psi \tag{55}$$
and
$$\lambda = \pm m - i(2/\mathrm{R}). \tag{56}$$

Therefore, the mass of the particle is determined only by the real part of λ. The imaginary part of λ has no direct physical meaning; it is merely a result of the fact that $\gamma^\mu\gamma^\nu L_{\mu\nu}$ does not satisfy the hermiticity condition.

Appendix 1.—To prove (26), we first establish the following equations:

(1) By differentiating

$$(\partial\xi/^\nu\partial x^\alpha)(\partial x^i/\partial\xi^\nu) = \delta_\alpha{}^i \tag{A.1}$$

with respect to x^i and by using

$$\left(\frac{\partial x^5}{\partial\xi^\mu}\right)\left(\frac{\partial x^i}{\partial\xi^\nu}\right)\left[\frac{\partial}{\partial x^i}\left(\frac{\partial\xi^\nu}{\partial x^5}\right)\right] = \frac{4\xi_\mu}{\rho^2}, \tag{A.2}$$

we find

$$\frac{\partial}{\partial x^i}\left(\frac{\partial x^i}{\partial\xi^\mu}\right) = -\left(\frac{\partial x^j}{\partial\xi^\mu}\right)\left(\frac{\partial x^i}{\partial\xi^\nu}\right)\left[\frac{\partial}{\partial x^i}\left(\frac{\partial\xi^\nu}{\partial x^j}\right)\right] - \left(\frac{4\xi_\mu}{\rho^2}\right). \tag{A.3}$$

(2) From

$$-\frac{\partial g}{\partial x^i} = -gg^{jk}\left(\frac{\partial g_{jk}}{\partial x^i}\right), \tag{A.4}$$

$$\left(\frac{\partial x^i}{\partial\xi^\mu}\right)\left(\frac{\partial\xi^\nu}{\partial x^i}\right) = \delta_\mu{}^\nu - \left(\frac{\xi_\mu\xi^\nu}{\rho^2}\right). \tag{A.5}$$

and

$$\xi^\mu\left(\frac{\partial x^i}{\partial\xi^\mu}\right) = 0, \tag{A.6}$$

it follows that

$$\frac{\partial}{\partial x^i}(-g)^{1/2} = (-g)^{1/2}\frac{\partial x^j}{\partial\xi^\nu}\frac{\partial}{\partial x^i}\left(\frac{\partial\xi^\nu}{\partial x^j}\right). \tag{A.7}$$

Equation (26) can be readily obtained by using (A.3) and (A.7).

Appendix 2.—We notice that

$$\alpha^i\gamma^5 + \gamma^5\alpha^i = 0.$$

Therefore, under the S-transformation,

$$S\beta^5S^{-1} = \gamma^5, \tag{A.8}$$

and equation (27) becomes

$$\alpha^i\frac{\partial}{\partial x^i}\phi + \alpha^i\Lambda_i\phi - \frac{2\gamma^5}{\mathrm{R}}\phi + m\phi = 0, \tag{A.9}$$

where
$$\Lambda_i = -\left(\frac{\partial S}{\partial x^i}\right)S^{-1} = S\left(\frac{\partial S^{-1}}{\partial x^i}\right). \tag{A.10}$$

It is useful to introduce the reciprocal base vectors $\epsilon_i{}^a$ such that

$$e_a{}^i\epsilon_j{}^a = \delta_j{}^i$$

and
$$e_a{}^i \epsilon_i{}^b = \delta_a{}^b. \tag{A.11}$$

Thus, we have
$$S^{-1}\gamma^a S = \epsilon_i{}^a \beta^i \tag{A.12}$$

and
$$S\gamma^\mu S^{-1} = \frac{\partial \xi^\mu}{\partial x^j}\alpha^j + \left(\frac{\xi^\mu}{\mathrm{R}}\right)\gamma^5. \tag{A.13}$$

Upon differentiating (A.12), we find

$$\alpha^j \Lambda_i - \Lambda_i \alpha^j = \frac{\partial \alpha^j}{\partial x^i} - \alpha^k\left(\frac{\partial \xi^\mu}{\partial x^k}\right)\left[\frac{\partial}{\partial x^i}\left(\frac{\partial x^j}{\partial \xi^\mu}\right)\right] - \left(\frac{\xi^\mu}{\mathrm{R}}\right)\left[\frac{\partial}{\partial x^i}\left(\frac{\partial x^j}{\partial \xi^\mu}\right)\right]\gamma^5. \tag{A.14}$$

Let us define

$$\Delta_i \equiv \frac{1}{2}\epsilon_j{}^a \eta_{ab}\left(\frac{\xi^\mu}{\mathrm{R}}\right)\left[\frac{\partial}{\partial x^i}\left(\frac{\partial x^j}{\partial \xi^\mu}\right)\right]\gamma^5\gamma^b \tag{A.15}$$

and
$$\Gamma_i \equiv \Lambda_i - \Delta_i.$$

The commutation between Γ_i and α^j is given by

$$\alpha^j \Gamma_i - \Gamma_i \alpha^j = \left(\frac{\partial \alpha^j}{\partial x^i}\right) - \alpha^k\left(\frac{\partial \xi^\mu}{\partial x^k}\right)\left[\frac{\partial}{\partial x^i}\left(\frac{\partial x^j}{\partial \xi^\mu}\right)\right]. \tag{A.16}$$

Equation (42) now follows by showing that

$$\alpha^i \Delta_i = (2\gamma^5/\mathrm{R}) \tag{A.17}$$

and
$$\Gamma^i{}_{jk} = -\left(\frac{\partial \xi^\mu}{\partial x^k}\right)\left[\frac{\partial}{\partial x^i}\left(\frac{\partial x^j}{\partial \xi^\mu}\right)\right]. \tag{A.18}$$

* This work was supported in part by the U. S. Atomic Energy Commission.

† On leave of absence from the Institute for Advanced Study, Princeton, New Jersey.

[1] Dirac, P. A. M., *Ann. Math.*, **36**, 657 (1935).

[2] Throughout this paper the natural units $\hbar = c = 1$ are used. All Greek indices such as $\mu, \nu, \ldots$ vary from 1 to 5, all Roman indices such as a, b, i, j, k, . . . vary from 1 to 4, and all repeated indices are to be summed over. In the 5-dimensional space, a covariant five-vector, say f_μ, is related to the contravariant five-vector f^μ, by $f^\mu = \eta^{\mu\nu} f_\nu$.

[3] See, e.g., Bade, W. L., and H. Jehle, *Rev. Mod. Phys.*, **25**, 714 (1953) for a review of the literature on covariant derivatives.

[4] Brill, D. R., and J. A. Wheeler, *Rev. Mod. Phys.*, **29**, 465 (1957), have discussed the motion of a neutrino in an arbitrary gravitational field by using the method of covariant derivative.

Reprinted from THE PHYSICAL REVIEW, Vol. 133, No. 6B, B1549–B1562, 23 March 1964
Printed in U. S. A.

Degenerate Systems and Mass Singularities*

T. D. LEE† AND M. NAUENBERG‡
Columbia University, New York, New York
(Received 24 October 1963)

For a system with degenerate energies, the power series expansions of the S-matrix elements may become singular. An elementary theorem in quantum mechanics is proved which shows that under certain general conditions such singularities do not appear in the power series expansions of the transition probabilities, provided these are averaged over an appropriate ensemble of degenerate states. Application of this theorem leads to the cancellations of mass singularities and infrared divergences in quantum electrodynamics. The question of whether a charged particle can have zero mass is studied.

I. INTRODUCTION

IN many cases it has been observed that the perturbation series expansion of the transition probabilities for a degenerate system often exhibits infinities which, however, can be cancelled by averaging over an appropriate ensemble of states. The well-known problem of infrared divergence[1,2] in electromagnetic theory is one such example. Another example is given by Kinoshita and Sirlin[3] in their calculation of the lowest order radiative correction to muon decay (or other decays through weak interactions). If the mass of the electron m_e is set mathematically to be zero, the partial decay rates of the muon contain ($ln\ m_e$) singularities, but the total decay rate remains finite. By using the detailed properties of Feynman graphs, Kinoshita[4] has also investigated the cancellations of such "mass singularities" for higher order diagrams.

As we shall show, the occurrence of such singularities and their cancellations are consequences of an elementary theorem in quantum mechanics which can be established without any explicit use of Feynman graphs, nor even the explicit form of the Hamiltonian.

Let us consider an arbitrary Hamiltonian (H_0+gH_1) which can be diagonalized by a unitary matrix U.

$$U^\dagger(H_0+gH_1)U=E\,, \tag{1}$$

where H_0 and E are both diagonal matrices and g is the interaction coupling constant. If the problem contains a continuum then $U=U_-$ or U_+ depending on whether incoming or outgoing scattered waves are used. The S matrix is given by

$$S=U_-^\dagger U_+\,, \tag{2}$$

where † indicates Hermitian conjugation. The corresponding transition probability from a state a to a state b is given by

$$\sum_{i,j}[(U_-)_{ib}^*(U_-)_{jb}][(U_+)_{ia}(U_+)_{ja}^*]\,. \tag{3}$$

For clarity, we assume the problem contains a certain parameter μ and the degeneracy occurs in the total Hamiltonian only when $\mu\rightarrow 0$. For $\mu\neq 0$, the (i,j)th matrix element of $U_\pm$ can be expanded in the familiar power series in g.

$$(U_\pm)_{ij}=\delta_{ij}+g(E_j-E_i\pm i\alpha)^{-1}(1-\delta_{ij})(H_1)_{ij}+0(g^2)\,, \tag{4}$$

where δ_{ij} is the matrix element of a unit matrix, α is a positive infinitesimal quantity, and E_i is the ith diagonal element of the matrix E. Furthermore, we assume that each term in the power series expansion is finite if there is no degeneracy. As the parameter $\mu\rightarrow 0$, the state of energy E_i becomes degenerate with other states which lie within a certain subset $D(E_i)$. Therefore, if some of the states i, j, a (or b) in (3) are in the same degenerate set, the power series expansion of the corresponding transition probability would contain infinites in the limit $\mu=0$. On the other hand, such infinities can be completely cancelled if we consider the power series expansion of the sum

$$\sum_{D(E_a)} U_{ia}U_{ja}^*\equiv T_{ij}(E_a)\,, \tag{5}$$

where the summation extends over all states a in the same degenerate subset $D(E_a)$ and U can be either U_+ or U_-. This can be easily verified by using (4) and neglecting second or higher order terms in g. In an equally elementary way, we shall establish in the following section a theorem which gives the general condition under which such cancellations can occur for every term in the power series expansion.

By applying this theorem to electrodynamics, we can derive the elimination of the "mass singularities" in the mathematical limit $m_e\rightarrow 0$ and the cancellations of the well-known infrared divergences. This will be done in Sec. III. The question of whether a spin-$\frac{1}{2}$ zero-mass particle can have an electric charge is discussed in the same section. It is shown that by altering the usual renormalization program and by limiting measurements only to the ensemble averages over the appropriate degenerate sets of both the initial and the

* This research was supported in part by the U. S. Atomic Energy Commission.
† Alfred P. Sloan Fellow.
‡ John S. Guggenheim Fellow.
[1] F. Bloch and H. Nordsieck, Phys. Rev. **52**. 54 (1937).
[2] For a recent article with an extensive bibliography see, for example, D. R. Yennie, S. C. Frautschi, and H. Suura, Ann. Phys. (N. Y.) **13**, 379 (1961).
[3] T. Kinoshita and A. Sirlin, Phys. Rev. **113**, 1652 (1959).
[4] T. Kinoshita, J. Math. Phys. **3**, 650 (1962); *Lectures in Theoretical Physics University of Colorado*, (Interscience Publishers, Inc., New York, 1961).

final states, finite results for any physical transition probabilities can be obtained in the power series expansion. An unusual feature is the necessity of including interference terms between certain graphs, some of which may contain disconnected parts. Another interesting result is that the limit $m_e \to 0$ does not correspond to a theory with two *uncoupled* two-component particles.

For clarity of presentation, a number of other related theorems are given in Appendices A, B, and C. The details of the cancellation of mass singularity for the bremsstrahlung process are given in Appendix D. In Appendix E, we include some further illustrations of the theorem for the case of a soluble model in field theory.

Throughout this paper the question of convergence of the power series is not discussed.

II. AN ELEMENTARY THEOREM

The general perturbation series of the U matrix can be derived by using the unitarity condition

$$U^\dagger U = 1 \tag{6}$$

and Eq. (1), which may be written as

$$[U,E] = (gH_1 + \Delta)U, \tag{7}$$

where

$$\Delta = H_0 - E \tag{8}$$

and U stands for *either* U_+ *or* U_-. The diagonal matrix Δ represents the negative of the energy shift introduced by the interaction gH_1. For example, in a field theory with no bound state, E is the free-particle Hamiltonian with physical masses, and H_0 is the same free-particle Hamiltonian but with bare masses.[5]

Let the formal power series expansions of Δ, U, T be

$$\Delta = \sum_1^\infty g^n \Delta_n , \tag{9}$$

$$U = \sum_0^\infty g^n U_n , \tag{10}$$

and

$$T(E_a) = \sum_0^\infty g^n T_n(E_a) , \tag{11}$$

where $[T(E_a)]_{ij}$ is defined by (5).

When the parameter $\mu \neq 0$, the eigenvalues E_i of the total Hamiltonian $(H_0 + gH_1)$ contain no degeneracy other than the usual continuum due to the infinite volume. The power series expansions of $U_\pm$ can be obtained by using (6) and (7). As $\mu \to 0$, degeneracy occurs, and these expansions may contain singular terms. The following theorem can be proved:

Theorem: If $\lim_{\mu\to 0}\Delta_n$ exists for all $n \leqq N$, then $\lim_{\mu\to 0}[T_n(E_a)]_{ij}$ exists for all $n \leqq (N+1)$ and for arbitrary states i and j.

Proof: We shall prove the theorem by induction. Assume that $\lim_{\mu\to 0}\Delta_n$ exists for all $n \leqq N$ and that

$$\lim_{\mu\to 0} [T_n(E_a)]_{ij} \quad \text{exists for all } i, j \text{ and } n \leqq M < (N+1) . \tag{12}$$

$[T_n(E_a)]$ is related to U_n by

$$[T_n(E_a)]_{ij} = \sum_{m=0}^{n} \sum_{D(E_a)} (U_m)_{ia} (U_{n-m})_{ja}^* , \tag{13}$$

where the first summation $\sum_{D(E_a)}$ extends over the subset $D(E_a)$ of all states a that are degenerate with E_a in the limit $\mu \to 0$. As discussed in the previous section, $\lim_{\mu\to 0}[T_1(E_a)]_{ij}$ always exists. Therefore, (12) holds for $M=1$. To show that $\lim_{\mu\to 0}[T_{M+1}(E_a)]_{ij}$ exists, we consider the following three cases:

(i) The state i lies outside the subset $D(E_a)$, and the state j may or may not lie outside $D(E_a)$. From (7), it follows that (since $E_i \neq E_a$):

$$(U_m)_{ia} = (E_a - E_i)^{-1} \left[\sum_k (H_1)_{ik} (U_{m-1})_{ka} + \sum_{l=1}^{m} (\Delta_l)_{ii} (U_{m-l})_{ia} \right] . \tag{14}$$

Substituting (14) into (13) and taking the limit $\mu \to 0$, we find for $n = M+1$

$$[T_{M+1}(E_a)]_{ij} = (E_a - E_i)^{-1} \left\{ \sum_k (H_1)_{ik} [T_M(E_a)]_{kj} + \sum_{l=1}^{M} (\Delta_l)_{ii} [T_{M+1-l}(E_a)]_{ij} \right\} , \tag{15}$$

where in the summation over l we have used the simple fact that $[T_0(E_a)]_{ij} = 0$ for the present case. Thus, $\lim_{\mu\to 0}[T_{M+1}(E_a)]_{ij}$ exists.

(ii) If the state j lies outside the degenerate set $D(E_a)$, but the state i may lie inside $D(E_a)$, the existence of $\lim_{\mu\to 0}[T_{M+1}(E_a)]_{ij}$ follows from the hermiticity relation

$$[T_{M+1}(E_a)]_{ij} = [T_{M+1}(E_a)]_{ji}^* \tag{16}$$

and (i).

(iii) If both i and j are within $D(E_a)$, then by using the unitarity of the $U_\pm$ matrix, we have (for $M \geq 0$)

$$[T_{M+1}(E_a)]_{ij} = - \sum_{m=0}^{M+1} \sum_b (U_m)_{ib} (U_{M+1-m})_{jb}^* , \tag{17}$$

where $\sum_b$ extends over all the states b not in $D(E_a)$. The

[5] Strictly speaking, there may also be a shift Δ_{vac} in the vacuum energy which, however, can be removed by considering, instead of (H_0+gH_1), the Hamiltonian $(H_0+gH_1-\Delta_{\text{vac}}\, I)$ where I is the unit matrix.

right-hand side can be written as a sum of $[T_{M+1}(E_a)]_{ij}$, where the states i and j are not degenerate with E_b. Case (iii) is then reduced to (i); therefore, $\lim_{\mu\to 0}[T_{M+1}(E_a)]_{ij}$ exists. The theorem is then proved by induction on M.

It is important to note that since $U_\pm$ is unitary, its matrix element cannot have a magnitude bigger than 1. Therefore, $\lim_{\mu\to 0}(U_\pm)_{ij}$ and $\lim_{\mu\to 0}[T(E_a)]_{ij}$ cannot be infinite. However, as $\mu\to 0$, infinities may occur in the power series expansion of $U_\pm$. The theorem states that such infinities do not occur in the power series of $[T(E_a)]_{ij}$.

Remarks

1. For a system that contains a continuum, the summation $\sum_{D(E_a)}$ in (13) represents the integration over all states that lie within the energy interval between $E_a-\epsilon$ and $E_a+\epsilon$, where $\epsilon\neq 0$ but can be chosen to be arbitrarily small.

2. Since $(\Delta_0)_{ii}=0$ and $(\Delta_1)_{ii}=-(H_1)_{ii}$, it follows that $\lim_{\mu\to 0}[T_n(E_a)]_{ij}$ exists for $n\leq 2$ provided $(H_1)_{ii}$ remains finite.

3. As stated above, the degenerate set $D(E_a)$ should contain all states whose energy is degenerate with E_a. In almost all problems, the theorem remains true if the subset $D(E_a)$ is substantially reduced. This can be most easily seen by considering a different problem in which H_1 is changed into H_1' where $(H'_1)_{jk}=(H_1)_{jk}$ if j and k lie in an arbitrarily chosen set S, and otherwise $(H_1')_{jk}=0$. Applying the theorem to this new problem we find the relevant degenerate set becomes the intersections $D\wedge S$ of the original $D(E_a)$ and S. For any given states i, j, and a, one can always choose the set S such that $(U_m)_{ia}$ and $(U_m)_{ja}$ remain unchanged for all $m\leq n$. Therefore, for the original problem we can replace $D(E_a)$ by $D\wedge S$ in (13) and the resulting sum remains finite in the limit $\mu\to 0$.

4. In the above proof we need only the expansion formulas (14) for those elements $(U_m)_{ia}$ where $E_i\neq E_a$. The complete recurrence formulas for all elements $(U_m)_{ij}$ and $(\Delta_m)_{ii}$ are given in Appendix A. These formulas will be useful for many explicit calculations.

5. In the above power series expansions, the energy denominators such as (E_a-E_i) in (15) refer to the total energy E_a and E_i. An alternative series can be developed in which the energy denominator is replaced by $(E_a{}^0-E_i{}^0)$ where $E_a{}^0$ and $E_i{}^0$ are the eigenvalues of H_0. A theorem can also be established for the existence of $[T_n(E_a{}^0)]_{ij}$ when H_0 becomes degenerate. (See Appendix B.)

6. For problems in field theory, each Feynman graph represents a part of the S matrix which is the product $U_-{}^\dagger U_+$. To obtain U_+ or U_- one may imagine that each of the Feynman graphs is cut into two by an arbitrarily drawn line. Rules can be derived[6] to represent U_+ and U_- as sums of the respective halves of all these cut graphs. Our theorem on $[T_n(E_a)]_{ij}$ refers to the existence of the corresponding sum of products of these cut graphs when degeneracy occurs.

[6] Cf., W. R. Frazer and L. Van Hove, Physica **24**, 137 (1958).

III. APPLICATIONS TO ELECTRODYNAMICS

We consider the pure electromagnetic interaction between electrons and photons. The matrix Δ is

$$\Delta=\delta m_e\int\bar{\psi}_e\psi_e d^3r\,, \tag{18}$$

where δm_e is the difference between the mechanical mass m_0 and the physical mass m_e of the electron, and ψ_e is the wave function operator of the electron. We assume there exists an ultraviolet cutoff in the theory. Problems related to the renormalization of the ultraviolet divergence will be discussed later. (See point 6 below.)

1. Electrodynamics contains degeneracy because photons have zero mass. This is the well-known infrared divergence.[1,2] For calculation purposes we can assign the parameter μ to be the fictitious mass of a neutral vector particle; in the limit $\mu=0$ this vector particle becomes the photon. It is well known that δm_e does not contain the infrared divergence. Therefore, $\lim_{\mu\to 0}\Delta_n$ exists for all n. For each given state a, the subset $D(E_a)$ consists of all other states which differ from a only in the number of infrared photons. The theorem proved in the previous section states that the power series expansion of ($U=U_+$ or U_-)

$$\sum_{D(E_a)} U_{ia}U_{ja}{}^*$$

does not contain infrared divergence.

2. Another application is one in which the parameter μ in theorem 2 is the physical mass m_e of the electron. For example, the state of an electron with a three-momentum $\mathbf{p}$ is degenerate with the state which consists of an electron with momentum $\mathbf{p}-\mathbf{k}$ and a photon with momentum $\mathbf{k}$ provided both m_e and the angle θ between $\mathbf{p}$ and $\mathbf{k}$ are zero. However, because of helicity conservation the matrix element between these two states for $m_e=0$ and small θ is proportional to θ, being zero if $\theta=0$. The transition amplitude for an electron to emit such a photon is proportional to the product of the matrix element divided by the energy denominator $[(p-k)^2+m^2]\cong E\omega[\theta^2+(m_e/E)^2]$, where E and ω are the magnitudes of the three vectors $\mathbf{p}$ and $\mathbf{k}$. The probability of such an emission with small θ is proportional to $\int\theta^2[\theta^2+(m_e/E)]^{-2}d(\cos\theta)$ which contains a $\ln\times(m_e/E)$ singularity as $m_e\to 0$. Similar considerations can be applied to states which consist of combinations of e^-, e^+, and γ moving in the same direction. The resulting singularities are called mass singularities.

It follows from either dimensional arguments or γ_5 invariance that as $m_e\to 0$ every order of the perturbation series for (δm_e) approaches zero. (A formal proof is given in Appendix C.) Therefore, $\lim_{\mu\to 0}\Delta_n = 0$ for all n. Our theorem states that summing over these degenerate states, there is no $(\ln m_e)$ singularity in $[T_n(E_a)]_{ij}$ or in the corresponding products of the cut graphs.

3. By using Eq. (2), the theorem can be readily applied to the S matrix. We consider first the case of mass singularity. It follows from the above discussions that the power series expansion of

$$\sum_{D(E_b)}\sum_{D(E_a)} |S_{ba}|^2 \tag{19}$$

contains no mass singularities, provided we sum over both subsets $D(E_a)$ and $D(E_b)$ which consist of the states that are degenerate (in the limit $m_e \to 0$) with the initial state a and the final state b, respectively.

To illustrate the use of (19) we consider the collision between an electron with a fixed external potential. The differential cross section $d\sigma_1$ which includes the lowest order radiative correction for such a collision without the emission of any hard photon (i.e., $\omega \geqq \epsilon$ where $\epsilon \neq 0$ but can be chosen to be arbitrarily small) is given by

$$d\sigma_1 = d\sigma_0\left\{1+\frac{e^2}{\pi}\left[2\left(\ln\frac{q^2}{m_e^2}-1\right)\ln\frac{\epsilon}{E}+\frac{3}{2}\ln\frac{q^2}{m_e^2}+0(1)\right]\right\}, \tag{20}$$

where e^2 = fine structure constant, E is the initial energy of the electron, q^2 is the square of the momentum transfer and $d\sigma_0$ is the differential cross-section without radiative correction and without the emission of soft photons. In (20) the infrared divergence has already been eliminated by including the contributions due to emissions of soft photons (i.e., $\omega = \leqq \epsilon$). The 0(1) term remains finite as $m_e \to 0$ and $(\epsilon/E) \to 0$. The contribution of vacuum polarization is not included in (20), but will be discussed in point 6 together with the problem of charge renormalization.

Let $\mathbf{p}$ and $\mathbf{p}'$ be, respectively, the initial and the final momentum of the electron, with $|\mathbf{p}| = |\mathbf{p}'| = E$. We define the set D (or D') to consist of all states in which an electron with momentum $\mathbf{p}-\mathbf{k}$ (or $\mathbf{p}'-\mathbf{k}'$) and a hard photon with momentum $\mathbf{k}$ (or $\mathbf{k}'$), where hard photons means $\omega = |\mathbf{k}| \geq \epsilon$ or $(|\mathbf{k}'| \geq \epsilon)$. In addition the angle between $\mathbf{k}$ and $\mathbf{p}$ (or $\mathbf{k}'$ and $\mathbf{p}'$) is less than or equal to δ which can be chosen to be arbitrarily small. In Fig. 1, the diagram (i) represents the collision process without the presence of any hard photons. Diagram (ii) represents the bremsstrahlung process where the initial state is an electron with momentum $\mathbf{p}$ and the final state is one in the degenerate set D'. Diagram (iii) represents the absorption process in which the initial state is one in the degenerate set D but the final state consists of a single electron with momentum $\mathbf{p}'$.

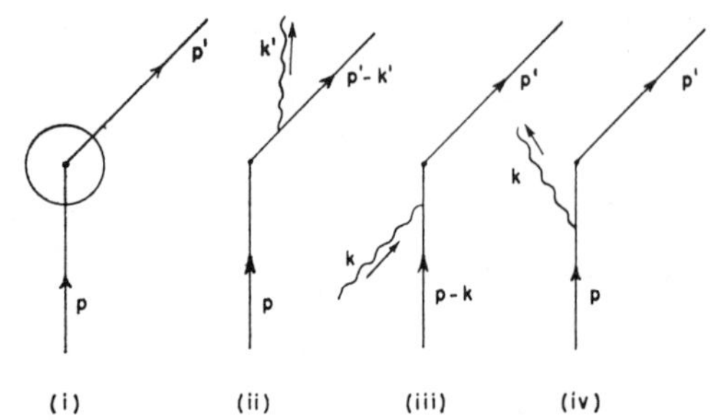

FIG. 1. Diagram (i) shows the elastic scattering of an electron in an external potential including radiative corrections. Diagrams (ii) and (iv) correspond to inelastic scattering with the emission of a single photon, while diagram (iii) illustrates the absorption of a photon in the initial state of the electron.

It can be readily verified that, after summing over their respective final and initial sets of states D' and D, neglecting terms which remain finite as $m_e \to 0$, the differential cross sections $d\sigma_2$ and $d\sigma_3$ for diagrams (ii) and (iii) are given by

$$d\sigma_2 = d\sigma_3 = d\sigma_0\frac{e^2}{\pi}\left[2\ln\frac{E\delta}{m_e}\right]\left[\ln\frac{E}{\epsilon}-\frac{3}{4}\right]. \tag{21}$$

Therefore, the sum $d\sigma_1 + d\sigma_2 + d\sigma_3$ contains no $(\ln m_e)$ singularity.

For practical calculations, (19) can be used to obtain the important radiative correction terms from the corresponding real emission and absorption processes, provided m_e is much smaller than all other values of energy and momentum transfer in the problem.

It is interesting to notice that for hard photons the mass singularity in the transition probability is removed only if the initial degenerate set $D(E_a)$ as well as the final degenerate set $D(E_b)$ are summed over in (19).

4. Another example is the decay of the muon. In this case, the initial state has no degeneracy (apart from the infrared photons) as $m_e \to 0$. Therefore, in (19) we need only to sum over the degenerate set of the final states. In particular, the power series expansion of the total decay rate of the muon does not contain any $(\ln m_e)$ singularity.[3]

For μ decays in which the momenta of ν_μ and $\bar{\nu}_e$ are fixed, the partial decay rate remains free from $(\ln m_e)$ singularity provided the appropriate degenerate states of γ, e^- and e^+ are summed over. This can be seen by applying remark 3 of Sec. II, where the set S contains only ν_μ and $\bar{\nu}_e$ with the given momenta.

5. For infrared photons, it can be shown that the power series of either

$$\sum_{D(E_a)} |S_{ba}|^2$$

or

$$\sum_{D(E_b)} |S_{ba}|^2 \tag{22}$$

already contains no infrared divergence. Since the problem of infrared divergence has been extensively studied in the literature[1,2], we will restrict ourselves only to a few remarks concerning the difference between the infrared divergence and the mass singularity in the limit $m_e \to 0$. For clarity, let us use the Bloch-Nordsieck approximation so that the electron currents $e\,j_\mu$ can be regarded as a static classical distribution. Let A_μ represent the electromagnetic field which consists of *only* soft photons ($\omega \leqq \epsilon$). The interaction Hamiltonian is given by

$$e\int j_\mu A_\mu d^3r\,. \tag{23}$$

For each given static classical distribution j_μ, the entire Hamiltonian for the soft photons can be diagonalized by a unitary matrix $U(j)$. Consider a problem in which the electron has an initial current distribution j_μ which becomes j_μ' after the collision with an external potential. Let q be the momentum transfer given to the external potential and $V(q)$ the corresponding matrix element. The S matrix (to first order in V) is then given by

$$S = U^\dagger(j')V(q)U(j)\,. \tag{24}$$

That the power series of (22) contains no infrared divergence can be easily established by using the unitarity relation (for arbitrary j_μ)

$$U(j)U^\dagger(j) = 1$$

and the fact that our Hilbert space consists of only soft photons which in turn form the complete degenerate set.

It is essential that in (24) the momentum transfer q and therefore the matrix element $V(q)$ are the same, irrespective of the number of soft photons that are emitted or absorbed by the electric current. In the case of hard photons, emissions and absorptions of different photons may drastically change the values of momentum transfer. For example, in Fig. 1 the momentum transfer given to the external potential in diagram (iv) is $\mathbf{p}-\mathbf{k}-\mathbf{p}'$, which differs from the momentum transfer $\mathbf{q}=\mathbf{p}-\mathbf{p}'$ in diagrams (i), (ii), and (iii). The final state of (iv) is not degenerate with the final states of the other three diagrams in Fig. 1. The mass singularity of (iv) is cancelled only if one includes diagrams listed in Fig. 3 which contain disconnected parts. (See Appendix D for the details.)

This situation is to be contrasted with the infrared problem. Let us suppose in Fig. 1, that diagram (i) represents the collision of an electron without any emission of photons (hard or soft) and the photons $\mathbf{k}$, $\mathbf{k}'$ in all other diagrams are soft photons ($|\mathbf{k}|$, $|\mathbf{k}'| \leqq \epsilon$). Then, according to (22), the contributions of (i) and (ii)+(iv) contains no infrared divergence.

6. In the above section we assume the existence of an ultraviolet cutoff λ in the theory. For electrodynamics, the ultraviolet divergences can be removed by a renormalization process. All above statements concerning the absence of singularities due to degeneracies are correct, provided the relations between the renormalized charge and the unrenormalized charge do not contain terms which become singular in the limit that degeneracy occurs. For the infrared divergence, this is the case. Therefore, all above statements about the removal of infrared divergence are also correct in terms of the renormalized charge.

The same, however, is not true for the limit $m_e=0$. This can be easily seen by recalling that the value of the observed electric charge e is usually determined by the well-known Thompson limit of electron-photon scattering which clearly does not exist if $m_e=0$. This difficulty can be overcome by defining the renormalized charge in a different way. Let $D_F'(k)$ be the complete propagator (including all radiative terms) of the photon. Instead of the usual Z_3, we may introduce[7] a Z_3' by requiring [$k^2=$ (4-momentum)^2]

$$k^2D_F'(k^2) = Z_3' \tag{25}$$

at $k^2 = -M^2 \neq 0$.

By applying our basic theorem, the power series expansion of $D_F'(-M^2)$ can be shown to be free of mass singularity. To see this, let us introduce a hypothetical neutral vector particle of mass M which has the same form of interaction with the electric field as that of the photon except the coupling constant is f instead of the bare charge e_0. The total interaction Hamiltonian now consists of this new interaction plus the usual electromagnetic interactions between the electrons and the photons. Consider the diagonal matrix element of the S matrix for the state which consists of only one such particle of mass M. The function $D_F'(-M^2)$ can be simply derived from the expansion of this matrix element to the first order in f but arbitrary orders in e_0^2. Since both the initial and the final states have no further degeneracy as $m_e \to 0$ (M remains finite), $D_F'(-M^2)$ can be regarded as a special case of (19) which, therefore, contains no mass singularity.

We now define the renormalized charge e' by

$$e' = Z_3'^{1/2}e_0\,. \tag{26}$$

In terms of e', Eq. (19) contains neither ultraviolet divergence nor infrared divergence nor ($\ln m_e$) singularity in the limit $m_e=0$.

7. The question whether a charged particle can have zero mass has been discussed in the literature.[7,8] Our results show that if we regard a spin-$\frac{1}{2}$ zero-mass particle as the limiting case of a nonzero mass particle as the mass $m \to 0$ and if we restrict ourselves only to measurements which consist of ensemble averages over the degenerate sets, then to each order of the perturbation series in terms of a new renormalized charge e', finite

[7] V. G. Vaks, Zh. Eksperim i Teor. Fiz. **40**, 792 (1961) [English transl.: Soviet Phys.—JETP **13**, 556 (1961)].

[8] K. M. Case and S. G. Gasiorowicz, Phys. Rev. **125**, 1055 (1962).

results can be obtained for these ensemble averages.[9] This, of course, does not mean that the entire sum of such power series exists. (Indeed, the fact that no such zero mass charged particle has been observed in nature leads us to suspect that perhaps the sum does not exist.) Nevertheless, in the limited sense of *ensemble averages* (over both initial and final states) and *power series calculations*, a theory with a zero mass charged particle can have a meaning. For problems in which all such zero mass particles are completely produced by massive particles, the initial state has no degeneracy. The single sum over the final set of degenerate states gives a finite result for the power series calculation.

The absence of any static limit makes the electrodynamics of a zero mass charged particle different from that of a finite mass charged particle even in the classical theory; e.g., there is no Coulomb's law. In the present form of quantum theory, several additional interesting features may be emphasized:

(i) Although there is no mass in the theory, a length M^{-1} is nevertheless introduced through the renormalization process.

(ii) For a spin-$\frac{1}{2}$ particle, the limiting process of starting with a finite mass m (therefore, a four-component theory) and then setting $m\rightarrow 0$ does not lead to two *uncoupled* two-component particles. This can be seen by considering the transition between such a charged particle with momentum $\mathbf{p}$ and, say, left-hand helicity to a right-hand helicity state with momentum $\mathbf{p}-\mathbf{k}$ through the emission of a photon with momentum $\mathbf{k}$ which makes a very small angle θ with $\mathbf{p}$. The matrix element for such a transition is proportional to (m/E) where $E=(m^2+|\mathbf{p}|^2)^{\frac{1}{2}}$. The transition probability for small values of θ and m is proportional to $(m/E)^2$ times the square of the energy denominator which is $\propto[\theta^2+(m/E)^2]$. The integrated transition probability is, therefore, proportional to $\int(m/E)^2[\theta^2+(m/E)^2]^{-2}\, d(\cos\theta)$ which remains nonzero in the limit $m\rightarrow 0$. This seems to indicate that a two-component theory of a zero mass spin-$\frac{1}{2}$ charged particle does not exist.

(iii) Another interesting feature is the necessity of averaging over the degenerate set for the *initial* states. Such averages require an ensemble in which states with different numbers of zero mass charged particles are populated with equal probability, provided they belong to the same degenerate set. However, the available phase space for states with N (unbound) particles is proportional to Ω^N where Ω is the volume of the entire system. It would appear that such an ensemble can never be realized. This dilemma can be resolved by considering the cases when these (initial) degenerate sets of states are themselves produced from certain finite numbers of nondegenerate states; e.g., these degenerate states are the decay products of some massive particles. These decay products can be regarded as wave packets extending over a finite volume V determined by the size of the experimental apparatus. The appropriate spatial part of the phase-space volume for the initial distribution of these degenerate states is, therefore, given by V and not Ω which is infinite.

Mathematically, this necessitates a certain change in carrying out the limit of infinite volume. We recall that Ω enters into all our formulas [e.g., Eqs. (13) and (14)] in two different ways. One is through the energy denominators such as $(E_a-E_i)_\Omega$ which should be set to its limiting value $(E_a-E_i)_{\Omega=\infty}$. The other is the trivial Ω dependence of the matrix elements $(H_1)_{ij}$, and in passing from a discrete momentum sum $\sum_k$ to the integration $(8\pi^3)^{-1}\Omega\int d^3k$. These latter ones can be combined to give factors $\Omega^m\int\pi d^3k_i\cdots$, which, apart from a three-dimensional δ function for over-all momentum conservation, should be replaced by $V^m\int\pi d^3k_i\cdots$, where V is to be regarded as an additional characteristic parameter of the initial ensemble.

It seems also possible that some of these Ω could have been replaced by different finite volumes, V_1, V_2, V_3, $\cdots$. The ensemble is then characterized by these parameters which, in turn, determine the relative populations of various degenerate systems containing different number of particles. However, we have not investigated these possibilities. Several additional interesting questions arise which have not been answered: Can the results of the limiting case $m\rightarrow 0$ be formulated by setting $m=0$ at the outset? Does the restriction to ensemble averages which seems to be necessary in the present case have a more fundamental bearing in the quantum theory of measurements? Is it possible to investigate the entire sum of the expansions, and thereby throw some light on the important question whether a zero mass charged particle can really exist or not?

ACKNOWLEDGMENTS

We wish to thank Professor L. Van Hove and Professor V. Weisskopf for the hospitality extended to us at CERN during the summer of 1963 when this paper was completed.

APPENDIX A

In this Appendix we will derive the complete recurrence formulas for the power series expansions of $(U_\pm)_{ij}$ and Δ_{ii}. As mentioned in Sec. II, these formulas can be derived by using Eq. (1). However, in order to make our discussion on Feynman graphs appear in a more transparent way, we will derive the power series expansion by using the time-dependent Schrödinger equation. This has an added advantage of deriving, in a natural way, the generalizations of the U matrices which approach $U_\pm$ in the limit $\alpha=0+$ [where α is introduced in Eq. (4)] and remain unitary for any nonzero α.

[9] It can be shown that the same conclusions hold for a spin-0 charged particle.

We choose an interaction representation in which the equation for an arbitrary dynamical variable $A(t)$ is given by

$$-i\frac{\partial A(t)}{\partial t}=[E,A(t)], \quad (A1)$$

where E is the diagonal matrix defined in (1). The solution of the Schrödinger equation for the state vector ψ in this representation is

$$\psi(t)=U(t,t_0)\psi(t_0), \quad (A2)$$

where

$$i\frac{\partial}{\partial t}U(t,t_0)=(gH_1+\Delta)U(t,t_0), \quad (A3)$$

$$U(t_0,t_0)=1, \quad (A4)$$

and

$$\Delta\equiv(H_0-E). \quad (A5)$$

For the S matrix we need the limit of $U(t,t_0)$ over an infinite time interval. To establish the existence of such limits, it is useful to multiply the coupling constant g by a slowly varying function in time, say, $\exp(-\alpha|t|)$, where α is a small real positive number. The limit $\alpha=0+$ corresponds to the physical situation. The interaction Hamiltonian $gH_1(t)+\Delta$ is then replaced by

$$H_{in}{}^{\alpha}(t)=gH_1(t)\exp[-\alpha|t|]+\Delta^{\alpha}(t), \quad (A6)$$

where $\Delta^{\alpha}(t)$ is related to the power series expansion of Δ,

$$\Delta=\sum_{n=1}^{\infty}g^n\Delta_n, \quad (A7)$$

by

$$\Delta^{\alpha}(t)=\sum_{n=1}^{\infty}g^n\Delta_n\exp(-n\alpha|t|). \quad (A8)$$

Let us define a unitary matrix $U^{\alpha}(t,-\infty)$ which satisfies

$$U^{\alpha}(-\infty,-\infty)=1 \quad (A9)$$

and

$$i(\partial/\partial t)U^{\alpha}(t,-\infty)=H_{in}{}^{\alpha}(t)U^{\alpha}(t,-\infty). \quad (A10)$$

For clarity, we consider first the case $t\leqq 0$. Expanding $U^{\alpha}(t,-\infty)$ in a power series

$$U^{\alpha}(t,-\infty)=\sum_{n=0}^{\infty}g^nU_n{}^{\alpha}(t,-\infty), \quad (A11)$$

the successive terms $U_n{}^{\alpha}(t,-\infty)$ satisfy

$$U_0{}^{\alpha}(t,-\infty)=1, \quad (A12)$$

and for $n\geq 1$

$$iU_n{}^{\alpha}(t,-\infty)=\int_{-\infty}^{t}H_1(t')\exp(\alpha t')U_{n-1}{}^{\alpha}(t',-\infty)dt'$$
$$+\sum_{m=1}^{n}\int_{-\infty}^{t}\Delta_m\exp(m\alpha t')U_{n-m}{}^{\alpha}(t',-\infty)dt'. \quad (A13)$$

The diagonal matrices Δ_m are independent of α and t. Their values can be uniquely determined by studying the above equations in the limit $\alpha=0+$.

Theorem A: If there is no degeneracy in the total energy E,

$$\lim_{\alpha\to 0+}U_n{}^{\alpha}(t,-\infty),\quad \lim_{\alpha\to 0+}\frac{\partial}{\partial\alpha}U_n{}^{\alpha}(t,-\infty),$$
$$\cdots\lim_{\alpha\to 0+}\frac{\partial^m}{\alpha\partial^m}U_n{}^{\alpha}(t,-\infty)$$

exist for all m and n.

Proof: By using (A13) and the time dependence of $H_1(t)$,

$$[H_1(t)]_{jk}=[H_1(0)]_{jk}\exp[i(E_j-E_k)t], \quad (A14)$$

the matrix elements of U_n are found to be

$$[U_n{}^{\alpha}(t,-\infty)]_{jk}=[U_n{}^{\alpha}(0,-\infty)]_{jk}\times\exp[i(E_j-E_k)t+n\alpha t] \quad (A15)$$

and

$$[U_n{}^{\alpha}(0,-\infty)]_{jk}=[E_k-E_j+in\alpha]^{-1}\times\Big\{\sum_l[H_1(0)]_{jl}[U_{n-1}{}^{\alpha}(0,-\infty)]_{lk}+\sum_{m=1}^{n}(\Delta_m)_{jj}\times[U_{n-m}{}^{\alpha}(0,-\infty)]_{jk}\Big\}. \quad (A16)$$

For $n=1$, the off-diagonal matrix element of U_1 is given by $(j\neq k)$

$$[U_1{}^{\alpha}(0,-\infty)]_{jk}=[E_k-E_j+i\alpha]^{-1}[H_1(0)]_{jk}, \quad (A17)$$

which clearly satisfies the theorem. In order that the theorem holds also for the diagonal element of $U_1{}^{\alpha}$, the matrix Δ_1 is determined to be

$$(\Delta_1)_{jj}=-[H_1(0)]_{jj}. \quad (A18)$$

Correspondingly,

$$[U_1{}^{\alpha}(0,-\infty)]_{jj}=0. \quad (A19)$$

Next, we assume that the theorem holds for all $n\leqq(N-1)$ and $m=1$. By considering the diagonal element $j=k$ in (A16), the matrix Δ_N is determined to be

$$(\Delta_N)_{jj}=-\sum_l[H_1(0)]_{jl}[U_{N-1}(0,-\infty)]_{lj}-\sum_{m=1}^{N-1}(\Delta_m)_{jj}(U_{N-m}(0,-\infty)]_{jj}, \quad (A20)$$

where $U_m(0,-\infty)=\lim_{\alpha\to 0+}U_m{}^{\alpha}(0,-\infty)$. It is easy to see that the theorem holds for all the off-diagonal matrix elements of $U_N{}^{\alpha}(0,-\infty)$. The diagonal matrix element is given by

$$[U_N{}^\alpha(0,-\infty)]_{jj}=(iN\alpha)^{-1}\Big\{\sum_l [H_1(0)]_{jl}$$

$$\times[U_{N-1}{}^\alpha(0,-\infty)-U_{N-1}(0-\infty)]_{lj}+\sum_{m=1}^{N-1}(\Delta_m)_{jj}$$

$$\times[U_{N-m}{}^\alpha(0,-\infty)-U_{N-m}(0,-\infty)]_{jj}\Big\}, \quad \text{(A21)}$$

which becomes, in the limit $\alpha=0$,

$$[U_N(0,-\infty)]_{jj}=(iN)^{-1}$$

$$\Big\{\sum_l [H_1(0)]_{jl}\frac{\partial}{\partial\alpha}[U_{N-1}{}^\alpha(0,-\infty)]_{lj}+\sum_{m=1}^{N-1}(\Delta_m)_{jj}$$

$$\times\frac{\partial}{\partial\alpha}[U_{N-m}{}^\alpha(0,-\infty)]_{jj}\Big\}_{\alpha=0+}. \quad \text{(A22)}$$

Similarly, one can obtain the limiting expressions for all the derivatives of $U_N{}^\alpha$ with respect to α. Theorem A is therefore proved by induction.

Remarks

1. In an entirely similar way, one can investigate the properties of the unitary matrix $U^\alpha(\infty,t)$ which satisfies

$$-i(\partial/\partial t)U^\alpha(\infty,t)=U^\alpha(\infty,t)H_{in}{}^\alpha(t) \quad \text{(A23)}$$

and

$$U^\alpha(\infty,\infty)=1.$$

Except for some minor changes, explicit power series solutions similar to (A16) can be obtained for $U^\alpha(\infty,t)$ for $t\geqq 0$. Furthermore,

$$\lim_{\alpha\to 0+} U_n{}^\alpha(\infty,t),\cdots \lim_{\alpha\to 0+}\frac{\partial^m}{\partial\alpha^m}U_n{}^\alpha(\infty,t)$$

exist for all n and m, where $U_n{}^\alpha(\infty,t_0)$ is given by the power series expansion

$$U^\alpha(\infty,t)=\sum_{n=0}^{\infty}g^nU_n{}^\alpha(\infty,t). \quad \text{(A24)}$$

2. Define

$$U(t,-\infty)=\sum_{\mu=0}^{\infty}\lim_{\alpha\to 0+}g^nU_n{}^\alpha(t,-\infty) \quad \text{(A25)}$$

and

$$U(\infty,t)=\sum_{n=0}^{\infty}\lim_{\alpha\to 0+}g^nU_n{}^\alpha(\infty,t). \quad \text{(A26)}$$

Both $U(t,-\infty)$ and $U(\infty,t)$ are particular solutions of (A3). It is clear that the previous restriction $t\leqq 0$ or $t\geqq 0$ can be extended to all t. Furthermore, by using (A10), (A15), and (A23), we find

$$U^\dagger(t,-\infty)[H_0+gH_1(t)]U(t,-\infty)=E$$

and

$$U(\infty,t)[H_0+gH_1(t)]U^\dagger(\infty,t)=E. \quad \text{(A27)}$$

3. For problems containing a continuum, it is convenient to introduce first a finite volume Ω for the system and then to take the limit $\Omega\to\infty$ *before* the limit $\alpha\to 0+$. The U_+ and U_- discussed in the previous sections [cf., Eq. (1)] are related to $U(t,t_0)$ by

$$U_+=U(0,-\infty)$$

and

$$U_-{}^\dagger=U(\infty,0).$$

APPENDIX B

We review the Raleigh-Schrödinger perturbation series expansion of the U matrix [Eq. (1)] in which the energy denominators refer to the eigenvalues $E_i{}^0$ of H_0. It is well known in the case that these eigenvalues are degenerate. There exists a special orthogonal set of degenerate eigenstates of H_0 for which the expansion of U remains valid. We will show, however, that the corresponding series expansion of $T(E_a{}^0)$, Eq. (5), exists in all representations.

We write (1) in the form

$$[U,H_0]=gH_1U+U\Delta, \quad \text{(B1)}$$

where $\Delta=H_0-E$. Substituting the formal power series expansion of U and Δ, Eqs. (9) and (10), in (B1) we obtain the recurrence relation

$$[U_n,H_0]=H_1U_{n-1}+\sum_{m=1}^{n}U_{n-m}\Delta_m. \quad \text{(B2)}$$

The solution of (B2) is subject to the unitarity condition

$$U_0=1,$$

and for $n\geqq 1$,

$$\sum_{m=0}^{n}U_m{}^\dagger U_{n-m}=\sum_{m=0}^{n}U_mU_{n-m}{}^\dagger=0. \quad \text{(B3)}$$

A convenient procedure to solve these equations is to introduce the slightly modified recurrence relation

$$[U_n{}^\alpha,H_0]+in\alpha U_n{}^\alpha=H_1U_{n-1}{}^\alpha+\sum_{m=1}^{n}U_{n-m}{}^\alpha\Delta_m, \quad \text{(B4)}$$

where α is a real parameter. The advantage of (B4) is that if a solution exists, it automatically satisfies the unitarity condition (B3), for all values of α. This can be readily verified by substituting (B4) in (B3). To obtain a solution of (B2) we then take the $\lim\alpha\to 0$. If there is a continuum, the limit taken from positive and from negative values of α differ in general. From the time-dependent formalism, Appendix A, we find that $\lim_{\alpha\to 0}U_\pm{}^\alpha=U_\pm$.

From (B4),

$$(U_n{}^\alpha)_{ij}=(E_j{}^0-E_i{}^0+in\alpha)^{-1}\Big\{\sum_k (H_1)_{ik}(U_{n-1}{}^\alpha)_{kj}+\sum_{m=1}^{n}\sum_k (U_{n-m}{}^\alpha)_{ik}(\Delta_m)_{kj}\Big\}, \quad \text{(B5)}$$

where $(\Delta_m)_{kj}=(\Delta_m)_{kk}\delta_{kj}$.

If the eigenvalues $E_i{}^0$ are not degenerate, the proof of the existence of $\lim_{\alpha\to0\pm}(U_n{}^\alpha)_{ij}$, $\lim_{\alpha\to0\pm}\partial/\partial\alpha(U_n{}^\alpha)_{ij}$ and higher derivates is the same as that given in Appendix A. We obtain

$$(\Delta_n)_{ii}=-\sum_k (H_1)_{ik}(U_{n-1})_{ki}-\sum_{m=1}^{n-1}(U_{n-m})_{ii}(\Delta_m)_{ii} \quad \text{(B6)}$$

and

$$(U_n)_{ii}=\lim_{\alpha\to0}\frac{1}{in}\Big\{\sum_k (H)_{ik}\frac{\partial}{\partial\alpha}(U_{n-1}{}^\alpha)_{ki}+\sum_{m=1}^{n-1}\frac{\partial}{\partial\alpha}(U_{n-m}{}^\alpha)_{ii}(\Delta_m)_{ii}\Big\}, \quad \text{(B7)}$$

where U_n stands for either $(U_+)_n$ or $(U_-)_n$, when $\alpha\to0^+$ or 0^-.

For degenerate eigenvalues $E_i{}^0$, it is clear that the $\lim\alpha\to0$ of (B5) in general does not exist. Suppose, however, that we allow Δ_n to have nondiagonal elements among degenerate states. To avoid confusion, we call this matrix Δ_n' and the corresponding solution of (B5), U_n'. Note that U' does not diagonalize the total Hamiltonian (H_0+gH_1). We can then prove the following theorem.

Theorem B: There exists a matrix Δ_n' which has nondiagonal elements only among degenerate states, for which

$$\lim_{\alpha\to0\pm}U_n{}'^\alpha,\quad \lim_{\alpha\to0\pm}\frac{\partial}{\partial\alpha}(U_n{}'^\alpha)\quad\cdots\quad\lim_{\alpha\to0\pm}\frac{\partial^m}{\partial\alpha^m}(U_n{}'^\alpha)$$

exists for all m and n.

Proof: For $n=1$

$$(U_1{}'^\alpha)_{ij}=\frac{1}{(E_j{}^0-E_i{}^0+i\alpha)}\{(H_1)_{ij}+(\Delta_1')_{ij}\}. \quad \text{(B8)}$$

Let

$$(\Delta_1')_{ij}=-(H_1)_{ij} \quad \text{(B9)}$$

for those i, j that satisfy $E_i{}^0=E_j{}^0$. Then

$$\lim_{\alpha\to0}(U_1{}'^\alpha)_{ij}=\frac{1}{(E_j{}^0-E_i{}^0)}(H_1)_{ij} \text{ provided } E_1{}^0\neq E_j{}^0,$$
$$=0 \text{ otherwise}. \quad \text{(B10)}$$

Now we assume the theorem holds for all $n\leq N-1$ and $m=1$. From (B5) we establish that when $E_i{}^0=E_j{}^0$,

$$(\Delta_N')_{ij}=-\Big\{\sum_k (H_1)_{ik}(U_{N-1}')_{kj}+\sum_{m=1}^{N-1}\sum_k (U_{N-m}')_{ik}(\Delta_m')_{kj}\Big\}$$

and when $E_i{}^0\neq E_j{}^0$

$$(\Delta_N')_{ij}=0. \quad \text{(B11)}$$

Hence, in the case $E_i{}^0=E_j{}^0$

$$(U_N{}'^\alpha)_{ij}=(iN\alpha)^{-1}\Big\{\sum_k (H_1)_{ik}(U_{N-1}{}'^\alpha-U_{N-1}')_{kj}+\sum_{m=1}^{N-1}\sum_k (U_{N-m}{}'^\alpha-U_{N-m}')_{ik}(\Delta_m')_{kj}\Big\}, \quad \text{(B12)}$$

which becomes in the $\lim\alpha\to0$

$$(U_N')_{ij}=(iN)^{-1}\Big\{\sum_k (H_1)_{ik}\frac{\partial}{\partial\alpha}(U_{N-1}{}'^\alpha)_{kj}+\sum_{m=1}^{N-1}\sum_k\frac{\partial}{\partial\alpha}(U_{N-m}{}'^\alpha)_{ik}(\Delta_m')_{kj}\Big\}_{\alpha=0}. \quad \text{(B13)}$$

When $E_i{}^0\neq E_j{}^0$ the $\lim_{\alpha\to0}U_N{}'^\alpha$ is obtained by setting $\alpha=0$ in (B5). Similar expressions hold for the derivative of $U_N{}'^\alpha$ with respect to α. Theorem B is then proved by induction.

From the unitarity of U' it can be readily established that Δ_N' is Hermitian. Let S be the unitary matrix which diagonalizes Δ', i.e.,

$$S^\dagger\Delta'S=\Delta, \quad \text{(B14)}$$

where Δ is diagonal. We note that S commutes with H_0. Hence $U=U'S$ is the unitary matrix which diagonalizes the total Hamiltonian.

Finally, since $S_{bc}=0$ when $E_b{}^0\neq E_c{}^0$, we have

$$[T(E_a{}^0)]_{ij}\equiv\sum_{D(E_a{}^0)}(U)_{ia}(U^\dagger)_{aj}$$
$$=\sum(U')_{ib}(S)_{ba}(S^\dagger)_{ac}(U'^\dagger)_{cj}$$
$$=\sum_{D(E_a{}^0)}(U')_{ia}(U'^\dagger)_{aj}. \quad \text{(B15)}$$

Substituting the expansion of U' in (B15), we obtain the power series expansion of $T(E_a{}^0)$.

Similarly to the treatment given in the text, we may consider the degeneracy in the eigenvalues $E_i{}^0$ of the unperturbed Hamiltonian H_0 as the limiting case when a parameter $\mu\to0$. For $\mu\neq0$, degeneracy does not occur, and the power series of $T(E_a{}^0,\mu)$ is obtained directly from the expansion of U, Eq. (B5). We have shown here that

$$\lim_{\mu\to0}T(E_a{}^0,\mu)=T(E_a{}^0)$$

has a power series expansion. However, this does not imply that

$$\lim_{\mu\to0}T_n(E_a{}^0,\mu)$$

exists, where $T_n(E_a{}^0,\mu)$ is the nth-order term in the expansion of $T(E_a{}^0,\mu)$. In general, the order of the summation in the power series expansion and the limit $\mu\to 0$ cannot be exchanged. This is to be contrasted with the case when the degeneracy occurs in the eigenvalues E_i of the total Hamiltonian, and the energy denominators in the expansion refer to E_i. The existence of the corresponding limit of $T_n(E_a,\mu)$ as $\mu\to 0$ is established in Sec. II.

APPENDIX C

To prove $(\delta m_e)=0$ in the limit $m_e=0$, we consider the wave function operator $\psi_m(x)$ of the electron with a bare mass m (physical mass m_e) in the Heisenberg representation. At any given time t, ψ_m, $\psi_m{}^\dagger$, and $\gamma_5\psi_{-m}$, $\psi_{-m}{}^\dagger\gamma_5$ obey the same anticommutation relations. Therefore, there exists a unitary matrix U which satisfies

$$U^\dagger\psi_m(x)U=\gamma_5\psi_{-m}(x) \tag{C1}$$

and

$$U^\dagger\bar{\psi}_m(x)U=-\bar{\psi}_{-m}(x)\gamma_5, \tag{C2}$$

where

$$\bar{\psi}_m(x)\equiv\psi_m{}^\dagger(x)\gamma_4. \tag{C3}$$

Throughout, the five γ_μ matrices are 4×4 Hermitian matrices that satisfy $\gamma_\mu\gamma_\nu+\gamma_\nu\gamma_\mu=2\delta_{\mu\nu}$. From (C1), (C2) and the equation of motion, it follows that U can be chosen to be independent of t and

$$U^\dagger H_m U=H_{-m}, \tag{C4}$$

where H_m and H_{-m} refer to the Hamiltonians in which the mechanical mass of the electron are, respectively, m and $-m$. Since high-frequency cutoff λ can be regarded as the result of some additional neutral vector fields which have a mass λ but with a wrong metric, all the above equations are also valid with a finite cutoff. Let $|\text{vac}\rangle_m$ refer to the vacuum state for H_m. We have, from (C4),

$$U^\dagger|\text{vac}\rangle_m=|\text{vac}\rangle_{-m}. \tag{C5}$$

Therefore, the propagator $S_m(x-y)$, which is defined to be the T product

$$T\langle\text{vac}|\psi_m(x)\bar{\psi}_m(y)|\text{vac}\rangle_m=T\langle\text{vac}|UU^\dagger\psi_m(x)\bar{\psi}_m(y)UU^\dagger|\text{vac}\rangle_m$$

satisfies

$$S_m(x-y)=-\gamma_5 S_{-m}(x-y)\gamma_5. \tag{C6}$$

The Fourier transform of $S_m{}^{-1}(x-y)$ may be written as

$$A_m+iB_m(\gamma\cdot p), \tag{C7}$$

where $(\gamma\cdot p)=\gamma_\mu p_\mu$, and $A_m B_m$ are functions of p^2. From (C6), we have

$$A_m=-A_{-m}\text{ and }B_m=B_{-m}. \tag{C8}$$

The physical mass m_e is given by

$$m_e=(A_m/B_m) \tag{C9}$$

at $p^2=-m_e^2$. As $m\to -m$, we have $m_e\to -m_e$ and $\delta m_e\to -\delta m_e$. In terms of m_e, δm_e is an odd function of m_e. Therefore, in the limit $m_e=0$, δm_e must also be zero.

APPENDIX D

We discuss in this section the cancellation of mass singularities in bremsstrahlung in the limit that the mass of the electron vanishes. The singularity appears here when a photon is emitted nearly parallel to the direction of the incident electron, before the electron scatters from the external potential [see Fig. 1, diagram (iv)].

In discussing the amplitude for processes containing nearly parallel electrons and photons, it is convenient to use helicity states. Let $u_\lambda(\mathbf{p})$ be the spin state of an electron with momentum $\mathbf{p}$ and helicity λ, where $\lambda=+(-)$ denotes spin parallel (antiparallel) to the momentum. Similarly, the four-vector $e_\eta(\mathbf{k})$ denotes the state of a photon with momentum $\mathbf{k}$ and helicity $\eta=\pm$. The Feynman amplitude corresponding to diagram (iv), Fig. 1, is given by

$$\frac{\bar{u}_{\lambda'}(\mathbf{p}')[\gamma\cdot V(q)][-(\gamma\cdot p)+(\gamma\cdot k')-im][\gamma\cdot e_\eta(\mathbf{k})]u_\lambda(\mathbf{p})}{(8EE'\omega)^{1/2}[(p-k)^2+m^2]}, \tag{D1}$$

where $p=(\mathbf{p},iE)$ and $p'=(\mathbf{p}',iE')$ are the initial and final four-momentum of the electron, $k=(\mathbf{k},i\omega)$ is the four-momentum of the emitted photon, $V(q)$ represents the external potential which depends on the four-momentum transfer $q=p'-p+k$, and $(\gamma\cdot a)\equiv\gamma_\mu a_\mu$ for any four-vector a.

If θ is the angle between $\mathbf{k}$ and $\mathbf{p}$, the denominator

$$(p-k)^2+m^2\cong\omega E\left(\frac{m^2}{E^2}+\theta^2\right)\text{ for }\theta\ll 1\text{ and }\frac{m}{E}\ll 1.$$

To evaluate the numerator, we note that in the limit $m\to 0$ we have

$$(\gamma\cdot k)(\gamma\cdot e_\beta)u_\beta\cong-\sqrt{2}\omega\theta u_\beta$$

for $\beta=+$ or $-$ and $\theta\ll 1$, while

$$(\gamma\cdot k)(\gamma\cdot e_-)u_+=(\gamma\cdot k)(\gamma\cdot e_+)u_-=0. \tag{D2}$$

For an incident electron with positive helicity, we obtain for (D1)

$$\frac{\bar{u}_{\lambda'}(\mathbf{p}')[i\gamma\cdot V(q)]u_+(\mathbf{p})\sqrt{2}\theta}{(8EE'\omega)^{1/2}\left(\omega E\left[\frac{m^2}{E^2}+\theta^2\right]\right)}\times\left\{\begin{matrix}E\\ E-\omega\end{matrix}\right\}, \tag{D3}$$

where the upper (lower) term in the curly bracket corresponds to a photon with positive (negative) helicity and θ, $(m/E)\ll 1$. In the case that the incident electron has negative helicity, the meaning of the upper and lower term in the bracket is interchanged.

The probability per unit frequency that the photon is emitted in the forward cone $0\leqq\theta\leqq\delta$, where $\delta\ll 1$, is then given by

$$\frac{|\bar{u}_{\lambda'}(\mathbf{p}')[-i\gamma\cdot V(q)]u_+(\mathbf{p})|^2}{(4\pi)^2E'E^3\omega}\left(\ln\frac{E\delta}{m}\right)\times\left\{\begin{matrix}E^2\\ (E-\omega)^2\end{matrix}\right\}. \quad \text{(D4)}$$

Equation (D4) becomes logarithmically divergent when $m\to 0$.

To obtain the contributions which cancel this singularity, we note that the singularity in (D4) is due to the degeneracy of the intermediate electron-photon state with the initial electron state [see diagram (iv), Fig. 1]. Since our basic theorem applies to any Hermitian Hamiltonian, we may consider a truncated electron-photon interaction which only couples these two states (see Sec. II, remark 3). Then the only other processes which have a degeneracy with respect to the initial state are given by diagrams (i) and (ii) of Fig. 2. According to (A16), the amplitude for diagram (ii) of Fig. 2 in the case $\mathbf{k}_1=\mathbf{k}$, $\theta_1\ll 1$ and $(m/E_1)\ll 1$, is given by

$$\frac{\bar{u}_{\lambda'}(\mathbf{p}')[-i\gamma\cdot V(q)]u_+(\mathbf{p}_1)\theta_1^2}{(8E'E_1\omega)^{1/2}(E_1+\omega)2i\alpha}[\omega E_1(m^2/E_1^2+\theta_1^2)+2i\alpha(E_1+\omega)]^{-1}\times\left\{\begin{matrix}(E_1+\omega)^2\\ E_1^2\end{matrix}\right\}, \quad \text{(D5)}$$

where θ_1 is the angle between $\mathbf{k}_1$ and $\mathbf{p}_1=\mathbf{p}-\mathbf{k}_1$, E and E_1 are, respectively, the initial and final energy of the electron. The interference of the processes illustrated on diagrams (i) and (ii) of Fig. 2 gives a second-order contribution which equals (D4) but with opposite sign. This result is obtained by using (D5) and integrating over the forward cone $\theta\leqq\theta\leqq\delta$, where θ is the angle between $\mathbf{k}$ and $\mathbf{p}$. [For $\theta_1\ll 1$, $(m/E_1)\ll 1$, and $k_1=k$, $\theta\cong(E_1\theta_1/E_1+\omega)$.] Hence the singularity is cancelled when adding these contributions.

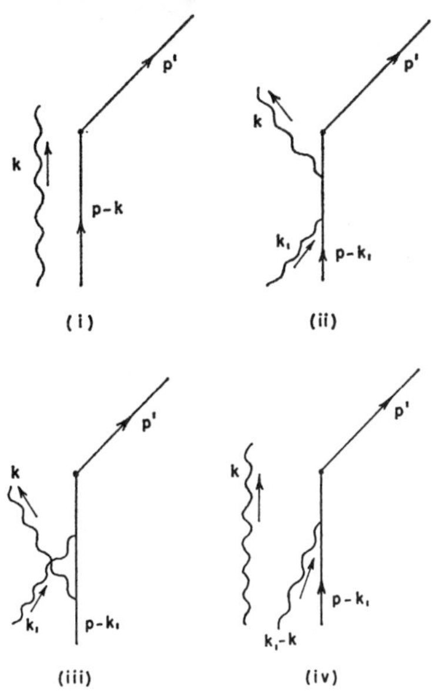

FIG. 2. Diagrams (i) to (iv) illustrate some of the processes in which the initial and the final states are degenerate with diagram (iv) in Fig. 1, provided $m_e=0$ and $\mathbf{p}$, $\mathbf{k}$, and $\mathbf{k}_1$ are parallel. By considering a truncated interaction Hamiltonian which has matrix elements only between states of a single electron and that of a single electron plus a photon, the mass singularity of diagram (iv) in Fig. 1 can be shown to be completely cancelled by the corresponding mass singularity in the sum of the first two diagrams (i) and (ii) in Fig. 2.

We note that the usual Feynman rules for diagram (ii) of Fig. 2 when $\mathbf{k}_1=\mathbf{k}_2$ lead to an amplitude different from Eq. (D5); in particular the energy-dependent factor in the curly bracket for the emission of photons with positive and negative helicity is given incorrectly, and a factor $\frac{1}{2}$ is missing.

In the case of the simple truncated electron-photon interaction discussed above, the process corresponding to diagram (iii), Fig. 2, for example, does not occur since it contains either two photons or electron-positron pairs in the intermediate state. It can be verified that this process also contains a mass singularity. Since for the complete Hamiltonian we must extend the interaction to include coupling to these states, other processes containing mass singularity will also occur, e.g., diagram (iv), Fig. 2. One finds that by adding the contributions of all of these processes the mass singularity is again canceled completely.

APPENDIX E

In this section, we consider a soluble model in field theory, which has been discussed in the literature.[10,11] Throughout this section, all unexplained notation is the same as that in Ref. 10. We will examine in particular the matrix elements of U, Eq. (1), between the V state and the $N+\theta$ scattering states in the limit $\mu=m_N+m_\theta-m_V\to 0$, where m_N, m_θ and m_V are the physical masses of the N, θ and V particles, respectively. We will show that in this limit, these matrix elements cannot be expanded in a power series in the unrenormalized coupling constant g, but that such an expansion does exist for the corresponding elements of the matrix $T(E_a)$ defined by Eq. (5). The power series expansion in terms of the renormalized coupling constant exists in the limit $\mu=0$ only after a certain change in the usual renormalization process. These modifications are similar to those used in Sec. III (point 6). Finally, to give a further illustration where the cancel-

[10] T. D. Lee, Phys. Rev. **95**, 1329 (1954).

[11] G. Källen and W. Pauli, Kgl. Danske Videnskab. Selskab, Mat. Fys. Medd. **30**, No. 7 (1955).

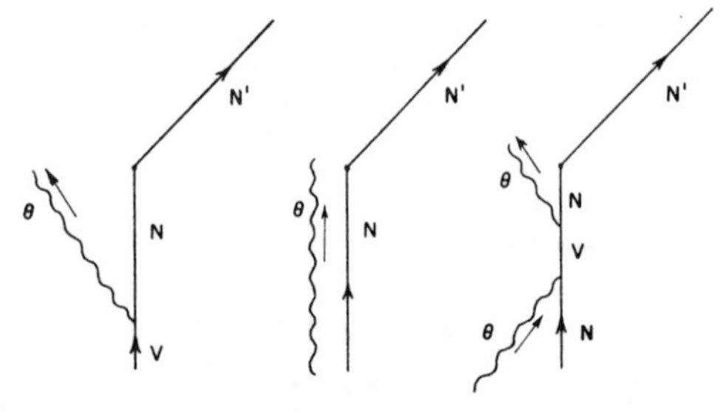

FIG. 3. Diagram (i) shows the emission of a θ particle by a V particle in an external field which transforms the N state into an N' state. Diagrams (ii) and (iii) illustrate the scattering process $N+\theta \to N+\theta'$ to zero and second order in g^2, respectively.

lation of singularities requires the inclusion of a disconnected diagram (see Fig. 3), we consider the emission of a θ by a V particle in an external field. In dealing with disconnected diagrams, a great deal of care must be taken to preserve unitarity in the expansion of the U and the S matrix, while cancelling singular terms. This requirement is satisfied by the explicit expansion formula given in Appendix A. In these expressions [e.g., Eq. (A16)] the unitarity of $U_\pm$ is maintained even for finite α, where α must be set equal to zero after the cancellation of singularities due to $\mu=0$.

To make the model finite, we assume for simplicity a sharp cutoff in the interaction at $\omega=\lambda$, where ω is the energy of the θ particle. In this case, there is a bound state G of the $N+\theta$ system which lies above the continuum, i.e., its mass $m_G>m_N+\lambda$. According to Ref. 10,

$$(V|U_+|V)=\left[1+g^2\sum_{\mathbf{k}}\frac{1}{2\omega\Omega(\omega+m_N-m_V)^2}\right]^{-1/2} \quad (E1)$$

$$(V|U_+|N\theta_{\mathbf{k}})=\frac{g}{(2\omega\Omega)^{1/2}}\left[(\omega+m_N-m_V)\left(1+g^2\sum_{\mathbf{k}'}\frac{1}{2\omega'\Omega(\omega'+m_N-m_V)(\omega'-\omega-i\alpha)}\right)\right]^{-1}, \quad (E2)$$

while the matrix element $(V|U_+|G)$ is obtained from (E1) by changing m_V to m_G. Here m_V and m_G are the two mass eigenvalues m of the equation

$$m_V{}^0-m=g^2\sum_{\mathbf{k}}\frac{1}{2\omega\Omega(\omega+m_N-m)}. \quad (E3)$$

In Eqs. (E1)–(E3), the sum is carried over momenta $\mathbf{k}$ such that $\omega\leqq\lambda$. For an infinite volume Ω,

$$\sum_{\mathbf{k}}\frac{1}{\Omega}\to\frac{1}{2\pi^2}\int_{m_\theta}^{\lambda}d\omega k\omega.$$

It is then clear that in the limit $\mu\to 0$, the integral in $(V|U_+|V)$, Eq. (E1), diverges. Therefore $(V|U_+|V)$ cannot be expanded in a power series in g^2 in this limit. This applies also to the matrix element $(V|U_+|N\theta_{\mathbf{k}})$, Eq. (E2), at the threshold energy $\omega=m_\theta$. Nevertheless, as an illustration of our basic theorem, we will prove by explicit calculations that the diagonal element of the matrix $T_+(E_V)$, Eq. (5), for the V state, does have a formal power series expansion in this limit. For clarity, we include a subscript $+$ in Eq. (5) to indicate that U_+ is being used. We have

$$(V|T_+(E_V)|V)=|(V|U_+|V)|^2+\sum_{D(E_V)}|(V|U_+|N\theta_{\mathbf{k}})|^2, \quad (E4)$$

where $D(E_V)$ is the set of $N\theta_{\mathbf{k}}$ states with $\omega\leqq m_\theta+\epsilon$, and ϵ is an arbitrarily small energy.

According to (E1), the nth-order contribution to $|(V|U_+|V)|^2$ is

$$\left[\frac{g^2}{4\pi^2}\int_{m_\theta}^{\lambda}\frac{d\omega k}{(\omega+m_N-m_V)^2}\right]^n \quad (E5)$$

which behaves as $\mu^{-n/2}$ in the limit $\mu\to 0$.

Now consider the second term in (E4) with $D(E_V)$ replaced by the complete set of $N\theta_{\mathbf{k}}$ states. From (E2) we have

$$\lim_{\Omega\to\infty}\sum_{\mathbf{k}}|(V|U_+|N\theta_{\mathbf{k}})|^2=\frac{g^2}{4\pi^2}\int_{m_\theta}^{\lambda}\frac{d\omega k f(\omega+i\alpha)f(\omega-i\alpha)}{(\omega+m_N-m_V)^2}, \quad (E6)$$

where

$$f(\omega)=\left[1+\frac{g^2}{4\pi^2}\int_{m_\theta}^{\lambda}\frac{d\omega' k'}{(\omega'+m_N-m_V)(\omega'-\omega)}\right]^{-1}. \quad (E7)$$

Using the relation

$$f(\omega+i\alpha)f(\omega-i\alpha)=\frac{2i\pi}{g^2k}(\omega+m_N-m_V)\times[f(\omega+i\alpha)-f(\omega-i\alpha)], \quad (E8)$$

(E6) is equal to

$$-\frac{1}{2\pi i}\int_C\frac{d\omega f(\omega)}{(\omega+m_N-m_V)}, \quad (E9)$$

where the path of integration C is a closed curve around the cut $m_\theta\leqq\omega\leqq\lambda$, which does not contain the poles of the integrand. To obtain the nth-order contribution, we now expand the right-hand side of (E7), in powers of g^2 and substitute the power series in (E9). The result for the nth-order term is

$$-\frac{1}{2\pi i}\int_C\frac{d\omega}{(\omega+m_N-m_V)}\times\left[-\frac{g^2}{4\pi^2}\int_{m_\theta}^{\lambda}\frac{d\omega' k'}{(\omega'+m_N-m_V)(\omega'-\omega)}\right]^n. \quad (E10)$$

In the limit $\mu \to 0$, the pole in the integrand at $\omega = m_V - m_N$ approaches the branch point at $\omega = m_\theta$ for $n \geqq 1$. The integral is therefore singular in this limit. We can readily separate out the singular part by deforming the contour; it corresponds precisely to the nth-order contribution to $|(V|U_+|V)|^2$, Eq. (E5), but with opposite sign. To complete the proof we note that

$$\sum_{D(E_V)} |(V|U_+|N\theta_{\mathbf{k}})|^2 = \sum_{\mathbf{k}} |(V|U_+|N\theta_k)|^2 - \sum_{\mathbf{k}}' |(V|U_+|N\theta_{\mathbf{k}})|^2, \quad \text{(E11)}$$

where on the right-hand side the sum $\sum_k$ extends over all $\mathbf{k}$ and the sum $\sum_k'$ extends only over those states $N\theta_k$ which are not in $D(E_V)$. The sum $\sum_{\mathbf{k}}'$ has a nonsingular power series in g^2 in the limit $\mu \to 0$.

As we have pointed out, this cancellation of singularities is essentially implied by the unitarity of U_+. In fact if we evaluate the integral in (E9) we obtain simply the normalization condition on the state $U_+|V)$

$$|(V|U_+|V)|^2 + \sum_{\mathbf{k}} |(V|U_+|N\theta_{\mathbf{k}})|^2 + |(V|U_+|G)|^2 = 1. \quad \text{(E12)}$$

In this connection it is interesting to note that for small g^2,

$$|(V|U_+|G)|^2 \cong g^{-2}(\lambda^2 - m_\theta^2)^{-1/2} 4\pi^2 \exp\left[-\frac{4\pi(\lambda + m_N - m_V)}{g^2(\lambda^2 - m_\theta^2)^{1/2}}\right]. \quad \text{(E13)}$$

Hence, the second term in (E12) does not have a strict power series expansion in g^2. Nevertheless, the difference between $1 - |(V|U_+|V)|^2$ and the series expansion generated by (E10) approaches zero as $g^2 \to 0$.

In order to eliminate the explicit dependence of the $N + \theta_{\mathbf{k}}$ scattering amplitude on the cutoff parameter λ, it is customary to introduce a renormalized coupling constant g_c related to the residue of the pole in this amplitude at $\omega = m_V - m_N$. By a subtraction in the integral in (E2) at $\omega_0 = m_V - m_N$, we obtain

$$(V|U_+|N\theta_{\mathbf{k}}) = \frac{Z_2^{1/2} g_c}{(2\omega\Omega)^{1/2}} (\omega + m_N - m_V)^{-1} \times \left[1 + \frac{g_c^2}{4\pi^2}(\omega + m_N - m_V) \times \int_{m_\theta}^{\lambda} \frac{d\omega' k'}{(\omega' + m_N - m_V)^2 (\omega' - \omega - i\alpha)}\right]^{-1}, \quad \text{(E14)}$$

where $g_c = Z_2^{1/2} g$, and $Z_2 \equiv |(V|U_+|V)|^2$ is given by (E1). The limit $\lambda \to \infty$ of (E14), keeping g_c fixed, exists. However, if we keep g_c fixed as $\mu \to 0$ then the integrand in the denominator of (E14) becomes singular for all ω, and $(V|U_+|N\theta_k)$ would vanish identically in this limit.

The way out of this difficulty is to define a new renormalized coupling constant g_c' by a subtraction in the integral in (E2) at $\omega_0 \neq m_V - m_N$. We then have

$$(V|U_+|N\theta_{\mathbf{k}}) = \frac{(Z')^{1/2} g_c'}{(2\omega\Omega)^{1/2}} \left[(\omega + m_N - m_V)\left(1 + \frac{g_c'^2}{4\pi^2}(\omega - \omega_0) \times \int_{m_\theta}^{\lambda} \frac{d\omega' k'}{(\omega' + m_N - m_V)(\omega' - \omega_0)(\omega' - \omega - i\alpha)}\right)\right]^{-1}, \quad \text{(E15)}$$

where $g_c' = (Z')^{1/2} g$ and

$$Z'\left[1 + \frac{g^2}{4\pi^2} \int_{m_\theta}^{\lambda} \frac{d\omega k}{(\omega + m_N - m_V)(\omega - \omega_0)}\right]^{-1}. \quad \text{(E16)}$$

It is easy to see that a power series expansion of $T(E)$, Eq. (5), exists also in $(g_c')^2$, since Z' is analytic in g^2 in the limit $\mu \to 0$.

As a last example, we consider a process which requires the inclusion of a disconnected diagram to cancel the mass singularity. For this purpose, we introduce a new field $\psi_{N'}$ in the model corresponding to a particle N' which interacts only with the N particle. The interaction Hamiltonian H' is time-dependent,

$$H' = f(t)(\psi_{N'}{}^\dagger \psi_N + \psi_N{}^\dagger \psi_{N'}), \quad \text{(E17)}$$

where $f(t)$ has a Fourier transform $\tilde{f}(\omega)$. The process of interest is the emission of a θ by a V particle: $V \to N' + \theta$. The transition amplitude to first order in H' is given by

$$-i\int_{-\infty}^{\infty} dt (N'\theta_{\mathbf{k}}|H'(t)U(t, -\infty)|V) = -i\tilde{f}(m_{N'} + \omega - m_V)(N\theta_{\mathbf{k}}|U_+|V). \quad \text{(E18)}$$

To first order in g, (E18) becomes

$$-i\tilde{f}(m_{N'} + \omega - m_V)\frac{g}{(2\omega\Omega)^{1/2}(m_V - m_N - \omega + i\alpha)} \quad \text{(E19)}$$

corresponding to the amplitude for diagram (i), Fig. 3.

The transition probability for emitting the θ in the energy range $m_\theta \leqq \omega \leqq \epsilon$, where $m_\theta \ll \epsilon \leqq \lambda$, is

$$\frac{g^2}{4\pi^2} \int_{m_\theta}^{\epsilon} \frac{d\omega k |\tilde{f}(m_{N'} + \omega - m_V)|^2}{(m_N + \omega - m_V)^2}. \quad \text{(E20)}$$

If $\tilde{f}(m_{N'} + \omega - m_V)$ is finite at $\omega = m_\theta$ this transition probability diverges in the limit $\mu \to 0$.

To cancel this singularity we must consider also the process $N + \theta_{\mathbf{k}_1} \to N' + \theta_{\mathbf{k}_2}$. The transition amplitude to first order in H' is

$$-i\tilde{f}(m_{N'}+\omega_2-m_N-\omega_1)(N\theta_{\mathbf{k}_2}|U_+|N\theta_{\mathbf{k}_1})\,. \quad \text{(E21)}$$

Evaluating (E21) to second order in g^2, we obtain according to Eqs. (A16) and (A17) given in Appendix A,

$$-i\tilde{f}(m_{N'}+\omega_2-m_N-\omega_1)\Bigg(\delta_{\mathbf{k}_2,\mathbf{k}_1}+\frac{g^2}{(4\omega_1\omega_2)^{1/2}\Omega}$$

$$\times[(\omega_2-\omega_1+2i\alpha)(m_N+\omega_1-m_V+i\alpha)]^{-1}\Bigg)\,. \quad \text{(E22)}$$

The first and second terms in (E22) are, respectively, the amplitudes for diagrams (ii) and (iii), Fig. 3. To order g^2, the contribution to the probability of finding the θ particle in the energy interval $m_\theta \leqq \omega \leqq \epsilon$ due to the interference between these amplitudes is given by

$$-\frac{g^2}{4\pi^2}|\tilde{f}(m_{N'}-m_N)|^2\int_{m_\theta}^{\epsilon}\frac{d\omega k}{(m_N+\omega-m_V)^2}\,. \quad \text{(E23)}$$

In Eq. (E23), we have summed over initial states of the θ particle in the same energy interval. Now if we add Eqs. (E20) and (E23) we see that the combined transition probability is finite in the limit $\mu\to 0$.

Remark

Note that Eq. (E22) differs from the usual Feynman amplitude in the factor 2 multiplying α. This factor can be neglected in the nondiagonal elements of the U matrix, but is essential here, since we are evaluating the interference term with the disconnected process, diagram (ii) Fig. 3, at $\mathbf{k}_1=\mathbf{k}_2$.

Reprinted from THE PHYSICAL REVIEW, Vol. 157, No. 5, 1376–1399, 25 May 1967
Printed in U. S. .A

Neutral Vector Mesons and the Hadronic Electromagnetic Current*

NORMAN M. KROLL
University of California, San Diego, La Jolla, California

AND

T. D. LEE
Columbia University, New York, New York

AND

BRUNO ZUMINO
New York University, New York, New York

(Received 6 January 1967)

The question of whether the entire hadronic electromagnetic current operator can be identical with a linear combination of the renormalized field operators for the known neutral vector mesons ρ^0, ϕ^0, and ω^0 is investigated in the context of a Lagrangian field theory. It is found that such an identity is completely consistent with gauge invariance, provided that these mesons are coupled only to conserved currents. The general renormalization problem of the strong interactions of these vector mesons is discussed. It is shown that the proposed identity between the hadronic electromagnetic current and the renormalized meson fields can be related to the possible identity between the unrenormalized currents generating the neutral vector mesons and those generating the photon; furthermore, this proposed identity leads to an exact relation between the entire $O(e^2)$ hadronic contribution to the photon propagator and the renormalized propagators of the neutral vector mesons, and such a relation implies, among other consequences, that to $O(e^2)$ and neglecting leptonic contributions, the ratio of the unrenormalized charge e_0 and the renormalized charge e is finite. Various experimental applications are given. In particular, the analysis of ϕ-ω mixing and their leptonic decay rates is made independently of the approximate validity of the SU_3 symmetry.

I. INTRODUCTION

THAT vector mesons might play a dominant role in the description of the electromagnetic interactions of hadrons was first suggested by the interpretation of the electromagnetic form factors of the nucleon.[1] Subsequently, the idea of vector dominance has been extended to apply to all electromagnetic interactions of hadrons.[2] It is evident from a study of the literature of this subject that it is, at the least, of great heuristic value to treat the vector mesons as elementary particles in this context. The utility of this kind of treatment can be substantially augmented by the inclusion of an explicit statement of the meaning of vector dominance (as described in dispersion theory) in the language of a local Lagrangian field theory. The statement which we propose for this purpose is the following: "To a very good approximation the entire hadronic electromagnetic current operator is *identical* with a linear combination of the known neutral vector-meson fields." The principal purpose of this paper is to exhibit a Lagrangian field theory in which the approximate identification referred to above becomes exact, and to examine its various theoretical implications and practical consequences; in this theory, the Maxwell equation can be written as

$$\partial F_{\mu\nu}/\partial x_\mu = e(\lambda_\rho \rho_\nu + \lambda_\phi \phi_\nu + \lambda_\omega \omega_\nu) - e(j_\nu)_{\text{lepton}}\,, \quad (1.1)$$

where

$$F_{\mu\nu} = \partial A_\nu/\partial x_\mu - \partial A_\mu/\partial x_\nu\,. \quad (1.2)$$

A_ν, ρ_ν, ϕ_ν, and ω_ν are, respectively, the field operators of the photon, the neutral ρ meson, the ϕ meson and the ω meson, λ_ρ, λ_ϕ, and λ_ω are constants, e is the charge of the electron, and $(j_\mu)_{\text{lepton}}$ is the current operator of the charged leptons.

The question of gauge invariance is studied in detail in Sec. II; it is shown that Eq. (1.1) is completely consistent with the requirement of gauge invariance, provided that the currents generating the three neutral vector-meson fields are all conserved currents.

In Secs. III and IV we examine the general renormalization problem of the strong interaction of a single massive vector meson which can be either stable or unstable. It is shown that if the unrenormalized current generating such a meson is conserved, then in the limit that the unrenormalized mass m_0 of the vector meson is infinite, the renormalized meson field becomes necessarily proportional to its unrenormalized current operator. In addition, it can be shown that the nonzero observed mass implies that the unrenormalized mass $m_0 \neq 0$, and m_0 must be ∞ if the theory is divergent. These considerations can be readily applied to the known neutral vector mesons. The proposed identity

* This research was supported in part by the U. S. Atomic Energy Commission and the National Science Foundation.

[1] Y. Nambu, Phys. Rev. **106**, 1366 (1957). W. R. Frazer and J. R. Fulco, *ibid.* **117**, 1603 (1960).

[2] M. Gell-Mann and F. Zachariasen, Phys. Rev. **124**, 953 (1961); M. Gell-Mann, *ibid.* **125**, 1067 (1962); Y. Nambu and J. J. Sakurai, Phys. Rev. Letters **8**, 79 (1962); M. Gell-Mann, D. Sharp, and W. G. Wagner, *ibid.* **8**, 261 (1962); G. Feldman and P. T. Mathews, Phys. Rev. **132**, 823 (1963); S. Bergman and S. Drell, *ibid.* **133**, B791 (1964); R. F. Dashen and D. H. Sharp, *ibid.* **133**, B1585 (1964); L. Stodolsky, *ibid.* **134**, B1099 (1964); G. Barton and B. G. Smith, Nuovo Cimento **36**, 436 (1965); R. Gatto, Ergeb. Exakt. Naturw., Vol. 39; M. Ross and L. Stodolsky, Phys. Rev. **149**, 1172 (1966); D. S. Beder, *ibid.* **149**, 1203 (1966).

between the hadronic electromagnetic current and the renormalized meson fields is, then, related to the possible identity between the unrenormalized currents generating the vector mesons and those generating the photon.

Throughout the paper, the renormalizability of the strong interactions of these neutral vector mesons is assumed. As is well known, the interaction between a neutral vector meson and a conserved vector current composed only of bilinear products of spin-$\frac{1}{2}$ and spin-0 fields with the minimal order of derivatives can be shown to be renormalizable in a perturbation series.

As we shall see, Eq. (1.1) implies that the isovector part of the hadronic electromagnetic current $(J_\mu{}^\gamma)$ is given by

$$(J_\mu{}^\gamma)_{\text{isovector}}=-(m_\rho{}^2/g_\rho)\rho_\mu, \qquad (1.3a)$$

where all symbols refer to the renormalized quantities; m_ρ is the observed mass of the neutral ρ meson, g_ρ is the renormalized coupling constant, etc. It will be shown that the unrenormalized mass $m_\rho{}^0$ of the neutral ρ meson must be greater than $2m_\pi$ where m_π is the observed pion mass. If the theory is divergent, then $m_\rho{}^0$ is ∞, and consequently (1.3a) becomes identical with an alternative proposal[3] that the unrenormalized isovector part of the hadronic electromagnetic current $(J_\mu{}^\gamma)^0$ is the same as the unrenormalized current $(J_\mu{}^\rho)^0$ which generates the ρ-meson field;

$$(J_\mu{}^\gamma)^0{}_{\text{isovector}}=(J_\mu{}^\rho)^0, \qquad (1.3b)$$

where both currents are normalized so that the spatial integrals of their fourth components are all equal to i times the z component isospin operator. The converse statement is also true: (1.3b) becomes identical with (1.3a), provided the unrenormalized mass $m_\rho{}^0=\infty$.

On the other hand, these two proposals would be different if the unrenormalized mass $m_\rho{}^0$ were finite. For example, (1.3a) implies that the isovector part of any electromagnetic form factor $F_{AB}{}^\gamma(q^2)$, which can be arbitrarily defined, for any real or virtual transition $A\rightarrow B+\gamma$ is related to the similarly defined form factor $F_{AB}{}^\rho(q^2)$ for the corresponding virtual or real transition $A\rightarrow B+\rho^0$, at the same 4-momentum transfer q_μ, by

$$[F_{AB}{}^\gamma(q^2)]_{\text{isovector}}=\frac{m_\rho{}^2}{m_\rho{}^2+q^2}F_{AB}{}^\rho(q^2), \qquad (1.4a)$$

where A and B can be any two hadronic systems. However, the alternative proposal (1.3b) leads to the identity

$$[F_{AB}{}^\gamma(q^2)]_{\text{isovector}} =\frac{m_\rho{}^2}{m_\rho{}^2+q^2}\left[1+\frac{q^2}{(m_\rho{}^0)^2}\right]F_{AB}{}^\rho(q^2), \qquad (1.4b)$$

[3] See, e.g., J. J. Sakurai, Ann. Phys. (N. Y.) **11**, 1 (1960); M. Gell-Mann and F. Zachariasen (Ref. 2).

which becomes the same as (1.4a) only if the unrenormalized mass $m_\rho{}^0$ becomes ∞. If $m_\rho{}^0$ were finite, then the alternative proposal (1.3b) implies that

$$[F_{AB}{}^\gamma(q^2)]_{\text{isovector}}=0 \quad \text{at} \quad q^2+(m_\rho{}^0)^2=0. \qquad (1.5)$$

Consequently, these two proposals (1.3a) and (1.3b) can be distinguished by examining experimentally the *zeros* of any isovector electromagnetic form factor in the timelike q^2 region.

Similar conclusions can be obtained for the ϕ meson and the ω meson. The renormalization problem of the actual ϕ-ω complex is slightly complicated because of their decays and because of their mixing. The general mathematical analysis is given in Sec. V and is independent of any assumption of the approximate validity of SU_3. This is based upon the fact that the renormalized hypercharge and baryon number currents can be defined independently of SU_3. The renormalized $\phi_\mu(x)$ and $\omega_\mu(x)$ fields are defined in such a way that if the ϕ and ω mesons were stable, then $\langle\text{vac}|\phi_\mu(x)|\omega\rangle=\langle\text{vac}|\omega_\mu(x)|\phi\rangle=0$; i.e., the matrix elements of $\phi_\mu(x)$ do not carry the ω-meson pole, and those of $\omega_\mu(x)$ do not carry the ϕ-meson pole. We find it most convenient to characterize ϕ-ω mixing by means of the resolution of the currents which act as sources of these fields in terms of the hypercharge and baryon number currents. Two angles, which we designate by θ_Y and θ_N, are in general necessary to describe this resolution. The isoscalar hadronic electromagnetic current becomes related to the renormalized fields ϕ_μ and ω_μ and the angle θ_Y by

$$(J_\mu{}^\gamma)_{\text{isoscalar}}=-\tfrac{1}{2}g_Y{}^{-1} \times[(\cos\theta_Y)m_\phi{}^2\phi_\mu-(\sin\theta_Y)m_\omega{}^2\omega_\mu], \qquad (1.6)$$

where g_Y is the renormalized hypercharge coupling constant.

In general, the two angles θ_Y and θ_N are different even to first order in the SU_3 symmetry-breaking interaction. Of course, in the limit of SU_3 symmetry, one must have $\theta_Y=\theta_N=0$. The actual values of θ_Y and θ_N depend on the nature of the SU_3 symmetry-breaking interactions for which a number of models can be made. We shall see, for example, that if one makes the *ad hoc* assumption that all SU_3 symmetry-breaking effects are due to the off-diagonal matrix elements of the "bare" mass matrix M_0 between $\phi_\nu{}^0$ and $\omega_\nu{}^0$, then

$$\theta_Y=\theta_N\neq 0. \qquad (1.7)$$

We call this model the "mass-mixing" model.[4] On the other hand, one may make the opposite assumption that the "bare" mass matrix M_0 is diagonal, but the

[4] The "mass-mixing" model is formally similar to the "particle-mixing" model considered by S. Coleman and H. J. Schnitzer, Phys. Rev. **134**, B863 (1964). We note that both this and the current-mixing model are consistent with the transversality of the vector mesons (i.e., the vector mesons are coupled to conserved currents). See also S. L. Glashow, Phys. Rev. Letters **11**, 48 (1963); J. J. Sakurai, Phys. Rev. **132**, 434 (1963).

SU_3 symmetry is broken by certain current operators terms. Such a model[5] is called the "current-mixing" model; in this model $\theta_Y \neq \theta_N$, but

$$m_\omega^2 \tan\theta_Y = m_\phi^2 \tan\theta_N. \tag{1.8}$$

Within the model, this relation between θ_Y and θ_N holds to all orders of the SU_3 symmetry-breaking interaction.

If one makes the further assumption that the SU_3 symmetry-breaking interaction transforms like the isoscalar member of a SU_3 octet,[6] and equates the observed masses of the nine vector mesons with the first-order perturbation expressions, then one finds $\theta_Y \cong 33°$, $\theta_N \cong 21°$ in the "current-mixing" model, but $\theta_Y = \theta_N \cong 32°$ in the "mass-mixing" model (or, $\theta_Y = \theta_N \cong 39°$ in a variation of the same "mass-mixing" model). The details of these special models are given in Sec. V 3. It is important to note that, independently of the dynamical model, the actual values of g_Y, θ_Y, and θ_N can be determined (at least, to a good approximation) by using the known rate of $\phi^0 \to K^+ + K^-$ and by measuring the leptonic decay rates of $\phi^0 \to l^+ + l^-$ and $\omega^0 \to l^+ + l^-$.

In Sec. VI, a discussion of the photon propagator is given. The entire $O(e^2)$ hadronic contribution to the photon propagation is expressed explicity in terms of the renormalized propagators of the vector mesons. From this expression, it follows that to $O(e^2)$ and neglecting leptonic contributions, the ratio of the unrenormalized charge e_0 and the renormalized charge e is finite. The upper limit of $(e_0/e)^2$ can be estimated, and we find

$$1 < (e_0/e)^2 < 1 + \tfrac{1}{4} m_\pi^{-2} e^2 \times [g_\rho^{-2} m_\rho^2 + \tfrac{1}{9} g_Y^{-2} (\cos^2\theta_Y m_\phi^2 + \sin^2\theta_Y m_\omega^2)] \cong 1.03.$$

Various applications of the proposed identity between the hadronic electromagnetic current and the vector-meson fields are discussed in Sec. VII. Some of these results have already been extensively studied in the literature; they are included in this paper, but with particular emphasis on the underlying assumptions and approximations used in the derivations.

In this paper, we consider only the usual hadronic electromagnetic current which is odd under the particle-antiparticle conjugation operator C determined by the strong interaction. The question whether there does, or does not, exist an additional $C = +1$ part of the hadronic electromagnetic current is not discussed.

II. GAUGE INVARIANCE

To simplify our discussion we will consider first only the isovector part of the hadronic electromagnetic current. The corresponding part of the Maxwell equation becomes, according to Eq. (1.1),

$$(\partial F_{\mu\nu}/\partial x_\mu)_{\text{isovector}} = e\lambda_\rho \rho_\nu. \tag{2.1}$$

[5] The "current-mixing" model is similar to the vector-mixing model considered by S. Coleman and H. J. Schnitzer (Ref. 4).

[6] M. Gell-Mann, Phys. Rev. **125**, 1067 (1962); S. Okubo, Progr. Theoret. Phys. (Kyoto) **27**, 949 (1962).

The complete Lagrangian can be written as (neglecting the weak interaction)

$$\mathfrak{L} = \mathfrak{L}_{\text{free}} + \mathfrak{L}_{\text{st}} + \mathfrak{L}_\gamma. \tag{2.2}$$

In this section we assume that the free Lagrangian $\mathfrak{L}_{\text{free}}$ and the strong interaction Lagrangian $\mathfrak{L}_{\text{st}}$ are already given, but demonstrate that for arbitrary $\mathfrak{L}_{\text{free}} + \mathfrak{L}_{\text{st}}$ a gauge-invariant Lagrangian $\mathfrak{L}_\gamma$ can be constructed which will yield Eq. (2.1) as part of its dynamical equations, provided that the neutral ρ-meson field is coupled only to a conserved current.

As a consequence of the strong interaction, the neutral ρ-meson field operator satisfies the dynamical equation

$$\partial G_{\mu\nu}{}^\rho/\partial x_\mu - m_\rho^2 \rho_\nu = g_\rho J_\nu{}^\rho + O(e), \tag{2.3}$$

where

$$G_{\mu\nu}{}^\rho = \frac{\partial}{\partial x_\mu}\rho_\nu - \frac{\partial}{\partial x_\nu}\rho_\mu. \tag{2.4}$$

m_ρ is the *observed* mass of ρ_μ and g_ρ is a *finite* coupling constant, depending on the normalization of the current $J_\mu{}^\rho$. Equation (2.3) is the "renormalized" field equation of the meson field. The term $O(e)$ shows that it is valid only if one neglects the electromagnetic interaction. [If one wishes, one may also regard Eq. (2.3) as the definition of $g_\rho J_\mu{}^\rho$.]

The current $J_\nu{}^\rho$ is assumed to be conserved:

$$\partial J_\nu{}^\rho/\partial x_\nu = 0;$$

therefore,

$$\partial \rho_\nu/\partial x_\nu = 0 \tag{2.5}$$

on account of Eq. (2.3). The detailed form of $J_\nu{}^\rho$ depends on the strong interaction Lagrangian and the renormalization process which will be discussed in the next section. Here, the discussion of gauge invariance can be made independently of the detailed form of $J_\nu{}^\rho$, provided that it is conserved. It can be readily shown that since $J_\nu{}^\rho$ is a conserved vector current and it transforms like the $I_z = 0$ member of an isotriplet under the isospin rotation, its spatial integral must be proportional to the z component I_z of the isospin operator. (See Appendix A.) For convenience, we shall adopt the normalization convention

$$-i\int J_4{}^\rho d^3r = I_z. \tag{2.6}$$

The isovector part of the hadronic electromagnetic interaction is assumed to be given by the Lagrangian density

$$(\mathfrak{L}_\gamma)_{\text{isovector}} = e(\lambda_\rho g_\rho/m_\rho^2) \times (J_\mu{}^\rho A_\mu + \tfrac{1}{2} g_\rho^{-1} G_{\mu\nu}{}^\rho F_{\mu\nu}) + O(e^2). \tag{2.7}$$

The term $O(e^2)$ depends on the derivatives of the charged fields in $J_\mu{}^\rho$. It is zero, if $J_\mu{}^\rho$ does not contain such derivatives; otherwise, it can be easily generated, say, by the usual minimal principle. Equation (2.7) gives a gauge invariant[7] Lagrangian $\mathfrak{L}_\gamma$. Upon varying with respect to A_μ, it gives

$$(\partial F_{\mu\nu}/\partial x_\mu)_{\text{isovector}}=e(J_\nu{}^\gamma)_{\text{isovector}},$$

where

$$(J_\nu{}^\gamma)_{\text{isovector}}=-\frac{\lambda_\rho g_\rho}{m_\rho{}^2}\left(J_\nu{}^\rho-g_\rho{}^{-1}\frac{\partial}{\partial x_\mu}G_{\mu\nu}{}^\rho\right)+O(e). \quad (2.8)$$

From Eqs. (2.3) and (2.8), it follows that

$$(J_\nu{}^\gamma)_{\text{isovector}}=\lambda_\rho\rho_\nu,$$

which is Eq. (2.1). By using the normalization condition, Eq. (2.6), and the fact that the total hadronic charge Q is given by

$$Q=I_z+\tfrac{1}{2}Y,$$

one finds

$$\lambda_\rho=-(m_\rho{}^2/g_\rho).$$

Thus,[8]

$$(J_\nu{}^\gamma)_{\text{isovector}}=J_\nu{}^\rho-g_\rho{}^{-1}(\partial G_{\mu\nu}{}^\rho/\partial x_\mu)=-(m_\rho{}^2/g_\rho)\rho_\mu. \quad (2.9)$$

Although we have established Eq. (2.9) only to the zeroth order in e, the inclusion of all orders in e is possible, but a full discussion will not be given in this paper. Furthermore we observe that the definition of g_ρ has not yet been completely given as it requires a specification of a normalization condition on ρ_μ, to be given in Sec. III. For the present, however, we point out that the combinations $g_\rho{}^{-1}G_{\mu\nu}{}^\rho$, $g_\rho{}^{-1}\rho_\mu$, and $\lambda_\rho g_\rho$ which appear in Eqs. (2.7) and (2.9) are all independent of this specification.

For any real or virtual photon process $A\rightarrow B+\gamma$, the isovector part of the matrix element of $J_\mu{}^\gamma$ is then related to the matrix element of $J_\mu{}^\rho$ for the corresponding virtual or real process $A\rightarrow B+\rho^0$ provided the 4-momentum q_λ of γ is the same as that of ρ^0. By using Eqs. (2.3), (2.5), and (2.9), we find

$$\langle B|J_\mu{}^\gamma(x)|A\rangle_{\text{isovector}}=\frac{m_\rho{}^2}{q^2+m_\rho{}^2}\langle B|J_\mu{}^\rho(x)|A\rangle, \quad (2.10)$$

where A and B can be any single- or multiple-particle states of the hadrons. Sometimes, it is convenient to express the matrix elements of $J_\mu{}^\gamma$ and $J_\mu{}^\rho$ in terms of a sum of form factors:

$$\langle B|J_\mu{}^\gamma(x)|A\rangle=\sum_i F_{AB}{}^\gamma(q^2)u_B{}^\dagger\Gamma_\mu{}^i u_A\exp(iq_\lambda x_\lambda),$$

and (2.11)

$$\langle B|J_\mu{}^\rho(x)|A\rangle=\sum_i F_{AB}{}^\rho(q^2)u_B{}^\dagger\Gamma_\mu{}^i u_A\exp(iq_\lambda x_\lambda),$$

where $u_B{}^\dagger\Gamma_\mu{}^i u_A$ denotes the appropriate choice of some spin-momentum functions. If A and B are, say, single nucleon states, then $\Gamma_\mu{}^i$ can be either the usual $\gamma_4\gamma_\mu$ or $\gamma_4\sigma_{\mu\nu}q_\nu$ and u_A, u_B the corresponding spinor functions. If A and B are multiparticle states, then $u_B{}^\dagger\Gamma_\mu{}^i u_A$ would depend not only on the spin-momentum variables but also on all other dynamical parameters that characterize the states A and B. The definitions of the form factors $F_{AB}{}^\gamma(q^2)$ and $F_{AB}{}^\rho(q^2)$, of course, depend on the explicit forms of $u_B{}^\dagger\Gamma_\mu{}^i u_A$ and therefore also on the index i. Equation (2.10) states that for whatever choice of such definitions, the form factor $F_{AB}{}^\rho(q^2)$ is related to the corresponding isovector part of $F_{AB}{}^\gamma(q^2)$ at the same q^2 by Eq. (1.4a):

$$[F_{AB}{}^\gamma(q^2)]_{\text{isovector}}=\frac{m_\rho{}^2}{q^2+m_\rho{}^2}F_{AB}{}^\rho(q^2). \quad (1.4a)$$

Thus, compared to $F_{AB}{}^\rho(q^2)$, the electromagnetic form factor $[F_{AB}{}^\gamma(q^2)]_{\text{isovector}}$ always vanishes more rapidly[9] at $|q^2|=\infty$.

Identical considerations can be applied to the isoscalar part. Let the dynamical equation of the ϕ and ω mesons be given by

$$\partial G_{\mu\nu}{}^\phi/\partial x_\mu-m_\phi{}^2\phi_\nu=\mathfrak{J}_\nu{}^\phi,$$

---

[7] Under the gauge transformation $A_\mu\rightarrow A_\mu+\partial\Lambda/\partial x_\mu$ the electromagnetic interaction Lagrangian density (2.7) plus (2.20) (which is given below) transforms like $\mathfrak{L}_\gamma\rightarrow\mathfrak{L}_\gamma-eJ_\mu\partial\Lambda/\partial x_\mu$, where $\mathfrak{L}_\gamma=(\mathfrak{L}_\gamma)_{\text{isovector}}+(\mathfrak{L}_\gamma)_{\text{isoscalar}}$ and $J_\mu=J_\mu{}^\rho+\frac{1}{2}Y_\mu$; for simplicity, all $O(e^2)$ terms are omitted. According to Eqs. (2.6) and (2.16), $-i\int J_4d^3r=Q=I_z+\frac{1}{2}Y$. The difference $J_\mu{}'$ between the current J_μ and the minimal electromagnetic current operator $(J_\mu)_{\min}$, which satisfies the same normalization condition $-i\int(J_4)_{\min}d^3r=Q$, may not be zero, but it is obvious that this difference $J_\mu{}'=J_\mu-(J_\mu)_{\min}$ must satisfy $[\partial J_\mu{}'/\partial x_\mu]=0$ and $\int J_4{}'d^3r=0$. An implicit assumption is made here that $[\partial J_\mu{}'/\partial x_\mu]=0$ is satisfied kinematically, *without* the use of dynamical equations. For example, $J_\mu{}'$ can be simply proportional to $(\partial G_{\mu\nu}{}^\rho/\partial x_\nu)$. Under the gauge transformation (of the second kind), the free Lagrangian and the strong interaction Lagrangian density transform like $(\mathfrak{L}_{\text{free}}+\mathfrak{L}_{\text{st}})\rightarrow(\mathfrak{L}_{\text{free}}+\mathfrak{L}_{\text{st}})+e(J_\mu)_{\min}(\partial\Lambda/\partial x_\mu)$. [If one wishes, one may also regard this as the definition of the minimal current $(J_\mu)_{\min}$.] Correspondingly, the total Lagrangian $\mathfrak{L}=\mathfrak{L}_{\text{free}}+\mathfrak{L}_{\text{st}}+\mathfrak{L}_\gamma$ transforms according to $\mathfrak{L}\rightarrow\mathfrak{L}+eJ_\mu{}'(\partial\Lambda/\partial x_\mu)$. The gauge invariance of the action integral $\int\mathfrak{L}d^4x$, and therefore also that of the equation of motion, is consequently guaranteed. For the ρ^0 meson, if we identify the unrenormalized current $(J_\mu{}^\rho)^0$, which enters in Eq. (3.2) below, as the minimal z-component isospin current, then the current $J_\mu{}^\rho$ is defined by Eq. (3.13), $J_\mu{}^\rho=(J_\mu{}^\rho)^0+(\text{constant})\partial G_{\mu\nu}{}^\rho/\partial x_\nu$; therefore, the corresponding difference, $J_\mu{}^\rho-(J_\mu{}^\rho)^0$, is $J_\mu{}'=(\text{constant})\ \partial G_{\mu\nu}{}^\rho/\partial x_\nu$, which does satisfy $(\partial J_\mu{}'/\partial x_\mu)=0$ in a purely kinematical way.

It is important to note that a single direct coupling $\rho_\mu A_\mu$ violates gauge invariance. This can be most easily seen by observing that such a term in the Lagrangian generates a non-gauge-invariant contribution to the ρ-meson current proportional to A_μ. This circumstance is associated with the fact that $(\partial\rho_\mu/\partial x_\mu)=0$ is *not* a kinematical identity, but is derived only after using the dynamical equation and the condition that $(\partial J_\mu{}^\rho/\partial x_\mu)=0$. This point has often been incorrectly stated in the literature. [See, e.g., M. Ross and L. Stodolsky (Ref. 2) and L. Stodolsky (Ref. 2).]

The same gauge-invariant Lagrangian densities (2.7) and (2.20) can also be cast into other different, but equivalent, forms. Some of these alternative forms are discussed in Appendix B.

[8] We emphasize that (2.7)–(2.9) are equations in renormalized quantities. The term $J_\mu{}^\rho A_\mu$ may be regarded as the "direct" photon-hadron coupling while the term $G_{\mu\nu}{}^\rho F_{\mu\nu}$ is the gauge-invariant photon-ρ meson mixing term discussed by Feldman and Matthews (Ref. 2). We note that Eq. (2.9) is achieved *not* via the introduction of a term of the form $\rho_\mu A_\mu$ in the Lagrangian, but by the assumption of a special relation between the "direct" source term and the mixing term. In the case of free photons, it is the direct source term rather than the mixing term which couples the photon to the hadrons, but nevertheless, Eq. (2.9) holds. This would seem to resolve the issue raised by Feldman and Matthews in their Ref. 14. A related resolution has been given by Barton and Smith (Ref. 2, Sec. 4.3) in the context of dispersion relations, with reference to the connection between $f_{\rho\pi\gamma}$ and $f_{\rho\pi\phi}$. See case 2 of Appendix B for further discussions.

[9] This property can be related to the Barton and Smith "boundary condition" [Ref. 2, Eq. (4.14)].

and

$$\partial G_{\mu\nu}{}^{\omega}/\partial x_\mu - m_\omega{}^2\omega_\nu = \mathcal{J}_\nu{}^{\omega}, \tag{2.12}$$

where ϕ_ν and ω_ν are, respectively, the "renormalized" field operators of the ϕ meson and the ω meson, m_ϕ and m_ω are the observed masses,

$$G_{\mu\nu}{}^{\phi} = \frac{\partial}{\partial x_\mu}\phi_\nu - \frac{\partial}{\partial x_\nu}\phi_\mu,$$

and

$$G_{\mu\nu}{}^{\omega} = \frac{\partial}{\partial x_\mu}\omega_\nu - \frac{\partial}{\partial x_\nu}\omega_\mu. \tag{2.13}$$

Both currents $\mathcal{J}_\nu{}^{\phi}$ and $\mathcal{J}_\nu{}^{\omega}$ are assumed to be conserved:

$$\partial \mathcal{J}_\nu{}^{\phi}/\partial x_\nu = \partial \mathcal{J}_\nu{}^{\omega}/\partial x_\nu = 0. \tag{2.14}$$

Consequently, as will also be proved in Appendix A, the hypercharge Y and the baryon number N must be linear functions of the spatial integrals of $\mathcal{J}_4{}^{\phi}$ and $\mathcal{J}_4{}^{\omega}$. Without any loss of generality, we may express this linear relationship in terms of four real constants g_Y, g_N, θ_Y, and θ_N:

$$g_Y Y = -i\int[\cos\theta_Y \mathcal{J}_4{}^{\phi} - \sin\theta_Y \mathcal{J}_4{}^{\omega}]d^3r,$$

and

$$g_N N = -i\int[\sin\theta_N \mathcal{J}_4{}^{\phi} + \cos\theta_N \mathcal{J}_4{}^{\omega}]d^3r. \tag{2.15}$$

It is convenient to define two conserved currents Y_μ and N_μ which satisfy the normalization conditions

$$-i\int Y_4 d^3r = Y, \tag{2.16}$$

and

$$-i\int N_4 d^3r = N; \tag{2.17}$$

these two currents are related to $\mathcal{J}_\mu{}^{\phi}$ and $\mathcal{J}_\mu{}^{\omega}$ by

$$g_Y Y_\mu = \cos\theta_Y \mathcal{J}_\mu{}^{\phi} - \sin\theta_Y \mathcal{J}_\mu{}^{\omega}$$

and

$$g_N N_\mu = \sin\theta_N \mathcal{J}_\mu{}^{\phi} + \cos\theta_N \mathcal{J}_\mu{}^{\omega}. \tag{2.18}$$

The inverse relations are

$$\mathcal{J}_\mu{}^{\phi} = [\cos(\theta_Y - \theta_N)]^{-1}[\cos\theta_N g_Y Y_\mu + \sin\theta_Y g_N N_\mu]$$

and

$$\mathcal{J}_\mu{}^{\omega} = [\cos(\theta_Y - \theta_N)]^{-1}[-\sin\theta_N g_Y Y_\mu + \cos\theta_Y g_N N_\mu]. \tag{2.19}$$

It is important to note that the above definitions of g_Y, g_N, θ_Y, and θ_N depend only on the conservation of Y and N, and are *independent* of any assumption concerning the approximate validity of SU_3 symmetry. The constants g_Y and g_N are the renormalized coupling constants, and the angles θ_Y and θ_N relate the currents $g_Y Y_\mu$ and $g_N N_\mu$ to $\mathcal{J}_\mu{}^{\phi}$ and $\mathcal{J}_\mu{}^{\omega}$. In general, these two angles are different, $\theta_Y \neq \theta_N$, even to the first order in the SU_3 symmetry-breaking interaction. Estimations of the actual values of θ_Y and θ_N can be made by making specific dynamical assumptions; these discussions will be given in Sec. V.

In order to identify the isoscalar part, $(J_\mu{}^{\gamma})_{\text{isoscalar}}$, of the hadronic electromagnetic current with a linear sum of the renormalized field operators ϕ_μ and ω_μ, we assume that the corresponding isoscalar part of the electromagnetic interaction is given by the following Lagrangian density:

$$(\mathcal{L}_\gamma)_{\text{isoscalar}} = -\tfrac{1}{2}e\{J_\mu{}^{Y}A_\mu + \tfrac{1}{2}g_Y{}^{-1} \times[\cos\theta_Y G_{\mu\nu}{}^{\phi} - \sin\theta_Y G_{\mu\nu}{}^{\omega}]F_{\mu\nu}\}. \tag{2.20}$$

As a consequence, the isoscalar part of the electromagnetic field is given by

$$(\partial F_{\mu\nu}/\partial x_\mu)_{\text{isoscalar}} = e(J_\nu{}^{\gamma})_{\text{isoscalar}}, \tag{2.21}$$

where $(J_\mu{}^{\gamma})_{\text{isoscalar}}$ is given by Eq. (1.6).

For any real or virtual photon process $A \to B+\gamma$, the isoscalar part of the matrix element of $J_\mu{}^{\gamma}$ is then related to the matrix elements of $\mathcal{J}_\mu{}^{\phi}$ and $\mathcal{J}_\mu{}^{\omega}$ for the corresponding virtual or real processes $A \to B+\phi^0$, and $A \to B+\omega^0$ at the same 4-momentum transfer. By using Eq. (1.6), we find

$$\langle B|J_\mu{}^{\gamma}|A\rangle_{\text{isoscalar}} = \tfrac{1}{2}g_Y{}^{-1}\left[\cos\theta_Y\left(\frac{m_\phi{}^2}{q^2+m_\phi{}^2}\right)\langle B|\mathcal{J}_\mu{}^{\phi}|A\rangle - \sin\theta_Y\left(\frac{m_\omega{}^2}{q^2+m_\omega{}^2}\right)\langle B|\mathcal{J}_\mu{}^{\omega}|A\rangle\right]. \tag{2.22}$$

The matrix elements of $\mathcal{J}_\mu{}^{\phi}$ and $\mathcal{J}_\mu{}^{\omega}$ are, in turn, related to those of Y_μ and N_μ through Eq. (2.19). We have, then, the following alternative expression:

$$2\langle B|J_\mu{}^{\gamma}|A\rangle_{\text{isoscalar}} = \left(\frac{m_\phi{}^2}{q^2+m_\phi{}^2}\right)[C_Y\langle B|Y_\mu|A\rangle + C_N\langle B|N_\mu|A\rangle] + \left(\frac{m_\omega{}^2}{q^2+m_\omega{}^2}\right)[(1-C_Y)\langle B|Y_\mu|A\rangle - C_N\langle B|N_\mu|A\rangle], \tag{2.23}$$

where

$$C_Y = [\cos(\theta_Y - \theta_N)]^{-1}\cos\theta_Y\cos\theta_N,$$

and

$$C_N = [g_Y\cos(\theta_Y - \theta_N)]^{-1}g_N\cos\theta_Y\sin\theta_Y. \tag{2.24}$$

Equation (2.23) shows that at $q^2=0$,

$$\langle B|J_\mu{}^{\gamma}|A\rangle_{\text{isoscalar}} = \tfrac{1}{2}\langle B|Y_\mu|A\rangle.$$

III. RENORMALIZATION OF THE NEUTRAL ρ MESON

In this section, we will discuss the renormalization problem of the *strong* interaction of the neutral ρ

meson, but only in the absence of the electromagnetic interaction.

Let us assume that the part of the Lagrangian density describing the neutral ρ meson and its strong interaction is given by

$$\mathcal{L}_\rho = -\tfrac{1}{4}(G_{\mu\nu}{}^0)^2 - \tfrac{1}{2}(m_\rho{}^0\rho_\mu{}^0)^2 - g_\rho{}^0\rho_\mu{}^0(J_\mu{}^\rho)^0, \quad (3.1)$$

where the superscript 0 denotes the unrenormalized quantities; thus, $\rho_\mu{}^0$ is the unrenormalized meson-field operator

$$G_{\mu\nu}{}^0 = \frac{\partial}{\partial x_\mu}\rho_\nu{}^0 - \frac{\partial}{\partial x_\nu}\rho_\mu{}^0,$$

$m_\rho{}^0$ is its unrenormalized mass, $(J_\mu{}^\rho)^0$ is the unrenormalized current and $g_\rho{}^0$ is the unrenormalized coupling constant. The equation of motion is given by

$$\partial G_{\mu\nu}{}^0/\partial x_\mu - (m_\rho{}^0)^2\rho_\nu{}^0 = g_\rho{}^0(J_\nu{}^\rho)^0. \quad (3.2)$$

In Eq. (3.1), we assume for simplicity that $(J_\nu{}^\rho)^0$ does not depend on $\rho_\mu{}^0$; otherwise, Eq. (3.1) has to be modified so that the equation of motion [Eq. (3.2)] remains valid.

The current operator $(J_\nu{}^\rho)^0$ transforms like the z component of an isospin triplet, and it is assumed to be conserved; i.e.,

$$\partial(J_\nu{}^\rho)^0/\partial x_\nu = 0. \quad (3.3)$$

Therefore, according to the general theorem established in Appendix A, the integral $\int(J_4{}^\rho)^0 d^3r$ is proportional to the observed z-component isospin operator I_z, and we may, without any loss of generality, choose

$$\int (J_4{}^\rho)^0 d^3r = iI_z. \quad (3.4)$$

Since the normalization of $\rho_\mu{}^0$ is fixed by the Lagrangian density (3.1) and the canonical rules, condition (3.4) defines the unrenormalized coupling constant $g_\rho{}^0$.

To obtain renormalized equations we first set

$$\rho_\mu{}^0 = Z^{1/2}\rho_\mu, \quad (3.5)$$

where Z is a wave-function renormalization constant. The Lagrangian density $\mathcal{L}_\rho$ given by Eq. (3.1) may be written in terms of ρ_μ in the form

$$\mathcal{L}_\rho = \mathcal{L}_{\text{free}} + \mathcal{L}_{\text{st}}, \quad (3.6)$$

where

$$\mathcal{L}_{\text{free}} = -\tfrac{1}{4}(G_{\mu\nu}{}^\rho)^2 - \tfrac{1}{2}m_\rho{}^2\rho_\mu{}^2, \quad (3.7)$$

$$\mathcal{L}_{\text{st}} = -\tfrac{1}{4}(Z-1)(G_{\mu\nu}{}^\rho)^2 - \tfrac{1}{2}[(m_\rho{}^0)^2 Z - m_\rho{}^2]\rho_\mu{}^2 - g_\rho{}^0 Z^{1/2}(J_\nu{}^\rho)^0\rho_\nu, \quad (3.8)$$

and m_ρ is the observed mass of the ρ meson. The above expression of $\mathcal{L}_\rho$ is convenient for a perturbation series expansion in terms of $\mathcal{L}_{\text{st}}$, since the free Lagrangian is constructed so that it would have the correct energy spectrum if the neutral ρ meson were stable.

The equation of motion of the renormalized field ρ_μ is given by

$$\frac{\partial G_{\mu\nu}}{\partial x_\mu} - m_\rho{}^2\rho_\nu = Z^{-1/2}g_\rho{}^0 Z_0(J_\nu{}^\rho)^0 + (1-Z_0)\frac{\partial G_{\mu\nu}}{\partial x_\mu} \equiv g_\rho J_\nu{}^\rho, \quad (3.9)$$

with

$$Z_0 = (m_\rho/m_\rho{}^0)^2, \quad (3.10)$$

and

$$J_\nu{}^\rho = \frac{g_\rho{}^0}{g_\rho}\frac{Z_0}{Z^{1/2}}(J_\nu{}^\rho)^0 + (1-Z_0)\frac{1}{g_\rho}\frac{\partial G_{\mu\nu}}{\partial x_\mu}.$$

The renormalized current $J_\nu{}^\rho$ is the same current as that used in Eq. (2.3). Its normalization condition [Eq. (2.6)] fixes the relation between g_ρ and Z; the relation being

$$Z_0 g_\rho{}^0 = Z^{1/2}g_\rho. \quad (3.11)$$

Hence,

$$\frac{\rho_\nu}{g_\rho} = \frac{1}{Z_0}\frac{\rho_\nu{}^0}{g_\rho{}^0}, \quad \frac{G_{\mu\nu}}{g_\rho} = \frac{1}{Z_0}\frac{G_{\mu\nu}{}^0}{g_\rho{}^0} \quad (3.12)$$

and

$$J_\nu{}^\rho = (J_\nu{}^\rho)^0 + (1-Z_0)\frac{1}{g_\rho}\frac{\partial G_{\mu\nu}}{\partial x_\mu}. \quad (3.13)$$

It follows from Eqs. (3.12) and (3.13) that the ratios ρ_ν/g_ρ and $G_{\mu\nu}/g_\rho$, and the renormalized current density $J_\nu{}^\rho$ are independent of Z as expected. The factor Z is, of course, of utility principally in connection with divergence difficulties and is introduced in order to make it possible to express the theory in terms of finite quantities. Apart from this requirement it can be chosen for convenience and has no physical consequences. Some convenient choices will be discussed later.

By using Eqs. (3.9), (3.10), and (3.13), one finds the important relation

$$(J_\nu{}^\rho)^0 = -\left(\frac{m_\rho{}^2}{g_\rho}\right)\rho_\nu + g_\rho{}^{-1}\left(\frac{m_\rho}{m_\rho{}^0}\right)^2\frac{\partial G_{\mu\nu}{}^\rho}{\partial x_\mu}. \quad (3.14)$$

While the precise value of $m_\rho{}^0$ depends on the form $(J_\nu{}^\rho)^0$ assumed in the particular strong-interaction theory, it can be shown that [see Eq. (4.7) in the next section] $m_\rho{}^0 > 2m_\pi$, and that

$$m_\rho{}^0 = \infty, \quad (3.15)$$

if the theory is divergent. Consequently, the right-hand side of Eq. (3.14) contains only finite quantities. In particular, if the theory is divergent, then $m_\rho{}^0 = \infty$, and

$$(J_\nu{}^\rho)^0 = -(m_\rho{}^2/g_\rho)\rho_\nu. \quad (3.16)$$

In this case, the unrenormalized current $(J_\nu{}^\rho)^0$ becomes proportional to the renormalized field operator ρ_ν. The renormalized current $J_\nu{}^\rho$ is, of course, different from the field operator ρ_ν.

The proportionality between the unrenormalized current operator and the renormalized field operator is a general consequence of a vector field interacting with a conserved current, provided that the unrenormalized mass $=\infty$. If the unrenormalized mass were finite, then one would have Eq. (3.14) instead of Eq. (3.16).

Next, we discuss the relation between the electromagnetic current $J_\nu{}^\gamma$ and the unrenormalized current $(J_\nu{}^\rho)^0$ of the ρ meson. If we assume Eq. (1.3a),

$$(J_\nu{}^\gamma)_{\text{isovector}}=-(m_\rho{}^2/g_\rho)\rho_\nu\,, \tag{1.3a}$$

then, by using Eq. (3.14), we find

$$(J_\nu{}^\gamma)_{\text{isovector}}=(J_\nu{}^\rho)^0-g_\rho{}^{-1}(m_\rho/m_\rho{}^0)^2(\partial G_{\mu\nu}{}^\rho/\partial x_\mu)\,, \tag{3.17}$$

which implies that in the case of an infinite unrenormalized mass $m_\rho{}^0$

$$(J_\nu{}^\gamma)_{\text{isovector}}=(J_\nu{}^\rho)^0. \tag{3.18}$$

Since we do not consider the renormalization problem of the electromagnetic interaction, there is no difference between $J_\nu{}^\gamma$ and the unrenormalized electromagnetic current operator $(J_\nu{}^\gamma)^0$.

There exists an alternative possibility in which one assumes that, instead of (1.3a),

$$(J_\nu{}^\gamma)^0{}_{\text{isovector}}=(J_\nu{}^\rho)^0\,, \tag{1.3b}$$

and therefore, neglecting higher-order electromagnetic corrections,

$$(J_\nu{}^\gamma)_{\text{isovector}}=-\left(\frac{m_\rho{}^2}{g_\rho}\right)\rho_\nu+g_\rho{}^{-1}\left(\frac{m_\rho}{m_\rho{}^0}\right)^2\frac{\partial G_{\mu\nu}{}^\rho}{\partial x_\mu}. \tag{3.19}$$

In the case of $m_\rho{}^0=\infty$, one has

$$(J_\nu{}^\gamma)_{\text{isovector}}=(-m_\rho{}^2/g_\rho)\rho_\mu. \tag{3.20}$$

It has already been mentioned in the introduction that these two views become identical if $m_\rho{}^0=\infty$; but if the unrenormalized mass $m_\rho{}^0$ turns out to be finite, then the proposal (1.3a) implies the form-factor relations given by Eq. (1.4a) while the proposal (1.3b) implies the alternative form-factor relation given by[10] Eq. (1.4b).

IV. THE ρ-MESON PROPAGATOR

Let us consider the usual spectral representation of the vacuum expectation value of the commutator $[\rho_\mu(x),\rho_\nu(0)]$ in the Heisenberg representation:

$$\langle[\rho_\mu(x),\rho_\nu(0)]\rangle_{\text{vacuum}}=\int\sigma_\rho(a)\left[\delta_{\mu\nu}-a^{-1}\frac{\partial^2}{\partial x_\mu\partial x_\nu}\right]\Delta_a(x)da\,, \tag{4.1}$$

where

$$\Delta_a(x)=-i(2\pi)^{-3}\int\omega^{-1}\sin\omega t\exp(i\mathbf{q}\cdot\mathbf{r})d^3q\,,$$

and

$$\omega=(\mathbf{q}^2+a)^{1/2}.$$

The function $\sigma_\rho(a)$ is related to the matrix elements of the spatial components of ρ_μ by

$$\sigma_\rho(a)=\tfrac{1}{3}\sum_\Gamma\delta(a-m_\Gamma{}^2)|\langle\Gamma|\boldsymbol{\rho}(0)|\text{vac}\rangle|^2\geqq0\,, \tag{4.2}$$

where the sum extends over all eigenstates $|\Gamma\rangle$ of the strong interaction Hamiltonian that satisfy

$$(\text{momentum})_\Gamma=0,\quad(\text{energy})_\Gamma=m_\Gamma,\quad\text{and}\quad(\text{spin})_\Gamma=1\,, \tag{4.3}$$

and the factor $\frac{1}{3}$ is due to the further sum over all three components of $\boldsymbol{\rho}(0)$. Throughout the paper, all boldface letters denote 3-vectors.

By using the Lagrangian given by Eq. (3.6), one finds that the components of the canonical momentum $\mathbf{\Pi}(x)$ of the field variable $\boldsymbol{\rho}(x)$ are given by

$$\Pi_j=-iZG_{j4}{}^\rho.$$

From the equation of motion, $\mathbf{\Pi}$ is also related to ρ_4 by

$$\rho_4=m_\rho{}^{-2}[i(Z_0/Z)\nabla\cdot\mathbf{\Pi}-g_\rho(J_4{}^\rho)^0]. \tag{4.4}$$

The comparison between the spectral representation [Eq. (4.1)] and the equal-time commutator between $\mathbf{\Pi}$ and $\boldsymbol{\rho}$ leads to the sum rule[11]

$$Z^{-1}=\int\sigma_\rho(a)da. \tag{4.5}$$

Similarly, by using the equal-time commutator between ρ_4 and $\boldsymbol{\rho}$, and assuming that $(J_4{}^\rho)^0$ commutes with $\boldsymbol{\rho}$ at equal time, one finds

$$Z^{-1}(m_\rho{}^0)^{-2}=(Z_0/Z)m_\rho{}^{-2}=\int a^{-1}\sigma_\rho(a)da\,, \tag{4.6}$$

where Z_0 is given by Eq. (3.10). For the physical ρ meson, the integrations in (4.5) and (4.6) extend from $4m_\pi{}^2$ to ∞, where m_π is the observed pion mass. By taking the ratio of (4.5) and (4.6), one obtains

$$(m_\rho{}^0)^2=\left[\int_{4m_\pi{}^2}^\infty a^{-1}\sigma_\rho(a)da\right]^{-1}\int_{4m_\pi{}^2}^\infty\sigma_\rho(a)da. \tag{4.7}$$

Since $\sigma_\rho\geqq0$, Eqs. (4.7), (4.5), and (4.6) imply, respectively,

$$(m_\rho{}^0)^2>4m_\pi{}^2,$$

$$Z^{-1}>0,\quad\text{and}\quad(Z_0/Z)>0.$$

In order that the theory be renormalizable [i.e., that the renormalized propagator given by Eq. (4.10) below exists], the behavior of $\sigma_\rho(a)$ as $a\to\infty$ must be such that

[10] Equation (1.4b) may be compared with Eq. (4.7) of the paper by Gell-Mann and Zachariasen (Ref. 2).

[11] Equations (4.5) and (4.6), together with Eq. (3.5), are the same as Eqs. (12) and (11), respectively, of K. Johnson, Nucl. Phys. **25**, 435 (1961).

$\int a^{-1}\sigma_\rho(a)da$ is convergent. Thus, (Z_0/Z) must be finite. On the other hand, the integral $\int \sigma_\rho(a)da$ need not be convergent. If it in fact diverges, then both Z^{-1} and the unrenormalized mass m_ρ^0 must be infinite. Consequently we establish (3.15).

Combining Eqs. (4.5) and (4.6) we find

$$1-Z_0=Z\int a^{-1}(a-m_\rho^2)\sigma_\rho(a)da. \tag{4.8}$$

For the hypothetical case of a stable ρ meson (i.e., if m_ρ were less than $2m_\pi$), Eq. (4.8) implies the inequality $0\leqq Z_0\leqq 1$, and hence $(m_\rho^0)^2\geqq m_\rho^2$ on account of Eq. (3.10).

The renormalized ρ-meson propagator

$$\hat{D}_{\mu\nu}(x)=-i(2\pi)^{-4}\int D_{\mu\nu}(q)\exp(iq_\lambda x_\lambda)d^4q \tag{4.9}$$

is given in terms of $\sigma_\rho(a)$ through the relation

$$D_{\mu\nu}(q)=\int\frac{\delta_{\mu\nu}+a^{-1}q_\mu q_\nu}{q^2+a-i\epsilon}\sigma_\rho(a)da, \tag{4.10}$$

where ϵ is a positive infinitesimal quantity, q_μ denotes the 4-momentum $(\mathbf{q},iq_0)$ and $q^2=\mathbf{q}^2-q_0^2$. Using (4.6), (4.1), and (4.2), one easily sees that $\hat{D}_{\mu\nu}(x)$ is related to the time-ordered product of the renormalized fields via

$$\hat{D}_{\mu\nu}(x)=\{T[\rho_\mu(x)\rho_\nu(0)]\}_{\text{vacuum}} -im_\rho^{-2}(Z_0/Z)\delta^4(x)\delta_{4\mu}\delta_{4\nu}. \tag{4.11}$$

In our notation, both d^4q and $\delta^4(x)$ are real. The last term in Eq. (4.11) cancels the noncovariant part in $\{T[\rho_\mu(x)\rho_\nu(0)]\}_{\text{vacuum}}$. The appearance of this noncovariant term is due to the fact that the commutator $[\rho_\mu(x),\rho_\nu(0)]$ does not vanish for $x_0=0$.

Because the ρ meson is an unstable particle, it is convenient to define the renormalized mass m_ρ in terms of the inverse propagator. Writing

$$D_{\mu\nu}(q)=\delta_{\mu\nu}F(q^2)+q_\mu q_\nu G(q^2), \tag{4.12}$$

and noting the relation

$$F+q^2G=(Z_0/Z)m_\rho^{-2}, \tag{4.13}$$

which follows from Eqs. (4.10) and (4.6), we obtain

$$D_{\mu\nu}^{-1}(q)=[F(q^2)]^{-1}[\delta_{\mu\nu}-q_\mu q_\nu Z_0^{-1}Zm_\rho^2G(q^2)]. \tag{4.14}$$

The renormalized mass is then defined as that value of $-q^2$ at which the real part of the $\delta_{\mu\nu}$ term of the inverse propagator vanishes; i.e., by the equation

$$\mathrm{Re}[1/F(-m_\rho^2)]=0. \tag{4.15}$$

For the physical ρ meson, $[F(-m_\rho^2)]^{-1}$ is pure imaginary; the same must also hold for $F(-m_\rho^2)$, and we find

$$\mathrm{Re}F(-m_\rho^2)=\int_{4m_\pi^2}^{\infty}\sigma_\rho(a)\mathcal{P}\left[\frac{1}{a-m_\rho^2}\right]da=0,$$

where $\mathcal{P}$ denotes the principal value. If, as is usually assumed, the imaginary part of $(1/F)$ as well as the derivative of the real part can be regarded as constant within the resonance width, then the width is given by the formula

$$\gamma m_\rho=[Z\pi\sigma(m_\rho^2)]^{-1}Z_1, \tag{4.16}$$

where we have written Z/Z_1 for the derivative of the real part. Thus

$$\frac{Z}{Z_1}\equiv\left[\frac{d}{dq^2}\mathrm{Re}\left(\frac{1}{F}\right)\right]_{q^2=-m_\rho^2}. \tag{4.17}$$

The definitions of m_ρ^2 and γ given by Eqs. (4.15) and (4.16) correspond to those given by a pole approximation for F in the resonance region of the form

$$F(q^2)\cong(Z_1/Z)(q^2+m_\rho^2-i\gamma m_\rho)^{-1}. \tag{4.18}$$

In the hypothetical case of a stable ρ meson, the renormalized mass may still be determined by Eq. (4.15), while the derivative designated in Eq. (4.17) determines the matrix element of the renormalized field ρ_μ between the vacuum state and the state of a ρ meson at rest; thus,

$$\langle\text{vac}|\boldsymbol{\rho}(0)|\rho\rangle=(2m_\rho)^{-1/2}(Z_1/Z)^{1/2}\mathbf{s}, \tag{4.19}$$

where $\mathbf{s}$ is the polarization vector $(\mathbf{s}^2=1)$. The spectral function $\sigma_\rho(a)$ contains a delta function of the form $(Z_1/Z)\delta(a-m_\rho^2)$ so that Eq. (4.6) implies

$$1=\frac{Z_1}{Z_0}+m_\rho^2\frac{Z}{Z_0}\int_b^\infty a^{-1}\sigma_\rho(a)da, \tag{4.20}$$

where b is the lower limit of the continuous spectrum in this hypothetical case. Hence one concludes

$$0<Z_1/Z_0<1. \tag{4.21}$$

So far, except for the requirement that the renormalized ρ-meson propagator $D_{\mu\nu}(q)$ should be free from divergence difficulties, the choice of the wave-function renormalization constant Z is completely arbitrary. It is clear that different choices of Z can differ from each other only by a finite positive multiplicative factor, and such different choices all lead to exactly the same physical result. On account of Eq. (4.19), the conventional choice for the renormalization constant is $Z=Z_1$. On the other hand, the fact that the ρ meson is unstable makes this choice less compelling, and we note that the choice $Z=Z_0$ simplifies a number of formulas. For the remainder of this section and in the applications of Sec. VII, we shall use the convention

$$Z=Z_0=(m_\rho/m_\rho^0)^2, \tag{4.22}$$

and therefore

$$\rho_\mu(x)=(m_\rho^0/m_\rho)\rho_\mu^0(x). \tag{4.23}$$

With this convention, it follows from Eq. (3.11) that

$$g_\rho=(m_\rho/m_\rho{}^0)g_\rho{}^0, \tag{4.24}$$

and from Eqs. (4.12) and (4.13) that

$$F(0)=m_\rho{}^{-2},$$

and

$$D_{\mu\nu}(q=0)=m_\rho{}^{-2}\delta_{\mu\nu}. \tag{4.25}$$

The width γ can also be explicitly expressed in terms of the transition matrix elements. Let us define a modified current operator $\mathcal{J}_\mu{}^\rho(x)$ whose matrix element between any two eigenstates A and B of the strong interaction Hamiltonian is given by[12]

$$\langle B|\mathcal{J}_\mu{}^\rho(x)|A\rangle=-[g_\rho D_{\mu\nu}(q)]^{-1}\langle B|\rho_\nu(x)|A\rangle, \tag{4.26}$$

where q_μ is the difference between the 4-momenta of the states A and B. This modified current operator $\mathcal{J}_\mu{}^\rho(x)$ does not satisfy either the usual Hermiticity condition or the locality condition. Nevertheless, since the integrated operator $\int\rho_4 d^3r$ is, according to Eqs. (3.4) and (3.14), $-i(g_\rho/m_\rho{}^2)I_z$, one has, by using Eqs. (4.25) and (4.26),

$$\int \mathcal{J}_4{}^\rho d^3r=iI_z. \tag{4.27}$$

The modified current operator $\mathcal{J}_\mu{}^\rho(x)$ is useful since the matrix elements of $J_\mu{}^\rho(x)$ must, by definition [Eq. (3.9)], satisfy

$$\langle B|J_\mu{}^\rho(x)|A\rangle=0 \quad\text{at}\quad q^2+m_\rho{}^2=0, \tag{4.28}$$

while those of $\mathcal{J}_\mu{}^\rho$ do not. By using Eqs. (4.2), (4.16), (4.18), and (4.26), we find that the partial decay width of the ρ^0 meson to the final channel Γ is given by

$$\text{rate}(\rho^0\to\Gamma)=\tfrac{2}{3}\pi g_\rho{}^2(Z_1/Z_0)|\langle\Gamma|\boldsymbol{\mathcal{J}}^\rho(0)|\text{vac}\rangle|^2(2m_\rho)^{-1}\times(\text{density of states})_\Gamma, \tag{4.29}$$

where $\boldsymbol{\mathcal{J}}^\rho$ denotes the spatial component of $\mathcal{J}_\mu{}^\rho(x)$. The total width γ is

$$\gamma=\sum \text{rate}(\rho^0\to\Gamma),$$

where the sum extends over all different channels Γ. The factor (Z_1/Z_0) is due to our choice [Eq. (4.22)] of the wave-function renormalization.

Equation (4.29) can be readily used for calculating the various decay rates of the neutral ρ meson. These applications will be discussed in Sec. VII. It is clear that all above discussions can be applied to any vector meson provided its current is conserved. The detailed treatment of the ϕ-ω mixing problem will be given in the following section.

[12] For single-particle states A and B, $\langle B|\mathcal{J}|A\rangle$ is, of course, related to the conventional vertex operator $\Gamma_\mu(p_B,p_A)$ via the formula

$$\langle B|\mathcal{J}_\mu(x)|A\rangle=u_B{}^\dagger\Gamma_\mu(p_B,p_A)u_A\exp(iq_\lambda x_\lambda),$$

where p_B, p_A denote, respectively, the 4-momenta of the states B and A, and $q=p_B-p_A$.

V. RENORMALIZATION OF THE ω AND φ MESONS AND THE MIXING PROBLEM

1. Renormalized Equations

Although the renormalization problem of the ω-ϕ complex is complicated by the possible mixing (i.e., virtual transition $\omega\rightleftarrows\phi$) between these two fields, it can nevertheless be carried out in a manner which parallels our treatment of the ρ meson. Let $\phi_\mu(x)$ and $\omega_\mu(x)$ denote the renormalized fields, and $\phi_\mu{}^0(x)$, $\omega_\mu{}^0(x)$ the corresponding unrenormalized fields. We write

$$\psi_\mu(x)=\begin{pmatrix}\phi_\mu(x)\\ \omega_\mu(x)\end{pmatrix} \tag{5.1}$$

and

$$\psi_\mu{}^0(x)=\begin{pmatrix}\phi_\mu{}^0(x)\\ \omega_\mu{}^0(x)\end{pmatrix}. \tag{5.2}$$

The part of the Lagrangian density describing the ϕ, ω mesons and their strong interactions can always be written, in analogy to Eq. (3.1) for the ρ-meson problem, as

$$\mathcal{L}_{\phi-\omega}=-\tfrac{1}{4}\tilde{\mathcal{G}}_{\mu\nu}{}^0K_0\mathcal{G}_{\mu\nu}{}^0-\tfrac{1}{2}\tilde{\psi}_\mu{}^0M_0{}^2\psi_\mu{}^0-\tilde{\psi}_\mu{}^0g_0J_\mu{}^0, \tag{5.3}$$

where

$$\mathcal{G}_{\mu\nu}{}^0=\begin{bmatrix}\dfrac{\partial}{\partial x_\mu}\phi_\nu{}^0-\dfrac{\partial}{\partial x_\nu}\phi_\mu{}^0\\[2ex] \dfrac{\partial}{\partial x_\mu}\omega_\nu{}^0-\dfrac{\partial}{\partial x_\nu}\omega_\mu{}^0\end{bmatrix},$$

and

$$J_\mu{}^0=\begin{pmatrix}Y_\mu{}^0\\ N_\mu{}^0\end{pmatrix}, \tag{5.4}$$

which denotes the unrenormalized current operators; g_0 is a (2×2) real matrix, K_0, $M_0{}^2$ are both (2×2) real symmetric positive-definite matrices, and $\sim$ (the tilde) denotes transposition. The currents $Y_\mu{}^0$ and $N_\mu{}^0$ are conserved; they satisfy

$$-i\int(Y_4{}^0)d^3r=\text{hypercharge}=Y$$

and

$$-i\int(N_4{}^0)d^3r=\text{baryon number}=N, \tag{5.5}$$

and will be referred to as the unrenormalized hypercharge current and baryon number current.

There is a certain arbitrariness in the matrices K_0, $M_0{}^2$, and g_0 so long as the normalization and orientation (in the internal space) of the fields $\psi_\mu{}^0$ is not specified. That is, by linearly transforming the fields,

$$\psi_\mu{}^0=A\psi_\mu{}', \tag{5.6}$$

where A is an arbitrary nonsingular real matrix, and

defining

$$\begin{aligned}\mathcal{G}_{\mu\nu}'&=A^{-1}\mathcal{G}_{\mu\nu}^0,\\ K_0'&=\tilde{A}K_0A,\\ M_0'^2&=\tilde{A}M_0^2A,\\ g_0'&=\tilde{A}g_0,\end{aligned} \tag{5.7}$$

the Lagrangian density (5.3) becomes

$$\mathcal{L}_{\phi-\omega}=-\tfrac{1}{4}\tilde{\mathcal{G}}_{\mu\nu}'K_0'\mathcal{G}_{\mu\nu}'-\tfrac{1}{2}\tilde{\psi}_\mu'M_0'^2\psi_\mu'-\tilde{\psi}_\mu'g_0'J_\mu^0.$$

One may therefore specify the normalization and orientation by imposing certain restrictions on the forms of K_0, M_0^2, and g_0. The conventional specification is the "canonical form," defined by requiring that K_0 be the unit matrix and M_0^2 be diagonal. It will, however, prove to be more convenient in connection with SU_3 considerations to use what we will refer to as the "aligned form." The aligned form is defined by requiring g_0 to be diagonal,

$$g_0=\begin{pmatrix} g_Y^0 & 0\\ 0 & g_N^0\end{pmatrix}, \tag{5.8}$$

and the diagonal elements of K_0 to be unity,

$$K_0=\begin{pmatrix} 1 & \kappa\\ \kappa & 1\end{pmatrix}, \tag{5.9}$$

where κ is a constant which may, or may not, be zero. Both the canonical and aligned forms are always possible and specify the normalization and orientation of ψ_μ^0 completely. We will for the present, however, leave the choice unspecified.

As in the case of the ρ meson, we assume that J_μ^0 does not depend upon ψ_μ^0. The equation of motion implied by Eq. (5.3) is then

$$K_0(\partial/\partial x_\mu)\mathcal{G}_{\mu\nu}^0-M_0^2\psi_\nu^0=g_0J_\nu^0. \tag{5.10}$$

By setting $g_0=0$ in (5.10), one easily sees that the squares of the mechanical (i.e., unrenormalized) masses of the ϕ and ω are the eigenvalues of the matrix M_{mech}^2,

$$M_{\text{mech}}^2\equiv K_0^{-1/2}M_0^2K_0^{-1/2}. \tag{5.11}$$

Under the transformation (5.7), we have

$$\begin{aligned}M_{\text{mech}}'^2&=K_0'^{-1/2}M_0'^2K_0'^{-1/2}\\ &=(K_0^{1/2}AK_0'^{-1/2})^{-1}M_{\text{mech}}^2(K_0^{1/2}AK_0'^{-1/2}),\end{aligned} \tag{5.12}$$

which is a similarity transformation. Hence, as expected, the unrenormalized masses are unaffected by the transformation.

To obtain renormalized equations we set

$$\psi_\mu^0=S\psi_\mu, \tag{5.13}$$

where the matrix S corresponds to the renormalization factor $Z^{1/2}$ introduced by Eq. (3.5) for the ρ meson. We may call S the renormalization matrix. Similarly to Eqs. (3.6) to (3.8), the Lagrangian density $\mathcal{L}_{\phi-\omega}$ given by Eq. (5.3) may be rewritten in terms of ψ_μ in the form

$$\mathcal{L}_{\phi-\omega}=\mathcal{L}_{\phi-\omega\text{ free}}+\mathcal{L}_{\phi-\omega\text{ st}}, \tag{5.14}$$

where

$$\mathcal{L}_{\phi-\omega\text{ free}}=-\tfrac{1}{4}\tilde{\mathcal{G}}_{\mu\nu}\mathcal{G}_{\mu\nu}-\tfrac{1}{2}\tilde{\psi}_\mu M^2\psi_\mu, \tag{5.15}$$

$$\begin{aligned}\mathcal{L}_{\phi-\omega\text{ st}}=&-\tfrac{1}{4}\tilde{\mathcal{G}}_{\mu\nu}(\tilde{S}K_0S-1)\mathcal{G}_{\mu\nu}\\ &-\tfrac{1}{2}\tilde{\psi}_\mu(\tilde{S}M_0^2S-M^2)\psi_\mu-\tilde{\psi}_\mu\tilde{S}g_0J_\mu^0,\end{aligned} \tag{5.16}$$

$$M^2=\begin{pmatrix} m_\phi^2 & 0\\ 0 & m_\omega^2\end{pmatrix}; \tag{5.17}$$

$\mathcal{G}_{\mu\nu}=S^{-1}\mathcal{G}_{\mu\nu}^0$, and m_ϕ^2, m_ω^2 are the observed masses of the designated mesons.

The equation of motion of the renormalized field ψ_μ can, in analogy to (3.9), be written as

$$\begin{aligned}\frac{\partial\mathcal{G}_{\mu\nu}}{\partial x_\mu}-M^2\psi_\nu&=(M^2S^{-1}M_0^{-2}g_0)J_\nu^0\\ &\quad+(1-M^2S^{-1}M_0^{-2}K_0S)\frac{\partial\mathcal{G}_{\mu\nu}}{\partial x_\mu}\\ &=\mathcal{J}_\nu\equiv gJ_\nu,\end{aligned} \tag{5.18}$$

where $\mathcal{J}_\nu$ and J_ν are related to the currents $\mathcal{J}_\nu^\phi$, $\mathcal{J}_\nu^\omega$, Y_ν, and N_ν introduced in Sec. II by

$$\mathcal{J}_\nu=\begin{pmatrix}\mathcal{J}_\nu^\phi\\ \mathcal{J}_\nu^\omega\end{pmatrix},\quad\text{and}\quad J_\nu=\begin{pmatrix}Y_\nu\\ N_\nu\end{pmatrix}.$$

We recall that Y_ν and N_ν satisfy the normalization Eqs. (2.16) and (2.17) and they represent, respectively, the renormalized hypercharge current and the renormalized baryon-number current. The matrix g is, according to Eq. (2.15),

$$g=T^{-1}g_D, \tag{5.19}$$

where T and g_D are, respectively, related to the angles θ_Y, θ_N and the renormalized coupling constants g_Y, g_N by

$$T=\begin{pmatrix}\cos\theta_Y & -\sin\theta_Y\\ \sin\theta_N & \cos\theta_N\end{pmatrix}, \tag{5.20}$$

and

$$g_D=\begin{pmatrix} g_Y & 0\\ 0 & g_N\end{pmatrix}. \tag{5.21}$$

Similarly to Eqs. (3.11) to (3.13), we have

$$M_0^{-2}g_0=SM^{-2}g. \tag{5.22}$$

Hence

$$\begin{aligned}g^{-1}M^2\psi_\mu&=g_0^{-1}M_0^2\psi_\mu^0=g_0'^{-1}M_0'^2\psi_\mu',\\ g^{-1}M^2\mathcal{G}_{\mu\nu}&=g_0^{-1}M_0^2\mathcal{G}_{\mu\nu}^0=g_0'^{-1}M_0'^2\mathcal{G}_{\mu\nu}',\end{aligned} \tag{5.23}$$

and

$$J_\nu=J_\nu^0+g^{-1}\frac{\partial\mathcal{G}_{\mu\nu}}{\partial x_\mu}-\bar{M}_{\text{mech}}^{-2}g^{-1}M^2\frac{\partial\mathcal{G}_{\mu\nu}}{\partial x_\mu}, \tag{5.24}$$

where

$$\bar{M}_{\text{mech}}^{-2}=(K_0^{-1/2}g_0)^{-1}M_{\text{mech}}^{-2}(K_0^{-1/2}g_0)$$
$$=(K_0'^{-1/2}g_0')^{-1}M_{\text{mech}}'^{-2}(K_0'^{-1/2}g_0'). \quad (5.25)$$

The matrix $\bar{M}_{\text{mech}}^{-2}$ is not necessarily symmetric, but it differs from M_{mech}^{-2} only by a similarity transformation and hence has the same eigenvalues. Equations (5.23) and (5.25) show that $g^{-1}M^2\psi_\mu$, $g^{-1}M^2\mathcal{G}_{\mu\nu}$, and $\bar{M}_{\text{mech}}^2$ are independent of the choice of the renormalization matrix S and also of the normalization and orientation conditions imposed on the unrenormalized fields. We shall refer to this property as *normalization-orientation invariance*. The current J_ν is, because of the presence of the term $g^{-1}(\partial\mathcal{G}_{\mu\nu}/\partial x_\mu)$, normalization-orientation invariant only for the hypothetical case of degenerate M^2. This is to be expected, since application of a linear transformation to both sides of Eq. (5.18) in general does not leave M^2 diagonal.

The isoscalar electromagnetic current is given by Eq. (1.6), which can also be written as

$$(J_\mu{}^\gamma)_{\text{isoscalar}}=-\tfrac{1}{2}(g^{-1}M^2\psi_\mu)_1, \quad (5.26)$$

and is therefore normalization-orientation invariant. The notation $(\ )_1$ means the upper component of the enclosed column matrix.

Equations (5.18) and (5.22) imply that, in analogy to Eq. (3.14) for the ρ-meson case

$$J_\nu{}^0=-g^{-1}M^2\psi_\nu+\bar{M}_{\text{mech}}^{-2}g^{-1}M^2(\partial\mathcal{G}_{\mu\nu}/\partial x_\mu). \quad (5.27)$$

For a divergent theory, the mechanical (i.e., unrenormalized) masses of the ϕ and ω fields are infinite, and hence $\bar{M}_{\text{mech}}^{-2}$ vanishes. Comparison with (5.26) and (5.27) then shows

$$(J_\nu{}^\gamma)_{\text{isoscalar}}=\tfrac{1}{2}(J_\nu{}^0)_1=\tfrac{1}{2}Y_\nu{}^0, \quad (5.28)$$

which plays the same role as Eq. (3.18) for the isovector current. Hence the alternative identifications of the isoscalar electromagnetic current with the *unrenormalized* hypercharge current on the one hand, and the renormalized ϕ-ω fields on the other become identical hypotheses for a divergent theory.

2. The ϕ-ω Propagator and Normalization-Orientation Conditions for the Renormalized Fields

Let us consider the spectral representation of the vacuum expectation value of the commutator $[\psi_\mu{}^0(x), \psi_\nu{}^0(0)]$ in the Heisenberg representation:

$$\langle[\psi_\mu{}^0(x),\tilde{\psi}_\nu{}^0(0)]\rangle_{\text{vacuum}}$$
$$=\int\sigma_{\phi\omega}{}^0(a)\left[\delta_{\mu\nu}-a^{-1}\frac{\partial^2}{\partial x_\mu\partial x_\nu}\right]\Delta_a(x)da. \quad (5.29)$$

The function $\Delta_a(x)$ has already been used in Eq. (4.1), and $\sigma_{\phi\omega}{}^0(a)$ is a (2×2) Hermitian matrix given by

$$\sigma_{\phi\omega}{}^0(a)=\tfrac{1}{3}\sum_\Gamma\delta(a-m_\Gamma{}^2)\langle\text{vac}|\boldsymbol{\psi}^0(0)|\Gamma\rangle$$
$$\times\langle\Gamma|\boldsymbol{\psi}^{0\dagger}(0)|\text{vac}\rangle, \quad (5.30)$$

where the boldface letters denote the space components and the sum extends over all $I=0$ eigenstates $|\Gamma\rangle$ of the strong interaction Hamiltonian which satisfy Eq. (4.3). From time-reversal invariance, or CPT invariance, it follows that $\sigma_{\phi\omega}{}^0(a)$ is also a real matrix. Similarly to the derivation of Eqs. (4.5) and (4.6), one can show that the following two sum rules hold for the unrenormalized spectral function $\sigma_{\phi\omega}{}^0(a)$:

$$\int\sigma_{\phi\omega}{}^0(a)da=K_0{}^{-1}, \quad (5.31)$$

and

$$\int a^{-1}\sigma_{\phi\omega}{}^0(a)da=M_0{}^{-2}. \quad (5.32)$$

For the realistic case of unstable ϕ meson and ω meson, both integrations extend from $9m_\pi{}^2$ to ∞. It follows from these two sum rules that the squares of the unrenormalized masses $(m_\phi{}^0)^2$ and $(m_\omega{}^0)^2$ (i.e., the eigenvalues of $M_{\text{mech}}{}^2$) satisfy

$$(m_\phi{}^0)^2>9m_\pi{}^2 \quad\text{and}\quad (m_\omega{}^0)^2>9m_\pi{}^2. \quad (5.33)$$

The above inequalities can be most easily established by choosing the canonical form $K_0=1$ and $M_0{}^2=M_{\text{mech}}{}^2$ diagonal.

We now discuss the question of a convenient specification of normalization and orientation conditions for the renormalized fields. It is sometimes convenient to specify normalization and orientation separately. We note that the renormalization matrix S is a real nonsingular matrix and that any arbitrary real nonsingular matrix can always be uniquely decomposed in the form

$$S=T_S{}^{-1}R, \quad (5.34)$$

where the matrix R is diagonal and positive,

$$R=\begin{pmatrix}R_\phi & 0\\ 0 & R_\omega\end{pmatrix}, \quad (5.35)$$

and

$$T_S=\begin{pmatrix}\cos\alpha_1 & -\sin\alpha_1\\ \sin\alpha_2 & \cos\alpha_2\end{pmatrix}, \quad (5.36)$$

with α_1, α_2 real. We shall refer to R as the normalization matrix and T_S as the orientation matrix of S. As an application of the above separation, we note that the renormalized current $J_\mu(x)$, defined by Eq. (5.18), is independent of the choice of the normalization matrix R. This can be verified by using Eq. (5.24) and noting that the only normalization-orientation noninvariant term in the right-hand side of (5.24) is

$$g^{-1}(\partial \mathcal{G}_{\mu\nu}/\partial x_\mu)=g_0^{-1}M_0{}^2SM^{-2}S^{-1}(\partial \mathcal{G}_{\mu\nu}{}^0/\partial x_\mu),$$

which is independent of the normalization matrix R since M^2 is diagonal by definition.

In the hypothetical case of a stable ϕ and ω meson, a convenient orientation may be defined by the requirement

$$\langle \text{vac}|\phi_\mu|\omega\rangle=\langle \text{vac}|\omega_\mu|\phi\rangle=0, \tag{5.37}$$

where $|\phi\rangle$ and $|\omega\rangle$ are the "physical" ϕ-meson and ω-meson states in the hypothetical case. Thus, the angles α_1 and α_2 in the orientation matrix T_S are given by

$$\tan\alpha_1=\langle \text{vac}|\phi_\mu{}^0|\omega\rangle/\langle \text{vac}|\omega_\mu{}^0|\omega\rangle,$$

and (5.38)

$$\tan\alpha_2=-\langle \text{vac}|\omega_\mu{}^0|\phi\rangle/\langle \text{vac}|\phi_\mu{}^0|\phi\rangle;$$

Equation (5.37) and the equation of motion (5.18) then imply

$$\langle \text{vac}|J_\mu|\phi\rangle=\langle \text{vac}|J_\mu|\omega\rangle=0, \tag{5.39}$$

and, therefore, the matrix elements $\langle A|J_\mu(0)|B\rangle$ are free of poles at $q^2=(p_A-p_B)^2=-m_\phi{}^2$ or $-m_\omega{}^2$ for arbitrary hadronic states A and B. The converse is also true; i.e., the requirement that $\langle A|J_\mu(0)|B\rangle$ be free of ϕ-ω poles implies Eq. (5.39) and, consequently, Eq. (5.37). The normalization matrix R can be fixed by specifying values for $\langle \text{vac}|\boldsymbol{\phi}|\phi\rangle$ and $\langle \text{vac}|\boldsymbol{\omega}|\omega\rangle$, the conventional choice, analogous to $Z=Z_1$ in Eq. (4.19), being

$$(2m_\phi)^{1/2}\mathbf{s}\cdot\langle \text{vac}|\boldsymbol{\phi}|\phi\rangle=(2m_\omega)^{1/2}\mathbf{s}\cdot\langle \text{vac}|\boldsymbol{\omega}|\omega\rangle=1, \tag{5.40}$$

where $|\phi\rangle$ and $|\omega\rangle$ refer to the ϕ-meson and ω-meson states in their respective rest systems and $\mathbf{s}$ is the polarization vector ($\mathbf{s}^2=1$).

Next, we turn to the realistic case of unstable ϕ and ω; it is convenient to discuss the normalization and orientation conventions with reference to the propagator. The unrenormalized propagator $\mathfrak{D}_{\mu\nu}{}^0(q)$ is related to $\sigma_{\phi\omega}{}^0(a)$, defined in Eq. (5.30), by

$$\mathfrak{D}_{\mu\nu}{}^0(q)=\int\frac{\delta_{\mu\nu}+a^{-1}q_\mu q_\nu}{q^2+a-i\epsilon}\sigma_{\phi\omega}{}^0(a)da. \tag{5.41}$$

Its inverse can be written as

$$(\mathfrak{D}_{\mu\nu}{}^0)^{-1}=M_0{}^2\delta_{\mu\nu}+(q^2\delta_{\mu\nu}-q_\mu q_\nu)\Pi_0(q^2), \tag{5.42}$$

where $\Pi_0(q^2)$ is a (2×2) matrix related to $\sigma_{\phi\omega}{}^0(a)$ by

$$\Pi_0(q^2)=\left[\int\frac{\sigma_{\phi\omega}{}^0(a)da}{q^2+a-i\epsilon}\right]^{-1}\times\left[\int\frac{\sigma_{\phi\omega}{}^0(a)}{a(q^2+a-i\epsilon)}da\right]M_0{}^2, \tag{5.43}$$

which, in view of Eq. (5.32), is also equal to

$$\Pi_0(q^2)=q^{-2}\left\{\left[\int\frac{\sigma_{\phi\omega}{}^0(a)da}{q^2+a-i\epsilon}\right]^{-1}-M_0{}^2\right\}. \tag{5.44}$$

From (5.43), we find that $\Pi_0(q^2)$ at $q^2=0$ is given by

$$\Pi_0(0)=M_0{}^{-2}\int a^{-2}\sigma_{\phi\omega}{}^0(a)da\,M_0{}^2,$$

and from (5.44) it follows that $\Pi_0(q^2)$, like $\sigma_{\phi\omega}{}^0(a)$, is a symmetric matrix.

The observed masses of the ϕ and ω mesons are determined by the two roots $q^2=-m_\phi{}^2$ and $q^2=-m_\omega{}^2$ of the equation

$$\det|M_0{}^2+q^2\,\text{Re}\Pi_0(q^2)|=0, \tag{5.45}$$

where $\text{Re}\Pi_0(q^2)$ denotes the real part of $\Pi_0(q^2)$.

At $q^2=-m_\omega{}^2$ and $q^2=-m_\phi{}^2$, the vanishing of the determinant requires that the matrix $M_0{}^2+q^2\Pi_0(q^2)$ must be, respectively, of the forms

$$M_0{}^2-m_\omega{}^2\,\text{Re}\Pi_0(-m_\omega{}^2)=N_\omega{}^2u(\omega)\tilde{u}(\omega),$$

and (5.46)

$$M_0{}^2-m_\phi{}^2\,\text{Re}\Pi_0(-m_\phi{}^2)=-N_\phi{}^2u(\phi)\tilde{u}(\phi),$$

where N_ω and N_ϕ are real numbers, and $u(\omega)$ and $u(\phi)$ are two real column [i.e., (2×1)] matrices, both normalized to unity,

$$\tilde{u}(\omega)u(\omega)=\tilde{u}(\phi)u(\phi)=1. \tag{5.47}$$

The difference in signs in the two equations in (5.46) corresponds to the assertion that the diagonal elements of the left-hand sides of these two equations are, respectively, nonnegative and nonpositive. This is rigorously true in the case of the pole approximation (which will be discussed in detail later), since $(q^2+m_\phi{}^2)>0$ at $q^2=-m_\omega{}^2$ but $(q^2+m_\omega{}^2)<0$ at $q^2=-m_\phi{}^2$. We regard these sign assignments as highly probable in the actual case. The treatment which follows can obviously be tailored to fit whatever signs actually occur.

A general orientation convention [that reduces to Eq. (5.37) in the hypothetical case of stable ϕ and ω mesons] can be obtained by choosing the orientation matrix T_S, defined by Eqs. (5.34) and (5.36), as

$$T_S=\begin{pmatrix}u_1(\omega) & u_2(\omega)\\ u_1(\phi) & u_2(\phi)\end{pmatrix}, \tag{5.48}$$

where $u_1(i)$, $u_2(i)$ are the two matrix elements of the (2×1) matrix $u(i)$, and $i=\phi$ or ω.

Let $\mathfrak{D}_{\mu\nu}{}^{-1}(q)$ be the inverse of the renormalized propagator, related to the inverse of the unrenormalized propagator by

$$\mathfrak{D}_{\mu\nu}{}^{-1}(q)=\tilde{S}[\mathfrak{D}_{\mu\nu}{}^0(q)]^{-1}S. \tag{5.49}$$

By using Eqs. (5.46) to (5.48), we find that, at the two

roots $q^2=-m_\phi^2$ and $q^2=-m_\omega^2$ of Eq. (5.45),

$$\tilde{S}[M_0^2+q^2\,\mathrm{Re}\Pi_0(q^2)]S=\begin{bmatrix}\frac{R_\phi^2N_\omega^2}{m_\phi^2-m_\omega^2}(q^2+m_\phi^2) & 0\\ 0 & \frac{R_\omega^2N_\phi^2}{m_\phi^2-m_\omega^2}(q^2+m_\omega^2)\end{bmatrix}. \tag{5.50}$$

Correspondingly, the real part of the renormalized propagator becomes

$$\mathrm{Re}[\mathfrak{D}_{\mu\nu}^{-1}(q)]=\begin{bmatrix}\frac{R_\phi^2N_\omega^2}{m_\phi^2-m_\omega^2}(q^2+m_\phi^2) & 0\\ 0 & \frac{R_\omega^2N_\phi^2}{m_\phi^2-m_\omega^2}(q^2+m_\omega^2)\end{bmatrix}\delta_{\mu\nu}+\cdots, \tag{5.51}$$

at the two roots $q^2=-m_\phi^2$ and $-m_\omega^2$ where $\cdots$ denotes the part of $\mathfrak{D}_{\mu\nu}^{-1}$ that is proportional to $q_\mu q_\nu$. [In the hypothetical case of the stable ϕ and ω mesons, Eq. (5.51) implies that, independently of the constants N_ϕ and N_ω the orientation condition, Eq. (5.37), is satisfied.]

So far as the renormalization problem is concerned, the only requirement on the renormalization matrix S is that the renormalized propagator

$$\mathfrak{D}_{\mu\nu}(q)=S^{-1}\mathfrak{D}_{\mu\nu}{}^0(q)\tilde{S}^{-1} \tag{5.52}$$

should be free from divergence difficulties. Thus, just as for the renormalization constant Z in the ρ-meson case, R_ϕ, or R_ω, is determined only up to an arbitrary finite multiplicative factor; all different choices of such finite multiplicative factors clearly lead to the same physical results. As we shall see below, a particularly convenient choice of R_ϕ and R_ω is to set simply

$$R_\phi^2N_\omega^2=R_\omega^2N_\phi^2=(m_\phi^2-m_\omega^2); \tag{5.53}$$

consequently, Eq. (5.51) becomes

$$\mathrm{Re}[\mathfrak{D}_{\mu\nu}^{-1}(q)]=\begin{pmatrix}q^2+m_\phi^2 & 0\\ 0 & q^2+m_\omega^2\end{pmatrix}\delta_{\mu\nu}+\cdots, \tag{5.54}$$

at the two roots $q^2=-m_\phi^2$ and $-m_\omega^2$.

For practical applications, it is reasonable to assume that in the resonance region from $q^2\cong-m_\omega^2$ to $q^2\cong-m_\phi^2$ the propagator is dominated by the two poles; i.e., only these two-pole contributions are included in $\mathfrak{D}_{\mu\nu}(q)$. [With this approximation the normalization choice (5.53) becomes, for the hypothetical case of stable ϕ-ω mesons, the same as the normalization condition given by Eq. (5.40)]. Thus, Eq. (5.54) should hold for the entire resonance region $m_\omega^2\lesssim-q^2\lesssim m_\phi^2$ and, in particular, the transformation matrix S satisfies

$$M^2=\tilde{S}M_0^2S=\begin{pmatrix}m_\phi^2 & 0\\ 0 & m_\omega^2\end{pmatrix}, \tag{5.55}$$

and

$$\tilde{S}\,\mathrm{Re}\Pi_0(-m_\phi^2)S=\tilde{S}\,\mathrm{Re}\Pi_0(-m_\omega^2)S=\begin{pmatrix}1 & 0\\ 0 & 1\end{pmatrix}. \tag{5.56}$$

We note that Eqs. (5.55) and (5.56) can also be derived by making a weaker approximation in which one neglects only the difference between the real part of $\Pi_0(q^2)$ at $q^2=-m_\omega$ and that at $q^2=-m_\phi^2$; i.e., we assume

$$\mathrm{Re}\Pi_0(-m_\phi^2)=\mathrm{Re}\Pi_0(-m_\omega^2). \tag{5.57}$$

[The imaginary parts of $\Pi_0(q^2)$ at $q^2=-m_\phi^2$ and $-m_\omega^2$ are clearly very different.] From Eq. (5.56), it follows that the pole approximation implies this weaker condition (5.57). Equation (5.57) is, in fact, equivalent to the pole approximation, for a hypothetical stable ϕ-ω system, but does not require the full extent of the pole approximation for the realistic case.

Finally, we wish to relate the renormalization matrix S with the matrices g_0 and g occurring in Eqs. (5.3) and (5.18), respectively. If in the resonance region, $m_\phi^2\gtrsim-q^2\gtrsim m_\omega^2$, the pole dominance is a good approximation, then by combining (5.22) and (5.55) one obtains

$$gg_0^{-1}=\tilde{S}. \tag{5.58}$$

If, in addition, the aligned form is chosen for the unrenormalized fields, then according to Eq. (5.8) g_0 is diagonal and, using (5.19) and (5.21), we have

$$S^{-1}=\tilde{T}\begin{pmatrix}g_Y^0/g_Y & 0\\ 0 & g_N^0/g_N\end{pmatrix}. \tag{5.59}$$

We recall that the matrix T depends on two angles θ_Y and θ_N which are introduced in Sec. II [Eq. (2.18)] to relate the currents $\mathcal{J}_\mu{}^\phi$ and $\mathcal{J}_\mu{}^\omega$ to g_YY_μ and g_NN_μ; Eqs. (5.59) and (5.13) show that the same matrix also transforms the unrenormalized field $\psi_\mu{}^0$, after multiplication by a diagonal matrix $g_0g_D^{-1}$, into the renormalized field ψ_μ. Note, however, that T is not equal to the T_S defined by Eqs. (5.34) and (5.36); these two matrices are related to each other by

$$R^{-1}T_S=\tilde{T}g_0g_D^{-1}, \tag{5.60}$$

where R, g_0, g_D are all diagonal matrices.

3. Special Models

Hitherto, our discussions have been completely general. In this section, we will discuss two specific models. As we shall see, the angles θ_Y and θ_N become related, though in different ways, in each of these two models. For the definitions of these models it is convenient to adopt the aligned form for the unrenormalized fields. Accordingly g_0 has the diagonal form (5.8) and K_0 the form (5.9). (In the limit of SU_3 symmetry, the aligned form also implies $K_0=1$ and M_0^2 diagonal. Therefore, in this limit the aligned form becomes the same as the canonical form.)

Current-Mixing Model

We consider first the special model in which the matrix M_0^2 is assumed to be diagonal:

$$M_0^2=\begin{pmatrix}(\mu_1^0)^2 & 0\\ 0 & (\mu_2^0)^2\end{pmatrix}. \tag{5.61}$$

The matrix K_0 may, or may not, be diagonal. If K_0 is also diagonal, then according to Eq. (5.9), K_0 is a unit matrix, and $(\mu_1^0)^2$, $(\mu_2^0)^2$ become, respectively, the (unrenormalized mass)^2 of the ϕ meson and the ω meson.

The SU_3 symmetry is assume *not* to be valid; as a result, the matrix $\Pi_0(q^2)$ is not diagonal. A likely mechanism is that in the absence of the SU_3 symmetry, the vacuum expectation value $\langle\text{vac}|Y_\mu^0(x)N_\nu^0(0)|\text{vac}\rangle$, for example, is no longer zero. Such mixed-current matrix elements can give rise to the off-diagonal matrix element of $\Pi_0(q^2)$, but *not* of M_0^2. For convenience, we call any model, in which M_0^2 is diagonal but $\Pi_0(q^2)$ is not diagonal, the "current-mixing" model. For our analysis, the precise mechanism of the SU_3 symmetry-breaking interaction is, however, immaterial.

Assuming the validity of (5.55), it is easy to verify that Eqs. (5.55) and (5.61) imply that S has the general form

$$S=\begin{pmatrix}(\mu_1^0)^{-1}m_\phi\cos\theta & -(\mu_1^0)^{-1}m_\omega\sin\theta\\ (\mu_2^0)^{-1}m_\phi\sin\theta & (\mu_2^0)^{-1}m_\omega\cos\theta\end{pmatrix}, \tag{5.62}$$

where the angle θ is real and its value depends on the matrix $\text{Re}\Pi_0(-m_\phi^2)=\text{Re}\Pi_0(-m_\omega^2)$. Upon comparing (5.62) with (5.59), we find

$$\left(\frac{g_Y^0}{g_Y}\right)^2=\left(\frac{\mu_1^0}{m_\phi}\right)^2\left(\cos^2\theta+\frac{m_\phi^2}{m_\omega^2}\sin^2\theta\right), \tag{5.63}$$

$$\left(\frac{g_N^0}{g_N}\right)^2=\left(\frac{\mu_2^0}{m_\omega}\right)^2\left(\cos^2\theta+\frac{m_\omega^2}{m_\phi^2}\sin^2\theta\right), \tag{5.64}$$

$$\tan\theta_Y=(m_\phi/m_\omega)\tan\theta, \tag{5.65}$$

$$\tan\theta_N=(m_\omega/m_\phi)\tan\theta, \tag{5.66}$$

and, consequently

$$m_\omega^2\tan\theta_Y=m_\phi^2\tan\theta_N. \tag{5.67}$$

It is important to note that these results hold to *all* orders in the SU_3 symmetry-breaking interaction, provided that the "current-mixing" model is valid.

An estimation of the values of θ_Y and θ_N can be made if one assumes that the SU_3 symmetry-breaking interaction transforms like the $I=0$ member of an octet under the SU_3 transformations. To simplify our discussions, we will further approximate $\text{Re}\Pi_0(q^2)$ by a constant matrix, as is the case in the pole approximation.

In the limit of SU_3 symmetry, the matrix $\Pi_0(q^2)$ must be a diagonal matrix. Let us denote its real part by

$$\lim_{SU_3\text{ sym}}\text{Re}\Pi_0=\begin{pmatrix}(\lambda_1^0)^2 & 0\\ 0 & (\lambda_2^0)^2\end{pmatrix}. \tag{5.68}$$

The renormalized masses of the octet and singlet vector mesons in the SU_3 symmetry limit are, respectively, given by

$$m_{\text{octet}}=(\mu_1^0/\lambda_1^0)\quad\text{and}\quad m_{\text{singlet}}=(\mu_2^0/\lambda_2^0). \tag{5.69}$$

The inclusion of the SU_3 symmetry-breaking interaction in the current-mixing model does not change M_0^2, but it alters $\text{Re}\Pi_0$ from (5.68) to

$$\text{Re}\Pi_0=\begin{pmatrix}(\lambda_1)^2 & \eta\\ \eta & (\lambda_2)^2\end{pmatrix}. \tag{5.70}$$

If the SU_3 symmetry-breaking interaction is assumed to transform like the $I=0$ member of an octet, then to the first order of such a symmetry-breaking interaction one must have

$$\lambda_2=\lambda_2^0. \tag{5.71}$$

Adopting the notations of Coleman and Schnitzer,[5] we define ϵ and β by

$$\lambda_1=\lambda_1^0(1+2\epsilon)^{1/2}\quad\text{and}\quad(\lambda_1^0\lambda_2^0)^{-1}\eta=\beta. \tag{5.72}$$

To first order in the SU_3 symmetry-breaking interaction, the observed masses of K^*, ρ, ϕ, and ω are related to the zeroth-order renormalized masses m_{octet}, m_{singlet} and the two parameters ϵ and β by

$$m_{K^*}=(1+\epsilon)^{-1/2}m_{\text{octet}}, \tag{5.73}$$

$$m_\rho=(1-2\epsilon)^{-1/2}m_{\text{octet}}, \tag{5.74}$$

$$m_\phi^{-2}+m_\omega^{-2}=m_{\text{singlet}}^{-2}+(1+2\epsilon)m_{\text{octet}}^{-2}, \tag{5.75}$$

and

$$(m_\phi m_\omega)^{-2}(m_{\text{octet}}m_{\text{singlet}})^2=1+2\epsilon-\beta^2. \tag{5.76}$$

Thus, one finds

$$m_{\text{octet}}=839\text{ MeV},\quad m_{\text{singlet}}=817\text{ MeV},\quad \epsilon=-0.115.$$

and

$$\beta = -0.18. \tag{5.77}$$

From Eqs. (5.75) and (5.76), it is clear that $\beta = +0.18$ is also a solution. However, since under the transformation $\phi_\mu^0 \to +\phi_\mu^0$, $\omega_\mu^0 \to -\omega_\mu^0$, β transforms $\to -\beta$, we may, without any loss of generality, choose β to be negative (so that $\tan\theta$ becomes positive). The angle θ in Eq. (5.62) is related to these parameters by

$$\tan\theta = (\beta m_\omega^2 m_{\text{singlet}})^{-1} m_{\text{octet}}(m_\omega^2 - m_{\text{singlet}}^2), \tag{5.78}$$

which together with Eqs. (5.65) and (5.66) yields

$$\theta \cong 26°,$$

$$\theta_Y \cong 33°,$$

and

$$\theta_N \cong 21°. \tag{5.79}$$

For convenience, we have chosen θ, and also θ_Y and θ_N, to be in the first quadrant, instead of the third quadrant. (Under $\phi_\mu \to -\phi_\mu$, $\omega_\mu \to -\omega_\mu$, but $\phi_\mu^0 \to +\phi_\mu^0$ and $\omega_\mu^0 \to +\omega_\mu^0$, one finds $\theta \to \theta + 180°$, $\theta_Y \to \theta_Y + 180°$ and $\theta_N \to \theta_N + 180°$.)

Mass-Mixing Model

Next, we consider a different model in which the ϕ-ω mixing is assumed to be due entirely to the off-diagonal matrix elements of M_0^2. In general, if the matrix M_0^2 is nondiagonal, then $\Pi_0(q^2)$ would also contain nonzero off-diagonal matrix elements. However, in the mass-mixing model, we make the *ad hoc* assumption that

$$\text{Re}\Pi_0(-m_\phi^2) = \text{Re}\Pi_0(-m_\omega^2) = \begin{pmatrix} (\lambda_1)^2 & 0 \\ 0 & (\lambda_2)^2 \end{pmatrix}, \tag{5.80}$$

but M_0^2 can be any arbitrary (2×2) real symmetric matrix. By using Eqs. (5.55) and (5.56), one finds that the matrix S is given by

$$S = \begin{pmatrix} \lambda_1^{-1} & 0 \\ 0 & \lambda_2^{-1} \end{pmatrix} U, \tag{5.81}$$

where U is the real orthogonal matrix which diagonalizes

$$\begin{pmatrix} \lambda_1^{-1} & 0 \\ 0 & \lambda_2^{-1} \end{pmatrix} M_0^2 \begin{pmatrix} \lambda_1^{-1} & 0 \\ 0 & \lambda_2^{-1} \end{pmatrix}. \tag{5.82}$$

By comparing (5.81) with (5.49), one derives in this mass-mixing model

$$\theta_Y = \theta_N, \tag{5.83}$$

$$(g_Y^0/g_Y) = \lambda_1, \tag{5.84}$$

and

$$(g_N^0/g_N) = \lambda_2. \tag{5.85}$$

The matrix M_0^2 is real symmetric matrix; therefore, it can be diagonalized by a real orthogonal matrix, say V:

$$M_0^2 = \tilde{V} \begin{pmatrix} (\mu_1)^2 & 0 \\ 0 & (\mu_2)^2 \end{pmatrix} V, \tag{5.86}$$

where

$$V = \begin{pmatrix} \cos a & \sin a \\ -\sin a & \cos a \end{pmatrix}, \tag{5.87}$$

and μ_1^2, μ_2^2 are the two eigenvalues of M_0^2. In the limit of SU_3 symmetry, $a=0$; therefore, to the first order in the SU_3 symmetry-breaking interaction, we can neglect a^2, and (5.82) becomes

$$\begin{pmatrix} (\mu_1/\lambda_1)^2 & \xi \\ \xi & (\mu_2/\lambda_2)^2 \end{pmatrix}, \tag{5.88}$$

where

$$\xi = (\lambda_1\lambda_2)^{-1}(\mu_1^2 - \mu_2^2) \sin a \cos a \cong (\lambda_1\lambda_2)^{-1}(\mu_1^2 - \mu_2^2) a.$$

The values of (μ_1/λ_1), (μ_2/λ_2), and ξ can be determined from the known vector-meson masses, if we make the following further dynamical assumptions:

(i) The eigenvalues of M_0^2 are *not* changed by the introduction of the SU_3 symmetry-breaking interaction; i.e., in the limit of SU_3 symmetry, one has from Eq. (5.86), $V=1$ and

$$\lim_{SU_3 \text{ sym}} M_0^2 = \begin{pmatrix} (\mu_1)^2 & 0 \\ 0 & (\mu_2)^2 \end{pmatrix}. \tag{5.89}$$

This assumption appears natural if one imagines the unrenormalized theory to be in the canonical form, instead of the aligned form which is used here. In the canonical form this theory is characterized by a nondiagonal g_0' given by $g_0' = V g_0$.

In addition, we approximate the matrix $\text{Re}\Pi_0(q^2)$ by a constant matrix, and denote its limiting form by

$$\lim_{SU_3 \text{ sym}} \text{Re}\Pi_0(q^2) = \begin{pmatrix} (\lambda_1^0)^2 & 0 \\ 0 & (\lambda_2^0)^2 \end{pmatrix}. \tag{5.90}$$

The renormalized octet and singlet masses in the SU_3 symmetry limit are, therefore, given respectively by

$$m_{\text{octet}} = (\mu_1/\lambda_1^0),$$

and

$$m_{\text{singlet}} = (\mu_2/\lambda_2^0). \tag{5.91}$$

(ii) The SU_2 symmetry-breaking interaction transforms like the isoscalar member of an octet. Thus, to first order in the SU_3 symmetry-breaking interaction,

$$\lambda_2^0 = \lambda_2. \tag{5.92}$$

The parameter λ_1 is assumed to be different from its SU_3 symmetry-limiting value λ_1^0. We write

$$\lambda_1 = \lambda_1^0(1+2\epsilon)^{1/2}. \tag{5.93}$$

The SU_3 symmetry-breaking interaction mixes ϕ-ω through the angle a, or the parameter ξ in (5.88), but the difference between the observed vector-meson masses and their zeroth-order masses m_{octet} and m_{singlet} depend on both ξ and ϵ. We have, as in Eqs. (5.73) and (5.74),

$$m_{K^*}=(1+\epsilon)^{-1/2}m_{\text{octet}}, \quad (5.94)$$

$$m_\rho=(1-2\epsilon)^{-1/2}m_{\text{octet}}, \quad (5.95)$$

but, instead of Eqs. (5.75) and (5.76),

$$m_\phi^2+m_\omega^2=(1+2\epsilon)^{-1}m_{\text{octet}}^2+m_{\text{singlet}}^2,$$

and (5.96)

$$m_\phi^2m_\omega^2=(1+2\epsilon)^{-1}(m_{\text{octet}}m_{\text{singlet}})^2-\xi^2.$$

From the known masses m_{K^*}, m_ρ, m_ϕ, and m_ω, we determine

$$m_{\text{octet}}=839\text{ MeV},$$

$$m_{\text{singlet}}=859\text{ MeV},$$

$$\epsilon=-0.115,$$

and

$$\xi=1.93\times10^5\,(\text{MeV})^2. \quad (5.97)$$

The angles θ_Y and θ_N are given by

$$\tan\theta_Y=\tan\theta_N=\xi^{-1}(m_{\text{singlet}}^2-m_\omega^2), \quad (5.98)$$

and therefore

$$\theta_Y=\theta_N\cong32°. \quad (5.99)$$

For convenience, we have chosen ξ positive and $\theta_Y=\theta_N$ in the first quadrant.

Mass-Mixing Model (A Variation)

The mass-mixing model implies $\theta_Y=\theta_N$, but the above estimation $\theta_Y=\theta_N\cong32°$ is based on further *ad hoc* assumptions (i) and (ii). In this section, we will give a variation of the same mass-mixing model in which all the above formulas (5.80)–(5.88) are assumed to remain applicable. The only change is that, instead of the previous additional assumption (i), one assumes[4]:

(i)′. The matrix $\text{Re}\Pi_0(q^2)$ is approximated by a constant matrix, and it is, for some unspecified dynamical reason, *not* changed by the SU_3 symmetry-breaking interaction. Thus, one may use Eq. (5.80), but set

$$\lambda_1=\lambda_1^0,$$

and (5.100)

$$\lambda_2=\lambda_2^0.$$

In place of Eq. (5.89), one may write

$$\lim_{SU_3\text{ sym}} M_0^2=\begin{pmatrix}(\mu_1^0)^2 & 0\\ 0 & (\mu_2^0)^2\end{pmatrix}. \quad (5.101)$$

Instead of by Eq. (5.91), the renormalized octet and singlet masses in the SU_3 symmetry limit are given by

$$m_{\text{octet}}=(\mu_1^0/\lambda_1^0)$$

and (5.102)

$$m_{\text{singlet}}=(\mu_2^0/\lambda_2^0).$$

We make the same assumption (ii) as in the previous case. By using Eqs. (5.86), (5.87), and neglecting $O(a^2)$, we find, in order to conform to the assumption (ii),

$$\mu_2=\mu_2^0. \quad (5.103)$$

The masses of K^*, ρ, ϕ, and ω are related to their SU_3-symmetry limits m_{octet}, m_{singlet}, and the ratio

$$(\mu_1/\mu_1^0)=1+2\delta,$$

by

$$m_{K^*}^2=m_{\text{octet}}^2(1+\delta), \quad (5.104)$$

$$m_\rho^2=m_{\text{octet}}^2(1-2\delta), \quad (5.105)$$

$$m_\phi^2+m_\omega^2=(1+2\delta)m_{\text{octet}}^2+m_{\text{singlet}}^2, \quad (5.106)$$

and

$$m_\phi^2m_\omega^2=(1+2\delta)(m_{\text{octet}}m_{\text{singlet}})^2-\xi^2, \quad (5.107)$$

where ξ is given by Eq. (5.88). Equation (5.98) remains applicable. By using the known vector-meson masses, one finds, in place of Eqs. (5.97) and (5.99),

$$m_{\text{octet}}=850\text{ MeV},$$

$$m_{\text{singlet}}=884\text{ MeV},$$

$$\delta=0.103,$$

$$\xi=2.07\times10^5\,(\text{MeV})^2,$$

and

$$\theta_Y=\theta_N\cong39°. \quad (5.108)$$

We note that Eqs. (5.104)–(5.107) reduce to Eqs. (5.94)–(5.96) if we set $\epsilon=-\delta$ and neglect $O(\epsilon^2)$. The difference between the two estimations [(5.99) and (5.108)] of $\theta_Y=\theta_N$ in the mass-mixing model lies, therefore, only in the higher-order terms of the SU_3 symmetry-breaking interaction.

VI. PHOTON PROPAGATOR

The hadronic contribution to the electromagnetic current may influence purely leptonic processes through its effect on the photon propagator. Indeed, discussions of contributions arising from this source to the anomalous magnetic moment of leptons[13] and to electron-positron scattering[14] have already appeared in the literature. We wish, however, to exhibit here the fact that our considerations imply an exact (in the strong interactions) connection between the order e^2 hadronic contributions to the photon propagator and the vector-meson propagators. Such a connection can be derived directly by considering the set S_γ of all e^2 order Feynman graphs for the photon propagator (to all orders in the strong interaction, but neglecting leptonic contribu-

[13] L. Durand III, Phys. Rev. **128**, 441 (1962); C. Bouchiat and L. Michel, J. Phys. Radium **22**, 121 (1961).

[14] R. Gatto, Nuovo Cimento **28**, 658 (1963).

tions) and the set S_{meson} of all Feynman graphs for the vector-meson propagators (to all orders in the strong interactor, but only zeroth order in e). By using the strong interaction Lagrangian (3.7), (3.8), (5.15), and (5.16), and the gauge-invariant electromagnetic interaction Lagrangian (2.7) and (2.20) [or, more simply, the alternative identities (1.3b) and (5.28)], it can be readily seen that each of the graphs in the set S_γ corresponds to a subset of graphs in the set S_{meson}. It is convenient to include also the free photon propagator in S_γ. One finds, then, that there exists a homomorphism between the set S_γ and the set S_{meson}. While it is straightforward to convert this homomorphism into algebraic relations, and to derive the results that are given by Eqs. (6.9) and (6.15) below, the detailed description of such a graphic procedure turns out to be somewhat unnecessarily complicated. Thus, we shall give, instead, a formal analytic proof in the following.

For convenience, we begin with the general expression first used by Källén,[15] and write for the vacuum expectation value of the photon commutator

$$\hat{K}_{\mu\nu}{}^{\gamma}(x)=\langle 0|[A_\mu(x),A_\nu(0)]|0\rangle$$

$$=-i\delta_{\mu\nu}\Delta_0(x)-L\frac{\partial^2}{\partial x_\mu \partial x_\nu}\Delta_0(x)+\int d^4y d^4z$$

$$\times\langle \text{vac}|[\mathcal{J}_\mu{}^\gamma(y),\mathcal{J}_\nu{}^\gamma(z)]|\text{vac}\rangle D_R(x-y)D_R(-z)\,, \quad (6.1)$$

where $\mathcal{J}_\mu{}^\gamma$ is the total electromagnetic current defined by

$$\partial F_{\mu\nu}/\partial x_\mu=-\mathcal{J}_\nu{}^\gamma\,, \quad (6.2)$$

$\Delta_0(x)$ is the same $\Delta_a(x)$ function given by Eq. (4.1) with $a=0$, and D_R is the retarded Green's function satisfying

$$\Box^2 D_R(x)=-\delta^4(x)\,. \quad (6.3)$$

The constant L is to be chosen so as to guarantee $\hat{K}_{\mu\nu}{}^\gamma=0$ for x spacelike. To order e^2 we may split $\hat{K}_{\mu\nu}{}^\gamma$ into a zeroth order plus a lepton part and a hadron part; thus

$$\hat{K}_{\mu\nu}{}^\gamma(x)=-i\delta_{\mu\nu}\Delta_0(x)+\hat{K}_{\mu\nu}{}^{\gamma l}(x)+\hat{K}_{\mu\nu}{}^{\gamma h}(x)\,, \quad (6.4)$$

with

$$\hat{K}_{\mu\nu}{}^{\gamma h}=e^2\int d^4yd^4z\langle\text{vac}|[J_\mu{}^\gamma(y),J_\nu{}^\gamma(z)]|\text{vac}\rangle$$

$$\times D_R(x-y)D_R(-z)-L_h\frac{\partial^2}{\partial x_\mu\partial x_\nu}\Delta_0(x)\,, \quad (6.5)$$

and $J_\nu{}^\gamma$ denotes the hadronic electromagnetic current. For simplicity of notation we include only the ρ contribution to $J_\nu{}^\gamma$ in the following discussion. The ϕ and ω contributions will be added to the final formula. Thus, applying Eq. (2.9), we write for the ρ contribution to $\hat{K}_{\mu\nu}{}^{\gamma h}$

$$\hat{K}_{\mu\nu}{}^{\gamma\rho}=\frac{e^2m_\rho{}^4}{g_\rho{}^2}\int d^4yd^4z\langle\text{vac}|[\rho_\mu(y),\rho_\nu(z)]|\text{vac}\rangle$$

$$\times D_R(x-y)D_R(-z)-L_\rho\frac{\partial^2}{\partial x_\mu\partial x_\nu}\Delta_0(x)$$

$$=\frac{e^2m_\rho{}^4}{g_\rho{}^2}\int d^4yd^4z\left[\delta_{\mu\nu}-a^{-1}\frac{\partial^2}{\partial x_\mu\partial x_\nu}\right]\Delta_a(y-z)\sigma_\rho(a)$$

$$\times D_R(x-y)D_R(-z)da-L_\rho\frac{\partial^2}{\partial x_\mu\partial x_\nu}\Delta_0(x)\,, \quad (6.6)$$

on account of Eq. (4.1).

Taking Fourier transforms, we find

$$K_{\mu\nu}{}^{\gamma\rho}(q)=i\int\hat{K}_{\mu\nu}{}^{\gamma\rho}(x)e^{-iq_\lambda x_\lambda}d^4x$$

$$=2\pi i\frac{e^2m_\rho{}^4}{g_\rho{}^2}\int_{4m_\pi{}^2}^\infty\frac{\sigma_\rho(a)}{a^2}da\ \epsilon(q_0)$$

$$\times\left[\delta_{\mu\nu}\delta(q^2+a)+\frac{q_\mu q_\nu}{a}(\delta(q^2+a)-\delta(q^2))\right], \quad (6.7)$$

where $\epsilon(q_0)=q_0/|q_0|$. The term in $\delta(q^2)$ is the contribution of the L_ρ term. The ρ contribution to the photon propagator

$$D_{\mu\nu}{}^\gamma(q)=q^{-2}\delta_{\mu\nu}+D_{\mu\nu}{}^{\gamma h}(q)+D_{\mu\nu}{}^{\gamma l}(q) \quad (6.8)$$

is obtained from (6.7) via the correspondence[16]

$$2\pi i\epsilon(q_0)\delta(q^2+a)\rightarrow 1/(q^2+a-i\epsilon)\,,$$

yielding the ρ contribution to $D_{\mu\nu}{}^{\gamma h}(q)$:

$$D_{\mu\nu}{}^{\gamma\rho}(q)=\frac{e^2m_\rho{}^4}{g_\rho{}^2}\left(\delta_{\mu\nu}-\frac{q_\mu q_\nu}{q^2}\right)\int da\frac{\sigma_\rho(a)}{a^2}\frac{1}{q^2+a-i\epsilon}\,. \quad (6.9)$$

It is of interest to note that

$$\lim_{q^2\to\infty}D_{\mu\nu}{}^{\gamma\rho}=\frac{e^2m_\rho{}^4}{g_\rho{}^2}\left(\delta_{\mu\nu}-\frac{q_\mu q_\nu}{q^2}\right)\frac{1}{q^2}\int da\frac{\sigma_\rho(a)}{a^2}\,, \quad (6.10)$$

so that the ρ contribution to the order e^2 part of the charge renormalization is given by[16a]

$$\frac{\delta(e_0{}^2)_\rho}{e^2}=\frac{e^2m_\rho{}^4}{g_\rho{}^2}\int_{4m_\pi{}^2}^\infty\frac{\sigma_\rho(a)}{a^2}da\,, \quad (6.11)$$

[15] G. Källén, Helv. Phys. Acta **25**, 417 (1952). The L used in this paper corresponds to M in Källén's notation.

[16] Some additional details and references to some of the relevant literature are given by L. M. Brown and F. Calogero, Phys. Rev. **120**, 653 (1960), Appendix.

[16a] *Note added in proof.* In writing (6.11) we use the ratio

$$\lim_{q^2\to\infty}q^2D^\gamma(q^2)/\lim_{q^2\to 0}q^2D^\gamma(q^2)$$

as the definition of $(e_0/e)^2$ where $D^\gamma(q^2)$ is the coefficient of $\delta_{\mu\nu}$ in $D_{\mu\nu}{}^\gamma$. An alternative definition of $(e_0/e)^2$ is given simply by the limit

$$\lim_{q^2\to 0}q^2[D^\gamma(q^2)]_{\text{unren}}$$

where $[D^\gamma(q^2)]_{\text{unren}}$ is the coefficient of $\delta_{\mu\nu}$ in the unrenormalized photon propagator $[D_{\mu\nu}{}^\gamma(q^2)]_{\text{unren}}$. It is interesting to note that for the present case these two definitions give the same result only if the unrenormalized mass $m_\rho{}^0$ is infinite. For details see T. D. Lee and B. Zumino [Nuovo Cimento (to be published)].

where e is the observed charge, and it is related to the unrenormalized charge e_0 by $e_0^2=e^2+(\delta e_0^2)_\rho+\cdots$. From Eq. (4.6) and $\sigma_\rho(a)\geq 0$, it follows that

$$0<\frac{\delta(e_0^2)_\rho}{e^2}<\frac{1}{4}\left(\frac{e^2m_\rho^4}{g_\rho^2m_\pi^2}\right)\int_{4m_\pi^2}^\infty \frac{\sigma_\rho(a)}{a}=\frac{1}{4}\frac{Z_0e^2m_\rho^2}{Zg_\rho^2m_\pi^2}. \quad (6.12)$$

We note that, on account of Eq. (3.11), the product Zg_ρ^2 is independent of the choice of the renormalization constant Z. For definiteness, we may adopt the renormalization convention $Z=Z_0$ given by Eq. (4.22). As we shall see in the next section, this choice leads to $(4\pi)^{-1}g_\rho^2\cong 2.3$. Thus, (6.12) becomes

$$\frac{\delta(e_0^2)_\rho}{e^2}<\frac{1}{4}\frac{e^2m_\rho^2}{g_\rho^2m_\pi^2}\cong 2.4\%. \quad (6.13)$$

$D_{\mu\nu}{}^{\gamma\rho}$ can be expressed explicitly in terms of the ρ-propagator $D_{\mu\nu}(q)$. Recalling Eqs. (4.10) and (4.12), we write

$$D_{\mu\nu}(q)=\int\frac{\delta_{\mu\nu}+a^{-1}q_\mu q_\nu}{q^2+a-i\epsilon}\sigma_\rho(a)da=\delta_{\mu\nu}F(q^2)+q_\mu q_\nu G(q^2),$$

and obtain for the ρ contribution to the photon propagator

$$D_{\mu\nu}{}^{\gamma\rho}(q)=(e^2m_\rho^4/g_\rho^2)\left(\delta_{\mu\nu}-\frac{q_\mu q_\nu}{q^2}\right)$$

$$\times\left\{\frac{1}{q^2-i\epsilon}[F(q^2)-F(0)-(q^2-i\epsilon)F'(0)]\frac{1}{q^2-i\epsilon}\right\}. \quad (6.14)$$

This expression is, for all physical processes, equivalent to that which would be derived by straightforward application of Feynman rules to the form of the theory given in Appendix B, Eq. (B8). The apparent photon mass squared appearing there, $(e''m_\rho{}^0/g_\rho{}^0)^2$, is equal to the term $(e^2m_\rho^4/g_\rho^2)F(0)$ of Eq. (6.14) to lowest order in e^2.

The complete expression for the hadronic contribution to the photon propagator, obtained from Eqs. (5.26) and (6.9), is

$$D_{\mu\nu}{}^{\gamma h}=e^2\left(\delta_{\mu\nu}-\frac{q_\mu q_\nu}{q^2}\right)\int\frac{da}{a^2(q^2+a-i\epsilon)}$$

$$\times\{g_\rho^{-2}m_\rho^4\sigma_\rho(a)+\tfrac{1}{4}[g^{-1}M^2\sigma_{\phi\omega}(a)M^2\tilde{g}^{-1}]_{11}\}, \quad (6.15)$$

where g and M^2 are given by Eqs. (5.19) and (5.17), respectively, the subscript 11 denotes the (1,1)th matrix element, and $\sigma_{\phi\omega}(a)$ is the renormalized spectral-function matrix, related to the unrenormalized spectral-function matrix $\sigma_{\phi\omega}{}^0$ of Eq. (5.30) by

$$\sigma_{\phi\omega}(a)=S^{-1}\sigma_{\phi\omega}{}^0\tilde{S}^{-1}, \quad (6.16)$$

and to the renormalized ϕ-ω propagator $\mathfrak{D}_{\mu\nu}$ by

$$\mathfrak{D}_{\mu\nu}(q)=\int_{9m_\pi^2}^\infty\frac{\delta_{\mu\nu}+a^{-1}q_\mu q_\nu}{q^2+a-i\epsilon}\sigma_{\phi\omega}(a)da. \quad (6.17)$$

To order e^2 and neglecting leptonic contributions, we find that the entire hadronic contribution to charge renormalization is finite; the unrenormalized charge e_0 is given by

$$(e_0/e)^2=1+e^2\int a^{-2}da$$

$$\times\{g_\rho^{-2}m_\rho^4\sigma_\rho(a)+\tfrac{1}{4}[g^{-1}M^2\sigma_{\phi\omega}(a)M^2\tilde{g}^{-1}]_{11}\}$$

$$<1+\tfrac{1}{4}(g_\rho m_\pi)^{-1}e^2m_\rho^4\int_{4m_\pi^2}^\infty a^{-1}\sigma_\rho(a)da$$

$$+\frac{1}{36}\left[g^{-1}M^2\int_{9m_\pi^2}^\infty a^{-1}\sigma_{\phi\omega}(a)daM^2\tilde{g}^{-1}\right]_{11}m_\pi^2. \quad (6.18)$$

The value of the above upper limit of $(e_0/e)^2$ can be estimated by using Eqs. (6.13), (5.32), (5.55), and the numerical values found in the next section. We find

$$1<(e_0/e)^2<1+\tfrac{1}{4}e^2m_\pi^{-2}$$

$$\times[g_\rho^{-2}m_\rho^2+\tfrac{1}{9}g_Y^{-2}(\cos^2\theta_Ym_\phi^2+\sin^2\theta_Ym_\omega^2)]$$

$$\cong 1.03. \quad (6.19)$$

The finiteness of the hadronic contribution to charge renormalization arises from the fact that the spectral representation of the photon propagator is a factor a^{-2} more convergent than that of the vector-meson propagator, and corresponds to the circumstance exhibited in Eqs. (2.10) and (2.22) that the matrix elements of the hadronic part of the electromagnetic current are more convergent, by a factor q^2, than the corresponding matrix elements for the vector-meson currents.

VII. APPLICATIONS

In this section, we list the various applications[17] of the above general discussions, and in particular the experimental implications that follow from the proposed identity between the hadronic current operator $J_\mu{}^\gamma$ and the vector-meson fields.

1. $\rho^0\to\pi^++\pi^-$

This decay depends only on the strong interaction. It has already been pointed out in Sec. IV, Eq. (4.28), that all matrix elements of the renormalized current $J_\mu{}^\rho(x)$ vanish at $q^2=-m_\rho^2$; therefore, it is useful to introduce a modified current operator $\hat{J}_\mu{}^\rho(x)$ defined by

[17] The various applications given in this section have almost all been discussed in one form or another in the literature. See Refs. 2, 4, and 21 for a partial list of these references.

Eq. (4.26). By using Eq. (4.29), one finds ($\hbar=c=1$)

$$\text{rate}(\rho^0\to\pi^++\pi^-)=(48\pi m_\rho^2)^{-1}(m_\rho^2-4m_\pi^2)^{3/2}\times[g_\rho\hat{F}_{\pi\pi}{}^\rho(-m_\rho^2)]^2(Z_1/Z_0)\,, \quad (7.1)$$

where $\hat{F}_{\pi\pi}{}^\rho$ is the vertex function (or, the $\pi\pi$-ρ form factor), defined by

$$\langle(\pi^+\pi^-)^{\text{in}}|\hat{\mathbf{J}}^\rho(0)|\text{vac}\rangle=\mathbf{s}(p_\pi/\omega_\pi)\hat{F}_{\pi\pi}{}^\rho(q^2)\,, \quad (7.2)$$

$|(\pi^+\pi^-)^{\text{in}}\rangle$ is the two-pion "incoming" p-wave eigenstate of the strong-interaction Hamiltonian in the center-of-mass system, $\mathbf{s}$ is the polarization vector (chosen to be real), ω_π is the pion energy$=\frac{1}{2}(-q^2)^{1/2}$, and p_π is its corresponding momentum.

If inelastic processes are neglected, then the state $|(\pi^+\pi^-)^{\text{in}}\rangle$ differs from the corresponding stationary state by a multiplicative factor $e^{-i\delta_p}$ where δ_p is the two-pion p-wave phase shift. In this approximation, on account of time-reversal invariance, the phase of the matrix element $\langle(\pi^+\pi^-)^{\text{in}}|\varrho(0)|\text{vac}\rangle$ is δ_p, or $\delta_p+180°$. If, in addition, the nonresonant background is neglected for q^2 near $-m_\rho^2$, the resonance factor $[q^2+m_\rho^2-i\gamma m_\rho]^{-1}$ in the ρ-propagator $D_{\mu\nu}(q^2)$ is proportional to

$$-i(e^{2i\delta_p}-1)=2e^{i\delta_p}\sin\delta_p\,.$$

By using Eq. (4.26), we see that $\langle(\pi^+\pi^-)^{\text{in}}|\hat{\mathbf{J}}^\rho(0)|\text{vac}\rangle$ and, consequently, $\hat{F}_{\pi\pi}{}^\rho(q^2)$ are real at $q^2=\text{real}\cong-m_\rho^2$. Furthermore, in the same region $\hat{F}_{\pi\pi}{}^\rho(q^2)$ is expected to be a relatively slowly varying function of q^2.

In Eq. (7.1), the factor (Z_1/Z_0) is due to our normalization convention [Eq. (4.22)]. The same convention also leads to Eq. (4.27), which implies that

$$\langle\pi^+|\hat{J}_4{}^\rho(0)|\pi^+\rangle=i\,; \quad (7.3)$$

therefore, the analytic continuation of the form factor $\hat{F}_{\pi\pi}{}^\rho(q^2)$ satisfies

$$\hat{F}_{\pi\pi}{}^\rho(q^2)=1 \quad \text{at} \quad q^2=0\,. \quad (7.4)$$

By using Eq. (7.1) and the experimental values that

$$\text{rate}(\rho^0\to\pi^++\pi^-)=115.5\pm8.2 \text{ MeV}$$
$$\text{and} \quad m_\rho=756.4\pm3.2 \text{ MeV}\,,$$

one finds

$$(4\pi)^{-1}[g_\rho\hat{F}_{\pi\pi}{}^\rho(-m_\rho^2)]^2(Z_1/Z_0)=2.28\pm0.16\,. \quad (7.5)$$

According to Eqs. (4.17), (4.22), and (4.25), the factor (Z_1/Z_0) depends on the variation of $(q^2+m_\rho^2)^{-1}\times\text{Re}F^{-1}(q^2)$ from $q^2=0$ to $q^2=-m_\rho^2$, where F is defined by Eq. (4.12). As we shall see, while both (Z_1/Z_0) and the value of $\hat{F}_{\pi\pi}{}^\rho(q^2)$ at $q^2=-m_\rho^2$ are not known, the product $(Z_1/Z_0)\hat{F}_{\pi\pi}{}^\rho(-m_\rho^2)$ can be determined by using (7.1) and the leptonic decay rates of ρ^0.

2. $\varrho^0\to l^++l^-$

By using Eqs. (1.3a), (4.19), and (4.22), it can be easily verified that

$$\text{rate}(\rho^0\to l^++l^-)=\tfrac{1}{3}\alpha^2[(g_\rho^2/4\pi)(Z_0/Z_1)]^{-1}\times m_\rho^{-2}(m_\rho^2-4m_l^2)^{1/2}(m_\rho^2+2m_l^2)\,, \quad (7.6)$$

where $l=e$ or μ and $\alpha=(137)^{-1}$. The present experimental value[18] of the branching ratio of $\rho^0\to\mu^++\mu^-$ is $(4.3\pm1.4)\times10^{-5}$. Thus, we find

$$(4\pi)^{-1}g_\rho^2(Z_0/Z_1)=(2.5\pm0.8)\,. \quad (7.7)$$

It is interesting to note that the rate of $\rho^0\to\pi^++\pi^-$ depends on $g_\rho^2(Z_1/Z_0)$, but the rate $\rho^0\to\mu^++\mu^-$ (or, e^++e^-) depends on $g_\rho^2(Z_0/Z_1)$. By taking the ratio of Eqs. (7.5) and (7.7), one finds

$$[\hat{F}_{\pi\pi}{}^\rho(-m_\rho^2)(Z_1/Z_0)]^2=0.9\pm0.3\,, \quad (7.8)$$

which is consistent with the approximation that both the vertex function and $(q^2+m_\rho^2)^{-1}\,\text{Re}F^{-1}(q^2)$ do not change much from $q^2=0$ to $q^2=-m_\rho^2$; i.e.,

$$\hat{F}_{\pi\pi}{}^\rho(q^2)\cong1 \quad \text{for} \quad -q^2\lesssim m_\rho^2\,, \quad (7.9)$$

and

$$(Z_1/Z_0)\cong1\,. \quad (7.10)$$

Under the same approximation, the decay $\rho\to\pi^++\pi^-$ determines the coupling constant g_ρ^2 to be

$$(4\pi)^{-1}g_\rho^2\cong2.3\,. \quad (7.11)$$

3. ϕ-ω Decays and the Determination of θ_Y and θ_N

From Eq. (2.12) it follows that any matrix element of the renormalized current operator $\mathcal{J}_\mu{}^\phi(x)$ vanishes at $q^2=-m_\phi^2$, and that of $\mathcal{J}_\mu{}^\omega(x)$ vanishes at $q^2=-m_\omega^2$. It is useful to define the modified current operator

$$\hat{\mathcal{J}}_\mu(x)=\begin{pmatrix}\hat{\mathcal{J}}_\mu{}^\phi(x)\\ \hat{\mathcal{J}}_\mu{}^\omega(x)\end{pmatrix} \quad (7.12)$$

by

$$\langle B|\hat{\mathcal{J}}_\mu(x)|A\rangle=-[\mathfrak{D}_{\mu\nu}(q)]^{-1}\langle B|\psi_\nu(x)|A\rangle\,, \quad (7.13)$$

where $\psi_\mu(x)$ is the renormalized ϕ-ω field operator given by Eq. (5.1), and $\mathfrak{D}_{\mu\nu}(q)$ is the renormalized ϕ-ω propagator given by Eq. (5.52). In analogy with Eq. (4.29), the decay rate of the ϕ meson into a state Γ can be expressed in terms of the matrix element $\langle\Gamma|\hat{\mathcal{J}}_\mu{}^\phi(0)|\text{vac}\rangle$, and the decay rate of the ω meson into a state Γ can be expressed in terms of the matrix element $\langle\Gamma|\hat{\mathcal{J}}_\mu{}^\omega(0)|\text{vac}\rangle$. The fact that these two decay rates involve $\hat{\mathcal{J}}_\mu{}^\phi$ and $\hat{\mathcal{J}}_\mu{}^\omega$ separately is a consequence of the orientation condition [Eq. (5.48)] that we have used.

In the following, we will assume the validity of the pole dominance approximation in the resonance region and choose the normalization convention given by Eq. (5.53). Thus, from Eqs. (5.55) and (5.56), it follows that, in the region from $q^2=-m_\phi^2$ to $-m_\omega^2$, the real part of the inverse of the renormalized ϕ-ω propagator is given by

$$\text{Re}[\mathfrak{D}_{\mu\nu}(q)]^{-1}=(q^2+M^2)\delta_{\mu\nu}-q_\mu q_\nu\,, \quad (7.14)$$

[18] A. Wehmann *et al.*, Phys. Rev. Letters **17**, 1113 (1966). See also R. Weinstein, in *Proceedings of the Thirteenth International Conference on High-Energy Physics, Berkeley, 1966* (University of California Press, Berkeley, 1967).

where

$$M^2=\begin{pmatrix} m_\phi^2 & 0 \\ 0 & m_\omega^2 \end{pmatrix}.$$

We recall that at $q^2=0$, $\mathfrak{D}_{\mu\nu}^{-1}(q)$ is always equal to $\tilde{S}M_0{}^2S\delta_{\mu\nu}$, and therefore

$$\mathfrak{D}_{\mu\nu}^{-1}(0)=M^2\delta_{\mu\nu}, \tag{7.15}$$

on account of Eq. (5.55), even though the pole approximation is only assumed for the resonance region $-m_\omega{}^2\geqq q^2\geqq -m_\phi{}^2$ (which does not include $q^2=0$). By using Eqs. (5.18) and (7.13), one finds that the spatial integral of $\mathfrak{J}_4(x)$ is related to the hypercharge Y and the baryon number N by

$$-i\int \mathfrak{J}_4(x)d^3r=g\begin{pmatrix} Y \\ N \end{pmatrix}, \tag{7.16}$$

which together with Eqs. (5.19), (5.20), and (5.21) gives

$$-i\int \mathfrak{J}_4{}^\phi(x)d^3r=[\cos(\theta_Y-\theta_N)]^{-1} \times[\cos\theta_N g_Y Y+\sin\theta_Y g_N N],$$

and (7.17)

$$-i\int \mathfrak{J}_4{}^\omega(x)d^3r=[\cos(\theta_Y-\theta_N)]^{-1} \times[-\sin\theta_N g_Y Y+\cos\theta_Y g_N N].$$

The decay rate of $\phi^0\rightarrow K^++K^-$, or $K_1{}^0+K_2{}^0$ depends on the vertex function $\hat{F}_{KK}{}^\phi(q^2)$ at $q^2=-m_\phi{}^2$; $\hat{F}_{KK}{}^\phi(q^2)$ is defined by

$$\langle(2K)^{\text{in}}|\mathfrak{J}_i{}^\phi(0)|\text{vac}\rangle =\sqrt{2}\mathbf{s}_i(p_K/\omega_K)\hat{F}_{KK}{}^\phi(q^2),\quad i=1,2,3, \tag{7.18}$$

where $|(2K)^{\text{in}}\rangle$ denotes the two-kaon $I=0$, $S=$strangeness$=0$, p-wave "incoming" eigenstate of the strong interaction Hamiltonian, $\mathbf{s}$ is the polarization vector (chosen to be real), ω_K is the kaon energy$=\frac{1}{2}(-q^2)^{1/2}$ and $p_K=(\omega_K{}^2-m_K{}^2)^{1/2}$. Just as in the case of $\hat{F}_{\pi\pi}{}^\rho(-m_\rho{}^2)$, the function $\hat{F}_{KK}{}^\phi(q^2)$ is real at $q^2=-m_\phi{}^2$, and $\hat{F}_{KK}{}^\phi(q^2)$ is expected to be a slowly varying function of q^2 near $q^2=-m_\phi{}^2$. From Eq. (7.17), we know that the diagonal matrix element of $\mathfrak{J}_4{}^\phi$ for a K^+ at rest is

$$\langle K^+|\mathfrak{J}_4{}^\phi(0)|K^+\rangle=i[\cos(\theta_Y-\theta_N)]^{-1}\cos\theta_N g_Y, \tag{7.19}$$

which implies that the analytic continuation of the same vertex function $\hat{F}_{KK}{}^\phi(q^2)$, at $q^2=0$, becomes

$$\hat{F}_{KK}{}^\phi(0)=[\cos(\theta_Y-\theta_N)]^{-1}\cos\theta_N g_Y. \tag{7.20}$$

The formulas for the rates $\phi^0\rightarrow K^++K^-$ (and $K_1{}^0+K_2{}^0$), $\phi^0\rightarrow l^++l^-$, and $\omega^0\rightarrow l^++l^-$ can be obtained by following the same derivations of Eqs. (7.1) and (7.6). In order to use these rates to determine the two mixing angles θ_Y and θ_N, we assume that the vertex function $\hat{F}_{KK}{}^\phi(q^2)$ at $q^2=-m_\phi{}^2$ can be approximated by its value at $q^2=0$, i.e.;

$$\hat{F}_{KK}{}^\phi(-m_\phi)\cong\hat{F}_{KK}{}^\phi(0) =[\cos(\theta_Y-\theta_N)]^{-1}\cos\theta_N g_Y. \tag{7.21}$$

By using Eqs. (7.14) and (7.21), one finds

$$\text{rate}(\phi^0\rightarrow K^++K^-)=(48\pi m_\phi{}^2)^{-1}(m_\phi{}^2-4m_K{}^2)^{3/2} \times[\cos(\theta_Y-\theta_N)]^{-2}g_Y{}^2\cos^2\theta_N, \tag{7.22}$$

where m_K is the mass of $K^\pm$. The same expression applies to rate$(\phi^0\rightarrow K_1{}^0+K_2{}^0)$, provided that m_K refers to the mass of the neutral K meson.

Similarly, by using Eqs. (1.6) and (5.40), one finds

$$\text{rate}(\phi^0\rightarrow l^++l^-)=(1/12)\alpha^2(g_Y{}^2/4\pi)^{-1}m_\phi{}^{-2} \times(m_\phi{}^2-4m_l{}^2)^{1/2}(m_\phi{}^2+2m_l{}^2)\cos^2\theta_Y, \tag{7.23}$$

and

$$\text{rate}(\omega^0\rightarrow l^++l^-)=(1/12)\alpha^2(g_Y{}^2/4\pi)^{-1}m_\omega{}^{-2} \times(m_\omega{}^2-4m_l{}^2)^{1/2}(m_\omega{}^2+2m_l{}^2)\sin^2\theta_Y. \tag{7.24}$$

From the known rate of $\phi^0\rightarrow K^++K^-=1.7\pm0.4$ MeV, one obtains

$$\left[\frac{g_Y\cos\theta_N}{\cos(\theta_Y-\theta_N)}\right]^2\frac{1}{4\pi}=1.4\pm0.3. \tag{7.25}$$

The coupling constant $g_Y{}^2$ and the two mixing angles θ_Y and θ_N can then be determined by measuring the leptonic decay rates of ϕ^0 and ω^0. These leptonic decay rates can also be calculated theoretically by assuming the particular values of θ_Y and θ_N determined in the Sec. V (3. *Special Models.*) The results are given in Table I.

TABLE I. The coupling constant $(g_Y{}^2/4\pi)$ and the leptonic decay rates are calculated by assuming the rate $(\phi^0\rightarrow K^++K^-)$ is 1.7 (MeV/$\hbar$). [In the limit of SU_3 symmetry, $(g_Y{}^2/4\pi)=\frac{3}{4}(g_\rho{}^2/4\pi)\cong1.7$.] To the accuracy given, there is no difference between the decay rates to μ pair and to e pair. The mean-square radii $R^2(K^0)$ and $R^2(\bar{K}^0)$ of the charge distributions of K^0 and $\bar{K}^0$ are calculated by using Eq. (7.42).

	Current-mixing model	Mass-mixing model	Mass-mixing model (variation)
θ_Y	33°	32°	39°
θ_N	21°	32°	39°
$(g_Y{}^2/4\pi)$	1.5	1.9	2.2
Rate $(\phi\rightarrow e^+e^-)$	2.2 (keV/$\hbar$)	1.7 (keV/$\hbar$)	1.2 (keV/$\hbar$)
Rate $(\omega\rightarrow e^+e^-)$	0.7 (keV/$\hbar$)	0.5 (keV/$\hbar$)	0.6 (keV/$\hbar$)
$R^2(K^0)$	-7.6×10^{-28} cm²	-7.0×10^{-28} cm²	-6.1×10^{-28} cm²
$R^2(\bar{K}^0)$	$+7.6\times10^{-28}$ cm²	$+7.0\times10^{-28}$ cm²	$+6.1\times10^{-28}$ cm²

4. Comparison Between $A\rightarrow B+\gamma$ and $A\rightarrow B+\rho^0$ (or, ϕ^0 and ω^0)

The identity (1.3a) between the hadronic electromagnetic current and the renormalized ρ-meson field implies that the isovector part of the electromagnetic form factor $F_{AB}{}^\gamma(q^2)$ of any real or virtual transition

$A \to B+\gamma$ is related to the corresponding form factor of $A \to B+\rho^0$ by Eq. (1.4a),

$$[F_{AB}{}^{\gamma}(q^2)]_{\text{isovector}} = \frac{m_\rho{}^2}{q^2+m_\rho{}^2} F_{AB}{}^{\rho}(q^2). \tag{1.4a}$$

Similarly, by using Eq. (2.22), one finds that the identity (1.6) implies that the isoscalar part of $F_{AB}{}^{\gamma}(q^2)$ is given by

$$[F_{AB}{}^{\gamma}(q^2)]_{\text{isoscalar}} = \tfrac{1}{2} g_Y{}^{-1} \left[\cos\theta_Y \left(\frac{m_\phi{}^2}{q^2+m_\phi{}^2} \right) F_{AB}{}^{\phi}(q^2) - \sin\theta_Y \left(\frac{m_\omega{}^2}{q^2+m_\omega{}^2} \right) F_{AB}{}^{\omega}(q^2) \right], \tag{7.26}$$

where $F_{AB}{}^{\phi}(q^2)$ and $F_{AB}{}^{\omega}(q^2)$ are related to the matrix elements of the renormalized currents $\mathcal{J}_\mu{}^\phi$ and $\mathcal{J}_\mu{}^\omega$ by

$$\langle B | \mathcal{J}_\mu{}^\phi(x) | A \rangle = \sum_i F_{AB}{}^{\phi}(q^2) u_B{}^\dagger \Gamma_\mu{}^i u_A \exp(iq_\lambda x_\lambda),$$
$$\langle B | \mathcal{J}_\mu{}^\omega(x) | A \rangle = \sum_i F_{AB}{}^{\omega}(q^2) u_B{}^\dagger \Gamma_\mu{}^i u_A \exp(iq_\lambda x_\lambda), \tag{7.27}$$

and $u_B{}^\dagger \Gamma_\mu{}^i u_A$ denotes the same spin-momentum function used in Eq. (2.11). In the following, we shall discuss these form-factor relations in three separate regions of q^2:

(i) At $q^2=0$, the validity of Eqs. (1.4a) and (7.26) follows directly from the properties that the spatial integrals of $J_4{}^\rho(x)$ and $[\cos\theta_Y\, \mathcal{J}_4{}^\phi(x) - \sin\theta_Y\, \mathcal{J}_4{}^\omega(x)]$ are, respectively, iI_z and $ig_Y Y$. [See Eqs. (2.6) and (2.15).]

(ii) Near the resonance, we have, on account of Eqs. (2.3) and (2.12), for any hadronic states A and B,

$$F_{AB}{}^{a}(q^2)=0 \quad \text{at} \quad q^2=-m_a{}^2, \tag{7.28}$$

where a can be either ρ, or ϕ, or ω. Although $(q^2+m_a{}^2)^{-1} \times F_{AB}{}^{a}(q^2)$ and, therefore, also Eqs. (1.4a) and (7.26) remain well defined at $q^2+m_a{}^2=0$, it is much more convenient to use the modified current operators $\hat{J}_\mu{}^\rho(x)$, $\hat{\mathcal{J}}_\mu{}^\phi(x)$, $\hat{\mathcal{J}}_\mu{}^\omega(x)$, and their related form factors $\hat{F}_{AB}{}^\rho(q^2)$, $\hat{F}_{AB}{}^\phi(q^2)$, and $\hat{F}_{AB}{}^\omega(q^2)$, instead of $J_\mu{}^\rho(x)$, $\mathcal{J}_\mu{}^\phi(x)$, and $\mathcal{J}_\mu{}^\omega(x)$ [or $F_{AB}{}^\rho(q^2)$, $F_{AB}{}^\phi(q^2)$, and $F_{AB}{}^\omega(q^2)$], in the q^2 region near the vector-meson resonances.

A direct consequence of the identity (1.3a) is that for any $|\Delta\mathbf{I}|=1$ transition at $q^2=-m_\rho{}^2$, the ratio between the rates of

$$A \to B+\rho^0 \tag{7.29}$$

and

$$A \to B+l^+ + l^- \quad (l=e, \text{ or } \mu) \tag{7.30}$$

is independent of the initial and final complexes A and B. This independence is supposed to be an *exact* one, provided that the higher-order radiative corrections are neglected; furthermore, in taking this ratio, one should use directly the observed rate of the $|\Delta\mathbf{I}|=1$ transition $A \to B+l^+ + l^-$ at $q^2=-m_\rho{}^2$ without any background subtraction.

Since the neutral ρ meson is unstable against the strong interaction, reaction (7.29) can only be observed by studying its decay products, such as 2π or 4π. Instead of (7.29), one may use, e.g.,

$$A \to B+\pi^+ + \pi^-. \tag{7.31}$$

The two-pion p-wave amplitude has the familiar resonance behavior $[q^2+m_\rho{}^2 - i\gamma m_\rho]^{-1}$ at $q^2 \cong -m_\rho{}^2$. By using only the resonant part of the two-pion amplitude, the ratio

$$r(q^2) \equiv \frac{\text{rate}[A \to B+(\pi^+ + \pi^-)_{p\text{-state}}]}{\text{rate}[A \to B+l^+ + l^-]} \tag{7.32}$$

can be measured, and it should be independent of A and B for any $|\Delta\mathbf{I}|=1$ transitions in the region $q^2 \cong -m_\rho{}^2$.

By following the same arguments that led to Eqs. (7.1) and (7.6), but without setting $q^2=-m_\rho{}^2$, one finds, for q^2 near $-m_\rho{}^2$,

$$r(q^2) = g_\rho{}^4 \left[\frac{(m_\rho{}^2 - 4m_\pi{}^2)^{3/2}}{16\alpha^2 (m_\rho{}^2 - 4m_l{}^2)^{1/2} (m_\rho{}^2 + 2m_l{}^2)} \right] \times \left[\frac{-q^2}{m_\rho{}^2} \right]^2 [\hat{F}_{\pi\pi}{}^\rho(q^2)]^2, \tag{7.33}$$

which is independent of A and B, and, in addition, is independent of the wave-function normalization factor (Z_1/Z_0). The functional dependence of the vertex function $\hat{F}_{\pi\pi}{}^\rho(q^2)$ at q^2 near $-m_\rho{}^2$ can be determined by measuring $r(q^2)$.

We note that the q^2-dependent factor in Eq. (7.33) must be expected to produce a shift[19] in the ρ peak observed in the process $A \to B+l^+ + l^-$ from that observed in $A \to B+\pi^+ + \pi^-$. The main shift may be expected to arise from the factor $(q^2)^2$; from this source alone the shift in the ρ peak is from $q^2=-m_\rho{}^2 \cong -(756 \text{ MeV})^2$ to

$$q^2 = -\tfrac{1}{4} m_\rho [3m_\rho + (m_\rho{}^2 - 8\gamma^2)^{1/2}] \cong -(745 \text{ MeV})^2. \tag{7.34}$$

Identical arguments can be applied to any $|\Delta\mathbf{I}|=0$ transition

$$A \to B+l^+ + l^-$$

by comparing its rate with that of

$$A \to B+\phi^0 \text{ (or } \omega^0)$$

at $q^2=-m_\phi{}^2$ (or $-m_\omega{}^2$). The rates of the latter reactions can be measured by, e.g., using the *resonant* part of $A \to B+K^+ + K^-$ for the ϕ meson and $A \to B+3\pi$ for the ω meson. The ratios of these rates to the corresponding lepton pair production rates are, again, independent of A and B, provided Eq. (1.6) holds.

(iii) For q^2 away from the vector-meson resonances, it is more convenient to use Eqs. (1.4a) and (7.26).

[19] One of us (N.M.K.) wishes to acknowledge a discussion of this point with M. Good and A. Silverman.

Although the form factors $F_{AB}{}^{\rho}(q^2)$, $F_{AB}{}^{\phi}(q^2)$, and $F_{AB}{}^{\omega}(q^2)$ are not known, one may assume some simple analytic functions for these vector-meson form factors. The phenomenological parameters contained in these functions can, then, be determined by using the experimental results on the electromagnetic form factor $F_{AB}{}^{\gamma}(q^2)$.

As an example, we may consider the special case $A=B$ and assume, for q^2 spacelike (or, for any q^2 away from the resonances),

$$F_{AB}{}^{\rho}(q^2)/F_{AB}{}^{\rho}(0)=[1+(q^2/\Lambda_1{}^2)]^{-1} \qquad (7.35)$$

and

$$\frac{F_{AB}{}^{\phi}(q^2)}{F_{AB}{}^{\phi}(0)}=\frac{F_{AB}{}^{\omega}(q^2)}{F_{AB}{}^{\omega}(0)}=\left[1+\left(\frac{q^2}{\Lambda_0{}^2}\right)\right]^{-1}, \qquad (7.36)$$

where Λ_1 and Λ_0 are phenomenological parameters characterizing the overall q^2 dependence of the $I=1$ and $I=0$ vertex functions. At $q^2=0$, the values of these form factors $F_{AA}{}^{\rho}(0)$, $F_{AA}{}^{\phi}(0)$, and $F_{AA}{}^{\omega}(0)$ are known; they can be readily determined by using (2.11), (7.27), and the identities

$$\int J_4{}^{\rho}(x)d^3r=iI_z,$$

$$\int \mathcal{g}_4{}^{\phi}(x)d^3r=i[\cos(\theta_Y-\theta_N)]^{-1} \times[\cos\theta_N g_Y Y+\sin\theta_Y g_N N], \qquad (7.37)$$

$$\int \mathcal{g}_4{}^{\omega}(x)d^3r=i[\cos(\theta_Y-\theta_N)]^{-1} \times[-\sin\theta_N g_Y Y+\cos\theta_Y g_N N].$$

Thus, the resulting electromagnetic form factor $F_{AA}{}^{\gamma}(q^2)$ in the spacelike q^2 region (or, for any q^2 away from the resonances) becomes dependent only on Λ_0, Λ_1, θ_Y, θ_N, and, if A has nonzero baryon number, the ratio (g_N/g_Y); among these, the angles θ_Y and θ_N can be either directly measured, or theoretically calculated by using special models.

Such a study for $A=B=$ single nucleon has been made by Massam and Zichichi[20]; they assumed $\theta_Y=\theta_N\cong 35°$ and found that the existing data in the spacelike q^2 region is consistent with $\Lambda_0\cong\Lambda_1\cong 1$ BeV.

At present, it is not possible to make a similar study for the timelike q^2 region away from the resonance. In this connection, we may recall the possibility of the alternative proposal [Eq. (1.3b)] which implies Eq. (1.4b) instead of Eq. (1.4a). Thus, it seems particularly interesting to investigate reactions such as $A\rightarrow B+l^++l^-$ for large $-q^2$ and see whether $[F_{AB}{}^{\gamma}(q^2)]_{\text{isovector}}$ can be zero at some $q^2=-(m_\rho{}^0)^2$. If $F_{AB}{}^{\gamma}(q^2)$ has a zero, then this could be regarded as a confirmation of Eq. (1.3b), and the value of the mechanical mass $m_\rho{}^0$ would become measurable. Otherwise, it is consistent with the assumption that $m_\rho{}^0$ is infinite, and the two different proposals (1.3a) and (1.3b) are the same.

[20] T. Massam and A. Zichichi, Nuovo Cimento **43**, 1137 (1966).

5. Electromagnetic Form Factor of K^0 and $\bar{K}^0$

Let $|K^0,p\rangle$ and $|\bar{K}^0,p\rangle$ denote, respectively, the state of a neutral K^0 and $\bar{K}^0$ with 4-momentum p_μ. From Lorentz invariance and current conservation, one has

$$\langle K^0,p'|J_\mu{}^{\gamma}(0)|K^0,p\rangle=-\langle\bar{K}^0,p'|J_\mu{}^{\gamma}|\bar{K}^0,p\rangle =\tfrac{1}{2}(\omega\omega')^{-1/2}(p'+p)_\mu F_{KK}{}^{\gamma}(q^2), \qquad (7.38)$$

where $i\omega$ and $i\omega'$ are, respectively, p_4 and p_4'. By using Eqs. (1.4a), (7.26), (7.37), and (7.38), one finds

$$F_{KK}{}^{\gamma}(q^2)=-\tfrac{1}{2}f_{KK}{}^{\rho}(q^2)\left(\frac{m_\rho{}^2}{q^2+m_\rho{}^2}\right) +\frac{1}{2}\frac{1}{\cos(\theta_Y-\theta_N)}\left[\frac{m_\phi{}^2\cos\theta_Y\cos\theta_N f_{KK}{}^{\phi}(q^2)}{q^2+m_\phi{}^2} +\frac{m_\omega{}^2\sin\theta_Y\sin\theta_N f_{KK}{}^{\omega}(q^2)}{q^2+m_\omega{}^2}\right], \qquad (7.39)$$

where

$$f_{KK}{}^{a}(q^2)=F_{KK}{}^{a}(q^2)/F_{KK}{}^{a}(0), \qquad (7.40)$$

and $a=\rho$, ϕ, and ω. At $q^2=0$, $f_{KK}{}^{a}(q^2)=1$.

The mean-square radius of the change distribution of K^0 is, by definition,

$$R^2(K^0)=-6(d/dq^2)F_{KK}{}^{\gamma}(q^2) \qquad (7.41)$$

at $q^2=0$; the corresponding mean-square radius of $\bar{K}^0$ is $R^2(\bar{K}^0)=-R^2(K^0)$. If the differences between the three derivatives $(d/dq^2)f_{KK}{}^{a}(q^2)$ at $q^2=0$ can be neglected, then one finds

$$R^2(K^0)=3\{-m_\rho{}^{-2}+[\cos(\theta_Y-\theta_N)]^{-1} \times[m_\phi{}^{-2}\cos\theta_Y\cos\theta_N+m_\omega{}^{-2}\sin\theta_Y\sin\theta_N]\}. \qquad (7.42)$$

[If one assumes (7.35) and (7.36), then this expression is valid, provided one neglects $(\Lambda_1{}^{-2}-\Lambda_0{}^{-2})$.] The numerical value of $R^2(K^0)$ can be estimated by using either the current-mixing model, or the mass-mixing model. The results, which are given in Table I, are about a factor of 30 smaller than the estimate given by Zeldovich.[21] While the existing experimental evidence in support of vector dominance seems substantial enough to make a value as large as that given by Zeldovich rather unlikely, nevertheless a measurement[22] of the charge radius of K^0 and $\bar{K}^0$ could constitute a

[21] Y. B. Zeldovich, Zh. Eksperim. i Teor. Fiz. **36**, 782 (1959) [English transl.: Soviet Phys.—JETP **9**, 984 (1959)]. Some earlier discussions have been given by G. Feinberg, Phys. Rev. **109**, 1381 (1958). A recent calculation by W. Frazer (private communication) gives results of the same order as those in Table I.

[22] One of us (N.M.K.) is grateful to O. Piccioni for a stimulating discussion of this point.

further relatively sensitive test of the theory, and it would also serve as a measure of the possible difference between the derivatives of the different strong interaction form factors $f_{KK}{}^{a}$ at $q^2=0$.

ACKNOWLEDGMENTS

This investigation was started during the summer of 1966 when we were all visitors at CERN; we wish to thank Professor L. Van Hove and Professor J. Prentki for the hospitality extended to us at CERN. We also wish to thank Professor J. Bernstein, Professor W. Frazer, and Professor G. C. Wick for discussions.

APPENDIX A: AN ELEMENTARY THEOREM

The current $J_\mu{}^\rho(x)$ is first introduced in Eq. (2.3). From its definition, we know that (i) $J_\mu{}^\rho(x)$ transforms like the $I_z=0$ member of an $I=1$ triplet, (ii) $J_\mu{}^\rho(x)$ commutes with the baryon number operator N, and (iii) $J_\mu{}^\rho(x)$ anticommutes with the particle antiparticle conjugation operator C. The following theorem can be easily established:

Theorem. If $J_\mu{}^\rho(x)$ is conserved under the strong interaction, then

$$-i\int J_4{}^\rho(x)d^3r=\lambda I_z\,, \tag{A1}$$

where λ is a constant.

Proof. Let $A_z=-i\int J_4{}^\rho d^3r$, and define

$$[I_j,A_k]=i\epsilon_{jkl}A_l\,, \tag{A2}$$

where the subscripts j, k, and l can be either x, or y, or z, and ϵ_{jkl} is the usual third-rank constant antisymmetric tensor. Since A_z and I_j commute with the strong interaction Hamiltonian $H_{\rm st}$, the other two components A_x and A_y must also commute with $H_{\rm st}$. We recall that the only single-particle eigenstate of $H_{\rm st}$ that is degenerate with $|\Lambda^0\rangle$ is $|\bar\Lambda^0\rangle$. The state $A_j|\Lambda^0\rangle$ must, therefore, be a linear function of $|\Lambda^0\rangle$ and $|\bar\Lambda^0\rangle$. From Eq. (A2) and $I_j|\Lambda^0\rangle=I_j|\bar\Lambda^0\rangle=0$, we find

$$A_j|\Lambda^0\rangle=A_j|\bar\Lambda^0\rangle=0. \tag{A3}$$

Similarly, we can show that in the sector of the single nucleon and single antinucleon states

$$\begin{bmatrix} p \\ n \\ \bar n \\ -\bar p \end{bmatrix}, \tag{A4}$$

the operator A_j must be of the form

$$\begin{pmatrix} a\sigma_j & b\sigma_j \\ c\sigma_j & d\sigma_j \end{pmatrix},$$

where σ_j is the usual (2×2) Pauli matrix, and a, b, c, d are constants. Now, $[A_z,N]=0$ implies $b=c=0$, and $\{A_z,C\}=0$ implies $a=d$. Thus, in the single nucleon-antinucleon sector,

$$A_z=a\begin{pmatrix} \sigma_z & 0 \\ 0 & \sigma_z \end{pmatrix}. \tag{A5}$$

Since all known hadrons are connected through $H_{\rm st}$ to some multiple-particle states of the nucleon, antinucleon, Λ^0, $\bar\Lambda^0$ system, the theorem is, then, proved by using Eqs. (A3) and (A5), and by setting $\lambda=2a$ in Eq. (A1).

In exactly the same way, we can also prove that the spatial integral of $(J_4{}^\rho)^0$, which satisfies Eqs. (3.2) and (3.3), must also be proportional to the operator I_z.

Similar considerations can also be applied to currents $\mathcal{J}_\mu{}^\phi$ and $\mathcal{J}_\mu{}^\omega$ which are defined by Eq. (2.12). Both currents (i) are isoscalars, (ii) commute with N, and (iii) anticommute with C. Let

$$Q_\alpha\equiv-i\int \mathcal{J}_4{}^\alpha d^3r\,, \tag{A6}$$

where $\alpha=\phi$ or ω. The conservation law of $\mathcal{J}_\mu{}^\alpha$ implies that $[Q_\alpha,H_{\rm st}]=0$. By using the above properties (i)–(iii), it is easy to see that Q_α must be of the form

$$a_\alpha\begin{pmatrix} 1 & 0 \\ 0 & -1 \end{pmatrix} \tag{A7}$$

in the sector

$$\begin{pmatrix} \Lambda^0 \\ \bar\Lambda^0 \end{pmatrix},$$

and Q_α is of the form

$$b_\alpha\begin{bmatrix} 1 & 0 & 0 & 0 \\ 0 & 1 & 0 & 0 \\ 0 & 0 & -1 & 0 \\ 0 & 0 & 0 & -1 \end{bmatrix} \tag{A8}$$

in the nucleon-antinucleon sector (A4) where a_α and b_α are constants. Since all known hadrons are connected through $H_{\rm st}$ to some multiple-particle states of the nucleon, antinucleon, Λ^0, $\bar\Lambda^0$ system, the operator Q_α must, therefore, be related to the hypercharge operator Y and the baryon number operator N by

$$Q_\alpha=a_\alpha N+(b_\alpha-a_\alpha)Y. \tag{A9}$$

This establishes Eq. (2.15).

APPENDIX B: ALTERNATIVE FORMS OF ρ-γ COUPLING

For simplicity, we will consider in this Appendix only the isovector part of the electromagnetic interaction. The Lagrangian density which includes both such an interaction and the strong interaction of the ρ^0 meson is given by

$$\mathfrak{L}=-\tfrac14(F_{\mu\nu}{}^0)^2-\tfrac14(G_{\mu\nu}{}^0)^2-\tfrac12(m_\rho{}^0\rho_\mu{}^0)^2 -(g_\rho{}^0\rho_\mu{}^0+e^0A_\mu{}^0)(J_\mu{}^\rho)^0-\tfrac12(e^0/g_\rho{}^0)G_{\mu\nu}{}^0F_{\mu\nu}{}^0\,, \tag{B1}$$

where the superscript zero denotes the unrenormalized quantities. If we neglect the renormalization problem of the electromagnetic interaction, then $e^0=e$, $A_\mu{}^0=A_\mu$, and $F_{\mu\nu}{}^0=F_{\mu\nu}$. The above Lagrangian density $\mathfrak{L}$ contains both the Lagrangian densities (2.7) and (3.1) which are used in Secs. II and III. By using the definition of $J_\mu{}^\rho$ [Eq. (3.13)] one can show that apart from a trivial partial integration,

$$\mathfrak{L}=-\tfrac{1}{4}(F_{\mu\nu}{}^0)^2+(2.7)+(3.1). \qquad \text{(B2)}$$

(1) We consider first the transformation

$$\rho_\mu{}^0=[1-(e^0/g_\rho{}^0)^2]^{-1/2}\rho_\mu{}',$$

and

$$A_\mu{}^0=-(e^0/g_\rho{}^0)[1-(e^0/g_\rho{}^0)^2]^{-1/2}\rho_\mu{}'+A_\mu{}'. \qquad \text{(B3)}$$

It is easy to see that in terms of $\rho_\mu{}'$ and $A_\mu{}'$, (B1) can be expressed in an alternative form

$$\mathfrak{L}=-\tfrac{1}{4}(F_{\mu\nu}{}')^2-\tfrac{1}{4}(G_{\mu\nu}{}')^2-\tfrac{1}{2}(m_\rho{}'\rho_\mu{}')^2 -(g'\rho_\mu{}'+e^0A_\mu{}')(J_\mu{}^\rho)^0, \qquad \text{(B4)}$$

where

$$m_\rho{}'=[1-(e^0/g_\rho{}^0)^2]^{-1/2}m_\rho{}^0, \qquad \text{(B5)}$$

$$g'=g_\rho{}^0[1-(e^0/g_\rho{}^0)^2]^{1/2}, \qquad \text{(B6)}$$

$$G_{\mu\nu}{}'=\frac{\partial}{\partial x_\mu}\rho_\nu{}'-\frac{\partial}{\partial x_\nu}\rho_\mu{}',$$

and

$$F_{\mu\nu}{}'=\frac{\partial}{\partial x_\mu}A_\nu{}'-\frac{\partial}{\partial x_\nu}A_\mu{}'.$$

According to Eq. (B4), the currents which generate the new (unrenormalized) fields $\rho_\mu{}'$ and $A_\mu{}'$ are, respectively, $g'(J_\mu{}^\rho)^0$ and $e^0(J_\mu{}^\rho)^0$. In contrast to Eq. (3.17) these two currents are now proportional to each other; therefore, apart from the coupling constants g' and e^0, they formally satisfy Eq. (1.3b), even though the Lagrangian (B1) is derived based on the identity (1.3a). In terms of these new field variables, the difference between the proposals (1.3a) and (1.3b) lies now in the form of other electromagnetic interactions which are not included in (B1). We note that if (1.3a) holds, then, for example, the lepton current j_μ interacts with $A_\mu{}'$ through the Lagrangian density

$$e^0j_\mu A_\mu{}^0=e^0j_\mu A_\mu{}'-[(e^0)^2/g']j_\mu\rho_\mu{}'.$$

(2) Next, we consider a different transformation:

$$\rho_\mu{}''=\rho_\mu{}^0+(e^0/g_\rho{}^0)A_\mu{}^0,$$

and

$$A_\mu{}''=[1-(e^0/g_\rho{}^0)^2]^{1/2}A_\mu{}^0. \qquad \text{(B7)}$$

The Lagrangian density (B1) now becomes

$$\mathfrak{L}=-\tfrac{1}{4}(F_{\mu\nu}{}'')^2-\tfrac{1}{4}(G_{\mu\nu}{}'')^2-\tfrac{1}{2}(m_\rho{}^0\rho_\mu{}'')^2-g_\rho{}^0\rho_\mu{}''(J_\mu{}^\rho)^0 +(m_\rho{}^0)^2(e''/g_\rho{}^0)\rho_\mu{}''A_\mu{}''-\tfrac{1}{2}(m_\rho{}^0)^2 \times(e''/g_\rho{}^0)^2(A_\mu{}'')^2, \qquad \text{(B8)}$$

where

$$e''=e^0[1-(e^0/g_\rho{}^0)^2]^{-1/2},$$

$$F_{\mu\nu}{}''=\frac{\partial}{\partial x_\mu}A_\nu{}''-\frac{\partial}{\partial x_\nu}A_\mu{}'', \qquad \text{(B9)}$$

and

$$G_{\mu\nu}{}''=\frac{\partial}{\partial x_\mu}\rho_\nu{}''-\frac{\partial}{\partial x_\nu}\rho_\mu{}''.$$

In terms of the transformed fields, a gauge transformation means that

$$A_\mu{}''\to A_\mu{}''+\partial\Lambda/\partial x_\mu$$

and

$$\rho_\mu{}''\to\rho_\mu{}''+(e''/g_\rho{}^0)\partial\Lambda/\partial x_\mu. \qquad \text{(B10)}$$

In Eq. (B8), the only photon-matter coupling is given by $(m_\rho{}^0)^2(e''/g_\rho{}^0)\rho_\mu{}''A_\mu{}''$. Such a term, by itself, clearly violates gauge invariance; but the combination

$$-\tfrac{1}{2}(m_\rho{}^0\rho_\mu{}'')^2+(m_\rho{}^0)^2(e''/g_\rho{}^0)\rho_\mu{}''A_\mu{}'' -\tfrac{1}{2}(m_\rho{}^0)^2(e''/g_\rho{}^0)^2(A_\mu{}'')^2 \qquad \text{(B11)}$$

is invariant under the gauge transformation (B10). Consequently, the Lagrangian density (B8) is also gauge-invariant. [Under the gauge transformation (B10), expression (B8) $\to$ (B8) $-e''(J_\mu{}^\rho)^0\partial\Lambda/\partial x_\mu$. This additional term $-e''(J_\mu{}^\rho)^0\partial\Lambda/\partial x_\mu$ is, as usual, canceled by a corresponding term generated by the free Lagrangian and the strong interaction Lagrangian of the charged particles under the same gauge transformation.]

In the language of Feynman graphs, (B8) shows that there is a direct ρ-photon coupling vertex given by $(m_\rho{}^0)^2(e''/g_\rho{}^0)\rho_\mu{}''A_\mu{}''$. The application of such vertices in the photon propagator would lead to a non-gauge-invariant and negative term for (photon mass)^2, which is, however, completely canceled by the additional term $-\tfrac{1}{2}(m_\rho{}^0)^2(e''/g_\rho{}^0)^2A_\mu{}^2$ in (B8).

ALGEBRA OF FIELDS*

T. D. Lee
Columbia University, New York, New York

and

Steven Weinberg†
University of California, Berkeley, California

and

Bruno Zumino
New York University, New York, New York

(Received 10 May 1967)

The algebra of currents[1] has proven remarkably powerful, but there are important commutation relations it leaves unspecified. In particular, we would like to know more about the Schwinger terms, and, more generally, about commutators involving space components of currents or their derivatives. We shall show in this note that the algebra of gauge fields can be used to rederive all results that can be derived from current algebra, but that it also makes definite statements about these other commutators. We will then use this additional information to complete and generalize the proof of the spectral-function sum rules[2] and will conclude with an argument in support of the view[3] that the hadronic electromagnetic current is to be identified with an appropriate linear combination of gauge fields.[4]

The Lagrangian for a set of massive gauge fields $\Phi_\alpha^{\ \nu}$ coupled to a general matter field ψ is[5]

$$\mathcal{L}=-\tfrac{1}{4}F_{\alpha\mu\nu}F_\alpha^{\ \mu\nu}-\tfrac{1}{2}m_0^{\ 2}\Phi_{\alpha\mu}\Phi_\alpha^{\ \mu}+\mathcal{L}_m(\psi,D_\mu\psi), \quad (1)$$

$$F_{\alpha\mu\nu}\equiv\partial_\mu\Phi_{\alpha\nu}-\partial_\nu\Phi_{\alpha\mu}+g_0C_{\alpha\beta\gamma}\Phi_{\beta\mu}\Phi_{\gamma\nu}, \quad (2)$$

$$D_\mu\psi\equiv\partial_\mu\psi+g_0T_\alpha\Phi_{\alpha\mu}\psi, \quad (3)$$

where $C_{\alpha\beta\gamma}$ are the structure constants[6] of a compact semisimple Lie algebra, and T_α is the matrix representation of this algebra on the ψ. The field equations for $\Phi_{\alpha\mu}$ may be written as

$$m_0^{\ 2}\Phi_\alpha^{\ \nu}=\partial_\mu F_\alpha^{\ \mu\nu}+g_0J_\alpha^{\ \nu}, \quad (4)$$

with $J_\alpha^{\ \nu}$ the usual current used in current algebra,

$$J_\alpha^{\ \nu}\equiv F_\beta^{\ \nu\mu}C_{\alpha\beta\gamma}\Phi_{\gamma\mu}+\textstyle\sum_\psi(\partial\mathcal{L}_m/\partial D_\nu\psi)T_\alpha\psi. \quad (5)$$

We note the following properties of the $\Phi_{\alpha\nu}$:

(A) From Eq. (4) we have immediately

$$\partial_\nu\Phi'_\alpha{}^\nu=\partial_\nu J_\alpha^{\ \nu}, \quad (6)$$

where

$$\Phi'_\alpha{}^\nu\equiv(m_0^{\ 2}/g_0)\Phi_\alpha^{\ \nu}. \quad (7)$$

Hence $\Phi'_\alpha{}^\nu$ has the same properties of conservation or partial conservation as does $J_\alpha^{\ \nu}$.

(B) Setting $\nu=0$ in Eq. (4), and using the canonical commutation relations for $\Phi_{\alpha i}$ and ψ, gives

$$[\Phi'_{\alpha0}(\vec{x},t),\Phi'_{\beta0}(\vec{y},t)]=iC_{\alpha\beta\gamma}\Phi'_{\gamma0}(\vec{x},t)\delta^3(\vec{x}-\vec{y}), \quad (8)$$

$$[\Phi'_{\alpha0}(\vec{x},t),\Phi'_{\beta i}(\vec{y},t)]=iC_{\alpha\beta\gamma}\Phi'_{\gamma i}(\vec{x},t)\delta^3(\vec{x}-\vec{y})+i(m_0^{\ 2}/g_0^{\ 2})\delta_{\alpha\beta}\partial_i\delta^3(\vec{x}-\vec{y}). \quad (9)$$

These commutation relations are precisely the same in form as those proposed[1] for the $J_{\alpha\nu}$, except that here the Schwinger term is an explicitly known c number.

(C) Using the canonical commutation relations of the $\Phi_{\alpha i}$ together with Eq. (9) gives

$$[\Phi'_{\alpha i}(\vec{x},t),\Phi'_{\beta j}(\vec{y},t)]=0, \tag{10}$$

$$[\partial_0\Phi'_{\alpha i}(\vec{x},t)-\partial_i\Phi'_{\alpha 0}(\vec{x},t),\Phi'_{\beta j}(\vec{y},t)]=-i(m_0^4/g_0^2)\delta_{\alpha\beta}\delta_{ij}\delta^3(\vec{x}-\vec{y})+iC_{\alpha\beta\gamma}\Phi'_{\gamma i}(\vec{x},t)\partial_j\delta^3(\vec{x}-\vec{y})$$
$$-i(g_0^2/m_0^2)C_{\alpha\delta\gamma}C_{\beta\delta\epsilon}\Phi'_{\gamma i}(\vec{x},t)\Phi'_{\epsilon j}(\vec{x},t)\delta^3(\vec{x}-\vec{y}). \tag{11}$$

We see that the properties (A) and (B) make the algebra of the fields $\Phi'_{\alpha\mu}$ at least powerful enough to allow for the rederivation of any strong-interaction result that can be derived from current algebra, while the precise value given for the Schwinger term in Eq. (9) and the additional commutators specified under (C) provide extra information not available from the algebra of currents.[7]

The most immediate consequences of this new information can be obtained by taking the vacuum expectation values of Eqs. (9) and (11). We find that[8]

$$\int\rho_{\alpha\beta}^{(1)}(\mu^2)\mu^{-2}d\mu^2+\int\rho_{\alpha\beta}^{(0)}(\mu^2)d\mu^2=(m_0^2/g_0^2)\delta_{\alpha\beta}, \tag{12}$$

$$\int\rho_{\alpha\beta}^{(1)}(\mu^2)d\mu^2=(m_0^4/g_0^2)\delta_{\alpha\beta}, \tag{13}$$

where $\rho_{\alpha\beta}^{(1)}$ and $\rho_{\alpha\beta}^{(0)}$ are the spin-1 and spin-0 spectral functions[9] of the $\Phi'_{\alpha\mu}$:

$$\langle\Phi'_{\alpha\mu}(x)\Phi'_{\beta\nu}(0)\rangle_0\equiv(2\pi)^{-3}\int d^4p\,\theta(p^0)e^{ip\cdot x}\{\rho_{\alpha\beta}^{(1)}(-p^2)[g_{\mu\nu}-p_\mu p_\nu/p^2]+\rho_{\alpha\beta}^{(0)}(-p^2)p_\mu p_\nu\}. \tag{14}$$

[In deriving Eq. (13) we have thrown away an infinite term arising from the vacuum expectation value of the last term in Eq. (11). This term does not have an unambiguous meaning, for $\langle\Phi'_{\gamma\mu}(x)\Phi'_{\gamma\nu}(x)\rangle_0$ is manifestly positive definite for any $\mu=\nu$, while Lorentz invariance requires that it have opposite sign for $\mu=\nu$ spacelike or timelike. By adopting a suitable regularization procedure, such as the ξ-limiting process,[10] one can show that the term in question does in fact vanish.]

If we are dealing with a chiral algebra of dimension $2n$, it is convenient to choose as gauge fields n vectors $v_{\alpha\mu}$ and n axial vectors $a_{\alpha\mu}$ and to express the commutation relations in terms of the fields

$$v'_\alpha{}^\nu\equiv(m_0^2/g_0)v_\alpha{}^\nu,\quad a'_\alpha{}^\nu\equiv(m_0^2/g_0)a_\alpha{}^\nu. \tag{15}$$

In particular, Eq. (9) takes the form

$$[v'_{\alpha 0}(\vec{x},t),v'_{\beta i}(\vec{y},t)]=[a'_{\alpha 0}(\vec{x},t),a'_{\beta i}(\vec{y},t)]=iC_{\alpha\beta\gamma}v'_{\gamma i}(\vec{x},t)\delta^3(\vec{x}-\vec{y})+i(m_0^2/g_0^2)\delta_{\alpha\beta}\partial_i\delta^3(\vec{x}-\vec{y}), \tag{16}$$

where $C_{\alpha\beta\gamma}$ are the structure constants of the algebra of the vectors, and from Eq. (11) one can derive the useful equality

$$[\partial_0v'_{\alpha i}(\vec{x},t)-\partial_iv'_{\alpha 0}(\vec{x},t),v'_{\beta j}(\vec{y},t)]=[\partial_0a'_{\alpha i}(\vec{x},t)-\partial_ia'_{\alpha 0}(\vec{x},t),a'_{\beta j}(\vec{y},t)]. \tag{17}$$

Until now we have used unrenormalized quantities, but the commutation relations and the sum rules can also be expressed in terms of renormalized quantities. We explain this in some detail for the case of the chiral SU(2)⊗SU(2), where the vector current is conserved and the axial-vector current is partially conserved. We may define the renormalized vector and axial-vector fields to be

$$(v_\alpha^{\nu})_{\text{ren}} = (m_0/m)v_\alpha^{\nu},$$

$$(a_\alpha^{\nu})_{\text{ren}} = (m_0/m)a_\alpha^{\nu} \tag{18}$$

and the renormalized coupling constant as

$$g = (m/m_0)g_0, \tag{19}$$

where m is the renormalized mass of the vector field.[11] It is very convenient, for our purpose, to choose for the axial-vector field the same renormalization factor as for the vector field. In terms of renormalized quantities, $v'_\alpha{}^{\nu}$ and $a'_\alpha{}^{\nu}$ become

$$v'_\alpha{}^{\nu} = (m^2/g)(v_\alpha^{\nu})_{\text{ren}},$$

$$a'_\alpha{}^{\nu} = (m^2/g)(a_\alpha^{\nu})_{\text{ren}}. \tag{20}$$

From Eq. (19) it is clear that the commutation relations given above, except for Eq. (11), can be expressed in terms of renormalized quantities only. This is true in particular for Eq. (16) and also for Eq. (17) as long as one only states the equality of the commutators.

If we take the vacuum expectation value of Eqs. (16) and (17) (and use $m_0^2/g_0^2 = m^2/g^2$) we obtain two sum rules relating the spectral functions for the renormalized vector and axial-vector fields. For the case of SU(2)⊗SU(2) these sum rules can be used to obtain the same successful relation between the masses of the A_1 and ρ mesons previously derived[2] using analogous sum rules from the chiral SU(2)⊗SU(2) current algebra. However, the current-algebra derivation suffers from the following deficiencies: (1) It assumes that the Schwinger terms of the currents are c numbers, while in simple models this is not true[12]; (2) it uses implicitly the Jacobi identity, which is not always allowed[13] for currents formed from limits of products of fields; and (3) it makes unjustified assumptions about the behavior of the currents at large momenta. All these difficulties are avoided by using the algebra of fields.[14] In addition, the current-algebra derivation only shows that the left-hand sides of the equations analogous to Eqs. (12) and (13) are proportional to $\delta_{\alpha\beta}$ but does not identify the coefficients. We see now that the coefficient m_0^2/g_0^2 is finite and equals m_ρ^2/g_ρ^2. The application of the methods of field algebra to Lie algebras other than SU(2)⊗SU(2) will be discussed in a separate note.

We have emphasized that, as far as the strong interactions are concerned, the algebra of fields is generally as useful as the algebra of currents, and sometimes more so. However, when we turn to electromagnetic- and weak-interaction calculations there arises a new question of great physical importance: Are the currents to which photons and leptons are coupled given by the hadronic currents J_α^{ν}, or by the hadronic gauge fields[3] Φ_α^{ν}, or by something else? The only place to find an answer is in calculations which are sensitive to those details of the commutation relations (such as the Schwinger terms) which distinguish between the algebra of currents and the algebra of fields. Most of the successful applications of current algebra have not been of this sort; they generally depend only on the conservation of the vector currents, the partial conservation of the axial currents, and the commutators of time components of currents, and therefore, according to Eqs. (6) and (8), these calculations could be performed in the algebra of fields with no change in results. The one noteworthy exception is a recent successful calculation[15] of the π^+-π^0 mass difference which depends upon the sum rules, Eqs. (12) and (13), for chiral SU(2)⊗SU(2) to eliminate both the quadratic and logarithmic divergences. We can conclude from this that the electromagnetic current belongs to an SU(2)⊗SU(2) multiplet of operators whose commutation relations are consistent with these sum rules. The only <u>known</u> example of such an operator algebra is afforded by the algebra of gauge fields.

We wish to thank S. Coleman, S. L. Glashow, K. Johnson, F. E. Low, H. Schnitzer, J. Schwinger, and J. Wess for discussions.

---

*This research was supported in part by the U. S. Air Force Office of Scientific Research, Office of Aerospace Research, U. S. Air Force; by the U. S. Atomic Energy Commission; and by the National Science Foundation.

†Presently Morris Loeb Lecturer at Harvard University, Cambridge, Massachusetts.

[1]M. Gell-Mann, Physics 1, 63 (1964).

[2]S. Weinberg, Phys. Rev. Letters 18, 507 (1967).

[3]N. M. Kroll, T. D. Lee, and Bruno Zumino, Phys. Rev. (to be published); T. D. Lee and Bruno Zumino, to be published.

[4]It is to be emphasized that this identity is assumed to hold independently of whether the unrenormalized

mass of the gauge fields is finite or infinite. There exists an alternative possibility in which such an identity holds only in the case of an infinite unrenormalized mass [see Ref. 3, and M. Gell-Mann and F. Zachariasen, Phys. Rev. 124, 953 (1961)]. However, if this alternative possibility is chosen, the hadronic currents satisfy the usual current algebra and not the field algebra given in this paper. Some discussion of the connection between currents and fields can also be found in J. Schwinger, Phys. Rev. 140, B158 (1965).

[5]C. N. Yang and R. L. Mills, Phys. Rev. 96, 191 (1954); R. Utiyama, Phys. Rev. 101, 1597 (1956); M. Gell-Mann and S. Glashow, Ann. Phys. (N.Y.) 15, 437 (1961). We are not assuming that $\mathcal{L}_m$ is invariant under the transformation $\delta\psi \propto T_\alpha\psi$, and in fact we have in mind a symmetry group like chiral SU(2) $\otimes$SU(2) which leaves neither $\mathcal{L}_m$ nor the vacuum invariant, although $\mathcal{L}_m$ may be assumed to have simple transformation properties corresponding to the partial conservation of currents and fields.

[6]We are using a canonical basis for the Lie algebra, in which the structure constants are totally antisymmetric and the invariant tensor is $\propto -\delta_{\alpha\beta}$.

[7]The commutation relation (10) shows that the space integrals of the vector and axial-vector octets do not generate the algebra U(6)$\otimes$U(6), but instead generate a nonsemisimple algebra having the space integrals of $\phi_{\alpha i}$ as its invariant Abelian subalgebra.

[8]These are a generalization to the non-Abelian case of sum rules for a single gauge field derived long ago by K. Johnson, Nucl. Phys. 25, 435 (1961), and D. G. Boulware and W. Gilbert, Phys. Rev. 126, 1563 (1962). While our work was in progress we received a preprint from R. Acharya, who also suggests using sum rules for spectral functions of gauge fields to derive the A_1-ρ mass ratio; he does not deal with the complications arising from the terms in the Lagrangian containing the structure constants of the non-Abelian Lie algebra.

[9]If the matter Lagrangian $\mathcal{L}_m$ is invariant under T_α then $J_\alpha{}^\nu$ and $\phi_\alpha{}^\nu$ will be conserved, so $\rho_{\alpha\beta}{}^{(0)}(\mu^2) = c_{\alpha\beta}\delta(\mu^2)$ for all β. If T_α is really a good symmetry of nature, we would not expect to find massless particles in the spectrum of $\Phi_\alpha{}^\nu$, so $c_{\alpha\beta}=0$. If T_α is broken by the vacuum, then such massless particles must appear, and $c_{\alpha\alpha}\neq 0$. If the matter Lagrangian is not invariant under T_α then $J_\alpha{}^\nu$ and $\Phi_\alpha{}^\nu$ will not be conserved, but they may be partially conserved, in which case $\rho_{\alpha\beta}{}^{(0)}$ will be dominated by low-lying spin-0 meson poles. It does not make a great deal of difference in using Eq. (12) whether we suppose that T_α is an exact symmetry of $\mathcal{L}_m$ broken by the vacuum, or an approximate symmetry of $\mathcal{L}_m$ which generates partially conserved currents.

[10]T. D. Lee and C. N. Yang, Phys. Rev. 128, 885 (1962); T. D. Lee, Phys. Rev. 128, 899 (1962).

[11]We sketch the renormalization procedure for a vector meson coupled to a conserved current. The renormalized and the unrenormalized propagators are related by $(\Delta_{\mu\nu})_{\rm unr} = Z(\Delta_{\mu\nu})_{\rm ren}$. At zero four-momentum $(\Delta_{\mu\nu})_{\rm unr} = m_0^{-2}g_{\mu\nu}$, as a consequence of current conservation. By assumption $(\Delta_{\mu\nu})_{\rm ren}$ is finite, and one can choose conveniently $Z=(m/m_0)^2$, where the physical mass m of the vector is defined as a zero of the real part of the inverse propagator. If the renormalized coupling constant g is defined at zero momentum transfer, one sees from the equation of motion for the vector field that it satisfies $g/g_0 = m/m_0$. This procedure can be appropriately modified to adapt it to the case of a vector meson coupled to a nonconserved current.

[12]D. Gross and R. Jackiw have pointed out in a private communication that the method used in Ref. 2 to derive the equality of V and A Schwinger terms does not work if the currents involve bilinear products of spin-0 fields, as in the σ model.

[13]K. Johnson and F. E. Low, Progr. Theoret. Phys. Suppl. 37-38, 74 (1966).

[14]Observe that it is not necessary to use the sum rule, Eq. (13), in its precise form which, as explained above, involves the dropping of an ambiguous divergent term. The equality of the integrals of the vector and the axial-vector spectral functions follows directly from Eq. (17), which is exact. The same remark can be made for the application described in T. Das, G. S. Guralnik, V. S. Mathur, F. E. Low, and J. E. Young, Phys. Rev. Letters 18, 759 (1967).

[15]Das, Guralnik, Mathur, Low, and Young, Ref. 14.

Reprinted from THE PHYSICAL REVIEW, Vol. 163, No. 5, 1667–1681, 25 November 1967
Printed in U. S. A.

Field-Current Identities and Algebra of Fields*

T. D. LEE
Columbia University, New York, New, York

AND

BRUNO ZUMINO
New York University, New York, New York
(Received 30 June 1967)

It is shown that the possible identity between the various hadronic current operators and the corresponding spin-1 meson field operators determines the general structure of the hadronic part of the total Lagrangian. In particular, the identity between the isovector electromagnetic current and the neutral ρ-meson field implies that the ρ dependence of the strong interaction must be the same as that in the Yang-Mills theory, except for the mass term of the ρ meson. The explicit form of the interaction Lagrangian makes possible a general study of the local equal-time commutators of the various hadronic current operators, including the effects of the electromagnetic interaction. Many of these electromagnetic correction terms depend only on the general requirement of gauge invariance, and are independent of whether the proposed field-current identities are valid or not. For example, the usual Schwinger term $\lambda(\partial/\partial r_j)\delta^3(r-r')$ in the commutator between the time component of any charged hadronic weak interaction current and the jth space component of its Hermitian conjugate should be replaced by $\lambda[(\partial/\partial r_j)+ieA_j]\delta^3(r-r')$, where A_j is the electromagnetic field operator. The contribution of such a correction term, i.e., $\lambda ieA_j\delta^3(r-r')$, remains present in the integrated form of the commutator. In the usual current algebra, λ is mathematically undefined. If field-current identities hold, then these current commutators are the same as the corresponding algebra of the field operators, and λ becomes a well-defined c number. Some speculative remarks concerning the possible extension of the algebra of fields to the lepton currents are presented.

1. INTRODUCTION

RECENTLY, it has been suggested[1] that the entire hadronic electromagnetic current operator is identical with a linear combination of the local field operators of the known neutral vector mesons, independent of whether the unrenormalized masses of these vector mesons are finite or infinite.[2] This identity is shown to be consistent with the requirement of gauge invariance; it gives a precise formulation of the idea of vector domi-

* This research was supported in part by the U. S. Atomic Energy Commission and the National Science Foundation.

[1] N. M. Kroll, T. D. Lee, and Bruno Zumino, Phys. Rev. **157**, 1376 (1967).

[2] There exists an alternative proposal in which the unrenormalized isovector part of the hadronic electromagnetic current is assumed to be the same operator as the unrenormalized current generating the neutral ρ-meson field. In such a case, the field-current identity holds only in the limit of an infinite unrenormalized meson mass. [See Refs. 1, 4, M. Gell-Mann and F. Zachariasen, Phys. Rev. **124**, 953 (1961).] It is important to note that in this alternative possibility, the products of the current operators would in general, be different from the products of the corresponding field operators even in the limit of infinite unrenormalized masses; thus, the hadronic current operators entering in the electromagnetic and the weak interactions satisfy the usual current algebra instead of the algebra of fields.

nance[3] in the language of a local Lagrangian field theory, and it leads, among other consequences, to the conclusion that, in the absence of the leptons, the entire hadronic contribution to the electric charge renormalization is finite.[1,4] The generalization that such field-current identities hold also for other observed hadronic currents, such as the various weak-interaction vector and axial-vector currents, implies that these observed hadronic currents satisfy the algebra of fields,[5] which is simpler than the usual current algebra.[6] It is found that all successful applications previously obtained from the current algebra can be rederived by using the algebra of fields; in addition, the usually troublesome Schwinger term[7] becomes a well-defined c number in the algebra of fields.

The purpose of this paper is to point out that these field-current identities also determine the general structure of both the strong-interaction and the hadronic part of the electromagnetic interaction. The explicit form of these interaction Lagrangians enables one to study systematically the electromagnetic corrections of the equal-time commutator of the various hadron currents. That some of these commutators must be affected by the electromagnetic interaction follows from general considerations of gauge invariance. For example, suppose that, in the *absence* of the electromagnetic interaction, the commutator

$$[V_4^{\rm wk}(r,t)_{S=0}, V_k^{\rm wk}(r',t)^\dagger{}_{S=0}] \tag{1.1}$$

is[8]

$$-2i\delta^3(r-r')J_k^\gamma(r',t)_{I=1}+\lambda(\partial/\partial r_k)\delta^3(r-r'), \tag{1.2}$$

where $(V_\mu^{\rm wk})_{S=0}$ is the hadronic strangeness-conserving part of the weak-interaction vector current, $(J_\mu^\gamma)_{I=1}$ is the hadronic isovector part of the electromagnetic current and λ is the coefficient of the Schwinger term. The *inclusion* of the electromagnetic interaction requires that the same commutator should be given by, instead of (1.2),

$$-2i\delta^3(r-r')(J_k^\gamma)_{I=1}+\lambda[(\partial/\partial r_k)+ieA_k]\delta^3(r-r'), \tag{1.3}$$

where A_μ is the electromagnetic field operator. We note that under the gauge transformation

$$A_\mu\to A_\mu-(\partial\Lambda/\partial x_\mu),$$
$$(V_\mu^{\rm wk})_{S=0}\to(V_\mu^{\rm wk})_{S=0}\exp(ie\Lambda),$$

the expression (1.1) acquires a multiplicative factor $\exp[ie\Lambda(r,t)-ie\Lambda(r',t)]$ while (1.3) acquires an additive term $-ie\lambda(\partial\Lambda/\partial r_k)\delta^3(r-r')$. It can be easily shown that these two changes are indeed identical by using the equality

$$\exp[ie\Lambda(r\ t)-ie\Lambda(r',t)](\partial/\partial r_j)\delta^3(r-r') = (\partial/\partial r_j)\delta^3(r-r')-ie(\partial\Lambda/\partial r_j)\delta^3(r-r'). \tag{1.4}$$

The contribution of this new term $ie\lambda A_k\delta^3(r-r')$ in (1.3) remains present even after one integrates the commutator (1.1) over d^3r, or d^3r'. In the usual current algebra λ is mathematically undefined, therefore, so would be the corresponding *integrated* equal-time current commutator. In the algebra of fields, λ is a well-defined c number, related to the renormalized mass m_ρ and the renormalized coupling constant g_ρ of the neutral ρ meson by

$$\lambda=2(m_\rho/g_\rho)^2. \tag{1.5}$$

In Sec. 2, we begin with the identity that $(J_\mu^\gamma)_{I=1}$ is related to the neutral ρ-meson field by

$$[J_\mu^\gamma(x)]_{I=1}=-(m_0^2/g_0)\rho_\mu^0(x) = -(m_\rho^2/g_\rho)[\rho_\mu^0(x)]_{\rm ren}, \tag{1.6}$$

where $(\rho_\mu^0)_{\rm ren}$ is the renormalized field operator, and m_0, g_0, $\rho_\mu^0(x)$ refer to the unrenormalized mass, the unrenormalized coupling constant, and the unrenormalized field operator of the neutral ρ meson.[9] Let $\boldsymbol{\varrho}_\mu(x)$ be the

[3] Y. Nambu, Phys. Rev. **106**, 1366 (1957); W. R. Frazer and J. R. Fulco, Phys. Rev. **117**, 1603 (1960); J. J. Sakurai, Ann. Phys. (N. Y.) **11**, 1 (1960); M. Gell-Mann and F. Zachariasen, Phys. Rev. **124**, 953 (1961); M. Gell-Mann, *ibid.* **125**, 1067 (1962); Y. Nambu and J. J. Sakurai, Phys. Rev. Letters **8**, 79 (1962); M. Gell-Mann, D. Sharp, and W. G. Wagner, *ibid.* **8**, 261 (1962); J. Schwinger, Phys. Rev. **140**, B158 (1965). For further discussions and applications of vector dominance ideas, see G. Feldman and P. Matthews, Phys. Rev. **132**, 823 (1963); S. Berman and S. Drell, *ibid.* **133**, B791 (1964); R. F. Dashen and D. H. Sharp, *ibid.* **133**, B1585 (1964); L. Stodolsky, *ibid.* **134**, B1099 (1964); G. Barton and B. G. Smith, Nuovo Cimento **36**, 436 (1965); R. Gatto, Ergeb. Exakt. Naturw. **39**, 106 (1965); M. Ross and L. Stodolsky, Phys. Rev. **149**, 1172 (1966); D. S. Beder, *ibid.* **149**, 1203 (1966); M. A. B. Bég and A. Pais, Phys. Rev. Letters **14**, 51 (1965).

While the approximation that the matrix elements of the hadronic vector current operator may be replaced by the corresponding elements of the vector-meson field operator has been discussed in many of these papers, in the context of a Lagrangian field theory the difficulty has always been the apparent violation of gauge invariance. [See, e.g., G. Feldman and P. Matthews (Ref. 3).] This difficulty is revolved in Ref. 1.

[4] T. D. Lee and Bruno Zumino (to be published).

[5] T. D. Lee, S. Weinberg, and Bruno Zumino, Phys. Rev. Letters **18**, 1029 (1967).

[6] M. Gell-Mann, Phys. Rev. **125**, 1067 (1962); Physics **1**, 63 (1964).

[7] T. Gotô and T. Imamura, Progr. Theoret. Phys. (Kyoto) **14**, 396 (1955); J. Schwinger, Phys. Rev. Letters **3**, 296 (1959). For an earlier discussion of a related term see R. Serber, Phys. Rev. **49**, 545 (1936).

[8] Throughout the paper, the subscript μ denotes the space-time index, $\mu=4$ is the time component, $x_4=it$, and $\mu=i$ (or j, or k) denotes the space component. The isospin index is represented by the superscript, e.g. a, which can be 0, 1, or 2, and $a=0$ denotes the neutral component. All boldface letters denote isovectors.

[9] We sketch the renormalization procedure for the ρ meson coupled to a conserved current. [For more details, see Ref. 1.] The renormalized and the unrenormalized propagators are related by $(D_{\mu\nu})_{\rm unr}=Z_\rho(D_{\mu\nu})_{\rm ren}$. At zero four-momentum $(D_{\mu\nu})_{\rm unr}=m_0^{-2}\delta_{\mu\nu}$, as a consequence of current conservation. By assumption $(D_{\mu\nu})_{\rm ren}$ is finite, and one can choose conveniently, $Z_\rho=(m_\rho/m_0)^2$, where the renormalized mass m_ρ of the ρ meson is defined as the zero of the real part of the inverse propagator. If the renormalized coupling constant g_ρ is defined at zero momentum transfer, one sees from the equation of motion of the ρ meson that it satisfies $g_\rho/g_0=m_\rho/m_0$. Since the current conservation holds more exactly for the neutral ρ meson, we will choose m_ρ and g_ρ to be those of the neutral ρ meson.

It is important to note that while the perturbation series of the interaction between a single neutral vector meson and a conserved current consisting of only spin-$\frac{1}{2}$ and spin-0 fields can be renormalized, this is not true for the perturbation series of the strong-interaction Lagrangian of the isovector ρ-meson system given by Eq. (2.6) following. Nevertheless, we shall *assume* in the same spirit as that used in the ξ-limiting process that the renormalized theory does exist [T. D. Lee and C. N. Yang, Phys. Rev. **128**, 885 (1962); T. D. Lee, *ibid.* **128**, 899 (1962)]. The failure of the perturbation series method implies only that the renormalized theory does not have a power-series expansion in terms of the square of the renormalized coupling constant g_ρ^2. For example, it may contain terms such as $\ln g_\rho$, or $g_\rho^{1/2}$, etc.

unrenormalized isovector ρ-meson field whose 0th component[8] is the neutral ρ-meson field $\rho_\mu^0(x)$. The identity (1.6) implies that under the strong interaction, the ρ-meson field satisfies the equation of field conservation

$$\frac{\partial}{\partial x_\mu}\varrho_\mu(x)=0. \tag{1.7}$$

As we shall see, this field-conservation equation determines the general form of the ρ-meson part of the strong interaction. The resulting Lagrangian $\mathcal{L}_{st}$ is identical in form to that in the Yang-Mills theory,[10] except for a mass term $-\frac{1}{2}(m_0\varrho_\mu)^2$. The presence of this mass term destroys the *local* isospin invariance which forms the starting point of the Yang-Mills theory. Such a local gauge invariance plays no role in the present paper; instead, we emphasize the importance of the field-conservation Eq. (1.7), which is a necessary consequence of the field-current identity (1.6).

The extensions of this Lagrangian to include the electromagnetic and the weak interactions are discussed in Sec. 3. The resulting algebra of fields, valid to all orders in e, is given in Sec. 4. The generalization to an arbitrary broken symmetry is discussed in Sec. 5. As an illustration of the general formalism, we give in Sec. 6 the special example of the $SU_3\times SU_3$ field algebra which is valid to all orders in e.

In Sec. 7, some speculative remarks are made concerning the lepton currents. It is pointed out that at present, our experimental information is completely consistent with the possibility that the leptonic part of the electromagnetic current (or, the weak interaction current) may also satisfy the algebra of fields. The experimental consequences of such a possibility are discussed.

It may be emphasized that the distinction between fields and sources (or, currents) has its origin in the study of the electromagnetic and gravitational fields. Both fields, having a zero rest mass and satisfying Bose statistics, do approach their respective classical limits at large distances. Therefore, there is a clear physical distinction between such fields and their sources. The same physical distinction does not exist for a massive boson field. The difference between whether a set of observed hadron, or lepton, current operators are identical with a corresponding set of massive meson fields, or not, lies only in the algebraic relations that such operators satisfy. Although throughout the paper all derivations are based on the usual local Lagrangian field theory, it is hoped that the resulting equal-time commutation relations can be of a more general nature, not necessarily depending on the validity of the local Lagrangian field theory.

[10] C. N. Yang and F. Mills, Phys. Rev. **96**, 191 (1954). A discussion of the implications of the field conservation equation for the Yang-Mills theory has been given by V. I. Ogievetskij and I. V. Polubarinov, Ann. Phys. (N. Y.) **25**, 358 (1963); and by Bruno Zumino, Acta Phys. Austriaca, Suppl. II, 212 (1966).

2. STRONG INTERACTION

For clarity, we consider first only the system of the ρ-meson fields. The free Lagrangian is

$$\mathcal{L}_{\text{free}}(\rho)=-\tfrac{1}{4}\mathbf{G}_{\mu\nu}^2-\tfrac{1}{2}(m_0\varrho_\nu)^2, \tag{2.1}$$

where

$$\mathbf{G}_{\mu\nu}=\frac{\partial}{\partial x_\mu}\varrho_\nu-\frac{\partial}{\partial x_\nu}\varrho_\mu. \tag{2.2}$$

In order that the field-current identity (1.6) holds, it follows from, e.g., Eq. (2.8) of Ref. 1 that, since the charged ρ meson does interact with the electromagnetic field, there must also exist a strong-interaction term between the neutral ρ meson and the charged ρ mesons. Such a term should be proportional to $(G_{\mu\nu}^1\rho_\nu^2-G_{\mu\nu}^2\rho_\nu^1)\rho_\mu^0$; from the isospin invariance, it becomes proportional to $(\mathbf{G}_{\mu\nu}\times\varrho_\nu)\cdot(\varrho_\mu)$. Thus, we assume the strong-interaction Lagrangian density of the ρ-meson system to be of the general form

$$\mathcal{L}_{\text{int}}(\rho)=\tfrac{1}{2}g_0\mathbf{G}_{\mu\nu}\cdot(\varrho_\mu\times\varrho_\nu)+F(\varrho_\mu), \tag{2.3}$$

where, for simplicity, F is assumed to depend *only* on ϱ_μ. This restriction is equivalent to the usual minimal principle in which the ρ-meson coupling is assumed to consist of only the minimal number of derivatives.

Theorem. If $\varrho_\mu(x)$ satisfies the field conservation equation

$$\partial\varrho_\mu(x)/\partial x_\mu=0, \tag{2.4}$$

then the function F must be of the form

$$F=-\tfrac{1}{4}g_0^2(\varrho_\mu\times\varrho_\nu)^2. \tag{2.5}$$

Therefore, the total Lagrangian density of the ρ-meson system is

$$\begin{aligned}\mathcal{L}_\rho&=\mathcal{L}_{\text{free}}(\rho)+\mathcal{L}_{\text{int}}(\rho),\\&=-\tfrac{1}{4}(\mathbf{f}_{\mu\nu})^2-\tfrac{1}{2}(m_0\varrho_\mu)^2,\end{aligned} \tag{2.6}$$

where

$$\mathbf{f}_{\mu\nu}=(\partial\varrho_\nu/\partial x_\mu)-(\partial\varrho_\mu/\partial x_\nu)-g_0(\varrho_\mu\times\varrho_\nu). \tag{2.7}$$

Proof. Let $\mathbf{S}_\mu(\rho)$ be the current operator, whose components are defined by

$$\begin{aligned}S_\mu{}^a(\rho)&=-\epsilon^{abc}\left[\partial\mathcal{L}(\rho)/\partial\left(\frac{\partial\rho_\nu{}^b}{\partial x_\mu}\right)\right]\rho_\nu{}^c,\\&=\epsilon^{abc}f_{\mu\nu}{}^b\rho_\nu{}^c,\end{aligned} \tag{2.8}$$

where the superscripts a, b, c denote the isospin indices and ϵ^{abc} is $+1$, -1, or 0 depending on whether abc is an even permutation of 012, an odd permutation, or otherwise. The isospin invariance of the strong interaction implies that in the absence of other fields

$$\frac{\partial}{\partial x_\mu}\mathbf{S}_\mu(\rho)=0. \tag{2.9}$$

From (2.1) and (2.3), the equation of the ρ-meson field is

$$\frac{\partial}{\partial x_\mu}\mathbf{f}_{\mu\nu}-m_0^2\varrho_\nu=g_0\mathbf{S}_\nu+g_0^2(\varrho_\nu\times\varrho_\mu)\times\varrho_\mu-(\partial F/\partial\varrho_\nu). \tag{2.10}$$

Let us define A^{ab} by

$$A^{ab}\rho_\nu{}^b=[-(\partial F/\partial \boldsymbol{\varrho}_\nu)+g_0{}^2(\boldsymbol{\varrho}_\nu\times\boldsymbol{\varrho}_\mu)\times\boldsymbol{\varrho}_\mu]^a. \quad (2.11)$$

This is always possible since the right-hand side is an isovector function of $\boldsymbol{\varrho}_\mu$, and it is also a space-time 4-vector. From Eqs. (2.4), (2.9), and (2.10), it follows that

$$\rho_\nu{}^b\frac{\partial}{\partial x_\nu}A^{ab}=0. \quad (2.12)$$

In this problem, the three spatial components $\boldsymbol{\varrho}_i$ ($i=1$ 2, 3) and their first time derivatives $(\partial\boldsymbol{\varrho}_i/\partial t)$ are independent variables. The time component $\boldsymbol{\varrho}_4$ is, on account of (2.10), a function of $(\partial\boldsymbol{\varrho}_i/\partial t)$ and $\boldsymbol{\varrho}_i$. Thus (in the c-number theory), at any given time the values of $\boldsymbol{\varrho}_i$ and $\boldsymbol{\varrho}_4$ can be arbitrarily chosen. In order that (2.12) hold at all times, one must have $(\partial A^{ab}/\partial x_\nu)=0$. Thus, $A^{ab}=\kappa\delta^{ab}$ where κ is a constant. It follows from Eq. (2.11) that F is given by

$$F=-\tfrac{1}{2}\kappa(\boldsymbol{\varrho}_\nu)^2-\tfrac{1}{4}g_0{}^2(\boldsymbol{\varrho}_\mu\times\boldsymbol{\varrho}_\nu)^2.$$

In this expression, the term $-\frac{1}{2}\kappa(\boldsymbol{\varrho}_\nu)^2$ should be combined with the mass term $-\frac{1}{2}m_0(\boldsymbol{\varrho}_\nu)^2$ in (2.1). The theorem is then proved.

We note that the first term $-\frac{1}{4}\mathbf{f}_{\mu\nu}{}^2$ in (2.6) is exactly the same as the Lagrangian in the Yang-Mills theory.[10] In contrast to the Yang-Mills theory, the second term $-\frac{1}{2}(m_0\boldsymbol{\varrho}_\nu)^2$ destroys the *local* isospin gauge invariance. The fact that $m_0\neq 0$ makes it possible to connect the field conservation equation with the equation of motion.

The above considerations can be readily extended to include other fields. In the absence of the electromagnetic and the weak interactions, the total Lagrangian $\mathcal{L}$ is known to be invariant under isospin transformations. If we assume the minimal principle, the field-conservation equation (2.4) then requires the sum of the free and the strong-interaction Lagrangian densities to be of the form

$$\mathcal{L}_{\text{free}}+\mathcal{L}_{\text{st}}=\mathcal{L}_\rho+\mathcal{L}_m(\psi,D_\nu\psi),$$

where $\mathcal{L}_m$ is invariant under the isospin rotation, $\mathcal{L}_\rho$ is given by (2.6), ψ represents all other matter fields of either half-integer or integer spin, and $D_\nu\psi$ is related to the matrix representation $-i\mathbf{T}$ of the isospin generator on ψ by

$$D_\nu\psi=\frac{\partial}{\partial x_\nu}\psi+g_0(\mathbf{T}\cdot\boldsymbol{\varrho}_\nu)\psi. \quad (2.13)$$

If ψ is an isospin $\frac{1}{2}$ field, then $\mathbf{T}=\frac{1}{2}i\boldsymbol{\tau}$, where $\boldsymbol{\tau}$ is the usual isospin Pauli matrix.

For the purpose of our subsequent discussions, the minimal principle is by no means necessary. The Lagrangian density can, then, assume a more general form

$$\mathcal{L}_{\text{free}}+\mathcal{L}_{\text{st}}=\mathcal{L}_\rho+\mathcal{L}_m(\psi,D_\nu\psi,\mathbf{f}_{\mu\nu}), \quad (2.14)$$

showing explicitly that $\mathcal{L}_m$ may also depend on $\mathbf{f}_{\mu\nu}$ which contains the derivatives of the ρ-meson field.

The ρ-meson equation becomes

$$\frac{\partial}{\partial x_\mu}\left(\mathbf{f}_{\mu\nu}-\frac{\partial\mathcal{L}_m}{\partial\mathbf{f}_{\mu\nu}}+\frac{\partial\mathcal{L}_m}{\partial\mathbf{f}_{\nu\mu}}\right)-m_0{}^2\boldsymbol{\varrho}_\nu=g_0\mathbf{S}_\nu, \quad (2.15)$$

where the ath isospin component of $\mathbf{S}_\nu$ is given by

$$S_\nu{}^a=-\epsilon^{abc}\left(f_{\mu\nu}{}^b-\frac{\partial\mathcal{L}_m}{\partial f_{\mu\nu}{}^b}+\frac{\partial\mathcal{L}_m}{\partial f_{\nu\mu}{}^b}\right)\rho_\mu{}^c$$
$$-\sum_\psi(\partial\mathcal{L}_m/\partial D_\nu\psi)T^a\psi. \quad (2.16)$$

Both the current $\mathbf{S}_\nu$ and the field $\boldsymbol{\varrho}_\nu$ are conserved

$$\frac{\partial}{\partial x_\nu}\mathbf{S}_\nu=\frac{\partial}{\partial x_\nu}\boldsymbol{\varrho}_\nu=0, \quad (2.17)$$

and the spatial integrals of both 4th components are proportional to the generators of the isospin group.

3. ELECTROMAGNETIC AND WEAK INTERACTIONS

We discuss first the explicit form of the electromagnetic-interaction Lagrangian which yields the field-current identity (1.6). By following the general procedures given in Ref. 1, one can easily construct such a gauge-invariant electromagnetic interaction from any hadronic free and strong-interaction Lagrangian density $(\mathcal{L}_{\text{free}}+\mathcal{L}_{\text{st}})$ of the form (2.14). It is convenient to separate $(\mathcal{L}_{\text{free}}+\mathcal{L}_{\text{st}})$ into two parts:

$$\mathcal{L}_{\text{free}}+\mathcal{L}_{\text{st}}=-\tfrac{1}{2}(m_0\boldsymbol{\varrho}_\mu)^2+\mathcal{L}_0, \quad (3.1)$$

where

$$\mathcal{L}_0=-\tfrac{1}{4}\mathbf{f}_{\mu\nu}{}^2+\mathcal{L}_m(\psi,D_\nu\psi,\mathbf{f}_{\mu\nu}). \quad (3.2)$$

The corresponding electromagnetic interaction can be generated by replacing $\boldsymbol{\varrho}_\mu$ and $(\partial\boldsymbol{\varrho}_\mu/\partial x_\nu)$ in $\mathcal{L}_0$ by $\hat{\rho}_\mu$ and $(\partial\hat{\rho}_\mu/\partial x_\nu)$, where the isospin components of $\hat{\rho}_\mu$ are related to those of $\boldsymbol{\varrho}_\mu$ by

$$\hat{\rho}_\mu{}^1=\rho_\mu{}^1,\quad \hat{\rho}_\mu{}^2=\rho_\mu{}^2,$$

and

$$\hat{\rho}_\mu{}^0=\rho_\mu{}^0+(e_0/g_0)A_\mu. \quad (3.3)$$

A_μ is the unrenormalized electromagnetic field and e_0 the unrenormalized charge of the electron. The resulting Lagrangian density $\mathcal{L}=\mathcal{L}_{\text{free}}+\mathcal{L}_{\text{st}}+(\mathcal{L}_\gamma)_{I=1}$ is then given by

$$\mathcal{L}=-\tfrac{1}{4}F_{\mu\nu}{}^2-\tfrac{1}{2}(m_0\boldsymbol{\varrho}_\mu)^2+\mathcal{L}_0{}', \quad (3.4)$$

where

$$\mathcal{L}_0{}'=\mathcal{L}_0(\boldsymbol{\varrho}_\mu\to\hat{\rho}_\mu)=-\tfrac{1}{4}\hat{f}_{\mu\nu}{}^2+\mathcal{L}_m(\psi,D_\nu{}'\psi,\hat{f}_{\mu\nu}), \quad (3.5)$$

$$\hat{f}_{\mu\nu}=(\partial\hat{\rho}_\nu/\partial x_\mu)-(\partial\hat{\rho}_\mu/\partial x_\nu)-g_0(\hat{\rho}_\mu\times\hat{\rho}_\nu), \quad (3.6)$$

$$D_\nu{}'\psi=(\partial\psi/\partial x_\nu)+g_0(\mathbf{T}\cdot\hat{\rho}_\nu)\psi, \quad (3.7)$$

and

$$F_{\mu\nu}=(\partial A_\nu/\partial x_\mu)-(\partial A_\mu/\partial x_\nu). \quad (3.8)$$

The Lagrangian density (3.4) is obviously gauge invariant. We note that from (3.4) both ϱ_ν and $(\partial F_{\mu\nu}/\partial x_\mu)$ are proportional to the variational derivative $(\delta\mathcal{L}_0'/\delta\hat{\rho}_\nu)$, which is defined to be $[(\partial\mathcal{L}_0'/\partial\hat{\rho}_\nu)-(\partial/\partial x_\mu)\partial\mathcal{L}_0'/\partial(\partial\hat{\rho}_\nu/\partial x_\mu)]$

$$m_0^2\varrho_\nu=\delta\mathcal{L}_0'/\delta\hat{\rho}_\nu, \quad (3.9)$$

and

$$\frac{\partial F_{\mu\nu}}{\partial x_\mu}=-\left(\frac{e_0}{g_0}\right)\frac{\delta\mathcal{L}_0'}{\delta\hat{\rho}_\nu{}^0}. \quad (3.10)$$

Thus, one has

$$\partial F_{\mu\nu}/\partial x_\mu=-(e_0m_0^2/g_0)\rho_\nu{}^0, \quad (3.11)$$

which gives the field-current identity (1.6).

For clarity, only the hadronic electromagnetic isovector current $[J_\mu{}^\gamma(x)]_{I=1}$ is included in the above Eqs. (3.4)–(3.11). It has been shown in Ref. 1 that the inclusion of the hadronic isoscalar electromagnetic current $[J_\mu{}^\gamma(x)]_{I=0}$ can be made in a similar way. In addition to the change $\varrho_\mu\to\hat{\rho}_\mu$ in the strong-interaction Lagrangian, one replaces also the unrenormalized field $\phi_\mu{}^0$ by $[\phi_\mu{}^0+\frac{1}{2}(e_0/g_Y{}^0)A_\mu]$ where $\phi_\mu{}^0$ is defined to be the field operator coupled to the hypercharge current with a coupling constant $g_Y{}^0$. As a result, $(J_\mu{}^\gamma)_{I=0}$ is proportional to $\phi_\mu{}^0$ which in turn is proportional to a linear combination of the renormalized fields $[\phi_\mu(x)]_{\text{ren}}$ and $[\omega_\mu(x)]_{\text{ren}}$ for the ϕ meson and the ω meson. We have, according to Eq. (1.6) of Ref. 1,

$$(J_\nu{}^\gamma)_{I=0}=-\tfrac{1}{2}g_Y{}^{-1}[\cos\theta_Y m_\phi{}^2(\phi_\mu)_{\text{ren}}-\sin\theta_Y m_\omega{}^2(\omega_\mu)_{\text{ren}}], \quad (3.12)$$

where g_Y is the renormalized coupling constant, θ_Y is one of the two mixing angles and m_ϕ, m_ω are the renormalized masses of the ϕ meson and the ω meson, respectively.

Next, we discuss the weak-interaction currents of the hadrons. The field-current identity (1.6) together with the isotriplet current hypothesis of Feynman and Gell-Mann[11] require that the strangeness conserving hadronic weak-interaction vector currents $(V_\mu{}^{\text{wk}})_{S=0}$ be proportional to the appropriate charge component of ϱ_μ.

$$(V_\mu{}^{\text{wk}})_{S=0}=-(m_0^2/g_0)(\rho_\mu{}^1-i\rho_\mu{}^2) \equiv-\sqrt{2}(m_0^2/g_0)\rho_\mu{}^+, \quad (3.13)$$

where $\rho_\mu{}^+$ is the field operator that annihilates (creates) the positively (negatively) charged ρ mesons. In terms of the renormalized mass m_ρ, the renormalized coupling constant g_ρ and the renormalized field $(\rho_\mu{}^+)_{\text{ren}}$, (3.13) can be written as

$$(V_\mu{}^{\text{wk}})_{S=0}=-\sqrt{2}(m_\rho{}^2/g_\rho)(\rho_\mu{}^+)_{\text{ren}}. \quad (3.14)$$

[11] R. P. Feynman and M. Gell-Mann, Phys. Rev. **109**, 193 (1958).

It is natural to extend the field-current identity also to other weak-interaction currents. For example, one would assume the strangeness-nonconserving hadronic weak-interaction current $(V_\mu{}^{\text{wk}})_{S=1}$ to be proportional to the field operator of the $K^*(890)$ meson, and the strangeness-conserving and nonconserving parts of the hadronic weak axial-vector currents $(A_\mu{}^{\text{wk}})_{S=0}$ and $(A_\mu{}^{\text{wk}})_{S=1}$ to be, respectively, proportional to the field operators of a charged axial-vector meson such as $A1$ and its corresponding SU_3 multiplet member, e.g., $K_A{}^*$. Such a generalization and some of its consequences have already been discussed in Ref. 5. One has then

$$\begin{aligned}[V_\mu{}^{\text{wk}}(x)]_{S=1}&=-\sqrt{2}(m_0{}^2/g_0)K_\mu{}^*(x)\\&=-\sqrt{2}(m_\rho{}^2/g_\rho)[K_\mu{}^*(x)]_{\text{ren}},\\ [A_\mu{}^{\text{wk}}(x)]_{S=0}&=-\sqrt{2}(m_0{}^2/g_0)A1_\mu(x)\\&=-\sqrt{2}(m_\rho{}^2/g_\rho)[A1_\mu(x)]_{\text{ren}},\\ [A_\mu{}^{\text{wk}}(x)]_{S=1}&=-\sqrt{2}(m_0{}^2/g_0)K_{A\mu}{}^*(x)\\&=-\sqrt{2}(m_\rho{}^2/g_\rho)[K_{A\mu}{}^*(x)]_{\text{ren}},\end{aligned} \quad (3.15)$$

where $K_\mu{}^*(x)$, $A1_\mu(x)$, and $K_{A\mu}{}^*(x)$ are the unrenormalized field operators of K^*, $A1$. and $K_A{}^*$ mesons. These are assumed to have the same unrenormalized coupling constant g_0 and the same unrenormalized mass m_0. [The general case that these mesons may have different unrenormalized masses is discussed in Sec. 5.] For convenience, we have chosen in (3.15) the same wave-function renormalization factor[9] $Z=Z_\rho=(m_\rho/m_0)^2$ for all these different meson fields, where m_ρ is the renormalized mass of the neutral ρ meson.

In the above expressions, all hadron currents are properly normalized, so that the semileptonic weak-interaction Lagrangian density is

$$2^{-1/2}GJ_\mu{}^{\text{wk}}j_\mu{}^{\text{wk}}+\text{H.c.}, \quad (3.16)$$

where

$$J_\mu{}^{\text{wk}}=\cos\theta_c(V_\mu{}^{\text{wk}}+A_\mu{}^{\text{wk}})_{S=0} +\sin\theta_c(V_\mu{}^{\text{wk}}+A_\mu{}^{\text{wk}})_{S=1}, \quad (3.17)$$

θ_c is the Cabibbo angle,[12] G is the Fermi constant for μ decay $\cong(10^{-5}/m_N{}^2)$, and $i_\mu{}^{wk}$ is the leptonic weak-interaction current.

4. SU_2 FIELD ALGEBRA (INCLUDING ELECTROMAGNETIC EFFECTS)

An important consequence of the field-current identities is that the observed hadronic electromagnetic and weak-interaction current operators should satisfy the same equal-time commutators as the corresponding fields. The details of these commutation relations which are called the field algebra have been analyzed in Ref. 5. The explicit Lagrangian density given in the previous section makes it possible to include in these algebraic relations also the necessary effects of the electromagnetic field.

[12] N. Cabibbo, Phys. Letters **10**, 513 (1963).

We begin with the Lagrangian density $\mathcal{L}$ given by (3.4) and, for convenience, adopt the Coulomb gauge. Let A_j^{tr} be the 3-vector which denotes the usual transverse electromagnetic vector potential; it satisfies

$$(\partial A_j^{\mathrm{tr}}/\partial r_j)=0. \tag{4.1}$$

In the Coulomb gauge, the 4-vector potential A_μ is given by

$$A_j=A_j^{\mathrm{tr}} \quad \text{and} \quad A_4=i\phi, \tag{4.2}$$

where, on account of (3.11), ϕ is the solution of the Laplace equation

$$\Delta\phi=i(e_0m_0^2/g_0)\rho_4^0.$$

Correspondingly, the electric field E_j can be written as $E_j=E_j^{\mathrm{tr}}+E_j^{\mathrm{long}}$, where

$$E_j^{\mathrm{tr}}=-(\partial A_j^{\mathrm{tr}}/\partial t) \quad \text{and} \quad E_j^{\mathrm{long}}=-(\partial\phi/\partial r_j). \tag{4.3}$$

To apply the canonical formalism, it is convenient to replace the $-\frac{1}{4}F_{\mu\nu}^2$ term in the Lagrangian density (3.4) by

$$\tfrac{1}{2}[(E_j^{\mathrm{tr}})^2+(E_j^{\mathrm{long}})^2-H_j^2], \tag{4.4}$$

where $H_j=\epsilon_{jkl}(\partial A_l^{\mathrm{tr}}/\partial r_k)$ denotes the magnetic field. It is clear that the spatial integral of (4.4) is the same as that of $-\frac{1}{4}F_{\mu\nu}^2$. The field algebra can then be most easily derived by choosing ρ_j^a, A_j^{tr}, and ψ as the generalized coordinates. By using (3.4), (3.5), and (4.4) one finds that their conjugate momenta are, respectively,

$$\hat{P}_j{}^a=i[f_{4j}{}^a-(\partial\mathcal{L}_m/\partial \dot{f}_{4j}{}^a)+(\partial\mathcal{L}_m/\partial \dot{f}_{j4}{}^a)],\quad \Pi_j^{\mathrm{tr}}=-E_j^{\mathrm{tr}},$$

and (4.5)

$$P_\psi=-i(\partial\mathcal{L}_m/\partial D_4'\psi).$$

It is useful to introduce the renormalization-independent field operators

$$\rho_\mu'^a(r,t)=(m_0^2/g_0)\rho_\mu{}^a(r,t)=(m_\rho^2/g_\rho)[\rho_\mu{}^a(r,t)]_{\mathrm{ren}}. \tag{4.6}$$

From the canonical commutation relations it follows that the field operator $\rho_\mu'^a$ satisfies the following equal-time commutation relations:

$$[\rho_i'^a(r,t),\rho_j'^b(r',t)]=0; \tag{4.7}$$

$$\begin{aligned}[\rho_4'^a(r,t),\rho_j'^b(r',t)]=\;&\epsilon^{abc}\delta^3(r-r')\rho_j'^c(r',t)\\&+(m_\rho/g_\rho)^2\delta^{ab}(\partial/\partial r_j)\delta^3(r-r')\\&+e\epsilon^{abc}(m_\rho/g_\rho)^2[A_j^{\mathrm{tr}}(r',t)]_{\mathrm{ren}},\end{aligned} \tag{4.8}$$

and

$$[\rho_4'^a(r,t),\rho_4'^b(r',t)]=\epsilon^{abc}\delta^3(r-r')\rho_4'^c(r',t). \tag{4.9}$$

In deriving these, we have made use of the fact that

$$\rho_4{}^a=m_0^{-2}[(\partial/\partial r_j)(i\hat{P}_j{}^a)-ig_0(-\epsilon^{abc}\hat{P}_j{}^b\rho_j{}^c+P_\psi T^a\psi)],$$

and the renormalized electromagnetic field $(A_j^{\mathrm{tr}})_{\mathrm{ren}}$ and that the renormalized charge e are related to the unrenormalized quantities A_j^{tr} and e_0 by

$$e(A_j^{\mathrm{rt}})_{\mathrm{ren}}=e_0A_j^{\mathrm{tr}}. \tag{4.10}$$

Besides (4.7)–(4.9), there are other commutation relations between the electromagnetic fields and $\rho_\mu{}^a$; these are given in Appendix A. Throughout the paper, e denotes the renormalized charge of the electron.

From the field-current identities (1.6) and (3.13), it follows that

$$(J_\mu{}^\gamma)_{I=1}=-\rho_\mu'^0,$$

and

$$(V_\mu{}^{\mathrm{wk}})_{S=0}=-\sqrt{2}\rho_\mu'^+=-(\rho_\mu'^1-i\rho_\mu'^2).$$

These observed hadronic current operators, therefore, satisfy the following field algebra (valid to all orders in e):

$$[J_i{}^\gamma(r,t),J_j{}^\gamma(r',t)]=[J_4{}^\gamma(r,t),J_4{}^\gamma(r',t)]=0, \tag{4.11}$$

$$[J_4{}^\gamma(r,t),J_j{}^\gamma(r',t)]=(m_\rho/g_\rho)^2(\partial/\partial r_j)\times\delta^3(r-r'), \tag{4.12}$$

$$[V_i{}^{\mathrm{wk}}(r,t),J_j{}^\gamma(r',t)]=0, \tag{4.13}$$

$$\begin{aligned}[V_4{}^{\mathrm{wk}}(r,t),J_\mu{}^\gamma(r',t)]&=-[J_4{}^\gamma(r,t),V_\mu{}^{\mathrm{wk}}(r',t)]\\&=i\delta^3(r-r')V_\mu{}^{\mathrm{wk}}(r',t),\end{aligned} \tag{4.14}$$

$$[V_i{}^{\mathrm{wk}}(r,t),V_j{}^{\mathrm{wk}}(r',t)]=[V_i{}^{\mathrm{wk}}(r,t),V_j{}^{\mathrm{wk}}(r',t)^\dagger]=0, \tag{4.15}$$

$$[V_4{}^{\mathrm{wk}}(r,t),V_\mu{}^{\mathrm{wk}}(r',t)]=0, \tag{4.16}$$

$$[V_4{}^{\mathrm{wk}}(r,t),V_4{}^{\mathrm{wk}}(r',t)^\dagger]=2i\delta^3(r-r')J_4{}^\gamma(r',t), \tag{4.17}$$

and

$$\begin{aligned}[V_4{}^{\mathrm{wk}}(r,t),V_j{}^{\mathrm{wk}}(r',t)^\dagger]=\;&-2i\delta^3(r-r')J_j{}^\gamma(r',t)\\&+2(m_\rho/g_\rho)^2\left(\frac{\partial}{\partial r_j}+ie[A_j^{\mathrm{tr}}(r',t)]_{\mathrm{ren}}\right)\delta^3(r-r'),\end{aligned} \tag{4.18}$$

where the dagger denotes Hermitian conjugation. For clarity, we have omitted the subscripts $I=1$ and $S=0$; in the above Eqs. (4.11)–(4.18), $J_\mu{}^\gamma$ stands for $(J_\mu{}^\gamma)_{I=1}$ and $V_\mu{}^{\mathrm{wk}}$ for $(V_\mu{}^{\mathrm{wk}})_{S=0}$.

Although for convenience we have adopted the Coulomb gauge in our derivation, it can be readily shown that the above Eqs. (4.11)–(4.18) are valid in any gauge, provided that in (4.18) one replaces $(A_j^{\mathrm{tr}})_{\mathrm{ren}}$ by the appropriate $(A_j)_{\mathrm{ren}}$.

All above formulas, except (4.18), are formally unchanged with the inclusion of the electromagnetic interaction. They differ from those of current algebra by having the Schwinger term finite and the commutators of all spatial current components zero.[13] As already noted in the introduction, the presence of a term proportional to $eA_j\delta^3(r-r')$ in the commutator $[V_4{}^{\mathrm{wk}}(r,t),V_j{}^{\mathrm{wk}}(r',t)^\dagger]$ is a general consequence of gauge invariance.

[13] In principle, since the matrix elements of these current operators are measurable, the validity of these commutation relations can be tested. A convenient way is to use, e.g., the appropriate sum rules for the high-energy neutrino processes developed by S. L. Adler [Phys. Rev. 143, 1144 (1965)]. At present, it is unclear whether the assumption of unsubtracted dispersion relations made in deriving these sum rules is justified or not.

The presence of the electromagnetic interaction destroys the isospin invariance. Nevertheless, one may still define the generator $\mathbf{I}$ of the isospin group by

$$\mathbf{I}=i\int \varrho_4' d^3r. \tag{4.19}$$

Through the field-current identities, its components I^a ($a=0, 1, 2$) are also given by

$$I^0=-i\int (J_4{}^\gamma)_{I=1}d^3r,$$

$$I^-=\tfrac{1}{2}(I^1-iI^2)=-i2^{-1/2}\int (V_4{}^{\rm wk})_{S=0}d^3r, \tag{4.20}$$

and

$$I^+=\tfrac{1}{2}(I^1+iI^2)=+i2^{-1/2}\int (V_4{}^{\rm wk})^\dagger{}_{S=0}d^3r.$$

The equal-time commutator of $I^a(t)$ is, on account of (4.9),

$$[I^a(t),I^b(t)]=\epsilon^{abc}I^c(t), \tag{4.21}$$

which is valid to all orders in e.

By using (4.7)–(4.9) and the additional canonical commutation relation

$$[\rho_4'^a(r,t),A_j{}^{\rm tr}(r',t)]=0, \tag{4.22}$$

one finds (valid to all orders in e)

$$[I^a(t),\rho_4{}^b(r,t)]=\epsilon^{abc}\rho_4{}^c(r,t), \tag{4.23}$$

but

$$[I^a(t),\hat{\rho}_j{}^b(r,t)]=\epsilon^{abc}\hat{\rho}_j{}^c(r,t). \tag{4.24}$$

Thus, under the isospin rotation, the components of ϱ_4, or the equivalent current operators

$$[V_4{}^{\rm wk}(r,t)]_{S=0},\quad [J_4{}^\gamma(r,t)]_{I=1},\quad \text{and}\quad [V_4{}^{\rm wk}(r,t)]^\dagger{}_{S=0}, \tag{4.25}$$

form an isotriplet, but for the spatial part it is those of $\hat{\rho}_j$, or

$$[V_j{}^{\rm wk}(r,t)]_{S=0},\quad [J_j{}^\gamma(r,t)]_{I=1}+e(m_\rho/g_\rho)^2[A_j{}^{\rm tr}(r,t)]_{\rm ren},\quad \text{and}\quad [V_j{}^{\rm wk}(r,t)]^\dagger{}_{S=0}, \tag{4.26}$$

that form an isotriplet.

5. BROKEN SYMMETRIES

As shown by Eq. (3.15), the generalization of the field-current identities to other weak-interaction currents requires the existence of both vector and axial-vector field operators. In place of the isospin symmetry, the appropriate symmetry group $\mathcal{G}$ becomes either

$$\mathcal{G}=SU_2\times SU_2,\quad \text{or}\quad SU_3,\quad \text{or}\quad SU_3\times SU_3, \tag{5.1}$$

or some other possibilities. All such symmetries are known to be broken or badly broken by the strong interaction. In the following, we assume the existence of N spin-1 fields $\phi_\mu{}^1$, $\phi_\mu{}^2$, $\cdots\phi_\mu{}^N$ consisting of both vector and axial-vector fields which include the three ρ-meson fields. The total number N of $\phi_\mu{}^a$ is the same as the number of generators of $\mathcal{G}$. For example, in the case $\mathcal{G}=SU_2\times SU_2$, one has $N=6$, and the set of spin-1 meson fields $\phi_\mu{}^a$ consists of three ρ-meson fields and three additional axial-vector meson fields. As we shall see, the existence of these spin-1 meson field operators makes possible a unified treatment of these broken symmetries.

It is useful to separate the total free and strong-interaction Lagrangian density of the hadrons into two parts

$$\mathcal{L}_{\rm free}+\mathcal{L}_{\rm st}=\mathcal{L}_\phi+\mathcal{L}_m, \tag{5.2}$$

where $\mathcal{L}_\phi$ depends *only* on the meson fields $\phi_\mu{}^a$, $\mathcal{L}_m$ depends on $\phi_\mu{}^a$ and also on other matter fields, represented by ψ. The infinitesimal transformations of the symmetry group $\mathcal{G}$ can be represented by

$$\phi_\mu{}^a\to\phi_\mu{}^a+C^{abc}(\delta\theta)^b\phi_\mu{}^c, \tag{5.3}$$

and

$$\psi\to\psi+T^a(\delta\theta)^a\psi, \tag{5.4}$$

where $\delta\theta^a$ is a set of infinitesimal numbers, C^{abc} is the totally antisymmetric structure constant of the symmetry group given by (5.1), and $-iT^a$ is the matrix representation of its Hermitian generators on ψ which satisfies

$$[T^a,T^b]=C^{abc}T^c. \tag{5.5}$$

In this section all superscripts a, b, or c, vary from 1 to N. [For SU_2 symmetry, $C^{abc}=-\epsilon^{abc}$, and T^a is related to the Pauli matrices τ^a by $T^a=\tfrac{1}{2}i\tau^a$ for the isospin-$\tfrac{1}{2}$ fields.]

We assume that the meson part $\mathcal{L}_\phi$ is *invariant* under the transformation (5.3), but $\mathcal{L}_m$ may violate the symmetry; i.e., under (5.3) and (5.4)

$$\mathcal{L}_\phi\to\mathcal{L}_\phi, \tag{5.6}$$

but

$$\mathcal{L}_m\to\mathcal{L}_m-P^a\delta\theta^a. \tag{5.7}$$

By assumption, the ρ-meson fields are included in the set $\phi_\mu{}^a$; therefore, the ρ-meson part of the Lagrangian, given by (2.6), is contained in $\mathcal{L}_\phi$. Thus, the invariance assumption of $\mathcal{L}_\phi$ under the larger group of transformations (5.3) requires that[14]

$$\mathcal{L}_\phi=-\tfrac{1}{4}(f_{\mu\nu}{}^a)^2-\tfrac{1}{2}m_0{}^2(\phi_\mu{}^a)^2, \tag{5.8}$$

where

$$f_{\mu\nu}{}^a=\frac{\partial}{\partial x_\mu}\phi_\nu{}^a-\frac{\partial}{\partial x_\nu}\phi_\mu{}^a+g_0C^{abc}\phi_\mu{}^b\phi_\nu{}^c. \tag{5.9}$$

In the case of broken symmetries, there is a great deal of arbitrariness in the symmetry-violating interaction

[14] Except for the mass term, $\mathcal{L}_\phi$ is the same as the Lagrangian of the generalized Yang-Mills theory discussed by R. Utiyama, Phys. Rev. **101**, 1597 (1956); M. Gell-Mann and S. Glashow, Ann. Phys. (N. Y.) **15**, 437 (1961).

$\mathcal{L}_m$. We will assume that, similar to (2.14), $\mathcal{L}_m$ is of the form

$$\mathcal{L}_m=\mathcal{L}_m(\psi,D_\nu\psi,f_{\mu\nu}{}^a); \quad (5.10)$$

i.e., $\mathcal{L}_m$ can be an arbitrary function of ψ, $f_{\mu\nu}{}^a$, and $D_\nu\psi$, where

$$D_\nu\psi=(\partial\psi/\partial x_\nu)+g_0T^a\phi_\nu{}^a\psi. \quad (5.11)$$

This symmetry-violating Lagrangian density (5.10) is quite general. Among others, it may contain symmetry-violating terms like

$$K^{ab}f_{\mu\nu}{}^af_{\mu\nu}{}^b,\quad M_{\mu\nu}{}^af_{\mu\nu}{}^a,\quad \text{etc.},$$

where K^{ab} can be an arbitrary set of constants, or it can be an arbitrary set of (space-time) scalar functions of ψ and $D_\nu\psi$, and $M_{\mu\nu}{}^a$ can be an arbitrary set of (space-time) tensor functions of ψ and $D_\nu\psi$.

It can be readily verified that, independent of the detailed structure of the symmetry-breaking form of $\mathcal{L}_m$, the canonical commutation rules imply that, in the absence of the electromagnetic interaction, $\phi_\mu{}^a$ satisfied the following algebra of fields[5]:

$$[\phi_i{}^a(r,t),\phi_j{}^b(r',t)]=0, \quad (5.12)$$

$$[\phi_4{}^a(r,t),\phi_4{}^b(r',t)] =-(g_0/m_0{}^2)C^{abc}\delta^3(r-r')\phi_4{}^c(r',t), \quad (5.13)$$

and

$$[\phi_4{}^a(r,t),\phi_j{}^b(r',t)]=-(g_0/m_0{}^2)C^{abc}\delta^3(r-r')\phi_j{}^c(r',t) +m_0{}^{-2}\delta^{ab}(\partial/\partial r_j)\delta^3(r-r'). \quad (5.14)$$

In deriving these, we have used the fact that the canonical momenta conjugate to $\phi_j{}^a$ and ψ, and are, respectively,

$$P_j{}^a=i[f_{4j}{}^a-(\partial\mathcal{L}_m/\partial f_{4j}{}^a)+(\partial\mathcal{L}_m/\partial f_{j4}{}^a)],$$

and

$$P_\psi=-i(\partial\mathcal{L}_m/\partial D_4\psi). \quad (5.15)$$

Through the equation of motion

$$\frac{\partial}{\partial x_\mu}\left[f_{\mu\nu}{}^a-\frac{\partial\mathcal{L}_m}{\partial f_{\mu\nu}{}^a}+\frac{\partial\mathcal{L}_m}{\partial f_{\nu\mu}{}^a}\right]-m_0{}^2\phi_\nu{}^a=g_0S_\nu{}^a, \quad (5.16)$$

where

$$S_\nu{}^a=C^{abc}[f_{\mu\nu}{}^b-(\partial\mathcal{L}_m/\partial f_{\mu\nu}{}^b)+(\partial\mathcal{L}_m/\partial f_{\nu\mu}{}^b)]\phi_\mu{}^c -\sum_\psi(\partial\mathcal{L}_m/\partial D_\nu\psi)T^a\psi, \quad (5.17)$$

the 4th component $\phi_4{}^a$ is given by

$$\phi_4{}^a=m_0{}^{-2}[(\partial/\partial r_j)(iP_j{}^a) -ig_0(C^{abc}P_j{}^b\phi_j{}^c-P_\psi T^a\psi)]. \quad (5.18)$$

The field-current identities (1.6), (3.14), and (3.15) imply that the various hadronic electromagnetic and weak-interaction current operators satisfy the same algebra of fields. From (5.16) and (5.17), it follows that

$$-(m_0{}^2/g_0)(\partial\phi_\nu{}^a/\partial x_\nu)=\partial S_\nu{}^a/\partial x_\nu=P^a, \quad (5.19)$$

where P^a is given by (5.7). The usual PCAC (partially conserved axial-vector current) approximation implies that for the divergence of the axial-vector isovector field $\phi_\mu{}^i$, the corresponding P^a can be approximated by a constant times the pion field; i.e.,

$$(\partial\phi_\mu{}^i/\partial x_\mu)\propto\pi^i(x),$$

where $\pi^i(x)$ denotes the pion fields and i represents the isospin index. [If one wishes, similar approximations can also be made for other pseudoscalar fields.]

If the following combination of derivatives of $\mathcal{L}_m$ commutes with $\phi_j{}^b$ at equal time:

$$[\partial\mathcal{L}_m/\partial f_{4i}{}^a(r,t)-\partial\mathcal{L}_m/\partial f_{i4}{}^a(r,t),\ \phi_j{}^b(r',t)]=0, \quad (5.20)$$

then one has, in addition,

$$[(\partial/\partial t)\phi_j{}^a(r,t)-i(\partial/\partial r_j)\phi_4{}^a(r,t),\ \phi_k{}^b(r',t)] =-i\delta^{ab}\delta^3(r-r')\delta_{jk}-iC^{abc}(g_0/m_0{}^2)\phi_j{}^c(r,t) \times(\partial/\partial r_k)\delta^3(r-r')+i(g_0/m_0)^2C^{adc}C^{dbe}\delta^3(r-r') \times\phi_j{}^c(r,t)\phi_k{}^e(r',t). \quad (5.21)$$

As an example of (5.20), $\mathcal{L}_m$ can be of the form

$$M_{\mu\nu}{}^a(\psi)f_{\mu\nu}{}^a+\mathcal{L}_m'(\psi,D_\nu\psi), \quad (5.22)$$

where $M_{\mu\nu}{}^a=\partial\mathcal{L}_m/\partial f_{\mu\nu}{}^a$ depends only on ψ.

We note that according to (5.8), the unrenormalized masses of $\phi_\mu{}^a$ are all equal. A simple way to break the symmetry is to introduce different unrenormalized masses for different $\phi_\mu{}^a$. Such a possibility can be easily incorporated in our scheme by introducing a symmetry-breaking term $(K_{ab}f_{\mu\nu}{}^af_{\mu\nu}{}^b)$ in $\mathcal{L}_m$. The unrenormalized masses of the mesons $\phi_\mu{}^a$ become, then, the different eigenvalues of the $(N\times N)$ matrix $m_0{}^2(1+K)^{-1}$, where $K=$matrix (K_{ab}). In this case, the commutation relations (5.12)–(5.14) remain valid, but (5.21) should be modified.

If the symmetry-breaking interaction $\mathcal{L}_m$ is independent of $f_{\mu\nu}{}^a$, but otherwise can be an arbitrary function of ψ and $D_\nu\psi$, then one has the following additional equal-time commutation relations

$$[(\partial/\partial t)\phi_j{}^a(r,t)-i(\partial/\partial r_j)\phi_4{}^a(r,t),\ \phi_4{}^b(r',t)] =-(g_0/m_0{}^2)C^{abc}\delta^3(r-r')((\partial\phi_j{}^a/\partial t)-i(\partial\phi_4{}^a/\partial r_j)) +i(g_0/m_0{}^2)C^{abc}\phi_4{}^c(r,t)(\partial/\partial r_j)\delta^3(r-r'), \quad (5.23)$$

and

$$[f_{4j}{}^a(r,t),f_{4k}{}^b(r',t)]=0. \quad (5.24)$$

By adding these algebraic relations to those already derived, one can also determine the equal-time commutator between

$$(\partial/\partial t)\phi_j{}^a(r,t)-i(\partial/\partial r_j)\phi_4{}^a(r,t),$$

and

$$(\partial/\partial t)\phi_k{}^b(r',t)-i(\partial/\partial r_k')\phi_4{}^b(r',t),$$

though its explicit form is somewhat lengthy.

6. $SU_3\times SU_3$ FIELD ALGEBRA (INCLUDING ELECTROMAGNETIC EFFECTS)

As an illustration of how the above considerations can be extended to include the electromagnetic interaction, we discuss in some detail the special case of

$$\mathcal{G}=SU_3\times SU_3. \tag{6.1}$$

[The general case is discussed in Appendix A.] We assume the existence of eight vector fields $v_\mu^1, v_\mu^2, \cdots, v_\mu^8$, and eight axial-vector fields $a_\mu^1, a_\mu^2, \cdots, a_\mu^8$, where the vector fields v_μ^α are related to the ρ-meson fields $\boldsymbol{\rho}_\mu$ and the K^{*+}, K^{*0} meson fields by

$$\begin{gathered} v_\mu^1=\rho_\mu^1,\quad v_\mu^2=\rho_\mu^2,\quad v_\mu^3=\rho_\mu^0,\\ 2^{-1/2}(v_\mu^4-iv_\mu^5)=K^{*+},\ \text{and}\ 2^{-1/2}(v_\mu^6-iv_\mu^7)=K^{*0}. \end{gathered} \tag{6.2}$$

In this case, it is convenient to represent the field tensors (5.9) by $V_{\mu\nu}^\alpha$ and $A_{\mu\nu}^\alpha$:

$$V_{\mu\nu}^\alpha=\frac{\partial}{\partial x_\mu}v_\nu^\alpha-\frac{\partial}{\partial x_\nu}v_\mu^\alpha-g_0f^{\alpha\beta\gamma}(v_\mu^\beta v_\nu^\gamma+a_\mu^\beta a_\nu^\gamma), \tag{6.3}$$

and

$$A_{\mu\nu}^\alpha=\frac{\partial}{\partial x_\mu}a_\nu^\alpha-\frac{\partial}{\partial x_\nu}a_\mu^\alpha-g_0f^{\alpha\beta\gamma}(v_\mu^\beta a_\nu^\gamma+a_\mu^\beta v_\nu^\gamma), \tag{6.4}$$

where

$$-f^{\alpha\beta\gamma}=\text{totally antisymmetric structure constant of the } SU_3 \text{ group}, \tag{6.5}$$

and α,β,γ vary independently from 1 to 8. According to (5.8) and (5.10), the free and the strong-interaction Lagrangian density of hadrons can be written as

$$\mathcal{L}_{\text{free}}+\mathcal{L}_{\text{st}}=-\tfrac{1}{2}m_0^2(v_\mu^\alpha)^2-\tfrac{1}{2}m_0^2(a_\mu^\alpha)^2+\mathcal{L}_0, \tag{6.6}$$

and

$$\mathcal{L}_0=-\tfrac{1}{4}(V_{\mu\nu}^\alpha)^2-\tfrac{1}{4}(A_{\mu\nu}^\alpha)^2+\mathcal{L}_m(\psi,D_\nu\psi,V_{\mu\nu}^\alpha,A_{\mu\nu}^\alpha), \tag{6.7}$$

where $D_\nu\psi$ is defined by (5.11).

Following the arguments given in Sec. 3, one finds that, similar to (3.4), the Lagrangian density, $\mathcal{L}=\mathcal{L}_{\text{free}}+\mathcal{L}_{\text{st}}+\mathcal{L}_\gamma$, is

$$\mathcal{L}=-\tfrac{1}{4}F_{\mu\nu}^2-\tfrac{1}{2}m_0^2(v_\mu^\alpha)^2-\tfrac{1}{2}m_0^2(a_\mu^\alpha)^2+\mathcal{L}_0'. \tag{6.8}$$

The function $\mathcal{L}_0'$ is the same as $\mathcal{L}_0$ except for the replacement of v_μ^α by $\hat{v}_\mu^\alpha$, and $\hat{v}_\mu^\alpha$ is given by

$$\begin{gathered} \hat{v}_\mu^3=v_\mu^3+(e_0/g_0)A_\mu,\\ \hat{v}_\mu^8=v_\mu^8+3^{-1/2}(e_0/g_0)A_\mu, \end{gathered} \tag{6.9}$$

and

$$\hat{v}_\mu^\alpha=v_\mu^\alpha\quad\text{for}\quad 3\neq\alpha\neq 8.$$

This replacement changes $D_\nu\psi$, $V_{\mu\nu}^\alpha$, and $A_{\mu\nu}^\alpha$ to $D_\nu'\psi$, $\hat{V}_{\mu\nu}^\alpha$, and $\hat{A}_{\mu\nu}^\alpha$, respectively, where

$$\hat{V}_{\mu\nu}^\alpha=\frac{\partial}{\partial x_\mu}\hat{v}_\nu^\alpha-\frac{\partial}{\partial x_\nu}\hat{v}_\mu^\alpha-g_0f^{\alpha\beta\gamma}(\hat{v}_\mu^\beta\hat{v}_\nu^\gamma+a_\mu^\beta a_\nu^\gamma), \tag{6.10}$$

and

$$\hat{A}_{\mu\nu}^\alpha=\frac{\partial}{\partial x_\mu}a_\nu^\alpha-\frac{\partial}{\partial x_\nu}a_\mu^\alpha-g_0f^{\alpha\beta\gamma}(\hat{v}_\mu^\beta a_\nu^\gamma+a_\mu^\beta\hat{v}_\nu^\gamma). \tag{6.11}$$

Correspondingly,

$$\begin{aligned} \mathcal{L}_0'=\mathcal{L}_0(v_\mu^\alpha\to\hat{v}_\mu^\alpha)=&-\tfrac{1}{4}(\hat{V}_{\mu\nu}^\alpha)^2-\tfrac{1}{4}(\hat{A}_{\mu\nu}^\alpha)^2\\ &+\mathcal{L}_m(\psi,D_\nu'\psi,\hat{V}_{\mu\nu}^\alpha,\hat{A}_{\mu\nu}^\alpha). \end{aligned} \tag{6.12}$$

To derive the field algebra valid to all orders in e, we follow the steps outlined in Sec. 4. In the Coulomb gauge, A_μ is given by

$$A_j=A_j^{\text{tr}}\quad\text{and}\quad A_4=i\phi, \tag{6.13}$$

where A_j^{tr} satisfies (4.1) and ϕ is the solution of the Laplace equation

$$\Delta\phi=ie_0(m_0^2/g_0)(v_4^3+3^{-1/2}v_4^8). \tag{6.14}$$

It is convenient to regard $\hat{v}_j^\alpha$, a_j^α, A_j^{tr}, and ψ as independent variables. Similar to (4.5), their conjugate momenta are, respectively,

$$\begin{aligned} (\hat{P}_v)_j^\alpha&=i[\hat{V}_{4j}^\alpha-(\partial\mathcal{L}_m/\partial\hat{V}_{4j}^\alpha)+(\partial\mathcal{L}_m/\partial\hat{V}_{j4}^\alpha)],\\ (\hat{P}_a)_j^\alpha&=i[\hat{A}_{4j}^\alpha-(\partial\mathcal{L}_m/\partial\hat{A}_{4j}^\alpha)+(\partial\mathcal{L}_m/\partial\hat{A}_{j4}^\alpha)],\\ \Pi_j^{\text{tr}}&=-E_j^{\text{tr}}=(\partial A_j^{\text{tr}}/\partial t), \end{aligned} \tag{6.15}$$

and

$$P_\psi=-i(\partial\mathcal{L}_m/\partial D_4'\psi).$$

By using the canonical commutation relations, the algebra of fields (5.12)–(5.14) can be easily generalized to include the electromagnetic effects. In writing down these expressions, it is convenient to use the renormalization-independent field operators introduced in Ref. 5. We define

$$\begin{aligned} v_\mu'^\alpha&=(m_0^2/g_0)v_\mu^\alpha=(m_\rho^2/g_\rho)(v_\mu^\alpha)_{\text{ren}},\\ a_\mu'^\alpha&=(m_0^2/g_0)a_\mu^\alpha=(m_\rho^2/g_\rho)(a_\mu^\alpha)_{\text{ren}}, \end{aligned} \tag{6.16}$$

and

$$\hat{v}_\mu'^\alpha=(m_0^2/g_0)\hat{v}_\mu^\alpha=(m_\rho^2/g_\rho)(\hat{v}_\mu^\alpha)_{\text{ren}},$$

where the subscript ren denotes the renormalized operator. For convenience, we have chosen the same wave-function renormalization for all v_μ^α and a_μ^α as that of the ρ meson; i.e.,

$$g_0v_\mu^\alpha=g_\rho(v_\mu^\alpha)_{\text{ren}},\quad g_0a_\mu^\alpha=g_\rho(a_\mu^\alpha)_{\text{ren}}, \tag{6.17}$$

where

$$(g_0/g_\rho)=(m_0/m_\rho).$$

Since, by definition, the renormalized electromagnetic field $(A_\mu)_{\text{ren}}$ and the renormalized charge e are related to the unrenormalized quantities A_μ and e_0 by

$$e(A_\mu)_{\text{ren}}=e_0A_\mu,$$

one finds from (6.9) that

$$\begin{aligned} (\hat{v}_\mu^3)_{\text{ren}}&=(v_\mu^3)_{\text{ren}}+(e/g_\rho)(A_\mu)_{\text{ren}},\\ (\hat{v}_\mu^8)_{\text{ren}}&=(v_\mu^8)_{\text{ren}}+3^{-1/2}(e/g_\rho)(A_\mu)_{\text{ren}}, \end{aligned} \tag{6.18}$$

and

$$(\hat{v}_\mu{}^\alpha)_{\rm ren}=(v_\mu{}^\alpha)_{\rm ren} \quad \text{for} \quad 3\neq\alpha\neq 8.$$

The equal-time commutators (5.12)–(5.14) become (valid to all orders in e)

$$[v_i'^\alpha(r,t),v_j'^\beta(r',t)]=[a_i'^\alpha(r,t),a_j'^\beta(r',t)] =[v_i'^\alpha(r,t),a_j'^\beta(r',t)]=0\,, \quad (6.19)$$

$$[v_4'^\alpha(r,t),v_4'^\beta(r',t)]=[a_4'^\alpha(r,t),a_4'^\beta(r',t)] =f^{\alpha\beta\gamma}\delta^3(r-r')v_4'^\gamma(r',t)\,, \quad (6.20)$$

$$[v_4'^\alpha(r,t),a_4'^\beta(r',t)]=f^{\alpha\beta\gamma}\delta^3(r-r')a_4'^\gamma(r',t)\,, \quad (6.21)$$

$$[v_4'^\alpha(r,t),v_j'^\beta(r',t)]=[a_4'^\alpha(r,t),a_j'^\beta(r',t)] =f^{\alpha\beta\gamma}\delta^3(r-r')\hat{v}_j'^\gamma+(m_\rho/g_\rho)^2 \times\delta^{\alpha\beta}(\partial/\partial r_j)\delta^3(r-r')\,, \quad (6.22)$$

and

$$[v_4'^\alpha(r,t),a_j'^\beta(r',t)]=[a_4'^\alpha(r,t),v_j'^\beta(r',t)] =f^{\alpha\beta\gamma}\delta^3(r-r')a_j'^\gamma(r',t)\,. \quad (6.23)$$

In the above expressions, the electromagnetic field enters explicitly only in (6.22) through $\hat{v}_j'^\gamma$. [Further commutation relations are given in Appendix A.]

In terms of $v_\mu'^\alpha$ and $a_\mu'^\alpha$, the observed hadronic electromagnetic current operators and the weak-interaction current operators become

$$(J_\mu{}^\gamma)_{I=1}=-v_\mu'^3\,, \quad (J_\mu{}^\gamma)_{I=0}=-3^{-1/2}v_\mu'^8\,,$$
$$(V_\mu{}^{\rm wk})_{S=0}=-(v_\mu'^1-iv_\mu'^2)\,,$$
$$(V_\mu{}^{\rm wk})_{S=1}=-(v_\mu'^4-iv_\mu'^5)\,, \quad (6.24)$$
$$(A_\mu{}^{\rm wk})_{S=0}=-(a_\mu'^1-ia_\mu'^2)\,,$$

and

$$(A_\mu{}^{\rm wk})_{S=1}=-(a_\mu'^4-ia_\mu'^5)\,.$$

These field-current identities imply that these hadron current operators satisfy the same equal-time commutation relations as those of the fields. Most of the commutators of these hadron currents do not explicitly depend on the electromagnetic field. Those which are modified by the presence of the electromagnetic fields are

$$[V_4{}^{\rm wk}(r,t)_{S=0},V_j{}^{\rm wk}(r',t)^\dagger{}_{S=0}] =[A_4{}^{\rm wk}(r,t)_{S=0},A_j{}^{\rm wk}(r',t)^\dagger{}_{S=0}] =-2i\delta^3(r-r')(J_j{}^\gamma)_{I=1} +2(m_\rho/g_\rho)^2\left[\frac{\partial}{\partial r_j}+ie(A_j)_{\rm ren}\right]\delta^3(r-r')\,, \quad (6.25)$$

and

$$[V_4{}^{\rm wk}(r,t)_{S=-1},V_j{}^{\rm wk}(r',t)^\dagger{}_{S=-1}] =[A_4{}^{\rm wk}(r,t)_{S=-1},A_j{}^{\rm wk}(r',t)^\dagger{}_{S=-1}] =-i\delta^3(r-r')[(J_j{}^\gamma)_{I=1}+3(J_j{}^\gamma)_{I=0}] +2(m_\rho/g_\rho)^2\left[\frac{\partial}{\partial r_j}+ie(A_j)_{\rm ren}\right]\delta^3(r-r')\,. \quad (6.26)$$

In both expressions the Schwinger term has a finite constant coefficient and the covariant derivative $[\partial/\partial r_j+ie(A_j)_{\rm ren}]$ occurs.

7. LEPTON CURRENTS

In this section, we will discuss some speculations concerning the lepton currents. Consider first the electromagnetic interaction. The hadronic part of the electromagnetic current $J_\nu{}^\gamma$ and the leptonic part of the electromagnetic current $j_\nu{}^\gamma$ are, by definition, related to the electromagnetic field $F_{\mu\nu}$ by

$$\partial F_{\mu\nu}/\partial x_\mu=e_0(J_\nu{}^\gamma+j_\nu{}^\gamma)\,. \quad (7.1)$$

Through the current field identities (1.6) and (3.12), the hadron currents satisfy the following field algebra:

$$[J_j{}^\gamma(r,t),J_k{}^\gamma(r',t)]=[J_4{}^\gamma(r,t),J_4{}^\gamma(r',t)]=0\,, \quad (7.2)$$

and

$$[J_4{}^\gamma(r,t),J_k{}^\gamma(r',t)]=\lambda_h(\partial/\partial r_j)\delta^3(r-r')\,, \quad (7.3)$$

where λ_h is a finite constant. [In the SU_3 field algebra, one has $\lambda_h=\frac{4}{3}(m_\rho/g_\rho)^2$.] Equations (7.2) and (7.3) reduce to (4.11) and (4.12), respectively, if only the isovector part is included in $J_\mu{}^\gamma$.

A natural question to ask is whether the lepton current operator $j_\mu{}^\gamma$ can satisfy a similar set of algebraic relations. As we shall see, our present experimental information is completely consistent with the proposal that the leptonic part of the electromagnetic current $j_\lambda{}^\gamma$ also satisfies the algebra of fields; i.e.,

$$[j_i{}^\gamma(r,t),j_k{}^\gamma(r',t)]=[j_4{}^\gamma(r,t),j_4{}^\gamma(r',t)]=0\,, \quad (7.4)$$

and

$$[j_4{}^\gamma(r,t),j_k{}^\gamma(r',t)]=\lambda_l(\partial/\partial r_k)\delta^3(r-r')\,, \quad (7.5)$$

where λ_l is a finite constant. To be sure, in the usual quantum electrodynamics, the operator $j_\lambda{}^\gamma$ is assumed to be equal to

$$s_\lambda=i\psi_e{}^\dagger\gamma_4\gamma_\lambda\psi_e+i\psi_\mu{}^\dagger\gamma_4\gamma_\lambda\psi_\mu\,, \quad (7.6)$$

where ψ_e and ψ_μ are the field operators of the electron and the μ meson; therefore, it cannot possibly satisfy the algebra of fields.

The feasibility of (7.4) and (7.5) can be demonstrated by considering a particular model in which one assumes the existence of a (hypothetical) neutral spin-1 boson field $B_\mu{}^0(x)$. There is a direct interaction between $B_\mu{}^0(x)$ and the charged lepton. This interaction can be represented by the Lagrangian density

$$\mathcal{L}_{\rm free}+\mathcal{L}_{\rm int}=-\tfrac{1}{2}(m_B{}^0B_\mu{}^0)^2-\frac{1}{4}\left(\frac{\partial}{\partial x_\mu}B_\nu{}^0-\frac{\partial}{\partial x_\nu}B_\mu{}^0\right)^2 -\sum_l\psi_l{}^\dagger\gamma_4\left[\gamma_\mu\left(\frac{\partial}{\partial x_\mu}-if_0B_\mu{}^0\right)+m_l{}^0\right]\psi_l\,, \quad (7.7)$$

where f_0 is the unrenormalized coupling constant, $m_B{}^0$, $m_l{}^0$ are, respectively, the unrenormalized masses of B^0

and the charged lepton l, $l=e$ or μ, and ψ_l is the field operator of the lepton.

Just as in (3.1), we may write

$$\mathcal{L}_{\text{free}}+\mathcal{L}_{\text{int}}=-\tfrac{1}{2}(m_B{}^0B_\mu{}^0)^2+\mathcal{L}_0. \qquad (7.8)$$

The electromagnetic interaction Lagrangian $\mathcal{L}_\gamma$ is then generated by replacing in $\mathcal{L}_0$ the operator $B_\mu{}^0(x)$ by

$$\hat{B}_\mu{}^0=B_\mu{}^0+(e_0/f_0)A_\mu, \qquad (7.9)$$

and $(\partial B_\mu{}^0/\partial x_\nu)$ by $(\partial\hat{B}_\mu{}^0/\partial x_\nu)$; i.e., in the absence of hadrons, one has

$$\mathcal{L}_{\text{free}}+\mathcal{L}_{\text{int}}+\mathcal{L}_\gamma=-\tfrac{1}{4}F_{\mu\nu}{}^2-\tfrac{1}{2}(m_B{}^0B_\mu{}^0)^2+\mathcal{L}_0', \qquad (7.10)$$

where

$$\mathcal{L}_0'=\mathcal{L}_0(B_\mu{}^0\rightarrow\hat{B}_\mu{}^0). \qquad (7.11)$$

It can be readily verified that in this model, if one includes also the hadron current $J_\nu{}^\gamma$, the Maxwell equation is

$$\frac{\partial}{\partial x_\mu}F_{\mu\nu}=-(e_0/f_0)(m_B{}^0)^2B_\nu{}^0+e_0J_\nu{}^\gamma. \qquad (7.12)$$

Comparison between (7.12) and (7.1) leads to the identity

$$\begin{aligned} j_\nu{}^\gamma(x)&=-[(m_B{}^0)^2/f_0]B_\nu{}^0(x)\\ &=-(m_B{}^2/f)[B_\nu{}^0(x)]_{\text{ren}}, \qquad (7.13)\end{aligned}$$

where $[B_\nu{}^0(x)]_{\text{ren}}$ is the renormalized field, related to the renormalized coupling constant f and the renormalized mass m_B by

$$(B_\nu{}^0)_{\text{ren}}=(m_B{}^0/m_B)B_\nu{}^0 \text{ and } (m_B{}^0/m_B)=(f_0/f). \qquad (7.14)$$

Consequently $j_\nu{}^\gamma(x)$ satisfies the field algebra (7.4) and (7.5), and the constant λ_l is given by

$$\lambda_l=(m_B{}^0/f_0)^2=(m_B/f)^2. \qquad (7.15)$$

In (7.10), there is a direct B^0-photon coupling through the replacement $B_\mu{}^0\rightarrow\hat{B}_\mu{}^0$ in

$$-\tfrac{1}{4}[(\partial B_\nu{}^0/\partial x_\mu)-(\partial B_\mu{}^0/\partial x_\nu)]^2.$$

As will be shown in Appendix B, this direct B^0-photon coupling can be removed by a canonical transformation. Furthermore, the physical consequences of the B^0 meson become particularly simple if one assumes that the unrenormalized theory is divergent; i.e., the unrenormalized mass $m_B{}^0\rightarrow\infty$. In this case, as a result of the canonical transformation, the B^0 meson becomes coupled *only* to $e^\pm$ and $\mu^\pm$ with the same renormalized coupling constant f given by (7.14). By using the results derived in Appendix B, one sees that the existence of the B^0 meson has essentially only the following two experimental consequences:

(i) Scattering between charged leptons (e.g., $e^\pm$ on $e^\pm$) can occur, besides through the usual photon exchanges, also through a virtual exchange of the B^0 meson. The lowest-order perturbation formulas for all photon-exchange amplitudes are the *same* as the usual ones. Since quantum electrodynamics has been tested by the colliding beam experiment[15] for 4-momentum transfer $|q^2|^{1/2}$ up to $\sim(\frac{1}{2}\text{ BeV})$, one should require, in the same range of q^2, the matrix element due to B^0 exchange $(f^2/4\pi)(q^2+m_B{}^2)^{-1}$ to be much smaller than that due to photon exchange (α/q^2). Thus, one expects

$$m_B{}^2\gg\alpha^{-1}(f^2/4\pi)(\tfrac{1}{2}\text{ BeV})^2\sim(f^2/4\pi)(5\text{ BeV})^2, \qquad (7.16)$$

which implies that $m_B\gg5$ BeV if $(f^2/4\pi)$ is $\sim O(1)$ and[15a] $m_B\gg\frac{1}{2}$ BeV if $(f^2/4\pi)\sim O(\alpha)$.

If $(f/m_B)^2$ is of the same order of magnitude as the weak-interaction coupling constant G [i.e., $(f^2/4\pi)\sim10^{-6}(m_B/m_N)^2$], it would be natural to identify B^0 as the neutral component of the (hypothetical) intermediate boson $W^\pm$ of the weak interaction. One may then expect m_B to have a comparable lower limit $\sim$2.5 BeV to that of $W^\pm$. [See remark (4) below.]

(ii) The B^0 meson can be produced through inelastic processes by scattering $\mu^\pm$ or $e^\pm$ on hadrons; e.g.,

$$\mu^\pm+p\rightarrow\mu^\pm+p+B^0, \qquad (7.17)$$

and

$$e^\pm+p\rightarrow e^\pm+p+B^0. \qquad (7.18)$$

Subsequently, the B^0 meson would decay into lepton pairs

$$B^0\rightarrow(e^++e^-), \text{ or } (\mu^++\mu^-). \qquad (7.19)$$

The B^0-meson production cross section $\sigma(l^\pm+p\rightarrow l^\pm+p+B^0)$ at an energy high above the threshold can be roughly estimated to be $\sim\alpha^{-1}(f^2/4\pi)$ times the cross section $\sigma(l^\pm+p\rightarrow l^\pm+p+\gamma)$ for a photon-emission process at the same incoming lepton energy and with a comparable 4-momentum transfer to the proton. The rates for the B^0 decay can be estimated by using the lowest-order perturbation formula. One finds, upon neglecting the lepton mass,

$$\begin{aligned}\text{Rate}(B^0\rightarrow\mu^++\mu^-)&=\text{Rate}(B^0\rightarrow e^++e^-)\\ &=\tfrac{1}{3}(f^2/4\pi)m_B. \qquad (7.20)\end{aligned}$$

From (i), one expects m_B to be $>(\sim1\text{ BeV})$. Thus, the effective way to investigate reactions (7.17) and (7.18) is to use either the high-energy electrons from,

---

[15] W. C. Barber, B. Gittelman, G. K. O'Neill, and B. Richter, Phys. Rev. Letters **16**, 1127 (1966).

[15a] *Note added in proof.* In the case $f^2/4\pi\sim O(\alpha)$, a stronger lower limit for m_B can be obtained by using the recent result for the anomalous magnetic moment of the negative μ-meson. Farley *et al.* (at the 1967 Stanford International Symposium on Electron-Photon Interactions) reported the experimental value

$$(g-2)_{\text{exp}}=(23332\pm10)\times10^{-7}.$$

The usual quantum electrodynamics prediction for the same quantity is

$$(g-2)_{\text{th}}=23312\times10^{-7}.$$

Assuming that the discrepancy is due to the existence of the B^0 meson, and using the relation

$$\delta g=(f^2/4\pi)(1/3\pi)(m_\mu/m_B)^2,$$

one finds $m_B\gtrsim1.6$ BeV, provided $(f^2/4\pi)\cong\alpha$. We wish to thank Professor L. Lederman for pointing this out to us.

e.g., SLAC, or the high-energy muons from, e.g., the AGS and CERN, provided $(f^2/4\pi)$ is not much smaller than α. At present, we are unaware of any such experiments.[16]

In the model, the existence of this hypothetical B^0 meson implies that j_ν^γ satisfies the algebra of fields. At present, the converse, which is perhaps the more interesting question, whether the algebraic relations (7.4) and (7.5) also imply the existence of a vector meson, like B^0, remains an open one. It is hoped that these algebraic relations and the generalized field-current identities may have a wider domain of applicability than the special local-field-theoretical model. In this connection, the following remarks, though based on the special model, may clarify some of the implications of these field-current identities.

(1) The Lagrangians (7.7) and (7.10) are both renormalizable in the usual sense. The mathematical problem is identical with the one discussed in Ref. 4 The unrenormalized mass $m_B{}^0$ cannot be zero; it can be infinite, if the unrenormalized theory is divergent. In the limit of an infinite $m_B{}^0$, the matrix element of s_λ between any two states $|a\rangle$ and $|b\rangle$ becomes identical with that of $j_\nu^\gamma(x)=-(m_B{}^2/f)[B_\nu{}^0(x)]_{\text{ren}}$; i.e.,

$$\lim_{m_B{}^0=\infty}\langle b|s_\lambda|a\rangle=\langle b|j_\nu{}^\gamma(x)|a\rangle$$
$$=-(m_B{}^2/f)\langle b|(B_\nu{}^0)_{\text{ren}}|a\rangle, \quad (7.21)$$

where s_λ is given by (7.6). Nevertheless, $j_\nu{}^\gamma$ satisfies the algebra of fields, but s_ν does not. [Similar conclusions also apply to the hadrons.[1,4]]

(2) By using the same arguments given in Refs. 1 and 4, one finds that, independently of whether the unrenormalized mass $m_B{}^0$ is finite or infinite, the unrenormalized photon propagator must be finite. As a result, the ratio between the unrenormalized charge e_0 and the renormalized charge e is also finite. We note that the Maxwell equation for the renormalized electromagnetic field tensor

$$(F_{\mu\nu})_{\text{ren}}=(e_0/e)(F_{\mu\nu}) \quad (7.22)$$

can be decomposed into a leptonic part

$$\left[\frac{\partial}{\partial x_\mu}(F_{\mu\nu})_{\text{ren}}\right]_{\text{lep}}=-(e_0/e)^2e(m_B{}^2/f)(B_\nu{}^0)_{\text{ren}}, \quad (7.23)$$

a hadronic $I=1$ part

$$\left[\frac{\partial}{\partial x_\mu}(F_{\mu\nu})_{\text{ren}}\right]_{I=1}=-(e_0/e)^2e(m_\rho{}^2/g)(\rho_\nu{}^0)_{\text{ren}}, \quad (7.24)$$

and a hadronic $I=0$ part

$$\left[\frac{\partial}{\partial x_\mu}(F_{\mu\nu})_{\text{ren}}\right]_{I=0}=-\tfrac{1}{2}(e_0/e)^2(e/g_Y)$$
$$\times[\cos\theta_Y m_\phi{}^2(\phi_\mu)_{\text{ren}}-\sin\theta_Y m_\omega{}^2(\omega_\mu)_{\text{ren}}]. \quad (7.25)$$

Except for $(e_0/e)^2$, only renormalized quantities occur in these expressions. The finiteness of $(e_0/e)^2$ and $(e_0/e)^2>1$ make possible the general expectation that the matrix elements of the renormalized field operators $(F_{\mu\nu})_{\text{ren}}$, $(\rho_\mu{}^0)_{\text{ren}}$, $\cdots$ between any two physical states should be finite.

The hadronic contribution to the photon propagator has already been discussed in Refs. 1 and 4. For the special case that

$$(f^2/4\pi)=O(\alpha^{1/2})\ll 1, \quad (7.26)$$

and, therefore,

$$\alpha(f^2/4\pi)^{-1}=O(\alpha^{1/2})\ll 1, \quad (7.27)$$

where $\alpha=(e^2/4\pi)\cong(137)^{-1}$, the leptonic contribution can be explicitly calculated.

From a straightforward perturbation calculation, one finds that, to first order in $(f^2/4\pi)$, the *renormalized* B^0-meson propagator is

$$(q^2+m_B{}^2)^{-1}(\delta_{\mu\nu}+m_B{}^{-2}q_\mu q_\nu)$$
$$\times[1+(q^2+m_B{}^2)^{-1}(f^2/4\pi)q^2(q^2\delta_{\mu\nu}-q_\mu q_\nu)F]. \quad (7.28)$$

The function F can be most conveniently expressed in terms of the usual spectral representation

$$F(q^2)=F_e(q^2)+F_\mu(q^2), \quad (7.29)$$

and

$$F_l(q^2)=-\frac{1}{3\pi}\int_0^1\frac{(3-2x)x^2(1-2x)dx}{m_l{}^2+q^2x(1-x)}, \quad (7.30)$$

where $l=e$ or μ, and m_l is the renormalized lepton mass. The unrenormalized photon propagator $(D_{\mu\nu}{}^\gamma)^0$ can be obtained by using Eq. (55) of Ref. 4. In the limit that the unrenormalized mass $m_B{}^0$ is infinite, one has, upon neglecting the hadronic contribution and higher-order terms in $\alpha(f^2/4\pi)^{-1}$, $(f^2/4\pi)$, and α,

$$(D_{\mu\nu}{}^\gamma)^0=q^{-2}(\delta_{\mu\nu}-q^{-2}q_\mu q_\nu)[1-\alpha(f^2/4\pi)^{-1}$$
$$\times(q^2+m_B{}^2)^{-1}m_B{}^2+\alpha(q^2+m_B{}^2)^{-2}m_B{}^4q^2F]. \quad (7.31)$$

Thus to the same order, the leptonic contribution to $(e_0/e)^2$ is

$$(e_0/e)^2=1+\alpha(f^2/4\pi)^{-1}. \quad (7.32)$$

From (7.31), one sees that the usual vacuum polarization term[17] $\alpha(15\pi)^{-1}q^2[m_e{}^{-2}+m_\mu{}^{-2}]$ is now changed to

$$\alpha q^2[(15\pi)^{-1}(m_e{}^{-2}+m_\mu{}^{-2})+(f^2/4\pi)^{-1}m_B{}^{-2}]. \quad (7.33)$$

(3) The possible presence of the B^0 meson leaves a

[16] We wish to thank Dr. R. Cool and Dr. W. Panofsky for discussions of these experimental problems.

[17] E. A. Uehling, Phys. Rev. 48, 55 (1935); R. Serber, *ibid.* 48, 49 (1935).

certain arbitrariness in the definition of the electromagnetic field A_μ. For example, one may define

$$A_\mu'=[1+(e_0/f_0)^2]^{1/2}A_\mu + (e_0/f_0)[1+(e_0/f_0)^2]^{-1/2}B_\mu^0. \quad (7.34)$$

Thus, by using the Lagrangian (7.10), one sees that, in the absence of hadrons, A_μ' satisfies

$$\frac{\partial}{\partial x_\mu}F_{\mu\nu}'=-e_1 s_\nu, \quad (7.35)$$

where s_ν is given by (7.6)

$$e_1=[1+(e_0/f_0)^2]^{-1/2}e_0, \quad (7.36)$$

and

$$F_{\mu\nu}'=(\partial A_\nu'/\partial x_\mu)-(\partial A_\mu'/\partial x_\nu).$$

If the hadron current J_ν^γ is included, then A_μ' satisfies

$$\frac{\partial}{\partial x_\mu}F_{\mu\nu}'=e_1(-s_\nu+J_\nu^\gamma). \quad (7.37)$$

[Further discussions will be given in Appendix B.] From (7.14) and the finiteness of $(e_0/e)^2$, it follows that in the limit of the unrenormalized mass $m_B^0=\infty$, (7.21) holds and $F_{\mu\nu}'=F_{\mu\nu}$.

(4) If the renormalized coupling constant $(f^2/4\pi)$ is of the same order as the semiweak coupling constant,

$$(f^2/4\pi)\sim 10^{-6}(m_B/m_N)^2, \quad (7.38)$$

where m_N is the nucleon mass, then it becomes natural to identify the B^0 meson as the neutral component of the usual charged weak-interaction intermediate boson $W^\pm$. One replaces the lepton current j_μ^{wk} in (3.16) by the $W^\pm$-meson field operator W_μ. The weak-interaction Lagrangian density (3.16) becomes, then,

$$fJ_\mu^{\text{wk}}W_\mu+\text{H.c.}, \quad (7.39)$$

where J_μ^{wk} is given by (3.17). In addition, there is a direct W-lepton coupling, in complete analogy with the direct B^0-lepton coupling in (7.7),

$$fs_\mu^{\text{wk}}W_\mu+\text{H.c.}, \quad (7.40)$$

where

$$s_\mu^{\text{wk}}=i\sum_l \psi_l^\dagger\gamma_4\gamma_\mu(1+\gamma_5)\psi_{\nu_l}, \quad (7.41)$$

and ψ_{ν_l} is the field operator of the neutrino ν_l.

A full investigation of these interesting, but hypothetical, possibilities clearly lies outside both the scope and the spirit of the present paper.

APPENDIX A

In this Appendix, we discuss the complete set of equal-time commutators of the local fields for the general case discussed in Sec. 5, but extended to include also the electromagnetic field. On account of (5.2) and (5.8), the strong-interaction Lagrangian density and the hadronic part of the free Lagrangian density can be written as

$$-\tfrac{1}{2}m_0^2(\phi_\nu^a)^2+\mathcal{L}_0(\psi,D_\nu\psi,f_{\mu\nu}^a), \quad (\text{A}.1)$$

where $\mathcal{L}_0$ is related to $\mathcal{L}_m$ of (5.10) by

$$\mathcal{L}_0=-\tfrac{1}{4}(f_{\mu\nu}^a)^2+\mathcal{L}_m. \quad (\text{A}.2)$$

The function $\mathcal{L}_0$ is invariant under isospin rotations; otherwise, it can be an arbitrary function of ψ, $D_\nu\psi$, and $f_{\mu\nu}^a$.

The total Lagrangian density $\mathcal{L}_{\text{free}}+\mathcal{L}_{\text{st}}+\mathcal{L}_\gamma$ can be derived from (A.1) by replacing in $\mathcal{L}_0$

$$\phi_\mu^a\to\hat{\phi}_\mu^a=\phi_\mu^a+(e_0/g_0)\xi^a A_\mu, \quad (\text{A}.3)$$

where ξ^a depends on the group $\mathcal{G}$. If $\mathcal{G}$ is the isospin SU_2 group, or the usual $SU_2\times SU_2$ group, then

$$\begin{aligned}\xi^a&=1 \quad \text{for} \quad \phi_\mu^a=\rho_\mu^0,\\ &=0 \quad \text{otherwise}.\end{aligned} \quad (\text{A}.4)$$

If $\mathcal{G}$ is the usual SU_3, or $SU_3\times SU_3$ group, then

$$\begin{aligned}\xi^a&=1 \quad &&\text{for} \quad \phi_\mu^a=v_\mu^3=\rho_\mu^0,\\ &=3^{-1/2} &&\text{for} \quad \phi_\mu^a=v_\mu^8,\\ &=0 &&\text{otherwise}.\end{aligned} \quad (\text{A}.5)$$

Through the replacement (A.3), the total Lagrangian density becomes

$$\mathcal{L}_{\text{free}}+\mathcal{L}_{\text{st}}+\mathcal{L}_\gamma=-\tfrac{1}{4}F_{\mu\nu}^2-\tfrac{1}{2}m_0^2(\phi_\nu^a)^2 -\tfrac{1}{4}(\hat{f}_{\mu\nu}^a)^2+\mathcal{L}_m(\psi,D_\nu'\psi,\hat{f}_{\mu\nu}^a), \quad (\text{A}.6)$$

where

$$D_\nu'\psi=(\partial\psi/\partial x_\nu)+g_0T^a\hat{\phi}_\nu^a\psi, \quad (\text{A}.7)$$

and

$$\hat{f}_{\mu\nu}^a=\frac{\partial}{\partial x_\mu}\hat{\phi}_\nu^a-\frac{\partial}{\partial x_\nu}\hat{\phi}_\mu^a+g_0C^{abc}\hat{\phi}_\mu^b\hat{\phi}_\nu^c. \quad (\text{A}.8)$$

From the Lagrangian density (A.6), one sees that the Maxwell equation takes the form

$$\frac{\partial}{\partial x_\mu}F_{\mu\nu}=-(e_0m_0^2/g_0)\xi^a\phi_\nu^a. \quad (\text{A}.9)$$

Just as in Secs. 4 and 6, we will, for convenience, adopt the Coulomb gauge, and regard $\hat{\phi}_j^a$, A_j^{tr}, and ψ as the generalized coordinates. Their conjugate momenta are, respectively,

$$\begin{aligned}\hat{P}_j^a&=i[\hat{f}_{4j}^a-(\partial\mathcal{L}_m/\partial\hat{f}_{4j}^a)+(\partial\mathcal{L}_m/\partial\hat{f}_{j4}^a)],\\ \Pi_j^{\text{tr}}&=-E_j^{\text{tr}}=(\partial A_j^{\text{tr}}/\partial t),\end{aligned} \quad (\text{A}.10)$$

and

$$P_\psi=-i(\partial\mathcal{L}_m/\partial D_4'\psi).$$

For $\mathcal{G}=SU_2$, (A.10) reduces to (4.5), and for $\mathcal{G}=SU_3\times SU_3$, (A.10) reduces to (6.15). From the equations of motion, it follows that

$$\phi_4^a=m_0^{-2}[(\partial/\partial r_j)(i\hat{P}_j^a) -ig_0(C^{abc}\hat{P}_j^b\hat{\phi}_j^c-P_\psi T^a\psi)]. \quad (\text{A}.11)$$

The totality of the canonical commutation relations between these generalized coordinates and their conjugate momenta determines the complete set of field algebra. The following equal-time commutators for $\phi_\mu{}^a(r,t)$ are valid to all orders in e^2:

$$[\phi_i{}^a(r,t),\phi_j{}^b(r',t)]=0, \quad (A.12)$$

$$[\phi_4{}^a(r,t),\phi_4{}^b(r',t)] = -(g_0/m_0{}^2)C^{abc}\phi_4{}^c(r',t)\delta^3(r-r'), \quad (A.13)$$

and

$$[\phi_4{}^a(r,t),\phi_j{}^b(r',t)] = -(g_0/m_0{}^2)C^{abc}\hat{\phi}_j{}^c(r',t)\delta^3(r-r') + m_0{}^{-2}\delta^{ab}(\partial/\partial r_j)\delta^3(r-r'). \quad (A.14)$$

We note that (A.12) and (A.13) are not affected by the presence of the electromagnetic field; they are identical with (5.12) and (5.13), respectively. Equation (A.14) differs from (5.14) by the presence of $\hat{\phi}_j{}^c$, instead of $\phi_j{}^c$, on the right-hand side. The above Eqs. (A.12)–(A.14) become (4.7)–(4.9) for $\mathcal{G}=SU_2$, and (6.19)–(6.23) for $\mathcal{G}=SU_3\times SU_3$.

In the Coulomb gauge, the transverse part of the vector potential $A_j{}^{\mathrm{tr}}$ and the corresponding transverse part of the electric field $E_j{}^{\mathrm{tr}}$ are canonical variables. They satisfy

$$[A_j{}^{\mathrm{tr}}(r,t),E_k{}^{\mathrm{tr}}(r',t)] = -i(\delta_{jk}-\partial_j\partial_k\nabla^{-2})\delta^3(r-r'), \quad (A.15)$$

$$[\phi_\mu{}^a(r,t),A_k{}^{\mathrm{tr}}(r',t)]=0, \quad (A.16)$$

$$[\phi_4{}^a(r,t),E_k{}^{\mathrm{tr}}(r',t)]=0, \quad (A.17)$$

$$[\phi_j{}^a(r,t),E_k{}^{\mathrm{tr}}(r',t)] = [\partial_t\phi_j{}^a(r,t)-i\partial_j\phi_4{}^a(r,t),A_k{}^{\mathrm{tr}}(r',t)] = i(e_0/g_0)\xi^a(\delta_{jk}-\partial_j\partial_k\nabla^{-2})\delta^3(r-r'), \quad (A.18)$$

and

$$[\partial_t\phi_j{}^a(r,t)-i\partial_j\phi_4{}^a(r,t),E_k{}^{\mathrm{tr}}(r',t)]=0, \quad (A.19)$$

where $\partial_t=(\partial/\partial t)$, $\partial_j=(\partial/\partial r_j)$, $\nabla^2=\partial_j{}^2$ and

$$\nabla^{-2}\delta^3(r-r') = -(4\pi)^{-1}[\sum_j(r_j-r_j')^2]^{-1/2}. \quad (A.20)$$

The magnetic field H_j is related to $A_j{}^{\mathrm{tr}}$ by

$$H_j=\epsilon_{jkl}\partial_k A_l{}^{\mathrm{tr}}. \quad (A.21)$$

The electric field E_j consists of two parts

$$E_j=E_j{}^{\mathrm{tr}}+E_j{}^{\mathrm{long}}, \quad (A.22)$$

in which $E_j{}^{\mathrm{long}}$ is not an independent variable, but is determined by $\phi_4{}^a$. One has, on account of (A.9),

$$\partial_j E_j{}^{\mathrm{long}} = -i(e_0 m_0{}^2/g_0)\xi^a\phi_4{}^a. \quad (A.23)$$

From (A.21)–(A.23) and the above commutation relations, it can be readily verified that

$$[\phi_\mu{}^a(r,t),H_k(r',t)]=0, \quad (A.24)$$

$$[\phi_4{}^a(r,t),E_k(r',t)] = -ie_0C^{abc}\xi^b\phi_4{}^c\partial_k\nabla^{-2}\delta^3(r-r'), \quad (A.25)$$

and

$$[\phi_j{}^a(r,t),E_k(r',t)] = i(e_0/g_0)(\xi^a\delta_{jk}-g_0C^{abc}\xi^b\phi_j{}^c\partial_k\nabla^{-2})\delta^3(r-r'). \quad (A.26)$$

All these commutation relations are valid independently of the detailed form of $\mathcal{L}_m$, and to all orders in e. It has been shown in Sec. 5 that, in the absence of the electromagnetic interaction, one has the additional commutation relations (5.21), if $\mathcal{L}_m$ satisfies (5.20), and also (5.23), if $\mathcal{L}_m$ is independent of $f_{\mu\nu}{}^a$. The generalization of (5.21) and (5.23) to include the electromagnetic effects can be derived in a straightforward, though somewhat tedious, way by using the above commutation relations together with the canonical ones;

$$[\hat{P}_j{}^a(r,t),\hat{\phi}_k{}^b(r',t)] = -i\delta_{jk}\delta^{ab}\delta^3(r-r'),$$
$$[\hat{P}_j{}^a(r,t),\hat{P}_k{}^b(r',t)]=0,$$

and the definitions of $\hat{P}_\nu{}^a$ and $\hat{\phi}_k{}^b$.

APPENDIX B

In this Appendix, we discuss some consequences of the (hypothetical) B^0 meson. For clarity, we consider a model containing the B^0 meson, the electromagnetic field, the electron and the muon, but in which the hadrons are represented only by the proton and the ρ^0 meson. The total Lagrangian density is

$$\mathcal{L} = -\tfrac{1}{4}F_{\mu\nu}{}^2 - \tfrac{1}{2}(m_B{}^0 B_\mu{}^0)^2 - \tfrac{1}{2}(m_\rho{}^0\rho_\mu{}^0)^2 - \frac{1}{4}\left[\frac{\partial\hat{B}_\nu{}^0}{\partial x_\mu} - \frac{\partial\hat{B}_\mu{}^0}{\partial x_\nu}\right]^2 - \frac{1}{4}\left[\frac{\partial\hat{\rho}_\nu{}^0}{\partial x_\mu} - \frac{\partial\hat{\rho}_\mu{}^0}{\partial x_\nu}\right]^2$$
$$-\sum_l \psi_l{}^\dagger\gamma_4\left[\gamma_\mu\left(\frac{\partial}{\partial x_\mu} - if_0\hat{B}_\mu{}^0\right)+m_l{}^0\right]\psi_l$$
$$-\psi_p{}^\dagger\gamma_4\left[\gamma_\mu\left(\frac{\partial}{\partial x_\mu}+ig_0\hat{\rho}_\mu{}^0\right)+m_p{}^0\right]\psi_p, \quad (B.1)$$

where

$$\hat{\rho}_\mu{}^0 = \rho_\mu{}^0 + (e_0/g_0)A_\mu, \quad (B.2)$$

and

$$\hat{B}_\mu{}^0 = B_\mu{}^0 + (e_0/f_0)A_\mu. \quad (B.3)$$

We find that (B.1) becomes the same as (7.10) if one neglects the hadrons p and ρ^0. From (B.1), the electromagnetic field satisfies

$$\frac{\partial}{\partial x_\mu}F_{\mu\nu} = -e_0[f_0{}^{-1}(m_B{}^0)^2B_\nu{}^0 + g_0{}^{-1}(m_\rho{}^0)^2\rho_\nu{}^0]. \quad (B.4)$$

It is convenient to define

$$s_\nu = \sum_l i\psi_l{}^\dagger\gamma_4\gamma_\nu\psi_l, \quad (B.5)$$

and

$$S_\nu = i\psi_p{}^\dagger\gamma_4\gamma_\nu\psi_p. \tag{B.6}$$

One can readily verify that both s_ν and S_ν are conserved. Similarly, both $\rho_\nu{}^0$ and $B_\nu{}^0$ fields satisfy the field-conservation equation. There exist in (B.1), besides the $(\rho_\nu{}^0 S_\nu)$ and the $(B_\nu{}^0 s_\nu)$ interactions, also a $\rho_\nu{}^0$-photon interaction and a $B_\nu{}^0$-photon interaction with coupling constants (e_0/g_0) and (e_0/f_0). These photon-meson direct couplings can be removed by the transformation

$$A_\mu \to A_\mu{}',$$

where $A_\mu{}'$ is defined by

$$A_\mu{}' = N^{-1}A_\mu + N[(e_0/g_0)\rho_\mu{}^0 + (e_0/f_0)B_\mu{}^0], \tag{B.7}$$

and

$$N = [1+(e_0/f_0)^2 + (e_0/g_0)^2]^{-1/2}. \tag{B.8}$$

In terms of $A_\mu{}'$, the Lagrangian density (B.1) becomes

$$\begin{aligned}\mathcal{L} = -\tfrac{1}{4}[F_{\mu\nu}{}'^2 + \tilde{\mathcal{G}}_{\mu\nu}{}^0 K_0 \mathcal{G}_{\mu\nu}{}^0] - \tfrac{1}{2}\tilde{\phi}_\mu{}^0 M_0{}^2 \phi_\mu{}^0 \\ + e_1(s_\mu - S_\mu)A_\mu{}' + \tilde{\phi}_\mu{}^0 G_0 \mathcal{J}_\mu \\ - \sum_l \psi_l{}^\dagger\gamma_4\left(\gamma_\mu \frac{\partial}{\partial x_\mu} + m_l{}^0\right)\psi_l \\ - \psi_p{}^\dagger\gamma_4\left(\gamma_\mu \frac{\partial}{\partial x_\mu} + m_p{}^0\right)\psi_p,\end{aligned} \tag{B.9}$$

where $\sim$ denotes the transpose of a matrix,

$$\begin{aligned} e_1 &= Ne_0, \\ F_{\mu\nu}{}' &= \frac{\partial}{\partial x_\mu}A_\nu{}' - \frac{\partial}{\partial x_\nu}A_\mu{}', \\ \phi_\mu{}^0 &= \begin{pmatrix}\rho_\mu{}^0 \\ B_\mu{}^0\end{pmatrix}, \quad \mathcal{J}_\mu = \begin{pmatrix} S_\mu \\ s_\mu\end{pmatrix}, \\ \mathcal{G}_{\mu\nu}{}^0 &= \frac{\partial}{\partial x_\mu}\phi_\nu{}^0 - \frac{\partial}{\partial x_\nu}\phi_\mu{}^0, \end{aligned} \tag{B.10}$$

K_0, M_0, and G_0 are all (2×2) matrices,

$$K_0 = N^2\begin{pmatrix} 1+(e_0/f_0)^2 & -e_0{}^2/(f_0g_0) \\ -e_0{}^2/(f_0g_0) & 1+(e_0/g_0)^2\end{pmatrix}, \tag{B.11}$$

$$M_0{}^2 = \begin{pmatrix}(m_\rho{}^0)^2 & 0 \\ 0 & (m_B{}^0)^2\end{pmatrix}, \tag{B.12}$$

and

$$G_0 = N^2\begin{pmatrix} -g_0[1+(e_0/f_0)^2] & -e_0{}^2/g_0 \\ e_0{}^2/f_0 & f_0[1+(e_0/g_0)^2]\end{pmatrix}. \tag{B.13}$$

One sees that the transformed electromagnetic field $A_\mu{}'$ satisfies

$$\frac{\partial}{\partial x_\mu}F_{\mu\nu}{}' = e_1(S_\nu - s_\nu). \tag{B.14}$$

The mathematical problem of the two coupled fields $\rho_\mu{}^0$ and $B_\mu{}^0$ is identical with the ϕ-ω mixing problem discussed in Ref. 1. Here we will only discuss the case that the unrenormalized theory is divergent. In this case, the unrenormalized masses $m_B{}^0$ and $m_\rho{}^0$ tend to infinity; i.e.,

$$m_B{}^0 \to \infty \quad \text{and} \quad m_\rho{}^0 \to \infty. \tag{B.15}$$

The renormalized coupling constants f and g are related to the unrenormalized ones by

$$(f/f_0) = (m_B/m_B{}^0) \quad \text{and} \quad (g/g_0) = (m_\rho/m_\rho{}^0). \tag{B.16}$$

In the limit (B.15), $(e_0/e)^2$ remains finite, but f_0 and g_0 both become infinity; therefore,

$$K_0 \to \begin{pmatrix}1 & 0 \\ 0 & 1\end{pmatrix}, \tag{B.17}$$

and

$$G_0 \to \begin{pmatrix}-g_0 & 0 \\ 0 & f_0\end{pmatrix}. \tag{B.18}$$

By substituting these limiting forms into (B.9), one finds that in the same limit the ρ^0 meson is coupled *only* to the hadron, while the B^0 meson is coupled *only* to the leptons. Equation (B.14) is *exactly the same* as the Maxwell equation in the usual electrodynamics without the B^0 meson. The experimental consequences of the possible existence of such a B^0 meson have already been discussed in Sec. 7.

7.B.1 Nuclear Physics B9 (1969) 209-243. North-Holland Publ. Comp., Amsterdam

NEGATIVE METRIC AND THE UNITARITY OF THE *S*-MATRIX

T. D. LEE *
CERN, Geneva

and

G. C. WICK
Columbia University, New York, N. Y. **

Received 22 November 1968

Abstract: It is shown that the introduction of a negative metric in quantum mechanics can lead to a unitary S-matrix, provided that all stable particle states are of positive square length. Such a possibility can be used to remove divergence difficulties in physics, while maintaining unitarity and relativistic invariance. For example, to the zeroth order of e^2, the usual photon propagator may be replaced by $(q^2)^{-1} - (q^2+m^2)^{-1}$, and to the zeroth order of the weak interaction constant G, the intermediate boson $W^{\pm}$ propagator may be given simply by $\delta_{\mu\nu}(q^2+m^2)^{-1}$, assuming that m is not too small. The general theoretical implications of such a theory with negative metric, but unitary S-matrix, are analyzed. Its various experimental consequences are discussed. Explicit examples are given for some soluble models.

1. INTRODUCTION

This paper is an attempt to revive the old idea that certain divergence difficulties of physics can be overcome by the use of an indefinite metric [1-3]. Specifically, our thesis is that, contrary to a widely held belief, a theory of this kind can give a *fully unitary* S-matrix, provided certain conditions are satisfied (see eq. (1.10) for a more precise formulation). A formal but fairly trivial proof of this statement for the general case is given below, and is illustrated in detail on some simple models (sects. 2 and 3). Later (sect. 5) we show why the interaction representation, as usually formulated, could lead to a different conclusion, and how this trouble can be avoided. Thus some foundation is laid for the hope that a more detailed formalism can be developed, leading to finite and physically acceptable results in problems now beset by incurable divergencies. Admittedly, non-

* On leave from Columbia University.

** This research was supported in part by the U.S. Atomic Energy Commission and the National Science Foundation.

unitarity of the S-matrix is not the only objection that has been raised against theories with an indefinite metric. Sect. 4 examines some questions of this kind.

Let us now recall the main features of the formalism. In such a theory, the Hamiltonian H is not Hermitian, but pseudo-Hermitian; i.e., it satisfies

$$\eta H^{\dagger} \eta = H , \tag{1.1}$$

where (and throughout the paper) the dagger sign denotes the usual Hermitian conjugation. The metric η is assumed to be a Hermitian matrix whose eigenvalues can always be chosen [2] to be either +1 or -1, and, therefore, one may set

$$\eta^2 = 1 . \tag{1.2}$$

As is well-known, from eq. (1.1) one can derive *in general* only a pseudo-unitary condition for the S-matrix:

$$S^{\dagger} \eta S = \eta , \tag{1.3}$$

which is, of course, unsatisfactory.

We wish to suggest the following way out of this difficulty. Suppose we start in the usual way with a zero-oder Hamiltonian describing 'bare' particles with various masses, spins, etc. Some of these particles will be 'normal' ($\eta = +1$) and some 'abnormal' ($\eta = -1$). Remember, however, that these are not real physical states, but mathematical fictions, out of which the real states are to be built. Suppose now that after interactions, mass renormalization terms, etc., are introduced, no *stable* abnormal particles are left. This seems perfectly possible (and in fact happens in the simple examples studied later) since we have plenty of parameters at our disposal to make it happen. Since the S-matrix is concerned only with asymptotic states which, by definition, consist of only stable particles such as p, e^-, γ, ν_e, ν_μ, $\bar{p}$, $e^+, \ldots$, and since these stable particles are all assumed to be of positive metric, eq. (1.3) then takes the customary, acceptable form

$$S^{\dagger} S = \text{the unit matrix } I . \tag{1.4}$$

Some comment is in order, to show that this is not a swindle. Clearly, the Hilbert space of the asymptotic states, all 'normal', must be a subspace of the original Hilbert space. This is not impossible; there are similar mathematical phenomena in the theory of scattering by a potential; furthermore, the examples of sects. 2 and 3 show that it can really happen. Does this mean that the negative metric does not really play any role, and could have been eliminated from the beginning? This does not seem likely, if one examines the way in which the abnormal 'bare' states are intertwined with the normal 'bare' states in the expansions of the physical scattering states. The negative metric is *not* a trivial device.

Amongst the possible applications of these ideas we have in mind, of course, the usual regularized photon propagator [3,4]

$$\frac{1}{q^2} - \frac{1}{q^2 + m^2} . \tag{1.5}$$

The values of the cut-off parameter m suggested by experiment are sufficiently high so that it does not give rise to a stable negative metric state. The same is true if one assumes the propagator of the intermediate boson W_μ in the weak interaction theory to be *‡

$$\frac{\delta_{\mu\nu}}{q^2 + m_W^2}, \tag{1.6}$$

instead of the usual

$$\frac{\delta_{\mu\nu}}{q^2 + m_W^2} + \frac{1}{m_W^2}\frac{q_\mu q_\nu}{q^2 + m_W^2} \tag{1.7}$$

The spin-0 part of such a W_μ is of negative metric. However, because of its large mass, there cannot be any stable W states. Furthermore, we may assume the interactions between W states to be of such a nature that there are also no stable 3W bound states, 5W bound states, etc. Thus, in all cases the S-matrix has only non-zero matrix elements between asymptotic states consisting of purely positive metric particles, and therefore it is also a unitary matrix.

In general, a non-Hermitian operator may have complex eigenvalues. We may classify the eigenstates of H into three categories: $|\mathrm{r}\rangle$, $|+, \mathrm{c}\rangle$ and $|-, \mathrm{c}\rangle$, defined by

$$H|\mathrm{r}\rangle = E_\mathrm{r}|\mathrm{r}\rangle, \tag{1.8}$$

where E_r is real,

$$H|\pm, \mathrm{c}\rangle = E_{\pm,\mathrm{c}}|\pm, \mathrm{c}\rangle, \tag{1.9}$$

where $E_{+,\mathrm{c}}$ and $E_{-,\mathrm{c}}$ are complex

$$\mathrm{Im}\, E_{+,\mathrm{c}} > 0 \text{ and } \mathrm{Im}\, E_{-,\mathrm{c}} < 0 .$$

The subscripts r and c denote the different eigenstates of H, where r refers to those with real eigenvalues and c those with complex eigenvalues; both consist of discrete and continuous parameters.

The precise mathematical statement of our basic requirement that there should be no stable negative metric is that *all eigenstates* $|\mathrm{r}\rangle$ *of real eigenvalues are of positive square length*; i.e., that

$$\langle \mathrm{r}|\eta|\mathrm{r}\rangle > 0 . \tag{1.10}$$

Since any such state $|\mathrm{r}\rangle$ has a purely oscillatory time dependence $\exp(-iE_\mathrm{r}t)$, if it were of a negative square length then it would correspond to a stable negative metric state. It is important to note that eq. (1.10) also rules out the pathological case that $|\mathrm{r}\rangle$ may be of zero length.

* This corresponds to the case $\xi = 1$ in the notations of Lee and Yang [5]. If one wishes, one can also assume the $W^\pm$ propagator to be $\delta_{\mu\nu}[(q^2 + m_W^2)^{-1} - (q^2 + \Lambda^2)^{-1}]$; in this case there is an additional spin-1 negative-metric field.

‡ See note added in proof.

From the pseudo-Hermiticity condition (1.1), it follows that

$$\langle +, c|\eta|r\rangle = \langle -, c|\eta|r\rangle = 0 \ , \tag{1.11}$$

and

$$\langle +, c'|\eta|+, c\rangle = \langle -, c'|\eta|-, c\rangle = 0 \ , \tag{1.12}$$

for all c, c' and r. Furthermore,

$$\langle r'|\eta|r\rangle = 0 \quad \text{if} \quad E_r \neq E_{r'} \ , \tag{1.13}$$

and

$$\langle +, c'|\eta|-, c\rangle = 0 \quad \text{if} \quad E_{+,c'} \neq E^*_{-,c} \ . \tag{1.14}$$

We shall assume, in analogy with ordinary quantum mechanics, that the eigenstates $|r\rangle$, $|+, c\rangle$ and $|-, c\rangle$ form a complete set of basic vectors. This, of course, cannot be proved in general for a non-Hermitian H (not even in a finite vector-space!). It seems, however, that some resemblance to the ordinary case must be preserved, in order to have a theory of physical interest. For this reason we assume, furthermore, that to each vector $|\ldots\rangle$ there is at least one vector (which can, of course, be the vector $|\ldots\rangle$ itself) that is not orthogonal to it, so that the usual expansion theorem holds, stating that a vector $|\zeta\rangle$ is determined by its 'projections' $\langle r|\eta|\zeta\rangle$ and $\langle \pm, c|\eta|\zeta\rangle$. Eq. (1.14) then shows that this is only possible if to each eigenvalue $E_{+,c}$ there is at least one complex conjugate eigenvalue of the $E_{-,c}$ type. The simplest assumption is that these eigenvalues occur in complex conjugate pairs. We shall write:

$$E_{+,c} = E^*_{-,c} \ , \tag{1.15}$$

and set the normalizations of $|\pm, c\rangle$ so that

$$\langle +, c'|\eta|-, c\rangle = \langle -, c'|\eta|+, c\rangle = I_{c'c} \tag{1.16}$$

where I is the unit matrix. Our basic requirement (1.10) enables us to normalize the states $|r\rangle$ in a similar form:

$$\langle r'|\eta|r\rangle = I_{r'r} \ . \tag{1.17}$$

The completeness theorem states that any vector $|\rangle$ can be expanded in terms of $|r\rangle$, $|+, c\rangle$ and $|-, c\rangle$:

$$|\rangle = \sum_r \langle r|\eta|\rangle|r\rangle + \sum_c \langle -, c|\eta|\rangle|+, c\rangle + \sum_c \langle +, c|\eta|\rangle|-, c\rangle \ . \tag{1.18}$$

Thus, a state of negative square length must be a coherent mixture ‡ of $|+, c\rangle$ and $|-, c\rangle$.

‡ This is to be contrasted with the free Hamiltonian H_0 of, say, the weakly interacting intermediate boson W_μ, with its propagator given by eq. (1.6). Its spin-0 part is of negative metric, and the corresponding eigenvalue m_W for such a particle at rest is *real* in the *absence* of the weak interaction H_{wk}. However, because of its large mass, W^+ is degenerate with the continuum states $e^+\nu_e, \mu^+\nu_\mu$, etc., which are all of positive metric. The matrix elements of H_{wk}, which connect the spin-0 part of $W^\pm$ to these continuum states, are those of an anti-Hermitian; i.e., $(H_{wk})_{ij} = -(H_{wk})^*_{ji}$. Consequently, H_{wk} removes the degeneracy, and changes the eigenvalues from real to complex. (See sect. 2 and 3 for further examples.)

For the continuum case, the states $|r\rangle$ can be chosen, just as in the usual theory, to be either $|r^{out}\rangle$, i.e., a plane wave plus outgoing waves only, or $|r^{in}\rangle$, i.e., a plane wave plus incoming waves. All above-mentioned ortho-normality relations hold separately for both $|r\rangle = |r^{out}\rangle$ and $|r\rangle = |r^{in}\rangle$. For example, eq. (1.17) can be written as

$$\langle r'^{in}|\eta|r^{in}\rangle = \langle r'^{out}|\eta|r^{out}\rangle = I_{r'r} \,. \tag{1.19}$$

The S-matrix is, by definition, given by

$$S_{r'r} = \langle r'^{in}|\eta|r^{out}\rangle \,. \tag{1.20}$$

Since both sets of states $\{|r^{in}\rangle, \ |+,c\rangle, \ |-,c\rangle\}$ and $\{|r^{out}\rangle, \ |+,c\rangle, \ |-,c\rangle\}$ are complete and satisfy the same ortho-normality relations, the matrix S connecting these two sets must be pseudo-unitary; i.e., eq. (1.3) holds. Upon substituting eqs. (1.19) and (1.20) into eq. (1.3), one proves the unitary condition (1.4).

In the above formal discussion, the existence of such a theory with pseudo-Hermitian Hamiltonian satisfying the stated basic condition (1.10) is implicitly assumed. To demonstrate its existence, we have investigated in some detail several soluble models *. These are discussed in sect. 2 and 3. In all these soluble examples, η has negative eigenvalues, and, therefore, the Hamiltonian is not Hermitian, but pseudo-Hermitian. Our basic condition of no stable negative metric states can be trivially satisfied, and the unitary of the resulting S-matrix can be explicitly demonstrated.

The presence of these eigenstates $|\pm, c\rangle$ with complex eigenvalues makes it necessary for the analytical continuation of the S-matrix to have poles and cuts on the physical sheet above the real axis. An interesting consequence is that, in the neighborhood of such a complex pole, the two-body scattering phase shift δ may decrease very rapidly with energy. In sect. 4, the motion of a wave packet is studied. It is shown that in spite of such a possibility there is no violation of the simple causality condition; i.e., conditions that can be stated in terms of only the average position of the wave packet.

Another consequence of these complex eigenvalues $E_{\pm,c}$ is that the operator $\exp(-iHt)$ becomes no longer bounded as $t \to \pm\infty$. Thus, the S-matrix can only be rigorously defined by using the eigenstates $|r^{out}\rangle$, or $|r^{in}\rangle$, of the total Hamiltonian H; the time-dependence of these states is simply $\exp(-iE_r t)$ where E_r is real. Furthermore, for self-consistency, we impose a *fundamental law* that in the physical world *only states of the form*

$$|t\rangle = \sum_r C_r \ |r\rangle \exp(-iE_r t) \tag{1.21}$$

can be realized, where C_r can be arbitrary constants. All physical colli-

* Since the completion of this paper, Dr. V. Glaser has kindly informed us that the question whether the use of indefinite metric could lead to a unitary S-matrix has been studied for the Nθ sector of the soluble model by Pauli [6], and for some other cases by Glaser [7] and Froissart [8]. Their conclusions on unitarity, though based on apparently different motivations and different approaches, are in agreement with ours. Compare also Heisenberg [9].

sion processes are described by forming the appropriate wave packets from these states. The result is the well-known expression (1.20) for the S-matrix. It should be emphasized that in imposing eq. (1.21) we have made another departure from the conventional formulation of quantum mechanics.

The usual alternative derivation of the S-matrix, based on the $t \to \pm\infty$ limits of the Green's function (or, the $U(t, t')$ matrix) of the time-dependent Schroedinger equation in the interaction representation, is *not* valid in the present theory. Nevertheless, it is shown in sect. 5 that, by introducing an artificial limiting process as $t \to \pm\infty$, both $|r^{\text{out}}\rangle$ and $|r^{\text{in}}\rangle$ states, and therefore also the same S-matrix, can be obtained from solutions of the time-dependent Schroedinger equation in the interaction representation. The usual Feynman rules [4] have to be modified, and used with care. In sect. 6, by using the soluble examples, several simple pitfalls are illustrated.

That the use of a negative metric can remove divergence difficulties is familiar to almost all physicists. However, it is usually believed that since the Lagrangian, or the Hamiltonian, is not Hermitian, the resulting S-matrix cannot be unitary. Our analysis shows that this belief is *without foundation*. While a Hermitian Hamiltonian does lead to a unitary S-matrix, the converse is not true. Thus, one could regard the use of negative metric as a possible physical reality. Besides being able to resolve divergence difficulties in physics, it has also other experimental consequences that can, in principle, be tested directly. For example, if, to the zeroth order in e^2, the photon propagator is indeed given by eq. (1.5) instead of the usual $(q^2)^{-1}$, then the corresponding heavy photon pole, called Γ^0, can be established by using, say, the colliding beam experiment

$$e^+e^- \to e^+e^- \, . \tag{1.22}$$

In eq. (1.5), the minus sign in front of $(q^2+m^2)^{-1}$ can be measured by comparing it with the sign of the usual $(q^2)^{-1}$ term. Such a minus sign would imply that, when e^2 corrections are included, the Γ^0 pole is above the real axis and on the physical sheet. (Incidentally, the partial width of Γ^0 to all lepton modes is $\approx$ 50 MeV if its mass m is $\approx$ 10 GeV.) Similar considerations can be extended to possible detection of the complex $W^\pm$ pole in weak interaction processes. The general theoretical and experimental implications of a theory with negative metric, but unitary S-matrix, are discussed in sect. 7.

2. A SIMPLE SOLUBLE MODEL

We consider the soluble model [10]

$$V \rightleftharpoons N\theta \, . \tag{2.1}$$

The Hamiltonian is given by

$$H = H_0 + H_1 \, , \tag{2.2}$$

where

$$H_o = m_V^o V^\dagger V + m_N N^\dagger N + \sum_k \omega_k a_k^\dagger a_k , \tag{2.3}$$

$$H_1 = g \sum_k (4\pi R \omega_k)^{-\frac{1}{2}} k u_k (V^\dagger N a_k - V N^\dagger a_k^\dagger) , \tag{2.4}$$

$$\omega_k = (\mu^2 + k^2)^{\frac{1}{2}} .$$

For convenience, we enclose the whole system inside a sphere of large radius R. The annihilation and creation operators a_k and $a_k^\dagger$ refer only to the S-wave part of the θ-meson field

$$\theta(r, t) = \sum_k (4\pi R \omega_k)^{-\frac{1}{2}} r^{-1} \sin(kr)(a_k + a_k^\dagger) + \text{p-wave} + \text{d-wave} + \cdots . \tag{2.5}$$

All summations over k extend over

$$k = (n\pi/R) , \tag{2.6}$$

where

$$n = 0, 1, 2, 3, \cdots .$$

At equal time,

$$[a_k, a_{k'}^\dagger] = \delta_{kk'} \tag{2.7}$$

and

$$[V, V^\dagger] = [N, N^\dagger] = 1 , \tag{2.8}$$

while all other commutators are zero. (If one wishes, eq. (2.8) can, of course, be changed into anticommutators, instead of commutators.) The parameters m_V^o, m_N^o and μ are all real and have their usual meanings. The form factor u_k and the coupling constant g are also both *real*. Therefore, the Hamiltonian is chosen to be *not* a Hermitian; instead, it satisfies the pseudo-Hermiticity relation

$$H = \eta H^\dagger \eta , \tag{2.9}$$

where the metric η is given by

$$\eta = (-1)^{V^\dagger V} . \tag{2.10}$$

Without any loss of generality, we adopt the choice (through the canonical transformation $\exp(-i m_N t)$)

$$m_N = 0 . \tag{2.11}$$

Throughout this and the following sections, we are concerned only with the question of unitarity of the S-matrix, and not with renormalization problems. Thus, the form factor u_k can be an arbitrary function, provided that it goes to zero faster than k^{-1} as $k \to \infty$, so that there is no divergence problem.

We consider first the Nθ sector. It is straightforward to verify that the eigenvalues E of the total Hamiltonian are roots of

$$h_R(E) = 0 \ , \tag{2.12}$$

where

$$h_R(E) = E - m_V^0 - g^2 \sum_k (4\pi R \omega_k)^{-1} k^2 u_k^2 (\omega_k - E)^{-1} \ . \tag{2.13}$$

As $R \to \infty$, the discrete roots for E real and $\geqslant \mu$ become a continuous distribution. Therefore, the limiting function

$$h_\infty(E) \equiv \lim_{R\to\infty} h_R(E) = E - m_V^0 - g^2 \int_0^\infty dk (4\pi^2 \omega_k)^{-1} k^2 u_k^2 (\omega_k - E)^{-1} \ , \tag{2.14}$$

has a cut along the real axis for $E \geqslant \mu$. Our basic condition that there should be no stable negative metric state implies that $h_\infty(E)$ has no poles on the real axis outside the cut; otherwise, as can be easily verified, there would be a stable negative metric state. Thus, we require

$$h_\infty(E) < 0 \text{ for } E \text{ real and} < \mu \ . \tag{2.15}$$

Since m_V^0, g and u_k can be arbitrary, this condition can be easily satisfied. For example, if $m_V^0 > \mu$, then eq. (2.15) holds. As a result of eq. (2.15), there is *no* ghost state‡.

As we shall see, the value of $h_\infty(E)$ immediately above the cut is related to the phase shift δ_p of Nθ scattering at energy ω_p by

$$e^{2i\delta_p} - 1 = i[h_\infty(\omega_p + i\epsilon)]^{-1} g^2 (2\pi)^{-1} p u_p^2 \ , \tag{2.16}$$

where $\epsilon = 0+$. By using eq. (2.14), one finds that the imaginary part of $h_\infty(\omega_p + i\epsilon)$ is negative, and is given by

$$\operatorname{Im} h_\infty(\omega_p + i\epsilon) = -(4\pi)^{-1} g^2 p u_p^2 \ . \tag{2.17}$$

Thus, there is a complex root,

$$E = E_+ \equiv m_V + \tfrac{1}{2} i\gamma \ , \tag{2.18}$$

of $h_\infty(E) = 0$ above the real axis on the physical sheet. Since

$$h_\infty(E^*) = h_\infty(E)^* \ , \tag{2.19}$$

its complex conjugate,

$$E = E_- = m_V - \tfrac{1}{2} i\gamma \ , \tag{2.20}$$

is also a root. The parameters m_V and γ are both real.

It is most instructive to study the emergence of these complex roots by considering the following example of a large but finite R: We note that for $g^2 = 0+$, the roots of eq. (2.12) are simply m_V^0 and all the ω_k. For definite-

‡ See Källen and Pauli [11]. For clarity, a ghost state is defined to be an eigenstate of the total Hamiltonian with a *real* eigenvalue, but is of *negative*, or *zero*, square length.

ness, let us first consider the case $m_V^0 > \mu$. The root m_V^0 is the only one that corresponds to a negative metric state, and it lies in the midst * of all the ω_k. At fixed R and m_V^0, as g^2 is slightly increased to a critical value g^2_{crit} which is $O(R^{-1})$, there will be a degeneracy between the eigenvalue of the negative metric eigenstate, and one of the numerous positive metric eigenstates. When g^2 is further increased, because the interaction Hamiltonian H_1 is anti-Hermitian (i.e., $H_1{}^\dagger = -H_1$), this degeneracy is removed by moving these two roots from the real axis into the complex plane. In the limit $R \to \infty$, $g^2_{crit} = O(R^{-1}) \to 0$, the roots on the real axis become the cut, and the two complex roots $E_\pm$ are given by eqs. (2.18) and (2.20). Both E_+ and E_- remain on the physical sheet.

Identical conclusions can be reached if eq. (2.15) holds, but $m_V^0 < \mu$. In this case, in order to satisfy eq. (2.15), g^2 must be bigger than a minimum value g^2_{min}. Now, for a large but finite R, by following the previous arguments, one can easily show that for $g^2 > g^2_{crit} = g^2_{min} + O(R^{-1})$, there must be two complex roots of $h_R(E) = 0$. In the limit of $R \to \infty$, the difference between g^2_{crit} and g^2_{min} goes to zero, and, as in the previous case, these two complex roots E_+ and E_- remain on the physical sheet.

Thus, in general, for R large but finite and $g^2 > g^2_{crit}$, the eigenvalues of H consist of two complex values E_+, E_- and a series of real values E_p given (see fig. 1) by:

$$E_p = (\mu^2 + p^2)^{\frac{1}{2}} \,, \tag{2.21}$$

where p is related to the phase shift δ_p by

$$pR = n\pi - \delta_p \,, \tag{2.22}$$

and $n = 0, 1, 2, \cdots$. Let $|+\rangle$, $|-\rangle$ and $|N\theta_p\rangle$ be the corresponding eigenstates of H. It can be readily verified that

$$|N\theta_p\rangle = C_p[V^\dagger + g(4\pi R)^{-1} \sum_k \omega_k^{-\frac{1}{2}} k u_k(\omega_k - E_p)^{-1} N^\dagger a_k^\dagger]|\text{vac}\rangle \,, \tag{2.23}$$

$$|\pm\rangle = C_\pm[V^\dagger + g(4\pi R)^{-1} \sum_k \omega_k^{-\frac{1}{2}} k u_k(\omega_k - E_\pm)^{-1} N^\dagger a_k^\dagger]|\text{vac}\rangle \,, \tag{2.24}$$

where

$$C_p = [-h'_R(E_p)]^{-\frac{1}{2}} \,, \qquad C_\pm = [-h'_R(E_\pm)]^{-\frac{1}{2}} \,, \tag{2.25}$$

and

$$h'_R(E) = \frac{d}{dE} h_R(E) \,. \tag{2.26}$$

From fig. 1, one sees that

$$-h'_R(E_p) > 0 \,, \tag{2.27}$$

for all E_p. One finds, in accordance with eqs. (1.11) - (1.17),

* We exclude the pathological case $u_k = 0$ for $\omega_k > \omega_{max}$, and m_V^0 is also $> \omega_{max}$.

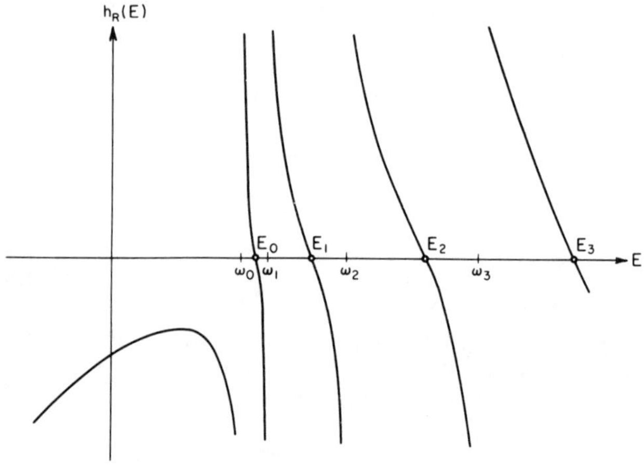

Fig. 1. Schematic diagram of $h_R(E)$ for R finite, E real and $g^2 > g^2_{\text{crit}}$. $\omega_n = [\mu^2 + (n\pi/R)^2]^{1/2}$ and E_n is the nth root of $h_R(E) = 0$.

$$\langle N\theta_p | \eta | N\theta_{p'} \rangle = \delta_{pp'} \tag{2.28}$$

where $\delta_{pp'}$ is the usual Kronecker δ-symbol,

$$\langle + | \eta | N\theta_p \rangle = \langle - | \eta | N\theta_p \rangle = 0 \ , \tag{2.29}$$

$$\langle + | \eta | + \rangle = \langle - | \eta | - \rangle = 0 \ , \tag{2.30}$$

$$\langle + | \eta | - \rangle = 1 \ . \tag{2.31}$$

Thus, all eigenstates with real eigenvalues are of square length +1, and those with complex eigenvalues are of zero length.

As $R \to \infty$, one replaces

$$\sum_k \text{ by } (R/\pi) \int dk \ ,$$

$$a_k \text{ by } (\pi/R)^{\frac{1}{2}} A_k \ , \tag{2.32}$$

where A_k satisfies, in place of the commutation relation (2.7),

$$[A_k, A_{k'}^\dagger] = \delta(k-k') \tag{2.33}$$

and $\delta(k-k')$ denotes the usual one-dimensional δ-function. The eigenstates $|\pm\rangle$ remain given by eq. (2.24), but those with real eigenvalues can be chosen to be either the outgoing states $|N\theta_p^{\text{out}}\rangle$, or the incoming states $|N\theta_p^{\text{in}}\rangle$. If one uses the eigenstate $|N\theta_p\rangle$, given by eq. (2.23), keeps E_p real, and lets $R \to \infty$, then $|N\theta_p\rangle$ becomes the stationary state $|N\theta_p^{\text{st}}\rangle$. The outgoing eigenstate is given by

$$|N\theta_p^{\text{out}}\rangle = K_p^{\text{out}} V^\dagger |\text{vac}\rangle + \int_0^\infty dk \ \phi_p^{\text{out}}(k) N^\dagger A_k^\dagger |\text{vac}\rangle \ , \tag{2.34}$$

where

$$K_p^{\text{out}} = [h_\infty(\omega_p + i\epsilon)]^{-1}(2\pi)^{-1}\,\omega_p^{-\frac{1}{2}}\,g p u_p\,, \tag{2.35}$$

$$\phi_p^{\text{out}}(k) = \delta(k-p) + (\omega_k - \omega_p - i\epsilon)^{-1}(2\pi)^{-1}\,\omega_k^{-\frac{1}{2}}\,g k u_k\,K_p^{\text{out}}\,, \tag{2.36}$$

and $\omega_p = (u^2 + p^2)^{\frac{1}{2}}$ is the corresponding eigenvalue. Similarly, the incoming eigenstate is given by

$$|\text{N}\theta_p^{\text{in}}\rangle = K_p^{\text{in}}\,V^\dagger|\text{vac}\rangle + \int_0^\infty dk\;\phi_p^{\text{in}}(k)\,N^\dagger A_k^\dagger|\text{vac}\rangle\,, \tag{2.37}$$

where

$$K_p^{\text{in}} = [h_\infty(\omega_p - i\epsilon)]^{-1}(2\pi)^{-1}\,\omega_p^{-\frac{1}{2}}\,g p u_p\,, \tag{2.38}$$

$$\phi_p^{\text{in}}(k) = \delta(k-p) + (\omega_k - \omega_p + i\epsilon)^{-1}(2\pi)^{-1}\,\omega_k^{-\frac{1}{2}}\,g k u_k\,K_p^{\text{in}}\,. \tag{2.39}$$

The states $|\text{N}\theta_p^{\text{out}}\rangle$ and $|\text{N}\theta_p^{\text{in}}\rangle$ separately satisfy the appropriate orthonormality relations for the continuum states:

$$\langle\text{N}\theta_{p'}^{\text{out}}|\,\eta\,|\text{N}\theta_p^{\text{out}}\rangle = \langle\text{N}\theta_{p'}^{\text{in}}|\,\eta\,|\text{N}\theta_p^{\text{in}}\rangle = \delta(p-p')\,, \tag{2.40}$$

$$\langle\pm|\,\eta\,|\text{N}\theta_p^{\text{out}}\rangle = \langle\pm|\,\eta\,|\text{N}\theta_p^{\text{in}}\rangle = 0\,. \tag{2.41}$$

The S-matrix can be derived by directly forming wave packets from the time-dependent solution

$$\exp(-i\omega_p t)\,|\text{N}\theta_p^{\text{out}}\rangle. \tag{2.42}$$

The asymptotic behavior of the Fourier transform of $\phi_p^{\text{out}}(k)$ determines that the phase shift δ_p is given by eq. (2.16), and that δ_p is real, on account of eq. (2.17). Thus, the unitarity of the S-matrix is established.

Of course, the same result follows also from eq. (1.20). We note that

$$K_p^{\text{out}} = \exp(2i\,\delta_p)K_p^{\text{in}}\,, \tag{2.43}$$

$$\phi_p^{\text{out}}(k) = \exp(2i\,\delta_p)\phi_p^{\text{in}}(k)\,, \tag{2.44}$$

$$|\text{N}\theta_p^{\text{out}}\rangle = \exp(2i\,\delta_p)\,|\text{N}\theta_p^{\text{in}}\rangle, \tag{2.45}$$

and, therefore, the S-matrix in the Nθ sector is given by

$$\langle p' | S | p \rangle = \langle \mathrm{N}\theta_{p'}^{\mathrm{in}} | \eta | \mathrm{N}\theta_{p}^{\mathrm{out}} \rangle = \delta(p' - p) \exp(2i\,\delta_p) \ . \tag{2.46}$$

In a similar, but more complicated, analysis, the unitarity condition of the S-matrix can also be explicitly demonstrated for the N$\theta\theta$ sector. In this case, the complex eigenvalues form continuous distributions. In order not to obscure the physics involved, details of the N$\theta\theta$ sector will be given in a separate paper. Instead, we will give in sect. 3 a generalized soluble model; it contains an N$\theta\theta'$ sector which also has continuous distributions of complex eigenvalues, but with a much simpler mathematical structure.

3. A GENERALIZED SOLUBLE MODEL

In this generalized soluble model *, there also exist, in addition to the $V(t)$, $N(t)$ and $\theta(r, t)$ fields introduced before, two additional fields: $V'(t)$ and $\theta'(r, t)$. Just as in the simple model the field $V(t)$ is of negative metric, but all other fields are of positive metric. Thus, the metric η remains given by eq. (2.10). The total Hamiltonian is given by

$$H = H_0 + H_1 \ , \tag{3.1}$$

where (set $m_{\mathrm{N}} = 0$),

$$H_0 = m_{\mathrm{V}}^{0}\, V^{\dagger} V + m_{\mathrm{V'}}^{0}\, V'^{\dagger} V' + \sum_k \omega_k a_k^{\dagger} a_k + \sum_k \omega_k' a_k'^{\dagger} a_k' \ , \tag{3.2}$$

$$H_1 = g \sum_k (4\pi R\,\omega_k)^{-\frac{1}{2}}\, k u_k (V^{\dagger} N a_k - V N^{\dagger} a_k^{\dagger})$$
$$+ g' \sum_k (4\pi R\,\omega_k')^{-\frac{1}{2}}\, k u_k' (V'^{\dagger} V a_k' - V' V^{\dagger} a_k'^{\dagger}) \ , \tag{3.3}$$

$$\omega_k = (\mu^2 + k^2)^{\frac{1}{2}} \ ,$$
$$\omega_k' = (\mu'^2 + k^2)^{\frac{1}{2}} \ .$$

Eqs. (2.5) - (2.8) remain valid. In addition, these new operators a_k', $a_k'^{\dagger}$ are related to $\theta'(r, t)$ by

$$\theta'(r, t) = \sum_k (4\pi R\,\omega_k)^{-\frac{1}{2}} r^{-1} \sin(kr)(a_k' + a_k'^{\dagger}) + \text{p-wave} + \text{d-wave} + \ldots \ , \tag{3.4}$$

and at equal time

$$[a_k', a_q'^{\dagger}] = \delta_{kq} \ , \tag{3.5}$$

$$[V', V'^{\dagger}] = 1 \ . \tag{3.6}$$

* This generalized model is essentially the same as the one studied by Ruijgrok and Van Hove [12], except $\theta \neq \theta'$ here. See also ref. [13].

The form factors u_k, u'_k and the parameters g, g', μ, μ', m^o_V, $m^o_{V'}$ are all real, so that

$$H_o^\dagger = H_o\,, \qquad H_1^\dagger = -H_1\,, \qquad H^\dagger = \eta H \eta\,. \tag{3.7}$$

In this generalized model, the only transitions are

$$V \rightleftharpoons N\theta\,, \tag{3.8}$$

with the coupling constant g, and

$$V' \rightleftharpoons V\theta'\,, \tag{3.9}$$

with the coupling constant g'. There are three conservation laws:

$$\mathcal{N}_{V'} + \mathcal{N}_V + \mathcal{N}_N = \text{constant}\,,$$

$$\mathcal{N}_N - \mathcal{N}_\theta = \text{constant}\,,$$

$$\mathcal{N}_{V'} + \mathcal{N}_{\theta'} = \text{constant}\,, \tag{3.10}$$

where $\mathcal{N}_{V'}$, $\mathcal{N}_V$, $\mathcal{N}_N$, $\mathcal{N}_\theta$ and $\mathcal{N}_{\theta'}$ are the total occupation numbers of V', V, N, θ and θ' respectively. All eigenstates of H with $\mathcal{N}_{V'} + \mathcal{N}_{\theta'} = 0$ are exactly the same as those in the simple model; for example, the solutions of the Nθ sector are given by the same expressions (2.23), (2.24), (2.34) and (2.37). In the following, we are *only* concerned with the case that there is no stable negative metric state in the Nθ sector; i.e., eq. (2.15) holds.

The scattering,

$$N\theta_s\,\theta'_t \to N\theta_k\,\theta'_q\,, \tag{3.11}$$

is of particular interest, since the Hilbert space spanned by all the eigenstates in the N$\theta\theta'$ sector contains the space spanned by the negative metric states $V'^\dagger a'^\dagger_q\,|\,\text{vac}\rangle$ which form a continuum. In eq. (3.11) we denote the initial momenta by s, t and final momenta by k, q. As we shall see, the absence of a stable negative metric state in the Nθ sector implies also that there is no stable negative metric state in the N$\theta\theta'$ sector. In the N$\theta\theta'$ sector, it can be readily verified that, for a finite R, the eigenvalues E of the total Hamiltonian are roots of

$$\mathcal{H}_R(E) = 0\,, \tag{3.12}$$

where

$$\mathcal{H}_R(E) = E - m^o_{V'} + g'^2 \sum_q (4\pi R\,\omega'_q)^{-1} q^2 u'^2_q [h_R(E - \omega'_q)]^{-1}\,, \tag{3.13}$$

the function $h_R(E)$ is given by eq. (2.13),

$$q = (l\pi/R)\,, \tag{3.14}$$

and the sum extends over $l = 0, 1, 2, \cdots$.

The analysis of eigenstates and eigenvalues in the N$\theta\theta'$ sector is straightforward; it is only slightly more complicated than that given in sect. 2. We give only the final results. For a finite R, all real roots of eq. (3.12) are sandwiched in between the two-parameter sequence

$$\omega_k + \omega'_q \ , \tag{3.15}$$

where k and q given by eqs. (2.6) and (3.14) respectively. It is convenient to label the successive real roots of eq. (3.12) by $E_{k,q}$. The lowest real root $E_{0,0}$ is always less than $(\mu+\mu')$. If the difference $(E_{0,0} - \mu - \mu')$ is $O(R^{-1})$, then in the limit $R \to \infty$, V' is unstable; otherwise, it is stable. One can also show that for a finite R,

$$\frac{d}{dE} \mathcal{H}_R(E) > 0 \tag{3.16}$$

at *every* $E = E_{k,q}$. From this inequality, it follows than, independently of whether V' is stable or not, *all eigenstates of H with real eigenvalues are of positive square length.*

In the complex plane, in the neighborhood (i.e., within a radius $O(R^{-1})$) of every

$$m_V \pm \tfrac{1}{2} i\gamma + \omega'_p \ , \tag{3.17}$$

there is a complex eigenvalue $E_{\pm,p}$ of H. The corresponding eigenstates $|\pm, p\rangle$ are all of zero length, and they satisfy eq. (1.16).

In the limit $R \to \infty$, $\mathcal{H}_R(E)$ becomes

$$\mathcal{H}_\infty(E) = E - m^o_{V'} + g'^2 \int_0^\infty dq(4\pi^2 \omega'_q)^{-1} q^2 u'^2_q [h_\infty(E - \omega'_q)]^{-1} \ , \tag{3.18}$$

which has a cut along the real axis for $E \geqslant (\mu+\mu')$. If V' is stable, then $[\mathcal{H}_R(E)]^{-1}$ has a pole on the real axis outside this cut; otherwise, it does not. In addition, the complex roots (3.17) in the limit $R \to \infty$ form two cuts which are parallel to the real axis from $m_V + \mu' + \frac{1}{2}i\gamma$ to $\infty + \frac{1}{2}i\gamma$ and from $m_V + \mu' - \frac{1}{2}i\gamma$ to $\infty - \frac{1}{2}i\gamma$.

As we shall see, the transition amplitude for N$\theta\theta'$ scattering, eq. (3.11), at a total energy

$$Q = \omega_s + \omega'_t \ , \tag{3.19}$$

is connected to the value of $\mathcal{H}_\infty(E)$ at

$$E = Q + i\epsilon \ , \tag{3.20}$$

where $\epsilon = 0+$. From eqs. (2.17) and (3.18), one finds that the imaginary part of $\mathcal{H}_\infty(Q+i\epsilon)$ is positive, and it is given by

$$\mathrm{Im}\ \mathcal{H}_\infty(Q+i\epsilon) = (gg')^2 \int dq\, |h_\infty(Q - \omega'_q + i\epsilon)|^{-2} (16\pi^3 \omega'_q)^{-1} q^2 u'^2_q k u^2_k \ , \tag{3.21}$$

where k and q are related by

$$\omega_k + \omega'_q = Q \ , \tag{3.22}$$

and the integration extands from $q = 0$ to $k = 0$. Thus, if V' is unstable, it corresponds to a pole on the unphysical sheet.

The S-matrix in the N$\theta\theta'$ sector can be most simply evaluated by using the explicit form of the eigenstates with real eigenvalues. These eigenstates can be chosen to be either the outgoing states $|N\theta_s\theta'^{\rm out}_t\rangle$, or the incoming states

$|N\theta_s\theta_t'^{in}\rangle$; they are given by

$$|N\theta_s\theta_t'^{out}\rangle = A_t'^{\dagger}\,|N\theta_s^{out}\rangle$$

$$+ K_{st}^{out}\{V'^{\dagger} - g'\int_0^\infty dq\,[2\pi\,\omega_q'^{\frac{1}{2}}\,h_\infty(Q-\omega_q'+i\epsilon)]^{-1}\,qu_q'\,V^{\dagger}A_q'^{\dagger}$$

$$- gg'\int_0^\infty dk\int_0^\infty dq\,[4\pi^2(\omega_k\,\omega_q')^{\frac{1}{2}}(\omega_k+\omega_q'-Q-i\epsilon)h_\infty(Q-\omega_q'+i\epsilon)]^{-1}$$

$$\times\, kq\,u_k\,u_q'\,N^{\dagger}\,A_k^{\dagger}A_q'^{\dagger}\}\,|\,\text{vac}\rangle\ , \tag{3.23}$$

where the eigenvalue Q is $\omega_s + \omega_t'$, the state $|N\theta_s^{out}\rangle$ is given by eq. (2.34),

$$K_{st}^{out} = [4\pi^2\,h_\infty(\omega_s+i\epsilon)]^{-1}[\,\mathcal{H}_\infty(Q+i\epsilon)]^{-1}(\omega_s\,\omega_t')^{-\frac{1}{2}}\,gg'\,s\,t\,u_s u_t'\ , \tag{3.24}$$

and, just as in eq. (2.32), we have made the replacement of

$$a_k \text{ by } (\pi/R)^{\frac{1}{2}}A_k\ ,$$

$$a_k' \text{ by } (\pi/R)^{\frac{1}{2}}A_k'\ . \tag{3.25}$$

Both A_k and A_k' separately satisfy the appropriate commutation relation (2.33). The incoming eigenstate $|N\theta_s\theta_t'^{in}\rangle$ is given by exactly the same expression as above, provided we replace in eqs. (3.23) and (3.24),

$$\text{all superscripts out by in, and all } i\epsilon \text{ by } -i\epsilon\ . \tag{3.26}$$

The matrix element of the S-matrix for the $N\theta\theta'$ scattering process (3.11) is given by

$$\langle k,q\,|S\,|\,s,t\rangle = \langle N\theta_k\,\theta_q'^{in}\,|\,\eta\,|\,N\theta_s\,\theta_t'^{out}\rangle\ . \tag{3.27}$$

It is convenient to factor out a δ-function due to the energy conservation. We define a matrix U_{ks} by

$$\langle k,q\,|S\,|\,s,t\rangle = \delta(\omega_k+\omega_q'-\omega_s-\omega_t')U_{ks}\ . \tag{3.28}$$

The unitarity of the S-matrix,

$$S^{\dagger}S = SS^{\dagger} = I\ , \tag{3.29}$$

is satisfied, if the matrix U_{ks} satisfies

$$\int_0^{k_{max}} U_{ks'}^{*}\,U_{ks}\,\rho(k)dk = [\rho(s)]^{-1}\,\delta(s-s')\ , \tag{3.30}$$

and

$$\int_0^{k_{max}} U_{s'k}^{*}\,U_{sk}\,\rho(k)dk = [\rho(s)]^{-1}\,\delta(s-s')\ , \tag{3.31}$$

where

$$\rho(k) = (\omega_q'/q) = [(Q-\omega_k)^2 - \mu'^2]^{-\frac{1}{2}}(Q-\omega_k) \ , \tag{3.32}$$

Q is given by eq. (3.19) and k_{max} corresponds to an energy ω_{max} of the θ-meson, determined by

$$Q = \omega_{max} + \mu' \ , \tag{3.33}$$

(i.e., θ' is at rest). The presence of $\rho(k)$ in these expressions is due to the fact that, in the N$\theta\theta'$ sector, the unit matrix I is given by

$$\langle k,q|I|s,t\rangle = \delta(k-s)\delta(q-t) = \delta(\omega_k + \omega_q' - \omega_s - \omega_t')\delta(k-s)[\rho(k)]^{-1} \ . \tag{3.34}$$

From the explicit forms of $|\mathrm{N}\theta_s\theta_t'^{\mathrm{out}}\rangle$ and $|\mathrm{N}\theta_s\theta_t'^{\mathrm{in}}\rangle$, and by using eq. (2.45), one finds that

$$U_{ks} = [\rho(k)]^{-1}\,\delta(k-s)\exp(2i\,\delta_k) - i[8\pi^3\,\mathcal{H}_\infty(Q+i\epsilon)]^{-1}$$

$$\times [h_\infty(\omega_k+i\epsilon)]^{-1}[h_\infty(\omega_s+i\epsilon)]^{-1}(\omega_k\,\omega_q'\,\omega_s\,\omega_t')^{-\frac{1}{2}}(gg')^2\,kqst\,u_k u_q' u_s u_t' \tag{3.35}$$

where δ_k is the phase shift of Nθ scattering given by eq. (2.16). It can then be verified that both eqs. (3.30) and (3.31) are valid, which *establishes* the *unitarity*. Furthermore, the outgoing and incoming states can be explicitly shown to be related through U_{ks} by

$$|\mathrm{N}\theta_s\theta_t'^{\mathrm{out}}\rangle = \int_0^{k_{max}} dk\,\rho(k)\,U_{ks}\,|\mathrm{N}\theta_k\theta_q'^{\mathrm{in}}\rangle \ , \tag{3.36}$$

where k and q are related by eq. (3.22), and k_{max} is given by eq. (3.33). By using eq. (3.28), this relation can also be written as

$$|\mathrm{N}\theta_s\theta_t'^{\mathrm{out}}\rangle = \int_0^\infty dk \int_0^\infty dq\,\langle k,q|S|s,t\rangle\,|\mathrm{N}\theta_k\theta_q'^{\mathrm{in}}\rangle \ . \tag{3.37}$$

4. CAUSALITY

In this section, we will study the question of causality * for a system with negative metric, but with stable particles all of positive metric. Although the S-matrix of such a system is unitary, the negative metric manifest itself in having complex poles (or cuts) above the real axis on the physical sheet, say at

$$m + \tfrac{1}{2}i\gamma \ , \tag{4.1}$$

for the analytic continuation of the relevant scattering amplitude. Such a pole could be sufficiently near the real axis to have observable effects. For example, at a physical (i.e., real) energy E near such a pole, the corre-

* This section is based mainly on results obtained by one of us (T.D.L.) and Zumino (unpublished).

sponding eigenvalue of the S-matrix could be dominated by the pole contribution, and it would be approximately given by

$$e^{2i\delta} \approx \frac{E - m + \frac{1}{2}i\gamma}{E - m - \frac{1}{2}i\gamma}. \tag{4.2}$$

At E near m, the phase shift δ decreases rapidly. The derivative $(\mathrm{d}\delta/\mathrm{d}E)$ is given by

$$\frac{\mathrm{d}\delta}{\mathrm{d}E} = -\tfrac{1}{2}[(E - m)^2 + \tfrac{1}{4}\gamma^2]^{-1}\gamma , \tag{4.3}$$

At $E = m$, $(\mathrm{d}\delta/\mathrm{d}E)$ becomes $-(2/\gamma)$ which can have a very large magnitude if γ is sufficiently small. It has been shown by Wigner [14] and others [15] that in the usual resonance scattering the derivative $(\mathrm{d}\delta/\mathrm{d}E)$ is positive and it corresponds to a time-delay in the motion of the wave packet due to the resonance. In the present case, this would be a time-advancement. Nevertheless, as we shall show, independently of how small γ may be, *no violation of causality can be concluded by studying the time dependence of the average position of the wave packet.*

For clarity, we omit possible spin and angular dependences, and assume the relevant channel is simply a two-body elastic S-wave scattering. In the center-of-mass system, the wave function at a large distance r can be written as

$$\psi_k = r^{-1}(e^{-ikr} + e^{2i\delta}\, e^{ikr})e^{-iEt} . \tag{4.4}$$

A wave packet $\psi(r,t)$ can be formed by coherently superposing ψ_k:

$$\psi(r,t) = \int_0^\infty \mathrm{d}k\, \psi_k(r,t) C_k , \tag{4.5}$$

where $|C_k|^2$ has a peak at $k = k_0$ and a width Δ; i.e.,

$$|C_k|^2 \ll 1 \text{ if } |k - k_0| \gg \Delta . \tag{4.6}$$

Throughout our discussion, we are only interested in the space-time region limited to

$$r \geqslant R , \tag{4.7}$$

and either

$$t < (-R/v) ,$$

or

$$t > (R/v) , \tag{4.8}$$

where v is the velocity, defined by

$$v = \left\langle \frac{\mathrm{d}E}{\mathrm{d}k} \right\rangle, \tag{4.9}$$

averaged over the wave packet, R is large enough so that the asymptotic expansion (4.4) holds, and furthermore

$$R \gg \gamma^{-1} . \tag{4.10}$$

For the specific applications we are interested in, γ^{-1} is $\lesssim 4 \times 10^{-13}$ cm for electrodynamics and $\lesssim 2 \times 10^{-11}$ cm for the weak interaction. (See eqs. (4.36) and (4.37) below.) It is convenient to decompose $\psi(r,t)$ into two parts:

$$\psi(r,t) = r^{-1} \exp(-ik_0 r - iE_0 t)\phi^{\text{in}} + r^{-1} \exp(ik_0 r - iE_0 t)\phi^{\text{out}} , \tag{4.11}$$

Where E_0 is the value of E at $k = k_0$,

$$\phi^{\text{in}}(r,t) = \int_0^\infty \exp[-i(k-k_0)r - i(E-E_0)t] C_k \, dk , \tag{4.12}$$

and

$$\phi^{\text{out}}(r,t) = \int_0^\infty \exp[2i\delta + i(k-k_0)r - i(E-E_0)t] C_k \, dk . \tag{4.13}$$

To simplify our discussion, we assume that the wave packet is well localized in r within a spatial width

$$O(\Delta^{-1}) \ll R .$$

The average radial position of the wave packet is given by

$$\int r \, |\psi|^2 \, d^3r = \langle r \rangle , \tag{4.14}$$

where ψ is assumed to be normalized; i.e.,

$$\int |\psi|^2 \, d^3r = 1 . \tag{4.15}$$

At $t < (-2R/v)$, the wave packet is well within the region $r \geqslant R$. (The factor 2 is introduced to avoid unnecessary complications at $r = R + O(\Delta^{-1})$ and $t = -R/v$.) Thus, $|\psi|^2$ is almost completely determined by $|r^{-1}\phi^{\text{in}}|^2$, and we can approximate

$$\langle r \rangle = 4\pi \int_0^\infty r \, |\phi^{\text{in}}(r,t)|^2 \, dr . \tag{4.16}$$

In this integration, we can use eq. (4.12) and make the further approximation by formally extending the integration to from $r = -\infty$ to $r = +\infty$. As we shall see, in the explicit example given below, these approximations amount to neglecting $\exp(-R\Delta)$ and $\exp(-R\gamma)$ as compared to unity. By partial integrations, one finds that

$$\langle r \rangle = v(t - t_0) \quad \text{for} \quad t < (-2R/v) , \tag{4.17}$$

where t_0 is given by

$$v t_0 = \frac{\int_0^\infty |C_k|^2 \frac{d\beta}{dk} \, dk}{\int_0^\infty |C_k|^2 \, dk} , \tag{4.18}$$

β = phase angle of C_k. Clearly, if one wishes, t_0 can be chosen to be zero by adjusting the origin of t.

Similarly, by using ϕ^{out}, one finds [14, 15] that

$$\langle r\rangle = v(t - t_0) + l \quad \text{for} \quad t > (2R/v)\ , \tag{4.19}$$

where

$$l = \frac{-2 \int_0^\infty |C_k|^2 \frac{d\delta}{dk}\, dk}{\int_0^\infty |C_k|^2\, dk}\ . \tag{4.20}$$

By comparing eqs. (4.17) and (4.19), one may interpret (l/v) as a time delay or a time advancement, depending on its sign.

For the present case, on account of eq. (4.3),

$$l = \frac{\int \gamma[(E-m)^2 + \frac{1}{4}\gamma^2]^{-1} \frac{dE}{dk} |C_k|^2\, dk}{\int |C_k|^2\, dk}\ . \tag{4.21}$$

It is easy to see that, independently of γ, there is an upper limit l_{max} for l:

$$l < l_{max} = O(\Delta^{-1})\ . \tag{4.22}$$

This follows from the simple fact that either (i) the wave packet width Δ in momentum space is $\ll \gamma$, then $|d\delta/dE| \leq (2/\gamma) \ll \Delta^{-1}$, or (ii) $\Delta = O(\gamma)$, then $|d\delta/dE| \leq (2/\gamma) = O(\Delta^{-1})$, or (iii) $\Delta \gg \gamma$, then since in the momentum space, the fraction of the wave packet within the level width γ is only $O(\gamma/\Delta)$ and for that fraction $(d\delta/dE)$ is $O(\gamma^{-1})$, the net pole contribution to l is the product

$$O(\gamma^{-1}) \times O(\gamma/\Delta) = O(\Delta^{-1})\ .$$

In addition, there is also a background contribution to l which is not included in eq. (4.21). In all these cases, as $O(\Delta^{-1})$ is the uncertainty of the wave packet position in the coordinate space, *one cannot*, by measuring only $\langle r\rangle$, *conclude whether there is any violation of causality or not*. Of course, as we shall see, what one should have done is not to measure the average position $\langle r\rangle$, but rather to seek for other more sensitive observables.

To measure the possible existence of complex poles such as eq. (4.1), the simplest method is to measure the *sign* of δ. For example, as will be further discussed in sect. 7, the spin =1, parity = -1 scattering state of

$$e^+e^- \to e^+e^-\ , \tag{4.23}$$

would exhibit a phase shift like eq. (4.2) at a center-of-mass energy E near m, if the photon propagator (to zeroth order in e^2) is given by

$$\delta_{\mu\nu}[(q^2)^{-1} - (q^2 + m^2)^{-1}]\ , \tag{4.24}$$

instead of the usual $\delta_{\mu\nu}(q^2)^{-1}$. A similar pole contribution would also occur in, say, the S-wave scattering of

$$\pi^{\pm}\pi^0 \to \pi^{\pm}\pi^0 , \tag{4.25}$$

if the charged intermediate boson $W^{\pm}$ propagator (to zeroth order in the weak interaction Fermi constant G) is given by

$$\delta_{\mu\nu}(q^2 + m_W^2)^{-1} + q_\mu q_\nu m_W^{-2}[(q^2 + m_W^2)^{-1} - (q^2 + m^2)^{-1}] , \tag{4.26}$$

instead of the usual propagator (1.7). (If $m = m_W$, then eq. (4.26) reduces to eq. (1.6).)

The possible existence of such a complex pole (4.1) can also be established, at least in principle, if one studies the shape of the wave packet. To illustrate such possibilities, let us consider again the simple example of a two-body S-wave elastic scattering, but keeping in mind that it also serves as a model for the realistic cases such as eqs. (4.23) or (4.25). In the realistic case, m is expected to be quite large, say, $\approx$ 10 GeV or larger. Thus, for example, in eq. (4.23) one may safely neglect the mass of $e^{\pm}$, and set

$$E = k , \tag{4.27}$$

where k denotes the magnitude of the momentum difference $\boldsymbol{p}(e^+) - \boldsymbol{p}(e^-)$ in the center-of-mass system, and its conjugate coordinate r is simply the magnitude of $\frac{1}{2}[\boldsymbol{r}(e^+) - \boldsymbol{r}(e^-)]$.

For simplicity, we assume

$$C_k = \tfrac{1}{2}(\Delta/\pi)^{\frac{3}{2}}[(k - m)^2 + \Delta^2]^{-1} ; \tag{4.28}$$

in addition, $e^{2i\delta}$ is assumed to be given by eq. (4.2). Throughout the following, we regard $(\Delta/m)^3$, $\exp(-\Delta R)$ and $\exp(-\gamma R)$ as extremely small; these parameters will be neglected when compared to unity. The smallness of $(\Delta/m)^3$ allows us to extend the k-integrals in eqs. (4.12) and (4.13) from $k = -\infty$ to $k = +\infty$. By using eqs. (4.2), (4.27) and (4.28) these integrals can be explicitly evaluated. One finds, for $r > R$,

$$\phi^{\text{in}}(r, t) = \tfrac{1}{2}(\Delta/\pi)^{\frac{1}{2}}\, e^{-\Delta|r + t|} , \tag{4.29}$$

and

$$\phi^{\text{out}}(r, t) = \tfrac{1}{2}(\Delta/\pi)^{\frac{1}{2}} \left(\frac{2\Delta - \gamma}{2\Delta + \gamma}\right) e^{-\Delta(t - r)} \quad \text{if } t > r , \tag{4.30}$$

but

$$\phi^{\text{out}}(r, t) = \tfrac{1}{2}(\Delta/\pi)^{\frac{1}{2}} \left[\left(\frac{2\Delta + \gamma}{2\Delta - \gamma}\right) e^{-\Delta(r - t)} - \frac{8\gamma\Delta}{4\Delta^2 - \gamma^2}\, e^{-\frac{1}{2}\gamma(r - t)}\right] \quad \text{if } t < r . \tag{4.31}$$

From these explicit expressions, one sees that, upon neglecting $\exp(-\Delta R)$ and $\exp(-\gamma R)$, $|\psi(r, t)|^2$ at $r > R$ is given by $[r^{-1}\phi^{\text{in}}(r, t)]^2$ if $t < -R$, and by $[r^{-1}\phi^{\text{out}}(r, t)]^2$ if $t > R$. Furthermore,

$$\int |\psi|^2 \, d^3r = 4\pi \int [\phi^{in}(r,t)]^2 \, dr = 1 \quad \text{for} \quad t < -R ,$$

$$= 4\pi \int [\phi^{out}(r,t)]^2 \, dr = 1 \quad \text{for} \quad t > R \quad , \qquad (4.32)$$

which is simply the unitarity condition. (In the limit $R \to \infty$, eq. (4.32) is exact.) The average position $\langle r \rangle$ is given explicitly by

$$\langle r \rangle = t \qquad \text{for} \quad t < -R ,$$

$$\langle r \rangle = t + \frac{4(4\Delta+\gamma)}{(2\Delta+\gamma)^2} \qquad \text{for} \quad t > R \qquad (4.33)$$

where the difference $|\langle r \rangle - t|$ is always $\leqslant 4\Delta^{-1}$. The tail $\exp[-\frac{1}{2}\gamma(r-t)]$ in eq. (4.31) is due to the complex pole (4.1). It could become the dominant term giving a clear 'time-advancement' effect, if one has a wave packet whose initial uncertainty Δ^{-1} in the coordinate space is much smaller than γ^{-1}, and if one can measure parameters that are sensitive to the tail $\exp[-\frac{1}{2}\gamma(r-t)]$.

For the specific applications we are interested in, the detection of such a tail is nearly impossible at present. We recall that the partial level width to the lepton modes e^+e^- and $\mu^+\mu^-$ for the complex heavy photon pole, called Γ^o, in reaction (4.23) is

$$\gamma_{\text{lepton modes}} = \tfrac{2}{3}\alpha m , \qquad (4.34)$$

where α is the fine-structure constant. For the corresponding complex pole due to the negative metric spin-0 part of the $W_\mu^\pm$ in reaction (4.25), its partial width to the lepton modes is zero if one neglects the lepton mass. Thus, for reaction (4.25) the corresponding width γ is determined essentially by the hadron modes; one may estimate

$$\gamma \approx (4\pi)^{-1} G m^3 , \qquad (4.35)$$

where G is the weak-interaction coupling constant. If m is ≈ 10 GeV, then eqs. (4.34) and (4.35) lead to, respectively,

$$\gamma > \gamma_{\text{lepton modes}} \approx 50 \text{ MeV}, \text{ or } \gamma^{-1} \lessapprox 4 \times 10^{-13} \text{ cm} \qquad (4.36)$$

for the heavy photon Γ^o-pole, and

$$\gamma \approx 1 \text{ MeV}, \text{ or } \gamma^{-1} \approx 2 \times 10^{-11} \text{ cm} \qquad (4.37)$$

for the spin-0 $W^\pm$-pole. From the known experimental results, one knows that m must be bigger than a few GeV. Therefore, it cannot be much smaller than 10 GeV, but it could also be much bigger than 10 GeV, and correspondingly, γ^{-1} may turn out to be much smaller than eqs. (4.36) and (4.37). At present, we do not know of any method that can produce a wave packet whose initial uncertainty Δ^{-1} in the coordinate space is of such a small magnitude.

5. INTERACTION REPRESENTATION

The total Hamiltonian H can be formally decomposed into two parts:

$$H = H_0 + H_1 \ . \tag{5.1}$$

In the interaction representation, H_0 is independent of time, and H_1 has the usual time dependence

$$H_1(t) = \exp(iH_0 t)\, H_1(0) \exp(-iH_0 t) \ . \tag{5.2}$$

To simplify our discussion, we shall assume that the metric η commutes with H_0,

$$[H_0, \eta] = 0 \ . \tag{5.3}$$

Furthermore, $H_1(t)$ is pseudo-Hermitian; i.e.,

$$\eta H_1(t) \eta = H_1(t)^\dagger \ , \tag{5.4}$$

which together with eq. (1.1) implies that H_0 is Hermitian.

The state vector $|t\rangle$ at time t is given by

$$|t\rangle = U(t, 0)\,|0\rangle \ , \tag{5.5}$$

and

$$U(t, 0) = \exp(iH_0 t) \exp[-iH(0)\,t] \ , \tag{5.6}$$

where $H(0)$ refers to the total Hamiltonian $H(t)$ at $t = 0$. The inverse of $U(t, 0)$ is

$$U(0, t) = \exp[iH(0)\,t] \exp(-iH_0 t) \ . \tag{5.7}$$

In the usual case of $\eta = +1$, although the mathematical limits $U(\infty, 0)$ and $U(0, -\infty)$ do not exist, one can introduce the adiabatic process and define *

$$U_\epsilon(t, 0) = \epsilon \int_0^t e^{-\epsilon\tau}\, e^{iH_0\tau}\, e^{-iH(0)\tau}\, d\tau \qquad \text{for} \qquad t > 0 \ , \tag{5.8}$$

and

$$U_\epsilon(0, t) = \epsilon \int_t^0 e^{\epsilon\tau}\, e^{iH(0)\tau}\, e^{-iH_0\tau}\, d\tau \qquad \text{for} \qquad t < 0 \ , \tag{5.9}$$

where $\epsilon = 0+$. The limits $U_\epsilon(\infty, 0)$ and $U\ (0, -\infty)$ do exist, provided that the spectra of H_0 and H are the same. These operators $U_\epsilon(\infty, 0)$ and $U_\epsilon(0, -\infty)$ connect the eigenstates of $H(0)$ with those of H_0, and their product gives an alternative form of the S-matrix in the interaction representation.

In the present case, η is not the unit matrix. Because of the complex eigenvalues $E_{\pm,c}$ of the total Hamiltonian $H(0)$, the limits $U_\epsilon(\infty, 0)$ and $U_\epsilon(0, \infty)$ *do not exist*. The usual interaction representation formalism of the S-matrix *is not valid*. Nevertheless, as we shall see, for purely computa-

* See, for example, Goldberger and Watson [16].

tional purposes, it is still useful to define $U_\epsilon(t,0)$ and $U_\epsilon(0,t)$; these operators, under some different rules, can lead to the correct S-matrix, which is defined by eq. (1.20).

Let us start with the general case discussed in sect. 1. We assume that to each continuum eigenstate $|r^{out}\rangle$ of the total Hamiltonian $H(0)$ there is a corresponding eigenstate $|r_o\rangle$ of H_o with the *same* real eigenvalue E_r; i.e., H_o contains the physical masses for all stable particles. (Such a correspondence is not assumed for other eigenstates of H and H_o.) Furthermore, the incoming part of $|r^{out}\rangle$ is the same as $|r_o\rangle$ in the asymptotic region in the coordinate space. By our previous assumption that all stable particles are of positive metric, similarly to eq. (1.17), these eigenstates of H_o satisfy

$$\langle r'_o|\eta|r_o\rangle = I_{r'r} , \tag{5.10}$$

where I is the unit matrix. (In the examples discussed in sects. 2 and 3, $|r_o\rangle$ corresponds to states such as $N^\dagger A_p^\dagger|vac\rangle$, $N^\dagger A_p^\dagger A_q'^\dagger|vac\rangle$, etc.) From these definitions, it follows then

$$|r^{out}\rangle = [1 - (H - E_r - i\epsilon)^{-1} H_1]|r_o\rangle , \tag{5.11}$$

and, similarly,

$$|r^{in}\rangle = [1 - (H - E_r + i\epsilon)^{-1} H_1]|r_o\rangle , \tag{5.12}$$

where H refers to $H(0)$ and $\epsilon = 0+$.

To see that the same relations can also be obtained through the use of $U_\epsilon(t,0)$ and $U_\epsilon(0,t)$, it is convenient to introduce the formal notion of 'equivalence'. We say two state vectors $|a\rangle$ and $|b\rangle$ are equivalent, if their difference

$$|\lambda\rangle = |a\rangle - |b\rangle \tag{5.13}$$

is a vector of zero length and if it is orthogonal to $|a\rangle$ (therefore also to be $|b\rangle$); i.e.,

$$\langle\lambda|\eta|\lambda\rangle = \langle\lambda|\eta|a\rangle = \langle\lambda|\eta|b\rangle = 0 . \tag{5.14}$$

By using the definition of $U_\epsilon(0,t)$, given by eq. (5.9), and the completeness theorem (1.18), one finds for $t < 0$

$$U_\epsilon(0,t)|r_o\rangle = |r^{out}\rangle + |\lambda^{out}\rangle , \tag{5.15}$$

where

$$|\lambda^{out}\rangle = \sum_{r'} C^r_{r'}(t)|r'^{out}\rangle + \sum_c C^r_{+,c}(t)|+,c\rangle + \sum_c C^r_{-,c}(t)|-,c\rangle , \tag{5.16}$$

$$C^r_{r'}(t) = i\epsilon(E_{r'} - E_r - i\epsilon)^{-1}\exp[i(E_{r'} - E_r)t]\exp(-\epsilon|t|)\quad\langle r'^{out}|\eta|r_o\rangle , \tag{5.17}$$

$$C^r_{+,c}(t) = i\epsilon(m_c - E_r + \tfrac{1}{2}i\gamma_c - i\epsilon)^{-1}\exp[i(m_c - E_r)t] \times\exp[(\tfrac{1}{2}\gamma_c - \epsilon)|t|]\langle -,c|\eta|r_o\rangle , \tag{5.18}$$

$$C^{\mathrm{r}}_{-,\mathrm{c}}(t) = i\epsilon(m_{\mathrm{c}} - E_{\mathrm{r}} - \tfrac{1}{2}i\gamma_{\mathrm{c}} - i\epsilon)^{-1} \exp[i(m_{\mathrm{c}} - E_{\mathrm{r}})t] \times \exp[(-\tfrac{1}{2}\gamma_{\mathrm{c}} - \epsilon)|t|]\langle +,\mathrm{c}|\eta|\mathrm{r_o}\rangle , \tag{5.19}$$

and

$$E_{\pm,\mathrm{c}} = m_{\mathrm{c}} \pm \tfrac{1}{2}i\gamma_{\mathrm{c}} . \tag{5.20}$$

From the orthonormality relations (1.11), (1.12), (1.16) and (1.17) it follows that

$$\langle\lambda^{\mathrm{out}}|\eta|\lambda^{\mathrm{out}}\rangle = \sum_{\mathrm{r'}} |C^{\mathrm{r}}_{\mathrm{r'}}(t)|^2 + \sum_{\mathrm{c}} [C^{\mathrm{r}*}_{+,\mathrm{c}}(t)C^{\mathrm{r}}_{-,\mathrm{c}}(t) + C^{\mathrm{r}*}_{-,\mathrm{c}}(t)C^{\mathrm{r}}_{+,\mathrm{c}}(t)] , \tag{5.21}$$

and

$$|\langle\lambda^{\mathrm{out}}|\eta|\mathrm{r}^{\mathrm{out}}\rangle|^2 = |C^{\mathrm{r}}_{\mathrm{r}}(t)|^2 . \tag{5.22}$$

Since every term on the right-hand sides of eqs. (5.21) and (5.22) has an $\exp(-2\epsilon|t|)$ dependence, in the limit $t = -\infty$

$$\lim_{t=-\infty} \langle\lambda^{\mathrm{out}}|\eta|\lambda^{\mathrm{out}}\rangle = 0 , \tag{5.23}$$

and

$$\lim_{t=-\infty} \langle\lambda^{\mathrm{out}}|\eta|\mathrm{r}^{\mathrm{out}}\rangle = 0 . \tag{5.24}$$

Thus, although $U_\epsilon(0,-\infty)$ does not exist, as $t \to -\infty$ the vector $U_\epsilon(0,t)\,|\mathrm{r_o}\rangle$ becomes 'equivalent' to $|\mathrm{r}^{\mathrm{out}}\rangle$. Furthermore, from eq. (5.16) one finds that

$$\lim_{t=-\infty} [U_\epsilon(0,t)|\mathrm{r_o}\rangle - \sum_{\mathrm{c}} C^{\mathrm{r}}_{+,\mathrm{c}}(t)|+,\mathrm{c}\rangle] = |\mathrm{r}^{\mathrm{out}}\rangle . \tag{5.25}$$

From a computational point of view eq. (5.25) is quite useful. Since $C^{\mathrm{r}}_{+,\mathrm{c}}(t)$ increases exponentially in time as $\exp(\frac{1}{2}\gamma_{\mathrm{c}}|t|)$, it can be easily separated out from $U\ (0,t)|\mathrm{r_o}\rangle$, and the remainder, in the limit $t = -\infty$, gives $|\mathrm{r}^{\mathrm{out}}\rangle$.

Similarly, one can write for $t > 0$

$$\langle\mathrm{r_o}|\eta U_\epsilon(t,0)\eta = \langle\mathrm{r}^{\mathrm{in}}| + \langle\lambda^{\mathrm{in}}| , \tag{5.26}$$

where

$$\langle\lambda^{\mathrm{in}}| = \sum_{\mathrm{r'}} D^{\mathrm{r}}_{\mathrm{r'}}(t)\langle\mathrm{r'}^{\mathrm{in}}| + \sum_{\mathrm{c}} D^{\mathrm{r}}_{+,\mathrm{c}}(t)\langle +,\mathrm{c}| + \sum_{\mathrm{c}} D^{\mathrm{r}}_{-,\mathrm{c}}(t)\langle -,\mathrm{c}| , \tag{5.27}$$

$$D^{\mathrm{r}}_{\mathrm{r'}}(t) = i\epsilon(E_{\mathrm{r'}} - E_{\mathrm{r}} - i\epsilon)^{-1} \exp[-i(E_{\mathrm{r'}} - E_{\mathrm{r}})t] \exp(-\epsilon t)\langle\mathrm{r_o}|\eta|\mathrm{r'}^{\mathrm{in}}\rangle , \tag{5.28}$$

$$D^{\mathrm{r}}_{+,\mathrm{c}}(t) = i\epsilon(m_{\mathrm{c}} - E_{\mathrm{r}} - \tfrac{1}{2}i\gamma_{\mathrm{c}} - i\epsilon)^{-1} \exp[-i(m_{\mathrm{c}} - E_{\mathrm{r}})t] \times \exp[-(\tfrac{1}{2}\gamma_{\mathrm{c}} + \epsilon)t]\langle\mathrm{r_o}|\eta|-,\mathrm{c}\rangle , \tag{5.29}$$

$$D^{\mathrm{r}}_{-,\mathrm{c}}(t) = i\epsilon(m_{\mathrm{c}} - E_{\mathrm{r}} + \tfrac{1}{2}i\gamma_{\mathrm{c}} - i\epsilon)^{-1}\exp[-i(m_{\mathrm{c}} - E_{\mathrm{r}})t]$$

$$\times \exp[(\tfrac{1}{2}\gamma_{\mathrm{c}} - \epsilon)t]\langle \mathrm{r_o}|\eta|+,\mathrm{c}\rangle \,. \qquad (5.30)$$

Similarly to eqs. (5.23) - (5.25), one finds

$$\lim_{t=\infty} \langle\lambda^{\mathrm{in}}|\eta|\lambda^{\mathrm{in}}\rangle = 0 \,, \qquad (5.31)$$

$$\lim_{t=\infty} \langle\lambda^{\mathrm{in}}|\eta|\mathrm{r}^{\mathrm{in}}\rangle = 0 \,, \qquad (5.32)$$

and

$$\lim_{t=\infty} [\langle \mathrm{r_o}|\eta\, U_\epsilon(t,0)\eta - \sum_{\mathrm{c}} D^{\mathrm{r}}_{-,\mathrm{c}}\langle -,\mathrm{c}|] = \langle \mathrm{r}^{\mathrm{in}}| \,. \qquad (5.33)$$

Thus, in the limit $t = \infty$, the state

$$\langle \mathrm{r_o}|\eta\, U_\epsilon(t,0)\eta \,, \qquad (5.34)$$

becomes 'equivalent' to $\langle \mathrm{r}^{\mathrm{in}}|$; furthermore, if we remove from eqs. (5.34) all its exponentially increasing terms in t, then the remainder gives $\langle \mathrm{r}^{\mathrm{in}}|$.

By using these expressions together with the definition of the S-matrix given by eq. (1.20), one derives

$$S_{\mathrm{r'r}} = \lim_{t=\infty} [\langle \mathrm{r'_o}|\eta\, U_\epsilon(t,0)\, U_\epsilon(0,t)\,|\mathrm{r_o}\rangle - \sum_{\mathrm{c}} D^{\mathrm{r'}}_{-,\mathrm{c}}(t) C^{\mathrm{r}}_{+,\mathrm{c}}(-t)] \,, \qquad (5.35)$$

which gives the formalism of the S-matrix in the interaction representation. We emphasize that $D^{\mathrm{r'}}_{-,\mathrm{c}}(t) C^{\mathrm{r}}_{+,\mathrm{c}}(-t)$ increases exponentially in time as $\exp(\gamma_{\mathrm{c}} t)$; its identification is unambiguous. After separating out all these exponentially increasing terms in t from

$$\langle \mathrm{r'_o}|\,\eta\, U_\epsilon(t,0)\, U_\epsilon(0,-t)\,|\mathrm{r_o}\rangle \,, \qquad (5.36)$$

the remainder, as $t \to \infty$, becomes the correct S-matrix.

In the present theory, the interest in this alternative expression of the S-matrix in the interaction representation is purely for computational purposes. The basic formulation of the S-matrix should always be eq. (1.20), which is derived from the motion of wave packets that can be constructed from the set of physical states (1.21). The state $|\mathrm{r_o}\rangle$ is *not* a physical state, and the notion of 'equivalence' does not seem to be a fundamental notion. That $U_\epsilon(0,t)|\mathrm{r_o}\rangle$ becomes equivalent to the physical state $|\mathrm{r}^{\mathrm{out}}\rangle$ as $t \to \infty$ is an interesting property, but it plays no important role in our discussion.

6. FEYNMAN DIAGRAM

It is useful to define

$$U_\epsilon(t,-t) \equiv U_\epsilon(t,0)\, U_\epsilon(0,-t) \,, \qquad (6.1)$$

where $t > 0$, and then decompose it into two parts, an exponentially increasing part in t, $U_\epsilon^{\exp}(t, -t)$, and a regular part $U_\epsilon^{\mathrm{reg}}(t, -t)$:

$$U_\epsilon(t, -t) = U_\epsilon^{\exp}(t, -t) + U_\epsilon^{\mathrm{reg}}(t, -t) \ . \tag{6.2}$$

Eq. (5.35) states that the S-matrix is given by

$$S_{\mathrm{r'r}} = \lim_{t=\infty} \langle \mathrm{r}_0' | \eta \, U_\epsilon^{\mathrm{reg}}(t, -t) | \mathrm{r}_0 \rangle \ . \tag{6.3}$$

One may formally expand $U_\epsilon(t, -t)$ as a power series of $H_1(t)$:

$$U_\epsilon(t, -t) = 1 + \sum_{n=1}^{\infty} (n!)^{-1} (-i)^n \int \prod \mathrm{d}t_i \, T[H_1(t_1) H_1(t_2) \dots H_1(t_n)] \ , \tag{6.4}$$

where T denotes the time-ordered product and the integration extends over each t_i varying separately from $-t$ to $+t$. For clarity, the ϵ-dependence is omitted on the right-hand side of eq. (6.4). The usual procedure [17] of converting time-ordered products to normal products can be directly applied; it leads to a set of Feynman-like diagrams expressed in terms of the appropriate space-time integrals. From these diagrams, in order to obtain the S-matrix, one has to subtract out the exponentially increasing part in t, and then take the limit $t = \infty$.

In the usual case (i.e., $\eta = 1$), it is most convenient to state Feynman rules in the momentum space. For the present case of indefinite metric, such rules must be used with great care. This is because the presence of exponentially increasing parts $U_\epsilon^{\exp}(t, -t)$ prevents one from directly transforming these space-time integrals in $U_\epsilon(t, -t)$ into the appropriate momentum-space integrals. We note that Fourier's theorem is not applicable to such exponentially increasing functions in t, but, as shown explicitly by eq. (6.3), neither are such functions relevant to the evaluation of the S-matrix.

To illustrate possible pitfalls, we shall return to the soluble cases, and give the following examples.

6.1. *How a naive, but incorrect, approach may sometimes lead to the correct answer*

We start with the simple Hamiltonian (2.2), and observe that in the interaction representation, on account of eq. (2.3),

$$V(t) = V(0) \exp(-i m_{\mathrm{V}}^{\mathrm{o}} t) \ ,$$

$$N(t) = N(0) \exp(-i m_{\mathrm{N}} t) \ ,$$

$$a_k(t) = a_k(0) \exp(-i \omega_k t) \ . \tag{6.5}$$

The free propagators $S_{\mathrm{V}}^{\mathrm{o}}(t)$, $S_{\mathrm{N}}(t)$ and $S_{\theta(k)}(t)$ of $V(t)$, $N(t)$ and $a_k(t)$ are, respectively, defined by

$$S_V^0(t) = T\langle \text{vac}\,|\eta\, V(t)V^\dagger(0)\,|\text{vac}\rangle\,,$$

$$S_N(t) = T\langle \text{vac}\,|\eta\, N(t)N^\dagger(0)\,|\text{vac}\rangle\,,$$

$$S_{\theta(k)}(t) = T\langle \text{vac}\,|\eta\, a_k(t)a_k^\dagger(0)\,|\text{vac}\rangle\,. \tag{6.6}$$

In the above notation, the superscript 0 denotes the free propagator; the N and θ propagators are without this superscript, since in this soluble model these free propagators are also the complete ones.

By using eq. (6.5), one finds that their Fourier transforms are

$$\tilde{S}_V^0(p_0) = i(p_0 - m_V^0 + i\epsilon)^{-1}\,,$$

$$\tilde{S}_N(p_0) = i(p_0 - m_N + i\epsilon)^{-1}\,,$$

$$\tilde{S}_{\theta(k)}(p_0) = i(p_0 - \omega_k + i\epsilon)^{-1}\,, \tag{6.7}$$

respectively, where $\epsilon = 0+$, and the Fourier transform $\tilde{F}(p_0)$ of any function $F(t)$ is defined by

$$F(t) = (2\pi)^{-1} \int_{-\infty}^{\infty} \tilde{F}(p_0)\mathrm{d}p_0 \exp(-i\, p_0 t)\,. \tag{6.8}$$

The vertex functions for $\mathrm{V} \rightleftharpoons \mathrm{N}\theta$ transitions are given by the relevant terms in $-iH_1$. From eq. (2.4), it follows that the vertex function is

$$i g(4\pi R\omega_k)^{-\frac{1}{2}}\, k u_k\,, \qquad \text{for} \qquad \mathrm{V} \to \mathrm{N} + \theta(k)\,,$$

$$-i g(4\pi R\omega_k)^{-\frac{1}{2}}\, k u_k\,, \qquad \text{for} \qquad \mathrm{N} + \theta(k) \to \mathrm{V}\,. \tag{6.9}$$

We will now 'blindly' apply the usual Feynman rules in momentum space to evaluate the formal power series (6.4); more specifically, we will (i) ignore possible exponentially increasing functions in t, and arbitrarily exchange the order of the power series sum with the relevant integrations in the Fourier integrals, and (ii) regard the resulting expression for eq. (6.4) as simply the S-matrix. Of course, both steps are incorrect. Yet, as we shall see, the final expression for Nθ scattering is, nevertheless, correct.

To carry out step (i), we first evaluate the V-propagator, defined by the sum of Feynman diagrams in fig. 2. By applying the usual Feynman rules in momentum space, the resulting V-propagator, labeled by a superscript F, is (as $R \to \infty$)

$$\tilde{S}_V^F(p_0) = i[p_0 - m_V^0 + i\epsilon + \Pi(p_0)]^{-1}\,, \tag{6.10}$$

where $\Pi(p_0)$ is given by the single Nθ loop diagram,

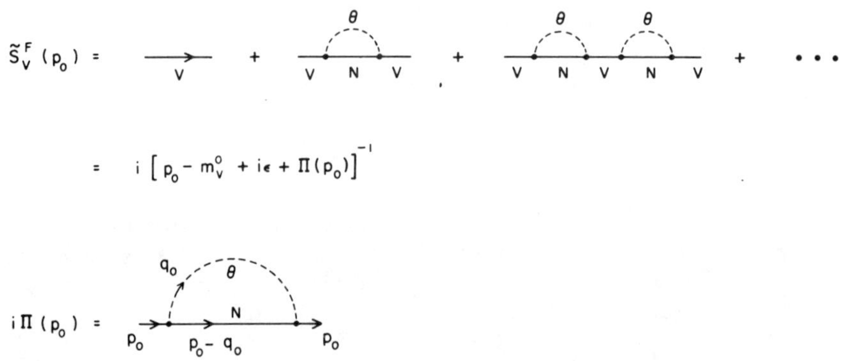

Fig. 2. Feynman diagrams for V propagator $\tilde{S}_V^F(p_0)$ in the momentum space.

$$\Pi(p_0) = i g^2 \int_0^\infty dk \int_{-\infty}^\infty dq_0 (p_0 - q_0 - m_N + i\epsilon)^{-1} (q_0 - \omega_k + i\epsilon)^{-1} (8\pi^3\, \omega_k)^{-1}\, k^2\, u_k^2$$

$$= g^2 \int_0^\infty dk (p_0 - \omega_k - m_N + i\epsilon)^{-1} (4\pi^2\, \omega_k)^{-1}\, k^2 u_k\,. \tag{6.11}$$

By following step (ii), one finds that the Nθ scattering phase shift at a total energy

$$p_0 = m_N + \omega_p\,,$$

is given by

$$e^{2i\delta_p} - 1 = g^2 (2\pi)^{-1}\, p u^2\, \tilde{S}_V^F(p_0)\,, \tag{6.12}$$

which is the same as eq. (2.16) since the Feynman propagator $\tilde{S}_V^F(p_0)$ in the momentum space is related to the function $h_\infty(E)$, defined by eq. (2.14), by

$$\tilde{S}_V^F(p_0) = i[h_\infty(p_0 + i\epsilon)]^{-1}\,, \tag{6.13}$$

after setting $m_N = 0$.

To understand why in the Nθ sector these two incorrect steps (i) and (ii) can lead to the correct answer, we evaluate the complete V-propagator $S_V(t)$ in the coordinate space. By definition, $S_V(t)$ is given by

$$S_V(t) = T\langle \text{vac}\,|\,\eta\, V(t)\, V^\dagger(0)\,|\,\text{vac}\rangle_{\text{Heisenberg}}\,, \tag{6.14}$$

where the subscript 'Heisenberg' denotes the Heisenberg representation. By using the explicit solutions of the eigenstates $|N\theta_p\rangle$ and $|\pm\rangle$ given by eqs. (2.23) and (2.24), one finds

$$S_V(t) = 0\,, \quad \text{if} \quad t < 0\,, \tag{6.15}$$

and

$$S_V(t)$$
$$= -\sum_p |C_p|^2 \exp(-iE_p t) - C_- C_+^* \exp(-iE_- t) - C_+ C_-^* \exp(-iE_+ t)\,, \quad \text{if } t > 0, \tag{6.16}$$

where $C_\pm$ and C_p are given by eq. (2.25). The integral representation of $S_V(t)$ is, as $R \to \infty$,

$$S_V(t) = i(2\pi)^{-1} \int_C [h_\infty(E)]^{-1} \exp(-iEt)\mathrm{d}E\,, \tag{6.17}$$

where the contour C starts from $-\infty + i\epsilon$, goes over the complex root E_+, and then ends at $\infty + i\epsilon$ (see fig. 3). One sees that, just as in eq. (6.2), $S_V(t)$ can be separated into two terms, an exponentially increasing function in t, $S_V^{\exp}(t)$, and a regular function S_V^{reg}:

$$S_V(t) = S_V^{\mathrm{reg}}(t) + S_V^{\exp}(t)\,, \tag{6.18}$$

where

$$S_V^{\mathrm{reg}}(t) = i(2\pi)^{-1} \int_{-\infty+i\epsilon}^{\infty+i\epsilon} [h_\infty(E)]^{-1} \exp(-iEt)\mathrm{d}E\,, \tag{6.19}$$

in which the integration is along a straight line slightly above the real axis, and

$$S_V^{\exp}(t) = i(2\pi)^{-1} \oint_{E_+} [h_\infty(E)]^{-1} \exp(-iEt)\mathrm{d}E\,, \tag{6.20}$$

in which the contour is along a small clockwise circle around the complex root $E = E_+$.

In the above usage of Feynman rules in the momentum space, step (i) does not lead to the entire $S_V(t)$. On the other hand, by comparing eqs. (6.13) and (6.19) one sees that $\tilde{S}_V^F(p_0)$ is precisely the Fourier transform of $S_V^{\mathrm{reg}}(t)$ which, according to eq. (6.3), should lead to the correct S-matrix.

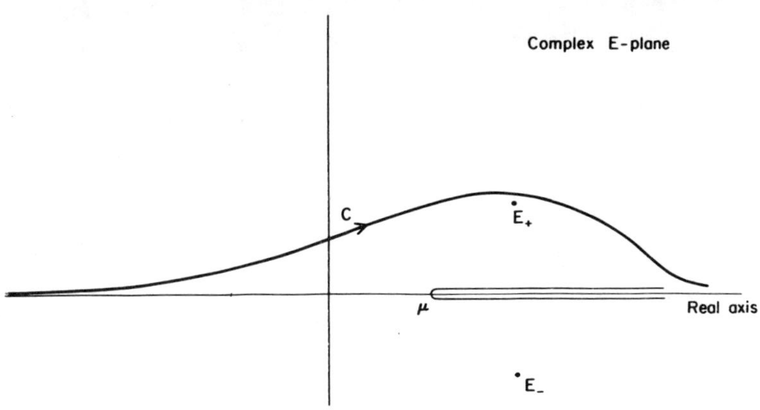

Fig. 3. Contour C for the integral representation of $S_V(t)$. (See eq. (6.17)).

6.2. *How incorrect procedure does lead to incorrect results*

Next, we will apply exactly the same above steps (i) and (ii) but to the $N\theta\theta'$ scattering of the generalized model discussed in sect. 3. As we shall see, unlike the previous example, this time the mistakes incurred in making these two incorrect steps do not cancel each other.

From the free Hamiltonian (3.2), one sees that the Fourier transforms of the free V' and θ'_k propagators, defined by

$$S^{o}_{V'}(t) = T\,\langle \text{vac}\,|\,\eta\, V'(t) V'^{\dagger}(0)\,|\,\text{vac}\rangle\,,$$

$$S_{\theta'(k)}(t) = T\,\langle \text{vac}\,|\,\eta\, a'_k(t) a'^{\dagger}_k(0)\,|\,\text{vac}\rangle\,. \tag{6.21}$$

are given, respectively, by

$$\tilde{S}^{o}_{V'}(p_o) = i(p_o - m^{o}_{V'} + i\epsilon)^{-1}\,,$$

$$\tilde{S}_{\theta'(k)}(p_o) = i(p_o - \omega'_k + i\epsilon)^{-1}\,. \tag{6.22}$$

The vertex function is, according to eq. (3.3),

$$i g'(4\pi R\,\omega'_k)^{-\frac{1}{2}}\, k u'_k \quad \text{for} \quad V' \to V + \theta'_k\,,$$

$$-i g'(4\pi R\,\omega'_k)^{-\frac{1}{2}}\, k u'_k \quad \text{for} \quad V + \theta'_k \to V'\,. \tag{6.23}$$

To follow step (i), we first evaluate the V' propagator, defined by the sum of Feynman diagrams given in fig. 4. By using the usual Feynman rules in momentum space, one finds that, similarly to eqs. (6.10) and (6.11), the resulting V' propagator is (as $R \to \infty$)

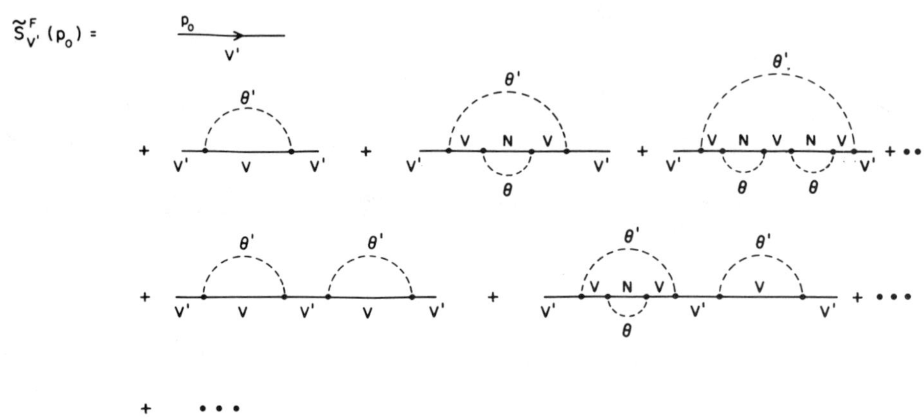

Fig. 4. Feynman diagrams for the V' propagator $\tilde{S}^{F}_{V'}(p_o)$ in the momentum space.

$$\tilde{S}^{F}_{V'}(p_0) = i[p_0 - m^0_V + i\epsilon + \Pi'^F(p_0)]^{-1} ,$$

and

$$\Pi'^F(p_0) = (g')^2 \int_0^\infty dk \int_{-\infty}^\infty dq_0 \tilde{S}^F_V(p_0 - q_0)(q_0 - \omega'_k + i\epsilon)(8\pi^3 \omega'_k)^{-1}(k u'_k)^2 . \quad (6.25)$$

The q_0 integration can be carried out by considering the closed contour bounded by the real axis and the semi-circle of radius ∞ below the real axis. By using eq. (6.13), one sees that the integrand

$$\tilde{S}^F_V(p_0 - q_0)(q_0 - \omega'_k + i\epsilon)^{-1} = i[h_\infty(p_0 - q_0 + i\epsilon)]^{-1}(q_0 - \omega'_k + i\epsilon)^{-1} , \quad (6.26)$$

has two poles inside this contour at

$$q_0 = \omega'_k - i\epsilon , \quad (6.27)$$

and

$$q_0 = p_0 - E_+ = p_0 - m_V - \tfrac{1}{2} i\gamma . \quad (6.28)$$

The q_0 integral in eq. (6.25) is then given simply by the appropriate residues at these two poles. It is straightforward to carry out step (ii). The resulting expression for the $N\theta\theta'$ scattering is, however, *incorrect.*

We recall that the correct V' propagator in coordinate space is given by

$$S_{V'}(t) = T\langle \text{vac} \,|\, \eta V'(t) V'^\dagger(0) \,|\, \text{vac}\rangle_{\text{Heisenberg}} \cdot \quad (6.29)$$

Similarly to eq. (6.18), we can decompose $S_{V'}(t)$ into an exponential part and a regular part:

$$S_{V'}(t) = S^{\text{reg}}_{V'}(t) + S^{\text{exp}}_{V'}(t) . \quad (6.30)$$

By using the explicit solutions given in sect. 3, we find

$$S^{\text{reg}}_{V'}(t) = i(2\pi)^{-1} \int_{-\infty}^\infty [\mathcal{H}_\infty(p_0 + i\epsilon)]^{-1} \exp(-ip_0 t) dp_0 , \quad (6.31)$$

where $\mathcal{H}_\infty$ is given by eq. (3.18). The Fourier transform of $S^{\text{reg}}_{V'}(t)$ is

$$\tilde{S}^{\text{reg}}_{V'}(p_0) = i[\mathcal{H}_\infty(p_0 + i\epsilon)]^{-1} , \quad (6.32)$$

which can also be written as

$$\tilde{S}^{\text{reg}}_{V'}(p_0) = i[p_0 - m^0_{V'} + i\epsilon + \Pi'(p_0)]^{-1} , \quad (6.33)$$

where, according to eq. (3.18),

$$\Pi'(p_0) = i(g')^2 \int_0^\infty dk \int_\Gamma dq_0 [h_\infty(p_0 - q_0)]^{-1}(q_0 - \omega'_k + i\epsilon)^{-1}(8\pi^3 \omega'_k)^{-1}(k u'_k)^2 , \quad (6.34)$$

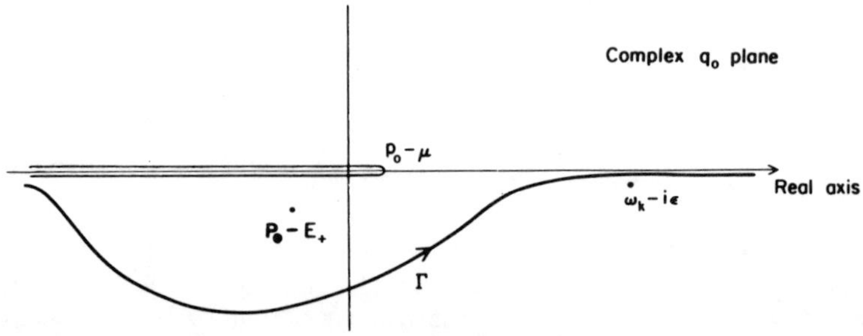

Fig. 5. Contour Γ for the integral (6.34).

and the contour Γ is given by fig. 5. By comparing eq. (6.25) with eq. (6.34), one sees that, unlike the previous case,

$$\tilde{S}^{\text{reg}}_{V'}(p_0) \neq \tilde{S}^{F}_{V'}(p_0) , \tag{6.35}$$

and therefore $\tilde{S}^{F}_{V'}(p_0)$ does not give the correct result. Rather, in eq. (6.25) one should detour the contour to Γ, as given by eq. (6.34), and avoid the contribution of the complex pole (6.28); the result would be $\tilde{S}^{\text{reg}}_{V'}(p_0)$, which can then be directly used to give the correct transition matrix element given by eqs. (3.28) and (3.35).

7. REMARKS

In this section, we will make a number of brief remarks concerning the various theoretical and experimental implications of the present theory.

(i) We recall that in this theory the Hamiltonian H is pseudo-Hermitian. All its eigenstates $|\mathrm{r}\rangle$ with real eigenvalues E_{r} are of positive square length. Furthermore, in the physical world at any time t, only states $|t\rangle$ of the form (1.21) can be realized,

$$|t\rangle = \sum C_{\mathrm{r}} |\mathrm{r}\rangle \exp(-iE_{\mathrm{r}} t) . \tag{7.1}$$

These states clearly satisfy the Schrödinger equation

$$i \frac{\mathrm{d}}{\mathrm{d}t} |t\rangle = H |t\rangle . \tag{7.2}$$

The expectation value of any operator, say a local operator $\bar{O}(x)$, is given by

$$\langle t | \eta \bar{O}(x) | t\rangle , \tag{7.3}$$

where

$$x_4 = it . \tag{7.4}$$

(ii) In such a relativistic local field theory, the CPT theorem may remain valid, provided that the usual CPT operator is further multiplied by η.

(iii) To understand some of the experimental consequences, let us consider the specific example of reaction (4.23),

$$e^+e^- \to e^+e^- \, , \tag{7.5}$$

and assume that the photon propagator (to the zeroth order in e^2) is given by

$$\delta_{\mu\nu}[(q^2)^{-1} - (q^2 + m^2)^{-1}] \, . \tag{7.6}$$

The inclusion of e^2 corrections would move the pole from the real axis to the complex plane. For simplicity, we shall call the complex spin = 1, parity = -1, heavy photon pole Γ^0. From the present (g - 2) experiment [18], one knows that the mass of Γ^0 must satisfy *

$$m \gtrsim 3.3 \text{ GeV} \, . \tag{7.7}$$

The existence of this pole can be experimentally established by studying the cross section of eq. (7.5). Since the width γ is, according to eq. (4.34), $>$ 50 MeV if m is $\approx$ 10 GeV, such a detection may be possible if m is not too high. The sign of the phase shift can be easily inferred from the negative sign of the coefficient of $(q^2 + m^2)^{-1}$ in eq. (7.6). Any measurement of the cross section that can establish the existence of the pole term $(q^2 + m^2)^{-1}$ would almost automatically include its interference with the usual $(q^2)^{-1}$ term, and, therefore, could be used to establish whether or not Γ^0 corresponds to a pole on the physical sheet above the real axis.

(iv) Other experimental consequences such as the motion of wave packets at small distances discussed in sect. 4 are extremely interesting, but difficult to observe. From a theoretical point of view, both the admission of negative metric η to the theory and the constraint (7.1) make a major departure from the usual foundation of quantum mechanics.

For clarity, let us again consider the simple two-body elastic S-wave scattering problem discussed in sect. 4. The wave function at a large distance r in the center-of-mass system is given by the asymptotic expansion (4.4)

$$\psi_k = r^{-1}(e^{-ikr} + e^{2i\delta} e^{ikr})e^{-iEt} \, , \tag{7.8}$$

and a wave packet is described by eq. (4.5)

$$\psi(r,t) = \int_0^\infty dk \; \psi_k(r,t) C_k \, . \tag{7.9}$$

At a large distance, $|\psi(r,t)|^2$ represents the usual relative probability of finding, at time t, the two particles at a distance r apart in the center-of-mass system. (To avoid confusion, we use the term 'probability' only if it satisfies the usual mathematical definition of being non-negative and having its total sum unity.) Once negative metric is used, it is obvious that there

* The presence of Γ^0 would give a change

$$\delta g = -(3\pi)^{-1} \alpha (\mu/m)^2 \, ,$$

where μ and m are, respectively, the muon mass and the real part of the complex Γ^0 pole $(m + \frac{1}{2} i \gamma)$. The present experimental value is

$$(\delta g)_{\text{exp}} = (g)_{\text{exp}} - (g)_{\text{usual theory}} = (108 \pm 62) \times 10^{-8} \, .$$

Hence, $m \gtrsim 3.3$ GeV; otherwise, the discrepancy would be more than three standard deviations.

exists a characteristic small distance a, within which the usual probability interpretation of $|\psi(r,t)|^2$ must fail. In order to maintain the usual probability interpretation of $|\psi(r,t)|^2$ for

$$r > a \ ,$$

we require that (i) the asymptotic expansion (7.8) should be valid for $r > a$, and furthermore (ii) at

$$r > a \text{ and } t < -r \ , \tag{7.10}$$

$\psi(r,t)$ is determined by ϕ^{in}, defined by eq. (4.12); i.e.,

$$|\psi(r,t)|^2 = |r^{-1}\phi^{in}(r,t)|^2 \ , \tag{7.11}$$

for a well-localized wave packet in the coordinate space. (For such a wave packet, one may neglect $\exp(-a\Delta)$ as compared to unity, where Δ is the width of the wave packet in the momentum space.) By using eqs. (4.29) and (4.31), one sees that, because of the term $\exp[-\frac{1}{2}\gamma(r-t)]$ in ϕ^{out}, eqs. (7.10) and (7.11) imply that

$$e^{-a\gamma} \ll 1 \ . \tag{7.12}$$

Thus, the characteristic distance a is determined not simply by the length for the validity of the asymptotic expansion (7.8), but also by γ^{-1}, which is usually the bigger one. The usual probability is not applicable for distances smaller than γ^{-1}. For the specific applications we have considered, γ^{-1} is sufficiently small. As given by eqs. (4.36) and (4.37), γ^{-1} is $\lessapprox 4 \times 10^{-13}$ cm if m is ≈ 10 GeV for Γ^0, and is $\approx 2 \times 10^{-11}$ cm if m is ≈ 10 GeV for $W^{\pm}$; therefore, this failure of probability interpretation is not in contradiction to our present experience. A challenging question is how to devise a practical experiment that can detect such a failure of probability interpretation if it indeed exists in nature.

(v) All stable particles p, e^-, γ, ν_e, ν_μ and their anti-particles must be of positive metric. All presently known metastable particles, such as $\mu^{\pm}$, Λ^0, $K^{\pm}$, etc., must also be of positive metric, since they all have clearly measurable lifetimes corresponding to appropriate poles on the unphysical sheet below the real axis. For resonances that have been observed in strong interaction processes, the signs of at least some of the phase shifts, e.g., N^*, have been measured, while those of the others may be inferred theoretically. All of these resonances also correspond to poles on the unphysical sheet below the real axis.

As we have seen, the proposals that the usual photon propagator may be replaced by eq. (7.6) and that the usual $W^{\pm}$ propagator may be replaced by eq. (4.26) can be used to remove divergence difficulties without violating unitarity, and such possibilities are within the general framework of the relativistic local field theory. While these proposals do imply some unusual phenomena such as the existence of complex singularities on the physical sheet, they are not, as yet, in contradiction with our present observations; their validity remains to be resolved by future experimental efforts.

The conventional formalism of quantum field theory with a positive definite metric would be completely satisfactory if it were not for serious divergence difficulties in electrodynamics, such as radiative corrections for weak decays and mass differences between hadrons within each isospin multiplet. For high order weak interaction processes these divergence difficulties become even more critical. Under the usual assumptions of current algebra, or field algebra, it appears that without some fundamental changes in quantum mechanics, these divergences can only be transferred* but not really removed. The possible existence of complex singularities such as Γ^0 and $W^{\pm}$ may provide a way out of such difficulties.

We wish to thank the members of the CERN Theoretical Study Division for their kind hospitality, and to thank many of our colleagues, especially Bruno Zumino, for helpful discussions.

NOTE ADDED IN PROOF

Dr. Haller has kindly informed us that the possibility of using eq. (1.6) as the W-propagator has also been considered by Haller, Landovitz and Goldberg [21]. However, their arguments are based on the incorrect assumption that the S-matrix remains given by the limit $t \to \infty$, $t' \to \infty$ of $U(t,t')$. As shown in sect. 5 of this paper, such a limit does not exist; in fact, it diverges exponentially in both t and t'.

REFERENCES

[1] P.A.M.Dirac, Proc. Roy. Soc. A180 (1942) 1.
[2] W.Pauli, Revs. Mod. Phys. 15 (1943) 175.
[3] W.Pauli and F.Villars, Revs. Mod. Phys. 21 (1949) 434.
[4] R.P. Feynman, Phys. Rev. 76 (1949) 769.
[5] T.D. Lee and C.N.Yang, Phys. Rev. 128 (1962) 885.
[6] W.Pauli, Proc. Int. Conf. on high energy physics at CERN (1958) p. 127.
[7] V.Glaser, private communication.
[8] M. Froissart, Suppl. Nuovo Cimento 14 (1959) 197.
[9] W.Heisenberg, Nucl. Phys. 4 (1957) 532.
[10] T.D. Lee, Phys. Rev. 95 (1954) 1329.
[11] G.Källen and W.Pauli, Mat. Fys. Medd. Dan. Vid. Selsk. 30 (1955) 7.
[12] T.W. Ruijgrok and L.Van Hove, Physica 22 (1956) 880.
[13] Fei Shian Chen-Cheung and C.M.Sommerfield, Phys. Rev. 152 (1966) 1401; Fei Shian Chen-Cheung, Phys. Rev. 152 (1966) 1407.
[14] E.P.Wigner, Phys. Rev. 98 (1955) 145.
[15] M. Froissart, M.L.Goldberger and K.M.Watson, Phys. Rev. 131 (1963) 2820.
[16] M.L.Goldberger and K.M.Watson, Collision theory (Wiley and Sons, New York, 1964) Ch. 5.
[17] G.C.Wick, Phys. Rev. 80 (1950) 268.
[18] Bailey et al., Proc. Int. Conf. on high energy physics, Vienna 1968, to be published.
[19] T.D. Lee, Phys. Rev. 171 (1968) 1731.
[20] M.Gell-Mann et al., to be published.
[21] K.Haller, L.F.Landovitz and I.Goldberg, Nuovo Cimento 48 (1967) 303.

* See Lee [19] for a method which can transform divergences in electrodynamics to those in weak interaction, and Gell-Mann et al. [20] for devices that can transfer divergences in 'non-diagonal' weak processes to those in 'diagonal' processes.

7.B.1 Nuclear Physics B10 (1969) 1-10. North-Holland Publ. Comp., Amsterdam

UNITARITY IN THE N$\theta\theta$ SECTOR OF SOLUBLE MODEL WITH INDEFINITE METRIC

T. D. LEE *
CERN, Geneva, Switzerland

and

G. C. WICK
Columbia University, New York, N. Y. **

Received 17 February 1969

Abstract: Assuming that the Nθ sector of a certain soluble model with an indefinite metric contains no stable V state, the unitarity condition of the S-matrix is explicitly verified in the N$\theta\theta$ sector.

1. INTRODUCTION

In a preceding paper [1], quoted as I in the following, we have promised a detailed study of the N$\theta\theta$ sector of the soluble model [2], by means of the explicit type of solution discovered by Amado [3] and investigated by many other authors [4] for other cases of the soluble model ***. The presence, in our case, of two complex roots of the eigenvalue condition in the Nθ sector requires certain modifications, but the method goes through all the same. Although the unitarity of the S-matrix has already been inferred from general considerations [1], it seems of some interest to verify it directly in a case which has been shown to give trouble [5] for different values of the parameters, i.e. when a stable V and a 'ghost' exist‡.

In order to avoid unnecessary repetitions, we shall refer as much as is practical to the earlier work of Pagnamenta [4]; the Hamiltonian and nota-

* On leave from Colombia University.

** This research was supported in part by the U.S. Atomic Energy Commission and the National Science Foundation.

*** The list of papers in ref. [4] could be much longer. See the last paper in the reference for further literature. Pagnamenta's two papers of ref. [4] will be quoted as AP1 and AP2 respectively.

‡ We recall that a 'ghost' state is defined to be an eigenstate of the total Hamiltonian with a real eigenvalue, but of a negative, or zero, square length. Our basic requirement, as given by eq. (3) below, that there should be no stable V excludes the existence of such a 'ghost' state, in contrast to the case considered by Källen and Pauli [5].

tions are taken over from eqs. (2.2) - (2.8) of our paper [1]. We go immediately to the limit $R = \infty$; see eqs. (2.32) and (2.33) of I. For a quicker comparison with Pagnamenta's formulae, we shall designate the function h_∞ in I, eq. (2.14), as $H(z)$ (z a complex value of the energy parameter E) and its limiting value on the upper rim of the cut on the real axis as $H^+(\omega) \equiv \equiv H(\omega + i0)$. Similarly $H^-(\omega) \equiv H(\omega - i0)$; unlike Pagnamenta we attach the $\pm$ designation to the *function*, not to some specific variable (a confusing habit, in our opinion). Thus we have:

$$H(z) = -m_V^0 + z - \frac{1}{\pi} \int_\mu^\infty \frac{\sigma(\omega)}{\omega - z} d\omega , \tag{1}$$

where

$$\sigma(\omega) = +(4\pi)^{-1} g^2 k u_k^2 ; \qquad k = (\omega^2 - \mu^2)^{\frac{1}{2}} ,$$

(compare eqs. (2.14) and (2.17) of I). Of course,

$$\operatorname{Im} H^+(\omega) = -\sigma(\omega) , \qquad (\omega > \mu) . \tag{2}$$

We emphasize that, in spite of superficial differences due to renormalizations and subtractions, adopted by Pagnamenta and other authors, the function $H(z)$ defined by the unsubtracted relation (1) corresponds exactly to $H(z)$ in AP1, 2.

As pointed out in I, one can easily see that if m_V^0 is such that on the uncut part of the real axis,

$$-\infty < \omega < \mu , \qquad H(\omega) < 0 , \tag{3}$$

then the equation $H(z) = 0$ has a complex root E_+ in the upper half-plane, eq. (2.18) of I, and a root $E_- = E_+^*$ in the lower half-plane. We shall need, in the following, a formula analogous to eq. (14) of AP1 for the reciprocal function $[H(z)]^{-1}$. In our case this function has simple complex poles at $z = E_\pm$, and is $O(z^{-1})$ at ∞. It is easy to establish, therefore, the Cauchy formula:

$$[H(z)]^{-1} = (z - E_+)^{-1} \Gamma_+ + (z - E_-)^{-1} \Gamma_- + \frac{1}{\pi} \int_\mu^\infty \operatorname{Im} [H^+(\omega)]^{-1} \frac{d\omega}{\omega - z} , \tag{4}$$

where

$$\Gamma_\pm^{-1} = H'(E_\pm) .$$

Of course

$$\operatorname{Im} [H^+(\omega)]^{-1} = \sigma(\omega) / |H^+(\omega)|^2 .$$

We now go over to the scattering problem.

2. THE $N\theta\theta$ SECTOR

Consider a state consisting of an incident wave $N^\dagger A_p^\dagger A_q^\dagger |\text{vac}\rangle$ plus additional terms which in the asymptotic region reduce to outgoing waves only. In the notation of I, this state is designated by $|N\theta_p\theta_q^{\text{out}}\rangle$. In view of I, eq. (2.11), the energy of this state is

$$E = \omega_p + \omega_q \,. \tag{5}$$

In the following, we use the abbreviations ω, ω', ω_1, ω_2 for ω_k, $\omega_{k'}$, ω_{k_1}, ω_{k_2}. The corresponding values of the form-factor u_k, etc. will be abbreviated likewise as u, u', u_1, u_2.

The state vector in question may be expanded:

$$|N\theta_p\theta_q^{\text{out}}\rangle = \int_\mu^\infty \mathrm{d}\omega\, \omega^{\frac{1}{2}}\, u\,\alpha(\omega)\, A_k^\dagger V^\dagger |\text{vac}\rangle$$
$$+ \int_\mu^\infty \mathrm{d}\omega_1 \int_\mu^\infty \mathrm{d}\omega_2\, \phi(\omega_1,\omega_2)\, A_{k_1}^\dagger A_{k_2}^\dagger N^\dagger |\text{vac}\rangle \,. \tag{6}$$

From the Schrödinger equation one obtains two relations between ϕ and α:

$$(E-\omega_1-\omega_2)\phi(\omega_1,\omega_2) = -(g/4\pi)u_1 u_2(\omega_1\omega_2)^{\frac{1}{2}}[\alpha(\omega_1)+\alpha(\omega_2)] \,, \tag{7}$$

and

$$(E - m_V^0 - \omega)\omega^{\frac{1}{2}} u\,\alpha(\omega) = (g/2\pi)\int_\mu^\infty \omega'^{-\frac{1}{2}} k' u' [\phi(\omega,\omega') + \phi(\omega',\omega)]\mathrm{d}\omega' \,. \tag{8}$$

The first relation may also be written as

$$\phi(\omega_1,\omega_2) = \delta(\omega_1-\omega_p)\delta(\omega_2-\omega_q)$$
$$-(g/4\pi)u_1 u_2(\omega_1\omega_2)^{\frac{1}{2}}(E+i0-\omega_1-\omega_2)^{-1}[\alpha(\omega_1)+\alpha(\omega_2)] \,. \tag{9}$$

Here the 'inhomogeneous' term on the right represents the incident wave, while the energy denominator in the second term is in the form corresponding to the assumed asymptotic behavior of the state (6). The second relation after eliminating ϕ by means of eq. (9) takes the form of an integral equation for $\alpha(\omega)$, which after some manipulations (compare, e.g., AP1, eqs. (58), (59), (60) and (II)) can be reduced to the following form. First introduce a new unknown function $\chi^-(\omega)$ through the substitution:

$$\tfrac{1}{2} g u_p u_q(\omega_p\omega_q)^{\frac{1}{2}} H^+(\omega_p)H^+(\omega_q)\alpha(\omega)$$
$$= \{\sigma(\omega_p)H^+(\omega_q)\delta(\omega-\omega_q) + (p \rightleftharpoons q)\} - \frac{1}{\pi}\sigma(\omega_p)\sigma(\omega_q)\chi^-(\omega) \,, \tag{10}$$

then χ^- is a solution of the integral equation

$$H^+(E-\omega)\chi^-(\omega) = \frac{1}{\pi}\int_\mu^\infty \frac{\sigma(\omega')\chi^-(\omega')d\omega'}{\omega'+\omega-E-i0} + \frac{H^+(\omega_q)}{\omega_p+i0-\omega} + \frac{H^+(\omega_p)}{\omega_q+i0-\omega}. \tag{11}$$

This is to be regarded as an integral equation for a function χ^- of a real variable ω, over the interval

$$\mu < \omega < \infty . \tag{12}$$

If $\chi^-(\omega')$ in the integrand on the right is regarded as known, however, one can regard the right-hand side of eq. (11) as the boundary value for $z \to \omega\text{-}i0$ of a function of a complex variable z. Since this is also true of the factor $H^+(E-\omega)$ on the left, we can set

$$\chi^-(\omega) = \chi(\omega - i0) , \tag{13}$$

where $\chi(z)$ is defined by

$$H(E-z)\chi(z) = \frac{1}{\pi}\int_\mu^\infty \frac{\sigma(\omega')\chi^-(\omega')}{\omega'+z-E}\,d\omega' + \frac{H^+(\omega_q)}{\omega_p - z} + \frac{H^+(\omega_p)}{\omega_q - z} . \tag{14}$$

One sees that $\chi(z)$ is meromorphic in the whole z-plane, except for a cut on the real axis, $z = \omega$ (real) such that

$$-\infty < \omega < E - \mu . \tag{15}$$

Notice that since $E = \omega_p + \omega_q > 2\mu$, the cut (15) overlaps the integration interval (12) which is also the cut for the function $H(z)$. All these circumstances have been discussed in detail by Pagnamenta, Sommerfield, etc.; therefore we summarize. One difference here is that the factor $H(E-z)$ on the left of eq. (14) has zeroes at

$$z = z_\pm \equiv E - E_\pm . \tag{16}$$

Hence $\chi(z)$ has two *poles* in the cut plane. In addition, of course, the last two terms in eq. (14) indicate that $\chi(z)$ has poles at $z = \omega_p$ and ω_q, but owing to eq. (5) these poles lie on the boundary of the cut plane

$$\mu < \omega_p, \omega_q < E - \mu .$$

If we approach one of these poles, for example ω_p, *from below* we see that $H(E-z)$ on the left tends to $H(E-\omega_p+i0) = H^+(\omega_q)$. As a result we calculate a residue -1 for $\chi(z)$ at these poles

$$\chi(z) = -(z-\omega_p)^{-1} - (z-\omega_q)^{-1} + \dots . \tag{17}$$

At first sight one obtains a different residue, however, if one approaches the poles *from above*; a more careful consideration shows, however, that in addition to the 'explicit' pole terms on the right-hand side of eq. (14), we must also examine the singularities which may arise from the integral term, when the integration path (the real interval from μ to $+\infty$) is 'pinched' between a pole of $\chi^-(\omega')$, say $\omega' = \omega_q + i0$, and the zero $\omega' = E - z$ of the denominator. This pinching occurs when z approaches $E - \omega_q \equiv \omega_p$ (respectively $E - \omega_p \equiv \omega_q$) *from above.* Thus while the residues calculated *from below,* eq. (17), require no correction, the different residues obtained from above turn out to be erroneous. The correction is easily calculated by means of eq. (1) (compare AP1) and the result is, quite simply, that eq. (17) holds above the cut as well.

We now notice with Pagnamenta that $\chi(z)$ may be determined from the Cauchy formula, in our case:

$$\chi(z) = \frac{R_+}{z - z_+} + \frac{R_-}{z - z_-} + \frac{1}{2\pi i}\int_{-\infty}^{E-\mu} \frac{\Delta\chi(\omega')d\omega'}{\omega' - z} , \tag{18}$$

if the residues R_+ and R_-, and the discontinuity across the cut (15),

$$\Delta\chi(\omega) = \chi(\omega + i0) - \chi(\omega - i0) , \tag{19}$$

are known. In order to evaluate the latter we follow more or less the procedure of AP2, eqs. (17) - (22), i.e. we calculate eq. (14) both above and below the cut and take the difference, thus:

$$H^-(E - \omega)\chi^+(\omega) - H^+(E - \omega)\chi^-(\omega)$$

$$= -2i\sigma(E - \omega)\chi^-(E - \omega) + 2\pi i\{H^+(\omega_q)\delta(\omega - \omega_p) + H^+(\omega_p)\delta(\omega - \omega_q)\} , \tag{20}$$

and replace $H^-(E - \omega) \to H^+(E - \omega) + 2i\sigma(E - \omega)$; we then divide by $H^+(E - \omega)$ obtaining

$$\Delta\chi(\omega) = -2i\sigma(E - \omega)[H^+(E - \omega)]^{-1}\{\chi^+(\omega) + \chi^-(E - \omega)\}$$

$$+ 2\pi i[\delta(\omega - \omega_p) + \delta(\omega - \omega_q)] . \tag{21}$$

The δ-terms when inserted into eq. (18) give, of course, the pole terms (17); they need no further attention. The key to the solution is the fact (compare AP1, 2 and Sommerfield's paper of ref. [4]) that the expression between braces $\{\ldots\}$ is the boundary value $M^+(\omega)$ of the function $M(z) = \chi(z) + \chi(E - z)$ for which one obtains the remarkably simple result

$$M(z) \equiv \chi(z) + \chi(E - z) = \frac{L}{H(z)H(E - z)} , \tag{22}$$

where L is a constant! To this end one shows that

$$L(z) = H(z)H(E - z)M(z) , \tag{23}$$

is regular analytic (holomorphic) in the whole z-plane. Since it is, rather obviously, bounded at ∞, it must be a constant, hence eq. (22). To see that $L(z)$ is holomorphic, one notices first that the poles (17) disappear when one forms the symmetric combination $\chi(z) + \chi(E - z)$. After that, there are still poles at $z_\pm = E - E_\pm$, see eq. (18), and at $E - z_\pm = E_\pm$. These poles, however, are zeroes of $H(E - z)$ and $H(z)$ respectively; the product (23), therefore, has no poles whatsoever. Finally, we must still show that $L(z)$ is continuous across the real axis; $M(z)$, of course, has both cuts (12) and (15), but if we manipulate eq. (20) in a different way, eliminating σ instead of H^-, we obtain:

$$H^-(E-\omega)[\chi^+(\omega)+\chi^-(E-\omega)] = H^+(E-\omega)\{\chi^-(\omega)+\chi^-(E-\omega)\}$$
$$+ 2\pi i\{H^+(\omega_q)\delta(\omega-\omega_p)+H^+(\omega_p)\delta(\omega-\omega_q)\}\,. \qquad (21')$$

We multiply on both sides by $H^+(\omega)$, whereupon the right-hand side is clearly seen to be invariant under the substitution $\omega \to E-\omega$, while the left-hand side becomes

$$H^+(\omega)H^-(E-\omega)M^+(\omega) \equiv L^+(\omega)\ .$$

Thus we get $L^+(\omega) = L^+(E-\omega)$. On the other hand, clearly $L(z) = L(E-z)$, hence $L^+(E-\omega) = L^-(\omega)$. Finally,

$$L^+(\omega) = L^-(\omega)\ . \qquad (24)$$

The proof of eq. (22) is thus completed.

One easily sees that this equation also allows one to express the residues $R_\pm$, eq. (18), in terms of the constant L

$$R_\pm = -L[H(z_\pm)H'(E_\pm)]^{-1}\ . \qquad (25)$$

When the results (21), (22) and (25) are inserted into eq. (18), one obtains, remembering eq. (2),

$$\chi(z) = (\omega_p-z)^{-1}+(\omega_q-z)^{-1}+L\{r_+(z_+-z)^{-1}+r_-(z_--z)^{-1}+I(E-z)\}\ , \qquad (26)$$

where

$$r_\pm = [H(z_\pm)H'(E_\pm)]^{-1}\ , \qquad (27)$$

and

$$I(z) = \frac{1}{\pi}\int_{-\infty}^{E-\mu}\frac{d\omega'}{E-z-\omega'}\,\frac{1}{H^+(\omega')}\,\mathrm{Im}\,\frac{1}{H^+(E-\omega')}$$
$$= \frac{1}{\pi}\int_{\mu}^{\infty}\frac{d\omega'}{\omega'-z}\,\frac{1}{H^+(E-\omega')}\,\mathrm{Im}\,\frac{1}{H^+(\omega')}\ . \qquad (28)$$

We still have to determine the constant L. To do this, one can (see AP1, eq. (31), etc.) substitute eq. (26) back into the integral equation, (11) or (14), and verify that the equation is satisfied for a certain value of L. We can avoid this rather tedious calculation, if we notice that, according to eq. (14), $\chi(z) = O(z^{-2})$ when $|z| \to \infty$. The terms of $O(z^{-1})$ in eq. (26) must therefore vanish. This gives the condition:

$$L[A(E)-r_+-r_-] = 2\ , \qquad (29)$$

where

$$A(E) = \frac{1}{\pi} \int_{\mu}^{\infty} \frac{d\omega}{H^+(E-\omega)} \operatorname{Im} \frac{1}{H^+(\omega)} . \tag{30}$$

Our solution is now complete.

3. THE S-MATRIX

The wave-function (9) could be labeled more properly $\phi(\omega_1\omega_2; \omega_p\omega_q)$; also $\alpha(\omega)$, of course, depends on ω_p and ω_q, although this is not indicated explicitly. The S-matrix element

$$\langle N\theta_1\theta_2^{\text{in}} | \eta | N\theta_p\theta_q^{\text{out}} \rangle = 2 \left(\frac{k_1 k_2 pq}{\omega_1\omega_2\omega_p\omega_q} \right)^{\frac{1}{2}} S(\omega_1\omega_2; \omega_p\omega_q) \tag{31}$$

can be calculated from the expansion (6). More rapidly $S(\omega_1\omega_2; \omega_p\omega_q)$ can be obtained from Møller's wave matrix $\phi(\omega_1\omega_2; \omega_p\omega_q)$. The usual recipe is: multiply ϕ by the square root of the final velocities $v_i = k_i\omega_i^{-1}$ ($i = 1, 2$) and divide by the square root of the initial velocities $p\omega_p^{-1}$ and $q\omega_q^{-1}$; then replace the denominator $(E + i0 - \omega_1 - \omega_2)^{-1}$ with $-2\pi i \delta(E - \omega_1 - \omega_2)$; in addition the 'unit operator' (the first term on the right of eq. (9)) must be symmetrized, as will automatically happen if we calculate eq. (31). Thus, remembering eq. (5),

$$S(\omega_1\omega_2; \omega_p\omega_q) = \Delta - 2\pi i \delta(\omega_1 + \omega_2 - \omega_p - \omega_q) T , \tag{32}$$

where

$$\Delta \equiv \Delta(\omega_1\omega_2; \omega_p\omega_q) = \tfrac{1}{2}[\delta(\omega_1 - \omega_p)\delta(\omega_2 - \omega_q) + \delta(\omega_1 - \omega_q)\delta(\omega_2 - \omega_p)] , \tag{33}$$

$$T \equiv T(\omega_1\omega_2; \omega_p\omega_q) = -(g/4\pi) u_1 u_2 [\omega_p\omega_q k_1 k_2/(pq)]^{\frac{1}{2}} [\alpha(\omega_1) + \alpha(\omega_2)] .$$

We decompose $T = T_a + T_b$ corresponding to the two terms on the right of eq. (10); then

$$\delta(\omega_1 + \omega_2 - \omega_p - \omega_q) T_a = -\frac{1}{\pi} \left[\frac{\sigma(\omega_p)}{H^+(\omega_p)} + \frac{\sigma(\omega_q)}{H^+(\omega_q)} \right] \Delta . \tag{34}$$

These terms, as one easily sees, correspond to disconnected diagrams in which one of the θ-particles goes through without interacting, while the other one is scattered elastically (see I, eq. (2.16)).

In order to calculate T_b we see from eqs. (9) and (10) that we need an expression for $\chi^-(\omega_1) + \chi^-(\omega_2)$. Since $\omega_2 = E - \omega_1$ (on the mass shell), this is just the expression between braces on the right-hand side of eq. (21'), while on the left side $\chi^+(\omega) + \chi^-(E-\omega) \equiv M^+(\omega)$. We calculate therefore easily, with the help of eq. (22):

$$\chi^-(\omega) + \chi^-(E-\omega) = -2\pi i[\delta(\omega - \omega_p) + \delta(\omega - \omega_q)] + \frac{L}{H^+(\omega)H^+(E-\omega)} . \tag{35}$$

Hence

$$\delta(E-\omega_1-\omega_2)T_b = -\frac{2i}{\pi}\frac{\sigma(\omega_p)\sigma(\omega_q)}{H^+(\omega_p)H^+(\omega_q)}\Delta$$

$$+\delta(\omega_1+\omega_2-\omega_p-\omega_q)\frac{1}{2\pi^2}[\sigma(\omega_1)\sigma(\omega_2)\sigma(\omega_p)\sigma(\omega_q)]^{\frac{1}{2}}\frac{L}{H^+(\omega_1)H^+(\omega_2)H^+(\omega_p)H^+(\omega_q)}. \tag{36}$$

Let us now put all the terms together. Notice the simplification

$$1+2i\,\sigma(\omega_p)/H^+(\omega_p)+2i\,\sigma(\omega_q)/H^+(\omega_q)-4\sigma(\omega_p)\sigma(\omega_q)/H^+(\omega_p)H^+(\omega_q)$$
$$= e^{2i(\delta_p+\delta_q)} \tag{37}$$

where δ_p is the phase shift in the Nθ sector

$$e^{2i\delta_p} = H^-(\omega_p)/H^+(\omega_p)\,. \tag{38}$$

Compare I, eq. (2.16). Then, with the abbreviations

$$\sigma_i = \sigma(\omega_i)\,, \qquad H_i = H^+(\omega_i)\,, \qquad (i = 1, 2, p, q)\,, \tag{39}$$

we have

$$S(\omega_1\omega_2;\omega_p\omega_q) = e^{i(\delta_p+\delta_q+\delta_1+\delta_2)}\{\Delta(\omega_1\omega_2;\omega_p\omega_q)$$

$$-i\pi^{-1}L\,\delta(\omega_1+\omega_2-\omega_p-\omega_q)[\sigma_1\sigma_2\sigma_p\sigma_q]^{\frac{1}{2}}|H_1H_2H_pH_q|^{-1}\}\,. \tag{40}$$

We now check unitarity

$$\int_\mu^\infty\int_\mu^\infty d\omega_1\,d\omega_2 S(\omega_1\omega_2;\omega_p\omega_q)S^*(\omega_1\omega_2;\omega_r\omega_s) = \Delta(\omega_r\omega_s;\omega_p\omega_q)$$

$$+e^{i(\delta_p+\delta_q-\delta_r-\delta_s)}[\sigma_p\sigma_q\sigma_r\sigma_s]^{\frac{1}{2}}|H_pH_qH_rH_s|^{-1}\pi^{-1}\delta(\omega_p+\omega_q-\omega_r-\omega_s)$$

$$\times\{-i(L-L^*)+|L|^2\frac{1}{\pi}\int\frac{\sigma_1\sigma_2}{|H_1|^2|H_2|^2}\delta(E-\omega_1-\omega_2)d\omega_1\,d\omega_2\}\,. \tag{41}$$

Now from eqs. (29) and (30) we see that the expression between braces, divided by $|L|^2$, is equal to

$$-\mathrm{Im}\,\frac{2}{L}+\frac{1}{\pi}\int_\mu^\infty\frac{\sigma(\omega)\sigma(E-\omega)}{|H^+(\omega)|^2|H^+(E-\omega)|^2}d\omega$$

$$= -\mathrm{Im}\,A(E)+\frac{1}{\pi}\int_\mu^\infty\mathrm{Im}\,\frac{1}{H^+(\omega)}\,\mathrm{Im}\,\frac{1}{H^+(E-\omega)}d\omega = 0\,. \tag{42}$$

The residues r_+ and r_- do not contribute because of the reality condition $H(z^*) = [H(z)]^*$. As a result we have proved the unitarity condition

$$\int d\omega_1 \, d\omega_2 \, S(\omega_1\omega_2; \omega_p\omega_q) S^*(\omega_1\omega_2; \omega_r\omega_s) = \tfrac{1}{2}\{(\omega_p - \omega_r)\delta(\omega_q - \omega_s) + (p \rightleftharpoons q)\} \, . \tag{43}$$

The proof is, of course, essentially the same as would apply to the situation considered in AP2, sect. V. The interest here was in showing that the existence of complex roots does not create any difficulties in extending unitarity to the situation envisaged here.

4. NON-EXISTENCE OF A BOUND STATE IN THE N$\theta\theta$ SECTOR

To complete our discussion of the N$\theta\theta$ sector of the soluble model, we shall prove that in the present case the *only* eigenstates with real eigenvalues in the N$\theta\theta$ sector are the scattering states $|N\theta_p\theta_q^{\text{out}}\rangle$ or $|N\theta_p\theta_q^{\text{in}}\rangle$. To establish this, we start from the time-dependent Schrödinger equation, but, instead of eq. (5), require that the energy value

$$E < 2\mu \, . \tag{44}$$

Such a state, if it did exist, would correspond to a bound state; however, as the following discussion will show, it in fact cannot exist. Formally, one may express such a state in terms of the right-hand side of eq. (6) where the functions $\phi(\omega_1, \omega_2)$ and $\alpha(\omega)$ still satisfy eqs. (7) and (8). For the bound state, instead of eq. (9), one has simply

$$\phi(\omega_1, \omega_2) = -(g/4\pi) u_1 u_2 (\omega_1\omega_2)^{\frac{1}{2}} (E - \omega_1 - \omega_2)^{-1} [\alpha(\omega_1) + \alpha(\omega_2)] \, . \tag{45}$$

After eliminating $\phi(\omega_1, \omega_2)$, the integral equation for $\alpha(\omega)$ becomes a homogeneous one which can be written as

$$\int_\mu^\infty K_{\omega\omega'} \, \alpha(\omega') d\omega' = 0 \, , \tag{46}$$

where

$$K_{\omega\omega'} = -\sigma(\omega) H(E - \omega) \delta(\omega - \omega') + \frac{1}{\pi} \frac{\sigma(\omega)\sigma(\omega')}{\omega + \omega' - E} \, . \tag{47}$$

We recall that $\sigma(\omega)$ is, by definition, real and positive. For $E < 2\mu$, because of eq. (3), $-H(E - \omega)$ is also real and positive. Thus, the matrix

$$K = (K_{\omega\omega'}) \, , \tag{48}$$

is real and symmetric. By using the identity

$$\int_\mu^\infty d\omega \int_\mu^\infty d\omega' \, \alpha^*(\omega)\alpha(\omega')\sigma(\omega)\sigma(\omega')(\omega + \omega' - E)^{-1} = \int_0^\infty d\lambda \, |\phi(\lambda)|^2 \, , \tag{49}$$

where

$$\phi(\lambda) = \int_{\mu}^{\infty} d\omega\ \alpha(\omega)\sigma(\omega) \exp[-\lambda(\omega - \tfrac{1}{2}E)] \ , \tag{50}$$

one sees that the matrix K is also positive definite. Consequently, eq. (46) has no solution except $\alpha(\omega) = 0$, which completes the proof.

REFERENCES

[1] T.D.Lee and G.C.Wick, Nucl. Phys. B9 (1969) 209.
[2] T.D.Lee, Phys. Rev. 95 (1954) 1329.
[3] R.D.Amado, Phys. Rev. 122 (1961) 697.
[4] A.Pagnamento, J. Math. Phys. 6 (1965) 955; 7 (1966) 356;
C.Sommerfield, J. Math. Phys. 6 (1965) 1170;
E.M.Kazes, J. Math. Phys. 6 (1965) 1172;
L.M.Scarfone, J. Math. Phys. 9 (1968) 246.
[5] G.Källen and W.Pauli, Mal. Fys. Medd. Dan. Vid. Selsk. 30 No. 7 (1955).

Reprinted from THE PHYSICAL REVIEW D, Vol. 2, No. 6, 1033–1048, 15 September 1970
Printed in U. S. A.

Finite Theory of Quantum Electrodynamics*

T. D. LEE AND G. C. WICK
Physics Department, Columbia University, New York, New York 10027
(Received 15 May 1970)

A new theory of quantum electrodynamics is presented, which is relativistically invariant, gauge invariant, unitary, and free of divergences. In this theory, mass renormalization, charge renormalization, and wave-function renormalization are all finite. Experimental consequences are discussed, and theoretical implications, especially those related to causality, are analyzed.

I. INTRODUCTION

IN spite of the spectacular success of the renormalized theory of quantum electrodynamics,[1] there remain unsatisfactory aspects concerning the inherent mathematical ambiguities in manipulating divergent expressions. Even apart from questions of mathematical rigor, there are serious difficulties when one tries to extend the renormalization process to hadronic systems. Consider, for example, the problem of the mass difference between $\pi^{\pm}$ and π^0. It seems reasonable to assume that this mass difference is due purely to the electromagnetic interaction. Yet, according to the usual theory, this mass difference is calculated to be infinite; furthermore, it can be shown that under rather general assumptions such an infinity, being closely connected with the equal-time commutator between the current operator and its derivatives, cannot be removed through strong interactions.[2] The same divergence difficulties exist for all observed (therefore finite) mass differences between different hadrons in the same isospin multiplet; all such mass differences are found to be infinite according to the usual theory of quantum electrodynamics. Similarly, by making the universality assumption about the weak interaction, one expects that at least the ratio between the radiative correction to the Fermi constant G_V in a β decay and that to the μ-decay constant G_μ should be finite. Again, it is infinite in the conventional theory.[3]

Now, the various fractional mass differences between different hadrons in the same isospin multiplet are certainly finite, and are all of the order of the fine structure constant α. By using the Cabibbo theory, the amount of the radiative correction to G_V/G_μ can be deduced from the observed rates of β decay and μ decay; it is, of course, finite, and also of the order of α. The fact that all these quantities are of the right order of magnitude strongly indicates that they are indeed due to second-order electromagnetic processes. It appears then that there must be fundamental changes in our basic formulation of quantum field theory, so that unrenormalized masses and unrenormalized coupling constants can become finite.

Recently, it has been found[4–6] that there exists a general class of field theories in which the S matrix is fully unitary, but the Lagrangian is not Hermitian. This makes it possible to construct relativistic local field theories which satisfy the unitarity requirement and are free from divergences. In particular, by replacing in the electromagnetic interaction the usual zero-mass photon field A_μ by a complex field

$$\phi_\mu = A_\mu + iB_\mu , \tag{1.1}$$

where B_μ is a massive boson field associated with a negative metric, it is possible to remove all infinities from the electromagnetic mass differences between hadrons, as well as those associated with radiative corrections to weak decays.

The purpose of the present paper is, on the one hand, to supply some further details of this new form of quantum electrodynamics, and on the other hand to extend a similar modification also to the fermion fields, so that the unrenormalized electric charge is also finite. The resulting theory is then completely free from divergent expressions.

In order to gain proper perspective, we shall first review briefly in Sec. II some elementary properties of quantum theories with indefinite metric, and then proceed to analyze the relations between the usual commutation (or, anticommutation) relations and the metric of the system. It is shown that, under rather general conditions, in the case of Fermi statistics the positive definiteness, or indefiniteness, of the metric is *uniquely* determined by the equal-time anticommutation relations. Thus, once the Lagrangian for a fermion system is given, the metric is completely specified (of course, up to a canonical transformation, connected

* Research supported in part by the U. S. Atomic Energy Commission.

[1] At present, quantum electrodynamics agrees with experiment better than ever: J. Aldins, S. J. Brodsky, A. J. Dufner, and T. Kinoshita, Phys. Rev. Letters **23**, 441 (1969); Phys. Rev. D **1**, 2378 (1970); T. Appelquist and S. J. Brodsky, Phys. Rev. Letters **24**, 562 (1970).

[2] J. D. Bjorken, Phys. Rev. **148**, 1467 (1966); G. C. Wick and B. Zumino, Phys. Letters **25B**, 479 (1967); I. S. Gerstein, B. W. Lee, H. T. Nieh, and H. J. Schnitzer, Phys. Rev. Letters **19**, 1064 (1967); D. G. Boulware and S. Deser, *ibid.* **20**, 1399 (1968).

[3] T. Kinoshita and A. Sirlin, Phys. Rev. **113**, 1652 (1958); E. S. Abers, R. E. Norton, and D. A. Dicus, Phys. Rev. Letters **18**, 676 (1967); E. S. Abers, D. A. Dicus, R. E. Norton, and H. R. Quinn, Phys. Rev. **167**, 1461 (1968). For a general discussion and more complete literature, see A. Sirlin, *ibid.* **176**, 1871 (1968).

[4] T. D. Lee and G. C. Wick, Nucl. Phys. **B9**, 209 (1969); **B10**, 1 (1969).

[5] T. D. Lee, in *Quanta*, edited by P. G. O. Freund, C. J. Goebel, and Y. Nambu (Chicago U. P., Chicago, 1970), p. 260.

[6] T. D. Lee, in *Proceedings of the Topical Conference on Weak Interactions, CERN, 1969*, edited by J. S. Bell (CERN, Geneva, 1969), p. 427.

with possible changes of basis vectors). For the case of Bose statistics, the usual commutation relations do not uniquely determine the metric; there remains a free choice whether the metric is positive definite, or indefinite.

In Sec. III, we give the details for the modified photon field $A_\mu + iB_\mu$. The interaction between the negative-metric B_μ field and the charged lepton and hadron pair states requires that the modified photon propagator have, besides the usual photon pole at

$$-(\text{four-momentum})^2 = 0, \tag{1.2}$$

also complex poles at

$$-(\text{four-momentum})^2 = (m_B \pm \tfrac{1}{2} i\gamma_B)^2, \tag{1.3}$$

where, to $O(\alpha)$, the partial width $(\gamma_B)_{\text{lepton}}$ due to charged lepton pairs can be calculated, and is given by

$$(\gamma_B)_{\text{lepton}} = \tfrac{2}{3}\alpha m_B. \tag{1.4}$$

As will be discussed, in order to maintain unitarity, it is necessary that the poles due to the negative-metric B_μ field are, indeed, *off* the real axis.

It is well known[7] that in order to regularize the usual charge renormalization, one cannot, because of gauge invariance, simply modify, say, the electron propagator by introducing convergence factors. This difficulty is resolved in Sec. IV. We introduce two new Dirac spinor fields; their charges are imaginary $\pm ie$, and their masses are complex

$$m_F \pm \tfrac{1}{2} i\gamma_F, \tag{1.5}$$

where m_F and γ_F are both real and not zero. Both fields obey Fermi statistics. As will be shown, in this theory, the metric is uniquely determined by the canonical anticommutation relations and is indefinite; the gauge invariance and, therefore, current conservation are fully satisfied by introducing the usual minimal electromagnetic interaction; the unitarity is maintained because of the complex masses; and the charge renormalization becomes finite since, because of their $\pm ie$ charges, the loop diagram due to these new fermion fields is of the opposite sign from the usual one.

The modified Feynman rule is given in Sec. V. Explicit evaluations of charge renormalization and photon propagator are carried out in Sec. VI. Experimental consequences of this new finite theory of quantum electrodynamics are discussed in Sec. VII, and the causality problem is analyzed in Sec. VIII.

II. INDEFINITE METRIC

In this section we collect, for the convenience of the reader, some well-known facts about vector spaces with an indefinite metric and self-adjoint operators,[8] and we discuss the relations between the usual canonical commutation (or anticommutation) rules and the general structure of metric that can be used in a quantum theory.

Let us consider a complex vector space $\mathcal{H}$; let $|x\rangle$, $|y\rangle, \ldots$ be vectors in $\mathcal{H}$, and let $|1\rangle, |2\rangle, \ldots$ be a basis, so that

$$|x\rangle = \sum_i x_i |i\rangle, \tag{2.1}$$

where $x_1, x_2, \ldots$ are complex numbers. We define the scalar product $\langle x|y\rangle$ to be the Hermitian form

$$\langle x|y\rangle = \sum_{i,j} \eta_{ij} x_i^* y_j, \tag{2.2}$$

where an asterisk denotes complex conjugation, and the matrix

$$\eta = (\eta_{ij}) \tag{2.3}$$

is assumed to be a *nonsingular Hermitian matrix*. One calls the metric indefinite if η is not a positive definite matrix. From (2.2), one has, of course,

$$\langle i|j\rangle = \eta_{ij}. \tag{2.4}$$

A linear operator A_{op} can be specified by the equations

$$A_{\text{op}}|i\rangle = \sum_j A_{ji}|j\rangle, \tag{2.5}$$

where the A_{ij}'s are complex numbers; the adjoint operator $\bar{A}_{\text{op}}$ is defined by the equation

$$\langle x|\bar{A}_{\text{op}}|y\rangle = \langle y|A_{\text{op}}|x\rangle^* \tag{2.6}$$

for all vectors $|x\rangle$ and $|y\rangle$, where $\langle x|A_{\text{op}}|y\rangle$ denotes the scalar product between $|x\rangle$ and $A_{\text{op}}|y\rangle$. According to (2.2) and (2.4), we have

$$\langle x|A_{\text{op}}|y\rangle = \sum_{i,j,k} x_i^* \eta_{ij} A_{jk} y_k. \tag{2.7}$$

The expectation value $\langle x|A_{\text{op}}|x\rangle$ of a self-adjoint operator, defined by

$$A_{\text{op}} = \bar{A}_{\text{op}}, \tag{2.8}$$

is, therefore, real for all vectors $|x\rangle$.

All these equations can, of course, be expressed in terms of the appropriate matrix notations. For example, Eqs. (2.2) and (2.7) can be simply written as

$$\langle x|y\rangle = \bar{x}y \tag{2.9}$$

and

$$\langle x|A_{\text{op}}|y\rangle = \bar{x}Ay, \tag{2.10}$$

where y is a column matrix whose ith matrix element is y_i, $\bar{x}$ is a row matrix whose ith matrix element is

$$\bar{x}_i = \sum_j x_j^* \eta_{ji}, \tag{2.11}$$

and A is the square matrix

$$A = (A_{ij}). \tag{2.12}$$

[7] W. Pauli and F. Villars, Rev. Mod. Phys. **21**, 434 (1949); see also R. P. Feynman, Phys. Rev. **76**, 749 (1949).

[8] As is well known, an indefinite metric in the space of state vectors was first used by P. A. M. Dirac, Proc. Roy. Soc. (London) **A180**, 1 (1942).

Note that the matrix element A_{ij} should not be confused with the scalar product between the basis vector $|i\rangle$ and the vector $A_{\mathrm{op}}|j\rangle$, which is given by

$$\langle i|A_{\mathrm{op}}|j\rangle=(\eta A)_{ij}. \tag{2.13}$$

It is worthwhile to point out that the well-known transformation law of the matrix η implies that η does not represent a *bona fide* operator. In fact, if one chooses new basis vectors $|1'\rangle, |2'\rangle, \ldots,$

$$|i'\rangle=\sum_j T_{ji}|j\rangle, \tag{2.14}$$

then, in terms of the new basis, (2.1) becomes

$$|x\rangle=\sum_i x_i|i\rangle=\sum_i x_i'|i'\rangle. \tag{2.15}$$

Thus $x_i=\sum_j T_{ij}x_j'$, or, symbolically,

$$x'=T^{-1}x, \tag{2.16}$$

where, again, x and x' are column matrices, and T^{-1} is the inverse of the matrix $T=(T_{ij})$. The transformed matrix A' for the operator A_{op}, defined by

$$A_{\mathrm{op}}|i'\rangle=\sum_j A_{ji}'|j'\rangle, \tag{2.17}$$

is given by

$$A'=(A_{ij}')=T^{-1}AT. \tag{2.18}$$

On the other hand, relative to the new basis, the metric is obviously given by the matrix $\eta'=(\eta_{ij}')$, where

$$\eta_{ij}'=\langle i'|j'\rangle, \tag{2.19}$$

so that

$$\eta'=T^\dagger\eta T. \tag{2.20}$$

In view of the difference between (2.18) and (2.20) it is worthwhile to make a clear distinction between η and an *operator*. For example, algebraic equations involving η are not in general independent of the choice of basis. *Bona fide* operators, of course, do not have this defect.[9] Similarly, the eigenvalues of an operator are basis independent. Those of η are not. This has its good side: One can not only transform η to a diagonal form, but also assume that the diagonal matrix elements are equal to ±1 (zero being excluded by the assumption that η is nonsingular). At any rate, the eigenvalues of η have no invariant meaning. The only invariant property of η is the "signature" which (in a finite dimensional space) is defined to be the number of its positive eigenvalues minus that of its negative eigenvalues.

We now turn to the general question of different classes of the metric η allowed in a quantum theory. [All η's related by the transformation (2.20) are said to belong to the same class.] Through the usual quantization rules, one has, to begin with, the appropriate commutation, or anticommutation, relations for a set of operators a and their adjoints $\bar{a}$; furthermore, one is often only interested in the case in which the matrix representation of the entire set of these operators a and $\bar{a}$ is irreducible. As we shall see, in such a case, these commutation, or anticommutation, relations then completely determine the possible classes of the metric η.

We note that in any basis, because of (2.6) and $\eta=\eta^\dagger$, the matrix representations of a and $\bar{a}$ are related by

$$\bar{a}=\eta^{-1}a^\dagger\eta, \tag{2.21}$$

and therefore the over-all sign of η cannot be determined by any set of algebraic equations between a and $\bar{a}$. Of course, such an over-all sign has *no* physical meaning, since it merely changes the entire S matrix by a minus sign; in any quantum theory, only the relative phase has no physical significance. Thus, it is convenient to take care of this trivial over-all sign choice by generalizing the definition of the same class of η to include also

$$\eta\rightarrow-\eta, \tag{2.22}$$

in addition to those related by (2.20); i.e.,

$$\eta\rightarrow T^\dagger\eta T. \tag{2.20'}$$

The mathematical details for the determination of different classes of η are given in Appendices A and B. Here, we shall discuss some of the main conclusions. Consider, for example, the case of a single harmonic oscillator obeying Fermi-Dirac statistics. We have

$$a^2=0 \quad\text{and}\quad \bar{a}^2=0. \tag{2.23}$$

There are two different possibilities for the anticommutation relation between a and $\bar{a}$:

$$\text{(i)}\quad a\bar{a}+\bar{a}a=1; \tag{2.24}$$

$$\text{(ii)}\quad a\bar{a}+\bar{a}a=-1. \tag{2.25}$$

As will be shown in Appendix A, in case (i), by using (2.22), the metric η can always be set to be positive definite; this corresponds to the usual positive-metric case, in which

$$\langle x|x\rangle \text{ is positive} \tag{2.26}$$

for all vectors $|x\rangle$, and therefore one can always choose

[9] For this reason, we have chosen the notation in which η is omitted in the scalar-product symbol, given by the left-hand side of (2.2) (being "absorbed" in the definition of $\langle x|$). Such a notation has, of course, already been used occasionally, for example, by W. Heisenberg [Nucl. Phys. **4**, 532 (1957)]. It differs from the more commonly used notation in the literature, for example, in Pauli [Rev. Mod. Phys. **15**, 175 (1943)], and also in our earlier papers on the subject (Refs. 4–6). The present notation of not explicitly displaying η has the advantage of avoiding unnecessary noncovariant equations.

Similarly, one notes that while the matrix representation $\bar{A}$ of the adjoint operator $\bar{A}_{\mathrm{op}}$ transforms in the same way as the matrix representation A of any operator A_{op}, the Hermitian conjugate $A^\dagger$ does not. Thus, it is always of considerable convenience to use only A and $\bar{A}$, but not $A^\dagger$.

In this connection, it is important to dispel the notion that certain noncovariant relations involving η, which occur in almost every treatment of the subject, are really indispensable. For example, as will be shown in the following, noncovariant equations such as $\eta a=-a\eta$ and $\eta=(-1)^{\bar{a}a}$, that are usually assumed in the theory of a harmonic oscillator with indefinite metric, can be completely replaced by the more general covariant condition (2.27).

the basis so that

$$\eta=1. \tag{2.26'}$$

In case (ii), there is also only a *single* class for the metric η; by using (2.22), one can always require, instead of (2.26), that

$$\langle x|(-1)^{\bar{a}a}|x\rangle \text{ be positive} \tag{2.27}$$

for all vectors $|x\rangle$. The metric η is therefore indefinite. If one wishes, one may choose a specific basis so that

$$\eta=(-1)^{\bar{a}a}. \tag{2.27'}$$

We note that (2.26) and (2.27) are valid for all bases, while (2.26′) and (2.27′) are, of course, basis dependent. In either case, the quantum of this oscillator is said to be of positive or negative metric, depending on whether case (i) or case (ii) holds.

In the case of a single harmonic oscillator obeying Bose-Einstein statistics, the commutation relation

$$a\bar{a}-\bar{a}a=-1 \tag{2.28}$$

can be reduced to

$$a\bar{a}-\bar{a}a=1 \tag{2.29}$$

by interchanging the roles of a and $\bar{a}$. As will be shown in Appendix B, this commutation relation limits the metric η to one of the following three classes.

(i) *Definite metric case.* In this case (2.26) holds and therefore (2.26′) is always a possible choice.

(ii) *Indefinite metric case (normal).* In this case (2.27) holds, and therefore (2.27′) is always a possible choice.

(iii) *Abnormal case.* In this case the metric η is indefinite; unlike the situation in cases (i) and (ii), the eigenvalues of $\bar{a}a$ are not integers, and they have neither upper nor lower bound.

In the following, we shall impose the condition that the operator $\bar{a}a$ should be bound either from above, or from below; this would ensure, at least for the free oscillator, a lower bound to the energy spectrum. This additional condition then rules out the abnormal case (iii). The class of metric is then uniquely determined by specifying whether η is definite or indefinite. The quantum of this oscillator is said to be of positive or negative metric, depending on whether η is definite or indefinite.

For physical applications of a quantum theory with indefinite metric, we choose the Hamiltonian H to be a self-adjoint operator. Let $|r\rangle$ denote any eigenvector of H with a real eigenvalue. We recall the elementary, but important, property[4] that the S matrix is unitary if

$$\langle r|r\rangle \text{ is positive} \tag{2.30}$$

for *all* such eigenvectors $|r\rangle$ with real eigenvalues. In such a theory, each physical observable is represented by a self-adjoint operator A, and each physical state $|\ \rangle$ must be a linear superposition of only eigenstates $|r\rangle$ belonging to the real eigenvalues. The expectation value of A is then given by the usual formula,

$$\langle A\rangle=\langle|A|\rangle/\langle\ |\ \rangle. \tag{2.31}$$

III. PHOTON FIELD

In this section we shall review briefly the basic formalism of the modified photon field

$$\phi_\mu=A_\mu+iB_\mu. \tag{3.1}$$

For clarity, we consider first a simple system consisting of three fields: A_μ, B_μ, and a charged lepton field ψ_l which can be either the usual electron field or the usual muon field. (The inclusion of other fermion fields of an unusual type will be discussed in Sec. IV.)

The Lagrangian density of this simple system is given by

$$\mathcal{L}_{\text{free}}(\phi)+\mathcal{L}(l,\phi), \tag{3.2}$$

where

$$\mathcal{L}_{\text{free}}(\phi)=-\tfrac{1}{4}(G_{\mu\nu}{}^2+F_{\mu\nu}{}^2)-\tfrac{1}{2}(m_B{}^0B_\mu)^2, \tag{3.3}$$

$$\mathcal{L}(l,\phi)=-\bar{\psi}_l\gamma_4\left[\gamma_\mu\left(\frac{\partial}{\partial x_\mu}-ie_0\phi_\mu\right)+m_l{}^0\right]\psi_l, \tag{3.4}$$

$$F_{\mu\nu}=\frac{\partial}{\partial x_\mu}A_\nu-\frac{\partial}{\partial x_\nu}A_\mu, \tag{3.5}$$

$$G_{\mu\nu}=\frac{\partial}{\partial x_\mu}B_\nu-\frac{\partial}{\partial x_\nu}B_\mu, \tag{3.6}$$

e_0 is the unrenormalized charge, and $m_B{}^0$ and $m_l{}^0$ are the unrenormalized masses of the B_μ and ψ_l fields. In the above expressions, $\bar{\psi}_l$ denotes the adjoint operator of ψ_l, and ϕ_μ is a self-adjoint vector field[10]; i.e.,

$$\bar{\phi}_\mu=+\phi_\mu \quad \text{for} \quad \mu\neq4$$

and

$$\bar{\phi}_4=-\phi_4.$$

Consequently, A_μ is a self-adjoint field,

$$\bar{A}_\mu=+A_\mu \quad \text{for} \quad \mu\neq4 \tag{3.7}$$

and

$$\bar{A}_4=-A_4, \tag{3.8}$$

but B_μ is not,[11]

$$\bar{B}_\mu=-B_\mu \quad \text{for} \quad \mu\neq4 \tag{3.9}$$

and

$$\bar{B}_4=+B_4. \tag{3.10}$$

The Hamiltonian H and its quantization can be carried out by following the usual canonical treatment. We may choose as generalized coordinates the lepton field ψ_l, the transverse electromagnetic vector potential

[10] Throughout the paper, the subscripts μ and ν denote the space-time indices; $\mu=4$ is the time component with $x_4=it$. All boldface letters denote three-dimensional vectors.

[11] If one wishes, one may replace iB_μ by B_μ', which is a self-adjoint vector field. Then, of course, the free Lagrangian for B_μ' appears in (3.3) with an unusual sign.

A_j^{tr} (in the Coulomb gauge[12]), and the spatial component B_j of the massive Boson field; their conjugate momenta are, respectively, $i\bar{\psi}_l$, $-E_j^{\text{tr}}$, and

$$\Pi_j = iG_{4j}. \quad (3.11)$$

At equal time, one has

$$\{\psi_l(\mathbf{r},t),\bar{\psi}_l(\mathbf{r}',t)\} = \delta^3(\mathbf{r}-\mathbf{r}'), \quad (3.12)$$

$$[-E_j^{\text{tr}}(\mathbf{r},t),A_k^{\text{tr}}(\mathbf{r}',t)] = -i(\delta_{jk}-\nabla^{-2}\nabla_j\nabla_k)\delta^3(\mathbf{r}-\mathbf{r}'), \quad (3.13)$$

and

$$[\Pi_j(\mathbf{r},t),B_k(\mathbf{r}',t)] = -i\delta_{jk}\delta^3(\mathbf{r}-\mathbf{r}'). \quad (3.14)$$

It is convenient to introduce the usual Fourier components of these fields. For example, we may write

$$\mathbf{B} = \sum_{\mathbf{k},t} (2\Omega\omega_k)^{-1/2}(a_{\mathbf{k}}{}^t e^{i\mathbf{k}\cdot\mathbf{r}} - \bar{a}_{\mathbf{k}}{}^t e^{-i\mathbf{k}\cdot\mathbf{r}})\hat{e}_{\mathbf{k}}{}^t + \sum_k (2\Omega\omega_k)^{-1/2}(a_{\mathbf{k}}{}^l e^{i\mathbf{k}\cdot\mathbf{r}} - \bar{a}_{\mathbf{k}}{}^l e^{-i\mathbf{k}\cdot\mathbf{r}})(\omega_k\hat{k}/m_B{}^0) \quad (3.15)$$

and

$$\mathbf{\Pi} = \sum_{\mathbf{k},t} (2\Omega\omega_k)^{-1/2}(a_{\mathbf{k}}{}^t e^{i\mathbf{k}\cdot\mathbf{r}} + \bar{a}_{\mathbf{k}}{}^t e^{-i\mathbf{k}\cdot\mathbf{r}})(-i\omega_k\hat{e}_{\mathbf{k}}{}^t) + \sum_{\mathbf{k}} (2\Omega\omega_k)^{-1/2}(a_{\mathbf{k}}{}^l e^{i\mathbf{k}\cdot\mathbf{r}} + \bar{a}_{\mathbf{k}}{}^l e^{-i\mathbf{k}\cdot\mathbf{r}})(-im_B{}^0\hat{k}), \quad (3.16)$$

where $\hat{e}_k{}^1$, $\hat{e}_k{}^2$, and $\hat{k} = |\mathbf{k}|^{-1}\mathbf{k}$ form a right-handed orthonormal set of unit vectors,

$$\omega_k = [\mathbf{k}^2 + (m_B{}^0)^2]^{1/2}, \quad (3.17)$$

and Ω is the volume of the system. The commutation relation (3.14) becomes then

$$[a_{\mathbf{k}}{}^\alpha, \bar{a}_{\mathbf{q}}{}^\beta] = -\delta_{\mathbf{kq}}\delta_{\alpha\beta}, \quad (3.18)$$

where α, β denote either $t=1$, 2, or l. As already discussed in the previous section, the anticommutation relation determines the metric of ψ_l to be positive; the commutation relations (3.13) and (3.14) limit the metric of these Boson fields to two possibilities: either positive or negative. We will now specify the metric of the system by requiring[13]

$$\langle\ |(-1)^{N_B}|\ \rangle \text{ to be positive} \quad (3.19)$$

for all vectors $|\ \rangle$, where

$$N_B \equiv -\sum_{\mathbf{k},t} \bar{a}_{\mathbf{k}}{}^t a_{\mathbf{k}}{}^t - \sum_{\mathbf{k}} \bar{a}_{\mathbf{k}}{}^l a_{\mathbf{k}}{}^l. \quad (3.20)$$

That is, A_μ is of positive metric and B_μ is of negative metric.

It is easy to verify that the total Hamiltonian H is self-adjoint. The equations of motion can be derived by using the usual Heisenberg equations, and the compatibility of the equal-time commutation relations (3.12)–(3.14) with relativistic invariance can be established following the usual arguments. For the unitarity property, we note that for the free field ($e_0=0$), our basic condition (2.30) is not satisfied since there exist negative-metric eigenstates of the free Hamiltonian with real eigenvalues, e.g., states with $N_B=$ odd integers. [For the free system, the condition (2.30) is, of course, irrelevant since the S matrix is the unit matrix.] In order to satisfy our basic condition (2.30), it is necessary to have $e_0\neq 0$. Consider, for example, the negative-metric state $|B\rangle$ of a single free B_μ quantum, i.e., $N_B=1$. For $m_B{}^0 > 2m_l{}^0$, such a state, in the limit $e_0=0$, is degenerate[14] with the positive-metric lepton pair states $|l^+l^-\rangle$. It can be readily verified that this degeneracy is removed in the presence of the interaction. Furthermore, the resulting eigenvalues are[15]

$$m_B \pm \tfrac{1}{2}i\gamma_B, \quad (3.21)$$

where γ_B and m_B are both real, denoting the width and the renormalized mass of the B_μ quantum. The corresponding eigenvectors $|\pm\rangle$ of these two complex eigenvalues consist of a coherent mixture of $|B\rangle$ and $|l^+l^-\rangle$; these two eigenvectors both have zero norm,

$$\langle +|+\rangle = \langle -|-\rangle = 0, \quad (3.22)$$

and can be normalized so that

$$\langle +|-\rangle = 1. \quad (3.23)$$

The remaining eigenstates $|r\rangle$ with real eigenvalues can then be shown to have positive norm, and therefore (2.30) is satisfied. Further details will be given in Secs. V and VI.

Since the propagator of the modified photon field $\phi_\mu = A_\mu + iB_\mu$ can be easily seen to be proportional to k^{-4} in the high-momentum limit as $k\to\infty$, one finds that, through a straightforward power counting, except for charge renormalization all higher-order processes of the Lagrangian (3.2) are finite. In order to render charge renormalization finite, additional new fermion fields have to be introduced; these will be discussed in Sec. IV.

---

[12] The Coulomb gauge is chosen here purely for convenience. In the Coulomb gauge, A_μ is given by $A_j = A_j{}^{\text{tr}}$ and $A_4 = i\phi$, where $A_j{}^{\text{tr}}$ satisfies $(\partial A_j{}^{\text{tr}}/\partial r_j)=0$ and ϕ is the solution of $\Delta\phi = -e_0\bar{\psi}_l\gamma_4\psi_l$. Correspondingly, the electric field is given by $E_j = E_j{}^{\text{tr}} + E_j{}^{\text{long}}$, where $E_j{}^{\text{tr}} = -(\partial A_j{}^{\text{tr}}/\partial t)$ and $E_j{}^{\text{long}} = -(\partial\phi/\partial r_j)$. In deriving the canonical momentum of $A_j{}^{\text{tr}}$ as $-E_j{}^{\text{tr}}$, we have, as usual, first set in the Lagrangian the spatial integral $\int E_j{}^{\text{tr}}E_j{}^{\text{long}}d^3r = 0$, and then taken its derivative with respect to $(\partial A_j{}^{\text{tr}}/\partial t)$.

[13] We emphasize that (3.19), as well as other equations in this section, are all basis independent. Of course, it follows from (3.19) that, if one wishes, one can always choose a specific basis so that $\eta = (-1)^{N_B}$.

[14] If $m_B{}^0 < 2m_l{}^0$, then, in the limit $e_0=0$, this negative-metric state is degenerate with the positive-metric three-photon state. The introduction of $e_0\neq 0$ would also remove this degeneracy and make it possible to satisfy (2.30).

[15] After the completion of this paper, Norman Kroll kindly drew our attention to the fact that in connection with his investigation on *ad hoc* modifications of quantum electrodynamics [Nuovo Cimento **45**, 65 (1966)], he has also noted the possibility of such complex poles and some of their consequences. The so-called "causality difficulty" mentioned in Kroll's paper, however, has been resolved in Ref. 4. (See also Sec. VIII of the present paper.)

IV. FERMION FIELDS

Let

$$\psi_F\equiv\begin{pmatrix}\psi_1\\ \psi_2\end{pmatrix} \tag{4.1}$$

denote a new fermion field, where ψ_1 and ψ_2 are both four-component Dirac fields, and

$$\bar\psi_F=(\bar\psi_1\,\bar\psi_2)$$

denotes the adjoint field, where the bar is used in the sense of Eq. (2.6). The Lagrangian density for a system consisting of A_μ, B_μ, the usual charged lepton fields ψ_e, ψ_μ, and this new fermion field ψ_F is given by

$$\mathcal{L}=\mathcal{L}_{\text{free}}(\phi)+\mathcal{L}(F,\phi)+\sum_{l=e,\mu}\mathcal{L}(l,\phi), \tag{4.2}$$

where $\mathcal{L}_{\text{free}}(\phi)$ and $\mathcal{L}(l,\phi)$ are given, respectively, by (3.3) and (3.4),

$$\mathcal{L}(F,\phi)\equiv-\bar\psi_F\tau_a\gamma_4\left[\gamma_\lambda\left(\frac{\partial}{\partial x_\lambda}+e_0\tau_b\phi_\lambda\right)+M_F{}^0\right]\psi_F, \tag{4.3}$$

and τ_a, τ_b are two of the usual (2×2) Pauli matrices which commute with all Dirac matrices γ_μ and satisfy

$$\tau_a{}^2=\tau_b{}^2=1, \tag{4.4}$$

$$\tau_a{}^\dagger=\tau_a,\quad \tau_b{}^\dagger=\tau_b, \tag{4.5}$$

and

$$\tau_a\tau_b+\tau_b\tau_a=0. \tag{4.6}$$

In (4.3), $M_F{}^0$ is also a (2×2) matrix, given by

$$M_F{}^0=m_F{}^0+i\tfrac{1}{2}\gamma_F{}^0\tau_b, \tag{4.7}$$

where $m_F{}^0$ and $\gamma_F{}^0$ are real numbers. The total Lagrangian density is therefore self-adjoint, because of (4.5) and (4.6). From (4.3), one finds that the conjugate momentum of ψ_F is $i\bar\psi_F\tau_a$. The canonical quantization rule gives then, in addition to (3.12)–(3.14),

$$\{\psi_F(\mathbf{r}',t),\bar\psi_F(\mathbf{r},t)\}=\tau_a\delta^3(\mathbf{r}-\mathbf{r}'). \tag{4.8}$$

By following the standard procedure of passing from Lagrangian to Hamiltonian, one finds the Heisenberg equation for the new fermion field,

$$\gamma_\lambda\left(\frac{\partial}{\partial x_\lambda}+e_0\tau_b\phi_\lambda\right)\psi_F+M_F{}^0\psi_F=0, \tag{4.9}$$

which implies

$$\frac{\partial\bar\psi_F}{\partial x_\lambda}\tau_a\gamma_4\gamma_\lambda-\bar\psi_F\tau_a\gamma_4(e_0\tau_b\gamma_\lambda\phi_\lambda+M_F{}^0)=0.$$

We have, therefore, two conservation laws:

$$\frac{\partial}{\partial x_\lambda}(\bar\psi_F\tau_a\tau_b\gamma_4\gamma_\lambda\psi_F)=0 \tag{4.10}$$

and

$$\frac{\partial}{\partial x_\lambda}(i\bar\psi_F\tau_a\gamma_4\gamma_\lambda\psi_F)=0, \tag{4.10'}$$

which correspond to the two similar conservation laws for the usual charged lepton fields,

$$\sum_{l=e,\mu}\frac{\partial}{\partial x_\lambda}(i\bar\psi_l\gamma_4\gamma_\lambda\psi_l)=0 \tag{4.11}$$

and

$$\frac{\partial}{\partial x_\lambda}(i\bar\psi_e\gamma_4\gamma_\lambda\psi_e-i\bar\psi_\mu\gamma_4\gamma_\lambda\psi_\mu)=0. \tag{4.11'}$$

The total electromagnetic current density j_μ is given by

$$j_\lambda\equiv e_0\bar\psi_F\tau_a\tau_b\gamma_4\gamma_\lambda\psi_F-ie_0\sum_{l=e,\mu}\bar\psi_l\gamma_4\gamma_\lambda\psi_l, \tag{4.12}$$

and it is clearly conserved:

$$\partial j_\mu/\partial x_\mu=0. \tag{4.13}$$

Under the gauge transformation

$$A_\mu\to A_\mu+\partial\Lambda/\partial x_\mu \tag{4.14}$$

and

$$B_\mu\to B_\mu, \tag{4.15}$$

the charged lepton field ψ_l ($l=e,\mu$) has the usual transformation

$$\psi_l\to e^{ie_0\Lambda}\psi_l, \tag{4.16}$$

and therefore

$$\bar\psi_l\to\bar\psi_l e^{-ie_0\Lambda};$$

the new fermion field ψ_F transforms according to

$$\psi_F\to e^{-e_0\tau_b\Lambda}\psi_F, \tag{4.17}$$

and therefore

$$\bar\psi_F\to\bar\psi_F e^{-e_0\tau_b\Lambda}.$$

By using (4.2), it can be readily verified that the theory is gauge invariant.

Just as in (3.19), we will now specify the metric of the system by requiring

$$\langle\,|(-1)^{N_B+N_F}|\,\rangle \text{ to be positive} \tag{4.18}$$

for all vectors $|\,\rangle$, where N_B is given by (3.20) and N_F is given by Eq. (C16) of Appendix C. By following the general discussions given in Sec. II, it is easy to see that the factor $(-1)^{N_F}$ is completely determined by the anticommutation relation (4.8), just as is the factor $(-1)^{\bar a a}$ in (2.27). The details are given in Appendix C. In the same appendix, the free Hamiltonian of the system is explicitly diagonalized. As will be shown there, the frequency spectrum of the free ψ_F field is given by

$$[\mathbf{k}^2+(m_F{}^0\pm\tfrac{1}{2}i\gamma_F{}^0)^2]^{1/2}. \tag{4.19}$$

We note that in the boson case, the bare mass $m_B{}^0$ of B_μ is assumed to be real; the width γ_B is acquired through the electromagnetic interaction between B_μ and the charged lepton pairs, and therefore it is not an independent parameter, as will be calculated in Sec. VI. On the other hand, in the fermion case, because of the conservation laws (4.10) and (4.10′), in order to ensure unitarity we have to assume a nonzero width $\gamma_F{}^0$ for the free ψ_F field.

V. FEYNMAN DIAGRAMS

In the present theory, because of complex singularities such as (3.21) and (4.19), the Green's function $U(t, -t)$ of the time-dependent Schrödinger equation in, say, the interaction representation diverges exponentially in the limit $t \rightarrow \infty$. On the other hand, the S matrix is, of course, well defined in terms of the eigenvectors of the total Hamiltonian with real eigenvalues; its matrix elements are given by

$$S_{r'r} = \langle r'^{\text{in}} | r^{\text{out}} \rangle , \tag{5.1}$$

where the superscript "out" (or "in") denotes states consisting of only plane waves and outgoing (or incoming) waves. Thus, one does not have the usual relation between the S matrix and the limit $U(t, -t)$ at $t = \infty$. Nevertheless, as pointed out in Ref. 4, it is possible to separate $U(t, -t)$ into a sum

$$U(t, -t) = U^{\text{reg}}(t, -t) + U^{\text{exp}}(t, -t) , \tag{5.2}$$

where as $t \rightarrow \infty$, $U^{\text{exp}}(t, -t)$ diverges exponentially, but

$$\lim_{t\rightarrow\infty} U^{\text{reg}}(t, -t) = S . \tag{5.3}$$

From (5.1) and (5.3), one obtains a set of modified Feynman rules which will be briefly reviewed in this section.

In general, in a quantum field theory with an indefinite metric, any S-matrix element can always be given by a sum over an appropriate set of Feynman graphs, just as in the usual theory with a definite metric. Each graph still stands for a multidimensional integral, integrated over a domain of some virtual four-momenta, which will be labeled collectively as k_μ; the integration over the space component $\mathbf{k}$ remains, as usual, over the entire real region, but that over the time component k_0 is now along a complex path. We note that, unlike the case in the usual theory, the integrand now has complex singularities. To obtain the correct integration path in the complex k_0 plane, it is convenient to first regard all imaginary parts γ_i of these complex singularities as independent parameters. For $\gamma_i = 0$, we have the usual Feynman rule: At fixed $\mathbf{k}$, the integration over k_0 is along the real axis, and the positions of the singularities of the integrand are determined by the usual $i\epsilon$ rule. Alternatively, one may regard (still for $\gamma_i = 0$) all singularities to be on the real axis, but the Feynman path C is slightly detoured so as to go either above or below the appropriate singularities. Now as γ_i increases from zero to its final value, these singularities will move continuously. We shall require the integration path C for dk_0 to be continuously

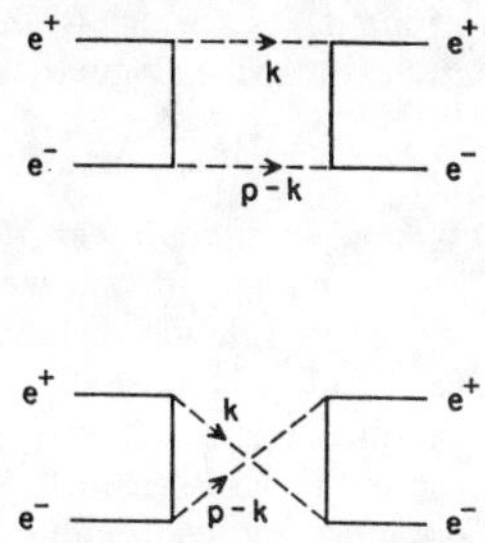

FIG. 1. Two examples of fourth-order e^+e^- scattering diagrams.

deformed in such a way that none of the singularities ever crosses C. It is clear that C is unambiguously defined only if there is no pinching along the path; i.e., if complex singularities on different sides of the path do not coalesce. As we shall see, such pinching occurs only over an integration domain of "zero" measure, and therefore does not contribute to the Feynman integral.

For clarity, let us consider some specific diagrams for, say, the elastic scattering of e^+e^- given in Fig. 1. The dashed line refers to the propagator of

$$\phi_\mu = A_\mu + iB_\mu ,$$

which, as will be explicitly calculated in Sec. VI, contains complex poles at

$$-(\text{four-momentum})^2 = M_B{}^2 \quad \text{and} \quad (M_B{}^*)^2 , \tag{5.4}$$

where

$$M_B = m_B + \tfrac{1}{2} i \gamma_B . \tag{5.5}$$

Thus, the integrand for either diagram in Fig. 1 has complex poles at, among others,

$$k_0 = (\mathbf{k}^2 + M_B{}^2)^{1/2} \tag{5.6}$$

and

$$k_0 = p_0 - [(\mathbf{p} - \mathbf{k})^2 + (M_B{}^*)^2]^{1/2} , \tag{5.7}$$

where p_μ denotes the total momentum of e^+ and e^-. According to the above "modified" Feynman rule, the pole (5.6) should be below the integration path C and the pole (5.7) should be above C. This is always possible except when these two poles coalesce, which occurs when

$$p_0 = (\mathbf{k}^2 + M_B{}^2)^{1/2} + [(\mathbf{p} - \mathbf{k})^2 + (M_B{}^*)^2]^{1/2} . \tag{5.8}$$

Except for the special center-of-mass reference frame, in any reference frame the total momentum $\mathbf{p}$ is not zero, and therefore Eq. (5.8) represents two independent conditions in order to satisfy both its imaginary and real parts:

$$|\mathbf{k}| = |\mathbf{p} - \mathbf{k}| \tag{5.9}$$

and

$$p_0 = \tfrac{1}{2} \operatorname{Re}(\mathbf{k}^2 + M_B{}^2)^{1/2} , \tag{5.10}$$

since p_0 denotes the total energy of e^+ and e^-, which must be real.

In $\mathbf{k}$ space, the points that satisfy these two conditions lie on a circle whose center is at $\frac{1}{2}|\mathbf{p}|$. For all other points, there is no pinching and the Feynman integral over k_0 is well defined; the result of this k_0 integration is a function of $\mathbf{k}$ which is singular at the above-mentioned circle, but with singularities no worse than

$$\{p_0-(\mathbf{k}^2+M_B^2)^{1/2}-[(\mathbf{p}-\mathbf{k})^2+(M_B^*)^2]^{1/2}\}^{-1}. \quad (5.11)$$

The subsequent integration over $\mathbf{k}$ is therefore absolutely convergent and well defined[16]; the ambiguity of the integrand over such a set of points of zero measure (a circle) in $\mathbf{k}$ space does not lead to any ambiguity in the final result. Therefore, in any system (except the center-of-mass system), the Feynman integral can be obtained by continuous deformation from the usual Feynman path. For the particular center-of-mass frame $\mathbf{p}=0$, we require the corresponding Feynman integral to be obtained as the limit $\mathbf{p}\rightarrow 0$.

The actual Feynman integration for the above diagrams can be carried out in exactly the same way as done in Sec. 5 of Ref. 5. The result is that, apart from the usual cut corresponding to the normal two-zero-mass–photon exchange, there is no additional imaginary part in the scattering amplitude due to the virtual two complex-mass heavy boson states. Consequently, unitarity holds. The usual condition of analyticity is, of course, violated.

The above discussions can be readily carried out for more complicated diagrams involving multiphoton exchanges. If the intermediate state consists of three or more particles, the equation for pinching always gives two independent conditions in any reference frame, including the center-of-mass frame. Take, for example, the case of three particles of energies, say,

$$\omega_k=(\mathbf{k}^2+M_B^2)^{1/2}, \quad \omega_q^*=[\mathbf{q}^2+(M_B^*)^2]^{1/2},$$

and $\nu_{p-k-q}=|\mathbf{p}-\mathbf{k}-\mathbf{q}|$. Just as in Eq. (5.8), the condition for pinching occurs when

$$p_0=\omega_k+\omega_q^*+\nu_{p-k-q}, \quad (5.12)$$

where $\mathbf{p}$ and p_0 denote the total momentum and energy of the system. Equation (5.12) implies two separate conditions,

$$|\mathbf{k}|=|\mathbf{q}| \quad (5.13)$$

and

$$\tfrac{1}{2}(p_0-\nu_{p-k-q})=\mathrm{Re}\,\omega_k, \quad (5.14)$$

independently of whether $\mathbf{p}=0$ or not. Just as in the above case of two-photon exchange, the points that satisfy these two conditions (5.13) and (5.14) give zero contribution to the Feynman integral.

In these Feynman integrals, since the domain of all three-momenta is kept real, but that of the fourth components is complex, the question of relativistic invariance naturally arises, especially for more complicated diagrams involving several complex masses. In this connection, there exists an alternative, but manifestly covariant, prescription given by Cutkosky, Landshoff, Olive, and Polkinghorne[17]; in their prescription, whenever there are a pair of regularized photon lines, one always first sets the complex masses in one line to be M_B and M_B^*, and in the other line M_B' and $M_B'^*$. The physical case $M_B=M_B'$ is obtained as the limit $M_B'\rightarrow M_B$. Relativistic invariance then becomes obvious; for example, in the two-photon-exchange diagrams discussed, for arbitrary $M_B\neq M_B'$, the Feynman integral is well defined in any system $\mathbf{p}\neq 0$ or $\mathbf{p}=0$. For simple diagrams, these two prescriptions give identical results. For more complicated diagrams there may be some differences,[17] in which case one should adopt the prescription of Cutkosky *et al.* with (whenever necessary) some further specifications as to the correct order of limits. For example, Cutkosky *et al.* found that, for the so-called double ice cream cone diagram (which corresponds to a diagram of order at least e^{10}), depending on the order of limits, their prescrrption leads to two different expressions for the S matrix, each being fully relativistic and unitary. The mathematical complexities involved in these diagrams of rather high order have prevented us from appreciating fully their arguments. Nevertheless, it seems reasonable to expect that such an ambiguity can be resolved by noting the Bose statistics nature of photons. All limits must be taken symmetrically with respect to different internal photon lines. This would then lead to a unique answer for the double ice cream cone diagram.

VI. PHOTON PROPAGATOR

As an explicit example, we shall evaluate the propagator D of the modified photon field

$$\phi_\mu=A_\mu+iB_\mu. \quad (6.1)$$

It is convenient to represent all propagators between the two fields A_μ and B_μ by a (2×2) matrix

$$\mathfrak{D}\equiv\begin{pmatrix} D_{AA} & D_{AB} \\ D_{BA} & D_{BB}\end{pmatrix}, \quad (6.2)$$

where the two subscripts denote, respectively, the initial and final states to be either A_μ or B_μ. The propagator for the coherent mixture $A_\mu+iB_\mu$ is then given by

$$D=\tilde{\xi}\mathfrak{D}\xi, \quad (6.3)$$

[16] To show this, it is only necessary to study the $\mathbf{k}$ integration near the singularities of (5.11). Let us use the cylindrical coordinates $\mathbf{k}=(\rho\cos\phi,\ \rho\sin\phi,\ z)$, where the z axis is parallel to $\mathbf{p}$. The singularities of (5.11) lie on the circle $z=\frac{1}{2}|\mathbf{p}|$ and $\rho=\rho_0$, where ρ_0 is its radius, determined by (5.10), i.e., $\rho_0=\frac{1}{2}\,\mathrm{Re}\,\omega_0$ and $\omega_0=(\rho_0^2+\frac{1}{4}|\mathbf{p}|^2+M_B^2)^{1/2}$. In the neighborhood of this circle, it is convenient to write $\rho=\rho_0+r\cos\theta$ and $z=\frac{1}{2}|\mathbf{p}|+r\sin\theta$. The region $r\leqq\epsilon\ll 1$ denotes, then, a doughnut which contains all singularities of (5.11); its cross section is a small circle of radius ϵ. Neglecting $O(\epsilon^2)$, one finds, inside this doughnut region, $(5.11)=[r(a\cos\theta+ib\sin\theta)]^{-1}$, where a and b are two real constants, different from zero, and given explicitly by $a=-\rho_0[\omega_0^{-1}+(\omega_0^*)^{-1}]$ and $b=i\frac{1}{2}\times[\omega_0^{-1}-(\omega_0^*)^{-1}]\,|\mathbf{p}|$. Since $d^3k=r\rho_0\,dr\,d\theta\,d\phi$, the $\mathbf{k}$ integration of (5.11) over this doughnut region is simply

$$2\pi\epsilon\rho_0\int_0^{2\pi} d\theta(a\cos\theta+ib\sin\theta)^{-1},$$

which goes to zero as $\epsilon\rightarrow 0$.

[17] R. E. Cutkosky, P. V. Landshoff, D. Olive, and J. C. Polkinghorne, Nucl. Phys. **B12**, 281 (1969). We wish to thank Dr. Cutkosky for a discussion on the double ice cream cone diagram.

where

$$\xi=\begin{pmatrix}1\\ i\end{pmatrix} \tag{6.4}$$

and $\tilde{\xi}$ is the transpose. For the free fields, we have

$$\mathfrak{D}=\mathfrak{D}_{\text{free}}=-i\begin{pmatrix} k^{-2}\delta_{\mu\nu}+(\cdots k_\mu k_\nu) & 0 \\ 0 & [k^2+(m_B{}^0)^2]^{-1}[\delta_{\mu\nu}+(m_B{}^0)^{-2}k_\mu k_\nu]\end{pmatrix}, \tag{6.5}$$

and therefore (6.3) becomes

$$D_{\text{free}}=\frac{-i}{k^2}\left(\frac{(m_B{}^0)^2}{k^2+(m_B{}^0)^2}\right)\delta_{\mu\nu}+(\cdots k_\mu k_\nu), \tag{6.6}$$

where $(\cdots k_\mu k_\nu)$ denotes gauge-dependent terms, which are always proportional to $k_\mu k_\nu$.

In Fig. 2, we introduce

$$i(k^2\delta_{\mu\nu}-k_\mu k_\nu)P \tag{6.7}$$

to represent the sum of all irreducible self-energy graphs in the propagator D_{AA}; because of the coherent mixture (6.1) that appears in all electromagnetic interactions, identical sets of irreducible self-energy graphs also appear in the other propagators D_{AB}, D_{BA}, and D_{BB}. Let Π be a (2×2) matrix, defined by

$$\Pi\equiv\xi\tilde{\xi}i(k^2\delta_{\mu\nu}-k_\mu k_\nu)P. \tag{6.8}$$

One, then, has

$$\mathfrak{D}=\mathfrak{D}_{\text{free}}+\mathfrak{D}_{\text{free}}\Pi\mathfrak{D}_{\text{free}}+\mathfrak{D}_{\text{free}}\Pi\mathfrak{D}_{\text{free}}\Pi\mathfrak{D}_{\text{free}}+\cdots, \tag{6.9}$$

which leads to

$$D=\frac{-i(m_B{}^0)^2\delta_{\mu\nu}}{k^2[k^2+(m_B{}^0)^2(1-P)]}+(\cdots k_\mu k_\nu). \tag{6.10}$$

In Fig. 2, the dashed line represents D itself. It is easy to see that the sum of the first four lowest-order graphs is finite, although each of them diverges logarithmically. With respect to the other graphs of higher order in α, since D is proportional to k^{-4} as $k\to\infty$, they are all finite, as may be shown by means of the "power counting theorem."[18] In this respect, the gauge invariance of our prescription is crucial, since it leads to the cancellation of terms that would otherwise lead to divergences.[19]

To calculate this sum, *we shall follow the modified Feynman rule given in Sec. V*. The result is

$$P(z)=-(\alpha/\pi)[\tfrac{2}{3}\ln(M_F M_F{}^*/m_e m_\mu)+f(z,m_e) +f(z,m_\mu)-f(z,M_F)-f(z,M_F{}^*)]+O(\alpha^2), \tag{6.11}$$

where $\alpha\cong(137)^{-1}$, $M_F=m_F+\frac{1}{2}i\gamma_F$,

$$z=-k^2, \tag{6.12}$$

$$f(z,\lambda)=(3z)^{-1}(4\lambda^2+2z)[1+(\tfrac{1}{4}z-\lambda^2)J(z,\lambda)], \tag{6.13}$$

$$J(z,\lambda)=\int_0^1\frac{dx}{\lambda^2-x(1-x)z}=\frac{2}{F}\ln\frac{F-z}{F+z}, \tag{6.14}$$

and

$$F=(z^2-4\lambda^2 z)^{1/2}. \tag{6.15}$$

We note that in (6.14) the integration path for x is along the real axis from 0 to 1; this follows from the modified Feynman rule, after some transformations and calculations. In (6.11), we have omitted the superscript 0 in all masses, since to $O(\alpha)$ it is immaterial whether one uses the bare masses or the observed masses.

For $\lambda=m_l$ ($l=e$ or μ), $J(z,m_l)$ has a branch point at $z=4m_l{}^2$, and we may choose the cut to be along the real axis from $z=4m_l{}^2$ to ∞. Outside the cut, the integral representation (6.14) is valid everywhere. The imaginary part of $f(z,m_l)$ along the cut is given by

$$\operatorname{Im}f(z,m_l)=\pm\pi\left(\frac{z+2m_l{}^2}{3z}\right)\left(\frac{z-4m_l{}^2}{z}\right)^{1/2}, \tag{6.16}$$

where the $+$ or $-$ sign depends on z being slightly above or below the cut. For $\lambda=M_F$, or $M_F{}^*$, $f(z,\lambda^2)$ is analytic everywhere except at the branch point $z=4M_F{}^2$, or $4(M_F{}^*)^2$. Furthermore, it follows from (6.14) that

$$f(z,\lambda)^*=f(z^*,\lambda^*) \tag{6.17}$$

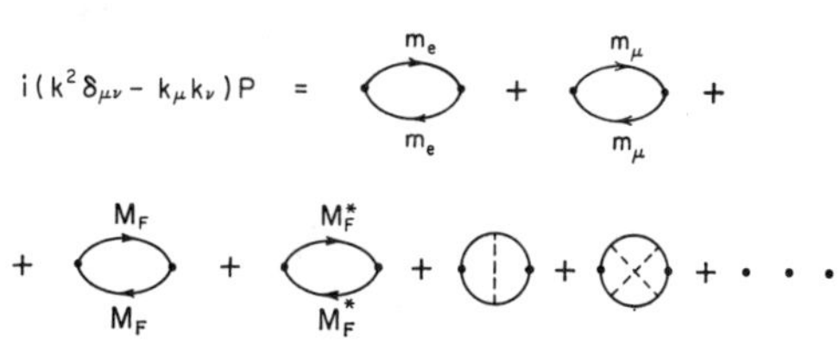

FIG. 2. Sum of irreducible photon self-energy graphs. The charges carried by the four fermion lines, which are labeled by their masses m_e, m_μ, M_F, and $M_F{}^*$, are, respectively, e_0, e_0, ie_0, and $-ie_0$.

[18] For a proof of this theorem, without postulating the validity of contour rotations in the energy variables, see W. Zimmermann, Commun. Math. Phys. 11, 1 (1968).

[19] We note that each of the first four graphs in Fig. 2 becomes logarithmically divergent (and therefore the sum becomes finite) only if the integration is performed after the gauge-invariant term $(k^2\delta_{\mu\nu}-k_\mu k_\nu)$ is factored out. Quite often, it is much more convenient to do the integration before any factoring. In such a case, in order to avoid any mathematical ambiguity, one may define $\Pi(m_e{}^2,m_\mu{}^2,M_F{}^2,M_F{}^{*2})$ to be the formal sum of all irreducible self-energy graphs, as before, but replace (6.8) by

$$i\xi\tilde{\xi}(k^2\delta_{\mu\nu}-k_\mu k_\nu)P=\lim_{\Lambda^2\to\infty}[\Pi(m_e{}^2,m_\mu{}^2,M_F{}^2,M_F{}^{*2}) -\Pi(\Lambda^2+m_e{}^2,\Lambda^2+m_\mu{}^2,\Lambda^2+M_F{}^2,\Lambda^2+M_F{}^{*2})].$$

By taking the limit $\Lambda\to\infty$ after the integration, one can then unambiguously arrive at (6.11).

and therefore $f(z,M_F)+f(z,M_F^*)$ is real along the entire real axis, in accordance with the unitarity condition. By using (6.10), (6.11), and (6.16), one finds that besides the usual photon pole $k^2=0$, D has poles at

$$-k^2=(m_B\pm\tfrac{1}{2}i\gamma_B)^2, \qquad (6.18)$$

where to $O(\alpha)$ and neglecting $(m_l/m_B)^2$ as compared to unity,

$$\gamma_B=\tfrac{2}{3}\alpha m_B \qquad (6.19)$$

and

$$m_B=(m_B{}^0)[1-\tfrac{1}{2}\,\mathrm{Re}P(m_B{}^2)]. \qquad (6.20)$$

The ratio of the renormalized charge e and the unrenormalized charge e_0 is, by definition, given by the behavior of D as $z\to 0$, i.e.,

$$D\to(-i/k^2)(e/e_0)^2 \quad \text{as } k^2\to 0. \qquad (6.21)$$

Since $f(z,\lambda^2)\to\frac{1}{9}$ as $z\to 0$, one finds for charge renormalization,

$$\left(\frac{e_0}{e}\right)^2=1+\frac{\alpha}{3\pi}\ln\left(\frac{M_FM_F{}^*}{m_em_\mu}\right)^2+O(\alpha^2), \qquad (6.22)$$

which, of course, is finite. As $z\to\infty$ and slightly above the real axis, one finds

$$\mathrm{Re}P\to(2\alpha/\pi z)(M_F{}^2+M_F{}^{*2}-m_e{}^2-m_\mu{}^2),$$
$$\mathrm{Im}P\to-\tfrac{2}{3}\alpha.$$

Therefore, one finds [after neglecting higher-order terms in both $O(\alpha/k^2)$ and $O(\alpha^2)$] that in the timelike region, as $-k^2\to\infty+i\epsilon$, where $\epsilon=0+$,

$$D\to-i(m_B{}^0)^2k^{-2}[k^2+(m_B{}^0)^2+\tfrac{2}{3}i\alpha(m_B{}^0)^2]^{-1}, \qquad (6.23)$$

while in the spacelike region, as $k^2\to\infty$,

$$D\to-i(m_B{}^0)^2k^{-2}[k^2+(m_B{}^0)^2]^{-1}. \qquad (6.24)$$

Except for the constant imaginary parts in (6.23), both limiting expressions are exactly the same as D_{free}.

VII. EXPERIMENTAL CONSEQUENCES

The following remarks concern some experimental consequences of the theory.

(1) The modification of the photon propagator from the usual $-ik^{-2}$ to $D(k)$, given by (6.10), in the present theory can be observed through any electromagnetic processes sensitive to the high k^2 behavior of the photon propagator, such as $e^\pm p$ and $\mu^\pm p$ scatterings for spacelike k^2, or pair productions of e^+e^- and $\mu^+\mu^-$ from any high-energy collision processes involving leptons and/or hadrons for timelike k^2. In the timelike region, for $-k^2$ near $m_B{}^2$, the transition probability, being proportional to $|D(k)|^2$, would exhibit a k^2 dependence identical to the standard Breit-Wigner resonance formula; on the other hand, the transition amplitude should have a phase that is of an opposite sign ($-90°$ instead of $+90°$). At $-k^2$ away from $m_B{}^2$, one may use the zeroth-order expression for $D(k)$:

$$D(k)=\frac{-im_B{}^2}{k^2(k^2+m_B{}^2)}\delta_{\mu\nu}+(\cdots k_\mu k_\nu). \qquad (7.1)$$

Consider, e.g., the differential cross section of

$$e^+e^-\to\mu^+\mu^- \qquad (7.2)$$

in the present theory; it differs from that in the usual quantum electrodynamics by a simple multiplicative factor

$$[m_B{}^2/(k^2+m_B{}^2)]^2. \qquad (7.3)$$

It is of interest to note that this factor can be rather substantial even if $-k^2$ is quite far from the resonance region. For example, for $m_B\sim 20$ GeV the width γ_B is only ~ 200 MeV; yet, at, say, $(-k^2)^{1/2}\sim 10$ GeV, the above factor is ~ 1.8.

(2) There should be a deviation in the gyromagnetic ratio g of the muon from the usual expression due to such a modification in the photon propagator:

$$\delta g=-(3\pi)^{-1}\alpha(m_\mu/m_B)^2. \qquad (7.4)$$

From the present experimental result,[20] one concludes that

$$m_B>5\ \text{GeV}. \qquad (7.5)$$

This limit is also consistent with the present high-energy experimental results[21] on

$$p+\text{uranium}\to p+\mu^++\mu^-+\cdots. \qquad (7.6)$$

(3) In principle, m_B can also be determined from the finite value of mass differences between hadrons in the same isospin multiplet and from radiative corrections to weak decays. In practice this is difficult, since all these terms depend on m_B only logarithmically, and none of these terms can be calculated accurately. For example, the mass difference Δm_π between $\pi^\pm$ and π^0 in the usual chiral $SU_2\times SU_2$ phenomenological Lagrangian method is infinite, while in the present theory, it is of course finite. By using the same approximation for the strong-interaction vertex, one finds

$$\Delta m_\pi=\frac{3\alpha m_\rho{}^2}{8\pi m_\pi}\left[2\ln 2+\tfrac{1}{8}(1+\delta^2)\frac{m_\pi{}^2}{m_\rho{}^2}\ln\frac{m_B{}^2}{m_\rho{}^2}+O\left(\frac{m_\pi{}^2}{m_\rho{}^2}\right)\right]$$
$$\cong 5\left[1+0.003(1+\delta^2)\ln\left(\frac{m_B}{m_\rho}\right)^2+O\left(\frac{m_\pi{}^2}{m_\rho{}^2}\right)\right]\text{MeV}, \qquad (7.7)$$

where the first term was first derived by Das *et al.*,[22] δ is the anomalous gyromagnetic ratio of the A_1 meson, and

[20] J. Bailey *et al.*, Phys. Letters **28B**, 287 (1968); see also E. Picasso, in *Proceedings of the Third International Conference on High-Energy Physics and Nuclear Structure* (Plenum, New York, 1970), p. 615.

[21] J. Christenson, G. Hicks, P. Limon, L. M. Lederman, B. Pope, and E. Zavattini, Phys. Rev. Letters (to be published).

[22] T. Das, G. S. Guralnik, V. S. Mathur, F. E. Low, and J. E. Young, Phys. Rev. Letters **18**, 759 (1967).

$O(m_\pi^2/m_\rho^2)$ denotes terms proportional to (m_π^2/m_ρ^2) but remaining finite even in the limit $m_B \to \infty$. The entire expression (7.7) is, of course, identical to those obtained by Gerstein *et al.*,[23] except for replacing their *ad hoc* cutoff parameter Λ by m_B. It is clear that once Δm_π is made finite, it becomes rather insensitive to the precise value of m_B.

For the radiative correction to weak decays, the dependence on m_B is again only logarithmical. The result is similar to those calculated[24] by using the charged intermediate vector boson $W^\pm$, except for replacing m_W by m_B. If such calculations involving strong-interaction vertices could be made accurate, then it would be possible to determine m_B from the observed value of the Cabibbo angle θ. This is, of course, far from the actual case. Assuming that $\theta = 0.22$, the best estimate[25] for m_W in the intermediate boson theory is $\ln(m_W/m_N) = 2.8 \pm 0.8$ if quark algebra is applicable and $\ln(m_W/m_N) = 3.5 \pm 1.0$ if field algebra is applicable. This leaves a large admissible range for m_B (assuming $m_B \sim m_W$) from about 7 to 90 GeV.

Thus, the best way to determine m_B is through the direct observation of possible deviations from the conventional quantum electrodynamics predictions at high energy.

(4) By using (6.10) and (6.11), one sees that both m_F and γ_F, and therefore also the finite value of charge renormalization $(e_0/e)^2$, can be determined by accurately measuring the photon propagator $D(k)$ at high k^2. However, these measurements are more difficult since it is then necessary to measure $D(k)$ at least to an accuracy comparable to, or better than, $O(\alpha)$.

(5) We remark that an attractive, but highly speculative, idea is to regard B_μ as the neutral component of the hypothetical charged intermediate boson field $W_\mu^\pm$ for the weak interaction, in which case one would expect $m_B \sim m_W$. The further speculation that *the semi-weak interaction coupling constant* $g^2/4\pi$ *is*, in fact, *the same as the fine structure constant* α leads to

$$m_B \sim m_W \sim (4\pi\alpha/G_V)^{1/2} \sim 100 \text{ GeV}. \qquad (7.8)$$

VIII. CAUSALITY

The presence of complex singularities near the physical region, but on the physical sheet, has an unusual effect on the propagation of wave packets in a collision process. As has been discussed in Ref. 4, because of the uncertainty principle, such unusual effects disappear if one studies *only* the average position of the wave packet; this then automatically removes all of the so-called "causality difficulties" or, more precisely, those difficulties that could be directly related to a classical description. In quantum physics, as we shall see, in the first place, there is no general agreement as to the precise meaning of causality. While in the present theory, unusual effects do occur if one analyzes the shape of the wave packet, such effects, though unusual, are not in contradiction with anything known at present about the physical world. Furthermore, it should be clear that such effects *cannot* ever lead to logical difficulties (i.e., self-contradictory predictions), since they are the mathematical consequences of a set of well-defined self-consistent equations. In the following, we shall briefly review these unusual effects and their connection with the causality question.

In most papers on causality, attempts have been made to transform the somewhat ill-defined problem of causality to that of relativistic invariance, which can be stated with precision. In the classical derivation of the dispersion formula for light waves,[26] one begins by assuming a sharp wave front for a physical signal in the theory; e.g., the signal is zero for the space-time region specified by, say, $t<0$ and $x<0$. The subsequent requirement that this wave front should not travel faster than the velocity of light leads to the well known classical dispersion formula which in fact rules out complex singularities such as those that appear in the present theory. We recall that in quantum field theory, it is not possible to construct such a sharp wave front for the incoming wave for all $t<0$, since that would require the superposition of plane waves of all frequencies ν, negative as well as positive. Even for the zero-mass photon field, a coherent mixture of many photon states necessarily covers *only* the *positive* energy range, and therefore $\nu \geqslant 0$, by using the time-dependent Schrödinger equation; for a massive field, the range of physically allowable frequencies is even smaller, since $\nu \geqslant h^{-1} \times (\text{mass}) > 0$. The impossibility of constructing a wave packet with a sharply defined front makes it impossible to apply such a classical argument.[27]

Another approach is to use local field theory. One requires the commutators between any two field operators at points separated by a spacelike distance to be zero. This requirement is *satisfied* in the present theory. The derivation[28] of the usual analyticity condition is not applicable to our theory because it assumes the energy spectrum to be real (or, the underlying Lagrangian to be Hermitian), which is obviously not true in

---

[23] I. S. Gerstein, B. W. Lee, H. T. Nieh, and H. J. Schnitzer, Phys. Rev. Letters **19**, 1064 (1967); **20**, 825 (1968).

[24] For a more recent discussion, see A. Sirlin, in *Proceedings of the Fourteenth International Conference on High-Energy Physics, Vienna, 1968*, edited by J. C. Prentki and J. Steinberger (CERN, Geneva, 1968). The fact that, in the intermediate boson theory of the weak interaction, the $O(\alpha)$ radiative correction to the ratio (G_V/G_μ) is finite in the conventional form of quantum electrodynamics has, of course, been known for quite some time [T. D. Lee, Phys. Rev. **128**, 899 (1962)].

[25] The values quoted are based on the recent calculations by A. Sirlin (private communication). We thank Dr. Sirlin for making these values available to us.

[26] R. Kronig, J. Opt. Soc. Am. **12**, 547 (1926); H. A. Kramers, Atti. Congr. Intern. Fis. Como **2**, 545 (1927); See also A. Sommerfeld, Ann. Phys. (Paris) **44**, 177 (1914); L. Brillouin, *ibid.* **44**, 203 (1914).

[27] N. G. Van Kampen, Phys. Rev. **89**, 1072 (1953); **91**, 1267 (1953).

[28] M. Gell-Mann, M. L. Goldberger, and W. Thirring, Phys. Rev. **95**, 1612 (1954).

the present theory. The violation of the usual analyticity condition is, however, totally consistent with the requirement of relativistic invariance.

Still another description of causality that has been used in the literature[29] is one deriving from the study of the average motion of a wave packet. For example, in the case of a simple S-wave elastic scattering, if the incoming wave packet has an average relative position $\langle r_{\rm in}\rangle = -vt$ for time $t \ll 0$, then the outgoing wave packet, for $t \gg 0$, has an average relative position

$$\langle r_{\rm out}\rangle = vt + l, \tag{8.1}$$

where

$$l = -2\langle d\delta/dk\rangle, \tag{8.2}$$

δ denotes the phase shift, and k is the relative momentum. A $-90°$ resonance, such as the one required by the B_μ field in our theory, would contribute a positive value for l, and therefore give rise to an advancement of the outgoing wave packet. However, it can be shown,[30] under very general conditions, that one has the inequality

$$l < l_{\rm max} = O(\Delta^{-1}), \tag{8.3}$$

where Δ denotes the momentum width of the *incoming* wave packet. Therefore, it is not possible to draw any strong conclusion concerning causality by studying the average position of a wave packet.

Rather unusual behavior of wave packets can, nevertheless, be demonstrated to exist in the present theory. In general, these unusual properties concern the detailed shape of the wave packet. Consider, for example, the elastic collision of e^+ and e^- in the center-of-mass system at the resonance energy m_B. For clarity, let us assume the radial dependence of the incoming wave to be given by

$$\phi^{\rm in}(r,t) \propto r^{-1} e^{-\Delta|r+t|} \tag{8.4}$$

at large relative distances r. As shown in Ref. 4, the presence of a complex pole, such as $(m_B + \frac{1}{2}i\gamma_B)$ in the photon propagator, implies that the outgoing wave has a radial dependence given by

$$\phi^{\rm out}(r,t) \propto r^{-1} e^{-\Delta(t-r)} \quad \text{for } t > r, \tag{8.5}$$

but

$$\phi^{\rm out}(r,t) \propto r^{-1}\left[e^{-\Delta(t-r)} - \frac{8\gamma\Delta}{(2\Delta+\gamma)^2} e^{-\frac{1}{2}\gamma(t-r)} \right] \quad \text{for } t < r, \tag{8.6}$$

where $\gamma = \gamma_B$, the first term $e^{-\Delta(t-r)}$ merely reproduces the shape of the initial wave, but the second term $e^{-\frac{1}{2}\gamma(t-r)}$ is quite unusual. Nevertheless, the presence of such a term is, of course, perfectly compatible with the requirement of relativistic invariance as well as with all existing experiments.

We note that at present in any high-energy experiment almost nothing is known concerning the *shape* of wave packets. Without some detailed knowledge of the shape of wave packets, one can study only the average positions $\langle r\rangle$, which, as mentioned above, are insensitive to the complex singularities. In order to see the unusual tail $e^{-\frac{1}{2}\gamma(t-r)}$, we may consider a measurement which can differentiate the time-advanced region, say $(t-r) > \tau$, from the time-retarded region $(t-r) < \tau$, where τ represents the experimental space-time resolution. At present, the best value of the time resolution in any high-energy experiment is $\tau \sim 10^{-10}$ sec. Assuming, for example, $m_B \sim 20$ GeV and therefore $\gamma_B \sim 200$ MeV, the full intensity of the unusual tail, integrated over the entire time-advanced region from $(t-r) = \tau$ to $(t-r) = \infty$, comes out to be

$$\sim \exp(-\gamma_B\tau) \sim \exp[-3\times 10^{13}]. \tag{8.7}$$

The smallness of this probability[31] makes it unlikely that we can realistically detect such an unusual effect in any near future (assuming that we can reach a center-of-mass energy $\sim$ the resonance energy m_B). Of course, in principle, this effect should be measurable.

In any quantum theory, what one really studies are only correlations between various events occurring at different space-time regions. The impossibility of constructing during $t < 0$ a sharp wave front for the incoming wave makes it also not possible to give a strict causal interpretation to such correlations. Thus, there does *not* exist a sharply defined causality principle. (We regard requirements such as the usual zero commutator of two local field operators separated by a spacelike distance as simply an expression of relativistic invariance and local canonical quantum field theory, but not of causality.[32]) The attribution as to which effect should be regarded as "noncausal," therefore, has a certain degree of arbitrariness, except in the classical limit.[33] It seems nevertheless appropriate to call the above described unusual tail $e^{-\frac{1}{2}\gamma(t-r)}$ in the outgoing wave packet "noncausal," although one must emphasize that there is *no* logical difficulty in having this particular

[29] E. P. Wigner, Phys. Rev. **98**, 145 (1955). Professor Wigner has kindly pointed out to us a different formulation of causality in his paper in *International School of Physics "Enrico Fermi," Course 29* (Academic, New York, 1964), p. 40. This formulation is not based on wave packets and is, therefore, not directly related to the point being made here. See also M. Froissart, M. L. Goldberger, and K. M. Watson, Phys. Rev. **131**, 2820 (1963); cf., however, H. M. Nussenzveig, Nuovo Cimento **20**, 694 (1961).

[30] See Ref. 4. The arguments used in Ref. 4 can be extended to the nonresonance region as well. The inequality (8.3) holds, provided the integral $\int (d\delta/dk)dk$, integrated over the entire momentum width of the incoming wave, is $\lesssim \pi$.

[31] The two-photon exchange processes can lead to a larger value, though still much too small to be detected at present. See T. D. Lee (Ref. 5) and Cutkosky *et al.* (Ref. 17).

[32] A similar opinion is also expressed by G. Wanders [Nuovo Cimento **14**, 168 (1959)], whose conclusions on causality seem to agree with ours also in other respects. General questions of causality have also been discussed by B. Ferretti, Nuovo Cimento **43**, 506 (1966); **43**, 516 (1966).

[33] Since a macroscopic body does not necessarily imply the validity of a classical limit (e.g., the superfluidity phenomenon), there also does not exist in quantum physics any general macrocausality principle that can be sharply defined without further qualifications.

kind of "noncausal" effect; the question of whether it indeed exists in nature can only be resolved by future experimentation.

ACKNOWLEDGMENTS

We wish to thank N. Christ, S. Coleman, R. Jackiw, and B. Zumino for helpful discussions.

APPENDIX A

Let us study anticommutation relations for one oscillator, assuming

$$a^2=\bar{a}^2=0,\quad a\bar{a}+\bar{a}a=\pm 1. \tag{A1}$$

The case with the + sign is, of course, well known, but we can study the two cases simultaneously if we define $N=\pm\bar{a}a$; then

$$N^2=N. \tag{A2}$$

As is well known, N satisfying (A2) can always be diagonalized. In fact, any vector ψ can always be decomposed into a sum of two vectors (one of which may be zero) $N\psi$ and $(1-N)\psi$ which are eigenvectors of N, corresponding to the eigenvalues 1 and 0, respectively. Both eigenvalues certainly exist; for if we assume a vector ψ with the eigenvalue 0, $N\psi=0$, then from $a\bar{a}\psi=\pm(\psi-N\psi)=\pm\psi$ we see that $\phi=\bar{a}\psi\neq 0$, since $a\phi=\pm\psi$; furthermore, $N\phi=\pm\bar{a}a\bar{a}\psi=\bar{a}\psi=\phi$, and therefore ϕ belongs to the eigenvalue 1. Conversely, if ϕ is an eigenvector belonging to the eigenvalue 1, $N\phi=\phi$, then $a\phi$ belongs to the eigenvalue 0, for $Na\phi=\pm\bar{a}a^2\phi=0$; notice again that $a\phi$ cannot be zero, since $N\phi=\pm\bar{a}a\phi=\phi\neq 0$. These calculations are well known from the ordinary case, where $\bar{a}=a^\dagger$ and the + sign holds in Eq. (A1). We have repeated them here, merely to remind the reader of the fact that the analysis makes no use whatsoever of the metric; it is based *entirely* on the algebraic structure of (A1). Pursuing the analysis further, one sees that the whole vector space $\mathcal{H}$ is the direct sum of two subspaces $\mathcal{H}=\mathcal{H}_0\oplus\mathcal{H}_1$, in one of which the operator $N=0$, while in the other it acts as the identity $N=1$. We have seen that if ψ belongs to the first subspace $\mathcal{H}_0$, then $\phi=\bar{a}\psi$ belongs to $\mathcal{H}_1$, and we have the relations

$$\begin{aligned}\bar{a}\psi&=\phi, & a\phi&=\pm\psi,\\ a\psi&=0, & \bar{a}\phi&=0.\end{aligned} \tag{A3}$$

The same relations are obtained if we start from any vector ϕ of $\mathcal{H}_1$, and define $\psi=\pm a\phi$. From this we see that $\bar{a}$ maps $\mathcal{H}_1$ into 0, and $\mathcal{H}_0$ into $\mathcal{H}_1$. Regarded as a mapping of $\mathcal{H}_0$ into $\mathcal{H}_1$, $\bar{a}$ has $\pm a$ as a left inverse. Likewise a maps $\mathcal{H}_0$ into 0, and $\mathcal{H}_1$ into $\mathcal{H}_0$; regarded as a mapping of $\mathcal{H}_1$ into $\mathcal{H}_0$, a has $\pm\bar{a}$ as a left inverse. It follows that $\mathcal{H}_0$ and $\mathcal{H}_1$ have the same dimensionality. Choosing bases $\psi_1,\ldots,\psi_n$ in $\mathcal{H}_0$ and $\phi_1,\ldots,\phi_n$ in $\mathcal{H}_1$ such that ψ_k and ϕ_k are related to each other in the same way as ψ and ϕ in (A3), a and $\bar{a}$ are reduced to the canonical form

$$a=\begin{pmatrix}0 & \pm I\\ 0 & 0\end{pmatrix},\quad \bar{a}=\begin{pmatrix}0 & 0\\ I & 0\end{pmatrix}, \tag{A4}$$

where I, of course, is the n-dimensional unit matrix. This may be obtained from the usual case, if in (A1) one identifies $\bar{a}$ with $\pm a^\dagger$. Notice again, however, that no use has been made of Hermiticity.

In order to determine uniquely the metric η, we shall impose the condition that *the set of matrices a and $\bar{a}$ should be irreducible.* It follows, then, from (A4) that both a and $\bar{a}$ are (2×2) matrices, i.e., there is only one ψ and one ϕ. To conform to the notations used in Sec. II, we shall denote ψ and ϕ by $|0\rangle$ and $|1\rangle$, respectively. Equation (A3) becomes

$$\begin{aligned}\bar{a}|0\rangle&\equiv|1\rangle, & a|1\rangle&\equiv\pm|0\rangle,\\ a|0\rangle&=0, & \bar{a}|1\rangle&=0.\end{aligned} \tag{A5}$$

Thus, one finds

$$\begin{aligned}&\langle 1|0\rangle=\pm\langle 1|a|1\rangle=\pm\langle 1|\bar{a}|1\rangle^*=0,\\ &\langle 0|1\rangle=0,\end{aligned} \tag{A6}$$

and

$$\langle 1|1\rangle=\langle 0|a\bar{a}|0\rangle=\pm\langle 0|0\rangle.$$

By using transformation (2.22), we can always choose $\langle 0|0\rangle$ to be +1, and therefore there are only two classes specified by (2.26) and (2.27).

The preceding result can be applied immediately to the case of ν "oscillators" with anticommuting variables $a_1, a_2, \ldots, a_\nu$. We assume

$$\{a_r,a_s\}=\{\bar{a}_r,\bar{a}_s\}=0, \tag{A7}$$

while

$$\{\bar{a}_r,a_s\}=\epsilon_r\delta_{rs}\quad(\epsilon_r=\pm 1). \tag{A8}$$

Then the above analysis can be applied immediately, say to the first pair of operators, a_1 and $\bar{a}_1$. We may in fact cut short all remaining calculations, if we notice that if we set $a_r^+=\epsilon_r\bar{a}_r$, the above equations have the usual form. The notation does not require that a_r^+ be Hermitian conjugate to a_r; it is simply another operator. The result is the customary one; if the set of matrices is assumed irreducible, the span has 2^ν dimensions. A basis may be chosen, in which the basis vectors are

$$|n_1,n_2,\ldots,n_\nu\rangle,$$

where each $n_k=0$ or 1. These vectors satisfy

$$a_k|0,0,\ldots,0\rangle=0, \tag{A9}$$

$$\epsilon_k\bar{a}_ka_k|n_1,n_2,\ldots,n_\nu\rangle=n_k|n_1,n_2,\ldots,n_\nu\rangle, \tag{A10}$$

and

$$|n_1,n_2,\ldots,n_\nu\rangle=(\bar{a}_1)^{n_1}(\bar{a}_2)^{n_2}\cdots(\bar{a}_\nu)^{n_\nu}|0,0,\ldots,0\rangle. \tag{A11}$$

We note in passing, that in a somewhat more general formulation (A8) could be replaced by

$$\{\bar{a}_r,a_s\}=K_{rs}, \tag{A12}$$

where the matrix $K=(K_{rs})$ is a *nonsingular Hermitian* matrix. This greater generality is only fictitious; a linear transformation $b_r=\sum_s T_{sr}a_s$, $\bar{b}_r=\sum_s T_{sr}^*\bar{a}_s\equiv\sum_s T_{rs}^\dagger\bar{a}_s$ will change the matrix K to $T^\dagger KT$. As is well known, one can always choose T so that (A12) reduces to (A8), with $\epsilon_r=\pm1$. Although trivial, this generalization is nevertheless useful in practice, as we shall see in Appendix C.

Just as in the case of a single oscillator, the metric η is completely determined, up to a real proportionality constant. By using (A9)–(A11), one easily sees that

$$\langle n_1',n_2',\ldots,n_\nu'|n_1,n_2,\ldots,n_\nu\rangle=\prod_{r=1}^{\nu}\epsilon_r^{n_r}\delta_{n_r'n_r}\times\langle 0,0,\ldots,0|0,0,\ldots,0\rangle. \quad (A13)$$

APPENDIX B

We examine here the consequences of the commutation relation

$$a\bar{a}-\bar{a}a=-1, \quad (B1)$$

which, by interchanging the roles of a and $\bar{a}$, is of course completely equivalent to the alternative form

$$a\bar{a}-\bar{a}a=+1.$$

Just as in the case for the anticommutation relation, we need the following assumption: (i) The algebra of matrices generated by a, $\bar{a}$ is irreducible, i.e., there is no invariant subspace. In addition, we assume (ii) there is at least one eigenvector $|\psi\rangle$ of the operator $\bar{a}a$:

$$\bar{a}a|\psi\rangle=\lambda|\psi\rangle. \quad (B2)$$

(For brevity, in this section, a vector is denoted by the symbols $|\psi\rangle$ or ψ as seems most convenient.)

The metric η, which is to be determined, is, as always, assumed to be Hermitian and nonsingular; therefore, there is no vector $\phi\neq0$ such that

$$\langle\ |\phi\rangle=0 \quad (B3)$$

for all vectors $|\ \rangle$.

The argument that follows is fairly trivial, as it relies on the customary construction of a string of eigenvectors of $\bar{a}a$,

$$\ldots,\ \bar{a}^2\psi,\ \bar{a}\psi,\ \psi,\ a\psi,\ a^2\psi,\ \ldots, \quad (B4)$$

corresponding to the eigenvalues

$$\ldots,\lambda-2,\ \lambda-1,\ \lambda,\ \lambda+1,\ \lambda+2,\ \ldots. \quad (B5)$$

The eigenvalue equations

$$\bar{a}a\phi_p=(\lambda+p)\phi_p,\quad \phi_p=a^p\psi$$

and

$$\bar{a}a\chi_p=(\lambda-p)\chi_p,\quad \chi_p=\bar{a}^p\psi \quad (B6)$$

follow, of course, from (B1) by recursion, with the qualification that the string may stop on the right at ϕ_p, if $a\phi_p=0$ (or on the left at χ_q, if $\bar{a}\chi_q=0$). In any event, by a well-known argument, the eigenvectors in (B4) form an independent system, their eigenvalues in (B5) being all different. In addition one easily sees that

$$a\phi_p=\phi_{p+1},\quad \bar{a}\phi_p=\bar{a}a\phi_{p-1}=(\lambda+p-1)\phi_{p-1},$$
$$\bar{a}\chi_p=\chi_{p+1},\quad a\chi_p=a\bar{a}\chi_{p-1}=(\lambda-p)\chi_{p-1}. \quad (B7)$$

All these equations hold for $p=1, 2, \ldots$ provided one defines $\phi_0=\chi_0=\psi$. Thus the string of vectors (B4) spans our invariant subspace, which according to our assumption (i) must be the whole space. Clearly the string may not stop on both sides, since (B1) cannot hold in a finite dimensional space (by the usual trace argument). We now have three possibilities.

(1) The string stops at the left: $\chi_{p+1}=0$ but $\chi_p\neq0$. $0=a\chi_{p+1}=(\lambda-p-1)\chi_p$; hence $\lambda=p+1$.

(2) The string stops at the right: $\phi_{p+1}=0$ but $\phi_p\neq0$ (where p may also be $=0$). But then from (B7), $0=\bar{a}\phi_{p+1}=(\lambda+p)\phi_p$; hence $\lambda=-p$.

(3) The string extends to infinity in both directions: In this case, as we shall see, λ is an arbitrary *real* number but not an integer.

It is now easy to see that in all three cases the metric of the space is completely determined, apart from a real proportionality constant. Two distinct vectors in (B4) are mutually orthogonal, since they belong to different eigenvalues of a self-adjoint operator. None of them can have zero norm, or we would have degeneracy, Eq. (B3). From (B6) and (B7) we have finally

$$\langle\phi_{p+1}|\phi_{p+1}\rangle=\langle\phi_p|\bar{a}a|\phi_p\rangle=(\lambda+p)\langle\phi_p|\phi_p\rangle$$

and

$$\langle\chi_{p+1}|\chi_{p+1}\rangle=\langle\chi_p|a\bar{a}|\chi_p\rangle=(\lambda-p-1)\langle\chi_p|\chi_p\rangle. \quad (B8)$$

These equations determine the norm of all vectors up to a common proportionality constant. They show that λ must be real. Furthermore, if a vector in the string (B4) has a negative eigenvalue, then the vector to the right of it (if it exists) has a square norm of the opposite sign. This never occurs in case (1), which is therefore recognized as the ordinary oscillator with positive-definite metric by setting $\bar{a}=b$ and $a=b^\dagger$. Case (2) is likewise recognized as Dirac's or Pauli's oscillator with indefinite metric. Choosing suitable normalization factors, one can reduce Eqs. (B7) to one of the standard forms commonly employed.

We shall not mention case (1) any further. In case (2) or (3) we may (by a suitable selection of the vector called ψ in the string) assume that

$$0\leqslant\lambda<1.$$

Case (2) corresponds to $\lambda=0$, in this case the vectors $\phi_1, \phi_2, \ldots$ do not exist, and the eigenvalues are $\ldots, -2, -1, 0$. From (B8), and by choosing

$$\langle\chi_0|\chi_0\rangle \text{ positive},$$

one finds (2.27), i.e.,

$$\langle\ |(-1)^{\bar{a}a}|\ \rangle \text{ positive}$$

for all vectors $|\ \rangle$.

In case (3), $0<\lambda<1$, we may define new basis vectors, setting

$$|\lambda\rangle=\psi,$$

$$|\lambda+p\rangle=[\lambda(\lambda+1)\cdots(\lambda+p-1)]^{-1/2}a^p|\lambda\rangle \tag{B9}$$

and

$$|\lambda-p\rangle=i^{-p}[(1-\lambda)(2-\lambda)\cdots(p-\lambda)]^{-1/2}\bar{a}^p|\lambda\rangle.$$

Then we get, denoting by $|\mu\rangle$ one of the new basis vectors ($\mu=\lambda$ or $\lambda\pm p$),

$$a|\mu\rangle=\mu^{1/2}|\mu+1\rangle$$

and

$$\bar{a}|\mu\rangle=(\mu-1)^{1/2}|\mu-1\rangle. \tag{B10}$$

Notice again that, when $\mu<0$, $\mu^{1/2}=+i|\mu|^{1/2}$ by definition. Thus, one finds

$$\langle\mu+1|\mu+1\rangle=(\mu/|\mu|)\langle\mu|\mu\rangle, \tag{B11}$$

i.e., all $\langle\mu|\mu\rangle$ have the same sign for $\mu>0$, but alternate signs for $\mu<0$.

APPENDIX C

In this appendix, we discuss the explicit diagonal form of the free Hamiltonian for the ψ_F field discussed in Sec. IV. According to (4.3), the free Lagrangian density is

$$\mathfrak{L}_{\text{free}}=-\bar{\psi}_F\tau_z\gamma_4\left(\gamma_\lambda\frac{\partial}{\partial x_\lambda}+M_F{}^0\right)\psi_F, \tag{C1}$$

where

$$M_F{}^0=m_F{}^0+\tfrac{1}{2}i\gamma_F{}^0\tau_x, \tag{C2}$$

and for convenience, we have set

$$\tau_a=\tau_z \quad\text{and}\quad \tau_b=\tau_x. \tag{C3}$$

The free Hamiltonian is then given by

$$H_{\text{free}}=\int d^3r\,\bar{\psi}_F\tau_z(-i\boldsymbol{\alpha}\cdot\nabla+\beta M_F{}^0)\psi_F, \tag{C4}$$

in which, as usual, $\gamma_4=\beta$ and $\gamma_j=-i\beta\alpha_j$ for $j=1, 2, 3$.

It is useful to introduce the c number 4-component spinor functions $u_\pm(\mathbf{p},s)$ and $v_\pm(\mathbf{p},s)$, defined by

$$(\boldsymbol{\alpha}\cdot\mathbf{p}+\beta M)u_+(\mathbf{p},s)=E_pu_+(\mathbf{p},s),$$
$$(\boldsymbol{\alpha}\cdot\mathbf{p}+\beta M^*)u_-(\mathbf{p},s)=E_p{}^*u_-(\mathbf{p},s), \tag{C5}$$
$$(\boldsymbol{\alpha}\cdot\mathbf{p}+\beta M)v_+(-\mathbf{p},s)=-E_pv_+(-\mathbf{p},s),$$

and

$$(\boldsymbol{\alpha}\cdot\mathbf{p}+\beta M^*)v_-(-\mathbf{p},s)=-E_p{}^*v_-(-\mathbf{p},s),$$

where

$$E_p=(\mathbf{p}^2+M^2)^{1/2}, \tag{C6}$$

$$M=m_F{}^0+\tfrac{1}{2}i\gamma_F{}^0, \tag{C7}$$

and $s=\pm1$ denotes the usual helicity. From (C5), it can be readily verified that the following orthogonality relations hold:

$$u_+{}^\dagger(\mathbf{p},s)v_-(-\mathbf{p},s')=v_+{}^\dagger(-\mathbf{p},s)u_-(\mathbf{p},s')=0. \tag{C8}$$

By a suitable choice of the normalization factors, one can set

$$\sum_s[u_+(\mathbf{p},s)u_-{}^\dagger(\mathbf{p},s)+v_+(-\mathbf{p},s)v_-{}^\dagger(-\mathbf{p},s)]=I, \tag{C9}$$

where I is the (4×4) unit matrix.

At any given time,

$$\psi_F=\begin{pmatrix}\psi_1\\ \psi_2\end{pmatrix}$$

can be expanded in terms of these spinors

$$\psi_1=\sum_{\mathbf{p},s}(2\Omega)^{-1/2}[a_+(\mathbf{p},s)u_+(\mathbf{p},s)+a_-(\mathbf{p},s)u_-(\mathbf{p},s)$$
$$+\bar{b}_-(-\mathbf{p},s)v_+(-\mathbf{p},s)+\bar{b}_+(-\mathbf{p},s)v_-(-\mathbf{p},s)]e^{i\mathbf{p}\cdot\mathbf{r}} \tag{C10}$$

and

$$\psi_2=\sum_{\mathbf{p},s}(2\Omega)^{-1/2}[a_+(\mathbf{p},s)u_+(\mathbf{p},s)-a_-(\mathbf{p},s)u_-(\mathbf{p},s)$$
$$+\bar{b}_-(-\mathbf{p},s)v_+(-\mathbf{p},s)-\bar{b}_+(-\mathbf{p},s)v_-(-\mathbf{p},s)]e^{i\mathbf{p}\cdot\mathbf{r}}, \tag{C11}$$

where Ω is the volume of the system. Correspondingly, the adjoint operators are given by

$$\bar{\psi}_1=\sum_{\mathbf{p},s}(2\Omega)^{-1/2}[\bar{a}_+(\mathbf{p},s)u_+{}^\dagger(\mathbf{p},s)+\bar{a}_-(\mathbf{p},s)u_-{}^\dagger(\mathbf{p},s)$$
$$+b_-(-\mathbf{p},s)v_+{}^\dagger(-\mathbf{p},s)+b_+(-\mathbf{p},s)v_-{}^\dagger(-\mathbf{p},s)]e^{-i\mathbf{p}\cdot\mathbf{r}} \tag{C12}$$

and

$$\bar{\psi}_2=\sum_{\mathbf{p},s}(2\Omega)^{-1/2}[\bar{a}_+(\mathbf{p},s)u_+{}^\dagger(\mathbf{p},s)-\bar{a}_-(\mathbf{p},s)u_-{}^\dagger(\mathbf{p},s)$$
$$+b_-(-\mathbf{p},s)v_+{}^\dagger(-\mathbf{p},s)-b_+(-\mathbf{p},s)v_-{}^\dagger(-\mathbf{p},s)]e^{-i\mathbf{p}\cdot\mathbf{r}}. \tag{C13}$$

By using (C9) and the quantization rule (4.8) (setting $\tau_a=\tau_z$), one finds

$$\{a_+(\mathbf{p},s),\bar{a}_-(\mathbf{p}',s')\}=\{b_+(\mathbf{p},s),\bar{b}_-(\mathbf{p}',s')\}$$
$$=\{a_-(\mathbf{p},s),\bar{a}_+(\mathbf{p}',s')\}=\{b_-(\mathbf{p},s),\bar{b}_+(\mathbf{p}',s')\}$$
$$=\delta_{ss'}\delta_{\mathbf{pp}'} \tag{C14}$$

and *all other equal-time anticommutators are zero.* These anticommutators are seen to correspond to the more general type of Eq. (A12), rather than the "diagonal" case of Eqs. (A7) and (A8) of Appendix A. The relations may be diagonalized by introducing new operators

$$a_1(\mathbf{p},s)=[a_+(\mathbf{p},s)+a_-(\mathbf{p},s)]/\sqrt{2},$$
$$a_2(\mathbf{p},s)=[a_+(\mathbf{p},s)-a_-(\mathbf{p},s)]/\sqrt{2}, \tag{C15}$$

and similar expressions for the b operators. Then

$$\{a_\rho(\mathbf{p},s),\bar{a}_\sigma(\mathbf{p}',s')\}=\{b_\rho(\mathbf{p},s),\bar{b}_\sigma(\mathbf{p}',s')\}=\epsilon_\rho\delta_{ss'}\delta_{\mathbf{pp}'}\delta_{\rho\sigma},\quad (C14')$$

where ρ, $\sigma=1$ or 2, and $\epsilon_1=1$, $\epsilon_2=-1$. All other equal-time anticommutators are zero. It then follows from the arguments given in Appendix A that the metric η is uniquely determined, up to the transformations (2.20′) and (2.22). We find, for a system consisting only of the fermion field ψ_F,

$$\langle\,|(-1)^{N_F}|\,\rangle \text{ to be positive}$$

for all vectors $|\,\rangle$, where

$$N_F=-\sum_{\mathbf{p},s}[\bar{a}_2(\mathbf{p},s)a_2(\mathbf{p},s)+\bar{b}_2(\mathbf{p},s)b_2(\mathbf{p},s)].\quad (C16)$$

In terms of the Fourier components $a_\pm$ and $b_\pm$, the free Hamiltonian becomes

$$H_{\text{free}}=\sum_{\mathbf{p},s}\{E_{\mathbf{p}}[\bar{a}_-(\mathbf{p},s)a_+(\mathbf{p},s)-b_+(\mathbf{p},s)\bar{b}_-(\mathbf{p},s)]+E_{\mathbf{p}}^*[\bar{a}_+(\mathbf{p},s)a_-(\mathbf{p},s)-b_-(\mathbf{p},s)\bar{b}_+(\mathbf{p},s)]\}.\quad (C17)$$

We remark that, if we follow the prescriptions of Appendix A, we will naturally arrive at a basis in which the operators $\bar{a}_\rho a_\rho$, $\bar{b}_\rho b_\rho$ are diagonal; in this basis $\bar{a}_-a_+$, etc., are not diagonal. We notice, however, that the metric is positive definite with respect to the degrees of freedom with $\rho=1$ and indefinite for $\rho=2$ [see Eq. (C14′)]. Furthermore, in the basis we have described, because of (A13), the matrix representation η of the metric, defined by (2.3) and (2.4), is diagonal, and the matrices representing a_1, a_2, b_1, b_2 (we omit the indices $\mathbf{p}$, s for simplicity) obey the rule

$$\bar{a}_\rho=\epsilon_\rho a_\rho^\dagger,\quad \bar{b}_\rho=\epsilon_\rho b_\rho^\dagger.\quad (C18)$$

In this basis, therefore,

$$\bar{a}_-=a_+^\dagger,\quad \bar{b}_-=b_+^\dagger,\qquad \bar{a}_+=a_-^\dagger,\quad \bar{b}_+=b_-^\dagger,\quad (C19)$$

and Eqs. (C14) take the standard form for fermion operators,

$$\{a_\lambda(\mathbf{p},s),a_\lambda^\dagger(\mathbf{p}',s')\}=\{b_\lambda(\mathbf{p},s),b_\lambda^\dagger(\mathbf{p}',s')\}=\delta_{\mathbf{pp}'}\delta_{ss'},\quad (C20)$$

where $\lambda=+$ or $-$ and all other equal-time anticommutators remain zero. Similarly, the free Hamiltonian becomes

$$H_{\text{free}}=\sum_{\mathbf{p},s}\{E_{\mathbf{p}}[a_+^\dagger(\mathbf{p},s)a_+(\mathbf{p},s)-b_+(\mathbf{p},s)b_+^\dagger(\mathbf{p},s)]+E_{\mathbf{p}}^*[a_-^\dagger(\mathbf{p},s)a_-(\mathbf{p},s)-b_-(\mathbf{p},s)b_-^\dagger(\mathbf{p},s)]\}.\quad (C21)$$

Now, technically we are still in the basis in which $\bar{a}_\rho a_\rho=\epsilon_\rho a_\rho^\dagger a_\rho$ and $\bar{b}_\rho b_\rho=\epsilon_\rho b_\rho^\dagger b_\rho$ are diagonal, and therefore H_{free} is not yet diagonal. However, through a familiar *unitary* transformation, one can easily transform the basis vectors into the eigenvectors of $a_+^\dagger a_+$, $a_-^\dagger a_-$, $b_+^\dagger b_+$, and $b_-^\dagger b_-$. Because this is a unitary transformation, the transformation law (2.20) governing the metric is the same as that for a *bona fide* operator. Therefore, in this new basis, Eqs. (C18)–(C21) remain valid and the free Hamiltonian (C21) is diagonal, though the matrix η is no longer diagonal.

It is useful, especially in the presence of interactions, to define

$$\psi_\pm=2^{-1/2}(\psi_1\pm\psi_2).\quad (C22)$$

The free Hamiltonian (C4) then becomes

$$H_{\text{free}}=\int[\bar{\psi}_-(-i\boldsymbol{\alpha}\cdot\nabla+\beta M)\psi_++\bar{\psi}_+(-i\boldsymbol{\alpha}\cdot\nabla+\beta M^*)\psi_-]d^3r.\quad (C23)$$

By using (4.3), one finds that the electromagnetic interaction is given by

$$H_{\text{int}}=\int e_0(\bar{\psi}_-\gamma_4\gamma_\mu\psi_+-\bar{\psi}_+\gamma_4\gamma_\mu\psi_-)\phi_\mu d^3r.\quad (C24)$$

Furthermore, according to (4.8),

$$\{\psi_+(\mathbf{r},t),\bar{\psi}_-(\mathbf{r}',t)\}=\{\psi_-(\mathbf{r},t),\bar{\psi}_+(\mathbf{r}',t)\}=\delta^3(\mathbf{r}-\mathbf{r}')\quad (C25)$$

and

$$\{\psi_+(\mathbf{r},t),\bar{\psi}_+(\mathbf{r}',t)\}=\{\psi_-(\mathbf{r},t),\bar{\psi}_-(\mathbf{r}',t)\}=0.\quad (C26)$$

From these relations and Eqs. (C10)–(C13), it is easy to see that ψ_+ annihilates "particles" of mass M and charge ie_0, and creates "antiparticles" of mass M and charge $-ie_0$, while $\bar{\psi}_-$ creates "particles" of mass M and charge ie_0, etc. By using (C8) and (C9), one can then proceed to derive the appropriate propagators and vertex functions for these fermion fields. The resulting expressions are just like the usual ones, except that the masses M and M^* are complex and the charges ie_0 and $-ie_0$ are imaginary.

Reprinted from THE PHYSICAL REVIEW D, Vol. 3, No. 4, 1046–1047, 15 February 1971
Printed in U. S. A.

Questions of Lorentz Invariance in Field Theories with Indefinite Metric*

T. D. LEE AND G. C. WICK
Columbia University, New York, New York 10027
(Received 1 December 1970)

It is pointed out that, contrary to the view expressed by Nakanishi, there is no real reason why one should rigidly follow the so-called "real space-momentum" prescription when "pinching" occurs in a field theory with indefinite metric. Since by adopting the modification introduced by Cutkosky *et al.*, a relativistic and unitary S matrix can be obtained, that should be the obvious choice.

THE paper by Nakanishi[1] makes a valid technical point about a suggestion made in one of our papers, but derives from it conclusions far beyond what the author has proved. The following comments seem all the more necessary, since one of us, in a certain sense, has been directly cited as a witness (albeit perhaps unintentionally) by the closing sentences of the above-mentioned paper. Here, then, is our opinion.

The question revolves around the proper definition of integration paths in Feynman integrals in a field theory with indefinite metric; since the usual "$i\epsilon$" prescription fails, one has to find a suitable generalization compatible with unitarity, relativistic invariance, etc. The problem was posed, but not solved, in our first two papers on the subject,[2] since no attempt was made there to discuss relativistic invariance. Later on, a simple solution, or so it seemed at the time, was suggested by one of us.[3] Let us refer to this, in view of later experience, as the "naive" prescription. Soon afterwards, it was pointed out in a paper by Cutkosky *et al.*[4] that, due to pinching, this naive prescription would lead to violations of relativistic invariance in some higher-order diagrams. A manifestly covariant, but more sophisticated, prescription for defining the Feynman integrals when pinching occurs in our theory was then described by CLOP in the same paper.

Section II of Nakanishi's paper discusses again the question of relativistic invariance of the naive prescription; it shows that noninvariance difficulties can occur in some simpler graphs than the ones pointed out earlier in the CLOP paper.[5] In our finite theory of quantum electrodynamics,[6] the simplest case that corresponds to Nakanishi's example would be some of the fourth-order e^+e^- scattering diagrams; for these diagrams, violation of relativistic invariance occurs only if one includes effects due to the square of the width of the B^0 quantum, which is in turn proportional to α^2. Previously, a comparison between the two above-mentioned prescriptions has been made[4] for the same type of diagrams, but only to first order in the width. To that order, these two prescriptions do give identical results. Owing to our incorrect impression that there was no discrepancy between the two prescriptions in these fourth-order diagrams, the naive form was allowed to persist in the discussion of these diagrams, given in Sec. V of our recent paper,[6] simply because, for pedagogical reasons, it seemed easier to present. This particular discussion should now, of course, be stricken out completely; whenever pinching occurs, one should follow the CLOP prescription. Fortunately, this has no effect on any other parts of that paper, nor does it affect any of the few actual calculations that have been performed so far, such as the vacuum-polarization calculation given in Sec. VI.

In the Introduction and in Sec. III of Nakanishi's paper, however, some very general conclusions are stated, for which no real foundation has been laid. The notion of a "nonanalytic barrier" with a "definite shape" following from the Hamiltonian or Lagrangian structure of the theory seems rather nebulous. True, for a relativistic field theory with complex masses, the contour deformation required by the CLOP prescription has not been derived from the Lagrangian formulation (though heuristic arguments can be given concerning their compatibility[7]). For that matter, neither has the naive "real space-momentum" prescription. It may be worthwhile to recall that this naive prescription was originally suggested by one of us, based only on an *ad hoc* application of the usual quantization rule to the Fourier components of a field with complex masses. Even for the

* Research supported in part by the U. S. Atomic Energy Commission.

[1] N. Nakanishi, this issue, Phys. Rev. D **3**, 811 (1971).

[2] T. D. Lee and G. C. Wick, Nucl. Phys. **B9**, 209 (1969); **B10**, 1 (1969).

[3] T. D. Lee, in *Quanta*, edited by P. G. O. Freund, C. J. Goebel and Y. Nambu (Chicago U. P., Chicago, 1970), p. 260.

[4] R. E. Cutkosky, P. V. Landshoff, D. Olive, and J. C. Polkinghorne, Nucl. Phys. **B12**, 281 (1969). This paper will be referred to as CLOP hereafter. In fairness to these authors, it should be pointed out that their involvement in the present controversy is entirely unintentional, the purposes of their investigation, as clearly stated in the introduction to their paper, being rather different from ours.

[5] The existence of difficulties was kindly mentioned to us by R. Cutkosky prior to the appearance of the CLOP paper (private communication). Nakanishi, however, in Ref. 6 of his paper, denies the validity of the CLOP counter-examples; although we do not quite follow his reasoning, we shall not argue this (by now somewhat academic) point any further.

[6] T. D. Lee and G. C. Wick, Phys. Rev. D **2**, 1033 (1970).

[7] See, e.g., lectures by T. D. Lee, in *Proceedings of the International School of Physics "Ettore Majorana," Erice, Italy, 1970*, edited by A. Zichichi (Academic, New York, to be published). The formal argument given there, which is based on Lagrangian field theory, leads automatically to the completely symmetrized limit of the CLOP prescription, and thereby resolves the ambiguity encountered in the CLOP paper for the so-called "double ice cream cone" diagram.

free field, these Fourier components (i.e., plane-wave solutions) would lead to an explicitly noncovariant representation in which the three-momentum is always real, but the energy complex. In the coordinate space, the equation of a free field with complex masses is, of course, manifestly covariant; however, its plane-wave solutions diverge exponentially in the asymptotic region. (One notes that, even for those solutions that diverge only along the time direction in a specific Lorentz frame, the same solutions viewed in other systems of reference would diverge in the asymptotic region along the spatial directions as well, thus violating the condition for the validity of the Fourier theorem.) The mathematical procedure of applying the usual quantization rules to these Fourier components must, therefore, be regarded as a purely formal one; its general validity is clearly questionable. Thus, both the naive and the CLOP prescriptions are, at present, merely recipes for evaluating the S matrix, connected only heuristically to the Lagrangian field theory.

Between these two prescriptions, the one that leads to a relativistically invariant unitary S matrix is clearly to be preferred.[8] The alternative raised by Nakanishi, *either* to sacrifice the Lorentz invariance *or* the Lagrangian field-theoretical formulation, appears to be an artificially created issue. His novel suggestion that there may be some merit in a "very slightly" noninvariant S matrix, obtained by rigidly adhering to the original naive prescription, is not, at any rate, a line of thought we would like to encourage.

[8] See also the remarks given near the end of Sec. V of Ref. 6.

PHYSICAL REVIEW D VOLUME 9, NUMBER 8 15 APRIL 1974

Vacuum stability and vacuum excitation in a spin-0 field theory*

T. D. Lee and G. C. Wick
Columbia University, New York, New York 10027
(Received 17 January 1974)

The theoretical possibility that in a limited domain in space the expectation value $\langle \phi(x) \rangle$ of a neutral spin-0 field may be abnormal (that is to say quite different from its normal vacuum expectation value) is investigated. It is shown that if the ϕ^3 coupling is sufficiently large, then such a configuration can be metastable, and its physical size may become substantially greater than the usual microscopic dimension in particle physics. Furthermore, independent of the strength of the ϕ^3 coupling, if $\phi(x)$ has sufficiently strong scalar interaction with the nucleon field, the state that has an abnormal $\langle \phi(x) \rangle$ inside a very heavy nucleus can become the minimum-energy state, at least within the tree approximation; in such a state, the "effective" nucleon mass inside the nucleus may be much lower than the normal value. Both possibilities may lead to physical systems that have not yet been observed.

I. INTRODUCTION

In a relativistic field theory, the vacuum state is defined to be the lowest energy level of the system. In analogy with other quantum-mechanical systems, however, a relativistic field may possess a degenerate lowest state. Perhaps the best known and simplest analogy is to Heisenberg's infinite ferromagnet, in which case the degeneracy of the ground state is due to rotational invariance. The assumption of a degeneracy of the vacuum state, connected with a symmetry group of the Lagrangian, obviously has some far-reaching consequences, the most alluring of which is the possibility to "understand" that puzzling aspect of particle physics, the existence of broken symmetries. As is well known, this has given rise to a host of interesting theoretical speculations.

Besides spontaneous symmetry breaking,[1] and other well-known consequences[2] related to it (Goldstone bosons, the Higgs phenomenon, etc.), the assumption of vacuum degeneracy, or near degeneracy, probably has other striking consequences, which have received little attention so far. We describe in the following an investigation of various questions which arise naturally out of the virtual existence, within a given dynamical scheme, of states which could play the same role as the observed vacuum state, but are nevertheless different from it. We shall see that, depending on the details of the theory and on the values of certain physical parameters, which are not too well known experimentally, there may or may not be consequences that are just as drastic as the already-known features of this kind of theory.

All the schemes so far considered in the literature have two assumptions in common:

(a) The Lagrangian of the system is invariant (or sometimes nearly invariant) under a certain group of transformations of the field variables.

(b) In the (observed) lowest state of the system, some of the field variables have expectation values which are not invariant under all transformations of the symmetry group. Because of (a) we must envisage the existence of other possible lowest states, or nearly lowest states, in which the expectation values of some of the fields are different; such states represent the abnormal vacuum states.

This is, of course, what is referred to in the literature as degeneracy of the vacuum; at the same time we are often reminded of the essential difference between this phenomenon and the common variety of degenerate ground state encountered in *finite* systems: In the latter case all the states of a degenerate multiplet have the same degree of physical reality; the system can easily be induced to make transitions from one substate to the others. On the other hand, only one vacuum state is realized in our world; all the others are unphysical.

On second thought, the difference is not as profound as it seems. For the sake of clarity, and at the cost of repeating familiar things, let us recall in somewhat loose terms what is really implied. In a field theory of this type, the system possesses several "equivalent" configurations of minimum potential energy; in the observed lowest state the system performs small zero-point oscillations about one of these configurations. When the system is excited the configuration will deviate more strongly, but in any event only locally, from the basic equilibrium configuration. Fundamentally the stability of the situation is attributed to the *infinite* nature of the system; owing to this, the system will never flip over as a whole from the normally observed minimum configuration to one of the others whose existence is required by the symmetry group. (As an example, the reader may recall what is usually said about the Heisenberg ferromagnet, spin waves, etc., and in particular the physical impossibility of rotating all the spins of an infinite ferromagnet simultaneously.)

Now in certain attempts at a sharp mathematical formulation of this state of affairs, it has even been asserted (perhaps on quite sound mathematical grounds) that in the limit of an infinite system one can construct a Hilbert space which contains only one "vacuum state," e.g., the observed one, and the excited states built upon it by local excitations. In this Hilbert space the physical quantities corresponding to local measurements are represented by well-defined operators; some global quantities such as the total energy or momentum are also represented, we hope, but the global generators of the group are not.

It may seem, at first sight, that in this way one has neatly thrown the abnormal degenerate vacuum states out the window, but physically it does not make so much difference, since in a certain sense they can reappear in the form of local excitations. In ferromagnetism the phenomenon is well known under the name of *domains* of magnetization. More generally we argue as follows: Suppose the configuration of the system flips over from the ordinary one to an abnormal equilibrium configuration, but only in a finite though large domain. As a volume effect, this will cost nothing; the difference in energy will be a relatively unimportant surface effect. In the case of a ferromagnet, for example, a very weak external field applied to a sufficiently large volume can easily cause the transition. Physical common sense suggests that any system with analogous features in the structure of the Lagrangian can exhibit similar phenomena under suitable circumstances. The absolute stability of the asymmetric vacuum state is therefore a relative thing.

In this paper we intend to investigate the general question of vacuum stability, and in particular to inquire whether it is experimentally possible in a limited domain in space to "excite" (flip) the ordinary vacuum to an abnormal one. As we shall see, our discussion can be readily extended to include also theories that have *no vacuum degeneracy*, but only other local minima in the field energy. For definiteness, we shall first consider the simple theory of a renormalizable spin-0 Hermitian field ϕ. The Lagrangian density is

$$\mathcal{L} = -\frac{1}{2}\left(\frac{\partial\phi}{\partial x_\mu}\right)^2 - U(\phi) + \text{counterterms}\,, \qquad (1.1)$$

where

$$U(\phi) = \tfrac{1}{2}a\phi^2 + (3!)^{-1}b\phi^3 + (4!)^{-1}c\phi^4\,, \qquad (1.2)$$

ϕ denotes the renormalized field operator, and a, b, c are the appropriately defined renormalized constants. As usual, the counterterms are for renormalization purposes; their precise definitions are given in Sec. III and in Appendix A. In $U(\phi)$, the constant c is assumed to be >0 so that the energy spectrum has a lower bound. Through the transformation $\phi(x) \to \phi(x) + \text{constant}$, one may always assume that for the vacuum state

$$\langle \text{vac} \mid \phi(x) \mid \text{vac} \rangle = 0\,. \qquad (1.3)$$

Thus, $U(\phi)$ does not contain a term linear in ϕ. [Note that in order to maintain (1.3) there is a linear term in the counterterms.] Furthermore, since the vacuum state is assumed to be the lowest-energy state, the constant a is also >0. For convenience, by using the transformation $\phi(x) \to -\phi(x)$, we may also choose the constant b to be $\geqslant 0$. As a result, but without any loss of generality, the three constants a, b, and c are all assumed to be positive.

To study the question whether there are other abnormal vacuum states, i.e., either degenerate or "excited" vacuumlike states, we find it convenient to first quantize the system in a box of a finite volume Ω with the periodic boundary condition, and then let $\Omega \to \infty$ in the end. A useful concept is to define an energy density function $\mathcal{E}(\overline{\phi})$:

$$\mathcal{E}(\overline{\phi}) \equiv \lim_{\Omega\to\infty} \Omega^{-1}(\text{minimum} \langle \;|H|\; \rangle)\,, \qquad (1.4)$$

where H is the total Hamiltonian and the minimum is taken among all states $|\;\rangle$ under the constraint

$$\Omega^{-1}\int \langle \;| \phi(x) |\; \rangle\, d^3r = \overline{\phi}\,. \qquad (1.5)$$

The value $\overline{\phi} = 0$ is, by definition, the minimum of $\mathcal{E}(\overline{\phi})$. Furthermore, it is convenient to adjust the constant part of the counterterms in (1.1) such that at the minimum $\overline{\phi} = 0$

$$\mathcal{E}(0) = 0\,. \qquad (1.6)$$

The question whether there are other, either degenerate or "excited," vacuumlike states then reduces simply to the investigation of the function $\mathcal{E}(\overline{\phi})$ for $\overline{\phi} \neq 0$, which turns out to have some rather interesting properties.

As will be shown in the next section, the dependence of $\mathcal{E}(\overline{\phi})$ on $\overline{\phi}$ bears a certain resemblance to the dependence of the Helmholtz free energy on the specific volume in thermodynamics. Just as in thermodynamics, when there is a phase transition, the Helmholtz free energy exhibits a straight-line dependence on the specific volume, its slope being the negative of the pressure; here, depending on the values of the renormalized constants a, b, and c, the function $\mathcal{E}(\bar{\phi})$ may also contain a straight section, say between $\phi_\alpha \le \bar{\phi} \le \phi_\beta$. The existence of such a straight section appears to be a general feature of the theory, provided that the ϕ^3 coupling constant b is sufficiently large. It exists even in the approximation of neglecting all loop diagrams; in such an approximation, one has $\mathcal{E}(\bar{\phi}) = U(\bar{\phi})$ *outside* the straight section, where U is given by (1.2). [Note that $U(\bar{\phi})$ does not contain any straight section.] Along the straight section $\phi_\alpha < \bar{\phi} < \phi_\beta$, the system actually comprises two phases, in analogy to the phase-transition phenomenon in thermodynamics. Outside the straight section, $\bar{\phi} < \phi_\alpha$ or $\bar{\phi} > \phi_\beta$, the system exists only in a single phase. The "true" vacuum state $\bar{\phi} = 0$ is included in the region $\bar{\phi} > \phi_\beta$, as illustrated in Fig. 1.

The inclusion of loop diagrams does not alter the general character of the energy density curve $\mathcal{E}(\bar{\phi})$. The explicit contributions of all one-loop and two-loop diagrams and some of the general properties of other multiloop diagrams are given in Sec. III. From these results, one expects that the function $\mathcal{E}(\bar{\phi})$ defined in either one of the two single-phase regions, say $\bar{\phi} < \phi_\alpha$, can be analytically continued beyond the point $\bar{\phi} = \phi_\alpha$ to the region $\bar{\phi} > \phi_\alpha$; its analytic continuation, called $\mathcal{E}_\alpha(\bar{\phi})$, is, of course, different from $\mathcal{E}(\bar{\phi})$ in the two-phase region. [This phenomenon is again in close analogy to the familiar gas-liquid transition in statistical mechanics; the analytic continuation of the gas (or liquid) phase is the supercooled gas (or superheated liquid) region, not the two-phase region.[3]] Similarly, one may analytically continue the function $\mathcal{E}(\bar{\phi})$, defined in the other single-phase region $\bar{\phi} > \phi_\beta$, to the region $\bar{\phi} \le \phi_\beta$ and call its analytic continuation $\mathcal{E}_\beta(\bar{\phi})$. In general, one expects the function $\mathcal{E}_\alpha(\bar{\phi})$ to have a minimum at

$$\bar{\phi} = \phi_{\text{vex}} , \tag{1.7}$$

where the subscript "vex" denotes the vacuum excitation state.

In the case of the degenerate vacuum, both the true vacuum state $\bar{\phi} = 0$ and the vacuum excitation state $\bar{\phi} = \phi_{\text{vex}}$ appear as the end points of the straight section $\phi_\alpha \le \bar{\phi} \le \phi_\beta$, i.e.,

$$\phi_\alpha = \phi_{\text{vex}} , \quad \phi_\beta = 0 ,$$

and because of (1.6)

$$\mathcal{E}(\phi_\alpha) = \mathcal{E}(\phi_\beta) = 0 . \tag{1.8}$$

From (1.2) one sees that if all loop diagrams are neglected, then the degeneracy occurs at

$$3ac = b^2 . \tag{1.9}$$

As we shall discuss in Sec. III, there is a simple and convenient way to define the renormalization constants so that (1.9) is the *exact* condition for degeneracy when all the loop diagrams are also included. Consequently, in order that the absolute minimum energy level is at $\bar{\phi} = 0$, we must have

$$3ac \ge b^2 ; \tag{1.10}$$

otherwise, the role of the states $\bar{\phi} = 0$ and $\bar{\phi} = \phi_{\text{vex}}$ will be interchanged.

In Sec. IV, we study the question of the lifetime of the system in the excited state $\bar{\phi} = \phi_{\text{vex}}$. We

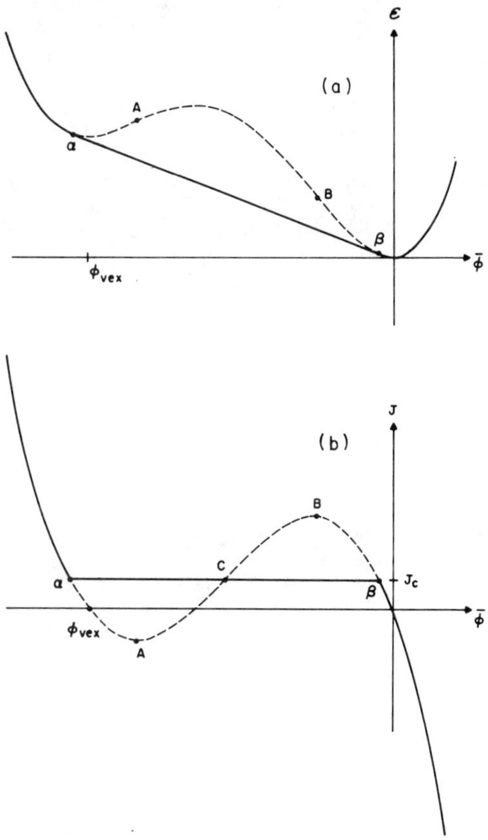

FIG. 1. Examples of graphs of $\mathcal{E}(\bar{\phi})$ and $J(\bar{\phi}) = -(d\mathcal{E}/d\bar{\phi})$ in the tree approximation. The two-phase region is between the points α and β; ϕ_α and ϕ_β are their abscissas. In (a), outside the interval $\phi_\alpha \le \bar{\phi} \le \phi_\beta$, $\mathcal{E}(\bar{\phi}) = U(\bar{\phi})$; inside the interval, the solid line refers to $\mathcal{E}$, and the dashed curve to U. In (b), the two areas $C\alpha A$ and $C\beta B$ (between the dashed curve and the solid line) are equal.

shall show that in the nondegenerate case ($3ac > b^2$), *as the volume* $\Omega \to \infty$, *the lifetime becomes zero.* On the other hand, there may exist metastable states which satisfy approximately

$$\langle \mid \phi(x) \mid \rangle = \phi_{\text{vex}} \tag{1.11}$$

in a *finite* volume L^3, where $L \gg m^{-1}$ and m^{-1} denotes the relevant microscopic length in the problem; m can be either $\sim O(b)$ or $O(a^{1/2})$. Outside the volume, except over a surface region of a volume $\sim O(L^2 m^{-1})$, one has $\langle \mid \phi(x) \mid \rangle = 0$. The excitation energy of such a state in its rest frame is given by

$$M_{\text{vex}} = L^3 \mathcal{E}_\alpha(\phi_{\text{vex}}) + O(L^2 m^3), \tag{1.12}$$

where $O(L^2 m^3)$ denotes the surface energy and $\mathcal{E}_\alpha(\bar{\phi})$ is the aforementioned analytic continuation of $\mathcal{E}(\bar{\phi})$. The lifetime τ of such a state is given by

$$\tau \gtrsim L, \tag{1.13}$$

provided $\ln(Lm)$ is not too large, though (Lm) must be $\gg 1$. Only in the special case of a vacuum degeneracy, i.e., $\mathcal{E}_\alpha(\phi_{\text{vex}}) = 0$, can the size L be arbitrarily large; its rest mass is determined completely by the surface energy. In general, the ratio of the width to the rest mass of such vacuum excitations in either the degenerate or the nondegenerate case is exceedingly small, given by

$$(\tau M_{\text{vex}})^{-1} \lesssim [L^4 \mathcal{E}_\alpha(\phi_{\text{vex}}) + O(L^3 m^3)]^{-1} \ll 1. \tag{1.14}$$

In Sec. V, we discuss the classical solutions corresponding to the vacuum excitations. The most interesting aspect of these solutions occurs when there is an extended external source. For definiteness, we may treat approximately the effect of a heavy nucleus as that of an "external source," assuming that there is a strong interaction $g\psi^\dagger \gamma_4 \psi \phi$ between the scalar field ϕ and the nucleon field ψ. As we shall see, within the tree approximation, if the surface energy can be neglected, then when g is sufficiently strong, or when the nuclear density is sufficiently high, the lowest-energy state becomes one in which the expectation value $\langle \phi(x) \rangle$ inside the nucleus can be quite different from its normal vacuum expectation value (which is zero, by our convention). Furthermore, inside the nucleus the "effective" mass of the nucleon becomes $m_N + g\langle \phi \rangle$, which can also be quite different from its normal value m_N.

A concrete example of such a strong interaction is given by the well-known σ model. This is examined in Sec. VI. It appears that, within the tree approximation, if the mass of the σ particle is $\lesssim$ a few GeV, there may well exist a new family of metastable, or even stable, superheavy nuclei.

By taking the zero pion mass limit, we can readily extend our discussion of the σ model to theories with Goldstone bosons; with some further minor modifications, it can also be applied to fields with Higgs mechanisms.

II. ENERGY DENSITY FUNCTION

To evaluate the energy density function $\mathcal{E}(\bar{\phi})$, defined by (1.4), we apply the standard Lagrangian-multiplier method to take into account the constraint (1.5). Let H_J be a new Hamiltonian, defined by

$$H_J \equiv H + J \int \phi(x)\, d^3r, \tag{2.1}$$

where J is the Lagrangian multiplier and H is the original Hamiltonian, which according to (1.1) is given by

$$H = \int [\tfrac{1}{2}\Pi^2 + \tfrac{1}{2}(\nabla\phi)^2 + U(\phi) + \text{counterterms}]\, d^3r, \tag{2.2}$$

where Π is the conjugate momentum of ϕ. Let the lowest eigenvalue of H_J be $\Omega\lambda_J$, i.e.,

$$H_J \mid \rangle = \Omega \lambda_J \mid \rangle. \tag{2.3}$$

By using (2.1), (1.4), and (1.5), we find the energy density function $\mathcal{E}(\bar{\phi})$ to be given by the Legendre transformation

$$\mathcal{E}(\bar{\phi}) = \lambda_J - J\bar{\phi}, \tag{2.4}$$

where

$$\bar{\phi} = \frac{\partial \lambda_J}{\partial J} \tag{2.5}$$

and

$$J = -\frac{\partial \mathcal{E}(\bar{\phi})}{\partial \bar{\phi}}. \tag{2.6}$$

To calculate λ_J, let us decompose

$$H_J = H_0 + H_1, \tag{2.7}$$

where

$$H_0 = \tfrac{1}{2} \int [\Pi^2 + (\nabla\phi)^2 + \tfrac{1}{2} a \phi^2]\, d^3r, \tag{2.8}$$

and regard H_1 as a perturbation. The power-series expansion of λ_J in terms of the constants J, b, and c can be readily derived. Following the treatment given by Coleman and Weinberg[4] (which is also formally analogous to some of the analysis developed in statistical mechanics and many-body problems[5]), we may regroup the perturbation-series expansion of λ_J into sums of tree diagrams, one-loop diagrams, two-loop diagrams, etc. The systematics of these loop diagrams will be given in the next section. Here we only discuss the tree approximation. It is not difficult to see[6] that in the tree approximation λ_J is given by the absolute minimum of

$$U_J(\phi) \equiv J\phi + U(\phi)$$
$$= J\phi + \tfrac{1}{2}a\phi^2 + (3!)^{-1}b\phi^3 + (4!)^{-1}c\phi^4, \qquad (2.9)$$

and $\phi = \overline{\phi}$ is the minimum point. (For completeness, a proof is given in Appendix A.) At $J=0$, one has $U_J = U$. Since we are interested in the case where the function $U(\phi)$ in the original Lagrangian (1.1) has more than one local minimum, the ϕ^3 coupling constant b cannot be too small:

$$b^2 > \tfrac{8}{3}ac. \qquad (2.10)$$

On the other hand, because of our convention that the absolute minimum of $U(\phi)$ should be at $\phi = 0$, we have

$$b^2 \leq 3ac. \qquad (2.11)$$

[The apparently narrow region defined by these two inequalities may be deceptive. Actually, only (2.10) is the relevant one. If $b^2 > 3ac$, then the absolute minimum of U is not at $\phi = 0$. By using the transformation $\phi \to \phi + \text{constant}$, this absolute minimum can be shifted back to $\phi = 0$. Under such a transformation, only the coupling constant c is invariant; the new constants a and b now satisfy $b^2 < 3ac$.]

Next, we consider the equation $\partial U_J/\partial\phi = 0$; i.e., on account of (2.9),

$$J = -\frac{\partial U}{\partial \phi} = -a\phi - \tfrac{1}{2}b\phi^2 - (3!)^{-1}c\phi^3, \qquad (2.12)$$

which at $J=0$ has three roots:

$$\phi = 0$$

and

$$\phi = \phi_\pm \equiv \frac{3}{2c}\left[-b \pm (b^2 - \tfrac{8}{3}ac)^{1/2}\right]. \qquad (2.13)$$

Among these, $\phi = 0$ is the absolute minimum of $U(\phi)$, $\phi = \phi_+$ is a local maximum, and $\phi = \phi_-$ is the other local minimum. As J increases, these two minima will move, and the corresponding values of $U(\phi)$ will also change. There is a critical value J_c at which these two minima become degenerate. As illustrated in Fig. 1, we may determine graphically the value $J = J_c$ by using Maxwell's rule of equal area. The absolute minimum $\phi = \overline{\phi}$ makes a sudden jump from $\overline{\phi} = \phi_B$ at $J = J_{c^-}$ to $\overline{\phi} = \phi_\alpha$ at $J = J_{c^+}$. By using (2.4) we find in the tree approximation

$$\mathcal{E}(\overline{\phi}) = U(\overline{\phi})$$

in the region

$$\overline{\phi} \geq \phi_B \text{ and } \overline{\phi} \leq \phi_\alpha. \qquad (2.14)$$

But in $\phi_\alpha \leq \overline{\phi} \leq \phi_B$, $\mathcal{E}(\overline{\phi})$ is a linear function of $\overline{\phi}$, which is simply the common tangent line of $U(\overline{\phi})$ at $\overline{\phi} = \phi_\alpha$ and ϕ_B.

Such behavior is analogous to the problem of phase transition in statistical mechanics. In the statistical analog, the roles of J, $\overline{\phi}$, $\mathcal{E}(\overline{\phi})$, and λ_J are replaced by those of pressure, specific volume, Helmholtz free energy density, and Gibbs free energy density, respectively. The straight section $\phi_\alpha \leq \overline{\phi} \leq \phi_B$ denotes the two-phase region. As already noted in the Introduction, the function $\mathcal{E}(\overline{\phi})$ in either one of the single-phase regions, $\overline{\phi} > \phi_B$ or $\overline{\phi} < \phi_\alpha$, can be analytically continued into the two-phase region. In the tree approximation, these two analytic continuations are identical and both lead to $U(\overline{\phi})$. This is again analogous to the Van der Waals approximation used in statistical mechanics. In statistical mechanics, the analytic continuations of the thermodynamical functions of the liquid and the gas phases are respectively those of the superheated liquid and the supercooled gas, which should be different functions, but they reduce to the same expression in the Van der Waals approximation.

In the present problem, except for the degenerate vacuum case, the energy density function $\mathcal{E}(\overline{\phi})$ has only one minimum at $\overline{\phi} = 0$, and that is the true vacuum state. On the other hand, if the ϕ^3 coupling constant b is not too small, the analytic continuation of $\mathcal{E}(\overline{\phi})$ is expected to have another minimum at $\overline{\phi} = \phi_{\text{vex}}$ which denotes the vacuum excitation. In the above, this property has been established in the tree approximation; as we shall see in the next section, if the coupling c is not too large, this property remains correct at least to every order in the loop expansion.

III. LOOP DIAGRAMS

The reduction of the perturbation-series expansion of $\mathcal{E}(\overline{\phi})$ into a sum of tree diagrams, one-loop diagrams, etc. has been given in Ref. 4. In this section we shall first briefly review the procedure, and then discuss some new properties.

A. Prototype diagrams

By using the free-field Hamiltonian H_0, defined by (2.8), as the unperturbed Hamiltonian, one can readily expand the energy density function $\mathcal{E}(\overline{\phi})$ as a power series in b, c, and $\overline{\phi}$. As will be shown in Appendix A, we may separate $\mathcal{E}(\overline{\phi})$ into a sum of tree diagrams and loop diagrams:

$$\mathcal{E}(\overline{\phi})=[\mathcal{E}(\overline{\phi})]_{\text{tree}}+\sum_{l=1}^{\infty}[\mathcal{E}(\overline{\phi})]_{l\text{-loop}}, \quad (3.1)$$

where $[\mathcal{E}(\overline{\phi})]_{l\text{-loop}}$ represents the summation over all one-particle irreducible scattering diagrams that have l loops and in which every external line carries a zero 4-momentum and contributes a factor $\overline{\phi}$ to the Feynman integral. For the tree diagrams (away from the two-phase region), one has

$$[\mathcal{E}(\overline{\phi})]_{\text{tree}}=U(\overline{\phi}), \quad (3.2)$$

where U is given by (1.2), provided that the renormalized constants a, b, and c in $U(\overline{\phi})$ are related to the appropriate scattering amplitudes at zero momentum. (See Sec. III B and Appendix A for further discussions of renormalization.)

For $l\neq 0$, it is useful to introduce $D(k)$, defined to be the propagator of the spin-0 particle moving in a given constant external field ϕ_{ext} whose value happens to be given by $\phi_{\text{ext}}=\overline{\phi}$. Thus, $D(k)$ is identical to the propagator of a free particle, but with its $(\text{mass})^2$ given by $(\partial^2 U/\partial\overline{\phi}^2)$; i.e.,

$$D(k)=-i[k^2+a(1+\Delta)]^{-1}, \quad (3.3)$$

where

$$\Delta=\overline{\phi}(b+\tfrac{1}{2}c\overline{\phi})/a. \quad (3.4)$$

Let us first consider the sum of all one-loop diagrams, and differentiate $[\mathcal{E}(\overline{\phi})]_{\text{one-loop}}$ with respect to a, but keeping b, c, and $\overline{\phi}$ fixed. We obtain

$$\frac{\partial}{\partial a}[\mathcal{E}(\overline{\phi})]_{\text{one-loop}} =\frac{1}{2}\int(2\pi)^{-4}d^4k[D(k)+\text{subtraction term}], \quad (3.5)$$

which can be readily established by first expanding both sides as a power series of $\overline{\phi}$, then noting that graphically the differentiation $\partial/\partial a$ on the one-loop diagram is just like cutting open one of its internal lines; this turns each loop diagram into a propagator diagram. Thus, diagram by diagram, both sides of (3.5) are equal. The subtraction term in (3.5) is needed to eliminate divergences. (The details of the subtraction term will be given in Sec. III B.) From Eq. (3.5), it follows that[4,5]

$$[\mathcal{E}(\overline{\phi})]_{\text{one-loop}}=\frac{i}{2}\int(2\pi)^{-4}d^4k \times\{\ln[iD(k)]+\text{subtraction term}\}. \quad (3.6)$$

Throughout the paper, $k^2=\vec{k}^2-k_0^2$ and d^4k is real.

It is straightforward to express the higher-order loop diagrams in terms of $D(k)$. In this way all external lines attached to a three-point vertex and all pairs of external lines attached to a four-point vertex are implicitly accounted for. We need only consider those l-loop diagrams, called *prototype diagrams*,[4,7] in which all external lines, if they exist, must be attached separately to different four-point vertices; i.e., every three-point vertex $b\phi^3$ connects only internal lines and every four-point vertex $c\phi^4$ connects at most one external line to the diagram. For any given $l>1$, there are only a finite number of such prototype diagrams. We shall evaluate these prototype diagrams according to the standard Feynman rule, except that each internal line gives a factor $D(k)$, not $-i(k^2+a)^{-1}$, to the Feynman integral. Otherwise, all the remaining factors in the Feynman integral are as usual, i.e., we assign factors b, c, and $\overline{\phi}$ respectively for a three-point vertex, a four-point vertex, and an external line. Except for the subtraction terms that are needed for renormalization purposes (and which will be discussed in Sec. III B), the function $[\mathcal{E}(\overline{\phi})]_{l\text{-loop}}$ for $l>1$ is simply given by the sum over the finite set of all different prototype l-loop diagrams. As an example, for $l=2$, there are only four different prototype diagrams; these are given by diagrams (i)–(iv) in Fig. 2. [Because of renormalization, one must combine these four diagrams together with diagrams (ii)′, (iii)′, and (iv)′ in Fig. 2. The explicit value of these two-loop diagrams is given in Sec. III C.]

B. Renormalization

In (3.6) the integral $\int d^4k\ln(iD)$ is quartically divergent; therefore three subtractions are needed to eliminate the infinities. The corresponding subtraction term should be at least a quadratic function in $\overline{\phi}$. However, it is entirely a matter of choice whether or not one should also subtract the *finite* $\overline{\phi}^3$ and $\overline{\phi}^4$ terms from the integral. Similar ambiguities also exist for higher-order loop diagrams. This problem is closely tied to the original freedom in defining the renormalized constants a, b, and c. Any finite loop-diagram contribution to the $\overline{\phi}^3$ and $\overline{\phi}^4$ terms can either be included in the renormalized constants [i.e., already included in the $b\overline{\phi}^3$ and $c\overline{\phi}^4$ terms in the original $U(\overline{\phi})$ function given by (1.2)] or not included. If they are included, then a corresponding subtraction is necessary in the relevant loop calculation to avoid double counting, but otherwise this is not necessary. As it turns out, there is a particularly convenient way to decide on which choice to make.

Let us first consider the special case of degenerate vacuum. If $3ac=b^2$, the function $U(\phi)$ in the original Lagrangian (1.1) is symmetric with respect to the transformation

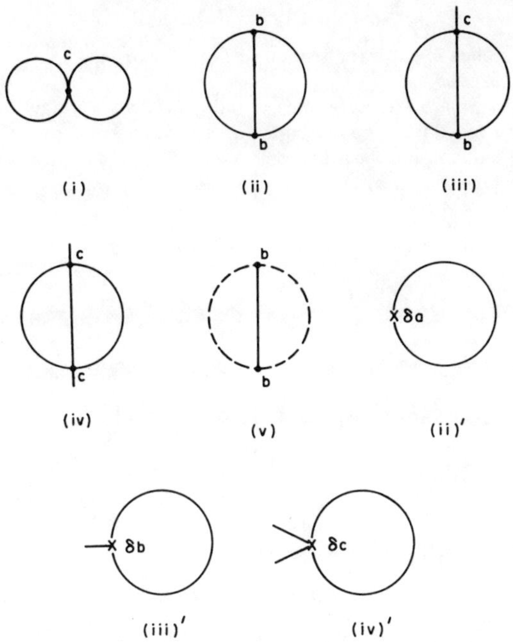

FIG. 2. Diagrams (i)–(iv) are examples of two-loop prototype diagrams. Each solid internal line carries a propagator factor D, given by (3.3). Diagrams (ii)′–(iv)′ are related to (ii)–(iv) through renormalization. In diagram (v), the dashed line carries a factor $-i(k^2+a)^{-1}$; hence (v) is not a prototype diagram.

$$\phi+\frac{b}{c}\rightarrow-\left(\phi+\frac{b}{c}\right). \tag{3.7}$$

It is clearly desirable that the symmetry should also be maintained by the counterterms; in that case the entire Lagrangian (1.1) is invariant under the same transformation, and consequently the vacuum degeneracy becomes an exact property. It is quite simple to show that the dependence of $[\mathcal{E}(\overline{\phi})]_{l\text{-loop}}$ on $\overline{\phi}$, except maybe for the subtraction terms, is completely through the variable Δ, given by (3.4). Since Δ is invariant under the transformation $(\overline{\phi}+b/c)\rightarrow-(\overline{\phi}+b/c)$, the same symmetry holds for $\mathcal{E}(\overline{\phi})$ if all these subtraction terms in the loop-diagram calculations are also functions of Δ. Because Δ is a quadratic function of $\overline{\phi}$ and because these subtraction terms should be at most quartic functions of $\overline{\phi}$, we require them to be *quadratic* functions of Δ.

Thus, in a power-series expansion in Δ,

$$[\mathcal{E}(\overline{\phi})]_{l\text{-loop}}=\alpha\Delta^3+\beta\Delta^4+\gamma\Delta^4+\cdots, \tag{3.8}$$

where $\alpha, \beta, \gamma, \ldots$ are constants. As a result, if $3ac=b^2$, the entire Lagrangian is symmetric under (3.7), and that implies a degenerate vacuum. In the following, the requirement (3.8) will be imposed also for the general case, even when there is no degeneracy.

With this requirement, and the convention that $\overline{\phi}=0$ denotes the true vacuum, we derive the inequality

$$b^2\leqslant 3ac, \tag{3.9}$$

which is the same as (2.11), but is now valid with the inclusion of all loop-diagram corrections, not just in the tree approximation.

C. Loop expansion

In order to understand the nature of the loop expansion, we establish first the following theorem:

Theorem 1. At any $l\geqslant 1$, $[\mathcal{E}(\overline{\phi})]_{l\text{-loop}}$ can be written in terms of l dimensionless functions $F_{l,1}$, $F_{l,2}, \ldots F_{l,l}$ which depend only on Δ:

$$[\mathcal{E}(\overline{\phi})]_{l\text{-loop}}=a^2\sum_{m=1}^{l}c^{m-1}[a^{-1}(b^2+2ac\Delta)]^{l-m}F_{l,m}(\Delta), \tag{3.10}$$

where Δ is given by (3.4).

Proof. Let us consider an l-loop prototype diagram with N three-point vertices, M four-point vertices, E external lines, and I internal lines. From the explicit Feynman rules given above, it follows that the corresponding Feynman integral for $\mathcal{E}(\overline{\phi})$ is of the form

$$b^N c^M\overline{\phi}^E f(a,\Delta). \tag{3.11}$$

Since the total number of loops is given by $l=I-N-M+1$ and since $(2I+E)$ is equal to $(3N+4M)$, we have

$$l=\tfrac{1}{2}N+M-\tfrac{1}{2}E+1. \tag{3.12}$$

The a dependence in (3.11) can be easily obtained from a simple dimensional analysis. Because Δ and c are both dimensionless, but a, b^2, $\overline{\phi}^2$, and $[\mathcal{E}(\overline{\phi})]^{1/2}$ are of the same dimension $(\text{mass})^2$, we obtain

$$f(a,\Delta)=a^{2-(N+E)/2}F(\Delta), \tag{3.13}$$

where F is dimensionless. For the special case of $E=0$ (i.e., those prototype diagrams with no external line $\overline{\phi}$), by using (3.11)–(3.13) we find that the Feynman integral of such a diagram is of the form

$$a^2c^M(b^2/a)^{l-M-1}F(\Delta). \tag{3.14}$$

Now, from the definition of prototype diagrams we see that any $E\neq 0$ prototype diagram can be transformed into an $E=0$ prototype diagram by simply replacing all four-point vertices that are attached to external lines by three-point vertices, but keeping all internal lines and other vertices unchanged. Formally, we may represent such a replacement by

$$c\bar{\phi}\phi_{\text{in}}^3 \to b\phi_{\text{in}}^3, \tag{3.15}$$

where ϕ_{in} denotes the appropriate internal line and $\bar{\phi}$ the external line. Thus, the sum over all different prototype diagrams that can be transformed into the same $E=0$ prototype diagram through (3.15) is equal to the Feynman integral of the $E=0$ diagram, provided we change $b\to(b+c\bar{\phi})$; therefore, (3.14) becomes

$$a^2c^M[a^{-1}(b^2+2ac\Delta)]^{l-M-1}F(\Delta).$$

Since M can vary from 0 to $l-1$, Theorem 1 is proved.

Remarks. According to (3.9), $b^2\leq 3ac$; we may regard the loop expansion as a power-series expansion in c, but treating Δ and (b^2/ac) [and therefore also $(b\bar{\phi}/a)$ and $(c\bar{\phi}^2/a)$] as $\lesssim O(1)$.

Theorem 2.

$$[\mathcal{E}(\bar{\phi})]_{\text{one-loop}}=\frac{a^2}{32\pi^2}[\tfrac{1}{2}(1+\Delta)^2\ln(1+\Delta)-\tfrac{1}{2}\Delta-\tfrac{3}{4}\Delta^2] \tag{3.16}$$

and

$$[\mathcal{E}(\bar{\phi})]_{\text{two-loop}}=\frac{a^2c}{2(32\pi^2)^2}[(1+\Delta)\ln(1+\Delta)-\Delta]^2+\frac{a(b^2+2ac\Delta)}{(32\pi^2)^2}\{\tfrac{1}{2}(1+\Delta)[\ln(1+\Delta)]^2-2(1+\Delta)\ln(1+\Delta)+2\Delta+\tfrac{1}{2}\Delta^2\}. \tag{3.17}$$

Proof. The evaluation of $[\mathcal{E}(\bar{\phi})]_{\text{one-loop}}$ follows readily from (3.6) and (3.8); the result is (3.16). (If $b=0$, that is in the pure ϕ^4 theory, the above expression for $[\mathcal{E}(\bar{\phi})]_{\text{one-loop}}$ reduces to the form derived by Coleman and Weinberg.[4]) The two-loop prototype diagrams are listed in Fig. 2. These diagrams can be calculated according to the general rules given in the previous sections. The calculation is somewhat involved because of renormalization. The details are given in Appendix B, and the result is (3.17).

The evaluation of higher-order loop diagrams is complicated partly because of the large number of diagrams and partly because of the renormalization procedures required to eliminate infinities. For simplicity, we shall consider the special case $c=0$. In such a case, there are only the $b\phi^3$ vertices, and the theory is superrenormalizable. The Feynman integrals of the majority of the l-loop diagrams are convergent. In the following theorem, we shall restrict our discussion to these convergent diagrams, or "primitively divergent" diagrams as in the case of $l=3$. (A primitively divergent diagram, as defined by Dyson,[8] is one whose Feynman integral, though divergent, becomes convergent when any one of its internal momenta is held fixed; here, the only example is in $l=3$.)

Theorem 3. If $c=0$ and if we include only convergent, or primitively divergent, diagrams, then

$$[\mathcal{E}(\bar{\phi})]_{\text{three-loop}}=(\text{constant})\,b^4[\ln(1+\Delta)-\Delta+\tfrac{1}{2}\Delta^2] \tag{3.18}$$

and for $l>3$

$$[\mathcal{E}(\bar{\phi})]_{l-\text{loop}}=(\text{constant})a^2(b^2/a)^{l-1}\times[(1+\Delta)^{3-l}-1+(l-3)\Delta-\tfrac{1}{2}(l-3)(l-2)\Delta^2]. \tag{3.19}$$

(This theorem is proved in Appendix C.)

Remarks. From Theorem 2 and Theorem 3, it follows that every term in the loop expansion is singular at $\Delta=-1$, i.e., at

$$\bar{\phi}=c^{-1}[-b\pm(b^2-2ac)^{1/2}], \tag{3.20}$$

which are the points of inflection A and B of the function $U(\bar{\phi})$, as illustrated in Fig. 1. This implies that the energy density function $\mathcal{E}(\bar{\phi})$ can be analytically continued from either one of the two single-phase regions, $\bar{\phi}<\phi_\alpha$ or $\bar{\phi}>\phi_\beta$, to the two-phase region. Let $\mathcal{E}_\alpha(\bar{\phi})$ denote the analytic continuation from $\bar{\phi}<\phi_\alpha$, and $\mathcal{E}_\beta(\bar{\phi})$ that from $\bar{\phi}>\phi_\beta$. If the loop expansion is used, then $\mathcal{E}_\alpha(\bar{\phi})$ has a singularity at A, and $\mathcal{E}_\beta(\bar{\phi})$ a singularity at B. [At $\Delta=-1$, the propagator $D(k)$ is that of a zero-mass particle. Thus, physically, it seems reasonable that there should be such singularities for these analytic continuations, independent of the loop expansion.] The true vacuum is at $\bar{\phi}=0$, and therefore it lies in the single phase region $\bar{\phi}>\phi_\beta$. The vacuum excitation $\bar{\phi}=\phi_{\text{vex}}$ denotes the minimum of the analytic continuation $\mathcal{E}_\alpha(\bar{\phi})$. From Fig. 1, one sees that the point $\bar{\phi}=\phi_{\text{vex}}$ lies in between $\phi=\phi_\alpha$ and the $\bar{\phi}$ corresponding to A.

In Fig. 3 we plot the modification of the J vs $\bar{\phi}$ curve due to the one-loop diagram for the special case $b^2=3ac$. Because of the symmetry under the transformation (3.7), the two-phase region, with the inclusion of the loop-diagram correction, remains given by

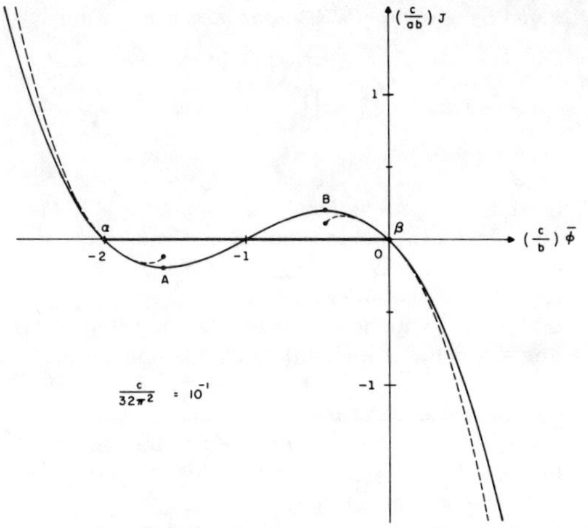

FIG. 3. J in units ab/c vs $\bar{\phi}$ in units b/c for the special case $b^2 = 3ac$ and $(32\pi^2)^{-1}c = 10^{-1}$. The solid curve denotes the tree approximation, and the dashed curve includes the one-loop approximation.

$$J = 0 \text{ and } -\frac{2b}{c} \leq \bar{\phi} \leq 0 . \tag{3.21}$$

It is convenient to introduce the dimensionless variables χ, V, and j, defined by

$$\bar{\phi} \equiv \frac{b}{c}(\chi - 1), \quad \mathcal{E} \equiv \frac{ab^2}{c^2} V, \quad J \equiv \frac{ab}{c} j , \tag{3.22}$$

and therefore $j = -\partial V/\partial\chi$. From (3.2) and (3.16), we have (for the special case $b^2 = 3ac$)

$$V_{\text{tree}} = \tfrac{1}{8}(1-\chi^2)^2 \tag{3.23}$$

and

$$V_{\text{one-loop}} = \tfrac{1}{3}\gamma\left[\tfrac{1}{2}(1+\Delta)^2 \ln(1+\Delta) - \tfrac{1}{2}\Delta - \tfrac{3}{4}\Delta^2\right] , \tag{3.24}$$

where

$$\Delta = \tfrac{3}{2}(\chi^2 - 1) \tag{3.25}$$

and

$$\gamma = \frac{c}{32\pi^2} . \tag{3.26}$$

In Fig. 3, for definiteness we assume arbitrarily

$$\gamma = 10^{-1} . \tag{3.27}$$

IV. STABILITY

In this section we discuss the stability problem if the system is not in the true vacuum state $\bar{\phi} = 0$, but in the vacuum excitation state $\bar{\phi} = \phi_{\text{vex}}$. As remarked before, only in the case of a degenerate vacuum do both $\bar{\phi} = 0$ and $\bar{\phi} = \phi_{\text{vex}}$ lie on the energy density curve $\mathcal{E}(\bar{\phi})$. In the nondegenerate case, while the true vacuum state $\bar{\phi} = 0$ is on the energy density curve $\mathcal{E}(\bar{\phi})$, the vacuum excitation state $\bar{\phi} = \phi_{\text{vex}}$ lies on the analytical continuation of $\mathcal{E}(\bar{\phi})$, denoted by $\mathcal{E}_\alpha(\bar{\phi})$, as illustrated in Fig. 1.

A. Nondegenerate case ($b^2 < 3ac$)

We assume that at time $t = 0$ the system is in the vacuum excitation state $|\,\rangle$ which satisfies

$$\langle\, |\phi(x)|\, \rangle = \phi_{\text{vex}} \tag{4.1}$$

at every point x in the volume Ω. For convenience, let us take Ω to be a cube, which will be divided into N smaller cubes, each of a linear size L; all adjacent cubes are separated by a distance δ. Hence,

$$\Omega = N(L + \delta)^3 , \tag{4.2}$$

where δ is of the order of the microscopic length of the problem, but L is much larger and may even be of a macroscopic dimension; e.g.,

$$\delta \sim O(a^{-1/2}) \text{ or } O(b^{-1})$$

and

$$L \gg \delta . \tag{4.3}$$

Let $p(t)$ be the probability that at a later time t the system is either in a state in which

$$\langle\, |\phi(x)|\, \rangle = \begin{cases} 0 \text{ in one of the cubes } L^3 \\ \text{arbitrary in the surface region} \sim O(L^2\delta) \\ \phi_{\text{vex}} \text{ outside} \end{cases} \tag{4.4}$$

or in states that differ from (4.4) by some additional high-energy quantum excitations inside the cube L^3 that has been singled out. In the nondegenerate vacuum case, one has $\mathcal{E}_\alpha(\phi_{\text{vex}}) > \mathcal{E}(0)$, where $\mathcal{E}_\alpha$ denotes the analytical continuation of $\mathcal{E}$. These states can have the same energy as the initial state, provided

$$L^3\mathcal{E}_\alpha(\phi_{\text{vex}}) = L^3\mathcal{E}(0) + \text{excitation energy} . \tag{4.5}$$

Since L is $\gg O(a^{-1/2})$ or $O(b^{-1})$, there is a large number of such states that satisfy (4.5); their entropy is proportional to L^3. Thus, by using the standard calculation of transition rates, one finds

$$p(t) = 1 - \exp(-\lambda_L t) , \tag{4.6}$$

where $\lambda_L \neq 0$, and at fixed L and δ the probability $p(t)$ is *independent* of N. As shown in Appendix D, a lower bound in λ_L can be easily estimated; we find for L sufficiently large

$$\ln\lambda_L > (-\kappa L^3), \tag{4.7}$$

where κ is positive-definite and depends only on the renormalized constants a, b, and c.

Since the N cubes are arranged to be physically separated from each other, they can be regarded as independent systems. For an initial state (4.1), the probability that at a later time t the system remains in the same state is

$$[1-p(t)]^N = \exp(-N\lambda_L t), \tag{4.8}$$

which, at a fixed L, approaches zero as N (and therefore Ω) becomes ∞. Thus, if the vacuum excitation state extends over an infinite volume, *its lifetime is zero.*

However, the lifetime of a vacuum excitation in a limited volume v is quite a different matter. Let us consider a finite volume v and a surface region s that surrounds v. The domain $v+s$ is, of course, inside the bigger volume Ω of the entire quantum system; for simplicity, one may assume Ω to be infinite. Let the vacuum excitation be described by the state $|\text{vex}\rangle$ which satisfies

$$\langle \text{vex}|\phi(x)|\text{vex}\rangle = \begin{cases} \phi_{\text{vex}} \text{ in } v \\ 0 \text{ outside } v+s \\ \text{arbitrary, though smooth,} \\ \text{inside } s\,. \end{cases} \tag{4.9}$$

Furthermore, we assume that in its rest system (i.e., $|\text{vex}\rangle$ of zero 3-momentum) the shape of v is one in which the linear dimension is $\sim O(v^{1/3})$ in all directions. Thus, because of (1.6), the rest mass of $|\text{vex}\rangle$ is

$$M_{\text{vex}} = v\mathcal{E}_\alpha(\phi_{\text{vex}}) + \text{surface energy}\,. \tag{4.10}$$

Such a state can decay through meson emissions. There are two dominant modes of decay: One is via the surface contraction, and the other is via the decay law (4.8), provided that v is sufficiently large. The latter resembles a "boiling" mechanism; we may first imagine that v is divided into n smaller volumes, $v = n(L+\delta)^3$, and then each smaller volume L^3 decays exponentially as $\exp(-\lambda_L t)$. Let τ_c and τ_b be, respectively, the time scales for surface contraction and for boiling. It is clear that

$$\tau_c \sim v^{1/3} \text{ and } \tau_b \sim (n\lambda_L)^{-1}\,. \tag{4.11}$$

For v small, the decay time is determined by τ_c, and for v sufficiently large by τ_b. To have a rough idea of the critical volume size when $\tau_c \sim \tau_b$, we may use the lower bound (4.7) as an estimate of λ_L. As shown in Appendix D, this lower bound is derived by using the WKB approximation; we may write

$$-\ln(\lambda_L\delta) \equiv P \sim 2L^3 \int [\,U(\phi) - U(\phi_{\text{vex}})]^{1/2} d\phi\,, \tag{4.12}$$

in which δ is, as before, $\sim O(a^{-1/2})$ or $O(b^{-1})$, $U(\phi)$ is given by (1.2), and the integral extends from ϕ_{vex} to ϕ_0, where $U(\phi_0) = U(\phi_{\text{vex}})$. Because $L \gg \delta$, we expect P to be quite large, and therefore at $\tau_b \sim \tau_c$ the critical volume v_c should also be rather large. For example, if we arbitrarily assume $L \sim 10\delta$, $\delta \sim 10^{-13}$ cm, and $P \sim 10^2$, then v_c is ~1 mm^3; the corresponding lifetime of the vacuum excitation state $|\text{vex}\rangle$ is ~3×10^{-12} sec. Since the theory is Lorentz-invariant, such a state can acquire a nonzero momentum; of course, its shape would then undergo a Lorentz contraction, and its lifetime a time dilatation.

B. Degenerate vacuum ($b^2 = 3ac$)

In this case, the system is invariant under the transformation

$$\bar{\phi} + \frac{b}{c} \to -\left(\bar{\phi} + \frac{b}{c}\right). \tag{4.13}$$

The states $\bar{\phi} = 0$ and $\bar{\phi} = -(2b/c)$ are therefore completely symmetrical with respect to each other. We observe that any classical path in the functional space $\phi(x)$ that connects these two states must pass through a potential barrier whose height is at least proportional to $\Omega^{2/3}$, where Ω is the volume of the entire system. The transition matrix element between these two states becomes zero as Ω approaches ∞. Consequently, in an infinite volume, the states $\bar{\phi} = 0$ and $\bar{\phi} = -(2b/c)$ are degenerate, and are both *stable*.

Next, we examine the lifetime of a vacuum excitation that extends over only a limited volume v (but Ω is still ∞). Let $|\text{vex}\rangle$ be such a vacuum excitation state defined by (4.9), where $\phi_{\text{vex}} = -(2b/c)$. In this case, the rest mass consists of only the surface energy, and the lifetime is determined completely by surface contraction. It is not possible to have "boiling" inside v, because of energy conservation. Near the surface, "boiling" is possible, but then there is no clear distinction between that and surface contraction.

In both the degenerate and the nondegenerate cases, we see that the vacuum excitation can, in principle, extend over a domain of macroscopic sizes. In the degenerate case, there is no limit to its size; the larger its dimension is, the bigger its mass but the smaller its width, and therefore the sharper the definition of the state. In the nondegenerate case, the same holds only if the "boiling" mechanism can be neglected, and that gives an upper limit to its size.

V. CLASSICAL SOLUTION

Some knowledge of the actual shape of the vacuum excitation state in space may be obtained by studying its classical solution; this is especially useful if its size may extend over a macroscopic region. For simplicity, we concentrate mainly on the degenerate case ($b^2 = 3ac$) in this section. With slight modifications, the method used below can be readily applied to the nondegenerate case ($b^2 \neq 3ac$) as well.

A. One spatial dimension

It is convenient to introduce the dimensionless variables

$$x = a^{-1/2}\xi ,\quad t = a^{-1/2}\tau , \quad \text{and} \quad \phi = \frac{b}{c}(\chi - 1) . \tag{5.1}$$

The wave equation for the degenerate case $b^2 = 3ac$ in a one-dimensional space becomes

$$-\frac{\partial^2\chi}{\partial\tau^2} + \frac{\partial^2\chi}{\partial\xi^2} + \tfrac{1}{2}\chi(1-\chi^2) = 0 . \tag{5.2}$$

We first examine the time-independent solution. From (5.2) it follows that if $\partial\chi/\partial\tau = 0$, then

$$\frac{dK}{d\xi} = 0 , \tag{5.3}$$

where

$$K \equiv \frac{1}{2}\left(\frac{d\chi}{d\xi}\right)^2 - \tfrac{1}{8}(1-\chi^2)^2 . \tag{5.4}$$

Thus, if we regard ξ as a fictitious "time," the problem becomes identical to one in elementary mechanics, in which there is a point particle at χ moving in a potential

$$W \equiv -\tfrac{1}{8}(1-\chi^2)^2 \tag{5.5}$$

and K is the total energy of the particle. The explicit solution $\chi = \chi(\xi)$ can then be readily obtained.

To illustrate the different types of solutions in this problem, we may consider, for example, the special case $K = 0$. The solutions are

$$\chi = \pm 1 \tag{5.6}$$

and

$$\chi = \pm\tanh\tfrac{1}{2}(\xi - \xi_0) , \tag{5.7}$$

where ξ_0 is a constant. In terms of the mechanical analog, (5.6) is the solution that the particle is at one of the two peaks of W, and (5.7) is the solution such that the particle goes from one peak to the other. In the field-theory problem, the two solutions in (5.6) represent simply the two degenerate vacuum states $\overline{\phi} = 0$ and $\overline{\phi} = -(2b/c)$. The solution in (5.7) gives the details of the transition from $\overline{\phi} = 0$ at, say, $x = +\infty$ to $\overline{\phi} = -(2b/c)$ at $x = -\infty$.

Through a Lorentz transformation, the solution (5.7) can be easily transformed to one in which the transition region moves with a velocity u. The explicit form is

$$\chi = \pm\tanh\theta , \tag{5.8}$$

where

$$\theta = \tfrac{1}{2}(1-u^2)^{-1/2}(\xi - u\tau + \text{constant}) .$$

B. Three-dimensional case

For simplicity, we consider only the spherically symmetrical solution. Again, we introduce the dimensionless variables

$$r = a^{-1/2}\rho ,\quad t = a^{-1/2}\tau , \quad \text{and} \quad \phi = \frac{b}{c}(\chi - 1) . \tag{5.9}$$

For the degenerate case ($b^2 = 3ac$), the wave equation becomes

$$-\frac{\partial^2\chi}{\partial\tau^2} + \frac{1}{\rho^2}\frac{\partial}{\partial\rho}\left(\rho^2\frac{\partial\chi}{\partial\rho}\right) + \tfrac{1}{2}\chi(1-\chi^2) = 0 . \tag{5.10}$$

For the time-independent solution $\partial\chi/\partial\tau = 0$, one has now, instead of (5.3),

$$\frac{dK}{d\rho} = -\frac{2}{\rho}\left(\frac{d\chi}{d\rho}\right)^2 , \tag{5.11}$$

where, as before,

$$K = \frac{1}{2}\left(\frac{d\chi}{d\rho}\right)^2 - \tfrac{1}{8}(1-\chi^2)^2 . \tag{5.12}$$

Again, we may consider the mechanical analog by regarding ρ as the "time" and χ as the "position" of a particle. The "potential" W is again given by (5.5). But now, because

$$\frac{dK}{d\rho} \leqslant 0 , \tag{5.13}$$

the particle is in a dissipative system, with a "time"-dependent frictional force. The motion of the particle can be discussed in the standard way[9] by plotting the $K = (\text{constant})$ contours in the phase space (with χ and $d\chi/d\rho$ as the coordinates). Since a regular solution at $\rho = 0$ implies that $\chi(0)$ is finite and $(d\chi/d\rho)_{\rho=0}$ is zero, at $\rho = 0$ the trajectory must begin at a point on the real axis (i.e., $d\chi/d\rho=0$) in the phase space. As ρ increases, because of (5.13), the value of K along the trajectory must keep on decreasing. From Fig. 4, one sees that the $K=0$ contour divides the entire phase space into one closed region $\mathcal{R}$ and four open regions. Thus, depending on the initial value $\chi(0)$, there are three types of solutions:

1. Stationary solution. If $\chi(0)=1$ or -1, then at all $\rho\geqslant 0$

$$\chi(\rho)=1 \text{ or } -1. \tag{5.14}$$

2. Runaway solution. For $\chi(0)>1$ or <-1, the trajectory in the phase space moves toward points at infinity as ρ increases.

3. Spiral solution. If $-1<\chi(0)<1$, the trajectory lies within the closed region $\mathcal{R}$ bounded by the $K=0$ contour. Inside $\mathcal{R}$, the minimum K is at the origin. As illustrated by the dashed curve in Fig. 4, a typical trajectory would begin at a point on the real axis at $\rho=0$, then spiral in, and eventually approach the origin as $\rho\to\infty$.

Returning to the field-theory problem, one sees that the two stable solutions given by (5.14) correspond to the two degenerate vacuum states $\overline{\phi}=0$ and $\overline{\phi}=-(2b/c)$. Both the runaway solution and the spiral solution have a field-energy content $\int d^3r[\frac{1}{2}(\nabla\phi)^2+U(\phi)]$ that is infinite. Thus, they are unphysical. This situation is quite different from the one-dimensional case; as shown in the previous section, there is a time-independent solution (5.7) in which χ is not a constant, and the solution has a finite field-energy. In three dimensions, a similar transition from $\chi\cong-1$ at, say, $\rho\ll R$ to $\chi=+1$ at $\rho\gg R$ gives rise to a surface energy which can always be reduced by decreasing R. Thus, such a solution cannot be stable (i.e., time-independent) as in the one-dimensional case.

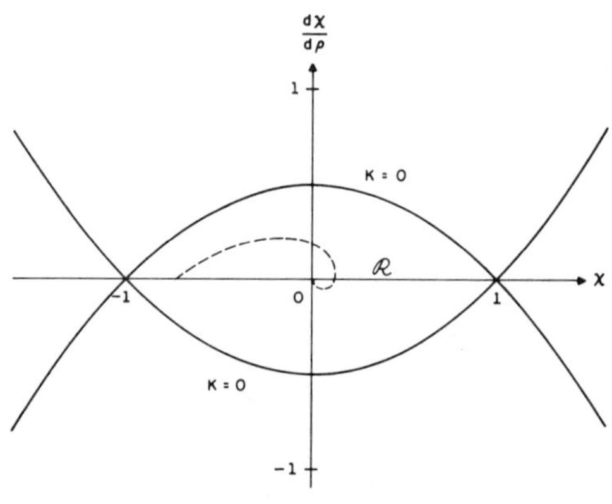

FIG. 4. Phase-space diagram for the mechanical analogy discussed in Sec. V B. Inside the region $\mathcal{R}$, the minimal K is $-\frac{1}{8}$ at the origin. The dashed curve illustrates a spiral solution.

C. Constant external source

It is therefore of interest to examine the three-dimensional time-independent classical solutions which may exist in the presence of an external source $J(x)$. For example, we may assume that the spin-0 field $\phi(x)$ is of parity +1 and interacts with a spin-$\frac{1}{2}$ nucleon field ψ through a scalar coupling $-g\psi^\dagger\gamma_4\psi\phi$. The Lagrangian density is given by

$$\mathcal{L}=-\frac{1}{2}\left(\frac{\partial\phi}{\partial x_\mu}\right)^2-U(\phi)-\psi^\dagger\gamma_4\left(\gamma_\mu\frac{\partial}{\partial x_\mu}+m_N\right)\psi$$
$$-g\psi^\dagger\gamma_4\psi\phi+\text{counterterms}, \tag{5.15}$$

where $U(\phi)$ is given by (1.2), m_N is the physical mass of the nucleon, $\psi^\dagger$ is the Hermitian conjugate of ψ, and g is the renormalized coupling constant. The wave equation is now of the form

$$\frac{\partial^2\phi}{\partial x_\mu{}^2}-\frac{dU}{d\phi}-J=0, \tag{5.16}$$

where (neglecting the counterterm)

$$J=g\psi^\dagger\gamma_4\psi.$$

In this section, we shall assume that in regions occupied by nuclear matter, the source J is a constant. Physically, we may assume either g weak or m_N large, so that

$$m_N{}^2\gg(g\phi_{\text{vex}})^2. \tag{5.17}$$

[The case $m_N{}^2\lesssim(g\phi_{\text{vex}})^2$ will be considered in the next section.] Thus, when ϕ changes from 0 to $O(\phi_{\text{vex}})$, the coupling term $g\phi\psi^\dagger\gamma_4\psi$ remains much smaller than the nucleon-mass term $m_N\psi^\dagger\gamma_4\psi$. The perturbation on ψ due to the variation of ϕ may therefore be neglected. So far as the classical solution is concerned, we may then regard $J(x)$ as a given function. For definiteness, we consider $J(x)$ to resemble the nucleon distribution in, say, a spherical heavy nucleus; it will be assumed to be time-independent and of the form

$$J(x)=\begin{cases}0 & \text{if } \rho>R,\\ (ab/c)j & \text{if } \rho<R,\end{cases}$$

where ρ is defined by (5.9); R and j are both dimensionless constants.

By using the dimensionless variables introduced in (5.9), we find that for the degenerate vacuum case ($b^2=3ac$), the time-independent spherically symmetric equation is, as before,

$$\frac{1}{\rho^2}\frac{d}{d\rho}\left(\rho^2\frac{d\chi}{d\rho}\right)+\tfrac{1}{2}\chi(1-\chi^2)=0 \tag{5.18}$$

in the outside region $\rho>R$; it is

$$\frac{1}{\rho^2}\frac{d}{d\rho}\left(\rho^2\frac{d\chi}{d\rho}\right)+\tfrac{1}{2}\chi(1-\chi^2)=j \tag{5.19}$$

in the inside region $\rho<R$. At $\rho=R$, the outside and inside solutions are joined together so that χ and $d\chi/d\rho$ are both continuous. The solution is then determined by requiring χ to be regular at the origin and at infinity.

The solutions that we are interested in are those in which R is large and χ is nearly a constant either inside or outside $\rho=R$; only near the boundary $\rho\cong R$ does χ have any significant variation. In order to have the "true" vacuum $\phi=0$ at infinity, we require that in the outside region $\chi\to 1$ as $\rho\to\infty$; the next term in the asymptotic expansion of χ is then exhibited in

$$\chi\to 1-\lambda\rho^{-1}e^{-\rho}, \tag{5.20}$$

where λ is a constant. In the inside region, we require that as $\rho\to 0$,

$$\chi\to\chi_0+\epsilon\rho^{-1}\sinh(\kappa\rho), \tag{5.21}$$

where $\epsilon\ll 1$, χ_0 satisfies

$$\chi_0(1-\chi_0^2)=2j \tag{5.22}$$

and

$$\kappa^2=\tfrac{1}{2}(3\chi_0^2-1). \tag{5.23}$$

It can be readily verified that in the outside region the asymptotic solution (5.20) satisfies the differential equation (5.18) to first order in $(\chi-1)$; similarly, in the inside region the corresponding limiting solution (5.21) satisfies (5.19) to first order in $(\chi-\chi_0)$. The exact determination of these parameters λ and ϵ in terms of j and R is rather involved, but some of the general characteristics can be derived without detailed calculations.

For $j^2<\frac{1}{27}$, Eq. (5.22) has three real roots, $\chi_0=\chi_\alpha,\chi_\beta,\chi_\gamma$, given by

$$\chi_\alpha=\frac{2}{\sqrt{3}}\cos\left(\frac{\delta}{3}+\frac{2\pi}{3}\right),$$
$$\chi_\beta=\frac{2}{\sqrt{3}}\cos\left(\frac{\delta}{3}\right), \tag{5.24}$$
$$\chi_\gamma=\frac{2}{\sqrt{3}}\cos\left(\frac{\delta}{3}+\frac{4\pi}{3}\right),$$

and

$$\cos\delta=-3\sqrt{3}\,j.$$

We choose $\pi\geq\delta\geq 0$, and therefore $\chi_\alpha\leq\chi_\gamma\leq\chi_\beta$. By following the same argument given in Sec. VB, one can show that for $j<0$ there is no solution which satisfies the desired boundary conditions (5.20) and (5.21). At $j=0$, the three roots are $\chi_\alpha=-1$, $\chi_\beta=+1$, and $\chi_\gamma=0$, but there is only one solution that satisfies the boundary conditions (5.20) and (5.21): $\chi(\rho)=1$ at all ρ.

At a fixed R, as j increases gradually from zero, the inside solution assumes (except near the surface $\rho=R$) the form (5.21) with $\chi_0=\chi_\beta$. Because of the continuity condition at $\rho=R$, the value of ϵ is $\sim O(e^{-\kappa R})$. For $R\gg 1$, which is the case of physical interest for the classical solution, ϵ is exceedingly small. Thus, $\chi\cong\chi_\beta<1$ near the origin. At larger ρ, the inside solution increases very slowly. It makes a rapid rise only when near the surface $\rho=R$. At the surface, it connects with the outside solution, and then approaches unity asymptotically as $\rho\to\infty$. According to (5.22), as j increases beyond $j=1/3\sqrt{3}$, the root $\chi_0=\chi_\beta$ ceases to exist, and therefore the solution disappears. Physically, this means that inside $\rho<R$, as j increases adiabatically from zero, the state shifts from $\phi(x)=0$ to $\phi(x)<0$, until j reaches the value at point B in Fig. 3. Beyond that, $\phi(x)$ has to make a jump to a completely different solution which represents the vacuum excitation state.

To obtain this other solution, let us first consider the case $R\gg 1$ and $j\ll 1$. We assume that the solution is approximately given by (5.21) in the region $\rho<(R-d)$ where $\chi_0=\chi_\alpha\cong -1$ and $d\sim O(1)$. In the region $\rho>(R+d)$, we assume that the solution is approximately given by (5.20). In the transition region $(R-d)<\rho<(R+d)$, we may neglect both R^{-1} and j as a zeroth approximation; thus, according to (5.7), we have

$$\chi\cong\tanh\tfrac{1}{2}(\rho-\rho_0) \tag{5.25}$$

where ρ_0 lies within the transition region. From the continuity condition, it follows that $\epsilon\sim O(e^{-R})$, and therefore $\chi\cong -1$ in the region $\rho<(R-d)$. Similarly, we find $\chi\cong +1$ in the region $\rho>R+d$. By multiplying (5.18) and (5.19) by $d\chi/d\rho$ and then integrating over all space, we find

$$j\delta\chi=2\int\rho^{-1}\left(\frac{d\chi}{d\rho}\right)^2 d\rho, \tag{5.26}$$

where

$$\delta\chi=\chi(R)-\chi(0). \tag{5.27}$$

In terms of the mechanical analog discussed in the previous section, (5.26) implies simply that the energy dissipated by the "frictional force" equals the work done by the "external force" j. To evaluate approximately the "energy dissipation," we need only to consider the transition region. By using (5.25), we find the right-hand side of (5.26) to be approximately given by $4/3R$. Since for $j\ll 1$ $\delta\lambda$ is <2, we derive the approximate condition

$$j>\frac{2}{3R} \tag{5.28}$$

in order to have the vacuum excitation solution inside $\rho = R$.

Next, we examine its field-energy content

$$4\pi \int H \rho^2 d\rho ,$$

where H is

$$H = \frac{1}{2}\left(\frac{d\chi}{d\rho}\right)^2 + \frac{1}{8}(1-\chi^2)^2 + \begin{cases} j\chi \text{ for } \rho < R , \\ 0 \text{ for } \rho > R . \end{cases}$$

By using the above solution, which is valid for $R \gg 1$ and $j \ll 1$, we find that to first order in j the integral of the Hamiltonian density H in the inside region, $\rho < R - d$, is given by

$$(4\pi)\int_0^{R-d} j\chi \rho^2 d\rho \cong -\frac{4\pi}{3} R^3 j .$$

The energy content in the transition region is approximately given by

$$4\pi R^2 \int_{R-d}^{R+d} \left[\frac{1}{2}\left(\frac{d\chi}{d\rho}\right)^2 + \frac{1}{8}(1-\chi^2)^2\right] d\rho \cong \frac{8\pi}{3} R^2 .$$

To the same order, we may neglect the energy content in the outside region $\rho > (R + d)$. The total field energy content is therefore

$$\frac{4\pi}{3} R^2 (2 - Rj) . \tag{5.29}$$

This is to be compared with the approximate energy content

$$\frac{4\pi}{3} R^3 j \tag{5.30}$$

of the other solution ($\chi \cong \chi_\beta \sim 1$ inside $\rho < R$). Thus, for $R \gg 1$, by comparing (5.29) with (5.30), we find that the vacuum excitation solution has a lower energy if $j > 1/R$.

To summarize: For $R \gg 1$, as j gradually increases from 0, the solution changes continuously from $\chi = 1$ everywhere to one in which $\chi \cong \chi_\beta \lesssim 1$ in the inside region $\rho \ll R$, but χ remains $\cong +1$ in the outside region $\rho \gg R$. As j becomes larger than $2/3R$, there appears another solution, called the vacuum excitation solution, in which $\chi \cong \chi_\alpha < -1$ for $\rho \ll R$, though χ is still $\cong +1$ for $\rho \gg R$. If j becomes $> 1/R$, then the vacuum excitation solution has a lower energy. When j exceeds $1/3\sqrt{3}$, the vacuum excitation becomes the only form of time-independent solution.

D. External source (free nucleon gas)

We now turn to the case in which the coupling constant g in (5.15) is assumed to be sufficiently strong that (5.17) may not hold. We recall that in the "true" vacuum, because of our convention (1.3), $\bar{\phi} = 0$; by definition, the nucleon mass is m_N. However, in states with $\bar{\phi} \neq 0$, the nucleon mass is $m_N + g\bar{\phi}$. In discussing the classical equation, if the solution $\phi(x)$ is slowly varying, we may expect $\phi(x)$ to replace locally the role of $\bar{\phi}$ in the quantum-mechanical treatment. Thus, the "effective" mass of the nucleon becomes $m_N + g\phi$, which in the present case may be quite different from m_N. For definiteness, let us again consider the example of a heavy nucleus. Inside the nucleus, we have

$$\langle \psi^\dagger \psi \rangle = n , \tag{5.31}$$

where n is the nucleon density, and $\langle\ \rangle$ denotes the expectation value. However, as we shall see, when g is strong (or relatively speaking, m_N not too large), in contrast with (5.17), $\langle \psi^\dagger \gamma_4 \psi \rangle \neq$ constant and must depend on ϕ.

To discuss the classical solution of the spin-0 field, we shall assume the nucleons to be approximately described by a degenerate Fermi distribution, characterized by a maximum Fermi momentum k_F. In the simple example of an equal number of protons and neutrons, k_F is given by

$$k_F = (\tfrac{3}{2}\pi^2 n)^{1/3} . \tag{5.32}$$

Since the classical solution $\phi(x)$ is expected to be slowly varying inside the nucleus, one may treat $m_N + g\phi(x)$ as the "effective" mass of the nucleon at x; the density of the kinetic energy of nucleons is therefore given by

$$\begin{aligned} U_N &= \frac{2}{\pi^2}\int_0^{k_F} k^2 (k^2 + M^2)^{1/2} dk \\ &= (2\pi^2)^{-1}\left[k_F (k_F{}^2 + M^2)^{1/2}(k_F{}^2 + \tfrac{1}{2}M^2) - \tfrac{1}{2}M^4 \ln \frac{k_F + (k_F{}^2 + M^2)^{1/2}}{M}\right], \end{aligned} \tag{5.33}$$

where $M^2 = (m_N + g\phi)^2$. The nuclear density n is determined both by the usual short-range nuclear forces (generated through the exchange of high-frequency virtual mesons) and by the long-range "classical" potential $\phi(x)$ (which, in the time-independent solution, is of zero frequency). In the following, we shall consider two models: (i) the free-gas model, to be discussed in this section, and (ii) the incompressible-fluid model, which will be discussed in Sec. VE. The actual physical situation should lie somewhere in between these two extreme possibilities.

Free-nucleon-gas model

In this model, we neglect all short-range nuclear forces, as well as the electromagnetic interaction between nucleons. The nucleons are treated as a free degenerate Fermi gas moving in a classical field $\phi(x)$. To derive the time-independent field equation, we consider the minimum of the field energy E, defined by

$$E \equiv \int [\tfrac{1}{2}(\nabla\phi)^2 + U_\phi + U_N] d^3r \tag{5.34}$$

but subject to the constraint that the total number of nucleons N is a constant, where for a system of equal number of neutrons and protons,

$$N = \frac{2}{3\pi^2}\int k_F{}^3 d^3r , \tag{5.35}$$

U_N is given by (5.33) and U_ϕ is given by (1.2); i.e.,

$$U_\phi = \tfrac{1}{2}a\phi^2 + (3!)^{-1}b\phi^3 + (4!)^{-1}c\phi^4 . \tag{5.36}$$

By setting, at constant k_F, the variational derivative of E with respect to ϕ equal to zero, we derive

$$-\nabla^2\phi + \frac{d}{d\phi}U_\phi + \left(\frac{\partial}{\partial\phi}U_N\right)_{k_F} = 0 . \tag{5.37}$$

Next, let us consider the variation of E with respect to k_F, at constant ϕ and under the constraint (5.35). By using the standard Lagrangian-multiplier method, we find that in order to have E minimum,

$$k_F{}^2[(k_F{}^2 + M^2)^{1/2} - \text{constant}] = 0 , \tag{5.38}$$

where the constant is the Lagrangian multiplier. Thus, at any point in space, either there is no nuclear matter, hence $k_F = 0$, or since $M = m_N + g\phi$, k_F is related to ϕ by

$$k_F{}^2 + (m_N + g\phi)^2 \equiv \omega^2 m_N{}^2 = \text{constant} , \tag{5.39}$$

which implies that the top energy of the degenerate Fermi sea is a constant. Together, (5.37) and (5.38) determine the classical time-independent equation for ϕ.

The most remarkable consequence of the above field equation is the possibility that it may have solutions in which the N nucleons can be bound together in a region of finite and nonzero volume, even though the nucleons are treated as free gas particles without any short-range forces. Furthermore, these solutions exhibit typical "saturation" properties; i.e., for N sufficiently large, the volume is proportional to N and the binding energy per nucleon is independent of N. In such solutions, the classical field $\phi \to 0$ at infinity, so that, in accordance with our convention (1.3), we have the usual vacuum at infinity. However, the constant ω in (5.39) is chosen to be <1, so that there can be a finite volume in space in which $g\phi$ is negative and $< -m_N(1-\omega)$. The nuclear matter will be confined in this volume, whose boundary is defined by

$$g\phi(x) = -m_N(1-\omega) < 0 . \tag{5.40}$$

As we shall see, if g is sufficiently large, one has

$$m_N + g\phi \cong 0 \tag{5.41}$$

inside the bound volume, except in a small region near the boundary; therefore, because of (5.39), inside the volume

$$k_F \cong \omega m_N ; \tag{5.42}$$

i.e., the nuclear density $n \cong 2(3\pi^2)^{-1}(\omega m_N)^3$ is also nearly a constant inside. Furthermore, because of (5.41), the "effective" mass of the nucleon is $\cong 0$. Thus, the field energy E for such a bound solution is given by

$$E = \left[\frac{(\omega m_N)^4}{2\pi^2} + U_\phi\left(-\frac{m_N}{g}\right)\right]\Omega_N + \text{surface energy} ,$$

where Ω_N denotes the volume of the bound solution and $U_\phi(-m_N/g)$ is the value of U_ϕ at $\phi = -m_N/g$. Because of (5.35), $k_F \propto \Omega_N{}^{-1/3}$. Therefore, if one *neglects* the surface energy, the minimum of E occurs at $(\partial E/\partial\Omega_N) = 0$; i.e.,

$$(\omega m_N)^4 = 6\pi^2 U_\phi(-m_N/g) . \tag{5.43}$$

By using (5.35), one finds

$$N = 2(3\pi^2)^{-1}(\omega m_N)^3\Omega_N + O(\Omega_N{}^{2/3}) .$$

The minimum energy E of the bound solution is given by

$$N^{-1}E = \omega m_N + O(N^{-1/3}) . \tag{5.44}$$

This is to be compared with the lowest energy Nm_N of the unbound solution [in which $\phi = 0$ and $k_F = 0$ everywhere, but one retains (5.35) by having the particles at infinity]. Now since, according to (5.36) $U_\phi = 0$ at $\phi = 0$, the bound solution has a *lower* energy than the unbound solution, provided g is sufficiently large that

$$6\pi^2 U_\phi(-m_N/g) < m_N{}^4 , \tag{5.45}$$

and therefore $\omega < 1$; in addition, N must be sufficiently large that the surface energy can be neglected. The binding energy per nucleon is $(1-\omega)m_N$.

[*Note.* One may wonder, as ϕ varies from 0 to a nonzero value, what happens to the energy change of the Dirac sea of negative energy states. Naively, one might expect the increment $\Delta\mathcal{E}$ in energy density to be given by the difference

$$d\equiv\Omega^{-1}\left\{\sum(p^2+m_N{}^2)^{1/2}-\sum[p^2+(m_N+g\phi)^2]^{1/2}\right\},$$

where Ω is the volume of the system. Since the summation extends over all negative energy states, the expression for d is ∞ if $\phi\neq 0$. Actually, $\Delta\mathcal{E}\neq d$. By following the argument given in Sec. III, it can be shown that $\Delta\mathcal{E}$ should be given by the appropriate sum of all nucleon loop diagrams. If we include only the one-nucleon-loop approximation, then $\Delta\mathcal{E}$ is equal to the above expression d, but only after a subtraction of a fourth-order polynomial in ϕ; its explicit expression, after summing over both proton and neutron, is

$$\Delta\mathcal{E}=-\frac{1}{8\pi^2}(m_N+g\phi)^4\ln\left[\frac{(m_N+g\phi)^2}{m_N{}^2}\right]$$
$$+\text{a finite fourth-order polynomial}.$$

The exact form of the polynomial depends on the definition of the renormalized constants a, b, and c, just as in the case of the boson loop discussed in Sec. III B. If one wishes, one may determine the polynomial by requiring that as $\phi\to 0$, $\Delta\mathcal{E}$ is $O(\phi^5)$; in that case

$$\Delta\mathcal{E}=-\frac{1}{8\pi^2}\{(m_N+g\phi)^4\ln[(m_N+g\phi)^2/m_N{}^2]$$
$$-2g\phi[m_N{}^3+\tfrac{7}{2}m_N{}^2g\phi+\tfrac{13}{3}m_N(g\phi)^2$$
$$+\tfrac{25}{12}(g\phi)^3]\},$$

which gives $\Delta\mathcal{E}=(16\pi^2)^{-1}m_N{}^4$ at $m_N+g\phi=0$. In a complete treatment, one should, of course, include $\Delta\mathcal{E}$ in $U_\phi(-m_N/g)$, together with other corrections due to higher-order fermion loops as well as all boson loops. (Note that the above expression gives $6\pi^2\Delta\mathcal{E}=\frac{3}{8}m_N{}^4<m_N{}^4$, therefore the inequality (5.45) is not violated.) In this section, however, our discussion is restricted to the semiclassical equation, without taking into account any loop correction.]

We emphasize that, unlike the other topics discussed in this paper, the existence of this rather unusual type of heavy "nucleus" is *independent* of the existence of another local minimum in U_ϕ (besides $\phi=0$); it may occur even if the ϕ^3 coupling $b=0$.

To illustrate more explicitly the details of such bound solutions, let us consider the degenerate vacuum case $b^2=3ac$. In addition, for simplicity we shall also assume

$$m_N=gb/c \tag{5.46}$$

so that both U_ϕ and U_N are symmetric under the transformation (3.7): $\phi+b/c\to-(\phi+b/c)$. By using the dimensionless variables introduced in (5.9), one has

$$r=a^{-1/2}\rho \quad\text{and}\quad \chi=1+(g\phi/m_N). \tag{5.47}$$

Let $\rho=R$ be the boundary of a spherically symmetric solution, representing a heavy nucleus. Because of (5.37)–(5.39), the corresponding time-independent equation is given by

$$\frac{1}{\rho^2}\frac{d}{d\rho}\left(\rho^2\frac{d\chi}{d\rho}\right)+\tfrac{1}{2}\chi(1-\chi^2)=0\,,\quad \rho>R \tag{5.48}$$

$$=j_N\,,\quad \rho<R \tag{5.49}$$

where

$$j_N=2\beta\chi\left\{(\omega^2-\chi^2)^{1/2}\omega-\tfrac{1}{2}\chi^2\ln\left[\frac{\omega+(\omega^2-\chi^2)^{1/2}}{\chi}\right]^2\right\}, \tag{5.50}$$

$$\beta=3g^4/2\pi^2c\,, \tag{5.51}$$

and

$$k_F/m_N=(\omega^2-\chi^2)^{1/2}. \tag{5.52}$$

The boundary $\rho=R$ is determined by

$$\chi(R)=\omega\,.$$

As $\rho\to\infty$, the asymptotic behavior of χ is given by (5.20);

$$\chi\to 1-\lambda\rho^{-1}e^{-\rho}\,, \tag{5.53}$$

where λ is a constant. As $\rho\to 0$, we have

$$\chi\to\epsilon\rho^{-1}\sinh\kappa\rho\,, \tag{5.54}$$

where

$$\kappa^2=2\beta\omega^2-\tfrac{1}{2}\,, \tag{5.55}$$

which is assumed to be >0. Both χ and $d\chi/d\rho$ are continuous at $\rho=R$. Therefore, the constant ϵ is $O(e^{-\kappa R})$; consequently, except when near the boundary, inside the nucleus $\chi\sim O(e^{-\kappa R})\cong 0$, provided that R is large.

Let χ_{out} and χ_{in} be, respectively, the solutions of (5.48) and (5.49) that satisfy the conditions (5.53) and (5.54). To study how these two solutions can be connected at $\rho=R$, we note that at $\rho\gg 1$,

$$\rho^{-2}\frac{\partial}{\partial\rho}\left(\rho^2\frac{\partial}{\partial\rho}\chi\right)\cong\frac{\partial^2\chi}{\partial\rho^2}\,.$$

Thus, when R is sufficiently large, (5.48) implies that at $\rho=R$ the outside solution χ_{out} satisfies

$$\frac{d\chi_{\text{out}}}{d\rho}=\tfrac{1}{2}(1-\chi_{\text{out}}{}^2)\,, \tag{5.56}$$

which can be easily derived by following the same steps leading to (5.4). Similarly, from (5.49) one concludes that in the transition region near the boundary,.

$$R\geqslant\rho\geqslant(R-d)\gg 1\,, \tag{5.57}$$

the inside solution $\chi = \chi_{in}$ satisfies

$$\frac{1}{2}\left(\frac{d\chi}{d\rho}\right)^2 - \tfrac{1}{8}(1-\chi^2)^2 - \int j_N \, d\chi = \text{constant}, \quad (5.58)$$

where

$$\int j_N \, d\chi = \tfrac{1}{6}\beta\Big[\omega(5\chi^2 - 2\omega^2)(\omega^2-\chi^2)^{1/2} - 3\chi^4 \ln\frac{\omega + (\omega^2-\chi^2)^{1/2}}{\chi}\Big]. \quad (5.59)$$

The width d of the transition region (5.57) is $\sim O(1)$; it is chosen such that at $\rho = R - d$, χ_{in} and $d\chi_{in}/d\rho$ are $\cong 0$. Since at $\rho = R$, $\chi_{in} = \omega$, (5.59) is zero; we obtain for R sufficiently large, at $\rho = R$,

$$\frac{d\chi_{in}}{d\rho} = [\tfrac{1}{12}(8\beta+3)\chi_{in}{}^2 - \tfrac{1}{2}]^{1/2}\chi_{in}. \quad (5.60)$$

The intersection of (5.56) and (5.60) determines χ and $d\chi/d\rho$ at $\rho = R$. We find

$$\chi(R) = \omega = \left(\frac{3}{8\beta}\right)^{1/4}, \quad (5.61)$$

provided that R is sufficiently large. This result, of course, agrees with (5.43). If we neglect the surface energy, then the binding energy per nucleon is

$$m_N(1-\omega). \quad (5.62)$$

As we shall see in Sec. VI, in the σ model the constant β is given by

$$\beta \cong \frac{g^2}{2\pi^2}\left(\frac{m_N}{m_\sigma}\right)^2.$$

[See Eq. (6.13).] This leads to a value $\beta \cong 10$ if $(4\pi)^{-1}g^2 \cong 15.7$ and $m_\sigma \cong m_N$. The corresponding value of ω is $\cong 0.44$. In Fig. 5, the two solid curves labeled "outside ($R=\infty$)" and "inside ($R=\infty$)" refer respectively to (5.56) and (5.60) with $\beta = 10$. These are to be compared with the dashed curves for $R = 10$, determined by the numerical solutions of (5.48) and (5.49). As a further illustration, the numerical solution of $\chi(\rho)$ is plotted in Fig. 6 for $R = 20$ and $\beta = 10$; the corresponding value of N is $\cong 210$ and that of ω is $\cong 0.46$, which is to be compared with the asymptotic value $\omega \cong 0.44$ if R is ∞.

E. External source (incompressible nucleon fluid)

In this model, we assume the short-range nuclear force to be so strong that the nuclear density n is a constant. The nuclear matter resembles an incompressible fluid. Thus, if we retain the approximation that the nucleon density is still related to the Fermi momentum k_F by (5.32) and that the kinetic energy of the nucleons remains given by

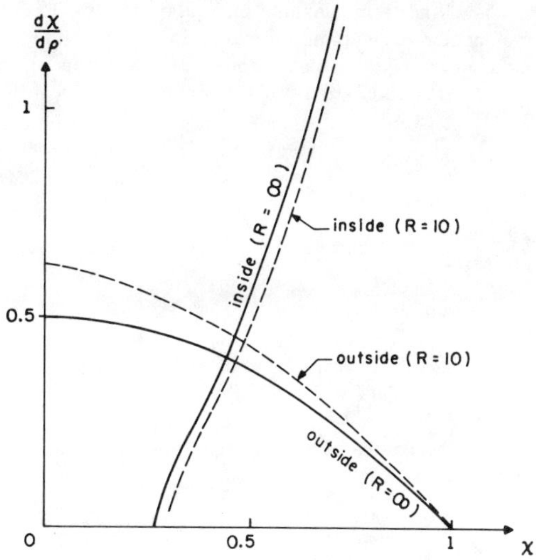

FIG. 5. χ vs $d\chi/d\rho$ at $\rho = R$. The "outside" curves refer to solutions of (5.48), integrating from $\rho = \infty$ to R. The "inside" curves refer to solutions of (5.49), integrating from $\rho = 0$ to R, with $\beta = 10$.

(5.33), then the time-independent equation for ϕ is

$$-\nabla^2\phi + \frac{d}{d\phi}U_\phi = 0 \quad (5.63)$$

outside the nucleus, and

$$-\nabla^2\phi + \frac{d}{d\phi}U_\phi + \left(\frac{\partial}{\partial\phi}U_N\right)_{k_F} = 0 \quad (5.64)$$

inside the nucleus, which is the same as (5.37), *except* that instead of (5.38) we have now

$$k_F = \text{constant}. \quad (5.65)$$

To illustrate the main feature of the model, let us consider again the degenerate vacuum case

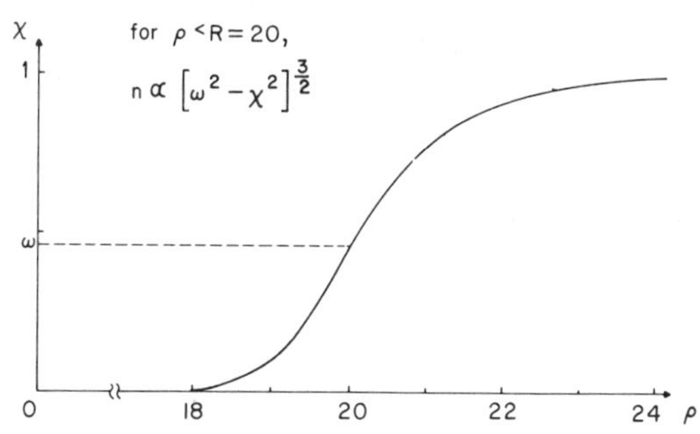

FIG. 6. Numerical solution of $\chi(\rho)$ in the free nucleon gas model. The total number of nucleons is $\cong 210$ and the top Fermi energy is $\omega m_N \cong 0.46 m_N$. The nuclear radius is $R = 20$, and the nucleon density n is zero outside the nucleus, but $\propto (\omega^2 - \chi^2)^{3/2}$ inside.

$b^2 = 3ac$, and let us assume that (5.46) holds. For the spherically symmetric case, in terms of the dimensionless variables ρ and χ introduced in (5.47), Eqs. (5.63) and (5.64) become

$$\frac{1}{\rho^2}\frac{d}{d\rho}\left(\rho^2\frac{d\chi}{d\rho}\right)+\tfrac{1}{2}\chi(1-\chi^2)=0\,,\quad \rho>R \tag{5.66}$$

$$=\frac{d}{d\chi}V_N\,,\quad \rho<R \tag{5.67}$$

where

$$V_N(\chi)=\beta\left\{\alpha(\alpha^2+\chi^2)^{1/2}(\alpha^2+\tfrac{1}{2}\chi^2)-\tfrac{1}{4}\chi^4\ln\left[\frac{\alpha+(\alpha^2+\chi^2)^{1/2}}{\chi}\right]^2\right\}, \tag{5.68}$$

$\rho = R$ is the radius of the nucleus, and α, β are both *constants*, given by

$$\alpha = k_F/m_N \quad\text{and}\quad \beta = 3g^4/2\pi^2 c\,. \tag{5.69}$$

The field energy of the system is given by

$$E = E_{\text{out}} + E_{\text{in}}\,, \tag{5.70}$$

where apart from a common multiplicative factor

$$E_{\text{out}} \propto \int_R^\infty \rho^2 d\rho\left[\frac{1}{2}\left(\frac{d\chi}{d\rho}\right)^2+\tfrac{1}{8}(1-\chi^2)^2\right], \tag{5.71}$$

$$E_{\text{in}} \propto \int_0^R \rho^2 d\rho\left[\frac{1}{2}\left(\frac{d\chi}{d\rho}\right)^2+V(\chi)\right], \tag{5.72}$$

and

$$V(\chi)=\tfrac{1}{8}(1-\chi^2)^2+V_N(\chi)-V_N(1)\,, \tag{5.73}$$

in which the constant term $-V_N(1)$ is arbitrarily added, such that for the true vacuum $\phi = 0$, $\chi = 1$, one has $V(1) = 0$. In contrast with the previous free-gas model, the nuclear radius R is predetermined by the given constant n and the given number of nucleons. By varying E_{out} and E_{in} independently, we derive the field equations (5.66) and (5.67).

Outside the nucleus, the solution has the same form as that in the previous section; e.g., the asymptotic solution remains given by (5.53) as $\rho\to\infty$. However, as will be analyzed, the solution χ inside the nucleus changes its character depending on the physical parameters. In the weak-coupling limit, as expected, the equation becomes identical to that in the constant-current model, discussed in Sec. V C. Similar behavior also occurs in the low-nucleon-density limit, even though the coupling constant g may be strong. But when the nucleon density is sufficiently high and g is strong, the solution resembles that in the free-gas model.

We first observe that as $g\to 0$, the minimum of V is at $\chi = \pm 1 + O(g)$; therefore, to zeroth order in g, the current

$$j \equiv \frac{dV_N}{d\chi} = 2\beta\{\alpha(\alpha^2+1)^{1/2} - \ln[\alpha+(\alpha^2+1)^{1/2}]\} \tag{5.74}$$

is a constant inside the nucleus. Equation (5.67) reduces to the previous equation, (5.19). Next, we consider the case where g is strong, but the nucleon density $n\to 0$, and therefore also $\alpha\to 0$. At $\chi = 0$, one has $dV/d\chi = 0$ and $d^2V/d\chi^2 = 2\beta\alpha^2 - \frac{1}{2}$; consequently as $\alpha \to 0$, the point $\chi = 0$ is a local *maximum* of V. The minimum of V remains at $\chi \cong \pm 1$; the solution then retains the character of the constant-current model. However, when α increases to

$$2\beta\alpha^2 > \tfrac{1}{2}\,,$$

the point $\chi = 0$ becomes a local minimum of V. When the nucleon density becomes sufficiently high, $\chi = 0$ becomes the absolute *minimum* of V. Thus, it resembles the free-gas model when g is strong and nuclear density is sufficiently high. (This is in contrast with the situation in the constant-current model, in which $\chi = 0$ is always the local maximum of the field energy.) The corresponding solution inside the nucleus can be readily obtained by using (5.67).

As $\rho \to 0$, the solution satisfies

$$\chi \to \epsilon\rho^{-1}\sinh(\kappa\rho)\,, \tag{5.75}$$

where

$$\kappa^2 = 2\beta\alpha^2 - \tfrac{1}{2}\,. \tag{5.76}$$

Because χ is continuous at $\rho = R$, and because outside the nucleus, according to (5.53), χ is $\leqslant 1$, one finds that ϵ is $\sim O(e^{-\kappa R})$. Thus, if R is sufficiently large, for the most part inside the nucleus, the value of χ *is near zero*. As ρ approaches R, χ begins to increase. If one neglects $O(R^{-1})$, then one has for ρ near R but inside the nucleus

$$\frac{1}{2}\left(\frac{d\chi}{d\rho}\right)^2 - V(\chi) \cong -V(0)\,; \tag{5.77}$$

for ρ near R and outside the nucleus, one has

$$\frac{1}{2}\left(\frac{d\chi}{d\rho}\right)^2 - \tfrac{1}{8}(1-\chi^2)^2 \cong 0\,. \tag{5.78}$$

Consequently, at $\rho = R$, χ satisfies

$$V_N(\chi) \cong \beta\alpha^4 + \tfrac{1}{8}\,. \tag{5.79}$$

In order for $\chi = 0$ to be the absolute minimum of V, we must have $V(0) < V(1)$, i.e.,

$$V_N(1) - \beta\alpha^4 > \tfrac{1}{8}\,. \tag{5.80}$$

If $\beta \gg 1$, this inequality can be satisfied for a relatively small α, and therefore also a relatively low nuclear density. Since for α small $V_N(1) \cong \frac{4}{3}\beta\alpha^3$, (5.80) can be satisfied if α is above a critical value α_c,

$$\alpha_c \approx \left(\frac{3}{32\beta}\right)^{1/3}, \tag{5.81}$$

provided that β is sufficiently large; the corresponding critical density is $\propto \alpha_c^3 \propto \beta^{-1}$.

The above discussions, after some minor changes, can be extended to cases where $b^2 \neq 3ac$ and $m_N \neq gb/c$.

VI. σ MODEL

It is not our purpose here to start a complete reinvestigation of the σ model[10] of strong interactions; such a project clearly lies outside the scope of the present paper. However, as we shall see, there are some rather new and interesting properties in the σ model when a sizable chunk of nuclear matter is present; these properties are closely related to those discussed above. In this section, we shall give only a brief survey of these new features. Our discussion will be restricted to the tree approximation.

The σ model consists of a spin-$\frac{1}{2}$ nucleon field ψ, a spin-0 (even-parity) field σ, and the usual pseudoscalar pion field $\vec{\pi}$. The Lagrangian density is given, apart from the counterterms for renormalization, by

$$\mathcal{L} = -\psi^\dagger \gamma_4 \gamma_\mu \frac{\partial}{\partial x_\mu} \psi - g\psi^\dagger \gamma_4[\sigma + i\vec{\pi}\cdot\vec{\tau}\gamma_5]\psi - \frac{1}{2}\left[\left(\frac{\partial\sigma}{\partial x_\mu}\right)^2 + \left(\frac{\partial\vec{\pi}}{\partial x_\mu}\right)^2\right] - U_\sigma, \tag{6.1}$$

where

$$U_\sigma = \tfrac{1}{4}\lambda^2[(\sigma^2+\vec{\pi}^2)-(\mu/\lambda)^2]^2 - C_\pi\sigma. \tag{6.2}$$

For convenience, we assume the parameters C_π, μ, and λ to be all positive. The minimum of the c-number function U_σ occurs at $\sigma=\sigma_0$ and $\vec{\pi}=0$, where σ_0 is $> (\mu/\lambda)$ and satisfies

$$C_\pi = \sigma_0(\lambda^2\sigma_0^2 - \mu^2). \tag{6.3}$$

In the tree approximation, the renormalized constants λ, μ, g, and C_π are related to the physical masses m_N, m_σ, and m_π of the particles by

$$m_N = g\sigma_0, \quad C_\pi = m_\pi^2\sigma_0, \quad m_\pi^2 = \lambda^2\sigma_0^2 - \mu^2, \tag{6.4}$$

and

$$m_\sigma^2 = 3\lambda^2\sigma_0^2 - \mu^2.$$

The vacuum state satisfies

$$\langle \text{vac}|\sigma(x)|\text{vac}\rangle = \sigma_0 \tag{6.5}$$

and $\langle \text{vac}|\vec{\pi}(x)|\text{vac}\rangle = 0$. In the σ model, the constant g is given by the well-known π-nucleon coupling, $(4\pi)^{-1}g^2 \cong 15.7$. The only unknown parameter is m_σ. However, from the absence of any 0+ resonance that has been positively identified experimentally, we may conclude m_σ is $\gg m_\pi$, and may perhaps be[11] $\sim O(m_N)$.

We note that if $\vec{\pi}=0$, then U_σ reduces to the form (1.2) with $\phi \propto (\sigma-\sigma_0)$. Owing to the smallness of m_π, and therefore also of C_π, the function U_σ has a local maximum at σ near zero and a local minimum, besides $\sigma=\sigma_0$, at σ near $-\sigma_0$. However, when $\vec{\pi}$ is now allowed to vary, this local minimum at σ near $-\sigma_0$ turns into a saddle point; it is connected to the absolute minimum point $\sigma=\sigma_0$ by a smooth path, $\sigma^2+\vec{\pi}^2 \cong \sigma_0^2$, without passing through any potential barrier. Thus, in the absence of nuclear matter, the σ model is quite different from the system discussed in the previous sections. On the other hand, when there is nuclear matter present in a certain region, then for a sufficiently large nuclear density and the region not too small, the σ model exhibits almost exactly the same property as that discussed in the previous sections.

It is convenient to introduce, similar to (5.9), the dimensionless variables

$$\rho \equiv \sqrt{2}\,\mu r \quad \text{and} \quad \chi \equiv \lambda\sigma/\mu; \tag{6.6}$$

i.e., on account of (6.4),

$$\rho = (m_\sigma^2 - 3m_\pi^2)^{1/2} r$$

and

$$\chi = (g\sigma/m_N)(m_\sigma^2 - m_\pi^2)^{1/2}(m_\sigma^2 - 3m_\pi^2)^{-1/2}.$$

For simplicity, let us consider a spherical nucleus of radius $\rho = R$. Furthermore, just as in Secs. V D and V E, we assume for the nucleons a degenerate Fermi distribution with a maximum Fermi momentum k_F, given by (5.32). By following exactly the same discussion given in the previous two sections, we find that outside the nucleus the classical time-independent spherically symmetric equation for σ (with $\vec{\pi}=0$) is

$$\frac{1}{\rho^2}\frac{d}{d\rho}\left(\rho^2\frac{d\chi}{d\rho}\right) - \frac{d}{d\chi}V_\sigma(\chi) = 0, \tag{6.7}$$

where

$$V_\sigma(\chi) = \tfrac{1}{8}(1-\chi^2)^2 - \eta\chi, \tag{6.8}$$

and where because of (6.4) η is given by

$$\eta = m_\pi^2(m_\sigma^2 - m_\pi^2)^{1/2}(m_\sigma^2 - 3m_\pi^2)^{-3/2} << 1. \quad (6.9)$$

Inside the nucleus, the corresponding equation is

$$\frac{1}{\rho^2}\frac{d}{d\rho}\left(\rho^2\frac{d\chi}{d\rho}\right) - \frac{d}{d\chi}V_\sigma(\chi) = j_N(\chi). \quad (6.10)$$

The function $j_N(\chi)$ depends on the nuclear model. Under the assumption of the free-gas model, we have $j_N = (j_N)_{\text{gas}}$ where, just as in (5.50),

$$(j_N)_{\text{gas}} = 2\beta\chi\left\{(\omega^2-\chi^2)^{1/2}\omega - \tfrac{1}{2}\chi^2 \ln\left[\frac{\omega + (\omega^2-\chi^2)^{1/2}}{\chi}\right]^2\right\}, \quad (6.11)$$

in which ω is a constant, related to the value of χ at $\rho = R$ by

$$\chi(R) = \omega, \quad (6.12)$$

and, because of (6.4),

$$\beta = \frac{g^2 m_N^2}{2\pi^2(m_\sigma^2 - m_\pi^2)}. \quad (6.13)$$

On the other hand, if we assume the incompressible-fluid model then $j_N = (j_N)_{\text{fluid}}$, where, just as in (5.67) and (5.68),

$$(j_N)_{\text{fluid}} = 2\beta\chi\left\{\alpha(\alpha^2+\chi^2)^{1/2} - \tfrac{1}{2}\chi^2 \ln\left[\frac{\alpha + (\alpha^2+\chi^2)^{1/2}}{\chi}\right]^2\right\}, \quad (6.14)$$

in which β is given by (6.13) and α is a constant related to the Fermi momentum k_F by

$$\alpha = \frac{k_F}{m_N}\left(\frac{m_\sigma^2 - m_\pi^2}{m_\sigma^2 - 3m_\pi^2}\right)^{1/2}. \quad (6.15)$$

In the limit $m_\pi \to 0$, one has $\eta \to 0$ and $\alpha \to m_N^{-1}k_F$; Eq. (6.10) reduces identically to either (5.49) or (5.67).

In the σ model, $(4\pi)^{-1}g^2 \cong 15.7$ and therefore (after neglecting m_π^2)

$$\beta \cong 10(m_N/m_\sigma)^2. \quad (6.16)$$

In the free nucleon gas model, by using (5.61) we find

$$\omega \cong 0.44(m_\sigma/m_N)^{1/2}. \quad (6.17)$$

If we neglect the surface energy, then according to (5.62) the binding energy per nucleon is $(1-\omega)m_N$. Thus, in this model, if m_σ is less than $\sim 5m_N$ there will be a new type of stable heavy nucleus, provided that the short-range nuclear force can be neglected and that the nucleon number is sufficiently large.

If we assume the incompressible-fluid model, then the field energy is given by (5.70)–(5.72), except that $V(\chi)$ is now

$$V(\chi) = V_\sigma(\chi) + V_N(\chi) - V_N(1), \quad (6.18)$$

where V_σ is given by (6.8), but V_N remains given by (5.68). The above expression reduces to (5.73) in the limit $m_\pi = 0$. As noted in Sec. V E, when the nucleon density is sufficiently high, the minimum energy state of a very heavy nucleus flips from the "normal" solution (in which χ is near unity and the nucleon mass $\cong m_N$) to an "abnormal" one, in which both χ and the effective nucleon mass are near 0. In order to produce the flip to the abnormal solution, (5.80) must be satisfied. By using (5.81) and (6.16), one finds that the critical density is approximately determined by

$$\alpha_c \equiv \left(\frac{k_F}{m_N}\right)_c \approx 0.21\left(\frac{m_\sigma}{m_N}\right)^{2/3}. \quad (6.19)$$

If $m_\sigma \cong m_N$, then the critical density is about the usual nuclear density

$$n_0 = \left[\frac{4\pi}{3}(1.3\times 10^{-13}\text{ cm})^3\right]^{-1}. \quad (6.20)$$

If $m_\sigma \neq m_N$, then the critical density n_c varies approximately as

$$n_c \approx n_0\left(\frac{m_\sigma}{m_N}\right)^2, \quad (6.21)$$

provided that m_σ is not too large.

In Fig. 7, the function $V(\chi)$ is plotted for $m_\sigma = m_N$ and the usual nuclear density $n = n_0$, with $m_\pi \neq 0$. From the plot one sees more explicitly that under these conditions, if the nucleus is sufficiently heavy (so that surface energy can be neglected), then, as expected, the abnormal solution has an energy comparable to but slightly higher than that of the normal solution. For $m_\sigma \gtrsim m_N$, one may produce the abnormal nuclear state by increasing the nuclear density through, say, high-energy collisions between very heavy nuclei. From Fig. 7 one observes that there is practically no potential barrier between the normal and the abnormal configurations once the critical density is reached; the corresponding production probability should, therefore, be not too small.

VII. REMARKS

In this paper we have investigated, among other things, the possibility that over a *limited* region in space the expectation value $\langle\phi\rangle$ of a spin-0 even-parity field $\phi(x)$ may be different from its "normal" vacuum expectation value (which can be chosen, by convention, to be zero). This investigation leads us to a study of several different physical problems, each containing some rather

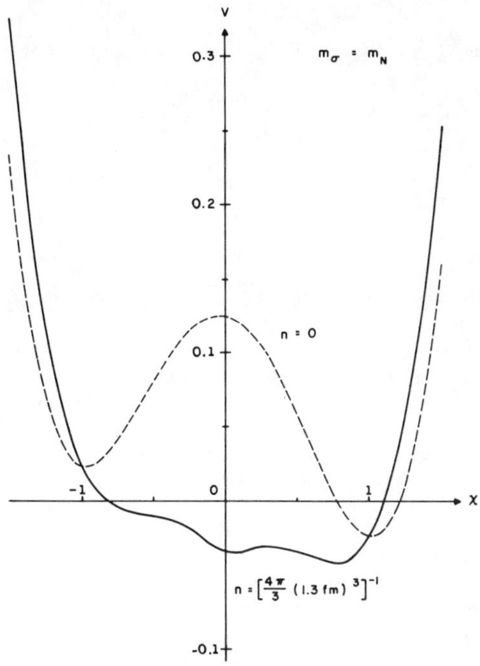

FIG. 7. $V(\chi)$ in the σ model for a nucleon density $n = [\frac{4}{3}\pi(1.3\times10^{-13}\text{ cm})^3]^{-1}$ and for $n = 0$. See Eq. (6.18) for the definition of $V(\chi)$.

interesting properties. However, not all of them have been fully examined in this paper.

If the spin-0 field has a strong interaction with some matter field, say the nucleon field with a large coupling g, then whenever there is a sizable bulk of nuclear matter present there is a tendency to have $\langle \phi(x)\rangle \cong -(m_N/g)$ in the region occupied by nuclear matter. This would reduce the "effective" nucleon mass to $\cong 0$, and thereby lower the kinetic energy of the nucleons. Within a certain range of the relevant physical parameters, this unusual solution may even become the lowest energy state. Thus, if such a strongly interacting scalar field does exist, there would be the possibility of a large class of "stable" or "metastable" superheavy nuclei, hitherto undiscovered.

As a mathematical model, such a possibility suggests also a possible extension to the bound-state description of a single nucleon, by replacing the role of the nucleus by a nucleon, and that of nucleons by a mixture of quarks plus a suitable quark-antiquark continuum. Since the effective quark mass might be near zero inside the bound state (though heavy outside), one could hope to resolve some of the present theoretical difficulties in such a description.

If the spin-0 field has a large ϕ^3 coupling constant b, then the function $U(\phi)$, defined by (1.2), can have another local minimum at $\phi = \phi_{\text{vex}} \neq 0$. In this case, even *without* the presence of nuclear matter, there could be the possibility of a pure vacuum excitation state, in which the expectation value $\langle \phi(x)\rangle \cong \phi_{\text{vex}}$ over an extended region in space. This leads naturally to the physical picture that the so-called vacuum actually more resembles a medium whose properties can be changed. If this is true, which of course we do not know at present, it must ultimately lead to rather striking physical consequences.

ACKNOWLEDGMENTS

We wish to thank N. H. Christ, A. H. Mueller, L. M. Lederman, and M. A. Ruderman for discussions.

APPENDIX A

In this appendix we give the details of the graphical representation of the energy density function $\mathcal{E}(\overline{\phi})$, which is defined by (2.4). It is convenient to introduce the unrenormalized field operator ϕ_0,

$$\phi_0 = Z^{1/2}\phi\,, \tag{A1}$$

where ϕ is the renormalized field operator, as before. The Lagrangian density (1.1) may be written as

$$\mathcal{L} = -\frac{1}{2}\left(\frac{\partial\phi_0}{\partial x_\mu}\right)^2 - \phi_0\delta J - \tfrac{1}{2}\phi_0^2(a+\delta a) - (3\,!)^{-1}\phi_0^3(b+\delta b) - (4\,!)^{-1}\phi_0^4(c+\delta c)\,, \tag{A2}$$

where δJ, δa, δb, and δc are counterterms; together with $(Z^{1/2}-1)$, these terms are needed to cancel the infinities.

The counterterm δJ is determined by requiring

$$\langle \text{vac}\,|\,\phi_0(x)\,|\,\text{vac}\rangle = 0\,. \tag{A3}$$

The precise definitions δa, δb, and δc will be given below [after Eq. (A23)].

From (2.1) and (2.2), the Hamiltonian H_J may be written as the sum of a zeroth-order term H_0 and a perturbation term H_1:

$$H_J = H_0 + H_1\,, \tag{A4}$$

$$H_0 = \tfrac{1}{2}\int [\Pi_0^{\,2} + (\nabla\phi_0)^2 + a_0\phi_0^{\,2}]\,d^3r\,, \tag{A5}$$

$$H_1 = \int [(J_0+\delta J)\phi_0 + (3\,!)^{-1}b_0\phi_0^{\,3} + (4\,!)^{-1}c_0\phi_0^{\,4}]\,d^3r\,, \tag{A6}$$

where Π_0 is the conjugate momentum of ϕ_0,

$$a_0 = a + \delta a\,,$$
$$b_0 = b + \delta b\,, \tag{A7}$$
$$c_0 = c + \delta c\,,$$

and the constant J_0 is related to J, introduced in (2.1), by

$$J_0 = JZ^{-1/2}\,. \tag{A8}$$

Since the counterterm δJ is determined by (A3), in which $|\text{vac}\rangle$ is defined to be the lowest-energy eigenstate of H_J with $J_0 = 0$, there should be a nonzero expectation value of $\phi(x)$ in the lowest-energy eigenstate $|\,\rangle$ of H_J when $J_0 \neq 0$. We define

$$\overline{\phi}_0 = \Omega^{-1}\int \langle\ |\phi_0(x)|\ \rangle\, d^3r\,. \tag{A9}$$

Both $\overline{\phi}_0$ and the corresponding lowest eigenvalue λ_J of $\Omega^{-1}H_J$ may be evaluated by regarding H_0 as the unperturbed Hamiltonian and H_1 as the perturbation. The perturbation series of λ_J is the sum of all connected Feynman graphs that have no external line. We may write

$$\lambda_J = (\lambda_J)_{\text{tree}} + (\lambda_J)_{\text{one-loop}} + (\lambda_J)_{\text{two-loop}} + \cdots, \tag{A10}$$

in which $(\lambda_J)_{\text{tree}}$ denotes the partial summation of all such diagrams that are trees and $(\lambda_J)_{l-\text{loop}}$ denotes the partial summation of all such diagrams that have l loops.

From (A6), (A9), and (2.3) one sees that, keeping a_0, b_0, and c_0 fixed,

$$\frac{\partial \lambda_J}{\partial J_0} = \overline{\phi}_0\,. \tag{A11}$$

We recall that according to (2.4)

$$\mathcal{E}(\overline{\phi}_0) = \lambda_J - J_0\overline{\phi}_0\,. \tag{A12}$$

Thus, keeping a_0, b_0, and c_0 fixed, we have

$$\frac{\partial \mathcal{E}}{\partial \overline{\phi}_0} = -J_0 \tag{A13}$$

and

$$\frac{\partial^2 \mathcal{E}}{\partial \overline{\phi}_0^{\,2}}\,\frac{\partial^2 \lambda_J}{\partial J_0^{\,2}} = -1\,. \tag{A14}$$

1. Tree diagrams

In Fig. 8, we list the sum $(\lambda_J)_{\text{tree}}$ of all the tree diagrams. In these diagrams, there is no external line. Every internal line carries a zero 4-momentum, so it gives to the Feynman amplitude a factor $-i(k^2 + a_0)^{-1}$ with $k = 0$. Every one-point vertex gives a factor $-iJ_0$, every three-point vertex a factor $-ib_0$, and every four-point vertex a factor $-ic_0$. From Fig. 8, it follows that, keeping a_0, b_0, and c_0 fixed, (A11) holds within the tree approximation, i.e., $(\partial\lambda_J/\partial J_0)_{\text{tree}} = \overline{\phi}_0$. Furthermore, in the same tree approximation, the full propagator of ϕ_0 at zero 4-momentum is simply $i(\partial^2\lambda_J/\partial J_0^{\,2})_{\text{tree}}$. Thus, one derives

$$\left(\frac{\partial^2 \lambda_J}{\partial J_0^{\,2}}\right)_{\text{tree}} = -(k^2 + a_0 + b_0\overline{\phi}_0 + \tfrac{1}{2}c_0\overline{\phi}_0^{\,2})^{-1}_{k=0}\,. \tag{A15}$$

Because of (A14), this leads to

$$\left(\frac{\partial^2 \mathcal{E}}{\partial \overline{\phi}_0^{\,2}}\right)_{\text{tree}} = a_0 + b_0\overline{\phi}_0 + \tfrac{1}{2}c_0\overline{\phi}_0^{\,2}\,. \tag{A16}$$

Again from Fig. 8, one sees that as $J_0 \to 0$, $(\lambda_J)_{\text{tree}} \to 0$ and $(\partial\lambda_J/\partial J_0)_{\text{tree}} \to 0$. Therefore, as $\overline{\phi}_0 \to 0$, one must have $(\mathcal{E})_{\text{tree}} \to 0$ and $(\partial\mathcal{E}/\partial\overline{\phi}_0)_{\text{tree}} \to 0$. Consequently,

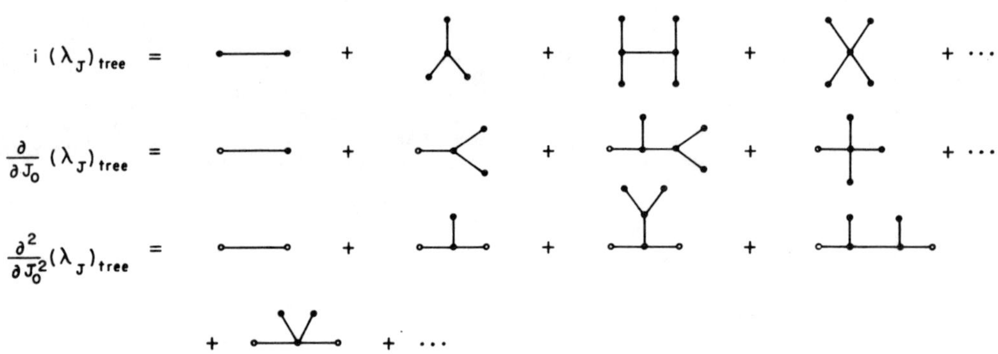

FIG. 8. Tree diagrams for λ_J and its derivatives. All lines carry zero 4-momentum. For the Feynman amplitude, there is a factor $-(i/a_0)$ to each line, $-iJ_0$ to each one-point vertex, $-ib_0$ to each three-point vertex, and $-ic_0$ to each four-point vertex. Each open circle denotes a differentiation with respect to $(-iJ_0)$.

$$[\mathcal{E}(\overline{\phi}_0)]_{\text{tree}} = \tfrac{1}{2}a_0\overline{\phi}_0^{\,2} + (3\,!)^{-1}b_0\overline{\phi}_0^{\,3} + (4\,!)^{-1}c_0\overline{\phi}_0^{\,4}. \tag{A17}$$

2. General expression

To find the general expression of $\mathcal{E}(\overline{\phi}_0)$, let us consider the scattering of n zero-momentum mesons whose interaction is given by the Lagrangian density (A2); n may vary from 2 to ∞. We *define* $[S(\overline{\phi}_0)]_{\text{loop}}$ to be the sum of all such one-particle irreducible scattering diagrams that are *not* trees; in these diagrams, each external line carries zero 4-momentum and gives a factor $\overline{\phi}_0$ to the Feynman integral. The corresponding factors for the internal line, the three-point vertex, and the four-point vertex are, respectively, $-i(k^2+a_0)^{-1}$, $-ib_0$, and $-ic_0$. (Note that there is no one-point vertex in these scattering graphs.) We shall now establish

$$\mathcal{E}(\overline{\phi}_0) = [\mathcal{E}(\overline{\phi}_0)]_{\text{tree}} + [iS(\overline{\phi}_0)]_{\text{loop}}\,. \tag{A18}$$

To prove this, we consider the sum (A10) and note that, similarly to (A15),

$$i\,\frac{\partial^2\lambda_J}{\partial J_0^{\,2}} = [\mathfrak{D}_J(k)]_{k=0}\,, \tag{A19}$$

where $\mathfrak{D}_J(k)$ is the full propagator of the meson field ϕ_0 in a theory in which the Hamiltonian is given by (A4). We may write

$$[i\mathfrak{D}_J(k)]^{-1} = k^2 + a_0 + i\Sigma(k)\,, \tag{A20}$$

where $\Sigma(k)$ is, by definition, the sum of all proper self-energy diagrams. Let us separate in $\Sigma(k)$ the J_0-dependent part $\Sigma_J(k)$ from the J_0-independent part $\Sigma_0(k)$:

$$\Sigma(k) = \Sigma_0(k) + \Sigma_J(k)\,, \tag{A21}$$

where as $J_0 \to 0$, $\Sigma_J(k) \to 0$ and therefore $\Sigma(k) \to \Sigma_0(k)$. According to (A6), the dependence on J_0 is completely due to the one-point vertex. Thus, every diagram in $\Sigma_J(k)$ is one-particle reducible—i.e., it is possible to separate every diagram in $\Sigma_J(k)$ into two disconnected parts by cutting an internal line open; one of these two disconnected parts contains the external momentum k_μ and the other does not. By repeating this cutting procedure and keeping only the part that contains k_μ, we can reduce each of these diagrams to a one-particle irreducible diagram in which there is no J_0 vertex, but besides the two external lines that carry k_μ we have also other zero-momentum external lines (as the remainder of the cutting). If we assign to each of these additional zero-momentum external lines a factor $\overline{\phi}_0$ to the Feynman amplitude, we find that $\Sigma_J(k)$, introduced in (A21), is equal to the summation over the set of all such different one-particle irreducible (proper self-energy) diagrams. In this set, for $k \neq 0$ every diagram has at least one zero-momentum external line. Among these diagrams, there are only two diagrams without any loop; these are simply $-ib_0\overline{\phi}_0$ and $-i\frac{1}{2}c_0\overline{\phi}_0^{\,2}$. The rest all have some loops.

Next, we note that for $k \neq 0$ the J_0-independent part $\Sigma_0(k)$, defined in (A20), consists of all one-particle irreducible proper self-energy diagrams that do not have any zero-momentum external line. Together, $\Sigma(k) = \Sigma_0(k) + \Sigma_J(k)$ is then the sum of all one-particle irreducible proper self-energy diagrams which may or may not have additional zero-momentum external lines. It is now straightforward to show that $[S(\overline{\phi}_0)]_{\text{loop}}$, defined above, is related to $\Sigma(k)$ at $k=0$ by

$$i\Sigma(0) = b_0\overline{\phi}_0 + \tfrac{1}{2}c_0\overline{\phi}_0^{\,2} + \frac{\partial^2}{\partial\overline{\phi}_0^{\,2}}[iS(\overline{\phi}_0)]_{\text{loop}}\,. \tag{A22}$$

By using (A14) and the boundary condition that at $\overline{\phi}_0 = 0$, both $\mathcal{E}$ and $(\partial\mathcal{E}/\partial\overline{\phi}_0)$ vanish. We establish (A18).

Equations (A17) and (A18) still differ from (3.1) and (3.2) by being expressed in terms of a_0, b_0, c_0, and $\overline{\phi}_0$ rather than the corresponding renormalized quantities. We note that whatever may be the precise definitions of these renormalized quantities, the counterterms δa, δb, δc, and $(Z^{1/2}-1)$ can always be expressed formally as sums over the appropriate set of diagrams in which only the renormalized quantities a, b, c, and $\overline{\phi}$ appear. Every one of these diagrams must have loops. By redefining "loop" to include also these loops in the counterterm, we derive (3.1) and (3.2).

3. Renormalized constants

To define the wave-function renormalization constant Z, we may follow the standard procedure: Set $J_0 = 0$. The full propagator of the ϕ_0 field becomes then

$$\mathfrak{D}_0(k) = -i[k^2 + a_0 + \Sigma_0(k)]^{-1}\,, \tag{A23}$$

where $\Sigma_0(k)$ is defined in (A21). Let $k^2 = -m_\phi^{\,2}$ be the zero of $[\mathfrak{D}_0(k)]^{-1}$. We require as $k^2 \to -m_\phi^{\,2}$

$$[\mathfrak{D}_0(k)]^{-1} \to iZ^{-1}(k^2 + m_\phi^{\,2})\,. \tag{A24}$$

Thus, Z is defined and m_ϕ is the physical mass of the meson. The renormalized constant a is defined by

$$\mathfrak{D}_0(k) \to -iZ/a \quad \text{as } k^2 \to 0\,. \tag{A25}$$

Consequently,

$$a = [a_0 + \Sigma_0(0)]\,Z\,. \tag{A26}$$

From (A18), (A22), and the fact that $\Sigma(k) \to \Sigma_0(k)$ as

$\bar{\phi}\to 0$, we obtain in the same limit, $\bar{\phi}\to 0$,

$$\mathcal{E}(\bar{\phi})\to \tfrac{1}{2}a\bar{\phi}^2+O(\bar{\phi}^3). \tag{A27}$$

We may expand the scattering amplitude $[S(\bar{\phi}_0)]_{\text{loop}}$ as a power series in $\bar{\phi}_0$:

$$[S(\bar{\phi}_0)]_{\text{loop}}=\sum_{n=2}^{\infty}(n!)^{-1}S_n\bar{\phi}_0^{\,n}, \tag{A28}$$

in which n denotes the number of external mesons in the scattering amplitude. From (A21), (A22), and (A26), it follows that

$$a=[a_0+iS_2]Z. \tag{A29}$$

Since S_2 contains a quadratically divergent Feynman integral, two counterterms δa and $(Z-1)$ are needed to render (A29) finite. The renormalized coupling constants b and c are related to the scattering amplitudes S_3 and S_4 by

$$b+\text{ finite term}=[b_0+iS_3]Z^{3/2} \tag{A30}$$

and

$$c+\text{ finite term}=[c_0+iS_4]Z^2. \tag{A31}$$

Since S_3 and Z both contain only logarithmically divergent integrals, one counterterm δb is sufficient to render (A30) finite; similarly, one counterterm δc is sufficient to render (A31) finite. The precise values of the finite terms in (A30) and (A31) are determined by imposing (3.8), as discussed in Sec. III B.

If one wishes, one may alter the above definition of Z by an arbitrary finite multiplicative factor, say $Z\to\zeta Z$, provided that $a\to\zeta a$, $b\to\zeta^{3/2}b$, $c\to\zeta^2 c$, and $\bar{\phi}\to\zeta^{-1/2}\bar{\phi}$; of course, the residue of the renormalized propagator at $k^2+m_\phi^{\,2}=0$ now becomes not 1 but ζ^{-1}.

APPENDIX B

According to the rules for the prototype diagram, given in Sec. III A, the Feynman amplitude for diagram (i) in Fig. 2 is given by

$$A_{(\text{i})}=\tfrac{1}{8}(-ic)I^2, \tag{B1}$$

where the factor $\frac{1}{8}$ denotes the inverse of the symmetry number, and

$$I=\frac{-i}{(2\pi)^4}\int\frac{d^4k}{k^2+a(1+\Delta)}-\text{constant}-\text{constant}\times\Delta, \tag{B2}$$

in which the two constants are determined by requiring I to be $O(\Delta^2)$ as $\Delta\to 0$. The integral (B2) can be readily evaluated. We find

$$I=(16\pi^2)^{-1}a[(1+\Delta)\ln(1+\Delta)-\Delta]. \tag{B3}$$

According to (A18), in order to obtain $\mathcal{E}(\bar{\phi})$ we should multiply the scattering amplitude (B1) by i; this gives the first term on the right-hand side of (3.17).

The evaluation of the prototype diagram (ii) in Fig. 2 is complicated, since it can be made finite only after we include also the diagram (ii)′. According to the rules given in Sec. III A, we find

$$A_{(\text{ii})}=\frac{-ib^2}{2(3!)(2\pi)^8}\int\frac{d^4k\,d^4q}{[k^2+a(1+\Delta)][q^2+a(1+\Delta)][(k+q)^2+a(1+\Delta)]}+\text{subtraction term}, \tag{B4}$$

$$A_{(\text{ii})'}=\frac{-\delta a}{2(2\pi)^4}\int\frac{d^4k}{k^2+a(1+\Delta)}+\text{ subtraction term}, \tag{B5}$$

and, to the lowest order,

$$\delta a=\frac{-ib^2}{2(2\pi)^4}\int\frac{d^4q}{(q^2+a)^2}, \tag{B6}$$

where, according to (3.8), the subtraction terms must be quadratic functions of Δ. Since both integrals in (B4) and (B5) are not primitively divergent, even with the subtraction terms included, (ii) and (ii)′ are still logarithmically divergent. It is convenient to introduce another diagram, diagram (v) in Fig. 2, in which the dashed line denotes the propagator $-i(k^2+a)^{-1}$. [The solid line remains $-i(k^2+a+a\Delta)^{-1}$.]

$$A_{(\text{v})}=\frac{-ib^2}{4(2\pi)^8}\int\frac{d^4k\,d^4q}{(k^2+a)(q^2+a)[(k+q)^2+a(1+\Delta)]}+\text{subtraction term}. \tag{B7}$$

Again, the subtraction term is assumed to be a quadratic function in Δ. We shall calculate first $A_{(\text{ii})'}+A_{(\text{v})}$ and $A_{(\text{ii})}-A_{(\text{v})}$ separately, and then sum these two terms together. By using the standard parametric representation of the Feynman integral, one can show that

$$A_{(\text{ii})'}+A_{(\text{v})}=\frac{iab^2}{4(16\pi^2)^2}\int\prod_1^3 dx_j\,\delta\Big(\sum_1^3 x_j-1\Big)F(x_2\Delta)[(x_1x_2+x_2x_3+x_1x_3)^{-2}-x_2^{-2}(x_1+x_3)^{-2}], \tag{B8}$$

where $x_j \geq 0$ and the function F is

$$F(y)=(1+y)\ln(1+y)-y-\tfrac{1}{2}y^2 . \tag{B9}$$

Similarly, we find

$$A_{(ii)}-A_{(v)}=\frac{iab^2}{12(16\pi^2)^2}\int\prod_1^3 dx_j\,\delta\left(\sum_1^3 x_j-1\right)(x_1x_2+x_2x_3+x_1x_3)^{-2}\left[F(\Delta)-\sum_{j=1}^3 F(x_j\Delta)\right]. \tag{B10}$$

Both expressions are now finite. It is straightforward to verify that

$$A_{(ii)}+A_{(ii)'}=\frac{iab^2}{4(16\pi^2)^2}\left[F(\Delta)-G(\Delta)\right], \tag{B11}$$

where

$$G(\Delta)=\tfrac{1}{2}(1+\Delta)[\ln(1+\Delta)]^2-(1+\Delta)\ln(1+\Delta)+\Delta . \tag{B12}$$

By using Theorem 1, one sees that diagrams (iii) + (iii)′ and (iv) + (iv)′ are related to (ii) + (ii)′ in a simple way. Their entire sum is given by (B11), provided one substitutes b^2 by $b^2+2ac\Delta$, but keeping a and Δ fixed. Equation (3.17) is then established, and this completes the proof of Theorem 2.

APPENDIX C

To establish Theorem 3 (stated in Sec. III C) we shall consider first the case $l>3$ and $c=0$. One can readily verify that in this case there is no primitively divergent prototype diagram; consequently, we need only consider the convergent ones. A typical example is given by diagram (i) in Fig. 9. Let I and V be, respectively, the number of internal lines and vertices in the diagram. We have

$$2I=3V,\quad l=I-V+1$$

and therefore

$$V=2(l-1). \tag{C1}$$

Since each vertex carries a factor b, the corresponding Feynman amplitude is proportional to b^V. The dimension of the energy density function is

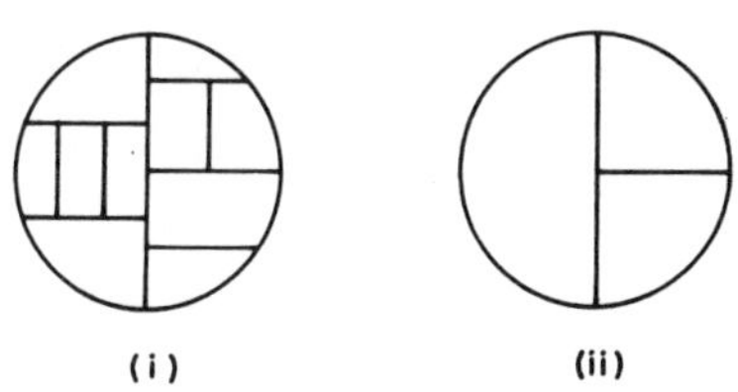

FIG. 9. Examples of prototype diagrams in a ϕ^3 theory. Diagram (i) is convergent, and diagram (ii) is primitively divergent.

$(\text{mass})^4$. Thus, from a simple dimensional consideration and by using (C1), one sees that the amplitude should be proportional to $a^2(b^2/a)^{l-1}$. Now, according to the rules for the prototype diagram given in Sec. III A, the parameter a appears only in the product $a(1+\Delta)$; this implies that the amplitude is proportional to

$$a^2(b^2/a)^{l-1}(1+\Delta)^{l-3} . \tag{C2}$$

Since the diagram is a convergent one, one finds the proportionality constant to be finite and independent of Δ. Equation (3.19) now follows because of (3.8).

Next, we consider the case $l=3$ and $c=0$. In this case, there is only one primitively divergent prototype diagram, given by (ii) in Fig. 9. By writing down explicitly the corresponding Feynman amplitude, one can readily derive (3.18). Theorem 3 is then proved.

APPENDIX D

In this appendix, we give an estimate of a lower bound for the decay rate λ_L, defined in (4.6). Let us expand the field operator $\phi(\vec{r},t)$ in terms of the Fourier series in the volume L^3:

$$\phi(\vec{r},t)=q_0+\sum_{\vec{k}\neq 0} q_k e^{i\vec{k}\cdot\vec{r}} . \tag{D1}$$

The Lagrangian for the system inside L^3 is

$$\int \mathcal{L}d^3r=\tfrac{1}{2}L^3\left[\left(\frac{dq_0}{dt}\right)^2-U(q_0)\right]+\cdots, \tag{D2}$$

where U is given by (1.2), and $\cdots$ is q_k-dependent $(\vec{k}\neq 0)$. The conjugate momentum of q_0 is

$$p_0=L^3\left(\frac{dq_0}{dt}\right) . \tag{D3}$$

Therefore, the Hamiltonian is

$$H=\tfrac{1}{2}[L^{-3}p_0^{\,2}+L^3U(q_0)]+\cdots . \tag{D4}$$

According to (4.1), at time $t=0$ the system is at $q_0=\phi_{\text{vex}}$, which is only a local minimum of $U(q_0)$. There is a potential barrier that separates this local minimum from the absolute minimum of $U(q_0)$, which is at $q_0=0$. To estimate the barrier-penetration probability, we shall use the WKB method for the q_0 degree of freedom, but suppress all other $\vec{k}\neq 0$ degrees of freedom (i.e., set $q_k = 0$ for $\vec{k}\neq 0$). The result is

$$\lambda_L \sim \omega \exp\left\{-2L^3 \int [U(q_0) - U(\phi_{vex})]^{1/2} dq_0\right\}, \tag{D5}$$

where

$$\omega^2 = a + b\phi_{vex} + \tfrac{1}{2}c\phi_{vex}^2 . \tag{D6}$$

In (D5), the integration is from $q_0 = \phi_{vex}$ to q_0', where $U(q_0') = U(\phi_{vex})$ and $\phi_{vex} < q_0' \leq 0$. Such an estimation of λ_L is obviously an underestimation, since by using the other $\vec{k}\neq 0$ degrees of freedom one can easily show that there are other paths leading from the local minimum $q_0 = \phi_{vex}$ to regions near the absolute minimum $q_0 = 0$, but passing through a *much lower potential barrier*.

---

*Research supported in part by the U. S. Atomic Energy Commission.

[1]For a history of this subject see Y. Nambu, Fields and Quanta 1, 33 (1970).

[2]J. Goldstone, Nuovo Cimento 19, 155 (1961); J. Goldstone, A. Salam, and S. Weinberg, Phys. Rev. 127, 965 (1962); P. W. Higgs, Phys. Lett. 12, 132 (1964); Phys. Rev. Lett. 13, 508 (1964); Phys. Rev. 145, 1156 (1966), and literature cited therein and in Ref. 1.

[3]T. D. Lee and C. N. Yang, Phys. Rev. 87, 404 (1952); 87, 417 (1952).

[4]S. Coleman and E. Weinberg, Phys. Rev. D 7, 1888 (1973).

[5]See, e.g., T. D. Lee and C. N. Yang, Phys. Rev. 117, 22 (1960).

[6]See, e.g., K. Symanzik, Commun. Math. Phys. 16, 48 (1970). Calculations very similar to those given here may be found in B. W. Lee, Nucl. Phys. B9, 649 (1969).

[7]Notions essentially identical to the "prototype diagram" have been used in many-body problems and statistical mechanics. See Ref. 5.

[8]F. J. Dyson, Phys. Rev. 75, 1736 (1949).

[9]See, e.g., R. Finkelstein, R. Le Levier, and M. Ruderman, Phys. Rev. 83, 326 (1951).

[10]J. Schwinger, Ann. Phys. (N.Y.) 2, 407 (1957); J. C. Polkinghorne, Nuovo Cimento 8, 179 (1958); 8, 781 (1958); M. Gell-Mann and M. Lévy, Nuovo Cimento 16, 53 (1960). For further references, see those given by B. W. Lee, *Chiral Dynamics* (Gordon and Breach, New York, 1972).

[11]For the present experimental status, see (e.g.) the review article by S. Protopopescu *et al.*, in *Experimental Meson Spectroscopy—1972*, edited by K.-W. Lai and A. H. Rosenfeld (A.I.P., New York, 1972).

PHYSICAL REVIEW D VOLUME 19, NUMBER 6 15 MARCH 1979

Feynman rules of quantum chromodynamics inside a hadron

T. D. Lee
Columbia University, New York, New York 10027
and Brookhaven National Laboratory, Upton, New York 11973

(Received 6 September 1978)

We start from quantum chromodynamics in a finite volume of linear size L and examine its color-dielectric constant κ_L, especially the limit κ_∞ as $L \to \infty$. By choosing as our standard $\kappa_L = 1$ when L = some hadron size R, we conclude that κ_∞ must be <1; furthermore, from the fact that a free quark has not been observed we can estimate an upper bound: $\kappa_\infty < 1.3 \times 10^{-2}\alpha$ where α is the fine-structure constant of QCD inside the hadron. A permanent quark confinement corresponds to the limit $\kappa_\infty = 0$. The hadrons are viewed as small domain structures (with color-dielectric constant = 1) immersed in a perfect, or nearly perfect, color-dia-electric medium, which is the vacuum. The Feynman rules of QCD inside the hadron are derived; they are found to depend on the color-dielectric constant κ_∞ of the vacuum that lies outside. We show that, when $\kappa_\infty \to 0$, the mass of any color-nonsinglet state becomes ∞, but for color-singlet states their masses and scattering amplitudes remain finite. These new Feynman rules also depend on the hadron size R. Only at high energy and large four-momentum transfer can such R dependence be neglected and, for color-singlet states, these new rules be reduced to the usual ones.

I. INTRODUCTION

In a recent paper,[1] it was emphasized that in order to give quantum chromodynamics[2] (QCD) a well-defined meaning, a convenient method is to first contain the whole system within a volume of size L^3. At a finite L, there is the usual perturbation series which is finite to every order of the renormalized coupling constant g. Assuming that the limit $L \to \infty$ exists, one expects the existence of a long-range order in the vacuum for an infinite volume. Because of relativistic invariance, such a long-range order must be a Lorentz scalar. It is then suggested that this long-range order can be expressed in terms of the (color) dielectric constant κ of the vacuum. In this approach,[3] a permanent quark confinement is simply viewed as the vacuum of an infinite volume being a perfect "dia-electric" substance with its dielectric constant $\kappa \to 0$, while the "vacuum" inside a hadron is normal ($\kappa = 1$). As we shall see, such a description leads to a set of Feynman rules that is quite different from the usual ones given in the literature. The propagator of the vector gauge field inside the hadron has an explicit dependence on the dielectric constant κ of the vacuum that lies outside the hadron. In the limit $\kappa \to 0$, the mass of any color-nonsinglet state, such as a single quark or a single vector gauge particle, becomes infinite. On the other hand, as will be shown in this paper, for color-singlet states, their masses and scattering amplitudes remain finite when $\kappa \to 0$.

In the following, we begin in Sec. II with a brief summary of what is currently known about the (color) dielectric constant of the vacuum. In order to take into account the long-range order of the vacuum and its long-wavelength fluctuations, we adopt the standard soliton description by introducing a scalar field to represent the dynamics of such collective motion. The related phenomenological Lagrangian is given in Sec. III. This enables us in Sec. IV to write down the appropriate Feynman rules and to analyze their properties. These new Feynman rules are different from the usual ones because they are applicable to configurations near a *soliton* (or bag) solution, while the usual ones are for perturbations near a *spatially homogeneous* solution, such as a pure infinite vacuum without the soliton (or bag). [Since in our picture all hadrons are solitons (or bags), it is difficult to have small perturbations around such a pure vacuum.] As we shall see, the Feynman rules derived here have an explicit dependence on the hadron size R. Only at very high energy and four-momentum transfer can one neglect the R dependence and thereby reduce these new rules to the usual ones as an approximation.

In a relativistic theory, the dielectric constant κ of the vacuum is always the inverse of its magnetic susceptibility μ because the velocity of light $c/(\kappa\mu)^{1/2}$ must equal c itself. Thus, $\kappa \ll 1$ means $\mu \gg 1$, and a zero dielectric constant is the same as an infinite magnetic susceptibility.[4] It is interesting to ask whether there exists a critical temperature at which the infinite-volume (QCD) system can undergo a phase transition in its dielectric constant (or magnetic susceptibility). This and other questions related to the long-range order, the scale determination of the hadron mass (or radius), and the experimental limit of the (color) dielectric constant are briefly discussed in Sec. V.

II. DIELECTRIC CONSTANT

For simplicity, let us consider in this section a pure QCD system consisting of only color gauge fields V_μ^a $(a=1,2,\ldots,8)$. We first contain the system within a finite volume L^3. So long as L is fixed, the (color) dielectric constant κ_L of the vacuum has no absolute meaning, since the transformation

$$V_\mu^a \to \kappa^{1/2} V_\mu^a\,, \qquad g \to \kappa^{-1/2} g \tag{1}$$

brings the covariant field derivative $V_{\mu\nu}^a \to \kappa^{1/2} V_{\mu\nu}^a$, and therefore the Lagrangian density

$$-\tfrac{1}{4} V_{\mu\nu}^a V_{\mu\nu}^a \to -\tfrac{1}{4}\kappa V_{\mu\nu}^a V_{\mu\nu}^a, \tag{2}$$

where g is the renormalized coupling constant, the subscripts μ, ν denote the space-time indices, and the superscript a is the color index. But when we consider two different volumes, say of sizes L and L', the ratio $\kappa_{L'}/\kappa_L$ is, of course, physically meaningful. Because of (1), this ratio is equal to the inverse of the corresponding ratio of the square of the renormalized coupling constants. By using the well-known properties of the β function,[5] one can readily derive

$$\kappa_{L'} < \kappa_L \text{ if } L' > L. \tag{3}$$

Furthermore, if

$$\beta(g) = 0 \text{ only at } g = 0, \tag{4}$$

then

$$\lim_{L\to\infty} \kappa_L = 0.$$

Throughout our discussions, we *assume* κ_L is a smooth function of L, and that the limit $L\to\infty$ exists. For completeness, a proof is given in Appendix A. [Although the mathematics given there is essentially identical to that used in deriving asymptotic freedom,[5] its application to the color dielectric constant focuses on a new aspect of the physics involved. As we shall see, this leads to the conclusion that domain structures (soliton solutions) should develop whenever there are quarks and antiquarks present; thereby one bypasses the usual difficulty of strong coupling in the so-called "infrared slavery."]

For small L, because g^2 is $\ll 1$, κ_L is clearly $\neq 0$. As a convention, we may define when $L=$ some hadronic radius R,

$$\kappa_R = 1. \tag{5}$$

Let

$$\kappa_\infty \equiv \lim_{L\to\infty} \kappa_L. \tag{6}$$

From (3) and (5), it follows that the dielectric constant of the vacuum for an infinite volume must be less than unity, i.e.,

$$\kappa_\infty < 1. \tag{7}$$

Furthermore, κ_∞ is zero if (4) holds. In the following, we assume

$$\kappa_\infty = 0 \tag{8a}$$

or at least

$$\kappa_\infty \ll 1. \tag{8b}$$

In the former, the vacuum for an infinite system is a perfect dia-electric, in the latter, a nearly perfect dia-electric. By following the arguments given in Sec. II of Ref. 1 (and as will also be shown in the following), one sees that the mass of any color-nonsinglet state diverges when $\kappa_\infty \to 0$. Thus, (8a) implies a permanent quark confinement, while (8b) implies a nearly permanent confinement.

At present, it is not known whether $\beta(g)=0$ has only a real root at $g=0$. From (7), we know that κ_∞ must be <1. But as yet, we are not able to decide by pure theoretical deduction how small κ_∞ actually is.[6] Now the fact that a single free quark has never been observed puts a lower limit on its mass m_q. As will be shown in Sec. V, if we set $m_q > 5$ GeV, which is a rather lenient lower bound of the free quark mass, then we can estimate an upper limit of κ_∞:

$$\kappa_\infty < 1.3\times 10^{-2}\alpha,$$

where $\alpha = (4\pi)^{-1} g_R{}^2$, with R chosen such that $\kappa_R = 1$, in accordance with our convention (5). Thus, we can regard that as an *experimental* determination κ_∞ must be $\ll 1$. On esthetic grounds, one may conjecture $\kappa_\infty = 0$. The otherwise puzzling phenomenon of quark confinement, or near confinement, now receives a natural "explanation".

III. SOLITON (OR BAG) MODEL

A. Phenomenological Lagrangian

In order to incorporate the long-range order of the vacuum and its long-wavelength fluctuations into the dynamics of the hadron, we adopt the soliton description[1] through the use of collective coordinates. The dielectric constant κ will now be represented by a scalar field σ. The simplest form is

$$\sigma \propto 1-\kappa. \tag{9}$$

If we assume (8a), then

$$\frac{\sigma}{\sigma_{vac}}=1-\kappa,$$

so that for an infinite volume, $\kappa=0$ and therefore $\sigma=\sigma_{vac}$. In the soliton picture, one has outside the hadron $\sigma=\sigma_{vac}$, but inside the hadron $\sigma\cong 0$ which gives $\kappa\cong 1$, consistent with our convention (5). The phenomenological Lagrangian density is assumed to be (see, however, the modification given in Sec. III D).

$$\mathcal{L}=-\tfrac{1}{4}\kappa V^a_{\mu\nu}V^a_{\mu\nu}-\psi^\dagger\gamma_4(\gamma_\mu D_\mu+f\sigma+m)\psi -\frac{1}{2}\left(\frac{\partial\sigma}{\partial x_\mu}\right)^2-U(\sigma)+\text{counterterms}, \tag{10}$$

where the superscript dagger denotes the Hermitian conjugation, $x_\mu=(\vec{r}, it)$,

$$V^a_{\mu\nu}=\frac{\partial}{\partial x_\mu}V^a_\nu-\frac{\partial}{\partial x_\nu}V^a_\mu+gC^{abc}V^b_\mu V^c_\nu,$$
$$D_\mu=\frac{\partial}{\partial x_\mu}-\tfrac{1}{2}ig\lambda^a V^a_\mu, \tag{11}$$

the λ^a's are the standard Gell-Mann matrices related to C^{abc} by the commutation relation

$$[\lambda^a,\lambda^b]=2iC^{abc}\lambda_c,$$

the γ_μ's are the usual Hermitian Dirac matrices, and the function $U(\sigma)$ has an absolute minimum at $\sigma=\sigma_{vac}$ and a local minimum at $\sigma=0$ with

$$U(\sigma_{vac})=0$$
and
$$U(0)\equiv p>0. \tag{12}$$

As we shall see, the detailed form of $U(\sigma)$ is not important. If one wishes, one may assume $U(\sigma)$ to be simply a quartic function of σ. In (10) and (11), ψ denotes the quark field which, besides being a color triplet, also has F flavors, m is the mass matrix for quarks inside the hadron, and g and f are both renormalizable coupling constants. Since σ is only a phenomenological field, describing the long-range collective effects of QCD, its short-wavelength components do not exist in reality. The counterterms in $\mathcal{L}$ are for renormalization; they consist *only* of those due to loop diagrams of the vector gauge field V^a_μ and the quark field ψ. In the following, we shall ignore all σ loops, i.e., σ will be approximated by a classical field.

In our phenomenological Lagrangian density, the σ field is coupled to the quarks in two ways: One is through the quark-V^a_μ coupling g and $-\frac{1}{4}\kappa V^a_{\mu\nu}V^a_{\mu\nu}$ which, because of (9), couples V^a_μ with σ, the other is a direct quark-σ coupling f. So far as the quark confinement problem is concerned, there is no need to have the direct quark-σ coupling; it suffices to have the vacuum be a perfect dia-electric ($\kappa_\infty=0$). The origin of the f coupling lies in the dichotomy that only inside the hadron is the coupling between the quark and the vector gauge field really g, which is relatively small; outside the hadron, because of (1), it is actually $g/\kappa_\infty^{1/2}$. For $\kappa_\infty\ll 1$, $g/\kappa_\infty^{1/2}$ becomes very large; on the one hand, this has the desirable effect of preventing the quarks from moving outside the hadron, but on the other hand it also presents a technical difficulty for a diagram analysis in powers of g^2. In (10) the direct quark-σ coupling f with

$$f\sigma_{vac}\gg \text{hadron mass} \tag{13}$$

is introduced phenomenologically to give a convenient alternative formulation of the same physical effect, but bypassing the above difficulty. As we shall see, because of (13) the f coupling restricts the quarks to staying always inside the hadron, and that enables us to take full advantage of the relative smallness of the quark-V^a_μ coupling g. With the f coupling we may now expand any physical observable, say the hadron mass M, in a power series of $\alpha\equiv(4\pi)^{-1}g^2$:

$$M=M_0+\alpha M_1+\alpha^2M_2+\cdots. \tag{14}$$

If f were zero, it would be difficult to derive the zeroth-order term M_0, since when $\alpha=0$ both g and $g/\kappa_\infty^{1/2}$ should be zero; hence, without the f coupling quarks would be unconfined. (See Sec. III D for further discussions.)

B. Zeroth-order approximation

With the direct quark-σ coupling f, we can now in the zeroth-order calculation neglect the exchange of V^a_μ. The description of the hadron reduces to that of a simple soliton model, consisting of the scalar σ field and the quark ψ field. The relevant part of the Lagrangian density (10) now consists simply of

$$-\psi^\dagger\gamma_4\left(\gamma_\mu\frac{\partial}{\partial x_\mu}+f\sigma+m\right)\psi-\frac{1}{2}\left(\frac{\partial\sigma}{\partial x_\mu}\right)^2-U(\sigma). \tag{15}$$

As mentioned before, since σ is only a phenomenological field that has no short-wavelength components, we shall neglect all σ-loop diagrams. The remaining σ diagrams are all tree diagrams, which correspond to the quasiclassical approximation that has been extensively studied in the literature.[7] Here we give only a summary of the results for light quark hadrons, with $m=0$. In this case, ψ and σ can be reduced to c-number functions which satisfy

$$(-i\vec{\alpha}\cdot\vec{\nabla}+f\beta\sigma)\psi=\epsilon\psi$$

and (16)

$$-\nabla^2\sigma + U'(\sigma) = -fN\psi^\dagger\beta\psi,$$

where $U'(\sigma) = dU/d\sigma$, $\vec{\alpha}$ and β are the usual Dirac matrices, $\epsilon > 0$, ψ satisfies $\int \psi^\dagger\psi d^3r = 1$, and N is the total number of quarks and antiquarks,

$$N = 2 \text{ for mesons}$$

and

$$N = 3 \text{ for baryons.}$$

As shown in Ref. 7, for $f\sigma_{vac} \gg$ hadron mass and under very general assumptions, the hadron acquires a well-defined surface $\mathcal{S}$. One has

$$\psi = 0$$

and

$$\sigma = \sigma_{vac} \text{ outside } \mathcal{S}; \qquad (17)$$

on the inner side of the surface $\mathcal{S}$,

$$-i\beta\vec{\alpha}\cdot\hat{n}\psi = \psi, \qquad (18)$$

where $\hat{n}$ is the unit vector normal to $\mathcal{S}$, and inside $\mathcal{S}$ the field σ is $\cong 0$, with its deviation from zero proportional to $\psi^\dagger\beta\psi$. For the $s_{1/2}$ orbit, $\mathcal{S}$ is a spherical surface of radius R. Inside $\mathcal{S}$, (16) can be further reduced, through scaling, to a simple system of two coupled first-order differential equations:

$$\frac{du}{d\rho} = (-1 + u^2 - v^2)v$$

and (19)

$$\frac{dv}{d\rho} + \frac{2v}{\rho} = (1 + u^2 - v^2)u,$$

where ρ is related to the radius r by $\rho = \epsilon r$,

$$\psi \propto \begin{pmatrix} u \\ i(\vec{\sigma}\cdot\vec{r}/r)v \end{pmatrix} \zeta,$$

$$\zeta = \begin{pmatrix} 0 \\ 1 \end{pmatrix} \text{ or } \begin{pmatrix} 1 \\ 0 \end{pmatrix}.$$

$\vec{\sigma}$ is the usual Pauli spin matrix, and the variables ρ, u, and v are all dimensionless. Although (19) does not contain any explicit parameters, its solutions form a one-parameter family. As $\rho \to 0$, one has $v \to 0$ and $u \to u(0)$. For every $u(0)$ between 0 and a critical value $u_c = 1.7419$, there is a solution of (19). The solution can be obtained by direct integration from $\rho = 0$ to $\rho = \rho_0$. At $\rho = \rho_0$, one has $u(\rho_0) = v(\rho_0)$ and therefore the boundary condition (18) is satisfied. The radius of the hadron is given by

$$R = \rho_0/\epsilon.$$

A convenient parameter to label these solutions can be either $u(0)$ or the integral

$$n \equiv \int (u^2 + v^2) d^3\rho$$

over the region $\rho \le \rho_0$. As the initial value $u(0) \to 0$, one has $n \to 0$, but as $u(0) \to u_c = 1.7419$, $n \to \infty$.

The physical description of the soliton solution then resembles that of a gas bubble (i.e., the hadron) inside a medium (i.e., the vacuum). The hadron mass is determined by three parameters:

$$p,\ s,\ \text{and } n, \qquad (20)$$

where p is defined by (12) which represents the pressure of the medium on the bubble, s is the surface tension which arises because σ changes from 0 to σ_{vac} across the soliton surface, and n determines the gas pressure inside the bubble which is due to both the kinetic energy of quarks and the excitation energy of σ. In either of the limits $n \to 0$ or $n \to \infty$ the hadron mass has the form, in the notation of (14),

$$M_0 = \frac{N\rho_0}{R} + \tfrac{1}{3}4\pi R^3 p + 4\pi R^2 s, \qquad (21)$$

where

$$\rho_0 = 2.0428 \text{ when } n \to 0$$

and (22)

$$\rho_0 = 1 \text{ when } n \to \infty.$$

The double limit $n \to 0$ and $s \to 0$ gives the MIT bag,[8,9] and the double limit $n \to \infty$ and $p \to 0$ gives the SLAC bag.[10]

When $n \to 0$, the field σ assumes a simple form:

$$\sigma = \sigma_{vac} \text{ outside } \mathcal{S}$$

and (23)

$$\sigma = 0 \text{ inside } \mathcal{S}.$$

For $n \ne 0$, though $\sigma = \sigma_{vac}$ outside $\mathcal{S}$, σ is only $\cong 0$ inside; as mentioned before, its deviation from 0 is proportional to $\psi^\dagger\beta\psi$.

C. Coulomb gauge

To simplify our discussions, in the following we shall regard $\sigma(x)$, and therefore also $\kappa(x)$, as a *given* function of $\vec{r}$. Let us consider a single hadron system so that its surface $\mathcal{S}$ is simply connected. We assume further that $\sigma(\vec{r})$ has the simple form given by (23); hence, because of (9)

$$\kappa(\vec{r}) = \kappa_\infty \to 0 \text{ outside } \mathcal{S}$$

and (24)

$$\kappa(\vec{r}) = 1 \text{ inside } \mathcal{S},$$

where the precise form of the surface $\mathcal{S}$ is to be determined by minimizing the energy of the hadron state under consideration, as is done in the above zeroth-order calculation. [It is quite straightforward to extend our analysis to the more complicated case in which $\kappa(\vec{r})$ is only $\cong 1$ inside $\mathcal{S}$, as in the general $n \neq 0$ solutions examined in the preceding section.]

From (10), it follows that V^a_μ satisfies (neglecting the counterterms)

$$\frac{\partial}{\partial x_\mu}(\kappa V^a_{\mu\nu}) = -gJ^a_\nu \,, \tag{25}$$

where

$$J^a_\nu \equiv j^a_\nu + \kappa C^{abc} V^b_\mu V^c_{\mu\nu} \tag{26}$$

and

$$j^a_\nu = \tfrac{1}{2} i \psi^\dagger \gamma_4 \gamma_\nu \lambda^a \psi. \tag{27}$$

It is convenient to introduce the three-vectors $\vec{E}^a$, $\vec{D}^a$, $\vec{B}^a$, and $\vec{H}^a$, as in the case of the usual electromagnetic field in a medium:

$$V^a_{4k} = iE^a_k, \quad V^a_{ij} = \epsilon_{ijk} B^a_k \,, \qquad \vec{D}^a = \kappa \vec{E}^a, \text{ and } \vec{H}^a = \kappa \vec{B}^a, \tag{28}$$

where the roman subscripts denote the space components, $\epsilon_{ijk} = +1$ if ijk is an even permutation of 123, -1 if it is an odd permutation, and 0 otherwise. Equation (25) becomes

$$\vec{\nabla}\cdot \vec{D}^a = gJ^a_0$$

and (29)

$$\vec{\nabla}\times\vec{H}^a - \dot{\vec{D}}^a = g\vec{J}^a,$$

where the dot denotes a time derivative, $J^a_\mu = (\vec{J}^a, iJ^a_0)$,

$$J^a_0 = \tfrac{1}{2}\psi^\dagger \lambda^a \psi - \kappa C^{abc} \vec{V}^b \cdot \vec{E}^c$$

and (30)

$$\vec{J}^a = \tfrac{1}{2}\psi^\dagger \vec{\alpha} \lambda^a \psi - \kappa C^{abc}(V^b_0 \vec{E}^c + \vec{V}^b \times \vec{B}^c).$$

In the Coulomb gauge, $\vec{V}^a$ is chosen to be divergence-free,

$$\vec{\nabla}\cdot\vec{V}^a = 0, \tag{31}$$

and it satisfies

$$\vec{V}^a \cdot \hat{n} = 0 \text{ on } \mathcal{S}, \tag{32}$$

where, as before, $\hat{n}$ is the normal of $\mathcal{S}$. Because the Lagrangian density (10) is locally gauge-invariant, one sees readily that (as will also be proved in Appendix B) both (31) and (32) can be satisfied. According to (24), $\kappa(\vec{r})$ has a finite discontinuity across $\mathcal{S}$, which together with (32) gives

$$\vec{V}^a \cdot \vec{\nabla}\kappa = 0. \tag{33}$$

The function V^a_0 will now be regarded as a functional of $\vec{V}^a$, $\dot{\vec{V}}^a$, and

$$j^a_0 = \tfrac{1}{2}\psi^\dagger \lambda^a \psi \tag{34}$$

through

$$-\vec{\nabla}\cdot(\kappa \vec{\nabla} V^a_0) = gI^a, \tag{35}$$

where on account of (28)–(33)

$$I^a = J^a_0 + \kappa C^{abc} \dot{\vec{V}}^b \cdot \vec{\nabla} V^c_0. \tag{36}$$

Because $\kappa C^{abc} \dot{\vec{V}}^b \cdot \vec{\nabla} V^c_0 = \vec{\nabla}\cdot(\kappa C^{abc} \dot{\vec{V}}^b V^c_0)$, the total "color" charge is

$$Q^a \equiv \int J^a_0 d^3r = \int I^a d^3r; \tag{37}$$

furthermore, it is a constant of motion. Let $G(\vec{r}, \vec{r}\,')$ be the Green's function defined by

$$-\vec{\nabla}\cdot[\kappa \vec{\nabla} G(\vec{r}, \vec{r}\,')] = \delta^3(\vec{r} - \vec{r}\,'), \tag{38}$$

where $\vec{\nabla}$ operates on $\vec{r}$. The solution V^a_0 of (35) satisfies

$$V^a_0(\vec{r}) = g\int G(\vec{r}, \vec{r}\,') I^a(\vec{r}\,') d^3r'. \tag{39}$$

Since I^a depends on V^a_0, the functional dependence of

$$V^a_0 = V^a_0(\vec{V}^b, \dot{\vec{V}}^b, j^b_0) \tag{40}$$

can be obtained from (36) and (39) by iteration.

From (24), one sees that as $\kappa_\infty \to 0$, the Green's function $G(\vec{r}, \vec{r}\,') \to \infty$ everywhere. We may expand $G(\vec{r}, \vec{r}\,')$ in powers of κ_∞:

$$G(\vec{r}, \vec{r}\,') = \kappa_\infty^{-1} G_0(\vec{r}, \vec{r}\,') + G_1(\vec{r}, \vec{r}\,') + \kappa_\infty G_2(\vec{r}, \vec{r}\,') + \cdots \,. \tag{41}$$

As shown in Appendix B, for arbitrary surface $\mathcal{S}$,

$$G_0(\vec{r}, \vec{r}\,') = \text{constant} \tag{42}$$

when both $\vec{r}$ and $\vec{r}\,'$ are inside $\mathcal{S}$; furthermore, $G_0(\vec{r}, \vec{r}\,')$ is independent of $\vec{r}$ when $\vec{r}$ is inside $\mathcal{S}$ (but $\vec{r}\,'$ may be outside). Likewise, owing to symmetry, $G_0(\vec{r}, \vec{r}\,')$ is independent of $\vec{r}\,'$ when $\vec{r}\,'$ is inside $\mathcal{S}$. Hence for a color-singlet hadron, because $Q^a = 0$ and $j^a_0 = \vec{V}^a = 0$ outside $\mathcal{S}$,

$$\lim_{\kappa_\infty \to 0} V^a_0(\vec{r}) = \text{finite}. \tag{43}$$

The Green's function $G(\vec{r}, \vec{r}\,')$ for a spherical surface of radius R can be readily derived (and is also given in Appendix B), from which one sees that, e.g., when $\vec{r}$ and $\vec{r}\,'$ are both inside the surface $G_0(\vec{r}, \vec{r}\,') = (4\pi R)^{-1}$ and

$$G_1(\vec{r},\vec{r}')=(4\pi)^{-1}\left\{\frac{1}{|\vec{r}-\vec{r}'|}+\frac{1}{R}\left[-1+\sum_{l=1}^{\infty}\frac{l+1}{l}\left(\frac{rr'}{R^2}\right)^l P_l(\cos\theta)\right]\right\}, \tag{44}$$

where the magnitudes of $\vec{r}$ and $\vec{r}'$ are r and r', respectively, the angle between them is θ, and the P_l's are the standard Legendre polynomials. [The sum in (44) can be easily carried out. See (B22) and (B23) of Appendix B.]

In the Coulomb gauge, ψ and $\vec{V}^a$ are regarded as independent generalized coordinates. Their conjugate momenta are, respectively, $i\psi^\dagger$ and

$$\Pi_i^a=-[\delta_{ij}-(\nabla_i\nabla_j/\nabla^2)]D_j^a\,. \tag{45}$$

It is useful to resolve $\vec{E}^a$ and $\vec{D}^a$ into their transverse and longitudinal components:

$$\vec{E}^a=\vec{E}^a_{\rm tr}+\vec{E}^a_{\rm long}$$

and (46)

$$\vec{D}^a=\vec{D}^a_{\rm tr}+\vec{D}^a_{\rm long}\,,$$

where $\vec{\nabla}\cdot\vec{D}^a_{\rm tr}=\vec{\nabla}\cdot\vec{E}^a_{\rm tr}=0$ and $\vec{\nabla}\times\vec{D}^a_{\rm long}=\vec{\nabla}\times\vec{E}^a_{\rm long}=0$. Equation (45) becomes

$$\vec{\Pi}^a=-\vec{D}^a_{\rm tr}. \tag{47}$$

By using (35) and (36), one can readily verify that

$$\int\vec{D}^a_{\rm long}\cdot\vec{E}^a_{\rm long}\,d^3r=g\int V_0^a(j_0^a+C^{abc}\vec{V}^b\cdot\vec{\Pi}^c)d^3r, \tag{48}$$

and therefore, through partial integrations, the Hamiltonian density can be set to be

$$\mathcal{H}=\tfrac{1}{2}\vec{D}^a_{\rm tr}\cdot\vec{E}^a_{\rm tr}+\tfrac{1}{2}\vec{D}^a_{\rm long}\cdot\vec{E}^a_{\rm long}+\tfrac{1}{2}\vec{B}^a\cdot\vec{H}^a$$
$$+\psi^\dagger\left(\vec{\alpha}\cdot\frac{1}{i}\vec{\nabla}+f\sigma+m\right)\psi-\vec{j}^a\cdot\vec{V}^a+\tfrac{1}{2}(\vec{\nabla}\sigma)^2+U(\sigma), \tag{49}$$

which is valid in the approximation that $\sigma(x)$ is a given time-independent classical field.

D. A modification

As mentioned in Sec. III A, the quark-V_μ^a coupling has the relatively small value g only inside the hadron; outside, it is $g/\kappa_\infty^{1/2}$ which $\to\infty$ when $\kappa_\infty\to 0$. In order to develop a series expansion in g, but not in $g/\kappa_\infty^{1/2}$, we introduce a direct quark-σ coupling f in the Lagrangian density (10). When $f\sigma_{\rm vac}\gg$ hadron mass, the quark field ψ becomes 0 outside the hadron surface $\mathcal{S}$, in accordance with (17); thereby we avoid the very large coupling $g/\kappa_\infty^{1/2}$. The same problem also arises between the vector gauge fields themselves because of self-coupling. Likewise, we would also like to restrict

$$V_\mu^a=0 \text{ outside } \mathcal{S}\,. \tag{50}$$

For simplicity, let us assume that, as in the preceding section, $\sigma(x)$ is given by (23). Within the context of a relativistic local theory, the above restriction can be achieved most simply by modifying the Lagrangian density from $\mathcal{L}$ to[11]

$$\mathcal{L}'=\mathcal{L}-\tfrac{1}{2}\kappa(f'\sigma)^2V_\mu^aV_\mu^a\,, \tag{51}$$

where, as before, κ and σ are related by (9) and $\mathcal{L}$ is given by (10). Like (13), this additional coupling f' satisfies

$$f'\sigma_{\rm vac}\equiv\mu\gg\text{hadron mass.} \tag{52}$$

The modified Lagrangian density $\mathcal{L}'$ is no longer locally gauge invariant, though it remains globally invariant. Hence the total "color" charge Q^a is still conserved. Furthermore, as we shall see, since

$$f'\sigma=0 \quad \text{inside } \mathcal{S}, \tag{53}$$

we retain almost all the physical consequences of the locally gauge-invariant Lagrangian density $\mathcal{L}$.

[One may wonder: Why not directly set in the Lagrangian density (10) $f=0$, or in (51) $f=f'=0$, and simply impose "$\psi=0$ and $V_\mu^a=0$ outside $\mathcal{S}$" as constraints? Although such an approach will be discussed in Sec. IV, in principle there are several advantages in starting from a Lorentz-invariant Lagrangian density. (i) By using its soliton solution, one can then *derive* these constraints. This way it ensures the self-consistency of the constraints, especially since V_0^a is not an independent variable. For example, without the f' coupling, even if $\psi=\vec{V}^a=0$ outside $\mathcal{S}$, according to (39) there is still a tail of nonzero V_0^a outside $\mathcal{S}$, and that might lead to a Van der Waals-type force between hadrons which would be in violation of experimental observations. However, in the classical limit, one can show that when $\kappa_\infty\to 0$ the long-range force between hadrons must vanish. (ii) A covariant Lagrangian theory allows one to study the motion of the hadron surface $\mathcal{S}$ and to ensure relativistic covariance.]

Instead of (25), V_μ^a now satisfies

$$\frac{\partial}{\partial x_\mu}(\kappa V_{\mu\nu}^a)-\kappa(f'\sigma)^2V_\nu^a=-gJ_\nu^a\,, \tag{54}$$

and instead of (35), V_0^a satisfies

$$-\vec{\nabla}\cdot(\kappa\vec{\nabla}V_0^a)+\kappa(f'\sigma)^2V_0^a=gI^a, \quad (55)$$

where J_ν^a and I^a remain given by (26) and (36), respectively. Because the local gauge invariance is broken in $\mathcal{L}'$, it is not possible to impose the divergence condition (31); all three components of $\vec{V}^a$ are independent variables. However, since $\dot{V}_0^a$ is absent in $\mathcal{L}'$, only ψ and $\vec{V}^a$ can be regarded as independent generalized coordinates; V_0^a remains a dependent variable. Just as in (39), we may write

$$V_0^a(\vec{r})=g\int G(\vec{r},\vec{r}\,')I^a(\vec{r}\,')d^3r', \quad (56)$$

where, instead of (38), $G(\vec{r},\vec{r}\,')$ is now defined by

$$[-\vec{\nabla}\cdot(\kappa\vec{\nabla})+\kappa(f'\sigma)^2]G(\vec{r},\vec{r}\,')=\delta^3(\vec{r}-\vec{r}\,'). \quad (57)$$

At any fixed $\mu\equiv f'\sigma_{\text{vac}}$, when $\kappa_\infty\to 0$, this new Green's function $G(\vec{r},\vec{r}\,')\to\infty$ everywhere, as before. We may again expand $G(\vec{r},\vec{r}\,')$ in powers of κ_∞:

$$G(\vec{r},\vec{r}\,')=\kappa_\infty^{-1}G_0(\vec{r},\vec{r}\,')+G_1(\vec{r},\vec{r}\,')+\kappa_\infty G_2(\vec{r},\vec{r}\,')+\cdots. \quad (58)$$

It will be proved in Appendix C that (42) and (43) remain valid, i.e.,

$$G_0(\vec{r},\vec{r}\,')=\text{constant} \quad (59)$$

when $\vec{r}$ and $\vec{r}\,'$ are both inside $\mathcal{S}$, and for a color-singlet hadron, $V_0^a(\vec{r})=\text{finite}$ when $\kappa_\infty\to 0$. Furthermore, for a spherical surface of radius R and for $\vec{r},\vec{r}\,'$ inside the sphere, $G_1(\vec{r},\vec{r}\,')$ remains given by (44), apart from an additive constant. Since for a color singlet the additive constant in $G(\vec{r},\vec{r}\,')$, or $G_1(\vec{r},\vec{r}\,')$, is of no importance, we may rewrite (44) as

$$G_1(\vec{r},\vec{r}\,')=(4\pi)^{-1}\left[\frac{1}{|\vec{r}-\vec{r}\,'|}+\sum_{l=1}^{\infty}\frac{l+1}{lR}\left(\frac{rr'}{R^2}\right)^l P_l(\cos\theta)\right]. \quad (60)$$

Outside $\mathcal{S}$, this new G_1 decreases exponentially; hence, it is quite different from the old G_1 which has only a power dependence on $\vec{r}$ and/or $\vec{r}\,'$. [For nonspherical surfaces, the new G_1 can be quite different from the old G_1 even inside $\mathcal{S}$. See expression (C17) of Appendix C.]

Inside $\mathcal{S}$, since $\sigma=0$ we have $\kappa=1$ and $\mathcal{L}'=\mathcal{L}$; hence, $\vec{D}^a=\vec{E}^a$ and our Lagrangian density is locally gauge invariant. We may again adopt the Coulomb gauge and require (31) and (32), i.e., $\vec{\nabla}\cdot\vec{V}^a=0$ inside the surface and $\hat{n}\cdot\vec{V}^a=0$ on the surface. As in (46), $\vec{E}^a$ may be decomposed into $\vec{E}^a=\vec{E}^a_{\text{tr}}+\vec{E}^a_{\text{long}}$:

$$\vec{E}^a_{\text{long}}=-\vec{\nabla}V_0^a+\vec{\nabla}\phi^a,$$
$$\vec{E}^a_{\text{tr}}=-\dot{\vec{V}}^a-gC^{abc}\vec{V}^bV_0^c-\vec{\nabla}\phi^a, \quad (61)$$

and

$$\nabla^2\phi^a=-g\vec{\nabla}\cdot(C^{abc}\vec{V}^bV_0^c)\,.$$

As will be shown in Appendix C, one can always choose ϕ^a such that

$$\vec{E}^a_{\text{tr}}\cdot\hat{n}=0 \quad \text{on } \mathcal{S}. \quad (62)$$

Equation (47) now becomes

$$\vec{\Pi}^a=-\vec{E}^a_{\text{tr}} \quad \text{inside } \mathcal{S}. \quad (63)$$

Outside the surface, when $f\sigma_{\text{vac}}\to\infty$ and $f'\sigma_{\text{vac}}\to\infty$, we have

$$\psi=\vec{V}^a=0. \quad (64)$$

Therefore, the modified Lagrangian (51) provides a self-consistent device to exclude the quark and the gauge fields from the outside region.

IV. FEYNMAN DIAGRAMS

A. Reduced Hamiltonian

To simplify our discussions, we again ignore the surface motion,[12] and assume σ and κ as given by (23) and (24), respectively. Inside the surface $\mathcal{S}$, the equations of motion reduce to

$$\frac{\partial}{\partial x_\mu}V^a_{\mu\nu}=-gJ^a_\nu$$

and

$$(\gamma_\mu D_\mu+m)\psi=0,$$

where D_μ and J_ν^a are given by (11) and (26). Outside $\mathcal{S}$, we impose the constraints $\psi=\vec{V}^a=0$, in accordance with (64). We shall adopt the Coulomb gauge: $\vec{\nabla}\cdot\vec{V}^a=0$ inside $\mathcal{S}$. In addition, ψ and $\vec{V}^a$ satisfy the boundary conditions (18) and (32), respectively.

By following the discussions given in the preceding section (and as will also be shown in Appendix C), the Hamiltonian of the system can be reduced to the form

$$H_0=\int_\Omega \mathcal{H}\,d^3r+sA, \quad (65)$$

where Ω refers to the volume within $\mathcal{S}$,

$$\mathcal{H}=\tfrac{1}{2}\vec{\Pi}^a\cdot\vec{\Pi}^a+\tfrac{1}{2}gV_0^a(j_0^a+C^{abc}\vec{V}^b\cdot\vec{\Pi}^c)+\tfrac{1}{2}\vec{B}^a\cdot\vec{B}^a+\psi^\dagger\left(\vec{\alpha}\cdot\frac{1}{i}\vec{\nabla}+m\right)\psi-\vec{j}^a\cdot\vec{V}^a+p, \quad (66)$$

p is given by (12), s is the surface tension defined in (20), and A is the area of $\mathcal{S}$. As before, $\vec{\Pi}^a$

$=-\vec{E}^a_{tr}$ is the conjugate momentum of $\vec{V}^a$, $j^a_0 = \frac{1}{2}\psi^\dagger\lambda^a\psi$, and $\vec{j}^a = \frac{1}{2}\psi^\dagger\vec{\alpha}\lambda^a\psi$. Equations (37) and (56) may now be written as

$$Q^a = \int_\Omega I^a d^3r = \int_\Omega (j^a_0 + C^{abc}\vec{V}^b\cdot\vec{\Pi}^c)d^3r, \tag{67}$$

$$V^a_0(\vec{r}) = g\int_\Omega G(\vec{r},\vec{r}\,')I^a(\vec{r}\,')d^3r', \tag{68}$$

and

$$I^a = j^a_0 + C^{abc}\vec{V}^b\cdot(-\vec{E}^c + \vec{\nabla}V^c_0),$$

where $G(\vec{r},\vec{r}\,')$ is given by (57).

When $\kappa_\infty \to 0$, one has $G(\vec{r},\vec{r}\,') \to \infty$. As shown in Appendix C, for color nonsinglets, $Q^a \neq 0$,

$$H_0 \to (2\kappa_\infty)^{-1}g^2G_0Q^aQ^a \to \infty, \tag{69}$$

where according to (59) G_0 is a constant. But for a color singlet, H_0 is *finite* and (65) becomes

$$H_0 = \int_\Omega \bar{\mathcal{H}}d^3r + sA, \tag{70}$$

where $\bar{\mathcal{H}}$ is the same as $\mathcal{H}$, provided V^a_0 is replaced by

$$\bar{V}^a_0(\vec{r}) \equiv g\int G_1(\vec{r},\vec{r}\,')I^a(\vec{r}\,')d^3r', \tag{71}$$

where G_1 is defined by (58). The explicit form of $\bar{\mathcal{H}}$ is

$$\bar{\mathcal{H}} = \tfrac{1}{2}\vec{\Pi}^a\cdot\vec{\Pi}^a + \tfrac{1}{2}g\bar{V}^a_0(j^a_0 + C^{abc}\vec{V}^b\cdot\vec{\Pi}^c) + \tfrac{1}{2}\vec{B}^a\cdot\vec{B}^a + \psi^\dagger\left(\vec{\alpha}\cdot\frac{1}{i}\vec{\nabla} + m\right)\psi + \vec{j}^a\cdot\vec{V}^a + p. \tag{72}$$

From (69) and (70), we conclude that when $\kappa_\infty \to 0$, the Hamiltonian operator is always divergent for color nonsinglets, but *for color singlets, it is always finite.*

B. Feynman rules

In the Coulomb gauge, the derivation from the Lagrangian to the Hamiltonian follows the standard canonical procedure. However, because the Lagrangian contains nonlinear terms which are $\dot{\vec{V}}^a$ dependent, there must be some additional action[13] which gives rise to new loop diagrams. The simplest derivation is to follow the path-integration method of Faddeev and Popov.[14,15] The result can be expressed in terms of an additional Hamiltonian

$$H_1 = -\tfrac{1}{2}i\,\mathrm{tr}\ln(1-\Delta^{-1}M) + i\,\mathrm{tr}\ln(1-\Delta^{-1}N), \tag{73}$$

where 1 is the unit matrix

$$\langle a,\vec{r}\,|1|\,b,\vec{\rho}\rangle = \delta^{ab}\delta^3(\vec{r}-\vec{\rho}),$$

and the matrices Δ, M, and N are given by

$$\langle a,\vec{r}\,|\Delta|\,b,\vec{\rho}\rangle = \delta^{ab}\vec{\nabla}^2\delta^3(\vec{r}-\vec{\rho}),$$

$$\langle a,\vec{r}\,|M|\,b,\vec{\rho}\rangle = gC^{abc}[\vec{V}^c(\vec{r}) + \vec{V}^c(\vec{\rho})]\cdot\vec{\nabla}\delta^3(\vec{r}-\vec{\rho}) + g^2C^{ace}C^{bde}\vec{V}^c(\vec{r})\cdot\vec{V}^d(\vec{\rho})\delta^3(\vec{r}-\vec{\rho}), \tag{74}$$

$$\langle a,\vec{r}\,|N|\,b,\vec{\rho}\rangle = gC^{abc}\vec{\nabla}\cdot[\vec{V}^c(\vec{r})\delta^3(\vec{r}-\vec{\rho})].$$

As before, $a, b, \ldots$ are the color indices, δ^{ab} is the usual Kronecker symbol, and $\vec{\nabla}$ acts on $\vec{r}$.

The Feynman diagrams can now be constructed by using the "effective Hamiltonian"

$$H_0 + H_1, \tag{75}$$

where H_0 and H_1 are given by (65) and (73), respectively. In the Coulomb gauge, the propagator of the gauge field between two space-time points x and x' consists of a longitudinal part $D^{ab}_{\rm long}(x,x')$ and a transverse part $D^{ab}_{\rm tr}(x,x')$. (i) The longitudinal propagator is instantaneous in time, given by

$$D^{ab}_{\rm long}(x,x') = \delta(t-t')\delta^{ab}G(\vec{r},\vec{r}\,'), \tag{76a}$$

where $G(\vec{r},\vec{r}\,')$ is defined by (57). As mentioned before, when $\kappa_\infty \to 0$, $G \to \infty$, and that leads to divergence whenever there are color-nonsinglet external lines. However, between color singlets all amplitudes remain finite when $\kappa_\infty \to 0$, and we may replace (76a) by

$$D^{ab}_{\rm long}(x,x') = \delta(t-t')\delta^{ab}G_1(\vec{r},\vec{r}\,'), \tag{76b}$$

where $G_1(\vec{r},\vec{r}\,')$ is defined by (58) and it is *independent* of κ_∞. Both G_1 and G are symmetric in $\vec{r}$ and $\vec{r}\,'$. For a spherical surface, G_1 is given by (44) or (60). For a general surface, according to (C17) of Appendix C, G_1 is determined by

$$-\nabla^2G_1(\vec{r},\vec{r}\,') = \delta^3(\vec{r}-\vec{r}\,') \quad \text{inside } \mathcal{S}$$

and, when $\vec{r}$ is on the surface, the inhomogeneous Neumann boundary condition

$$-\hat{n}\cdot\vec{\nabla}G_1(\vec{r},\vec{r}\,') = A^{-1} \text{ on } \mathcal{S}, \tag{76c}$$

where $\vec{r}\,'$ is inside $\mathcal{S}$, A is the area of $\mathcal{S}$, and, as before, $\hat{n}$ is the unit normal vector of $\mathcal{S}$. (G_1 is determined up to an additive constant, which is immaterial for color-singlet states.)

To derive the transverse part of the gauge field propagator $D^{ab}_{\rm tr}(x,x')$, we need the various radiation modes in a cavity of surface $\mathcal{S}$ (when $g=0$). The boundary condition of the radiation field can be derived most simply by integrating respectively the top equation of (29) over a small volume across $\mathcal{S}$ and the bottom equation over a small loop. This leads to the conclusion that the normal component of $\vec{D}$ and the tangential component of $\vec{H}$ should be both continuous across $\mathcal{S}$. Since both are zero outside $\mathcal{S}$, they must remain zero[16] on the inner side of $\mathcal{S}$. Now inside $\mathcal{S}$, $\kappa=1$; for the radiation field and when $g=0$, $\vec{D}^a = -\dot{\vec{V}}^a$ and $\vec{H}^a = \vec{\nabla}\times\vec{V}^a$. Therefore, we have the boundary condition

$$\hat{n}\cdot\vec{\dot{V}}^a=0$$

and (77)

$$\hat{n}\times(\vec{\nabla}\times\vec{V}^a)=0 \text{ on } S,$$

where, as before, $\hat{n}$ is the normal vector. (See Appendix C for further discussions.)

In the special case that S is a spherical surface of radius R, the usual TE and TB modes give the complete solution. These radiation modes can be expressed in terms of the scalar solution of

$$\nabla^2\phi+k^2\phi=0,$$

which may in turn be written as

$$\phi^{\lambda}_{k,l,m}(\vec{r})=\text{constant}\times j_l(kr)Y_{l,m}(\alpha,\beta), \tag{78}$$

where r, α, β are the spherical coordinates of $\vec{r}$, $Y_{l,m}$ is the usual spherical harmonics, j_l is the spherical Bessel function, $k>0$, and $\lambda=E$ or B depending on the TEor TB mode. For the TE mode

$$\vec{V}^E_{k,l,m}(\vec{r})=\vec{\nabla}\times(\vec{r}\phi^E_{k,l,m}) \tag{79}$$

and k is determined by

$$\frac{d}{dr}[rj_l(kr)]=0 \quad \text{at } r=R.$$

For the TB mode

$$\vec{V}^B_{k,l,m}(\vec{r})=\vec{\nabla}\times[\vec{\nabla}\times(\vec{r}\phi^B_{k,l,m})] \tag{80}$$

and k is determined by

$$j_l(kR)=0.$$

The constant in (78) is chosen so that the integral of $|\vec{V}^{\lambda}_{k,l,m}|^2$ over the volume $r\leq R$ is unity. At $r=R$, in either mode $\vec{V}^{\lambda}_{k,l,m}$ has only tangential components while $\vec{\nabla}\times\vec{V}^{i}_{k,l,m}$ has only a normal component, in accordance with (77).

The Feynman propagator $D^{ab}_{\text{tr}}(x,x')$ of the radiation field is given by

$$[D^{ab}_{\text{tr}}(x,x')]_{ij}=\delta^{ab}\sum_{\lambda,k,l,m}(2k)^{-1}[\vec{V}^{\lambda}_{k,l,m}(\vec{r})]_i \times[\vec{V}^{\lambda}_{k,l,m}(r')]^*_j e^{\mp ik(t-t')}, \tag{81}$$

where the minus sign in the exponent is for $t>t'$ and the plus sign for $t<t'$.

To derive the propagator of the quark field, we need the complete set of c-number solutions $(\chi_n)_\pm$ of the Dirac equation in a cavity:

$$\left(\vec{\alpha}\cdot\frac{1}{i}\vec{\nabla}+\beta m\right)(\chi_n)_\pm=\pm\epsilon_n(\chi_n)_\pm\ ,$$

where ϵ_n is >0 and $(\chi_n)_\pm$ satisfies the boundary condition (18). The Feynman propagator $S(x,x')$ of the quark field is given by, for $t>t'$,

$$S(x,x')=\sum_n(\chi_n(\vec{r}))_+(\chi^\dagger_n(\vec{r}')\beta)_+e^{-i\epsilon_n(t-t')}, \tag{82a}$$

and for $t<t'$,

$$S(x,x')=-\sum_n(\chi_n(\vec{r}))_-(\chi^\dagger_n(\vec{r}')\beta)_-e^{i\epsilon_n(t-t')}. \tag{82b}$$

The various g-dependent terms in H_0+H_1, defined by (75), give directly the great variety of vertices in this problem. These vertices, together with the propagators D^{ab}_{long}, D^{ab}_{tr}, and S given above, complete our discussion of Feynman rules.

We emphasize that the quark-vector and vector-vector interactions are local in character, while the color singletness of the hadron state is a global property. This is why in (76a) there is an explicit dependence on κ_∞ in the propagator $D^{ab}_{\text{long}}(x,x')$. When $\kappa_\infty\to 0$, our Feynman rules explicitly forbid the appearance of external color nonsinglets, such as free quarks or free vector gauge particles. The propagators D^{ab}_{long}, D^{ab}_{tr}, and S also depend on the linear size R of the hadron surface. Only at high energy and large four-momentum transfer can one neglect the effect of hadron surfaces; in that case, one recovers the "usual" Feynman rules used in the current literature on QCD.

V. REMARKS

A. Phase transition

In our picture, at low temperature, an infinite-volume QCD exists only in a *single* phase. A good analogy is to think of an infinite ferromagnet below the Curie temperature, which has a long-range order and also exists in a single phase. The hadrons are then analogous to some small-domain structures within the infinite ferromagnet. Another analogy that has been used frequently in the soliton (or bag) model is to regard the vacuum as a liquid and hadrons as bubbles. It seems reasonable to assume that there should exist a critical temperature at which the infinite-volume QCD system can undergo a phase transition in its dielectric constant κ (or magnetic susceptibility $\mu=1/\kappa$). The value of the critical temperature and the nature of the phase transition depend sensitively on the excitation curve of hadron spectroscopy. If one follows Chodos *et al.*[8] and approximates the excitéd hadron system as noninteracting bubbles of an ideal relativistic gas, then not only does there exist a critical temperature T_c, but T_c is also the maximum temperature that the system can attain.[8,17] However, as the size l of such bubbles grows, its dielectric constant κ_l should decrease, and therefore the effective coupling $g/\kappa_l^{1/2}$ would increase, which makes the quarks behave less and less like an ideal gas. Furthermore, as the bubbles increase in size, their mutual interactions should become more important. Hence, it remains an interesting open question whether the idea of a maximum temperature is really correct.

B. Long-range order

Let us leave our problem for the moment and consider a ferromagnet of volume L^3 at a nonzero temperature below the Curie temperature. The magnetization per unit volume $M(H, L)$ under an external magnetic field has the following familiar properties: At any finite volume, $M(H,L)$ is an analytic function of H and $\lim_{H\to 0} M(H,L)=0$. Thus,

$$\lim_{L\to\infty}\lim_{H\to 0} M(H,L)=0. \tag{83}$$

On the other hand, if we take the infinite-volume limit first and then the $H\to 0$ limit, a finite nonzero magnetization results:

$$\lim_{H\to 0}\lim_{L\to\infty} M(H,L)\neq 0; \tag{84}$$

this is why we usually say the infinite ferromagnet carries a long-range order, which is characterized by its magnetization and given by the double limit (84).

In our problem, we may take the ratio of the (color) dielectric constant between two volumes of sizes l^3 and L^3 with $l<L$. By using (A2), (A5), (A7), and (A16) in Appendix A, we find, after neglecting $O(\alpha^2)$,

$$\kappa_L/\kappa_l=\left[1+\frac{11}{2\pi}\alpha\ln(L/l)\right]^{-1}, \tag{85}$$

where $\alpha=(4\pi)^{-1}g_l^{\,2}$. The approximation of neglecting $O(\alpha^2)$ is, of course, not a good one, but some insight into the long-range order in QCD may be obtained even in such a crude approximation. Let us keep the small size l always fixed. At any finite L, we have, according to (85), $\lim_{\alpha\to 0}(\kappa_L/\kappa_l)=1$ and therefore

$$\lim_{L\to\infty}\lim_{\alpha\to 0}[1-(\kappa_L/\kappa_l)]=0. \tag{86}$$

On the other hand, $\lim_{L\to\infty}(\kappa_L/\kappa_l)=0$ at any $\alpha\neq 0$. Hence,

$$\lim_{\alpha\to 0}\lim_{L\to\infty}[1-(\kappa_L/\kappa_l)]=1 \tag{87}$$

in analogy with (83) and (84). Hence, we may regard the long-range order in QCD to be characterized by the perfect dia-electric property of its vacuum when the volume becomes infinite. (See Ref. 1 for further discussions.)

C. Scale of hadron radius

Let us consider a QCD system in the limit of quark mass $=0$ inside the hadron. At first sight, since there seems to be no mass scale in the problem, it may be difficult to see how the physical hadron radius can possibly emerge. To explain this apparent contradiction, let us introduce, as in Appendix A, an ultraviolet momentum cutoff Λ and denote g_0 as the unrenormalized coupling constant. Inside a hadron of radius R, let α be the relevant fine-structure constant of the color gauge field. Clearly, α must be a function of $R\Lambda$ and g_0:

$$\alpha=f(R\Lambda,g_0).$$

When $\Lambda\to\infty$, one must vary g_0 accordingly so that the limiting function

$$\alpha(R)\equiv\lim_{\Lambda\to\infty} f(R\Lambda,g_0) \tag{88}$$

exists. This limiting function must contain a length scale since R has a dimension but α does not. (In a realistic theory, because the quark masses inside the hadron are not really zero, there are additional mass scales.)

In a quasiclassical soliton (or bag) model, the actual value of R is determined by minimizing the hadron energy spectrum, of which the lowest level is the pion. Its mass in a quasiclassical calculation is

$$M_\pi\cong\tfrac{1}{3}4\pi pR^3+4\pi sR^2+\frac{2\rho_0}{R}-\frac{\alpha(R)}{R}\eta_0, \tag{89}$$

where p, s, ρ_0 are given by (21) and (22), and η_0 is a positive number. As discussed in Ref. 1, since the minimum of $M_\pi^{\,2}=0$, $\alpha(R)$ should be near a critical value α_c; when $\alpha(R)=\alpha_c$, $M_\pi^{\,2}=0$, and that in turn determines the pion radius R.

D. Experimental upper limit of κ_∞

As mentioned in Sec. II, according to (7) κ_∞ must be <1; but as yet, no one has been able to determine through pure theoretical deduction just how small κ_∞ actually is. Within our picture, the fact that a single free quark has never been observed sets a lower limit on the free quark mass m_q. As shown in Appendix D,

$$\kappa_\infty\leq(0.11)\alpha(m_p/m_q)^{4/3}, \tag{90}$$

where m_p is the proton mass and $\alpha=g^2/4\pi$ is the fine structure for hadrons. Even without any detailed search of the literature, we may set $m_q>5$ GeV, from which we determine an *experimental* upper limit of κ_∞:

$$\kappa_\infty<1.3\times10^{-2}\alpha. \tag{91}$$

ACKNOWLEDGMENTS

I wish to thank N. H. Christ and A. H. Mueller for many very helpful discussions. The kind hospitality of the Physics Department of BNL during the summer of 1978 is gratefully acknowledged. This research was supported in part by the U. S. Department of Energy.

APPENDIX A

The proof of properties (3) and (4) concerning the (color) dielectric constant of the vacuum, which are stated in Sec. II, will be given in this appendix. As in Sec. II, let us consider a pure QCD system consisting of only color gauge fields V_μ^a. For clarity, we introduce an *ultraviolet* momentum cutoff Λ, and denote g_0 as the unrenormalized coupling constant. The whole system is then enclosed in a finite volume L^3. Let κ_L and g_L be, respectively, the (color) dielectric constant of the vacuum and the renormalized coupling constant. Both κ_L and g_L are functions of g_0, L, and Λ. From dimensional considerations, they must depend only on g_0 and the product $L\Lambda$, i.e., $\kappa_L = F(L\Lambda, g_0)$ and

$$g_L = G(L\Lambda, g_0). \tag{A1}$$

Let us consider two different volumes l^3 and L^3, but with the same Λ and g_0. From (1), it follows that

$$\frac{\kappa_L}{\kappa_l} = \left(\frac{g_l}{g_L}\right)^2, \tag{A2}$$

where, as in (A1),

$$g_l = G(l\Lambda, g_0) \tag{A3}$$

or its inverse $g_0 = g_0(l\Lambda, g_l)$. Eliminating g_0 between (A1) and (A3), we may express g_L in terms of $L\Lambda$, $l\Lambda$, and g_l:

$$g_L = G(L\Lambda, g_0(l\Lambda, g_l)). \tag{A4}$$

Since the theory is a renormalizable one, the limit $\Lambda \to \infty$ of (A4) should exist; in this limit,

$$g_L = g\left(\frac{l}{L}, g_l\right) \equiv \lim_{\Lambda\to\infty} G(L\Lambda, g_0(l\Lambda, g_l)),$$

which may be written as

$$g_L = g(\lambda, g_l), \tag{A5}$$

where

$$\lambda = l/L. \tag{A6}$$

At $\lambda = 1$, (A5) becomes

$$g_l = g(1, g_l). \tag{A7}$$

Because $g_L = G(L\Lambda, g_0)$ is independent of l, one has $(\partial g_L/\partial l)_{L,\Lambda,g_0} = 0$; hence, $g(\lambda, g_l)$ satisfies the familiar renormalization-group equation[18] (but without the usual γ and δ functions)

$$[-(\partial/\partial \ln\lambda) + \beta(\partial/\partial g_l)]g(\lambda, g_l) = 0, \tag{A8}$$

where

$$\beta = -\partial g_l/\partial \ln l. \tag{A9}$$

So far, the finite volume is just a device for the infrared cutoff. Presumably, our physical result should not be sensitive to the precise form in which this infrared cutoff is introduced. In order to derive the dependence of β on g_l, we assume l^{-1} = the usual momentum value M chosen for renormalization.[19] The β function as a power series of g_l is then given by[19,20]

$$\beta(g_l) = -11(16\pi^2)^{-1}g_l^3 - 102(16\pi^2)^{-2}g_l^5 + O(g_l^7) \tag{A10}$$

for a pure SU_3 gauge field system. The solution of (A8) has the standard form

$$g(\lambda, g_l) = f(z), \tag{A11}$$

where

$$z = \ln\lambda + \int_{g_1}^{g_l} \frac{dg'}{\beta(g')} \tag{A12}$$

and g_1 is an arbitrary constant, which will be chosen so that

$$0 < g_1 \ll 1$$

and

$$\beta(g') < 0 \quad \text{for} \quad 0 < g' \leq g_1. \tag{A13}$$

Let $f^{-1}(z)$ be the inverse of $f(z)$, i.e., $f^{-1}(f(z)) = z$. From (A11), one has $z = f^{-1}(g(\lambda, g_l))$, which at $\lambda = 1$, because of (A7), becomes $z = f^{-1}(g_l)$. Now z is also given by (A12). By setting $\lambda = 1$, one derives

$$f^{-1}(g_l) = \int_{g_1}^{g_l} \frac{dg'}{\beta(g')}. \tag{A14}$$

For g_l sufficiently small, the $O(g_l^5)$ term in (A10) may be neglected. Hence,

$$f^{-1}(g_l) \cong \tfrac{8}{11}\pi^2(g_l^{-2} - g_1^{-2}); \tag{A15}$$

its inverse function is

$$f(z) \cong [(8\pi^2)^{-1}11z + g_1^{-2}]^{-1/2},$$

and therefore

$$g(\lambda, g_l) \cong \left[\frac{(8\pi^2/11)g_l^2}{g_l^2\ln\lambda + (8\pi^2/11)}\right]^{1/2}. \tag{A16}$$

Since $g(\lambda, g_l)$ is an even function of g_l, we need only consider positive values of g_l. From (A14), we see that (i) the physical value of g_l can vary from 0 to ∞ if $\beta(g') = 0$ only at $g' = 0$, otherwise (ii) g_l can only vary between 0 and $\bar{g}$ where $\bar{g}$ is the smallest positive-definitive root of $\beta(\bar{g}) = 0$. When $g_l = \bar{g}$, the integral in (A14) diverges. In either case, setting $g_l = x$, one has in the physical region $df^{-1}(x)/dx = 1/\beta(x) < 0$; hence, the inverse function $f(z)$ satisfies

$$\frac{df(z)}{dz} < 0. \tag{A17}$$

By using (A11) and (A12), one finds

$$\frac{\partial}{\partial\lambda} g(\lambda, g_l) < 0. \tag{A18}$$

Since $\lambda = l/L$, this means for $L > l$, $g_L > g_l$ and therefore, because of (A2), which is $\kappa_L < \kappa_l$, the inequality (3) of Sec. II.

According to (A14) and (A15), as g_l varies from 0 to g_1, $f^{-1}(g_l)$ decreases from ∞ to 0. When g_l increases to $> g_1$, $f^{-1}(g_l)$ becomes negative. In the above case (i), as $g_l \to \infty$ we have either $f^{-1}(\infty) = -A$ = finite or $f^{-1}(\infty) = -\infty$; in the former the inverse function $f(z) \to \infty$ as $z \to -A+$, which means that because of (A11) and (A12), at any fixed g_l, $g_L = g(\lambda, g_l)$ becomes singular at a finite $\lambda = l/L$, in violation of our assumption that κ_L (therefore, also g_L) is a smooth function of L when $L \to \infty$. Consequently, we are left only with the latter: $f^{-1}(\infty) = -\infty$ or $f(-\infty) = \infty$. Since according to (A12), $z \to -\infty$ means $\lambda = l/L \to 0$, we conclude that in case (i) as $L \to \infty$, $g_L \to \infty$, and therefore $\kappa_L \to 0$. Statement (4) of Sec. II is then established.

In case (ii), $f^{-1}(g_l) \to -\infty$ as $g_l \to \bar{g}-$. Thus, $f(z) \to \bar{g}$ as $z \to -\infty$, and that gives $\lim_{L\to\infty} g_L = \bar{g}$. Adopting convention (5) of Sec. II, we find

$$\kappa_\infty = (g_R/\bar{g})^2, \tag{A19}$$

where, as before, R is some hadron radius chosen to set the scale $\kappa_R = 1$, and $\bar{g}$ is the smallest positive-definite root of $\beta(\bar{g}) = 0$.

APPENDIX B

In this appendix, we shall prove a number of technical points mentioned in Sec. III C concerning the Coulomb gauge. As we shall see, all of these proofs are quite elementary.

1. Conditions (31) and (32)

Suppose $\vec{V}^a$ satisfies neither (31) nor (32). Under an infinitesimal gauge transformation, $\vec{V}^a$ becomes $\vec{V}^a(\theta)$ where

$$\vec{V}^a(\theta) = \vec{V}^a + C^{abc}\theta^b\vec{V}^c - g^{-1}\vec{\nabla}\theta^a \tag{B1}$$

and θ^a is an infinitesimal. We first show that it is always possible to choose θ^a such that

$$\nabla^2\theta^a - g\vec{\nabla}\cdot(C^{abc}\theta^b\vec{V}^c) = g\epsilon\vec{\nabla}\cdot\vec{V}^a \tag{B2}$$

and therefore

$$\vec{\nabla}\cdot\vec{V}^a(\theta) = (1-\epsilon)\vec{\nabla}\cdot\vec{V}^a, \tag{B3}$$

where $\epsilon = 0+$. This can be readily established by expanding

$$\theta^a = g\theta_0^a + g^2\theta_1^a + g^3\theta_2^a + \cdots \tag{B4}$$

and noting that the inhomogeneous Laplace equations

$$\nabla^2\theta_0{}^2 = \epsilon\vec{\nabla}\cdot\vec{V}^a \tag{B5}$$

and

$$\nabla^2\theta_i^a = C^{abc}\vec{\nabla}\cdot(\theta_{i-1}^b\vec{V}^c) \text{ for } i \geq 1 \tag{B6}$$

all have solutions. By using (B3) and varying ϵ continuously, we can transform $\vec{\nabla}\cdot\vec{V}^a$ from any value to 0. Hence, (31) is satisfied.

Next, we assume $\vec{V}^a$ to satisfy (31) but not (32), i.e., $\vec{\nabla}\cdot\vec{V}^a = 0$, but $\hat{n}\cdot\vec{V}^a \neq 0$ on $\mathcal{S}$. Again, we consider an infinitesimal gauge transformation $\vec{V}^a \to \vec{V}^a(\theta)$, but now we want to choose θ^a such that (31) remains satisfied,

$$\vec{\nabla}\cdot\vec{V}^a(\theta) = 0,$$

and, in addition (B7)

$$\hat{n}\cdot\vec{V}^a(\theta) = (1-\epsilon)\hat{n}\cdot\vec{V}^a \text{ on } \mathcal{S},$$

where $\epsilon = 0+$. By using (B4), we note that in order to achieve (B7) we must have

$$\nabla^2\theta_0^a = 0 \tag{B8}$$

and

$$\hat{n}\cdot\vec{\nabla}\theta_0^a = \epsilon\hat{n}\cdot\vec{V}^a \text{ on } \mathcal{S}. \tag{B9}$$

Furthermore, besides (B6), we also need (for $i \geq 1$)

$$\hat{n}\cdot\vec{\nabla}\theta_i^a = C^{abc}\theta_{i-1}^b(\hat{n}\cdot\vec{V}^c) \text{ on } \mathcal{S}. \tag{B10}$$

For simplicity, let us first assume $\mathcal{S}$ to be a spherical surface of radius R. The general solution of (B8) can be written in terms of the spherical coordinates (r, α, β)

$$\theta_0^a = \sum A_{lm}^a r^l Y_{lm}(\alpha, \beta),$$

where Y_{lm} is the usual spherical harmonics. Let

$$\hat{n}\cdot\vec{V}^a = \sum B_{lm}^a Y_{lm}(\alpha,\beta) \text{ at } r = R,$$

where

$$B_{0,0}^a \propto \int_{r=R} \hat{n}\cdot\vec{V}^a R^2 \sin\alpha\, d\alpha\, d\beta$$

$$= \int_{r\leq R} \vec{\nabla}\cdot\vec{V}^a d^3r = 0.$$

By choosing

$$A_{lm}^a = \epsilon B_{lm}^a/(lR^{l-1}) \text{ for } l \geq 1$$

and leaving $A_{0,0}^a$ arbitrary, the above θ_0^a satisfies both (B8) and (B9). Let $\bar{\theta}_1^a$ by any particular solution of

$$\nabla^2\bar{\theta}_1^a = C^{abc}\vec{\nabla}\cdot(\theta_0^b\vec{V}^c)$$

and

$\alpha_1^a \equiv \theta_1^a - \overline{\theta}_1^a$.

In order to have (B6) and (B10) for $i=1$, α_1^a must satisfy

$$\nabla^2 \alpha_1^a = 0$$

and

$$\hat{n} \cdot \vec{\nabla} \alpha_1^a = \hat{n} \cdot (C^{abc} \theta_0^b \vec{V}^c - \vec{\nabla} \overline{\theta}_1^a),$$

which have exactly the same form as (B8) and (B9). Consequently, α_1^a exists and so does θ_1^a. Likewise, we can construct $\theta_2^a, \theta_3^a, \ldots$, and therefore (B7) is satisfied. By varying ϵ continuously, we can transform $\hat{n} \cdot \vec{V}^a$ from any value to 0 on S. (When S is not a spherical surface, by choosing the appropriate curvilinear coordinates, we can apply identical arguments as those given above.) Conditions (31) and (32) are then established.

2. Equations (42)-(44)

Consider a simple problem in classical electrostatics: A unit point charge is placed at $\vec{r} = \vec{r}_0$ inside a simply connected surface S. The dielectric constant is $\kappa = 1$ inside S, but $\kappa = \kappa_\infty \to 0$ outside. The electrostatic potential $V(\vec{r})$ in this problem is the Green's function $G(\vec{r}, \vec{r}_0)$ defined by (38). Let $V(\vec{r}) = V_{in}(\vec{r})$ or $V_{out}(\vec{r})$ depending on whether $\vec{r}$ is inside or outside S. Hence,

$$-\nabla^2 V_{out}(\vec{r}) = 0$$

and (B11)

$$-\nabla^2 V_{in}(\vec{r}) = \delta^3(\vec{r} - \vec{r}_0).$$

The boundary conditions at the surface S are

$$V_{in}(\vec{r}) = V_{out}(\vec{r})$$

and (B12)

$$\hat{n} \cdot \vec{\nabla} V_{in}(\vec{r}) = \kappa_\infty \hat{n} \cdot \vec{\nabla} V_{out}(\vec{r}),$$

where, as before, $\hat{n}$ is the normal vector of S. We now expand V_α ($\alpha =$ in or out) in powers of κ_∞:

$$V_\alpha = \kappa_\infty^{-1} V_\alpha^{(0)} + V_\alpha^{(1)} + \kappa_\infty V_\alpha^{(2)} + \cdots . \tag{B13}$$

Equation (B11) becomes

$$-\nabla^2 V_{in}^{(0)} = 0, \quad -\nabla^2 V_{in}^{(1)} = \delta^3(\vec{r} - \vec{r}_0),$$

$$-\nabla^2 V_{in}^{(l)} = 0 \text{ for } l \geq 2$$

and (B14)

$$-\nabla^2 V_{out}^{(m)} = 0 \text{ for all } m.$$

The boundary condition (B12) requires that at the surface S

$$V_{in}^{(l)} = V_{out}^{(l)} \text{ for all } l,$$

$$\hat{n} \cdot \vec{\nabla} V_{in}^{(0)} = 0, \tag{B15}$$

and

$$\hat{n} \cdot \vec{\nabla} V_{in}^{(m)} = \hat{n} \cdot \vec{\nabla} V_{out}^{(m-1)} \text{ for } m \geq 1.$$

Let $\vec{E}_\alpha^{(n)}$ be the electrostatic field associated with $V_\alpha^{(n)}$:

$$\vec{E}_\alpha^{(n)} \equiv -\vec{\nabla} V_\alpha^{(n)} . \tag{B16}$$

Now $\vec{E}_{in}^{(0)}$ is, by definition, irrotational; from (B14) it is also divergence-free, and from (B15) its normal component is zero on S. Hence, $\vec{E}_{in}^{(0)} = 0$ and therefore

$$V_{in}^{(0)}(\vec{r}) = \text{constant}. \tag{B17}$$

So far, $\vec{r}_0$ is assumed to be inside S. When $\vec{r}$ is also inside S, the Green's function $G(\vec{r}, \vec{r}_0) = V_{in}(\vec{r})$ and $G_0(\vec{r}, \vec{r}_0) = V_{in}^{(0)}(\vec{r})$ where G_0 is defined by (41). Thus, (B17) shows that when $\vec{r}$ and $\vec{r}_0$ are both inside S, $G_0(\vec{r}, \vec{r}_0)$ is independent of $\vec{r}$; by symmetry $G_0(\vec{r}, \vec{r}_0)$ is also independent of $\vec{r}_0$, and therefore (42) follows. By following exactly the same argument, but for $\vec{r}_0$ outside S, we find $G_0(\vec{r}, \vec{r}_0)$ to be independent of $\vec{r}$, so long as $\vec{r}$ is inside S. These properties may be summarized as follows:

$$G_0(\vec{r}, \vec{r}\,') \text{ is independent of } \vec{r} \text{ (or } \vec{r}\,'\text{) whenever } \vec{r} \text{ (or } \vec{r}\,'\text{) is inside } S. \tag{B18}$$

To determine the constant in (B17), we have to examine $V_{out}^{(0)}$. Since according to (B14) and (B15), $-\nabla^2 V_{out}^{(0)} = 0$ outside S and $V_{out}^{(0)} = V_{in}^{(0)} =$ constant on S, one sees that $V_{out}^{(0)}(\vec{r})$ is the same as the electrostatic potential in free space outside a *perfect conductor* of surface S. Let $\sigma(\vec{r})$ be the surface charge density of that problem. We have $-\vec{\nabla} V_{out}^{(0)} = \hat{n}\sigma$ on S. From (B14) and (B15) we see that

$$-\nabla^2 V_{in}^{(1)} = \delta^3(\vec{r} - \vec{r}_0)$$

and (B19)

$$-\hat{n} \cdot \vec{\nabla} V_{in}^{(1)} = \sigma \text{ on } S.$$

Hence,

$$\int \sigma\, d^2S = -\int_\Omega \nabla^2 V_{\rm in}^{(1)} d^3r = 1$$

where d^2S is the surface element of $\mathcal{S}$ and Ω is the volume inside $\mathcal{S}$. From (41), it follows that $G_1(\vec{r},\vec{r}_0)=V_{\rm in}^{(1)}(\vec{r})$ when $\vec{r}$ is inside $\mathcal{S}$. Consequently, inside $\mathcal{S}$, $G_1(\vec{r},\vec{r}_0)$ is the electrostatic potential generated by a positive unit charge at $\vec{r}=\vec{r}_0$ with the boundary condition (B19) on $\mathcal{S}$.

If $\psi=\vec{V}^a=0$ outside $\mathcal{S}$, then $j_0^a=J_0^a=I^a=0$ also outside $\mathcal{S}$. Hence, according to (37), (39), (41), and (B18), as $\kappa_\infty\to 0$,

$$V_0^a(\vec{r})=\kappa_\infty^{-1} g G_0(\vec{r},\vec{R})Q^a + g\int G_1(\vec{r},\vec{r}\,')I^a(\vec{r}\,')d^3r' + O(\kappa_\infty), \quad \text{(B20)}$$

where $\vec{R}$ is any position vector on $\mathcal{S}$. According to (B18), $G_0(\vec{r},\vec{r}\,')=G_0(\vec{r},\vec{R})$ whenever $\vec{r}\,'$ is inside $\mathcal{S}$. Now, for a color singlet $Q^a=0$, $V_0^a(\vec{r})$ is finite as $\kappa_\infty\to 0$. Equation (43) then follows.

To derive (44), we assume $\mathcal{S}$ to be a spherical surface of radius R. As before, a unit charge is placed at $\vec{r}=\vec{r}_0$ inside $\mathcal{S}$. It is straightforward to show that for $r<R$

$$G(\vec{r},\vec{r}_0)=V_{\rm in}(\vec{r})=(4\pi)^{-1}\left[\frac{1}{|\vec{r}-\vec{r}_0|}+\sum_{l=0}^{\infty}\frac{(l+1)(1-\kappa_\infty)}{l+(l+1)\kappa_\infty}\frac{(rr_0)^l}{R^{2l+1}}P_l(\cos\theta)\right] \quad \text{(B21)}$$

and for $r>R$

$$G(\vec{r},\vec{r}_0)=V_{\rm out}(\vec{r})=(4\pi r)^{-1}\sum_{l=0}^{\infty}\frac{2l+1}{l+(l+1)\kappa_\infty}\left(\frac{r_0}{r}\right)^l P_l(\cos\theta),$$

where r and r_0 are, respectively, the magnitudes of $\vec{r}$ and $\vec{r}_0$, θ is the angle between them, and the P_l's are Legendre polynomials. Taking the limit $\kappa_\infty\to 0$ and by using (41), we obtain (44).

As $\kappa_\infty\to 0$, an alternative form of (44) can be given in terms of image charges. We observe that for $r<R$, (B21) becomes

$$G(\vec{r},\vec{r}_0)=V_{\rm in}(\vec{r})=(4\pi)^{-1}\left[\frac{1-2\kappa_\infty}{\kappa_\infty R}+\frac{1}{|\vec{r}-\vec{r}_0|}+\frac{Rr_0}{|r_0^2\vec{r}-R^2\vec{r}_0|}+\frac{r_0}{R}\int_{(R/r_0)^2}^{\infty}dx\left(\frac{1}{|\vec{r}-x\vec{r}_0|}-\frac{1}{xr_0}\right)\right]+O(\kappa_\infty) \quad \text{(B22)}$$

and for $r>R$

$$G(\vec{r},\vec{r}_0)=V_{\rm out}(\vec{r})=(4\pi)^{-1}\left[\frac{1-2\kappa_\infty}{\kappa_\infty r}+\frac{2}{|\vec{r}-\vec{r}_0|}+\int_0^1\frac{dx}{x}\left(\frac{1}{|\vec{r}-x\vec{r}_0|}-\frac{1}{r}\right)\right]+O(\kappa_\infty).$$

Thus, as $\kappa_\infty\to 0$, $V_{\rm in}(\vec{r})$, apart from an additive constant, is the electrostatic potential due to a unit charge at $\vec{r}=\vec{r}_0$ inside the sphere, an image point charge of magnitude (R/r_0) at $\vec{r}=R^2\vec{r}_0/r_0^2$ outside the sphere, and a continuous line distribution, also outside the sphere, along the $\vec{r}_0$ direction from $\vec{r}=R^2\vec{r}_0/r_0^2$ to ∞; $V_{\rm out}(\vec{r})$ is that due to two point charges, one of magnitude $1-2\kappa_\infty$ at $\vec{r}=0$ and the other of magnitude $2\kappa_\infty$ at $\vec{r}=\vec{r}_0$, plus a continuous line distribution of image charges along $\vec{r}_0$ from $\vec{r}=0$ to $\vec{r}_0$, all inside the sphere. The integrals in (B22) can be readily integrated. We find for $r<R$

$$G(\vec{r},\vec{r}_0)=V_{\rm in}(\vec{r})=(4\pi)^{-1}\left\{\frac{1-2\kappa_\infty}{\kappa_\infty R}+\frac{1}{|\vec{r}-\vec{r}_0|}+\frac{Rr_0}{|r_0^2\vec{r}-R^2\vec{r}_0|} - R^{-1}\ln\left[\frac{1}{2}-\frac{rr_0}{2R^2}\cos\theta+\frac{1}{2R^2}(R^4-2R^2rr_0\cos\theta+r^2r_0^2)^{1/2}\right]\right\}+O(\kappa_\infty) \quad \text{(B23)}$$

and for $r>R$

$$G(\vec{r},\vec{r}_0)=V_{\rm out}(\vec{r})=(4\pi)^{-1}\left\{\frac{1-2\kappa_\infty}{\kappa_\infty r}+\frac{2}{|\vec{r}-\vec{r}_0|}-r^{-1}\ln\left[\frac{1}{2}-\frac{r_0}{2r}\cos\theta+\frac{1}{2r}(r^2-2rr_0\cos\theta+r_0^2)^{1/2}\right]\right\}+O(\kappa_\infty).$$

APPENDIX C

Several properties of the modified Lagrangian density $\mathcal{L}'$ and the related Green's function $G(\vec{r},\vec{r}\,')$ are stated in Sec. III D and Sec. IV; these will be established in this appendix.

1. *Equations (59) and (60).* The proof is essentially identical to that given in Appendix B 2. Let

$V(\vec{r})$ be the solution of

$$[-\vec{\nabla}\cdot(\kappa\vec{\nabla})+\kappa(f'\sigma)^2]V(\vec{r})=\delta^3(\vec{r}-\vec{r}_0) \tag{C1}$$

that equals the Green's function $G(\vec{r},\vec{r}_0)$, defined by (57),

$$G(\vec{r},\vec{r}_0)=V(\vec{r}). \tag{C2}$$

Denote $V(\vec{r})=V_{in}(\vec{r})$ or $V_{out}(\vec{r})$ depending on whether $\vec{r}$ is inside or outside the surface S. For $\vec{r}_0$ inside S, because of (23), (24), and (52), (C1) becomes

$$(\nabla^2-\mu^2)V_{out}(\vec{r})=0$$

and (C3)

$$-\nabla^2 V_{in}(\vec{r})=\delta^3(\vec{r}-\vec{r}_0)$$

with the boundary conditions at the surface S:

$$V_{in}(\vec{r})=V_{out}(\vec{r})$$

and (C4)

$$\hat{n}\cdot\vec{\nabla}V_{in}(\vec{r})=\kappa_\infty\hat{n}\cdot\vec{\nabla}V_{out}(\vec{r}),$$

where, as in (B12), $\hat{n}$ is the normal vector of S. Just as in (B13), we may expand V_α (α = in or out) in powers of κ_∞:

$$V_\alpha=\kappa_\infty^{-1}V_\alpha^{(0)}+V_\alpha^{(1)}+\kappa_\infty V_\alpha^{(2)}+\cdots, \tag{C5}$$

where because of (C3)

$$-\nabla^2 V_{in}^{(0)}=0,\quad -\nabla^2 V_{in}^{(1)}=\delta^3(\vec{r}-\vec{r}_0),$$

$$-\nabla^2 V_{in}^{(l)}=0,\ \text{for } l\geq 2 \tag{C6}$$

and

$$(\nabla^2-\mu^2)V_{out}^{(m)}=0,\ \text{for all } m.$$

The boundary condition (C4) becomes

$$V_{in}^{(l)}=V_{out}^{(l)},\quad \text{for all } l$$

$$\hat{n}\cdot\vec{\nabla}V_{in}^{(0)}=0, \tag{C7}$$

$$\hat{n}\cdot\vec{\nabla}V_{in}^{(m)}=\hat{n}\cdot\vec{\nabla}V_{out}^{(m-1)},\ \text{for } m\geq 1$$

at the surface S.

Since $V_{in}^{(0)}$ satisfies exactly the same equations here as in Appendix B 2, by following identical arguments as those given after (B15), one establishes, as in (B17),

$$V_{in}^{(0)}(\vec{r})=\text{constant}, \tag{C8}$$

and, as in (B18),

$G_0(\vec{r},\vec{r}')$ is independent of $\vec{r}$ (or $\vec{r}'$) whenever

$\vec{r}$ (or $\vec{r}'$) is inside S. (C9)

Hence, (59) follows.

If $\psi=\vec{V}^a=0$ outside S, then $j_0^a=J_0^a=I^a=0$ also outside S. Hence, (56) reduces to (B20), i.e.,

$$V_0^a(\vec{r})=\kappa_\infty^{-1}gG_0(\vec{r},\vec{R})Q^a$$

$$+g\int G_1(\vec{r},\vec{r}')I^a(r')d^3r'+O(\kappa_\infty), \tag{C10}$$

where, as before, $\vec{R}$ is any position vector on S. According to (C9), $G_0(\vec{r},\vec{r}')=G_0(\vec{r},\vec{R})$ whenever $\vec{r}'$ is inside S. For a color singlet, $Q^a=0$; hence (C10) becomes

$$V_0^a(\vec{r})=g\int G_1(\vec{r},\vec{r}')I^a(\vec{r}')d^3r'+O(\kappa_\infty), \tag{C11}$$

which is finite when $\kappa_\infty\to 0$.

To derive (60), let us assume S to be a spherical surface of radius R. By using (C6)–(C8), one sees that $V_{out}^{(0)}$ is given by

$$V_{out}^{(0)}=qe^{-\mu r}/r, \tag{C12}$$

where q is a constant. The function $V_{in}^{(1)}$ satisfies

$$-\nabla^2 V_{in}^{(1)}=\delta^3(\vec{r}-\vec{r}_0)$$

and (C13)

$$-\hat{n}\cdot\vec{\nabla}V_{in}^{(1)}\equiv\sigma=\text{constant on S}.$$

Because of the Gauss theorem, $\sigma=(4\pi R^2)^{-1}$ when S is a spherical surface of radius R. Since $\hat{n}\cdot\vec{\nabla}V_{in}^{(1)}=\hat{n}\cdot\vec{\nabla}V_{out}^{(0)}$ on S, we determine the constant in (C12): $q=(4\pi)^{-1}(1+\mu R)^{-1}\exp(\mu R)$. For $r<R$,

$$G_0(\vec{r},\vec{r}_0)=V_{in}^{(0)}(\vec{r})=(4\pi R)^{-1}(1+\mu R)^{-1} \tag{C14}$$

and for $r>R$

$$G_0(\vec{r},\vec{r}_0)=V_{out}^{(0)}(\vec{r})=(4\pi r)^{-1}(1+\mu R)^{-1}e^{-\mu(r-R)}. \tag{C15}$$

By setting $\mu=0$, we recover the previous form of $G_0(\vec{r},\vec{r}_0)$ discussed in Appendix B.

In the present case of a spherical surface, (C13) is identical to (B19); the solution $V_{in}^{(1)}$ must also be the same, apart from an additive constant which is not determined by (C13). Thus, (60) is established.

For an arbitrary surface and assuming $\mu\gg$ hadron mass, one has

$$\hat{n}\cdot\vec{\nabla}V_{out}^{(0)}\cong-\mu V_{out}^{(0)}\quad\text{on S}. \tag{C16}$$

In this approximation, since $\hat{n}\cdot\vec{\nabla}V_{out}^{(0)}=\hat{n}\cdot\vec{\nabla}V_{in}^{(1)}$ and $V_{out}^{(0)}=V_{in}^{(0)}$ = constant on S, the normal electric field defined by (C13) is a constant for any surface. Hence for $\vec{r}$ and $\vec{r}_0$ both inside S, $G_1(\vec{r},\vec{r}_0)=V_{in}^{(1)}(\vec{r})$ is the electrostatic potential generated by a positive unit charge at $\vec{r}=\vec{r}_0$, with the normal component of its electric field being a constant on the surface. (C17)

We note that for nonspherical surfaces, even inside $\mathcal{S}$ the Green's function $G_1(\vec{r}, \vec{r}_0)$ determined by (C17) can be quite different from that derived in Appendix B; however, their difference must be of the form $f(\vec{r}) + f(\vec{r}_0)$ due to the reciprocity relation. [Compare (C17) with the derivation following (B19) in Appendix B.] As mentioned earlier, outside $\mathcal{S}$ the $G_1(\vec{r}, \vec{r}_0)$ function of Appendix B has a long tail, which leads to a Van der Waals-type force between hadrons, in violation of experimental results.

2. Equation (62). Inside $\mathcal{S}$, since $\sigma = 0$ and $\kappa = 1$, the Lagrangian density is locally gauge invariant. In the Coulomb gauge, according to (31) and (32), $\vec{\nabla} \cdot \vec{V}^a = 0$ inside $\mathcal{S}$ and $\hat{n} \cdot \vec{V}^a = 0$ on $\mathcal{S}$.

To derive (62), we first show that it is always possible to choose a solution ϕ^a of the last equation of (61),

$$\nabla^2 \phi^a = -g \vec{\nabla} \cdot (C^{abc} \vec{V}^b V_0^c), \tag{C18}$$

such that

$$\hat{n} \cdot \vec{\nabla} \phi^a = 0 \quad \text{on } \mathcal{S}. \tag{C19}$$

To see this, let us assume that $\phi^a = \phi_0^a$ is a solution of (C18) inside $\mathcal{S}$, i.e.,

$$\nabla^2 \phi_0^a = -g \vec{\nabla} \cdot (C^{abc} \vec{V}^b V_0^c) \tag{C20}$$

but $\hat{n} \cdot \vec{\nabla} \phi_0^a \neq 0$ on $\mathcal{S}$. For simplicity, let $\mathcal{S}$ be a spherical surface of radius R and let (r, α, β) be the spherical coordinates of $\vec{r}$. Expand

$$\hat{n} \cdot \vec{\nabla} \phi_0^a = \sum_{l,m} A_{lm}^a Y_{lm}(\alpha, \beta) \quad \text{at } r = R, \tag{C21}$$

where the Y_{lm}'s are the spherical harmonics. Because of (C20) and (32),

$$A_{00}^a \propto \int_\Omega \nabla^2 \phi_0^a d^3 r = 0, \tag{C22}$$

where Ω denotes the volume inside $\mathcal{S}$. Now choose ϕ_1^a to be a solution of $\nabla^2 \phi_1^a = 0$ inside $\mathcal{S}$. Hence

$$\phi_1^a = \sum_{l,m} B_{lm}^a r^l Y_{lm}(\alpha, \beta). \tag{C23}$$

We shall set

$$B_{lm}^a = -A_{lm}^a / l R^{l-1} \text{ for all } l \neq 0; \tag{C24}$$

therefore, for arbitrary B_{00}^a, $\hat{n} \cdot \vec{\nabla}(\phi_0^a + \phi_1^a) = 0$ on $\mathcal{S}$. Thus, $\phi^a \equiv \phi_0^a + \phi_1^a$ satisfies (C18) and (C19). Equation (62) now follows.

3. Reduced Hamiltonian (65). In this section, we assume

$$\psi = \vec{V}^a = 0 \quad \text{outside } \mathcal{S}, \tag{C25}$$

in accordance with (64). By using (10) and (51), we find the total Lagrangian $L \equiv \int \mathcal{L}' d^3 r$ can be decomposed into three parts:

$$L = \int_\Omega \mathcal{L}_{\text{in}} d^3 r - sA + \int_{\text{out}} \mathcal{L}_{\text{out}} d^3 r, \tag{C26}$$

in which we omit the counterterms in (10), A is the area of $\mathcal{S}$, s is the surface tension defined in (20) [due to the integration of $\frac{1}{2}(\nabla\sigma)^2 + U(\sigma)$ over the surface], Ω refers to the volume inside $\mathcal{S}$ and "out" the volume outside,

$$\mathcal{L}_{\text{in}} = -\tfrac{1}{4} V_{\mu\nu}^a V_{\mu\nu}^a - \psi^\dagger \gamma_4 (\gamma_\mu D_\mu + m) \psi - p, \tag{C27}$$

where p is given by (12), and

$$\mathcal{L}_{\text{out}} = \tfrac{1}{2} \kappa_\infty [(\vec{\nabla} V_0^a)^2 + \mu^2 (V_0^a)^2]. \tag{C28}$$

In the Coulomb gauge, the conjugate momentum of $\vec{V}^a$ is

$$\vec{\Pi}^a = -\vec{E}_{\text{tr}}^a, \tag{C29}$$

where $\vec{E}_{\text{tr}}^a$ is given by (61) and it satisfies (62). The Hamiltonian is given by

$$H = \int_\Omega (\vec{\Pi}^a \cdot \dot{\vec{V}}^a + i\psi^\dagger \dot{\psi}) d^3 r - L. \tag{C30}$$

Through partial integrations, we find

$$\int_\Omega \vec{E}_{\text{long}}^a \cdot \vec{E}_{\text{long}}^a d^3 r = \int V_0^a \hat{n} \cdot \vec{\nabla} V_0^a d^2 S + g \int_\Omega V_0^a (j_0^a + C^{abc} \vec{V}^b \cdot \vec{\Pi}^c) d^3 r \tag{C31}$$

and

$$\int_\Omega (\vec{\Pi}^a \cdot \dot{\vec{V}}^a + i\psi^\dagger \dot{\psi} - \mathcal{L}_{\text{in}}) d^3 r = \int_\Omega \mathcal{H} \, d^3 r - \tfrac{1}{2} \int (V_0^a \hat{n} \cdot \vec{\nabla} V_0^a)_{\text{in}} d^2 S, \tag{C32}$$

where $\mathcal{H}$ is given by (66), $d^2 S$ is the surface element of $\mathcal{S}$, and $(\,)_{\text{in}}$ refers to the value of the inner side of $\mathcal{S}$. Now, outside $\mathcal{S}$, because of (C25), Eq. (55) becomes $(\nabla^2 - \mu^2) V_0^a = 0$. Hence,

$$\int_{\text{out}} \mathcal{L}_{\text{out}} d^3 r = -\tfrac{1}{2} \kappa_\infty \int (V_0^a \hat{n} \cdot \vec{\nabla} V_0^a)_{\text{out}} d^2 S. \tag{C33}$$

Furthermore, at the surface $(V_0^a)_{\rm out}=(V_0^a)_{\rm in}$ and $\kappa_\infty(\hat n\cdot\vec\nabla V_0^a)_{\rm out}=(\hat n\cdot\vec\nabla V_0^a)_{\rm in}$. Hence, H is the same as H_0 given by (65).

4. "Radiation field" inside the hadron. The Lagrangian density $\mathcal{L}'$ given by (51) is not locally gauge invariant. As mentioned before, inside the hadron, $\sigma=0$ and therefore $\mathcal{L}'=\mathcal{L}$ given by (10), which is locally gauge invariant. Outside, $\sigma=\sigma_{\rm vac}\neq 0$. Hence, local gauge invariance is broken. So long as we restrict ourselves to the inside region, we can perform local gauge transformations. However, in order to make a connection with the outside solution, it is sometimes necessary to transform the inside solution to a specific gauge; only then can one satisfy the continuity condition at the surface.

As an example, let us consider the free radiation fields $\vec V^E_{k,l,m}$ and $\vec V^B_{k,l,m}$ of the TE and TB modes discussed in Sec. IV B. For these free field solutions, we may set $g=0$. Equation (54) becomes

$$\frac{\partial}{\partial x_\alpha}V_{\alpha\beta}=0 \quad \text{inside } S$$

and (C34)

$$\frac{\partial}{\partial x_\alpha}V_{\alpha\beta}-\mu^2V_\beta=0 \quad \text{outside } S,$$

where for simplicity the color index is omitted and since $g=0$, $V_{\alpha\beta}=(\partial V_\beta/\partial x_\alpha)-(\partial V_\alpha/\partial x_\beta)$. Assume S to be a spherical surface of radius R. For $r<R$, the solution for the TE, or TB, mode may be written as ($\lambda=E$ or B)

$$\vec V^\lambda(x)=\vec V^\lambda_{k,l,m}(\vec r)e^{\pm ikt}$$

and (C35)

$$V_0^\lambda(x)=0,$$

where $\vec V^\lambda_{k,l,m}(\vec r)$ is given by (79) or (80). In spherical coordinates $\vec r=(r,\theta,\phi)$, since $\vec\nabla\cdot\vec V^\lambda=0$ we have

$$\frac{\partial V_r^\lambda}{\partial r}+\frac{1}{r}\frac{\partial V_\theta^\lambda}{\partial\theta}+\frac{1}{r\sin\theta}\frac{\partial V_\phi^\lambda}{\partial\phi}=0 \qquad \text{(C36)}$$

and at $r=R-$, because of (77),

$$V_r^\lambda=0. \qquad \text{(C37)}$$

In the outside region, the lower equation in (C34) gives

$$\frac{\partial V_\beta}{\partial x_\beta}=0; \qquad \text{(C38)}$$

for $\mu\gg R^{-1}$, its solution near the surface ($r=R+$) must be of the form

$$V_\alpha(r,\theta,\phi,t)=(R/r)V_\alpha(R,\theta,\phi,t)e^{-M(r-R)}, \qquad \text{(C39)}$$

where

$$M=\mu\{1+O[(\mu R)^{-2}]\}.$$

In order to connect the outside solution with the inside solution, we set the outside solution V_α at $r=R+$,

$$V_\theta=V_\theta^\lambda,\quad V_\phi=V_\phi^\lambda,$$

$$V_0=V_0^\lambda=0,$$

and (C40)

$$V_r=-\frac{1}{\mu}\left(\frac{\partial V_r^\lambda}{\partial r}\right)_R,$$

where V_α^λ is the inside solution given by (C35), and $(\partial V_r^\lambda/\partial r)_R$ refers to its value at $r=R$. The last term in (C40) is needed so that at $r=R+$, neglecting $O(1/\mu R)$, $\partial V_\alpha/\partial x_\alpha=0$. To the zeroth order in $(\mu R)^{-1}$, (C40) and (C35) give the same field at $r=R$. Hence, to that order the continuity condition at the surface is satisfied.

It is of interest to examine the first-order correction. To the first order in $(\mu R)^{-1}$, we must modify the inside solution (C35) by a local gauge transformation (which is allowed since the inside Lagrangian density is locally gauge invariant):

$$\vec V=\vec V^\lambda+\mu^{-1}\vec\nabla\chi$$

and

$$V_0=-\mu^{-1}\dot\chi=\mp\mu^{-1}(ik)\chi, \qquad \text{(C41)}$$

where χ is analytic inside the sphere ($r\leq R$) and it should satisfy the following boundary conditions at the surface:

$$\chi(R,\theta,\phi,t)=0$$

and

$$\left(\frac{\partial\chi}{\partial r}\right)_R=-\left(\frac{\partial V_r^\lambda}{\partial r}\right)_R. \qquad \text{(C42)}$$

Consequently, (C40) and (C41) satisfy the continuity condition at $r=R$. [That such an χ exists can be easily seen by assuming, e.g., $\chi\propto r^l(r^2-R^2)\times Y_{lm}(\theta,\phi)$ for the appropriate TE or TB solution.]

We note further that as $\kappa_\infty\to 0$ the integral $\kappa_\infty(f'\sigma)^2\vec V^a\cdot\vec V^a$ over the outside region becomes zero. Hence, the constraint $\vec V^a=0$ outside S is derived by using the Lagrangian (51) provided that we take the limit $\kappa_\infty\to 0$ first and then $f'\sigma_{\rm vac}\equiv\mu\to\infty$.

APPENDIX D

In this appendix, we give the derivation of (90), which leads to the experimental upper limit of κ_∞ discussed in Sec. V D. From (69), the mass of a free quark is given approximately by

$$m_q\cong\frac{2g^2G_0}{3\kappa_\infty}+\frac{4\pi}{3}pR^3, \qquad \text{(D1)}$$

where p is defined in (12). Here, for simplicity, we neglect the surface tension s. According to (C14)

$$G_0=(4\pi R)^{-1}(1+\mu R)^{-1}. \quad \text{(D2)}$$

We discuss the two limiting cases: $\mu \ll (4\pi p/m_q)^{1/3}$ and $\mu \gg (4\pi p/m_q)^{1/3}$.

1. $\mu \ll (4\pi p/m_q)^{1/3}$. In this case, $G_0 \cong (4\pi R)^{-1}$. By setting $dm_q/dR=0$, we find

$$m_q=\frac{4}{3}\left(\frac{2\alpha}{3\kappa_\infty}\right)^{3/4}(4\pi p)^{1/4}, \quad \text{(D3)}$$

where $\alpha=(4\pi)^{-1}g^2$. Likewise, the proton mass is

$$m_p=\tfrac{4}{3}(3\rho_0)^{3/4}(4\pi p)^{1/4}, \quad \text{(D4)}$$

where ρ_0 is given by (22). Hence,

$$\kappa_\infty=\frac{2\alpha}{9\rho_0}\left(\frac{m_p}{m_q}\right)^{4/3}. \quad \text{(D5)}$$

Taking $\rho_0=2.0428$, we obtain

$$\kappa_\infty=(0.109)\alpha(m_p/m_q)^{4/3}. \quad \text{(D6)}$$

2. $\mu \gg (4\pi p/m_q)^{1/3}$. In this case, $G_0 \cong (4\pi\mu R^2)^{-1}$ and that gives

$$m_q=\frac{5}{6}\left(\frac{4\alpha}{3\kappa_\infty\mu}\right)^{3/5}(4\pi p)^{2/5}. \quad \text{(D7)}$$

By setting $\rho_0=2.0428$, we find, instead of (D6),

$$\kappa_\infty=(0.012)\alpha(m_p/m_q)^{5/3}(m_p/\mu). \quad \text{(D8)}$$

Since (D8) always gives a smaller value than (D6), we establish (90).

---

[1]R. Friedberg and T. D. Lee, Phys. Rev. D 18, 2623 (1978).

[2]For various references on QCD, see, e.g., the review article by W. Marciano and H. Pagels, Phys. Rep. 36C, 139 (1978).

[3]Ideas very similar to Ref. 1 have been discussed in an earlier paper by J. Kogut and L. Susskind, Phys. Rev. D 9, 3501 (1974). However, their Feynman rules are quite different from those discussed in this paper. See also G. 't Hooft, in *Proceedings of the Marseille Conference on Recent Progress in Lagrangian Field Theory and Applications*, edited by C. Korthals-Altes *et al.* (CNRS, Marseille, 1974), and the review article by P. Hasenfratz and J. Kuti, Phys. Rep. 40C, 75 (1978).

[4]Cf. H. Pagels and E. Tomboulis, Princeton University report (unpublished).

[5]H. D. Politzer, Phys. Rev. Lett. 30, 1346 (1973); D. Gross and F. Wilczek, *ibid.* 30, 1343 (1973); G. 't Hooft, Nucl. Phys. B61, 455 (1973); B62, 444 (1973), first presented in some brief remarks at the Marseille meeting, 1972 (unpublished).

[6]Cf., however, some recent attempts by C. G. Callan, R. F. Dashen, and D. J. Gross, Institute for Advanced Study reports (unpublished).

[7]R. Friedberg and T. D. Lee, Phys. Rev. D 16, 1096 (1977).

[8]A. Chodos, R. L. Jaffe, K. Johnson, C. B. Thorn, and V. F. Weisskopf, Phys. Rev. D 9, 3471 (1974); A. Chodos, R. L. Jaffe, K. Johnson, and C. B. Thorn, *ibid.* 10, 2599 (1974); K. Johnson, Acta Phys. Polon. B 6, 865 (1975); T. DeGrand, R. L. Jaffe, K. Johnson, and J. Kiskis, Phys. Rev. D 12, 2060 (1975).

[9]M. Creutz, Phys. Rev. D 10, 1749 (1974); M. Creutz and K. S. Soh, *ibid.* 12, 443 (1975). See also P. Hasenfratz and J. Kuti, Ref. 3, for discussions on a modified MIT bag with a surface tension.

[10]W. A. Bardeen, M. S. Chanowitz, S. D. Drell, M. Weinstein, and T.-M. Yan, Phys. Rev. D 11, 1094 (1975).

[11]If we do not assume (23), then, in general, σ is only $\cong 0$ inside the hadron (as in any of the $n \neq 0$ solutions discussed in Sec. II B, which includes the SLAC bag). In such a case, we must assume $f'/f \to 0$, so that $f'\sigma \to 0$ inside the hadron, even though $f\sigma$ may be of the order of the hadron mass; therefore, there is always local gauge invariance inside. Outside, both $f'\sigma_{\text{vac}}$ and $f\sigma_{\text{vac}}$ are assumed to $\to \infty$.

[12]In order to analyze the surface motion, it is useful to start from the covariant Lagrangian density (51) and to study the motion of its soliton solutions. For examples of surface motion analysis see, e.g., R. F. Dashen, B. Hasslacher, and A. Neveu, Phys. Rev. D 11, 3424 (1975); J. Goldstone and R. Jackiw, *ibid.* 11, 1486 (1975); J. L. Gervais and B. Sakita, *ibid.* 11, 2943 (1975); N. H. Christ and T. D. Lee, *ibid.* 12, 1606 (1975); C. Rebbi, *ibid.* 12, 2407 (1975). For further references, see the review article by L. D. Faddeev and V. E. Korepin, Phys. Rep. 42C, 1 (1978).

[13]T. D. Lee and C. N. Yang, Phys. Rev. 128, 885 (1962).

[14]L. D. Faddeev and V. N. Popov, Phys. Lett. 25B, 29 (1967).

[15]See the review article by E. S. Abers and B. W. Lee, Phys. Rep. 9C, 1 (1973).

[16]This is the same boundary condition used in the MIT bag model (Ref. 8) but applied here only to the radiation field. In deriving the Feynman rules, a straightforward application of the MIT boundary condition to the longitudinal part of the vector field might encounter difficulties because of the Gauss theorem. Such difficulties are resolved here by explicitly constructing the longitudinal propagator (76a) or (76b).

[17]R. Hagedorn, Nuovo Cimento Suppl. 3, 147 (1965); Nuovo Cimento 52A, 1336 (1967); K. Huang and S. Weinberg, Phys. Rev. Lett. 25, 895 (1970).

[18]M. Gell-Mann and F. E. Low, Phys. Rev. 95, 1300 (1954); C. G. Callan, Phys. Rev. D 2, 1541 (1970); K. Symanzik, Commun. Math. Phys. 18, 227 (1970).

[19]D. J. Gross and F. Wilczek, Phys. Rev. Lett. 30, 1343 (1973); H. D. Politzer, *ibid.* 30, 1346 (1973).

[20]W. E. Caswell, Phys. Rev. Lett. 33, 244 (1974); D. R. T. Jones, Nucl. Phys. B75, 531 (1974).

PHYSICAL REVIEW D VOLUME 22, NUMBER 4 15 AUGUST 1980

Operator ordering and Feynman rules in gauge theories

N. H. Christ and T. D. Lee
Columbia University, New York, New York 10027
(Received 10 April 1980)

The ordering of operators in the Yang-Mills Hamiltonian is determined for the $V_0 = 0$ gauge and for a general noncovariant gauge $\chi(V_i) = 0$, with χ a linear function of the spatial components of the gauge field V_μ. We show that a Cartesian ordering of the $V_0 = 0$ gauge Hamiltonian defines a quantum theory equivalent to that of the usual, covariant-gauge Feynman rules. However, a straightforward change of variables reduces this $V_0 = 0$ gauge Hamiltonian to a $\chi(V_i) = 0$ gauge Hamiltonian with an unconventional operator ordering. The resulting Hamiltonian theory, when translated into Feynman graphs, is shown to imply new nonlocal interactions, even in the familiar Coulomb gauge.

I. INTRODUCTION

Gauge theories have a wide range of applications in physics, from quantum electrodynamics (QED) to quantum chromodynamics (QCD) and the unifying theory of weak and electromagnetic interactions. Because the gauge transformations contain arbitrary functions of time, the usual canonical quantization procedure can only be carried out in a specific gauge. It is natural to inquire whether there are rules to ensure that the quantum theories in different gauges are indeed the same. As we shall see, this question is closely connected with the ordering problem of operators, especially in the non-Abelian Yang-Mills theory because of the intrinsic nonlinear nature of the interaction.

In this paper, we will show that the Yang-Mills theory has a Cartesian realization in the $V_0=0$ gauge; in this gauge the naive ordering of operators in the Hamiltonian is correct. We may then go from the $V_0=0$ gauge to other gauges such as the axial gauge, Coulomb gauge, or covariant gauges via either the operator or the path-integration formalism, thereby resolving whatever ambiguities may arise in these gauges.

The fact that the Euler-Lagrange equations of motion may allow arbitrary functions of time is by no means restricted to relativistic field theories. For convenience of nomenclature, all such theories will be referred to as gauge theories. In the next section, we give one of the simplest examples of such a system.

In Sec. III we review the operator formulation of the Yang-Mills theory in the $V_0=0$ gauge and establish our notation. Next, in Sec. IV, we exploit the residual symmetry of the $V_0=0$ gauge to make a change of coordinates from the Cartesian basis provided by the gauge potentials $V_i(x)$ to new variables $\phi_a(x)$ and $A_i(x)$: $\phi_a(x)$ are pure gauge variables while the gauge potentials $A_i(x)$ obey the constraint $\chi(A_i)=0$. When expressed in terms of these new variables, the gauge-invariant sector of the theory is recognized as identical to the Yang-Mills theory formulated in a noncovariant, $\chi(A_i)=0$ gauge with a specific ordering of operators in the Hamiltonian. This discussion is in precise analogy with the treatment of the simple mechanical system given in Sec. II. As shown in the Appendix, many of our formulas are also identical to those of rigid-body rotation, when one changes from the laboratory frame to the rotating-body frame.

In Sec. V we justify our assertion that the $V_0=0$ gauge provides a Cartesian realization of the quantum Yang-Mills theory: Using functional integration we demonstrate the equivalence of that theory and the usual, covariant-gauge Feynman rules.

The Hamiltonian formulation of the quantum Yang-Mills theory obtained in Sec. IV is not a convenient one for weak-field perturbation theory. In Sec. VI, this Hamiltonian operator theory is translated into the Lagrangian, path-integral language with careful attention paid to the question of operator ordering. This path-integral description implies new nonlocal interactions, called $\mathcal{V}_1+\mathcal{V}_2$ in our paper, that must be added to the usual Feynman rules. Finally, in the conclusion we show explicitly how to equate our Hamiltonian to that obtained by Schwinger for the Coulomb gauge.[1-3] Although the $\mathcal{V}_2$ term is new, the $\mathcal{V}_1$ term was derived by Schwinger in 1962; both have been left out in the conventional treatment[4] of Coulomb-gauge Feynman rules.

II. A SIMPLE MECHANICAL EXAMPLE

Let us consider a point particle in a three-dimensional space at position $\vec{r}$. Its Lagrangian is

$$L=\tfrac{1}{2}(\dot{\vec{r}}-\dot{\vec{q}}\times\vec{r})^2-V(r)\,, \tag{2.1}$$

where $\vec{q}$ is another coordinate vector, but $\vec{q}$ is absent in L. As usual, the dot denotes the time derivative and $r=|\vec{r}|$. From (2.1), one sees immediately that L is invariant under the transformation

$$\vec{r}\to\vec{r}+\vec{\epsilon}\times\vec{r}$$

and (2.2)

$$\vec{q}\to\vec{q}+\dot{\vec{\epsilon}}+\vec{\epsilon}\times\vec{q}$$

where $\vec{\epsilon}=\vec{\epsilon}(t)$ can be an arbitrary infinitesimal vector function of time t. Except for the $\frac{1}{2}(\vec{q}\times\vec{r})^2$ term, this would be the problem of a nonrelativistic charged particle moving in a central potential and under the influence of an external magnetic field.

We may further simplify the problem by imposing the constraint that $\vec{r}$ lies in the (x,y) plane and $\vec{q}=\hat{z}q$, where $\hat{z}$ is the unit vector along the z axis. Equation (2.1) becomes then

$$L=\tfrac{1}{2}(\dot{x}^2+\dot{y}^2)-(x\dot{y}-y\dot{x})q+\tfrac{1}{2}q^2r^2-V(r)\,, \qquad (2.3)$$

where x and y are the *Cartesian* coordinates of $\vec{r}$ which is now a two-dimensional vector. In terms of the polar coordinates $x=r\cos\theta$ and $y=r\sin\theta$, (2.3) can be written as

$$L=\tfrac{1}{2}[\dot{r}^2+r^2(\dot{\theta}-q)^2]-V(r) \qquad (2.4)$$

and (2.2) is simply the Abelian group of transformations

$$\theta\to\theta+\epsilon(t)$$

and (2.5)

$$q\to q+\dot{\epsilon}(t)\,,$$

where $\epsilon(t)$ can now be any finite function of t.

The invariance group of this simple example shares with the gauge groups of QED or QCD the special feature that its elements contain arbitrary functions of t. Consequently, the canonical procedure from the Lagrangian to the Hamiltonian and to quantization requires a specific choice of the "gauge." The Lagrange equations of motion can, of course, be written down without specifying the gauge. We find in polar coordinates

$$\dot{\theta}-q=0 \qquad (2.6)$$

and

$$\ddot{r}+\frac{dV}{dr}=0\,. \qquad (2.7)$$

A. $q=0$ gauge

Because of (2.5), any orbit $\vec{r}=\vec{r}(t)$ and $q=q(t)$ can be transformed to one in which $q=0$ at all times. In this gauge, $L=\frac{1}{2}\dot{\vec{r}}^2-V(r)$, the momentum $\vec{p}$ is $\dot{\vec{r}}$ and the Hamiltonian H is $\frac{1}{2}\vec{p}^2+V(r)$. Thus, in quantum mechanics, $\vec{p}=-i\vec{\nabla}$ and

$$H=-\tfrac{1}{2}\nabla^2+V(r)\,. \qquad (2.8)$$

The angular momentum operator

$$l=-i\left(x\frac{\partial}{\partial y}-y\frac{\partial}{\partial x}\right) \qquad (2.9)$$

commutes with H. To be consistent with the equation of motion (2.6), only eigenstates of H with $l=0$ should be accepted. These eigenstates are all θ independent, and that leads to

$$H=-\frac{1}{2r}\frac{d}{dr}\left(r\frac{d}{dr}\right)+V(r)\,. \qquad (2.10)$$

B. $y=0$ gauge

From (2.5), we see that any orbit $\vec{r}=\vec{r}(t)$ and $q=q(t)$ can also be transformed to one with $y=0$ at all times, albeit there are two branches: x can be >0 or <0, with $x=0$ being the point that the Jacobian of the transformation is zero. In the $y=0$ gauge, $r^2=x^2$ and the Lagrangian (2.3) becomes

$$L=\tfrac{1}{2}\dot{x}^2+\tfrac{1}{2}q^2x^2-V(x)\,, \qquad (2.11)$$

where, for definiteness, we choose the branch of positive x. Since the above L does not contain $\dot{q}$, we may follow the standard procedure to eliminate q through $\partial L/\partial q=0$, which in the present example is simply $x^2q=0$. Hence, (2.11) becomes

$$L=\tfrac{1}{2}\dot{x}^2-V(x)\,. \qquad (2.12)$$

The conjugate momentum p is $\dot{x}$ and the classical Hamiltonian is

$$H=\tfrac{1}{2}p^2+V(x)\,. \qquad (2.13)$$

In passing over to quantum mechanics, in order that the spectrum of this Hamiltonian be identical to that of (2.10), it is important not to treat x in the $y=0$ gauge as a Cartesian coordinate; in (2.13) p^2 is the operator

$$-\frac{1}{x}\frac{d}{dx}\left(x\frac{d}{dx}\right)$$

and not $-d^2/dx^2$.

The $q=0$ gauge is the analog of the $V_0=0$ gauge in QED or QCD; the $y=0$ gauge corresponds to either the usual axial gauge or the Coulomb gauge. In this simple example, the choice of which coordinate in what guage is Cartesian can be determined by the one who makes up the problem. In the Yang-Mills theory, however, one is guided by the requirement of relativistic invariance. As we shall see, that leads to the choice of the $V_0=0$ gauge as the starting point.

Before leaving this example, we note that if one wishes, one may choose a more general gauge in which an arbitrary function $\chi(x,y,q)=0$, provided that any point in the (x,y,q) space can be transformed onto the surface $\chi(x,y,q)=0$ through the gauge transformation (2.5). Since this mechanical example is such a simple one, we shall refrain from exhibiting further details, except to remark that in passing over to the path-integration formalism, it is useful to absorb a factor of $r^{1/2}$ into the state vectors so that H, given by (2.10), becomes

$$\bar{H} \equiv r^{1/2} H r^{-1/2} = -\frac{1}{2}\frac{d^2}{dr^2} + V(r) - \frac{1}{8r^2}\,, \tag{2.14}$$

which changes the volume element from rdr to dr and adds to the potential $V(r)$ a new term $-(8r^2)^{-1}$.

III. $V_0 = 0$ GAUGE

For definiteness, let us consider an SU_N gauge theory consisting of a spin-$\frac{1}{2}$ fermion field ψ which belongs to the N-dimensional representation of the SU_N group and a spin-1 gauge field V_μ^l with $l = 1, 2, \ldots, M$ and $M = N^2 - 1$. The Lagrangian density is

$$\mathcal{L} = -\tfrac{1}{4} V_{\mu\nu}^l V_{\mu\nu}^l - \psi^\dagger \gamma_4 \gamma_\mu D_\mu \psi\,, \tag{3.1}$$

where all repeated indices are summed over, a dagger denotes Hermitian conjugation,

$$V_{\mu\nu}^l = \frac{\partial}{\partial x_\mu} V_\nu^l - \frac{\partial}{\partial x_\nu} V_\mu^l + g f^{lmn} V_\mu^m V_\nu^n\,,$$
$$D_\mu = \frac{\partial}{\partial x_\mu} - i g T^l V_\mu^l\,, \tag{3.2}$$

$x_\mu = (\vec{r}, it)$, the γ_μ's are the 4×4 Hermitian Dirac matrices, and the T^l's are $N \times N$ matrices that satisfy

$$T^l = T^{l\dagger}, \quad \mathrm{Tr}(T^l T^m) \propto \delta^{lm}$$
and
$$[T^l, T^m] = i f^{lmn} T^n \tag{3.3}$$

with f^{lmn} the antisymmetric structure constants of the group algebra. For simplicity, the fermion mass has been set equal to zero. The electric and magnetic fields E_i^l and B_i^l are given by the usual expressions

$$E_i^l = i V_{i4}^l = -\dot{V}_i^l - \nabla_i V_0^l + g f^{lmn} V_0^m V_i^n$$
and
$$\epsilon_{ijk} B_k^l = V_{ij}^l = \nabla_i V_j^l - \nabla_j V_i^l + g f^{lmn} V_i^m V_j^n\,,$$

where $V_0^l = -iV_4^l$ and the subscripts i, j, k denote the space indices which vary from 1 to 3. The Lagrangian equations of motion are

$$\gamma_\mu D_\mu \psi = 0 \tag{3.4}$$

and

$$\frac{\partial}{\partial x_\mu} V_{\mu\nu}^l + g(f^{lmn} V_\mu^m V_{\mu\nu}^n + I_\nu^l) = 0\,, \tag{3.5}$$

where

$$I_\nu^l = i\psi^\dagger \gamma_4 \gamma_\nu T^l \psi\,.$$

It is convenient to introduce the matrix function

$$V_\mu \equiv T^l V_\mu^l\,. \tag{3.6}$$

The Lagrangian density (3.1) is invariant under the SU_N transformation

$$V_\mu \to u V_\mu u^\dagger + \frac{i}{g} u \frac{\partial u^\dagger}{\partial x_\mu} \tag{3.7}$$

and

$$\psi \to u\psi\,, \tag{3.8}$$

where $u = u(x)$ is any $N \times N$ unitary matrix function of x with $\det u = 1$. From any configuration $V_\mu = F_\mu(\vec{r}, t)$, we may choose $u^\dagger$ to be the following time-ordered function:

$$u^\dagger(\vec{r}, t) = T \exp\left[-i \int_0^t g F_0(\vec{r}, t')\right]\,, \tag{3.9}$$

where $F_0 = -iF_4$ and T is the time-ordering operator. Hence, $u^\dagger$ satisfies

$$\frac{\partial u^\dagger}{\partial t} = -igF_0 u^\dagger\,. \tag{3.10}$$

The transformation (3.7) then brings V_μ from the configuration $F_\mu(x)$ to the gauge in which

$$V_0(x) = 0\,. \tag{3.11}$$

In the $V_0 = 0$ gauge, $E_i^l = -\dot{V}_i^l$ and (3.1) becomes

$$\mathcal{L} = \tfrac{1}{2}(\dot{V}_i^l \dot{V}_i^l - B_i^l B_i^l) - \psi^\dagger \gamma_4 \gamma_\mu D_\mu \psi\,. \tag{3.12}$$

The conjugate momentum of V_i^l is simply

$$\Pi_i^l = \dot{V}_i^l\,, \tag{3.13}$$

and the Hamiltonian density is

$$\mathcal{H} = \tfrac{1}{2}(\Pi_i^l \Pi_i^l + B_i^l B_i^l) - g I_i^l V_i^l + \psi^\dagger \gamma_4 \gamma_i \nabla_i \psi\,. \tag{3.14}$$

The usual canonical quantization procedure leads to

$$[V_i^l(\vec{r}, t), \Pi_j^m(\vec{r}\,', t)] = i\delta_{ij}\delta^{lm}\delta^3(\vec{r} - \vec{r}\,')\,, \tag{3.15}$$
$$\{\psi(\vec{r}, t), \psi^\dagger(\vec{r}\,', t)\} = \delta^3(\vec{r} - \vec{r}\,')\,. \tag{3.16}$$

The equal-time commutators between the V_i's and between the Π_i's are zero; likewise, the equal-time anticommutators between the ψ's and between the $\psi^\dagger$'s are also zero.

Remnants of the original gauge transformations (3.7) and (3.8) remain important. In accordance with (3.6), we denote

$$V_i = T^l V_i^l$$
and
$$\Pi_i = T^l \Pi_i^l\,. \tag{3.17}$$

It can be readily verified that the Hamiltonian density $\mathcal{H}$ and the commutation relations are invariant under a time-dependent SU_N transformation

$$V_i \to uV_iu^\dagger - \frac{i}{g}(\nabla_i u)u^\dagger ,$$
$$\Pi_i \to u\Pi_i u^\dagger , \tag{3.18}$$

and

$$\psi \to u\psi ,$$

where $u = u(\vec{r})$ can be any $N\times N$ unitary matrix function of $\vec{r}$ with $\det u = 1$. Since the $u(\vec{r})$'s are time-independent, the invariance group $\{u(\vec{r})\}$ is generated by the $\vec{r}$-dependent operators $\mathcal{G}^l$ which are conserved:

$$\mathcal{G}^l \equiv J^l + \psi^\dagger T^l \psi , \tag{3.19}$$

where

$$J^l \equiv \frac{1}{g} D_i^{lm}\Pi_i^m$$

and (3.20)

$$D_i^{lm} = \delta^{lm}\nabla_i - gf^{lmn}V_i^n .$$

It is straightforward to verify

$$[J^l(\vec{r},t), J^m(\vec{r}\,',t)] = if^{lmn}\delta^3(\vec{r}-\vec{r}\,')J^n(\vec{r},t) , \tag{3.21}$$

$$[\mathcal{G}^l(\vec{r},t), \mathcal{G}^m(\vec{r}\,',t)] = if^{lmn}\delta^3(\vec{r}-\vec{r}\,')\mathcal{G}^n(\vec{r},t) , \tag{3.22}$$

$$[\mathcal{G}^l(\vec{r},t), \psi(\vec{r}\,',t)] = -\delta^3(\vec{r}-\vec{r}\,')T^l\psi(\vec{r},t) , \tag{3.23}$$

$$[\mathcal{G}^l(\vec{r},t), \Pi_i^m(\vec{r}\,',t)] = if^{lmn}\delta^3(\vec{r}-\vec{r}\,')\Pi_i^n(\vec{r},t) , \tag{3.24}$$

and

$$[\mathcal{G}^l(\vec{r},t), V_i^m(\vec{r}\,',t)] = if^{lmn}\delta^3(\vec{r}-\vec{r}\,')V_i^n(\vec{r},t) - \frac{1}{g}\delta^{lm}\nabla_i\delta^3(\vec{r}-\vec{r}\,') , \tag{3.25}$$

where ∇_i is the differential operator with respect to $\vec{r}$. Consequently, $\mathcal{G}^l$ commutes with the Hamiltonian $H = \int \mathcal{H}\, d^3r$ and

$$\dot{\mathcal{G}}^l(\vec{r},t) = i[H, \mathcal{G}^l(\vec{r},t)] = 0 . \tag{3.26}$$

By commuting H with ψ, we obtain the equation of motion (3.4) for ψ. Likewise, by commuting H with V_i^l and Π_i^l, we derive (3.5) for $\nu = i$ which can be 1, 2, or 3 but not 4. We note that $i\mathcal{G}^l/g$ is identical to the left-hand side of (3.5) when $\nu = 4$. Thus, in order to be consistent with all the Lagrangian equations of motion, in the $V_0 = 0$ gauge we require all state vectors $|\;\rangle$ to satisfy

$$\mathcal{G}^l|\;\rangle = 0 . \tag{3.27}$$

In the Schrödinger picture the operators $V_i^l = V_i^l(\vec{r})$ and $\Pi_i^l = \Pi_i^l(\vec{r})$ are all t independent. The state vector in the V_i^l representation is the functional

$$\Psi(V_i) \equiv \langle V_i^l|\;\rangle . \tag{3.28}$$

In this representation, the Hamiltonian H is

$$H = \mathcal{K} + \mathcal{V} , \tag{3.29}$$

where

$$\mathcal{K} = -\tfrac{1}{2}\int \frac{\delta}{\delta V_i^l(\vec{r})}\frac{\delta}{\delta V_i^l(\vec{r})}\, d^3r \tag{3.30}$$

and

$$\mathcal{V} + \int (\tfrac{1}{2}B_i^l B_i^l - gI_i^l V_i^l + \psi^\dagger\gamma_4\gamma_i\nabla_i\psi)\, d^3r . \tag{3.31}$$

In Sec. IV we shall see how the introduction of curvilinear coordinates can be used to eliminate the constraint (3.27), in complete analogy to the passage from (2.8) to (2.10) in the simple mechanical example discussed in the previous section.

IV. NONCOVARIANT GAUGES

Let us start from the $V_0 = 0$ gauge quantum theory of Sec. III and show how to reach other noncovariant gauges such as the axial or Coulomb gauges. For notational clarity the gauge field in these other gauges will be referred to as $A_\mu = T^l A_\mu^l$. The spatial components of A_μ obey a gauge condition

$$\chi(A_i) = 0 . \tag{4.1}$$

Among possible choices for χ one has

$$\chi(A_i) = \begin{cases} A_3 & \text{in axial gauge ,} \\ \nabla_i A_i & \text{in Coulomb gauge .} \end{cases} \tag{4.2}$$

For simplicity we will treat χ as a linear homogeneous functional of A_i. As in (4.2), we assume $\chi = T^l\chi^l$ to be an $N\times N$ Hermitian matrix with zero trace. Thus at any given space-time point, (4.1) expresses $M = N^2 - 1$ conditions:

$$0 = \chi^l(A_i, \vec{r}) \equiv \int d^3r'\langle \vec{r}, l\,|\,\Gamma_j\,|\,\vec{r}\,', m\rangle A_j^m(\vec{r}\,', t) , \tag{4.3}$$

where the matrix element of Γ_j is real, and as before, the parameters l and m can vary from 1 to M. In addition, we will assume that for every field configuration $V_i(\vec{r},t)$ in the $V_0 = 0$ gauge there exists a unique gauge-transformation matrix $u(\vec{r},t)$ such that[5]

$$V_i = uA_iu^{-1} + \frac{i}{g}u\nabla_i u^{-1} . \tag{4.4}$$

If we view the $N\times N$ matrix u as a function $u(\phi_a)$ of M group parameters ϕ_a, $1 \le a \le M$, then the gauge transformation $u(\vec{r},t)$ in turn specifies M functions $\phi_a(\vec{r},t)$ such that

$$u(\vec{r},t) = u(\phi_a(\vec{r},t)) . \tag{4.5}$$

Equation (4.4) can thus be viewed as expressing the gauge field $V_i(\vec{r},t)$ in terms of the curvilinear coordinates $A_i(\vec{r},t)$, $\phi_a(\vec{r},t)$. For example, if our group is SU_2, then $M=3$ and we might choose the ϕ_a's to be simply the three Euler angles. (See the Appendix for details.)

Because of (3.30), the V_i^l's can be regarded as the Cartesian coordinates. We recall that in any coordinate transformation from a set of Cartesian coordinates $q_1, q_2, \ldots$ to one of curvilinear coordinates $Q_1(q_a), Q_2(q_a), \ldots$, the standard kinetic energy term in the Lagrangian may be written as

$$K \equiv \tfrac{1}{2}\sum_\alpha \dot{q}_\alpha^{\,2} = \tfrac{1}{2}\sum_{\alpha,\beta} \dot{Q}_\alpha M_{\alpha\beta}\dot{Q}_\beta . \tag{4.6}$$

In the quantum Hamiltonian, the corresponding operator in the coordinate representation is

$$\mathcal{K} \equiv -\tfrac{1}{2}\sum_\alpha \frac{\partial^2}{\partial q_\alpha^{\,2}} = -\tfrac{1}{2}\sum_{\alpha,\beta} |M|^{-1/2}\frac{\partial}{\partial Q_\alpha}\left(M^{-1}{}_{\alpha\beta}|M|^{1/2}\frac{\partial}{\partial Q_\beta}\right), \tag{4.7}$$

where

$$M_{\alpha\beta} = \sum_\gamma \frac{\partial q_\gamma}{\partial Q_\alpha}\,\frac{\partial q_\gamma}{\partial Q_\beta}, \quad M^{-1}{}_{\alpha\beta} = \sum_\gamma \frac{\partial Q_\alpha}{\partial q_\gamma}\,\frac{\partial Q_\beta}{\partial q_\gamma}, \tag{4.8}$$

and $|M| = \det M$. Hence,

$$|M|^{-1/2} = \det\left(\frac{\partial Q_\alpha}{\partial q_\beta}\right) \tag{4.9}$$

in which $\partial Q_\alpha/\partial q_\beta$ is the (α,β)th element of the matrix.

Applied to our case, (3.30) becomes

$$\mathcal{K} = \tfrac{1}{2}|M|^{-1/2}\int d^3r \int d^3r' [p_a(\vec{r})|M|^{1/2}M^{-1}(\vec{r},\vec{r}')_{ab}p_b(\vec{r}') + p_a(\vec{r})|M|^{1/2}M^{-1}(\vec{r},\vec{r}')^l_{ai}P^l_i(\vec{r}')$$
$$+ P^l_i(\vec{r})|M|^{1/2}M^{-1}(\vec{r},\vec{r}')^l_{ia}p_a(\vec{r}') + P^l_i(\vec{r})|M|^{1/2}M^{-1}(\vec{r},\vec{r}')^{lm}_{ij}P^m_j(\vec{r}')], \tag{4.10}$$

where p_a and P_i^l are the momenta conjugate to ϕ_a and A_i^l. In the coordinate representation of the Schrödinger picture, $\phi_a(\vec{r})$ and $A_i^l(\vec{r})$ are all t independent. Clearly,

$$p_a(\vec{r}) = -i\frac{\delta}{\delta\phi_a(\vec{r})} . \tag{4.11}$$

However, because of the constraints (4.3) obeyed by $A_i^l(\vec{r})$, its conjugate momentum must be defined more carefully. We can expand $A_i^l(\vec{r})$ in terms of a complete set of real orthonormal functions $f_i^l(\vec{r})_N$, each obeying the same set of constraints

$$\int d^3r' \langle \vec{r}, l|\Gamma_j|\vec{r}', m\rangle f_j^m(\vec{r}')_N = 0 \tag{4.12}$$

for all N, $\vec{r}$, and l. The expansion is

$$A_i^l(\vec{r}) = \sum_N Q_N f_i^l(\vec{r})_N . \tag{4.13}$$

The Q_N's may now be viewed as the independent generalized coordinates. The momentum $P_i^l(\vec{r})$ is then given by

$$P_i^l(\vec{r}) = \sum_N f_i^l(\vec{r})_N\left(-i\frac{\partial}{\partial Q_N}\right) . \tag{4.14}$$

Thus, $P_i^l(\vec{r})$ also obeys the constraints

$$\chi^l(P_i,\vec{r}) = \int d^3r' \langle \vec{r}, l|\Gamma_j|\vec{r}', m\rangle P_j^m(\vec{r}') = 0. \tag{4.15}$$

The following construction for the functions $f_i^l(\vec{r})_N$ will be useful. We may view $\langle \vec{r}, l|\Gamma_i^\dagger|\vec{r}', l'\rangle$ as a transformation matrix which maps a vector $|\xi\rangle$ in the functional space $h \equiv \{|\vec{r}', l'\rangle\}$ to a vector $|\bar{\xi}\rangle$ in a larger space $\bar{h} \equiv \{|i, \vec{r}, l\rangle\}$, where $|\bar{\xi}\rangle$ is given by

$$\langle i, \vec{r}, l|\bar{\xi}\rangle \equiv \langle \vec{r}, l|\Gamma_i^\dagger|\xi\rangle . \tag{4.16}$$

Let $|N\rangle$ be a vector in $\bar{h}$ that is orthogonal to all such $|\bar{\xi}\rangle$'s. If we denote

$$f_i^l(\vec{r})_N \equiv \langle i, \vec{r}, l|N\rangle , \tag{4.17}$$

then since

$$\langle \bar{\xi}|N\rangle = \int \langle \xi|\Gamma_i|\vec{r}, l\rangle f_i^l(\vec{r})_N d^3r = 0 \tag{4.18}$$

for all the $|\xi\rangle$ vectors in h, (4.12) follows.

The inverse of the mass matrix M appearing in (4.10) has been divided into four blocks with

$$M^{-1}(\vec{r},\vec{r}')_{ab} = \int d^3r'' \frac{\delta\phi_a(\vec{r})}{\delta V_k^n(\vec{r}'')}\,\frac{\delta\phi_b(\vec{r}')}{\delta V_k^n(\vec{r}'')} ,$$

$$M^{-1}(\vec{r},\vec{r}')^l_{ai} = M^{-1}(\vec{r}',\vec{r})^l_{ia} = \int d^3r'' \frac{\delta\phi_a(\vec{r})}{\delta V_k^n(\vec{r}'')}\,\frac{\delta A_i^l(\vec{r}')}{\delta V_k^n(\vec{r}'')} , \tag{4.19}$$

$$M^{-1}(\vec{r},\vec{r}')^{lm}_{ij} = \int d^3r'' \frac{\delta A_i^l(\vec{r})}{\delta V_k^n(\vec{r}'')}\,\frac{\delta A_j^m(\vec{r}')}{\delta V_k^n(\vec{r}'')} .$$

Next, let us determine these functional derivatives in terms of the gauge-fixing function $\chi(A_i,\vec{r})$. Solving Eq. (4.4) for A_i, and considering small variations of A_i, ϕ_a, and V_i we obtain

$$\delta A_i = \delta u^{-1}V_i u + u^{-1}\delta V_i u + u^{-1}V_i\delta u + \frac{i}{g}\delta u^{-1}\nabla_i u + \frac{i}{g}u^{-1}\nabla_i\delta u . \tag{4.20}$$

By using (3.3), we may convert the above equation into the form

$$\delta A_i^l = U^{ml}\delta V_i^m + \frac{1}{g}\mathfrak{D}_i^{lm}(\lambda_a^m \delta\phi_a), \tag{4.21}$$

where λ_a^l is defined by

$$iu^{-1}\frac{\partial u}{\partial \phi_a} \equiv \lambda_a \equiv T^l \lambda_a^l, \tag{4.22}$$

U^{ml} satisfies

$$u^{-1}T^m u = U^{ml}T^l, \tag{4.23}$$

and $\mathfrak{D}_i^{lm}$ represents the covariant-derivative operator containing the field A_i^n

$$\mathfrak{D}_i^{lm} = \delta^{lm}\nabla_i - gf^{lmn}A_i^n. \tag{4.24}$$

Because of the first two equations in (3.3), the matrix $U = (U^{ml})$ is real and orthogonal, and therefore also unitary.

The requirement that $\chi(A_i + \delta A_i, \vec{r})$ should also vanish then relates δV_i and $\delta\phi_a$:

$$0 = \int d^3r' l\langle \vec{r}, |\Gamma_i|\vec{r}', m\rangle \Big[U^{nm}(\vec{r}')\delta V_i^n(\vec{r}') + \frac{1}{g}\mathfrak{D}_i^{mn}(\lambda_a^n(\vec{r}')\delta\phi_a(\vec{r}'))\Big], \tag{4.25}$$

which can be solved for $\delta\phi_a(\vec{r})$, thereby determining

$$\frac{\delta\phi_a(\vec{r})}{\delta V_i^l(\vec{r}')} = -g\lambda^{-1}(\vec{r})_a^m \langle \vec{r}, m|(\Gamma_j\mathfrak{D}_j)^{-1}\Gamma_i|\vec{r}, n\rangle U^{ln}(\vec{r}'), \tag{4.26}$$

where $(\Gamma_j\mathfrak{D}_j)^{-1}$ is the inverse of the matrix $\Gamma_j\mathfrak{D}_j$ whose matrix elements are

$$\langle \vec{r}, l|\Gamma_j\mathfrak{D}_j|\vec{r}', l'\rangle \equiv \int d^3r'' \langle \vec{r}, l|\Gamma_j|\vec{r}'', m\rangle \times \langle \vec{r}'', m|\mathfrak{D}_j|\vec{r}', l'\rangle, \tag{4.27}$$

with $\mathfrak{D}_j$ as the antisymmetric matrix, defined by

$$\langle \vec{r}, l|\mathfrak{D}_j|\vec{r}', m\rangle = \mathfrak{D}_j^{lm}(\vec{r})\delta^3(\vec{r} - \vec{r}'). \tag{4.28}$$

The matrix $\lambda^{-1}(\vec{r})$ is the inverse of the $M\times M$ matrix $\lambda(\vec{r}) \equiv (\lambda_a^n(\vec{r}))$, with $(\lambda^{-1})_a^m \lambda_a^n = \delta^{mn}$. If (4.26) is used in (4.21), we can also obtain

$$\frac{\delta A_i^m(\vec{r})}{\delta V_j^l(\vec{r}')} = U^{lm}(\vec{r})\delta_{ij}\delta^3(\vec{r} - \vec{r}') - \langle \vec{r}, m|\mathfrak{D}_i(\Gamma_k\mathfrak{D}_k)^{-1}\Gamma_j|\vec{r}', n\rangle U^{ln}(\vec{r}'). \tag{4.29}$$

Note that (4.29) automatically obeys

$$\int d^3r \langle \vec{r}'', l''|\Gamma_i|\vec{r}, m\rangle \frac{\delta A_i^m(\vec{r})}{\delta V_j^l(\vec{r}')} = 0. \tag{4.30}$$

The determinant $|M|$ in (4.10) can now be determined explicitly. From (4.9), we see that $|M|$ is invariant under an orthogonal transformation among the Cartesian coordinates q_α. By using the notation of (4.13) and (4.16), we may consider the variations

$$\delta A_i^l(\vec{r}) = \sum_N f_i^l(\vec{r})_N \delta Q_N \tag{4.31}$$

and

$$\delta V_i^l(\vec{r}') = U^{lm}[\delta \bar{V}_i^m(\vec{r}')_I + \delta \bar{V}_i^m(\vec{r}')_{II}], \tag{4.32}$$

with

$$\delta \bar{V}_i^l(\vec{r}')_I \equiv \sum_\xi \langle \vec{r}', l|\Gamma_i^\dagger(\Gamma_j\Gamma_j^\dagger)^{-1/2}|\xi\rangle \delta q_\xi \tag{4.33}$$

and

$$\delta \bar{V}_i^l(\vec{r}')_{II} = \sum_N f_i^l(\vec{r}')_N \delta q_N, \tag{4.34}$$

where ξ runs over any complete set of orthonormal basis vectors $|\xi\rangle$ in h. Hence, the matrix $(\partial Q_\alpha/\partial q_\beta)$ in (4.9) now takes a 2×2 block form. We find

$$|M|^{-1/2} = \det\begin{bmatrix} -g\lambda^{-1}(\Gamma_i\mathfrak{D}_i)^{-1}(\Gamma_j\Gamma_j^\dagger)^{1/2} & 0 \\ X & 1 \end{bmatrix} \tag{4.35}$$

where X is a rectangular matrix whose matrix elements are

$$X_{N\xi} = \frac{\partial Q_N}{\partial q_\xi} = -\int f_i^l(\vec{r})_N \langle \vec{r}, l|\mathfrak{D}_i(\Gamma_k\mathfrak{D}_k)^{-1}(\Gamma_j\Gamma_j^\dagger)^{1/2}|\xi\rangle d^3r.$$

From (4.35), we see that

$$|M|^{1/2} = \det(-g^{-1}(\Gamma_j\Gamma_j^\dagger)^{-1/2}\Gamma_i\mathfrak{D}_i\lambda) = \text{const}\,\mathcal{J}|\lambda|, \tag{4.36}$$

where $|\lambda| = \prod_{\vec{r}} \det\lambda(\vec{r})$, $\mathcal{J}$ is the Faddeev-Popov determinant

$$\mathcal{J} = \det|\Gamma_i\mathfrak{D}_i|, \tag{4.37}$$

and const $= \det|-g^{-1}(\Gamma_j\Gamma_j^\dagger)^{-1/2}|$ which is independent of both $A_i(\vec{r})$ and $\phi_a(\vec{r})$.

Our partial derivative formulas (4.26) and (4.29) may also be substituted into the expression (4.10) for $\mathcal{K}$ yielding

$$\mathcal{K} = \tfrac{1}{2}\mathcal{J}^{-1}\int d^3r P_i^l(\vec{r})\mathcal{J}P_i^l(\vec{r}) + \tfrac{1}{2}\mathcal{J}^{-1}|\lambda|^{-1}\int\int d^3r\, d^3r' [-P_i^l(\vec{r})\mathfrak{D}_i - gp_a(\vec{r})(\lambda^{-1})_a^l]\mathcal{J} \times (\Gamma_k\mathfrak{D}_k)^{-1}\Gamma_j\Gamma_j^\dagger(\mathfrak{D}_{k'}^\dagger\Gamma_{k'}^\dagger)^{-1}|\lambda|[\mathfrak{D}_{i'}P_{i'}^{l'}(\vec{r}') - g(\lambda^{-1})_{a'}^{l'}p_{a'}(\vec{r}')], \tag{4.38}$$

where in the double integral the matrix element $\langle \vec{r}, l|O|\vec{r}', l'\rangle$ between any pair of momenta is simply written as O. The above formula appears quite similar to the familiar Coulomb-gauge formula, $P_i^l(\vec{r})$ replacing the transverse momenta, the quantity in square brackets the charge density, and the central matrix element the Coulomb Green's function. However, the analogy is not yet completely precise because of the presence of the angle variables $\phi_a(\vec{r})$, their conjugate momenta $p_a(\vec{r})$, and the matrix λ.

In order to eliminate the angular dependence, it is useful to first construct the group generators in terms of the p_a's. Let us introduce the following parallel expression to (4.22):

$$i\frac{\partial u}{\partial \phi_a}u^{-1} \equiv \Lambda_a \equiv T^l \Lambda_a^l \,. \tag{4.39}$$

Because of (4.23), Λ_a^l and λ_a^m are related by

$$\Lambda_a^l = U^{lm}\lambda_a^m \,, \tag{4.40}$$

which implies

$$\det\Lambda = \det\lambda \,. \tag{4.41}$$

Since u is unitary and T^l Hermitian, both Λ_a^l and λ_a^l are real. We define two sets of operators $\{j^l(\vec{r})\}$ and $\{J^l(\vec{r})\}$:

$$p_a \equiv \lambda_a^l j^l \equiv \Lambda_a^l J^l \,. \tag{4.42}$$

Hence,

$$\begin{aligned} &j^l = (\lambda^{-1})_a^l p_a, \quad J^l = (\Lambda^{-1})_a^l p_a \,, \\ &j^l = U^{ml}J^m \text{ and } J^l = U^{lm}j^m \,. \end{aligned} \tag{4.43}$$

On account of (4.5) and (4.11), the $\vec{r}$ dependence of j^l, J^l, λ_a^l, and Λ_a^l is entirely through their dependence on $\phi_a(\vec{r})$ and $p_a(\vec{r})$. By using (3.3) and by differentiating (4.22) and (4.39) with respect to ϕ_b, we can verify that

$$[j^l(\vec{r}), j^m(\vec{r}')] = -if^{lmn}j^n(\vec{r})\delta^3(\vec{r}-\vec{r}') \tag{4.44}$$

and

$$[J^l(\vec{r}), J^m(\vec{r}')] = if^{lmn}J^n(\vec{r})\delta^3(\vec{r}-\vec{r}') \,. \tag{4.45}$$

Similarly, through differentiation of (4.23), it follows that

$$\begin{aligned} &[J^l(\vec{r}), U^{mn}(\vec{r}')] = if^{lms}U^{sn}(\vec{r})\delta^3(\vec{r}-\vec{r}') \,, \\ &[j^l(\vec{r}), U^{mn}(\vec{r}')] = -if^{lns}U^{ms}(\vec{r})\delta^3(\vec{r}-\vec{r}') \,, \end{aligned} \tag{4.46}$$

which together with (4.43)–(4.45) lead to

$$[J^l(\vec{r}), j^m(\vec{r}')] = 0 \,. \tag{4.47}$$

Furthermore, by using

$$|\lambda|^{-1}\frac{\partial}{\partial\phi_a}|\lambda| = (\lambda^{-1})_b^m \frac{\partial}{\partial\phi_a}\lambda_b^m$$

and by differentiating (4.22), we can derive, after some manipulation, a useful commutation relation

$$[p_a, (\lambda^{-1})_a^l|\lambda|] = 0 \,. \tag{4.48}$$

Thus, for arbitrary functions $f(\phi_a)$ and $g(\phi_a)$, j^l satisfies the Hermiticity condition

$$\int f^* j^l g\, d\tau_\phi = \left(\int g^* j^l f\, d\tau_\phi\right)^* , \tag{4.49}$$

where $d\tau_\phi = |\lambda|\Pi d\phi_a$. Likewise, J^l is also Hermitian. That there should exist two sets of operators $\{J_a^l(\vec{r})\}$ and $\{-j_a^l(\vec{r})\}$, both satisfying the same group algebra and mutually commuting, has a simple geometrical meaning. In the case of the SU_2 group, this situation is identical to the familiar problem of rigid-body rotation, with J^l as the angular momentum operator in the laboratory frame and j^l that in the body frame. The details will be given in the Appendix.

We shall now show that the operator J^l defined by (4.43) is equal to $g^{-1}D_i^{lm}\Pi_i^m$, given by (3.20). From (4.4), (4.23), and (4.39) we find

$$V_i^l = U^{lm}A_i^m - \frac{1}{g}\Lambda_a^l \nabla_i \phi_a \,, \tag{4.50}$$

which together with (3.20) and (4.24) leads to

$$D_i^{lm}U^{mn} = U^{lm}\mathfrak{D}_i^{mn} \,. \tag{4.51}$$

The operator $D_i^{lm}\Pi_i^m$ can be written as a linear combination of the momenta p_a and P_i^l:

$$\begin{aligned} D_i^{lm}(\vec{r})\Pi_i^m(\vec{r}) &= -iD_i^{lm}(\vec{r})\frac{\delta}{\delta V_i^m(\vec{r})} \\ &= \int d^3r' \left[D_i^{lm}(\vec{r})\frac{\delta\phi_a(\vec{r}')}{\delta V_i^m(\vec{r})}p_a(\vec{r}') \right. \\ &\qquad \left. + D_i^{lm}(\vec{r})\frac{\delta A_j^n(\vec{r}')}{\delta V_i^m(\vec{r})}P_j^n(\vec{r}') \right] . \end{aligned} \tag{4.52}$$

The substitution of Eqs. (4.26) and (4.29) simplifies this considerably. For example, by using the transpose of (4.51), we find

$$\begin{aligned} D_i^{lm}(\vec{r})\frac{\delta A_j^n(\vec{r}')}{\delta V_i^m(\vec{r})} &= \langle \vec{r}', n|(U^\dagger\delta_{ij} - \mathfrak{D}_j(\Gamma_k\mathfrak{D}_k)^{-1}\Gamma_i U^\dagger)D_i|\vec{r}, l\rangle \\ &= -\langle \vec{r}', n|(\delta_{ij} - \mathfrak{D}_j(\Gamma_k\mathfrak{D}_k)^{-1}\Gamma_i)\mathfrak{D}_i U^\dagger|\vec{r}, l\rangle \\ &= 0 \end{aligned} \tag{4.53}$$

in which the matrix $\langle \vec{r}, l|U|\vec{r}', m\rangle = U^{lm}(\vec{r})\delta^3(\vec{r}-\vec{r}')$ is real, and $\langle \vec{r}, l|D_i|\vec{r}', m\rangle = D_i^{lm}(\vec{r})\delta^3(\vec{r}-\vec{r}')$ is antisymmetric, like $\mathfrak{D}_i$. Likewise,

$$D_i^{lm}(\vec{r})\frac{\delta\phi_a(\vec{r}')}{\delta V_i^m(\vec{r})} = g\lambda^{-1}(\vec{r})_a^m U^{lm}(\vec{r})\delta^3(\vec{r}-\vec{r}') \,. \tag{4.54}$$

Consequently, (4.52) becomes

$$D_i^{lm}\Pi_i^m = g(\Lambda^{-1})_a^l p_a = gJ^l \,, \tag{4.55}$$

which establishes the identity between the two definitions of J^l, (3.20) and (4.43).

Let us finally examine the requirement (3.27) that the physical subspace of the $V_0 = 0$ gauge Hilbert space be annihilated by the operator $\mathcal{G}^l = J^l + \psi^\dagger T^l \psi$. Under the transformation (4.4), the state vector (3.28) becomes a functional of A_i and ϕ_a. Equation (3.27) can be written as

$$\left[\frac{\delta}{\delta\phi_a(r)} + i\Lambda^l_a \psi^\dagger T^l \psi\right]\Psi[A_i, \phi_a] = 0\,, \tag{4.56}$$

an equation which is easily solved:

$$\Psi[A_i, \phi_a] = \mathfrak{U}(\phi_a)\tilde{\Psi}(A_i)'. \tag{4.57}$$

Here $\tilde{\Psi}(A_i)$ is any state vector depending on the vector potential A_i and the fermionic degrees of freedom, but independent of ϕ_a. $\mathfrak{U}(\phi_a)$ is the unitary transformation acting on the fermionic degrees of freedom which represents the group element $u(\phi_a)$. Since the generators of $\mathfrak{U}(\phi_a)$ are simply $i\psi^\dagger T^l \psi$, $1 \le l \le N$, the equation

$$\frac{\partial \mathfrak{U}}{\partial \phi_a}\mathfrak{U}^{-1} = -i\Lambda^l_a \psi^\dagger T^l \psi \tag{4.58}$$

is the Hilbert-space analog of (4.39) and this equation implies directly that the state $\Psi[A_i, \phi_a]$ defined in Eq. (4.57) does indeed satisfy Gauss' law, Eq. (4.56). [In the SU_2 case, if we represent u in terms of the Euler angles a, b, c as in (A1) of the Appendix

$$u = e^{-i\tau_z b/2} e^{-i\tau_y a/2} e^{-i\tau_z c/2}\,,$$

then the corresponding $\mathfrak{U}$ is given by

$$\mathfrak{U} = e^{-i\mathcal{T}_z b/2} e^{-i\mathcal{T}_y a/2} e^{-i\mathcal{T}_z c/2}\,,$$

where $\mathcal{T}_i = \int d^3r\, \psi^\dagger \frac{1}{2}\tau_i \psi$].

If the Hamiltonian H is applied to this gauge-invariant subspace of the Hilbert space, it is convenient to work directly with the ϕ-independent state vectors $\tilde{\Psi}(A)$ and to absorb the unitary operator $\mathfrak{U}(\phi_a)$ into H. The resulting Hamiltonian,

$$\tilde{H} = \mathfrak{U}(\phi)^{-1} H \mathfrak{U}(\phi)\,, \tag{4.59}$$

is then extremely simple. For example, (4.58) and (4.40) imply that under conjugation by $\mathfrak{U}$, the combination $(\lambda^{-1})^m_a p_a$ appearing in $\mathcal{K}$ becomes

$$\begin{aligned}(\lambda^{-1})^m_a \mathfrak{U}^{-1} p_a \mathfrak{U} &= -(\lambda^{-1})^m_a \Lambda^l_a \mathfrak{U}^{-1}\psi^\dagger T^l \psi \mathfrak{U} \\ &= -(\lambda^{-1})^m_a \Lambda^l_a (U^{-1})^{l'l} \psi^\dagger T^{l'} \psi \\ &= -\psi^\dagger T^m \psi\,.\end{aligned} \tag{4.60}$$

Likewise,

$$\mathfrak{U}^{-1}\psi^\dagger \alpha_j(-i\nabla_j - gV^l_j T^l)\psi \mathfrak{U} = \psi^\dagger \alpha_j(-i\nabla_j - gA^l_j T^l)\psi\,. \tag{4.61}$$

Thus, from (3.29), (4.38), and (4.59) we find

$$\begin{aligned}\tilde{H}(P,A) = {}& \tfrac{1}{2}\mathcal{J}^{-1}\int d^3r\, P^l_i(\vec{r})\mathcal{J}P^l_i(\vec{r}) \\ &+ \tfrac{1}{2}\mathcal{J}^{-1}\int d^3r \int d^3r' \{[-P^l_i(\vec{r})\mathfrak{D}_i + g\psi^\dagger T^l\psi]\mathcal{J} \times (\Gamma_k \mathfrak{D}_k)^{-1}\Gamma_j \Gamma^\dagger_j (\mathfrak{D}^\dagger_{k'}\Gamma^\dagger_{k'})^{-1}[\mathfrak{D}_{i'}P^{l'}_{i'}(\vec{r}') + g\psi^\dagger T^{l'}\psi]\} \\ &+ \int d^3r [\tfrac{1}{2}B^l_j B^l_j + \psi^\dagger \alpha_j(-i\nabla_j - gA^l_j T^l)\psi]\,,\end{aligned} \tag{4.62}$$

where, as in (4.37), $\mathcal{J} = \det|\Gamma_k \mathfrak{D}_k|$. The angle variables have now been completely eliminated from the problem. The Yang-Mills quantum theory, when restricted to the states $\tilde{\Psi}(A)$, becomes very close to the canonical theory that would have been naively proposed for the gauge $\chi[A]$. However, the precise ordering of the operators in (4.62), in particular, the appearance of the Jacobian $\mathcal{J}$, is not the conventional one. In fact, as will be shown in Sec. VI, the operator ordering in (4.62) yields additional vertices in the Feynman rules. Although for the case of the Coulomb gauge, the operator ordering implied by (4.62) can be shown to be identical to that proposed by Schwinger[1] as will be discussed in Sec. VII, the derivation presented above appears to be particularly simple and clear, the kinetic energy in (4.62) being essentially the familiar formula (4.7) for the Laplacian in curvilinear coordinates.

Let us conclude this section by specializing our resulting Hamiltonian operator in (4.62) to the Coulomb gauge for the group SU_2. In that case the matrix Γ_j becomes

$$\langle \vec{r}, l \mid \Gamma_j \mid \vec{r}', m\rangle = \nabla_j \delta^3(\vec{r} - \vec{r}')\delta^{lm}\,. \tag{4.63}$$

The operator $\Gamma_k \mathfrak{D}_k$ becomes

$$\Gamma_k \mathfrak{D}_k = \nabla_k \mathfrak{D}_k = \nabla_k(\nabla_k \delta^{lm} - g\epsilon^{lmn}A^n_k) \tag{4.64}$$

so that the Jacobian $\mathcal{J}$ of (4.37) can be recognized as the familiar, Coulomb-gauge, Faddeev-Popov determinant. Equation (4.62) becomes

$$\tilde{H}=\int d^3r\left[\tfrac{1}{2}\mathcal{J}^{-1}\vec{\Pi}_i^T\circ\mathcal{J}\vec{\Pi}_i^T+\tfrac{1}{2}\vec{B}_i^{\,2}+\psi^\dagger\alpha_i(-i\nabla_i)-\tfrac{1}{2}g\vec{\tau}\cdot\vec{A}_i\psi\right]$$
$$+\tfrac{1}{2}g^2\int d^3r\,d^3r'\,\mathcal{J}^{-1}\{[-\vec{\Pi}_i^T(\vec{r})\times\vec{A}_i(\vec{r})+\tfrac{1}{2}\psi^\dagger(\vec{r})\vec{\tau}\psi(\vec{r})]^m$$
$$\cdot\langle\vec{r},m|(\nabla_j\mathcal{D}_j)^{-1}(-\nabla^2)(\nabla_{j'}\mathcal{D}_{j'})^{-1}|\vec{r}',m'\rangle\mathcal{J}\cdot[\vec{A}_k(\vec{r}')\times\vec{\Pi}_k(\vec{r}')+\tfrac{1}{2}\psi^\dagger(\vec{r}')\vec{\tau}\psi(\vec{r}')]^{m'}\}, \quad (4.65)$$

where $T^l=\frac{1}{2}\tau^l$, all SU_2 vectors are indicated by arrows, and P_i^l is replaced by the more familiar notation $(\vec{\Pi}_i^T)^l$. Both $\vec{A}_i$ and $\vec{\Pi}_i^T$ are transverse. Note that the matrix $\langle\vec{r},m|(\Gamma_k\mathcal{D}_k)^{-1}(\Gamma_j\Gamma_j^\dagger)(\mathcal{D}_k^\dagger\Gamma_k^\dagger)^{-1}|\vec{r}',m'\rangle$ has reduced to precisely the usual non-Abelian, Coulomb, Green's function.

V. COVARIANT GAUGES

We will now show that the Cartesian operator ordering of the $V_0=0$ gauge Hamiltonian (3.29)–(3.31) defines a quantum theory identical to the theory determined by the usual covariant-gauge Feynman rules. The transformation of the $V_0=0$ gauge quantum theory of Sec. III into a covariant gauge is most easily done using the Feynman path-integral formalism.[6]

For the case of a single coordinate q and conjugate momentum p the path integral for the Schrödinger Green's function applied to a state $|\ \rangle$ is obtained by writing

$$\langle q'|e^{-iH(t'-t)}|\ \rangle=\lim_{\mathfrak{N}\to\infty}\prod_{n=1}^{\mathfrak{N}}\int dq(n)\langle q(n+1)|[1-i\epsilon H(p,q)]|q(n)\rangle\langle q(1)|\ \rangle, \quad (5.1)$$

where $q(n)$ denotes the coordinate q at time $t_n=t+(n-1)\epsilon$ with $\epsilon=(t'-t)/\mathfrak{N}$ and $q(\mathfrak{N}+1)=q'$. Each matrix element in the product on the right-hand side of (5.1) is represented by

$$\langle q(n+1)|1-i\epsilon H(p,q)|q(n)\rangle=\int\frac{dp(n)}{2\pi}e^{ip(n)[q(n+1)-q(n)]}\left[1-i\epsilon H\left(p(n),\frac{q(n)+q(n+1)}{2}\right)\right], \quad (5.2)$$

so that (5.1) becomes

$$\langle q'|e^{-iH(t'-t)}|\ \rangle=\lim_{\mathfrak{N}\to\infty}\prod_{n=1}^{\mathfrak{N}}\int\frac{dq(n)dp(n)}{2\pi}\exp\left\{i\epsilon\left[\frac{p(n)[q(n+1)-q(n)]}{\epsilon}-H\left(p(n),\frac{q(n+1)+q(n)}{2}\right)\right]\right\}\langle q(1)|\ \rangle, \quad (5.3)$$

the familiar Hamiltonian path integral for a one-dimensional problem. Our substitution of the variable $\frac{1}{2}[q(n+1)+q(n)]$ for the operator q in $H(p,q)$ of (5.1) corresponds to the Weyl-ordered form of the Hamiltonian. As can be readily verified, for a classical Hamiltonian $p^2f(q)+pu(q)+v(q)$, the Weyl form of the quantum-mechanical Hamiltonian is

$$H=\tfrac{1}{4}[p^2f(q)+2pf(q)p+f(q)p^2]+\tfrac{1}{2}[pu(q)+u(q)p]+v(q), \quad (5.4)$$

where q and p are operators. By substituting this expression into the left-hand side of (5.2), one sees that it leads precisely to the right-hand side.

Equation (5.3) can be immediately generalized to represent the $V_0=0$ gauge Schrödinger Green's function by a path integral for the gauge theory without fermions:

$$\langle V'|e^{-iH(t'-t)}|\rangle=\lim_{\mathfrak{N}\to\infty}\int\prod_{n=1}^{\mathfrak{N}}d[V(n)]d[\Pi(n)]$$
$$\times\left\{\exp\left[\sum_{n=1}^{\mathfrak{N}}i\epsilon\int d^3r\,\mathrm{tr}[\Pi_i(n)[V_i(n+1)-V_i(n)]\epsilon^{-1}-\tfrac{1}{2}\Pi_i(n)\Pi_i(n)-\tfrac{1}{2}B_i(n)B_i(n)]\right]\right\}\Psi[V_i(1)], \quad (5.5)$$

where $\Psi[V_i]$ is introduced by (3.28), $V_i(n)$ and $\Pi_i(n)$ refer to the matrix form (3.17) of the fields $V_i(\vec{r},t_n)$ and $\Pi_i(\vec{r},t_n)$, with $t_n=t+(n-1)\epsilon$. For simplicity, the position dependence is suppressed, $V_i(\mathfrak{N}+1)=V_i'$, and the differentials $d[V(n)]$ and $d[\Pi(n)]$ stand, respectively, for the products

$\prod_{i,l,\vec{r}} dV_i^l(\vec{r},t_n)$ and $\prod_{i,l,\vec{r}} d\Pi_i^l(\vec{r},t_n)$.

To include fermions in the path-integral description, it is necessary to introduce a representation of the fermion Hilbert space as a space of polynomials of generators $\hat{q}_1, \hat{q}_2, \ldots$ of a Grassmann algebra[7]

$$\{\hat{q}_\alpha, \hat{q}_\beta\} = 0$$

for all α and $\beta = 1, 2, \ldots$. Their differentials $d\hat{q}_\alpha$ and derivative operators $\partial/\partial\hat{q}_\alpha$ satisfy

$$\left\{\frac{\partial}{\partial\hat{q}_\alpha}, \hat{q}_\beta\right\} = \frac{\partial\hat{q}_\beta}{\partial\hat{q}_\alpha} = \delta_{\alpha\beta} \tag{5.6}$$

and

$$\{d\hat{q}_\alpha, \hat{q}_\beta\} = \{d\hat{q}_\alpha, d\hat{q}_\beta\} = 0 . \tag{5.7}$$

In addition, there are the usual integration rules

$$\int d\hat{q}_\alpha = 0 \text{ and } \int \hat{q}_\alpha d\hat{q}_\beta = \delta_{\alpha\beta} . \tag{5.8}$$

Let

$$\{\psi_\alpha(\vec{r})\} \tag{5.9}$$

be a complete orthonormal set of c-number single-particle spinor functions. By introducing for each $\psi_\alpha(\vec{r})$ a Grassmann generator $\hat{q}_\alpha$, we can represent a multiparticle state vector $|\ \rangle$ with the probability amplitude $C_r(\alpha_1, \ldots, \alpha_r)$ for the states $\psi_{\alpha_1}, \ldots, \psi_{\alpha_r}$ to be occupied by the polynomial

$$\langle q|\rangle \equiv \sum_{r=0}^{\infty} \sum_{\alpha_1<\alpha_2<\cdots<\alpha_r} C_r(\alpha_1, \ldots, \alpha_r) \times \prod_{\beta \neq \alpha_i} \hat{q}_\beta \tag{5.10}$$

in which, for definiteness, the product of the $\hat{q}_\beta$'s is arranged in the order of increasing β. Equation (5.10) is the $\hat{q}$ representation of the bra vector $|\ \rangle$; its ket vector $\langle\ |$ in the $\hat{q}$ representation then assumes the form

$$\langle\ |\hat{q}\rangle \equiv \sum_{r=0}^{\infty} \sum_{\alpha_1<\alpha_2<\cdots<\alpha_r} \pm C_r^*(\alpha_1, \ldots, \alpha_r)\hat{q}_{\alpha_1}\hat{q}_{\alpha_2}\cdots\hat{q}_{\alpha_r} , \tag{5.11}$$

where the $\pm$ sign is determined by the normalization condition

$$\langle\ |\ \rangle = \int \langle\ |q\rangle \prod_\alpha d\hat{q}_\alpha \langle q|\ \rangle = \sum_{r=0}^{\infty} \sum_{\alpha_1<\alpha_2<\cdots<\alpha_r} |C_r(\alpha_1, \ldots, \alpha_r)|^2 . \tag{5.12}$$

With the above definition, a multiplication $\hat{q}_\alpha$ onto $\langle \hat{q}|\ \rangle$ becomes the annihilation operator for the αth state, while the differentiation $\partial/\partial\hat{q}_\alpha$ is the corresponding creation operator.

The "Fourier transform" of any polynomial $\psi(\hat{q})$ is given by

$$\phi(\overline{q}) \equiv \int \exp\left(\sum_\alpha \overline{q}_\alpha \hat{q}_\alpha\right)\psi(\hat{q})\prod_\alpha d\hat{q}_\alpha , \tag{5.13}$$

where the $\overline{q}_\alpha$'s anticommute with each other as well as with the $\hat{q}_\alpha$'s. By using (5.6) and (5.8), we see that

$$\overline{q}_\beta \phi(\overline{q}) = \int \exp\left(\sum_\alpha \overline{q}_\alpha \hat{q}_\alpha\right)\frac{\partial\psi(\hat{q})}{\partial\hat{q}_\beta}\prod_\alpha d\hat{q}_\alpha , \tag{5.14}$$

and therefore the usual partial-integration rule holds. The "δ function" is defined by

$$\begin{aligned}\langle \hat{q}|\hat{q}'\rangle &\equiv \delta(\hat{q}-\hat{q}') \\ &\equiv \int \prod_\alpha d\overline{q}_\alpha \exp\sum_\alpha \overline{q}_\alpha(\hat{q}_\alpha - \hat{q}'_\alpha) \\ &= \prod_\alpha (\hat{q}'_\alpha - \hat{q}_\alpha)\end{aligned} \tag{5.15}$$

since, for arbitrary $\psi(\hat{q})$,

$$\psi(\hat{q}) = \int \delta(\hat{q}-\hat{q}')\tilde{\prod_\alpha} d\hat{q}'_\alpha \psi(\hat{q}') , \tag{5.16}$$

where the order of product $\tilde{\prod}_\alpha$ is the transpose of that in $\prod_\alpha$; i.e., if in (5.15) $\prod_\alpha d\overline{q}_\alpha = d\overline{q}_1 d\overline{q}_2 \cdots$, then $\delta(\hat{q}-\hat{q}') = (\hat{q}'_1 - \hat{q}_1)(\hat{q}'_2 - \hat{q}_2)\cdots$ in the same order, but in (5.16) $\tilde{\prod}_\alpha d\hat{q}'_\alpha = \cdots d\hat{q}'_2 d\hat{q}'_1$ in the transposed order. Hence, the inverse transform of (5.13) is

$$\psi(\hat{q}) = \int \exp\left(\sum_\alpha \hat{q}_\alpha \overline{q}_\alpha\right)\phi(\overline{q})\tilde{\prod_\alpha} d\overline{q}_\alpha . \tag{5.17}$$

As an illustration, we first examine the case of a single mode with only one generator $\hat{q}$. The corresponding basis vectors of the Hilbert space can be chosen to be $|l\rangle$, where l = occupation number = 0 or 1. In accordance with (5.10)–(5.12), we may write

$$\begin{aligned}&\langle \hat{q}|0\rangle = \hat{q}, \quad \langle \hat{q}|1\rangle = 1 , \\ &\langle 0|\hat{q}\rangle = -1, \quad \langle 1|\hat{q}\rangle = \hat{q} .\end{aligned} \tag{5.18}$$

Therefore, we have the usual orthonormality relation

$$\langle l|l'\rangle = \int \langle l|\hat{q}\rangle d\hat{q} \langle \hat{q}|l'\rangle = \delta_{ll'}$$

and the completeness condition

$$\sum_{l=0,1} \langle \hat{q}'|l\rangle\langle l|\hat{q}\rangle = -\hat{q}' + \hat{q} = \delta(\hat{q}'-\hat{q}) . \tag{5.19}$$

If a Hamiltonian H has $|\,l\rangle$ as its eigenvector with eigenvalue $E_l = lE$, then the matrix representation of e^{-iHt} is

$$\langle \hat{q}'|\, e^{-iHt}\,|\,\hat{q}\rangle = \sum_{l=0,1} \langle \hat{q}'|\,l\rangle\, e^{-iE_l t}\langle l\,|\,\hat{q}\rangle = -\hat{q}' + e^{-iEt}\hat{q}\,. \tag{5.20}$$

Returning to the general case, with the help of these equations we can easily write down the fermionic analog of (5.2):

$$\langle \hat{q}(n+1)|\, 1 - i\epsilon H\,|\,\hat{q}(n)\rangle = \int \prod_\alpha d\bar{q}_\alpha(n) \exp\Big[\sum_\beta \bar{q}_\beta(n)(\hat{q}_\beta(n+1) - \hat{q}_\beta(n))\Big] \times \Big[1 + i\epsilon \sum_{\gamma,\gamma'} \bar{q}_\gamma(n) H_{\gamma\gamma'} \hat{q}_{\gamma'}(n)\Big]\,, \tag{5.21}$$

where in our case, because of (3.1),

$$H_{\gamma\gamma'} = -i \int d^3r\, \psi_\gamma^\dagger(\vec{r})\, \alpha_j D_j\, \psi_{\gamma'}(\vec{r}) \tag{5.22}$$

with $\psi_\gamma(\vec{r})$ as a member of the c-number spinor set (5.9). Thus, if we define the anticommuting functions

$$\begin{aligned} \psi(n) &\equiv \psi(\vec{r}, t_n) \equiv \sum_\alpha \psi_\alpha(\vec{r})\hat{q}_\alpha(n)\,, \\ \bar{\psi}(n) &\equiv \bar{\psi}(\vec{r}, t_n) \equiv -\sum_\alpha \bar{\psi}_\alpha(\vec{r})\bar{q}_\alpha(n)\,, \end{aligned} \tag{5.23}$$

where $\bar{\psi}_\alpha = \psi_\alpha^\dagger \gamma_4$, $t_n = t + (n-1)\epsilon$ as before, and write $\int \prod_\alpha dq_\alpha(n)$ and $\int \prod_\alpha d\bar{q}_\alpha(n)$ as $\int d[\psi(n)]$ and $\int d[\bar{\psi}(n)]$, respectively, the fermionic Green's function becomes

$$\langle \psi'|\, e^{-iH(t'-t)}\,|\,\rangle = \lim_{\mathfrak{N}\to\infty} \int \prod_{n=1}^{\mathfrak{N}} d[\bar{\psi}(n)]\, d[\psi(n)] \times \Big\{\exp\Big[\sum_{n=1}^{\mathfrak{N}} \int d^3r\, [-\bar{\psi}(n)\gamma_4(\psi(n+1) - \psi(n)) - i\epsilon\bar{\psi}(n)\gamma_j D_j \psi(n)]\Big]\Big\}\, \Psi[\psi(1)]\,, \tag{5.24}$$

where $\Psi[\psi(1)] = \langle \psi(1)|\,\rangle$. Equations (5.5) and (5.24) can be combined to give a path-integral expression for the Yang-Mills and fermion Green's function. By using the Cartesian form $\Pi_i(n)\,\Pi_i(n)$ for the kinetic energy in (5.5), with no additional terms we have explicitly incorporated the operator ordering of the $V_0 = 0$ gauge Hamiltonian (3.29)–(3.31).

We can now transform the $V_0 = 0$ gauge quantum theory defined by (5.5) and (5.24) into a noncovariant gauge, $\chi(A_i) = 0$, by a simple change of variables. Of course, we must be careful when transforming $V(n+1) - V(n)$ and $\psi(n+1) - \psi(n)$ to keep all terms in the exponent through order ϵ if we intend the transformed theory to be equivalent to the original one. Such a procedure is completely equivalent to the operator manipulations of Sec. IV and leads to the same result.[8]

In this section we wish to use the path-integral formulation to transform to a covariant gauge. This can be done most economically if we introduce into the integrand of (5.5) the factor

$$1 = \int \prod_{n=1}^{\mathfrak{N}} d[\phi_a(n)]\, \mathfrak{J}\left(\exp\left\{\sum_{n=1}^{\mathfrak{N}} \int d^3r\, \frac{-i\epsilon}{2\alpha}\, \mathrm{tr}\Big[\nabla_i A_i(\vec{r}, t_n) + \frac{1}{\epsilon}(A_0(\vec{r}, t_{n+1}) - A_0(\vec{r}, t_n))\Big]^2\right\}\right), \tag{5.25}$$

where $A_i(n) = A_i(\vec{r}, t_n)$ and $A_0(n) = A_0(\vec{r}, t_n)$ are Hermitian matrices, defined by

$$A_i(n) = u(n)^{-1} V_i(n) u(n) + \frac{i}{g}\, u(n)^{-1}\nabla_i u(n) \tag{5.26}$$

and

$$A_0(n) = -\frac{i}{g\epsilon} \ln[u(n)^{-1} u(n+1)] = -\frac{i}{g\epsilon}\, u(n)^{-1}(u(n+1) - u(n)) + \frac{i}{2g\epsilon}\,[u(n)^{-1}(u(n+1) - u(n))]^2 + \cdots\,, \tag{5.27}$$

with

$$u(n) = u(\phi_a(n)) \tag{5.28}$$

the $N\times N$ unitary matrix function of the M group parameters $\phi_a(n)=\phi_a(\vec{r},t_n)$ in (4.5), and the differential $d[\phi_a(n)]=\Pi_{a,\vec{r}}\, d\phi_a(\vec{r},t_n)$. Except for a numerical factor, the Jacobian $|\mathfrak{J}|$ is given by

$$|\mathfrak{J}|=\det\langle\vec{r},t_n|\mathfrak{J}|\,\vec{r}\,',t_n\rangle$$
$$=\det\frac{\delta}{\delta\phi_a(n)}\left[\nabla_i A_i(n')+\frac{(A_0(n'+1)-A_0(n'))}{\epsilon}\right]. \tag{5.29}$$

In order to match the final configuration of (5.5), we set $A_i(\mathfrak{N}+1)=V_i(\mathfrak{N}+1)=V_i'$, i.e.,

$$u(\mathfrak{N}+1)=1. \tag{5.30}$$

Furthermore, since the integrand of (5.25) is invariant under $A_0(n)\to A_0(n)+\text{constant}$ for all $n=1,2,\ldots,\mathfrak{N}+1$, we may choose $A_0(\mathfrak{N}+1)=0$.

We now examine the region where the summand in the exponent of (5.25) is of order 1. Because in the summand the coefficient of $[A_0(n+1)-A_0(n)]^2$ is proportional to ϵ^{-1}, we expect only those configurations with $A_0(n+1)-A_0(n)$ of order $\epsilon^{1/2}$ to contribute to the integral over $\phi_a(n)$. Thus since $A_0(\mathfrak{N}+1)=0$, $A_0(n)$ should be of order $(\mathfrak{N}-n)^{1/2}\epsilon^{1/2}=O(1)$ and, from (5.27), $u(n+1)-u(n)$ is only of order ϵ.

If we perform $\int d[\Pi(n)]$ in (5.5) we find, up to a numerical factor,

$$\int d[\Pi(n)]\exp\left(i\epsilon\sum_n\int d^3r\,\mathrm{tr}\{\Pi_i(n)[V_i(n+1)-V_i(n)]\epsilon^{-1}-\tfrac{1}{2}\Pi_i(n)\Pi_i(n)\}\right)$$
$$=\exp\left(i\tfrac{1}{2}\sum_n\int d^3r\,\mathrm{tr}[V_i(n+1)-V_i(n)]^2\epsilon^{-1}\right). \tag{5.31}$$

Next we can use (5.26) to express this exponent in terms of A_i and u. A particularly symmetrical form is obtained if the quantity in square brackets in (5.31) is conjugated with the matrix $[u(n)^{-1}u(n+1)]^{1/2}u(n+1)^{-1}$:

$$\begin{aligned}\epsilon^{-1}\mathrm{tr}[V_i(n+1)-V_i(n)]^2=\epsilon^{-1}\mathrm{tr}\Big\{&[u(n)^{-1}u(n+1)]^{1/2}A_i(n+1)[u(n)^{-1}u(n+1)]^{-1/2}\\&+\frac{i}{g}[u(n)^{-1}u(n+1)]^{1/2}\nabla_i[u(n)^{-1}u(n+1)]^{-1/2}\\&-[u(n)^{-1}u(n+1)]^{-1/2}A_i(n)[u(n)^{-1}u(n+1)]^{1/2}\\&-\frac{i}{g}[u(n)^{-1}u(n+1)]^{-1/2}\nabla_i[u(n)^{-1}u(n+1)]^{1/2}\Big\}^2.\end{aligned} \tag{5.32}$$

Using (5.27) to replace $u(n)^{-1}u(n+1)$ by a function of $A_0(n)$,

$$u(n)^{-1}u(n+1)=1+ig\epsilon A_0(n)+O(\epsilon^2),$$

we can expand[9] the right-hand side of Eq. (5.32) through order ϵ:

$$\epsilon^{-1}\mathrm{tr}[V_i(n+1)-V_i(n)]^2=\epsilon\,\mathrm{tr}\{[A_i(n+1)-A_i(n)]\epsilon^{-1}+ig[A_0(n),A_i(n)]+\nabla_iA_0(n)\}^2+O(\epsilon^{3/2}) \tag{5.33}$$

where we treat $A_i(n+1)-A_i(n)$ as of order $\epsilon^{1/2}$. Thus, if we change integration variables from $V_i(n)$, $\psi(n)$, and $\bar{\psi}(n)$ to the unitarily equivalent set $A_i(n)$,

$$\psi'(n)=u(n)^{-1}\psi(n)$$

and (5.34)

$$\bar{\psi}'(n)=\bar{\psi}(n)u(n),$$

the exponentials in (5.5), (5.24), and (5.25) combine to give the usual covariant-gauge action up to terms vanishing with ϵ:

$$\begin{aligned}&i\epsilon\sum_n\int d^3r\,\mathrm{tr}\Big\{\tfrac{1}{2}[(A_i(n+1)-A_i(n))\epsilon^{-1}+ig[A_0(n),A_i(n)]+\nabla_iA_0(n)]^2\\&\qquad-\tfrac{1}{2}[B_i(n)]^2-\frac{1}{2a}[\nabla_iA_i(n)+(A_0(n+1)-A_0(n))\epsilon^{-1}]^2\Big\},\\&-i\epsilon\sum_n\int d^3r\Big\{\bar{\psi}'(n)\Big[\frac{\gamma_4(\psi'(n+1)-\psi'(n))}{i\epsilon}+gA_0(n)\gamma_4\psi'(n+1)+\gamma_i(\nabla_i-igA_i(n))\psi'(n)\Big]\Big\}.\end{aligned} \tag{5.35}$$

Finally, we must use Eq. (5.27) to change the integration variable $\phi_a(n)$ to $A_0^m(n)$ where $T^m A_0^m(n) = A_0(n)$. The resulting Jacobian $\det[\delta\phi_{a'}(n')/\delta A_0^m(n)]|\bar{\mathcal{G}}|$ can then be worked out explicitly. By using (5.27), we see that the matrix $\delta A_0^m(n)/\delta\phi_a(n')$ can be resolved into two terms:

$$\frac{\delta A_0^m(n)}{\delta\phi_a(n')}=\frac{1}{g\epsilon}(\delta_{nn'}\langle a|G(n)|m\rangle + \delta_{n+1n'}\langle a|G'(n)|m\rangle), \tag{5.36}$$

where, on account of (4.22),

$$\langle a|G(n)|m\rangle = \lambda_a^m(n) - \tfrac{1}{2}g\epsilon f^{klm}\lambda_a^k(n)A_0^l(n) + O(\epsilon^2). \tag{5.37}$$

Because of the boundary condition (5.30), $A_0^m(\mathfrak{N})$ depends only on $u(\mathfrak{N})$. Hence, we have

$$\langle a|G'(\mathfrak{N})|m\rangle = 0,$$

and therefore

$$\det\frac{\delta A_0^m(n)}{\delta\phi_a(n')} = \prod_{n=1}^{\mathfrak{N}} \det\frac{\langle a|G(n)|m\rangle}{g\epsilon}. \tag{5.38}$$

From (5.37), it follows that

$$\begin{aligned}\det\langle a|G(n)|m\rangle &= [1-\tfrac{1}{2}g\epsilon f^{klm}\lambda_a^k(n)A_0^l(n)(\lambda^{-1}(n))_a^m] \times \det\lambda(n) + O(\epsilon^2)\\ &= \det\lambda(n) + O(\epsilon^2),\end{aligned} \tag{5.39}$$

where $\lambda(n)$ stands for the $M\times M$ matrix $\lambda_a^m(n)$. Since $\mathfrak{N} = O(\epsilon^{-1})$, when $\epsilon\to 0$ we can drop the $O(\epsilon^2)$ term in (5.39) and derive

$$\det\frac{\delta\phi_a(n')}{\delta A_0^m(n)} = \text{const}\times\left[\prod_{n=1}^{\mathfrak{N}}\det\lambda(n)\right]^{-1}. \tag{5.40}$$

By following steps very similar to those that led to (4.36), we may express $|\bar{\mathcal{G}}|$ of (5.29) as

$$|\bar{\mathcal{G}}| = \text{const}\times\mathcal{G}_c\times\prod_n \det\lambda(n), \tag{5.41}$$

which when multiplied by (5.40) gives, in the limit $\epsilon\to 0$, a constant times the usual, covariant-gauge Faddeev-Popov determinant

$$\mathcal{G}_c = \det\{\partial_\mu \mathfrak{D}_\mu\}. \tag{5.42}$$

Consequently, apart from a constant multiplicative factor, the original $V_0 = 0$ gauge Schrödinger Green's function has been transformed precisely into the usual covariant-gauge expression

$$\langle V'|e^{-iH(t'-t)}|\rangle = \int d[A_\mu]d[\psi']d[\bar\psi']\mathcal{G}_c \exp\left\{i\epsilon\int d^4x\,\text{tr}\left[-\tfrac{1}{4}F_{\mu\nu}{}^2 - \bar\psi'\gamma_\mu\mathfrak{D}_\mu\psi' - \frac{1}{2\alpha}(\partial_\mu A_\mu)^2\right]\right\}, \tag{5.43}$$

where

$$d[A_\mu]d[\psi']d[\bar\psi'] = \lim_{n\to\infty}\prod_{n=1}^{\mathfrak{N}} d[A_\mu(n)]d[\psi'(n)]d[\bar\psi'(n)].$$

VI. NONCOVARIANT-GAUGE FEYNMAN RULES

The quantum Yang-Mills theory in the gauge $\chi(A_i) = 0$ is completely specified by the Hamiltonian operator (4.62) derived in Sec. IV. However, if this Hamiltonian is divided into free and interacting pieces,

$$H_{\text{free}} + H_{\text{int}}, \tag{6.1}$$

the resulting Dyson-Wick perturbation theory in H_{int} will be complicated because of the quadratic dependence of H_{int} on the canonical momenta P_i^l. In the interaction representation the P_i^l become $\dot{A}_i^l$, the Wick contraction of $\dot{A}_i^l(x)$ and $\dot{A}_j^l(y)$ contains noncovariant $\delta(x_0-y_0)$ terms, these δ-function terms can be systematically resummed and the result represented as new, nonpolynomial additions[10] to H_{int}. The vertices implied by this modified H'_{int}, when joined with "naive" Feynman propagators, then provide the correct Feynman perturbation series for the original quantum theory.

In this section we will derive these simplified Feynman rules for the noncovariant-gauge Hamiltonian (4.62). In practice, these Feynman rules are most easily obtained using functional integration.[6] We first represent the Schrödinger Green's function by an integral over classical trajectories in phase space as in (5.1). Instead of immediately evaluating the Gaussian integral over P_i^l, we next replace the term in H_{int} quadratic in P_i^l by a Gaussian integral with respect to a new variable A_0^l coupling linearly with P_i^l. Finally, the integration over P_i^l is carried out leaving a Lagrangian functional integral over A_i^l and A_0^l which, if the original Hamiltonian was Weyl ordered, directly specifies the proper Feynman rules.

The first step in this procedure is the most difficult. We must rearrange the operators in the Hamiltonian (4.62) into Weyl order. The Weyl ordering of the Laplacian (4.7) in curvilinear coordinates,

$$\mathcal{K} = -\tfrac{1}{2}\sum_{\alpha,\beta}|M|^{-1/2}\frac{\partial}{\partial Q_\alpha}M^{-1}{}_{\alpha\beta}|M|^{1/2}\frac{\partial}{\partial Q_\beta}, \tag{4.7}$$

is quite straightforward. The result is particularly simple if we extract a factor of $|M|^{-1/4}$ from the state vectors:

$$|M|^{1/4}\mathcal{K}|M|^{-1/4}=-\frac{1}{2}\left\{\frac{1}{4}\frac{\partial}{\partial Q_\alpha}\frac{\partial}{\partial Q_\beta}M^{-1}{}_{\alpha\beta}+\frac{1}{2}\frac{\partial}{\partial Q_\alpha}M^{-1}{}_{\alpha\beta}\frac{\partial}{\partial Q_\beta}+\frac{1}{4}M^{-1}{}_{\alpha\beta}\frac{\partial}{\partial Q_\alpha}\frac{\partial}{\partial Q_\beta}\right\}+\frac{1}{8}\left(\frac{\partial}{\partial Q_\alpha}\frac{\partial Q_\beta}{\partial q_\gamma}\right)\left(\frac{\partial}{\partial Q_\beta}\frac{\partial Q_\alpha}{\partial q_\gamma}\right), \tag{6.2}$$

in which the differentiations appearing inside the curly brackets are arranged in Weyl order, as in (5.4).

However, complications arise because what we are interested in is the Weyl ordering of the Hamiltonian (4.62) which, having been simplified by restriction to gauge-invariant states, (4.57), no longer has the form (4.7). In order to apply this simple formula to the case at hand, we must return to the form (4.38) for the gauge-theory Hamiltonian in which the angle variables ϕ_a and their conjugate momenta p_a appear. Equation (6.2) then gives a Weyl-ordered form for H, from which ϕ_a and p_a must again be eliminated as follows: The terms in curly brackets in (6.2) which contain derivatives with respect to ϕ_a are

$$\begin{aligned}-\tfrac{1}{2}\int d^3r\int d^3r'\Big\{&\langle\vec{r},l|(\Gamma_k\mathfrak{D}_k)^{-1}\Gamma_j\Gamma_j^\dagger(\mathfrak{D}_{k'}^\dagger\Gamma_{k'}^\dagger)^{-1}|\vec{r},l'\rangle g^2\\
&\times\left[\frac{1}{4}\frac{\delta}{\delta\phi_a(\vec{r})}\frac{\delta}{\delta\phi_{a'}(\vec{r}\,')}\lambda^{-1}(\vec{r})^l_a\lambda^{-1}(\vec{r}\,')^{l'}_{a'}+\frac{1}{2}\frac{\delta}{\delta\phi_a(\vec{r})}\lambda^{-1}(\vec{r})^l_a\lambda^{-1}(\vec{r}\,')^{l'}_{a'}\frac{\delta}{\delta\phi_{a'}(\vec{r}\,')}\right.\\
&\left.+\tfrac{1}{4}\lambda^{-1}(\vec{r})^l_a\lambda^{-1}(\vec{r}\,')^{l'}_{a'}\frac{\delta}{\delta\phi_a(\vec{r})}\frac{\delta}{\delta\phi_{a'}(\vec{r}\,')}\right]\\
&+ig[P^l_i(\vec{r})\langle\vec{r},l|\mathfrak{D}_i(\Gamma_k\mathfrak{D}_k)^{-1}\Gamma_j\Gamma_j^\dagger(\mathfrak{D}_{k'}^\dagger\Gamma_{k'}^\dagger)^{-1}|\vec{r}\,',l'\rangle+\langle\vec{r},l|\mathfrak{D}_i(\Gamma_k\mathfrak{D}_k)^{-1}\Gamma_j\Gamma_j^\dagger(\mathfrak{D}_{k'}^\dagger\Gamma_{k'}^\dagger)^{-1}|\vec{r}\,',l'\rangle P^l_i(\vec{r})]\\
&\times\left[\frac{1}{2}\frac{\delta}{\delta\phi_{a'}(\vec{r}\,')}\lambda^{-1}(\vec{r}\,')^{l'}_{a'}+\tfrac{1}{2}\lambda^{-1}(\vec{r}\,')^{l'}_{a'}\frac{\delta}{\delta\phi_{a'}(\vec{r}\,')}\right]\Big\}.\end{aligned}\tag{6.3}$$

If this operator is applied to the state

$$|\lambda|^{1/2}\mathfrak{U}(\phi_a)\tilde{\Psi}(A_i), \tag{6.4}$$

the result is

$$\begin{aligned}-\tfrac{1}{2}|\lambda|^{1/2}\mathfrak{U}(\phi_a)\int d^3r\int d^3r'\Big\{&\langle\vec{r},l|(\Gamma_k\mathfrak{D}_k)^{-1}\Gamma_j\Gamma_j^\dagger(\mathfrak{D}_{k'}^\dagger\Gamma_{k'}^\dagger)^{-1}|\vec{r}\,',l\rangle g^2\\
&\times\left[-\psi^\dagger(\vec{r}\,')T^{l'}\psi(\vec{r}\,')\psi^\dagger(\vec{r})T^l\psi(\vec{r})+\frac{1}{4}\left(\frac{\delta}{\delta\phi_a(\vec{r})}\lambda^{-1}(\vec{r}\,')^{l'}_{a'}\right)\left(\frac{\delta}{\delta\phi_{a'}(\vec{r}\,')}\lambda^{-1}(\vec{r})^l_a\right)\right]\\
&+g[P^l_i(\vec{r})\langle\vec{r},l|\mathfrak{D}_i(\Gamma_k\mathfrak{D}_k)^{-1}\Gamma_j\Gamma_j^\dagger(\mathfrak{D}_{k'}^\dagger\Gamma_{k'}^\dagger)^{-1}|\vec{r}\,',l\rangle\\
&+\langle\vec{r},l|\mathfrak{D}_i(\Gamma_k\mathfrak{D}_k)^{-1}\Gamma_j\Gamma_j^\dagger(\mathfrak{D}_{k'}^\dagger\Gamma_{k'}^\dagger)^{-1}|\vec{r}\,',l'\rangle P^l_i(\vec{r}\,')]\psi^\dagger(\vec{r}\,')T^{l'}\psi(\vec{r}\,')\Big\}\tilde{\Psi}(A_i).\end{aligned}\tag{6.5}$$

In obtaining this expression we have made repeated use of the identities (4.48) and (4.60). The state (6.4) above is simply (4.57) multiplied by $|\lambda|^{1/2}$, the ϕ_a-dependent part of the factor $|M|^{1/4}$ introduced in (6.2).

Next let us examine the remainder term in (6.2):

$$\begin{aligned}\frac{1}{8}\left(\frac{\partial}{\partial Q_\alpha}\frac{\partial Q_\beta}{\partial q_\gamma}\right)\left(\frac{\partial}{\partial Q_\beta}\frac{\partial Q_\alpha}{\partial q_\gamma}\right)\\
=\tfrac{1}{8}\int d^3r\int d^3r'\int d^3r''\Big\{&-[P^l_i(\vec{r}),\langle\vec{r}\,',l'|\mathfrak{D}_{i'}(\Gamma_{k'}\mathfrak{D}_{k'})^{-1}\Gamma_j|\vec{r}\,'',l''\rangle][P^{l'}_{i'}(\vec{r}\,'),\langle\vec{r},l|\mathfrak{D}_i(\Gamma_k\mathfrak{D}_k)^{-1}\Gamma_j|\vec{r}\,'',l''\rangle]\\
&-2ig[P^l_i(\vec{r}),\lambda^{-1}(\vec{r}\,')^{l'}_{a'}\langle\vec{r}\,',l'|(\Gamma_{k'}\mathfrak{D}_{k'})^{-1}\Gamma_j|\vec{r}\,'',n'\rangle U^{l''n'}(\vec{r}\,'')]\\
&\times\left(\frac{\delta}{\delta\phi_{a'}(\vec{r}\,')}\langle\vec{r},l|\delta_{ij}-\mathfrak{D}_i(\Gamma_k\mathfrak{D}_k)^{-1}\Gamma_j|\vec{r}\,'',n\rangle U^{l''n}(\vec{r}\,'')\right)\\
&+g^2\left(\frac{\delta}{\delta\phi_a(\vec{r})}\lambda^{-1}(\vec{r}\,')^{l'}_{a'}\langle\vec{r}\,',l'|(\Gamma_{k'}\mathfrak{D}_{k'})^{-1}\Gamma_j|\vec{r}\,'',n'\rangle U^{l''n'}(\vec{r}\,'')\right)\\
&\times\left(\frac{\delta}{\delta\phi_{a'}(\vec{r}\,')}\lambda^{-1}(\vec{r})^l_a\langle\vec{r},l|(\Gamma_k\mathfrak{D}_k)^{-1}\Gamma_j|\vec{r}\,'',n\rangle U^{l''n}(\vec{r}\,'')\right)\Big\}.\end{aligned}\tag{6.6}$$

The last term in (6.6) can be evaluated quite explicitly if we use the commutation relation (4.44) in the form

$$\lambda^{-1}(\vec{r})^l_a \frac{\delta\lambda^{-1}(\vec{r}')^{l'}_b}{\delta\phi_a(\vec{r})} - \lambda^{-1}(\vec{r}')^{l'}_{a'} \frac{\delta\lambda^{-1}(\vec{r})^l_b}{\delta\phi_{a'}(\vec{r}')} = f^{ll'n}\lambda^{-1}(\vec{r})^n_b \delta^3(\vec{r}-\vec{r}') \tag{6.7}$$

with the result

$$\tfrac{1}{8}g^2 \int d^3r \int d^3r' \Big\{ \langle \vec{r}, l | (\Gamma_k \mathfrak{D}_k)^{-1}\Gamma_j \Gamma_j^\dagger (\mathfrak{D}_{k'}^\dagger \Gamma_{k'}^\dagger)^{-1} | \vec{r}', l' \rangle \left(\frac{\delta}{\delta\phi_a(\vec{r})} \lambda^{-1}(\vec{r}')^{l'}_{a'} \right)\left(\frac{\delta}{\delta\phi_{a'}(\vec{r}')} \lambda^{-1}(\vec{r})^l_a \right)$$
$$ - \langle \vec{r}, l' | (\Gamma_k \mathfrak{D}_k)^{-1}\Gamma_j t^l - t^l (\Gamma_k \mathfrak{D}_k)^{-1}\Gamma_j | \vec{r}, n \rangle \langle \vec{r}, l | (\Gamma_{k'} \mathfrak{D}_{k'})^{-1}\Gamma_j t^{l'} | \vec{r}, n \rangle \delta^3(\vec{r}-\vec{r}') \Big\}, \tag{6.8}$$

where $(t^l)_{mn} = -if^{lmn}$ is the adjoint representation analog of the generator T^l. The ϕ-dependent term in this result cancels the ϕ dependence of (6.5) so the combination of (6.5) and (6.8) will not depend on ϕ.

The second term in (6.6), linear in $P^l_i(\vec{r})$, is also not hard to simplify. Because the vector $|\xi\rangle$ defined by

$$\langle i, \vec{r}, l | \xi \rangle \equiv \langle \vec{r}, l | [\delta_{ij} - \mathfrak{D}_i (\Gamma_k \mathfrak{D}_k)^{-1}\Gamma_j)] | \vec{r}'', n \rangle \tag{6.9}$$

is automatically orthogonal to $|\bar{\xi}\rangle$,

$$\langle i, \vec{r}, l | \bar{\xi} \rangle = \langle \vec{r}, l | \Gamma_i^\dagger | \xi \rangle$$

given by (4.16), we can replace

$$P^l_i(\vec{r}) = \sum_N f^l_i(\vec{r})_N \left(-i \frac{\partial}{\partial Q_N} \right)$$

by $-i\delta/\delta A^l_i(\vec{r})$ so that the second term becomes

$$\tfrac{1}{4}g^2 \int d^3r \int d^3r' \langle \vec{r}', l' | (\Gamma_k \mathfrak{D}_k)^{-1}\Gamma_i t^l | \vec{r}, m \rangle \langle \vec{r}, m | (\Gamma_{k'} \mathfrak{D}_{k'})^{-1}\Gamma_j | \vec{r}', n \rangle \langle \vec{r}, l | (\delta_{ij} - \mathfrak{D}_i (\Gamma_{k''} \mathfrak{D}_{k''})^{-1}\Gamma_j) t^{l'} | \vec{r}', n \rangle. \tag{6.10}$$

This result can be combined with the second, ϕ-independent term in (6.8) to yield

$$\mathcal{V}_1(A) = \tfrac{1}{8}g^2 \int d^3r \langle \vec{r}, l' | (\Gamma_k \mathfrak{D}_k)^{-1}\Gamma_j | \vec{r}, l \rangle \langle \vec{r}, m | (\Gamma_{k'} \mathfrak{D}_{k'})^{-1}\Gamma_j (t^{l'}t^l + t^l t^{l'}) | \vec{r}, m \rangle$$
$$ + \tfrac{1}{4}g^2 \int d^3r \int d^3r' \langle \vec{r}', l' | (\Gamma_k \mathfrak{D}_k)^{-1}\Gamma_i t^l | \vec{r}, m \rangle \langle \vec{r}, m | (\Gamma_{k'} \mathfrak{D}_{k'})^{-1}\Gamma_j t^{l'} | \vec{r}', n \rangle \langle \vec{r}, l | \mathfrak{D}_i (\Gamma_{k''} \mathfrak{D}_{k''})^{-1}\Gamma_j | \vec{r}', n \rangle ,$$

which, on account of the identity, valid for arbitrary states $|X\rangle$, $|Y\rangle$, and $|Z\rangle$,

$$\int d^3r f^{abc} \{ \langle \vec{r}, a | \mathfrak{D}_i | X \rangle \langle \vec{r}, b | Y \rangle \langle \vec{r}, c | Z \rangle + \langle \vec{r}, a | X \rangle \langle \vec{r}, b | \mathfrak{D}_i | Y \rangle \langle \vec{r}, c | Z \rangle + \langle \vec{r}, a | X \rangle \langle \vec{r}, b | Y \rangle \langle \vec{r}, c | \mathfrak{D}_i | Z \rangle \} = 0 \tag{6.11}$$

can be further reduced to

$$\mathcal{V}_1(A) = \tfrac{1}{8}g^2 \int d^3r \langle \vec{r}, l' | (\Gamma_k \mathfrak{D}_k)^{-1}\Gamma_j | \vec{r}, l \rangle \times \langle \vec{r}, m | (\Gamma_{k'} \mathfrak{D}_{k'})^{-1}\Gamma_j t^{l'} t^l | \vec{r}, m \rangle . \tag{6.12}$$

In order to simplify the first term in (6.6), we observe that

$$-i \frac{\delta}{\delta A^{l'}_{i'}(\vec{r}')} \langle \vec{r}, l | \mathfrak{D}_i (\Gamma_k \mathfrak{D}_k)^{-1}\Gamma_j | \vec{r}'', l'' \rangle$$
$$ = g \langle \vec{r}, l | [-\delta_{ii'} + \mathfrak{D}_i (\Gamma_k \mathfrak{D}_k)^{-1}\Gamma_{i'}] t^l | \vec{r}', n \rangle \times \langle \vec{r}', n | (\Gamma_{k'} \mathfrak{D}_{k'})^{-1}\Gamma_j | \vec{r}'', l'' \rangle$$

and therefore

$$[P^{l'}_{i'}(\vec{r}'), \langle \vec{r}, l | \mathfrak{D}_i (\Gamma_k \Gamma_k^\dagger)^{-1}\Gamma_j | \vec{r}'', l'' \rangle] , \tag{6.13}$$

regarded as vectors in the functional space $\{|i, \vec{r}, l\rangle\}$, are both orthogonal to $\langle i, \vec{r}, l | \bar{\xi} \rangle$ given by (4.16). Since in the operator identity

$$P^l_i(\vec{r}) = -i \frac{\delta}{\delta A^l_i(\vec{r})} + i \int d^3\bar{r} \langle \vec{r}, l | \Gamma_i^\dagger (\Gamma_{\bar{k}} \Gamma_{\bar{k}}^\dagger)^{-1} \Gamma_{\bar{i}} | \bar{r}, \bar{l} \rangle \frac{\delta}{\delta A^{\bar{l}}_{\bar{i}}(\bar{r})}$$

the second term is parallel to $\langle i, \vec{r}, l | \bar{\xi} \rangle$ and consequently orthogonal to (6.13), we can simplify the first term in (6.6) by replacing $P^l_i(\vec{r})$ by $-i\delta/\delta A^l_i(\vec{r})$, and $P^{l'}_{i'}(\vec{r}')$ by $-i\delta/\delta A^{l'}_{i'}(\vec{r}')$. Thus, the first term in (6.6) becomes

$$\mathcal{V}_2(A) = \tfrac{1}{8}g^2 \int d^3r \int d^3r' \{ \langle \vec{r}', l' | \delta_{i'i} - \mathfrak{D}_{i'} (\Gamma_k \mathfrak{D}_k)^{-1}\Gamma_i | \vec{r}, n \rangle \langle \vec{r}, l | \delta_{ii'} - \mathfrak{D}_i (\Gamma_{k'} \mathfrak{D}_{k'})^{-1}\Gamma_{i'} | \vec{r}', n' \rangle \times \langle \vec{r}, n | t^l (\Gamma_{\bar{k}} \mathfrak{D}_{\bar{k}})^{-1}\Gamma_j \Gamma_j^\dagger (\mathfrak{D}_{\bar{k}'}^\dagger \Gamma_{\bar{k}'}^\dagger)^{-1} t^{l'} | \vec{r}', n' \rangle \} . \tag{6.14}$$

The results (6.5), (6.8), (6.12), and (6.14) can now be combined, giving a Weyl-ordered form for the $\chi(A_i) = 0$ gauge Hamiltonian

$$
\begin{aligned}
\bar{H}(P,A) &\equiv \mathcal{J}^{1/2}\tilde{H}(P,a)\mathcal{J}^{-1/2} \\
&= \tfrac{1}{2}\int d^3r\, P_i^l(\vec{r})P_i^l(\vec{r}) + \mathcal{V}_1(A) + \mathcal{V}_2(A) \\
&\quad + \tfrac{1}{2}\int d^3r \int d^3r' \{[-P_i^l(\vec{r})\mathfrak{D}_i + g\psi^\dagger T^l\psi](\Gamma_k\mathfrak{D}_k)^{-1}\Gamma_j\Gamma_j^\dagger(\mathfrak{D}_{k'}^\dagger\Gamma_{k'}^\dagger)^{-1}[\mathfrak{D}_{i'}P_{i'}^{l'}(\vec{r}\,') + g\psi^\dagger T^{l'}\psi]\}_W \\
&\quad + \int d^3r[\tfrac{1}{2}B_j^lB_j^l + \psi^\dagger\alpha_j(-i\nabla_j - gA_j^lT^l)\psi],
\end{aligned}
\tag{6.15}
$$

where $\mathcal{J}$ and $\tilde{H}(P,A)$ are given by (4.37) and (4.62), respectively, and $\{\;\}_W$ indicates that the enclosed operators are to be Weyl-ordered, i.e., arranged in the same order as those inside the curly brackets in (6.2). Thus, with the substitution

$$\tilde{\Psi}(A) = \mathcal{J}^{-1/2}\bar{\Psi}(A), \tag{6.16}$$

the time-dependent Schrödinger equation $i\dot{\tilde{\Psi}}(A) = \tilde{H}(P,A)\tilde{\Psi}(A)$ becomes

$$i\dot{\bar{\Psi}}(A) = \bar{H}(P,A)\bar{\Psi}(A). \tag{6.17}$$

As was discussed in Sec. V, the Schrödinger Green's function, applied to the state $\bar{\Psi}(A)$, can be written as

$$
\begin{aligned}
\langle A'|e^{-i\bar{H}(t'-t)}|\rangle = \lim_{\mathfrak{N}\to\infty} \int \prod_{n=1}^{\mathfrak{N}} d[A(n)]\, d[P(n)]\, d[\bar{\psi}(n)]\, d[\psi(n)] \\
\times\Bigg(\exp\Bigg\{i\epsilon\sum_{n=1}^{\mathfrak{N}} \int d^3r\Big[P_i^l(n)(A_i^l(n+1) - A_i^l(n))\epsilon^{-1} \\
- \bar{H}\Big(P(n), \frac{A(n+1)+A(n)}{2}\Big)\Big]\Bigg\}\Bigg)\bar{\Psi}[A_i(1)],
\end{aligned}
\tag{6.18}
$$

where all notations are the same as those in (5.5) and (5.24), except for the replacements of $V(n)$ and $\Pi(n)$ by $A(n)$ and $P(n)$; hence, $A_i(\mathfrak{N}+1) = A_i'$, $d[A(n)] = \Pi_{i,l,\vec{r}}\, dA_i^l(\vec{r},t_n)$, $A_i^l(n) = A_i^l(\vec{r},t_n)$, $\bar{\psi}(n) = \psi^\dagger(\vec{r},t_n)\gamma_4$, ... The quadratic interaction of the momenta $P_i^l(\vec{r})$ can be removed by introducing a Gaussian integral over a new variable $A_0^l(n) = A_0^l(\vec{r},t_n)$:

$$
\begin{aligned}
&\exp\Big\{-i\epsilon\tfrac{1}{2}\sum_{n=1}^{\mathfrak{N}} \int d^3r\int d^3r'[-P_i^l(\vec{r})\mathfrak{D}_i + g\psi^\dagger T^l\psi](\Gamma_k\mathfrak{D}_k)^{-1}\Gamma_j\Gamma_j^\dagger(\mathfrak{D}_{k'}^\dagger\Gamma_{k'}^\dagger)^{-1}(\mathfrak{D}_{i'}P_{i'}^{l'}(\vec{r}\,') + g\psi^\dagger T^{l'}\psi]\Big\} \\
&\quad = \int\prod_{n=1}^{\mathfrak{N}} d[A_0(n)]\,\mathcal{J}\exp\Big\{i\epsilon\sum_{n=1}^{\mathfrak{N}}\Big[\int d^3r[P_i^l(n)\mathfrak{D}_iA_0^l(n) - g\psi^\dagger(n)T^l\psi(n)A_0^l(n)] \\
&\qquad + \tfrac{1}{2}\int d^3r\int d^3r'A_0^l(\vec{r},n)\langle l,\vec{r}|\mathfrak{D}_k^\dagger\Gamma_k^\dagger(\Gamma_j\Gamma_j^\dagger)^{-1}\Gamma_{k'}\mathfrak{D}_{k'}|l',\vec{r}\,'\rangle A_0^{l'}(\vec{r}\,',n)\Big]\Big\},
\end{aligned}
\tag{6.19}
$$

where, apart from a constant factor, $\mathcal{J} = \det|\Gamma_k\mathfrak{D}_k|$ as in (4.37), which differs from

$$[\det(\mathfrak{D}_k^\dagger\Gamma_k^\dagger(\Gamma_j\Gamma_j^\dagger)^{-1}\Gamma_{k'}\mathfrak{D}_{k'})]^{1/2}$$

only by another A_i^l-independent factor. One should note that just as $\bar{H}$ in (6.18) is evaluated at the symmetric point $\frac{1}{2}[A(n+1)+A(n)]$, the same rule must also apply to the operator $\mathfrak{D}_i$ wherever it appears in (6.19). Finally, we can perform the P_i^l integrations. Because P_i obeys the constraint (4.15), $\Gamma_jP_j = 0$, we have

$$
\begin{aligned}
&\int d[P(n)]\exp\Big(i\epsilon\int d^3r\Big\{P_i^l(n)[(A_i^l(n+1) - A_i^l(n))\epsilon^{-1} + \mathfrak{D}_iA_0^l(n)] - \tfrac{1}{2}P_i^l(n)P_i^l(n)\Big\}\Big) \\
&\quad = \exp\Big\{i\epsilon\Big[\int d^3r\,\tfrac{1}{2}[(A_i^l(n+1) - A_i^l(n))\epsilon^{-1} + \mathfrak{D}_iA_0^l(n)]^2 \\
&\qquad - \tfrac{1}{2}\int d^3r\,d^3r'A_0^l(\vec{r})\langle l,\vec{r}|\mathfrak{D}_k^\dagger\Gamma_k^\dagger(\Gamma_j\Gamma_j^\dagger)^{-1}\Gamma_{k'}\mathfrak{D}_{k'}|l',\vec{r}\,'\rangle A_0^{l'}(\vec{r}\,')\Big]\Big\}.
\end{aligned}
\tag{6.20}
$$

Substituting (6.19) into (6.18), applying (6.20) to the P_i^l integrations and taking the limit $\epsilon\to 0$, we find

$$\langle A' | e^{-iH(t'-t)} | \rangle = \int d[A_\mu] \mathcal{J} d[\bar{\psi}(n)] d[\psi(n)] \exp\left(i \int_t^{t'} dt_1 \left\{ -\mathcal{V}_1 - \mathcal{V}_2 + \int d^3 r [-\tfrac{1}{4} F^l_{\mu\nu} F^l_{\mu\nu} - \bar{\psi} \gamma^\mu D_\mu \psi] \right\} \right). \quad (6.21)$$

Perturbation expansion of (6.21) gives the usual Feynman rules for the gauge $\chi(A_i) = 0$ with the addition of the new term $\mathcal{V}_1 + \mathcal{V}_2$.

In order to justify this statement, we must recall the connection between occurrence of the symmetric combination $\frac{1}{2}[A(n+1) + A(n)]$ in our definition of (6.18) and (6.19) and the usual Feynman rules. This connection can be most easily illustrated by considering the case of a simple one-dimensional harmonic oscillator with frequency ω. Let the "contraction" between any two functions $F(t)$ and $G(t)$ of the coordinate $q(t)$ and the velocity dq/dt be given by the standard expression

$$F(t)^{\cdot} G(t')^{\cdot} \equiv \frac{\int d[q] F(t) G(t') \exp(-i \int L(q, dq/dt) dt)}{\int d[q] \exp(-i \int L(q, dq/dt) dt)} . \quad (6.22)$$

With our convention of replacing $q(t)$ by the symmetric form $\frac{1}{2}[q(n+1) + q(n)]$ and dq/dt by the antisymmetric form $[q(n+1) - q(n)]\epsilon^{-1}$, and then taking the limit $\epsilon \to 0$, it is straightforward to verify that

$$q(t)^{\cdot} q(t')^{\cdot} = D(t-t') = \frac{i}{2\pi} \int \frac{e^{-ik_0(t-t')}}{k_0^2 - \omega^2} dk_0, \quad (6.23)$$

$$\left[\frac{dq(t)}{dt}\right]^{\cdot} q(t')^{\cdot} = \frac{d}{dt} D(t-t'), \quad (6.24)$$

and

$$\left[\frac{dq(t)}{dt}\right]^{\cdot} \left[\frac{dq(t')}{dt'}\right]^{\cdot} = -\frac{d^2}{dt^2} D(t-t')$$

$$= \omega^2 D(t-t') + i\delta(t-t'). \quad (6.25)$$

These equations correspond to the usual "covariant" Feynman propagators.

We note that in (6.22) when $t = t'$, the product $F(t)G(t)$ for $F = dq/dt$ and $G = q$ is, because of our choice,

$$\frac{1}{2\epsilon}[q(n+1) - q(n)][q(n+1) + q(n)]$$

$$= \frac{1}{2\epsilon}[q(n+1)^2 - q(n)^2],$$

which, after the $dq(n)$ and $dq(n+1)$ integrations, clearly gives zero, in accordance with the right-hand side of (6.24),

$$\frac{d}{dt} D(0) = \frac{1}{2\pi} \int \frac{k_0}{k_0^2 - \omega^2} dk_0 = 0 .$$

For $F = G = dq/dt$, the factor $F(t)G(t)$ is

$$\frac{1}{\epsilon^2}[q(n+1) - q(n)]^2$$

whose integrated value is $\frac{1}{2}\omega + i\epsilon^{-1}$, which in the limit $\epsilon \to 0$ agrees with the right-hand side of (6.25). This is to be contrasted with the usual Wick definition: the vacuum expectation value of $t([dq(t)/dt][dq(t')/dt'])$, or $T(p(t)p(t'))$, which is finite when $t = t'$ and differs from (6.25) by $i\delta(t-t')$. It is this difference that, when summed directly in the Dyson-Wick perturbation formalism, leads to the Jacobian[10] or Faddeev-Popov ghost[11] term.

Our result (6.21) implies that for the gauge $\chi(A_i) = 0$, in addition to the Jacobian $e^{\ln \mathcal{J}}$, the usual Feynman rules must be augmented by the potential-like term $\mathcal{V}_1 + \mathcal{V}_2$ given in (6.12) and (6.14). Although the derivation is somewhat lengthy, the physical origin of both terms is the same as that of $r^{1/2}$ and $-(8r^2)^{1/2}$ in (2.14), for the simple mechanical example.

The Jacobian is usually converted into an additional term $-i\delta(0)\ln\mathcal{J}$ in the Lagrangian. In contrast, the new $\mathcal{V}_1$ and $\mathcal{V}_2$ terms have very different characteristics; they are both real and without the $\delta(0)$ factor. When expanded in power series of A^l_i, each can be written as

$$\mathcal{V}_\alpha = O(g^4 A^2) + O(g^5 A^3) + O(g^6 A^4) + \cdots , \quad (6.26)$$

where $\alpha = 1$ or 2. [There is an $O(g^2)$ constant term in $\mathcal{V}_2$, which can be dropped.] For example, in the Coulomb gauge and for SU_2, to order g^4 we have

$$\mathcal{V}_1(A) = -\frac{g^4}{4} \int d^3r\, d^3r'\, d^3r''$$

$$\times K_{ij}(\vec{r} - \vec{r}') K_{ik}(\vec{r} - \vec{r}'') \vec{A}_j(\vec{r}') \cdot \vec{A}_k(\vec{r}''), \quad (6.27)$$

where

$$K_{ij}(\vec{\rho}) = \left[\tfrac{1}{3}\delta_{ij}\delta^3(\vec{\rho}) - \frac{1}{4\pi\rho^5}(3\rho_i\rho_j - \rho^2\delta_{ij})\right]\frac{1}{4\pi\rho}, \quad (6.28)$$

where $\rho = |\vec{\rho}|$ and the components of $\vec{A}_j$ are A^l_j. The corresponding expression for $\mathcal{V}_2(A)$ is similar, but more complicated. In (6.28), the singular term $\propto \delta^3(\vec{\rho})/\rho$ gives rise to infinites which are presumably relevant for the cancellation of divergences from the usual two-loop g^4-Feynman graphs; the remainder is finite when $\rho \neq 0$ and leads to new nonlocal interactions between the gauge fields which could be of physical importance.

VII. CONCLUSION

The $\chi(A_i) = 0$ gauge generating function $Z(J_\mu)$ implied by the discussion of Sec. VI is

$$Z(J_\mu)=\int d[A_\mu]\,\delta(\chi(A_l))\times\exp\Big(i\int d^4x\,\{-\tfrac{1}{4}F^l_{\mu\nu}F^l_{\mu\nu}+J^l_\mu A^l_\mu - i\,\delta(0)\,\langle x,l\,|\ln(\Gamma_k\mathfrak{D}_k)|\,x,l\rangle -\mathcal{V}_1(A)-\mathcal{V}_2(A)\}\Big). \qquad (7.1)$$

The first two terms in the exponent make up the usual classical action which, if a factor of g is absorbed into A_μ, is of order g^{-2}. The third Faddeev-Popov term is of order g^0 and although peculiar to the functional-integral description, makes an essential contribution to the Feynman rules. The final two terms, of order g^2, have been the object of the discussion in this paper, and are given explicity by (6.12) and (6.14). If Planck's constant is distinguished from one, then these three classes of terms are of order $\hbar^{-1}$, $\hbar^0$, and $\hbar$, respectively.

As we have seen, the extra terms $\mathcal{V}_1$ and $\mathcal{V}_2$ are precisely those which arise when the term in the quantum Hamiltonian $\bar{H}$, quadratic in the conjugate momenta $P^l_i(\vec{r})$, is Weyl ordered. In the pure Yang-Mills case, Eq. (6.15) can be written as

$$\bar{H}(P,A)=\tfrac{1}{2}\int d^3r\{[E^l_i(\vec{r})]^2+[B^l_i(\vec{r})]^2\}_W+\mathcal{V}_1+\mathcal{V}_2\,, \qquad (7.2)$$

where

$$E^l_i(\vec{r})=P^l_i(\vec{r})+\int d^3r'\langle\vec{r},l\,|\Gamma^\dagger_i(\mathfrak{D}^\dagger_k\Gamma^\dagger_k)^{-1}\mathfrak{D}_j|\,\vec{r}\,',l'\rangle\times P^{l'}_j(\vec{r}\,')\,, \qquad (7.3)$$

and, as before, the subscript W denotes Weyl ordering. For the case of the Coulomb gauge, our result (7.2) can be directly compared with that of Schwinger.[1,2] In our notation, Schwinger's Hamiltonian takes the form

$$\tfrac{1}{2}\int d^3r\{[E^l_i(\vec{r})_W]^2+[B^l_i(\vec{r})]^2\}+\mathcal{V}_1\,, \qquad (7.4)$$

where the operator $E^l_i(\vec{r})$ defined by (7.3) is first to be Weyl ordered and then squared. The equality of the expression (7.4) and $\bar{H}$ in (7.2) follows from the identity

$$-\frac{1}{8}\left[\frac{\partial}{\partial Q_N}F^\gamma_N+F^\gamma_N\frac{\partial}{\partial Q_N}\right]^2=-\frac{1}{8}\left[\frac{\partial}{\partial Q_N}\frac{\partial}{\partial Q_N}F^\gamma_NF^\gamma_{N'}+2\frac{\partial}{\partial Q_N}F^\gamma_NF^\gamma_{N'}\frac{\partial}{\partial Q_{N'}}+F^\gamma_NF^\gamma_{N'}\frac{\partial}{\partial Q_N}\frac{\partial}{\partial Q_{N'}}\right]+\frac{1}{8}\frac{\partial F^\gamma_N}{\partial Q_{N'}}\frac{\partial F^\gamma_{N'}}{\partial Q_N}\,, \qquad (7.5)$$

in which we may let Q_N be that given by (4.13) and F^γ_N be

$$\frac{\delta Q_N}{\delta V^l_j(\vec{r})}=\int d^3r' f^{l'}_i(\vec{r}\,')_N\langle\vec{r}\,',l\,|\delta_{ij}-\mathfrak{D}_i(\Gamma_k\mathfrak{D}_k)^{-1}\Gamma_j|\,\vec{r},l\rangle\,. \qquad (7.6)$$

The last term on the right-hand side of (7.5) can then be written as

$$\int d^3r\frac{\partial}{\partial Q_N}\left[\frac{\delta Q_{N'}}{\delta V^l_j(\vec{r})}\right]\frac{\partial}{\partial Q_{N'}}\left[\frac{\delta Q_N}{\delta V^l_j(\vec{r})}\right], \qquad (7.7)$$

which is precisely the first term inside the curly brackets in (6.6), the quantity defined as $\mathcal{V}_2$. Thus, $[E^l_i(\vec{r})_W]^2$ equals $[E^l_i(\vec{r})^2]_W+\mathcal{V}_2$ and the Hamiltonians (7.2) and (7.4) are identical.

Although the detailed arguments used in this paper to deduce the Hamiltonian operator $\bar{H}$ differ significantly from the method used by Schwinger, there is a close relationship between the initial physical assumptions. While Schwinger[1] determines $\bar{H}$ so that the Lorentz group generators obey the proper commutation relations, we show that $\bar{H}$ is equivalent to a $V_0=0$ gauge Hamiltonian with Cartesian operator ordering. Of course, the Lorentz group generators are gauge invariant and can be easily seen to obey the proper commutation relations when expressed in the $V_0=0$ gauge. Hence, these two methods lead to the same Hamiltonian operator. However, it is the Weyl-ordered $\bar{H}$ of (7.2) given in this paper that must be used to deduce the Feynman rules.

Note added in proof. We wish to thank J.-L. Gervais, I. Muzinich, and T. N. Tudron for informing us of the following papers which have also discussed the transformation from the $V_0=0$ gauge to the Coulomb gauge along lines similar to those in Sec. IV of our paper: V. N. Gribov, lecture at the 12th Winter School of the Leningrad Nuclear Physics Institute, 1977 (unpublished); J.-L. Gervais and B. Sakita, Phys. Rev. D 18, 453 (1978); M. Creutz, I. Muzinich, and T. N. Tudron, *ibid.* 19, 531 (1979); T. N. Tudron, Syracuse University Report No. SU-4217-156 (unpublished). Finally, we are indebted to R. Marmelius for bringing the work of R. Utiyama and J. Sakamoto [Prog. Theor. Phys. 55, 1631 (1976)] to our attention. These authors discuss the Coulomb-gauge operator-ordering problem from a viewpoint quite similar to ours. However, their technique for solving the Gauss constraint is significantly different from ours and the result seems more complex.

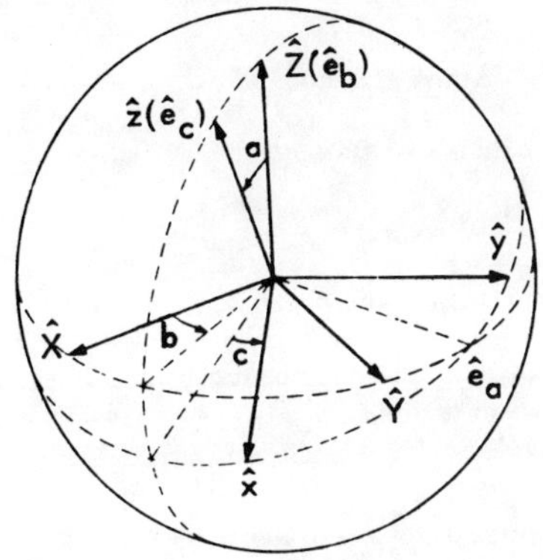

FIG. 1. The basis vectors of Σ_{lab} are $\hat{X}$, $\hat{Y}$, $\hat{Z}$ and those of Σ_{body} are $\hat{x}$, $\hat{y}$, $\hat{z}$. The rotation sequence "*cab*" takes Σ_{lab} to Σ_{body}: first an angle b around $\hat{Z}(\hat{e}_b)$, then an angle a around $\hat{e}_a$, and finally an angle c around $\hat{z}(\hat{e}_c)$.

ACKNOWLEDGMENTS

We thank Edward Yao and Charles Sommerfield for bringing the work of J. Schwinger and R. P. Treat to our attention. This research was supported in part by the U. S. Department of Energy.

APPENDIX

In this appendix, we consider the special case of SU_2 and give a simple physical interpretation of the operators j^l, J^l and their commutation relations (4.44)–(4.47). There are three group parameters ϕ_a, which may be chosen to be the Eulerian angles a, b, and c. Thus, the u matrix of (4.5) can be written as

$$u = e^{-i\tau_z b/2} e^{-i\tau_y a/2} e^{-i\tau_z c/2}, \tag{A1}$$

where τ_x, τ_y, τ_z are the usual Pauli matrices.

We recall that the origin of Eulerian angles a, b, c lies in the description of a rigid-body rotation. As shown in Fig. 1, there are two reference systems, the laboratory frame Σ_{lab} and the body frame Σ_{body}. Each frame is defined by a basis of three orthogonal unit vectors: $\hat{X}$, $\hat{Y}$, $\hat{Z}$ for Σ_{lab} and $\hat{x}$, $\hat{y}$, $\hat{z}$ for Σ_{body}. To go from Σ_{lab} to Σ_{body}, we first rotate an angle b along

$$\hat{Z} \equiv \hat{e}_b,$$

which moves the y axis from $\hat{Y}$ to $\hat{e}_a$; then an angle a along $\hat{e}_a$ which rotates the z axis from $\hat{Z}$ to $\hat{z}$; finally an angle c along

$$\hat{z} \equiv \hat{e}_c .$$

Consider now a point P in space, whose coordinates in Σ_{lab} and Σ_{body} are, respectively, X, Y, Z and x, y, z. Let us define

$$R \equiv \tfrac{1}{2}(\tau_x X + \tau_y Y + \tau_z Z)$$

and

$$r \equiv \tfrac{1}{2}(\tau_x x + \tau_y y + \tau_z z) .$$

From the definition of the Eulerian angles, it follows that

$$R = uru^\dagger ,$$

where u is given by (A1). (Here, R plays the role of V_i in the gauge-field case, and r the role of A_i.)

We keep Σ_{lab} fixed, and consider the rotation of Σ_{body} by changing a, b, and c. The angular velocity vector is

$$\dot{a}\hat{e}_a + \dot{b}\hat{e}_b + \dot{c}\hat{e}_c , \tag{A2}$$

where the dot denotes a time derivative. Let us refer to the components of (A2) in Σ_{body} and Σ_{lab} as ω^l and Ω^l, respectively. It is useful to define

$$iu^\dagger \dot{u} \equiv \omega = \tfrac{1}{2}\vec{\tau}\cdot\vec{\omega} \tag{A3}$$

and

$$i\dot{u}u^\dagger \equiv \Omega = \tfrac{1}{2}\vec{\tau}\cdot\Omega . \tag{A4}$$

Then, the matrices ω and Ω are related by

$$\Omega = u\omega u^\dagger , \tag{A5}$$

and the components of $\vec{\omega}$ and $\vec{\Omega}$ are the aforementioned ω^l and Ω^l, related by

$$\Omega^l = U^{lm}\omega^m \tag{A6}$$

with U^{lm} given by (4.23). The quantities λ^l_a and Λ^l_a, defined by (4.22) and (4.39), are related to ω^l and Ω^l by

$$\omega^l = \lambda^l_a \dot{\phi}_a \quad \text{and} \quad \Omega^l = \Lambda^l_a \dot{\phi}_a . \tag{A7}$$

The Lagrangian L of a rigid body with no external forces is a function only of ω^1, ω^2, and ω^3:

$$L = L(\omega^l) .$$

Through (A2), L is also a function of a, b, c and $\dot{a}, \dot{b}, \dot{c}$. The conjugate momenta of a, b, and c are given, respectively, by

$$p_a = \frac{\partial\omega^l}{\partial\dot{a}}\frac{\partial L}{\partial\omega^l}, \quad p_b = \frac{\partial\omega^l}{\partial\dot{b}}\frac{\partial L}{\partial\omega^l} \quad \text{and} \quad p_c = \frac{\partial\omega^l}{\partial\dot{c}}\frac{\partial L}{\partial\omega^l} . \tag{A8}$$

A comparison between (A8) and (4.43) gives

$$j^l = \frac{\partial L}{\partial\omega^l} , \tag{A9}$$

which is simply the component of the angular momentum vector in Σ_{body}. The same vector viewed in Σ_{lab} carries the component J^l:

$$\hat{x}j^1 + \hat{y}j^2 + \hat{z}j^3 = \hat{X}J^1 + \hat{Y}J^2 + \hat{Z}J^3 , \tag{A10}$$

or $J^l = U^{lm}j^m$, as in (4.43).

In the quantum theory,

$$\vec{J} \equiv \hat{X}J^1 + \hat{Y}J^2 + \hat{Z}J^3 \tag{A11}$$

is the rotational operator. Its components in $\Sigma_{\rm lab}$ satisfy the usual commutation relation

$$[J^l, J^m] = i\epsilon^{lmn}J^n . \qquad \text{(A12)}$$

On the other hand, its components the j^l's in $\Sigma_{\rm body}$ do not. From (A10), we have $j^1 = \hat{x}\cdot\vec{J}$ and similar expressions for J^2 and J^3. Since $\hat{x}$, $\hat{y}$, and $\hat{z}$ are unit vectors fixed in the body frame, under a rotation they transform like $\vec{J}$; all rotate like vectors. Hence, the scalar products $\hat{x}\cdot\vec{J}$, $\hat{y}\cdot\vec{J}$, and $\hat{z}\cdot\vec{J}$ are invariants, and that gives

$$[J^l, j^m] = 0 . \qquad \text{(A13)}$$

Likewise, we can show that

$$[j^l, j^m] = -i\epsilon^{lmn}j^n . \qquad \text{(A14)}$$

Equations (4.44)–(4.47) are simply generalizations of these standard commutation rules, but for the gauge field.

It is important to note that in the generalized Laplacian, $\mathcal{K}$ of (4.38), the angular momentum (i.e., charge) operator that appears naturally is j^l. Since $j^l = U^{ml}J^m$, the Gauss theorem (3.27) takes on the form

$$j^l|\rangle = -U^{ml}\psi^\dagger T^l\psi|\rangle , \qquad \text{(A15)}$$

which is the alternative expression for (4.60).

In terms of a, b, and c the components of $\vec{\omega}$ and $\vec{\Omega}$ are given by

$$\omega^1 = \dot{a}\sin c - \dot{b}\sin a\cos c ,$$
$$\omega^2 = \dot{a}\cos c + \dot{b}\sin a\sin c , \qquad \text{(A16)}$$
$$\omega^3 = \dot{b}\cos a + \dot{c}$$

and

$$\Omega^1 = -\dot{a}\sin b + \dot{c}\sin a\cos b ,$$
$$\Omega^2 = \dot{a}\cos b + \dot{c}\sin a\sin b , \qquad \text{(A17)}$$
$$\Omega^3 = \dot{b} + \dot{c}\cos a .$$

The matrices λ and Λ can be obtained by differentiating these expressions with respect to $\dot{a}$, $\dot{b}$, and $\dot{c}$. Likewise, the components j^l and J^l can be expressed in terms of $p_a = -i\partial/\partial a$, $p_b = -i\partial/\partial b$, and $p_c = -i\partial/\partial c$:

$$j^1 = \sin c\, p_a - \frac{\cos c}{\sin a}p_b + \cos c\frac{\cos a}{\sin a}p_c ,$$
$$j^2 = \cos c\, p_a + \frac{\sin c}{\sin a}p_b - \sin c\frac{\cos a}{\sin a}p_c , \qquad \text{(A18)}$$
$$j^3 = p_c ,$$

and

$$J^1 = -\sin b\, p_a - \cos b\frac{\cos a}{\sin a}p_b = \frac{\cos b}{\sin a}p_c ,$$
$$J^2 = \cos b\, p_a - \sin b\frac{\cos a}{\sin a}p_b + \frac{\sin b}{\sin a}p_c , \qquad \text{(A19)}$$
$$J^3 = p_b .$$

By using (A18) and (A19), one can also verify the commutation relations (A12)–(A14) directly.

---

[1]J. Schwinger, Phys. Rev. 127, 324 (1962).

[2]J. Schwinger, Phys. Rev. 130, 406 (1963).

[3]R. P. Treat, Phys. Rev. D 12, 3145 (1975).

[4]These terms do not appear, for example, in the Coulomb gauge, Hamiltonian quantum mechanics of Eq. (13.24) in E. S. Abers and B. W. Lee, Phys. Rep. 9C, 1 (1973).

[5]With appropriate boundary conditions, the existence of such a unique gauge transformation can be proven for the axial gauge. However, for the Coulomb gauge, $U(r,t)$ may either fail to exist if we require continuity in time [N. H. Christ, A. H. Guth, and E. J. Weinberg, Nucl. Phys. B114, 61 (1976)] or fail to be unique [V. N. Gribov, lecture at the 12th Winter School of the Leningrad Nuclear Physics Institute, 1977 (unpublished)]. Of course, the Coulomb gauge may well provide a satisfactory description of processes which do not involve these nonperturbative field configurations.

[6]Application of the path-integral formalism to gauge theories appears in E. S. Abers and B. W. Lee, Phys. Rep. 9C, 1 (1973), and references therein. The influence of Hamiltonian operator ordering on the path-integral formalism has been discussed by B. S. DeWitt, Rev. Mod. Phys. 29, 377 (1957); S. F. Edwards and Y. V. Gulyaev, Proc. R. Soc. London A279, 229 (1964); D. W. McLaughlin and L. S. Schulman, J. Math. Phys. 12, 2520 (1971); F. A. Berezin, Theor. Math. Phys. (USSR) 6, 141 (1971); K. S. Chang, J. Math. Phys. 13, 1723 (1972); M. Mizrahi, *ibid.* 16, 2201 (1975); J.-L. Gervais and A. Jevicki, Nucl. Phys. B110, 93 (1976).

[7]F. A. Berezin, *The Method of Second Quantization* (Academic, New York, 1966).

[8]This point has been stressed in the context of soliton quantization by J.-L. Gervais and A. Jevicki, Nucl. Phys. B110, 93 (1976).

[9]It is in this step that a similar, path-integral treatment of the noncovariant $\chi(A_i) = 0$ gauge generates additional terms. In that case, the gauge-fixing term in curly brackets in (5.25) is replaced by $\delta[\chi(A_i)]$ so that $[A_0(n+1) - A_0(n)]^2/\epsilon$ no longer appears in the exponent and $A_0(n)$ becomes of order $\epsilon^{-1/2}$, being constrained only by its appearance in (5.33). Consequently, in contrast to the present covariant case, there will be additional A_0^2 and A_0^3 terms that can enter (5.33) in order ϵ. These terms then lead to the nonlocal potentials $\mathcal{V}_1$ and $\mathcal{V}_2$ of Sec. VI.

[10]T. D. Lee and C. N. Yang, Phys. Rev. 128, 885 (1962).

[11]V. N. Popov and L. D. Faddeev, Kiev ITP report (unpublished); L. D. Faddeev and V. N. Popov, Phys. Lett. 25B, 29 (1967); R. P. Feynman, Acta. Phys. Pol. 24, 697 (1963).

Reprinted from
TRANSACTIONS OF THE NEW YORK ACADEMY OF SCIENCES
Series II Volume 40 Pages 111-123
September 15, 1980 27767

IS THE PHYSICAL VACUUM A MEDIUM?*

T. D. Lee

Department of Physics
Columbia University
New York, New York 10027

WHAT IS A VACUUM?

In the nineteenth century, to understand how the electromagnetic force, and later the electromagnetic wave, could be transmitted from one place to another, the vacuum was viewed as a medium, called aether. In his note 3075 on experimental research, Faraday wrote[1]:

> For my own part, considering the relation of a vacuum to the magnetic force and the general character of magnetic phenomena external to the magnet, I am more inclined to the notion that in the transmission of the force there is such an action, external to the magnet, than that the effects are merely attraction and repulsion at a distance. Such an action may be a function of the aether; for it is not at all unlikely that, if there be an aether, it should have other uses than simply the conveyance of radiations.

However, since at that time the nonrelativistic Newtonian mechanics was the only one available, the vacuum was thought to provide an absolute frame that could be distinguished from other moving frames by measuring the velocity of light. As is well known, this situation led to the downfall of aether and the rise of relativity.

Now we know that the vacuum is Lorentz invariant, which means that just by running around and changing the reference system, we are not going to alter the vacuum. Lorentz invariance is not everything, however. We may still ask: What is this vacuum state? In general, the vacuum is simply the lowest energy state of a Lorentz-invariant theory. Hence, the vacuum can be as complicated as any spin-0 field $\phi(x)$ at the zero 4-momentum limit:

$$\text{vacuum} \sim \phi \text{ at 4-momentum } k_\mu = 0. \tag{1}$$

Like a spin-0 field, perhaps the vacuum state can also carry quantum numbers, such as the isospin **I**, the parity P, and the strangeness S.

In quantum electrodynamics (QED), the complexity of the vacuum can be calculated in terms of the familiar power series in $\alpha \simeq 1/137$. In the following, we will be particularly interested in the vacuum structure beyond QED. Furthermore, to answer the question in the title of this note, we must inquire whether the properties of the vacuum, like those of any medium, can actually be changed under suitable conditions. If the answer is an affirmative one, it would be physically meaningful to regard the vacuum as a medium. Our analysis will be based primarily on two observations: missing symmetry and quark confinement. These two observations are certainly among the most remarkable phenomena in modern physics.

*Supported in part by a contract with the U.S. Department of Energy.

0028-7113/80/0040-0111 $1.75/1 © 1980, NYAS

Missing Symmetry

Symmetry principles have played an important part since the beginning of physics. Apart from their intrinsic aesthetic beauty, symmetry considerations have provided an extremely powerful and useful tool in our effort to understand nature. Gradually, they have become the backbone of our theoretic formulation of physical laws. Yet, especially over the past two decades, with few exceptions, most of the symmetries used in physics have been found to be broken. If we consider symmetry quantum numbers, such as the isospin **I**, the strangeness S, the parity P, . . . , we find

$$\frac{d}{dt}\left\{\begin{matrix} \mathbf{I} \\ \mathrm{S} \\ \mathrm{P} \\ \mathrm{C} \\ \mathrm{CP} \\ \cdot \\ \cdot \\ \cdot \end{matrix}\right\}_{\text{matter}} \neq 0. \tag{2}$$

Aesthetically, this may appear disturbing. Why should nature abandon perfect symmetry? Physically, this also seems mysterious. What happens to these missing quantum numbers? Where do they go? Can it be that matter alone does not form a closed system? If we also include the vacuum, perhaps symmetry can be restored

$$\frac{d}{dt}\left\{\begin{matrix} \mathbf{I} \\ \mathrm{S} \\ \mathrm{P} \\ \mathrm{C} \\ \mathrm{CP} \\ \cdot \\ \cdot \\ \cdot \end{matrix}\right\}_{\text{matter + vacuum}} = 0. \tag{3}$$

This is, of course, the basic idea underlying the general heading "spontaneous symmetry breaking." In such a scheme, one assumes that there exists some spin-0 field ϕ that can carry the missing quantum number and whose vacuum expectation value is not zero:

$$\phi_{\text{vac}} \equiv \langle \text{vac} | \phi | \text{vac} \rangle \neq 0. \tag{4}$$

Consequently, the observed asymmetry can be attributed entirely to the state vector of our universe, not to the physical law. On the other hand, unless we have other links that connect matter with the vacuum, how can we be sure that this idea is right and not merely a tautology?

A way out of this dilemma is to realize that in Equation 1, the restriction $k_\mu = 0$ for the vacuum state is only a mathematical idealization. After all, it is probable that the universe does have a finite radius, and k_μ is therefore never strictly zero. In regard to the microscopic system of particle physics, there is little difference between $k_\mu = 0$ and k_μ nearly 0; the latter corresponds to a state that varies only very slowly over a large space-time extension. This means that if the idea expressed by Equation 4 is correct, under suitable conditions, we must be able to produce excitations, or domain structures, in the vacuum. In such an excited state, there exists a volume Ω whose size

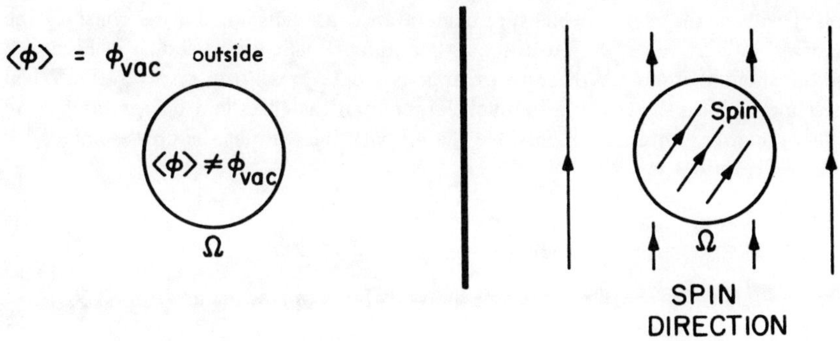

FIGURE 1. Domain structures in a ferromagnet versus in the vacuum.

is much greater than the relevant microscopic dimension; inside Ω, we have $\langle\phi(x)\rangle \neq \phi_{\text{vac}}$, but outside Ω, $\langle\phi(x)\rangle = \phi_{\text{vac}}$.

How can we produce such a change in $\langle\phi(x)\rangle$? The problem is analogous to the formation of "domain structures" in a ferromagnet. We may draw the analog:

$$\langle\phi(x)\rangle \leftrightarrow \text{spin}$$

$$J = \text{matter source} \leftrightarrow \text{magnetic field},$$

as shown in Figure 1.

In the case of a very large ferromagnet, because its spin interacts linearly with the magnetic field, a domain structure can be created by applying an external magnetic field over a large volume. Similarly, by applying over a large volume any matter source J that has a linear interaction with $\phi(x)$, one may hope to create[2] a domain structure in $\langle\phi(x)\rangle$. If the missing symmetry is due to $\phi_{\text{vac}} \neq 0$, by changing $\langle\phi(x)\rangle$, we may alter the dynamics of symmetry change inside Ω. Of course, to perform a realistic experiment to reclaim all our missing symmetry is not easy. However, perhaps one should ask whether such domain structures in the form of vacuum excitations have already existed in nature. After all, there do exist numerous sources of matter in this universe. This leads us to discuss our second phenomenon: "quark confinement."

Quark Model

There is by now a fairly large body of evidence supporting the validity of the quark[3] model of hadrons. The problem of quark confinement manifests itself in a most

FIGURE 2. Diagrams for $e^+e^- \rightarrow \mu^+\mu^-$, or $q\bar{q}$.

explicit way in the R-value measurement in an e^+e^- collision. Let us consider the diagrams for $e^+e^- \rightarrow \mu^+\mu^-$ and the quark–antiquark pair $q\bar{q}$, as shown in Figure 2. The threshold energies for these two reactions are $2m_\mu$ and $2m_q$, respectively. When the center-of-mass energy E_{cm} is much larger than the threshold energy, and if we *neglect* the strong interaction between the $q\bar{q}$ pair, the ratio of the cross sections for these two reactions is

$$\frac{\sigma(e^+e^- \rightarrow q\bar{q})}{\sigma(e^+e^- \rightarrow \mu^+\mu^-)} = 3Q_q^2, \tag{5}$$

where q stands for the flavor of the quark, which can be u, d, s, . . . , Q_q is its

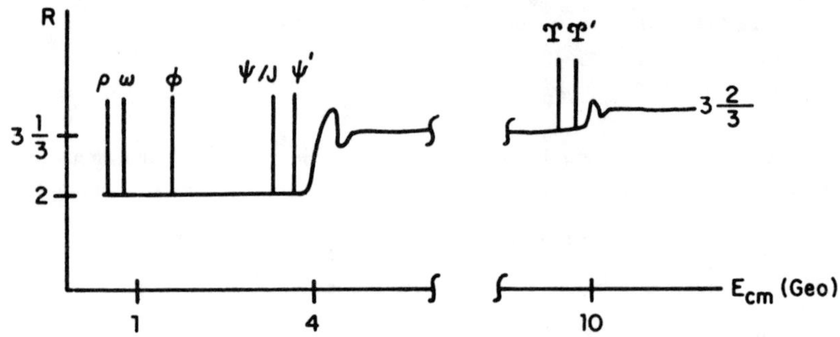

FIGURE 3. Schematic drawing of R.

corresponding charge in units of e, and the factor 3 is due to the three colors of each flavor. We define

$$R \equiv \frac{\sum_q \sigma(e^+e^- \rightarrow q\bar{q} \rightarrow \text{hadrons})}{\sigma(e^+e^- \rightarrow \mu^+\mu^-)},$$

where the sum extends over all quarks whose mass m_q is $<\frac{1}{2} E_{cm}$. From Equation 5, we have

$$R \simeq 3 \sum_q Q_q^2.$$

The experimental result[4] is consistent with the theoretical expectation, drawn schematically in Figure 3. When E_{cm} is $<2m_c$ but >1 GeV, which is assumed to be much larger than $2m_u$, $2m_d$, and $2m_s$, we have

$$R \simeq 3\left[\left(\frac{2}{3}\right)^2 + \left(\frac{1}{3}\right)^2 + \left(\frac{1}{3}\right)^2\right] = 2,$$

since the charges for u, d, and s are, respectively, $\frac{2}{3}$, $-\frac{1}{3}$, and $-\frac{1}{3}$. When E_{cm} is between $2m_c$ and $2m_b$, we may write $R = 2 + \Delta R$, where $\Delta R \simeq 3 \cdot (\frac{2}{3})^2 = \frac{4}{3}$, because the charge of the c quark is $\frac{2}{3}$. When E_{cm} is increased to above $2m_b$, the R value should increase by another amount $\Delta R \simeq 3 \cdot (1/3)^2 = 1/3$. The experimental confirmation of these theoretical expectations strongly supports the following conclusions:

Quarks have fractional charges: $Q_q = 2/3, -1/3, -1/3, 2/3$, and $-1/3$, respectively, for q = u, d, s, c, and b quarks.

Each flavor variety has three colors,[5] and the final states are symmetric with respect to color permutations; that is, the observed final hadron states are color singlets.

Quark "masses" are quite small.

Except near resonances, strong interactions between quarks and antiquarks can be neglected[6] (at least in the calculations of total cross sections). The critical question, then, is: Why in the final state do we not "see" these quarks in their free form? This is the well-known problem of quark confinement.[7]

Quantum Chromodynamics

In quantum chromodynamics (QCD), there are quark fields ψ_q^a, where q, the flavor index, = u, d, s, . . . , and a, denoting the color, = 1, 2, 3. In addition, we have the gauge fields V^i, where i = 1, 2, . . . 8, representing the color index of the gauge-field quanta. The theory is a renormalizable one, so there is no difficulty with respect to divergences in the ultraviolet region. However, there are complications in the infrared region. To give QCD a well-defined meaning, we first contain the whole system within a volume of size L^3 to avoid infrared difficulties. Let g_L be the renormalized coupling constant in the long-wavelength limit, momentum $k \sim L^{-1}$. It is possible to prove under rather general assumptions and valid to all orders in (coupling)^2 that

$$g_L > g_l \qquad \text{if } L > l. \tag{6}$$

This result is connected with the familiar "asymptotic freedom" property[8] of the theory: when $l \to 0$, g_l decreases to 0. Since there is no scale in the theory (assuming that all quarks are of zero mass), this implies that when l increases, g_l must also increase, which leads to Equation 6. The difficulty lies in the infrared limit. When $L \to \infty$, it seems likely that g_L may $\to \infty$, or at least $\gg 1$. Since the true physical system is one with $L = \infty$, we may always be in the ultrastrong coupling limit. In the next section, we shall see how this difficulty can be resolved by regarding the vacuum as a color dielectric medium.

Color Dielectric Constant

Let us introduce κ_L, which is called the color dielectric constant of the vacuum in a volume L^3. As a convention, we shall adopt a standard renormalized coupling constant g, defined by

$$g = g_l \text{ when } l = \text{some arbitrarily chosen length, say the proton radius.} \tag{7}$$

The dielectric constant κ_L will then be defined as

$$g_L{}^2 = \frac{g^2}{\kappa_L}. \tag{8}$$

Consequently, in accordance with Equation 7,

$$\kappa_l = 1 \qquad \text{when } l = \text{proton radius.} \tag{9}$$

Equation 6 now implies that

$$\kappa_L < \kappa_l \qquad \text{if } L > l. \tag{10}$$

To the second order,[8] we have

$$\frac{\kappa_L}{\kappa_l} = \frac{1}{1 + \dfrac{1}{2\pi}\dfrac{g^2}{4\pi}\left(11 - \dfrac{2}{3}n_f\right)\ln\dfrac{L}{l} + 0(g^4)}, \tag{11}$$

where n_f is the number of quark flavor varieties (assumed to be less than 17). With the convention of Equation 9, we find for the vacuum in an infinite volume $L = \infty$,

$$\kappa_\infty < 1, \tag{12}$$

which is valid to all orders in g^2. The ultrastrong coupling difficulty mentioned before corresponds to

$$\kappa_\infty = 0, \tag{13}$$

or

$$\kappa_\infty \ll 1. \tag{14}$$

In the former, we call the vacuum in QCD a perfect color diaelectric medium, in the latter a nearly perfect color diaelectric medium.

A Hypothetic Problem in Classical Electromagnetism

In QED, our usual convention is to set the dielectric constant of the vacuum state $\kappa_{vac} = 1$. It is then possible to prove that all physical media have their dielectric constants $\kappa \geqslant 1$. This can be seen most easily by using the familiar formula

$$\mathbf{D} = \mathbf{E} + 4\pi\mathbf{P}, \tag{15}$$

where **D** is the displacement vector, **E** the electric field, and **P** the polarization vector. Since under **E**, all atoms have their polarization **P** in the same direction as **E**, so as to produce a screening effect, we have $\kappa > 1$.

In this section, we shall consider a hypothetic problem. Let us imagine that in classical electromagnetism, but without the quantum theory of atoms, there could be a medium with a dielectric constant.

$$\kappa \equiv \kappa_{med} \ll 1, \quad \text{or} \quad \simeq 0; \tag{16}$$

that is, this hypothetic medium is antiscreening. Now, suppose we place a small charge distribution ϵ in the medium. As we shall show, no matter how small ϵ is, the medium will develop a hole surrounding the charge. Inside the hole, we have $\kappa = 1$, but outside $\kappa = \kappa_{med}$, as shown in Figure 4. To see this, let us assume that such a hole is formed. Because of the antiscreening nature of the medium, the induced charge on the inner surface of the hole is of the same sign as ϵ. Consequently, if we want to reduce the size of the hole, we must perform work to overcome the repulsion between ϵ and the induced charge. That work is infinite if the hole is to be eliminated in toto. Hence, the hole will not disappear. (This situation is completely different for a normal medium whose dielectric constant is >1. Imagine that a similar hole was created. The corresponding induced charge would have a sign opposite to that of ϵ. The hole would automatically shrink to 0, resulting in the usual Coulomb field distribution around the charge. The medium will remain homogeneous if ϵ is sufficiently small.) Here, no

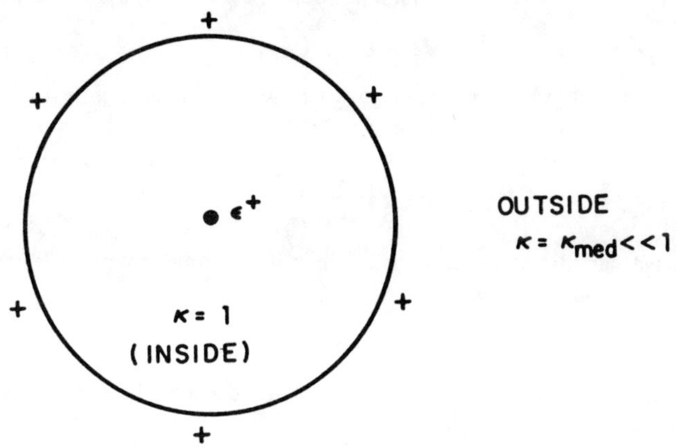

FIGURE 4. Effect of a charge in a hypothetical diaelectric medium in classical electromagnetism.

matter how small ϵ is, when $\kappa_{med} \rightarrow 0$, inhomogeneity in the form of holes must occur in the medium.

We may estimate the radius R of such a hole. Let D_{out} and E_{out} be, respectively, the normal components of **D** and **E** outside the hole when $r = R$. Similarly, let D_{in} and E_{in} be the corresponding components inside the hole when $r = R$. For a spherical hole, we have

$$D_{in} = E_{in} = D_{out} = \frac{\epsilon}{R^2}$$

and

$$E_{out} = \frac{\epsilon}{\kappa_{med} R^2},$$

where ϵ is the total charge of this small charge distribution, which is assumed to have a nonzero extension. The electric energy inside the hole is finite and independent of κ_{med}. The electric energy outside the hole is given by the volume integral of $\mathbf{D} \cdot \mathbf{E}$, which is $\propto \kappa_{med}^{-1}$. Therefore, when $\kappa_{med} \to 0$, the total electric energy U_{el} is

$$U_{el} \sim \frac{1}{2} \frac{\epsilon^2}{\kappa_{med} R}. \tag{17}$$

In addition, there is the energy U_{hole} needed to create such a hole. The amount U_{hole} is a function of R. When R is large, U_{hole} should be proportional to the volume plus a term proportional to the surface, and so on. We may write

$$U_{hole} = \frac{4\pi}{3} R^3 p + 4\pi R^2 s + \cdots, \tag{18}$$

where p, s, ... are positive constants. The total energy of the system is given by

$$M = U_{el} + U_{hole}, \tag{19}$$

which is drawn schematically in Figure 5. Its minimum, as determined by $dM/dR = 0$, gives (in the approximation $U_{hole} \simeq (4\pi/3)R^3 p$)

$$M \equiv M_\epsilon \sim \frac{4}{3}\left(\frac{\epsilon^2}{2\kappa_{med}}\right)^{3/4} (4\pi p)^{1/4},$$

from which we can draw the conclusion that if the total charge $\epsilon \neq 0$, $R \neq 0$. Furthermore, the total energy

$$M_\epsilon \to \infty \quad \text{when} \quad \kappa_{med} \to 0. \tag{20}$$

It is not difficult to show that the same conclusion can also be reached without the approximation $U_{hole} \simeq (4\pi/3)R^3 p$.

Next, we replace the single charge distribution by a dipole distribution, that is, two small charge distributions of total charge ϵ^+ and ϵ^-. It is easy to see that when κ_{med} is sufficiently small, the minimal energy state again requires the formation of a hole around these two charges. As before, inside the hole $\kappa = 1$, and outside $\kappa = \kappa_{med}$. When

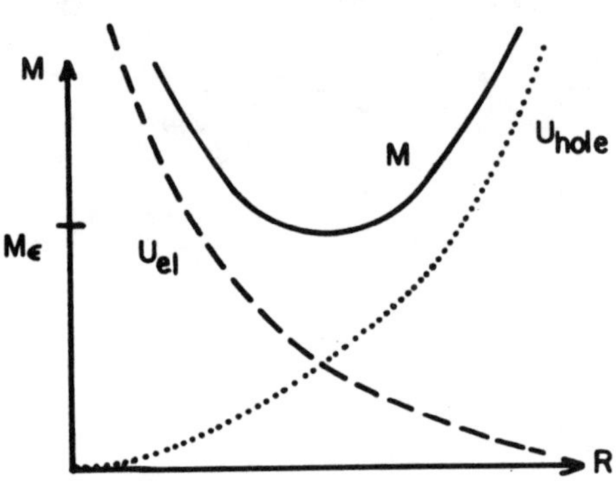

FIGURE 5. Total energy $M = U_{el} + U_{hole}$ of the system shown in FIGURE 4.

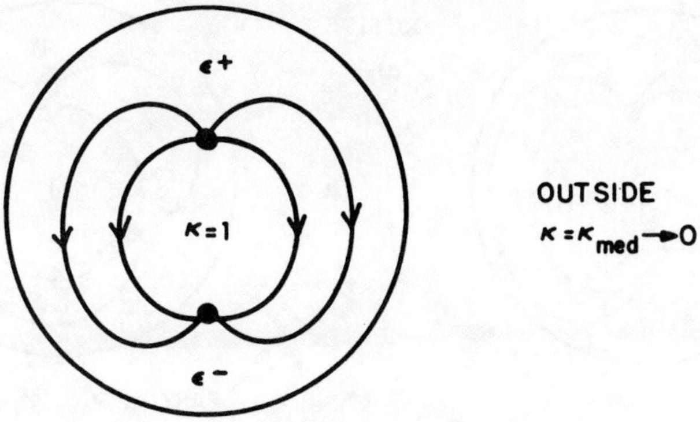

Figure 6. Distribution of the electric field for a dipole placed inside the hypothetical diaelectric medium.

$\kappa_{med} \to 0$, at the surface of the hole, the electric field inside should be parallel to the surface so that **D** is 0 outside, as shown in Figure 6. Thus, U_{el} remains finite, as does the total energy $M_{\epsilon^+\epsilon^-}$:

$$M_{\epsilon^+\epsilon^-} = \text{finite when } \kappa_{med} \to 0. \qquad \textbf{(21)}$$

If we try to separate the two charges ϵ^+ and ϵ^- by an infinite distance, because of Equation 20 the work required must also be infinite. This is the analog to "quark confinement" in our hypothetical problem.

Quark Confinement

We now return to the corresponding problem in QCD, discussed under Quantum Chromodynamics and Color Dielectric Constant. The vacuum is assumed to be a perfect (or nearly perfect) color diaelectric medium. By following the same argument given above, we see that whenever there are quarks or antiquarks there must be inhomogeneity in the space surrounding these particles. We may call these either bags,[9,10] or domain structures, or soliton solutions.[11] Inside the "bag," the dielectric constant κ is 1. But outside, $\kappa = \kappa_\infty$ which is 0, or $\ll 1$. If the total "color" is nonzero, the mass of such a bag would be infinite when $\kappa_\infty \to 0$, in analogy with Equation 20. However, just as in Equation 21, the bag mass remains finite when $\kappa_\infty \to 0$, if inside the bag one has a color singlet, such as

$$\text{meson: } \bar{q}^a q^a, \quad \text{or baryon: } \epsilon_{abc} q^a q^b q^c,$$

where a, b, and c are color indexes that can vary from 1 to 3 (see Figure 7). From this one can derive that the work required to separate the quarks to a large distance r is approximately proportional to r. The quark confinement is then "explained" by the assumption that the vacuum in QCD is a perfect (or nearly perfect) color diaelectric.

Of course, at present it is not known that the mass M_q of a truly free quark is indeed infinite. However, the fact that no free quark has been observed so far in the

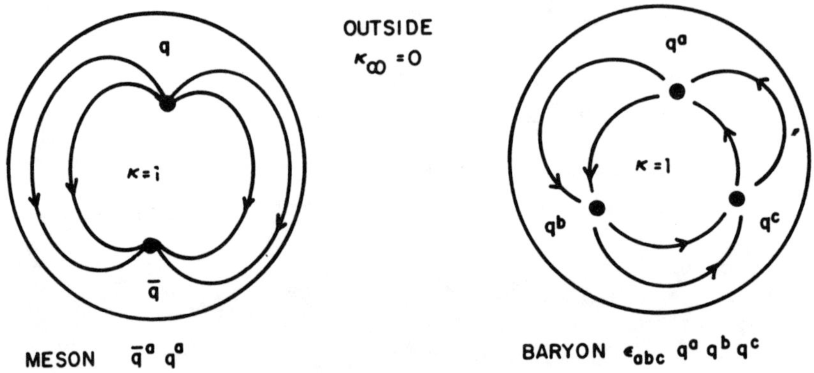

FIGURE 7. Color electric fields inside hadrons.

final state of any high-energy collision sets a lower bound $M_q > 5$ GeV for q = u, d, and s. It can be shown that this lower bound implies [12]

$$\kappa_\infty < 0.013\left(\frac{g^2}{4\pi}\right), \tag{22}$$

where g is defined by Equation 7.

One may wonder in what sense we have bypassed the difficulty of ultrastrong coupling, mentioned under QUANTUM CHROMODYNAMICS. To understand this, let us consider other problems in physics in which interactions are also ultrastrong and yet there are no mathematical difficulties. A good example is the interaction potential $V(r)$ between two helium atoms. As shown in Figure 8, for $r < a$, the potential V is $\simeq$ infinite; for $r > a$, V is small. Because the strong potential is of a repulsive nature, the two helium atoms will avoid the strong-interaction region, and therefore this strong interaction does not present any difficulty. The usual method is to replace the strong repulsive part of V by a hard-sphere potential, which leads to a new boundary value

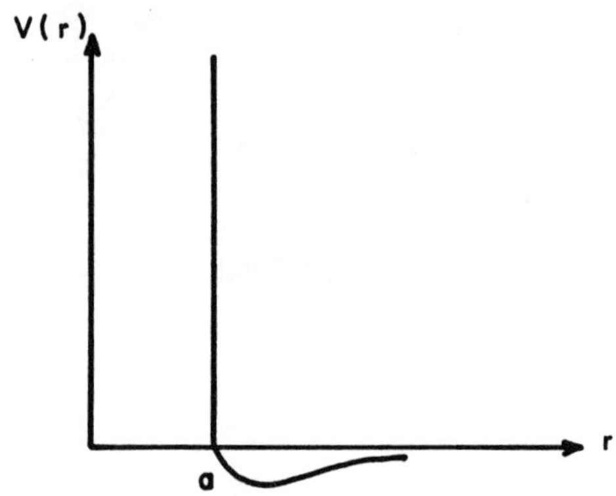

FIGURE 8. Potential V between two helium atoms.

problem: the distance r between the two helium atoms is restricted to a region $r \geqslant a$ in which the potential $V(r)$ is small and therefore can be regarded as a perturbation.

From Figure 7, we see that when $\kappa_\infty \longrightarrow 0$, that is, $g_\infty{}^2 = g^2/\kappa_\infty \longrightarrow \infty$, the quarks are confined to a domain in which $\kappa = 1$. Thus, the ultrastrong region, $g_\infty \longrightarrow \infty$, exerts a repulsion against the quarks and antiquarks. Just as in the above problem of helium–helium interaction, the particles automatically stay away from the ultra-strong-interaction region. Because of Lorentz invariance, the corresponding boundary-value problem in the present case requires the relativistic soliton solution, as has been given in the literature.[11,12] When one expands around these soliton solutions, only the coupling g inside the soliton is the relevant parameter, and that resolves the difficulty. (It is only when one attempts to make a simple-minded plane-wave expansion around a homogeneous background that one has to use g_∞ as the expansion parameter. Exactly the same kind of difficulty would appear if one were to force a similar plane-wave expansion in the aforementioned helium–helium problem.)

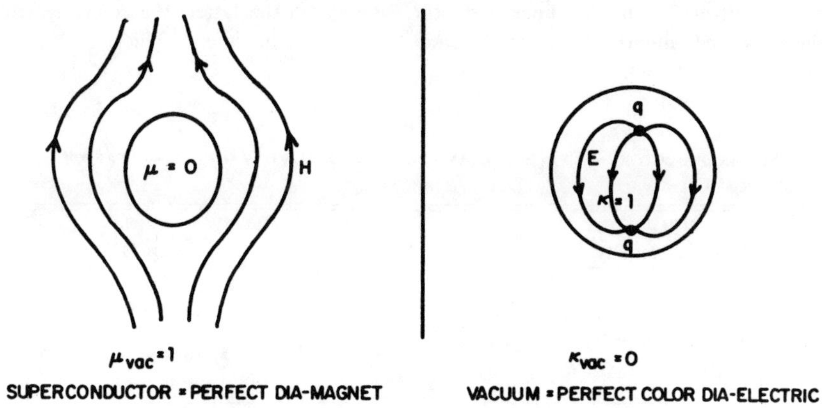

FIGURE 9. Superconductivity in QED versus quark confinement in QCD.

Because QCD is asymptotically free, the quarks inside the bag behave approximately like free particles, and their effective masses can be relatively small.

Since our theory is relativistic invariant, the velocity of light $c/(\kappa\mu)^{1/2}$ in the vacuum must remain c. Hence,

$$\kappa\mu = 1, \tag{23}$$

where κ is the color dielectric constant and μ is the corresponding "magnetic" susceptibility. When $\kappa = \kappa_\infty \longrightarrow 0$, one must have $\mu = \mu_\infty \longrightarrow \infty$.

With Equation 23, we can consider κ to be a Lorentz-invariant quantity; it plays the same part as the ϕ-field discussed under WHAT IS A VACUUM? and MISSING SYMMETRY. It may be useful to introduce a phenomenological spin-0 field ϕ that depends linearly on κ:

$$\phi \propto 1 - \kappa. \tag{24}$$

All hadrons can be viewed as domain structures in the physical vacuum. Inside the hadron, we have $\kappa = 1$, and therefore $\phi = 0$. Outside the hadron, κ is κ_∞ and ϕ is ϕ_{vac}. In the case that $\kappa_\infty = 0$, Equation 24 becomes simply $\phi = \phi_{vac}(1 - \kappa)$.

REMARKS

As mentioned before, in QED there does not exist any diaelectric medium at zero frequency; however, there do exist dispersion laws that exhibit diaelectric natures at some nonzero frequencies.[13] That can lead to physical effects somewhat similar to those discussed above.

Another analog is the comparison between the superconductivity in QED and the quark confinement in QCD. When we switch from QED to QCD, we replace the magnetic field **H** by the color electric field **E**, the superconductor by the QCD vacuum, and the QED vacuum by the interior of the hadron; therefore in FIGURE 9, the inside by the outside and the outside by the inside. In the former, the magnetic field is expelled outward from the superconductor, whereas in the latter, the color electric field is pushed into the bag, which leads to quark confinement. This situation is summarized in TABLE 1.

TABLE 1

SUPERCONDUCTOR = PERFECT DIAMAGNET IN QED *vs.* VACUUM = PERFECT DIAELECTRIC IN QCD

QED Superconductivity		QCD
H	↔	**E**
$\mu_{inside} = 0$	↔	$\kappa_{vacuum} = 0$
$\mu_{vacuum} = 1$	↔	$\kappa_{inside} = 1$
Inside	↔	outside
Outside	↔	inside

Quark confinement is a large-scale phenomenon. Therefore, at least on the phenomenologic level, it should be understandable through a quasiclassical macroscopic theory, much like the London–Landau theory for superfluidity. In this paper, we suggest that the use of the color dielectric constant of the vacuum serves such a purpose.

By viewing the vacuum as a Lorentz-invariant medium, we would expect that many of its parameters, such as the color dielectric constant and the average values of some symmetry numbers can be subjected to changes. We may, by suitably arranging either the matter density or the energy content over a relatively large volume, alter the properties of the vacuum. Thereby, it may lead us to other striking phenomena, hitherto completely unknown to us.

REFERENCES

1. Quoted by MAXWELL, J. C. 1954. A Treatise on Electricity and Magnetism. 3rd edit.: 432. Dover Publications. New York, N.Y.

2. LEE, T. D. & G. C. WICK. 1974. Phys. Rev. D **9:** 2291.
3. ZWEIG, G. 1964. CERN Report. Unpublished; GELL-MANN, M. 1964. Phys. Lett. **8:** 214.
4. Proceedings of the 1979 International Symposium on Lepton and Photon Interactions at High Energies, Fermilab.
5. GREENBERG, O. W. 1964. Phys. Rev. Lett. **13:** 598.
6. FEYNMAN, R. P. 1969. Phys. Rev. Lett. **23:** 1415.
7. MANDELSTAM, S. 1976. Phys. Rep. **23C:** 245; NAMBU, Y. Ibid.: 250; SUSSKIND, L. & J. KOGUT. Ibid.: 348; & K. G. WILSON. Ibid.: 331.
8. POLITZER, H. D. 1973. Phys. Rev. Lett. **30:** 1346; GROSS, D. & F. WILCZEK. 1973. Phys. Rev. Lett. **30:** 1343; t'HOOFT, G. 1973. Phys. Rev. B **61:** 455; 1973. Phys. Rev. B **62:** 444.
9. A. CHODOS, R. L. JAFFE, K. JOHNSON, C. B. THORN & V. F. WEISSKOPF. 1974. Phys. Rev. D **9:** 3471.
10. BARDEEN, W. A., M. S. CHANOWITZ, S. D. DRELL, M. WEINSTEIN & T.-M. YAN. 1975. Phys. Rev. D **11:** 1094.
11. FRIEDBERG, R. & T. D. LEE. 1978. Phys. Rev. D **18:** 2623.
12. LEE, T. D. 1979. Phys. Rev. D **19:** 1802.
13. JAVAN, A. & P. L. KELLEY. 1966. J. Quantum Electron. **2:** 470; HAPPER, W. & A. C. TAM. 1977. *In* Laser Spectroscopy III. J. L. Hall & J. L. Carlsten, Eds.: 333. Springer-Verlag. Berlin, Heidelberg, New York.

VI

Symmetry Principles

Commentary

In paper [14], the question is raised of whether the observed baryon conservation law is connected to a local gauge transformation, similar to that connected to electric charge conservation. This would lead to a vector field coupling to baryon number in the way that the electromagnetic field couples to electric charge. The Eotvos experiment on the variation between chemical elements of the ratio of inertial to gravitational mass shows that any such coupling must be exceedingly weak, if it exists at all. This raises strong doubts about the existence of the local gauge transformation for baryon number, and foreshadows the later suggestion, by Sakharov and others, that baryon number conservation is only an approximate symmetry.

In paper [20], it is shown that the combined invariance of the strong interactions under charge conjugation and isospin implies a multiplicative conservation law for systems with zero baryon number and hypercharge. The conserved quantum number G is -1 for all pions, so there cannot be strong transitions in which the pion number changes from odd to even. This classification of states according to G proved very useful in later work on pion-nucleon interactions and on weak interactions. In paper [41], the G quantum number is used to analyze annihilations of nucleon and antinucleon into a system containing K, $\overline{K}$ and pions. Correlations among the decay products are derived.

By early 1956, it had become clear as the result of analyses by R. Dalitz and others that a problem existed with respect to the decays of K mesons. Decays occurred into both two and three pions, which would not be allowed for a single, spin zero meson, if parity were conserved in the decay. Following efforts to understand the situation through a variety of hypotheses, Lee and Yang decided to analyze the evidence for parity conservation in various particle interactions. The outcome was a historic series of papers, published over a period of a few months in late 1956 and early 1957. The first result was paper [21], in which it is shown that whereas there is good evidence for parity *(P)* conservation in strong and electromagnetic interactions, there was no such evidence for weak interactions. The critical point is made that in order to test parity invariance in any interaction, it is necessary to observe a pseudoscalar correlation. Specific tests to do this are proposed for the processes of beta decay, pi-mu-electron decay, and hyperon decay. It is also pointed out that no evidence exists for charge conjugation *(C)* or time reversal *(T)*, invariance of weak interactions.

Evidence was soon provided by experiments, carried out by Wu and collaborators, by Lederman, Garwin and collaborators, and by Telegdi and collaborators that these possibilities were very real, and that weak interactions did not satisfy P or C invariance. These early experiments seemed to show that CP invariance was satisfied. These results immediately raised the question of what theory correctly describes weak decay processes.

Paper [24] describes a two-component theory of massless neutrinos, first discussed by Hermann Weyl, in which the neutrino has only a single spin state, and the antineutrino has opposite spin to the neutrino. It is shown that in this theory, *P* and *C* are not conserved, but that *T* and *CP* may be. This theory is equivalent to a four-component theory in which the neutrino field always is multiplied by the operator $1 + \gamma_5$. The paper goes on to discuss the implications of this theory for beta decay and pi-mu-electron decay processes, which turned out to be correct. At the time the paper was written, it was thought that the beta decay interaction included a tensor coupling. Later experiments showed that the coupling was instead an axial vector coupling, and it was straightforward to extend the two-component theory to that case.

Paper [25] discusses the implications of the *TCP* theorem, proved earlier by Pauli and Luders for parity conserving interactions. It shows that *TCP* invariance implies equality of lifetimes for particle and antiparticle, without the need for assuming *C* or *CP* invariance. It analyzes the consequences for experiment of various assumptions about *T*, *C* and *P* non-invariance, and shows, for example, that if *TP* is conserved, then in the absence of final state interactions, a correlation proportional to the pseudoscalar (spin • momentum) cannot be observed. It also discusses the decays of neutral *K* mesons, and shows that *TCP* invariance alone implies the existence of two distinct lifetimes for these decays, but that further assumptions, such as *CP* invariance, are necessary to imply equality of various partial decay rates. This analysis became directly relevant seven years later, when *CP* non-invariance of neutral *K*-meson decays was discovered experimentally.

Paper [28] analyzes parity non-conservation in hyperon decay. It shows that an angular asymmetry of the decay pion with respect to the production plane would indicate parity non-conservation, and that this asymmetry could be large. Later experiments indeed detected such a large asymmetry in hyperon decays.

Paper [32] is based on lectures Lee delivered at Brookhaven National Laboratory in January 1957. It is a general survey of weak interactions as of that time, following the Wu, Lederman-Garwin, and Telegdi experiments. It was used for a long time afterwards as a working manual for weak interaction experiments and their theoretical analyses.

In 1964 Fitch, Cronin and collaborators made the experimental discovery that in the decay of neutral *K* mesons, *CP* was not conserved. Later experiments showed that these decays also were not *T* invariant, although in each case the amount by which the invariance was violated was quite small. In a series of papers, Lee and collaborators explored possible mechanisms for these violations.

In paper [73], it was proposed that the weak interaction Hamiltonian actually is *CP* invariant, and the observed violation is an environmental effect, due to the existence of a long range vector field present in space, which acts oppositely on *K* and $\overline{K}$, producing an energy difference between them. As a consequence of this, the amount of *CP* violation would depend on the energy of the decaying meson. Later experiments showed that this possibility is not realized.

Paper [78] analyzes a variety of phenomenological models of *CP* violation, in which the violating Hamiltonian has different strengths and different strangeness selection rules. The effect of these Hamiltonians on neutral *K* meson decays is discussed.

In paper [80], it is hypothesized that the electromagnetic interaction of strongly interacting particles (hadrons) is *C* violating, and that this is the origin of the observed *CP* violation in neutral *K* decays. Various tests of this hypothesis are proposed. Later experiments showed that this hypothesis is not satisfied in nature.

In paper [146], a model is proposed in which *T* violation, and *CP* violation occur through spontaneous symmetry breaking in a theory with a *T* invariant Hamiltonian. In this paper, the model involves a complex scalar field, carrying a conserved quantum number, and interacting with a gauge field. This scalar field can be identified with the Higgs field in some versions of the unified theory of electromagnetic and weak interactions. It is shown that the *T* violation at low energies should be small in this model. Paper [147] extends the idea of spontaneously broken *T* violation to the interaction of a single real scalar field, while paper [149] discusses more detailed calculations of *T* violating effects in these models. The models in which *T* violation is a spontaneously broken symmetry remain one of the viable approaches to explaining this still mysterious aspect of physics.

In paper [54], the internal symmetries of the strong interactions are studied group theoretically. A criterion that any group must satisfy in order to describe observed properties of these symmetries is proposed.

Paper [93] discusses the relations between *C, P* and *T* symmetries, and internal particle symmetries that are satisfied in local field theories.

Reprinted from THE PHYSICAL REVIEW, Vol. 98, No. 5, 1501, June 1, 1955
Printed in U. S. A.

Conservation of Heavy Particles and Generalized Gauge Transformations

T. D. LEE, *Columbia University, New York, New York*

AND

C. N. YANG, *Institute for Advanced Study, Princeton, New Jersey*

(Received March 2, 1955)

The possibility of a heavy-particle gauge transformation is discussed.

THE conservation laws of nature fall into two distinct categories: those that are related to invariance under space-time displacements and rotations, and those that are not. In the former category there are the conservation laws of momentum, energy, and angular momentum. In the latter category we find the conservation laws of electric charge, of heavy particles, and the approximate conservation laws of isotopic spin, and perhaps others.[1] We notice that the best known within this second category, the conservation of electric charge, is related to invariance under gauge transformations,[2] which expresses the nonmeasurability of the phase of the complex wave function of a charged particle.

We want to ask here whether similar gauge invariances should be related to all conservation laws of the second category. This question has been discussed in connection with the conservation of isotopic spin by Yang and Mills.[3] We wish here to discuss the problem in connection with the conservation of heavy particles.

If we take the conservation of heavy particles to mean invariance under the transformation

$$\psi_N \to e^{i\alpha}\psi_N, \quad \psi_P \to e^{i\alpha}\psi_P, \tag{1}$$

for the wave function of the heavy particles (neutrons and protons), a general gauge transformation (heavy-particle gauge transformation) is a transformation like (1) with the phase α an arbitrary function of space-time. Invariance under such a transformation means that the relative phase of the wave function of a heavy particle at two different space-time points is not measurable.

Such a gauge transformation is formally completely identical with the electromagnetic gauge transformation. Invariance under such a transformation therefore necessitates the existence of a neutral vector massless field coupled to all heavy particles. A nucleon would have a "heavy-particle charge" of $+\eta$ in such a field and an antinucleon would have a "heavy-particle charge" of $-\eta$. The force between two massive bodies therefore would contain a contribution from the Coulomb-like repulsion between such "heavy-particle charges." The total force including the gravitational attraction is:

$$\text{Force} = -G(M_1M_2/R^2) + \eta^2(A_1A_2/R^2). \tag{2}$$

Here M_1, M_2, A_1, and A_2 are the inertia masses and mass numbers of the two bodies. There should also be a magnetic-dipole-like interaction between individual nuclei because the nucleons are in constant motion in a nucleus. But in a macroscopic object the nuclear spins average out so that (2) is correct unless the two bodies are spinning at high speeds.

Now the packing fraction of various atoms differ so that M/A varies fractional-wise from substance to substance by $\sim 10^{-3}$. This means that the observed gravitation mass [which contains a contribution from the η^2 term in (2)] divided by the inertia mass would vary fractional-wise from substance to substance by $10^{-3}\eta^2/G(M_P)^2$, where M_P is the mass of the proton[4] Very careful measurements by Eötvös and co-workers. have shown this variation to be $<10^{-8}$. Therefore

$$\eta^2/G(M_P)^2 < 10^{-5}.$$

It may be remarked that since the packing fraction differs most between hydrogen and, say, carbon, Eötvös' experiment could yield a more sensitive detection of η^2 by a factor of 10 if repeated with a comparison of hydrogen and carbon.

The assumption that leads to the above line of reasoning and the force expression (2) is that the phase factor α in (1) should be space-time-dependent. It should be noticed that in addition the assumption has also been made that the transformation that generates the conservation of heavy particles is of the specific form (1).

We wish to thank Dr. J. Robert Oppenheimer for an interesting discussion.

[1] See M. Gell-Mann and A. Pais, Proceedings of the Glasgow Conference, July, 1954 (to be published).

[2] W. Pauli, Revs. Modern Phys. **13**, 203 (1941).

[3] C. N. Yang and R. L. Mills, Phys. Rev. **96**, 191 (1954).

[4] Eötvös, Pekár, and Fekete, Ann. Physik **68**, 11 (1922).

Reprinted from THE PHYSICAL REVIEW, Vol. 100, No. 3, 932–933, November 1, 1955
Printed in U. S. A.

Speculations on Heavy Mesons*

T. D. LEE AND J. OREAR
Department of Physics, Columbia University, New York, New York
(Received August 22, 1955)

FROM the analysis of Dalitz[1] it is becoming necessary to assume that the $K_{\pi 3}^{\pm}$ and $K_{\pi 2}^{\pm}$ are due to at least two different charged particles. Recent experimental evidence[2] encourages the hypothesis that the observable lifetimes are exactly the same (at least after travel times$>10^{-9}$ sec). As a solution to this possible dilemma we propose that there exist two heavy mesons, the $\theta^{\pm}$ and the $\tau^{\pm}$, where $\theta^{\pm}\rightarrow\pi^{\pm}+\pi^0$ and $\tau^{\pm}\rightarrow\pi^{\pm}+2\pi$. We propose further that the heavier of these two has a lifetime of 10^{-8} sec with a significant branching ratio for gamma decay to the lighter one. Finally, if we assign a lifetime of order 10^{-9} sec or less to the lighter one, we achieve a situation where both particles can have different lifetimes but in most experiments appear to have exactly the same lifetime. Because of Dalitz's results[1] we require the τ to be a particle which cannot decay into two pions because of spin and parity conservation. However, both τ and θ may contribute to the $K_{\mu 2}$, $K_{\mu 3}$, K_{e2}, and K_{e3} while the θ may also contribute to a part of the $K_{\pi 3}$ (unless the spin and parity of the θ are 0+). We now discuss all the cascade possibilities.

Case I. θ and τ both have zero spin (single-gamma decay forbidden).—

$$\theta^{\pm}\rightarrow\tau^{\pm}+2\gamma, \tag{1}$$

or

$$\tau^{\pm}\rightarrow\theta^{\pm}+2\gamma. \tag{2}$$

With our present knowledge it is not possible to calculate accurately how large a mass difference is needed to give a two-gamma decay time of 10^{-8} sec. Our estimates show that a mass difference as small as 10 Mev is reasonable if one assumes that the θ and τ are strongly coupled through nucleon-hyperon intermediate states.

Case II. θ and τ do not both have zero spin.—

(a) $M_{\theta}>M_{\tau}$.

$$\theta^{\pm}\rightarrow\tau^{\pm}+\gamma. \tag{3}$$

For magnetic dipole radiation, a mass difference $\lesssim 100$ kev is required. Electric dipole requires just a few kev. A larger (and intuitively more reasonable) mass difference is permissible if we can manufacture a selection rule forbidding the strong coupling between θ and τ. The most convenient selection rule in the framework of the Gell-Mann scheme[3] is conservation of strangeness, S.[4] Because of the existence of the θ^0, the θ must have its usual assignment

$$(S)_{\theta^{\pm}}=\pm 1 \quad\text{or}\quad I_{\theta}=\tfrac{1}{2}. \tag{4}$$

The only possibility left for the τ is

$$(S)_{\tau^{\pm}}=\pm 2 \quad\text{or}\quad I_{\tau}=0. \tag{5}$$

Two important consequences of this assignment are that the τ^0 should not exist, in contrast to the existence of θ^0 and $\bar{\theta}^0$, and that a τ^+ cannot be produced in association with a single Λ^0 or Σ. However, it may be produced by the following reactions:

$$\pi^-+p\rightarrow\Xi^-+\tau^+, \tag{6}$$

$$p+n\rightarrow\Lambda^0+\Lambda^0+\tau^+. \tag{7}$$

(b) $M_{\tau}>M_{\theta}$. From the production experiments, the K-mesons must be produced abundantly together with Λ^0 and Σ. Since in this case the $\tau^{\pm}$ is to be the longer-lived particle, its strangeness must be ± 1. Consequently in this case the conservation-of-strangeness selection rule is not possible and the mass difference must not be greater than $\sim$100 kev unless the gamma radiation is of higher order than dipole.

The possible existence of a $\tau^{\pm}$ with a strangeness different from the $\theta^{\pm}$ permits the $\tau^{\pm}$ to have a lifetime very much shorter than the $\theta^{\pm}$, since they would then not be strongly coupled through intermediate nucleon-hyperon states. One is now tempted to speculate further whether the decay of the $\tau^{\pm}$ may be connected in some way with the universal Fermi interaction. We have explored the possibility of regarding the universal Fermi interaction as a secondary process via a virtual emission of $\tau^{\pm}$. In order to give the desired scalar and tensor coupling for the Fermi interaction, the $\tau^{\pm}$ has to be described by a pseudovector field,[5] ϕ_{ρ} $(\rho=1,2,3,4)$, coupled to the fermions by

$$H=g_{AB}\bar{\psi}_A\gamma_{\mu}\gamma_{\nu}\psi_B\frac{\partial\phi_{\rho}}{\partial x_{\lambda}}\epsilon_{\mu\nu\rho\lambda}+g_{AB}'\bar{\psi}_A\psi_B\phi_{\rho}^2, \tag{8}$$

where (A,B) represents the (p,n), (μ,ν), and (e,ν) groups, and $\epsilon_{\mu\nu\rho\lambda}$ is the isotopic antisymmetric tensor. The interaction (8) would give the decay of $\tau^{\pm}$ into 3π, $\mu+\nu$, and $e+\nu$ with comparable rates. If one sets $g_{pn}\sim g_{\mu\nu}\sim g_{e\nu}$, then from the strength of the Fermi interactions one obtains for the $\tau^{\pm}$ lifetime $\sim 10^{-16}$ sec.

We wish to thank the theoretical group at the Radiation Laboratory, University of California for several communications which have contributed to the above discussion.

* Supported by the joint program of the U. S. Atomic Energy Commission and the Office of Naval Research.

[1] R. Dalitz, Proceedings of the Fifth Annual Rochester Conference on High-Energy Physics (Interscience Publishers, Inc., New York, 1955), and private communication.

[2] Harris, Orear, and Taylor, preceding Letter [Phys. Rev. **100**, 932 (1955)].

[3] M. Gell-Mann, Phys. Rev. **92**, 833 (1953).

[4] Conservation of strangeness is mathematically equivalent to conservation of the z component of isotopic spin, I_z. For mesons, $S=2(Q-I_z)$ where Q is the charge.

[5] According to the latest analysis of Dalitz, the $K_{\pi 3}$ data cannot exclude a (1+) particle, although (0−) gives a better fit. All spin and parity configurations where two-pion decay is also possible are strongly excluded for spins less than six.

T. D. LEE and C. N. YANG
1956, Aprile
Il Nuovo Cimento
Serie X, Vol. 3, pag. 749-753

Charge Conjugation, a New Quantum Number G, and Selection Rules Concerning a Nucleon-Antinucleon System.

T. D. LEE
Columbia University - New York, N.Y.

C. N. YANG
Institute for Advanced Study - Princeton, N.J.

(ricevuto il 30 Gennaio 1956)

Summary. (*). — A new quantum number is introduced which leads to selection rules concerning transitions between states with heavy particle number = 0. They are applied to the system nucleon-antinucleon; it is also shown that they can be extended to include the conservation of strangeness.

(*) *Editor's care.*

Whenever several conservation laws operate for the same system it is oftentimes possible to obtain new quantum numbers and new selection rules. For the pion-nucleon-antinucleon system in strong interaction one has besides the usual spin parity conservation laws the additional conservation laws of isotopic spin, charge conjugation and heavy particle number. We shall see that these conservation laws together do lead to a new quantum number for all systems with heavy particle number = 0. Application of this result to a nucleon-antinucleon system gives the selection rules tabulated in Tables I and II.

We shall be concerned with these operators:

$$I_1,\, I_2,\, I_3 = \text{isotopic spin operators}\,,$$

$$G = C\exp\,[i\pi I_2]\,,$$

$$\text{and}\quad N = \text{number of heavy particles.}$$

TABLE I. – *Selection Rules for* $\overline{p} + \text{n} \to m\pi$.

State	Spin parity	I	G	$\pi^- + \pi^0$	$2\pi^- + \pi^+$	$\pi^- + 2\pi^0$	$2\pi^- + \pi^+ + \pi^0$	$\pi^- + 3\pi^0$	$3\pi^- + 2\pi^+$	$2\pi^- + \pi^+ + 2\pi^0$	$\pi^- + 4\pi^0$
$^1S_0$	0^-	1	—	×			—	—			
$^3S_1$	1^-	1	+		—	—			—	—	—
$^1P_1$	1^+	1	+	×	—	—			—	—	—
$^3P_0$	0^+	1	—	—	×	×	—	—			
$^3P_1$	1^+	1	—	×			—	—			
$^3P_2$	2^+	1	—	—			—	—			

× means strictly forbidden and — means forbidden so far as the isotopic spin is a good quantum number.

Here C is the charge conjugation operator. The advantage of using the operator G instead of C is that G commutes with I_1, I_2, and I_3, while C does not.

To see this one need only write down in the rest system the explicit form for these operators between the four states [1] with the same z component of spatial spin:

$$\begin{pmatrix} p \\ n \\ \overline{n} \\ -\overline{p} \end{pmatrix}.$$

They are

$$I_1 = \begin{pmatrix} 0 & 1 & 0 & 0 \\ 1 & 0 & 0 & 0 \\ 0 & 0 & 0 & 1 \\ 0 & 0 & 1 & 0 \end{pmatrix}, \quad I_2 = \begin{pmatrix} 0 & -i & 0 & 0 \\ i & 0 & 0 & 0 \\ 0 & 0 & 0 & -i \\ 0 & 0 & i & 0 \end{pmatrix}, \quad I_3 = \begin{pmatrix} 1 & 0 & 0 & 0 \\ 0 & -1 & 0 & 0 \\ 0 & 0 & 1 & 0 \\ 0 & 0 & 0 & -1 \end{pmatrix},$$

$$G = \begin{pmatrix} 0 & 0 & 1 & 0 \\ 0 & 0 & 0 & 1 \\ -1 & 0 & 0 & 0 \\ 0 & -1 & 0 & 0 \end{pmatrix}, \quad N = \begin{pmatrix} 1 & 0 & 0 & 0 \\ 0 & 1 & 0 & 0 \\ 0 & 0 & -1 & 0 \\ 0 & 0 & 0 & -1 \end{pmatrix}.$$

(1) Notice the minus sign in front of the $\overline{p}$ state in the convention here. The definition of $\overline{n}$ and $\overline{p}$ as the antineutron and antiproton states is here so chosen that they are identically related to the neutron and proton states. In other words, charge conjugation is here defined in the same way for the neutron and the proton.

TABLE II. – *Selection Rules for* $\overline{p}+p \to m\pi$ *or* $\overline{n}+n \to m\pi$.

State	Spin parity	C	I	G	$2\pi^0$	$\pi^+ + \pi^-$	$3\pi^0$	$\pi^+ + \pi^- + \pi^0$	$4\pi^0$	$\pi^+ + \pi^- + 2\pi^0$	$2\pi^+ + 2\pi^-$	$5\pi^0$	$\pi^+ + \pi^- + 3\pi^0$	$2\pi^+ + 2\pi^- + \pi^0$
$^1S_0$	0^-	+	0	+	×	×	—	—				—	—	—
			1	—	×	×			—	—	—			
$^3S_1$	1^-	—	0	—	×	—	×		×	—	—	×		
			1	+	×		×	—	×			×	—	—
$^1P_1$	1^+	—	0	—	×	×	×		×	—	—	×		
			1	+	×	×	×	—	×			×	—	—
$^3P_0$	0^+	+	0	+			×	×				—	—	—
			1	—	—	—	×	×	—	—	—			
$^3P_1$	1^+	+	0	+	×	×	—	—				—	—	—
			1	—	×	×			—	—	—			
$^3P_2$	2^+	+	0	+			—	—				—	—	—
			1	—	—	—			—	—	—			

× means strictly forbidden and — means forbidden so far as the isotopic spin is a good quantum number.

For a multiple particle system, I and N are additive and G multiplicative. One therefore has *in general*

$$[I_i, G] = [I_i, N] = 0\,, \tag{1}$$

$$GN + NG = 0\,, \tag{2}$$

$$G^2 = (-1)^N\,. \tag{3}$$

Besides G is unitary by definition and N has integral eigenvalues. Solving the relations (1) to (3) one obtains the following:

(1) For a state with a definite value $\neq 0$ for N, the quantum numbers are N, I^2 and I_3. The operation of G on the system gives a state with the same I^2, I_3 but changes the sign of N.

(2) For a state with $N = 0$ the quantum numbers are $N = 0$, I^2, I_3 and $G = \pm 1$. Besides, *all components of the same I-multiplet have the same value for G*. The pions obviously belong to this category, with $I^2 = 1(1+1)$ and $I_3 = \pm 1$ or 0. To find their value for G we notice that a π^0 has matrix elements connecting it to the ${}^1S_0$ state of a proton-antiproton system. Such a system has $C = +1$ [2]. Now π^0 has a total I-spin equal to 1. Therefore $\exp[i\pi I_2]$ for π^0 is -1. Hence by definition $G = -1$ for π^0. *Therefore $G = -1$ for all pions, charged or neutral.*

The existence of the quantum number G leads to interesting selection rules [3] concerning transitions between states with $N = 0$. In particular one sees that *an even number of pions cannot by strong interaction go into an odd number of pions.* When applied to a system consisting of a nucleon and an antinucleon one obtains the selection rules tabulated in Tables I and II.

The above consideration can easily be extended to include the conservation [4] of strangeness S. (S is defined to be $2Q - 2I_3 - N$). The commutation relations between S and the other operators are

$$[N, S] = [S, I_i] = 0\,,$$

$$GS + SG = 0\,,$$

---

(2) L. Wolfenstein and D. G. Ravenhall: *Phys. Rev.*, **88**, 279 (1952); A. Pais and R. Jost: *Phys. Rev.*, **87**, 871 (1952).

(3) Most of these results have been stated in the literature in various forms. See A. Pais and R. Jost: *Phys. Rev.*, **87**, 871 (1952); L. Michel: *Nuovo Cimento*, **10**, 319 (1953); D. Amati and B. Vitale: *Nuovo Cimento*, **2**, 719 (1955).

(4) M. Gell-Mann: *Phys. Rev.* (in press); T. Nakano and K. Nishijima: *Prog. Theor. Phys.*, **10**, 581 (1954).

and

$$G^2 = (-1)^{N+S}.$$

Solving these commutation relations one obtains results similar to (1) and (2) discussed above, except that S is now an additional quantum number. G is a good quantum number $= \pm 1$ in this case only if both N and S are zero.

RIASSUNTO (*)

Si introduce un nuovo numero quantico G che porta a regole di selezione riguardanti le transizioni fra stati con un numero N di particelle pesanti nullo. Tali regole sono poi applicate al sistema nucleone-antinucleone e viene anche mostrato che esse possono essere estese in modo da comprendere la conservazione della « stranezza » S.

(*) *A cura della Redazione.*

Reprinted from THE PHYSICAL REVIEW, Vol. 104, No. 1, 254–258, October 1, 1956
Printed in U. S. A.

Question of Parity Conservation in Weak Interactions*

T. D. LEE, *Columbia University, New York, New York*

AND

C. N. YANG,† *Brookhaven National Laboratory, Upton, New York*

(Received June 22, 1956)

The question of parity conservation in β decays and in hyperon and meson decays is examined. Possible experiments are suggested which might test parity conservation in these interactions.

RECENT experimental data indicate closely identical masses[1] and lifetimes[2] of the $\theta^+(\equiv K_{\pi 2}{}^+)$ and the $\tau^+(\equiv K_{\pi 3}{}^+)$ mesons. On the other hand, analyses[3] of the decay products of τ^+ strongly suggest on the grounds of angular momentum and parity conservation that the τ^+ and θ^+ are not the same particle. This poses a rather puzzling situation that has been extensively discussed.[4]

One way out of the difficulty is to assume that parity is not strictly conserved, so that θ^+ and τ^+ are two different decay modes of the same particle, which necessarily has a single mass value and a single lifetime. We wish to analyze this possibility in the present paper against the background of the existing experimental evidence of parity conservation. It will become clear that existing experiments do indicate parity conservation in strong and electromagnetic interactions to a high degree of accuracy, but that for the weak interactions (i.e., decay interactions for the mesons and hyperons, and various Fermi interactions) parity conservation is so far only an extrapolated hypothesis unsupported by experimental evidence. (One might even say that the present $\theta-\tau$ puzzle may be taken as an indication that parity conservation is violated in weak interactions. This argument is, however, not to be taken seriously because of the paucity of our present knowledge concerning the nature of the strange particles. It supplies rather an incentive for an examination of the question of parity conservation.) To decide unequivocally whether parity is conserved in weak interactions, one must perform an experiment to determine whether weak interactions differentiate the right from the left. Some such possible experiments will be discussed.

PRESENT EXPERIMENTAL LIMIT ON PARITY NONCONSERVATION

If parity is not strictly conserved, all atomic and nuclear states become mixtures consisting mainly of the state they are usually assigned, together with small percentages of states possessing the opposite parity. The fractional weight of the latter will be called $\mathfrak{F}^2$. It is a quantity that characterizes the degree of violation of parity conservation.

The existence of parity selection rules which work well in atomic and nuclear physics is a clear indication that the degree of mixing, $\mathfrak{F}^2$, cannot be large. From such considerations one can impose the limit $\mathfrak{F}^2 \lesssim (r/\lambda)^2$, which for atomic spectroscopy is, in most cases, $\sim 10^{-6}$. In general a less accurate limit obtains for nuclear spectroscopy.

Parity nonconservation implies the existence of interactions which mix parities. The strength of such interactions compared to the usual interactions will in general be characterized by $\mathfrak{F}$, so that the mixing will be of the order $\mathfrak{F}^2$. The presence of such interactions would affect angular distributions in nuclear reactions. As we shall see, however, the accuracy of these experiments is not good. The limit on $\mathfrak{F}^2$ obtained is not better than $\mathfrak{F}^2<10^{-4}$.

To give an illustration, let us examine the polarization experiments, since they are closely analogous to some experiments to be discussed later. A proton beam polarized in a direction z perpendicular to its momentum was scattered by nuclei. The scattered intensities were compared[5] in two directions A and B related to each other by a reflection in the x–y plane, and were found to be identical to within $\sim 1\%$. If the scattering originates from an ordinary parity-conserving interaction plus a parity-nonconserving interaction (e.g., $\boldsymbol{\sigma}\cdot\mathbf{r}$), then the scattering amplitudes in the directions A and B are in the proportion $(1+\mathfrak{F})/(1-\mathfrak{F})$, where $\mathfrak{F}$ represents the ratio of the strengths of the two kinds of interactions in the scattering. The experimental result therefore requires $\mathfrak{F}<10^{-2}$, or $\mathfrak{F}^2<10^{-4}$.

The violation of parity conservation would lead to an electric dipole moment for all systems. The magnitude of the moment is

$$\text{moment} \sim e\mathfrak{F}\times(\text{dimension of system}). \qquad (1)$$

* Work supported in part by the U. S. Atomic Energy Commission.

† Permanent address: Institute for Advanced Study, Princeton, New Jersey.

[1] Whitehead, Stork, Perkins, Peterson, and Birge, Bull. Am. Phys. Soc. Ser. II, **1**, 184 (1956); Barkas, Heckman, and Smith, Bull. Am. Phys. Soc. Ser. II, **1**, 184 (1956).

[2] Harris, Orear, and Taylor, Phys. Rev. **100**, 932 (1955); V. Fitch and K. Motley, Phys. Rev. **101**, 496 (1956); Alvarez, Crawford, Good, and Stevenson, Phys. Rev. **101**, 503 (1956).

[3] R. Dalitz, Phil. Mag. **44**, 1068 (1953); E. Fabri, Nuovo cimento **11**, 479 (1954). See Orear, Harris, and Taylor [Phys. Rev. **102**, 1676 (1956)] for recent experimental results.

[4] See, e.g., *Report of the Sixth Annual Rochester Conference on High Energy Physics* (Interscience Publishers, Inc., New York, to be published).

[5] See, e.g., Chamberlain, Segrè, Tripp, and Ypsilantis, Phys. Rev. **93**, 1430 (1954).

The presence of such electric dipole moments would have interesting consequences. For example, if the proton has an electric dipole moment $\cong e\times(10^{-16}$ cm), the perturbation caused by the presence of the neighboring $2p$ state of the hydrogen atom would shift the energy of the $2s$ state by about 1 Mc/sec. This would be inconsistent with the present theoretical interpretations of the Lamb shift. Another example is found in the electron-neutron interaction. An electric dipole moment for the neutron $\cong e\times(10^{-18}$ cm) is the upper limit allowable by the present experiments.

By far the most accurate measurement of the electric dipole moment was made by Purcell, Ramsey, and Smith. They gave[6] an upper limit for the electric dipole moment of the neutron of $e\times(5\times10^{-20}$ cm). This value sets the upper limit for $\mathfrak{F}^2$ as $\mathfrak{F}^2<3\times10^{-13}$, which is also the most accurate verification of the conservation of parity in strong and electromagnetic interactions. We shall see, however, that even this high degree of accuracy is not sufficient to supply an experimental proof of parity conservation in the weak interactions. For such a proof an accuracy of $\mathfrak{F}^2<10^{-24}$ is necessary.

QUESTION OF PARITY CONSERVATION IN β DECAY

At first sight it might appear that the numerous experiments related to β decay would provide a verification that the weak β interaction does conserve parity. We have examined this question in detail and found this to be not so. (See Appendix.) We start by writing down the five usual types of couplings. In addition to these we introduce the five types of couplings that conserve angular momentum but do not conserve parity. It is then apparent that the classification of β decays into allowed transitions, first forbidden, etc., proceeds exactly as usual. (The mixing of parity of the *nuclear states* would not measurably affect these selection rules. This phenomenon belongs to the discussions of the last section.) The following phenomena are then examined: allowed spectra, unique forbidden spectra, forbidden spectra with allowed shape, β-neutrino correlation, and $\beta-\gamma$ correlation. It is found that these experiments have no bearing on the question of parity conservation of the β-decay interactions. This comes about because in all of these phenomena no interference terms exist between the parity-conserving and parity-nonconserving interactions. In other words, the calculations always result in terms proportional to $|C|^2$ plus terms proportional to $|C'|^2$. Here C and C' are, respectively, the coupling constants for the usual parity-conserving interactions (a sum of five terms) and the parity-nonconserving interactions (also a sum of five terms.) Furthermore, it is well known[7] that without measuring the spin of the neutrino we cannot distinguish the couplings C from the couplings C' (provided the mass of the neutrino is zero). The experimental results concerning the above-named phenomena, which constitute the bulk of our present knowledge about β decay, therefore cannot decide the degree of mixing of the C' type interactions with the usual type.

The reason for the absence of interference terms CC' is actually quite obvious. Such terms can only occur as a pseudoscalar formed out of the experimentally measured quantities. For example, if three momenta $\mathbf{p}_1$, $\mathbf{p}_2$, $\mathbf{p}_3$ are measured, the term $CC'\mathbf{p}_1\cdot(\mathbf{p}_2\times\mathbf{p}_3)$ may occur. Or if a momentum $\mathbf{p}$ and a spin $\boldsymbol{\sigma}$ are measured, the term $CC'\mathbf{p}\cdot\boldsymbol{\sigma}$ may occur. In all the β-decay phenomena mentioned above, no such pseudoscalars can be formed out of the measured quantities.

POSSIBLE EXPERIMENTAL TESTS OF PARITY CONSERVATION IN β DECAYS

The above discussion also suggests the kind of experiments that could detect the possible interference between C and C' and consequently could establish whether parity conservation is violated in β decay. A relatively simple possibility is to measure the angular distribution of the electrons coming from β decays of oriented nuclei. If θ is the angle between the orientation of the parent nucleus and the momentum of the electron, an asymmetry of distribution between θ and $180°-\theta$ constitutes an unequivocal proof that parity is not conserved in β decay.

To be more specific, let us consider the allowed β transition of any oriented nucleus, say Co^{60}. The angular distribution of the β radiation is of the form (see Appendix):

$$I(\theta)d\theta=(\text{constant})(1+\alpha\cos\theta)\sin\theta d\theta, \qquad (2)$$

where α is proportional to the interference term CC'. If $\alpha\neq0$, one would then have a positive proof of parity nonconservation in β decay. The quantity α can be obtained by measuring the fractional asymmetry between $\theta<90°$ and $\theta>90°$; i.e.,

$$\alpha=2\left[\int_0^{\pi/2}I(\theta)d\theta-\int_{\pi/2}^{\pi}I(\theta)d\theta\right]\Big/\int_0^{\pi}I(\theta)d\theta.$$

It is noteworthy that in this case the presence of the magnetic field used for orienting the nuclei would automatically cause a spatial separation between the electrons emitted with $\theta<90°$ and those with $\theta>90°$. Thus, this experiment may prove to be quite feasible.

It appears at first sight that in the study of γ-radiation distribution from β-decay products of oriented nuclei one can form a pseudoscalar from the spin of the oriented nucleus and the γ-ray momentum $\mathbf{p}_\gamma$. Thus it may seem to offer another possible experimental test of parity conservation. Unfortunately, the nuclear levels have definite parities, and electromagnetic inter-

[6] E. M. Purcell and N. F. Ramsey, Phys. Rev. **78**, 807 (1950); Smith *et al.* as quoted in N. F. Ramsey, *Molecular Beams* (Oxford University Press, London, 1956).
[7] C. N. Yang and J. Tiomno, Phys. Rev. **79**, 495 (1950).

actions conserve parity. (Any small mixing of parities characterized by $\mathfrak{F}^2<3\times10^{-15}$ would not affect the arguments here.) Consequently the γ rays carry away definite parities. Thus the observed probability function must be an even function of $\mathbf{p}_\gamma$. This property eliminates the possibility of forming a pseudoscalar quantity. It is therefore not possible to use such experiments as a test of parity conservation.

In β-γ-γ' triple correlation experiments one can, by some rather similar but more complicated reasoning, prove that a measurement of the three momenta cannot supply any information on the question of parity conservation in β decay.

In β-γ correlation experiments the nature of the polarization of the γ ray could provide a test. To be more specific, let us consider the polarization state of γ rays emitted parallel to the β ray. If parity conservation holds for β decay, the γ ray will be unpolarized. On the other hand, if parity conservation is violated in β decay, the γ ray will in general be polarized. However, this polarization must be circular in nature and therefore may not lend itself to easy experimental detection. (The usual ways of measuring polarization through Compton effect, photoelectric effect, and photodisintegration of the deuteron are all incapable of detecting circular polarization. This is because circular polarization is specified by an axial vector parallel to the direction of propagation. From the observed momenta in these detection techniques such an axial vector cannot be formed.) For other directions of γ-ray propagation, elliptical polarization will result if parity is not conserved. This effect will thus be more difficult to detect.

QUESTION OF PARITY CONSERVATION IN MESON AND HYPERON DECAYS

If the weak interactions, such as the β-decay interactions or the decay interactions of mesons and hyperons, do not conserve parity, parity mixing will occur in all interactions by means of second-order processes. To examine this effect let us consider, for example, the decay of the Λ^0:

$$\Lambda^0\rightarrow p+\pi^-.$$

The assumption that parity is not conserved in this decay implies that the Λ^0 exists virtually in states of opposite parities. It could therefore possess an electric dipole moment of a magnitude

$$\text{moment}\sim e\mathcal{G}^2\times(\text{dimension of }\Lambda^0), \qquad (3)$$

where $\mathcal{G}$ is the coupling strength of the decay interaction of the Λ^0. ($\mathcal{G}^2\lesssim10^{-12}$.) The electric dipole moment of the Λ^0 is therefore $\lesssim e\times(10^{-25}\text{ cm})$.

Clearly the proton would have an electric dipole moment of the same order of magnitude. The existence of such a small electric dipole moment is, as we have seen, completely consistent with the present experimental information. Another way of putting this is to observe that by comparing Eq. (3) with Eq. (1), one has

$$\mathfrak{F}\sim\mathcal{G}^2.$$

Since all the weak interaction including β interactions are characterized by coupling strengths $\mathcal{G}^2<10^{-12}$, a violation of parity in weak interactions would introduce a parity mixing characterized by an $\mathfrak{F}^2<10^{-24}$. This is outside the present limit of experimental knowledge, as we have discussed before.

If the weak interactions violate parity conservation, parity would be defined and measured in strong and electromagnetic interactions only, just as strangeness is. Furthermore it is important to notice that with the conservation of strangeness, as with every conservation law, there is an element of arbitrariness introduced into the parity of all systems. The parity of all strange particles would be defined only up to a factor of $(-1)^S$, where S is the strangeness. The parity of the Λ^0 (relative to the nucleons) is therefore a matter of definition. But once this is defined, the parity of other strange particles would be measurable from the strong interactions.

POSSIBLE EXPERIMENTAL TESTS OF PARITY CONSERVATION IN MESON AND HYPERON DECAYS

To have a sensitive unequivocal test of whether parity is conserved in weak interactions, one must decide whether the weak interactions differentiate between the right and the left. This is possible only if one produces interference between states of opposite parities. The mere observation of two decay products of opposite parities originating from a "particle" cannot provide conclusive evidence that parity is not conserved. Such indeed is the state of affairs of the present $\theta-\tau$ puzzle.

As we have discussed before, these interference terms are possible only if the observed quantities can form a pseudoscalar such as $\mathbf{p}_1\cdot(\mathbf{p}_2\times\mathbf{p}_3)$. The observation of Λ^0 decays in association with their production does provide such a possible pseudoscalar and hence a possible test of whether parity is conserved in the Λ^0 decay interaction. Let us consider the experiment

$$\pi^-+p\rightarrow\Lambda^0+\theta^0,\quad \Lambda^0\rightarrow p+\pi^-. \qquad (4)$$

Let $\mathbf{p}_{\text{in}}$, $\mathbf{p}_\Lambda$, and $\mathbf{p}_{\text{out}}$ be, respectively, the momenta in the laboratory system of the incoming pion, the Λ^0, and the decay pion. We define a parameter R as the projection of $\mathbf{p}_{\text{out}}$ in the direction of $\mathbf{p}_{\text{in}}\times\mathbf{p}_\Lambda$. The value of R ranges from approximately -100 Mev/c to approximately $+100$ Mev/c. Switching from a right-handed convention for vector products (which we use) to a left-handed convention means a switch of the sign of R. Parity conservation in the weak decay interaction of Λ^0 can therefore be experimentally checked by investigating whether $+R$ and $-R$ have equal probabilities of occurrence.

To see more clearly the meaning of the parameter R, one transforms $\mathbf{p}_{out}(\rightarrow\mathbf{p}')$ into the center-of-mass system of Λ^0. The new vector $\mathbf{p}'$ has a constant magnitude $\cong$100 Mev/c. The frequency distribution of this vector $\mathbf{p}'$ can then be plotted on a spherical surface. Taking the z axis for this sphere to be in the direction of $\mathbf{p}_{in}\times\mathbf{p}_\Lambda$, one can prove the following two symmetries:

(*a*) The frequency distribution on the sphere remains unchanged under a rotation through 180° around the z axis. This symmetry follows from parity conservation in the strong reaction producing the Λ^0. It does not depend on the nature of the weak interaction.

(*b*) If parity is conserved in the decay interaction of Λ^0, the frequency distribution on the sphere is unchanged under a reflection with respect to the production plane of Λ^0.

To prove statement (*a*), one need only consider the invariance of the production process under a reflection with respect to the production plane defined by $\mathbf{p}_{in}$ and $\mathbf{p}_\Lambda$. This reflection is the resultant of an inversion and a rotation through 180° around the z axis (which is normal to the production plane). The state of polarization of Λ^0 is thus invariant under a 180° rotation around the z axis, leading to the stated symmetry.[8]

Statement (*b*) follows[8] directly from the assumption that the weak interaction as well as the strong interaction conserves parity. A reflection with respect to the production plane must then leave the whole process invariant.

The frequency distribution of R is just the projection of the distribution on the sphere onto the z axis. An asymmetry between $+R$ and $-R$ therefore implies parity nonconservation in Λ^0 decay. However, if the spin of Λ^0 is unpolarized, no asymmetry[9] can obtain even if parity is not conserved in Λ^0 decay. To obtain a polarized Λ^0 beam, the experiment is therefore best done at a definite nonforward angle of production of Λ^0 and at a definite incoming energy.

The above discussions apply also to any other strange particle decay if (1) the particle has a nonvanishing spin and (2) it decays into two particles at least one of which has a nonvanishing spin, or it decays into three or more particles. Thus the above considerations can be applied also to the decays of $\Sigma^\pm$ and *maybe* also to $K_{\mu 2}^\pm$, $K_{\mu 3}^\pm$ and $K_{\pi 3}^\pm$ ($\equiv\tau^\pm$).

In the decay processes

$$\pi\rightarrow\mu+\nu, \tag{5}$$

$$\mu\rightarrow e+\nu+\nu, \tag{6}$$

starting from a π meson at rest, one could study the distribution of the angle θ between the μ-meson momentum and the electron momentum, the latter being in the center-of-mass system of the μ meson. If parity is conserved in neither (5) nor (6), the distribution will not in general be identical for θ and $\pi-\theta$. To understand this, consider first the orientation of the muon spin. If (5) violates parity conservation, the muon would be in general polarized in its direction of motion. In the subsequent decay (6), the angular distribution problem with respect to θ is therefore closely similar to the angular distribution problem of β rays from oriented nuclei, which we have discussed before. (Entirely similar considerations can be applied to $\Xi^-\rightarrow\Lambda^0+\pi^-$ and $\Lambda^0\rightarrow p+\pi^-$.)

REMARKS

If parity conservation is violated in hyperon decay, the decay products will have mixed parities. This, however, does not affect the arguments of Adair[10] and of Treiman[11] concerning the relationship between the spin of the hyperons and the angular distribution of their decay products in certain special cases.[12]

One may question whether the other conservation laws of physics could also be violated in the weak interactions. Upon examining this question, one finds that the conservations of the number of heavy particles, of electric charge, of energy, and of momentum all appear to be inviolate in the weak interactions. The same cannot be said of the conservation of angular momentum, and of parity. Nor can it be said of the invariance under time reversal. It might appear at first sight that the equality of the life times of $\pi^\pm$ and of those of $\mu^\pm$ furnish proofs of the invariance under charge conjugation of the weak interactions. A closer examination of this problem reveals, however, that this is not so. In fact, the equality of the life times of a charged particle and its charge conjugate against decay through a weak interaction (to the lowest order of the strength of the weak interaction) can be shown to follow from the invariance under proper Lorentz transformations (i.e., Lorentz transformation with neither space nor time inversion). One has therefore at present no experimental proof of the invariance under charge conjugation of the weak interactions. In the present paper, only the question of parity nonconservation is discussed.

[8] This proof for statement (*a*) is correct only if Λ^0 exists as a single particle with a definite parity in the strong interactions, (as discussed in the last section); i.e. if Λ^0 does not exist as two degenerate states Λ_1^0 and Λ_2^0 of opposite parity, as has been suggested [T. D. Lee and C. N. Yang, Phys. Rev. **102**, 290 (1956)]. [It is to be emphasized, that if parity is indeed not conserved in the weak interactions, there would be (at present) no necessity to introduce the complication of two degenerate states of opposite parity at all.] On the other hand, statement (*b*) is correct even if Λ^0 exists as two degenerate states Λ_1^0 and Λ_2^0 of opposite parity. *To summarize, violation of the symmetry stated in* (*a*) *implies the existence of the parity doublets* Λ_1^0 *and* Λ_2^0 *with a mass difference less than their widths. Violation of the symmetry stated in* (*b*) *implies the nonconservation of parity in* Λ *decay*. See also footnote 12 and T. D. Lee and C. N. Yang, Phys. Rev. (to be published).

[9] Also the interference may accidentally be absent if the relative phase between the two parities in the decay product is 90°. This, however, cannot be the case if time-reversal invariance is preserved in the decay process.

[10] R. K. Adair, Phys. Rev. **100**, 1540 (1955).

[11] S. B. Treiman, Phys. Rev. **101**, 1216 (1956).

[12] The existence of Λ_1^0 and Λ_2^0 of opposite parity may affect these relationships. This is similar to the violation of symmetry (*a*) discussed in footnote 8. See T. D. Lee and C. N. Yang, Phys. Rev. (to be published).

The conservation of parity is usually accepted without questions concerning its possible limit of validity being asked. There is actually no *a priori* reason why its violation is undesirable. As is well known, its violation implies the existence of a right-left asymmetry. We have seen in the above some possible experimental tests of this asymmetry. These experiments test whether the present elementary particles exhibit asymmetrical behavior with respect to the right and the left. If such asymmetry is indeed found, the question could still be raised whether there could not exist corresponding elementary particles exhibiting opposite asymmetry such that in the broader sense there will still be over-all right-left symmetry. If this is the case, it should be pointed out, there must exist two kinds of protons p_R and p_L, the right-handed one and the left-handed one. Furthermore, at the present time the protons in the laboratory must be predominantly of one kind in order to produce the supposedly observed asymmetry, and also to give rise to the observed Fermi-Dirac statistical character of the proton. This means that the free oscillation period between them must be longer than the age of the universe. They could therefore both be regarded as stable particles. Furthermore, the numbers of p_R and p_L must be separately conserved. However, the interaction between them is not necessarily weak. For example, p_R and p_L could interact with the same electromagnetic field and perhaps the same pion field. They could then be separately pair-produced, giving rise to interesting observational possibilities.

In such a picture the supposedly observed right-and-left asymmetry is therefore ascribed not to a basic non-invariance under inversion, but to a cosmologically local preponderance of, say, p_R over p_L, a situation not unlike that of the preponderance of the positive proton over the negative. Speculations along these lines are extremely interesting, but are quite beyond the scope of this note.

The authors wish to thank M. Goldhaber, J. R. Oppenheimer, J. Steinberger, and C. S. Wu for interesting discussions and comments. They also wish to thank R. Oehme for an interesting communication.

APPENDIX

If parity is not conserved in β decay, the most general form of Hamiltonian can be written as

$$\begin{aligned} H_{\text{int}}= & (\psi_p^\dagger\gamma_4\psi_n)(C_S\psi_e^\dagger\gamma_4\psi_\nu+C_S'\psi_e^\dagger\gamma_4\gamma_5\psi_\nu) \\ & +(\psi_p^\dagger\gamma_4\gamma_\mu\psi_n)(C_V\psi_e^\dagger\gamma_4\gamma_\mu\psi_\nu+C_V'\psi_e^\dagger\gamma_4\gamma_\mu\gamma_5\psi_\nu) \\ & +\tfrac{1}{2}(\psi_p^\dagger\gamma_4\sigma_{\lambda\mu}\psi_n)(C_T\psi_e^\dagger\gamma_4\sigma_{\lambda\mu}\psi_\nu \\ & +C_T'\psi_e^\dagger\gamma_4\sigma_{\lambda\mu}\gamma_5\psi_\nu)+(\psi_p^\dagger\gamma_4\gamma_\mu\gamma_5\psi_n) \\ & \times(-C_A\psi_e^\dagger\gamma_4\gamma_\mu\gamma_5\psi_\nu-C_A'\psi_e^\dagger\gamma_4\gamma_\mu\psi_\nu) \\ & +(\psi_p^\dagger\gamma_4\gamma_5\psi_n)(C_P\psi_e^\dagger\gamma_4\gamma_5\psi_\nu+C_P'\psi_e^\dagger\gamma_4\psi_\nu), \end{aligned} \quad (A.1)$$

where $\sigma_{\lambda\mu}=-\frac{1}{2}i(\gamma_\lambda\gamma_\mu-\gamma_\mu\gamma_\lambda)$ and $\gamma_5=\gamma_1\gamma_2\gamma_3\gamma_4$. The ten constants C and C' are all real if time-reversal invariance is preserved in β decay. This however, will not be assumed in the following.

Calculation with this interaction proceeds exactly as usual. One obtains, e.g., for the energy and angle distribution of the electron in an allowed transition

$$N(W,\theta)dW\sin\theta d\theta=\frac{\xi}{4\pi^3}F(Z,W)pW(W_0-W)^2 \times\left(1+\frac{ap}{W}\cos\theta+\frac{b}{W}\right)dW\sin\theta d\theta, \quad (A.2)$$

where

$$\xi=(|C_S|^2+|C_V|^2+|C_S'|^2+|C_V'|^2)|M_{\text{F.}}|^2 +(|C_T|^2+|C_A|^2+|C_T'|^2+|C_A'|^2)|M_{\text{G.T.}}|^2, \quad (A.3)$$

$$a\xi=\tfrac{1}{3}(|C_T|^2-|C_A|^2+|C_T'|^2-|C_A'|^2)|M_{\text{G.T.}}|^2 -(|C_S|^2-|C_V|^2+|C_S'|^2-|C_V'|^2)|M_{\text{F.}}|^2, \quad (A.4)$$

$$\begin{aligned} b\xi= & \gamma[(C_S^*C_V+C_SC_V^*)+(C_S'^*C_V'+C_S'C_V'^*)]|M_{\text{F.}}|^2 \\ & +\gamma[(C_T^*C_A+C_A^*C_T)+(C_T'^*C_A'+C_A'^*C_T')] \times|M_{\text{G.T.}}|^2. \end{aligned} \quad (A.5)$$

In the above expression all unexplained notations are identical with the standard notations. (See, e.g., the article by Rose.[13])

The above expression does not contain any interference terms between the parity-conserving part of the interactions and the parity-nonconserving ones. It is in fact directly obtainable by replacing in the usual expression the quantity $|C_S|^2$ by $|C_S|^2+|C_S'|^2$, and $C_SC_V^*$ by $C_SC_V^*+C_S'C_V'^*$, etc. This rule also holds in general, except for the cases where a pseudoscalar can be formed out of the measured quantities, as discussed in the text.

When a pseudoscalar can be formed, for example, in the β decay of oriented nuclei, interference terms would be present, as explicitly displayed in Eq. (2). In an allowed transition $J\rightarrow J-1$ (no), the quantity α is given by

$$\alpha=\beta\langle J_z\rangle/J,$$

$$\beta=\text{Re}\left[C_TC_T'^*-C_AC_A'^*+i\frac{Ze^2}{\hbar cp}(C_AC_T'^*+C_A'C_T^*)\right] \times|M_{\text{G.T.}}|^2\frac{v_e}{c}\frac{2}{\xi+(\xi b/W)}, \quad (A.6)$$

where $M_{\text{G.T.}}$, ξ, and b are defined in Eqs. (A.3)–(A.5), v_e is the velocity of the electron, and $\langle J_z\rangle$ is the average spin component of the initial nucleus. For an allowed transition $J\rightarrow J+1$ (no), α is given by

$$\alpha=-\beta\langle J_z\rangle/(J+1). \quad (A.7)$$

The effect of the Coulomb field is included in all the above considerations.

[13] M. E. Rose, in *Beta- and Gamma-Ray Spectroscopy* (Interscience Publishers, Inc., New York, 1955), pp. 271–291.

Errata

Question of Parity Conservation in Weak Interactions, T. D. LEE AND C. N. YANG [Phys. Rev. 104, 254 (1956)]. Equation (A.4) in the Appendix of this paper should be amended to read as follows:

$$a\xi=-(|C_S|^2-|C_V|^2+|C_S'|^2-|C_V'|^2)|M_{\text{F.}}|^2$$

$$+\tfrac{1}{3}(|C_T|^2-|C_A|^2+|C_T'|^2-|C_A'|^2)|M_{\text{G.T.}}|^2$$

$$+2\,\text{Re}\left\{i\frac{Ze^2}{\hbar cp}\left[(C_SC_V^*+C_S'C_V'^*)|M_{\text{F.}}|^2\right.\right.$$

$$\left.\left.-\tfrac{1}{3}(C_TC_A^*+C_T'C_A'^*)|M_{\text{G.T.}}|^2\right]\right\}. \quad \text{(A.4)}$$

This change does not affect the text of the paper, nor does it affect the other parts of the Appendix.

The authors are indebted to Dr. R. B. Curtis and to Dr. M. Morita[1] for pointing out to them the error in the original Eq. (A.4).

[1] M. Morita, Progr. Theoret. Phys. Japan **10**, 364 (1953).

Reprinted from THE PHYSICAL REVIEW, Vol. 105, No. 5, 1671–1675, March 1, 1957
Printed in U. S. A.

Parity Nonconservation and a Two-Component Theory of the Neutrino

T. D. LEE, *Columbia University, New York, New York*

AND

C. N. YANG, *Institute for Advanced Study, Princeton, New Jersey*

(Received January 10, 1957; revised manuscript received January 17, 1957)

A two-component theory of the neutrino is discussed. The theory is possible only if parity is not conserved in interactions involving the neutrino. Various experimental implications are analyzed. Some general remarks concerning nonconservation are made.

RECENTLY the question has been raised[1,2] as to whether the weak interactions are invariant under space inversion, charge conjugation, and time reversal. It was pointed out that although these invariances are generally held to be valid for all interactions, experimental proof has so far only extended to cover the strong interactions. (We group here the electromagnetic interactions with the strong interactions.) To test the possible violation of these invariance laws in the weak interactions, a number of experiments were proposed. One of these is to study the angular distribution of the β ray coming from the decay of oriented nuclei. We have been informed by Wu[3] that such an experiment is in progress. The preliminary results indicate a large asymmetry with respect to the spin direction of the oriented nuclei. Since the spin is an axial vector, its observed correlation with the β-ray momentum (which is a polar vector) can be understood only in terms of a violation of the law of space inversion invariance in β decay.

In view of this information and especially in view of the large asymmetry found, we wish to examine here a possible theory of the neutrino different from the conventionally accepted one. In this theory for a given momentum p the neutrino has only *one* spin state, the spin being always parallel to p. The spin and momentum of the neutrino together therefore automatically define the sense of the screw.

In this theory the mass of the neutrino must be zero, and its wave function need only have two components instead of the usual four. That such a relativistic theory is possible is well known.[4] It was, however, always rejected because of its intrinsic violation of space inversion invariance, a reason which is now no longer valid. (In fact, as we shall see later, in such a theory the violation of space inversion invariance attains a maximum.)

In Sec. 1 we describe this two-component theory of the neutrino. It is then shown in Sec. 2 that this theory is mathematically equivalent to a familiar four-component neutrino formalism for which all parity-conserving and parity-nonconserving Fermi couplings C and C' (as defined in the appendix of reference 1) are always related in the following manner: $C_S=C_S'$, $C_V=C_V'$, etc. or $C_S=-C_S'$, $C_V=-C_V'$, etc. Sections 3 to 8 are devoted to the physical consequences of the theory that can be put to experimental test. In the last section some general remarks about nonconservation are made.

I. NEUTRINO FIELD

1. Consider first the Dirac equation for a free spin-$\frac{1}{2}$ particle with zero mass. Because of the absence of the mass term, one needs only three anticommuting Hermitian matrices. Thus the neutrino can be represented by a spinor function φ_ν which has only two components.[4] The Dirac equation for φ_ν can be written as ($\hbar=c=1$)

$$\boldsymbol{\sigma}\cdot\mathbf{p}\varphi_\nu=i\partial\varphi_\nu/\partial t, \tag{1}$$

where σ_1, σ_2, σ_3 are the usual 2×2 Pauli matrices. The relativistic invariance of this equation for proper Lorentz transformations (i.e., Lorentz transformations without space inversion and time inversion) is well known. In particular, for the space rotations through an angle θ around, say, the z axis, the wave function transforms in the following way:

$$\varphi\rightarrow\exp(-i\sigma_3\theta/2)\phi. \tag{2}$$

The σ matrices are therefore the spin matrices for the neutrino. For a state with a definite momentum $\mathbf{p}$, the energy and the spin along $\mathbf{p}$ are given, respectively, by

$$H=(\boldsymbol{\sigma}\cdot\mathbf{p}),$$
$$\sigma_p=(\boldsymbol{\sigma}\cdot\mathbf{p})/|\mathbf{p}|.$$

They are therefore related by

$$H=|\mathbf{p}|\sigma_p. \tag{3}$$

In the c-number theory, for a given momentum, the particle has therefore two states: a state with positive energy, and with $\frac{1}{2}$ as the spin component along $\mathbf{p}$, and a state with negative energy and with $-\frac{1}{2}$ as the spin component along $\mathbf{p}$.

It is easy to see that in a hole theory of such particles, *the spin of a neutrino* (defined to be a particle in the

---

[1] T. D. Lee and C. N. Yang, Phys. Rev. **104**, 254 (1956).

[2] Lee, Oehme, and Yang, Phys. Rev. (to be published).

[3] Wu, Ambler, Hayward, Hoppes, and Hudson. We wish to thank Professor C. S. Wu for informing us of the progress of the experiment.

[4] See, e.g., W. Pauli, *Handbuch der Physik* (Verlag Julius Springer, Berlin, 1933), Vol. 24, 226–227.

positive-energy state) *is always parallel to its momentum while the spin of an antineutrino* (defined to be a hole in the negative-energy state) *is always antiparallel to its momentum* (i.e., the momentum of the antineutrino). Many of the experimental implications discussed in later sections are direct consequences of this correlation between the spin and the momentum of a neutrino. We have remarked in the introduction that such a correlation defines automatically the sense of a screw. With the usual (right-handed) conventions which we adopt throughout this paper, the spin and the velocity of the neutrino represent the spiral motion of a right-handed screw while the spin and the velocity of the antineutrino represent the spiral motion of a left-handed screw.

We shall now discuss some general properties[5] of this neutrino field:

(A) In this theory it is clear that the neutrino state and the antineutrino state cannot be the same. A Majorana theory for such a neutrino is therefore impossible.

(B) The mass of the neutrino and the antineutrino in this theory is necessarily zero. This is true for the physical mass even with the inclusion of all interactions. To see this, one need only observe that all the one-particle *physical* states consisting of one neutrino (or one antineutrino) must belong to a representation of the inhomogeneous proper Lorentz group identical with the representation to which the free neutrino states discussed above belong. For such a representation to exist at all, the mass must be zero.

(C) That the theory does not conserve parity is well known. We see it also in the following way: Under a space inversion P, one inverts the momentum of a neutrino but not its spin direction. Since in this theory the two are always parallel, the operator P applied to a neutrino state leads to a nonexisting state. Consequently the theory is not invariant under space inversion.

(D) By the same reasoning one concludes that the theory is also not invariant under charge conjugation C which changes a particle into its antiparticle but does not change its spin direction or momentum.

(E) It is possible, however, for the theory to be invariant under the operation CP, as this operation changes a neutrino into an antineutrino and simultaneously reverses its momentum while keeping the spin direction fixed. By the Lüders-Pauli theorem[6] it follows that the theory can be invariant under time reversal T.

For the free neutrino field, as described by (1), one can prove that the theory is indeed invariant under time reversal and under CP.

2. We shall in this section indicate how one can use the conventional four-component formalism of the neutrino (with violation of parity conservation) and obtain the same results as the present theory.

We start from Eq. (1) and enlarge the matrices by the following definitions (1 represents a 2×2 unit matrix):

$$\boldsymbol{\alpha}\equiv\begin{pmatrix}\boldsymbol{\sigma} & 0\\ 0 & -\boldsymbol{\sigma}\end{pmatrix},\quad \beta\equiv\begin{pmatrix}0 & 1\\ 1 & 0\end{pmatrix},\tag{4}$$

$$\psi_\nu\equiv\begin{pmatrix}\phi\\ 0\end{pmatrix},\tag{5a}$$

$$\boldsymbol{\gamma}\equiv -i\beta\boldsymbol{\alpha},\quad \gamma_4\equiv\beta,\quad \gamma_5\equiv\gamma_1\gamma_2\gamma_3\gamma_4=\begin{pmatrix}-1 & 0\\ 0 & 1\end{pmatrix}.\tag{6}$$

An immediate consequence of these definitions is

$$\gamma_5\psi_\nu=-\psi_\nu.\tag{7a}$$

The free neutrino part of the Lagrangian is, as usual,

$$L_\nu=\psi_\nu^\dagger\gamma_4\left(\gamma_\mu\frac{\partial}{\partial x_\mu}\right)\psi_\nu,\tag{8}$$

where $\psi_\nu^\dagger$ = Hermitian conjugate of ψ_ν. The most general interaction Lagrangian not containing derivatives for the process

$$n\rightarrow p+e+\bar{\nu}\tag{9a}$$

is exactly as usual; namely, it is the sum of the usual S, V, T, A, and P couplings:

$$+L_{\text{int}}=-H_{\text{int}}=\sum\left[-2C_i(\psi_p^\dagger O_i\psi_n)(\psi_e^\dagger O_i\psi_\nu)\right],\tag{10a}$$

where i runs over S, V, T, A, and P and

$$\begin{aligned}O_S&=\gamma_4,\\ O_V&=\gamma_4\gamma_\mu,\\ O_T&=-\frac{1}{2\sqrt{2}}i\gamma_4(\gamma_\lambda\gamma_\mu-\gamma_\mu\gamma_\lambda),\\ O_A&=-i\gamma_4\gamma_\mu\gamma_5,\\ O_P&=\gamma_4\gamma_5.\end{aligned}\tag{11}$$

It is not difficult to prove that Eqs. (5a) and (7a) are consistent with a relativistic theory even in the presence of the interaction (10a). Another way of proving this is to start from the conventional theory of the neutrino with the interaction Hamiltonian given in (A.1) of reference 1 and observe that when

$$C_S=-C_S',\quad C_V=-C_V',\ \text{etc.}\tag{12a}$$

the neutrino field ψ_ν there always appears in interactions in the combination $(1-\gamma_5)\psi_\nu$. In the explicit representa-

[5] We have received a manuscript from Professor A. Salam on a theory of the neutrino similar to the one discussed in the present paper. He specifically discussed points (A) and (B) that we discuss here. He also gave the Michel parameter for the μ decay that agrees with the ones obtained below in Sec. 6.

[6] G. Lüders, Kgl. Dansk Videnskab. Selskab, Mat.-fys. Medd. **28**, No. 5 (1954); W. Pauli, *Niels Bohr and the Development of Physics* (Pergamon Press, London, 1955).

tion that we have adopted above, this means that only the first two components of ψ_ν contribute to the interaction. *All calculations using the conventional theory of the neutrino with the Hamiltonian* (A.1) *of reference* 1 *concerning* β *decay therefore gives the same result as the present theory if we take the choice of constants* (12a). There exists, however, the possibility that in the decay of the neutron a neutrino[7] is emitted:

$$n\rightarrow p+e+\nu. \tag{9b}$$

The corresponding general form (not including derivatives of the fields) of the Hamiltonian is

$$H_{\text{int}}=\sum[2C_i(\psi_p{}^\dagger O_i\psi_n)(\psi_e{}^\dagger O_i\psi_\nu{}')], \tag{10b}$$

where O_i has been defined in Eq. (11). The field $\psi_\nu{}'$ is a four-component spinor defined in terms of the two-component neutrino field ϕ by

$$\psi_\nu{}'=\begin{pmatrix}0\\ \sigma_2\varphi^\dagger\end{pmatrix}. \tag{5b}$$

From Eq. (6), we see that

$$\gamma_5\psi_\nu{}'=+\psi_\nu{}'. \tag{7b}$$

It can be shown that (5b) and (7b) are consistent with a relativistic theory even in the presence of interaction (10b). *It can also be proved that one can use again the Hamiltonian* (A.1) *of reference* 1 *for the conventional theory of the neutrino with the choice of the coupling constants*

$$C_S=C_S{}',\quad C_V=C_V{}',\ \text{etc.} \tag{12b}$$

and obtain the same result as the present theory.

The two possible choices (12a) and (12b) depend on whether, in the β decay of the neutron, process (9a) or (9b) prevails, i.e., whether a neutrino[7] or an antineutrino is emitted. We shall see in Sec. 3 that experimentally it will be easy to decide which of the two choices is appropriate (if the theory is correct). [We do not consider the possibility here of the simultaneous presence of (9a) and (9b), since the double beta decay process does not seem to be observed experimentally.]

II. EXPERIMENTAL IMPLICATIONS

3. We consider in this section the experiment of the β decay of oriented nuclei already discussed in reference 1, and currently being carried out.[3] For the present theory, according to Eqs. (12a) or (12b), Eq. (A.6) reduces to

$$\beta=\mp\left(\frac{v_e}{c}\right)\frac{|C_T|^2-|C_A|^2-(2Ze^2/\hbar cp)\,\text{Im}(C_AC_T{}^*)}{|C_T|^2+|C_A|^2}. \tag{13}$$

The choice of the $\mp$ sign depends on whether

$$n\rightarrow p+e+\bar{\nu}\quad(\bar{\nu}=\text{left-handed screw}),$$

or

$$n\rightarrow p+e+\nu\quad(\nu=\text{right-handed screw}). \tag{14}$$

In writing down (13) the Fierz interference terms has been set equal to zero, which is in conformity with the experimental results,[8] and which implies [see Eq. (A.5) of reference 1]:

$$\text{Real part of } C_AC_T{}^*=0. \tag{15}$$

By measuring the momentum dependence of the asymmetry parameter β, one can test whether the present theory is correct.

It is interesting to notice that for a positron emitter the asymmetry parameter has the opposite sign. This is a direct consequence of the fact that in positron and electron emission, the neutrino and antineutrino emitted have opposite spirality.

4. An experiment such as the one being carried out by Cowan and collaborators[9] measures the cross section for neutrino absorption, which can be calculated in both the present theory and the usual theory. Now one determines the magnitude of the β-coupling constants to give the observed lifetimes of nuclei against β decay. The calculated value of the cross section turns out then to be *twice as great* in the present theory as in the usual theory. This follows from the following simple reasoning: The neutrino flux is an experimental quantity independent of the theory. If the neutrinos in a given direction have only one spin state instead of the usual two, by a detailed balancing argument they must have twice the cross section for absorption as the usual ones.

5. In the decay of $\pi^\mp$ mesons at rest, let us consider the component of angular momentum along the direction of $\mathbf{p}_\mu$, the momentum of the μ meson. The orbital angular momentum contributes nothing to this component. The μ spin component is therefore completely determined (irrespective of its total spin) by the spin component of the ν or $\bar{\nu}$. There are then two possibilities:

$$\text{(A)}\quad \begin{aligned}&\pi^+\rightarrow\mu^++\nu,\quad(\mu^+\text{ spin along }\mathbf{p}_\mu)=+\tfrac{1}{2},\\ &\pi^-\rightarrow\mu^-+\bar{\nu},\quad(\mu^-\text{ spin along }\mathbf{p}_\mu)=-\tfrac{1}{2};\end{aligned} \tag{16}$$

or

$$\text{(B)}\quad \begin{aligned}&\pi^+\rightarrow\mu^++\bar{\nu},\quad(\mu^+\text{ spin along }\mathbf{p}_\mu)=-\tfrac{1}{2},\\ &\pi^-\rightarrow\mu^-+\nu,\quad(\mu^-\text{ spin along }\mathbf{p}_\mu)=+\tfrac{1}{2}.\end{aligned} \tag{17}$$

In each case the μ mesons with fixed $\mathbf{p}_\mu$ form a polarized beam. (It was pointed out in reference 1 that if parity is not conserved in the decay of π mesons, the μ mesons would in general be polarized.) Furthermore, the polarization is now complete (i.e., in a pure state). If this theory of the neutrino is correct, then the $\pi-\mu$ decay is a perfect polarizer of the μ meson, offering a

[7] The neutrino as defined in Sec. 1 is a particle with spin parallel to its momentum representing a right-hand screw. Similarly, the antineutrino as defined there is a particle with spin antiparallel to its momentum representing a left-handed screw. We use this definition throughout the present paper.

[8] See, e.g., R. Sherr and R. H. Miller, Phys. Rev. **93**, 1076 (1954).

[9] See C. L. Cowan, Jr. *et al.*, Science **124**, 103 (1956).

natural way to measure the spin and the magnetic moment of the μ meson. (It turns out that the $\mu - e$ decay may serve as a good analyzer, as we shall discuss in the next section.)

The choice of the two possibilities (16) and (17) will be further discussed in Sec. 7.

6. For the $\mu^- - e^-$ decay the process can be

$$\mu^- \to e^- + \nu + \bar{\nu}, \tag{18}$$

or

$$\mu^- \to e^- + 2\nu, \tag{19}$$

or

$$\mu^- \to e^- + 2\bar{\nu}. \tag{20}$$

Consider process (18) first. The decay coupling can be written with the notations defined in Eq. (11). (We assume no derivitive coupling.)

$$H_{\text{int}} = \sum_{i=V,A} f_i(\psi_e{}^\dagger O_i \psi_\mu)(\psi_\nu{}^\dagger O_i \psi_\nu). \tag{21}$$

It is easy to see that in the present theory, where ψ_ν satisfies (7a), the S-, T-, and P-type couplings do not exist. We have assumed in writing down (21) that the spin of the μ meson is $\frac{1}{2}$. For a μ^- at rest with spin completely polarized, the normalized electron distribution is given by

$$dN = 2x^2[(3-2x) + \xi \cos\theta(1-2x)]dx d\Omega_e(4\pi)^{-1}, \tag{22}$$

where p = electron momentum, $x = p/$maximum electron momentum, θ = angle between electron momentum and the spin direction of the μ, Ω_e = solid angle of electron momentum, and

$$\xi = [|f_V|^2 + |f_A|^2]^{-1}[f_V f_A{}^* + f_A f_V{}^*]. \tag{23}$$

The mass of the electron is neglected in this calculation. The decay probability per unit time is ($\hbar = c = 1$):

$$\lambda = M^5[|f_A|^2 + |f_V|^2]/(3\times 2^8 \pi^3), \tag{24}$$

where M is the mass of the μ meson. The spectrum (22) for a nonpolarized μ meson,

$$dN = 2x^2[3-2x]dx d\Omega_e(4\pi)^{-1}, \tag{25}$$

is characterized[5] by a Michel[10] parameter $\rho = \frac{3}{4}$, which is consistent with known[11] experimental results.

One sees that for not too small values of ξ, the spectrum (22) is sensitive to $\cos\theta$, especially in the region of large momentum for the electrons. Therefore the $\mu - e$ decay may turn out to be a very good analyzer of the μ-meson spin.

An analysis of the so-called universality of the Fermi couplings is easier in this theory because there are fewer coupling constants, and also because $\pi - \mu - e$ decay measurements would supply information concerning the parameter ξ of (23).

If process (19) or (20) prevails, the spectrum becomes

$$dN = 12x^2(1-x)dx[1 + \eta \cos\theta]d\Omega_e(4\pi)^{-1}. \tag{26}$$

This is characterized[5] by a Michel parameter[10] $\rho = 0$ which is not consistent with experiments.[11] One therefore concludes that (18) is the correct process.

A general theorem concerning the relationship between μ^+ and μ^- decays will be stated in Sec. 9.

7. If experiments should show that in the decay of the π meson, process (16) prevails, and in the β-decay process (9a) prevails, then one would say that the ν (the right-handed screw), the μ^-, and the e^- are light particles, and there is a conservation of light particles. If processes (17) and (9b) prevail, one would say that the $\bar{\nu}$ (the left-handed screw), the μ^-, and the e^- are light particles, and there is a conservation of light particles. Similar concepts have been discussed before.[12]

We have already seen in Sec. 3 that the sign of β in Eq. (13) determines whether

$$n \to p + e + \bar{\nu} \tag{9a}$$

or

$$n \to p + e + \nu \tag{9b}$$

is the process for β decay. To decide whether

$$\pi^+ \to \mu^+ + \nu, \quad (\mu^+ \text{ spin along } \mathbf{p}_\mu) = \tfrac{1}{2} \tag{16}$$

or

$$\pi^+ \to \mu^+ + \bar{\nu}, \quad (\mu^+ \text{ spin along } \mathbf{p}_\mu) = -\tfrac{1}{2} \tag{17}$$

one will have to determine the spin of μ^+ along its direction of motion.

8. The $\pi - \mu - e$ type experiment discussed in Secs. 6 and 7 can be done with the $K_{\mu 2} - \mu - e$ decays. The analysis is dependent on the spin of $K_{\mu 2}$. If this spin is not zero, the polarization of the μ meson is not necessarily complete. The degree of polarization can be experimentally found by a comparison of the angular distribution of the electrons in $\pi - \mu - e$ decay and in $K - \mu - e$ decay.

Another interesting experiment is to measure the momentum and polarization of the electron emitted in a β decay. A polarization of the electron results only if parity is not conserved; a measurement of this polarization is a measurement of a quantity similar to the parameter β in Eq. (13). The polarization in such a case will be along the direction of the momentum of the electron. Polarization along other directions can result if the momentum of the recoil nucleus is also determined. Theoretical considerations of such possibilities are being made by Dr. R. R. Lewis.

GENERAL REMARKS

9. Some general remarks concerning the conservation and nonconservation of the parity P, the charge conjugation C, and the time reversal T will be made in this section. Except for the last paragraph, no assump-

[10] L. Michel, Proc. Phys. Soc. (London) **A63**, 514 (1950).
[11] See, e.g., Sargent *et al.*, Phys. Rev. **99**, 885 (1955).
[12] E. J. Konopinski and H. M. Mahmoud, Phys. Rev. **92**, 1045 (1953).

tion that the neutrino is a two-component wave is made.

Since the preliminary result of the oriented nucleus experiment that there is a strong asymmetry, Eq. (A.6) of reference 1 shows that not only parity, but also charge conjugation is not conserved[2] in β decay. A measurement of the velocity dependence of the asymmetry parameter could supply[2] some information concerning time reversal invariance or noninvariance. If the $\pi-\mu-e$ decay should show any forward-backward asymmetry (as discussed in reference 1, and further analyzed above in Sec. 6 for the two-component neutrino theory), it can be shown from theorem 2 of reference 2 that charge conjugation invariance must be violated in both the $\pi-\mu$ and $\mu-e$ decays.

It is, however, easy to show from the Lüders-Pauli theorem[6] that even if C, T, and P are all not conserved, a stable particle ($e^{\pm}$ or $p^{\pm}$, or a deuteron, etc.) must have *exactly* the same mass as its antiparticle.

One can also prove that even if C, T, and P are all not conserved, the e^+ angular distribution in $\pi^+-\mu^+-e^+$ decay is exactly the same as the e^- angular distribution in $\pi^--\mu^--e^-$ decay. The only difference in the two cases is that the average spin of μ^+ along $\mathbf{p}_\mu$ is the opposite of that of μ^- along $\mathbf{p}_\mu$. (The decays are here assumed to occur in free space from $\pi^{\pm}$ at rest.)

It is further obvious from the Lüders-Pauli theorem[6] that if time reversal invariance is not violated, the operation CP is conserved. This means that the left-right asymmetry that is found in a laboratory is always exactly opposite to that found in the antilaboratory.

Should it further turn out that the two-component theory of the neutrino described above is correct, one would have a natural understanding of the violation of parity conservation in processes involving the neutrino. An understanding of the $\theta-\tau$ puzzle presents now a problem on a new level because no neutrinos are involved in the decay of $K_{\pi 2}$ and $K_{\pi 3}$. Perhaps this means that a more fundamental theoretical question should be investigated: the origin of all weak interactions. Perhaps the strange particles belong to strange representations of the Lorentz group. (Nature seems to make use of simple but odd representations.) It is also interesting to note that the massless electromagnetic field is the cause of the breakdown of the conservation of isotopic spin. The similarity to the massless two-component neutrino field that introduces the nonconservation of parity may not be accidental.

Reprinted from THE PHYSICAL REVIEW, Vol. 106, No. 2, 340–345, April 15, 1957
Printed in U. S. A.

Remarks on Possible Noninvariance under Time Reversal and Charge Conjugation*

T. D. LEE, *Columbia University, New York, New York*

AND

REINHARD OEHME AND C. N. YANG, *Institute for Advanced Study, Princeton, New Jersey*
(Received January 7, 1957)

Interrelations between the nonconservation properties of parity, time reversal, and charge conjugation are discussed. The results are stated in two theorems. The experimental implications for the K-$\bar{K}$ complex are discussed in the last section.

IN a recent paper[1] the question has been raised as to whether the weak interactions are invariant under a space inversion. It was also pointed out there that similar to the situation for space inversion there exists at present no experimental proof that weak interactions are invariant under charge conjugation. Consequently the absolute invariance under charge conjugation is also an open question.

The present note is devoted to a study of questions concerning the invariance under charge conjugation C, and under time reversal T (which is defined to be the Wigner time[2] reversal. It *does not* switch a particle into its antiparticle; nor does it change the sign of the spatial coordinates).

1. CPT THEOREM

For the discussion of the experimental consequences of possible nonconservation of P, C, and/or T, a theorem[3] which we shall call the CPT theorem proves very important.

To understand the meaning of the theorem one recalls first that the operations P and C in any many-particle system (with possibilities of creation and annihilation) are represented by unitary operators that operate on the state vectors. The operation T, on the other hand, is represented[2] by the *operator* of complex conjugation *multiplied* by a unitary operator. In the Schrödinger representation the transformation of a second quantized spin 0 field described by $\varphi(r)$ and $\pi(r)$ under these operations[4] can be brought into the following form:

$$\begin{aligned} P\phi(r)P^{-1}&=\eta_P\phi(-r), & P\pi(r)P^{-1}&=\eta_P{}^*\pi(-r),\\ C\phi(r)C^{-1}&=\eta_C\phi^\dagger(r), & C\pi(r)C^{-1}&=\eta_C{}^*\pi^\dagger(r),\\ T\phi(r)T^{-1}&=\eta_T\phi(r), & T\pi(r)T^{-1}&=-\eta_T{}^*\pi(r),\end{aligned}\tag{1}$$

where † means Hermitian conjugate and the phases η_P, η_C, and η_T have absolute values equal to 1. For the spin $\frac{1}{2}$ field $\psi(r)$, the transformations are

$$\begin{aligned} P\psi(r)P^{-1}&=\eta_P\gamma_4\psi(-r),\\ C\psi(r)C^{-1}&=\eta_C\psi^\dagger(r),\\ T\psi(r)T^{-1}&=\eta_T\gamma_1\gamma_2\gamma_3\psi(r),\end{aligned}\tag{2}$$

where the γ matrices are so chosen that γ_1, γ_2, and γ_3 are real and γ_4 is pure imaginary. The phases η_P, η_C, and η_T also have absolute values equal to unity. The transformation properties of fields of higher spin are similar.

From the CPT theorem one concludes that for any local Hermitian Hamiltonian H which is invariant under proper Lorentz transformations (i.e., Lorentz transformations that involve neither space nor time inversions), there always exists a choice of the phases η_C, η_P, and η_T for the various fields (usually in more than one way) with the following properties: (a) H commutes with the product of the operators P, C, and T taken in any order; and (b) if this choice of phases does not make H commute with P, then no other choice does, and the theory is not invariant under space inversion. (Of course, if this choice of phases makes H commute with P, then the theory is invariant under space inversion.) The same holds for C and T.

We shall illustrate this theorem by an example where H is invariant under proper Lorentz transformations. Let ψ_p, ψ_n, ψ_e, and ψ_ν be the fields describing the proton, the neutron, the electron, and the neutrino. The neutrino is assumed to be a non-Majorana particle with a nonvanishing mass. Consider

$$\begin{aligned} H=H_{\text{free}}+\int\{&g_1(\psi_p{}^\dagger\gamma_4\psi_n)(\psi_e{}^\dagger\gamma_4\psi_\nu)\\ &+g_2(\psi_p{}^\dagger\gamma_4\psi_n)(\psi_e{}^\dagger\gamma_4\gamma_5\psi_\nu)\\ &+g_3(\psi_p{}^\dagger\gamma_4\gamma_5\psi_n)(\psi_e{}^\dagger\gamma_4\gamma_5\psi_\nu)\\ &+g_4(\psi_p{}^\dagger\gamma_4\gamma_5\psi_n)(\psi_e{}^\dagger\gamma_4\psi_\nu)\\ &+\text{Hermitian conjugate}\}d^3(r),\end{aligned}\tag{3}$$

* *Note added in proof.*—This paper was written in December, 1956, before parity nonconservation was experimentally established.

[1] T. D. Lee and C. N. Yang, Phys. Rev. **104**, 254 (1956).

[2] E. P. Wigner, Gött. Nachr., Math. Naturw. Kl. (1932), p. 546.

[3] See W. Pauli's article in *Niels Bohr and the Development of Physics* (Pergamon Press, London, 1955). G. Lüders, Kgl. Danske Videnskab. Selskab, Mat.-fys. Medd. **28**, No. 5 (1954). J. Schwinger, Phys. Rev. **91**, 720, 723 (1953); **94**, 1366, formula (54) and p. 1576, discussions after formula (208). We are indebted to Professor Pauli for informing us of the work of Schwinger.

[4] We discuss here only the usual "type" of fields. The possibility of the existence of unusual "types" has been pointed out by Wigner. [An account of these unusual types has been given by L. Michel and A. S. Wightman, Princeton University lecture notes (unpublished).] An examination of these unusual "types" would be an important task if space-time conservation laws should indeed be found to break down for the weak interactions.

where $\gamma_5=\gamma_1\gamma_2\gamma_3\gamma_4$. This example is a special case of an example considered by Pauli.[3]

Writing H as

$$H=H(g_1,g_2,g_3,g_4),$$

one easily proves that

$$PHP^{-1}=H(g_1\eta_P,\ -g_2\eta_P,\ g_3\eta_P,\ -g_4\eta_P), \tag{4}$$

$$CHC^{-1}=H(g_1^*\eta_C^*,\ -g_2^*\eta_C^*,\ g_3^*\eta_C^*,\ -g_4^*\eta_C^*), \tag{5}$$

$$THT^{-1}=H(g_1^*\eta_T,\ g_2^*\eta_T,\ g_3^*\eta_T,\ g_4^*\eta_T). \tag{6}$$

In deriving these formulas, use has been made of the fact that $TgT^{-1}=g^*$ and $T\gamma_iT^{-1}=\gamma_i^*$. The phases η_P, η_C, and η_T are products of the respective phases of the four interacting fields. They are given by

$$\begin{aligned}\eta_P&=\eta_P^*(p)\eta_P(n)\eta_P^*(e)\eta_P(\nu),\\ \eta_C&=\eta_C^*(p)\eta_C(n)\eta_C^*(e)\eta_C(\nu),\\ \eta_T&=\eta_T^*(p)\eta_T(n)\eta_T^*(e)\eta_T(\nu).\end{aligned} \tag{7}$$

Using (4), (5), and (6), one can calculate the commutation relation between H and the six operators TCP, TPC, $\cdots$, PCT. It is found that with suitable choices of the phases η, the Hamiltonian H commutes with all of the six, as required by the CPT theorem. In fact the conditions on the phases η are simply

$$\eta_P=\eta_C\eta_T=\pm1. \tag{8}$$

It follows from the CPT theorem that, if one of the three operators P, C, and T is not conserved, at least one other must also be not conserved. It is of course also possible that all three are separately not conserved. In the example above, by assigning suitable values to the coupling constants g, one can construct examples for all the five possibilities of conservation or nonconservation of P, C, and T. These examples are displayed in Table I.

TABLE I. Examples of theories with various possible nonconservation properties.

Value of coupling constants	Conserved operators	Nonconserved operators
g_1=real, g_3=real, $g_2=g_4=0$	P, C, T	$\cdots$
g_1=real, g_3=complex, $g_2=g_4=0$	P, CT, TC	C, T
g_1=real, g_2=imaginary, $g_3=g_4=0$	C, PT, TP	P, T
g_1=real, g_2=real, $g_3=g_4=0$	T, CP, PC	C, P
g_1=real, g_2=complex, $g_3=g_4=0$	PCT, *and* permutations	P, C, T

2. LIFETIME OF CHARGE CONJUGATE PARTICLES AGAINST WEAK DECAY

Consider now a Hamiltonian

$$H=H_{\text{strong}}+H_{\text{weak}},$$

where both terms are invariant under a proper Lorentz transformation. In all subsequent discussions we shall assume that H_{strong} is invariant under C, P, and T. The phases η of the fields are defined (up to, possibly, some arbitrary factors) by this invariance; i.e., by the requirements that

$$CH_{\text{strong}}C^{-1}=H_{\text{strong}},\ \text{etc.}$$

On the other hand, the weak interactions may violate the invariance of C, P, and T. One can prove the following theorem.

Theorem 1.—If a particle A decays through the interaction H_{weak}, and if the particle and its antiparticle $\bar{A}$ do not decay into the same final products (as e.g. when A is charged), then to the lowest order of H_{weak} the lifetimes of A and $\bar{A}$ are the same, even if H_{weak} is *not* invariant under charge conjugation.

Proof.—Consider the case that particle A has spin zero. [The proof for the general case follows along the same lines.] Then the final states B and $\bar{B}$ in the decays

$$A\rightarrow B,\quad \bar{A}\rightarrow\bar{B}$$

also have spin zero. Using the identity

$$\langle\psi_1|\psi_2\rangle^*=\langle T\psi_1|T\psi_2\rangle, \tag{9}$$

one obtains

$$\begin{aligned}\langle B|H_{\text{weak}}|A\rangle^*&=\langle TB|TH_{\text{weak}}T^{-1}|TA\rangle\\ &=\langle TB|C^{-1}P^{-1}H_{\text{weak}}PC|TA\rangle,\end{aligned}$$

by the CPT theorem. Consider first the case that H_{weak} commutes (or anticommutes) with P. Then

$$\langle B|H_{\text{weak}}|A\rangle^*=\pm\langle TB|C^{-1}H_{\text{weak}}C|TA\rangle.$$

For a spinless system,

$$|TA\rangle=|A\rangle,\quad |TB\rangle=|B\rangle. \tag{10}$$

Hence

$$\begin{aligned}\langle B|H_{\text{weak}}|A\rangle^*&=\pm\langle B|C^{-1}H_{\text{weak}}C|A\rangle\\ &=\pm\langle CB|H_{\text{weak}}|CA\rangle=\pm\langle\bar{B}|H_{\text{weak}}|\bar{A}\rangle.\end{aligned} \tag{11}$$

This shows that the lifetimes of A and $\bar{A}$ are the same. If H_{weak} does not commute with P, we write

$$H_{\text{weak}}=H_1+H_2, \tag{12}$$

where

$$\begin{aligned}H_1&=\tfrac{1}{2}[H_{\text{weak}}+PH_{\text{weak}}P^{-1}],\\ H_2&=\tfrac{1}{2}[H_{\text{weak}}-PH_{\text{weak}}P^{-1}].\end{aligned} \tag{13}$$

Then

$$PH_1P^{-1}=H_1, \tag{14}$$

$$PH_2P^{-1}=-H_2. \tag{15}$$

The decays of A through H_1 and through H_2 lead to states B_1 and B_2 with opposite parities. They are orthogonal to the order considered, and hence they contribute independently without interference to the decay rate of A. The lifetimes of A and $\bar{A}$ are therefore again the same.

A consequence of this theorem has already been mentioned in a previous paper[1]: The identity of the experimental lifetimes of $\pi^\pm$ and of $\mu^\pm$ does not constitute a proof that charge conjugation invariance holds for the weak interactions.

For a discussion of a case where A and $\bar{A}$ may decay into the same final channels, see Sec. 4.

3. DEPENDENCE OF INTERFERENCE EFFECTS ON CONSERVATION OF C AND T

One would like to ask what are the experimentally detectable manifestations of a weak nonconservation of P, C, or T? For the nonconservation of parity, the answer is clearly to be sought in experiments to differentiate the right-handed screw from the left-handed. Some such experiments have been discussed before.[1]

If parity is indeed not strictly conserved, some of these experiments could *also reveal whether C and/or T are or are not conserved.* To illustrate this let us consider the experiment of the angular distribution of β decay from oriented nuclei. The degree of asymmetry was given in the appendix of reference 1 as proportional to

$$\mathrm{Re}\left[C_T C_T'^* - C_A C_A'^* + i\frac{Ze^2}{\hbar c p}(C_A C_T'^* + C_A' C_T^*)\right]. \quad (16)$$

Applying the arguments of Sec. 1 we recognize that the two terms in (16) are present or absent depending on whether C or T are not conserved. To be more specific: The first term vanishes if C is strictly conserved, the second term vanishes if T is strictly conserved. If this experiment shows any asymmetry, the p dependence and the Z dependence of the asymmetry could therefore reveal whether C and/or T are nonconserved. (The existence of any asymmetry rules out the possibility that both C and T are conserved, a conclusion we already drew on general grounds in Sec. 1.)

We notice that if C is strictly conserved, the asymmetry discussed above vanishes in the absence of the Coulomb distortion of the electron wave function. In fact, when C is conserved the asymmetry is directly dependent on the existence of a difference of the Coulomb phase shifts for opposite parities. It turns out that this is a consequence of a general theorem which we state and prove below:

Theorem 2.—If, in addition to the assumptions stated in Sec. 2 concerning H_{strong} and H_{weak}, we assume that H is strictly invariant under charge conjugation, (i.e., $[H,C]=0$) and if the decay products in the final state B are *free* particles, then to the lowest order of H_{weak} there is no interference between the parity-conserving and the parity-nonconserving parts of H in the decay of A, provided the interference is sought for in experiments measuring a term of the form $\boldsymbol{\sigma}\cdot\mathbf{p}^1$.

Proof.—We again illustrate the proof by considering the case that A is spinless. The general proof follows along the same lines. We perform the decomposition of H_{weak} as in Eqs. (12)–(15). The final state B consists of two states B_1 and B_2 of opposite parities reached from A through H_1 and H_2, respectively. Clearly H_1 commutes with C, and also, by the CPT theorem, commutes with CPT. Hence using identity (9) and Eq. (10) one obtains

$$\begin{aligned}\langle B_1|H_1|A\rangle^* &= \langle TB_1|TH_1T^{-1}|TA\rangle\\ &= \langle B_1|TH_1T^{-1}|A\rangle\\ &= \langle B_1|P^{-1}C^{-1}H_1CP|A\rangle\\ &= \langle B_1|P^{-1}H_1P|A\rangle = \langle B_1|H_1|A\rangle.\end{aligned}$$

Thus $\langle B_1|H_1|A\rangle$ is real. Similarly one easily proves that $\langle B_2|H_2|A\rangle$ is pure imaginary.

In the above the states B_1 and B_2 are taken as stationary states of H_{strong} consisting of standing waves. [Otherwise Eq. (10) does not hold.] Transition amplitudes into them have a relative phase factor which is, according to the above, pure imaginary. The observed final states are equal to these amplitudes multiplied by the *outgoing* part of the stationary states B_1 and B_2. Such outgoing parts always have real relative amplitudes if the stationary states B_1 and B_2 represent free particles. The theorem now follows immediately.

Using this theorem, one concludes that if any left-right asymmetry of the form $\boldsymbol{\sigma}\cdot\mathbf{p}$ is found, the part of this asymmetry that is independent of the distortion of the final-state wave functions can arise only if charge conjugation symmetry breaks down for the weak interactions. In particular, in decays where there is no strong final-state interactions, as, e.g., in $\pi\to\mu+\nu$ and $\mu\to e+\nu+\nu$ decays, the detection[1] of parity nonconservation through the observation of $\boldsymbol{\sigma}\cdot\mathbf{p}$ becomes impossible if C is strictly conserved.

4. K^0, $\bar{K}^0$ DECAY MODES

The existence of the particle $\bar{K}^0$ and some properties of its decay were predicted[5] and discussed under the assumption that charge conjugation is strictly conserved. We wish to discuss in this section the decay of K^0 and $\bar{K}^0$ under the assumption that C, P, and T are conserved for the strong interactions, but are not necessarily conserved in the weak decay interactions.

In the first place, the conservation of strangeness with respect to the strong interactions still requires that two particles K^0 and $\bar{K}^0$ with opposite strangeness exist. To understand their decay processes it is interesting to consider the charge conjugation symmetrical and antisymmetrical combinations introduced in reference 5 (compare, however, footnote 11):

$$K_1=\frac{1}{\sqrt{2}}(K^0+\bar{K}^0),\quad K_2=\frac{1}{\sqrt{2}}(K^0-\bar{K}^0). \quad (17)$$

Unlike the situation in reference 5, if C is now not invariant in the decay process, K_1 and K_2 can decay into the same final states:

$$\begin{aligned}&K_1\to\pi^++\pi^-,\\ &K_2\to\pi^++\pi^-,\\ &K_1\to\pi^\pm+e^\mp+\nu,\\ &K_2\to\pi^\pm+e^\mp+\nu,\\ &K_1\to\pi^++\pi^-+\pi^0,\\ &K_2\to\pi^++\pi^-+\pi^0.\end{aligned}$$

[5] M. Gell-Mann and A. Pais, Phys. Rev. 97, 1387 (1955).

Interference effects would therefore set in in these decay processes. The questions that one would like to ask are then: what the the lifetimes of the particles K^0 and $\bar{K}^0$? What are the branching ratios into various decay modes?

These questions can be answered by using a Weisskopf-Wigner type of treatment[6] of the time-dependent Schrödinger equation. We write the time-dependent amplitudes of the particles[7] K^0 and $\bar{K}^0$ as $a(t)$ and $b(t)$. The various channels of decay are denoted by j. $F_j(\omega)e^{-i\omega t}$ represents the amplitude of the decay product in the channel j with the energy ω.[8] [We choose units such that $\hbar=1$.] The zero of energy is taken to be the rest energy of K. The Schrödinger equations are then

$$i\frac{da}{dt}=\sum_{j,\omega} H_{aj}(\omega)F_j(\omega)e^{-i\omega t}, \tag{18}$$

$$i\frac{db}{dt}=\sum_{j,\omega} H_{bj}(\omega)F_j(\omega)e^{-i\omega t}, \tag{19}$$

$$i\frac{\partial F_j(\omega)}{\partial t}=e^{i\omega t}[H_{ja}(\omega)a+H_{jb}(\omega)b], \tag{20}$$

where $H_{ja}=H_{aj}^*$ are the matrix elements. The Weisskopf-Wigner treatment consists of first assuming an exponential time dependence for a and b, and then in sums over ω neglecting the variation of the matrix elements with ω in the interval $|\omega|\lesssim$uncertainty of energy of the original state. Using this treatment, one obtains

$$\begin{pmatrix} a \\ b \end{pmatrix}=\psi e^{-\frac{1}{2}\lambda t}. \tag{21}$$

The amplitude ψ and the decay constant λ are given by the eigenequation

$$\Gamma\psi=\lambda\psi. \tag{22}$$

Γ is a 2×2 Hermitian matrix with matrix elements given by

$$\begin{aligned} \Gamma_{11}=\Gamma_{22}&=\sum_j \Gamma_{aj}=\sum_j \Gamma_{bj}, \\ \Gamma_{12}&=\sum_j (\Gamma_{aj}\Gamma_{bj})^{\frac{1}{2}}e^{i\delta_j}=\Gamma_{21}^*, \end{aligned} \tag{23}$$

where

$$\begin{aligned} \Gamma_{aj}&=2\pi|H_{aj}|^2 \text{ (density of states per unit } d\omega)_{\omega=0}, \\ \Gamma_{bj}&=2\pi|H_{bj}|^2 \text{ (density of states per unit } d\omega)_{\omega=0}, \end{aligned} \tag{24}$$

and

$$e^{i\delta_j}=(\text{phase of } H_{aj}H_{bj}^{-1})_{\omega=0}. \tag{25}$$

[6] V. F. Weisskopf and E. P. Wigner, Z. Physik **63**, 54 (1930); **65**, 18 (1930).

[7] In this paper we assume that the K^0 particle (strangeness $=+1$) is a single state.

[8] Each channel j represents a possible decay state that is an eigenstate of H_{strong}. Thus it has a definite spin, charge, and parity.

In the foregoing derivation, use has been made of Eq. (11) which leads to

$$H_{aj}^*=\pm H_{bj'},$$

where j' is the charge conjugate channel of j (which may or may not be the same as j). It is important to notice that this equation is a consequence of the *CPT* theorem.

The two eigenvalues λ_+, λ_- of (22) correspond to the two decay lifetimes. The general solution is a linear superposition of two solutions $\psi_\pm$ of the form (21), each of which is characterized by a pure exponential decay. Since the 2×2 matrix Γ is Hermitian, the two solutions represent linear orthogonal combinations of the states K and $\bar{K}$.

In writing down Eqs. (18) and (19) we did not include a slight difference of mass in the form of a mass operation M for the states of the K particle. This restriction can be easily removed by adding to the right-hand sides of (18) and (19) the terms $\frac{1}{2}(M_{11}a+M_{12}b)$ and $\frac{1}{2}(M_{21}a+M_{22}b)$, respectively. The mathematical treatment is very similar to the above simple case except that we have now

$$-d\psi/dt=(\Gamma+iM)\psi, \tag{26}$$

where Γ is the same Hermitian matrix given by Eq. (23). (iM) is an anti-Hermitian matrix representing the effects of the mass shifts. By using Eq. (11) one can show that, similar to Eq. (23), M is a Hermitian matrix with

$$M_{11}=M_{22}. \tag{27}$$

Equation (26) can now be readily solved. Its eigenstates, defined by

$$(\Gamma+iM)\psi_\pm=\lambda_\pm\psi_\pm,$$

are

$$\psi_\pm=\begin{pmatrix} p \\ \pm q \end{pmatrix}(|p|^2+|q|^2)^{-\frac{1}{2}}, \tag{28}$$

with the corresponding time constants

$$\lambda_\pm=\Gamma_{11}+iM_{11}\pm(pq); \tag{29}$$

where p and q are two complex numbers given by

$$p^2=\Gamma_{12}+iM_{12}, \quad q^2=\Gamma_{21}+iM_{21}=\Gamma_{12}^*+iM_{12}^*. \tag{30}$$

If at $t=0$ a K particle is produced, then at a later time the state function ψ can be expressed in terms of these two eigenstates $\psi_\pm$ as

$$\psi(t)=\left(\frac{1}{2p}\right)(|p|^2+|q|^2)^{\frac{1}{2}}[\psi_+e^{-\frac{1}{2}\lambda_+t}+\psi_-e^{-\frac{1}{2}\lambda_-t}]. \tag{31}$$

It is convenient to separate the real and imaginary parts of $\lambda_\pm$. Without loss of generality, we may write

$$\lambda_+=\gamma_+, \quad \lambda_-=\gamma_-+2i\Delta, \tag{32}$$

where γ_+, γ_- are two real numbers representing the reciprocal lifetimes of the short-lived ones and the long-lived ones respectively and Δ is the mass difference between these two eigenstates. One notices that *these two eigenstates* ψ_+ *and* ψ_- *do not in general represent the states* K_1 *and* K_2 introduced in (17). *In fact they even may not be orthogonal to each other* (see footnote 11).

A general discussion of the decay processes is rather involved. We shall make only the following remarks:

(*A*) The fractional number of K mesons that decay at time t after its production is given by

$$N(t)dt=-d[\psi^\dagger\psi]. \tag{33}$$

Using (26), one easily shows that

$$-\frac{d}{dt}[\psi^\dagger\psi]=\psi^\dagger\Gamma\psi.$$

By using (28)–(31), Eq. (33) becomes

$$N(t)=\tfrac{1}{2}(1+\alpha)^{-1}\{\gamma_+e^{-\gamma_+t}+\gamma_-e^{-\gamma_-t}+\alpha e^{-\frac{1}{2}(\gamma_++\gamma_-)t}\times[(\gamma_++\gamma_-)\cos\Delta t-2\Delta\sin\Delta t]\}, \tag{34}$$

where

$$\alpha=\psi_+{}^\dagger\psi_-=[|p|^2-|q|^2][|p|^2+|q|^2]^{-1} \tag{35}$$

is a real number representing the nonorthogonality of these two eigenstates. The four real numbers γ_+, γ_-, Δ, and α characterize the decay of the K particle. They satisfy the inequalities

$$\gamma_\pm\geqq 0,\quad \alpha^2\leqq\frac{4\gamma_+\gamma_-}{(\gamma_++\gamma_-)^2+4\Delta^2}, \tag{36}$$

which follow from the fact that Γ is a positive Hermitian matrix. These conditions also insure that $N(t)\geqq 0$ for all t.

Experimentally $N(t)$ is measurable. From $N(t)$ one can in principle determine all four constants γ_+, γ_-, Δ, and α. Indications from presently existing experiments[9] show that probably $\gamma_+/\gamma_->100$. Equation (36) then shows that $\alpha^2<4\gamma_-/\gamma_+<0.04$.

(*B*) The above discussion also leads easily to a determination of the branching ratio of the long-lived component (and the short-lived component) into the various decay modes. If charge conjugation is conserved, the long-lived component is an eigenstate of charge conjugation.[5] Consequently its decay into charge conjugate channels such as $\pi^+e^-\nu$ and $\pi^-e^+\bar{\nu}$ must be equally probable, as is well known. If charge conjugation is not strictly conserved, decays into $\pi^+e^-\nu$ and $\pi^-e^+\nu$ may have different probabilities for the long-lived component.

A more complete discussion of the charge asymmetry of the decay of the long-lived K^0 will be given in the appendix. We mention here only that Lederman[10] has kindly informed us that experimental work in this direction is in progress. It is important to notice that if the experiments should yield a large asymmetry, and a small α (as mentioned above), Eq. (A7) would impose very strict conditions on the relative magnitudes of the amplitudes f_1, g_1, f_2 and g_2. (To see this roughly we need only examine the limiting case $\alpha=0$ discussed below.)

(*C*) If $\alpha=0$, the two eigenstates are orthogonal.[11] Also $|p|=|q|$. [See (35).] This is the case if the mass matrix is negligible. In this case $\psi_\pm$ are both 1:1 superpositions of the particle K^0 and $\bar{K}^0$. The fraction of particles decaying in dt, namely $N(t)dt$, becomes the sum of two pure exponentials by (34). Furthermore (A7) shows that the decays of the long-lived component into charge conjugate channels such as $\pi^++e^-+\nu$ and $\pi^-+e^++\bar{\nu}$ are equally probable.

One of us (Reinhard Oehme) would like to express his gratitude to Professor Robert Oppenheimer for his kind hospitality at the Institute for Advanced Study.

APPENDIX

In this appendix we shall show the interrelationship between the parameters p, q and the branching ratio for the decay of, say, the long-lived component of K particle into various charge conjugate states.

Consider first the following decay channel of the K particle

$$K\to e^-+\pi^++\nu. \tag{A1}$$

The final product may be in states with either parity $=+1$ or parity $=-1$. Let us denote the matrix elements for the decay process into these two types of states by f_1 and f_2. Similarly, we denote the matrix elements for

$$K\to e^++\pi^-+\bar{\nu}, \tag{A2}$$

with the final state having parity $=+1$ and parity $=-1$, by g_1 and g_2.

By using the CPT theorem and Eq. (11), the corresponding matrix elements for the decay of $\bar{K}$,

$$\bar{K}\to e^++\pi^-+\bar{\nu}, \tag{A3}$$

are related to that of (A1). These elements are f_1^* and $-f_2^*$. Similarly the matrix elements for

$$\bar{K}\to e^-+\pi^++\nu \tag{A4}$$

are g_1^* and $-g_2^*$. Let ψ_+ represent the long-lived component K_+ of the K particle. The matrix elements for

[9] K. Lande *et al.*, Phys. Rev. **103**, 1901 (1956).

[10] L. Lederman (private communication).

[11] We recall that since the strangeness S is conserved in the strong interaction, the phase η_c of a K particle ($S=+1$) under charge conjugation is not fixed by the strong interactions. If the weak interaction is not invariant under charge conjugation, the phase η_c is defined only up to a factor $e^{iS\theta}$. If ψ_+ is orthogonal to ψ_-, there exists however a most convenient choice which makes $\psi_\pm$ identical with the K_1, K_2 defined in (17).

the decay of K_+,

$$K_+\to e^-+\pi^++\nu, \tag{A5}$$

into the two different final parity states are proportional to $pf_1+qg_1^*$ and $pf_2-qg_2^*$, respectively, while the corresponding elements for

$$K_+\to e^++\pi^-+\bar{\nu} \tag{A6}$$

are proportional to $pg_1+qf_1^*$ and $pg_2-qf_2^*$. The branching ratio r for the decay of K_+ into $e^-+\pi^++\nu$ and $e^++\pi^-+\bar{\nu}$ is, therefore,

$$r=\frac{|pf_1+qg_1^*|^2+|pf_2-qg_2^*|^2}{|pg_1+qf_1^*|^2+|pg_2-qf_2^*|^2}. \tag{A7}$$

Session VII

WEAK INTERACTIONS

Chairman: C. N. Yang

LEE: Introductory survey.

In addition to the strong and electromagnetic interactions, there exists a large class of interactions characterized by coupling constants of order 10^{-13}. This class of weak interactions can be divided into two distinct groups. The first group contains processes not involving neutrinos, while the second group consists of processes in which neutrinos participate. Particles involved in the first group also participate in the strong interactions, and therefore each of these particles has a well-defined value of isotopic spin I and its z — component I_z. The weak interactions in the first group are further characterized by the selection rule $\Delta I_z = \pm \frac{1}{2}$, which of course is identical with non-conservation of strangeness. The interactions in the second group involve leptons that do not participate in any strong interactions and, in contrast with the particles of the first group, no useful assignment of isotopic spin to these leptons has been found. In spite of this difference, all weak interactions seem to have striking features in common, such as the similarity in the strengths of the coupling constants and the recently observed violation of conservation of parity and charge conjugation. While non-conservation of parity and charge conjugation have been proved only for reactions in the second group, there are strong suspicions that this is true also for the reactions in the first group - I allude, for instance, to the $\tau - \theta$ puzzle.

I shall first review briefly the present experimental status of the violation of conservation laws, together with the immediate theoretical implications. Next I shall develop some further theoretical considerations and speculations, and the particular points of view I shall adopt are fully shared by Prof. Yang.

I. The Experimental Situation

The first experiment demonstrating the non-conservation of parity and charge conjugation was performed by the Columbia and NBS groups, and involved the beta decay of polarized Co^{60} nuclei according to the reaction

$$Co^{60}\,(\text{polarized}) \longrightarrow Ni^{60} + e^- + \nu .$$

The alignment of the Co^{60} nuclei at low temperature was achieved by a magnetic field, and the electrons were found to be emitted preferentially in a direction antiparallel to the nuclear spin. From this asymmetry one can immediately deduce that parity is not conserved in the process, without invoking any detailed theory of beta decay. In fact, the direction of the magnetizing current loop together with the

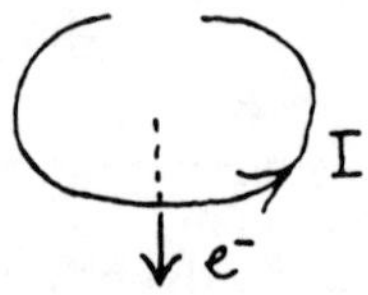

preferential direction of the emitted electrons can be used directly to establish the difference between a right-handed coordinate system and a left-handed coordinate system. Furthermore, from the observed magnitude of the asymmetry one can conclude that charge conjugation as well as parity is not conserved in this reaction. The transition is of pure Gamow-Teller type, so that only tensor coupling is involved. If C_T is the coupling constant for the part of the interaction which commutes with parity, and C_T' is the coupling constant for the part which anticommutes with parity, then the relationship

$$C_T' \simeq -C_T$$

would account for the observed asymmetry.

The second confirmation of non-conservation of parity and charge conjugation came from the measurement of the angular distribution of electrons from $\pi - \mu - e$ decay; the experiments were done at Columbia and Chicago. The pion decays into a muon, which then decays into an electron:

$$\pi^{\pm} \longrightarrow \mu^{\pm} + \nu \quad (\text{or } \bar{\nu})$$

$$\mu^{\pm} \longrightarrow e^{\pm} + \nu + \bar{\nu} \; .$$

The angular distribution of the electrons with respect to the muon momentum was found to be asymmetrical, the electrons being preferentially emitted in a direction antiparallel to the muon momentum. The angular distribution is, apart from the depolarization effect, the same for both signs of the electric charge.

Following the initial experiments, a large amount of work was done on the longitudinal polarization of the emitted electrons, on the circular polarization of the associated gamma rays together with beta-gamma correlations, on the beta angular distributions from other nuclei such as Co^{58} and Co^{56}, and on $\pi - \mu - e$ decay with the muons stopped in many kinds of materials. All results seem to confirm rather conclusively the essential findings of the earlier experiments.

II. The Two-Component Theory of the Neutrino

The quantitative data on the violation of the conservation laws of parity and charge conjugation can be explained in a simple and appealing way by use of the two-component theory of the neutrino. This theory and its applications are by now well known, and I shall make only a few remarks about it.

(a) The two-component theory of the neutrino can be expressed, as a matter of convenience, in terms of the usual four-component Dirac theory of the neutrino, by imposing a subsidiary condition on the neutrino field ψ_ν :

$$\gamma_5 \psi_\nu = - \psi_\nu .$$

This ensures that the mass of such particles will be zero.

(b) In the two-component theory, the neutrino (which is the particle of the theory) will always have its spin parallel to its momentum. If we use this structure to define a right-handed coordinate system, then the neutrino always has right-handed spirality and the antineutrino always has left-handed spirality:

Neutrino: right-handed spirality (ν)

Anti-neutrino: left-handed spirality ($\bar{\nu}$)

For a given momentum, there exist only these two states.

(c) Let us describe the decay of Co^{60} in terms of these definitions. Since the electrons are seen to be emitted preferentially in a direction antiparallel to the spin of the Co^{60} nucleus, we conclude that $C' = -C$, where C and C' stand for the coupling constants of the parity conserving and non-conserving parts of the interaction in the conventional four-component theory. If we interpret this result in terms of the two-component theory we are led to the conclusion that the emitted neutrino particle has left-handed spirality, i.e. it actually is an anti-neutrino:

$$Co^{60} \longrightarrow Ni^{60} + e^- + \bar{\nu}.$$

The neutron decay process therefore is

$$n \longrightarrow p + e^- + \bar{\nu}.$$

Similarly, when we examine π-μ-e decay in the light of the two-component theory, we find that both a neutrino and an anti-neutrino must be emitted in the μ decay, not two neutrinos or two anti-neutrinos:

$$\mu^\pm \longrightarrow e^\pm + \nu + \bar{\nu}.$$

This can be shown by computing the Michel parameter ρ from the electron spectrum in the two cases. In the former case we get $\rho = .75$, while in the latter case we get $\rho = 0$ which is certainly ruled out by the experimental evidence. The best experimental value to date is $\rho = 0.68$. While the value of .75 is somewhat higher than the experimental one, the theory nevertheless predicts an electron angular distribution and an energy dependence of that distribution which are in reasonable quantitative agreement with experiment.

(d) Another very interesting conclusion to be drawn from the two-component theory concerns the expected longitudinal polarization of the electrons. The electron-neutrino coupling term (not involving derivatives) has the general form $\psi_e^\dagger O_i \psi_\nu$, where O_i may be the scalar, tensor, vector, etc. operator. Using the subsidiary condition we may write

$$\psi_e^\dagger O_i \psi_\nu = \frac{1}{2} \psi_e^\dagger O_i (1-\gamma_5) \psi_\nu$$

$$= \frac{1}{2} \psi_e^\dagger (1 \pm \gamma_5) O_i \psi_\nu \begin{cases} + \text{ for } S, T, P \\ - \text{ for } V, A \end{cases}$$

One can easily show that this interaction leads to a longitudinal polarization of the electrons given by

$$\langle \vec{\sigma} \cdot \hat{p} \rangle_{e^-} = \begin{cases} -\frac{v}{c} & \text{for } S, T, P \\ +\frac{v}{c} & \text{for } V, A \end{cases}$$

Thus if the interaction is scalar, tensor, or pseudoscalar, the emitted electron will have left-handed spirality; if the interaction is vector or axial vector, then it will have right-handed spirality. Now we know that beta decay occurs predominantly through scalar and tensor interactions, and consequently we expect the electron to have left-handed spirality. For positrons, of course, the opposite is true. These predictions have been verified experimentally for Gamow-Teller transitions.

(e) Even though the two-component theory describes the experimental phenomena very well, we may ask whether it could also account for the other conceivable reaction

$$n \longrightarrow p + e^- + \nu$$

where the spirality is opposite to the observed one. Indeed, the mere use of two-component theories does not exclude such possibilities. However, if both reactions exist at the same time we would have a non-zero rate for double beta decay, and the observed asymmetries would not attain their maximum possible values. In fact, as we shall hear from Prof. Case tomorrow, by using a less restricted Majorana form of the two-component theory we can extend it to get a non-zero neutrino rest mass. Yet experimentally we find that the mass is small, that the rate of the double beta process is also small, and that the asymmetry in the angular distribution is large. We therefore are led to ask whether an underlying principle is behind all this. Indeed, the evidence points very strongly to the law of conservation of leptons. This law and the two-component theory together imply three things:

1. The mass of the neutrino is zero.
2. The rate of the double beta process is zero.
3. Conservation of parity is violated, and the resulting asymmetries can attain their maximum values.

These statements seem to be in good agreement with experiment, and therefore we next turn our attention to the lepton conservation law.

III. The Law of Conservation of Leptons

This law states that whenever a light fermion is emitted, another one is annihilated or an anti-fermion is emitted. We thus may assign a leptonic number to light fermions and anti-fermions, the signs being opposite in the two cases, and are led to the conservation of leptonic number.

By inspecting the beta decay process we find that the leptonic number for the neutrino is the same as that for the electron, and the decay of the muon tells us that μ^- and electron have the same leptonic number. The lepton conservation leads to several immediate theoretical predictions.

We consider the decay of the negative pion,

$$\pi^- \longrightarrow \mu^- + \bar{\nu} ,$$

and draw a diagram showing the spiralities of the emitted particles:

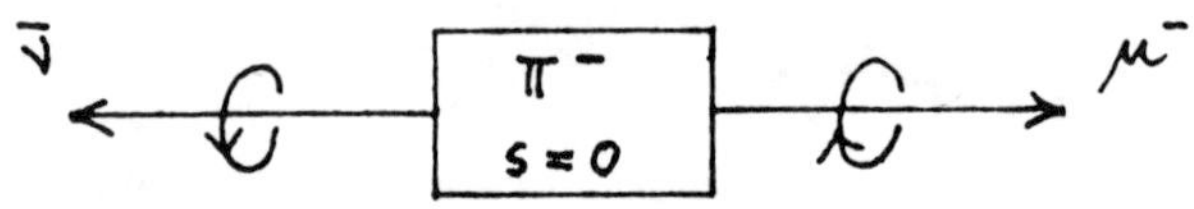

Here a μ^- is created, and hence an antineutrino must also come out. Since the latter has left-handed spirality, and since the pion has spin zero, the μ^- emitted must have left-handed spirality to conserve total angular momentum. Next we draw a similar diagram for the subsequent decay of the muon, $\mu^- \longrightarrow e^- + \nu + \bar{\nu}$:

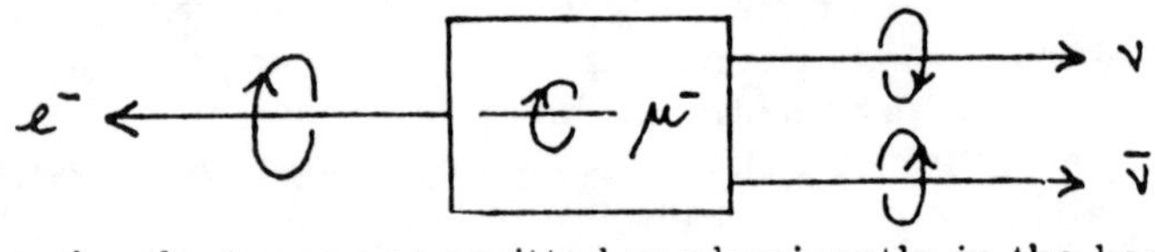

Since the electrons are emitted predominantly in the backward direction with respect to the muon momentum, and since we are merely interested in getting correct signs, we have drawn for simplicity only the high energy electron coming out in the backward direction, with the neutrino pair proceeding in the forward direction. The spiralities in the neutrino pair cancel, and conservation of angular momentum then tells us that the electron has the same spin component as the muon. Thus the e^- emitted in μ^- decay has right-handed spirality. Both of these predictions may of course be confirmed experimentally.

Let us assume that the above is correct, and look for further implications. The decay electron has right-handed spirality, and if in the interaction the electron and neutrino fields are coupled together, then by the arguments of Section II(d) the coupling must be vector or axial vector. In this case the vector and axial vector interactions are exactly the same, since the extra γ_5's operating on the ψ_ν in each of the (e, ν) and (μ, ν) factors simply produce compensating changes in sign. Thus the coupling contains vector, but not scalar or pseudoscalar or tensor parts. However, an antisymmetrical tensor of second rank is equivalent to a six-vector, and by virtue of the peculiar properties of the two-component neutrino three of its space components precisely cancel three of its time components. Consequently we may either say that there exists no tensor interaction, or that there is a tensor interaction with a non-vanishing coupling constant but what is multiplied by that coupling constant happens to be zero.

IV. The Universal Fermi Interaction

We next turn to the Universal Fermi Interaction, which is an attempt to gain a more unified understanding of certain of the weak interactions. We draw the famous triangle representing the interactions of interest:

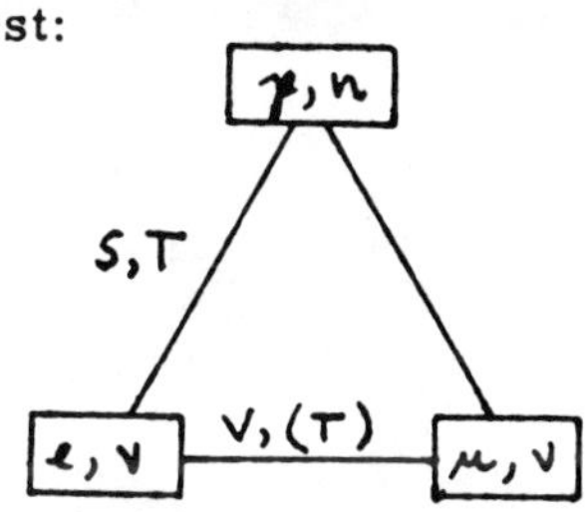

Beta decay information tells us that the interaction between (p, n) and (e, ν) is scalar and tensor, while the two-component neutrino theory plus the law of conservation of leptons implies that the coupling between (e, ν) and (μ, ν) is vector. This means that the Universal Fermi Interaction cannot be realized in the way we have expressed it. If all these coupling types turn out to be experimentally correct, we prefer to think that the similarity in coupling constants cannot be accounted for in terms of such a limited scheme. Rather it is a universal feature of all weak interactions, and not just those involving leptons. Nevertheless, at this moment it is very desirable to recheck even the old beta interactions to see whether the coupling is really scalar, a point to which we shall return later.

We next consider the third leg of the triangle, the muon capture process:

$$\mu^- + p \longrightarrow n + \nu .$$

This process used to be hard to analyze because we only observed its rate, but the fact that the emitted neutrino carries away angular momentum can be used to gain more detailed information. The law of conservation of leptons tells us that a neutrino is emitted, not an anti-neutrino. For convenience let us assume that the captured muon is 100% polarized. Then the neutron angular distribution will not be isotropic, but will take the form $1 - \cos\theta$ for scalar and vector coupling, and $1 + \frac{1}{3}\cos\theta$ for tensor and axial vector coupling, where θ is the angle between the spin of the muon and the momentum of the neutron. Hence, by measuring this angular distribution we can get quite definite information on the coupling types for the third leg, and this is particularly important in view of the apparent coupling discrepancy in the other two legs. Further information on the nature of this process can be obtained by observing the bremsstrahlung emitted during the reaction

$$\mu^- + p \longrightarrow n + \nu + \gamma,$$

measuring the angular distribution together with the circular polarization of the gamma rays. The conclusions to be drawn from these considerations depend on further experiments. The present data seem fairly well accounted for by the two-component neutrino theory, and we may say that in a modest way we have some understanding of these phenomena. The same cannot be said, however, with regard to the decay of K mesons and hyperons, the topic to which we turn next.

V. The Decay of K-Mesons and Hyperons

(a) We first consider the $\tau - \theta$ puzzle. Once we have observed the non-conservation of parity and charge conjugation in some weak interactions, we are free to conjecture that these violations extend to the entire class. Therefore the $\tau - \theta$ duality can be explained in a simple way, perhaps even the correct way, by just saying that the θ particle is identical to the τ particle. In the decay of K-mesons into two or three pions we measure at most two independent momenta, going to the center of mass system in the three pion case. If the K-meson has no spin, then the parity conserving and non-conserving modes of decay can never interfere because we cannot form a pseudoscalar out of two momentum vectors. Consequently the maximum possible evidence that parity is not conserved in K decay is precisely what we have seen: a spinless particle decaying into two or three pions, the masses and lifetimes for the two cases being the same.

Nevertheless, several attempts have been made to base a theory on the idea that only reactions involving neutrinos violate parity conservation, and that reactions not involving neutrinos do not violate parity conservation. Under this assumption the Dalitz-Fabry plot tells us that the θ particle is not the same as the τ particle. We would then expect to have two particles with different lifetimes, contradictory to what seems to be observed. Thus there is definite experimental evidence against the idea that the K-meson exists in a parity doublet, and in fact there is no theoretical necessity for such a proposal. Assuming spinless particles, we then have in the case of τ and θ decay the maximum possible information on non-conservation of parity.

(b) We may hope to gain additional information from hyperon decay, since hyperons, being fermions, certainly have spin. Let us consider the production of Λ^0 :

$$\pi^- + p \longrightarrow \Lambda^0 + K^0 .$$

This is a strong interaction which we may regard as a polarizer of the Λ^0 . We then use the subsequent decay of the Λ^0 , which is a weak interaction, as an analyzer of this polarization:

$$\Lambda^0 \longrightarrow p + \pi^- .$$

In fact, for this decay to serve as an analyzer at all, it must necessarily violate the conservation law of parity. However, since the polarizer is a strong interaction and since the analyzer involves slow particles, it is possible that neither polarizer nor analyzer are as effective as in the case of the leptons in π-μ-e decay, where the polarizer is known to be nearly 100% effective and the analyzer is also extremely effective. To illustrate, let us assume the spins of the Λ^0 and K^0 to be 1/2 and 0, respectively. If we carry out the production near threshold so that only S and P-waves are involved, and if the production angular distribution has the form $(1 \pm \cos\theta)^2$, then the polarization of the Λ^0 is zero for all angles. Here θ is the angle between the incoming pion momentum and the outgoing hyperon momentum in the center of mass system. Observation of course does not substantiate that this is the actual angular distribution. It is, however, the most peaked distribution in the forward or backward direction possible for S and P-waves, and may not be too far from the actual one. Here we would have a case in which the polarizer is not a very effective one. To obtain definite conclusions on the non-conservation of parity it is imperative to have good knowledge about polarizations. For that reason it will probably be very useful to investigate such production and decay processes in a bubble chamber near threshold energies.

(c) The θ_1^0 and θ_2^0 particles were introduced on the assumption of invariance under charge conjugation. Since this assumption is invalid for at least some of the weak interactions, we shall examine the implications of this violation on their expected behavior. From the strong production reactions we infer that there should exist two states (i.e. two particles) of θ^0 type, with strangeness + 1 and - 1. The existence of two states implies in general the existence of two lifetimes, and we wish to find out how the short-lived and the long-lived particles behave. Let us first assume that charge conjugation is not conserved, but that charge conjugation times parity is conserved. We immediately conclude that the short-lived particle can decay into two pions but the long-lived one cannot. The situation remains the same as under the original assumption, and evidently we cannot see a violation of charge conjugation invariance unless we see a violation of parity. If charge conjugation times parity is not conserved, the long-lived particle can decay into two pions but the 2π to 3π branching ratio, based on phase space arguments, would be quite small and the 2π mode might escape experimental detection. The striking behavior of θ_1^0 and θ_2^0 thus happens to be an insensitive test body for non-conservation of C or CP.

VI. Time Reversal Invariance

The question of whether CP is or is not conserved brings us to the question of invariance under time reversal. According to the famous TCP theorem, if T is invariant then CP is invariant, and if T is not invariant then CP is not invariant. We have seen that θ_1^0 and θ_2^0 do not furnish a sensitive test for T invariance. Fortunately more sensitive tests exist in beta decay, and I shall mention only a few of these.

One method, involving the type of experiment already done, measures the beta angular distribution from polarized nuclei, determining $\vec{J}\cdot\vec{p}_e$ for a ($J \rightarrow J$, no) transition. This provides information about the real part of $|M_F||M_{GT}|(C_s C_T'^* + C_s' C_T^*)$. The imaginary part of this quantity can be obtained in a beta-gamma cascade

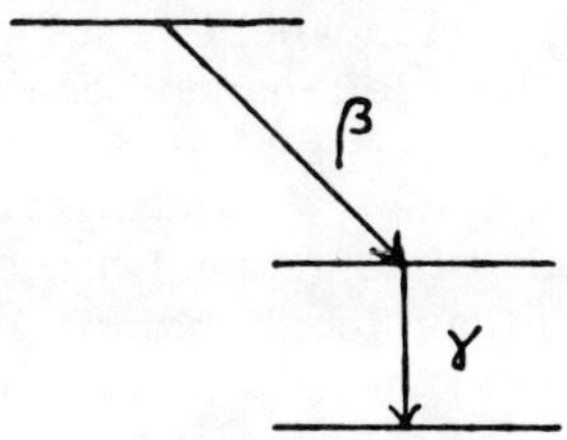

by measuring $\vec{J}\cdot(\vec{p}_e\times\vec{p}_\gamma)(\vec{J}\cdot\vec{p}_\gamma)$ or $\vec{J}\cdot(\vec{p}_e\times\vec{p}_\nu)$
A measurement of both the real and the imaginary parts for the same element would yield a great deal of information. Even the measurement of $\vec{J}\cdot(\vec{p}_e\times\vec{p}_\nu)$ alone will give much information. The real part of $|M_F||M_{GT}|(C_s C_T'^* + C_s' C_T^*)$ has been measured by the Leyden and NBS groups for Co^{58}, and was found to be zero. The ratio $|M_F|/|M_{GT}|$ had been determined previously, and neither M_F nor M_{GT} was found to vanish. Since C_s seems to be non-zero, and since certainly C_T is non-zero, it follows that the quantity in question is purely imaginary. This implies violation of time reversal invariance. But this argument is based on much experimental information obtained at different times, and it seems desirable to check all these results. Of course it would be extremely useful to measure the imaginary part directly, as outlined above. The question of the existence of the scalar interaction needs careful re-examination, and to that end a measurement of the spirality of the emitted positron is suggested. The answers, bearing on the existence of a Universal Fermi Interaction and on time reversal invariance, should be forthcoming in the near future.

VII. The Mach Principle

If it turns out (not to our great surprise, perhaps) that T is not invariant, what are the implications? This question leads us to a brief discussion of the Mach Principle which states that the laws of physics should not depend upon the choice of any particular geometrical coordinate system. If we believe in this principle then the present information on asymmetries can still be accounted for in two ways, depending on whether T is invariant or not. If T is invariant then CP is invariant, and when we perform mirror reflections we must simultaneously change particles to antiparticles to preserve the overall symmetry of space and time. We nevertheless have lost one symmetry property compared to the earlier situation. If T is not invariant then CP is not invariant, and changing particles into antiparticles does not help. But a firm believer in Mach's Principle might conjecture, for instance, that there exist two different kinds of nucleons. One kind of proton would be a right-handed one (p_R) and the other would be a left-handed one (p_L). They have opposite spirality. Our world consists almost 100% of one kind, which accounts for the observed asymmetry. But overall space symmetry can still be maintained by taking into account the p_L which do not exist in any appreciable quantity in our local cosmological region. Thus P is kept invariant, and from the TCP theorem we know that TC will be invariant, even though C is not invariant. Thus we can still have overall time reversal symmetry by changing particles to anti-particles.

The second possibility is certainly not so simple or attractive as the first.

In conclusion, let me emphasize again the curious nature of all weak interactions. In spite of their great diversity, they display striking similarities in the strengths of coupling constants, in the peculiar selection rule on I_a , and in the non-conservation of parity and charge conjugation. Perhaps all these are merely different aspects of a single and unifying principle underlying all weak interactions.

Reprinted from
Rochester Conference 1957
Interscience Publisher's, Section VII, pp. 1–12

Reprinted from THE PHYSICAL REVIEW, Vol. 106, No. 6, 1367–1369, June 15, 1957
Printed in U. S. A.

Possible Detection of Parity Nonconservation in Hyperon Decay*

T. D. LEE, J. STEINBERGER,
Columbia University, New York, New York

AND

G. FEINBERG, P. K. KABIR, AND C. N. YANG,
Institute for Advanced Study, Princeton, New Jersey

(Received May 2, 1957)

RECENTLY, various experiments[1–3] established the nonconservation of parity in β decay, π decay, and μ decay. The purpose of this note is to emphasize that, in view of these developments, experiments on hyperon production and decay in $(\pi+p)$ collisions of the type done by various groups using bubble chambers,[4] seem now to be especially important for a clarification of the following related questions: (i) whether parity conservation is violated in hyperon decays[5] and (ii) whether parity doublets exist.[6]

A detailed analysis concerning the possible detection of parity doublets exists in the literature.[6] In the following we shall make a phenomenological study of the problem of possible detection of parity nonconservation in hyperon decay under the assumption that there exist no parity doublets for either K mesons or hyperons.[7]

To make the analysis unambiguous and to draw conclusions that are relatively definite, it is necessary that one knows something about the polarization of the hyperons produced. It seems that a good plan is to study hyperon production and decay near threshold.

Production and decay of Σ^-. For example, let us consider the production of Σ^- from (π^-+p) collisions:

$$\pi^-+p\to\Sigma^-+K^+. \tag{1}$$

It is perhaps worthwhile to try to do the experiments at laboratory kinetic energies of the pion of, say, 955 Mev and 1 Bev, corresponding to center-of-mass total kinetic energies of the Σ^-+K^+ system of 30 Mev and 60 Mev. At these energies one hopes that only s and p waves are produced in the Σ^-+K^+ system.

It is then easy to see that the differential production cross section per unit solid angle $d\Omega$ (in the center-of-mass system of production) of the Σ^- produced is given by

$$I(\theta)=|a+b\cos\theta|^2+|c|^2\sin^2\theta, \tag{2}$$

where a can be chosen as real and b and c are complex numbers. We use the following notations:

$$\begin{aligned}\mathbf{p}_{\text{in}}&=\text{momentum of the incoming }\pi^-,\\ \mathbf{p}_\Sigma&=\text{momentum of the }\Sigma^-\text{ produced,}\\ \theta&=\text{angle between }\mathbf{p}_{\text{in}}\text{ and }\mathbf{p}_\Sigma.\end{aligned} \tag{3}$$

In (3) both $\mathbf{p}_{\text{in}}$ and $\mathbf{p}_\Sigma$ are measured in the center-of-mass system of production. The polarization of the Σ^- produced at the angle θ is always in the direction of $\mathbf{p}_{\text{in}}\times\mathbf{p}_\Sigma$ and has the magnitude

$$P(\theta)=[I(\theta)]^{-1}2\sin\theta\times\text{Im}[c^*(a+b\cos\theta)], \tag{4}$$

where $P(\theta)$ is defined to be the average spin of the Σ^- in units of $\frac{1}{2}\hbar$. In Eq. (4) the *assumptions have been made that the spin of* Σ^- *is* $\frac{1}{2}$ *and that the spin of* K^+ *is* 0.

If parity is not conserved in the decay of Σ^-, the polarization $P(\theta)$ can be measured by using the decay process of Σ^-,

$$\Sigma^-\to n+\pi^-, \tag{5}$$

as an analyzer. Let R be the projection of the momentum of the decay pion in the direction of $\mathbf{p}_{\text{in}}\times\mathbf{p}_\Sigma$. The distribution function for R at an angle θ of production is given by

$$W(\theta,\xi)d\Omega d\xi=I(\theta)d\Omega\times\tfrac{1}{2}[1+\alpha p(\theta)\xi]\alpha\xi, \tag{6}$$

where

$$\xi=R/(\text{maximum value of }R)\cong R/(100\text{ Mev}/c).$$

In terms of the coefficients a, b, and c, defined in Eq. (2), $W(\theta,\xi)$ can be written as

$$\begin{aligned}W(\theta,\xi)d\Omega d\xi=&[|a+b\cos\theta|^2+|c|^2\sin^2\theta]d\Omega\times\tfrac{1}{2}d\xi\\ &+\alpha\sin\theta\,\text{Im}[c^*(a+b\cos\theta)]d\Omega\times\xi d\xi.\end{aligned} \tag{7}$$

The existence of a nonvanishing α would constitute an unambiguous proof of parity nonconservation in Σ^- decay. In such a case the final state of $(n+\pi^-)$ in process (5) would be a mixture of $s_{\frac{1}{2}}$ and $p_{\frac{1}{2}}$ states with amplitudes, say, A and B respectively. The asymmetry parameter α is related to these amplitudes by

$$\alpha=2\,\text{Re}(A^*B)/(|A|^2+|B|^2). \tag{8}$$

If time reversal leaves invariant the decay process of Σ^-, then[8]

$$\alpha=\pm\frac{2|A|\times|B|}{|A|^2+|B|^2}\cos(\delta_p-\delta_s), \tag{9}$$

where δ_p and δ_s are, respectively, the phase shifts of $(n+\pi^-)$ scattering in the $p_{\frac{1}{2}}$ and $s_{\frac{1}{2}}$ states at about 117 Mev in their center-of-mass system. If the decay interaction is invariant under charge conjugation, then[8]

$$\alpha=\pm\frac{2|A|\times|B|}{|A|^2+|B|^2}\sin(\delta_p-\delta_s). \tag{10}$$

The following remarks are useful concerning the measurements of α and $p(\theta)$.

1. The polarization $P(\theta)$ may sometimes be very small. E.g., if (1) gives

$$I(\theta)=(1+\cos\theta)^2,\quad\text{or}\quad I(\theta)=(1-\cos\theta)^2, \tag{11}$$

then $P(\theta)=0$ identically.

2. At production energies *near the threshold*, the variations of the quantities a, b, and c, introduced in Eq. (2), with respect to p_Σ are given by

$$\begin{aligned} a&=a_0(p_\Sigma)^{\frac{1}{2}}, \\ b&=b_0(p_\Sigma)^{\frac{3}{2}}\exp(i\chi_b), \\ c&=c_0(p_\Sigma)^{\frac{3}{2}}\exp(i\chi_c), \end{aligned} \tag{12}$$

where a_0, b_0, c_0, χ_b, and χ_c are all *real* constants independent of p_Σ. Thus by selecting two or three energy values near threshold, it is possible to determine a_0, b_0, c_0, and χ_b from the angular and energy dependence of $I(\theta)$ alone.

If the energy dependence of the cross section should be not representable by (12), one would have an indication that resonance effects might be important in the π^-+p system near the threshold for Σ^- production.

3. If $\chi_b\neq 0$, then by comparing the coefficients of the $\sin\theta$ and $\sin\theta\cos\theta$ terms in $W(\theta,\xi)$, the phase χ_c cal also be determined.

4. From the values of these five real constants, a_0, b_0, c_0, χ_b, and χ_c, the asymmetry parameter α can then be deduced from $W(\theta,\xi)$ [Eq. (7)].

5. If

$$|\alpha|>|\sin(\delta_p-\delta_s)|,$$

then from Eq. (10) both invariance under charge conjugation and conservation of parity do not hold in the decay of Σ^-.

Since the phase shifts in the $J=\frac{1}{2}$ states are all small, the conclusion is essentially that any appreciable asymmetry with respect to the sign of ξ in $W(\theta,\xi)$ is an indication that conservation of parity and invariance under charge conjugation do not hold in the decay of Σ^-.

Production and decays of other hyperons. The foregoing analysis can also be applied to the productions and decays of other hyperons. We consider, for definiteness, the following processes concerning Λ^0:

$$\pi+p\rightarrow\Lambda^0+K^0, \tag{13}$$

and

$$\Lambda^0\rightarrow p+\pi^-. \tag{14}$$

All the previous formulas for $I(\theta)$, $p(\theta)$, $W(\theta,\xi)$, and α [i.e., Eqs. (2), (4), (6), (7), and (8)] remain unchanged. The only difference is that in Eq. (8) the amplitudes A and B of s- and p-wave final states in the decay process of Λ^0 are now each a mixture of two isotopic spin states. These amplitudes can be written as

$$\begin{aligned} A&=(\tfrac{2}{3})^{\frac{1}{2}}A_{\frac{1}{2}}+(\tfrac{1}{3})^{\frac{1}{2}}A_{\frac{3}{2}}, \\ B&=(\tfrac{2}{3})^{\frac{1}{2}}B_{\frac{1}{2}}+(\tfrac{1}{3})^{\frac{1}{2}}B_{\frac{3}{2}}, \end{aligned} \tag{15}$$

where $A_{\frac{1}{2}}$, $B_{\frac{1}{2}}$ are, respectively, the s- and p-wave amplitudes for final states with the total isotopic spin value $I=\frac{1}{2}$, and $A_{\frac{3}{2}}$, $B_{\frac{3}{2}}$ the corresponding amplitudes for states with $I=\frac{3}{2}$. In place of Eqs. (9) and (10) we have now the following conditions for invariance under time reversal and charge conjugation:

If the decay process is invariant under time reversal, then we can choose[8]

$$\begin{aligned} A_{\frac{1}{2}}&=|A_{\frac{1}{2}}|e^{i\delta_1}, \\ A_{\frac{3}{2}}&=\pm|A_{\frac{3}{2}}|e^{i\delta_3}, \\ B_{\frac{1}{2}}&=\pm|B_{\frac{1}{2}}|e^{i\delta_{11}}, \\ B_{\frac{3}{2}}&=\pm|B_{\frac{3}{2}}|e^{i\delta_{31}}. \end{aligned} \tag{16}$$

On the other hand, if the decay process is invariant under charge conjugation operation, then these amplitudes are[8]

$$\begin{aligned} A_{\frac{1}{2}}&=|A_{\frac{1}{2}}|e^{i\delta_1}, \\ A_{\frac{3}{2}}&=\pm|A_{\frac{3}{2}}|e^{i\delta_3}, \\ B_{\frac{1}{2}}&=\pm i|B_{\frac{1}{2}}|e^{i\delta_{1,1}}, \\ B_{\frac{3}{2}}&=\pm i|B_{\frac{3}{2}}|e^{i\delta_{3,1}}. \end{aligned} \tag{17}$$

The phase shifts δ are the usual pion-nucleon scattering phase shifts at 37-Mev total kinetic energy:

$$\begin{aligned} \delta_1&=\text{phase shift for } s \text{ waves, } I=\tfrac{1}{2},\ J=\tfrac{1}{2}, \\ \delta_3&=\text{phase shift for } s \text{ waves, } I=\tfrac{3}{2},\ J=\tfrac{1}{2}, \\ \delta_{\lambda\mu}&=\text{phase shift for } p \text{ waves, } I=\lambda/2,\ J=\mu/2. \end{aligned}$$

All these phase shifts are small at 37-Mev total kinetic energy. Therefore any appreciable asymmetry in $W(\theta,\xi)$ with respect to the sign of ξ is an indication that conservation of parity and invariance under charge conjugation do not hold in the decay of Λ^0.

A measurement of the branching ratio in the decay processes

$$\Lambda^0\rightarrow p+\pi^-, \tag{18}$$

$$\Lambda^0\rightarrow n+\pi^0, \tag{19}$$

and a measurement of the distribution function $W(\theta,\xi)$ for process (19) would lead to additional information concerning the amplitudes $A_{\frac{1}{2}}$, $B_{\frac{1}{2}}$, $A_{\frac{3}{2}}$, and $B_{\frac{3}{2}}$.

* Work supported in part by the U. S. Atomic Energy Commission.

[1] Wu, Ambler, Hayward, Hoppes, and Hudson, Phys. Rev. **105**, 1413 (1957).

[2] Garwin, Lederman, and Weinrich, Phys. Rev. **105**, 1415 (1957).

[3] J. Friedman and V. Telegdi, Phys. Rev. **105**, 1681 (1957).

[4] Budde, Chretien, Leitner, Samios, Schwartz, and Steinberger, Phys. Rev. **103**, 1827 (1956); D. A. Glaser (private communication); R. Adair (private communication).

[5] T. D. Lee and C. N. Yang, Phys. Rev. **104**, 254 (1956).

[6] T. D. Lee and C. N. Yang, Phys. Rev. **104**, 822 (1956).

[7] In view of the recent experimental developments (references 1, 2, and 3) there appears to be at present no *theoretical* necessity to introduce the complication of parity doublets. (See footnote 8 in reference 5.)

[8] Lee, Oehme, and Yang, Phys. Rev. **106**, 340 (1957). See the proof of Theorem 2.

ELEMENTARY PARTICLES and WEAK INTERACTIONS

by

T. D. Lee and C. N. Yang

BROOKHAVEN NATIONAL LABORATORY

Associated Universities, Inc.
under contract with the
United States Atomic Energy Commission

Foreword

The substance of these notes comes from published and unpublished research of the undersigned. In form and presentation these notes follow largely a series of six lectures given by one of us (T.D. Lee) at Brookhaven National Laboratory during January 1957. The lecture notes were originally edited by Drs. L.C.L. Yuan, B.H. McCormick, W. Chinowsky, and R.K. Adair, to all of whom grateful acknowledgment is hereby made. Dr. Yuan is especially to be thanked for the time and advice he generously contributed in the process of the later expansion and changes in the notes.

T.D. Lee
C.N. Yang

Table of Contents

I.	General review	1
II.	The θ-τ puzzle	10
III.	Experimental limits on the validity of parity conservation	16
IV.	Some general discussions on the consequences due to possible non-invariance under *P*, *C*, and *T*	21
V.	Various possible experimental tests on invariance under *P*, *C*, and *T*	28
VI.	A two-component theory of the neutrino	42
VII.	Possible law of conservation of leptons and the universal Fermi interactions	54
VIII.	Time reversal invariance and Mach's principle	60

I. GENERAL REVIEW

1. Introduction

In these discussions we shall try to review some general patterns of the interactions between various elementary particles and to study some general questions concerning the symmetry properties of these particles. The first natural question that one would like to ask is, what precisely constitutes an elementary particle? Suppose a new particle is observed, how do we know that it is an elementary particle and not merely a composite system consisting of some already known elementary particles? The answer is that we do not know. Nevertheless, the term "elementary particle" is well defined in a negative sense. We believe we understand what is meant by an atom, a molecule, and a nucleus. Any small particle that is not an atom, not a molecule, and not a nucleus (except the hydrogen nucleus) is called an elementary particle.

At first sight these so-called elementary particles form a very inhomogeneous group. The mass of a particle may be large (as $M_{\Xi} \cong 1321$ Mev) or may be very small (as the mass of $\gamma = 0$). The lifetime may be very short (as the lifetime of $\Sigma^0 \cong 10^{-20}$ sec) or may be very long (as in the case of a proton which has infinite lifetime). The particle may be electrically charged or may be neutral. Except for a very few of them, we do not know the spin and parity of these particles. Again except for those between a very few of these particles, we do not know their interactions. In fact there is still confusion as to the identity of the particles that have already been observed. Some of these particles that look very different may turn out to be the same particles. Some of them that look very similar may indeed be several different particles. In spite of this almost complete lack of understanding, some general patterns and general rules have been found. These properties will be our main topics of discussion.

2. Review of the Classifications of Interactions

We find that the interactions between these elementary particles fall into four distinct groups:

(i) Strong interactions. This group includes the forces responsible for the production and the scattering of nucleons, pions, hyperons, and K-mesons. It is characterized by a coupling constant of the order of magnitude 1 ($f^2/\hbar c \cong 1$).

(ii) Electromagnetic interactions. These are characterized by the coupling constant, $e^2/\hbar c = 1/137$.

(iii) Weak decay interactions. These are characterized by a dimensionless coupling constant, $\mathcal{G}^2/\hbar c \cong 10^{-14}$.

[(iv) Gravitational interactions. The gravitational interaction can be characterized by a dimensionless coupling constant $Gm^2/\hbar c \cong 2 \times 10^{-39}$ with G as Newton's gravitational constant and m chosen to be the mass of the proton. The gravitational interaction will not be further discussed.]

Among these three interactions only the electromagnetic interaction is well understood. About the other two groups, the strong interactions and the weak decay interactions, we know really very little. Nevertheless, we believe they possess certain symmetry properties.

3. Invariance Properties and Conservation Laws that are USUALLY Accepted as Exact

We shall now list those symmetry properties and conservation laws that were, before the end of 1956, generally accepted to be valid for all three groups of interactions. These are:

(i) Conservation of energy and momentum. This follows from the invariance under translations in the four-dimensional space. The infinitesimal translations in space are represented by the energy-momentum operators P_μ. Thus, the homogeneity of space implies the conservation of energy and momentum.

(ii) Invariance under the proper Lorentz transformation. A proper Lorentz transformation is a Lorentz transformation without either space inversion or time reversal. This invariance implies, among other things, the conservation of angular momentum.

(iii) (?) Invariance under space inversion P (or, the conservation of parity). P is a transformation which changes $\mathbf{r} \rightarrow -\mathbf{r}$; $t \rightarrow +t$; and particles $\rightarrow$ particles.

(iv) (?) Invariance under time reversal T (i.e., $\mathbf{r} \rightarrow +\mathbf{r}$; $t \rightarrow -t$; and particle $\rightarrow$ particle).

(v) (?) Invariance under charge conjugation C. The charge conjugation operator C changes a particle to its antiparticle, but leaves $\mathbf{r} \rightarrow +\mathbf{r}$ and $t \rightarrow +t$.

(vi) Conservation of charge Q. This conservation law is related to the invariance under gauge transformation.

(vii) Conservation of heavy particle number N.

Because of our belief that these invariance principles and conservation laws are valid for all interactions, strong, electromagnetic, and weak interactions, the most important characteristics of an elementary particle are its mass, spin, parity, charge, and heavy particle number. The question of its detailed dynamical behavior, such as scattering cross sections, production cross sections, decay modes, and lifetimes, is usually studied under specific assumptions about these intrinsic characteristics of the particle.

However, recent experiments[1,2] have led to a different picture of these symmetry principles. In particular, they show that the invariance under space inversion and charge conjugation are not valid in certain weak interactions. We shall return to these questions in detail in our later discussions.

---

[1]C.S. Wu, E. Ambler, R.W. Hayward, D.D. Hoppes, and R.P. Hudson, *Phys. Rev.* **105**, 1413 (1957).

[2]R.L. Garwin, L.M. Lederman, and M. Weinrich, *Phys. Rev.* **105**, 1415 (1957); J.I. Friedman and V.L. Telegdi, *Phys. Rev.* **105**, 1681 (1957).

4. Invariance Properties That Are Approximately True

It has been found that in addition to the above symmetry principles there are some further approximate invariance principles. These are the conservation of isotopic spin I and the conservation of I_z (or strangeness S).[3,4] We shall discuss the properties of these approximate conservation laws.

A. PION-NUCLEON SYSTEM

We review briefly the concept and the validity of isotopic spin in a system consisting of pions and nucleons. We describe the proton and the neutron as a spinor in the isotopic spin space:

	I	I_z	
p	½	½	(1.1)
n	½	$-$½ .	

The pions are considered as forming a vector in the isotopic spin space according to the assignments:

	I	I_z	
π^+	1	1	
π^0	1	0	(1.2)
π^-	1	-1 .	

The total isotopic spin of a pion-nucleon system is given by the quantum mechanical sum of the isotopic spin vectors of all particles. If the total isotopic spin **I** is conserved, then all physical observables are left invariant under a rotation in the isotopic spin space. We find that for the *strong interactions* this is indeed true.

Role of the Electromagnetic Field

From the assignments (1.1) and (1.2) we find the following empirical relation between the charge Q and I_z *for a pion-nucleon system:*

$$Q=I_z+(N/2) \tag{1.3}$$

where N is the heavy particle number ($N=1$ for p and n; $N=0$ for π). Because of this relationship the electromagnetic field destroys the invariance under arbitrary isotopic rotations except along the z axis. Thus the total isotopic spin quantum number I is not conserved in the electromagnetic interaction, while I_z is still conserved. To see this in detail let us represent the field of the nucleon by a two-component spinor function

$$\psi=\begin{pmatrix}\psi_p\\ \psi_n\end{pmatrix}.$$

---

[3] M. Gell-Mann, *Phys. Rev.* **93**, 933 (1953).
[4] K. Nishijima, *Progr. Theoret. Phys. (Japan)* **12**, 107 (1954).

The corresponding isotopic spin operator for a nucleon is then given by

$$\mathbf{I}_{\text{nucleon}}=(\tfrac{1}{2})\int\psi^{\dagger}\gamma_4\boldsymbol{\tau}\psi d^3r \tag{1.4}$$

with τ representing the 2×2 Pauli matrices. The electromagnetic field is described by the Hamiltonian

$$H_{\gamma}=-\int j_{\mu}A_{\mu}d^3r \tag{1.5}$$

where A_{μ} is the electromagnetic potential and j_{μ} is the electric current given by

$$j_{\mu}=(\tfrac{1}{2})ie[\psi^{\dagger}\gamma_4\gamma_{\mu}(1+\tau_3)\psi]\,. \tag{1.6}$$

We therefore immediately find that I_z commutes with H_{γ},

$$[I_z,H_{\gamma}]=0\,, \tag{1.7}$$

which means that I_z is conserved in the electromagnetic interactions. On the other hand, I_x, I_y, and I^2 $(=I_x^2+I_y^2+I_z^2)$ all do not commute with H_{γ}. In fact, under a rotation in the isotopic spin space the noncommuting part in H_{γ} behaves like the third component of an isotopic spin vector. Consequently, we find that the electromagnetic interactions can violate conservation of I by

$$\Delta I=0,\ \pm1\,, \tag{1.8}$$

whereas I_z is still conserved, i.e.,

$$\Delta I_z=0\,. \tag{1.9}$$

Similarly, we can apply the above arguments to any pion-nucleon system and obtain selection rules identical to Equations (1.8) and (1.9). The breakdown of the total isotopic spin by the electromagnetic fields accounts, at least in a qualitative sense, for the mass difference between the neutron and the proton ($\Delta m\cong2.5m_e$) and the mass difference between $\pi^{\pm}$ and π^0 ($\Delta m\cong9m_e$).

B. CONSERVATION OF *I* SPIN IN SYSTEMS INVOLVING OTHER ELEMENTARY PARTICLES

It is natural (in fact, it is almost necessary) to assume that I spin is conserved for *all strong interactions*. Let us, as an example, consider the following strong interaction:

$$\pi^-+p\rightarrow\Lambda^0+K^0\,. \tag{1.10}$$

Suppose, for the sake of argument, that this reaction does not conserve the isotopic spin vector **I**. By going through the virtual processes of emission and absorption of Λ^0+K^0, we would find that isotopic spin is not a good quantum number for the π^-+p system. Since the reaction (1.10) is a strong interaction, the violation of isotopic spin conservation in the pion-nucleon system will be strong. From our previous discussions we know that this is not the case. Consequently, reaction (1.10) is expected to conserve the isotopic spin. Similarly, we expect that the conservation law of isotopic spin should be valid for all strong interactions.

Accepting, then, that isotopic spin is conserved in all strong interactions, what must be the isotopic spin assignment of each of the other elementary particles in the various strong reactions, and what will be the relationship between I_z and Q for these new particles? Consider a particle A which is involved in the strong interactions. As the strong interactions conserve isotopic spin, we must assign to the particle A an isotopic spin quantum number I and its z component I_z. By a rotation in the I spin space, we generate $2I + 1$ states that are degenerate with respect to the strong interactions. We shall show that these isotopic spin multiplets are expected to be states of charges differing from each other by unity.

Let us consider the charge Q of one of these states. It is easy to see that if Q is a function of I_z then it must be a linear function of I_z. This follows from the fact that in a system of these particles A, the total charge Q and the total I_z of the system are given, respectively, by the linear sum of the Q and I_z of the individual particles. Consequently Q can at most be a linear function of I_z, i.e.,

$$Q = \alpha_A I_z + (Y_A/2) \tag{1.11}$$

where α_A and Y_A are quantities independent of I_z.

We want to demonstrate that the constant α_A must be 1. Let us consider a strong interaction of the form

$$\text{pions} + \text{nucleons} \to A + B\,. \tag{1.12}$$

From Equation (1.3) and the assumption that I_z is conserved in reaction (1.12), we have

$$(Q)_{\text{total}} = (I_z)_{\text{total}} + [(N)_{\text{total}}/2] = Q_A + Q_B. \tag{1.13}$$

For simplicity, suppose that

$$I_A = I_B = ½\,,$$

and consider first the state where the z components of both I spins are up $(\uparrow_A)(\uparrow_B)$. We have then $(I_z)_{\text{total}} = 1$. The total charge Q is

$$Q = Q(\uparrow_A) + Q(\uparrow_B) \tag{1.14}$$

where $Q(\uparrow_A)$ and $Q(\uparrow_B)$ are the charges corresponding to the states $\uparrow_A$ and $\uparrow_B$. Since the total I spin is conserved, we must have also the corresponding state with $(I_z)_{\text{total}} = 0$, which is $(1/\sqrt{2})(\uparrow_A\downarrow_B + \downarrow_A\uparrow_B)$. Furthermore, by considering the left-hand side of the reaction (1.12) and by using Equation (1.13) we conclude that the corresponding total charge for this state, $(I_z)_{\text{total}} = 0$, must be $Q - 1$ with Q given by Equation (1.14). Hence we have

$$Q(\uparrow_A) + (Q(\downarrow_B) = Q(\downarrow_A) + Q(\uparrow_B) = Q - 1\,.$$

By comparing with (1.11), we have

$$\alpha_A = \alpha_B = 1\,.$$

In a similar way, this result $\alpha_A = \alpha_B = 1$ can be generalized to particles of I values other than ½. It therefore follows that for all the strange particles[5] which interact strongly with the pion-nucleon system, we must have the relation,

[5]The term "strange particle" is again defined in a negative sense. It applies to any elementary particle which participates in the strong interactions but is not a nucleon or a pion.

$$Q = I_z + (Y/2) \qquad (1.15)$$

where Y is a quantity independent of I_z.

From this relation, Equation (1.15), there follow a number of consequences which we list below:

(i) In all these discussions we shall assume that *there does not exist a doubly charged particle.* Consequently from (1.15), I is limited to the values

$$I = 0, \tfrac{1}{2}, 1 \qquad (1.16)$$

for a single particle.

(ii) If we assume as consistent with experimental evidence that there are no charged particles of about the same mass as the Λ^0, then we must have

$$I_{\Lambda^0} = 0\,. \qquad (1.17)$$

(iii) Consider again the reaction

$$\pi^- + p \rightleftarrows \Lambda^0 + K^0\,. \qquad (1.10)$$

The total isotopic spin of the left-hand side is $(I)_{\text{total}} = \frac{1}{2}$ or $3/2$. By using the above properties (i) and (ii) we conclude

$$I_K = \tfrac{1}{2}\,, \qquad (1.18)$$

and for the I_z component

$$(I_z)_{K^0} = -\tfrac{1}{2}\,. \qquad (1.19)$$

From Equation (1.15) the corresponding particle with $I_z = +\frac{1}{2}$ must be a positively charged particle K^+. By applying the charge conjugation operator $\mathfrak{C}$ to K^+, K^0, we generate two other K-particles, K^- and $\bar{K}^0$, with

$$I_{\bar{K}} = \tfrac{1}{2}\,,\ (I_z)_{\bar{K}^0} = \tfrac{1}{2}, \quad \text{and} \quad (I_z)_{K^-} = -\tfrac{1}{2}\,. \qquad (1.20)$$

Thus we uniquely determine the I spin of the K-particles and deduce that there must be at least four such particles.

(iv) In a manner entirely analogous to the above, we can use the reaction, say,

$$\pi^- + p \rightarrow \Sigma^- + K^+\,. \qquad (1.21)$$

By comparing the I_z value on both sides we conclude that

$$(I_z)_{\Sigma^-} = -1\,.$$

Assuming that there is no doubly charged particle we have

$$(I)_{\Sigma} = 1\,, \qquad (1.22)$$

which implies a triplet $\Sigma^{\pm}$ and Σ^0.

Role of Electromagnetic Interactions

Now since we have determined the relationship between charge and I_z, Equation (1.15), for all the strange particles, we can apply the same argument concerning the role of electromagnetic interaction as that used in the pion-neucleon system. In an entirely similar way it can be shown that the electromagnetic interaction conserves I_z, but not the total isotopic spin I. The selection rule is

$$\Delta I = 0, \ \pm 1 \tag{1.23}$$

and

$$\Delta I_z = 0 \,. \tag{1.24}$$

Consequently, for example, Σ^0 is unstable against γ emission:

$$\Sigma^0 \rightarrow \Lambda^0 + \gamma \,. \tag{1.25}$$

Since I_z is conserved in the electromagnetic interaction as well as in the strong interactions and since the charge Q and the heavy particle number N are known to be conserved in all these interactions, any linear combinations of these three quantities will be conserved in both the strong and the electromagnetic interactions. In particular, it is useful to define

$$S/2 \equiv Q - I_z - (N/2), \quad \text{and} \quad Y \equiv S + N \,. \tag{1.26}$$

Because of the conservation of I_z, it follows that both S and Y are conserved in the strong as well as the electromagnetic interactions. The quantity S is called the strangeness quantum number.[3,4] From the assignment of I_z for various particles we find

$$\begin{aligned} &(S)_{\pi} = (S)_p = (S)_n = 0 \,, \\ &(S)_{K^+} = (S)_{K^0} = +1 \,, \\ &(S)_{K^-} = (S)_{\bar{K}^0} = -1 \,, \\ &(S)_{\Lambda^0} = (S)_{\Sigma^\pm} = (S)_{\Sigma^0} = -1 \,. \end{aligned} \tag{1.27}$$

Both S and Y do not vary with respect to different I_z components in the same I multiplets.

Let us consider now the assignment of I to the cascade particle Ξ^-. The production of the Ξ^- with two neutral K-mesons has been observed.[6] However, we do not know whether these K-mesons are K^0's or $\bar{K}^0$'s. Since S is conserved in production processes, it follows that the S of Ξ^- is either zero or ± 2 depending on the nature of these K-mesons. For Ξ^-, the charge Q is -1 and the heavy particle number N is $+1$. From Equation (1.26) we have

$$(I_z)_{\Xi^-} = -(3/2) - (\tfrac{1}{2})(S) \,.$$

In order that there be no doubly charged particle observed, we must have $|I_z| \leqslant 1$. Consequently, the isotopic spin assignment to Ξ^- is

$$(I_z)_{\Xi^-} = -\tfrac{1}{2} \quad \text{and} \quad I_{\Xi} = \tfrac{1}{2} \,. \tag{1.28}$$

Thus there should exist another Ξ^0 with $(I_z)_{\Xi^0} = +\tfrac{1}{2}$.

In Table 1 are listed various quantum numbers together with the masses and lifetimes for these particles.

---

[6] J.D. Sorrels, R.B. Leighton, and C.D. Anderson, *Phys. Rev.* **100**, 1457 (1955).

Table 1

Particle	Lifetime (sec)	Mass (Mev)	Spin, Parity	I	I_z	(S)
Ξ^-	$(4.6<\tau<200)\times10^{-10}$	1321 (±3.5)		½	−½	−2
Σ^-	1.6×10^{-10}	1196.65(±0.35)		1	−1	−1
Σ^+	0.69×10^{-10}	1189.7 (±0.25)			+1	−1
Σ^0	$(<1)\times10^{-11}$	1188.65$(^{+3}_{-2})$			0	−1
Λ^0	3×10^{-10}	1115		0	0	−1
p		938	½	½	½	0
n	10^3	939	½		−½	0
K^+	1.2×10^{-8}	494		½	½	1
K^0	0.95×10^{-10}	494			−½	1
$\overline{K}^0$	0.95×10^{-10}	494		½	½	−1
K^-	1.2×10^{-8}	494			−½	−1
$\pi^\pm$	2.6×10^{-8}	140	0^-	1	±1	0
π^0	10^{-15}	135	0^-		0	0
$\mu^\pm$	2.2×10^{-6}	106	½			
$e^\pm$		0.51	½			
$\nu, \bar{\nu}$		0	½			
γ		0	1			

C. WEAK INTERACTIONS

If we examine the weak interactions in detail, we find that they are divided distinctly into two groups:

(i) Those which are characterized by a nonconservation of I_z with

$$\Delta I_z = \pm \tfrac{1}{2}\,. \tag{1.29}$$

The neutrinos are not involved in these reactions. As examples of these reactions we have the decay of all hyperons, the $K_{\pi 2}$, the $K_{\pi 3}$, etc.

(ii) Those which involve neutrinos, such as the β decay, the μ decay, the π decay, the $K_{\mu 2}$, etc.

These two groups seem to have completely different characteristics. Yet they share a remarkable common feature which is that the strengths of the coupling constants for these two groups seem to be about the same. Of course, we do not really know how to calculate these coupling constants, because only for very few of these weak interactions, like β decay and μ decay, do we know something about the detailed dynamics of the decay reaction. In most of the decays we do not know how the various fields are coupled. Consequently, we can only estimate these coupling constants in a very crude way. For cases where the detailed interaction is unknown, we use the formula (with $\hbar = c = 1$)

$$1/\tau = 2\pi \mathcal{G}^2 (1/R)^2 \rho_E \tag{1.30}$$

where ρ_E is the number of final states per unit energy,

$$\rho_E = \frac{8\pi^3}{\Omega} \int \prod_{i=1}^{n} \left(\frac{1}{8\pi^3} d^3p_i d^3r_i \right) \delta^3(\textstyle\sum \mathbf{p}_i)\,\delta(\sum E_i - E)\,;$$

Table 2

	Lifetime (sec)	$\mathcal{G}^2\times10^{14}$
$\Lambda^0\to p+\pi^-$	3×10^{-10}	1.6
$\Sigma^-\to n+\pi^-$	1.6×10^{-10}	1.2
$K^0\to2\pi$	0.95×10^{-10}	2.8
$\pi^\pm\to\mu^\pm+\nu$	2.6×10^{-8}	0.3
$\mu^\pm\to e^\pm+\nu+\bar{\nu}$	2.2×10^{-6}	2
$K^\pm\to\mu^\pm+\nu$	1.2×10^{-8}	0.02

R represents a characteristic length of these decays; and $\Omega=(4\pi/3)R^3$. The total number of the particles in the decay product is n, and $\mathbf{p}_i$, E_i are their corresponding momenta and energies. In Table 2 are listed the various lifetimes and the corresponding coupling constants for several of the decay interactions in both group (i) and group (ii). In all these reactions we use

$$R=\hbar/m_\pi c \tag{1.31}$$

for simplicity. These results of course have significance only in their crude order of magnitude. For the purpose of comparison we include in Table 2 also μ decay calculated in the same way, even though we do know a great deal about its coupling.[7]

We observe from Table 2 the remarkable fact that although the lifetimes of these particles vary over a wide range from 10^{-10} sec to 10^{-6} sec the corresponding coupling constants $\mathcal{G}^2$ seem to be much more stable.

On the other hand, as remarked before, these decay interactions are separated into two distinct groups. In the first group, (i), the neutrinos are not involved; instead there is a nonconservation of I_z. In the other group, (ii), every reaction contains some neutrinos. Furthermore, these reactions are between many particles for which isotopic spin seems to play no role. The fact that they share approximately the same strength in coupling constants does suggest strongly a deep common origin for all weak interactions. As we shall see later, these weak interactions *may* share another significant feature, namely the violation of invariance under space inversion and charge conjugation.

[7]The coupling constants for β decay and μ^- capture are not included in Table 1. It is well known that they have the same order of magnitude as that for μ decay. See, e.g., E. Fermi, *Elementary Particles,* Yale University Press, 1951.

II. THE θ-τ PUZZLE

Among the various interesting phenomena concerning elementary particles, we would like to discuss specifically first the θ-τ puzzle, because it was due to this puzzle that a re-examination of the experimental basis of our various supposedly exact conservation laws was made. In Table 3 are listed the recently measured values of the mass, abundance, and lifetime of the various decay modes of charged K-mesons.[8]

Table 3

Type	Abundance	Mass of K from primary particle	Mass of K from secondary particles	Lifetime
τ	5.56±0.41	966.3±2.1	966.1±0.7	$(1.19\pm0.05)\times10^{-8}$
τ'	2.15±0.47	967.7±4		
$K_{\mu 2}$	58.2 ±3.0	967.2±2.2	965.8±2.4	$(1.24\pm0.02)\times10^{-8}$
$K_{\pi 2}$	28.9 ±2.7	966.7±2.0	962.8±1.8	$(1.21\pm0.02)\times10^{-8}$
$K_{\mu 3}$	2.83±0.95	969 ±5		$(0.88\pm0.23)\times10^{-8}$
K_{e3}	3.23±1.30			$(1.44\pm0.46)\times10^{-8}$

We see that the masses are extremely close to each other and the lifetimes agree within the experimental error of $\sim\pm5\%$. About three years ago, Dalitz pointed out that by plotting the angular and energy distribution of the three π-mesons from the decay of the τ-meson ($\equiv K_{\pi 3}$), it is possible to determine the spin and parity of the τ. *In the following discussion of the θ-τ puzzle, we shall assume that both spin and parity are absolutely conserved.*

1. Review of the Spin-Parity Determination of θ and τ (Dalitz's Analysis[9])

Let us consider first a θ-meson. The θ-meson is defined to be a K-particle which can decay, among other modes, into two π-mesons, e.g.,

$$\theta^+ \to \pi^+ + \pi^0 . \tag{2.1}$$

Assuming that both the spin and parity are conserved in reaction (2.1), the parity

[8] R.W. Birge, D.H. Perkins, J.R. Peterson, D.H. Stork, and M.N. Whitehead, *Nuovo cimento* **4**, 834 (1956); V. Fitch and R. Motley, *Phys. Rev.* **101**, 496 (1956); L.W. Alvarez, F.S. Crawford, M.L. Good, and M.L. Stevenson, *Phys. Rev.* **101**, 503 (1956); J. Orear, G. Harris, and S. Taylor, *Phys. Rev.* **104**, 1463 (1956).

[9] R. Dalitz, *Phil. Mag.* **44**, 1068 (1953); *Phys. Rev.* **94**, 1046 (1954); E. Fabri, *Nuovo cimento* **11**, 479 (1954).

of the θ is uniquely determined by its spin value. Let J be the spin of the θ-particle. Because each π-meson is a pseudoscalar, we have

$$P_\theta=(-1)^J \tag{2.2}$$

where $P_\theta=$ parity of the θ-meson. Thus, the spin-parity assignment of θ can only be $(0+)$, $(1-)$, $(2+)$, etc.

On the other hand, the possible spin-parity assignments for a τ-meson are quite different. The τ-meson is defined to be a K-meson which can decay, among other modes, into three π-mesons, e.g.,

$$\tau^+\rightarrow\pi^+ + \pi^+ + \pi^- . \tag{2.3}$$

The final decay state of a τ-meson is characterized by two momenta in its center-of-mass system. We may choose these two momenta to be:

(i) The relative momentum, **p**, between the two π^+ mesons.

(ii) The momentum, **k**, of the π^- in the center-of-mass system of the τ-meson. (It may differ by a factor 3/2, if the momentum is chosen to be that of the π^- with respect to the center-of-mass system of the two π^+ system.)

Let **J** be the spin of the τ-meson; then

$$\mathbf{J}=\mathbf{L}_p+\mathbf{L}_k \tag{2.4}$$

where $\mathbf{L}_k$ and $\mathbf{L}_p$ are respectively the orbital angular momentum of the π^- and the relative angular momentum of the two π^+. If the spin of the τ is zero, then the final state of the three pions must consist of states with $L_p{}^2=L_k{}^2$. Hence the parity of the final state must be -1. However, if the spin of τ is not zero then the parity of the final state can be either $+1$ or -1. Consequently the spin-parity assignments for τ are $(0-)$, $(1\pm)$, $(2\pm)$, etc.

It is easy to see the following simple conclusions:

(i) If the spin of τ is zero and if parity is conserved in the decay, then $\theta\neq\tau$.

(ii) If there exists a zero energy pion (π^+ or π^-) in the 3π state of τ decay and if parity is conserved in the decay, then $\theta\neq\tau$.

(iii) If there exists a zero energy π^- in the 3π state of the τ decay then the spin of τ must be even. If, further, parity is conserved, the parity of τ must be odd.

Of course, in reality it is not possible to observe a zero energy pion. But quite a few low energy ($\sim$⅓ Mev) π^- and π^+ have been observed. Thus, even without detailed statistical analysis, it is to be expected that *most probably* the spin of τ is even, and that if parity is conserved in the decay process then τ and θ are two different particles. To evaluate the exact meaning of the likelihood it is necessary to perform a detailed statistical analysis.

Let us characterize the system by an angle θ and a parameter ε defined to be

$$\theta=\angle(\mathbf{p},\mathbf{k}) \quad \text{and} \quad \varepsilon=(k/k_{\max})^2 . \tag{2.5}$$

If ψ $(\mathbf{p},\mathbf{k})$ represents the final state wave function of the 3π in the τ decay then the probability distribution function $P(\theta,\varepsilon)$ is

$$P(\theta,\varepsilon)\propto|\psi(\mathbf{p},\mathbf{k})|^2\sqrt{\varepsilon(1-\varepsilon)}\,d\varepsilon\,d(\cos\theta) . \tag{2.6}$$

Experimentally with an energy value ranging from ~⅓ Mev to ~50 Mev and with a total of ~1000 events, Dalitz and others find that

$$|\psi|^2=1 \tag{2.7}$$

is an extremely good fit to these data. From the above conclusion (ii) we see that if τ is the same as θ, then $|\psi|^2$ must be zero if, say, $\varepsilon=0$. In fact, it is possible to prove the following rigorous statements concerning the behavior of ψ at various limiting regions.

(iv) If τ is the same as θ and if parity is conserved in the τ decay, then

$$\text{as } \varepsilon\to 0,\ |\psi|^2\to \varepsilon^n \quad (n\geqslant 1)\ ;$$

$$\text{as } \varepsilon\to 1,\ |\psi|^2\to (1-\varepsilon)^m \quad (m\geqslant 2)\ ;$$

$$\text{and as } \theta\to 0,\ |\psi|^2\to \sin^2\theta\ . \tag{2.8}$$

To prove (iv), we consider first the simple case that the spin of τ is 1. From Equation (2.2) we see that if τ is the same as θ then the parity of τ is -1. Consequently the orbital parity due to **k** and **p** must be $+1$. Thus, ψ should be an axial vector. Furthermore, ψ must be an even function of **p** because of the Bose-statistical property of the two π^+-mesons. Hence it is easy to deduce that

$$\psi=(\mathbf{k}\times\mathbf{p})(\mathbf{k}\cdot\mathbf{p})f[k^2,(\mathbf{k}\cdot\mathbf{p})^2,p^2]\ .$$

Or,

$$|\psi|^2\propto\varepsilon^2(1-\varepsilon)^2\sin^2\theta\cos^2\theta|f|^2\ . \tag{2.9}$$

The function f is expected to be a regular function of its arguments. (This is the case, e.g., if the decay of τ can be represented by a local field theory with decay interaction involving derivative couplings to any high, but finite, orders.) Hence we see that the wave function ψ satisfies the conditions given by Equation (2.8). Similarly, by going over the same type of arguments of forming various tensor functions with 2 vectors, **k** and **p**, one can easily prove (2.8) for any other spin values.

From (2.9) we see that it is very difficult to pick a function f such that in all observed regions in (θ,ε) space $|\psi|^2\cong 1$ so as to be compatible with experimental results. In fact, there are good arguments to expect the function f to be near 1. The reason is, as pointed out by Dalitz, that these pions are very low in energy, and an expansion of f into powers of k^2 and p^2 may be legitimate,

$$f=1+0(k^2R^2)+0(p^2R^2)+\ \ldots\ . \tag{2.10}$$

From dimensional grounds we expect R to be some length that characterizes the τ-meson. For low energy pions we may neglect the terms $0(k^2R^2)$ and $0(p^2R^2)$ and we have

$$f\cong 1\ . \tag{2.11}$$

By using the condition $f\cong 1$ in the distribution function ψ, we can calculate the probability that one obtains $|\psi|^2=1$ from experimental events if ψ is actually given by Equation (2.9). This probability will be extremely small ($\lesssim 10^{-30}$).[10] Similar conclusions can be obtained for other spin values (except for very high ones). Thus, we conclude that *it is extremely improbable that τ and θ are the same particle* (under, of course, the framework that parity is conserved). The most probable spin-parity assignments of τ are $0-$ and $2-$.

One may question how good is the approximation $f \cong 1$. The average kinetic energy of π in the τ decay is $\sim$26 Mev. If R is taken to be $\hbar / m_\pi c$ where m_π is the pion mass, then

$$(k^2R^2) \cong \tfrac{1}{3},$$

which is not very small. However the region where the distribution function $|\psi|^2$ [Equation (2.8)] deviates most from the experimental situation is precisely the region (e.g., $k \to 0$ or $p \to 0$) where the relevant expansion parameter is small. Consequently, one expects that the conclusions of Dalitz are statistically very significant. Thus, there seem to be two particles of different spin-parity values. The difficulty is, then, why should they have, within experimental error, the same mass and the same lifetime? This is the famous θ-τ puzzle.

2. Previous Attempts to Solve the θ-τ Puzzle

Mainly for historical reasons we shall review some attempts that have been made to solve the θ-τ puzzle. Although the Dalitz analysis was made more than three years ago, at that time various different masses of K-mesons were reported. In fact there were indications that there might be a big mass difference between the various K-particles. Also, the statistics of Dalitz's analysis at that time were not very good. No one was greatly alarmed that there probably existed a K-particle (τ-meson) which was different from the θ-particle. There were too many K-particles anyway. However, about the spring of 1955, the situation was changed. The experimental mass values gradually converged, though still with large probable error. The Dalitz plot had more points and seemed to indicate convincingly that the τ is a $(0-)$-particle. Also, at about the same time people started to measure the lifetime. Before the lifetime measurements were done, some physicists speculated about what would happen if the lifetimes turned out to be the same. At first sight it seemed that, if the lifetimes of $K_{\pi 2}$ and $K_{\pi 3}$ are measured to be the same, then this evidence would be used to argue that θ must be the same particle as τ and that the conclusion of Dalitz's analysis is probably a manifestation of some statistical fluctuations. Later the experiments on lifetimes indeed showed that the observed decay constants for $K_{\pi 2}$ and $K_{\pi 3}$ are about the same.[8]

Nevertheless, with all the available evidence at that time (1955) it was not difficult to find schemes which could be made compatible with (i) the results of Dalitz's analysis, (ii) the aparent identity of lifetimes, and (iii) the approximate equality of masses. In describing these attempts we shall assume that the conclusions concerning Dalitz's analysis are correct and that τ and θ are two different particles.

---

[10]See, e.g., *Proc. Sixth Ann. Rochester Conf.*, Interscience, New York, 1956.

A. APPARENT LIFETIME EQUALITY

One such hypothesis, proposed to explain the apparent identity in lifetimes, is the so-called cascade process.[11] The idea is that τ and θ are two different particles with two different lifetimes, say, 10^{-8} sec and 10^{-10} sec. (On phase space arguments, one expects $K_{\pi 3}{}^{+}$ to live longer than $K_{\pi 2}{}^{+}$.) The long-lived one, say τ, is assumed to have a heavier mass. In addition to its other decay modes, τ decays into the light one, θ, by γ radiations. Under the experimental conditions in these lifetime measurements, only the long-lived K-mesons would be observed. Thus this cascade process can account for their apparent equality of lifetimes. If the spin-parities of τ and θ are $0-$ and $0+$, then the cascade process is

$$\tau \rightarrow \theta + 2\gamma\,. \tag{2.11}$$

In order to make the branching ratio correct a mass difference $m_\tau - m_\theta \cong 10$ Mev is required.

B. MASS DEGENERACY[12]

By taking analogy with isotopic spin invariance, a mass degeneracy means possibly a new symmetry property. This, for example, is the case of mass degeneracy between a proton and a neutron. The only difference is that we now have degeneracy between states with different parities, instead of between states with different charges. If one regards this mass degeneracy between τ and θ to be not an accident, then it means, just as in the case for isotopic spin, that the strong interaction must be invariant under a new symmetry operator. This operator, denoted by C_p, when acting on a θ-particle changes it into a τ, and when acting on τ converts it into a θ. The operator C_p is called "parity conjugation" by analogy with "charge conjugation."

$$C_p|\theta\rangle = |\tau\rangle, \quad C_p|\tau\rangle = |\theta\rangle\,. \tag{2.12}$$

The approximate mass degeneracy now follows from the assumption that C_p commutes with the strong part of the Hamiltonian

$$[C_p, H_{\text{strong}}] = 0\,. \tag{2.13}$$

Because of Equation (2.13) it follows that all strong interactions should be invariant under the operation of C_p. In particular, let us take an example, say,

$$\pi^- + p \rightarrow \Lambda_1{}^0 + \theta^0\,. \tag{2.14}$$

Under the operation C_p, there is no change in $\pi^- + p$ but θ^0 becomes τ^0. Therefore the Λ^0 must be a parity doublet, which we shall denote by $\Lambda_1{}^0$ and $\Lambda_2{}^0$. Equation (2.14) becomes

$$\pi^- + p \rightarrow \Lambda_2{}^0 + \tau^0 \tag{2.15}$$

under C_p.

We conclude therefore that there must be a parity doubling not only of the K-mesons but also of hyperons. There will be two Λ's of opposite parity and two Σ^+'s

[11] T.D. Lee and J. Orear, *Phys. Rev.* **100**, 932 (1956).
[12] T.D. Lee and C.N. Yang, *Phys. Rev.* **102**, 290 (1956).

of opposite parity, etc. In fact, there must be a parity doublet for all strange particles with odd strangeness quantum number. The commutation relation between C_p and the H_γ (electromagnetic interaction) is not known. A natural choice, of course, is to take advantage of the above explanation for the lifetimes and to assume that C_p does not commute with H_γ. This can introduce a large mass difference and make the cascade process possible. Combining these two possibilities, it seemed that one could have an explanation for the θ-τ puzzle. However, within half a year,[8] the mass measurements have been greatly improved, with the result that the mass difference can at most be ~ 1 or 2 Mev, which makes process (2.11) very unlikely. A direct search for such γ-rays was performed[13] and it also led to negative results.

In the early part of 1956 it seemed that the true sloution of the θ-τ puzzle might in fact lie in something quite different. Thus, an investigation of the experimental basis of the law of conservation of parity was made.[14] We shall discuss now in some detail the conclusions reached through such an investigation.

---

[13]L. Alvarez, *Proc. Sixth Ann. Rochester Conf.*, Interscience, New York, 1956.
[14]T.D. Lee and C.N. Yang, *Phys. Rev.* **104**, 254 (1956).

III. EXPERIMENTAL LIMITS ON THE VALIDITY OF PARITY CONSERVATION

In this section we shall try to discuss the limit on the validity of parity conservation in various fields of physics. If parity is not a rigorously conserved quantum number then eigenstates ψ of the entire Hamiltonian are, in general, not eigenstates of the parity operator. Thus we expect

$$\psi = \psi_p + F\psi_{-p} \tag{3.1}$$

where ψ_p and ψ_{-p} are of opposite parity and

$$F = \text{probability amplitude for parity mixing.} \tag{3.2}$$

It is useful to know from the various evidence in atomic and nuclear physics exactly what is the upper limit one can impose on the magnitude of F.

1. Atomic Spectroscopy

From the various parity selection rules concerning the radiative transitions for an atomic system, we find an upper limit for F:

$$|F|^2_{\text{atom}} < \left(\frac{r}{\lambda}\right)^2_{\text{atom}} \cong 10^{-6} \tag{3.3}$$

for a typical atomic transition. In (3.3), r is the radius of the atom and λ the wavelength of the radiation. In principle, by studying transitions involving photons of long wavelengths it is possible to make this upper limit much smaller than 10^{-6}. (It may not be impossible to reach the limit $|F|^2_{\text{atom}} < 10^{-12}$.)

2. Nuclear Spectroscopy

While the above condition sets a limit on parity nonconservation in atomic interactions, the same limit cannot be applied directly to nuclear interactions. Nevertheless, by using the various parity selection rules in nuclear spectroscopy, say β decay, it is possible to put a corresponding limit for the nuclear system,

$$|F|^2_{\text{nucleus}} < \left(\frac{r}{\lambda}\right)^2_{\text{nucleus}} \cong 10^{-4}. \tag{3.4}$$

3. Nuclear Reactions

The measurement by Chamberlain et al.[15] on the double scattering of protons offers a very direct test of parity conservation. In this experiment, a beam of incom-

[15] O. Chamberlain, E. Segrè, R. Tripp, C. Wiegand, and T. Ypsilantis, *Phys. Rev.* **93**, 1430 (1954).

ing protons with momentum $\mathbf{p}_1$ is scattered by a first target into momentum $\mathbf{p}_2$ and is further scattered by a second target into momentum $\mathbf{p}_3$. If parity is conserved, the cross section should be independent of such a quantity as $(\mathbf{p}_1 \times \mathbf{p}_2) \cdot \mathbf{p}_3$. The measurement shows the absence of such terms, giving directly an upper limit on F. From the measurements one can conclude

$$|F|^2 < 10^{-4} . \tag{3.5}$$

4. Static Electric Dipole Moment

If parity is conserved then the static electric dipole moment of any eigenstate of the Hamiltonian must be zero. Thus a measurement on the absence of such electric dipole moment for elementary particles gives also an upper limit of F. Smith, Purcell, and Ramsey[16] have measured the electric dipole moment of the neutron and found it to be smaller than 5×10^{-20} cm $\times$ electronic charge. If one takes the natural size of the neutron to be 10^{-13} cm, this gives a very severe limit on F:

$$|F|^2 < 3 \times 10^{-13} . \tag{3.6}$$

This limit applies directly to the structure of the neutron.

However, it is possible to show that *if time-reversal invariance holds then the static electric dipole moment must still be zero even though parity may not be conserved.* This is so because if parity is not conserved then the wave function ψ is a mixture of states with opposite parities ψ_p and ψ_{-p} as indicated by Equation (3.1). But if time reversal is invariant then ψ_p and ψ_{-p} will be 90° out of phase. Thus they cannot contribute to the diagonal element of the electric dipole moment which is a real quantity. We shall give a formal proof of the impossibility of having a static electric dipole if time reversal is invariant.

Consider a particle A with spin $\mathbf{J}$. The state function $|A\rangle_m$ describes the particle A at rest with $J_z = m$. The time reversal operator T is represented by

$$T = U_T \times \text{complex conjugation} \tag{3.7}$$

where U_T is a unitary operator. If invariance under time reversal is assumed, then

$$T|A\rangle_m = U_T|A^*\rangle_m = e^{i\delta m}|A\rangle_{-m} \tag{3.8}$$

where * represents a pure complex conjugation, and $e^{i\delta m}$ is a possible phase factor. Let $\mathbf{D}$ be the electric dipole moment,

$$\mathbf{D} = \sum e_i \mathbf{r}_i . \tag{3.9}$$

The average value of $\mathbf{D}$ must be proportional to that of $\mathbf{J}$.

$$\langle A|\mathbf{D}|A\rangle_m = K\langle A|\mathbf{J}|A\rangle_m \tag{3.10}$$

where K is a real numerical constant. If we take the complex conjugation of both sides in Equation (3.10) and replace 1 by $U_T^\dagger U_T$, Equation (3.10) becomes

---

[16] E.M. Purcell and N.F. Ramsey, *Phys. Rev.* **78**, 807 (1950); Smith et al. as quoted in N.F. Ramsey, *Molecular Beams,* Oxford University Press, 1956.

$$\langle A^*|U_T{}^\dagger U_T\mathbf{D}^* U_T{}^\dagger U_T|A^*\rangle_m = K^*\langle A^*|U_T{}^\dagger U_T\mathbf{J}^* U_T{}^\dagger U_T|A^*\rangle_m .$$

From (3.8) and the properties that

$$T\mathbf{J}T^{-1} = U_T\mathbf{J}^* U_T{}^\dagger = -\mathbf{J} \quad \text{and} \quad T\mathbf{r}T^{-1} = U_T\mathbf{r}^* U_T{}^\dagger = +\mathbf{r}\,, \tag{3.11}$$

we have

$$\langle A|\mathbf{D}|A\rangle_{-m} = -K\langle A|\mathbf{J}|A\rangle_{-m} . \tag{3.12}$$

Comparison of (3.10) and (3.12) shows that

$$\langle A|\mathbf{D}|A\rangle_m = 0 . \tag{3.13}$$

Thus, the static electric dipole moment must be zero if time reversal is invariant.[17] However, the matrix elements of **r** between two *different* states of dominantly the *same parity* are, in general, not zero if time reversal invariance holds but parity is not conserved. (Here the term "same parity" refers to that part determined by the strong interactions.)

Thus, if time reversal invariance holds, the most severe limit on parity nonconservation is given by spectroscopic evidence and experiments of the nuclear double scattering type. These limits are already quite strong and thus demand that the strong interactions and the electromagnetic interactions both must conserve parity. However, these limits throw no light on the invariance properties of the weak interactions.

5. β Decay Experiments

Prior to the recent experiments on β angular distribution from polarized nuclei and on the longitudinal polarization of β-rays, there already existed an immense body of knowledge in the field of β decay. These previous experiments[18] consist of (i) spectra (allowed, forbidden, etc.) and *ft* values, (ii) β-neutrino correlation, (iii) β-γ correlation, (iv) polarized nuclei and the angular distribution of secondary γ-rays, and (v) β-γ-γ′ angular correlation. We shall show that *these experiments* (i) *to* (v) *are not relevant so far as the question of parity conservation in weak interactions is concerned.* They neither prove nor disprove the conservation of parity in β decay.

The most general form of the interaction Hamiltonian for nonconservation of parity is

$$H_{\text{int}} = \sum_i (\psi_p{}^\dagger O_i\psi_n)(C_i\psi_e{}^\dagger O_i\psi_\nu + C_i'\psi_e{}^\dagger O_i\gamma_5\psi_\nu)\,, \quad i=S, T, V, A, P\,, \tag{3.14}$$

where $O_S=\gamma_4$, $O_V=\gamma_4\gamma_\mu$, $O_T=-[i/(2\sqrt{2})]\gamma_4(\gamma_\lambda\gamma_\mu-\gamma_\mu\gamma_\lambda)$, $O_A=-i\gamma_4\gamma_\mu\gamma_5$, and $O_P=\gamma_4\gamma_5$. In Equation (3.14) derivative couplings are not included. The conclusions [Equations (3.18) and (3.20) below] are correct even if there are such derivative coupling terms. We have now ten complex coupling constants. Any observed quantity will be related to the sum of the absolute squares of certain matrix elements.

---

[17]By the *CPT* theorem (see Chapter IV) the same is true if *C·P* is invariant. That the static dipole moment must be zero if *C·P* is invariant was first pointed out by L. Landau [*Nuclear Phys.* **3**, 127 (1957)].

[18]See, e.g., K. Siegbahn, *Beta- and Gamma-Ray Spectroscopy,* Interscience, New York, 1955.

$$\sum|M|^2=(\sum f_{ij}C_i^*C_j+\text{c.c.})+(\sum f_{ij}'C_i^*C_j'+\text{c.c.})+(\sum g_{ij}C_i^*C_j'+\text{c.c.}) \tag{3.15}$$

where f_{ij}, f_{ij}' and g_{ij} are certain functions of the measured momenta and spins. It is well known that, as the neutrino has zero mass, it satisfies not only the familiar Dirac equation

$$\left(\gamma_\mu \frac{\partial}{\partial x_\mu}\right)\psi_\nu=0\,, \tag{3.16}$$

but also the equation

$$\gamma_\mu\left(\frac{\partial}{\partial x_\mu}\right)\gamma_5\psi_\nu=0\,. \tag{3.17}$$

Thus, one has[19]

$$f_{ij}=f_{ij}'\,. \tag{3.18}$$

Furthermore, we can show that the g_{ij} must be pseudoscalar quantities. To see this, let us consider the following *formal* transformation:

$$C_i\to C_i\,,\quad C_i'\to -C_i' \tag{3.19}$$

together with $\mathbf{r}\to-\mathbf{r}$; $\mathbf{p}\to-\mathbf{p}$; and spin $\mathbf{s}\to+\mathbf{s}$. This formal mathematical transformation leaves the Hamiltonian H_{int} invariant. Thus it must also leave Equation (3.15) invariant. It then follows that under this formal transformation the interference terms g_{ij} must transform as

$$g_{ij}(\mathbf{p},\mathbf{s},\ \ldots\ )\to g_{ij}(-\mathbf{p},+\mathbf{s},\ \ldots\ )=\ -g_{ij}(\mathbf{p},\mathbf{s},\ \ldots\ ). \tag{3.20}$$

Consequently g_{ij} transform like pseudoscalar quantities. This means that in order to answer the question of parity conservation it is necessary to observe, experimentally, a pseudoscalar quantity. To observe a pseudoscalar quantity one must measure at least three linear momenta $\mathbf{p}_1$, $\mathbf{p}_2$, $\mathbf{p}_3$ or a spin $\mathbf{s}$ and a momentum $\mathbf{p}$ so as to form quantities like $(\mathbf{p}_1\times\mathbf{p}_2)\cdot\mathbf{p}_2$ or $\mathbf{s}\cdot\mathbf{p}$, etc. In the experiments on the spectra, β-ν correlation and β-γ correlation, it is clear that no such pseudoscalar quantity can be formed. With the parity nonconserving Hamiltonian Equation (3.14), the theoretical results for these experiments are identical with that of the conventional parity conserving Hamiltonian provided one replaces $C_i^*C_j$ in the old formulas:

$$C_i^*C_j\to(C_i^*C_j+C_i'^*C_j')\,. \tag{3.21}$$

In the measurement of polarized nuclei and the angular distribution of the secondary γ-ray it is possible to form pseudoscalar terms like

$$(\mathbf{s}\cdot\mathbf{p}_\gamma)\,. \tag{3.22}$$

However, since γ interaction conserves parity and since the multipole γ radiation in nuclear transitions has very accurately defined parities, the observed angular distribution must be invariant under the transformation

[19]It is important to note that in general (3.15) is invariant under the mathematical transformation $C_i\to C_i'$ and $C_i'\to C_i$. This property can serve as a good check for the correctness of various expressions [cf. Equations (5.3), (5.11), etc.]. We wish to thank Dr. Pauli for a communication on the usefulness of this transformation.

$$\mathbf{p}_\gamma \to -\mathbf{p}_\gamma \,. \tag{3.23}$$

Thus terms of the form (3.22) cannot exist.[20]

In the case of β-γ-γ' angular correlation measurements, one can easily form pseudoscalar quantities that are also invariant under transformation (3.23), such as

$$\mathbf{P}_e \cdot (\mathbf{p}_\gamma \times \mathbf{p}_{\gamma'}) \, (\mathbf{p}_\gamma \cdot \mathbf{p}_{\gamma'}) \,. \tag{3.24}$$

These terms cannot be ruled out by using the parity conservation property of the γ radiation. However, if time reversal is invariant for the strong interactions (including γ interactions), then such terms must all vanish. [This follows immediately from Equation (4.29)]. Consequently the absence of such terms can be used to prove the invariance of time reversal for strong interactions and electromagnetic interactions but not for the invariance properties of the weak interactions. We can summarize as follows: The previous accurate measurements of β decay (i) to (v) do not throw any light whatsoever upon a possible nonconservation of parity in the decay process. In order to detect possible parity nonconservation terms we must try to perform other experiments such as to measure $\mathbf{s} \cdot \mathbf{p}_e$, etc.

---

[20]The nonexistence of terms like $\mathbf{s} \cdot \mathbf{p}_\gamma$ and $\mathbf{p}_e \cdot (\mathbf{p}_\gamma \times \mathbf{p}_{\gamma'})$ can also serve as evidence for the parity conservation for the strong interactions (including γ interactions) in a nuclear system, but not for weak interactions.

IV. SOME GENERAL DISCUSSIONS ON THE CONSEQUENCES DUE TO POSSIBLE NON-INVARIANCE UNDER *P*, *C*, AND *T*

1. The *CPT* Theorem.[21] Equality of Mass and Lifetime Between a Particle and its Antiparticle[22]

Before we discuss in detail the various tests on the conservation of parity P, charge conjugation C, and time reversal T in weak interactions, it is useful to recall a general theorem concerning the interrelationship between these operators C, P, T and proper Lorentz invariance.

The *CPT* Theorem: If a local Lagrangian theory (which may contain derivative couplings to any high but finite orders) is invariant under the proper Lorentz transformations, it is invariant under the product of CPT (and its permutation PCT, etc.) although the theory may not be separately invariant under each one of these three operators C, P, and T.

It follows from this theorem that if P is not conserved in the weak interactions, then at least one of the other invariances C or T should not be conserved. In the following discussions we shall assume that the general framework of field theory, under which the CPT theorem is proved, is valid. At first sight it seems that the observed equality of lifetimes in the decay of π^+, π^- and the similar equality for the μ^+, μ^- may already form a proof that C is conserved in weak decays. As we shall see, the equality of the masses and lifetimes for a particle and its antiparticle follows directly from proper Lorentz invariance and the CPT theorem. It does not prove at all that C is invariant. We shall state these consequences of the CPT theorem in the form of two theorems. Let H be the complete Hamiltonian which may be separated into two parts

$$H = H_s + H_w \tag{4.1}$$

where H_s represents the strong interactions together with the γ interactions, and H_w the weak interactions. We assume that both H_s and H_w are invariant under the proper Lorentz transformation. Consequently, H_s and H_w are both invariant under the compound operation of PCT, i.e.,

$$PCTHT^{-1}C^{-1}P^{-1} = PCU_T H^* U_T^{\dagger} C^{\dagger} P^{\dagger} = H \tag{4.2}$$

where P, C, U_T are all unitary operators [cf. Equation (3.7)]. We shall further assume that H_s is invariant under the separate operation of each one of these three operators C, P, and T while H_w may or may not be invariant under C, P, and T separately. (The operators C, P, T are defined by using the invariance properties of H_s.)

[21]See W. Pauli's article in *Niels Bohr and the Development of Physics*, Pergamon, London, 1955; J. Schwinger, *Phys. Rev.* **91**, 720, 723 (1953); **94**, 1366 (1953); G. Lüders, *Kgl. Danske Videnskab. Selskab, Mat.-fys. Medd.* **28**, No. 5 (1954).

[22]T.D. Lee, R. Ochme, and C.N. Yang, *Phys. Rev.* **106**, 340 (1957).

Theorem 1:[23] If α is a stable particle, then

$$M_\alpha = M_{\bar{\alpha}} \tag{4.3}$$

where M_α and $M_{\bar{\alpha}}$ are the masses of α and its antiparticle $\bar{\alpha}$. Equation (4.3) is valid to all orders in H_w.

Proof: Consider a particle α at rest.

$$H|\alpha\rangle = M_\alpha|\alpha\rangle \quad \text{or} \quad H^*|\alpha^*\rangle = M_\alpha|\alpha^*\rangle \tag{4.4}$$

and

$$PCU_T H^*|\alpha^*\rangle = M_\alpha \cdot PCU_T|\alpha^*\rangle\,. \tag{4.5}$$

Let

$$|\bar{\alpha}\rangle \equiv PCU_T|\alpha^*\rangle\,. \tag{4.6}$$

We then have

$$H|\bar{\alpha}\rangle = M_\alpha|\bar{\alpha}\rangle \tag{4.7}$$

by using Equation (4.2). From the definitions of P, C, T [see Chapter I, section 3, and Equation (3.8)] we know that if $|\alpha\rangle$ represents a particle at rest with spin J and its z component J_z, then $|\alpha\rangle$, defined by Equation (4.6), represents its antiparticle state also at rest and with its spin along z component $-J_z$. Theorem 1 follows immediately from Equation (4.7).

Remarks: What we have proved is actually more than just the equality of masses. By taking $|\alpha\rangle$ to be any eigenstate of the total Hamiltonian H we can generate another state $|\bar{\alpha}\rangle$ by (4.6) which has the same eigenvalue. Thus the complete energy spectrum for a group of particles is identical with that of a corresponding group of antiparticles. By considering the energy spectrum of a particle, say a proton p, in a magnetic field, one can prove the equality of the magnitude of magnetic moment of p and $\bar{p}$ (again to all orders of H_w).

Theorem 2: Consider the decay of A via H_w,

$$A \to B \quad \text{and} \quad \bar{A} \to \bar{B} \tag{4.8}$$

(with the states $B \neq \bar{B}$);[24] then, to the lowest order in H_w,

$$(\text{lifetime of } A) = (\text{lifetime of } \bar{A}). \tag{4.9}$$

Proof: Since we are only interested in the lowest order in H_w, the states $|A\rangle_m$, $|B\rangle_m$, $|\bar{A}\rangle_m$, $|\bar{B}\rangle_m$ can be taken to be eigenstates of H_s with z component spin $J_z = m$. Furthermore, since H_s is invariant under T and C, we have [cf. Equation (3.8)]

$$|\bar{A}\rangle_m = C|A\rangle_m\,, \quad |\bar{B}\rangle_m = C|B\rangle_m\,; \tag{4.10}$$

$$T|A\rangle_m = e^{i\delta_m(A)}|A\rangle_{-m}\,, \quad T|B\rangle_m = e^{i\delta_m(B)}|B\rangle_{-m}\,; \tag{4.11}$$

and similar equations for $|\bar{A}\rangle_m$ and $|\bar{B}\rangle_m$ under time reversal. The states $|B\rangle_m$ and $|\bar{B}\rangle_m$ are taken to be stationary states consisting of standing waves (including all the strong interactions).

---

[23] T.D. Lee and C.N. Yang, *Phys. Rev.* **105**, 1671 (1957).

[24] This condition $B \neq \bar{B}$ applies to cases, e.g., where A and $\bar{A}$ have opposite charges $\pm Q$ or opposite heavy particle number $\pm N$, etc. General discussions concerning cases where B may be the same as $\bar{B}$ (e.g., in the decay of K^0 and $\bar{K}^0$) will be found in Chapter V, section 9.

We separate

$$H_w = H_+ + H_- \quad \text{with} \quad H_\pm \equiv \frac{1}{2}[H_w \pm P H_w P^\dagger]\,. \tag{4.12}$$

Hence

$$P H_\pm P^\dagger = \pm H_\pm\,. \tag{4.13}$$

The matrix elements $\langle B|H_\pm|A\rangle$ are related to $\langle \bar{B}|H_\pm|\bar{A}\rangle$ by the *CPT* theorem,

$$\langle B|H_\pm|A\rangle_m{}^* = \langle B|T^{-1}TH_\pm T^{-1}T|A\rangle_m$$

$$= \langle B|TH_\pm T^{-1}|A\rangle_{-m} e^{i\theta} = \langle B|C^\dagger P^\dagger H_\pm PC|A\rangle_{-m} e^{i\theta} = \pm \langle \bar{B}|H_\pm|\bar{A}\rangle_{-m} e^{i\theta} \tag{4.14}$$

where $\theta = \delta_m(A) - \delta_m(B)$. In the expression for the lifetime of A, no pseudoscalar quantities can be formed. From arguments similar to those used in the previous chapter, we conclude that there can be no interference term between H_+ and H_-. Thus we have

$$(\text{lifetime})_A = \sum_B (|\langle B|H_+|A\rangle_m|^2 + |\langle B|H_-|A\rangle_m|^2)\cdot(\text{phase space})_B\,. \tag{4.15}$$

Equation (4.15) is clearly independent of the m value. From (4.14) we have

$$(\text{lifetime of } A) = (\text{lifetime of } \bar{A}).$$

From theorems 1 and 2, we conclude that *the equality of the masses and lifetimes of a particle and its antiparticle cannot be used as evidence for the invariance under the charge conjugation operator C. Rather, it may serve as evidence for the validity of the CPT theorem.* Indeed, as we shall see later, the operator C is not conserved, at least in some of the weak interactions.

2. General Remarks Concerning Invariance or Non-invariance Under Time Reversal

Consider the weak decay of a particle

$$A \to B \tag{4.16}$$

through the weak interaction H_w. Let H_w be written as a sum of many terms (with C_i as coupling constants),

$$H_w = \sum C_i H_i\,, \tag{4.17}$$

such that under a time reversal operation T

$$T H_w T^{-1} = \sum C_i{}^* H_i \quad \text{and} \quad T H_i T^{-1} = H_i\,. \tag{4.18}$$

Thus, if T is invariant then C_i are all real and vice versa. The proof or disproof concerning the invariance under time reversal, then, rests completely on the possibility of measuring the relative phases between these C_i. We consider first case A.

A. CASE IN WHICH THERE ARE NO STRONG INTERACTIONS BETWEEN VARIOUS DECAY PRODUCTS IN THE FINAL STATES

In this case the final state is given as

$$\psi=\sum_{\mathbf{p},\mathbf{s}}\langle\mathbf{p},\mathbf{s}\,|\,H_w\,|\,A\rangle\cdot|\,\mathbf{p},\mathbf{s}\,\rangle \tag{4.19}$$

where $|\,\mathbf{p},\mathbf{s}\,\rangle$ represents free particle states of momenta $\mathbf{p}_1$, $\mathbf{p}_2$ ... and spins $\mathbf{s}_1$, $\mathbf{s}_2$ ... in the final state B. By assumption $|\mathbf{p},\mathbf{s}\rangle$ is an eigenstate of H_s. From (4.17) we can write

$$\psi=\sum C_i M_i(\mathbf{p},\mathbf{s})\cdot|\mathbf{p},\mathbf{s}\rangle \tag{4.20}$$

where

$$M_i(\mathbf{p},\mathbf{s})\equiv\langle\mathbf{p},\mathbf{s}|H_i|A\rangle\,. \tag{4.21}$$

Using (4.18) and the property [cf. Equation (3.11)]

$$T|\mathbf{p},\mathbf{s}\rangle=|-\mathbf{p},-\mathbf{s}\rangle\,, \tag{4.22}$$

we have

$$M_i^*(\mathbf{p},\mathbf{s})=M_i(-\mathbf{p},-\mathbf{s})\,. \tag{4.23}$$

Now let us consider the measurement of an observable O which is a function of the momenta $\mathbf{p}_i$ and spins $\mathbf{s}_i$ of some of the particles in the final state. Using (4.20), we obtain

$$\langle\psi|O|\psi\rangle=\sum_{i,j}C_i^*C_jO_{ij} \tag{4.24}$$

where

$$O_{ij}\equiv\sum_{\mathbf{p},\mathbf{s}}M_i^*(\mathbf{p},\mathbf{s})M_j(\mathbf{p},\mathbf{s})\,\langle\mathbf{p},\mathbf{s}|O|\mathbf{p},\mathbf{s}\rangle. \tag{4.25}$$

It is clear that

$$O_{ij}^*=O_{ji}\,.$$

Let us separate various observables O into even and odd functions $\mathbf{p}$ and $\mathbf{s}$ (see Table 4). We denote by O_+ the even functions and O_- the odd functions. Hence

$$\langle\mathbf{p},\mathbf{s}|O_\pm|\mathbf{p},\mathbf{s}\rangle=\pm\langle-\mathbf{p},-\mathbf{s}|O_\pm|-\mathbf{p},-\mathbf{s}\rangle\,, \tag{4.26}$$

and

$$TO_\pm T^{-1}=\pm O_\pm\,. \tag{4.27}$$

Using (4.25), (4.22), and (4.27) we have

$$(O_+)_{ij}=(O_+)_{ji}=\text{real}\quad\text{and}\quad(O_-)_{ij}=-(O_-)_{ji}=\text{imaginary}. \tag{4.28}$$

Table 4

Some Examples of O_+ and O_-

O_+	O_-
$\mathbf{p}_1\cdot\mathbf{p}_2$	$\mathbf{s}\cdot(\mathbf{p}_1\times\mathbf{p}_2)$
$\mathbf{s}\cdot\mathbf{p}$	$\mathbf{p}_1\cdot(\mathbf{p}_2\times\mathbf{p}_3)$
$\mathbf{s}_1\cdot\mathbf{s}_2$	$\mathbf{s}_1\cdot(\mathbf{s}_2\times\mathbf{s}_3)$

Thus, the diagonal elements of $O_\pm$ are related to the real and imaginary parts of $C_i^*C_j$, i.e.,

$$\langle\psi|O_+|\psi\rangle=\sum(O_+)_{ij}(C_i^*C_j+C_iC_j^*) \quad \text{and} \quad \langle\psi|O_-|\psi\rangle=\sum(O_-)_{ij}(C_i^*C_j-C_iC_j^*)\,. \tag{4.29}$$

Consequently, we reach the conclusion that *if in the final state there are no strong interactions between the various decay products then the existence of any observables of the form O_- serves as a proof that H_w is not invariant under time reversal.*

Next we consider case B.

B. CASE IN WHICH THERE ARE STRONG INTERACTIONS BETWEEN VARIOUS DECAY PRODUCTS IN THE FINAL STATE

In this case $|\mathbf{p},\mathbf{s}\rangle$ is not an eigenstate of the strong Hamiltonian H_s. Let $|B\rangle$ be an eigenstate of H_s. Furthermore state $|B\rangle$ is chosen to be a stationary state. Since H_s is invariant under T, we have [cf. Equation (3.8)]

$$T|B\rangle_m=|B\rangle_{-m} \tag{4.30}$$

where $J_z=m$. For simplicity we have absorbed the phase $e^{i\delta_m}$ in the definition of $|B\rangle_{-m}$. The final state in the momentum representation is now

$$|\psi\rangle=\sum_{B,\mathbf{p},\mathbf{s}}\langle\mathbf{p},\mathbf{s}|B^{\mathrm{out}}\rangle\langle B|H_w|A\rangle\cdot|\mathbf{p},\mathbf{s}\rangle \tag{4.31}$$

where the sum also extends over various final states B. The state $|B^{\mathrm{out}}\rangle$ is the outgoing wave part of the stationary states $|B\rangle$. The state $|B^{\mathrm{out}}\rangle$ contains a phase factor $e^{i\eta_B}$ with η_B as the phase shift due to the final state interactions. Thus, by using (4.30), (4.22), and (4.18) we can write Equation (4.31) as

$$|\psi\rangle=\sum C_i e^{i\eta_B}M_{iB}(\mathbf{p},\mathbf{s})\cdot|\mathbf{p},\mathbf{s}\rangle \tag{4.32}$$

with

$$M_{iB}^*(\mathbf{p},\mathbf{s})=M_{iB}(-\mathbf{p},-\mathbf{s})\,. \tag{4.33}$$

Similar to (4.24), we have

$$\langle\psi|O|\psi\rangle=\sum C_i^*C_j e^{-i\eta_B+i\eta_{B'}}O_{iB,jB'} \tag{4.34}$$

and

$$O_{iB,jB'}=\sum_{\mathbf{p},\mathbf{s}}M_{iB}^*(\mathbf{p},\mathbf{s})M_{jB'}(\mathbf{p},\mathbf{s})\langle\mathbf{p},\mathbf{s}|O|\mathbf{p},\mathbf{s}\rangle\,. \tag{4.35}$$

Again, as in case A, we separate O into two different types, O_+ and O_-. Similar to (4.29), their matrix elements are respectively

$$\langle\psi|O_+|\psi\rangle=\sum(O_+)_{iB,jB'}[(C_i^*C_j+C_iC_j^*)\cos(\eta_B-\eta_{B'})+ i(C_i^*C_j-C_iC_j^*)\sin(\eta_B-\eta_{B'})] \tag{4.36}$$

and

$$\langle\psi|O_-|\psi\rangle=\sum(O_-)_{iB,jB'}[(C_i^*C_j-C_iC_j^*)\cos(\eta_B-\eta_{B'})+ i(C_i^*C_j+C_iC_j^*)\sin(\eta_B-\eta_{B'})]\,. \tag{4.37}$$

Also, as in the case of (4.28) we can prove

$$(O_+)_{iB,jB'}=\text{real} \quad \text{and} \quad (O_-)_{iB,jB'}=\text{imaginary}\,. \tag{4.38}$$

It is easy to see that if there are no final state interactions ($\eta_B=\eta_{B'}=\ \ldots\ =0$) then Equation (4.37) reduces to (4.29). In the present case *measurements on both O_+ and O_- may serve as tests on time reversal.* The terms $(C_i{}^*C_j-C_j{}^*C_i)\sin(\eta_B-\eta_{B'})$ in O_+ and the terms $(C_i{}^*C_j-C_j{}^*C_i)\cos(\eta_B-\eta_{B'})$ in O_- can be used to detect possible violation of time reversal invariance. As examples we list, for β decay, terms of the form $(\mathbf{p}_e\cdot\mathbf{p}_\nu)Z$, $(\mathbf{s}\cdot\mathbf{p}_e)Z$, and $\mathbf{s}\cdot(\mathbf{p}_e\times\mathbf{p}_\nu)$. (For detailed discussions see Chapter V, sections 1 to 3.)

3. Remark on Invariance Under Charge Conjugation

In this section we shall make only a general remark concerning invariance under charge conjugation. If we separate H_w into the parity conserving part H_1 and the parity nonconserving part H_2 [cf. Equation (4.12)],

$$H_w=H_1+H_2 \tag{4.39}$$

with

$$PH_1P^\dagger=H_1 \quad \text{and} \quad PH_2P^\dagger=-H_2\,. \tag{4.40}$$

Furthermore, we decompose H_1 and H_2 into the sum of various terms as in Equation (4.17),

$$H_1=\sum_i(C_1)_i(H_1)_i, \quad H_2=\sum_i(C_2)_i(H_2)_i \tag{4.41}$$

with

$$T(H_\alpha)_iT^{-1}=(H_\alpha)_i \quad (\alpha=1,2)\,. \tag{4.42}$$

Thus, *if T is invariant, $(C_1)_i$ and $(C_2)_i$ must all be real.* From the CPT theorem it follows now that *if charge conjugation C is conserved then $(C_1)_i$ must be real while $(C_2)_i$ must be imaginary.*

4. Summary

We summarize the above discussions [cf. Equations (4.36) and (4.37)].

(i) Let O_+ be a scalar and be even in $\mathbf{p}$ and $\mathbf{s}$ (e.g., $\mathbf{p}_1\cdot\mathbf{p}_2$). The observation of such a term in the form

$$\langle O_+\rangle_{\text{scalar}}=\sum O_{iB,jB'}(C_i{}^*C_j+C_iC_j{}^*)\cos(\eta_B-\eta_{B'}) \tag{4.43}$$

does not violate the invariance under P, nor under C, nor under T; while the presence of a term in the form

$$\langle O_+\rangle_{\text{scalar}}=\sum iO_{iB,jB'}(C_i{}^*C_j-C_iC_j{}^*)\sin(\eta_B-\eta_{B'}) \tag{4.44}$$

violates the invariance under C and T, but not under P [e.g., the $\mathbf{p}_e\cdot\mathbf{p}_\nu)Z$ term in Equation (5.3)].

(ii)[22] Let O_+ be a pseudoscalar but still be even in $\mathbf{p}$ and $\mathbf{s}$ (e.g., $\mathbf{s}\cdot\mathbf{p}$). The observation of such a term in the form

$$\langle O_+\rangle_{ps}=\sum O_{iB,jB'}(C_i^*C_j+C_iC_j^*)\cos(\eta_B-\eta_{B'}) \tag{4.45}$$

violates the invariance under P and C, but not under T [e.g., the Z independent term of $\mathbf{s}\cdot\mathbf{p}_e$ in Equation (5.11)]; while the presence of a term in the form

$$\langle O_+\rangle_{ps}=\sum iO_{iB,jB'}(C_i^*C_j-C_iC_j^*)\sin(\eta_B-\eta_{B'}) \tag{4.46}$$

violates the invariance under P and T, but not under C [e.g., the $(\mathbf{s}\cdot\mathbf{p}_e)Z$ term in Equation (5.11)].

(iii) Let O_- be a scalar and be odd in **p** and **s** [e.g., $\mathbf{s}\cdot(\mathbf{p}_1\times\mathbf{p}_2)$]. The observation of such a term in the form

$$\langle O_-\rangle_{\text{scalar}}=\sum O_{iB,jB'}(C_i^*C_j-C_iC_j^*)\cos(\eta_B-\eta_{B'}) \tag{4.47}$$

violates the invariance under T and C, but not under P; while the existence of a term in the form

$$\langle O_-\rangle_{\text{scalar}}=\sum iO_{iB,jB'}(C_i^*C_j+C_iC_j^*)\sin(\eta_B-\eta_{B'}) \tag{4.48}$$

does not violate the invariance under P, not under C, nor under T. A particular example of (4.48) is the existence of the usual production of polarized nuclei by a single scattering. Because of time reversal invariance it follows that such polarization is zero in the Born approximation.

(iv) Let O_- be a pseudoscalar and be odd in **p** and **s** [e.g., $\mathbf{p}_1\cdot(\mathbf{p}_2\times\mathbf{p}_3)$]. The observation of such a term in the form

$$\langle O_-\rangle_{ps}=\sum O_{iB,jB'}(C_i^*C_j-C_iC_j^*)\cos(\eta_B-\eta_{B'}) \tag{4.49}$$

violates the invariance under P and T, but not under C; while the existence of a term like

$$\langle O_-\rangle_{ps}=\sum iO_{iB,jB'}(C_i^*C_j+C_iC_j^*)\sin(\eta_B-\eta_{B'}) \tag{4.50}$$

violates the invariance under P and C, but not under T.

V. VARIOUS POSSIBLE EXPERIMENTAL TESTS[25] ON INVARIANCE UNDER *P*, *C*, AND *T* IN WEAK INTERACTIONS

1. Electron-Neutrino Correlation in β Decay

We take for the β decay Hamiltonian

$$H=\sum(\psi_p{}^{\dagger}O_i\psi_n)[C_i(\psi_e{}^{\dagger}O_i\psi_\nu)+C_i'(\psi_e{}^{\dagger}O_i\gamma_5\psi_\nu)] \tag{5.1}$$

where i runs over the usual S, V, T, A, and P types of interactions with

$$O_S=\gamma_4,\quad O_V=\gamma_4\gamma_\mu,\quad O_T=-[i/(2\sqrt{2})]\,\gamma_4(\gamma_\lambda\gamma_\mu-\gamma_\mu\gamma_\lambda),$$
$$O_A=-i\gamma_4\gamma_\mu\gamma_5,\quad\text{and}\quad O_P=\gamma_4\gamma_5\,. \tag{5.2}$$

The angular energy distribution of the electron for the allowed transition is given by[26]

$$N(W,\theta)dW\sin\theta\,d\theta=\frac{\xi}{4\pi^3}F(Z,W)\,pW(W_0-W)^2$$
$$\times\left[1+\frac{p}{W}\cos\theta\,(a+a'\frac{Ze^2}{\hbar cp})+\frac{b}{W}\right]\sin\theta\,d\theta \tag{5.3}$$

where

$$\xi=(|C_S|^2+|C_V|^2+|C_S'|^2+|C_V'|^2)|M_F|^2+$$
$$(|C_T|^2+|C_A|^2+|C_T'|^2+|C_A'|^2)|M_{GT}|^2\,, \tag{5.4}$$

$$a\xi=(1/3)(|C_T|^2-|C_A|^2+|C_T'|^2-|C_A'|^2)|M_{GT}|^2-$$
$$(|C_S|^2-|C_V|^2+|C_S'|^2-|C_V'|^2)|M_F|^2\,, \tag{5.6}$$

$$a'\xi=(i/3)(C_AC_T{}^*-C_TC_A{}^*+C_A'C_T'^*-C_T'C_A'^*)|M_{GT}|^2+$$
$$i(C_V{}^*C_S-C_S{}^*C_V+C_V'^*C_S'-C_S'^*C_V')|M_F|^2\,, \tag{5.7}$$

$$b\xi=\gamma(C_S{}^*C_V+C_SC_V{}^*+C_S'^*C_V'+C_S'C_V'^*)|M_F|^2+$$
$$\gamma(C_T{}^*C_A+C_A{}^*C_T+C_T'^*C_A'+C_A'^*C_T')|M_{GT}|^2\,. \tag{5.8}$$

---

[25] Since January 1957 a large number of experiments have been performed to test the non-conservation properties of P and C in weak decays. For a review of these numerous experimental works, see, e.g., *Proc. Seventh Ann. Rochester Conf.*, Interscience, New York, 1957.

[26] Cf. Equation (A.4) of reference 14. The existence of the term $a'Ze^2/\hbar cp$ was pointed out by M. Morita and R. Curtis.

In the above expressions, all unexplained notations are identical with the standard notations.[27] Equations (5.4) to (5.8) can be obtained directly by applying the general rules, Equation (3.21), to the corresponding "old" expressions calculated previously under the assumption that parity is conserved.

We notice that the term containing

$$a'Z(\mathbf{p}_e \cdot \mathbf{p}_\nu) \tag{5.9}$$

in (5.3) is of the form described by Equation (4.44) with the phase shifts due to the Coulomb effect ($\mathbf{p}_e$, $\mathbf{p}_\nu$ are respectively the momenta of e and ν). Thus, the presence of this term would violate the invariance under C and T.

2. Experiments with β Decay from Polarized Nuclei[14]

We consider first the experiment on angular distribution for allowed β transition from polarized nuclei. Let θ be the angle between $\mathbf{p}_e$ and direction of the spin $\mathbf{J}$ of the polarized nuclei. The angular distribution is, in general, of the form

$$1 + \alpha \cos \theta\,. \tag{5.10}$$

The corresponding expression α for $\mathbf{J} \rightarrow \mathbf{J} - 1$(no) transition is

$$\alpha = \beta \frac{\langle J_z \rangle}{J}$$

with

$$\beta = Re\left[\pm (C_T C_T'^* - C_A C_A'^*) + i\frac{Ze^2}{\hbar c p}(C_A C_T'^* + C_A' C_T^*)\right] \frac{v_e}{c} \frac{2}{\xi + (\xi b/W)} |M_{GT}|^2\,. \tag{5.11}$$

For $J \rightarrow J+1$(no), α is given by

$$\alpha = -\beta J_z/(J+1)\,. \tag{5.12}$$

For $J \rightarrow J$(no), the corresponding α is[28]

$$\alpha = \beta \frac{J_z}{J(J+1)} + \beta' \frac{J_z}{\sqrt{J(J+1)}} \tag{5.13}$$

where

$$\beta' = Re\Big[(C_S^* C_T' + C_S'^* C_T - C_V^* C_A' - C_V'^* C_A) \pm$$

$$i\frac{Ze^2}{\hbar c p}(C_S^* C_A' + C_S'^* C_A - C_V^* C_T' - C_V'^* C_T)\Big] \frac{v_e}{c} \frac{2 M_F M_{GT}}{\xi + (\xi b/W)}\,.$$

---

[27] See, e.g. , the article by M.E. Rose in *Beta- and Gamma-Ray Spectroscopy*, Interscience, New York, 1955, pp. 271-91.

[28] Equation (5.13) has been independently calculated by many authors: M. Morita (private communication); R. Curtis and R. Lewis (private communication); J.D. Jackson, S.B. Treiman, and H.W. Wyld (*Phys. Rev.*, in press). In deriving Equation (5.13) we assume that the strong interactions are invariant under time reversal. Consequently the nuclear matrix elements M_F and M_{GT} are real quantities. If the strong interactions are not invariant under time reversal then the product ($M_F M_{GT}$) should be replaced by ($M_F^* M_{GT}$) and set inside the square brackets together with the coupling constants. It may be emphasized that the condition $\beta' = 0$ can be used as evidence of invariance under time reversal for both the weak and the strong interactions.

In Equations (5.11) to (5.13) the upper signs are for e^- emission and the lower signs for e^+ emission.

The detection of $\alpha \neq 0$ gives definite proof that P is not conserved. We notice further that in the expression for β [Equation (5.11)] there are two different terms of the forms of $\mathbf{J} \cdot \mathbf{p}_e$ and $Z\mathbf{J} \cdot \mathbf{p}_e$. Comparing with Equations (4.45) and (4.46) we see that the presence of the first term violates the invariance of P and C while the presence of the second term violates the invariance of P and T.

The first conclusive experimental evidence on the nonconservation of parity in weak interactions was done by Wu, Ambler, Hayward, Hoppes, and Hudson[1] using polarized Co^{60}. The decay scheme for Co^{60} is

$$Co^{60} \rightarrow Ni^{60} + e^- + \bar{\nu} \tag{5.14}$$

with $J=5 \rightarrow J=4$(no) and $v_e/c \cong 0.65$. Wu et al. obtained for the asymmetry parameter β,

$$\beta \cong -0.7 \,. \tag{5.15}$$

Since from the He^6 recoil experiment[29] we know that

$$\frac{|C_A|^2 + |C_A'|^2}{|C_T|^2 + |C_T'|^2} < \frac{1}{3} \tag{5.16}$$

and since for Ni^{60}

$$Ze^2/\hbar c = 0.2 \,,$$

the second term in the expression for β [Equation (5.11)] has an upper limit

$$\left| \frac{2(v_e/c) Re[i(Ze^2/\hbar c p)(C_A C_T'^* + C_A' C_T^*)]}{|C_A|^2 + |C_A'|^2 + |C_T|^2 + |C_T'|^2} \right| < 0.23 \,. \tag{5.17}$$

Thus from the observed magnitude [Equation (5.15)] we conclude that *in β decay both C and P are not conserved.* By more accurate measurement to study the Z dependence oı p dependence of the asymmetry parameter β and β' one can obtain information concerning time reversal invariance.

3. Other β Decay Experiments

There are other experiments in β decay which can serve as tests for possible non-invariance under P, C, and T. We list the following:

A. β-γ CORRELATION AND THE CIRCULAR POLARIZATION OF THE γ-RAY[14]

Consider a successive β,γ decay scheme

$$A \rightarrow B^* + e^{\pm} + \nu \quad \text{and} \quad B^* \rightarrow B + \gamma \tag{5.18}$$

[29] B.M. Rustad and S.L. Ruby, *Phys. Rev.* **97**, 991 (1955).

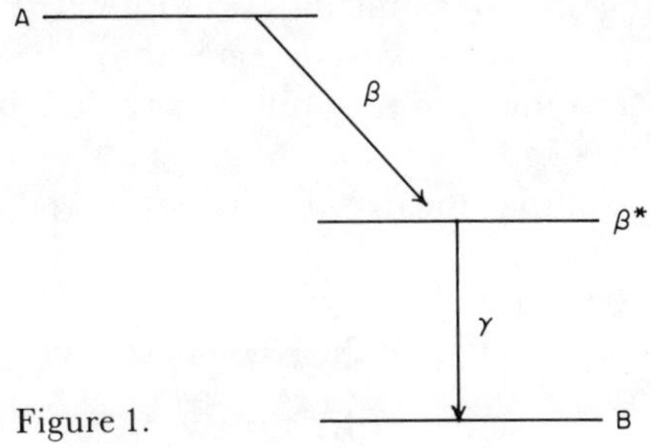

Figure 1.

in which the initial nucleus A is not polarized (see Figure 1). Because of the nonconservation of parity, the intermediate nucleus B^* will be polarized along the momentum $\mathbf{p}_e$ of the β-ray. The polarization of the B^* can then be detected by measuring the direction of the γ-ray with respect to $\mathbf{p}_e$ and the state of circular polarization of the γ-ray. The magnitude of the polarization of B^* can be easily deduced from Equation (5.10).

B. POLARIZATION OF $e^{\pm}$

By measuring the longitudinal polarization of the electron $(\boldsymbol{\sigma}\cdot\mathbf{p}_e)$ from unpolarized nuclei one can also obtain a test of parity conservation.[30] By a detailed study of its possible Z dependence it is also possible to obtain information on time reversal invariance. (See Chapter VI, section 2, for a more detailed discussion.)

C. POLARIZED NUCLEI AND RECOIL EXPERIMENTS[31]

From the experiments of Wu et al., it is proven that parity and charge conjugation are not conserved in β decay. The next important question is, of course, on the invariance under time reversal T. (By CPT theorem this is equivalent to the invariance under $C\cdot P$.) In the above mentioned experiments, by a careful study of Z dependent terms such as

$$(\mathbf{J}\cdot\mathbf{p}_e)Z \quad \text{and} \quad (\mathbf{p}_e\cdot\mathbf{p}_\nu)Z\,, \tag{5.19}$$

information on T invariance can be obtained. While their presence would prove that we do not have invariance under time reversal, their absence may be due to other reasons. For example, in Equations (5.7) and (5.11), if

$$C_A=C_A'=0 \quad \text{and} \quad C_V=C_V'=0$$

[30] T.D. Lee and C.N. Yang, *Phys. Rev.* **105**, 1671 (1957); L. Landau, *Nuclear Phys.* **3**, 127 (1957); J.D. Jackson, S.B. Treiman, and H.W. Wyld, *Phys. Rev.* **106**, 517 (1957); R.B. Curtis and R.R. Lewis, *Phys. Rev.* **107**, 543 (1957). For a summary of the various measurements on the longitudinal polarization of β-particles see, e.g., *Proc. Seventh Ann. Rochester Conf.*, Interscience, New York 1957.

[31] J.D. Jackson, S.B. Treiman, and H.W. Wyld, *Phys. Rev.* **106**, 517 (1957).

then these terms [Equation (5.19)] would also vanish. A critical test is then to measure possible time reversal nonconservation terms that are due to the interference terms between C_S and C_T. One such possibility has been pointed out by Jackson, Trieman, and Wyld.[31]

Let us consider, for example, the neutron decay from polarized neutrons

$$n \to p+e^-+\bar{\nu}\,. \tag{5.20}$$

A simultaneous measurement of $\mathbf{p}_e$ and the recoil proton can give a measurement of

$$\sigma_n \cdot (\mathbf{p}_e \times \mathbf{p}_\nu)\,. \tag{5.21}$$

From the previous general argument, Equation (4.47), we see that *the Z independent term of this quantity violates the invariance under T and C.* The coefficient of this term is proportional to

$$Im(C_S C_T^* - C_V C_A^* + C_S' C_T'^* - C_V' C_A'^*)\,. \tag{5.22}$$

4. π Decay[14,23]

Consider the reaction for π^+ decay

$$\pi^+ \to \mu^+ + \nu\,. \tag{5.23}$$

The parity nonconservation for this reaction can be established by measuring the longitudinal polarization of the μ-meson. If the μ-meson has spin ½, its polarization state is described by a density matrix

$$1+A_+ \boldsymbol{\sigma}_\mu \cdot \hat{\mathbf{p}}_\mu \tag{5.24}$$

where $\hat{\mathbf{p}}_\mu$ is a unit vector along the momentum of μ. Assuming that μ and ν have no strong interactions between them, from Equation (4.45) we see that *the presence of A_+ violates the conservation laws of P and C.*

The density matrix for the polarization of the μ^- from the corresponding π^- decay,

$$\pi^- \to \mu^- + \bar{\nu}\,, \tag{5.25}$$

is

$$1+A_- \boldsymbol{\sigma}_\mu \cdot \hat{\mathbf{p}}_\mu\,. \tag{5.26}$$

By *CPT* theorem, the final result must be invariant under the operation of *C·P·T*. Since there are no final state interactions, we see that

$$A_+ = -A_-\,. \tag{5.27}$$

5. μ Decay[14,23]

The easiest way to analyze the spin state of the $\mu^\pm$-meson from π decay is to use the possible parity nonconservation terms in μ decay

$$\mu^\pm \to e^\pm + \nu + \bar{\nu}. \tag{5.28}$$

If the μ-meson is polarized with spin $\boldsymbol{\sigma}_\mu$, the angular distribution of $e^\pm$ would be of the form

$$1+B_\pm \boldsymbol{\sigma}_\mu \cdot \hat{\mathbf{p}}_e \tag{5.29}$$

where $\hat{\mathbf{p}}_e$ is a unit vector along the momentum of e. Similar to (5.27), we have [32]

$$B_+ = -B_- . \tag{5.30}$$

Combining (5.24), (5.26), and (5.29), in the decays

$$\pi^\pm \rightarrow \mu^\pm \rightarrow e^\pm$$

the angular correlation between $\mathbf{p}_\mu$ and $\mathbf{p}_e$ is of the form

$$1+\alpha_\pm \hat{\mathbf{p}}_\mu \cdot \hat{\mathbf{p}}_e \tag{5.31}$$

with $\mathbf{p}_\mu$ measured in the rest system of $\pi^\pm$ and $\mathbf{p}_e$ in the rest system of $\mu^\pm$. Furthermore, from (5.27) and (5.30), we have

$$\alpha_+ = \alpha_- . \tag{5.32}$$

The observation of $\alpha_\pm$ *proves that P and C are not conserved in both π decay and μ decay.*

Experiments[2] on π-μ-e decays give for α_+

$$\alpha_+ = -0.26 \pm 0.02 \tag{5.33}$$

with μ^+ stopped in carbon and

$$\alpha_+ \cong -0.16 \tag{5.34}$$

with μ^+ stopped in emulsion. The difference between these two numbers is, of course, due to the large depolarization effect in emulsion.

The longitudinal polarization of the μ-meson from π decay offers a natural possibility for measuring the magnetic moment of the μ-meson. This was first measured by Garwin et al.[2] Their result gives a g value[33] for μ^+,

$$g = +2.00 \pm 0.10 . \tag{5.35}$$

This value strongly indicates that the spin of μ is ½. A more accurate measurement of this is of particular interest because it may give a very severe test of validity electrodynamics at a much smaller distance ($\sim 10^{-13}$ cm) than that tested by previous experiments with electrons.

6. $K^\pm$ Decay

We consider first the decay of $K_{\mu 2}$

$$K^\pm \rightarrow \mu^\pm + \nu \text{ (or } \bar{\nu}) . \tag{5.36}$$

As in the π-μ-e decay, one would expect that here again the μ-meson could be polarized along its direction of motion, and then the μ-e decay would be an analyzer

[32] If in the μ^+ decay two neutrinos are emitted, $\mu^+ \rightarrow e^+ + 2\nu$, then the corresponding μ^- decay is $\mu^- \rightarrow e^- + 2\bar{\nu}$. In this case Equation (5.30) is still correct.

[33] More recently T. Coffin et. al. [*Phys. Rev.* **106**, 1108 (1957)] measured the g value for μ^+ using the magnetic resonance technique. They found $g_{\mu^+} = +2.0064 \pm 0.0048$.

of the polarization through measurements of the distribution of $\mathbf{p}_\mu \cdot \mathbf{p}_e$. We will show later, in a special two-component theory of the neutrino, that this electron distribution is identical with the distribution from the π-μ-e decay if the K-meson spin is zero. Therefore in this special theory the K-μ-e decay distribution may offer a way to obtain some information about the spin of the K-meson[23] (cf. Chapter VII, section 1).

Next we consider the decays of $K_{\pi 2}$ and $K_{\pi 3}$:

$$K^+ \rightarrow \pi^+ + \pi^0 \tag{5.37}$$

and

$$K^+ \rightarrow \pi^- + 2\pi^+ . \tag{5.38}$$

In the decay of $K_{\pi 2}$ only one independent momentum can be measured in the rest system of K, and in the decay of $K_{\pi 3}$ only two independent momenta can be measured. In neither case can one form a pseudoscalar quantity out of these observed momenta. Thus, if the K-meson has zero spin (or if it is unpolarized) it is impossible to observe any interference term between the even and odd states in the decay of $K_{\pi 2}$ and $K_{\pi 3}$. *The strongest evidence for nonconservation of parity in this case is precisely the present θ-τ problem*, namely, *the K-meson can decay into a 2π system and a 3π system with the same lifetime and same mass.*

The determination of the spin-parity of the K-meson through the angular energy distribution of 3π mode (Dalitz's analysis) can still be used. The distribution function is now given by

$$[|\psi_{J,+}(\mathbf{k},\mathbf{p})|^2 + \alpha|\psi_{J,-}(\mathbf{k},\mathbf{p})|^2]\sqrt{\varepsilon(1-\varepsilon)}\; d\varepsilon \; d(\cos\theta) \tag{5.39}$$

where $\psi_{J,+}$ and $\psi_{J,-}$ are the wave functions for a 3π system with total angular momentum J and parity $+1$ and -1 respectively [cf. Equation (2.6)]. The constant α is a real positive number. It is to be expected that $J=0$ or 2 would still be favorable.

7. Λ^0 Decay and Σ Decay

Information concerning parity nonconservation can be obtained by studying the decay of hyperons.[14] We consider as an example the following reactions:

$$\pi^- + p \rightarrow \Sigma^- + K^+ \tag{5.40}$$

and

$$\Sigma^- \rightarrow n + \pi^- . \tag{5.41}$$

The first reaction (5.40) can be thought of as the polarizer of Σ^-. Since it is a strong interaction, in order to produce a polarized Σ^- it is necessary to measure the direction of two momenta, say $\mathbf{p}_{\text{in}}$, the momentum of the incoming pion, and $\mathbf{p}_\Sigma$, the momentum of Σ^-. The spin $\mathbf{s}_\Sigma$ of the Σ^- then will be polarized along the direction $(\mathbf{p}_{\text{in}} \times \mathbf{p}_\Sigma)$. In the subsequent decay (5.41), if in the angular distribution of $\mathbf{p}_{\text{out}}$ (the momentum of the decay pion in the rest system of Σ^-) a term of the form $\mathbf{s}_\Sigma \cdot \mathbf{p}_{\text{out}}$ is observed, then parity is not conserved. Because what one measures is a quantity of the form $\mathbf{p}_{\text{out}} \cdot (\mathbf{p}_{\text{in}} \times \mathbf{p}_\Sigma)$, it is necessary to exclude cases where $\mathbf{p}_{\text{in}}$ is parallel to $\mathbf{p}_\Sigma$.

We shall give a more detailed consideration of processes (5.40) and (5.41) by further assuming that (i) the spin of Σ is ½, (ii) the spin of K is zero, and (iii) in the production process only s and p waves are important.[34] Under these assumptions, the differential production cross section per unit solid angle $d\Omega$ (in the center-of-mass system of production) of the Σ^- produced is given by

$$I(\theta)=|a+b\cos\theta|^2+|c|^2\sin^2\theta \tag{5.42}$$

where $\theta=\angle(\mathbf{p}_{in},\mathbf{p}_\Sigma)$.

The corresponding polarization $P(\theta)$ is

$$P(\theta)=[I(\theta)]^{-1}2\sin\theta\cdot Im[c^*(a+b\cos\theta)] \tag{5.43}$$

where $P(\theta)$ is defined to be the average spin of the Σ^- in units of $(½)\hbar$. It is important to notice that if $I(\theta)=(1\pm\cos\theta)^2$ then $P(\theta)=0$ at all angles.

Because of the possible nonconservation of parity in the decay process (5.41) the polarization $P(\theta)$ can be measured. Let R be the projection of $\mathbf{p}_{out}$ in the direction of $\mathbf{p}_{in}\times\mathbf{p}_\Sigma$. The distribution function for R at an angle θ of production is given by

$$W(\theta,\xi)d\Omega d\xi=I(\theta)d\Omega\cdot(½)\,[1+\alpha p(\theta)\xi]\alpha\xi \tag{5.44}$$

where

$$\xi=R/(\text{maximum value of } R)\cong R/(100\text{ Mev/c})\,.$$

In terms of the coefficients a, b, and c, defined in (5.42), $W(\theta,\xi)$ can be written as

$$W(\theta,\xi)d\xi d\Omega=(|a+b\cos\theta|^2+|c|^2\sin^2\theta)(½)d\xi d\Omega+$$
$$\alpha\sin\theta\, Im[c^*(a+b\cos\theta)]\xi d\xi d\Omega\,. \tag{5.45}$$

The existence of a nonvanishing α constitutes an unambiguous proof of parity nonconservation in Σ^- decay. In such a case the final state of $(n+\pi^-)$ in process (5.41) is a mixture of $s_{1/2}$ and $p_{1/2}$ states with amplitudes, say, A and B respectively. The asymmetry parameter α is related to these amplitudes by

$$\alpha=\frac{2Re(A^*B)}{|A|^2+|B|^2}\,. \tag{5.46}$$

If time reversal leaves invariant the decay process of Σ^- then (cf. Chapter IV, section 2)

$$\alpha=\pm\frac{2|A|\cdot|B|}{|A|^2+|B|^2}\cos(\delta_p-\delta_s)\,, \tag{5.47}$$

where δ_p and δ_s are, respectively, the phase shifts of $(n+\pi^-)$ scattering in the $p_{1/2}$ and $s_{1/2}$ states at about 117 Mev in their center-of-mass system. If the decay interaction is invariant under charge conjugation then (cf. Chapter IV, section 3)

$$\alpha=\pm\frac{2|A|\cdot|B|}{|A|^2+|B|^2}\sin(\delta_p-\delta_s)\,. \tag{5.48}$$

[34] T.D. Lee, J. Steinberger, G. Feinberg, P.K. Kabir, and C.N. Yang, *Phys. Rev.* **106**, 1367 (1957).

Thus a large observed value of α means that both C and P are not conserved in the decay of Σ^-. Similar considerations can easily be applied to the productions and decays of other hyperons.

8. Ξ Decay

Another example of a possible test of parity nonconservation is through the decay of the Ξ-particle.[14]

$$\Xi^- \rightarrow \Lambda^0 + \pi^- \tag{5.49}$$

and

$$\Lambda^0 \rightarrow p + \pi^- . \tag{5.50}$$

As in the case of π-μ-e decay, if parity is not conserved in both (5.49) and (5.50), then the distribution of the momentum of the Λ^0, $\mathbf{p}_\Lambda$, and that of the pion, $\mathbf{p}_\pi$, may contain odd powers of

$$(\mathbf{p}_\Lambda \cdot \mathbf{p}_\pi) \tag{5.51}$$

where $\mathbf{p}_\pi$ is measured in the rest system of Λ^0 and $\mathbf{p}_\Lambda$ in the rest system of Ξ^-.

We shall illustrate the calculation of such a distribution by considering the special case that the spins of Λ^0 and Ξ^- are both ½. Let ϕ_Ξ be the initial spin state of Ξ^- at rest. The wave function of the corresponding Λ^0 in the decay (5.49) can be written as

$$\phi_\Lambda = M_\Lambda \cdot \phi_\Xi \tag{5.52}$$

where

$$M_\Lambda = A + B\boldsymbol{\sigma} \cdot \hat{\mathbf{p}}_\Lambda$$

with $\hat{\mathbf{p}}_\Lambda$ a unit vector along $\mathbf{p}_\Lambda$ and A, B the relative probability amplitudes of the two final states of opposite parity ($|A|^2 + |B|^2 = 1$). The subsequent wave function of p in (5.50) is

$$\phi_p = M_p \cdot \phi_\Lambda \tag{5.53}$$

where

$$M_p = a + b\boldsymbol{\sigma} \cdot \hat{\mathbf{p}}_\pi , \quad (|a|^2 + |b|^2 = 1)$$

with $\hat{\mathbf{p}}_\pi$ a unit vector along $\mathbf{p}_\pi$.

The final distribution of $\mathbf{p}_\pi$ and $\mathbf{p}_\Lambda$ for an unpolarized Ξ is

$$(½)\sum |\phi_p|^2 = (½)\,\mathrm{tr.}\,[M_\Lambda{}^\dagger M_p{}^\dagger M_p M_\Lambda] = 1 + \alpha \cos\theta \tag{5.54}$$

where

$$\theta = \angle\, (\mathbf{p}_\Lambda, \mathbf{p}_\pi) \quad \text{and} \quad \alpha = (a^* b + b^* a)\,(A^* B + B^* A) .$$

In Equation (5.54), the sum extends over the two initial spin states of ϕ_Ξ $(=\uparrow,\downarrow)$. Thus the observation of a nonvanishing value for α shows that P is not conserved in both Ξ^- decay and Λ^0 decay.

9. K^0, $\bar{K}^0$ Decay[22]

As a final example we consider the decay of neutral K-mesons. The existence of two neutral K-particles of different lifetimes and many of their properties were predicted and discussed[35] under the assumption that the decay of the K-meson is strictly invariant under charge conjugation C. Although there is as yet no explicit proof that C is not conserved in the decay of the K-particle, the various recent experiments[25] on parity and charge conjugation nonconservations in other weak interactions do give a strong indication that C is probably not invariant in K decay. As we shall see, the curious behavior of K^0 and $\bar{K}^0$ turns out to be remarkably insensitive to any possible nonconservation of P, C, or T. From the strong production processes we know that there must exist two different states, K^0 and $\bar{K}^0$, of opposite strangeness quantum number [Equation (1.27)]. Thus, independent of any assumption about the invariance or non-invariance properties, there should exist, in general, two lifetimes for their decay. We shall discuss the K-$\bar{K}$ decay in some detail under the following two possibilities.

A. THE DECAY PROCESSES ARE INVARIANT UNDER T

First we consider the various consequences if under time reversal T the decay processes are invariant. It follows, then, from the CPT theorem that $C{\cdot}P$ is conserved, although C may not be conserved. We define the states $|K_1\rangle$ and $|K_2\rangle$ by

$$|K_1\rangle=(1/\sqrt{2})\,(|K^0\rangle+C{\cdot}P|K^0\rangle) \quad \text{and} \quad |K_2\rangle=(1/\sqrt{2})\,(|K^0\rangle-C{\cdot}P|K^0\rangle). \tag{5.55}$$

Thus, $|K_1\rangle$ and $|K_2\rangle$ are eigenstates of $C{\cdot}P$ with eigenvalues $+1$ and -1 respectively. If $C{\cdot}P$ is strictly invariant, $|K_1\rangle$ and $|K_2\rangle$ each will decay exponentially in time. The lifetime of $|K_1\rangle$ is, in general, different from that of $|K_2\rangle$. Furthermore

$$K_1{}^0 \rightarrow \pi^+ + \pi^- \tag{5.56}$$

in addition to its other possible modes of decay. [For example, if $(\text{spin})_{K^0}=0$, 2, 4 . . . , then $K_1{}^0$ also $\rightarrow 2\pi^0$.] On the other hand $K_2{}^0$ cannot decay into a 2π system. This follows from the fact that a 2π system with total spin J is *always* an eigenstate of $C{\cdot}P$ with eigenvalue $+1$. From phase space arguments, it is expected that $K_1{}^0$ has a shorter lifetime.

In order to detect any difference between the present case and the case when C is conserved it is necessary to measure a pseudoscalar quantity. For example, in the decay of the long-lived K-particle, $K_2{}^0$,

$$K_2{}^0 \rightarrow \pi^+ + e^- + \bar{\nu}\,, \tag{5.57}$$

the electron may be longitudinally polarized and its polarization $(\boldsymbol{\sigma}\cdot\mathbf{p})_{e^-}$ may be measured. The corresponding quantity $(\boldsymbol{\sigma}\cdot\mathbf{p})_{e^+}$ for the positron in the decay,

$$K_2{}^0 \rightarrow \pi^- + e^+ + \nu\,, \tag{5.58}$$

is expected to be of opposite sign from that of $(\boldsymbol{\sigma}\cdot\mathbf{p})_{e^-}$. However in a measurement which does not involve a pseudoscalar quantity, the nonconservation of C can not be tested. For example, let r be the branching ratio

[35] M. Gell-Mann and A. Pais, *Phys. Rev.* **97**, 1387 (1955).

$$r \equiv \frac{\text{rate of } K_2^0 \to \pi^+ + e^- + \bar{\nu}}{\text{rate of } K_2^0 \to \pi^- + e^+ + \nu}. \tag{5.59}$$

We should expect $r=1$ in the present case.

B. K^0 AND $\bar{K}^0$ DECAY PROCESSES MAY NOT BE INVARIANT UNDER T

Next we consider the more general case that under time reversal T the decay processes may not be invariant. We define $|\bar{K}^0\rangle$ to be

$$|\bar{K}^0\rangle = C \cdot |K^0\rangle \tag{5.60}$$

where the charge conjugation operator C is defined through the strong interaction. Suppose at $t=0$ a K-particle is produced. At a later time its wave function can be described as

$$\psi(t) = a(t)|K^0\rangle + b(t)|\bar{K}^0\rangle\,, \tag{5.61}$$

or simply as

$$\psi(t) = \begin{pmatrix} a(t) \\ b(t) \end{pmatrix}. \tag{5.62}$$

The differential equation for $\psi(t)$ can be written as

$$-d\psi/dt = (\Gamma + iM)\psi\,, \tag{5.63}$$

where

$$\Gamma_{11} = \Gamma_{22} = \sum_j \Gamma_{aj} = \sum_j \Gamma_{bj}\,, \quad \Gamma_{12} = \Gamma_{21}{}^* = \sum_j (\Gamma_{aj}\Gamma_{bj})^{1/2} e^{i\delta_j} \tag{5.64}$$

with

$\Gamma_{aj} = 2\pi |H_{aj}|^2 \times$(density of states per unit energy);

$\Gamma_{bj} = 2\pi |H_{bj}|^2 \times$(density of states per unit energy); and

$e^{i\delta_j} =$ phase of (H_{aj}/H_{bj});

where H_{aj} and H_{bj} are the matrix elements of the Hamiltonian for the decay processes, a and b refer to states $|K^0\rangle$ and $|\bar{K}^0\rangle$ respectively, and j refers to any possible decay state (consisting of standing waves)that is an eigenstate of H_{strong}. Thus state $|j\rangle$ has a definite spin, charge, and parity. That $\Gamma_{11} = \Gamma_{22}$ follows from

$$H_{aj}{}^* = \pm H_{b\bar{j}}\,, \tag{5.65}$$

with $|\bar{j}\rangle = C|j\rangle$. Equation (5.65) is an immediate consequence of the CPT theorem [cf. Equation (4.14)].

Similarly, we have for the mass operator M,

$$M^\dagger = M \quad \text{and} \quad M_{11} = M_{22}\,. \tag{5.66}$$

Equation (5.63) can now be readily solved. Its eigenstates, defined by

$$(\Gamma + iM)\psi_\pm = \lambda_\pm \psi_\pm\,,$$

are

$$\psi_\pm = \begin{pmatrix} p \\ \pm q \end{pmatrix} (|p|^2 + |q|^2)^{-1/2}\,, \tag{5.67}$$

with the corresponding time constants

$$\lambda_\pm = \Gamma_{11} + iM_{11} \pm (pq)\,, \tag{5.68}$$

where p and q are two complex numbers given by

$$p^2 = \Gamma_{12} + iM_{12} \quad \text{and} \quad q^2 = \Gamma_{21} + iM_{21} = \Gamma_{12}{}^* + iM_{12}{}^*\,. \tag{5.69}$$

If at $t=0$ a K^0-particle is produced, then at a later time the state function ψ can be expressed in terms of these two eigenstates $\psi_\pm$ as

$$\psi(t) = [1/(2p)](|p|^2 + |q|^2)^{1/2}(\psi_+ e^{-\lambda_+ t/2} + \psi_- e^{-\lambda_- t/2})\,. \tag{5.70}$$

It is convenient to separate the real and imaginary parts of $\lambda_\pm$. Without loss of generality we may write

$$\lambda_+ = \gamma_+ \quad \text{and} \quad \lambda_- = \gamma_- + 2i\Delta\,, \tag{5.71}$$

where γ_+, γ_- are two real numbers representing the reciprocal lifetimes of the short-lived ones and the long-lived ones respectively, and Δ is the mass difference between these two eigenstates. One notices that *these two eigenstates* ψ_+ *and* ψ_- *do not in general represent the states* K_1 *and* K_2 introduced in (5.55). *In fact they may not be orthogonal to each other.*

The fractional number of K-mesons that decay at time t after production is given by

$$N(t)dt = -d(\psi^\dagger\psi)\,. \tag{5.72}$$

Using (5.63) one easily shows that

$$-\frac{d}{dt}(\psi^\dagger\psi) = \psi^\dagger\Gamma\psi\,.$$

By using Equations (5.67) to (5.70), Equation (5.72) becomes

$$N(t) = (\tfrac{1}{2})(1+\alpha)^{-1}\{\gamma_+ e^{-\gamma_+ t} + \gamma_- e^{-\gamma_- t} + \alpha e^{-(\gamma_+ + \gamma_-)t/2}[(\gamma_+ + \gamma_-)\cos\Delta t - 2\Delta\sin\Delta t]\}\,, \tag{5.73}$$

where

$$\alpha = \psi_+{}^\dagger\psi_- = (|p|^2 - |q|^2)(|p|^2 + |q|^2)^{-1} \tag{5.74}$$

is a real number representing the non-orthogonality of these two eigenstates. The four real numbers γ_+, γ_-, Δ, and α characterize the decay of the K-particle. They satisfy the inequalities

$$\gamma_\pm \geqslant 0 \quad \text{and} \quad \alpha^2 \leqslant \frac{4\gamma_+\gamma_-}{(\gamma_+ + \gamma_-)^2 + 4\Delta^2} \tag{5.75}$$

which follow from the fact that Γ is a positive Hermitian matrix. These conditions also insure that $N(t) \geqslant 0$ for all t.

Experimentally $N(t)$ is measurable. From $N(t)$ one can in principle determine all four constants γ_+, γ_-, Δ, and α. Indications from existing experiments[36] show that probably $(\gamma_+/\gamma_-) > 100$. Equation (5.75) then shows that $\alpha^2 < 4(\gamma_-/\gamma_+) < 0.04$.

The above discussion also leads easily to a determination of the branching ratio of the long-lived component (and the short-lived component) into the various decay

[36] K. Lande, E.T. Booth, J. Impeduglia, and L.M. Lederman, *Phys. Rev.* **103**, 1901 (1956).

modes. If time reversal is invariant the long-lived component is an eigenstate of $C\cdot P$. As discussed in part A above, its decay into charge conjugate channels such as $\pi^+ + e^- + \bar{\nu}$ and $\pi^- + e^+ + \nu$ must be equally probable. If time reversal is not strictly invariant, decays into $\pi^+ + e^- + \bar{\nu}$ and $\pi^- + e^+ + \nu$ may have different probabilities for the long-lived component.

We consider first the following decay channel of the K-particle:

$$K^0 \to e^- + \pi^+ + \bar{\nu}\,. \tag{5.76}$$

The final product may be in states with either parity $= +1$ or parity $= -1$. Let us denote the matrix elements for the decay process into these two types of states by f_1 and f_2. Similarly, we denote the matrix elements for

$$K^0 \to e^+ + \pi^- + \nu \tag{5.77}$$

with the final state having parity $= +1$ and parity $= -1$ by g_1 and g_2.

By using the CPT theorem and Equation (4.14), the corresponding matrix elements for the decay of $\bar{K}$,

$$\bar{K}^0 \to e^+ + \pi^- + \nu\,, \tag{5.78}$$

are related to that of (5.76). These elements are f_1^* and $-f_2^*$. Similarly the matrix elements for

$$\bar{K}^0 \to e^- + \pi^+ + \bar{\nu} \tag{5.79}$$

are g_1^* and $-g_2^*$. Let ψ_+ represent the long-lived component K_+ of the K-particle. The matrix elements for the decay of K_+,

$$K_+ \to e^- + \pi^+ + \bar{\nu}\,, \tag{5.80}$$

into the two different final parity states are proportional to $pf_1 + qg_1^*$ and $pf_2 - qg_2^*$, respectively, while the corresponding elements for

$$K_+ \to e^+ + \pi^- + \nu \tag{5.81}$$

are proportional to $pg_1 + qf_1^*$ and $pg_2 - qf_2^*$. The branching ratio r for the decay of K_+ into $e^- + \pi^+ + \bar{\nu}$ and $e^+ + \pi^- + \nu$ is therefore,

$$r = \frac{|pf_1 + qg_1^*|^2 + |pf_2 - qg_2^*|^2}{|pg_1 + qf_1^*|^2 + |pg_2 - qf_2^*|^2}\,. \tag{5.82}$$

A detection of $\gamma \neq 0$ would establish the non-invariance of K^0 decay under time reversal. However, from (5.82), we see that if $|p| = |q|$ then $r = 0$. Also, if $\alpha = 0$, the two eigenstates ψ_+, ψ_- are orthogonal and $|p| = |q|$. (This is the case if the mass operator M is negligible.) In this case the decays of long-lived components into charge conjugate channels such as $\pi^+ + e^- + \nu$ and $\pi^- + e^+ + \bar{\nu}$ are equally probable. Furthermore we recall that because the strong interaction conserves I_z, the behavior of $|K^0\rangle$ under a charge conjugation operator cannot be determined within a phase factor $e^{iI_z\theta}$. If ψ_+ is orthogonal to ψ_-, by choosing this phase factor to be that of (p/q) these two eigenstates ψ_+ and ψ_- can be made to be identical with $|K_1\rangle$ and $|K_2\rangle$, Equation (5.55). From experimental results on the two lifetimes of ψ_+ and ψ_- we know that $\alpha^2 < 0.04$. Thus the branching ratio r may be quite small even though time reversal may not be conserved.

As a final remark we note that because of the largeness of phase space volume for 2π decay it might be expected that in the Γ matrix [Equation (5.63)] only the matrix elements due to the decay into a certain 2π mode are of dominant importance. If in the calculation of the mass operator M the virtual processes via the same 2π mode give also the dominant contributions, then we should expect without further assumptions that the lifetime ratio γ_+/γ_- should be large; that the short-lived one ψ_- and the long-lived one ψ_+ should be almost identical with K_1 and K_2 respectively; and that ψ_- should decay mostly into that 2π mode while ψ_+ should decay dominantly to other modes such as $\pi^+ + e^- + \bar{\nu}$, etc.

VI. A TWO-COMPONENT THEORY OF THE NEUTRINO[37]

The various experimental results on nonconservation of parity and charge conjugation in β decay, π decay, and μ decay can be expressed in a particularly simple and appealing way by using a two-component theory of the neutrino.

1. The Neutrino Field

Consider first the Dirac equation for a free spin ½ particle with zero mass. Because of the absence of the mass term one needs only three anticommuting Hermitian matrices. Thus the neutrino can be represented by a spinor function ϕ_ν which has only two components. The Dirac equation for ϕ_ν can be written as ($\hbar=c=1$)

$$(\boldsymbol{\sigma}\cdot\mathbf{p})\phi_\nu=i\dot{\phi}_\nu \tag{6.1}$$

where σ_1, σ_2, σ_3 are the usual 2×2 Pauli matrices. The relativistic invariance of this equation for proper Lorentz transformations (i.e., Lorentz transformations without space inversion and time inversion) is well known. In particular for the space rotations through an angle θ around, say, the z axis, the wave function transforms in the following way:

$$\phi_\nu \rightarrow \exp(-i\sigma_3\theta/2)\,\phi_\nu\,.$$

The σ matrices are therefore the spin matrices for the neutrino. For a state with a definite momentum $\mathbf{p}$, the energy and the spin along $\mathbf{p}$ (defined to be the helicity $\mathcal{H}$) are given respectively by

$$H=(\boldsymbol{\sigma}\cdot\mathbf{p}) \quad \text{and} \quad \mathcal{H}=(\boldsymbol{\sigma}\cdot\mathbf{p})/|\mathbf{p}|\,. \tag{6.2}$$

They are therfore related by

$$H=|\mathbf{p}|\,\sigma_p\,. \tag{6.3}$$

In the c number theory, for a given momentum the particle therefore has two states: a state with positive energy, and with ½ as the spin component along $\mathbf{p}$, and a state with negative energy and with $-$½ as the spin component along $\mathbf{p}$.

It is easy to see that in a hole theory of such particles, *the spin of a neutrino* (defined to be a particle in the positive energy state) *is always parallel to its momentum*

[37]The possibility of a two-component relativistic theory of a spin ½ particle was first discussed by H. Weyl [*Z. Physik* **56**, 330 (1929)]. However, in such a theory parity is not manifestly conserved; therefore, in the past it was always rejected. (Cf. W. Pauli, *Handbuch der Physik*, Verlag Julius Springer, Berlin, 1933, Vol. 24, pp. 226-7.) The possible use of this two-component theory for expressing the nonconservation property of parity in neutrino processes was independently proposed and considered by T.D. Lee and C.N. Yang, *Phys. Rev.* **105**, 1671 (1957); A. Salam, *Nuovo cimento* **5**, 299 (1957); and L. Landau, *Nuclear Phys.* **3**, 127 (1957).

while the spin of an antineutrino (defined to be a hole in the negative energy state) *is always antiparallel to its momentum* (i.e., the momentum of the antineutrino). Many of the experimental implications on nonconservation of parity and charge conjugation can be directly related to this correlation between the spin and the momentum of a neutrino. With the usual (right-handed) conventions which we adopt throughout this paper, the spin and the velocity of the neutrino represent the spiral motion of a right-handed screw ($\mathcal{H}_\nu = +1$), while the spin and the velocity of the antineutrino represent the spiral motion of a left-handed screw ($\mathcal{H}_{\bar{\nu}} = -1$).

We shall now discuss some general properties of this neutrino field under the further assumption that the law of conservation of leptons is valid.[38]

A. The mass of the neutrino and the antineutrino in this theory is necessarily zero. This is true for the physical mass even with the inclusion of all interactions. To see this, one need only observe that all the one-particle physical states consisting of one neutrino (or one antineutrino) must belong to a representation of the inhomogeneous proper Lorentz group identical with the representation to which the free neutrino states discussed above belong. For such a representation to exist at all the mass must be zero.

B. That the use of such a theory fits naturally into the nonconservation property of parity is well known. We see it also in the following way: Under a space inversion P, one inverts the momentum of a neutrino but not its spin direction. Since in this theory the two are always parallel, the operator P applied to a neutrino state leads to a nonexisting state. Consequently the theory is not invariant under space inversion.

C. By the same reasoning one concludes that the theory is also not invariant under charge conjugation C which changes a particle into its antiparticle but does not change its spin direction or momentum.

D. It is possible, however, for the theory to be invariant under the operation CP, as this operation changes a neutrino into an antineutrino and simultaneously reverses its momentum while keeping the spin direction fixed. By the CPT theorem it follows that the theory can be invariant under time reversal T.

For the free neutrino field, as described by (6.1), one can prove that the theory is indeed invariant under time reversal and under CP.

2. β Decay

To discuss the β decay phenomena with the two-component neutrino theory and to compare the present results with those calculated previously, it is convenient to indicate how one can use the conventional four-component formalism of the neutrino (with nonconservation of parity) and obtain the same results as the present theory.

We start from Equation (6.1) and enlarge the matrices by the following definitions: ($1 \equiv 2 \times 2$ unit matrix)

[38]If the law of conservation of leptons is not valid then the mass of a two-component neutrino is, in general, not zero. In a special case even parity may still be conserved. See footnote 39 and K.M. Case, *Phys. Rev.* **107**, 307 (1957). Cf. also A. Salam, *Nuovo cimento* **5**, 299 (1957).

$$\boldsymbol{\alpha} \equiv \begin{pmatrix} \boldsymbol{\sigma} & 0 \\ 0 & -\boldsymbol{\sigma} \end{pmatrix}, \quad \beta \equiv \begin{pmatrix} 0 & 1 \\ 1 & 0 \end{pmatrix}, \tag{6.4}$$

$$\psi_\nu = \begin{pmatrix} \phi_\nu \\ 0 \end{pmatrix}, \tag{6.5a}$$

$$\boldsymbol{\gamma} \equiv -i\beta\boldsymbol{\alpha}, \quad \gamma_4 \equiv \beta, \quad \gamma_5 \equiv \gamma_1\gamma_2\gamma_3\gamma_4 = \begin{pmatrix} -1 & 0 \\ 0 & 1 \end{pmatrix}. \tag{6.6}$$

An immediate consequence of these definitions is

$$\gamma_5\psi_\nu = -\psi_\nu \,. \tag{6.7a}$$

The free neutrino part of the Lagrangian is, as usual,

$$L_\nu = \psi_\nu{}^\dagger\gamma_4\left(\gamma_\mu\frac{\partial}{\partial x_\mu}\right)\psi_\nu\,, \tag{6.8}$$

where $\psi_\nu{}^\dagger$ = Hermitian conjugate of ψ_ν. The most general interaction Lagrangian with no derivative couplings for the process

$$n \to p + e^- + \bar{\nu} \tag{6.9a}$$

is exactly as usual; namely, it is the sum of the usual S, V, T, A, and P couplings:

$$+L_{\text{int}} = -H_{\text{int}} = \sum -\mathfrak{G}_i(\psi_p{}^\dagger O_i\psi_n)(\psi_e{}^\dagger O_i\psi_\nu) \tag{6.10a}$$

where i runs over S, V, T, A, and P; and O_S, O_V, etc. are given in Equation (5.2).

It is not difficult to prove that Equations (6.5a) and (6.7a) are consistent with a relativistical theory even in the presence of the interaction (6.10a). Another way of proving this is to start from the conventional theory of the neutrino with the interaction Hamiltonian given in Equation (5.1) and observe that when

$$C_S = -C_S' = (1/2)\,\mathfrak{G}_S \quad \text{and} \quad C_V = -C_V' = (1/2)\,\mathfrak{G}_V, \quad \text{etc.}, \tag{6.11a}$$

the neutrino field ψ_ν always appears in interactions in the combinations $(1-\gamma_5)\psi_\nu$. In the explicit representation we have adopted above this means that only the first two components of ψ_ν contribute to the interaction. *All calculations using the conventional theory of the neutrino with the Hamiltonian Equation* (5.1) *concerning β decay therefore give the same result as the present theory if we take the choice of constants* (6.11a). There exists, however, the possibility that in the decay of the neutron a neutrino (defined to be a right-handed screw) is emitted,

$$n \to p + e + \nu\,. \tag{6.9b}$$

The corresponding general form (not including derivatives of the fields) of the Hamiltonian is

$$H_{\text{int}} = \sum \mathfrak{G}_i{}'(\psi_p{}^\dagger O_i\psi_n)(\psi_e{}^\dagger O_i\psi_\nu{}') \tag{6.10b}$$

where O_i has been defined in Equation (5.2). The field $\psi_\nu{}'$ is a four-component spinor defined in terms of the two-component neutrino field ϕ_ν by

$$\psi_\nu{}' = \begin{pmatrix} 0 \\ \sigma_2\,\phi_\nu{}^\dagger \end{pmatrix}. \tag{6.5b}$$

From Equation (6.6) we see that

$$\gamma_5 \psi_\nu' = +\psi_\nu' . \tag{6.7b}$$

It can be shown that (6.5b) and (6.7b) are consistent with a relativistic theory even in the presence of interaction (6.10b). *It can also be proved that one can use again the Hamiltonian Equation* (5.1) *for the conventional theory of the neutrino with the choice of the coupling constants*

$$(1/2)\,\mathfrak{G}_S' = C_S = C_S', \quad (1/2)\,\mathfrak{G}_V' = C_V = C_V', \quad \text{etc.} \tag{6.11b}$$

and obtain the same result as the present theory.

The two possible choices (6.11a) and (6.11b) depend on whether, in the β decay of the neutron process, (6.9a) or (6.9b) prevails, i.e., whether an antineutrino ($\mathcal{H}_{\bar\nu} = -1$, a left-handed screw) or a neutrino ($\mathcal{H}_\nu = +1$, a right-handed screw) is emitted. We shall see that experimentally it will be easy to decide which of the two choices is appropriate. [In this section we do not consider the possibility of the simultaneous presence of both (6.9a) and (6.9b).][39]

A. ANGULAR DISTRIBUTION OF β-RAYS FROM POLARIZED NUCLEI

The experimental results of Wu et al. show that in the decay of

$$\mathrm{Co}^{60} \to \mathrm{Ni}^{60} + e^- + \bar\nu, \quad J=5 \underset{(\mathrm{no})}{\to} J=4 , \tag{6.12}$$

the angular distribution of e^- is

$$1 + \alpha \cos\theta \tag{6.13}$$

with

$$\alpha \cong -0.7 \frac{\langle J_z \rangle}{J} \tag{6.14}$$

and

$$\theta = \angle\, (\mathbf{J}, \mathbf{p}_e) .$$

This result is consistent with the coupling constants assignment

$$C_T' = -C_T, \quad C_A' = -C_A \cong 0 . \tag{6.15}$$

By comparing (6.15) with (6.11a) and (6.11b) we see that in the β^- decay a $\bar\nu$ (with a left-hand helicity, $\mathcal{H}_{\bar\nu} = -1$) is emitted, while in the β^+ decay a ν (with a right-hand helicity, $\mathcal{H}_\nu = +1$) is emitted.

[39]If the conservation law of leptons is valid then (6.9a) and (6.9b) cannot be simultaneously present. It is of interest to note that if reactions (6.9a) and (6.9b) were simultaneously present with exactly equal amplitude, the angular distribution of β-rays from polarized nuclei, for example, would appear to be symmetrical. In such a case the theory is identical with the conventional Majorana theory of the neutrino with a parity conserving Hamiltonian. This, of course, is not the case that occurs in nature. The general relationship between the present two-component theory and the Majorana theory with a parity nonconserving Hamiltonian and, possibly, nonzero mass has recently been investigated by K. Case, *Phys. Rev.* **107**, 307 (1957).

B. LONGITUDINAL POLARIZATION OF β-RAYS

As remarked before, if parity is not conserved the electrons can be longitudinally polarized.[30] We shall see that with the two-component theory of the neutrino all β emitters can be used as almost perfect polarizers for the β-rays. The degree of polarization depends on v/c of the β-rays. Let us consider a coupling term between e^- and ν,

$$\psi_e^\dagger O_i \psi_\nu ,$$

with O_i given in Equation (5.2). By using Equation (6.7a), together with the commutation relations

$$O_i \gamma_5 = \pm \gamma_5 O_i$$

with the plus sign for $i=V, A$, and the minus sign for $i=S, T, P$, we obtain

$$\begin{aligned} \psi_e^\dagger O_i \psi_\nu &= \psi_R^\dagger O_i \psi_\nu \ \text{ if } i=V, A ; \\ &= \psi_L^\dagger O_i \psi_\nu \ \text{ if } i=S, T, P ; \end{aligned} \tag{6.16}$$

where

$$\psi_R = (\tfrac{1}{2})(1-\gamma_5)\psi_\nu \quad \text{and} \quad \psi_L = (\tfrac{1}{2})(1+\gamma_5)\psi_\nu . \tag{6.17}$$

If the electron is extremely relativistic, $v/c=1$, then ψ_R and ψ_L are both eigenstates of the free Hamiltonian of the electron, provided the mass term is neglected. In this case if the coupling (6.16) is S and/or T and/or P the electron will be 100% polarized with its spin antiparallel to its momentum (i.e., with a left-handed helicity), while if the coupling is V and/or A the electron is also 100% polarized but with its spin parallel to its momentum (i.e., with a right-handed helicity).

In general for relativistic or nonrelativistic electrons the term $(1\mp\gamma_5)/2$ acts as a projection operator for the longitudinal polarization. The helicity $\mathcal{H}$ can be calculated as

$$\mathcal{H}_{e^-} = \langle \boldsymbol{\sigma} \cdot \hat{\mathbf{p}}_e \rangle_{\text{Av}} = \pm v/c \tag{6.18}$$

where $\hat{\mathbf{p}}_e$ is a unit vector along the momentum of e^-. In Equation (6.18) the plus sign is for V, A couplings and the minus sign for S, T, P couplings. Thus we see that *in*

$$n \rightarrow p + e^- + \bar{\nu}$$

if vector and axial vector couplings are absent the electrons will be longitudinally polarized with

$$\mathcal{H}_{e^-} = -v/c . \tag{6.19}$$

Equation (6.19) is true for any β emitter independent of whether the nuclei are polarized or not, and is independent of whether the β decay is allowed or forbidden. The possible deviation from (6.19) can then be used as a measure for the strengths of vector and axial vector coupling constants. In deriving Equations (6.18) and (6.19) we have not included possible depolarization effects due to a Coulomb field. However,

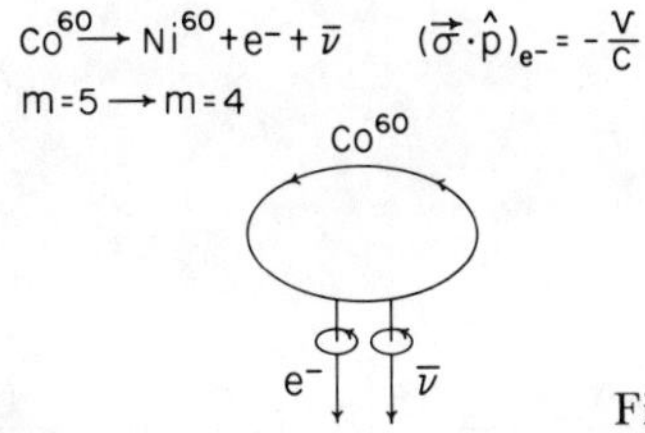

Figure 2.

if $v/c = 1$ then it is easy to see that these results of longitudinal polarization cannot be effected by any Coulomb field.[40]

Similarly for any β^+ emission,

$$p \to n + e^+ + \nu \, ,$$

the positions are longitudinally polarized with

$$\mathcal{H}_{e^+} = +v/c \tag{6.20}$$

provided V and A couplings are zero.

Because of these properties it is possible to understand in a simple way that in the decay of Co^{60}, Equation (6.12), if $\bar{\nu}$ is emitted the electrons are emitted predominantly antiparallel to the spin direction of Co^{60} [i.e., $\alpha < 0$ in Equation (6.14)]. To understand the sign of α, let us neglect the A coupling, and consider the special case with

$$(J_z)_{Co^{60}} = 5 \to (J_z)_{Ni^{60}} = 4$$

in the Co^{60} decay. The e^- and $\bar{\nu}$ emitted are both left-handed particles (i.e., spin antiparallel to momentum). Take the particular case that $\mathbf{p}_e$, $\mathbf{p}_{\bar{\nu}}$ are all parallel to the $\pm z$ axes. In this case the orbital angular momentum along the z axis is zero. In order to conserve J_z, both e^- and $\bar{\nu}$ must be emitted predominantly parallel to each other with both of their momenta $\mathbf{p}_e$ and $\mathbf{p}_{\bar{\nu}}$ along the $-z$ axis (cf. Figure 2). By using (6.19), it is easy to see that in this case with $C_A = 0$ the relative probabilities are

$$1-(v/c) \quad \text{for } \cos\theta = +1$$

and

$$1+(v/c) \quad \text{for } \cos\theta = -1 \, .$$

According to Equations (5.11) and (6.11a) the complete formula of α for

---

[40]This may be proved in the following way. Consider the motion of a charged particle in a coulomb field U,

$$(\boldsymbol{\alpha}\cdot\mathbf{p} + \beta m + U)\psi = E\psi \, .$$

Except for the βm term the Hamiltonian commutes with $(1 \pm \gamma_5)$. Thus if the final expression for $\langle \boldsymbol{\sigma}\cdot\hat{\mathbf{p}} \rangle$ is expanded in a power series of (m/E), the term with zeroth power in (m/E) is independent of the presence of U; i.e.,

$$\langle \boldsymbol{\sigma}\cdot\hat{\mathbf{p}} \rangle_{e^-} = +1 + o(m/E)$$

for V, A couplings and

$$\langle \boldsymbol{\sigma}\cdot\hat{\mathbf{p}} \rangle_{e^+} = -1 + o(m/E)$$

for S, A, P couplings.

$$J+1 \rightarrow J(\text{no}) \tag{6.21}$$

is

$$\alpha = +\frac{J_z}{J}\beta\,,$$

$$\beta = -\frac{v}{c}\,\frac{|C_T|^2 - |C_A|^2 + (2Ze^2/\hbar cp)Im(C_A C_T^*)}{|C_T|^2 + |C_A|^2}\,. \tag{6.22}$$

In (6.22) the Fierz term is put to be zero. The minus sign in the coefficient of $|C_A|^2$ is of course due to the fact that e^- is emitted with a right-hand helicity via the axial vector coupling.

By a similar argument, for the corresponding simple case of β^- decay in

$$J \rightarrow J+1(\text{no}) \tag{6.23}$$

the $(e^-, \bar{\nu})$ system should carry a $(\Delta J_z) = -1$. Thus the electrons are expected to be emitted predominantlyparallel to $\mathbf{p}_{\bar{\nu}}$ and both of these momenta are parallel to the spin direction of the nuclei. The general formula for (6.23) is

$$\alpha = -\frac{\langle J_z \rangle}{(J+1)}\beta \tag{6.24}$$

with β given by (6.22). The difference of sign in (6.24) and (6.21) is completely expected.

In the decay

$$J \rightarrow J(\text{no})\,, \tag{6.25}$$

let us first consider the contribution due to the tensor coupling in the Gamow-Teller matrix element alone. Even if one takes the special case of the completely polarized nuclei $J_z = J$ the change of z component nuclear angular momentum can be 0 or 1. In the first case of $\Delta J_z = 0$ no asymmetry is present. In the second case, the e^- and $\bar{\nu}$ should carry a $\Delta J_z = +1$ which causes the asymmetry parameter α to be negative, $\alpha < 0$. Thus we expect in the transition (6.25) if $M_F = 0$ and $C_A = 0$ the sign of α should be negative but with its magnitude greatly reduced. The general formula for $J \rightarrow J(\text{no})$ is

$$\alpha = +\frac{\langle J_z \rangle}{J(J+1)}\beta + \frac{\langle J_z \rangle}{\sqrt{J(J+1)}}\beta'\,,$$

$$\beta' = -Re\left[C_S^* C_T - C_V^* C_A + i\frac{Ze^2}{\hbar cp}(C_S^* C_A - C_V^* C_T)\right]\frac{v_e}{c}\,\frac{4M_F \cdot M_{GT}}{\xi + (\xi b/W)} \tag{6.26}$$

with ξ and b given in Equations (5.4) and (5.8).

Similarly it is easy to see that for β^+ decay the asymmetry parameter due to tensor coupling (or due to the axial vector coupling term alone) must change its sign. The complete formulas can be easily obtained from Equations (5.11) to (5.13).

3. Capture Cross Sections for the Neutrino

An experiment such as the one being carried out by Cowan et al.[41] measures the cross section for neutrino absorption, which can be calculated in both the present theory and the usual theory. Now one determines the magnitude of the β coupling constants to give the observed lifetimes of nuclei against β decay. The calculated value of the cross section turns out then to be *twice as great* in the present theory as in the usual theory using the four-component theory and conservation of parity. This follows from the following simple reasoning: The neutrino flux is an experimental quantity independent of the theory. If the neutrinos in a given direction have only one spin state instead of the usual two, by a detailed balancing argument they must have a cross section for absorption twice as great as the usual ones. Actually from the experiments of Wu et al.[1] one expects the neutrino emitted to be longitudinally polarized. Thus an increment of cross section is expected if we use the hermiticity property of the H_{weak}. This effect should be present even if the neutrino is described by a four-component theory with parity nonconservation.

4. π Decay

In the decay of π-mesons at rest let us consider the component of angular momentum along the direction $\mathbf{p}_\mu$, the momentum of the μ-meson. The orbital angular momentum contributes nothing to this component. The μ spin component is therefore completely determined (irrespective of its total spin) by the spin component of the ν or $\bar{\nu}$. There are then two possibilities: (i)

$$\pi^+ \to \mu^+ + \nu, \quad \mu^+ \text{ spin along } \mathbf{p}_\mu = +\tfrac{1}{2},$$

$$\pi^- \to \mu^- + \bar{\nu}, \quad \mu^- \text{ spin along } \mathbf{p}_\mu = -\tfrac{1}{2}; \tag{6.27}$$

or (ii)

$$\pi^+ \to \mu^+ + \bar{\nu}, \quad \mu^+ \text{ spin along } \mathbf{p}_\mu = -\tfrac{1}{2},$$

$$\pi^- \to \mu^- + \nu, \quad \mu^- \text{ spin along } \mathbf{p}_\mu = +\tfrac{1}{2}. \tag{6.28}$$

In each case the μ-mesons with fixed p_μ form a polarized beam. Furthermore, the polarization is complete (i.e., in a pure state). In this theory of the neutrino the π-μ decay is then a perfect polarizer of the μ-meson, and offers a natural way to measure the spin and the magnetic moment of the μ-meson. (It turns out that the μ-e decay can serve as a good analyser, as we shall discuss in the next section.)

The choice of the two possibilities (6.27) and (6.28) will be further discussed in Chapter VII.

---

[41] C.L. Cowan Jr., F. Reines, F.B. Harrison, H.W. Kruse, and A.D. McGuire, *Science* **124**, 103 (1956).

5. μ Decay

For the μ^--e^- decay the process can be

$$\mu^- \to e^- + \nu + \bar{\nu}\,, \tag{6.29}$$

or

$$\mu^- \to e^- + 2\nu\,, \tag{6.30}$$

or

$$\mu^- \to e^- + 2\bar{\nu}. \tag{6.31}$$

Consider process (6.29) first. The decay interaction (without derivative coupling) can be written with the notations defined in Equation (5.2):[42]

$$H_{\text{int}} = \sum_i G_i(\psi_e{}^\dagger O_i \psi_\nu)(\psi_\nu{}^\dagger O_i \psi_\mu). \tag{6.32}$$

We have assumed in writing (6.32) that the spin of the μ-meson is ½.

Becuse of the subsidiary condition (6.7a) satisfied by the neutrino field ψ_ν, the S-coupling term in the Hamiltonian (6.32) gives a result identical to that of the P-coupling term; the V-coupling term is the same as the A-coupling term; and the T-coupling term is identically zero. Thus, in (6.32) there are only two independent constants,

$$g_1 \equiv G_S - G_P \quad \text{and} \quad g_2 \equiv G_V + G_A\,. \tag{6.33}$$

The electron (or positron) emitted from a μ-meson decay at rest will be longitundinally polarized with a helicity $\mathcal{H}$ given by

$$\mathcal{H}_{e^-} = -\xi \quad \text{and} \quad \mathcal{H}_{e^+} = +\xi \tag{6.34}$$

where

$$\xi = (|g_1|^2 - 4|g_2|^2)(|g_1|^2 + 4|g_2|^2)^{-1} \tag{6.35}$$

and the helicity $\mathcal{H}$ is defined by

$$\mathcal{H} = (\boldsymbol{\sigma}\cdot\mathbf{p})/|\mathbf{p}| \tag{6.2}$$

with $\boldsymbol{\sigma}$ and $\mathbf{p}$ the spin and momentum vectors of the electron (or positron). Equation (6.34) is independent of the polarization state of the μ-meson.

For a μ^- at rest with spin completely polarized, the normalized electron distribution is given by

---

[42]In reference 23 the Hamiltonian for μ decay is written as

$$H = \sum_{i=V,A} f_i(\psi_e{}^\dagger O_i \psi_\mu)(\psi_\nu{}^\dagger O_i \psi_\nu)\,.$$

The f_S and f_A are related to G_i of (6.32) by

$$(½)(G_S - G_P) = f_A + f_V \quad \text{and} \quad G_V + G_A = f_A - f_V\,.$$

It is also possible to write the Hamiltonian for μ decay in the form

$$H = \sum_i G_i'(\psi_e{}^\dagger O_i \psi_\nu')(\psi_\nu'{}^\dagger O_i \psi_\mu)$$

where ψ_ν' is given by (6.56). The G_i' are related to G_i by

$$(½)(G_S - G_P) = G_V' + G_A' \quad \text{and} \quad G_V + G_A = -(½)(G_S' - G_P')\,.$$

$$(dN)_{e^-}=2x^2[(3-2x)+\xi \cos\theta(1-2x)]\, dx\, d\Omega(4\pi)^{-1} \tag{6.36}$$

where

$x=p/$(maximum electron momentum),

$\theta=$ angle between the electron momentum and the spin direction,

$d\Omega=$ solid angle of electron momentum,

and ξ is given by (6.35). The corresponding distribution of e^+ from a completely polarized μ^+ at rest is

$$(dN)_{e^+}=2x^2[(3-2x)-\xi \cos\theta(1-2x)]dx\, d\Omega(4\pi)^{-1}\,. \tag{6.37}$$

The decay probability per unit time is ($\hbar=c=1$)

$$\lambda=m_\mu{}^5(|g_1|^2+4|g_2|^2)(3\times 2^{11}\pi^3)\,. \tag{6.38}$$

The energy spectrum of the electron (or positron) is

$$dN=2x^2(3-2x)dx\,. \tag{6.39}$$

The spectrum (6.39) is characterized by a Michel parameter[43] $\rho=¾$ which is not inconsistent with but *seems* to be slightly higher than the present experimental value[44] $\rho=0.68$.

Integration of (6.36) and (6.37) over the energy of $e^\pm$ gives for the over-all angular distribution

$$(dN)_{e\pm}=[1\pm(⅓)\xi]d\Omega(4\pi)^{-1}\,. \tag{6.40}$$

The mass of the electron (or positron) is neglected in all the above formulas (6.34) to (6.40).

That in (6.37) the angular distribution depends sensitively on the energy of the electron can be understood in a simple way. By using (6.16) to (6.18), we expect for the helicity of e^- [cf. (6.35)]

$$\mathcal{H}_{e^-}=-1 \quad \text{if only } g_1\neq 0$$

and

$$\mathcal{H}_{e^+}=+1 \quad \text{if only } g_2\neq 0. \tag{6.41}$$

Consequently, there is no interference term between g_1 and g_2.

Let us first discuss the case that only $g_1\neq 0$ ($g_2=0$). The e^- emitted is of left-hand helicity. Consider the special case that $\mathbf{p}_e$, $\mathbf{p}_\nu$, $\mathbf{p}_{\bar\nu}$, and $\boldsymbol{\sigma}_\mu$ are all along the $+z$ or $-z$ axis. From conservation of angular momentum along the z axis it is easy to show that if $\boldsymbol{\sigma}_\mu // +z$ axis then for $x=1$,

$$\mathbf{p}_e // -z \text{ axis} \tag{6.42}$$

and for $x<½$,

$$\mathbf{p}_e // +z \text{ axis} \quad (g_1\neq 0,\ g_2=0). \tag{6.43}$$

---

[43]L. Michel, *Proc. Phys. Soc. (London)* **A63**, 514 (1950).

[44]C.P. Sargent, M. Rinehart, L.M. Lederman, and K.C. Rogers, *Phys. Rev.* **99**, 885 (1955). These authors give $\rho=0.68\pm0.10$. More recent measurements by L. Rosenson (*Phys. Rev.*, in press) and by K. Crowe, *Bull. Am. Phys. Soc.* **2**, 206 (1957), give the same value for ρ but with a smaller error. A slight deviation of ρ value from 0.75 may indicate that there is a possible "non-local" effect in the μ decay Lagrangian.

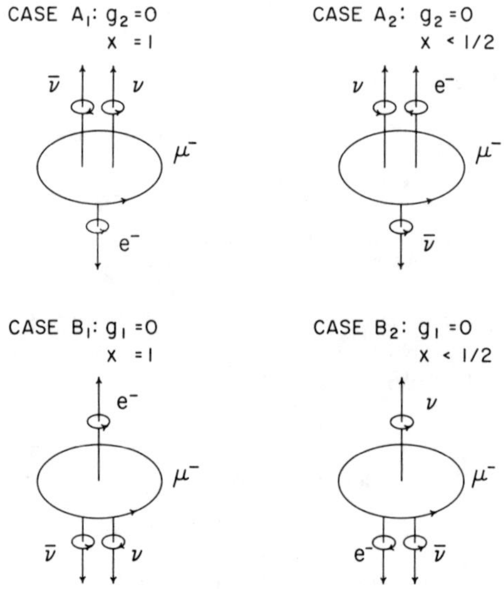

Figure 3.

This can be seen directly by inspection of Figure 3 (case A_1 and case A_2). Furthermore for $x=1$ all three momenta $\mathbf{p}_e$, $\mathbf{p}_\nu$, and $\mathbf{p}_{\bar{\nu}}$ must be colinear. Thus, from (6.42), the corresponding angular distribution must be of the form $1-\cos\theta$ (with $x=1$ and $g_2=0$), which means that the asymmetry is maximum. For other values of $x(<1)$ these three momenta may not be colinear and the asymmetry does not attain its maximum value. A comparison between (6.42) and (6.43) explains the energy dependence $(1-2x)$ in the $\cos\theta$ term in (6.37). In an entirely similar way one can apply the above considerations to the case that $g_1=0$ and $g_2\neq 0$ (cases B_1 and B_2 of Figure 3).

Experimentally, the angular distribution of $e^\pm$ has been measured[2,25] with respect to the momentum of the μ-meson from π decay. The experimental results seem to agree quite well with the distribution function (6.37). The results for μ^+ stopped in carbon were discussed in Chapter V, section 5. From Equation (5.33), it can be concluded that the parameter ξ, (6.35), must lie within the limits

$$1\geqslant|\xi|>0.78\,. \tag{6.44}$$

The algebraic sign of ξ depends on whether in π decay the helicity of μ is $+1$ or -1. Equation (6.44) gives only a lower limit of ξ. The actual value depends on the degree of depolarization of μ^+. The precise value and sign of ξ can be obtained more directly by a measurement of the helicity for $e^\pm$ from μ decay, Equation (6.34). This point will be further discussed in connection with the law of conservation of leptons in Chapter VII.

If in the $\mu^\pm$ decay process

$$\mu\rightarrow e+2\nu \tag{6.30}$$

or

$$\mu\rightarrow e+2\bar{\nu} \tag{6.31}$$

prevails, the corresponding energy spectrum is characterized by a Michel parameter[43] $\rho=0$ which is not consistent with experiments.[44]

6. Remarks

In the above sections we have seen that the various effects due to nonconservation of parity and charge conjugation in β decay, π decay, and μ decay can be conveniently described by the use of a two-component theory of the neutrino. Nevertheless, as we have noticed before (cf. footnote 39) the mere use of a two-component theory does not preclude, for example, the possibility that in addition to

$$n \rightarrow p+e^{-}+\bar{\nu} \tag{6.9a}$$

we may also have

$$n \rightarrow p+e^{-}+\nu\,. \tag{6.9b}$$

From the experimental results on the slowness of the rate for double β decay processes and the largeness of asymmetry in the β angular distribution from polarized nuclei, we know that reaction (6.9b), if it exists at all, must be described by a much smaller coupling constant than that of reaction (6.9a). Recently K. Case[45] was able to show that by using the Majorana theory with a Hamiltonian which does not conserve parity it is possible to generalize the equation of motion [Equation (6.1)] for the neutrino field and to construct a two-component theory with, possibly,

$$m_\nu \neq 0\,.$$

In the special case that $m_\nu=0$, his generalization reduces to the present two-component theory discussed in this chapter. However, in the general case, if the mass of the neutrino $m_\nu \neq 0$ then the rate of double β decay process cannot attain its minimum value and the asymmetry of the β angular distribution from polarized nuclei cannot reach its maximum value. Yet experimentally, the mass of the neutrino $m_\nu \cong 0$, the rate of double β decay process $\cong 0$, and the asymmetry of the β angular distribution $\cong$ its maximum value. These three facts seem to be strongly suggestive of the possible existence of a law of conservation of leptons.[46] It can be shown easily that *the existence of a conservation law of leptons together with the use of a two-component theory of the neutrino necessitates* (i) $m_\nu=0$, (ii) *rate of double β decay = its minimum value,* and (iii) *parity must be nonconserved and the observed asymmetry due to such nonconservation can attain its maximum value.* In the next chapter we shall analyze in some detail the various consequences of such a conservation law of leptons.

---

[45] K. Case, *Phys. Rev.* **107**, 307 (1957).

[46] The concept of a possible conservation law of leptons was first considered by E. Konopinski and H.M. Mahmond, *Phys. Rev.* **92**, 1045 (1953). Some discussions on analyzing the conservation law of leptons together with the use of a two-component theory of the neutrino are given in T.D. Lee and C.N. Yang, *Phys. Rev.* **105**, 1671 (1957).

VII. POSSIBLE LAW OF CONSERVATION OF LEPTONS AND THE UNIVERSAL FERMI INTERACTIONS

1. Law of Conservation of Leptons[46]

The law of conservation of leptons states that to each particle it is possible to assign a leptonic number l ($l \neq 0$ for leptons) and the sum of leptonic numbers must be conserved in all reactions. From our previous discussions we know that β decay and μ decay are represented by[47]

$$n \rightarrow p + e^- + \bar{\nu} \tag{7.1}$$

and

$$\mu^- \rightarrow e^- + \nu + \bar{\nu} \,. \tag{7.2}$$

Consequently, the assignments of l must be chosen as

$$\begin{aligned} &l = \text{same (say, } l = -1) && \text{for } e^-, \nu, \text{ and } \mu^- , \\ &l = +1 && \text{for } e^+, \bar{\nu}, \text{ and } \mu^+ , \text{ and} \\ &l = 0 && \text{for } \pi, \gamma, K, \text{ and all heavy particles.}^{48} \end{aligned} \tag{7.3}$$

The conservation law of leptons then necessitates for the π decay

$$\pi^- \rightarrow \mu^- + \bar{\nu} \quad \text{and} \quad \pi^+ \rightarrow \mu^+ + \nu \,. \tag{7.4}$$

Equation (7.4) implies that the helicities $\mathcal{H}$ of $\mu^\pm$ in the rest system of π-mesons are

$$\mathcal{H}_{\mu^-} = -1 \quad \text{and} \quad \mathcal{H}_{\mu^+} = +1 \tag{7.5}$$

where $\mathcal{H}$ is defined to be

$$\mathcal{H} = (\boldsymbol{\sigma} \cdot \mathbf{p})/|\mathbf{p}| \tag{7.6}$$

with $\boldsymbol{\sigma}$, $\mathbf{p}$ the spin and momentum vectors. The measurement of the helicities of μ^- and μ^+ from $\pi^\pm$ decay can serve as a test for the validity of the conservation law of leptons.

Next we consider the helicities of $e^\pm$ in the μ decay. As we have discussed before, the helicities of $e^\pm$ measured in the rest system of the μ-meson are given by [see (6.34) and (6.35)]

$$\mathcal{H}_{e^-} = -\xi \quad \text{and} \quad \mathcal{H}_{e^+} = +\xi \tag{7.7}$$

---

[47]Throughout this chapter we use the two-component theory for the neutrino.

[48]That the leptonic number l for all heavy particles can be chosen as zero is evident. Both pion and photon can be created singly, thus their leptonic numbers must both be zero. If the number of leptons is absolutely conserved, then, because of the existence of decay modes $K_{\pi 2}$ and $K_{\pi 3}$, the leptonic number l for K-mesons must also be zero.

where ξ is related to the coupling constants for μ decay by (6.35) and (6.33). By comparing the theoretical angular distribution for e with the observed value we have already found that

$$1 \geqslant |\xi| > 0.78 \,. \tag{6.44}$$

Now if the law of conservation of leptons is correct, by using (7.5) we conclude that

$$-1 \leqslant \xi < -0.78 \,. \tag{7.8}$$

The minus sign for ξ is a consequence of the conservation law of leptons. Thus the measurement of the helicities of e^- and e^+ from μ decay can also serve as a test of the validity of the conservation law of leptons.

The conservation law of leptons also has a direct effect on the helicities of $\mu^\pm$ from $K_{\mu 2}{}^\pm$ decay. From the assignments of l in (7.3), the $K_{\mu 2}{}^\pm$ decay should be described by the reactions

$$K_{\mu 2}{}^- \rightarrow \mu^- + \bar{\nu} \quad \text{and} \quad K_{\mu 2}{}^+ \rightarrow \mu^+ + \nu \,. \tag{7.9}$$

Thus if the spin of K is zero the helicities of $\mu^\pm$ in the rest system of K are respectively[49]

$$\mathcal{H}_{\mu} = -1 \quad \text{and} \quad \mathcal{H}_{\mu^+} = +1 \,. \tag{7.10}$$

2. Universal Fermi Interactions

In this section we shall analyze the possibility of the so-called universal Fermi interactions by comparing the Hamiltonians for μ decay, β decay, and μ capture processes.[50] We represent these three processes by[51]

$$H_1 = \sum G_i (\psi_e{}^\dagger O_i \psi_\nu)(\psi_\nu{}^\dagger O_i \psi_\mu) + \text{hermitian conjugate} \,,$$

$$H_2 = \sum \mathfrak{G}_i (\psi_p{}^\dagger O_i \psi_n)(\psi_e{}^\dagger O_i \psi_\nu) + \text{hermitian conjugate} \,,$$

and

$$H_3 = \sum \mathcal{G}_i (\psi_\nu{}^\dagger O_i \psi_\mu)(\psi_n{}^\dagger O_i \psi_p) + \text{hermitian conjugate} \,, \tag{7.11}$$

respectively.

From (7.8) and the lifetime of the μ-meson we have for the coupling constants G_i of μ decay (cf. Chapter VI, section 5)

---

[49]The application of the conservation law of leptons to $K_{\mu 2}$ decay is of interest because it gives a result opposite to that obtained by using the "attribute rule" proposed by R.G. Sachs [*Phys. Rev.* **99**, 1573 (1955)]. Cf. also W.G. Holladay, *Phys. Rev.*, in press.

[50]For references to previous works on the various Fermi interactions see, e.g., E. Fermi, *Elementary Particles,* Yale University Press, 1951; L. Michel, *Revs. Mod. Phys.* **29**, 159 (1957).

[51]In the present discussion of universal Fermi interactions we group the spinor fields ψ_e, ψ_ν, and ψ_μ in a particular form which is compatible with the idea of the conservation law of leptons. Had we grouped, for example, the μ decay interaction H in a different way (cf. footnote 42), many of the following conclusions about the similarity between the β decay coupling constants and the μ decay coupling constants would be changed. These changes can be easily obtained by using the relationship between G_i and G_i' given in footnote 42. (Compare especially G_i' with $\mathfrak{G}_i$.)

$$\eta \equiv \frac{|G_V+G_A|^2}{|G_S-G_P|^2} > 2.3 \tag{7.12}$$

and

$$|G_S-G_P|^2+4|G_V+G_A|^2=(7.5\times 10^{-49}\ \text{erg-cm}^3)^2 . \tag{7.13}$$

The right-hand side of (7.12) is only a lower limit. The actual value of η is probably much larger than this limit. For example, $\eta \cong 5$ if in the measurement of π-μ-e decay 90% of the μ-mesons stopped in carbon are polarized and $\eta \to \infty$ if only 80% of the stopped μ-mesons are polarized.

The corresponding values for the various coupling constants for β decay have been extensively studied.[52] These results may be summarized[53] as follows:[52]

(i) From the absence of a Fierz term it can be concluded that

$$\frac{|\mathfrak{G}_S^* \mathfrak{G}_V + \mathfrak{G}_V^* \mathfrak{G}_S|}{|\mathfrak{G}_S^* \mathfrak{G}_S + \mathfrak{G}_V^* \mathfrak{G}_V|} = 0.00 \pm 0.10 \tag{7.14}$$

and

$$\frac{|\mathfrak{G}_A^* \mathfrak{G}_T + \mathfrak{G}_T^* \mathfrak{G}_A|}{|\mathfrak{G}_A^* \mathfrak{G}_A + \mathfrak{G}_T^* \mathfrak{G}_T|} = 0.00 \pm 0.04 . \tag{7.15}$$

(ii) From the β-ν angular correlation experiment on He^6 it can be concluded that

$$\left|\frac{\mathfrak{G}_A}{\mathfrak{G}_T}\right|^2 < \frac{1}{3} . \tag{7.16}$$

(iii) From the β-ν angular correlation experiment on Ne^{19} it can be concluded that

$$\left|\frac{\mathfrak{G}_V}{\mathfrak{G}_S}\right|^2 < 1.4 \quad \text{if} \quad \mathfrak{G}_A \cong 0 \tag{7.17}$$

and

$$\left|\frac{\mathfrak{G}_V}{\mathfrak{G}_S}\right|^2 < 3 \quad \text{if} \quad \left|\frac{\mathfrak{G}_A}{\mathfrak{G}_T}\right|^2 = \frac{1}{3} . \tag{7.18}$$

(iv) From the β-X diagram it can be concluded that

$$\frac{|\mathfrak{G}_S|^2+|\mathfrak{G}_V|^2}{|\mathfrak{G}_A|^2+|\mathfrak{G}_T|^2} = 0.79 . \tag{7.19}$$

(v) From the *ft* value of O^{14} and (7.19), the absolute magnitudes of $|\mathfrak{G}_S|^2+|\mathfrak{G}_V|^2$ and $|\mathfrak{G}_A|^2+|\mathfrak{G}_T|^2$ are

$$|\mathfrak{G}_S|^2+|\mathfrak{G}_V|^2=(2.0\times 10^{-49}\ \text{erg-cm}^3)^2 \tag{7.20}$$

and

$$|\mathfrak{G}_A|^2+|\mathfrak{G}_T|^2=(2.5\times 10^{-49}\ \text{erg-cm}^3)^2 . \tag{7.21}$$

It should be noted that $\mathfrak{G}_i$ is related to C_i and C_i' in (5.1) by [cf. (6.11a)]

$$\mathfrak{G}_i = 2C_i = -2C_i' . \tag{7.22}$$

---

[52] For detailed references on these experimental works see, e.g., K. Siegbahn, *Beta- and Gamma-Ray Spectroscopy*, Interscience, New York, 1955; L. Michel, *Revs. Mod. Phys.* **29**, 159 (1957).

[53] These results, (7.14) to (7.21), on various $\mathfrak{G}_i$ for β decay were compiled by C.S. Wu. We are grateful to Professor Wu for her permission to quote these results here.

(vi) Thus, if the β decay interaction is invariant under time reversal T, then

$$\mathfrak{G}_V \cong 0 \quad \text{and} \quad \mathfrak{G}_A \cong 0 . \tag{7.23}$$

By comparing $\mathfrak{G}_i$ [(7.20), (7.21), and (7.23)] with G_i [(7.12) and (7.13)] we see that the β decay coupling constants $\mathfrak{G}_i$ are very different from the μ decay coupling constants G_i. In this case the idea of universal Fermi interactions seems to be difficult to maintain.[51]

(vii) If β decay interaction is not invariant under time reversal then the limits on $\mathfrak{G}_i$ are those given by (7.20), (7.21), and the inequalities (7.17) and (7.18). In this case the β decay coupling constants may not be incompatible with the μ decay coupling constants.

In conclusion we wish to remark that to give a definitive statement on the universal Fermi interaction it is necessary to obtain a much sharper limit on $|\mathfrak{G}_V/\mathfrak{G}_S|^2$ than that in (7.17) and (7.18). Because of the nonconservation of parity this can now be obtained by a direct measurement on the helicities of $e^\pm$ from any β transition that has a large Fermi matrix element M_F [cf. (6.18)]. As we shall see in the next section, because of the nonconservation of parity it is also possible to measure the various $\mathcal{G}$ for μ^- capture processes.

3. μ^- Capture Process[54]

In this section we shall study in detail the capture of μ^- by nuclei,

$$\mu^- + p \to n + \nu . \tag{7.24}$$

The Hamiltonian for this process is given by

$$H_3 = \sum_i \mathcal{G}_i (\psi_n^\dagger O_i \psi_p)(\psi_\nu^\dagger O_i \psi_\mu) \tag{7.25}$$

where ψ_ν is the two-component neutrino field. We list the following results for capture of μ^- in hydrogen.

(i) The rate for process (7.24) in hydrogen is

$$(1/\tau)_{\text{cap}} = p_\nu^2 \xi / 2\pi^2 a^3 \tag{7.26}$$

where

$$\xi = |\mathcal{G}_S + \mathcal{G}_V|^2 + 3|\mathcal{G}_A + \mathcal{G}_T|^2 , \tag{7.27}$$

p_ν is the momentum of the neutrino and a is the Bohr radius of the μ^--mesic atom.

(ii) If in the capture process (7.24) the μ^- is completely polarized, the angular distribution of the neutron is of the form

$$1 + \alpha \cos\theta_1 \tag{7.28}$$

where

$$\theta_1 = \angle\, (\sigma_\mu, \mathbf{p}_n) \tag{7.29}$$

representing the angle between the spin of the μ^--meson and the momentum of the neutron. The asymmetry parameter α is given by

---

[54]K. Huang, C.N. Yang, and T.D. Lee, *Phys. Rev.*, in press.

$$\alpha\xi = -|\mathcal{G}_S + \mathcal{G}_V|^2 + |\mathcal{G}_A + \mathcal{G}_T|^2 . \tag{7.30}$$

(iii) The transition probability $(1/\tau)_{\rm rad}$ for the radiative capture process of μ^- in hydrogen,

$$\mu^- + p \to n + \nu + \gamma , \tag{7.31}$$

is

$$(1/\tau)_{\rm rad} = \frac{e^2\eta}{6\pi\hbar c\xi}(1/\tau)_{\rm cap} \tag{7.32}$$

where

$$\eta = |\mathcal{G}_S|^2 + |\mathcal{G}_V|^2 + 3|\mathcal{G}_A|^2 + 3|\mathcal{G}_T|^2 . \tag{7.33}$$

(iv) In process (7.31) the γ-ray is circularly polarized. The polarization parameter β may be defined as

$$\beta = \frac{N_R - N_L}{N_R + N_L} \tag{7.34}$$

where N_R and N_L are respectively the number of right-handed and left-handed γ-rays. The parameter β is given by

$$\beta\eta = |\mathcal{G}_S|^2 - |\mathcal{G}_V|^2 - 3|\mathcal{G}_A|^2 + 3|\mathcal{G}_T|^2 . \tag{7.35}$$

Equation (7.35) is independent of either the state of polarization of the captured μ^--meson or the energy of the γ-rays.

(v) For the radiative capture of a 100% polarized μ^- the angular distribution of the γ-ray is of the form

$$1 + \beta \cos\theta_2 \tag{7.36}$$

where β is given by Equation (7.34) and

$$\theta_2 = \angle\, (\boldsymbol{\sigma}_\mu, \mathbf{p}_\gamma) \tag{7.37}$$

where $\mathbf{p}_\gamma$ is the momentum of the γ-ray.

(vi) In the radiative capture process (7.31) the angular distribution of the γ-ray with respect to the momentum of the neutrino $\mathbf{p}_\nu$ is

$$1 + \gamma \cos\theta_3 \tag{7.38}$$

where

$$\theta_3 = \angle\, (\mathbf{p}_\nu, \mathbf{p}_\gamma)$$

and

$$\gamma\eta = -|\mathcal{G}_S|^2 + |\mathcal{G}_V|^2 - |\mathcal{G}_A|^2 + |\mathcal{G}_T|^2 . \tag{7.39}$$

In all the above expressions we neglect v/c terms for the nucleon wave function and we replace ψ_μ by its value at the origin.

(vii) In (7.24) we assume that the law of conservation of leptons is valid. Otherwise instead of (7.24) and (7.31) we may have

$$\mu^- + p \to n + \nu \tag{7.40}$$

and

$$\mu^- + p \to n + \bar{\nu} + \gamma . \tag{7.41}$$

In the measurements of $(1/\tau)_{\mathrm{cap}}$, $(1/\tau)_{\mathrm{rad}}$, θ_1, θ_2, and θ_3 for reactions (7.40) and (7.41) the corresponding parameters ξ, η, α, β, γ are replaced by ξ' η', α', β', and γ' where

$$\xi'=\xi, \quad \eta'=\eta, \quad \alpha'=-\alpha, \quad \beta'=-\beta, \quad \text{and} \quad \gamma'=-\gamma\,. \tag{7.42}$$

(viii) If the μ^- capture process is invariant under time reversal then all $\mathcal{G}_i$ are real. The measurement of α, β, γ, η, ξ affords a complete determination of the four coupling constants $\mathcal{G}_V$, $\mathcal{G}_S$, $\mathcal{G}_A$, $\mathcal{G}_T$ plus a check on the validity of the conservation law of leptons. (It may also give a test on time reversal invariance.)

The above considerations can be extended to μ^- capture in heavy nuclei. However, the uncertainty in the nuclear matrix elements would make the corresponding formulas less definite.

VIII. TIME REVERSAL INVARIANCE AND MACH'S PRINCIPLE

From various recent measurements we know that P as well as C are not conserved, at least in some of the weak interactions. There remains unanswered a very fundamental question which is whether $C{\cdot}P$ is invariant or not. Or, by the CPT theorem we may ask whether T (time reversal) is invariant or not. If it turns out that T may or may not be conserved, what will then be the implication? This leads us naturally to a discussion of Mach's principle.

Should we follow the spirit of Mach's principle, we would believe that the laws of physics cannot depend on the geometrical coordinate system that we happen to choose. There should exist no absolute system. The present asymmetry may then be made compatible with this interpretation of Mach's principle in two ways.

(i) If T is invariant, then $C{\cdot}P$ is invariant. The right-left symmetry in space is retained by changing particle to antiparticle as we change from a right-handed system to a left-handed system.[55]

(ii) If, experimentally, the weak interactionsare found to be not invariant under T, the over-all symmetry may still bemaintained by conjecturing the existence of two different kinds of elementary particles with the same masses, charges, and spins, but exhibiting opposite asymmetries under a space inversion. In such a picture, the observed right-left asymmetry is ascribed not to a basic non-invariance under space inversion but to a cosmologically local preponderance of one kind of the elementary particles over the other kind. Consequently, in this broader sense P is still conserved. By the CPT theorem, all interactions are also invariant under $C \cdot T$. Thus, the time reversal symmetry can also be retained by changing particles to antiparticles as we reverse the chronological order of any sequence of events.

If this is the case, then there must exist two types of protons, p_R and p_L, the right-handed one and the left-handed one. At the present time, the protons in the laboratory must be predominantly of only one kind, say p_R, which accounts for the observed asymmetry and the observed Fermi-Dirac statistical characters of the protons. This means that the free oscillation period between p_R and p_L must be longer than the age of the universe. They could, therefore, both be regarded as stable particles. It is reasonable to assume that there exists only one kind of electromagnetic field. Thus, in an experiment on pair productions by γ radiation we expect the same cross sections for

$$\gamma \rightarrow p_R + \bar{p}_R \qquad (8.1)$$

[55]C.N. Yang, International Congress on Theoretical Physics, Seattle, Sept. 1956 [*Revs. Mod. Phys.* **29**, 231 (1957)]; T.D. Lee and C.N. Yang, *Phys. Rev.* **105**, 1671 (1957). This possibility was also independently considered and particularly emphasized by L. Landau, *Nuclear Phys.* **3**, 127 (1957), and by E.P. Wigner, *Bull. Am. Phys. Soc.* **2**, 36 (1957).

and for

$$\gamma \to p_L + \bar{p}_L\,. \tag{8.2}$$

The $\bar{p}_L$ would appear to be a stable negative particle with a mass equal to that of a proton. The detection of such particles, if produced, may not be difficult.

Reprinted from THE PHYSICAL REVIEW, Vol. 112, No. 5, 1796–1798, December 1, 1958
Printed in U. S. A.

Decay Modes of a $(\theta+\bar{\theta})$ System*

M. GOLDHABER, T. D. LEE,† AND C. N. YANG‡
Brookhaven National Laboratory, Upton, New York

(Received August 12, 1958)

Relations among the different states of a system consisting of a θ meson, a $\bar{\theta}$ meson, and a certain number of pions are discussed. Theorems concerning certain correlations between the production probabilities for charged and uncharged θ mesons and the decay modes ($\theta_1{}^0$ and $\theta_2{}^0$) are obtained.

I

THE purpose of this note is to study the decay of a system which consists of a θ meson, a $\bar{\theta}$ meson, and a certain number of pions. Such a system can be produced, e.g., in a collision between an antinucleon and a nucleon. We shall show that by the combined use of the isotopic-spin rotation operator and the charge conjugation operator there exist some interesting correlations, not only in production but also between some of the decay modes of the θ and $\bar{\theta}$. To be specific, let us first consider the reaction

$$\bar{p}+n\rightarrow\theta+\bar{\theta}+m\pi. \tag{1}$$

We shall denote the final state of the m pions by D_m which represents both a specific charge distribution and a fixed momentum distribution of the pions. The charges and the decay modes[1] of the θ and $\bar{\theta}$ mesons are of interest here. Let us define $P_{ij}(D_m)$ to be[2] the probability of observing the m pions with a distribution D_m together with a θ_i of momentum $\mathbf{k}_a$ and a θ_j of momentum $\mathbf{k}_b$ where i (j) runs over $+$, $-$, 1, or 2 representing the cases in which[3] θ_i (θ_j) is θ^+, θ^-, $\theta_1{}^0$, or $\theta_2{}^0$.

We assume that the two decay modes 1 and 2 are given by

$$\theta_1=\frac{1}{\sqrt{2}}(\theta^0+\bar{\theta}^0),\quad \theta_2=\frac{-i}{\sqrt{2}}(\theta^0-\bar{\theta}^0);$$

i.e., (2)

$$\theta^0=\frac{1}{\sqrt{2}}(\theta_1+i\theta_2),\quad \bar{\theta}^0=\frac{1}{\sqrt{2}}(\theta_1-i\theta_2).$$

It is well known that this decomposition corresponds to the two experimental decay modes θ_1 and θ_2 if time-reversal invariance holds in the decay.

II

To study the dependence of P_{ij} on i and j, we consider all the states of a pair of θ and $\bar{\theta}$ of given momenta $\mathbf{k}_a$ $\mathbf{k}_b$. There are eight possible states altogether.

$$\begin{aligned}&\text{Charge}=1: && (+\bar{0}),\ (\bar{0}+);\\ &\text{Charge}=-1: && (-0),\ (0-);\\ &\text{Charge}=0: && (+\,-),\ (-\,+),\ (0\bar{0}),\ (\bar{0}0).\end{aligned} \tag{3}$$

Here we adopt the obvious notation, e.g., $(+\bar{0})$ means the state with a θ^+ meson having momentum $\mathbf{k}_a$ and a $\bar{\theta}^0$ meson with momentum $\mathbf{k}_b$. For the states with charge $=+1$, we have, by (2),

$$\begin{aligned}(+\bar{0})&=\frac{1}{\sqrt{2}}(+1)-\frac{i}{\sqrt{2}}(+2),\\ (\bar{0}+)&=\frac{1}{\sqrt{2}}(1+)-\frac{i}{\sqrt{2}}(2+).\end{aligned} \tag{4}$$

A general state of charge $+1$ is a superposition of these two states. One sees that in any superposition the decay modes $(+1)$ and $(+2)$ always have the same probability. Also the modes $(1+)$ and $(2+)$ have the same probability. Similar considerations can be extended to the states with charge $=0$ and -1. One easily proves in this way the following theorem:

Theorem 1.—

$$P_{\pm1}(D_m)=P_{\pm2}(D_m), \tag{5}$$

$$P_{1\pm}(D_m)=P_{2\pm}(D_m), \tag{6}$$

$$P_{11}(D_m)=P_{22}(D_m), \tag{7}$$

$$P_{12}(D_m)=P_{21}(D_m). \tag{8}$$

In proving this theorem the only assumptions are that the decomposition (2) holds and that the total strangeness of the pair $\theta\bar{\theta}$ is zero (so that, e.g., the pair $\theta^+\theta^0$ is excluded).

III

It is possible to obtain more identities if one uses the conservation of isotopic spin and of the charge conjugation operator in the strong interactions. To do this, one has to study the transformation of the eight states (3) under an isotopic spin rotation and/or charge conjugation. The problem is identical to that of the transformation of the states describing a nucleon-antinucleon

* Work performed under the auspices of the U. S. Atomic Energy Commission.

† Permanent address: Columbia University, New York, New York.

‡ Permanent address: Institute for Advanced Study, Princeton, New Jersey.

[1] M. Gell-Mann and A. Pais, Phys. Rev. **97**, 1387 (1955).

[2] If θ should have nonzero spin, then $P_{ij}(D_m)$ should be defined as the probability of finding θ_i with momentum k_a and spin S_a and θ_j with momentum k_b and spin S_b.

[3] Throughout this paper, a $\theta^\pm$ meson is identical with a $K^\pm$ meson, including all the different decay modes of $K^\pm$.

TABLE I. Eigenfunctions for which strangeness $S=0$. Notations are explained in the text.

$Q=I_3$	I	G	Wave function: Form a	Wave function: Form b
1	1	−1	$2^{-\frac{1}{2}}[(+\bar{0})+(\bar{0}+)]$	$=\frac{1}{2}[(+1)-i(+2)+(1+)-i(2+)]$
1	1	1	$2^{-\frac{1}{2}}[(+\bar{0})-(\bar{0}+)]$	$=\frac{1}{2}[(+1)-i(+2)-(1+)+i(2+)]$
−1	1	−1	$-2^{-\frac{1}{2}}[(0-)+(-0)]$	$=-\frac{1}{2}[(1-)+i(2-)+(-1)+i(-2)]$
−1	1	1	$-2^{-\frac{1}{2}}[(0-)-(-0)]$	$=-\frac{1}{2}[(1-)+i(2-)-(-1)-i(-2)]$
0	1	−1	$\frac{1}{2}[-(+-)+(0\bar{0})+(\bar{0}0)-(-+)]$	$=\frac{1}{2}[-(+-)-(-+)+(11)+(22)]$
0	1	1	$\frac{1}{2}[-(+-)+(0\bar{0})-(\bar{0}0)+(-+)]$	$=\frac{1}{2}[-(+-)+(-+)-i(12)+i(21)]$
0	0	−1	$\frac{1}{2}[-(+-)-(0\bar{0})+(\bar{0}0)+(-+)]$	$=\frac{1}{2}[-(+-)+(-+)+i(12)-i(21)]$
0	0	1	$\frac{1}{2}[-(+-)-(0\bar{0})-(\bar{0}0)-(-+)]$	$=\frac{1}{2}[-(+-)-(-+)-(11)-(22)]$

pair.[4,5] Following the arguments used in reference 4 for studying the latter problem, we first consider the four states[6] of a single θ or a single $\bar{\theta}$:

$$\begin{pmatrix} \theta^+ \\ \theta^0 \\ \bar{\theta}^0 \\ -\theta^- \end{pmatrix}. \tag{9}$$

The isotopic spin operators which operate on these states are

$$I_1=\tfrac{1}{2}\begin{pmatrix} 0 & 1 & 0 & 0 \\ 1 & 0 & 0 & 0 \\ 0 & 0 & 0 & 1 \\ 0 & 0 & 1 & 0 \end{pmatrix},\quad I_2=\tfrac{1}{2}\begin{pmatrix} 0 & -i & 0 & 0 \\ i & 0 & 0 & 0 \\ 0 & 0 & 0 & -i \\ 0 & 0 & i & 0 \end{pmatrix},$$

$$I_3=\tfrac{1}{2}\begin{pmatrix} 1 & 0 & 0 & 0 \\ 0 & -1 & 0 & 0 \\ 0 & 0 & 1 & 0 \\ 0 & 0 & 0 & -1 \end{pmatrix}. \tag{10}$$

Instead of discussing the charge conjugation operator C, we define, as in reference 4,

$$G=C\exp[i\pi I_2], \tag{11}$$

and obtain

$$G=\begin{pmatrix} 0 & 0 & 1 & 0 \\ 0 & 0 & 0 & 1 \\ -1 & 0 & 0 & 0 \\ 0 & -1 & 0 & 0 \end{pmatrix}. \tag{12}$$

The operator G commutes with I. The strangeness number operator is diagonal in this representation,

$$S=\begin{pmatrix} 1 & 0 & 0 & 0 \\ 0 & 1 & 0 & 0 \\ 0 & 0 & -1 & 0 \\ 0 & 0 & 0 & -1 \end{pmatrix}. \tag{13}$$

We notice that

$$SG+GS=0. \tag{14}$$

[4] T. D. Lee and C. N. Yang, Nuovo cimento 3, 749 (1956).

[5] See also L. Michel, Nuovo cimento 10, 319 (1953); A. Pais and R. Jost, Phys. Rev. 87, 871 (1952); D. Amati and B. Vitale, Nuovo cimento 2, 719 (1955); C. Goebel, Phys. Rev. 103, 258 (1956).

[6] Note the minus sign in front of the θ^- state. The convention used here is similar to that discussed in footnote 1 of reference 4.

For a system of two particles with momenta $\mathbf{k}_a$ and $\mathbf{k}_b$, each described by a four-component wave function (9), the operators I and S are additive, whereas G is multiplicative. Of the sixteen states of the two-particle system, eight belong to $S=0$ and are given in (3). For these states, the charge

$$Q=I_3 \tag{15}$$

and I^2, G, and S all commute and can be simultaneously diagonalized. The eight eigenstates can be easily constructed and are displayed in Table I in the column headed by "Form a." In the next column "Form b," these eigenfunctions are rewritten by the use of (2) so that their decay modes are explicitly exhibited.

In reaction (1) the process in general goes through many channels of I and G which produce interference effects. However, the interference between different G values disappears if we *average* over the collisions $(\bar{p}n)$ and $(n\bar{p})$, where the states are defined in a manner similar to Eq. (3). To see this, one writes

$$(\bar{p}n)=\tfrac{1}{2}[(\bar{p}n)+(n\bar{p})]+\tfrac{1}{2}[(\bar{p}n)-(n\bar{p})],$$
$$(n\bar{p})=\tfrac{1}{2}[(\bar{p}n)+(n\bar{p})]-\tfrac{1}{2}[(\bar{p}n)-(n\bar{p})].$$

The term $(\bar{p}n)+(n\bar{p})$ belongs to $G=-1$ and the term $(\bar{p}n)-(n\bar{p})$ to $G=+1$. The interference terms between $G=\pm 1$ therefore have opposite signs in $(\bar{p}n)$ and $(n\bar{p})$ and cancel exactly upon taking the average. In other words, after the average one may consider (1) as going through incoherent noninterfering channels with definite G values. Since the pions are[4] eigenstates of G, for any given distribution D_m of the pions, there are therefore no interferences between the states of the $(\theta\bar{\theta})$ system with different G values. Table I shows then that for charge $Q=1$, the first two rows cannot interfere and one obtains, in addition to (5) and (6), the identity

$$P_{+1}(D_m)=P_{1+}(D_m).$$

These considerations are easily extended to states with $Q=0$ and -1.

Theorem 2.—Consider $\bar{p}+n$ and $n+\bar{p}$ which are related by the substitution of a $\bar{p}$ by an n with equal momentum and spin and vice versa. After summation over these two initial states, the partial cross sections satisfy

$$P_{\pm 1}(D_m)=P_{1\pm}(D_m)=P_{\pm 2}(D_m)=P_{2\pm}(D_m), \quad (16)$$

$$P_{11}(D_m)=P_{22}(D_m), \quad (17)$$

$$P_{12}(D_m)=P_{21}(D_m), \quad (18)$$

and

$$P_{11}(D_m)+P_{22}(D_m)+P_{12}(D_m)+P_{21}(D_m) = P_{+-}(D_m)+P_{-+}(D_m). \quad (19)$$

We thus have the interesting result that in those collisions where θ^0 and $\bar{\theta}^0$ are produced, their decay modes (θ_1^0 and θ_2^0) are in general not independent of each other but are related in a manner given by Eqs. (17)–(19).

In the capture of an antiproton by a neutron, both at rest, it is clear that theorem 2 applies.

A simple consequence of (16)–(19) is that in a collision $\bar{p}+n$, one has

$$\tfrac{1}{2}(n_+ + n_-)=n_1=n_2, \quad (20)$$

where n_i is the total number of θ_i produced ($i=1, 2, +,$ or $-$) integrated over all angles.

IV

It is easy to see that the same identities (theorems 1 and 2) are also valid for the reaction

$$\bar{n}+p \rightarrow \theta+\bar{\theta}+m\pi. \quad (1')$$

Furthermore, for both reactions (1) and (1′) additional equalities and inequalities may be obtained when m takes on some specific small values. In all cases, we found it convenient to use the eigenstates tabulated in Table I. They are useful also in discussing

$$p+\bar{p} \rightarrow \theta+\bar{\theta}+m\pi,$$

and

$$n+\bar{n} \rightarrow \theta+\bar{\theta}+m\pi.$$

Reprinted from THE PHYSICAL REVIEW, Vol. 122, No. 6, 1954–1961, June 15, 1961
Printed in U. S. A.

Some Considerations on Global Symmetry

T. D. LEE AND C. N. YANG
Institute for Advanced Study, Princeton, New Jersey
(Received February 3, 1961)

If the recently discovered Y^* state is related to the $T=\frac{3}{2}$, $J=\frac{3}{2}$ resonance in πp scattering, global symmetry considerations should become relevant. In this paper, global symmetry is discussed with a view to understanding its group structure. Also discussed is a possibility of reconciling the conflict, pointed out by Pais, between certain experimental results and global symmetry. The partial widths of the Y^* state are calculated and also those of the companion excited states Z^* and Ξ^*. A generalization of the quantum number G is discussed.

1. INTRODUCTION

RECENT experiments[1,2] have established the existence of an excited state $Y^{*\pm}$ in the $\Lambda+\pi^{\pm}$ system. The spin and parity of the state are not yet measured. As discussed in reference 1, the state shows certain resemblance to the $J=\frac{3}{2}$, $T=\frac{3}{2}$, p-state resonance N^* of the $p+\pi$ system, and the resemblance is reminiscent of the concept of global symmetry.

In this paper we proceed along this line of thinking and assume that indeed the $Y^*=\Lambda+\pi^{\pm}$ resonance is in the $J=\frac{3}{2}$, p state, and that the resonance is related to the $J=\frac{3}{2}$, $T=\frac{3}{2}$, p-state resonance N^* of the $p+\pi$ system by global symmetry. To analyze this relation it is necessary to know the quantum numbers of various states with respect to the global symmetry operations. It is therefore important to know the structure of the global symmetry group. Now global symmetry[3–6] means some symmetry, larger than isotopic spin invariance, that describes an approximate analogy between the various baryons. But in the literature its group property has not been fully discussed. We shall in this paper formulate in mathematical terms the requirements that global symmetry must satisfy. It appears that the simplest group $\mathcal{G}_0$ satisfying these requirements can be generated by three independent 2-dimensional unitary unimodular transformations together with a discrete transformation. Adopting this group as the global symmetry group we then try to assign quantum numbers to the various particles and the resonance states Y^*, N^*.

Certain approximate relations are then written down between the widths of Y^* and N^*, and also for the various partial widths of Y^*. Companion resonance states Z^* and Ξ^* are also discussed.

It has been pointed out by Pais[5] that any kind of global symmetry is in conflict with certain experimental facts. We suggest in Sec. 6 that if global symmetry is needed to understand the resonant state Y^*, a way to resolve Pais' conflict is to have the K mesons as a mixture of states which have different quantum numbers under the global symmetry transformations. The interactions between each of these states and other particles could still predominantly satisfy global symmetry. Such a picture, while not completely satisfactory, does offer a possible consistent scheme incorporating global symmetry that leads to useful experimental information.

It should be emphasized that much of our results about widths have already been discussed in the literature from the viewpoint of symmetry considerations.[7,8] Furthermore a detailed calculation using a specific dynamical model has been performed by Amati, Stanghellini, and Vitale[7] for the states Y^* and Z^*. Various discussions on the global symmetry group properties have also existed in the literature.[3–6] The present paper is written not in the spirit of presenting something entirely new, nor even in that of presenting something which we believe to be necessarily relevant[9] to physical facts. But if a similarity between Y^* and N^* exists, an analysis along the present line would be useful.

For completeness we include in Sec. 8 a discussion of charge conjugation invariance, together with a generalization of the quantum number G. Also included are some remarks in Sec. 9 concerning a global symmetry that does not put Ξ and nucleons in the same multiplet.

2. REQUIREMENTS ON THE GLOBAL SYMMETRY GROUP

The global symmetry group must by definition contain the isotopic spin group and the strangeness group [defined by the operators $\exp(iS\theta)$ or $\exp(i(S+N)\theta)$, where S=strangeness and N=baryon number; the strangeness group commutes with the isotopic spin group]. It has an 8×8 unitary representation to which the 8 baryons belong. [Cf. Sec. 9 for the case of a symmetry between N, Σ, and Λ only.] In order that the

[1] M. Alston, L. W. Alvarez, P. Eberhard, M. L. Good, W. Graziano, H. K. Ticho, and S. G. Wojcicki, Phys. Rev. Letters **5**, 520 (1960).

[2] M. Ferro-Luzzi, J. P. Berge, J. Kirz, J. J. Murray, A. H. Rosenfeld, and M. Watson, Bull. Am. Phys. Soc. **5**, 509 (1960); H. J. Martin, W. Chinowsky, L. B. Leipuner, F. T. Shively, and R. K. Adair, Bull. Am. Phys. Soc. **6**, 40 (1961).

[3] J. Schwinger, Phys. Rev. **104**, 1164 (1956); Ann. Phys. **2**, 407 (1957).

[4] M. Gell-Mann, Phys. Rev. **106**, 1296 (1957).

[5] A. Pais, Phys. Rev. **110**, 574 (1958); **110**, 1480 (1958); **112**, 624 (1958); **122**, 317 (1961). See also A. Pais, *Proceedings of the Fifth Annual Rochester Conference on High-Energy Physics* (Interscience Publishers, Inc., New York, 1955).

[6] J. Tiomno, Nuovo cimento **6**, 69 (1957); R. E. Behrends, Nuovo cimento **11**, 424 (1959). See also G. Feinberg and F. Gürsey, Phys. Rev. **114**, 1153 (1959).

[7] D. Amati, A. Stanghellini, and B. Vitale, Nuovo cimento **13**, 1143 (1959); Phys. Rev. Letters **5**, 524 (1960).

[8] Ph. Meyer, J. Prentki, and Y. Yamaguchi, Phys. Rev. Letters **5**, 442 (1960).

[9] See the discussion of R. H. Dalitz and S. Tuan, Phys. Rev. Letters **2**, 425 (1959), Ann. Phys. **10**, 307 (1960). See also M. Ross and G. Shaw, Phys. Rev. Letters **5**, 578 (1960).

8 baryons be analogous to each other under the global symmetry group, the representation must be irreducible. For the isotopic spin rotation subgroup the 8×8 representation breaks up into two doublets (N and Ξ), one triplet (Σ), and one singlet (Λ). For the strangeness subgroup this breakup must conform with the usual assignments of $S+N=+1, -1, 0, 0$ for N, Ξ, Σ, and Λ, respectively.

To state the above requirements explicitly we introduce as usual the states

$$\begin{aligned} Y^+ &= \Sigma^+, \\ Y^0 &= \frac{1}{\sqrt{2}}(\Sigma^0+\Lambda^0), \\ Z^0 &= \frac{1}{\sqrt{2}}(\Sigma^0-\Lambda^0), \\ Z^- &= \Sigma^-, \end{aligned} \tag{1}$$

and write

$$N\equiv\begin{Vmatrix} p \\ n \end{Vmatrix},\quad Y\equiv\begin{Vmatrix} Y^+ \\ Y^0 \end{Vmatrix},\quad Z\equiv\begin{Vmatrix} Z^0 \\ Z^- \end{Vmatrix},\quad \Xi\equiv\begin{Vmatrix} \Xi^0 \\ \Xi^- \end{Vmatrix}, \tag{2}$$

$$B\equiv\begin{Vmatrix} N \\ \Xi \\ Y \\ Z \end{Vmatrix}\equiv\begin{Vmatrix} p \\ n \\ \Xi^0 \\ \Xi^- \\ Y^+ \\ Y^0 \\ Z^0 \\ Z^- \end{Vmatrix}. \tag{3}$$

It is now useful to introduce (operating on the column matrix B) the following three sets of operators, each set satisfying the commutation relations for angular momenta:

$$(\mathcal{L}_1,\mathcal{L}_2,\mathcal{L}_3):\quad \mathcal{L}_i\equiv\tfrac{1}{2}\begin{Vmatrix} \sigma_i & 0 & 0 & 0 \\ 0 & \sigma_i & 0 & 0 \\ 0 & 0 & \sigma_i & 0 \\ 0 & 0 & 0 & \sigma_i \end{Vmatrix},\quad (i=1, 2, 3), \tag{4}$$

$$(\mathfrak{M}_1,\mathfrak{M}_2,\mathfrak{M}_3):\quad \mathfrak{M}_1\equiv\tfrac{1}{2}\begin{Vmatrix} 0 & 0 & 0 & 0 \\ 0 & 0 & 0 & 0 \\ 0 & 0 & 0 & I \\ 0 & 0 & I & 0 \end{Vmatrix},\quad \mathfrak{M}_2\equiv\tfrac{1}{2}i\begin{Vmatrix} 0 & 0 & 0 & 0 \\ 0 & 0 & 0 & 0 \\ 0 & 0 & 0 & -I \\ 0 & 0 & I & 0 \end{Vmatrix},\quad \mathfrak{M}_3\equiv\tfrac{1}{2}\begin{Vmatrix} 0 & 0 & 0 & 0 \\ 0 & 0 & 0 & 0 \\ 0 & 0 & I & 0 \\ 0 & 0 & 0 & -I \end{Vmatrix}, \tag{5}$$

$$(\mathfrak{N}_1,\mathfrak{N}_2,\mathfrak{N}_3):\quad \mathfrak{N}_1\equiv\tfrac{1}{2}\begin{Vmatrix} 0 & I & 0 & 0 \\ I & 0 & 0 & 0 \\ 0 & 0 & 0 & 0 \\ 0 & 0 & 0 & 0 \end{Vmatrix},\quad \mathfrak{N}_2\equiv\tfrac{1}{2}i\begin{Vmatrix} 0 & -I & 0 & 0 \\ I & 0 & 0 & 0 \\ 0 & 0 & 0 & 0 \\ 0 & 0 & 0 & 0 \end{Vmatrix},\quad \mathfrak{N}_3\equiv\tfrac{1}{2}\begin{Vmatrix} I & 0 & 0 & 0 \\ 0 & -I & 0 & 0 \\ 0 & 0 & 0 & 0 \\ 0 & 0 & 0 & 0 \end{Vmatrix}, \tag{6}$$

where

$$\sigma_1\equiv\begin{Vmatrix} 0 & 1 \\ 1 & 0 \end{Vmatrix},\quad \sigma_2\equiv\begin{Vmatrix} 0 & -i \\ i & 0 \end{Vmatrix},\quad \sigma_3\equiv\begin{Vmatrix} 1 & 0 \\ 0 & -1 \end{Vmatrix},$$

and

$$I=\begin{Vmatrix} 1 & 0 \\ 0 & 1 \end{Vmatrix}.$$

Clearly $\mathcal{L}_i$, $\mathfrak{M}_j$, $\mathfrak{N}_k$ all commute for any i, j, k.

The physical observables, Q=charge, S=strangeness, N=baryon number, T_1, T_2, T_3=isotopic spin are related to $\mathcal{L}$, $\mathfrak{M}$, and $\mathfrak{N}$ by

$$T_i=\mathcal{L}_i+\mathfrak{M}_i, \tag{7}$$

$$\tfrac{1}{2}(S+N)=\mathfrak{N}_3, \tag{8}$$

$$Q=T_3+\mathfrak{N}_3=\mathcal{L}_3+\mathfrak{M}_3+\mathfrak{N}_3. \tag{9}$$

The 8×8 representation must (i) contain $\exp(iT_1\theta_1)$, $\exp(iT_2\theta_2)$, $\exp(iT_3\theta_3)$, $\exp(i\mathfrak{N}_3\theta_4)$. In other words, T_1, T_2, T_3, $\mathfrak{N}_3$ are among the infinitesimal generators of the representation. Furthermore (ii) the 8×8 representation is irreducible.

Many possible 8×8 representations of groups can be found that satisfy (i) and (ii). Since we are not interested in unnecessary additional symmetries, we take the group to be *isomorphic* with the representation, and shall identify a group element with its 8×8 representation. The simplest such possibility, $\mathcal{G}_0$, is discussed in the following sections. Other possibilities are discussed in Appendix A.

3. THE GROUP $\mathcal{G}_0$

The group contains arbitrary $\mathcal{L}$ transformations [i.e., $U_\mathcal{L}\equiv\exp i(l_1\mathcal{L}_1+l_2\mathcal{L}_2+l_3\mathcal{L}_3)$, where $l_1l_2l_3$ are real numbers], arbitrary $\mathfrak{M}$ transformations [$\exp i(m_1\mathfrak{M}_1+m_2\mathfrak{M}_2+m_3\mathfrak{M}_3)$, which mixes Y and Z], and arbitrary $\mathfrak{N}$ transformations [$\exp i(n_1\mathfrak{N}_1+n_2\mathfrak{N}_2+n_3\mathfrak{N}_3)$, which mixes N and Ξ]. The product of all these is reducible since no mixing of N, Ξ with Y, Z has been introduced. To effect such a mixing we introduce the discrete element[10]

$$R\equiv\begin{Vmatrix} 0 & 0 & I & 0 \\ 0 & 0 & 0 & I \\ I & 0 & 0 & 0 \\ 0 & I & 0 & 0 \end{Vmatrix}, \tag{10}$$

[10] This discrete operator has been discussed by various authors. See reference 5.

TABLE I. The irreducible representations of the group $\mathcal{G}_0$. The quantity $\chi_{\mathcal{M}}(U)$ is the character of the 2×2 matrix U for the irreducible representations of dimension $2\mathcal{M}+1$.

Representation	Dimension	Character for elements (11)	Character for elements (12)
$(\mathcal{L}\mathcal{M}\mathcal{M}+1)$	$(2\mathcal{L}+1)(2\mathcal{M}+1)^2$	$\chi_{\mathcal{L}}(U_{\mathcal{L}})\chi_{\mathcal{M}}(U)\chi_{\mathcal{M}}(U')$	$\chi_{\mathcal{L}}(U_{\mathcal{L}})\chi_{\mathcal{M}}(UU')$
$(\mathcal{L}\mathcal{M}\mathcal{M}-1)$	$(2\mathcal{L}+1)(2\mathcal{M}+1)^2$	$\chi_{\mathcal{L}}(U_{\mathcal{L}})\chi_{\mathcal{M}}(U)\chi_{\mathcal{M}}(U')$	$-\chi_{\mathcal{L}}(U_{\mathcal{L}})\chi_{\mathcal{M}}(UU')$
$(\mathcal{L}\mathcal{M}\mathcal{N})$ $\mathcal{M}>\mathcal{N}$	$2(2\mathcal{L}+1)(2\mathcal{M}+1)(2\mathcal{N}+1)$	$\chi_{\mathcal{L}}(U_{\mathcal{L}})[\chi_{\mathcal{M}}(U)\chi_{\mathcal{N}}(U')$ $+\chi_{\mathcal{M}}(U')\chi_{\mathcal{N}}(U)]$	0

and other necessary elements to form a group. The elements of the group are then the 8×8 matrices of the form

$$U_{\mathcal{L}}\begin{Vmatrix} aI & bI & 0 & 0 \\ -b^*I & a^*I & 0 & 0 \\ 0 & 0 & a'I & b'I \\ 0 & 0 & -b'^*I & a'^*I \end{Vmatrix}, \tag{11}$$

and those of the form

$$U_{\mathcal{L}}\begin{Vmatrix} 0 & 0 & aI & bI \\ 0 & 0 & -b^*I & a^*I \\ a'I & b'I & 0 & 0 \\ -b'^*I & a'^*I & 0 & 0 \end{Vmatrix}. \tag{12}$$

Here

$$\begin{Vmatrix} a & b \\ -b^* & a^* \end{Vmatrix} \equiv U = \text{arbitrary } 2\times2 \text{ unimodular unitary matrix,}$$

and (13)

$$\begin{Vmatrix} a' & b' \\ -b'^* & a'^* \end{Vmatrix} \equiv U' = \text{arbitrary } 2\times2 \text{ unimodular unitary matrix.}$$

This group will be called $\mathcal{G}_0$. It has an invariant subgroup (11) which is the direct product[11] of three SU_2, and the quotient of the group by this invariant subgroup is the two-element group. [It is, however, not the direct product of the two-element group with the invariant subgroup.]

In Appendix A we shall give a few other possible groups satisfying conditions (i) and (ii).

[11] We use SU_n to denote the group of unitary $n\times n$ matrices with determinant unity. Strictly speaking, the invariant subgroup (11) is only locally identical with $SU_2\times SU_2\times SU_2$. If one changes the signs of $U_{\mathcal{L}}$ and a, b, a', b' at the same time, (11) is unchanged. Hence (11) has a 1 to 2 homomorphism with $SU_2\times SU_2\times SU_2$. In other words, $SU_2\times SU_2\times SU_2$ is the covering group of (11). Similarly $\mathcal{G}_0$, which is defined to be the group of matrices of the form (11) and (12), has a covering group which we shall call $\mathcal{G}_0'$. $\mathcal{G}_0'$ has a 2 to 1 homomorphism with $\mathcal{G}_0$. It has an invariant subgroup $SU_2\times SU_2\times SU_2$. The relationship between $\mathcal{G}_0'$ and $\mathcal{G}_0$ is entirely similar to the familiar relationship between SU_2 and O_3, the group of 3×3 rotations. To simplify matters we shall not dwell on the difference between $\mathcal{G}_0$ and $\mathcal{G}_0'$ in the text. The irreducible representations in Table I are actually representations of $\mathcal{G}_0'$. Those representations $(\mathcal{L}\mathcal{M}\mathcal{M}\lambda)$ with $\mathcal{L}+2\mathcal{M}=$integer, and those representations $(\mathcal{L}\mathcal{M}\mathcal{N})$, $\mathcal{M}>\mathcal{N}$, with $\mathcal{L}+\mathcal{M}+\mathcal{N}=$integer are also representations of $\mathcal{G}_0$. (The others are not single-valued representations of $\mathcal{G}_0$.) Since the baryons belong to a representation of this type, and all known particles have transition elements to a collection of baryons and antibaryons, only representations of the same type (i.e., single-valued representations of $\mathcal{G}_0$) enter into the discussion of known particles.

4. IRREDUCIBLE REPRESENTATION OF $\mathcal{G}_0$

For any representation of $\mathcal{G}_0$, the infinitesimal operators $\mathcal{L}_1$, $\mathcal{L}_2$, $\mathcal{L}_3$, $\mathcal{M}_1$, $\mathcal{M}_2$, $\mathcal{M}_3$, $\mathcal{N}_1$, $\mathcal{N}_2$, $\mathcal{N}_3$ form three sets of commuting angular momenta. One can diagonalize $\mathcal{L}^2$, $\mathcal{L}_3$, $\mathcal{M}^2$, $\mathcal{M}_3$, $\mathcal{N}^2$, $\mathcal{N}_3$ simultaneously. Now

$$\begin{aligned} R\mathcal{L}_iR^{-1}&=\mathcal{L}_i,\\ R\mathcal{M}_iR^{-1}&=\mathcal{N}_i,\\ R\mathcal{N}_iR^{-1}&=\mathcal{M}_i.\end{aligned} \tag{14}$$

Hence in any representation the set of eigenvalues of $\mathcal{M}^2$ must be the same as those of $\mathcal{N}^2$. One has therefore two kinds of irreducible representations:

(α) $\mathcal{L}^2$, $\mathcal{M}^2$, $\mathcal{N}^2$ have unique eigenvalues $\mathfrak{L}(\mathfrak{L}+1)$, $\mathfrak{M}(\mathfrak{M}+1)$, $\mathfrak{M}(\mathfrak{M}+1)$, respectively. Since R commutes with $\mathcal{M}_3+\mathcal{N}_3$, the state with $\mathcal{M}_3=\mathcal{N}_3=\mathfrak{M}$ is an eigenstate of R. The eigenvalue λ can be ±1. We denote this representation by the symbol $(\mathfrak{L}\mathfrak{M}\mathfrak{M}\lambda)$, where $2\mathfrak{L}=$integer$\geqq0$, $2\mathfrak{M}=$integer$\geqq0$, $\lambda=\pm1$.

The states of this representation are designated by $\mathcal{L}_3$, $\mathcal{M}_3$, $\mathcal{N}_3$, each running in integral steps between and including $\pm\mathfrak{L}$, $\pm\mathfrak{M}$, $\pm\mathfrak{M}$ respectively. The operator R switches the indices $\mathcal{M}_3$ and $\mathcal{N}_3$ for a state:

$$R|\mathcal{M}_3=a,\quad \mathcal{N}_3=b\rangle=\lambda|\mathcal{M}_3=b,\quad \mathcal{N}_3=a\rangle. \tag{15}$$

(β) $\mathcal{L}^2$ has a unique eigenvalue $\mathfrak{L}(\mathfrak{L}+1)$. $\mathcal{M}^2$ and $\mathcal{N}^2$ each has two eigenvalues $\mathfrak{M}(\mathfrak{M}+1)$ and $\mathfrak{N}(\mathfrak{N}+1)$ where $\mathfrak{M}\neq\mathfrak{N}$. We denote this representation by the symbol $(\mathfrak{L}\mathfrak{M}\mathfrak{N})$, where $2\mathfrak{L}$, $2\mathfrak{M}$, and $2\mathfrak{N}$ are integers $\geqq0$ and $\mathfrak{M}>\mathfrak{N}$.

The states of this representation are states for which

$$\{\mathcal{M}^2=\mathfrak{M}(\mathfrak{M}+1),\quad \mathcal{M}_3=-\mathfrak{M},-\mathfrak{M}+1,\cdots+\mathfrak{M},$$
$$\text{while } \mathcal{N}^2=\mathfrak{N}(\mathfrak{N}+1),\quad \mathcal{N}_3=-\mathfrak{N},-\mathfrak{N}+1,\cdots+\mathfrak{N}\};$$

and

$$\{\mathcal{M}^2=\mathfrak{N}(\mathfrak{N}+1),\quad \mathcal{M}_3=-\mathfrak{N},-\mathfrak{N}+1,\cdots+\mathfrak{N},$$
$$\text{while } \mathcal{N}^2=\mathfrak{M}(\mathfrak{M}+1),\quad \mathcal{N}_3=-\mathfrak{M},-\mathfrak{M}+1,\cdots+\mathfrak{M}\}.$$

The operator R switches the states between these two sets. In a suitable representation, R satisfies

$$R|\mathfrak{M}\mathfrak{M}_3\mathfrak{N}\mathfrak{N}_3\rangle=|\mathfrak{N}\mathfrak{N}_3\mathfrak{M}\mathfrak{M}_3\rangle.$$

The dimensions of the irreducible representations are tabulated in Table I. Also given are the characters of the representations. From the characters the decomposition of the direct product of two representations can be easily

TABLE II. Quantum number assignments for particles and excited states. $T=\mathfrak{L}+\mathfrak{M}$=isotopic spin. P_K=parity of K meson. The quantum number G_1 is explained in Sec. 8. In this table only the $(0,\frac{1}{2},\frac{1}{2},\lambda_K)$ part is listed for the K mesons.

Particle	Representation	$\mathfrak{L}$	$\mathfrak{L}_3$	$\mathfrak{M}$	$\mathfrak{M}_3$	$\mathfrak{N}$	$\mathfrak{N}_3$	G_1	T	R
p, n	$(\frac{1}{2},\frac{1}{2},0)$	$\frac{1}{2}$	$\pm\frac{1}{2}$	0	0	$\frac{1}{2}$	$\frac{1}{2}$		$\frac{1}{2}$	
Ξ^0, Ξ^-		$\frac{1}{2}$	$\pm\frac{1}{2}$	0	0	$\frac{1}{2}$	$-\frac{1}{2}$		$\frac{1}{2}$	
Y^+, Y^0		$\frac{1}{2}$	$\pm\frac{1}{2}$	$\frac{1}{2}$	$\frac{1}{2}$	0	0		1, 0	
Z^0, Z^-		$\frac{1}{2}$	$\pm\frac{1}{2}$	$\frac{1}{2}$	$-\frac{1}{2}$	0	0		1, 0	
π^+	$(1,0,0,\lambda_\pi)$	1	1	0	0	0	0	-1	1	λ_π
π^0	$\lambda_\pi=\pm1$	1	0	0	0	0	0	-1	1	λ_π
π^-		1	-1	0	0	0	0	-1	1	λ_π
K^+	$(0,\frac{1}{2},\frac{1}{2},\lambda_K)$	0	0	$\frac{1}{2}$	$\frac{1}{2}$	$\frac{1}{2}$	$\frac{1}{2}$	$-\lambda_K$	$\frac{1}{2}$	λ_K
K^0	$\lambda_K=\pm1$	0	0	$\frac{1}{2}$	$-\frac{1}{2}$	$\frac{1}{2}$	$\frac{1}{2}$	$-\lambda_K$	$\frac{1}{2}$	K_1^0: $\lambda_K P_K$
$\bar{K}^0$		0	0	$\frac{1}{2}$	$\frac{1}{2}$	$\frac{1}{2}$	$-\frac{1}{2}$	$-\lambda_K$	$\frac{1}{2}$	K_2^0: $-\lambda_K P_K$
K^-		0	0	$\frac{1}{2}$	$-\frac{1}{2}$	$\frac{1}{2}$	$-\frac{1}{2}$	$-\lambda_K$	$\frac{1}{2}$	λ_K
N^*	$(\frac{3}{2},\frac{1}{2},0)$	$\frac{3}{2}$	...	0	0	$\frac{1}{2}$	$\frac{1}{2}$		$\frac{3}{2}$	
Ξ^*		$\frac{3}{2}$	...	0	0	$\frac{1}{2}$	$-\frac{1}{2}$		$\frac{3}{2}$	
Y^*		$\frac{3}{2}$	...	$\frac{1}{2}$	...	0	0		1	
Z^*		$\frac{3}{2}$	...	$\frac{1}{2}$	...	0	0		2	

found in the standard way. (Except for the quantum number λ, it can also be found by the usual vector sum rule for $\mathfrak{L}$, $\mathfrak{M}$, and $\mathfrak{N}$ separately.)

5. QUANTUM NUMBERS

To assign quantum numbers to the states we first notice that Eqs. (7)–(9) give the isotopic spin, the strangeness, and the charge in terms of these quantum numbers.

The 8 baryons clearly belong to the representation $(\frac{1}{2},\frac{1}{2},0)$. It seems natural that the pions should be assigned to the 3×3 representation $(1,0,0,\lambda_\pi)$. The two possibilities $\lambda_\pi=\pm1$ are, of course, physically different, and differentiable. It seems natural to assign the K mesons to the representation $(0,\frac{1}{2},\frac{1}{2},\lambda_K)$ with again the two possibilities $\lambda_K=\pm1$. These assignments are tabulated in Table II.

For the state $N^*=\pi+p$ we notice that $\pi+p$ always belongs to either $(\frac{3}{2},\frac{1}{2},0)$ $T=\frac{3}{2}$, or $(\frac{1}{2},\frac{1}{2},0)$ $T=\frac{1}{2}$. But N^* has a total $T=\frac{3}{2}$. Hence it belongs to $(\frac{3}{2},\frac{1}{2},0)$. The natural assumption is therefore that Y^* is in the same multiplet structure $(\frac{3}{2},\frac{1}{2},0)$, as indicated in Table II. Since Y^* can go into $\Lambda+\pi$, its isotopic spin $T=1$. The multiplet $(\frac{3}{2},\frac{1}{2},0)$ also contains a $T=2$ state[1,4,7] which will be called Z^*. In addition to the 3 Y^* states and 5 Z^* states there should also be 4 Ξ^* states with $T=\frac{3}{2}$.

6. BREAKDOWN OF GLOBAL SYMMETRY

Even if global symmetry has any valid basis, there must be relatively strong interactions that violate it. One manifestation of this violation lies in the mass difference between the hyperons. Another manifestation was first pointed out by Pais,[5] who showed that the following reactions

$$\pi^++p\rightarrow\Sigma^++K^+, \tag{16}$$

$$K^++n\rightarrow K^0+p, \tag{17}$$

and many others violate global symmetry. For the group $\mathcal{G}_0$ discussed above, the conservation of $\mathfrak{L}_3$ is violated by both (16) and (17).

In face of these difficulties, does global symmetry have any validity at all? And if it has, does it ever produce useful physical information?

It would be difficult to answer these questions. But if the answers to the above questions are affirmative, presumably the baryons and the states N^*, Y^* allow more directly the application of global symmetry than reactions (16) and (17). For example, if the global-symmetry-destroying interactions produce relatively little mixing for the baryons, pions and Y^*, N^*, but produce large mixing for the K mesons, then apparent violation of global symmetry is not unnatural for (16) and (17). While the mixing may be the multiplet $(0,\frac{1}{2},\frac{1}{2})$ with any multiplet possessing a $T=\frac{1}{2}$, $\mathfrak{N}=\frac{1}{2}$ component, it seems that the mixing of $(0,\frac{1}{2},\frac{1}{2})$ with $(1,\frac{1}{2},\frac{1}{2})$ is the simplest possibility.

In this view, then, the usual interactions (baryons, pions, and K mesons) are regarded as predominantly globally symmetrical. The globally unsymmetrical interactions give rise to, among others, two effects: (a) mass splitting of the states within each multiplet. (b) a strong mixing for the K meson of a $(1,\frac{1}{2},\frac{1}{2})$ $T=\frac{1}{2}$ component with the $(0,\frac{1}{2},\frac{1}{2})$ state. The globally unsymmetrical interactions may, for example, have a very small range, so that the two effects (a) and (b) are the only ones that one need consider as causing global unsymmetry in the zeroth approximation. The influence of global unsymmetry is then quite limited in scope, though not in magnitude, and one can derive consequences that can be checked with experimental information. [This is true only insofar as one does not probe into the *very small* range where the strong unsymmetric force is assumed to be effective. It may be instructive to recall the well-known symmetry between $e^\pm$ and $\mu^\pm$. In that case, the asymmetry between these two particles

TABLE III. Phase-space factor Ω, projection weight w, and relative partial widths of resonance levels. The weights w are calculated from the quantum number assignments. The phase space factor Ω is computed from experimental resonance energies for N^* and Y^*, and from an assumed energy spectrum for Z^* and Ξ^*. The partial widths of other disintegration processes, such as $Z^{*+}\rightarrow\pi^+ +\Sigma^0$ etc., can be inferred from the table through a simple isotopic spin rotation.

Particle	Total energy (Mev)	Disintegration products	Ω (Mev³)	w	Computed relative partial width
$(N^*)^{++}$	1237	π^++p	9.7×10^6	1	1
Y^{*+}	1385	$\pi^++\Lambda$	7.3×10^6	$\frac{2}{3}$	0.5
		$\Sigma^++\pi^0$	1.6×10^6	$\frac{1}{6}$	0.03
		$\Sigma^0+\pi^+$	1.6×10^6	$\frac{1}{6}$	0.03
Z^{*++}	~1539(?)	$\pi^++\Sigma^+$	~$18(?)\times10^6$	1	~1.9(?)
Ξ^{*+}	~1637(?)	$\pi^++\Xi^0$	~$15(?)\times10^6$	1	~1.5(?)

seems to be completely characterized by their large mass difference which, presumably, is also generated by some unsymmetrical forces, strong in magnitude but remarkably limited in its symmetry destroying effects.]

If one asks how does it happen that only the K particles have a strong mixing, a possible answer could be that without the globally unsymmetrical interaction two multiplets $(1,\frac{1}{2},\frac{1}{2})$ and $(0,\frac{1}{2},\frac{1}{2})$ happen to lie relatively close together and, therefore, result in large mixings for the states with $T=\frac{1}{2}$. It would then be reasonable to expect the existence of other excited states K^*.

In such a picture the K meson is a mixture of $(0,\frac{1}{2},\frac{1}{2})$ and $(1,\frac{1}{2},\frac{1}{2})$, each of which, in interacting with the other particles, still predominantly *satisfies* global symmetry. Thus, e.g., in reactions such as K^-+baryon with multiple pion productions, one can apply the global symmetry arguments and obtain equalities and inequalities between the various related processes.

7. POSITIONS AND WIDTHS OF N^*, Y^*, Z^*, AND Ξ^*

The discussion of Sec. 6 suggests a zeroth order calculation of the partial widths of $N^*Y^*Z^*$ and Ξ^*, for the processes tabulated in Table III. These processes represent a transition from a $(\frac{3}{2},\frac{1}{2},0)$ multiplet to a product of a $(\frac{1}{2},\frac{1}{2},0)$ multiplet and a $(1,0,0,\lambda_\pi)$ multiplet. The decomposition therefore yields unique weights w which are related to the squares of the appropriate transition amplitude. The calculation of these weights from the usual tables of Clebsch-Gordan coefficients is straightforward and the result is tabulated in Table III. Besides these weights due to the projection of the initial state on final states, there is also a phase-space-potential-barrier factor Ω for the p-wave state. We take it to be given by

$$\Omega=q^3E_B/(E_B+E_\pi), \tag{18}$$

where $q=$ momentum of pion in the rest system of the resonance state, $E_B=$ total energy of the final baryon, and $E_\pi=$ total energy of the final pion. In the approximation that other effects due to global asymmetrical interactions are neglected, the partial widths of each resonance level are proportional to the appropriate products of w and Ω. To calculate Ω one needs the excitation energy of the resonance states. For N^* and Y^* we take the values in reference 1. To guess at the energies of Z^* and Ξ^* we write the total energy E of the excited state in the multiplet $(\frac{3}{2},\frac{1}{2},0)$ in the form

$$E(\equiv E_B+E_\pi)$$
$$=E_{N^*}+\alpha'\mathcal{L}\cdot\mathfrak{M}+\beta'(\mathfrak{N}^2-\tfrac{3}{4})+\gamma'(\mathfrak{N}_3-\tfrac{1}{2}), \tag{19}$$

where α', β' and γ' are constants. Similarly, for the 8 baryons in the multiplet $(\frac{1}{2},\frac{1}{2},0)$ we have an analogous expression

$$E=E_N+\alpha\mathcal{L}\cdot\mathfrak{M}+\beta(\mathfrak{N}^2-\tfrac{3}{4})+\gamma(\mathfrak{N}_3-\tfrac{1}{2}), \tag{20}$$

where

$$\alpha=E_\Sigma-E_\Lambda\cong77\text{ Mev},$$
$$\beta=\tfrac{4}{3}[\tfrac{1}{2}(E_\Xi+E_N)-\tfrac{1}{4}(3E_\Sigma+E_\Lambda)]\cong-59\text{ Mev}, \tag{21}$$
$$\gamma=-(E_\Xi-E_N)\cong-380\text{ Mev}.$$

One sees that by taking[12]

$$\alpha'=\alpha,\quad \beta'=\beta,\quad \gamma'\cong-400\text{ Mev}, \tag{22}$$

one obtains the experimental resonance energy for Y^*. With this choice the resonance energies for Z^* and Ξ^* can be computed and are tabulated in Table III, with the corresponding phase space factors Ω.

The last column of Table III shows a smaller total width for Y^* than N^* [in the ratio of approximately 0.56:1], and shows a very small branching ratio of $Y^*\rightarrow\Sigma+\pi$. Both of these are in general agreement with experimental information.[1]

8. CHARGE CONJUGATION INVARIANCE

With the inclusion of the unitary operator C, representing charge conjugation, the symmetry group is enlarged. The irreducible representations become larger in general, corresponding to, e.g., the fact that a particle and its antiparticle have the same mass. To study the combined group generated from $\mathcal{G}_0$, C and the baryon number gauge transformation $\exp(i\theta N)$, we start from their commutation relations. For the sake of clarity we shall formulate this discussion in theorems. We shall also only deal with particles and states that have transition matrix elements into n baryons and antibaryons, $n=1, 2, 3, \cdots$.

Theorem 1.

$$G_1\equiv C\exp[i\pi(\mathcal{L}_2+\mathfrak{M}_2+\mathfrak{N}_2)] \tag{23}$$

commutes with all elements of the group $\mathcal{G}_0$.

Proof: For a single baryon-antibaryon the explicit representations of C, $\mathcal{L}$, $\mathfrak{M}$, $\mathfrak{N}$, R are given in Appendix B. The theorem follows from a straightforward explicit

[12] A similar guess on the masses of excited levels has been made by A. Pais (private communication). See also reference 7.

computation of the commutators. For other states the theorem follows because G_1, C, $\exp(i\pi\mathcal{L}_2)$, $\exp(i\pi\mathfrak{M}_2)$, and $\exp(i\pi\mathfrak{N}_2)$ are all multiplicative for a collection of particles.

Theorem 2.

$$G_1N+NG_1=0,\quad [N,\mathcal{G}_0]_-=0,\quad G_1^2=1. \tag{24}$$

Proof: This theorem follows directly from the explicit representation of Appendix B.

Theorem 3. The full group generated by $\mathcal{G}_0$, G_1 and $\exp(iN\theta)$ is the direct product group $\mathcal{G}_0\times O_2^{\pm}$, where $O_2^{\pm}$ is the group of all 2×2 real orthogonal matrices with determinant$=\pm1$.

The proof of this theorem is again straightforward.

The irreducible representations of $O_2^{\pm}$ are either (a) 2×2 in size, in which $N=+\alpha$ and $N=-\alpha$ [α=integer] each occurs once, representing physically a pair of particles, and G_1 switches the two states; or (b) the representation is of dimension 1×1 in which $N=0$ and either $G_1=+1$ or $G_1=-1$.

For a state with $N=0$, the operator G_1 is *one and the same numerical constant* ($=\pm1$) *for all states in a multiplet of* $\mathcal{G}_0$. By (23), C brings one state into another in the same multiplet of $\mathcal{G}_0$.

Theorem 4. For the pions, $G_1=-1$.

Proof: The pions are eigenfunctions of $\mathfrak{N}$ with eigenvalue $\mathfrak{N}=0$. Hence[13] $G_1=G=-1$.

Theorem 5. For the $(0,\frac{1}{2},\frac{1}{2},\lambda)$, $N=0$ representation, if the total angular momentum $J=0$, and the system has transition matrix elements into a baryon-antibaryon pair, then

$$G_1\lambda=-1. \tag{25}$$

For the $(1,\frac{1}{2},\frac{1}{2},\lambda)$, $N=0$ representation under the same assumption,

$$G_1\lambda=1. \tag{26}$$

Proof: In the notation of Appendix B, we have four possible states for the baryon-antibaryon pair that belong to $(0,\frac{1}{2},\frac{1}{2})$, with $\mathfrak{M}_3=\frac{1}{2}$, $\mathfrak{N}_3=\frac{1}{2}$:

$$\{p\bar{Z}^0\}+\{n\bar{Z}^+\}, \tag{27}$$

$$\{\bar{Z}^0p\}+\{\bar{Z}^+n\}, \tag{28}$$

$$\{Y^+\bar{\Xi}^0\}+\{Y^0\bar{\Xi}^+\}, \tag{29}$$

and

$$\{\bar{\Xi}^0Y^+\}+\{\bar{\Xi}^+Y^0\}, \tag{30}$$

where each curly bracket represents a state, for which the first symbol inside specifies the state of particle a and the second symbol that of particle b. Consider a [c number] product wave function of an orbital part, a spin part, and a charge part [depending on the other quantum numbers, $\mathcal{L}$, $\mathcal{L}_3$, $\mathfrak{M}$, $\mathfrak{M}_3$, etc.] of the two particles. For a state $J=0$, the product of the first two parts is antisymmetry in the interchange $a\leftrightarrow b$. Hence the charge part is symmetric since the entire wave function must be antisymmetric. Now under $a\leftrightarrow b$,

$$(27)\leftrightarrow(28),\quad (29)\leftrightarrow(30).$$

[13] T. D. Lee and C. N. Yang, Nuovo cimento 3, 749 (1956).

Hence the states are either

$$(27)+(28)\quad\text{or}\quad(29)+(30), \tag{31}$$

or superpositions of these two. By using the explicit matrices listed in Appendix B it can be directly verified that under

$$R:\quad (27)\leftrightarrow(29),\qquad (28)\leftrightarrow(30),$$

$$G_1:\quad (27)\leftrightarrow-(30),\quad (28)\leftrightarrow-(29).$$

Hence under RG_1 both wave functions in (31) remain themselves but change sign. Thus (25) is proved. A similar proof holds for (26).

Applied to the K mesons, if the K meson admixture $(1,\frac{1}{2},\frac{1}{2})$ can be neglected, Theorem 5 states that

$$(G_1)_K=-\lambda_K.$$

While all the four states $\mathfrak{M}_3=\pm\frac{1}{2}$ and $\mathfrak{N}_3=\pm\frac{1}{2}$ are eigenstates of G_1 with eigenvalues $-\lambda_K$, only two: $\mathfrak{M}_3=\mathfrak{N}_3=\pm\frac{1}{2}$ are eigenstates of R with eigenvalues λ_K.

If, further, one assumes that time reversal invariance holds, then the two states K_1^0 and K_2^0 have simple behaviors under R. To see this, we notice that $(G_1)_K$ is a numerical constant. Hence (23) shows that

$$C|\mathfrak{M}_2=\tfrac{1}{2},\,\mathfrak{N}_2=-\tfrac{1}{2}\rangle=-G_1|\mathfrak{M}_2=-\tfrac{1}{2},\,\mathfrak{N}_2=\tfrac{1}{2}\rangle.$$

If P is the parity operator, we have

$$CP|\mathfrak{M}_2=\tfrac{1}{2},\,\mathfrak{N}_2=-\tfrac{1}{2}\rangle=-G_1P|\mathfrak{M}_2=-\tfrac{1}{2},\,\mathfrak{N}_2=\tfrac{1}{2}\rangle.$$

Now K_1^0, (K_2^0) is an eigenstate of CP with eigenvalue $+1$, (-1). Hence

$$|K_{1,2}^0\rangle=|\mathfrak{M}_2=\tfrac{1}{2},\,\mathfrak{N}_2=-\tfrac{1}{2}\rangle\mp(G_1P)|\mathfrak{M}_2=-\tfrac{1}{2},\,\mathfrak{N}_2=\tfrac{1}{2}\rangle.$$

One obtains with the use of (15):

$$R|K_{1,2}^0\rangle=\mp(G_1P)|K_{1,2}^0\rangle=\pm(\lambda_KP_K)|K_{1,2}^0\rangle,$$

where P_K is the parity of K^0 [with respect to, say, $\bar{\Lambda}n$]. Thus we obtain the entries in Table II for K_1^0, K_2^0 under R.

9. REMARKS

For a symmetry to exist between Y^* and N^* it is not necessary that all 8 baryons be brought into global symmetry. For example, one could have a symmetry between N, Y, and Z without Ξ. A simple symmetry group in such a case[11] is $SU_2\times SU_3$ which has more parameters than $\mathcal{G}_0$. The irreducible representations in such a case can be written down and an analysis like the above for $\mathcal{G}_0$ can be made. There also would be a companion Z^* $T=2$ state together with N^* and Y^*. The weight factors w for N^*, Y^*, Z^* remain the same as in Table III.

APPENDIX A

We give here several examples other than $\mathcal{G}_0$ satisfying conditions (i) and (ii) of Sec. 2.

(A) The group $\mathcal{G}_1$. The group is isomorphic with $SU_2\times SU_4$ where SU_2 consists of the $\mathcal{L}$ transformations

$U_{\mathcal{L}}=\exp i(l_1\mathcal{L}_1+l_2\mathcal{L}_2+l_3\mathcal{L}_3)$ and SU_4 consists of all the unimodular unitary transformations between N, Y, Z, Ξ. Clearly R is an element of SU_4. Hence $\mathcal{G}_0$ is a subgroup of $\mathcal{G}_1$. The usual $g_1=g_2=g_3=g_4$ case[4,5] [where g_1, g_2, g_3, g_4 are, respectively, the coupling constants between pion and $\bar{N}N$, $\bar{Y}Y$, $\bar{Z}Z$, $\bar{\Xi}\Xi$] has this larger symmetry $\mathcal{G}_1$ rather than $\mathcal{G}_0$. Notice that if the coupling constants $g_1=-g_2=-g_3=g_4$, the symmetry group is the smaller $\mathcal{G}_0$, with the pion assignment $(1,0,0,\lambda_\pi)$ where $\lambda_\pi=-1$.

(B) The group O_7'. The group is generated by the invariant subgroup $SU_2\times SU_2\times SU_2$ of $\mathcal{G}_0$, together with the elements

$$\exp(i\theta)\begin{Vmatrix}0&0&\sigma_1&0\\0&0&0&\sigma_1\\\sigma_1&0&0&0\\0&\sigma_1&0&0\end{Vmatrix}. \tag{32}$$

To find the group we introduce the matrices

$$\rho_1=\begin{Vmatrix}0&0&I&0\\0&0&0&I\\I&0&0&0\\0&I&0&0\end{Vmatrix},\quad \rho_2=i\begin{Vmatrix}0&0&-I&0\\0&0&0&-I\\I&0&0&0\\0&I&0&0\end{Vmatrix},\quad \rho_3=\begin{Vmatrix}I&0&0&0\\0&I&0&0\\0&0&-I&0\\0&0&0&-I\end{Vmatrix},$$

$$\tau_1=\begin{Vmatrix}0&I&0&0\\I&0&0&0\\0&0&0&I\\0&0&I&0\end{Vmatrix},\quad \tau_2=i\begin{Vmatrix}0&-I&0&0\\I&0&0&0\\0&0&0&-I\\0&0&I&0\end{Vmatrix},\quad \tau_3=\begin{Vmatrix}I&0&0&0\\0&-I&0&0\\0&0&I&0\\0&0&0&-I\end{Vmatrix}.$$

The infinitesimal generators of the invariant subgroup $SU_2\times SU_2\times SU_2$ of $\mathcal{G}_0$ are

$$\sigma_1,\ \sigma_2,\ \sigma_3,\quad (1+\rho_3)\tau_1,\quad (1+\rho_3)\tau_2,\quad (1+\rho_3)\tau_3,$$
$$(1-\rho_3)\tau_1,\quad (1-\rho_3)\tau_2,\quad (1-\rho_3)\tau_3, \tag{33}$$

or

$$\sigma_i,\ \tau_i,\ \rho_3\tau_i.$$

The infinitesimal generator for (32) is

$$\rho_1\sigma_1.$$

Taking the commutator of this generator with those listed in (33) one obtains additional generators. Altogether by repeatedly taking commutators one obtains the following 21 infinitesimal generators:

$$\sigma_i,\ \tau_i,\ \rho_1\sigma_i,\ \rho_3\tau_i,\ \rho_2\sigma_i\tau_j. \tag{34}$$

Taking further commutators gives rise to no new independent generators. The group obtained from these 21 generators is O_7' which has a 2-to-1 homomorphism with the group of 7×7 proper real orthogonal matrices O_7, as already discussed by various authors[6,3] in the literature. To see this we define seven anticommuting Hermitian matrices

$$\gamma_1=\rho_3\sigma_1,\quad \gamma_2=\rho_3\sigma_2,\quad \gamma_3=\rho_3\sigma_3,\quad \gamma_4=\rho_1\tau_1,$$
$$\gamma_5=\rho_1\tau_2,\quad \gamma_6=\rho_1\tau_3,\quad \gamma_7=\rho_2. \tag{35}$$

Then

$$\gamma_\mu\gamma_\nu+\gamma_\nu\gamma_\mu=2\delta_{\mu\nu}, \tag{36}$$

$$\gamma_1\gamma_2\gamma_3\gamma_4\gamma_5\gamma_6\gamma_7=-i. \tag{37}$$

The 21 infinitesimal generators (34) are then $i\gamma_\mu\gamma_\nu$ $(\mu\neq\nu)$. The group generated by (34) is therefore of the form

$$\exp(\textstyle\sum a_{\mu\nu}\gamma_\mu\gamma_\nu),$$

where $a_{\mu\nu}$ are real numbers. Now

$$[\exp(\textstyle\sum a_{\mu\nu}\gamma_\mu\gamma_\nu)]\gamma_i[\exp(-\sum a_{\mu\nu}\gamma_\mu\gamma_\nu)]=\sum_{j=1}^{7} b_{ij}\gamma_j, \tag{38}$$

where $\|b_{ij}\|$ is a 7×7 real orthogonal matrix with determinant$=1$. It is easy to prove that, conversely, for every such $\|b_{ij}\|$, there exist two sets of real $a_{\mu\nu}$'s satisfying (38). The group has thus a 2-to-1 homomorphism with O_7. [T. A. Tarski has pointed out to us that O_7' is called spin (7) in the standard language.]

In terms of the γ's, the element R of $\mathcal{G}_0$ is

$$R=\rho_1=-i\gamma_4\gamma_5\gamma_6=\gamma_1\gamma_2\gamma_3\gamma_7$$
$$=[\exp(\pi\gamma_1\gamma_2/2)][\exp(\pi\gamma_7\gamma_3/2)]. \tag{39}$$

Thus R is an element of O_7', hence $\mathcal{G}_0$ is a subgroup of O_7'.

Both of the above two groups contain $\mathcal{G}_0$ as a subgroup. There exist also groups that satisfy conditions (i) and (ii) but do not contain $\mathcal{G}_0$ as a subgroup.

(C) The group SU_3. It was pointed out to us by Speiser and Tarski[14] that the group[11] SU_3 has an irreducible 8×8 representation which satisfies both conditions (i) and (ii). However, in this case it seems impossible to incorporate π mesons and K mesons without introducing more new bosons. It is clear that $\mathcal{G}_0$ is not a subgroup of SU_3. The full implications and consequences of such possibilities still need to be investigated.

APPENDIX B

We give in this Appendix explicit matrices for G_1, R, $\mathcal{L}$, $\mathfrak{M}$, $\mathfrak{N}$, and N between the 16 states that describe a

[14] D. R. Speiser and J. A. Tarski (private communication).

single baryon or antibaryon:

$$\begin{Vmatrix} B \\ B' \end{Vmatrix},$$

where

$$B=\begin{Vmatrix} p \\ n \\ \Xi^0 \\ \Xi^- \\ Y^+ \\ Y^0 \\ Z^0 \\ Z^- \end{Vmatrix} \quad\text{and}\quad B'=\begin{Vmatrix} \bar{\Xi}^+ \\ -\bar{\Xi}^0 \\ -\bar{n} \\ \bar{p} \\ \bar{Z}^+ \\ -\bar{Z}^0 \\ -\bar{Y}^0 \\ \bar{Y}^- \end{Vmatrix}.$$

The antibaryon states[15] are defined such that under the charge conjugation operation all baryon states p, n, Ξ^0, Ξ^-, Y^+, Y^0, Z^0, Z^- are transformed in an *identical* way into their respective antibaryon states $\bar{p}$, $\bar{n}$, $\bar{\Xi}^0$, $\bar{\Xi}^+$, $\bar{Y}^-$, $\bar{Y}^0$, $\bar{Z}^0$, and $\bar{Z}^+$. The minus signs in B' are so chosen that the matrices G_1, R, $\mathcal{L}$, $\mathfrak{M}$, $\mathfrak{N}$, and N are given by

$$G_1=\begin{Vmatrix} &&&& I&0&0&0 \\ &&&& 0&I&0&0 \\ &&&& 0&0&I&0 \\ &&&& 0&0&0&I \\ I&0&0&0 &&&& \\ 0&I&0&0 &&&& \\ 0&0&I&0 &&&& \\ 0&0&0&I &&&& \end{Vmatrix},$$

$$R=\begin{Vmatrix} 0&0&I&0 &&&& \\ 0&0&0&I &&&& \\ I&0&0&0 &&&& \\ 0&I&0&0 &&&& \\ &&&& 0&0&I&0 \\ &&&& 0&0&0&I \\ &&&& I&0&0&0 \\ &&&& 0&I&0&0 \end{Vmatrix},$$

$$N=\begin{Vmatrix} I&0&0&0 &&&& \\ 0&I&0&0 &&&& \\ 0&0&I&0 &&&& \\ 0&0&0&I &&&& \\ &&&& -I&0&0&0 \\ &&&& 0&-I&0&0 \\ &&&& 0&0&-I&0 \\ &&&& 0&0&0&-I \end{Vmatrix},$$

$$\mathcal{L}_i=\tfrac{1}{2}\begin{Vmatrix} \sigma_i&0&0&0 &&&& \\ 0&\sigma_i&0&0 &&&& \\ 0&0&\sigma_i&0 &&&& \\ 0&0&0&\sigma_i &&&& \\ &&&& \sigma_i&0&0&0 \\ &&&& 0&\sigma_i&0&0 \\ &&&& 0&0&\sigma_i&0 \\ &&&& 0&0&0&\sigma_i \end{Vmatrix}, \quad (i=1, 2, 3)$$

$$\mathfrak{M}_1=\tfrac{1}{2}\begin{Vmatrix} 0&0&0&0 &&&& \\ 0&0&0&0 &&&& \\ 0&0&0&I &&&& \\ 0&0&I&0 &&&& \\ &&&& 0&0&0&0 \\ &&&& 0&0&0&0 \\ &&&& 0&0&0&I \\ &&&& 0&0&I&0 \end{Vmatrix},$$

$$\mathfrak{M}_2=\tfrac{1}{2}i\begin{Vmatrix} 0&0&0&0 &&&& \\ 0&0&0&0 &&&& \\ 0&0&0&-I &&&& \\ 0&0&I&0 &&&& \\ &&&& 0&0&0&0 \\ &&&& 0&0&0&0 \\ &&&& 0&0&0&-I \\ &&&& 0&0&I&0 \end{Vmatrix},$$

$$\mathfrak{M}_3=\tfrac{1}{2}\begin{Vmatrix} 0&0&0&0 &&&& \\ 0&0&0&0 &&&& \\ 0&0&I&0 &&&& \\ 0&0&0&-I &&&& \\ &&&& 0&0&0&0 \\ &&&& 0&0&0&0 \\ &&&& 0&0&I&0 \\ &&&& 0&0&0&-I \end{Vmatrix},$$

$$\mathfrak{N}_1=\tfrac{1}{2}\begin{Vmatrix} 0&I&0&0 &&&& \\ I&0&0&0 &&&& \\ 0&0&0&0 &&&& \\ 0&0&0&0 &&&& \\ &&&& 0&I&0&0 \\ &&&& I&0&0&0 \\ &&&& 0&0&0&0 \\ &&&& 0&0&0&0 \end{Vmatrix},$$

$$\mathfrak{N}_2=\tfrac{1}{2}i\begin{Vmatrix} 0&-I&0&0 &&&& \\ I&0&0&0 &&&& \\ 0&0&0&0 &&&& \\ 0&0&0&0 &&&& \\ &&&& 0&-I&0&0 \\ &&&& I&0&0&0 \\ &&&& 0&0&0&0 \\ &&&& 0&0&0&0 \end{Vmatrix},$$

and

$$\mathfrak{N}_3=\tfrac{1}{2}\begin{Vmatrix} I&0&0&0 &&&& \\ 0&-I&0&0 &&&& \\ 0&0&0&0 &&&& \\ 0&0&0&0 &&&& \\ &&&& I&0&0&0 \\ &&&& 0&-I&0&0 \\ &&&& 0&0&0&0 \\ &&&& 0&0&0&0 \end{Vmatrix},$$

where $\sigma_i (i=1, 2, 3)$ are the 2×2 Pauli spin matrices and I is the 2×2 unit matrix. All empty places in the above matrices are zeroes.

[15] We use the notation that, e.g., $\bar{Y}^-$ is the antiparticle of Y^+ and is negatively charged.

CP INVARIANCE AND THE 2π DECAY MODE OF THE K_2^0 *

J. BERNSTEIN **, N. CABIBBO and T. D. LEE ***
CERN, Geneva, Switzerland

Received 20 August 1964

In a recent experiment Christenson et al. [1] have found definite indications that the long lived component of the K^0, the $K^0{}_2$, decays into two π mesons

$$K_2^0 \rightarrow 2\pi . \qquad (1)$$

The most direct interpretation of their experiment is to assume that CP invariance is violated [2] in this decay and hence in the weak interactions in general. However, it is important to realize that this experiment involves an entirely new domain of phenomena, those involving extremely small energies. The experiment of Christenson et al. is capable of detecting energy differences of the order of 10^{-8} eV. No other experiment in particle physics, up to now, has exhibited such sensitivity to small energies. On the other hand, none of the other experiments [3] on weak interactions that might have shown CP violation has done so. Therefore it is not unnatural to try to explain the new experiments in some other way. In this note we shall attempt to reconcile the two assumptions:

i) CP invariance holds for *all* interactions:
ii) the long lived component of the K^0, in (local) vacuum, or at least in what is usually called "vacuum", decays into 2π's.

Up to now vacuum is particle physics has been taken to mean the absence of neighbouring material objects, although not necessarily that of long range fields such as gravitational or electromagnetic fields. These conventional fields, which are produced by distant bodies such as Milky Way and the other galaxies, are sufficiently well understood so that one can assume, with some confidence, that their effects on the K^0, $\bar{K}^0$ system can be neglected, either because they are too small (we have in mind the possible electromagnetic conversion between $K^0{}_1$ and $K^0{}_2$ due to the difference in the charge radius of K^0 and $\bar{K}^0$) or because they have the same effect on particles and antiparticles (e.g., gravitational forces).

In this note we postulate that there is a new long range, extremely weak, field which can be neglected so long as one does not measure energies as small as those measured in the experiment of Christensen et al. This field, we assume, produces a potential energy which is equal in magnitude and opposite in sign for K^0 and $\bar{K}^0$, say $\frac{1}{2}V$ for K^0 but $-\frac{1}{2}V$ for $\bar{K}^0$. This new interaction is CP conserving and if we could maintain a given K^0 and $\bar{K}^0$ system, but transform the environment into its CP conjugate state, then this potential energy would reverse sign †.

* Work supported in part by the U.S. Atomic Energy Commission. the Army Research Office (Durham) and the National Science Foundation.

** Permanent address: New York University, New York, N.Y., U.S.A.
*** Permanent address: Columbia University, New York, N.Y., U.S.A.

† During the completion of this paper, we learnt of a preprint by J. S. Bell and J. K. Perring, who have in-

As we shall see, this possibility leads to unique predictions concerning other "apparent" *CP* violating results which are in general quite *different* from the alternative possibility [2] of assuming *CP* non-invariance of weak interactions. Our analysis does not depend on the precise nature of the new interaction except that it must produce a potential energy difference V between K^0 and $\bar{K}^0$. [To give an example, we may envisage that this potential energy difference is transmitted by a four-vector field similar to the electromagnetic interaction. It can be generated by any strongly interacting particle and has an amplitude proportional to, say, I_z, or $\frac{1}{2}(N+S)$, of the particle where I_z = z component of isotopic spin and N = baryon number, S = strangeness. (We leave open the question of whether it can have leptonic couplings)].

Whatever the origin of this interaction with macroscopic bodies, the fact that it gives rise to a static potential energy implies that so far as energies of the order of 10^{-8} eV can be neglected in the *decays of all particles other than neutral* K *mesons, there are no observable "apparent" CP non-invariant effects*. For the neutral K meson, all other 'apparent" *CP* non-invariant effects are completely determined by the parameter V.

We first discuss the problem of a K meson in local isolation (i.e., in the absence of any neighbouring matter but taking into account the effects of the new long range interaction with distant objects). Let $\psi(t)$ be the wave function of a K meson in the laboratory system. In what follows, we regard t as the independent variable and the co-ordinate $\boldsymbol{r}$ of the K meson as a dependent variable given by $(\boldsymbol{r}_0 + \boldsymbol{v}t)$ where $\boldsymbol{v}$ is the K meson velocity and $\boldsymbol{r}_0$ is its initial co-ordinate. The relative velocity of the laboratory and the distant macroscopic bodies will be neglected compared to the velocity of the K in the laboratory. We may write:

$$\psi(t) = a(t)|K^0\rangle + b(t)|\bar{K}^0\rangle\,, \tag{2}$$

or simply

$$\psi(t) = \begin{pmatrix} a(t) \\ b(t) \end{pmatrix} \tag{3}$$

where

$$|\bar{K}^0\rangle \equiv CP|K^0\rangle\,. \tag{4}$$

The time development of $\psi(t)$ can be determined from

$$i\frac{d\psi}{dt} = (\eta + \xi\,\sigma_x + \zeta\,\sigma_z)\psi \tag{5}$$

where

$$\sigma_x = \begin{pmatrix} 0 & 1 \\ 1 & 0 \end{pmatrix}, \quad \sigma_z = \begin{pmatrix} 1 & 0 \\ 0 & -1 \end{pmatrix}, \tag{6}$$

and, for a locally isolated system, η, ξ, ζ, are given by ($\hbar = c = 1$)

$$\eta = \eta_0 = \frac{1}{2\gamma}[(m_1 + m_2) - i\tfrac{1}{2}(\Gamma_1 + \Gamma_2)]\,, \tag{7}$$

$$\xi = \xi_0 = \frac{1}{2\gamma}[(m_1 - m_2) - i\tfrac{1}{2}(\Gamma_1 - \Gamma_2)] \tag{8}$$

and

$$\zeta = \zeta_0 = \tfrac{1}{2}V\,. \tag{9}$$

The subscript o indicates that these are values for the K meson in a locally isolated system, and,

$$\gamma = (1 - v^2)^{-\frac{1}{2}} \tag{10}$$

where v = velocity of the K meson in the laboratory. In the above equations, $\zeta_0\sigma_z$ is the effect of the new interaction, m_1, Γ_1 and m_2, Γ_2 are the mass and width of $K^0{}_1$ and $K^0{}_2$ if $V = 0$. The solution of eq. (5) is well known. There are two eigensolutions $\psi_1(t)$ and $\psi_2(t)$ which are given by

$$\psi_1 = (1 + |\epsilon|^2)^{-\frac{1}{2}}[\psi_+ + \epsilon\psi_-]\,e^{i\lambda_1 t}$$

and

$$\psi_2 = (1 + |\epsilon|^2)^{-\frac{1}{2}}[\psi_- - \epsilon\psi_+]\,e^{i\lambda_2 t} \tag{11}$$

where

$$\psi_\pm \equiv \frac{1}{\sqrt{2}}\begin{pmatrix} 1 \\ \pm 1 \end{pmatrix}, \tag{12}$$

$$\epsilon = \zeta^{-1}[-\xi + (\xi^2 + \zeta^2)^{\frac{1}{2}}]\,, \tag{13}$$

$$\lambda_1 = \eta + (\xi^2 + \zeta^2)^{\frac{1}{2}} \tag{14}$$

and

$$\lambda_2 = \eta - (\xi^2 + \zeta^2)^{\frac{1}{2}}\,. \tag{15}$$

Using eqs. (7)-(9) and assuming small V, we find

$$\epsilon = \epsilon_0 \cong \tfrac{1}{2}\gamma V[(m_1 - m_2) - i\tfrac{1}{2}(\Gamma_1 - \Gamma_2)]^{-1}\,.$$

The value ϵ_0 for the experiment of Christenson et al. is found to be

$$|\epsilon_0| \cong 2 \times 10^{-3}\,. \tag{16}$$

It is important to note that if V is due to the average of the fourth component of a vector field, then V is independent of the energy of the K meson. Thus *the parameter* ϵ_0 *depends linearly on* γ. (On the other hand, V is proportional to γ^{-1} if it is due to a spin 0 field, and is porportional to γ if it is due to a spin 2 field. However, in the frame-

dependently considered the possibility of a long range vector interaction in connection with the experiment of Christenson et al.

work of a simple local field theory, both spin 0 and spin 2 fields give equal static potential for a particle and its antiparticle. Therefore it seems more attractive to assume that this new interaction is due to a vector field with an extremely small mass.)

The solution, eq. (11), also determines other "apparent" CP non-invariant effects in the K decay. For example, the ratio

$$r_i \equiv \frac{\text{rate }(K_i^0 \to \pi^- + l^+ + \nu_l)}{\text{rate }(K_i^0 \to \pi^+ + l^- + \bar{\nu}_l)}$$

can be different from 1, where i = 1 or 2, and l = e or μ. Assuming that the $\Delta Q = \Delta S$ rule holds, we find

$$r_1 = \left|\frac{1+\epsilon_0}{1-\epsilon_0}\right|^2$$

and

$$r_2 = \left|\frac{1-\epsilon_0}{1+\epsilon_0}\right|^2 \qquad (17)$$

which gives $|r_i - 1| \simeq 6\times 10^{-3}$ for the ϵ_0 given by eq. (16).

Next, we discuss the problem of a K meson in a medium. Let n and n' denote, respectively, the complex indices of refraction of K^0 and $\bar{K}^0$ in the medium [4]. Eq. (5) remains valid where, instead of eqs. (7)-(9), the parameters η, ξ, ζ are now given by

$$\eta = \frac{1}{2\gamma}\left[(m_1+m_2) - i\tfrac{1}{2}(\Gamma_1+\Gamma_2)\right] - \tfrac{1}{2}(n+n'-2)\,kv\,, \qquad (18)$$

$$\xi = \frac{1}{2\gamma}\left[(m_1-m_2) - i\tfrac{1}{2}(\Gamma_1-\Gamma_2)\right]\,. \qquad (19)$$

$$\zeta = \tfrac{1}{2}\left[V - (n-n')\,kv\right]\,, \qquad (20)$$

and k is the momentum of the K meson.

The solution, eq. (11), also remains valid. For $|\zeta| \ll |\xi|$, the parameter ϵ is given approximately by

$$\epsilon \cong \tfrac{1}{2}\gamma\left[V - (n-n')\,kv\right]\left[(m_1-m_2) - i\tfrac{1}{2}(\Gamma_1-\Gamma_2)\right]^{-1}. \qquad (21)$$

The absolute magnitude of ϵ in a medium can be obtained by measuring the corresponding rate of reaction (1). The interference term between V and $(n-n')\,kv$ in the expression for $|\epsilon|$ can also determine the sign of V.

It is interesting to speculate on the possible nature of such a long range interaction. As mentioned in the preceding discussion, a simple possibility is to make the ad hoc assumption of the existence of a four-vector field which interacts with all particles. For definiteness, let us assume that the effective "charge" for all strongly interacting particles is

$$f \times I_z \qquad (22)$$

or $f \times \frac{1}{2}(N+S)$, where f is a dimensionless coupling constant. The potential energy difference V between K^0 and $\bar{K}^0$ due to the Milky Way is simply given by *

$$V \approx \tfrac{1}{2} f^2 (M_g/M_p)\, R^{-1} \qquad (23)$$

where M_g = mass of the galaxy, m_p = mass of proton, and R = effective radius of the galaxy, defined so that $G M_g R^{-1}$ = gravitational potential on earth due to the galaxy, G is the gravitation constant. The factor $\frac{1}{2}$ is due to the isospin value of p. It seems reasonable to assume that effects from very distant galaxies are small, and the main contribution is due only to the Milky Way. By taking $M_g \approx 10^{11} \times$ solar mass, $R \sim$ 10 kpc and $V \sim 5\times 10^{-4}\ \text{cm}^{-1}$, we find

$$f^2 \sim 2.5 \times 10^{-49}\,. \qquad (24)$$

Such a long range force would also give rise to a slight difference between the observed "gravitational" mass and inertial mass **. The current experimental upper limit [6] on f^2 is

$$f^2 <\sim 2\times 10^{-45} \qquad (25)$$

which is certainly compatible with eq. (24).

Two of us (J. B. and T. D. L.) wish to thank Dr. S. F. Tuan for discussions, and Professors V. F. Weisskopf and L. Van Hove for the hospitality extended to us at CERN.

References

1) J.H. Christenson, J.W. Cronin, V.L. Fitch and R. Turlay, Phys. Rev. Letters 13 (1964) 138.
2) N. Cabibbo, Physics Letters, 12 (1964) 137. R.G. Sachs, to be published.
3) M.T. Burgy, V.E. Krohn, T.B. Novey, G.R. Ringo and V.L. Telegdi, Phys. Rev. Letters 1 (1958) 324; see also: J.W. Cronin and O.E. Overseth, Proc. 1962 Int. Conf. on high energy physics, CERN, p. 453.
4) M. L. Good, Phys. Rev. 106 (1957) 591.
5) T.D. Lee and C.N. Yang, Phys. Rev. 98 (1955) 1501.
6) R.H. Dicke, Phys. Rev. 126 (1962) 1580.

* We implicitly assume that the mass of this vector field is smaller than, or comparable to, R^{-1}.

** The reasoning here is very similar to that given by Lee and Yang [5].

Symmetry Properties of Leptons in the Zero-Mass Limit (*).

T. D. LEE (**)

CERN - Geneva

(ricevuto il 4 Settembre 1964)

Summary. — The symmetry properties of all known leptons are studied for those processes in which their mass differences can be neglected. It is found that in this approximation both the electromagnetic interaction H_{elect} and the weak interaction H_{weak} are invariant under a $U_2 \times U_2$ group of transformations. Thus, in the zero-mass limit, all lepton systems are characterized by six quantum numbers (in addition to charge, momentum, etc.). The electromagnetic interaction is known to be further invariant under a discrete symmetry operator C (charge conjugation), or P (parity). It is proposed that the weak interaction is also invariant under a discrete symmetry operator, called D, which exchanges charged leptons and neutral leptons. The C symmetry is violated by H_{elect}, and the D symmetry is violated by H_{weak} and by the mass differences. Applications of these symmetry properties to high-energy lepton reactions are discussed.

1. – Introduction.

It is well known that there exists a remarkable symmetry between muons and electrons if their mass difference $(m_\mu - m_e)$ is neglected. It has been shown by FEINBERG and GÜRSEY (1) that in this approximation the free Hamiltonian, the electromagnetic interaction and the weak interaction are all invariant under a U_2 group of transformations among the leptons. The purpose of this note is to make a more complete study of the symmetry properties of leptons by neglecting the mass differences between all leptons. As will be shown in

(*) Work supported in part by the U.S. Atomic Energy Commission.

(**) Permanent address: Columbia University, New York.

(1) G. FEINBERG and F. GÜRSEY: *Phys. Rev.*, **128**, 378 (1962).

945

the next Section, in the zero-mass limit all known interactions are *invariant* under a $U_2 \times U_2$ group. Thus, all lepton systems are characterized by six quantum numbers (in addition to charge, momentum, etc.). The application of this symmetry property leads to a number of relations between different leptonic processes which are valid to all orders in the electromagnetic and the weak interactions, provided the zero-mass limit is a good approximation.

The question of zero-mass limit in electrodynamics has been investigated in the literature (2,3). If radiative corrections are included, the power series expansions of most differential cross-sections contain logarithmic singularities as the mass of the charged particle $\to 0$. The underlying reason for having such a mass singularity is the appearance of degenerate states in the zero-mass limit. Similarly to the infra-red divergence, these singularities can be completely removed for every term in the power series expansion of any transition probability, provided it is averaged over an appropriate ensemble of degenerate states. For these averages, the zero-mass limit becomes a good approximation if

$$\frac{m_l}{E} \ll 1 \tag{1}$$

is satisfied, where m_l is the mass of the charged lepton and E is, for most processes, simply the energy of the lepton. As discussed in refs. (2) and (3), for any given Feynman graph the evaluation of its mass singularity is a relatively simple task. Thus, the zero-mass symmetry properties can also be applied to each individual graph which contains mass singularity, provided that the singular part is calculated separately and eq. (1) is satisfied.

At present, there exists no fundamental theory of weak interactions. From all existing weak reactions, it appears that, so far as the leptons are concerned, only the combination

$$j_\lambda = i\psi_e^\dagger \gamma_4 \gamma_\lambda (1+\gamma_5)\psi_{\nu_e} + i\psi_\mu^\dagger \gamma_4 \gamma_\lambda (1+\gamma_5)\psi_{\nu_\mu} \tag{2}$$

occurs in the weak interaction Hamiltonian H_{weak}, where ψ_e, ψ_μ, ψ_{ν_e}, ψ_{ν_μ} denote, respectively, the field operators for the leptons e^-, μ^-, ν_e, ν_μ, and $\gamma_1, \dots \gamma_5$ are the usual five anticommuting Hermitian matrices. In the following, we consider the general case

$$H_{\text{weak}} = \text{arbitrary function of } j_\lambda \text{ and } j_\lambda^\dagger , \tag{3}$$

and make no further restriction about its specific form.

---

(2) T. D. LEE and M. NAUENBERG: *Phys. Rev.*, **133**, B 1549 (1964).
(3) T. KINOSHITA: *Journ. Math. Phys.*, **3**, 650 (1962).

Lacking a complete theory, it is not possible to make definite statements concerning the properties of mass-singularity in weak interactions. However, from the general discussions given in ref. (2), and the form of H_{weak} given by eq. (3), it is expected that to arbitrary high orders in weak interactions there also should exist no mass singularity in the appropriate ensemble averages, provided the ultra-violet divergence difficulties are solved. Furthermore, since the symmetry property is valid to all orders in all interactions, its applications to high-energy leptonic processes may serve as a test of our basic assumption eq. (3).

2. – Symmetry properties.

The Hamiltonian density H of the known leptons can be written as

$$H = H_{\text{free}} + H_{\text{elect}} + H_{\text{weak}} . \tag{4}$$

In order to study the zero-mass limit, it is useful to decompose the free particle Hamiltonian density H_{free} into two terms

$$H_{\text{free}} = H_0 + H_m , \tag{5}$$

where

$$H_m = m_e \psi_e^\dagger \gamma_4 \psi_e + m_\mu \psi_\mu^\dagger \gamma_4 \psi_\mu . \tag{6}$$

The weak interaction part of the Hamiltonian density is given by eq. (3), and the electromagnetic part is given by

$$H_{\text{elect}} = e J_\lambda A_\lambda , \tag{7}$$

where

$$J_\lambda = i\psi_e^\dagger \gamma_4 \gamma_\lambda \psi_e + i\psi_\mu^\dagger \gamma_4 \gamma_\lambda \psi_\mu , \tag{8}$$

and A_λ describes the electromagnetic field.

In eq. (2), the neutrino fields can be chosen to satisfy

$$\gamma_5 \psi_{\nu_l} = \psi_{\nu_l} , \tag{9}$$

where $l = \text{e}$ or μ. Similarly, the charged lepton fields ψ_l can be decomposed into

$$\psi_l = \psi_{l,\text{L}} + \psi_{l,\text{R}} , \tag{10}$$

947

where

$$\psi_{l,\mathrm{L}} = \tfrac{1}{2}(1+\gamma_5)\psi_l$$

and

$$\psi_{l,\mathrm{R}} = \tfrac{1}{2}(1-\gamma_5)\psi_l \, .$$

Theorem. – The Hamiltonian density $(H_0 + H_{\text{elect}} + H_{\text{weak}})$ is invariant under a $U_2 \times U_2$ group which transforms

$$\text{(11)} \qquad \begin{cases} \varphi_\mathrm{L} \equiv \begin{pmatrix} \psi_{\mathrm{e,L}} \\ \psi_{\mu,\mathrm{L}} \end{pmatrix} \to u\, \varphi_\mathrm{L}, \\ \varphi_\nu \equiv \begin{pmatrix} \psi_{\nu_\mathrm{e}} \\ \psi_{\nu_\mu} \end{pmatrix} \to u \varphi_\nu \, , \end{cases}$$

and

$$\text{(12)} \qquad \varphi_\mathrm{R} \equiv \begin{pmatrix} \psi_{\mathrm{e,R}} \\ \psi_{\mu,\mathrm{R}} \end{pmatrix} \to v \varphi_\mathrm{R} \, ,$$

but leaves the field operators of all other particles invariant. The matrices u and v are two arbitrary (2×2) unitary matrices.

This theorem can be easily proved by noticing that both J_λ and j_λ are invariant under these two independent U_2 transformations. The special case $u = v$ has been discussed by FEINBERG and GÜRSEY (1).

From this theorem, it follows that $(H_0 + H_{\text{elect}} + H_{\text{weak}})$ commutes with the following eight Hermitian operators:

$$\text{(13)} \qquad N_\mathrm{L} = \int (\varphi_\mathrm{L}^\dagger \varphi_\mathrm{L} + \varphi_\nu^\dagger \varphi_\nu)\, \mathrm{d}^3 r \, ,$$

$$\text{(14)} \qquad N_\mathrm{R} = \int \varphi_\mathrm{R}^\dagger \varphi_\mathrm{R} \, \mathrm{d}^3 r \, ,$$

$$\text{(15)} \qquad L_i = \tfrac{1}{2} \int (\varphi_\mathrm{L}^\dagger \sigma_i \varphi_\mathrm{L} + \varphi_\nu^\dagger \sigma_i \varphi_\nu)\, \mathrm{d}^3 r$$

and

$$\text{(16)} \qquad R_i = \tfrac{1}{2} \int \varphi_\mathrm{R}^\dagger \sigma_i \varphi_\mathrm{R} \, \mathrm{d}^3 r \, ,$$

where the σ_i $(i = 1, 2, 3)$ are the usual three anticommuting (2×2) Pauli matrices. These operators obey the following commutation relations:

$$\text{(17)} \qquad [L_a, L_b] = i \varepsilon_{abc} L_c \, ,$$

$$\text{(18)} \qquad [R_a, R_b] = i \varepsilon_{abc} R_c \, ,$$

where $\varepsilon_{abc} = +1$ or -1, depending on whether (a, b, c) is an even or odd permutation of $(1, 2, 3)$; otherwise, $\varepsilon_{abc} = 0$. All other commutators between these eight operators are zero. In the zero-mass limit, all lepton systems are characterized by six quantum numbers (in addition to charge, momentum, etc.):

$$N_L,\ N_R,\ L,\ L_3,\ R \text{ and } R_3\,, \tag{19}$$

where N_L, N_R, L_3 and R_3 denote the eigenvalues of the corresponding operators, $L(L+1)$ and $R(R+1)$ are, respectively, the eigenvalue of the operators $\boldsymbol{L}^2$ and $\boldsymbol{R}^2$. All nonleptons are also eigenstates of these operators but with eigenvalues $= 0$.

The mass term H_m breaks the symmetry. The full Hamiltonian H commutes only with

$$N \equiv N_L + N_R \tag{20}$$

and

$$M \equiv L_3 + R_3\,, \tag{21}$$

which are the two familiar leptonic numbers that remain conserved in the presence of the mass term. From the discussion given in the introduction, we expect the zero-mass limit to be important for any high-energy leptonic processes. The use of these operators $\boldsymbol{L}$, $\boldsymbol{R}$, etc., gives a number of relations between the amplitudes for different leptonic processes.

3. – Applications.

It is convenient to classify the twelve single-lepton states into two $L = \frac{1}{2}$ doublets with $N_L = 1$ $(N_R = R = 0)$

$$\begin{pmatrix} e_L \\ \mu_L \end{pmatrix} \quad \text{and} \quad \begin{pmatrix} \nu_e \\ \nu_\mu \end{pmatrix}, \tag{22}$$

two $L = \frac{1}{2}$ doublets with $N_L = -1$ $(N_R = R = 0)$

$$\begin{pmatrix} \bar{\mu}_L \\ -\bar{e}_L \end{pmatrix} \quad \text{and} \quad \begin{pmatrix} \bar{\nu}_\mu \\ -\bar{\nu}_e \end{pmatrix}, \tag{23}$$

one $R = \frac{1}{2}$ doublet with $N_R = 1$ $(N_L = L = 0)$

$$\begin{pmatrix} e_R \\ \mu_R \end{pmatrix} \tag{24}$$

and one $R=\frac{1}{2}$ doublet with $N_R=-1$ $(N_L=L=0)$

$$\begin{pmatrix} \bar{\mu}_R \\ -\bar{e}_R \end{pmatrix} \tag{25}$$

where l_s ($l=$ e or μ) denotes the charged lepton l^- with helicity s and $\bar{l}_s$ denotes its antiparticle (*i.e.*, one with opposite charge and opposite helicity). A particularly useful operator is

$$S=\exp[i\pi(L_2+R_2)], \tag{26}$$

which exchanges the upper and the lower members of the doublets.

For simple reactions, the use of this exchange operator and the four conservation laws of N_L, N_R, L_3, R_3 are sufficient to obtain almost all physically interesting results. For example, let us consider either the reaction

$$l_1(\boldsymbol{q})+A\to l_2(\boldsymbol{p})+B, \tag{27}$$

or

$$A+B\to l_1(\boldsymbol{q})+l_2(\boldsymbol{p})+C, \tag{28}$$

where A, B, C denote arbitrary systems of nonleptons and $l_1(\boldsymbol{q})$ [or $l_2(\boldsymbol{p})$] can be any one of the twelve single lepton states with momentum $\boldsymbol{q}$ (or $\boldsymbol{p}$).

Let $a[l_1;l_2]$ be the-zero mass limit of the amplitude for reaction (27). The leptons l_1 and l_2 must have the same quantum numbers N_L, N_R, L, L_3 and R, R_3; otherwise, $a[l_1;l_2]=0$. The use of S relates the amplitude for $l_1=\uparrow$ with that for $l_1=\downarrow$ member of the same doublet ([4]). Thus, we find

$$a[l_1(\uparrow);l_2(\uparrow)]=a[l_1(\downarrow);l_2(\downarrow)]. \tag{29}$$

For example, if $l_1(\uparrow)=\nu_e$ and $l_2(\uparrow)=e_L$ then eq. (29) becomes

$$a[\nu_e;e_L]=a[\nu_\mu;\mu_L]. \tag{30}$$

Similar expressions can easily be written down for all other choices of $l_1(\uparrow)$ and $l_2(\uparrow)$.

([4]) Throughout this Section, we use $l_i(\uparrow)$ and $l_i(\downarrow)$ to denote, respectively, the two $L_3=+\frac{1}{2}$ and $L_3=-\frac{1}{2}$ members (or, $R_3=+\frac{1}{2}$ and $R_3=-\frac{1}{2}$ members) of any $L=\frac{1}{2}$ doublet (or, $R=\frac{1}{2}$ doublet).

950

In an identical way, we can analyse reaction (28). Let $b[l_1, l_2]$ be the zero-mass limit of its amplitude. The two leptons must have the same L and R, but their N_L, N_R, L_3 and R_3 must be of opposite signs; otherwise, $b[l_1, l_2]=0$. The further use of S leads to the identity ([4]):

$$b[l_1(\uparrow), l_2(\downarrow)] = -b[l_1(\downarrow), l_2(\uparrow)]\,. \tag{31}$$

For example, if $l_1(\uparrow)=e_L$ and $l_2(\downarrow)=-\bar{e}_L$, eq. (31) becomes

$$b[e_L, \bar{e}_L] = b[\mu_L, \bar{\mu}_L]\,. \tag{32}$$

Next, we discuss the reaction

$$l(\boldsymbol{q}) + A \to l_1(\boldsymbol{p}) + l_2(\boldsymbol{p}') + l_3(\boldsymbol{p}'') + B\,, \tag{33}$$

where A and B are two arbitrary systems of nonleptons. Let the amplitude of reaction (33) in the zero-mass limit be $c[l; l_1, l_2, l_3]$. For any given initial state l, the four conservation laws of N_L, N_R, L_3, R_3 limit the number of possible choices of l_1, l_2, l_3 for which the amplitude is not zero. These choices can be further classified into two classes:

α) the leptons l, l_1, l_2, l_3 all have the same quantum number $L=\frac{1}{2}$, or all have $R=\frac{1}{2}$. By using the properties of $\boldsymbol{L}$ and $\boldsymbol{R}$, we find

$$\begin{aligned} c_\alpha[l(\uparrow); l_1(\uparrow), l_2(\uparrow), l_3(\downarrow)] + c_\alpha[l(\uparrow); l_1(\uparrow), l_2(\downarrow), l_3(\uparrow)] + \\ + c_\alpha[l(\uparrow); l_1(\downarrow), l_2(\uparrow), l_3(\uparrow)] = 0\,, \end{aligned} \tag{34}$$

where the subscript α indicates that eq. (34) is valid only for case α). The use of S leads to

$$c[l(\uparrow); l_1(\uparrow), l_2(\uparrow), l_3(\downarrow)] = -c[l(\downarrow); l_1(\downarrow), l_2(\downarrow), l_3(\uparrow)]\,,$$

$$c[l(\uparrow); l_1(\uparrow), l_2(\downarrow), l_3(\uparrow)] = -c[l(\downarrow); l_1(\downarrow), l_2(\uparrow), l_3(\downarrow)]\,,$$

and

$$c[l(\uparrow); l_1(\downarrow), l_2(\uparrow), l_3(\uparrow)] = -c[l(\downarrow); l_1(\uparrow), l_2(\downarrow), l_3(\downarrow)]\,, \tag{35}$$

β) Two of the leptons, say, l and l_1 have the same quantum number $L=\frac{1}{2}$ and other two l_2 and l_3 have $R=\frac{1}{2}$. (Or, l, l_1 have $R=\frac{1}{2}$ and l_2, l_3 have $L=\frac{1}{2}$). In this case we have, instead of eq. (34).

$$c_\beta[l(\uparrow); l_1(\uparrow), l_2(\uparrow), l_3(\downarrow)] = -c_\beta[l(\uparrow); l_1(\uparrow), l_2(\downarrow), l_3(\uparrow)]\,, \tag{36}$$

951

where the subscript β indicates that it holds only for case β). The equations listed in (35) remain valid.

As an example, the complete set of identities for $l = e_L$ is given below (arranged in descending order of practical importance):

(37) $$c[e_L; e_L, e_L, \bar{e}_L] = c[e_L; e_L, \mu_L, \bar{\mu}_L] + c[e_L; \mu_L, e_L, \bar{\mu}_L],$$

(38) $$c[e_L; e_L, e_R, \bar{e}_R] = c[e_L; e_L, \mu_R, \bar{\mu}_R],$$

(39) $$c[e_L; e_L, e_L, \bar{\nu}_e] = c[e_L; e_L, \mu_L, \bar{\nu}_\mu] + c[e_L; \mu_L, e_L, \bar{\nu}_\mu],$$

(40) $$c[e_L; e_L, \nu_e, \bar{e}_L] = c[e_L; e_L, \nu_\mu, \bar{\mu}_L] + c[e_L; \mu_L, \nu_e, \bar{\mu}_L],$$

(41) $$c[e_L; \nu_e, e_R, \bar{e}_R] = c[e_L; \nu_\mu, \mu_R, \bar{\mu}_R],$$

(42) $$c[e_L; e_L, \nu_e, \bar{\nu}_e] = c[e_L; e_L, \nu_\mu, \bar{\nu}_\mu] + c[e_L; \mu_L, \nu_e, \bar{\nu}_\mu],$$

(43) $$c[e_L; \nu_e, \nu_e, \bar{e}_L] = c[e_L; \nu_e, \nu_\mu, \bar{\mu}_L] + c[e_L; \nu_\mu, \nu_e, \bar{\mu}_L],$$

(44) $$c[e_L; \nu_e, \nu_e, \bar{\nu}_e] = c[e_L; \nu_e, \nu_\mu, \bar{\nu}_\mu] + c[e_L; \nu_\mu, \nu_e, \bar{\nu}_\mu],$$

and all other unrelated amplitudes $c[e_L; l_1, l_2, l_3] = 0$.

These identities are valid for arbitrary nonleptonic systems A and B. For different identities, A and B can (and sometimes must) be different. All above identities are also valid for any given momentum distribution. The different momentum distributions are related by the antisymmetric relation

(45) $$c[l; l_1(\boldsymbol{p}), l_2(\boldsymbol{p}'), l_3(\boldsymbol{p}'')] = -c[l; l_2(\boldsymbol{p}'), l_1(\boldsymbol{p}), l_3(\boldsymbol{p}'')] = \\ = -c[l; l_1(\boldsymbol{p}), l_3(\boldsymbol{p}''), l_2(\boldsymbol{p}')].$$

In eqs. (34)–(44), all amplitudes $c[l; l_1, l_2, l_3]$ refer to that for the momentum distribution given by (33). Thus *e.g.*, $c[e_L; e_L, \mu_L, \bar{\mu}_L]$ denotes $c[e_L; e_L(\boldsymbol{p}), \mu_L(\boldsymbol{p}'), \bar{\mu}_L(\boldsymbol{p}'')]$, which for $\boldsymbol{p} \neq \boldsymbol{p}'$ is not the same as

$$-c[e_L; \mu_L, e_L, \bar{\mu}_L] = -c[e_L; \mu_L(\boldsymbol{p}), e_L(\boldsymbol{p}'), \bar{\mu}_L(\boldsymbol{p}'')].$$

As a final example, we discuss the reaction

(46) $$l_1(\boldsymbol{q}) + l_2(-\boldsymbol{q}) \to l_3(\boldsymbol{p}) + l_4(\boldsymbol{p}') + A,$$

where A is, again, any system of nonleptons. Let $d[l_1, l_2, l_3, l_4]$ be the zero-mass limit of the amplitude for reaction (46). As before, the conservation laws of N_L, N_R, L_3 and R_3 limit the different choices of lepton states. These states can be further classified as follows:

952

α) The four leptons l_1, l_2, l_3, l_4 all have the same quantum number $L = \frac{1}{2}$. or all have $R = \frac{1}{2}$. By using the symmetry properties, we have

$$d_\alpha[(l_1(\uparrow), l_2(\downarrow); l_3(\uparrow), l_4(\downarrow)] + d_\alpha[l_1(\uparrow)\, l_2(\downarrow); l_3(\downarrow), l_4(\uparrow)] - \\ - d_\alpha[l_1(\uparrow), l_2(\uparrow); l_3(\uparrow), l_4(\uparrow)] = 0\,. \tag{47}$$

For example, when $l_1(\uparrow) = l_3(\uparrow) = e_L$ and $l_2(\uparrow) = l_4(\uparrow) = \bar{\mu}_L$, eq. (47) becomes

$$d[e_L, \bar{e}_L; e_L, \bar{e}_L] - d[e_L, \bar{e}_L; \mu_L, \bar{\mu}_L] - d[e_L, \bar{\mu}_L; e_L, \bar{\mu}_L] = 0\,. \tag{48}$$

β) The two initial leptons l_1, l_2 have the same quantum number $L = \frac{1}{2}$, but the two final ones l_3, l_4 both have $R = \frac{1}{2}$; or, the initial leptons both have $R = \frac{1}{2}$ and the final ones both have $L = \frac{1}{2}$. In this case, we have

$$d_\beta[l_1(\uparrow), l_2(\downarrow); l_3(\uparrow), l_4(\downarrow)] = - d_\beta[l_1(\uparrow), l_2(\downarrow); l_3(\downarrow), l_4(\uparrow)]\,. \tag{49}$$

γ) One of the initial leptons, say l_1, has $L = \frac{1}{2}$ and the other one l_2 has $R = \frac{1}{2}$. Thus, one of the final leptons, say l_3, must have $L = \frac{1}{2}$ and the other one l_4 has $R = \frac{1}{2}$. We find

$$d_\gamma[l_1(\uparrow), l_2(\uparrow); l_3(\uparrow), l_4(\uparrow)] = - d_\gamma[l_1(\uparrow), l_2(\downarrow); l_3(\uparrow), l_4(\downarrow)]\,. \tag{50}$$

Here the subscripts α, β, γ indicate the cases in which these equations are valid.

The application of S leads to the following identities which are valid for all cases:

$$d[l_1(\uparrow), l_2(\uparrow); l_3(\uparrow), l_4(\uparrow)] = d[l_1(\downarrow), l_2(\downarrow); l_3(\downarrow), l_4(\downarrow)]\,,$$

$$d[l_1(\uparrow), l_2(\downarrow); l_3(\uparrow), l_4(\downarrow)] = d[l_1(\downarrow), l_2(\uparrow); l_3(\downarrow), l_4(\uparrow)]\,,$$

and

$$d[l_1(\uparrow), l_2(\downarrow); l_3(\downarrow), l_4(\uparrow)] = d[l_1(\downarrow), l_2(\uparrow); l_3(\uparrow), l_4(\downarrow)]\,.$$

All these relations are valid to all orders of H_{elect} and H_{weak}. They can be applied to any Feynman graph (or sum of graphs) for a high-energy leptonic process if it does not contain mass singularity or if its mass singularity has already been separated out.

4. – Discrete symmetries.

Throughout this paper, we study only those symmetry properties of leptons that are independent of space and time. In the zero-mass limit, there are twelve

953

field operators: ψ_{ν_e}, ψ_{ν_μ}, $\psi_{e,L}$, $\psi_{e,R}$, $\psi_{\mu,L}$, $\psi_{\mu,R}$ and their conjugate operators $\psi^c_{\nu_e}$, $\psi^c_{\nu_\mu}$, ..., $\psi^c_{\nu_{\mu,R}}$. In the representation γ_1, γ_2, γ_3 are real and γ_4, γ_5 are imaginary, the conjugate field operator of ψ_l is related to its Hermitian conjugate $\psi_l^\dagger$ by

$$(\psi_l^c)_a = (\psi_l^\dagger)_a , \tag{51}$$

where $a = 1, 2, 3, 4$, and $l = (\nu_e)$, or (ν_μ), or (e, L), ..., or (μ, R). The six operators ψ_{ν_e}, ψ_{ν_μ}, $\psi_{e,L}$, $\psi_{\mu,L}$ and $\psi^c_{e,R}$, $\psi^c_{\mu,R}$ satisfy the same set of anticommutation relations (which is *different* from that satisfied by $\psi^c_{\nu_e}$, $\psi^c_{\nu_\mu}$, $\psi^c_{e,L}$, $\psi^c_{\mu,L}$ and $\psi_{e,R}$, $\psi_{\mu,R}$). The quantization rules and H_0 are invariant under a U_6 group of transformations among these six field operators at the same space time point. This U_6 symmetry is broken by the interactions of leptons.

It can be shown that the only subgroup of U_6 which leaves the currents J_λ and j_λ invariant is the $U_2 \times U_2$ group given by eqs. (11) and (12), according to which these six operators are grouped into three doublets, φ_L, φ_ν and φ^c_R: We know that the electromagnetic interaction is invariant under charge conjugation C which connects φ_L with φ^c_R,

$$C\varphi_L C^\dagger = \varphi^c_R \tag{52}$$

and

$$C^2 = 1 . \tag{53}$$

The electromagnetic current changes its sign under C, but

$$CH_{\text{elect}} C^\dagger = H_{\text{elect}} . \tag{54}$$

The charge conjugation symmetry is broken by the weak interaction. The similarity in the roles of the three doublets φ_L, φ_ν and φ^c_R leads one to introduce another discrete operator D, defined by

$$D\varphi_L D^\dagger = \varphi_\nu \tag{55}$$

and

$$D^2 = 1 .$$

The leptonic current j_λ transforms into j^*_λ under D; *i.e.*,

$$Dj_\lambda D^\dagger = j^*_\lambda , \tag{56}$$

where $j^*_\lambda = j^\dagger_\lambda$ for $\lambda \neq 4$, but $-j^*_\lambda = j^\dagger_4$ for $\lambda = 4$. It is natural to impose the condition

$$DH_{\text{weak}} D^\dagger = H_{\text{weak}} . \tag{57}$$

The D symmetry is broken by the electromagnetic interaction.

954

The symmetry property of C requires that all (leptonic or nonleptonic) electromagnetic currents should change sign under C. Similarly, D symmetry (eq. (57)) requires that all nonleptonic currents in weak interactions ([5]) should obey a similar transformation under D as that given by eq. (56).

The D symmetry leads to identities which, *e.g.*, equate the amplitude for reaction

$$\nu_e + A \rightleftarrows e_L + B \tag{58}$$

with that for

$$e_L + A^D \rightleftarrows \nu_e + B^D , \tag{59}$$

in the zero-mass limit, where A and B are arbitrary systems of nonleptons, and A^D, B^D are their respective D-conjugate system. For example, if $A = n$, $B = p + \pi^0$, then $A^D = p$ and $B^D = n + \pi^0$. The consequence of D symmetry ([6]) is valid to all orders of H_{weak}, but is violated by H_{elect} and by the mass differences.

* * *

I wish to thank Professors V. F. WEISSKOPF and L. VAN HOVE for the hospitality extended to me at CERN.

([5]) This implies a symmetry requirement also for the strongly interacting particles, which can be satisfied in many models of strong interactions. A full discussion will be given elsewhere.

([6]) To lowest order in weak interaction and for leptonic interaction with nonstrange particles, D symmetry is the same as the $|\Delta I| = 1$ rule. See T. D. LEE and C. N. YANG: *Phys. Rev.*, **119**, 1410 (1960).

RIASSUNTO (*)

Si studiano le proprietà di simmetria di tutti i leptoni noti per quei processi in cui si possono trascurare le loro differenze di massa. Si trova che in questa approssimazione sia l'interazione elettromagnetica H_{elect} che l'interazione debole H_{weak} sono invarianti rispetto a un gruppo di trasformazioni $U_2 \times U_2$. Così, nel limite di massa zero, tutti i sistemi leptonici sono caratterizzati da sei numeri quantici (oltre alla carica, l'impulso ecc.). È noto, inoltre, che l'interazione elettromagnetica è invariante rispetto a un operatore di simmetria discreto C (coniugazione della carica) o P (parità). Si postula che l'interazione debole sia anch'essa invariante rispetto ad un operatore di simmetria discreto, D, che scambia i leptoni carichi coi leptoni neutri. La simmetria C è violata da H_{elect} e la simmetria D da H_{weak} e dalle differenze di massa. Si discutono le applicazioni di queste proprietà di simmetria alle reazioni dei leptoni di alta energia.

(*) *Traduzione a cura della Redazione.*

955

Reprinted from
Nuova Cimento **35**, 945–955

Reprinted from THE PHYSICAL REVIEW, Vol. 137, No. 6B, B1621–B1627, 22 March 1965
Printed in U. S. A.

Hypercharge Conservation, *CP* Invariance and the Possible Existence of a Zero-Mass Zero-Spin Field*

T. D. LEE
Department of Physics, Columbia University, New York, New York

(Received 12 November 1964)

It is shown that by introducing a neutral zero-mass zero-spin field ϕ, the Lagrangian density of all interactions (including weak interactions) can be made invariant under the hypercharge gauge transformation. Consequently, there exists a hypercharge current density J_μ that is absolutely conserved. The current density J_μ is related to the usual hypercharge current density j_μ of all the presently known particles by

$$J_\mu = j_\mu - \lambda^{-1}(\partial\phi/\partial x_\mu),$$

where λ is a coupling parameter. The same Lagrangian density is invariant under *CP* transformation, time-reversal transformation, and Lorentz transformation. It turns out that if the conserved quantity $\int J_4 d^3r \neq 0$, then there exists an energy difference between any hypercharged particle and its antiparticle with the same momentum. Such an energy difference would induce decays such as $K_2^0 \to 2\pi$, and the decay rate is proportional to the square of the K-meson energy.

1. INTRODUCTION

IT is well-known that the strong and electromagnetic interactions conserve the hypercharge Y and are invariant under the hypercharge gauge transformation

$$\psi_a(\mathbf{r},t) \to e^{iY_a\theta}\psi_a(\mathbf{r},t), \tag{1}$$

where θ is an arbitrary constant, $Y_a(=0, \text{ or } \pm1, \cdots)$ is the hypercharge of the particle a, and ψ_a is its field operator. The weak interaction allows processes such as

$$\Lambda^0 \to p + \pi^-, \text{ etc.} \tag{2}$$

Since $Y_\Lambda = Y_\pi = 0$ and $Y_p = 1$, it follows that reaction (2) violates Y conservation, and thus the weak interaction is usually regarded to be noninvariant under the hypercharge gauge transformation.

Recently, Christenson, Cronin, Fitch, and Turlay[1] observed that the long-lived component of the neutral K-meson, the K_2^0, decays into two π mesons

$$K_2^0 \to \pi^+ + \pi^- \tag{3}$$

which apparently indicates that *CP* is also not conserved in the weak interaction. The *CP* invariance is connected with the transformation

$$\psi_a(\mathbf{r},t) \to \eta_a\bar{\psi}_a(-\mathbf{r},t), \tag{4}$$

where η_a is a phase factor and $\bar{\psi}_a$ is the field operator for the antiparticle of a. The experiment by Christenson *et al.* seems to imply that the weak interaction might not be invariant under the transformation (4).

The purpose of this paper is to point out that by introducing a neutral field, ϕ which is associated with a zero-rest mass and zero-spin particle, it is possible to preserve both the hypercharge gauge invariance and the *CP* invariance for the weak interaction, and at the same time allow all the observed weak reactions such as (2) and (3).

To see how this can be achieved we may consider, for example, reaction (2) and represent the relevant part of the usual weak-interaction Lagrangian density by (in the absence of ϕ)

$$G[\bar{p}\Lambda\pi] + \text{H.c.}, \tag{5}$$

where G is the appropriate weak-coupling constant and

$$[\bar{p}\Lambda\pi] = \psi_p^\dagger\gamma_4\gamma_\mu(1+b\gamma_5)\psi_\Lambda\frac{\partial\psi_\pi}{\partial x_\mu};$$

the $\gamma_1, \gamma_2, \cdots \gamma_5$ are the usual Dirac matrices, b = constant, and the dagger denotes Hermitian conjugation. The expression (5) is not invariant under the hypercharge gauge transformation. In order to maintain the gauge invariance, we replace (5) by

$$G[\bar{p}\Lambda\pi]\exp(-i\lambda\phi) + \text{H.c.} \tag{6}$$

and assume that under the hypercharge gauge transformation (1), the field ϕ transforms according to

$$\phi \to \phi - \lambda^{-1}\theta, \tag{7}$$

where the parameter λ is real and has the dimension of a length. The gauge invariance requires that ϕ have zero mass. Under the *CP* transformation, we assume that

$$\phi(\mathbf{r},t) \to -\phi(-\mathbf{r},t). \tag{8}$$

The new Lagrangian density (6) is, then, invariant under the Lorentz transformation, the hypercharge gauge transformation, and the *CP* transformation.

As a consequence of the hypercharge gauge invariance, there exists a conserved hypercharge current density

$$J_\mu \equiv j_\mu - \lambda^{-1}(\partial\phi/\partial x_\mu) \tag{9}$$

which satisfies the conservation law

$$(\partial J_\mu/\partial x_\mu) = 0, \tag{10}$$

where j_μ is the usual hypercharge current density due to all the presently known particles. For example, in

* This research was supported in part by the U. S. Atomic Energy Commission.

[1] J. H. Christenson, J. W. Cronin, V. L. Fitch, and R. Turlay, Phys. Rev. Letters 13, 138 (1964).

the case of fermions such as p, n, Ξ, etc., we have

$$j_\mu = i \sum_a Y_a \psi_a^\dagger \gamma_4 \gamma_\mu \psi_a . \tag{11}$$

Equation (10) implies that

$$\int J_4 d^3r \equiv iQ \tag{12}$$

is absolutely conserved, and its value is determined by the initial condition of the system. Decays such as $\Lambda^0 \to p + \pi^-$ do not violate the conservation of hypercharge current J_μ, provided we also include in the definition of J_μ the contribution of the ϕ field.

It will be shown that if the system is in any state with $Q \neq 0$, there must exist a difference of energy between any hypercharge particle and its antiparticle of the same momentum. For K^0 and $\bar{K}^0$, this energy difference is given by

$$V = -2\lambda^2 (Q/\Omega) , \tag{13}$$

where Ω is the volume of the system and is, presumably, the same as the volume of the "universe." Under CP transformation, J_4 changes sign. The CP invariance implies that, for every eigenstate (of the entire system) with $Q = Q_0$, there exists another eigenstate of the same energy but with $Q = -Q_0$. The existence of this energy difference V between K^0 and $\bar{K}^0$ in a given state of the system, therefore, does not contradict the requirement of CP invariance.

It has been pointed out[2] that, independent of the precise nature of the mechanism, if such an energy difference V exists between K^0 and $\bar{K}^0$ then the decay $K^0 \to 2\pi$ can occur, and its rate is proportional to

$$|\epsilon|^2 , \tag{14}$$

where ($\hbar = c = 1$)

$$\epsilon \cong \tfrac{1}{2} \gamma V [(m_1 - m_2) - i\tfrac{1}{2}(\Gamma_1 - \Gamma_2)]^{-1} ; \tag{15}$$

m_i, Γ_i are the rest mass and the decay width of K_i^0 ($i = 1, 2$),

$$\gamma = (1 - v^2)^{-1/2} , \tag{16}$$

and v is the velocity of K_2^0 with respect to the rest system of the "universe." Since in the present theory, the energy difference V is independent of γ, the observed rate for the 2π decay of K_2^0 must be proportional to γ^2.

Identical considerations can, of course, be applied to other quantities such as strangeness, or I_z. The hypercharge is used only as a representative of any such quantities that could be conserved by introducing a zero-mass and zero-spin field. Throughout this paper we follow the method of canonical formalism of quantum-field theory. The effect of general relativity is not discussed.

[2] J. Bernstein, N. Cabibbo, and T. D. Lee, Phys. Letters 12, 146 (1964).

2. LAGRANGIAN AND HAMILTONIAN

In general, we can write the usual weak-interaction Lagrangian density in the absence of the ϕ field as

$$L_{\text{weak}}(0) = G \sum L_{\Delta Y} , \tag{17}$$

where the subscript ΔY denotes the amount of hypercharge violation and the sum extends over $\Delta Y = 0$ and ± 1. The weak-interaction Lagrangian density in the presence of the ϕ field is given by

$$L_{\text{weak}} = G \sum (L_{\Delta Y}) \exp[-i\lambda(\Delta Y)\phi] . \tag{18}$$

The Lagrangian density of the entire system can be written as

$$L = L_\phi + L_\psi + L_{\text{int}} , \tag{19}$$

in which the free-field Lagrangian density is given by L_ϕ and L_ψ, where

$$L_\phi = -\tfrac{1}{2}(\partial\phi/\partial x_\mu)^2 , \tag{20}$$

$$L_\psi = -\sum_a \psi_a^\dagger \gamma_4 (\gamma_\mu (\partial/\partial x_\mu) + m_a) \psi_a , \tag{21}$$

and the interaction Lagrangian density is given by

$$L_{\text{int}} = L_{\text{strong}} + L_{\text{elect}} + L_{\text{weak}} , \tag{22}$$

where L_{strong} and L_{elect} are, respectively, the Lagrangian density for the strong interaction and the electromagnetic interaction, and L_{weak} is given by Eq. (18). For simplicity, we include only the expression for the free Fermions in Eq. (21). All repeated indices are to be summed over.

The Lagrangian density L is invariant under the Lorentz transformation, the time-reversal transformation, the CP transformation and the hypercharge gauge transformation. By using L, the equation of motion is found to be

$$(\partial^2\phi/\partial x_\mu^2) = -(\partial/\partial\phi) L_{\text{weak}} \tag{23}$$

and

$$\gamma_4 \left(\gamma_\mu \frac{\partial}{\partial x_\mu} + m_a \right) \psi_a = \frac{\partial}{\partial \psi_a^\dagger} L_{\text{int}} - \frac{\partial}{\partial x_\mu} \left[\frac{\partial}{\partial \psi_{a,\mu}^\dagger} L_{\text{int}} \right] , \tag{24}$$

where

$$\psi_{a,\mu} = (\partial/\partial x_\mu)\psi_a \quad \text{and} \quad \psi_{a,\mu}^\dagger = (\partial/\partial x_\mu)\psi_a^\dagger .$$

The hypercharge current density J_μ is related to the Lagrangian density by

$$J_\mu = -i \sum_a Y_a \left[\frac{\partial L}{\partial \psi_{a,\mu}} \psi_a - \psi_a^\dagger \frac{\partial L}{\partial \psi_{a,\mu}^\dagger} \right] + \lambda^{-1} \frac{\partial L}{\partial \phi_{,\mu}} , \tag{25}$$

where

$$\phi_{,\mu} = \partial\phi/\partial x_\mu .$$

From the equations of motion and the gauge-invariance property of L, the conservation law [Eq. (10)]

$$\partial J_\mu / \partial x_\mu = 0$$

follows. An alternative form of J_μ is given by Eq. (9)

$$J_\mu = j_\mu - \lambda^{-1}(\partial\phi/\partial x_\mu),$$

where

$$j_\mu = -i\sum_a Y_a\left[\frac{\partial L}{\partial\psi_{a,\mu}}\psi_a - \psi_a^\dagger\frac{\partial L}{\partial\psi_{a,\mu}^\dagger}\right]. \quad (26)$$

The usual canonical formalism can be directly applied to the present problem. The conjugate momentum of the ϕ field is given by

$$\pi = \partial\phi/\partial t. \quad (27)$$

The corresponding Hamiltonian density can be written as

$$H = H_\phi + H_\psi + H_{\text{int}}, \quad (28)$$

where

$$H_\phi = \tfrac{1}{2}\pi^2 + \tfrac{1}{2}(\nabla\phi)^2, \quad (29)$$

$$H_\psi = \sum_a \psi_a^\dagger\left[\boldsymbol{\alpha}\cdot\frac{1}{i}\nabla + \beta m_a\right]\psi_a, \quad (30)$$

and

$$H_{\text{int}} = H_{\text{strong}} + H_{\text{elect}} + H_{\text{weak}}. \quad (31)$$

The matrices $\boldsymbol{\alpha}$ and β are the usual Dirac matrices. The H_{strong}, H_{elect}, and H_{weak} describe, respectively, the strong, the electromagnetic, and the weak interaction which includes the modification due to the ϕ field.

For simplicity, we have assumed in Eq. (30) that except for ϕ all other fields are Fermion fields. To simplify further our discussions, we consider, in the following, only the case that L_{int} does not contain derivatives of ψ_a. (All our conclusions can, of course, be applied to any other case.) Thus,

$$H_{\text{weak}} = -L_{\text{weak}} \quad (32)$$

and j_μ is given by Eq. (11). Both π and ϕ are Hermitian operators which satisfy

$$[\pi(\mathbf{r},t),\phi(\mathbf{r}',t)] = -i\delta^3(\mathbf{r}-\mathbf{r}'), \quad (33)$$

and the ψ_a obeys the usual anticommutation relation.

3. A CANONICAL TRANSFORMATION

To exhibit more explicitly the constant of motion implied by the conservation of hypercharge current, we consider the canonical transformation[3,4] generated by the unitary operator

$$U \equiv \exp\left[i\lambda\int\phi\rho d^3r\right], \quad (34)$$

where

$$\rho = -ij_4 = \textstyle\sum_a Y_a\psi_a^\dagger\psi_a \quad (35)$$

and is the hypercharge density of the ψ_a's. Under the canonical transformation,

$$U\psi_a U^\dagger = \psi_a\exp(-i\lambda Y_a\phi), \quad (36)$$

$$U\phi U^\dagger = \phi, \quad (37)$$

and

$$U\pi U^\dagger = \pi - \lambda\rho. \quad (38)$$

The Hamiltonian density is transformed into

$$H_U = UHU^\dagger = H_\phi + H_\psi + H_{\text{strong}} + H_{\text{elect}} + H_{\text{weak}}(0) - \lambda[\rho\pi + \mathbf{j}\cdot\nabla\phi] + \tfrac{1}{2}\lambda^2\rho^2, \quad (39)$$

where $\mathbf{j}$ is the spatial part of the 4-vector j_μ, the terms H_ϕ, H_ψ, H_{strong}, and H_{elect} are the same as that given by Eqs. (29)–(31), but $H_{\text{weak}}(0)$ is related to Eq. (17) by

$$H_{\text{weak}}(0) = -L_{\text{weak}}(0), \quad (40)$$

and is independent of ϕ. In H_U, the coupling between ϕ and ψ_a is of a derivative type with λ as the coupling constant.

To understand the meaning of conservation of hypercharge in this new representation, it is convenient to consider the Fourier expansions of ϕ and π in a volume Ω:

$$\phi(\mathbf{r}) = \Omega^{-1/2}[\phi_0 + \sum_{\mathbf{k}\neq 0}\phi_{\mathbf{k}}\exp(i\mathbf{k}\cdot\mathbf{r})] \quad (41)$$

and

$$\pi(\mathbf{r}) = \Omega^{-1/2}[\pi_0 + \sum_{\mathbf{k}\neq 0}\pi_{\mathbf{k}}\exp(-i\mathbf{k}\cdot\mathbf{r})], \quad (42)$$

where

$$\pi_{\mathbf{k}} = \pi_{-\mathbf{k}}^\dagger, \quad \phi_{\mathbf{k}} = \phi_{-\mathbf{k}}^\dagger,$$

and the commutation relations

$$[\pi_{\mathbf{k}},\phi_{\mathbf{k}'}] = -i\delta_{\mathbf{k}\mathbf{k}'}$$

are satisfied for all $\mathbf{k}$ and $\mathbf{k}'$. The transformed Hamiltonian H_U is independent of ϕ_0. Thus,

$$d\pi_0/dt = 0. \quad (43)$$

The conservation of hypercharge, after the canonical transformation, becomes simply

$$\pi_0 = \text{constant}, \quad (44)$$

which is a quantum number characterizing the particular state of the system. Under CP transformation, any state with $\pi_0 \neq 0$ is transformed into another state which has a π_0 of the opposite sign but with the same magnitude. The π_0 is related to the Q, introduced in Eq. (12), by $\pi_0 = \Omega^{-1/2}(\lambda Q)$.

It is important to notice that the hypercharge gauge transformation becomes a totally trivial operation after the canonical transformation. This may at first sight seem rather strange, but it is actually a general feature of such gauge invariance. In the Appendix, we give a simple example to illustrate further the same property.

4. APPARENT *CP* NONINVARIANCE

In this section we discuss the various consequences for the state which has a $\pi_0 \neq 0$. From Eq. (39), we

[3] Similar canonical transformations have been used by F. J. Dyson, Phys. Rev. **73**, 929 (1948), and R. J. Glauber, *ibid.* **84**, 395 (1951).

[4] The present considerations are, however, suggested by some results on the (physically unrelated) polaron problem by T. D. Lee, F. Low, and D. Pines, Phys. Rev. **90**, 297 (1953). See the discussions given in the Appendix.

notice that, after the canonical transformation, the Hamiltonian can be written as

$$\int H_U d^3r = -\Omega^{-1/2}\lambda\pi_0 \int \rho d^3r + \int H' d^3r, \quad (45)$$

where

$$H' = H_\phi + H_\psi + H_{\text{strong}} + H_{\text{elect}} + H_{\text{weak}}(0) + H_1 \quad (46)$$

and

$$H_1 = -\lambda[\rho\pi_1 + \mathbf{j}\cdot\nabla\phi_1] + \tfrac{1}{2}\lambda^2\rho^2. \quad (47)$$

In Eq. (47), the operators ϕ_1 and π_1 are defined to be

$$\phi_1 = \sum_{\mathbf{k}\neq 0} \Omega^{-1/2}\phi_{\mathbf{k}} \exp(i\mathbf{k}\cdot\mathbf{r})$$

and

$$\pi_1 = \sum_{\mathbf{k}\neq 0} \Omega^{-1/2}\pi_{\mathbf{k}} \exp(-i\mathbf{k}\cdot\mathbf{r}). \quad (48)$$

The first term on the right-hand side of Eq. (45) shows that there is an energy difference V between any hypercharge particle and its antiparticle of the same momentum. For particles with $Y=1$, this potential energy difference is given by

$$V = -2\Omega^{-1/2}\lambda\pi_0. \quad (49)$$

As mentioned in the Introduction, the existence of such an energy difference implies that

$$K_2^0 \to 2\pi$$

can occur and the analysis made in Ref. 2 is applicable. In the following, we list the various consequences for the weak decays. Most of these results have already been stated in Ref. 2.

(i). In a vacuum (i.e., in the absence of any neighboring matter), the two states $|K_1^0\rangle$ and $|K_2^0\rangle$, each of which has a single lifetime, are related to $|K^0\rangle$ and its CP conjugate state $|\bar{K}^0\rangle$ by

$$|K_1^0\rangle = [2(1+|\epsilon|^2)]^{-1} \times[(1+\epsilon)|K^0\rangle + (1-\epsilon)|\bar{K}^0\rangle]e^{-i\lambda_1 t} \quad (50)$$

and

$$|K_2^0\rangle = [2(1+|\epsilon|^2)]^{-1} \times[(1-\epsilon)|K^0\rangle - (1+\epsilon)|\bar{K}^0\rangle]e^{-i\lambda_2 t}, \quad (51)$$

where $|K^0\rangle$ and $|\bar{K}^0\rangle$ are time-independent,

$$\lambda_1 = \eta + (\xi^2+\zeta^2)^{1/2}, \quad (52)$$

$$\lambda_2 = \eta - (\xi^2+\zeta^2)^{1/2}, \quad (53)$$

$$\epsilon = \zeta^{-1}[-\xi + (\xi^2+\zeta^2)^{1/2}], \quad (54)$$

$$\eta = (2\gamma)^{-1}[(m_1^0+m_2^0) - i\tfrac{1}{2}(\Gamma_1^0+\Gamma_2^0)], \quad (55)$$

$$\xi = (2\gamma)^{-1}[(m_1^0-m_2^0) - i\tfrac{1}{2}(\Gamma_1^0-\Gamma_2^0)], \quad (56)$$

and

$$\zeta = \tfrac{1}{2}V. \quad (57)$$

In Eqs. (55) and (56), the m_1^0, Γ_1^0 and m_2^0, Γ_2^0 are the mass and width of K_1^0 and K_2^0 in the absence of V, and γ is given by Eq. (16). Since $V\neq 0$, the actual mass and width of K_1^0 and K_2^0 are determined by λ_1 and λ_2. We may define

$$\lambda_j \equiv \gamma^{-1}[m_j - i\tfrac{1}{2}\Gamma_j], \quad (58)$$

where $j=1, 2$. The "observed mass" m_1, m_2 and the "observed width" Γ_1 and Γ_2 depend on γ. If $|\epsilon|\ll 1$ (which corresponds to $\gamma \ll 10^3$), then

$$(m_1 - i\tfrac{1}{2}\Gamma_1) \cong (m_1^0 - i\tfrac{1}{2}\Gamma_1^0) + \tfrac{1}{4}\gamma^2 V^2[\Delta m^0 - i\tfrac{1}{2}\Delta\Gamma^0]^{-1} \quad (59)$$

and

$$(m_2 - i\tfrac{1}{2}\Gamma_2) \cong (m_2^0 - i\tfrac{1}{2}\Gamma_2^0) - \tfrac{1}{4}\gamma^2 V^2[\Delta m^0 - i\tfrac{1}{2}\Delta\Gamma^0]^{-1}, \quad (60)$$

where

$$\Delta m^0 = m_1^0 - m_2^0, \quad (61)$$

and

$$\Delta\Gamma^0 = \Gamma_1^0 - \Gamma_2^0. \quad (62)$$

For large values of γ ($\gg 10^3$),

$$\lambda_1 \cong \tfrac{1}{2}V + \frac{1}{2\gamma}[(m_1^0+m_2^0) - i\tfrac{1}{2}(\Gamma_1^0+\Gamma_2^0)] \quad (63)$$

and

$$\lambda_2 \cong -\tfrac{1}{2}V + \frac{1}{2\gamma}[(m_1^0+m_2^0) - i\tfrac{1}{2}(\Gamma_1^0+\Gamma_2^0)]. \quad (64)$$

Thus, the lifetimes of K_1^0 and K_2^0 become the same at the extremely high-energy limit.

(ii) From Eqs. (50), (51), and the CP invariance property of the theory, we find (in vacuum)

$$\frac{\text{Rate}(K_2^0 \to \pi^+ + \pi^-)}{\text{Rate}(K_1^0 \to \pi^+ + \pi^-)} = \frac{\text{Rate}(K_2^0 \to \pi^0 + \pi^0)}{\text{Rate}(K_1^0 \to \pi^0 + \pi^0)} = |\epsilon|^2. \quad (65)$$

The result of Christenson *et al.*[1] shows that at $\gamma \cong 2$, $|\epsilon| \cong 2\times 10^{-3}$. This corresponds to a

$$|V| \cong 2\times 10^{-8}\ \text{eV} \quad (66)$$

For $\gamma \ll 10^3$, the parameter $|\epsilon|^2$ varies according to γ^2; for $\gamma \gg 10^3$, the parameter $|\epsilon|^2 \cong 1$.

For the 3π decay mode, if we neglect the high angular momentum states in the final $(\pi^- + \pi^+ + \pi^0)$ system, then

$$\frac{\text{Rate}(K_1^0 \to \pi^+ + \pi^- + \pi^0)}{\text{Rate}(K_2^0 \to \pi^+ + \pi^- + \pi^0)} = \frac{\text{Rate}(K_1^0 \to \pi^0 + \pi^0 + \pi^0)}{\text{Rate}(K_2^0 \to \pi^0 + \pi^0 + \pi^0)} = |\epsilon|^2. \quad (67)$$

If we assume the $\Delta Q = \Delta S$ rule for the leptonic decay modes, then

$$\text{Rate}(K_1^0 \to \pi^- + l^+ + \nu_l) = \text{Rate}(K_2^0 \to \pi^+ + l^- + \bar{\nu}_l), \quad (68)$$

$$\text{Rate}(K_1^0 \to \pi^+ + l^- + \bar{\nu}_l) = \text{Rate}(K_2^0 \to \pi^- + l^+ + \nu_l), \quad (69)$$

and

$$\frac{\text{Rate}(K_1{}^0\to\pi^-+l^++\nu_l)}{\text{Rate}(K_1{}^0\to\pi^++l^-+\bar{\nu}_l)}=\left|\frac{1+\epsilon}{1-\epsilon}\right|^2, \tag{70}$$

where $l=e$ or μ.

(iii) All above results [Eqs. (50)–(54) and (65)–(70)] are applicable to $K_1{}^0$ and $K_2{}^0$ in a medium, provided we replace Eqs. (55)–(57) by

$$\eta=(2\gamma)^{-1}[(m_1{}^0+m_2{}^0)-i\tfrac{1}{2}(\Gamma_1{}^0+\Gamma_2{}^0)]-\tfrac{1}{2}(n+n'-2)kv, \tag{71}$$

$$\xi=(2\gamma)^{-1}[(m_1{}^0-m_2{}^0)-i\tfrac{1}{2}(\Gamma_1{}^0-\Gamma_2{}^0)], \tag{72}$$

and

$$\zeta=\tfrac{1}{2}[V-(n-n')kv], \tag{73}$$

where k is the momentum of the K meson and n, n' are, respectively the complex index of refraction of K^0 and $\bar{K}^0$ in the medium.

The sign of V can be determined either by observing the interference term between the decay of $K_1{}^0$ with that of $K_2{}^0$ in vacuum, or by studying the decay rates of $K_1{}^0$ or $K_2{}^0$ in a medium.

In all above formulas, the γ is measured with respect to the rest system of the entire volume Ω. On the other hand, we can also measure the value of γ with respect to earth by direct means. Combining these two measurements, it would become possible to measure the "absolute" velocity of the earth.

(iv) In the decays of all other particles, no apparent CP noninvariant term can be observed if an energy difference of the order of 10^{-8} eV between the particle and its antiparticle can be neglected.

5. EMISSION AND ABSORPTION OF ϕ QUANTA

From Eq. (47) we find that the coupling between ϕ and ψ_a is of a derivative type and that the coupling parameter, λ has the dimension of a length. It is convenient to introduce a dimensionless coupling constant, f:

$$f=\lambda m_N, \tag{74}$$

where m_N=mass of the nucleon. To study the emission and absorption processes of ϕ quanta with $\mathbf{k}\neq0$, we briefly describe the rules for Feynman graphs.

Let us start with H_U [Eq. (45)] and use the interaction representation. From Eq. (48), it follows that as $\Omega\to\infty$, the propagator of ϕ_1 is given by[5]

$$\phi_1{}^\cdot(x)\phi_1{}^\cdot(y)=(16\pi^4)^{-1}\int(ik^2)^{-1}\exp[ik_\lambda(x-y)_\lambda]d^4k \tag{75}$$

$$\equiv\tfrac{1}{2}D_F(x-y),$$

where $k^2=k_\lambda k_\lambda$ and d^4k is real. It is well known that the second time derivative of the right-hand side of Eq. (75) differs from the corresponding contraction of the time derivatives of ϕ_1 by a $\delta^4(x-y)$ function which, in the evaluation of the S matrix, contributes terms that are exactly canceled by the corresponding terms generated by the $\frac{1}{2}\lambda^2\rho^2$ in H_1 [Eq. (47)]. Therefore, in deriving the Feynman rules we can[6] ignore the $\delta^4(x-y)$ function, but regard the contraction of the derivatives of the ϕ_1 as given by

$$\left[\frac{\partial}{\partial x_\mu}\phi_1(x)\right]^\cdot\left[\frac{\partial}{\partial y_\nu}\phi_1(y)\right]^\cdot=-\frac{\partial^2}{\partial x_\mu\partial x_\nu}[\tfrac{1}{2}D_F(x-y)] \tag{76}$$

and, at the same time, replace Eq. (47) by

$$H_1=-\frac{f}{m_N}j_\mu\frac{\partial}{\partial x_\mu}\phi_1(x). \tag{77}$$

In the limit that the weak-coupling constant $G\to0$, the current j_μ is conserved. From Eq. (77), it follows that the emission and absorption rate of any ϕ quanta with $\mathbf{k}\neq0$ must be zero if $G=0$. Therefore, the actual emission and absorption probability for each ϕ quantum is proportional to G^2f^2. The same conclusion can also be derived in a more transparent way by using the original Hamiltonian, H given by Eq. (28).

The ϕ quanta can be emitted whenever there is a violation of the conservation of j_μ such as, e.g.,

$$\Lambda^0\to p+\pi^-. \tag{78}$$

The rate for the process

$$\Lambda^0\to p+\pi^-+\phi \tag{79}$$

is (to lowest order in G^2 and f^2) proportional to

$$f^2|M|^2(16\pi^3Em_N{}^2)^{-1}d^3k, \tag{80}$$

where M is the matrix element for the $\Lambda^0\to p+\pi^-$ vertex in the ϕ emission, k is the momentum of ϕ and $E=|k|$. For the emission of soft ϕ quantum, we can regard M as independent of k. The branching ratio is given by

$$B_1\equiv\frac{\text{Rate}(\Lambda^0\to p+\pi^-+\phi)}{\text{Rate}(\Lambda^0\to p+\pi^-)}=f^2E_m{}^2(8\pi^2m_N{}^2)^{-1}, \tag{81}$$

where E_m is the maximum energy of the soft ϕ quantum emitted. Similarly, the rate for

$$\Lambda^0\to p+\pi^-+\phi_1+\phi_2+\cdots+\phi_N \tag{82}$$

is (to lowest order in G^2 and f^2) proportional to

$$(n!)^{-1}f^{2N}|M|^2\prod_i(16\pi^3E_im_N{}^2)^{-1}d^3k_i, \tag{83}$$

where k_i is the momentum of ϕ_i and $E_i=|k_i|$. For soft

[5] See G. C. Wick, Phys. Rev. **80**, 268 (1950) for the definition of $\phi^\cdot(x)\phi^\cdot(y)$.

[6] For a detailed proof of this well-known result see, for example, the Appendix in T. D. Lee and C. N. Yang, Phys. Rev. **128**, 885 (1962).

quanta emission, the corresponding branching ratio

$$B_n \equiv \frac{\text{Rate}(\Lambda^0 \to p+\pi^- + n\phi)}{\text{Rate}(\Lambda^0 \to p+\pi^-)} \tag{84}$$

is given by

$$B_n = [(2n)!n!]^{-1}[f^2 E_m{}^2(4\pi^2 m_N{}^2)^{-1}]^n, \tag{85}$$

where E_m is the maximum value of the *total* energy given to the n ϕ-quanta system.

Identical formulas can be applied to the ϕ-emission probability associated with any $\Delta Y \neq 0$ weak reactions. From existing data, we know that

$$f^2 \ll 1. \tag{86}$$

Indeed, it seems that $(4\pi)^{-1}f^2$ can be as big as the fine structure constant without violating any known observations.

For ϕ quantum of sufficiently high energy, reactions such as $\phi+p \to \Lambda^0+\pi^+$ can occur. The corresponding cross section is expected to be $\sim G^2 f^2$. If the energy of ϕ is $\lesssim 175$ MeV, then to the lowest order of G^2 (but to arbitrary orders in f^2 and e^2) the ϕ quantum cannot be absorbed by a nucleon at rest. Thus, the mean free path of ϕ in a medium is, in general, much longer than that of the neutrino.

So far, we consider the case that ϕ is related to the hypercharge gauge transformation (or strangeness-gauge transformation). Therefore, for weak decays which satisfy $\Delta Y=0$, the emission probability for ϕ is zero. On the other hand, if ϕ is connected with the gauge transformation associated with the z component of isotopic spin, I_z, then reactions such as

$$n \to p+e^- + \bar{\nu}_e + \phi \tag{87}$$

or

$$\phi + p \to n + l^+ + \nu_l \tag{88}$$

become possible ($l=e$ or μ).

6. REMARKS

(i). The possibility that our system (or universe) is in a state with $\pi_0 \neq 0$ evokes many questions concerning its nature. From Eqs. (49), (66), and (74), we notice that the energy density due to π_0 is given by

$$\tfrac{1}{2}\Omega^{-1}\pi_0{}^2 = (8f^2)^{-1}V^2 m_N{}^2 > 10^7 \text{ BeV/cm}^3. \tag{89}$$

The total energy in the $\mathbf{k}=0$ mode is proportional to the entire volume Ω of the system (which is much greater than the rest-mass energy of the baryons in the universe). The corresponding energy contained in any other mode in the ground state of the system is the well-known zero-point energy $\frac{1}{2}|\mathbf{k}|$. The zero-point energy density contained in all the $\mathbf{k}\neq 0$ modes diverges in the ultraviolet region. If we introduce a momentum cutoff $k_{\max} \gg m_N$, the energy density in all the $\mathbf{k}\neq 0$ states is

$$(8\pi^2)^{-1}k_{\max}{}^4 \gg 10^{34}(\tfrac{1}{2}\Omega^{-1}\pi_0{}^2). \tag{90}$$

Although the fractional energy contained in the $\mathbf{k}=0$ mode is extremely small, the fact that its average value is of a macroscopic nature gives rise to the many striking physical effects discussed in this paper.

The absorption and scattering of the ϕ quantum in matter have some similarities with neutrinos. Thus, it is of some interest to compare the energy density contained in the $\mathbf{k}=0$ mode of the ϕ field with the degenerate neutrino energy density of the universe, which is, of course, not known. If we make the arbitrary assumption that these two energy densities should be comparable, then the Fermi momentum of the degenerate neutrinos would be ~ 10 eV. As has been discussed by Weinberg,[7] such a high value of Fermi momentum is, theoretically, not impossible. Similarly, an energy density of the magnitude given by Eq. (89) may also be possible for the ϕ field in some oscillatory cosmological models.

(ii). The present theory resembles the Bose-Einstein condensation of a system of Bose particles. In either case, the $\mathbf{k}=0$ mode acquires a macroscopic value for its occupation number. Such solutions are intrinsically different in both mathematical content and physical manifestation from the usual ones. It is possible that the existence of such eigensolutions is a general feature of any boson system. In this sense, the present theory may also serve as a simple model for a general class of field theories.

In the case of Bose gas, it is known that there are quasistationary states which correspond to the flow of superfluid. The simplest way is to divide the entire volume into many smaller but macroscopic boxes, and to allow variations between different boxes. The same can also be done in the present case by allowing the value for π_0 to change gradually from one box to the next one. In this way, we can construct a macroscopic field $\pi_0(x)$. The solution $\pi_0(x)\neq$constant is not an eigenstate of the Hamiltonian of the entire system, but it might be regarded as a quasistationary solution. Such a macroscopic field, if it exists, should have some cosmological influences, and the present value of π_0 as determined by the observed $K_2{}^0 \to 2\pi$ rate may, then, fit into a general picture concerning the evolution of our universe.

(iii). An entirely different mechanism has been suggested[2,8] to account for the 2π decay mode of $K_2{}^0$ and, at the same time, to maintain *CP* invariance by introducing a long-range spin-1 field V_μ interacting with the hypercharge (or I_z) current density j_μ of the presently known particles.

Since j_μ is not conserved, V_μ must have a mass m.

---

[7] S. Weinberg, Phys. Rev. **128**, 1457 (1962).

[8] J. S. Bell and J. K. Perring, Phys. Rev. Letters **13**, 348 (1964). (There is an unpublished note by F. Gürsey and A. Pais in which they considered the possible existence of a pseudoscalar field in connection with the experiment by Christenson *et al.* The proposal by Gürsey and Pais seems to be totally different from the theory discussed in this paper and from the suggestion made by Bernstein, Cabbibo and Lee, and by Bell and Perring.)

It has been pointed out by Weinberg[9] that the mass, m cannot be too small. In order to account for the observed rate of $K_2^0 \to 2\pi$, the experimental absence of real emission of such V_μ quanta and the present experimental accuracy[10] of the equality between the observed gravitational mass and the inertial mass, the Compton wavelength m^{-1} of V_μ should be about (or less than) the radius of the earth. The corresponding (coupling constant)^2 is about (or larger than) 10^{-46}. Thus, such a field, if it exists, could also be detected by improving the present accuracy of the Eötvos-type experiment by another order of magnitude. In contrast, the theory discussed in this paper does not give any such observable effect in the Eötvos-type experiment.

If the vector field V_μ exists, its predictions on all apparent CP noninvariant phenomena are the same as that discussed in the present paper. These two different theories can be differentiated either by improving the present accuracy on the equality between the gravitational mass and inertial mass or by observing the real emission of either the V_μ or the ϕ quantum and determining its spin.

Another variation on the same theoretical idea is to couple V_μ with J_μ [Eq. (9)]. Since J_μ is conserved, V_μ could then also have zero mass. While such a model has some appeal because of its possible connection with the gauge invariance of the second kind, it does appear at the present time to be too speculative to warrant a full discussion.

ACKNOWLEDGMENTS

I wish to thank J. Bernstein, G. Feinberg, R. Serber, and G. C. Wick for discussions.

APPENDIX

In this Appendix we give a simple example of the interaction between an electron and the phonon field in a solid to illustrate the interplay between the gauge transformation, the canonical transformation, and the apparent asymmetry phenomenon.[4] Let the Hamiltonian be given by

$$H=\frac{p^2}{2m}+\sum \omega_{\mathbf{k}}(\alpha_{\mathbf{k}}{}^\dagger\alpha_{\mathbf{k}}+\tfrac{1}{2}) +\sum [V_{\mathbf{k}}\alpha_{\mathbf{k}}e^{i\mathbf{k}\cdot\mathbf{r}}+V_{\mathbf{k}}{}^*\alpha_{\mathbf{k}}{}^\dagger e^{-i\mathbf{k}\cdot\mathbf{r}}], \quad \text{(A1)}$$

where $\mathbf{r}$, $\mathbf{p}$ are the coordinate and momentum of the electron, $\alpha_{\mathbf{k}}$ and $\alpha_{\mathbf{k}}{}^\dagger$ are the annihilation and creation operators of the phonon, $\omega_{\mathbf{k}}$ is its frequency and $V_{\mathbf{k}}$ is the interaction form factor. The Hamiltonian (A1) is invariant under the transformation

$$\alpha_{\mathbf{k}}\to\alpha_{\mathbf{k}}e^{i\mathbf{k}\cdot\mathbf{d}}$$

and

$$\mathbf{r}\to\mathbf{r}-\mathbf{d}. \quad \text{(A2)}$$

[9] S. Weinberg, Phys. Rev. Letters **13**, 495 (1964).
[10] R. H. Dicke, Phys. Rev. **126**, 1580 (1962).

This invariance is connected with the conservation law that the total-momentum operator

$$\mathbf{p}+\sum_{\mathbf{k}}\alpha_{\mathbf{k}}{}^\dagger\alpha_{\mathbf{k}}\mathbf{k} \quad \text{(A3)}$$

commutes with H.

We may introduce a canonical transformation generated by a unitary matrix

$$U\equiv\exp[i\sum\alpha_{\mathbf{k}}{}^\dagger\alpha_{\mathbf{k}}\mathbf{k}\cdot\mathbf{r}] \quad \text{(A4)}$$

which transforms

$$U\alpha_{\mathbf{k}}U^\dagger=\alpha_{\mathbf{k}}e^{-i\mathbf{k}\cdot\mathbf{r}},$$

$$U\mathbf{r}U^\dagger=\mathbf{r},$$

and

$$U\mathbf{p}U^\dagger=\mathbf{p}-\sum\alpha_{\mathbf{k}}{}^\dagger\alpha_{\mathbf{k}}\mathbf{k}. \quad \text{(A5)}$$

Thus, the transformed Hamiltonian becomes

$$UHU^\dagger=\frac{1}{2m}(\mathbf{p}-\sum\alpha_{\mathbf{k}}{}^\dagger\alpha_{\mathbf{k}}\mathbf{k})^2+\sum\omega_{\mathbf{k}}(\alpha_{\mathbf{k}}{}^\dagger\alpha_{\mathbf{k}}+\tfrac{1}{2}) +\sum[V_{\mathbf{k}}\alpha_{\mathbf{k}}+V_{\mathbf{k}}{}^*\alpha_{\mathbf{k}}{}^\dagger]. \quad \text{(A6)}$$

In the transformed system, $UHU^\dagger$ is independent of $\mathbf{r}$; therefore $\dot{\mathbf{p}}=0$, and the conservation of momentum becomes simply

$$\mathbf{p}=\text{constant}. \quad \text{(A7)}$$

Furthermore, after the canonical transformation, the gauge transformation (A2) becomes an identity transformation for the transformed $\alpha_{\mathbf{k}}$.

This simple example shares all the features of the theory discussed in the paper. Indeed, if the system (after the canonical transformation) is in a state with

$$\mathbf{p}\neq0,$$

the first term on the right hand side of (A6) contains a part which is of the form

$$-\frac{1}{m}\sum(\mathbf{p}\cdot\mathbf{k})\alpha_{\mathbf{k}}{}^\dagger\alpha_{\mathbf{k}}. \quad \text{(A8)}$$

This implies that the energy of a phonon with momentum $\mathbf{k}$ is different from that of $-\mathbf{k}$.

We may carry the analogy further and imagine that for some observational reasons it is easy to detect the phonons, but the existence of the extra electron is not known. Thus, it would appear that in this solid phonons are created and annihilated spontaneously and there is an apparent nonconservation of momentum. By postulating the existence of an additional electron it is possible to construct the Hamiltonian, Eq. (A1), which is invariant under the gauge transformation (A2), thus conserving momentum, and which is invariant under the discrete space inversion transformation $\mathbf{k}\to-\mathbf{k}$. However, if the system is in a state with a nonzero value of total momentum, there would exist an apparent noninvariance under space inversion as well, which is in complete analogy with the apparent CP noninvariance discussed in the paper.

Reprinted from THE PHYSICAL REVIEW, Vol. 138, No. 6B, B1490–B1496, 21 June 1965
Printed in U. S. A.

Analysis of *CP*-Noninvariant Interactions and the K_1^0, K_2^0 System*

T. D. LEE
Department of Physics, Columbia University, New York, New York

AND

L. WOLFENSTEIN†
CERN, Geneva, Switzerland

(Received 10 February 1965)

In order to explain the recent observation of $K_2^0 \to \pi^+ + \pi^-$ by Christenson *et al.*, we assume that the relevant interaction Hamiltonian can be separated into two parts: $H_G + H_F$, where H_G is the usual *CP*-conserving and *T*-invariant weak interaction and H_F is a new interaction which is not invariant under *CP*. Depending on the selection rule of H_F with respect to the strangeness quantum number *S*, this new interaction H_F can be much stronger than H_G, or somewhat weaker than H_G, or very much weaker than H_G. The properties of these three different classes of possibilities are discussed and some possible experimental tests are suggested. The system of K_1^0 and K_2^0 is analyzed for these different possibilities without assuming the validity of *CPT* invariance. Efforts are made to separate out the various experimental consequences that follow from different symmetry requirements which may be satisfied by this new *CP*-noninvariant interaction H_F.

I. GENERAL DISCUSSIONS

RECENTLY, Christenson *et al.*[1] observed that the long-lived component of the K^0, the K_2^0, decays into two π mesons with a rate given by

$$|\epsilon| \equiv [\text{Rate}(K_2^0 \to \pi^+ + \pi^-)/\text{Rate}(K_1^0 \to \pi^+ + \pi^-)]^{1/2} \cong 2.2\times 10^{-3}. \quad (1)$$

Throughout this paper we assume that these reactions can occur in a vacuum which has a definite $CP=+1$ value.[2] The operators *C* and *P* denote, respectively, the charge conjugation and the space inversion which are defined by the strong and the electromagnetic interactions. Thus, the final state $(\pi^+ + \pi^-)$ must have a definite $CP=+1$ value. Since out of the states K^0 and $\bar{K}^0$ only one initial state with $CP=+1$ can be constructed, the observation that both K_1^0 and K_2^0 can decay into $(\pi^+ + \pi^-)$ shows that *CP* cannot be conserved in the $K_{\pi 2}$ decay. If *CPT* invariance is assumed, then the time-reversal invariance must also be violated in the $K_{\pi 2}^0$ decay. From the magnitude of the observed value $|\epsilon|$ and the already existing experimental limits on the possible *CP* violation in other weak processes,[3] it appears that the amount of either *CP* violation or the time-reversal violation, at least in these particular reactions, is relatively small.

We assume that it is possible to separate the relevant interaction Hamiltonian into two parts:

$$H_G + H_F, \quad (2)$$

where H_G is the usual weak interaction which satisfies the $\Delta S \neq \pm 2$ rule and is invariant under *CP* and *T* separately, where *S* is the strangeness quantum number and *T* is the time reversal operator. The remainder H_F is a new interaction which does not conserve *CP*, nor is it invariant under *T*. For order-of-magnitude estimations, we may regard H_G and H_F as characterized, respectively, by the coupling constants *G* and *F* where *G* can be either the Fermi coupling constant G_V in β decay, or the usual strangeness-nonconserving weak coupling constant $\sim \frac{1}{4} G_V$, and *F* is a new coupling constant which has the same dimension as G_V.

Clearly, with only one observed *CP*-violating reaction, a great variety of different *CP*-noninvariant Hamiltonian H_F can be proposed. It is important to notice that, at present, not only is the detailed form of H_F unknown, but even the approximate order of magnitude of its coupling constant *F* is limited only within an extremely wide range by the existing experiments. As we shall see, there are three very different classes of possibilities depending on the particular selection rule that H_F satisfies with respect to the strangeness quantum number *S*. These three classes are listed in Table I. Some specific properties of these three classes of possibilities will be discussed in the following:

(1) If H_F satisfies the $\Delta S = 0$ rule and if it does not contain any leptonic operators, then (F/G) can be $\approx 10^3$. Thus, H_F is much stronger than the usual weak interaction H_G. The dimensionless coupling constant of H_F is given by (within a factor ~ 10)

$$(4\pi)^{-1} F m_p^2 \approx |\epsilon| \cong 2.2\times 10^{-3}, \quad (3)$$

* This research was supported in part by the U. S. Atomic Energy Commission.

† National Science Foundation Postdoctoral Fellow on leave from the Carnegie Institute of Technology, Pittsburgh, Pennsylvania.

[1] J. H. Christenson, J. W. Cronin, V. L. Fitch, and R. Turlay, Phys. Rev. Letters 13, 138 (1964).

[2] The alternative possibility has been considered elsewhere. See, e.g., J. Bernstein, N. Cabibbo, and T. D. Lee, Phys. Letters 12, 146 (1964); J. S. Bell and J. K. Perring, Phys. Rev. Letters 13, 348 (1964); T. D. Lee, Phys. Rev. (to be published); F. Gürsey and A. Pais (unpublished).

[3] M. T. Burgy, V. E. Krohn, T. B. Novey, G. R. Ringo and V. L. Telegdi, Phys. Rev. Letters 1, 324 (1958). See, also, J. W. Cronin and O. E. Overseth, in *Proceedings of the 1962 Conference on High Energy Physics at CERN*, edited by F. J. M. Farley and M. E. Meyer (North-Holland Publishing Company, Amsterdam, 1963), p. 453.

TABLE I. Three different classes of CP-noninvariant interaction H_F.

F/G (within a factor $\sim$10)	Conditions satisfied by H_F	Remarks
$\sim 10^3$	$\Delta S=0$ transitions only P is conserved C invariance violated T invariance violated	Both strong and weak processes have C-noninvariant and T-noninvariant amplitudes which are $\sim\|\epsilon\|$ times the C-, T-invariant amplitudes.
$\sim 10^{-2}$–10^{-3}	$\Delta S=\pm1$ transitions allowed $\Delta S=\pm2$ transitions forbidden CP invariance violated	Only weak processes have CP-violating amplitudes.
$\sim 10^{-9}$	$\Delta S=\pm2$ transitions allowed CP invariance violated	Neglecting terms of $O(10^{-9})$, but keeping those of $O(\epsilon)$, only $K_{1,2}{}^0$ decays violate CP invariance. These violations are due entirely to the mass operator of the $K_{1,2}{}^0$ system.

where m_p is the mass of the proton. Assuming that CPT invariance holds,[4,5] the observation of the $(\pi^+ + \pi^-)$ decay mode of $K_2{}^0$ and the present limit[6] on the electric dipole moment of the neutron require that H_F must conserve P but violate C invariance and T invariance. The question whether H_F may or may not contain terms that violate the conservation of isotopic spin $\mathbf{I}$ can be left open. We note that if H_F satisfies the $|\Delta\mathbf{I}|=0$ rule, then it is convenient to regard H_F as a small C-, or T-, noninvariant part of the usual strong interaction; otherwise, it is more convenient to regard H_F as a completely new interaction independent of the existing strong interactions.

The decay of $K_2{}^0$ to 2π, as well as other CP, or T, violating weak processes, can occur through the second-order expression H_GH_F, which contains terms that violate the strangeness conservation, P invariance, C invariance, PC invariance, and T invariance. The possible existence of such a H_F can also be tested by examining the limits of T invariance and C invariance in the strong interactions. If Eq. (3) holds, then all nuclear matrix elements could have a small admixture of T-noninvariant amplitude which is $\sim|\epsilon|$ times the T invariant amplitude. Such a T-noninvariant amplitude can be observed by studying, e.g., the E2, M1 interference terms in an appropriate nuclear transition.[7]

Another possible test of such a relatively strong H_F was pointed out by Friedberg, Lee, and Schwartz.[8] The decay of η^0 is known to violate the G parity. In particular, the decay mode

$$\eta^0 \rightarrow 3\pi \tag{4}$$

can occur[9] via the virtual electromagnetic interactions to a final (3π) state with $I=1$ and $C=+1$. It can also decay, now, through H_F to a final state with $C=-1$ and $I=0$, or 2, with a relative amplitude $\sim r(\epsilon/\alpha)$, where α is the fine structure constant and r is a reduction factor depending on the final-state pion interactions and the angular momentum barrier effects. The approximate order of magnitude of r is expected to be $\lesssim 1$, if H_F contains a $|\Delta\mathbf{I}|=2$ part; otherwise, it could be much smaller than 1. The interference between the $C=+1$ amplitudes can result in an asymmetry between π^+ and π^- in the Dalitz plot of the final (π^+,π^-,π^0) distribution. Because of the known strong pion interactions in the final states, such an asymmetry does not violate CPT invariance. The detection of such an asymmetry would be an unequivocal proof of violation of C invariance in the η^0 decay; it would also establish that the C-noninvariant interaction H_F is much stronger than the usual weak interaction H_G.

(2) If H_F allows the $\Delta S=\pm1$ transitions, but forbids the $\Delta S=\pm2$ transitions, then F must be of the same order of magnitude as, or somewhat smaller than, G. (The H_F may contain, in addition, also some $\Delta S=0$ transition amplitudes.) If the observed slow rate[1] of $K_2{}^0 \rightarrow \pi^+ + \pi^-$ is regarded as a typical example of CP noninvariance, then F should be smaller than G and, within a factor $\sim$10, is given by

$$F \approx 10^{-2}G\text{–}10^{-3}G. \tag{5}$$

Such a H_F can be easily described by assigning some appropriate phase factors to the various coupling constants in the usual weak interactions. The general observable effects in such a case have already been discussed in the literature.[10] Recently, in connection

[4] The general theoretical arguments for expecting CPT invariance have been given by W. Pauli, *Niels Bohr and the Development of Physics* (Pergamon Press, Inc., London, 1955) and J. Schwinger, Phys. Rev. **91**, 720, 732 (1953); **94**, 1366 (1953). See, also, G. Lüders, Kgl. Danske Videnskab. Selskab Mat. Fys. Medd. **28**, No. 5 (1954).

[5] If H_F is not invariant under CPT, then there would be a mass difference between any strongly interacting particle and its antiparticle; e.g., $\langle K^0|H_F|K^0\rangle$ could be different from $\langle \bar{K}^0|H_F|\bar{K}^0\rangle$. Such a difference is clearly ruled out if F is $\approx 10^3\, G$.

[6] E. M. Purcell and N. F. Ramsey, Phys. Rev. **78**, 807 (1950); N. F. Ramsey, *Molecular Beams* (Oxford University Press, London, 1956).

[7] See, e.g., B. Jacobsohn and E. Henley, Phys. Rev. **113**, 225 (1959).

[8] R. Friedberg, T. D. Lee, and M. Schwartz (unpublished).

[9] If H_F contains a C-conserving but $|\Delta\mathbf{I}|=1$, or 3, part then η^0 can also decay via H_F to a 3π final state with $C=+1$ and $I=1$, or 3.

[10] See, e.g., T. D. Lee and C. N. Yang, Brookhaven National Laboratory BNL 443 (T-91), 1957, Chaps. IV and V (unpublished).

with the observation[1] of Christenson *et al.*, many specific models have also been proposed.[11] In this paper, we will not investigate further this well-studied case.

(3) The final possibility is that H_F may be very much weaker[12] than H_G; i.e.,

$$F \approx 10^{-9} G \tag{6}$$

or

$$F \approx 10^{-15} m_p^{-2}. \tag{7}$$

In this case, H_F must allow a $\Delta S = \pm 2$ transition and, in particular, the matrix element

$$\langle \bar{K}^0 | H_F | K^0 \rangle \neq 0. \tag{8}$$

It will be shown that, in this case, to the accuracy of keeping $|4\pi F/G^2 m_p^2| \approx |\epsilon|$, but neglecting $(F/G) \sim 10^{-9}$ as compared to 1, the CP-violating effects in all known weak processes are *already completely determined*, provided the validity of the CPT theorem[4] is assumed; otherwise, an additional new real parameter is needed. We note that if one neglects $F/G \sim 10^{-9}$, then in all known weak processes, except those involving K_1^0 and K_2^0, there is no observable CP-violating effect. For the K_1^0 and K_2^0 states, terms of the order F/G^2 can occur, but only through the mass operator. Assuming the validity of the CPT theorem, the eigenstates K_1^0 and K_2^0, each of which has a definite lifetime, are given by[13]

$$K_1^0 = (1+|\epsilon|^2)^{-1/2}[(1+\epsilon)K^0 + (1-\epsilon)\bar{K}^0] \tag{9}$$

and

$$K_2^0 = (1+|\epsilon|^2)^{-1/2}[(1+\epsilon)K^0 - (1-\epsilon)\bar{K}^0], \tag{10}$$

where ϵ is a complex parameter,

$$\epsilon = |\epsilon| e^{i\delta}. \tag{11}$$

The magnitude of ϵ is proportional to F/G^2. As will be shown in Sec. III, to the accuracy of neglecting F/G, the parameter ϵ has already been measured. The absolute magnitude of ϵ is given by Eq. (1), and the phase δ is given by

$$\tan\delta \cong -2[(m_1 - m_2)/(\gamma_1 - \gamma_2)], \tag{12}$$

where m_i and γ_i are the observed mass and width of K_i^0 ($i=1$ or 2). To the same accuracy, this parameter ϵ determines all the CP violating effects in $K_{1,2}^0$ decays.

Within the present experimental techniques, while it is possible to measure weak-interaction amplitudes to the accuracy of $|\epsilon| \sim 2\times 10^{-3}$, it seems rather unlikely that any known weak-interaction amplitude can be determined to the accuracy of $F/G \sim 10^{-9}$. Thus, in the case that H_F is very much weaker than H_G, we may consider a simple *phenomenological model*, in which, except for the mass operator of the K_1^0, K_2^0 system, all matrix elements are assumed to conserve CP and to be invariant under T. All CP-violating phenomena in this model are due to the CP noninvariance of the mass operator of the K_1^0 and K_2^0 states. The further assumption of CPT invariance requires all these CP violations to be described by one single phenomenological parameter ϵ, which has already been measured [except that Eq. (12) has two discrete solutions δ and $(\pi+\delta)$]. If the underlying CP-violating interaction H_F does satisfy the general conditions given by Eqs. (7) and (8), then to the assumed accuracy of neglecting F/G but keeping F/G^2, all observable effects of such an extremely weak interaction are identical with this simple phenomenological model. If CPT invariance is violated, an additional real parameter η is needed to characterize the mass operator. To the same assumed accuracy, all observable CP-violating or CPT-violating effects of H_F contain only this single unknown parameter η.

In the above, we summarize the three different possibilities of the CP-noninvariant interaction H_F. The remaining part of the paper concerns, mainly, the $K_{1,2}^0$ system. In Sec. II, we review the general properties of the $K_{1,2}^0$ system, which are valid for all three classes of H_F, and without assuming the validity of the CPT theorem. Applications to the simple phenomenological model, in which only the mass operator violates CP invariance, are given in Sec. III, if CPT invariance holds, and in Sec. IV, if CPT invariance is also violated.

II. GENERAL DESCRIPTION OF K_1^0 AND K_2^0 STATES

In this section, we review the general description of the K_1^0 and K_2^0 system which is valid for *any* interaction Hamiltonian and *without* the assumption of CPT invariance.[14] We consider a coherent mixture of K^0 and $\bar{K}^0$ whose amplitude at time t is described by the wave function

$$\psi(t) = a(t)|K^0\rangle + b(t)|\bar{K}^0\rangle \tag{13}$$

or simply as

$$\psi(t) = \begin{pmatrix} a(t) \\ b(t) \end{pmatrix}, \tag{14}$$

where t is measured in the rest system of the K meson, and the state $|\bar{K}^0\rangle$ is defined to be

$$|\bar{K}^0\rangle = CP|K^0\rangle. \tag{15}$$

The CP operator can only be determined by the strong and the electromagnetic interactions up to a phase factor

$$e^{iS\chi}, \tag{16}$$

where S is the strangeness operator and χ an arbitrary real number. However, since H_G is, by definition, the usual CP-conserving weak interaction, and since it must allow the $\Delta S = \pm 1$ transitions, this phase factor

[11] R. G. Sachs, Phys. Rev. Letters **13**, 286 (1964); N. Cabibbo, Phys. Letters **12**, 137 (1964); T. N. Truong, Phys. Rev. Letters **13**, 358 (1964); S. Glashow, *ibid.* **14**, 35 (1965).

[12] L. Wolfenstein, Phys. Rev. Letters **13**, 562 (1964). Our estimate of the order of magnitude of F is somewhat lower than in this reference. We obtain it from $(Fm_p^2/4\pi) \approx |\epsilon| (G_V m_p^2/4\pi)^2 (\frac{1}{4})^2$. A similar result is obtained from $(Fm_p^2/4\pi) \approx |\epsilon| \times |(m_1 - m_2)/m_K|$.

[13] T. D. Lee, R. Oehme, and C. N. Yang, Phys. Rev. **105**, 1671 (1957).

[14] Cf. R. Sachs, Ann. Phys. (N. Y.) **22**, 239 (1963). Sachs' notation is related to ours by $r = e^{i\phi}$, $s = \cot(\theta/2)$. He uses a different normalization condition for the eigenstates, however.

χ can be fixed by H_G. In the following, we will use this choice of phase factor and regard the CP operator as completely determined.

By using the Weisskopf-Wigner method[15] of solving the time-dependent Schrödinger equation, the differential equation for $\psi(t)$ is given by

$$-d\psi/dt=(\Gamma+iM)\psi\,, \tag{17}$$

where Γ and M are two (2×2) Hermitian matrices,

$$\Gamma=\Gamma^\dagger \tag{18}$$

and

$$M=M^\dagger. \tag{19}$$

The matrix elements of Γ are related to that of the relevant perturbation Hamiltonian $\mathcal{H}$ by (neglecting $\mathcal{H}^3$)

$$\Gamma_{11}=2\pi\sum_j \rho_j|\langle j|\mathcal{H}|K^0\rangle|^2, \tag{20}$$

$$\Gamma_{22}=2\pi\sum_j \rho_j|\langle j|\mathcal{H}|K^0\rangle|^2, \tag{21}$$

and

$$\Gamma_{21}=\Gamma_{12}^*=2\pi\sum_j \rho_j\langle \bar{K}^0|\mathcal{H}|j\rangle\langle j|\mathcal{H}|K^0\rangle, \tag{22}$$

where the sum extends over all possible decay channels j, and ρ_j denotes the density of states per unit energy in the jth channel. The matrix elements of M are given by (neglecting $\mathcal{H}^3$)

$$M_{11}=m_K{}^0+\langle K^0|\mathcal{H}|K^0\rangle +\sum_\alpha \mathcal{P}[|\langle\alpha|\mathcal{H}|K^0\rangle|^2/(m_\alpha{}^0-m_K{}^0)], \tag{23}$$

$$M_{22}=m_K{}^0+\langle \bar{K}^0|\mathcal{H}|\bar{K}^0\rangle +\sum_\alpha \mathcal{P}[|\langle\alpha|\mathcal{H}|\bar{K}^0\rangle|^2/(m_\alpha{}^0-m_K{}^0)], \tag{24}$$

and

$$M_{21}=M_{12}^*=\langle \bar{K}^0|\mathcal{H}|K^0\rangle +\sum_\alpha \mathcal{P}[\langle \bar{K}^0|\mathcal{H}|\alpha\rangle\langle\alpha|\mathcal{H}|K^0\rangle/(m_\alpha{}^0-m_K{}^0)], \tag{25}$$

where $\mathcal{P}$ stands for the principal value. In the case that H_F is much stronger than H_G (i.e., $F\approx10^3G$), the perturbation Hamiltonian $\mathcal{H}$ in all the above equations is given by

$$\mathcal{H}=H_G; \tag{26}$$

the sum α stands for the appropriate sums and integrations over all (unperturbed) eigenstates of the strong interaction, the electromagnetic interaction, and the new interaction H_F; and $m_\alpha{}^0$, $m_K{}^0$ are the corresponding (unperturbed) eigenvalues of the states α and K^0 (or $\bar{K}^0$). In the other two cases ($F\lesssim G$ or $F\ll G$), the perturbation Hamiltonian is given by

$$\mathcal{H}=H_G+H_F, \tag{27}$$

the states α, K^0 (or $\bar{K}^0$) are eigenstates of the strong and electromagnetic interactions only, and $m_\alpha{}^0$, $m_K{}^0$ are the corresponding unperturbed eigenvalues.

We may express $(\Gamma+iM)$ in terms of the usual Pauli matrices σ_1, σ_2, σ_3:

$$(\Gamma+iM)=D+i(E_1\sigma_1+E_2\sigma_2+E_3\sigma_3), \tag{28}$$

where

$$\sigma_1=\begin{pmatrix}0&1\\1&0\end{pmatrix},\quad \sigma_2=\begin{pmatrix}0&-i\\i&0\end{pmatrix},\quad \sigma_3=\begin{pmatrix}1&0\\0&-1\end{pmatrix}$$

and D, E_1, E_2, and E_3 are four complex numbers. It is convenient to introduce the *complex* variables E, θ, and ϕ:

$$E=(E_1^2+E_2^2+E_3^2)^{1/2}, \tag{29}$$

$$E_3=E\cos\theta, \tag{30}$$

$$E_2=E\sin\theta\sin\phi, \tag{31}$$

and

$$E_1=E\sin\theta\cos\phi. \tag{32}$$

The eigenstates $K_1{}^0$ and $K_2{}^0$, defined by

$$(\Gamma+iM)|K_j{}^0\rangle=[\tfrac{1}{2}\gamma_j+im_j]|K_j{}^0\rangle, \tag{33}$$

where $j=1$ or 2 can be readily obtained; these solutions are identical in form to that of a spin-$\frac{1}{2}$ magnet in a magnetic field, and are given by

$$|K_1{}^0\rangle=N_1\begin{pmatrix}\cos\frac{1}{2}\theta\\ e^{i\phi}\sin\frac{1}{2}\theta\end{pmatrix}, \tag{34}$$

and

$$|K_2{}^0\rangle=N_2\begin{pmatrix}\sin\frac{1}{2}\theta\\ -e^{i\phi}\cos\frac{1}{2}\theta\end{pmatrix}, \tag{35}$$

where

$$N_1=[|\cos\tfrac{1}{2}\theta|^2+|e^{i\phi}\sin\tfrac{1}{2}\theta|^2]^{-1/2} \tag{36}$$

and

$$N_2=[|\sin\tfrac{1}{2}\theta|^2+|e^{i\phi}\cos\tfrac{1}{2}\theta|^2]^{-1/2}. \tag{37}$$

The observed mass m_j and width γ_j of $K_j{}^0$ are related to D and E by

$$D=\tfrac{1}{4}(\gamma_1+\gamma_2)+\tfrac{1}{2}i(m_1+m_2) \tag{38}$$

and

$$iE=\tfrac{1}{4}(\gamma_1-\gamma_2)+\tfrac{1}{2}i(m_1-m_2). \tag{39}$$

A useful parameter is

$$\alpha=\langle K_1{}^0|K_2{}^0\rangle =N_1N_2[(\cos\tfrac{1}{2}\theta)^*(\sin\tfrac{1}{2}\theta) -(e^{i\phi}\sin\tfrac{1}{2}\theta)^*(e^{i\phi}\cos\tfrac{1}{2}\theta)]. \tag{40}$$

From the property that Γ is a positive-definite Hermitian matrix, it follows that (independent of the validity of the CPT theorem)

$$|\alpha|^2\leqq\gamma_1\gamma_2/[(m_1-m_2)^2+\tfrac{1}{4}(\gamma_1+\gamma_2)^2], \tag{41}$$

which is $\sim1.3\times10^{-3}$, if $|m_1-m_2|$ is set to be about equal to γ_1.

[15] V. F. Weisskopf and E. P. Wigner, Z. Physik **63**, 54 (1930); **65**, 18 (1930).

The states K^0 and $\bar{K}^0$ can be expressed as linear combinations of the eigenstates K_1^0 and K_2^0:

$$|K^0\rangle = N_1^{-1}\cos\tfrac{1}{2}\theta|K_1^0\rangle + N_2^{-1}\sin\tfrac{1}{2}\theta|K_2^0\rangle \quad (42)$$

and

$$e^{i\phi}|\bar{K}^0\rangle = N_1^{-1}\sin\tfrac{1}{2}\theta|K_1^0\rangle - N_2^{-1}\cos\tfrac{1}{2}\theta|K_2^0\rangle. \quad (43)$$

Let us consider a neutral K-meson beam which, at $t=0$, is in a pure K^0 state (or a pure $\bar{K}^0$ state). The fractional number of all K mesons decaying per unit time at time t is given by the familiar expression

$$\begin{aligned}\mathfrak{N}_\pm(t) &= A_\pm\gamma_1\exp(-\gamma_1 t) + B_\pm\gamma_2\exp(-\gamma_2 t)\\ &+\{C_\pm[\tfrac{1}{2}(\gamma_1+\gamma_2) - i(m_1-m_2)]\\ &\times\exp[-\tfrac{1}{2}(\gamma_1+\gamma_2)t + i(m_1-m_2)t] + \text{c.c.}\}, \quad (44)\end{aligned}$$

where the plus (or minus) sign in the subscripts denotes the case that the neutral K meson is, at $t=0$, in a K^0 state (or a $\bar{K}^0$ state). The coefficients A_+, A_-, B_+, B_- are real and C_+, C_- are complex. These eight real coefficients depend only on two complex numbers θ and ϕ:

$$A_+ = N_1^{-2}|\cos\tfrac{1}{2}\theta|^2, \qquad B_+ = N_2^{-2}|\sin\tfrac{1}{2}\theta|^2,$$

$$A_- = N_1^{-2}|e^{-i\phi}\sin\tfrac{1}{2}\theta|^2, \quad B_- = N_2^{-2}|e^{-i\phi}\cos\tfrac{1}{2}\theta|^2,$$

$$C_+ = (N_1N_2)^{-1}(\cos\tfrac{1}{2}\theta)^*(\sin\tfrac{1}{2}\theta)\alpha,$$

and

$$C_- = -(N_1N_2)^{-1}(e^{-i\phi}\sin\tfrac{1}{2}\theta)^*(e^{-i\phi}\cos\tfrac{1}{2}\theta)\alpha. \quad (45)$$

The integrated value of $N_\pm(t)$ over all t must be 1; i.e.,

$$A_\pm + B_\pm + C_\pm + C_\pm^* = 1. \quad (46)$$

The constants $A_\pm$, $B_\pm$, and $C_\pm$ can be directly measured by studying the time development of such a K-meson beam. These measurements can be used to test the various symmetry or symmetry-violation properties of the K_1^0, K_2^0 system.

Remarks

(1) In general, the (2×2) matrix $(\Gamma + iM)$ depends on eight real parameters, of which four parameters can be determined by the observed widths γ_1, γ_2 and masses m_1, m_2 of K_1^0 and K_2^0. The remaining four parameters can be chosen to be the real and the imaginary parts of θ and ϕ:

$$\theta = \theta_R + i\theta_I \quad (47)$$

and

$$\phi = \phi_R + i\phi_I. \quad (48)$$

Three of these, θ_R, θ_I, and ϕ_I, can be determined by measuring $A_\pm$, $B_\pm$, $C_\pm$ and by using Eq. (45). The real part of ϕ depends on the relative phase between K^0 and $\bar{K}^0$, which, as remarked earlier, is determined only if we use the usual CP-conserving weak interaction H_G and the definition of $\bar{K}^0$ given by Eq. (15).[16]

(2) If CPT invariance holds, then[13]

$$\Gamma_{11} = \Gamma_{22} \quad (49)$$

and

$$M_{11} = M_{22}. \quad (50)$$

By using Eq. (28), it follows that

$$E_3 = 0. \quad (51)$$

Thus,

$$\theta = \tfrac{1}{2}\pi, \quad (52)$$

$$N_1 = N_2 = [\tfrac{1}{2}(1+|e^{i\phi}|^2)]^{-1/2}, \quad (53)$$

and α is real. The validity of the CPT theorem can be tested by measuring the constants $A_\pm$, $B_\pm$, $C_\pm$ in Eq. (44) and observing that

$$A_+ = B_+ = A_-e^{-2\phi_I} = B_-e^{-2\phi_I} \quad (54)$$

and

$$C_+ = -C_-e^{-2\phi_I} \text{ is real}. \quad (55)$$

If CPT invariance holds, the eigenstates K_1^0 and K_2^0 can also be expressed in the forms given by Eqs. (9) and (10) which depend only on one complex parameter

$$\epsilon = (1-e^{i\phi})/(1+e^{i\phi}). \quad (56)$$

(3) *Independent* of CPT invariance, if T invariance holds, then

$$\Gamma_{12}^*/\Gamma_{12} = M_{12}^*/M_{12}. \quad (57)$$

Thus, the imaginary part of ϕ must be zero; i.e.,

$$\phi_I = 0. \quad (58)$$

By using Eqs. (36), (37), and (45), we find

$$N_1 = N_2,$$

α is pure imaginary,

$$A_\pm = B_\mp, \quad (59)$$

and

$$C_+ = C_-^*. \quad (60)$$

Equations (59) and (60) can be used as tests of T invariance (without any assumption concerning the invariance of CPT, or CP, or C, or P).

(4) Independent of CPT invariance, if F/G is $\sim10^{-9}$, then by using Eqs. (20)–(25) and Eq. (27), it is easy to see that to the accuracy of keeping F/G^2, but neglecting F/G, all CP violations and CPT violations are due to the mass operator M. As already discussed in the previous section, to this assumed accuracy, such a theory is identical with the simple phenomenological model in which, except the mass operator M of the K_1^0, K_2^0 system, all other transition matrix elements satisfy the consequences of CP invariance and T invariance. Thus

$$\Gamma_{12} = \Gamma_{21} \text{ is real and } \Gamma_{11} = \Gamma_{22},$$

[16] For the simple phenomenological model, the effects of H_G and H_F are clearly separated and ϕ_R can be determined from experiments, as discussed in Sec. III and IV. For the other classes, this is less simple. Sometimes, it may be more convenient to define the relative phase between K^0 and $\bar{K}^0$ such that the amplitudes for K^0 and $\bar{K}^0$ going to the $I=0$ two-pion state are relatively real. [A method of determining ϕ_R by using this convention has been discussed by T. T. Wu and C. N. Yang, Phys. Rev. Letters 13, 380 (1964).] The two definitions of the relative phase between K^0 and $\bar{K}^0$ are the same for the simple phenomenological model.

which correspond to

$$E_2, \quad \text{real} \tag{61}$$

and

$$E_3, \quad \text{real}. \tag{62}$$

The further consequences of this simple phenomenological model will be studied in the next two sections.

III. CONSEQUENCES OF THE SIMPLE PHENOMENOLOGICAL MODEL AND CPT INVARIANCE

In the case of the simple phenomenological model, if CPT invariance holds, the eigenstates K_1^0 and K_2^0 are given by Eqs. (9) and (10). By using Eqs. (31), (39), (52), and (56), the parameter ϵ is given by

$$2\epsilon/(1-\epsilon^2)=(E_2/|E|)e^{i\beta}, \tag{63}$$

where the phase β is defined by

$$iE=|E|e^{-i\beta}=\tfrac{1}{4}(\gamma_1-\gamma_2)+\tfrac{1}{2}i(m_1-m_2). \tag{64}$$

According to Eq. (61), E_2 is real. Thus, the phase δ and the magnitude $|\epsilon|$ of the parameter ϵ satisfy

$$\tan\delta=\frac{\sin\beta-|\epsilon|^2\sin(2\delta+\beta)}{\cos\beta-|\epsilon|^2\cos(2\delta+\beta)}. \tag{65}$$

In this model, all the transition matrix elements for the decays of K^0 and $\bar{K}^0$ into any continuum states, such as 2π, 3π, etc., satisfy the consequences of CP invariance. The magnitude $|\epsilon|$ is, therefore, given by Eq. (1). Consequently, both $|\epsilon|$ and δ have been experimentally measured. If we neglect $|\epsilon|^2$ in Eq. (65), the phase δ is given by Eq. (12):

$$\tan\delta\cong\tan\beta=-2(m_1-m_2)/(\gamma_1-\gamma_2),$$

which has always two solutions, $\delta=\beta$ and $\delta=\pi+\beta$. There are, at present, some discrepancies between the existing measurements[17] of the mass difference (m_1-m_2); therefore, the exact value of β, or δ, is not yet definite.

The following is a list of some CP-violating effects in this simple model under the assumption of CPT invariance[18]:

(1)
$$\frac{\text{Rate}(K_2^0\to 2\pi^0)}{\text{Rate}(K_1^0\to 2\pi^0)}=\frac{\text{Rate}(K_2^0\to \pi^++\pi^-)}{\text{Rate}(K_1^0\to \pi^++\pi^-)}=|\epsilon|^2, \tag{66}$$

(2)
$$\frac{\text{Rate}(K_1^0\to 3\pi^0)}{\text{Rate}(K_2^0\to 3\pi^0)}=|\epsilon|^2. \tag{67}$$

The decay of $K_1^0\to\pi^++\pi^-+\pi^0$ can occur with the final 3π system in either a $CP=-1$ state (which shows CP violation) or a $CP=+1$ state (which is compatible with CP conservation). The amplitude of $K_1^0\to\pi^++\pi^-+\pi^0$ in a final $CP=-1$ state is ϵ times the corresponding amplitude of K_2^0 decay. The amplitude of $K_1^0\to\pi^++\pi^-+\pi^0$ in a final $CP=+1$ state does not violate CP conservation, but is reduced from the corresponding $CP=-1$ amplitude of K_2^0 decay only by the centrifugal barrier effects.

(3) For leptonic decays, we assume that the $\Delta Q=\Delta S$ rule holds for the transition matrix elements between the K^0, $\bar{K}^0$ states and any leptonic states [i.e., the $\Delta Q=\Delta S$ rule holds for H_G].

$$\text{Rate}(K_1^0\to\pi^++l^-+\bar{\nu}_l)=\text{Rate}(K_2^0\to\pi^++l^-+\bar{\nu}_l), \tag{68}$$

$$\text{Rate}(K_1^0\to\pi^-+l^++\nu_l)=\text{Rate}(K_2^0\to\pi^-+l^++\nu_l), \tag{69}$$

and

$$\frac{\text{Rate}(K_1^0\to\pi^-+l^++\nu_l)}{\text{Rate}(K_1^0\to\pi^++l^-+\bar{\nu}_l)}=\left|\frac{1+\epsilon}{1-\epsilon}\right|^2\cong1+4|\epsilon|\cos\delta, \tag{70}$$

where $l=e$ or μ. Identical relations also hold for K_{l4} decays, provided we replace π^+ and π^- in the above formulas by $(\pi^++\pi^0)$ and $(\pi^-+\pi^0)$, respectively.

(4) By using Eqs. (52) and (56), the coefficients $A_\pm$, $B_\pm$, and $C_\pm$ in Eq. (44) are given by

$$A_\pm=B_\pm=(1+|\epsilon|^2)/2|1\pm\epsilon|^2\cong\tfrac{1}{2}[1\mp2|\epsilon|\cos\delta] \tag{71}$$

and

$$C_\pm=\pm(\epsilon+\epsilon^*)/2|1\pm\epsilon|^2\cong\pm|\epsilon|\cos\delta. \tag{72}$$

IV. CONSEQUENCES OF THE SIMPLE MODEL WITHOUT CPT INVARIANCE

In this section, we list the CP-violating and CPT-violating phenomena for the same simple phenomenological model, but without assuming CPT invariance. According to Eqs. (61) and (62), these violations depend only on two real parameters E_2 and E_3. The observed rate of $K_2^0\to\pi^++\pi^-$ gives one relations between these two parameters. The whole problem, therefore, con-

[17] The experimental value of the mass difference between K_1^0 and K_2^0 varies from 0.50 γ_1 to 1.9 γ_1. See B. Aubert *et al.*, Phys. Letters **10**, 215 (1964); U. Camerini *et al.*, Phys. Rev. **128**, 362 (1962); J. H. Christenson *et al.*, *Conference on Fundamental Aspects of Weak Interactions*, BNL, New York (1963); J. H. Christenson, Tech. Rept. 34, Princeton University (1964); V. L. Fitch, Nuovo Cimento **22**, 1160 (1961); M. L. Good, *et al.*, Phys. Rev. **124**, 1223 (1961). In Ref. 12, the lowest of these values was used to obtain a numerical value for the ratio in Eq. (70).

[18] While the equalities given by Eqs. (66)–(72) can be used to test the validity of the simple model (i.e., $F\approx10^{-9}G$), it is important to notice that some of these results are also valid in other cases. For example, if we consider the case $F\approx10^{-2}G$ and assume that H_F and H_G both satisfy the $|\Delta\mathbf{I}|=\frac{1}{2}$ rule for the strangeness-nonconserving amplitudes. The $|\Delta\mathbf{I}|=\frac{3}{2}$ amplitude is, then, due to the combined effect of electromagnetic interaction and either H_G or H_F. Thus, neglecting $O(\epsilon\alpha)$, the equality

$$\frac{\text{Rate}(K_2^0\to\pi^++\pi^-)}{\text{Rate}(K_1^0\to\pi^++\pi^-)}=\frac{\text{Rate}(K_2^0\to2\pi^0)}{\text{Rate}(K_1^0\to2\pi^0)}$$

remains valid. Assuming CPT invariance and the $\Delta Q=\Delta S$ rule one finds, in addition, for this case that $|\epsilon|$ and δ are given within 10% by Eqs. (1) and (12). Thus, except for Eq. (67), Eqs. (66) and (68)–(72) also hold to a fair degree of approximation for the case $F\approx10^{-2}G$, provided that H_F and H_G both satisfy the $|\Delta\mathbf{I}|=\frac{1}{2}$ rule. [The results for δ can be most conveniently obtained by using Eqs. (20) and (23) of Wu and Yang (Ref. 16).]

tains only one real parameter which has not yet been measured.

It is convenient to represent the K_1^0 and K_2^0 states by

$$|K_1^0\rangle=[2(1+|\epsilon_1|^2)]^{-1/2}\begin{pmatrix}1+\epsilon_1\\1-\epsilon_1\end{pmatrix} \tag{73}$$

and

$$|K_2^0\rangle=[2(1+|\epsilon_2|^2)]^{-1/2}\begin{pmatrix}1+\epsilon_2\\-(1-\epsilon_2)\end{pmatrix}, \tag{74}$$

where ϵ_1 and ϵ_2 are related to θ and ϕ by

$$(1-\epsilon_1)/(1+\epsilon_1)=e^{i\phi}\tan\tfrac{1}{2}\theta \tag{75}$$

and

$$(1-\epsilon_2)/(1+\epsilon_2)=e^{i\phi}\cot\tfrac{1}{2}\theta, \tag{76}$$

which can, in turn, be expressed in terms of the real parameters E_2 and E_3.

We note that

$$\frac{\text{Rate}(K_2^0\to\pi^++\pi^-)}{\text{Rate}(K_1^0\to\pi^++\pi^-)}=\frac{\text{Rate}(K_2^0\to 2\pi^0)}{\text{Rate}(K_1^0\to 2\pi^0)}=\frac{1+|\epsilon_1|^2}{1+|\epsilon_2|^2}|\epsilon_2|^2 \tag{77}$$

and

$$\frac{\text{Rate}(K_1^0\to 3\pi^0)}{\text{Rate}(K_2^0\to 3\pi^0)}=\frac{1+|\epsilon_2|^2}{1+|\epsilon_1|^2}|\epsilon_1|^2. \tag{78}$$

From Eq. (77) and the experimental result[1] of Christenson *et al.*, it follows that

$$\frac{|\epsilon_2|^2}{1+|\epsilon_2|^2}\leqq|\epsilon|^2\cong 4.8\times10^{-6},$$

or

$$|\epsilon_2|\leqq|\epsilon|\times[1-|\epsilon|^2]^{-1/2}\cong 2.2\times10^{-3}, \tag{79}$$

which, together with the inequality (41), implies that

$$|\epsilon_1|\leqq[\sim 3.6\times10^{-2}]. \tag{80}$$

Thus, $|\epsilon_1|$ and $|\epsilon_2|$ must both be small numbers, and $|\epsilon_1|^2$, $|\epsilon_2|^2$ and $|\epsilon_1\epsilon_2|$ may be neglected as compared to 1. In this *approximation*, all formulas become considerably simpler. The results are given below:

(1) Let

$$\epsilon_j=|\epsilon_j|\exp(i\delta_j); \tag{81}$$

then

$$|\epsilon_1|=|\epsilon_2|=|\epsilon| \tag{82}$$

and

$$\tfrac{1}{2}(\delta_1+\delta_2)=\delta, \tag{83}$$

where

$$\delta=\beta \quad\text{or}\quad (\pi+\beta). \tag{84}$$

The constants $|\epsilon|$ and β have both been measured, and are given, respectively, by Eqs. (1) and (64).

The entire problem contains only one unknown real parameter which may be chosen to be

$$\tfrac{1}{2}(\delta_1-\delta_2)=\eta. \tag{85}$$

(2) The coefficients $A_\pm$, $B_\pm$, and $C_\pm$ in Eq. (44) are given by

$$A_\pm=\tfrac{1}{2}[1\mp2|\epsilon|\cos(\delta-\eta)], \tag{86}$$

$$B_\pm=\tfrac{1}{2}[1\mp2|\epsilon|\cos(\delta+\eta)], \tag{87}$$

and

$$C_\pm=\pm|\epsilon|(\cos\delta)e^{-i\eta}. \tag{88}$$

(3) For the K^0 decay into pions, Eqs. (66) and (67) remain valid. For the K^0 decay into leptons, if $\Delta Q=\Delta S$ rule holds, we have

$$\text{Rate}(K_j^0\to\pi^-+l^++\nu_l)/\text{Rate}(K_j^0\to\pi^++l^-+\bar{\nu}_l)=1+4|\epsilon|\cos\delta_j, \tag{89}$$

where $l=e$ or μ and $j=1$ or 2. The same formula holds for K_{l4} decays, provided $\pi^\mp$ is replaced by $(\pi^\mp+\pi^0)$.

(4) If T invariance holds (but not CPT invariance), then by using Eqs. (58), (75), and (76), and since $\cos\delta\neq0$, we find

$$\eta=\tfrac{1}{2}\pi. \tag{90}$$

(5) It has been suggested[2] that the observation of $K_2^0\to\pi^++\pi^-$ can be compatible with CP invariance, provided there exists an energy difference ΔE between the K^0 and the $\bar{K}^0$ states due to some cosmological background. The apparent CP noninvariance is attributed not to the basic violation of the interaction, but is regarded as the result of the initial (or boundary) condition of our universe which gives rise to this background field. As remarked in the paper by Bernstein *et al.*,[2] the amount of this energy difference ΔE is proportional to E^{2J}, where J=spin of the cosmological background field (or, rank of the corresponding field tensor) and E is the energy of the K meson measured in the rest system of the cosmological background field.

From a phenomenological point of view, except for this energy dependence of ΔE, the description of the K_1^0 and K_2^0 system in such a cosmological background field is the same as the simple model (without CPT invariance) considered in this paper, provided

$$\epsilon_2=-\epsilon_1; \tag{91}$$

i.e.,

$$\eta=\tfrac{1}{2}\pi,$$

which is identical with Eq. (90). If $J=0$, then there is no difference between the cosmological possibility with CP invariance and the case of an extremely weak CP-noninvariant, CPT-noninvariant, but T-invariant H_F with $F\sim10^{-9}G$, unless either the measurement of the appropriate transition amplitudes can be improved to include terms of the order of 10^{-9}, of the observation of a real emission of the cosmological field quantum can become feasible.

ACKNOWLEDGMENT

One of us (L.W.) wishes to thank Professor L. Van Hove and Professor V. F. Weisskopf for the hospitality of CERN.

Reprinted from THE PHYSICAL REVIEW, Vol. 139, No. 6B, B1650–B1659, 20 September 1965
Printed in U. S. A.

Possible C, T Noninvariance in the Electromagnetic Interaction*

J. BERNSTEIN
New York University, New York, New York

AND

G. FEINBERG AND T. D. LEE
Columbia University, New York, New York

(Received 12 May 1965)

The experimental foundation of the various discrete symmetry properties for the strong and the electromagnetic interactions is reviewed. It is found that there is strong evidence that both interactions are invariant under P and CPT, and good evidence that the strong interaction is invariant under C and T. However, at present, there is a complete lack of evidence that the electromagnetic interaction of the strongly interacting particles is invariant under C and T. Possible experiments that can test such C, T invariance or noninvariance are discussed. It is pointed out that if the electromagnetic interaction of the strongly interacting particles has strong violations of C, T invariance, then, through second-order processes, C-violating and CP-violating effects can be observed in the usual strong and weak processes, respectively. In particular, the decay $K_2^0 \to \pi^+ + \pi^-$ may occur with an amplitude $\sim(\alpha/\pi)$ times that of $K_1^0 \to \pi^+ + \pi^-$.

I. INTRODUCTION

SINCE the discovery[1] that

$$K_2^0 \to \pi^+ + \pi^-, \tag{1}$$

in apparent violation of CP invariance, it has become clear that the experimental foundation of all discrete symmetries should be re-examined for all the interactions. Such a study will be made in this paper for both the strong and electromagnetic interactions. We shall see that there is strong evidence supporting the premise that both interactions conserve P and CPT to a great accuracy; furthermore, the strong interaction is invariant under T to a fair degree of accuracy ($\sim 2\%$), and through CPT and P invariance the same accuracy applies also to C conservation in the strong interaction. The situation is, however, very different for the electromagnetic interactions. As will become clear, at present there exists *no* evidence that the electromagnetic interactions of the *strongly interacting particles* are invariant under C and T. Indeed, all existing experimental results are compatible with the possibility of a very large violation of C and T in the electromagnetic interaction of these strongly interacting particles. Throughout this paper, the three operators C, P, and T denote, respectively, the usual particle-antiparticle conjugation, space inversion, and time reversal.

For definiteness, let us write the nonleptonic part of the electromagnetic current operator $\mathcal{J}_\mu$ as

$$\mathcal{J}_\mu = J_\mu + K_\mu, \tag{2}$$

where

$$CJ_\mu C^{-1} = -J_\mu \tag{3}$$

and

$$CK_\mu C^{-1} = +K_\mu. \tag{4}$$

Both J_μ and K_μ are assumed to have the same transformation properties under CPT, and both are vector currents under P (in the absence of the weak interaction). The condition of C, or T invariance is

$$K_\mu = 0. \tag{5}$$

We note that if $K_\mu \neq 0$ and if some of its matrix elements are comparable in magnitude to those of J_μ (i.e., large C, T violation), then, through virtual electromagnetic processes, all strong reactions may violate C and T to order of α where $\alpha = (137)^{-1}$.

It has been suggested recently,[2,3] in connection with reaction (1), that the violation of CP invariance is not due to the usual weak interaction, but, rather, is due to the possible existence of a new C, T noninvariant interaction called H_F, whose coupling constant F is much stronger than the Fermi constant G_V of the usual weak interaction $H_{\rm wk}$. It is estimated that

$$F \sim 10^3 G_V; \tag{6}$$

or, the dimensionless constant (Fm_p^2) is given by

$$Fm_p^2 \sim 10^{-2}, \tag{7}$$

where m_p is the mass of the proton. The usual weak interaction violates C invariance and P invariance, but it is assumed to be invariant under CP and T. The new interaction H_F is assumed to violate C invariance and T invariance, but it conserves strangeness, and is invariant under CT and P. The decay $K_1^0 \to 2\pi$ can occur through

---

* This research was supported in part by the U. S. Atomic Energy Commission and the National Science Foundation.

[1] J. H. Christenson, J. W. Cronin, V. L. Fitch, and R. Turlay, Phys. Rev. Letters 13, 138 (1964).

[2] T. D. Lee and L. Wolfenstein, Phys. Rev. 138, B1490 (1965).

[3] L. B. Okun (unpublished report). See also J. Prentki and M. Veltman, Phys. Letters 15, 88 (1965), in which they consider the possibility that the C, T-noninvariant interaction is simply the usual SU_3-violating, but SU_2-conserving part of the strong interaction. This possibility seems to encounter several difficulties, especially in view of the present accuracy ($\sim 2\%$ in relative amplitude) of T invariance in many nuclear reactions (see Ref. 12). In all such reactions, large violations of SU_3 symmetry may occur because of the difference between the Compton wavelengths of π and K. It is also difficult to see how such a strong violation of C, T invariance manages to produce only a very small CP-noninvariant amplitude in the decay $K_2^0 \to \pi^+ + \pi^-$.

H_{wk} alone, but reaction (1) can occur only through the second-order CP-noninvariant term $H_F H_{wk}$; thus, its amplitude is much smaller than that of $K_1^0 \to 2\pi$.

If such an H_F exists, then, at least to order (Fm_p^2), an "effective" C, T-noninvariant current K_μ will appear. However, total lack of evidence for C, T invariance for the nonleptonic part of the electromagnetic interaction suggests another more interesting possibility. The electromagnetic interaction may itself show a large violation of C and T, and the "new" interaction H_F may well be simply a manifestation of the second-order electromagnetic effects. This explains why the magnitude of (Fm_p^2) is $\sim\alpha$.

While these are merely some hypothetical possibilities, they do reflect our present state of ignorance and should, therefore, provide an incentive for a critical study of the various experimental implications of possible C, T violations in the electromagnetic interactions of strongly interacting particles. To set a sensitive limit on the magnitude of possible C noninvariances, experiments should be done to measure the rates of

$$\eta^0 \to \pi^0 + e^+ + e^-,$$

or

$$\phi^0 \to \omega^0(\text{or } \rho^0) + \gamma,$$

and to study the possible $\pi^\pm$ asymmetry in

$$\omega^0(\text{or } \eta^0) \to \pi^+ + \pi^- + \gamma,$$

etc. For a direct limit on possible T noninvariance in the electromagnetic interactions, a reaction such as

$$\Sigma^0 \to \Lambda^0 + e^+ + e^-$$

is important. A systematic phenomenological discussion of these and similar reactions will be given in the subsequent sections.

The electromagnetic interaction is usually thought to be invariant under C, P, and T separately. There exists a so-called "minimal" principle, which states that if, in the absence of the electromagnetic field A_μ, the Lagrangian $\mathcal{L}$ is known, then the replacement of $(\partial/\partial x_\mu)$ by $(\partial/\partial x_\mu - ieA_\mu)$ changes $\mathcal{L}$ to another Lagrangian $\mathcal{L}_A$, which contains all the dependence on the electromagnetic field. A consequence of this "minimal" principle is that if $\mathcal{L}$ is invariant under T, so must $\mathcal{L}_A$ be, since $\partial/\partial x_\mu$ and $(\partial/\partial x_\mu - ieA_\mu)$ transform in the same way under T.

It must be pointed out that the validity of such a "minimal" principle for the strongly interacting particles is, as yet, unclear, and it will not be assumed in this paper.[3a] Throughout our discussions, we assume that the electromagnetic interactions of the leptons are invariant[4] under C, P, and T separately, but the electromagnetic interactions of the nonleptons can have strong violations of C, T invariance.

II. PRESENT EXPERIMENTAL LIMITS ON C, P, T INVARIANCES FOR STRONG INTERACTIONS AND ELECTROMAGNETIC INTERACTIONS

Since the discovery[5,6] of the nonconservation of parity in weak interactions,[7] there have been many experiments to search for possible P-violating effects in nuclear reactions. These experiments[8] establish that the magnitudes of the P-nonconserving amplitudes are smaller than that of the corresponding P-conserving amplitude by, at least, a factor $\sim 10^{-5}$, which is about the order of magnitude of the dimensionless weak coupling constant $G_V m_p^2$. Thus, we should regard both strong and electromagnetic interactions to be P conserving (neglecting the weak-coupling constant).

Invariance under CPT implies[9] mass and lifetime equalities between any particle and its antiparticle. In view of the possibly large violations of C and CP invariances, we may use such equalities as evidence for CPT invariance. Among such equalities, the most accurate one is that between K^0 and $\bar{K}^0$:

$$\langle K^0 | H_{st} + H_\gamma + H_{wk} | K^0 \rangle = \langle \bar{K}^0 | H_{st} + H_\gamma + H_{wk} | \bar{K}^0 \rangle, \quad (8)$$

where H_{st} and H_γ are, respectively, the Hamiltonians for the strong and the electromagnetic interactions. From the experimental mass difference[10] Δm between K_1^0 and K_2^0, we conclude that Eq. (8) holds to $\sim |\Delta m/m_K| \sim 10^{-14}$. In the following, we take H_{st} and H_γ to be invariant under P and CPT; therefore, to the same great accuracy, both interactions are also invariant under CT.

The present experiments[11] on the polarization and

[3a] *Note added in proof.* It turns out that the principle of minimal electromagnetic interaction is compatible with the violation of C invariance where C is the particle-antiparticle conjugation operator determined by the strong interaction. For a complete discussion, see T. D. Lee, Phys. Rev. (to be published).

[4] The very precise experimental results on $(g-2)$ and Lamb shift show that the electromagnetic interactions of both e and μ are accurately described by the usual form $ie\psi^\dagger\gamma_4\gamma_\mu\psi A_\mu$ which is invariant under C, P, and T separately.

[5] C. S. Wu, E. Ambler, R. W. Hayward, D. D. Hoppes, and R. P. Hudson, Phys. Rev. **105**, 1413 (1957).

[6] R. L. Garwin, L. M. Lederman, and M. Weinrich, Phys. Rev. **105**, 1415 (1957); J. I. Friedman and V. L. Telegdi, *ibid.* **105**, 1681 (1957).

[7] T. D. Lee and C. N. Yang, Phys. Rev. **104**, 254 (1956).

[8] We list but a few of such relatively recent experiments: F. Boehm and E. Kankeleit, University of California Report No. Calt-63-13 (unpublished); Yu G. Abov, P. A. Krupchitsky, and Yu. A. Oratovsky, *Proceedings of the International Congress on Nuclear Physics, Paris, 1964* (to be published); L. Grodzins and F. Genovese, Phys. Rev. **121**, 228 (1961); R. E. Segel *et al.*, Phys. Rev. **123**, 1382 (1961); D. E. Alburger *et al.*, Phil. Mag. **6**, 171 (1961); R. Haas, L. B. Leipuner, and R. K. Adair, Phys. Rev. **116**, 1221 (1959); F. Boehm and U. Hauser, Nucl. Phys. **14**, 615 (1959); D. A. Bromley *et al.*, Phys. Rev. **114**, 758 (1959).

[9] T. D. Lee, R. Oehme, and C. N. Yang, Phys. Rev. **106**, 340 (1957).

[10] B. Aubert *et al.*, Phys. Letters **10**, 215 (1964); U. Camerini *et al.*, Phys. Rev. **128**, 362 (1962); J. H. Christenson *et al.*, Conference on Fundamental Aspects of Weak Interactions (Brookhaven National Laboratory, New York, 1963) (unpublished); J. H. Christenson, Princeton University, Technical Report 34, 1964 (unpublished); V. L. Fitch, Nuovo Cimento **22**, 1160 (1961); R. H. Good *et al.*, Phys. Rev. **124**, 1223 (1961).

[11] A. Abashian and E. M. Hafner, Phys. Rev. Letters **1**, 255 (1958).

angular asymmetry in the p-p scattering give an upper limit on the magnitude of the T-noninvariant amplitude of not more than a few percent of the T-invariant amplitude. A corresponding upper limit of $\sim 2\%$ has also been obtained from the existing experiments on reciprocity relations[12] in low-energy nuclear reactions. Thus, the strong interaction has been found to be invariant under the time-reversal operation to within a few percent. By using CT invariance, we can conclude that the strong interaction is also invariant under C to within a few percent.

The question of C, T invariance in the electromagnetic interaction for the strongly interacting particles is, however, a completely open one. We note that the electromagnetic interaction of a single physical nucleon is characterized by the matrix element $\langle N'|\mathcal{J}_\mu(x)|N\rangle$, which, by invariance under the continuous Lorentz transformations and space reflection, must be of the form

$$\langle N'|\mathcal{J}_\mu(x)|N\rangle = ieU_{N'}{}^\dagger\gamma_4[\gamma_\mu F_1+i(N_\mu'+N_\mu)F_2+(N_\mu'-N_\mu)F_3]U_N \times\exp[-i(N_\lambda'-N_\lambda)x_\lambda], \quad (9)$$

where N_μ and N_μ' are, respectively, the 4-momenta of the initial and the final single physical nucleon states $|N\rangle$ and $|N'\rangle$, U_N and U_N' are the solutions of the free Dirac equations with N_μ and N_μ' as their respective 4-momenta. The γ_1, γ_2, γ_3, γ_4 are the usual Dirac matrices and F_1, F_2, F_3 are functions of the square of the 4-momentum transfer

$$q^2=(N_\mu'-N_\mu)^2 \quad (10)$$

only. [We use for a 4-vector $N_\mu=(\mathbf{N},iN_0)$, where N_0 is real.] From Hermiticity, we find F_1, F_2, and F_3 to be all real. If T invariance holds, then F_1 and F_2 are real, but F_3 is pure imaginary; thus,

$$F_3\neq 0 \quad (11)$$

implies T noninvariance. However, because of the conservation of current, the T-noninvariant term F_3 actually vanishes on the nucleon mass shells.[13] The electromagnetic form factors of the nucleon can be measured by studying the scattering of

$$l^\pm+N\rightarrow l^\pm+N, \quad (12)$$

where $l=e$ or μ, in which the nucleons *are* on their mass shells. Thus, neglecting higher order electromagnetic effects, no information concerning the possible C, T noninvariance in the operator $\mathcal{J}_\mu$ can be obtained from reaction (12).

The electromagnetic property of a nucleus is determined by the corresponding properties of its constituents. It is a good approximation (i) to regard the nucleus as a collection of physical nucleons interacting through their strong nuclear forces, and (ii) to neglect other terms which can occur in the nucleon-photon vertex when the nucleons are off the mass shell. Within these approximations, and neglecting higher order radiative processes, there are still no T-noninvariant terms which contribute to any nuclear γ transition. The accuracy of these two approximations may be crudely estimated to be $\sim O(q^2/m_\pi^2)$ for (i), at least for T-violating effects, and $\sim O(V/m_N)$ for (ii), where V is the nuclear potential energy and m_N, m_π are, respectively, the masses of the nucleon and the pion. Thus, to the accuracy of $\sim$a few percent, no information concerning C, T invariance of H_γ can be obtained by studying the nuclear photoprocesses; rather, such processes can only be used to establish more firmly the C, T invariance of the strong interaction.

If H_γ strongly violates C, T invariance, then, through higher order effects, the reciprocity relation for any strong reaction may be violated with a fractional difference of the order of 10^{-2}–10^{-3}. All nuclear matrix elements in either γ or β transitions may also contain a small admixture of T-noninvariant amplitude which is $\sim(10^{-2}$–$10^{-3})$ times the T-invariant amplitude. Such a T-noninvariant amplitude can be observed by measuring the relative phase[14] between G_V and G_A in β decay, or by studying, e.g., the $E2$, $M1$ interference term through the detection of a $[\mathbf{k}_\beta\cdot(\mathbf{k}_\gamma\times\mathbf{k}_\gamma')](\mathbf{k}_\gamma\cdot\mathbf{k}_\gamma')$ term in an appropriate β-γ-γ transition.[15]

Reciprocity relations such as $\gamma+A\leftrightarrows B+C$ can be used to test possible T noninvariance in H_γ, where A, B, and C are any strongly interacting particle states. However, because of Hermiticity, a detailed spin-momentum analysis is necessary to test the reciprocity relation. As possible examples for such reciprocity tests, we may list $\gamma+n\leftrightarrows p+\pi^-$ or $\gamma+p\leftrightarrows n+\pi^+$.

Other direct tests of the C, T invariance properties of H_γ can be obtained by studying the relevant decays of mesons and hyperons. These will be analyzed in detail in the subsequent sections.

III. DECAYS OF THE PSEUDOSCALAR MESONS

Since H_{st} does conserve C, its eigenstates such as π^0 and η^0 are also eigenstates of C. In view of the possibility that H_γ may not be invariant under C, we cannot use the γ decay modes to determine the C values of these mesons. For definiteness, we assume that *virtual*

[12] L. Rosen and J. E. Brolley, Jr., Phys. Rev. Letters **2**, 98 (1959); D. Bodansky *et al.*, Phys. Rev. Letters **2**, 101 (1959).

[13] We would like to thank Professor R. Serber and Professor G. C. Wick for pointing this out, and for enlightening discussions.

[14] M. T. Burgy *et al.*, Phys. Rev. Letters **1**, 324 (1958). It should be emphasized that, independent of T invariance, (G_A/G_V) must be real if the nonleptonic current $(V_\mu+A_\mu)$ in the weak interaction satisfies the charge symmetry condition; i.e., $[V_\mu+A_\mu]^*=-[\exp(i\pi I_y)][V_\mu+A_\mu][\exp(-i\pi I_y)]$, where I_y is the y component of the isospin operator, * denotes the Hermitian conjugation if $\mu\neq 4$ and (-1) times the Hermitian conjugation if $\mu=4$.

[15] For a detailed discussion, see E. M. Henley and B. A. Jacobsohn, Phys. Rev. **113**, 225 (1959); B. A. Jacobsohn and E. M. Henley, *ibid.* **113**, 234 (1959).

transitions such as

$$N+\bar{N}\leftrightarrows\pi^0 \tag{13}$$

can occur through H_{st}; thus,

$$C_{\pi^0}=+1. \tag{14}$$

A further confirmation of Eq. (14) is given by Eq. (46) below. (The validity of many of our subsequent conclusions on tests of C, T invariance depends, however, only on the relative C values of different mesons states.)

It is convenient to decompose the $C=\mp1$ currents J_μ and K_μ into various components which have different transformation properties under the isotopic spin rotations:

$$J_\mu=J_\mu{}^s+J_\mu{}^v \tag{15}$$

and

$$K_\mu=K_\mu{}^s+K_\mu{}^v+\cdots, \tag{16}$$

where the superscripts s and v indicate that these currents transform as isoscalar ($I=0$) or isovector ($I=1$), respectively.[15a] If

$$K_\mu\neq0, \tag{17}$$

then H_γ is not invariant under either C or T. A conclusive test of C, T noninvariance in H_γ would be the discovery of any process whose existence implies Eq. (17).

1. π^0 Decay

To first order in $\mathcal{J}_\mu$, the decay $\pi^0\to e^++e^-$ is forbidden because of current conservation. To second order in $\mathcal{J}_\mu$, the decays

$$\pi^0\to2\gamma, \tag{18}$$

$$\pi^0\to\gamma+e^++e^-, \tag{19}$$

and

$$\pi^0\to2e^++2e^-, \tag{20}$$

contain only matrix elements of $J_\mu\cdot J_\nu$ and $K_\mu\cdot K_\nu$. Therefore, these reactions do not give direct tests of possible C, T noninvariance.

The decay

$$\pi^0\to3\gamma \tag{21}$$

can occur only if C invariance is violated. Assuming that $K_\mu\neq0$, its rate can be estimated. We find

$$\frac{\text{Rate}(\pi^0\to3\gamma)}{\text{Rate}(\pi^0\to2\gamma)}\sim e^2\frac{(\text{phase space})_{3\gamma}}{(\text{phase space})_{2\gamma}}(kR)^6\sim3\times10^{-6}, \tag{22}$$

where $(e^2/4\pi)=\alpha$, k is the average momentum of the γ ray, the ratio of phase space is $\sim(4\pi^2)^{-1}(m_\pi R)^2$, and R is taken to be $\sim m_\pi{}^{-1}$. The present experimental limit[16] is only

$$\left[\frac{\text{Rate}(\pi^0\to3\gamma)}{\text{Rate}(\pi^0\to2\gamma)}\right]_{\text{expt}}<3.8\times10^{-4}. \tag{23}$$

Thus, pion decays seem to be particularly insensitive to any possible violation of C, T in H_γ.

2. η^0 Decays

From the decays

$$\rho^0\to\pi^++\pi^-, \tag{24}$$

and

$$\rho^0\not\to\eta^0+\pi^0, \tag{25}$$

we may conclude

$$C_\rho=-1, \tag{26}$$

and

$$C_\rho=-C_\eta C_{\pi^0}. \tag{27}$$

Thus, it follows that

$$C_\eta=C_{\pi^0}. \tag{28}$$

The same conclusion would follow if we assume virtual transitions such as $N+\bar{N}\leftrightarrows\eta^0$ to be allowed.

The η^0 can only decay through H_γ. The decay

$$\eta^0\to\pi^0+e^++e^- \tag{29}$$

through a single-photon intermediate state violates C invariance, and it can occur only if the isovector part $K_\mu{}^v\neq0$. The general matrix element of $\mathcal{J}_\mu$ for this transition is given by

$$\begin{aligned}\langle\pi^0|\mathcal{J}_\mu(x)|\eta^0\rangle&=\langle\pi^0|K_\mu{}^v(x)|\eta^0\rangle\\&=[f_1(\eta_\mu+\pi_\mu)+f_2(\eta_\mu-\pi_\mu)]\\&\quad\times[4m_\eta\omega_\pi]^{-1/2}\exp[i(\eta_\lambda-\pi_\lambda)x_\lambda],\end{aligned} \tag{30}$$

where η_μ and π_μ are, respectively, the 4-momenta of the initial η^0 and the final π^0 states, m_η is the mass of η^0, ω_π is the energy of the pion, and f_1 and f_2 are form factors depending only on the square of the 4-momentum transfer,

$$q^2=(\eta_\lambda-\pi_\lambda)^2. \tag{31}$$

Conservation of current requires that

$$f_1(m_\eta{}^2-m_\pi{}^2)=q^2f_2, \tag{32}$$

where m_π is the mass of π^0. To obtain an estimate of the decay rate, we assume[17]

$$f_1=-\tfrac{1}{6}e\langle r^2\rangle q^2, \tag{33}$$

[15a] *Note added in proof.* From a phenomenological point of view, one should also decompose $K_\mu=(K_\mu)_1+(K_\mu)_8+\cdots$, where $(K_\mu)_1$ and $(K_\mu)_8$ transform, respectively, like a unitary singlet and a member of the unitary octet under the SU_3 group of transformations. If $K_\mu=(K_\mu)_1+(K_\mu)_8$ only, then, in the limit of perfect SU_3 symmetry, $\eta^0\not\to\pi^0+e^++e^-$ and there is no T-noninvariant effect in $\Sigma^0\to\Lambda^0+e^++e^-$. If $K_\mu=(K_\mu)_1$ only [and, consequently, K_μ is an isoscalar], then there are many additional forbidden processes such as $\phi^0\not\to\omega^0+\gamma$, $\phi^0\not\to\rho^0+\gamma$, $\omega^0\not\to\rho^0+\gamma$ and the absence of π^+, π^- asymmetries in $\omega^0\to\pi^++\pi^-+\pi^0$ (or γ).

[16] D. Cline and R. M. Dowd, Phys. Rev. Letters **14**, 530 (1965).

[17] Such a form factor, after eliminating the photon propagator, gives an "effective" C, T-noninvariant point interaction:

$$\langle r^2\rangle\tfrac{1}{6}e^2[\phi_\eta(\partial/\partial x_\mu)\phi_\pi-\phi_\pi(\partial/\partial x_\mu)\phi_\eta](i\psi_e{}^\dagger\gamma_4\gamma_\mu\psi_e),$$

where ϕ_π, ϕ_η, and ψ_e are the operators for π^0, η^0, and e^-. In this paper, we assume that there does not exist any additional direct strangeness conserving interaction between e^+, e^- and the strongly interacting particles. For a discussion on the limits of such point interactions, see G. Feinberg and M. Goldhaber, Proc. Natl. Acad. Sci. **45**, 1301 (1959).

where $\langle r^2 \rangle$ is the average of (radius)^2 of the "mixed" charge distribution between η^0 and π^0. The decay rate is given by [neglecting the mass of $e^\pm$]

$$\text{Rate}(\eta^0 \to \pi^0+e^++e^-) = (1728\pi)^{-1}\alpha^2[\langle r^2\rangle m_\eta^2]^2 m_\eta \times[(1-\epsilon^2)(1-8\epsilon+\epsilon^2)-12\epsilon^2 \ln\epsilon], \quad (34)$$

where

$$\epsilon = (m_\pi/m_\eta)^2. \quad (35)$$

The corresponding spectrum of $e^\pm$ is, apart from a normalization constant, given by

$$[4k_+k_- - 2m_\eta(k_++k_-)+(m_\eta^2-m_\pi^2)]dk_+dk_-, \quad (36)$$

where $k_\pm$ is the energy of $e^\pm$. The rate of $\eta^0 \to 2\gamma$ may be estimated by using SU_3 symmetry:

$$\text{Rate}(\eta^0 \to 2\gamma) \sim \tfrac{1}{3}(m_\eta/m_\pi)^3 \times \text{Rate}(\pi^0 \to 2\gamma). \quad (37)$$

Combining Eqs. (34) and (37), we find the ratio

$$[\text{Rate}(\eta^0 \to \pi^0+e^++e^-)/\text{Rate}(\eta^0 \to 2\gamma)] \sim 0.04[\langle r^2\rangle m_\eta^2]^2, \quad (38)$$

which is ~ 1 if $\langle r^2 \rangle$ is set arbitrarily to be the same as the mean-square radius of the proton.

If reaction (29) is observed, then it clearly indicates the C noninvariance in η^0 decay, and it shows that such violation also does not conserve the isospin I.

Another possibility is to study the spectrum of π^+ and π^- in the decays

$$\eta^0 \to \pi^++\pi^-+\pi^0 \quad (39)$$

and

$$\eta^0 \to \pi^++\pi^-+\gamma. \quad (40)$$

Let $dN(E_+,E_-)$ be the number of events in which the energies of π^+ and π^- are between $E^\pm$ and $E^\pm+dE^\pm$. In either of these two reactions, the observation of

$$dN(E_+=E_1,E_-=E_2) \neq dN(E_+=E_2,E_-=E_1) \quad (41)$$

for any energies E_1 and E_2 is an unequivocal proof of C noninvariance.

A possible $\pi^\pm$ asymmetry in reaction (39) has been discussed elsewhere[2,18] under the assumption that a C, T noninvariant H_F exists whose coupling constant is given by Eq. (7). If H_γ strongly violates C and T, then H_F represents simply the second-order electromagnetic effect. We note that if K_μ contains only an isoscalar part $K_\mu{}^s$, then H_F satisfies the $|\Delta\mathbf{I}|<2$ rule; otherwise, H_F may contain a $|\Delta I|=2$ and $C=-1$ term.

Any $\pi^\pm$ asymmetry in reaction (40) must be due to the existence of the $I=0$ (or 2) part of K_μ; in addition, there must be strong pion interactions. To analyze such asymmetries, we may make a partial-wave analysis of the (2π) system for reaction (40) and the decay

$$\eta^0 \to 2\pi^0+\gamma \quad (42)$$

which violates C conservation. As an illustration, only the lowest possible angular-momentum states will be kept. The (2π) can be either in an $I=1$ p state produced by $J_\mu{}^v$, or in an $I=0$ d state produced by $K_\mu{}^s$. The amplitudes for these two states can be represented, respectively, by $\sqrt{2}a(\mathbf{p}\cdot\mathbf{H})$ and $i\sqrt{3}b(\mathbf{p}\cdot\mathbf{H})(\mathbf{p}\cdot\mathbf{k})$, where $\mathbf{p}$ is the relative momentum of the two pions, $\mathbf{k}$ is the momentum of the photon, and $\mathbf{H}$ is its magnetic field. The energy distribution of reaction (40) is, then, given by the invariant phase space times

$$k^2[p^2-(\mathbf{p}\cdot\hat{k})^2] \times[|a|^2+|b|^2(\mathbf{p}\cdot\mathbf{k})^2+i(a^*b-b^*a)(\mathbf{p}\cdot\mathbf{k})], \quad (43)$$

where $\mathbf{p}$ ranges over the entire momentum space and $\hat{k}$ is a unit vector along $\mathbf{k}$; the corresponding density for reaction (42) is given by

$$|b|^2[p^2-(\mathbf{p}\cdot\hat{k})^2][\mathbf{p}\cdot\mathbf{k}]^2k^2, \quad (44)$$

where $\mathbf{p}$ ranges only over half of the momentum space. From CPT invariance, we find that the relative phase between a and b is given by $(\delta_p-\delta_d)$ or $(\pi+\delta_p-\delta_d)$, where δ_p is the strong-interaction phase shift of the 2-pion system in the $I=1$ p state and δ_d is the same in the $I=0$ d state. Thus, the asymmetry, if it exists, can also be used to obtain information concerning the pion interactions.

3. Weak Decays

We consider first the $K_1{}^0$-$K_2{}^0$ system. To make our analysis definite, we will assume here that H_{wk} is invariant under both T and CP, and H_γ has a large C, T noninvariant part (i.e., $K_\mu \neq 0$). To zeroth order in $\mathcal{G}_\mu$, there is no CP-violating effect. Thus,

$$CP=+1, \quad \text{for } K_1{}^0$$

and

$$CP=-1, \quad \text{for } K_2{}^0. \quad (45)$$

If we use Eq. (45) and note that the three pions in the decay $K_2{}^0 \to 3\pi$ are produced predominantly in s states, it follows, then, that

$$(CP)_{\pi^0}=-1. \quad (46)$$

Thus, $C_{\pi^0}=+1$ which confirms Eq. (14).

To first order in $\mathcal{G}_\mu$, since H_{wk} does not conserve C, the decay

$$K_i{}^0 \to \pi^0+e^++e^- \quad (47)$$

can occur via either J_μ or K_μ, where $i=1$ or 2. For either decay, there is only one form factor, due to current conservation; therefore, it does not yield any direct information concerning possible T or CP noninvariance. On the other hand, the decay

$$K_i{}^0 \to \pi^++\pi^-+\gamma \quad (48)$$

can be used to test possible CP noninvariance. Let $\boldsymbol{\varepsilon}\cdot\mathbf{M}_J{}^i(\mathbf{p},\mathbf{k})$ and $\boldsymbol{\varepsilon}\cdot\mathbf{M}_K{}^i(\mathbf{p},\mathbf{k})$ be, respectively, the transition amplitudes due to $(J_\mu A_\mu)H_{\text{wk}}$ and $(K_\mu A_\mu)H_{\text{wk}}$, where A_μ is the electromagnetic field operator, $\mathbf{p}$ is the relative momentum between π^+ and π^-, $\mathbf{k}$ is the

[18] T. D. Lee, Phys. Rev. **139**, B1415 (1965).

momentum of γ, and $\boldsymbol{\varepsilon}$ is its polarization vector. By applying CP, we find

$$\mathbf{M}_J{}^i(\mathbf{p},+\mathbf{k})=\pm\mathbf{M}_J{}^i(\mathbf{p},-\mathbf{k}) \tag{49}$$

and

$$\mathbf{M}_K{}^i(\mathbf{p},+\mathbf{k})=\mp\mathbf{M}_K{}^i(\mathbf{p},-\mathbf{k}), \tag{50}$$

where the upper signs are for $i=1$, and the lower signs for $i=2$. The probability distribution, after summing over the two polarization directions of γ, is proportional to

$$\sum_{\alpha,\beta}[(\mathbf{M}_\alpha{}^i)^*\cdot\mathbf{M}_\beta{}^i-(\mathbf{M}_\alpha{}^i\cdot\hat{k})^*(\mathbf{M}_\beta{}^i\cdot\hat{k})], \tag{51}$$

where the sum extends over α (and β)$=J$ and K, and $\hat{k}$ is a unit vector along $\mathbf{k}$. By using Eqs. (49) and (50), we find that in expression (51) the interference term between $\mathbf{M}_J{}^i$ and $\mathbf{M}_K{}^i$ is an odd function in $\mathbf{k}$, and, therefore, it is also an odd function in $\mathbf{p}$; the other terms are even in $\mathbf{p}$. Thus, the assumption that $K_\mu\neq0$ can result in a $\pi^\pm$ asymmetry, similar to that given by Eq. (41):

$$dN(E_+=E_1,E_-=E_2)\neq dN(E_+=E_2,E_-=E_1).$$

Conversely, the observation of such an asymmetry gives also a direct proof of CP noninvariance in the K_i^0 decays.

To second order in $\mathcal{J}_\mu$, decays such as $K_2^0\to2\pi$ and $K_1^0\to3\pi^0$ can occur. The smallness of the observed rate of reaction (1) is attributed to the smallness of the fine-structure constant. To the same order, explicit CP- and T-noninvariant effects can also be observed in K_{l3} and K_{l4} decays.

For the weak decays of charged mesons, to second order in $\mathcal{J}_\mu$, there exist explicit T-noninvariant observables, such as $\boldsymbol{\sigma}_\mu\cdot(\mathbf{p}_\mu\times\mathbf{p}_\pi)$ in the $K_{l3}{}^\pm$ decay, etc. The T-noninvariant amplitudes are, however, smaller than the corresponding T-invariant amplitudes by a factor $\sim O(\alpha)$. In the radiative decay $\pi^\pm$ (or $K^\pm$)$\to\mu^\pm$ +neutrino$+\gamma$, we may consider only the inner-bremsstrahlung process. Since both the lepton current and the weak interaction are assumed to be invariant under T, no T-noninvariant term such as $\boldsymbol{\sigma}_\mu\cdot(\mathbf{p}_\mu\times\mathbf{p}_\gamma)$ can be observed.

From CPT invariance, we have (neglecting higher order terms in H_{wk}, but valid to all orders in H_γ)

$$\sum_n[\text{Rate}(K^+\to n\pi)+\text{Rate}(K^+\to n\pi+\gamma)]$$
$$=\sum_n[\text{Rate}(K^-\to n\pi)+\text{Rate}(K^-\to n\pi+\gamma)], \tag{52}$$

$$\sum_n[\text{Rate}(K^+\to n\pi+l^++\nu_l)+\text{Rate}(K^+\to n\pi+l^++\nu_l+\gamma)]$$
$$=\sum_n[\text{Rate}(K^-\to n\pi+l^-+\bar{\nu}_l)+\text{Rate}(K^-\to n\pi+l^-+\bar{\nu}_l+\gamma)], \tag{53}$$

$$\text{Rate}(K^+\to l^++\nu_l)+\text{Rate}(K^+\to l^++\nu_l+\gamma)$$
$$=\text{Rate}(K^-\to l^-+\bar{\nu}_l)+\text{Rate}(K^-\to l^-+\bar{\nu}_l+\gamma), \tag{54}$$

where $l=e$ or μ, and γ stands for any numbers of photons and (l^+,l^-) pairs. For decays that are allowed by H_{wk}, from CP invariance, each individual partial decay rate of K^+ is equal to that of K^-, provided virtual effects of H_γ are neglected. Thus, e.g., neglecting $O(\alpha)$, we obtain

$$\text{Rate}(K^+\to\pi^++2\pi^0)=\text{Rate}(K^-\to\pi^-+2\pi^0), \tag{55}$$

$$\text{Rate}(K^+\to2\pi^++\pi^0)=\text{Rate}(K^-\to2\pi^-+\pi^0), \tag{56}$$

etc. The decay $K^\pm\to\pi^\pm+\pi^0$ is forbidden by H_{wk}, if H_{wk} rigorously satisfies the $|\Delta\mathbf{I}|=\frac{1}{2}$ selection rule. In this case, the $K_{2\pi}{}^\pm$ decay can only occur through $H_{\text{wk}}H_\gamma H_\gamma$ which violates CP invariance. However, by using Eqs. (52)–(56) and the experimental limits on radiative decays, we find that the equality

$$\text{Rate}(K^+\to\pi^++\pi^0)=\text{Rate}(K^-\to\pi^-+\pi^0) \tag{57}$$

should hold to an accuracy $\sim O(10^{-2})$. Thus, any lack of difference between the $K_{2\pi}{}^+$ and $K_{2\pi}{}^-$ branching ratios does not reflect the T invariance or T non-invariance of H_γ.

IV. DECAYS OF THE VECTOR MESONS

The vector mesons all decay through H_{st}. If H_γ is not invariant under C and T, then, through second-order processes, C, T-noninvariant effects, such as $\pi^\pm$ asymmetry in $\omega^0\to\pi^++\pi^-+\pi^0$, $\phi^0\to2K_1^0$, $\rho^0\to\eta^0+\pi^0$, etc., can occur. The C, T-noninvariant amplitudes are smaller than the corresponding C, T-invariant amplitudes by a factor $\sim O(\alpha)$. In the following, we will discuss in detail two types of C, T-noninvariant processes that can occur to the first order in $\mathcal{J}_\mu$.

1. $A^0(1-)\to B^0(1-)+\gamma$

In this class, there are

$$\phi^0\to\omega^0+\gamma, \tag{58}$$

$$\phi^0\to\rho^0+\gamma, \tag{59}$$

and

$$\omega^0\to\rho^0+\gamma. \tag{60}$$

We observe that by using Eq. (45) and the decay $\phi^0\to K_1^0+K_2^0$, it follows that

$$C_\phi=-1. \tag{61}$$

From the fact that the π^+,π^- in the decay $\omega^0\to\pi^+ +\pi^-+\pi^0$ are produced in an antisymmetric state, or $\omega^0\nrightarrow3\pi^0$, we may conclude

$$C_\omega=-C_{\pi^0}=-1. \tag{62}$$

The C_ρ is, according to Eq. (26), also -1. Decays such as $\omega^0\to\pi^0+\gamma$ or $\omega^0\to\pi^0+e^++e^-$ conserve C. However, reactions (58)–(60) all violate C invariance. Thus, only the current K_μ can contribute to these decays. Furthermore, reaction (58) can occur only if $K_\mu{}^s\neq0$ and reactions (59) and (60) can occur only if $K_\mu{}^v\neq0$.

By using invariance under space reflection and the continuous Lorentz transformations, the general matrix element of $\mathcal{J}_\nu$ between any two physical (1−) states $|A\rangle$ and $|B\rangle$ is given by

$$\langle B|\mathcal{J}_\lambda|A\rangle = ie\varphi_\mu'[(p'+p)_\lambda(F_1\delta_{\mu\nu}+F_2q_\mu q_\nu) +F_3\delta_{\lambda\nu}q_\mu+F_4\delta_{\lambda\mu}q_\nu+q_\lambda(F_5\delta_{\mu\nu}+F_6q_\mu q_\nu)]\varphi_\nu, \quad (63)$$

where p_λ, p_λ' are, respectively, the 4-momenta of the states $|A\rangle$ and $|B\rangle$, and $q_\lambda = p_\lambda' - p_\lambda$. The polarizations of the initial and final states are given by φ_μ' and φ_ν, respectively, which satisfy the normalization conditions

$$\varphi_\nu^2 = (2\omega_A)^{-1} \quad \text{and} \quad \varphi_\nu'^2 = (2\omega_B)^{-1}, \quad (64)$$

where ω_A and ω_B are the energies of A and B. From current conservation, we have

$$(m_A^2 - m_B^2)F_1 + q^2F_5 = 0 \quad (65)$$

and

$$(m_A^2 - m_B^2)F_2 + q^2F_6 + F_3 + F_4 = 0, \quad (66)$$

where m_A and m_B are, respectively, the masses of A and B. For the radiative decay

$$A \to B + \gamma, \quad (67)$$

the longitudinal part $q_\lambda(F_5\delta_{\mu\nu}+F_6q_\mu q_\nu)$ cannot contribute. Assuming that all form factors are regular at $q^2=0$ and dropping the longitudinal part, we find, at $q^2=0$, $\langle B|\mathcal{J}_\lambda|A\rangle$ becomes

$$ie\varphi_\mu'[a(p'+p)_\lambda q_\mu q_\nu + b(\delta_{\lambda\nu}q_\mu - \delta_{\lambda\mu}q_\nu) -\tfrac{1}{2}a(m_A^2-m_B^2)(\delta_{\lambda\nu}q_\mu+\delta_{\lambda\mu}q_\nu)]\varphi_\nu, \quad (68)$$

where a and b are related to these form factors by

$$a = F_2(0) \quad (69)$$

and

$$b = \tfrac{1}{2}[F_3(0) - F_4(0)]. \quad (70)$$

[If $m_B = m_A = m$, then as $q_\lambda \to 0$, expression (68) corresponds to a system of two spin-1 neutral particles with a "mixed" magnetic moment $=(eb/2m)\times$spin and a "mixed" quadrupole moment $=(e/m^2)(2am^2-b)$. If A and B are the same particle, then $\langle A|\mathcal{J}_\lambda|A\rangle = 0$ by CPT invariance and Hermiticity.] The rate for reaction (67) is given by

$$\text{Rate}(A \to B+\gamma) = (24)^{-1}\alpha m_A^{-5}m_B^{-2}(m_A^2+m_B^2) \times (m_A^2-m_B^2)^3 g^2, \quad (71)$$

where

$$g^2 = \tfrac{1}{4}(m_A^2-m_B^2)^2|a|^2 + |b|^2 -\tfrac{1}{2}(m_A^2+m_B^2)^{-1}(m_A^2-m_B^2)^2(a^*b+ab^*). \quad (72)$$

If H_γ violates C, T invariance strongly and if K_μ^s exists, then the parameter g^2 for $\varphi^0 \to \omega^0+\gamma$ may be ≈ 1, and the corresponding branching ratio becomes $\approx 1.9\%$. Similarly, if K_μ^v exists and if the corresponding g^2 is taken to be ≈ 1, then the branching ratio for reaction (59) is $\approx 2.4\%$. The reaction (60) has an extremely small branching ratio, so that it is unlikely to be of any practical use.

If H_γ does strongly violate C, T invariance, observations of reaction (58) and (59), though difficult, may become feasible. The detections of these decay modes are unambiguous proofs of C noninvariance.

2. $A^0(1-) \to \pi^+ + \pi^- + \gamma$

There are three such decays:

$$\phi^0 \to \pi^+ + \pi^- + \gamma, \quad (73)$$

$$\omega^0 \to \pi^+ + \pi^- + \gamma, \quad (74)$$

and

$$\rho^0 \to \pi^+ + \pi^- + \gamma. \quad (75)$$

If H_γ is not invariant under C, then there would be an asymmetry in the π^+ and π^- distribution in these decays [see Eq. (41)],

$$dN(E_+=E_1, E_-=E_2) \neq dN(E_+=E_2, E_-=E_1).$$

Such an asymmetry can occur in reactions (73) and (74) if K_μ^v exists, and in reaction (75) if K_μ^s exists.

The analysis of these asymmetries is similar to that in the case of η^0 decay. We make a partial-wave analysis of the (2π) system for these reactions together with

$$A^0 \to 2\pi^0 + \gamma, \quad (76)$$

where A^0 stands for either one of the three mesons: ϕ^0, ω^0, and ρ^0. Reaction (76) does not violate C invariance.

As an illustration of the form of asymmetry that might be observed in these decays, we keep only the lowest possible angular-momentum states. The (2π) system can be either in a $I=0$ s state via J_μ, or in a $I=1$ p state via K_μ. The amplitudes for these two states can be represented by $\sqrt{3}a\mathbf{s}\cdot\mathbf{E}$ and $i\sqrt{2}b\mathbf{s}\cdot(\mathbf{p}\times\mathbf{H})$, respectively, where $\mathbf{s}$ is the polarization vector of A^0, $\mathbf{p}$ is the relative momentum of the two pions, and $\mathbf{E}$ and $\mathbf{H}$ are, respectively, the electric and magnetic field of the γ. The complex constants a and b are two phenomenological parameters; they are comparable in magnitude, if the C, T violation is a large one. The pion spectrum, after summing and averaging over the polarizations of A^0 and γ, is given by the invariant phase space times

$$k^2\{|a|^2 + \tfrac{1}{2}|b|^2[p^2 + (\mathbf{p}\cdot\hat{k})^2] - i(a^*b - ab^*)(\mathbf{p}\cdot\hat{k})\} \quad (77)$$

for reaction $A^0 \to \pi^+ + \pi^- + \gamma$, and the same phase space times

$$k^2|a|^2 \quad (78)$$

for reaction (76), where $\hat{k}$ is a unit vector along the momentum direction of γ. In the evaluation of the phase space $\mathbf{p}$ ranges over the entire momentum space in (77), but only over half of the momentum space in (78). From CPT invariance, the relative phase between a and b is given by $(\delta_s - \delta_p)$ or $(\delta_s - \delta_p + \pi)$, where δ_s and δ_p are, respectively, the appropriate average phase shifts of the (2π) system in the $I=0$ s state and the $I=1$ p state.

Before leaving the meson systems, we remark that the identical analysis can be applied to the annihilation of the nucleon and antinucleon system at rest; e.g.,

$$p+\bar{p}\rightarrow\pi^++\pi^-+\gamma. \quad (79)$$

The initial ${}^1S_0$ state has the identical symmetry properties as a mixture of η^0 and π^0, and the ${}^3S_1$ state has the same properties as a mixture of ω^0 (or ϕ^0) and ρ^0.

V. $\Sigma^0\rightarrow\Lambda^0+e^++e^-$

For a baryon system, tests of possible T noninvariance in H_γ become more accessible than that of C noninvariance. A suitable case is[19]

$$\Sigma^0\rightarrow\Lambda^0+e^++e^-. \quad (80)$$

Let $\boldsymbol{\sigma}_\Sigma$, $\boldsymbol{\sigma}_\Lambda$, $\mathbf{p}$, $\mathbf{k}_+$, and $\mathbf{k}_-$ denote, respectively, the Pauli spin matrix of Σ, spin matrix of Λ, momentum of Λ^0, momentum of e^+, and momentum of e^-. Since the lepton currents conserve C, P, and T, the distribution, as well as any T-noninvariant term, must be symmetric with respect to the exchange between $\mathbf{k}_+$ and $\mathbf{k}_-$ in the one-photon approximation. We define the normal vector of the decay plane to be parallel to

$$\mathbf{N}\equiv\hat{p}\times(\hat{k}_++\hat{k}_-), \quad (81)$$

where $\hat{p}$, $\hat{k}_+$, and $\hat{k}_-$ are unit vectors along $\mathbf{p}$, $\mathbf{k}_+$, and $\mathbf{k}_-$. Neglecting the possibility of any spin measurements for $e^\pm$, the only T-noninvariant, but P-invariant, observables are

$$\boldsymbol{\sigma}_\Sigma\cdot\mathbf{N} \quad\text{and}\quad \boldsymbol{\sigma}_\Lambda\cdot\mathbf{N}, \quad (82)$$

which are even functions with respect to the exchange between e^+ and e^-. If either one of these averages $\langle\boldsymbol{\sigma}_\Sigma\cdot\mathbf{N}\rangle$ and $\langle\boldsymbol{\sigma}_\Lambda\cdot\mathbf{N}\rangle$ is not zero, then it clearly proves T noninvariance, and consequently, also C noninvariance in this decay.

Sometimes, in order not to lose any information contained in the limited statistics available, it may be helpful to use the detailed form of the spin-momentum distribution for an experimental analysis, rather than a simple over-all average. The theoretical results of these distribution functions for reaction (80) are given below.

The matrix element of $\mathcal{J}_\mu$ is given by

$$\begin{aligned}\langle\Lambda|\mathcal{J}_\mu|\Sigma\rangle&=\langle\Lambda|J_\mu{}^v+K_\mu{}^v|\Sigma\rangle\\&=ieU_\Lambda{}^\dagger\gamma_4[\gamma_\mu F+i(\Lambda_\mu+\Sigma_\mu)F'\\&\qquad+i(\Lambda_\mu-\Sigma_\mu)F'']U_\Sigma,\end{aligned} \quad (83)$$

where Λ_μ, Σ_μ are, respectively, the 4-momenta of the states $|\Lambda\rangle$ and $|\Sigma\rangle$, U_Σ and U_Λ are spinor solutions of the free-particle Dirac equations with the same 4-momenta as the physical Σ^0 and Λ^0. If T invariance holds, then F, F', and F'' are all real functions of q^2, where

$$q^2=(\Lambda_\mu-\Sigma_\mu)^2. \quad (84)$$

In the following, we will, however, assume these functions to be complex so as to violate T invariance.

From current conservation, these functions satisfy the following relation

$$F=(m_\Sigma+m_\Lambda)F'+(m_\Sigma-m_\Lambda)^{-1}q^2F'', \quad (85)$$

where m_Λ and m_Σ are the masses of Λ^0 and Σ^0, respectively. Thus, there are, in general, only two independent complex form factors. Let F_0 and F_0' be the values of F and F' at $q^2=0$. We have

$$F_0=(m_\Sigma+m_\Lambda)F_0' \quad (86)$$

which may be regarded as the "mixed" gyromagnetic ratio between Σ^0 and Λ^0. The value of F_0 is related to the decay $\Sigma^0\rightarrow\Lambda^0+\gamma$:

$$\begin{aligned}\text{Rate}(\Sigma^0\rightarrow\Lambda^0+\gamma)=\tfrac{1}{2}\alpha|F_0|^2m_\Sigma{}^{-3}(m_\Sigma-m_\Lambda)^3\\\times(m_\Sigma+m_\Lambda).\end{aligned} \quad (87)$$

The initial Σ^0 may be produced in any reaction, or any combination of reactions. The spin state of Σ^0 in its rest system is completely characterized by a (2×2) density matrix:

$$D_\Sigma=\tfrac{1}{2}(1+\boldsymbol{\sigma}_\Sigma\cdot\mathbf{s}_\Sigma), \quad (88)$$

where $\mathbf{s}_\Sigma$ is a real vector whose direction and magnitude determine the average spin direction and polarization of the initial Σ^0.

By using Eqs. (83) and (88), it is straightforward to evaluate the resulting (2×2) density matrix D_Λ for the Λ^0. It is convenient to introduce G which is a linear function of F and F':

$$G=(m_\Lambda+m_\Sigma)F-2m_\Sigma(m_\Lambda+E)F', \quad (89)$$

where

$$E=(p^2+m_\Lambda{}^2)^{1/2} \quad (90)$$

and

$$p=|\mathbf{p}|. \quad (91)$$

In terms of the form factors G and F, the density matrix for the Λ^0 is given by (neglecting the masses of $e^\pm$, but without any nonrelativistic approximation)

$$\begin{aligned}D_\Lambda=GG^*(1+\hat{k}_+\cdot\hat{k}_-)+2FF^*p^2[1-(\hat{p}\cdot\hat{k}_+)(\hat{p}\cdot\hat{k}_-)]\\+i(GF^*-G^*F)p\mathbf{N}\cdot(\mathbf{s}_\Sigma+\boldsymbol{\sigma}_\Lambda)+D_\Lambda',\end{aligned} \quad (92)$$

where $\mathbf{N}$ is defined in Eq. (81). We note that the coefficients of $\mathbf{N}\cdot\mathbf{s}_\Sigma$ and $\mathbf{N}\cdot\boldsymbol{\sigma}_\Lambda$ are the same, and can be nonzero only if G and F are not relatively real, which violates the condition of T invariance.

The term D_Λ' contains all the spin-spin correlations; it is, therefore, zero if we average over either the initial spin direction $\mathbf{s}_\Sigma$ or the final spin direction of Λ^0. The detailed form of D_Λ' is somewhat complicated, and is given by

$$D_\Lambda'=\boldsymbol{\sigma}_\Lambda\cdot\mathbf{S}_\Lambda', \quad (93)$$

[19] The rate for this decay has been analyzed by G. Feinberg, Phys. Rev. **109**, 1019 (1958); G. Feldman and T. Fulton, Nucl. Phys. **8**, 106 (1958).

where

$$\mathbf{S}_\Lambda'=\mathbf{s}_\Sigma\{GG^*(1+\hat{k}_+\cdot\hat{k}_-)+2FF^*[p^2(\hat{k}_+\cdot\hat{k}_-)-(\mathbf{p}\cdot\hat{k}_+)(\mathbf{p}\cdot\hat{k}_-)]\}$$
$$+\mathbf{p}\{-(GF^*+FG^*)[\mathbf{s}_\Sigma\cdot(\hat{k}_++\hat{k}_-)]+2FF^*[-(\mathbf{s}_\Sigma\cdot\mathbf{p})(1+\hat{k}_+\cdot\hat{k}_-)+(\mathbf{p}\cdot\hat{k}_+)(\mathbf{s}_\Sigma\cdot\hat{k}_-)+(\mathbf{p}\cdot\hat{k}_-)(\mathbf{s}_\Sigma\cdot\hat{k}_+)]\}$$
$$+\hat{k}_+\{(GF^*+FG^*)(\mathbf{p}\cdot\mathbf{s}_\Sigma)+2FF^*[-p^2(\mathbf{s}_\Sigma\cdot\hat{k}_-)+(\mathbf{p}\cdot\hat{k}_-)(\mathbf{p}\cdot\mathbf{s}_\Sigma)]\}$$
$$+\hat{k}_-\{(GF^*+FG^*)(\mathbf{p}\cdot\mathbf{s}_\Sigma)+2FF^*[-p^2(\mathbf{s}_\Sigma\cdot\hat{k}_+)+(\mathbf{p}\cdot\hat{k}_+)(\mathbf{p}\cdot\mathbf{s}_\Sigma)]\}. \quad (94)$$

Combining Eqs. (92) and (94), we may write

$$D_\Lambda=T+\boldsymbol{\sigma}_\Lambda\cdot\mathbf{S}_\Lambda, \quad (95)$$

where

$$\mathbf{S}_\Lambda=i(GF^*-FG^*)p\mathbf{N}+\mathbf{S}_\Lambda' \quad (96)$$

and

$$T=\tfrac{1}{2}\,\mathrm{tr}D_\Lambda, \quad (97)$$

which is related to the decay rate by

$$\mathrm{Rate}(\Sigma_0\to\Lambda^0+e^++e^-)$$
$$=\alpha^2\int[4\pi^3E(m_\Lambda+E)]^{-1}(q^2)^{-2}T \times d^3k_+d^3k_-\delta(k_++k_-+E-m_\Sigma). \quad (98)$$

In the above, all momenta are measured in the rest system of the Σ^0. Let us first choose the z axis to be parallel to $\mathbf{p}$, and then make a Lorentz transformation in the (z,t) subspace to a particular rest system of Λ^0. In this system, the spin direction of Λ^0 is parallel to $\mathbf{S}_\Lambda$ and its magnitude is $|T^{-1}\mathbf{S}_\Lambda|$.

In this decay, the range of $|q^2|$ is $\leqq(m_\Sigma-m_\Lambda)^2$; thus, we may expand G and F as power series in q^2. By using

$$q^2=2Em_\Sigma-(m_\Sigma^2+m_\Lambda^2) \quad (99)$$

and Eqs. (86) and (89), we find $G(q^2=0)=0$. For small q^2, G and F are given by

$$F=F_0+O[q^2] \quad (100)$$

and

$$G=(dG/dq^2)_0q^2+O[(q^2)^2]. \quad (101)$$

Substituting these expressions into Eq. (92), we find that if $\mathbf{s}_\Sigma=0$, the approximate amount of Λ^0 polarization along $\mathbf{N}$ is given by

$$2\sin\phi\left|F_0^{-1}\left(\frac{dG}{dq^2}\right)_0\right|\times\frac{k_+k_-(k_+-k_-)\sin\theta}{k_+^2+k_-^2-k_+k_-(1-\cos\theta)}, \quad (102)$$

where $k_\pm=|\mathbf{k}_\pm|$, θ is the angle between $\mathbf{k}_+$ and $\mathbf{k}_-$,

$$\cos\theta=(2k_+k_-)^{-1}\times[(m_\Sigma^2-m_\Lambda^2)-2m_\Sigma(k_++k_-)+2k_+k_-], \quad (103)$$

ϕ is the relative phase between the two parameters $(dG/dq^2)_0$ and F_0, and ϕ is 0 or π if T invariance holds. The dimension of $|(dG/dq^2)_0/F_0|$ is the same as m_Σ^{-1}. Assuming that H_γ strongly violates T invariance, and using only simple dimensional considerations, we may expect an over-all average of $\langle\mathbf{N}\cdot\boldsymbol{\sigma}_\Lambda\rangle$ to be about

$$\frac{m_\Sigma(m_\Sigma-m_\Lambda)\langle r^2\rangle}{\ln[(m_\Sigma-m_\Lambda)^2/q_{\min}^2]}, \quad (104)$$

where $q_{\min}^2$ is the minimum value of q^2 among the events and we have taken $(dG/dq^2)_0\sim\frac{1}{6}iF_0\langle r^2\rangle(m_\Sigma+m_\Lambda)$. Clearly, it is more favorable to take events with relatively large q^2, but this must be balanced against the small number of such events.

VI. COMPARISON BETWEEN BRANCHING RATIOS OF PARTICLES AND ANTIPARTICLES

We consider first the weak decays, and assume that H_{wk} conserves CP, but H_γ can have strong violations of C and T. Thus, the partial weak-decay rates of any particle are the same as that of its antiparticle, provided H_γ is neglected. For example, the equality

$$\mathrm{Rate}(\Sigma^+\to n+\pi^+)=\mathrm{Rate}(\bar{\Sigma}^-\to\bar{n}+\pi^-) \quad (105)$$

holds to $O(\alpha)$. Violation of the equality, Eq. (105), depends not only on H_γ being noninvariant under C and T, but also on the strong interactions in the final states. A more sensitive test is to compare the partial radiative decay rates. If H_γ violates C, T invariance strongly, then the radiative decay rate of a particle may be quite different from that of its antiparticle, provided the radiation is not just the inner-bremsstrahlung of the lepton currents. The observation of, say,

$$\mathrm{Rate}(\Sigma^+\to n+\pi^++\gamma)\neq\mathrm{Rate}(\bar{\Sigma}^-\to\bar{n}+\pi^-+\gamma) \quad (106)$$

is a conclusive proof of CP violation.

These considerations can be applied to any weak decays, and the particular case of K-meson decays has already been studied in Sec. III.3.

Next, we make a few brief remarks concerning strong reactions. The strong interaction H_{st} conserves C, P, T separately. Since H_γ may be strongly noninvariant under C, T, the differential cross section of

$$A+B\to\gamma+C+D+\cdots \quad (107)$$

can be very different from that of its C-conjugate process

$$\bar{A}+\bar{B}\to\gamma+\bar{C}+\bar{D}+\cdots, \quad (108)$$

where $A, B, C, D\cdots$ are any strongly interacting particle states and $\bar{A}$, $\bar{B}$, $\cdots$ are the corresponding antiparticle states. The most convenient case is probably the annihilation of p by $\bar{p}$; i.e.,

$$A=p \quad \text{and} \quad B=\bar{p}. \tag{109}$$

Another possible test is to compare the branching ratios, or detailed distributions, of the radiative decays of any resonance with that of its C-conjugate state. If H_γ violates C, T invariance, then the distribution and branching ratio of, say,

$$N^{*0} \rightarrow p+\pi^- +\gamma \tag{110}$$

may be different from that of

$$\bar{N}^{*0} \rightarrow \bar{p}+\pi^+ +\gamma. \tag{111}$$

From CPT invariance, such a difference can occur only if the strong final-state interaction is not neglected.

ACKNOWLEDGMENTS

We wish to thank Professor P. Franzini, Professor M. Schwartz, and Professor C. S. Wu for discussions.

AN ELEMENTARY DISCUSSION OF POSSIBLE NON-INVARIANCE UNDER *T*, *CP* AND *CPT* IN HYPERON DECAYS *

T. D. LEE

Department of Physics, Columbia University, New York, N.Y.

(*Received April 20, 1965*)

1. INTRODUCTION

The question whether the weak interactions are invariant under the time reversal operation *T*, or *CP* the product of the charge conjugation *C* and the space inversion *P*, has been raised [1] in the early days when the possibility of non-conservation of parity was being studied. After the discoveries [2, 3] that parity is not conserved, several experiments were performed to test the time reversal invariance in weak interactions. It was found that within the experimental accuracy [4], of about a few % in relative amplitudes, time reversal invariance holds in the β-decay, and to a much lesser accuracy, the same holds [5] for the Λ^0 decay. If *CPT* invariance [6] is assumed, to the same degree of experimental accuracy *CP* is also converved in these decays.

Recently, Christenson *et al.* observed [7] that the long-lived component of the neutral K^0 meson can decay into $(\pi^+ + \pi^-)$, thus suggesting that *CP* invariance is violated in the $K^0_{2\pi}$ decay. The observed non-invariant amplitude is quite small, being only $\sim 2 \times 10^{-3}$ relative to the corresponding *CP* conserving amplitude. If *CPT* invariance is assumed, then the same experiment implies that time reversal invariance is also violated.

The experiments which established parity non-conservation usually consist of directly observing a right-left asymmetric effect from an, otherwise, initially right-left symmetric state. The conclusions that space inversion symmetry is violated in these experiments can be reached without any theoretical assumptions. The same is also true for the violation of charge conjugation symmetry. It is important to note

* This research was supported in part by the U.S. Atomic Energy Commission.

5

that in all these weak interaction experiments, which pertain to testing the time reversal invariance or non-invariance, not a single one consists of comparing a reaction with its time reversal process. The relations between these experimental observations and time reversal symmetry are obtained through indirect theoretical reasoning, and some of these conclusions are valid only under additional assumptions such as *CPT* invariance. Similar criticism applies also to many of the existing tests of *CP* invariance.

It seems, therefore, desirable to review the underlying theoretical arguments of some of these tests, and to separate out the various implications of different symmetry requirements. With this motive, we will analyze in this note the simple example of the decay of a spin $\frac{1}{2}$ hyperon, say,

$$\Lambda^0 \rightarrow \mathrm{N}+\pi \tag{1}$$

where N stands for either p or n and π represents the corresponding π^- or π^0. The consequences of possible non-invariance under *T*, *CP* and *CPT* are derived in Section II. As is well known, the time reversal invariance in the hyperon decay means [8] that the relative phase between the final $s_{\frac{1}{2}}$ and $p_{\frac{1}{2}}$ amplitudes is determined by the corresponding strong interaction phase shifts. In Section III, the same result is obtained by an alternative proof which is based only on the reciprocity relation between different reaction rates, without the explicit use of the time reversal operator [9] *T*. A simple example is given in Section IV which illustrates the difference between the consequences of time reversal invariance in quantum mechanics and that in classical mechanics, and which emphasizes again that the present tests of time reversal invariance concern only the reciprocity relations between various differential cross-sections, rather than the detailed time reversal operation *T*.

Throughout these discussions we assume that the amplitude of reaction (1) can be represented by the corresponding matrix element of a Hermitian operator [10] H_{weak}. The validity of the local field theory, or *CPT* invariance, is not assumed.

2. Λ^0 DECAY

In the decay of Λ^0, the final $(\mathrm{N}+\pi)$ system can be in either a $s_{\frac{1}{2}}$ or a

$p_{\frac{1}{2}}$ spin-orbital state. Let these two transition amplitudes be denoted by $A_s(I)$ and $A_p(I)$ where $I = \frac{1}{2}$, or $\frac{3}{2}$, is the total iso-spin of the final state. The relative phase $\phi(I)$ is given by

$$\frac{A_s(I)}{A_p(I)} = \left|\frac{A_s(I)}{A_p(I)}\right| e^{i\phi(I)}. \tag{2}$$

Similarly, in the decay of the anti-lambda,

$$\bar{\Lambda}^0 \to \bar{\mathrm{N}} + \pi, \tag{3}$$

the corresponding $s_{\frac{1}{2}}$ and $p_{\frac{1}{2}}$ transition amplitudes are $\bar{A}_s(I)$ and $\bar{A}_p(I)$, and their relative phase $\bar{\phi}(I)$ is given by

$$\frac{\bar{A}_s(I)}{\bar{A}_p(I)} = \left|\frac{\bar{A}_s(I)}{\bar{A}_p(I)}\right| e^{i\bar{\phi}(I)}. \tag{4}$$

The following theorem states the separate consequences of the invariance requirements under T, CP and CPT for the Λ^0 and $\bar{\Lambda}^0$ decays. Throughout the present paper, we neglect the effects of electromagnetic interactions and assume that the strong interaction is separately invariant under T, C and P.

Theorem

1. If T invariance holds then, independent of CP invariance,

$$\phi(I) = \begin{cases} \delta_s(I) - \delta_p(I), & \text{or} \\ \delta_s(I) - \delta_p(I) - \pi \end{cases} \tag{5}$$

and

$$\bar{\phi}(I) = \begin{cases} \delta_s(I) - \delta_p(I), & \text{or} \\ \delta_s(I) - \delta_p(I) - \pi \end{cases} \tag{6}$$

where $\delta_s(I)$ and $\delta_p(I)$ are, respectively, the $s_{\frac{1}{2}}$ and $p_{\frac{1}{2}}$ phase shifts due to the strong interactions of the $(\mathrm{N}+\pi)$ system with a total iso-spin I.

2. If CP invariance holds then, independent of T invariance,

$$A_s(I) = -\bar{A}_s(I) \tag{7}$$

$$A_p(I) = +\bar{A}_p(I) \tag{8}$$

and, consequently,

$$\phi(I) = \bar{\phi}(I) + \pi. \tag{9}$$

For convenience, we chose the anti-particle states $\bar{\Lambda}^0$ and $\bar{\text{N}}$ to be identically related to their respective particle states Λ^0 and N through the *CP* operation.

3. If *CPT* invariance holds then, independent of either *T* invariance or *CP* invariance,

$$|A_s(I)| = |\bar{A}_s(I)|, \tag{10}$$

$$|A_p(I)| = |\bar{A}_p(I)| \tag{11}$$

and

$$\tfrac{1}{2}[\phi(I)+\bar{\phi}(I)] = \begin{cases}[\delta_s(I)-\delta_p(I)]+\tfrac{1}{2}\pi, & \text{or} \\ [\delta_s(I)-\delta_p(I)]-\tfrac{1}{2}\pi. \end{cases} \tag{12}$$

Some of these results, e.g. Eqs. (5) and (6), are well known and have already been proved in the literature [8]. For pedagogical reasons, a formal proof of this theorem is given below.

Proof. We consider the rest system of Λ^0. Let $|(\text{N}\pi)^{\text{st}}_{I,\text{s}}\rangle$ and $|(\text{N}\pi)^{\text{st}}_{I,\text{p}}\rangle$ be, respectively, the stationary wave eigen-states of the strongly interacting $(\text{N}+\pi)$ system in the $\text{s}_{\frac{1}{2}}$ and $\text{p}_{\frac{1}{2}}$ orbits and with a total iso-spin $I = \frac{1}{2}$ or $\frac{3}{2}$. The corresponding incoming wave states $|(\text{N}\pi)^{\text{in}}_{I,\text{s}}\rangle$ and $|(\text{N}\pi)^{\text{in}}_{I,\text{p}}\rangle$ are related to these stationary states by

$$|(\text{N}\pi)^{\text{in}}_{I,l}\rangle = \text{e}^{-\text{i}\delta_l(I)}|(\text{N}\pi)^{\text{st}}_{I,l}\rangle \tag{13}$$

where l = s or p. The transition amplitude $A_l(I)$ is given by

$$A_l(I) = \langle(\text{N}\pi)^{\text{in}}_{I,l}|H_{\text{weak}}|\Lambda^0\rangle \tag{14}$$

$$= \text{e}^{+\text{i}\delta_l(I)}\langle(\text{N}\pi)^{\text{st}}_{I,l}|H_{\text{weak}}|\Lambda^0\rangle \tag{15}$$

where $|\Lambda^0\rangle$ is the physical Λ^0 state. The time reversal operator T is represented [11] by the joint operation of a complex conjugation times a unitary operator U_T. Since the strong interaction is invariant under T, all its stationary eigen-states $|j, m\rangle$ which have zero total momentum can be chosen to transform under T as

$$T|j, m\rangle = U_T|j, m\rangle^* = (-1)^m|j, -m\rangle, \tag{16}$$

where j is the total angular momentum quantum number and m its z-component. Both $|\Lambda^0\rangle$ and $|(\text{N}\pi)^{\text{st}}_{I,l}\rangle$ satisfy Eq. (16). If H_{weak} is invariant under the time reversal operation, then

$$TH_{\text{weak}}T^{-1} = U_T H^*_{\text{weak}} U^\dagger_T = H_{\text{weak}}, \tag{17}$$

and, as a consequence, $\langle(\mathrm{N}\pi)^{\mathrm{st}}_{I,l}|H_{\mathrm{weak}}|\Lambda^0\rangle$ are real. Thus, Eq. (5) can be obtained by using Eq. (15).

For the decay of $\bar{\Lambda}^0$, we may denote the corresponding incoming wave eigen-state of the strongly interacting $(\bar{\mathrm{N}}+\pi)$ system by $|(\bar{\mathrm{N}}\pi)^{\mathrm{in}}_{I,l}\rangle$. Since the strong interaction is invariant under C, Eq. (13) implies that the $|(\bar{\mathrm{N}}\pi)^{\mathrm{in}}_{I,l}\rangle$ state is also related to the stationary state $|(\bar{\mathrm{N}}\pi)^{\mathrm{st}}_{I,l}\rangle$ by

$$|(\bar{\mathrm{N}}\pi)^{\mathrm{in}}_{I,l}\rangle = \mathrm{e}^{-i\delta_l(I)}|(\bar{\mathrm{N}}\pi)^{\mathrm{st}}_{I,l}\rangle. \tag{18}$$

Equation (6) can be derived by using the relation

$$\bar{A}_l(I) = \langle(\bar{\mathrm{N}}\pi)^{\mathrm{in}}_{I,l}|H_{\mathrm{weak}}|\bar{\Lambda}^0\rangle. \tag{19}$$

To establish the consequences of CP invariance, we may choose

$$|\bar{\Lambda}^0\rangle = CP|\Lambda^0\rangle, \tag{20}$$

$$|(\bar{\mathrm{N}}\pi)^{\alpha}_{I,\mathrm{p}}\rangle = +CP|(\mathrm{N}\pi)^{\alpha}_{I,\mathrm{p}}\rangle \tag{21}$$

and

$$|(\bar{\mathrm{N}}\pi)^{\alpha}_{I,\mathrm{s}}\rangle = -CP|(\mathrm{N}\pi)^{\alpha}_{I,\mathrm{s}}\rangle \tag{22}$$

where α = stationary or incoming. Equations (7)–(9) are the direct consequences of the assumption that H_{weak} is invariant under CP, i.e.

$$CPH_{\mathrm{weak}}P^{-1}C^{-1} = H_{\mathrm{weak}}. \tag{23}$$

If H_{weak} is invariant under CPT, then

$$CPTH_{\mathrm{weak}}T^{-1}P^{-1}C^{-1} = H_{\mathrm{weak}}. \tag{24}$$

Equations (10), (11) and (12) follow immediately by using Eqs (14)–(16) and (18)–(22). A special consequence of Eqs (10) and (11) is that CPT invariance implies [1] the equality of life time between Λ^0 and $\bar{\Lambda}^0$.

We note that Eqs (5) and (6) are consequences of Eqs (7)–(12) [i.e. T invariance is a consequence of CP invariance and CPT invariance], Eqs (7)–(9) are consequences of Eqs (5), (6) and (10)–(12) [i.e. T invariance and CPT invariance imply CP invariance], and that Eqs (10)–(12) are consequences of Eqs (5)–(9) [i.e. T invariance and CP invariance imply CPT invariances].

The absolute magnitudes and the relative phases of these transition amplitudes can be directly measured by studying the decay rates and the spin directions [5, 12] for reactions (1) and (3). As shown in [12], if in reaction (1) the initial Λ^0 is at rest and is completely polarized along the unit vector $\hat{S}_\Lambda$, then at any given momentum direction $\hat{k}$ the final nucleon is also completely polarized, and its spin direction $\hat{S}_N$ (measured in its own rest system) is given by

$$\hat{S}_N = [1-\alpha\cos\theta]^{-1}[(-\alpha+\cos\theta)\hat{k}+\beta(\hat{k}\times\hat{S}_\Lambda)+\gamma(\hat{k}\times\hat{S}_\Lambda)\times\hat{k}] \tag{25}$$

where $\hat{k}$ and $\hat{S}_N$ are unit vectors, $\cos\theta = \hat{k}\cdot\hat{S}_\Lambda$,

$$\alpha = 2\,\mathrm{Re}\,(A_s^* A_p)/[|A_s|^2+|A_p|^2], \tag{26}$$

$$\beta = -2\,\mathrm{Im}\,(A_s^* A_p)/[|A_s|^2+|A_p|^2], \tag{27}$$

and

$$\gamma = [|A_s|^2-|A_p|^2]/[|A_s|^2+|A_p|^2]. \tag{28}$$

The A_s and A_p are, respectively, the corresponding $s_{\frac{1}{2}}$ and $p_{\frac{1}{2}}$ amplitudes which are linearly related to the $A_s(I)$ and $A_p(I)$ by using the appropriate Clebsch-Gordon coefficients depending on whether the nucleon is p or n. The measurements of the rates and the parameters α, β, γ for the decays of Λ^0 and $\bar{\Lambda}^0$ give direct tests of *T*, or *CP*, or *CPT*, invariance in these reactions.

Among these tests, the ones for *CP* invariance consist of directly comparing two *CP* conjugate processes; their physical implications are, therefore, self-evident. The tests for *T* invariance and *CPT* invariance are less intuitively obvious. For this reason, an alternative proof of Eqs (5), (6) and (10)–(12), based on reciprocity relations, will be given in the next section.

3. RECIPROCITY

In order to make clear the consequence of time reversal symmetry, reaction (1)

$$\Lambda^0 \to N+\pi$$

should be considered together with its reversed process

$$N+\pi \to \Lambda^0. \tag{29}$$

Let $\langle\hat{k}, \hat{S}_N|M|\hat{S}_\Lambda\rangle$ and $\langle\hat{S}'_\Lambda|M|\hat{k}', \hat{S}'_N\rangle$ be, respectively, the transition matrix elements of reactions (1) and (29) where $\hat{S}_\Lambda$, $\hat{S}_N$, $\hat{k}$ are the same unit vectors as those used in Eq. (25) and $\hat{S}'_\Lambda$, $\hat{S}'_N$, $\hat{k}'$ are the corresponding unit vectors for reaction (29).

If time reversal symmetry holds, then reaction *rates* of (1) and (29) are related by the reciprocity relation which states that for arbitrary directions $\hat{k}$, $\hat{S}_N$ and $\hat{S}_\Lambda$,

$$|\langle \hat{k}, \hat{S}_N|M|\hat{S}_\Lambda\rangle| = |\langle -\hat{S}_\Lambda|M|-\hat{k}, -\hat{S}_N\rangle|. \tag{30}$$

This reciprocity relation deals directly with observations. It is important to note that the usual T invariance implies not only the transition probabilities but also the transition amplitudes satisfying the reciprocity relation [13, 14]. In this section, we will show that the previously proved consequences of T invariance can be derived by using *only the reciprocity relations between the relevant reaction rates.*

The reciprocity relation, Eq. (30), holds for the $(N+\pi)$ system in any isotopic spin state I. To first order in H_{weak} (but all orders in the strong interaction), the transition matrix elements for reactions (1) and (29) are given, respectively, by

$$\langle \hat{k}, \hat{S}_N|M|\hat{S}_\Lambda\rangle = \langle (\hat{k}, \hat{S}_N)_I^{in}|H_{weak}|\hat{S}_\Lambda\rangle \tag{31}$$

and

$$\langle \hat{S}'_\Lambda|M|\hat{k}', \hat{S}'_N\rangle = \langle \hat{S}'_\Lambda|H_{weak}|(\hat{k}', \hat{S}'_N)_I^{out}\rangle \tag{32}$$

where $|\hat{S}_\Lambda\rangle$ and $|\hat{S}'_\Lambda\rangle$ are the same physical $|\Lambda^0\rangle$ state used in Eq. (14), but with the Λ^0 spin polarized along $\hat{S}_\Lambda$ and $\hat{S}'_\Lambda$ respectively. The $|(\hat{k}, \hat{S}_N)_I^{out}\rangle$, (or $|(\hat{k}, \hat{S}_N)_I^{in}\rangle$) state is the outgoing (or incoming) eigenstate of the strong interaction for the $(N+\pi)$ system and, in the coordinate representation, it has an asymptotic form that consists of a plane wave with momentum $\boldsymbol{k}$ and spin $\hat{S}_N$ plus the appropriate outgoing (or incoming) waves. These eigen-states can be expanded in terms of the spherical waves used in the previous section. For example, in the non-relativistic limit, the explicit asymptotic forms of these expansions in the coordinate representation are given by

$$\begin{aligned}\langle \boldsymbol{r}|(\hat{k}, \hat{S}_N)_I^{out(in)}\rangle \to\ & e^{\pm i\delta_s(I)}U_N(kr)^{-1}\sin[kr+\delta_s(I)]+\\ &+ie^{\pm i\delta_p(I)}(\boldsymbol{\sigma}\cdot\boldsymbol{r})(\boldsymbol{\sigma}\cdot\hat{k})U_N(kr^2)^{-1}\sin[kr-\tfrac{1}{2}\pi+\delta_p(I)]+\dots\end{aligned} \tag{33}$$

as the relative distance $r = |\boldsymbol{r}| \to \infty$, where the $+$ signs in the exponents are for the outgoing state and the $-$ signs are for the incoming state. The components of the vector $\boldsymbol{\sigma}$ are the usual Pauli matrices, and U_N is a Pauli spinor which satisffies

$$(\boldsymbol{\sigma} \cdot \hat{S}_N)U_N = U_N. \tag{34}$$

Equation (31) and the rotational symmetry property of H_{weak} allow us to write

$$\langle \hat{k}, \hat{S}_N | M | \hat{S}_\Lambda \rangle = U_N^\dagger [a_s(I) e^{i\delta_s(I)} + a_p(I) e^{i\delta_p(I)} (\boldsymbol{\sigma} \cdot \hat{k})] U_\Lambda \tag{35}$$

where † denotes the Hermitian conjugation and the spinor U_Λ satisfies

$$(\boldsymbol{\sigma} \cdot \hat{S}_\Lambda) U_\Lambda = U_\Lambda. \tag{36}$$

The $a_s(I)$ and $a_p(I)$ are related to the $A_s(I)$ and $A_p(I)$ of the previous section by

$$A_l(I) = a_l(I) e^{i\delta_l(I)} \tag{37}$$

where $l =$ s or p. By using the assumed Hermiticity property of H_{weak}, the explicit form of $|(\hat{k}, \hat{S}_N)_I^{out}\rangle$ and Eq. (32), we find that the corresponding matrix element for reaction (29) is given by

$$\langle \hat{S}'_\Lambda | M | \hat{k}', \hat{S}'_N \rangle = U'^\dagger_\Lambda [a_s^*(I) e^{i\delta_s(I)} + a_p^*(I) e^{i\delta_p(I)} (\boldsymbol{\sigma} \cdot \hat{k}')] U'_N \tag{38}$$

where U'_N and U'_Λ are the spinors whose spin directions are $\hat{S}'_N$ and S'_Λ respectively. Substituting Eqs (35) and (38) into Eq. (30), we find that if the reciprocity relation is satisfied then

$$\frac{a_p^*(I)}{a_p(I)} = \frac{a_s^*(I)}{a_s(I)} \tag{39}$$

which gives Eq. (5).

Equations (35) and (38) also determine directly the transition amplitude of the resonant scattering

$$N + \pi \to \Lambda^0 \to N + \pi. \tag{40}$$

At the resonant energy, the amplitude of (40) is proportional to

$$\begin{aligned} &\langle \hat{k}, \hat{S}_N | M | \hat{k}', \hat{S}'_N \rangle \\ &\quad = U_N^\dagger [a_s e^{i\delta_s} + a_p e^{i\delta_p} (\boldsymbol{\sigma} \cdot \hat{k})][a_s^* e^{i\delta_s} + a_p^* e^{i\delta_p} (\boldsymbol{\sigma} \cdot \hat{k}')] U'_N \end{aligned} \tag{41}$$

where $\hat{S}'_N$ and $\hat{k}'$ are, respectively, the spin and momentum directions of the initial nucleon and $\hat{S}_N$, $\hat{k}$ are that of the final nucleon. In Eq. (41), we suppress the explicit I-dependence in a_s and a_p. The reciprocity relation between the transition probabilities for the resonant scattering is given by

$$|\langle \hat{k}, \hat{S}_N|M|\hat{k}', \hat{S}'_N\rangle| = |\langle -\hat{k}', -\hat{S}'_N|M|-\hat{k}, -\hat{S}_N\rangle|, \tag{42}$$

which can also be used to derive Eq. (39), or Eq. (5). For example, let us consider the simple case of a backward resonant scattering, i.e.

$$\hat{k} = -\hat{k}'. \tag{43}$$

Equation (41) becomes simply

$$\langle \hat{k}, \hat{S}_N|M|-\hat{k}, \hat{S}'_N\rangle = U_N^\dagger[C+D(\boldsymbol{\sigma}\cdot\hat{k})]U'_N, \tag{44}$$

where

$$C = |a_s|^2 e^{2i\delta_s} - |a_p|^2 e^{2i\delta_p} \tag{45}$$

and

$$D = (a_s^* a_p - a_p^* a_s)e^{i(\delta_s+\delta_p)}. \tag{46}$$

The transition probability for the resonant scattering from, say, $\hat{S}'_N = \hat{k}$ to $\hat{S}_N = \hat{k}$ is proportional to $(C+D)$, and the corresponding probability for the reversed process from $\hat{S}'_N = -\hat{k}$ to $\hat{S}_N = -\hat{k}$ is proportional to $(C-D)$. Thus, if the reciprocity relation, Eq. (42), holds, D must be zero, which gives another derivation of Eq. (5).

In the same way, by comparing the reaction rates between (3) and

$$\overline{\mathrm{N}}+\pi \rightarrow \bar{\Lambda}^0, \tag{47}$$

we can derive Eq. (6) without explicitly using the anti-unitary operator T; similarly, Eqs (10)–(12) can be derived by comparing the reaction rates between (1) and (47).

In this simple case of Λ^0 and $\bar{\Lambda}^0$ decays, we have shown that all the consequences of T invariance and CPT invariance can be derived by using only reciprocity relations between the various differential cross-sections. The same can also be established for other presently proposed tests of T invariance and CPT invariance in weak interactions [15].

4. DISCUSSIONS

In the decay of $\Lambda^0 \rightarrow \mathrm{N}+\pi$, if the initial Λ^0 is completely polarized along $\hat{S}_\Lambda$, then at any given momentum direction $\hat{k}$ the final nucleon must also be completely polarized along $\hat{S}_N$ which is given by Eq. (25). Let us now consider the reversed reaction $\mathrm{N}+\pi \rightarrow \Lambda^0$, where the initial polarization direction $\hat{S}'_{\mathrm{N}}$ and the incident momentum direction $\hat{k}'$ are given by

$$\hat{S}'_{\mathrm{N}} = -\hat{S}_{\mathrm{N}} \quad \text{and} \quad \hat{k}' = -\hat{k}. \tag{48}$$

We note that had the system obeyed *classical* mechanics, then time reversal invariance would imply that the final Λ^0 in the reversed reaction must be completely polarized along the reversed direction $\hat{S}'_{\Lambda}$ where

$$\hat{S}'_{\Lambda} = -\hat{S}_{\Lambda}. \tag{49}$$

For the quantum mechanical system, while the final Λ^0 does remain completely polarized, its direction $\hat{S}'_{\Lambda}$ is, in general, *different* from $-\hat{S}_{\Lambda}$.

To demonstrate this, we may consider the special case

$$A_{\mathrm{s}}(I) = -A_{\mathrm{p}}(I) \tag{50}$$

and neglect $\delta_{\mathrm{s}}(I)$ and $\delta_{\mathrm{p}}(I)$. Eq. (5) is, then, satisfied. The final nucleon in the decay $\Lambda^0 \to \mathrm{N}+\pi$ is now always polarized along $\hat{S}_{\mathrm{N}} = -\hat{k}$ while the final Λ^0 in the reversed reaction $\mathrm{N}+\pi \to \Lambda^0$ is always polarized along $\hat{S}'_{\Lambda} = -\hat{k}'$ which can be very different from $-\hat{S}_{\Lambda}$.

The time reversal operator T in quantum mechanics relates the solution of the Schroedinger Equation $\psi(t)$ at a time t with that at $-t$. In the decay $\Lambda^0 \to \mathrm{N}+\pi^-$, the final state $\psi(t=\infty)$ is a coherent mixture of $\mathrm{s}_{\frac{1}{2}}$ and $\mathrm{p}_{\frac{1}{2}}$ waves which, of course, can also be expanded as another coherent mixture of $|(\hat{k}, \hat{S}_{\mathrm{N}})_I^{\mathrm{in}}\rangle$ states. Under T, the state $|(\hat{k}, \hat{S}_{\mathrm{N}})_I^{\mathrm{in}}\rangle$ becomes $|(-\hat{k}, -\hat{S}_{\mathrm{N}})_I^{\mathrm{out}}\rangle$, and $T\psi(t=\infty)$ becomes a corresponding *coherent mixture* of $|(-\hat{k}, -\hat{S}_{\mathrm{N}})_I^{\mathrm{out}}\rangle$ states which is, obviously, very different from a single $|(-\hat{k}, -\hat{S}_{\mathrm{N}})_I^{\mathrm{out}}\rangle$ state. However, as stated in Eq. (48), it is precisely this single $|(-\hat{k}, -\hat{S}_{\mathrm{N}})_I^{\mathrm{out}}\rangle$ state that is being used as the initial state in the reversed reaction $\mathrm{N}+\pi \to \Lambda^0$. On the other hand, the reciprocity relation, Eq. (30), does equate the transition probabilities between the Λ^0 decay and the reversed reaction whose initial state is given by Eq. (48).

The above simple example merely illustrates once again these elementary aspects of quantum mechanics. It also illustrates that while the mathematical operation of the anti-unitary operator T deals with the symmetry between the solution ψ at a time t and that at $-t$, the direct experimental test of such symmetry properties usually does not go beyond the reciprocity relations between various reaction rates. In connection with the recent observation [7] of Christenson *et al.*, while we can at least envisage theoretically the possibility that, in some distant future, it may become possible to test directly the relevant reciprocity relations [16], it seems virtually impossible to ever construct the desired coherent time reversed state $T\psi(t = \infty)$ for a direct testing of the symmetry (or, violation of symmetry) properties of the time reversal operation.

I wish to thank Professors G. Feinberg, R. Serber and G. C. Wick for several enjoyable discussions.

REFERENCES

1) T. D. Lee, R. Oehme and C. N. Yang, Phys. Rev. **106** (1957) 340.
2) C. S. Wu, E. Ambler, R. W. Hayward, D. D. Hoppes and R. P. Hudson, Phys. Rev. **105** (1957) 1413.
3) R. Garwin, L. M. Lederman and M. Weinrich, Phys. Rev. **105** (1957) 1415; J. I. Friedman and V. L. Telegdi, Phys. Rev. **105** (1957) 1681.
4) M. T. Burgy, V. E. Krohn, T. B. Novey, G. R. Rings and V. L. Telegdi, Phys. Rev. **110** (1958) 1214.
5) J. W. Cronin and O. E. Overseth, Proc. Int. Conf. on High Energy Physics, (CERN, 1962), p. 453.
6) W. Pauli, Niels Bohr and the Development of Physics, (Pergamon Press, London, 1955);
J. Schwinger, Phys. Rev. **91** (1953) 720, 723; **94** (1953) 1366;
G. Lüders, Kgl. Danske Videnskab. Selskab, Mat.-fys. Medd. **28** (1954) No. 5.
7) J. H. Christenson, J. W. Cronin, V. L. Fitch and R. Turlay, Phys. Rev. Letters **13** (1964) 138.
8) T. D. Lee and C. N. Yang, Elementary Particles and Weak Interactions, (Brookhaven National Laboratory, 1957), p. 34.
9) The usual T invariance preserves the magnitude $|\langle\psi|\phi\rangle| = |T\psi|T\phi\rangle|$ for all ψ and ϕ in the Hilbert space, while in this note the reciprocity relation between various reaction rates refers specifically only to those ψ and ϕ which represent asymptotically the appropriate initial and final systems in which every particle has a definite momentum and a definite spin. Thus, by itself, the T invariance appears to be a stronger mathematical condition.
10) If, as proposed by Lee and Wolfenstein (Phys. Rev. to be published), the coupling constant of the T non-invariant interaction, called H_F, is $\sim 10^3$ times that of the usual T invariant weak interaction, called H_G, then the relevant operator

H_{weak} whould be the sum of H_G plus the second order term due to $H_F H_G$; otherwise, $H_{\text{weak}} = H_G + H_F$.

11) E. P. Wigner, Gött. Nachr., Math. Naturw. Kl. (1932) 546; Group Theory, (Academic Press, New York and London, 1959), Chapter 26.
12) T. D. Lee and C. N. Yang, Phys. Rev. **108** (1957) 1645.
13) J. A. Wheeler, Phys. Rev. **52** (1937) 1107.
14) See also J. M. Blatt and V. F. Weisskopf, Theoretical Nuclear Physics, (John Wiley and Co., 1952), p. 528.
15) The mass equality of p and $\bar{\text{p}}$, which is valid to all orders in H_{weak} if *CPT* invariance holds, can be derived by using the reciprocity relation between, say, $\gamma + \text{p} \to \gamma + \text{p}$ and its *CPT* conjugate process $\gamma + \bar{\text{p}} \to \gamma + \bar{\text{p}}$. The same reciprocity relation leads also to the well known identities between the electromagnetic properties of p and $\bar{\text{p}}$.
16) Such tests may become feasible in the immediate future if the coupling constant F of the T non-invariant interaction H_F turns out to be $\sim 10^3$ times the Fermi coupling constant G of the ususal weak interaction. [See reference [10]]. In such a case, all strong reactions can violate the reciprocity relation by a fractional difference $\sim (10^{-2} - 10^{-3})$ between the relevant reaction rates. The current experimental accuracy of reciprocity relation in strong interactions is $\sim 2\,\%$ as determined by L. Rosen and J. E. Brolley, Jr., Phys. Rev. Letters **2** (1959) 98 for the reactions $\text{p} + \text{t} \leftrightarrows \text{d} + \text{d}$. This accuracy is compatible with $F \sim 10^3 G$, but it also implies that F cannot be much bigger than $10^3 G$. [Cf., however, the discussion by J. Prentki and M. Veltman, Physics Letters **15** (1965) 88 in which a different view is attempted.]

Preludes in Theoretical Physics
North-Holland Publishing Co. (1966), pp. 5–16

WEAK INTERACTIONS AND QUESTIONS OF C, P, T NON-INVARIANCES[†]

T. D. Lee
Columbia University, New York, N. Y.

A talk given at the Oxford International Conference on Elementary Particles, September, 1965.

[†] This research was supported in part by the U. S. Atomic Energy Commission.

This talk will be divided into two parts: the first part consists of a review of the present status of weak interactions and the second part discusses some recent theoretical speculations about C , P , T non-invariances.

I. Weak Interactions

All known weak interactions can be divided into three groups:

$$\mathcal{L}_{wk} = \mathcal{L}_{\ell\ell} + \mathcal{L}_{\ell h} + \mathcal{L}_{hh} \tag{1}$$

where the subscript ℓ denotes the leptons and h the hadrons. The three terms $\mathcal{L}_{\ell\ell}$, $\mathcal{L}_{\ell h}$ and $\mathcal{L}_{hh}$ describe, respectively, the weak reactions which involve only leptons, leptons and hadrons, and only hadrons. Among these, the μ-decay part of $\mathcal{L}_{\ell\ell}$ is known explicitly and is given by

$$(\mathcal{L}_{\ell\ell})_{\mu\text{-decay}} = \frac{g_\mu}{\sqrt{2}} \, (j_\lambda)^*_\mu \, (j_\lambda)_e + \text{h.c.} \tag{2}$$

where

$$(j_\lambda)_\ell = i\,\psi_\ell^\dagger \, \gamma_4 \, \gamma_\lambda \, (1+\gamma_5)\, \psi_{\nu_\ell} \ , \tag{3}$$

$\ell = e$ or μ , and $(j_\lambda)^*$ is the hermitian conjugate of (j_λ) times $+1$, or -1 , depending on whether $\lambda \neq 4$, or $\lambda = 4$.

This (current X current) form of (V, A) interaction[1,2] can be generalized to include all leptonic parts of the weak interaction:

$$\frac{g_\mu}{\sqrt{2}} \, \mathcal{J}_\lambda^* \, \mathcal{J}_\lambda = \mathcal{L}_{\ell\ell} + \mathcal{L}_{\ell h} + \cdots \tag{4}$$

where

$$\mathcal{J}_\lambda = j_\lambda + J_\lambda + S_\lambda \ , \tag{5}$$

$$j_\lambda = (j_\lambda)_\mu + (j_\lambda)_e , \tag{6}$$

J_λ, S_λ are, respectively, the strangeness conserving and the strangeness non-conserving parts of the hadron currents, and the $\cdots$ denotes part of the non-leptonic interaction which, because of the observed $|\Delta \underset{\sim}{I}| = \frac{1}{2}$ rule, does not appear to contain the complete $\mathcal{L}_{hh}$.

All known weak reactions are consistent with the following selection rules:[4]

Accuracy	Selection Rule
Exact	$\Delta Q = \Delta N = \Delta L_e = \Delta L_\mu = 0$
$\sim 10^{-6}$	$\lvert\Delta S\rvert \leqq 1$, $\Delta Q_h = \Delta S$
$\sim \alpha$	$\lvert\Delta \underset{\sim}{I}\rvert \leqq 1$ Iso-triplet: V_λ, $(J_\lambda^{el})_{I=1}$, V_λ^*
$\sim 10^{-1}$	$\mathcal{L}_{\ell h} \sim SU_3$ octet $\mathcal{L}_{hh} \sim SU_3$ octet (?)

The accuracy represents only the theoretical expectations. In many cases (e.g., the $\Delta Q_h = \Delta S$ rule), the theoretical expectations often far exceed the actual experimental verifications.

In the above table, Q, N and S are, respectively, the total charge, the baryon number and the strangeness, Q_h is the total charge of the hadrons (in units of e), L_e and L_μ are the two leptonic numbers defined by

$$L_\ell = \begin{cases} +1 & \text{for } \ell^- \text{ and } \nu_\ell \\ -1 & \text{for } \ell^+ \text{ and } \bar{\nu}_\ell \\ 0 & \text{otherwise} \end{cases} \tag{7}$$

For all $\Delta S \neq 0$ processes, the iso-spin $\underset{\sim}{I}$ cannot be conserved. The $|\Delta \underset{\sim}{I}| \leq 1$ rule, therefore, implies the usual $|\Delta \underset{\sim}{I}| = \frac{1}{2}$ rule for all $\Delta S \neq 0$ reactions; it also implies the usual $|\Delta \underset{\sim}{I}| = 1$ rule for the $\Delta S = 0$ leptonic processes. The V_λ refers to the vector part of the strangeness conserving current J_λ, and $(J_\lambda^{el})_{I=1}$ is proportional to the iso-vector part of the electromagnetic current of the hadrons. Under an iso-spin rotation V_λ, $(J_\lambda^{el})_{I=1}$ and V_λ^* transform like the three members of an $I = 1$ triplet. From this iso-triplet property it follows that the vector current is conserved; i.e.

$$\frac{\partial V_\lambda}{\partial x_\lambda} = 0 . \tag{8}$$

In the limit of SU_3 symmetry, the currents J_λ and S_λ are assumed to be proportional to members of a single octet current operator $(O_\lambda)^i_j$:

$$J_\lambda \propto (O_\lambda)^1_2$$

and

$$S_\lambda \propto (O_\lambda)^1_3 . \tag{9}$$

Furthermore, it is assumed that the same proportionality constant holds for both the vector and the axial-vector part of J_λ and S_λ. Let us denote the proportionality constant by $\tan\theta$; then, by definition,

$$\tan\theta = \frac{g_S}{g_V} \tag{10}$$

where g_V is the Fermi constant in the usual β-decay (i.e. $n \to p + e^- + \nu_e$) and

g_S is the corresponding vector coupling constant in the strangeness non-conserving decays. It is found that the angle θ, called the Cabibbo angle, is given by

$$|\theta| \cong .24$$

if π_{e3} and K_{e3} decay rates are used, and

$$|\theta| \cong .26$$

if a least square fit is used to include all existing experiments.[5]

I. 2. Coupling Constants and Renormalization

A. Radiative Corrections of (g_V/g_μ)

It has been suggested that[3,6]

$$g_V^2 + g_S^2 = g_\mu^2 . \tag{11}$$

Thus, the Cabibbo angle θ is also given by

$$\cos\theta = \frac{g_V}{g_\mu} . \tag{12}$$

From Eq. (10), the value of θ has already been determined. By using Eq. (12) and θ, the ratio (g_V/g_μ) can be calculated. One finds

$$\left(\frac{g_\mu - g_V}{g_\mu}\right)_{th} \cong \begin{cases} 2.9\% & \text{if } |\theta| \cong .24 \\ 3.4\% & \text{if } |\theta| \cong .26 \end{cases} \tag{13}$$

Now, without radiative correction, the observed value is[7]

$$\left[\frac{g_\mu - g_V}{g_\mu}\right]_{uncor.} = 1.19\% . \tag{14}$$

Since the vector current is conserved, there is no correction due to the strong interaction. The electromagnetic interaction, however, does introduce a correction term.

We may write

$$\left[\frac{g_\mu - g_V}{g_\mu}\right]_{cor.} = 1.19\% + r \qquad (15)$$

where r is the radiative correction. The radiative correction for the μ-decay is straightforward; that of the β-decay is complicated because of the strong interaction. In the following, we give only the result for the "bare" nucleon.

In the Fermi theory, r is divergent:[8]

$$(r)_{Fermi} = \frac{\alpha}{4\pi}\left[\pi^2 - \frac{25}{4} + 6(\ln\frac{\lambda}{m_p}) + 3(\ln\frac{m_p}{2E_m}) - 2.85\right] \qquad (16)$$

where m_p is the mass of the proton, E_m is the maximum β energy and λ is the high energy cut-off parameter. The presence of the strong interaction might provide such a cut-off.

If we assume the existence of an intermediate boson $W^\pm$, then r becomes finite, and it is given by[9]

$$(r)_W = \frac{\alpha}{4\pi}\left[\pi^2 - \frac{25}{4} + 6\ln(\frac{m_W}{m_p}) + 3\ln(\frac{m_p}{2E_m}) - 2.85\right] - \frac{3}{10}\left(\frac{m_\mu}{m_W}\right)^2 \qquad (17)$$

where m_W is the mass of $W^\pm$. This finite result can be obtained by using only the lowest order perturbation method (without using the ξ-limiting process). Eq. (17) is, again, derived only for the bare nucleon. [If the cut-off parameter λ due to the strong interaction is $> m_W$, then Eq. (17) may be applicable to a physical nucleon.]

From the high energy neutrino experiments, we know that

$$m_W > 2 \text{ Gev},$$

and, therefore,

$$\left[\frac{g_\mu - g_V}{g_\mu}\right]_{cor} > 2.4\% \tag{18}$$

which is consistent with Cabibbo's result, Eq. (13).

We may also assume that Cabibbo's result is correct and use Eq. (13) and Eq. (17) [or Eq. (16)] to obtain an estimation of m_W [or λ]. We find

$$\begin{aligned} m_W \text{ [or } \lambda] &\cong 8\, m_p && \text{if } |\theta| \cong .24 \\ & 30\, m_p && \text{if } |\theta| \cong .26 \end{aligned} \tag{19}$$

In the latter case, and if $W^{\pm}$ exists, the dimensionless weak interaction coupling constant becomes comparable to that of the electromagnetic interaction. Of course, Eq. (19) can only be regarded as, at best, indicative.

The significance of the equality $g_V^2 + g_S^2 = g_\mu^2$ can be best illustrated[3] by using a quark model,[10] or any one of the triplet models for the hadrons. Let ψ_1, ψ_2, ψ_3 be the field operators for the three quarks. The current $\mathcal{J}_\lambda$ becomes simply

$$\mathcal{J}_\lambda = \psi_2'^{\dagger}\gamma_4\gamma_\lambda(1+\gamma_5)\psi_1 + \psi_e^{\dagger}\gamma_4\gamma_\lambda(1+\gamma_5)\psi_{\nu_e} + \psi_\mu^{\dagger}\gamma_4\gamma_\lambda(1+\gamma_5)\psi_{\nu_\mu} \tag{20}$$

where $\psi_2' = \cos\theta\,\psi_2 + \sin\theta\,\psi_3$. Eq. (20) shows clearly the symmetry between the leptons and the hadrons. However, the extension of such symmetry to the non-leptonic weak interaction is still unclear.

B. Renormalization Constant of (g_A/g_V)

Recently, Adler and Weissberger[11] have obtained an approximate expression or the renormalized (g_A/g_V) in β-decay. They started from the assumption that the axial vector current A_λ in β-decay satisfies the commutation relation[12]

$$[\int A_4^* d^3 r , \int A_4 d^3 r] = 2 I_z \tag{21}$$

where I_z is the z-component of the iso-spin operator. Such commutation rule follows, e.g., if we use the expression of $\mathcal{J}_\lambda$ given by Eq. (20). By taking into account only the part of $\frac{\partial A_\mu}{\partial x_\mu}$ that is proportional to the single pion field operator[13] ϕ_π, Adler and Weissberger derived the following expression for (g_A/g_V) :

$$\left|\frac{g_A}{g_V}\right| = \left\{1 + \frac{4m_N^2}{\pi f_{\pi N}^2} \int_{m_N+m_\pi}^{\infty} \frac{W\,dW}{W^2 - m_N^2} [\sigma_{tot}^-(W) - \sigma_{tot}^+(W)]\right\}^{-\frac{1}{2}} \tag{22}$$

where $(f_{\pi N}^2/4\pi) \cong 14.6$ is the pion-nucleon coupling constant, m_N is the mass of the nucleon and $\sigma_{tot}^{\pm}(W)$ is the total cross-section for scattering of a <u>zero-mass</u> $\pi^\pm$ on a proton at a center-of-mass energy W. The evaluation of Eq. (21) involves some off-mass-shell extrapolations of the observed scattering cross-sections. Their results are

$$\left|\frac{g_A}{g_V}\right| \cong \begin{cases} 1.24 & \text{Adler} \\ 1.16 & \text{Weissberger} \end{cases} \tag{23}$$

which agree quite well with the observed value $\left(\frac{g_A}{g_V}\right)_{exp} = -1.18 \pm .02$.

Similar calculations have been extended by Amati, Bouchiat and Nuyts[14] and by Levinson and Muzinich[15] to include the leptonic decays of the hyperons.

I. 3. Symmetry Properties of the Leptons

As is well known, except for the mass difference between μ and e, all known interactions, electromagnetic and weak, are symmetric with respect to the exchange

$$\mu \rightleftarrows e$$

and $$\nu_\mu \neq \nu_e \ . \tag{24}$$

This suggests that it may be meaningful to regard the masses m_μ and m_e as not due to the weak interaction [and, perhaps, also not due to the electromagnetic interaction].

In this section, we will study the symmetry properties of $H_{free} + H_{weak}$ in the limit

$$m_\mu = m_e = 0 \ . \tag{25}$$

We may decompose the lepton field operators

$$\psi_\ell = \ell_L + \ell_R \tag{26}$$

where

$$\ell_L = \tfrac{1}{2}(1 + \gamma_5)\psi_\ell \ ,$$

$$\ell_R = \tfrac{1}{2}(1 - \gamma_5)\psi_\ell \ , \tag{27}$$

and $\ell = e$ and μ .

We note that the leptonic weak interaction current j_λ and the leptonic electromagnetic current j_λ^{el} are both invariant under a $(U_2 \times U_2)$ group:[16, 17]

$$\begin{pmatrix} e_L \\ \mu_L \end{pmatrix} \rightarrow u \begin{pmatrix} e_L \\ \mu_L \end{pmatrix}$$

$$\begin{pmatrix} \nu_e \\ \nu_\mu \end{pmatrix} \rightarrow u \begin{pmatrix} \nu_e \\ \nu_\mu \end{pmatrix}$$

$$\begin{pmatrix} e_R \\ \mu_R \end{pmatrix} \rightarrow v \begin{pmatrix} \nu_R \\ \mu_R \end{pmatrix} \tag{28}$$

where u and v are two arbitrary (2 X 2) unitary matrices.

Furthermore, there exists a discrete transformation[17] D under which

$$D \mathcal{J}_\mu D^{-1} = \mathcal{J}_\mu^* . \tag{29}$$

The D-transformation exchanges

$$\nu_e \rightleftarrows e_L$$

$$\nu_\mu \rightleftarrows \mu_L$$

and

$$\psi_1 \rightleftarrows \psi_2' \tag{30}$$

where ψ_1 and ψ_2 are the quark fields used in Eq. (20). The transformation properties of the known hadrons under D can be readily obtained by using Eq. (30). [The existence of D-symmetry, of course, does not depend on the quark model.] Thus, the leptonic part of the weak interaction is symmetric under D and the $U_2 \times U_2$ group of transformations.

The D-symmetry can be easily extended to the non-leptonic part of the weak interaction as well. The weak interaction is, then, invariant under a group $\mathcal{G}_{wk}$ of transformations where $\mathcal{G}_{wk}$ is the extension of the $U_2 \times U_2$ group by using the discrete element D.

The $U_2 \times U_2$ group of symmetry transformations is violated by the masses m_μ and m_e, and the D-symmetry is violated by the masses of the leptons, by the electromagnetic interaction and by the SU_3 violating part of the strong interactions. However, both symmetries could be valid to all orders of the weak interaction.

The operator D changes the weak interaction current $\mathcal{J}_\mu$ to its (hermitian) conjugate current $\mathcal{J}_\mu^*$; therefore, the operator D has a rule in the weak interaction which is quite similar to that of the charge conjugation operator in the electromagnetic interaction.

II. Questions of C, P, T Non-invariances[18]

II. 1. Two π Decay Mode of K_2^o

As you all know, it was discovered last year by Christenson, Cronin, Fitch and Turlay[19] that the long-lived component K_2^o of the neutral K meson has a 2π decay mode

$$K_2^o \rightarrow \pi^+ + \pi^- . \tag{31}$$

Since both the particle anti-particle conjugation operator C and the space inversion P (in the center of mass system) interchange π^+ and π^-, the final (2π) system in this decay must be of $CP = +1$. On the other hand, the same long-lived component K_2^o also decays into 3π,

$$K_2^o \rightarrow \pi^+ + \pi^- + \pi^o \tag{32}$$

in which the π^+ and π^- are observed to be predominantly in the S state. Thus, the final 3π state is (or, at least, predominantly is) of $CP = -1$.

Now, K_2^o is, by definition, one with a definite lifetime. The fact that it can decay into different states with opposite CP values shows conclusively that CP is not conserved in the K_2^o decays. The CP violation is, therefore, established independently of the detailed theory of the K_1^o, K_2^o system.

The physical K_2^o state and its decay processes are determined by the total Hamiltonian H. The observed CP non-invariance implies that

$$[H, CP] \neq 0 \tag{33}$$

where

$$H = H_{st} + H_\gamma + H_{wk} . \tag{34}$$

From this single experiment it is, of course, not possible to conclude which part of H is responsible for this CP violation; whether it is due to the strong interaction[20] H_{st}, or the electromagnetic interaction[21, 22] H_{γ}, or the weak interaction H_{wk} or a combination of different interactions. There is also the possibility of a sub-weak interaction,[23] which however will not be discussed here.

The amplitude of this new decay mode is very small:

$$\epsilon = \left| \frac{\text{Rate } (K_2^o \to \pi^+ + \pi^-)}{\text{Rate } (K_1^o \to \pi^+ + \pi^-)} \right|^{\frac{1}{2}} \cong 2.2 \times 10^{-3} \qquad (35)$$

where K_1^o refers to the short-lived component of K meson. The smallness of the violation parameter ϵ is one of the striking features of this new discovery.

II. 2. Experimental Foundation of C , P , T Invariances

It is well known that parity is not conserved in the weak interaction. Since the discovery of the non-conservation of parity in weak interactions, there have been many experiments to search for possible P violating effects in nuclear reactions.[24] These experiments establish that the magnitudes of the P non-conserving amplitudes are smaller than that of the corresponding P conserving amplitudes by, at least, a factor $\sim 10^{-5}$, which is about the order of magnitude of the dimensionless weak coupling constant $g_V m_p^2$. Thus, we should regard both strong and electromagnetic interactions to be P conserving.

Invariance under CPT implies[25] mass and lifetime equalities between any particle and its anti-particle. In view of the possible large violations of C and CP invariances we may use such equalities as evidence for CPT invariance. Among such equalities, the most accurate one is that between K^o and $\bar{K}^o$:

$$< K^o \mid H_{st} + H_\gamma + H_{wk} \mid K^o > \; = \; < \bar{K}^o \mid H_{st} + H_\gamma + H_{wk} \mid \bar{K}^o > \qquad (36)$$

From the experimental mass difference[26] Δm between K_1^o and K_2^o, we conclude that Eq.(36) holds to $\sim | \Delta m/m_K | \sim 10^{-14}$. Thus, we should regard H_{st} and H_γ to be both invariant under CPT to an accuracy $\sim 10^{-14}$ and $\sim 10^{-12}$, respectively. Eq. (36) also shows that the $\Delta S = 0$ non-leptonic part of H_{wk} conserves CPT to an accuracy $\sim 10^{-8}$. For the other parts of H_{wk}, evidence for the CPT invariance is given by the upper limits on the lifetime difference $\Delta\tau$ between that of a particle and its anti-particle:[27]

$$\left| \frac{\Delta\tau}{\tau} \right| < \begin{cases} 10^{-3} & \text{for } \mu^\pm \\ 8 \times 10^{-2} & \text{for } \pi^\pm \\ 15 \times 10^{-2} & \text{for } K^\pm . \end{cases} \qquad (37)$$

There exist many good evidences which show that within an accuracy of about $\sim 10^{-2}$ in relative amplitudes the strong interaction is invariant under the particle anti-particle conjugation C ; i.e. $[H_{st}, C] = 0$.

Among these experimental evidences, we may mention the recent study by the Columbia group[28] on the equality between the energy distributions of π^+ and π^- in the annihilation of $\bar{p}$ and p,

$$\bar{p} + p \rightarrow \pi^+ + \pi^- + \ldots \qquad (38)$$

which places an upper limit on the C non-invariant amplitude to be not more than ~1% of the C invariant amplitude. A similar upper limit of ~2% is obtained by studying the energy distributions of K^+ and K^- in the same $(\bar{p} + p)$ annihilation experiment. Further evidence of C invariance of the strong interaction comes from the smallness of the CP violation parameter ϵ in K_2^o decay. Additional supporting evidence can also be obtained from the p-p scattering experiments[29] and from the experiments on reciprocity relations in nuclear reactions.[30]

There exists also some evidence that, within certain experimental accuracies, the weak interaction is also invariant under CP or T :

For the $\Delta S \neq 0$ non-leptonic weak interaction we can use the smallness of ϵ to conclude that within an accuracy of $\sim 10^{-2}$ it is invariant under CP ; otherwise, the observed amplitude of $K_2^o \rightarrow \pi^+ + \pi^-$ would be much larger.

For the $\Delta S \neq 0$ leptonic weak interaction, the experimental limits on CP invariance are, at present, unclear. The detailed experimental status was summarized by Dr. Steinberger this morning.

For the $\Delta S = 0$, leptonic weak reactions, the relative phase ϕ between the β-decay coupling constants g_A and g_V has been measured, and is found to be[31]

$$\phi = 180^o \pm 8^o \tag{39}$$

which is consistent with T invariance.[32] If "CPT" invariance is assumed, to the same accuracy we also have CP invariance.

In contrast, there exists at present <u>no</u> evidence[21,33] that H_γ of the strongly interacting particles is, or is not, invariant under C or T .

(a) Nucleon Form Factor

The most extensively studied problem concerning the electromagnetic property of the strongly interacting particles is the nucleon form factor. From proper Lorentz invariance and space inversion, the matrix element of the electromagnetic current $\mathcal{J}_\mu^{el}$ is given by

$$< N' | \mathcal{J}_\mu^{el} | N > = i U_N'^{\dagger} \gamma_4 [\gamma_\mu F_1 + \sigma_{\mu\nu} q_\nu F_2 + q_\mu F_3] U_N \ . \tag{40}$$

where q_μ is the 4-momentum transfer. From Hermiticity, we have F_1, F_2, F_3 to be all real. From T invariance, F_1, F_2 are real but F_3 is imaginary. Hence, $F_3 = 0$ if T invariance holds. However, it follows from

$$\frac{\partial \mathcal{J}_\mu^{el}}{\partial x_\mu} = 0 , \tag{41}$$

$F_3 = 0$ already. Thus, no information on T invariance can be obtained from $e^- + p \rightleftarrows e^- + p$ experiment.

(b) Nuclear γ Transition

The nuclear electromagnetic property is determined by that of its constituent nucleons and the strong nuclear forces. Independent of whether H_γ is, or is not, invariant under T, there should be no T violating effect in nuclear γ transitions within an accuracy $\sim 1\%$, provided that H_{st} is invariant under T.

An extensive study made in collaboration with Bernstein and Feinberg shows that at present there exists no direct or indirect evidence that H_γ of the strongly interacting particles is or is not invariant under C.

The following table summarizes these results:

	H_{st}	H_γ	H_{wk}
CPT	√	√	√
P	√	√	X
C	√	?	X
T or CP	√	?	√ (?)

The conclusion is quite clear: to explain the $K_2^o \to 2\pi$ we may invoke a small

CP non-invariant amplitude in H_{st}, or in H_{wk} (if the observed parameter ϵ is a typical example of the CP violation). There is no difficulty in doing this, but one is only puzzled by the presence of such a small violation.

An alternative, and, perhaps, much more attractive, possibility is to assume that H_{st} and H_{wk} are invariant under CP, but H_{γ} has a large violation of C invariance. In this case, all strong and weak processes can have a small C and CP non-invariant amplitude through virtual emission and absorption of γ. In particular, K_2^o can decay into 2π with a fractional amplitude

$$\epsilon \cong \frac{\alpha}{\pi} \quad . \tag{42}$$

On the other hand, it has been well established that, at least for the leptons, the electromagnetic interaction H_{γ} is invariant under the charge conjugation operator, which shall be denoted by C_{γ}. The possible non-invariance of H_{γ} under the particle anti-particle conjugation operation C, naturally, leads one to inquire whether C is, or is not, the same as C_{γ}.

II. 3. A Unifying View of Discrete Symmetry Violations

We note that if the identity of a particle could be taken for granted, then it would be possible to define T, the pure time reversal, and P, the pure space inversion, unambiguously. However, the distinguishability between different particles depends on their interactions, and degeneracies occur if certain interactions are absent. It is, therefore, not possible to give a unique definition of P and T without any reference to some specific interactions.

To illustrate this point, let us consider the following two examples of weak decays.

(i)

$$K^+ \rightarrow \begin{cases} \pi^+ + \pi^0 \\ \pi^+ + \pi^+ + \pi^- \end{cases} . \tag{43}$$

This is the well known $\theta - \tau$ puzzle, based on which the original theoretical proposal of the non-conservation of parity was first suggested. However, if we could switch off the strong and the electromagnetic interactions, then from reaction (43) alone, it would not be possible to conclude any parity non-conservation. If $H_{st} = H_\gamma = 0$, we could conclude from reaction (43) that the parity conservation holds and that the parities for the pions in S-states satisfy

$$P(\pi^+ \pi^0) = P(\pi^+ \pi^+ \pi^-) . \tag{44}$$

Now, if H_{st} and H_γ are turned on, it would, then, be discovered that Eq. (44) is violated. Thus, inasmuch as we regard H_{wk} as violating the conservation of parity where parity $(= P_{st} = P_\gamma)$ is determined by H_{st} and H_γ, we could obtain the same result by regarding H_{wk} to be parity conserving, and H_{st} and H_γ to be parity violating where parity $(= P_{wk})$ is determined by the weak interaction.

(ii) Another example of parity non-conservation is the $\pi_{\mu 2}$ decays:

$$\begin{aligned} \pi^+ &\rightarrow \mu^+ + \nu_\mu(L) , \\ \pi^- &\rightarrow \mu^- + \bar{\nu}_\mu(R) \end{aligned} \tag{45}$$

where L and R denote the helicities.

In the absence of the electromagnetic interaction we might regard the physical pions as

$$\pi_1 \equiv \left(\frac{1}{\sqrt{2}}\right) (\pi^+ + \pi^-) \tag{46}$$

and

$$\pi_2 \equiv (\frac{1}{\sqrt{2}})(\pi^+ - \pi^-) \ . \tag{47}$$

The particle π_1, or π_2, decays with equal probability to the left-handed ν_μ and to the right-handed $\bar{\nu}_\mu$, without any apparent violation of parity. Again, by considering only this weak reaction it is not possible to conclude that H_{wk} is not parity conserving.

Indeed, all known weak reactions are consistent with the assumption that H_{wk} is invariant under P_{wk}, T_{wk} and C_{wk}. A simple choice[34] would be to define C_{wk} to be the operator D, given by Eq. (29) [which conjugates the weak interaction current γ_μ], $T_{wk} = T_{st}$, and $P_{wk} = D^{-1} C_{st} P_{st}$ where C_{st}, T_{st} and P_{st} are determined by the strong interaction.

We now make the fundamental assumption: Each interaction[35] H_i (i = strong, γ and weak) is separately invariant under its own P_i, T_i and C_i. This assumption is consistent with all the existing experiments. From the experimental evidences discussed in section II.2, we conclude that

$$C_{st} T_{st} P_{st} = C_\gamma T_\gamma P_\gamma = C_{wk} T_{wk} P_{wk} \tag{48}$$

and

$$P_{st} = P_\gamma \neq P_{wk} \ . \tag{49}$$

Since reaction (38) is a strong process, the particle anti-particle conjugation operator C must be the same as C_{st}; i.e.

$$C_{st} = C \ ;$$

furthermore, under C_{st}

$$C_{st} | p > = | \bar{p} > \ ,$$

$$C_{st} | n > = | \bar{n} > \ , \tag{50}$$

and, consequently,

$$C_{st} \mid \pi^{+} > \;=\; \mid \pi^{-} > \;, \text{ etc.}$$

[The C_{st} may, for example, be regarded as the baryon number conjugation, and can, in principle, be different from the charge conjugation C_{γ} .]

The possible C non-invariance of H_{γ} means simply the possible mis-match between the particle anti-particle conjugation C_{st} and the charge conjugation C_{γ} ; i.e.

$$C_{st} \neq C_{\gamma} \quad . \tag{51}$$

In this case, we may set

$$T_{wk} = T_{st} \neq T_{\gamma} \tag{52}$$

and, consequently, $C_{wk} P_{wk} = C_{st} P_{st} \neq C_{\gamma} P_{\gamma}$. The decay $K_2^o \rightarrow \pi^{+} + \pi^{-}$ can be attributed to the C_{st} non-conservation of H_{γ} , or the C_{γ} non-conservation of H_{st} .

The following table illustrates these possibilities:

H_{st}	C_{st}	T_{st}	P_{st}
	$\neq$	$\neq$	$=$
H_{γ}	C_{γ}	T_{γ}	P_{γ}
	$\neq$	$\neq$	$\neq$
H_{wk}	C_{wk}	T_{wk}	P_{wk}
		$=$	
		T_{st}	

The mis-matches between the various operators, $C_{st} \neq C_{\gamma} \neq C_{wk} \neq C_{st}$, $P_{st} = P_{\gamma} \neq P_{wk}$ and $T_{wk} = T_{st} \neq T_{\gamma}$, give rise to all the non-conservations of C , P and T .

Remarks.

1. We emphasize that the existing experimental evidences require that H_{st} be invariant, within an accuracy $\sim 10^{-2}$, under C_{st}, T_{st} and P_{st}; the presently used form of H_{wk} is invariant under C_{wk}, T_{wk} and P_{wk}; if we assume that H_γ consists of only the minimal electromagnetic interaction of the spin 0 and $\frac{1}{2}$ particles, then H_γ is automatically invariant under C_γ, T_γ and P_γ.

2. It is interesting to note that our usual T and P often refer to T_γ and P_γ. This is because at large distances, due to its long-range character, the electromagnetic force predominates. Thus, in all collision processes, the asymptotic conditions are determined by the physical masses and the electromagnetic properties of the incoming and outgoing particles. It is, therefore, convenient to identify $T = T_\gamma$ and $P = P_\gamma$, since $(H_{free} + H_\gamma)$ is invariant under both T_γ and P_γ. On the other hand, the internal structures of the non-leptons are determined mainly by H_{st}, which leads us to identify the particle anti-particle conjugation operator C with C_{st}. However, such identifications may lead to unnecessary confusion if $C_{st} \neq C_\gamma$. The H_{st} and H_γ are, then, not invariant under $C_{st} T_\gamma P_\gamma$, which would be identified as CTP; yet, the "CTP" Theorem remains satisfied, since all these interactions are invariant under $C_{st} T_{st} P_{st} = C_\gamma T_\gamma P_\gamma = C_{wk} T_{wk} P_{wk}$

3. The leptons have no strong interaction; their electromagnetic interaction is, therefore, invariant under $C = C_\gamma$, $P = P_\gamma$ and $T = T_\gamma$ separately.

4. As remarked before, the definition of the discrete space-time transformation "P" and "T" cannot be given without any reference to a specific interaction. Such ambiguity, however, does not exist for the continuous inhomogeneous Lorentz transformations. This may be regarded as one of the reasons why the invariance under the continuous space-time transformations holds better than the discrete space-time transformations.

II. 4. Classification of C_{st} Non-invariant Electromagnetic Interaction

Let $\mathcal{J}_\lambda^{el}$ be the total electromagnetic current which can be decomposed to a leptonic part $(\mathcal{J}_\lambda^{el})_\ell$ and a non-leptonic (i.e.,hadronic) part $(\mathcal{J}_\lambda^{el})_h$:

$$(\mathcal{J}_\lambda^{el}) = (\mathcal{J}_\lambda^{el})_\ell + (\mathcal{J}_\lambda^{el})_h \quad . \tag{53}$$

The operator C_{st} applies only to the hadrons. Let us define

$$2I_\lambda = (\mathcal{J}_\lambda^{el})_h - C_{st}(\mathcal{J}_\lambda^{el})_h C_{st}^{-1} \tag{54}$$

and

$$2K_\lambda = (\mathcal{J}_\lambda^{el})_h + C_{st}(\mathcal{J}_\lambda^{el})_h C_{st}^{-1} \quad . \tag{55}$$

Thus, we can write

$$(\mathcal{J}_\lambda^{el})_h = I_\lambda + K_\lambda \tag{56}$$

where

$$C_{st} I_\lambda C_{st}^{-1} = -I_\lambda \tag{57}$$

and

$$C_{st} K_\lambda C_{st}^{-1} = +K_\lambda \quad . \tag{58}$$

By definition, all electromagnetic currents change sign under C_γ; i.e.

$$C_\gamma (\mathcal{J}_\lambda^{el})_\alpha C_\gamma^{-1} = -(\mathcal{J}_\lambda^{el})_\alpha \tag{59}$$

where $\alpha = \ell$ or h . A mis-match between C_{st} and C_γ means that

$$K_\mu \neq 0 \tag{60}$$

and vice versa.

We note that, since H_{st} conserves both $(\mathcal{J}_\mu^{el})_h$ and C_{st}, it must also conserve K_μ . Thus, the electric charge associated with K_μ is conserved by H_{st} ; i.e.

$$[Q_K, H_{st}] = 0 \tag{61}$$

where $$Q_K = -i \int K_4 \, d^3 r \quad . \tag{62}$$

All possible C_{st} non-invariant H_γ can, thus, be classified, according to whether the operator Q_K is zero or not, into the following two classes:

(A) $Q_K = 0$ identically.

As an example, we may choose

$$K_\mu = i\lambda \frac{\partial}{\partial x_\nu} \left(\phi_\mu \omega_\nu - \phi_\nu \omega_\mu \right) \tag{63}$$

where λ is a real constant and ϕ_μ, ω_μ are the field operators for the spin 1 particles ϕ^o and ω^o respectively. Such a K_λ is connected with the mixed magnetic moment current, and it gives rise to the decay

$$\phi^o \rightarrow \omega^o + \gamma \; . \tag{64}$$

Its 4th component is the divergence of a 3-vector which leads to $Q_K = 0$.

From the strong reactions such as $\phi^o \rightarrow K_1^o + K_2^o$ and $\omega^o \rightarrow \pi^+ + \pi^- + \pi^o$, we determine that

$$C_{st}(\phi^o) = C_{st}(\omega^o) = -1 \; . \tag{65}$$

Consequently,

$$C_{st} K_\mu C_{st}^{-1} = +K_\mu \; . \tag{66}$$

However, from Eq. (64) and the requirement that H_γ is invariant under C_γ, we find

$$C_\gamma(\phi^o) = -C_\gamma(\omega^o) \tag{67}$$

and

$$C_\gamma K_\mu C_\gamma^{-1} = -K_\mu \; . \tag{68}$$

In this example, K_μ transforms like a unitary octet under the SU_3 group of transformations.

(B) $Q_K \neq 0$.

The simplest example of this class of electromagnetic interactions is to assume the existence of a new strongly interacting charged, but $C_{st} = 1$, particle $a^{\pm}$ which is of $I = N = S = 0$. We assume that the strong interaction between $a^{\pm}$ and the known particles is of such a form that

$$C_{st} \, | \, a^+ > \; = \; | \, a^+ > \tag{69}$$

and

$$T_{st} \, | \, a^+_{\underset{\sim}{k}, \underset{\sim}{s}} > \; = \; \eta \, | \, a^-_{-\underset{\sim}{k}, -\underset{\sim}{s}} > \tag{70}$$

where η is a phase factor. The existence of the $a^{\pm}$ particle with such a strong interaction, then, necessitates the mis-match between C_{st} and C_{γ} and between T_{st} and T_{γ}.

In this case, the current K_{μ} transforms like a singlet under either the SU_3 or the iso-spin transformations. [In footnote 36, it is proved that under very general conditions, the current K_{μ} of these $a^{\pm}$-particles remains a unitary singlet, even if the a's are not themselves unitary singlets.]

Among the various possible tests of the possible C_{st} and T_{st} non-invariances of H_{γ} we list a few examples:

(i) There can be asymmetries between the energy distributions of π^+ and π^- in, say,

$$\eta^o \rightarrow \pi^+ + \pi^- + \gamma \; . \tag{71}$$

Other decays such as $\eta^o \rightarrow \pi^+ + \pi^- + \pi^o$ and the decays of η' can also be used[37] to test the C_{st} and T_{st} invariances of H_{γ}.

(ii) There can be correlations such as

$$\underset{\sim}{J}_N \cdot (\underset{\sim}{k} \times \underset{\sim}{k}') \quad \text{and} \quad \underset{\sim}{J}_{N^*} \cdot (\underset{\sim}{k} \times \underset{\sim}{k}') \tag{72}$$

in the single photon exchange approximation of the inelastic reaction

$$e^- + N \rightarrow e^- + N^* \tag{73}$$

where $\underset{\sim}{k}$, $\underset{\sim}{k}'$ are, respectively, the initial and the final momenta of e^-, and $\underset{\sim}{J}_N$, $\underset{\sim}{J}_{N^*}$ are, respectively, the spin vectors of N and N^*.

(iii) In[38]

$$\pi^- + p \rightarrow n + e^+ + e^- \tag{74}$$

there can be correlation, in the single photon exchange approximation, between the spin of the neutron and the vector $\underset{\sim}{k}_+ \times \underset{\sim}{k}_-$ where $\underset{\sim}{k}_\pm$ are the momentum of $\underset{\sim}{e}^\pm$.

(iv) If the particle $a^\pm$ exists, then it can be produced strongly; e.g.,

$$\bar{p} + p \rightarrow a^+ + a^- + \pi^+ + \pi^- + \cdots \ . \tag{75}$$

Under C_{st}, $p \rightarrow \bar{p}$, $\bar{p} \rightarrow p$, $\pi^\pm \rightarrow \pi^\mp$ but $a^+ \rightarrow a^+$ and $a^- \rightarrow a^-$. Thus, there can be large asymmetry between the energy distributions of a^+ and that of a^-; i.e.

$$dN\,(E_{a^+} = E_1\ ,\quad E_{a^-} = E_2) \neq dN\,(E_{a^+} = E_2\ ,\quad E_{a^-} = E_1)\ . \tag{76}$$

Quite likely, the $a^\pm$, if it exists, may decay through the weak interaction. If we assume that its weak interaction satisfies both the D-symmetry and the octet current hypothesis, then many of its branching ratios become uniquely determined. For example, if the $a^\pm$ is a Boson (denoted by $a^\pm_B$), we find, in the limit of SU_3 symmetry and assuming that $a^+ \rightarrow +a^-$ under $D = C_{wk}$,

$$\text{Rate}\ (a^+_B \rightarrow \eta^o + \ell^+ + \nu_\ell) \propto \frac{1}{6}\,(3\cos^2\theta - 1)^2$$

$$\text{Rate}\ (a^+_B \rightarrow K^o + \ell^+ + \nu_\ell) \propto \sin^2\theta\ \cos^2\theta$$

$$\text{Rate}\ (a^+_B \rightarrow \bar{K}^o + \ell^+ + \nu_\ell) \propto \sin^2\theta\ \cos^2\theta$$

and

$$\text{Rate}\ (a^+_B \rightarrow \pi^o + \ell^+ + \nu_\ell) \propto \frac{1}{2}\ \sin^4\theta \tag{77}$$

426

with the same common proportional constant. The angle θ is the Cabibbo angle. Identical results can be derived if the $a^{\pm}$ is a Fermion (denoted by $a_F^{\pm}$) by, respectively, replacing η^o, K^o, $\bar{K}^o$, π^o by Λ^o, n, Ξ^o and Σ^o (or by $\bar{\Lambda}^o$, $\bar{n}$, $\bar{\Xi}^o$, $\bar{\Sigma}^o$).[39] The fact that we have not, as yet, observed the $a^{\pm}$ leads to a lower limit ~1 or 1.5 Gev for its mass[40] m_a, depending on whether it is a Boson or a Fermion.

In conclusion, we emphasize that the possibility that H_γ has large violation of C_{st} is only a theoretical speculation, since it is based on just one experimental result: $\epsilon \cong (\alpha/\pi)$. Nevertheless, the fact that there exists, at present, no evidence that H_γ is, or is not, invariant under C_{st} and T_{st} should provide sufficient incentive for further experimental efforts in this direction.

Our concept of "C" has undergone several different evolutionary stages. It started with the charged conjugation operator C_γ determined by the electromagnetic interaction of the electron. Later, the operator C_γ was extended to the strongly interacting particles and became the particle anti-particle conjugation operator C_{st}. So long as everything is conserved, we may set these two conjugation operators equal. However, we now know that there must be non-conservations; it becomes, then, necessary to make the logical distinction between these two operators, and it is important to investigate which conjugation defined by what interaction is violated by what other interaction.

References

1. R. Feynman and M. Gell-Mann, Phys. Rev. 109, 193 (1958).
2. R. Marshak and G. Sudarshan, Padua-Venice International Conference, 1957; Phys. Rev. 109, 1860 (1958). J. J. Sakurai, Nuovo Cimento 7, 649 (1958).
3. N. Cabibbo, Phys. Rev. Letters 10, 513 (1963).
4. The references for the various selection rules are quite numerous. We give only a partial list of the theoretical papers. For the $|\Delta S| \leq 1$ rule, see L. B. Okun and B.M. Pontecorvo, J. Exp. Theor. Phys. (U.S.S.R.) 32, 1587 (1957) [translation, Soviet Phys. J.E.T.P. 5, 1297 (1957)]. For the $\Delta Q_h = \Delta S$ rule and the iso-triplet hypothesis of the vector current, see R. Feynman and M. Gell-Mann (reference 1). For the $|\Delta \underset{\sim}{I}| = \frac{1}{2}$ rule, see M. Gell-Mann and A. Pais, Proceedings of the International Conference on High Energy Physics, Pergamon Press, London (1955); R. Marshak, S. Okubo, E. C. G. Sudarshan, W. B. Teutsch and S. Weinberg, Phys. Rev. 112, 665 (1958). For the $|\Delta \underset{\sim}{I}| = 1$ rule see T. D. Lee and C. N. Yang, Phys. Rev. 126, 2239 (1962). For the SU_3 transformation properties of the weak interaction see N. Cabibbo (reference 3).
5. Willis et al., Phys. Rev. Letters 13, 291 (1964).
6. M. Gell-Mann and M. Levy, Nuovo Cimento 16, 705 (1960).
7. The $(g_\mu)_{\text{uncor.}}$ is related to the observed lifetime τ_μ and the mass m_μ of the muon by $\tau_\mu^{-1} = (g_\mu)^2_{\text{uncor.}} m_\mu^5 (192\pi^3)^{-1}$. From the values $m_\mu = (206.768 \pm 0.003)\, m_e$ and $\tau_\mu = (2.198 \pm 0.001) \times 10^{-6}$ sec. [Charpak et al., Phys. Letters 1, 16 (1962); Farley et al., Proc. of International Conference on High Energy Physics, CERN, Geneva (1962), p. 415], we find

$(g_\mu)_{uncor.} = (1.4320 \pm .0011) \times 10^{-49}$ erg. cm.^{3} The value of $(g_V)_{uncor.}$ can be determined from the various β-decay ft values, it was found that

$(g_V)_{uncor.} = (1.4149 \pm 0.0022) \times 10^{-49}$ erg. cm.^{3} [See C. S. Wu, Lecture Notes of International School of Physics "Enrico Fermi," Varenna, Italy (1964) for the detailed experimental references.]

8. T. Kinoshita and A. Sirlin, Phys. Rev. 113, 1652 (1959). S. M. Berman, Phys. Rev. 112, 267 (1958).
9. T. D. Lee, Phys. Rev. 128, 899 (1962).
10. M. Gell-Mann, Phys. Letters 8, 214 (1964); G. Zweig, CERN preprint (1964).
11. S. L. Adler, Phys. Rev. Letters 14, 1051 (1965). W. I. Weissberger, Phys. Rev. Letters 14, 1047 (1965). See also S. Fubini and G. Furlan (preprint).
12. M. Gell-Mann, Phys. Rev. 125, 1067 (1962); Physics 1, 63 (1964).
13. M. Gell-Mann and M. Levy, loc. cit.
14. D. Amati, C. Bouchiat and J. Nuyts (preprint).
15. C. A. Levinson and I. J. Muzinich (preprint).
16. F. Gürsey and G. Feinberg, Phys. Rev. 128, 378 (1962). [Cf. also the appendix in R. E. Marshak and S. Okubo, Nuovo Cimento 19, 1227 (1961).]
17. T. D. Lee, Nuovo Cimento 35, 945 (1965).
18. The content of this part of the lecture follows closely the discussions given by the following series of papers: T. D. Lee and L. Wolfenstein, "An Analysis of CP Non-invariant Interactions and the K_1^o, K_2^o System," Phys. Rev. 138, B1490 (1965); J. Bernstein, G. Feinberg and T. D. Lee, "Possible C, T Non-invariance in the Electromagnetic Interaction," Phys. Rev. 139, B1650 (1965); T. D. Lee, "Minimal Electromagnetic Interaction and C, T Non-invariance," Phys. Rev. (to be published);

T. D. Lee, "Classification of All C Non-invariant Electromagnetic Interactions and the Possible Existence of a Charged, But C = 1 , Particle," Phys. Rev. (to be published).

19. J. H. Christenson, J. W. Cronin, V. L. Fitch and R. Turlay, Phys. Rev. Letters 13, 138 (1964). See also A. Abashian et al., Phys. Rev. Letters 13, 243 (1964).

20. T. D. Lee and L. Wolfenstein, loc. cit. L. B. Okun (preprint). J. Prentki and M. Veltman, Phys. Letters 15, 88 (1965).

21. J. Bernstein, G. Feinberg and T. D Lee, loc. cit.

22. Cf. also S. Barshay, Phys. Letters 17, 78 (1965).

23. L. Wolfenstein, Phys. Rev. Letters 13, 562 (1964). T. D. Lee and L. Wolfenstein, loc. cit.

24. We list but a few of the relatively recent such experiments: F. Boehm and E. Kankeleit, Calt-63-13 (preprint); Yu. G. Abov, P. A. Krupchitsky and Yu. A. Oratovsky, Comptes Rendus du Cong. Inter. de Phys. Nucl., Paris (1964); L. Grodzins and F. Genovese, Phys. Rev. 121, 228 (1961); R. E. Segel et al., Phys. Rev. 123, 1382 (1961); D. E. Alburger et al., Phil. Mag. 6, 171 (1961); R. Haas, L. B. Leipuner, R. K. Adair, Phys. Rev. 116, 1221 (1959); F. Boehm and U. Hauser, Nuclear Physics 14, 615 (1959); D. A. Bromley et al., Phys. Rev. 114, 758 (1959).

25. T. D. Lee, R. Oehme and C. N. Yang, Phys. Rev. 105, 1671 (1957).

26. B. Aubert et al., Phys. Letters 10, 215 (1964); U. Camerini et al., Phys. Rev. 128, 362 (1962); J. H. Christenson et al., Conference on Fundamental Aspects of Weak Interactions, BNL, New York (1963); J. H. Christenson, Technical Report 34, Princeton University (1964); V. L. Fitch, Nuovo Cimento 22, 1160 (1961); M. L. Good et al., Phys. Rev. 124, 1223 (1961).

27. $\mu^{\pm}$ lifetime:

R. A. Lundby, Phys. Rev. 125, 1686 (1962).

M. Eckhause, Carnegie Inst. Tech. Report 9286 (1962).

Farley et al., Proc. International Conference on High Energy Physics, Geneva (1962).

Meyer et al., Nevis Report 114, Columbia University (1963).

$\pi^{\pm}$ lifetime:

Jakobson, Schulz and Steinberger, Phys. Rev. 81, 894 (1951).

Durbin, Loar and Havens, Phys. Rev. 88, 179 (1952).

Anderson et al., Phys. Rev. 119, 2050 (1960).

Ashkin et al., Nuovo Cimento 16, 490 (1960).

A. Barkas and A. Rosenfeld, Proc. International Conference on High Energy Physics, Rochester (1960).

$K^{\pm}$ lifetime:

P. Nordin, Jr., Phys. Rev. 123, 2168 (1961).

Bhowmik et al., Nuovo Cimento 20, 857 (1961).

Barkas et al., Phys. Rev. 124, 1209 (1961).

Boyarski et al., Phys. Rev. 128, 2398 (1962).

F. Crawford, Proc. International Conference on High Energy Physics, Rochester (1957).

28. C. Baltay, N. Barash, P. Franzini, N. Gelfand, L. Kirsch, G. Lütjens, J. C. Severiens, J. Steinberger, D. Tycko and D. Zanello, Phys. Rev. Letters 15, 591 (1965).

29. A. Abashian and E. M. Hafner, Phys. Rev. Letters 1, 255 (1958). Hwang et al., Phys. Rev. 119, 352 (1960).

30. L. Rosen and J. E. Brolley, Jr., Phys. Rev. Letters 2, 98 (1959). D. Bodansky et al., Phys. Rev. Letters 2, 101 (1959). Huang, Ophel, Thorndike and Wilson, Phys. Rev. 119, 352 (1960).

31. M. T. Burgy et al., Phys. Rev. Letters 1, 324 (1958).

32. It should be emphasized that, independent of T invariance, (g_A/g_V) must be real if the strangeness conserving current J_μ satisfies the charge symmetry condition $J_\mu^* = -[\exp(i\pi I_y)]\, J_\mu\, [\exp(-i\pi I_y)]$ where I_y is the y-component of the iso-spin operator.

33. Recently, there have been several new experiments testing the C_{st} invariance of H_γ : L. R. Price and F. S. Crawford, Phys. Rev. Letters 15, 123 (1965); J. S. Lindsay and G. A. Smith, Phys. Rev. Letters 15, 221 (1965); A. Rittenberg and G. R. Kalbfleisch, Phys. Rev. Letters 15, 556 (1965); C. Alff et al. (this conference). Unfortunately, the conclusions that can be drawn from these experiments are still extremely limited. In particular, if the $C_{st} = +1$ current K_μ transforms either like a unitary octet or like a unitary singlet, or like an iso-scalar; then in the limit of SU_3 symmetry, or iso-spin rotation symmetry, $\eta^o \not\rightarrow \pi^o + e^+ + e^-$, in the single photon exchange approximation; similarly, there is no time reversal non-invariant effect in $\Sigma^o \rightarrow \Lambda^o + e^+ + e^-$, etc. [See the theoretical discussions by N. Cabibbo, Phys. Rev. Letters 14, 965 (1965); T. D. Lee, loc. cit.; G. Feinberg, Phys. Rev. (to be published).]

34. We note that this simple solution satisfies (i) $C_{wk}\, T_{wk}\, P_{wk} = C_{st}\, T_{st}\, P_{st}$, (ii) $C_{wk}^2 = 1$ and (iii) both H_{st} and H_γ violate C_{wk} and P_{wk} invariances. If we want to maintain these three conditions, then the general solution of C_{wk}, T_{wk} and P_{wk} can be obtained by multiplying this special form of C_{wk}, T_{wk}

and P_{wk} by arbitrary members of the group $\mathcal{G}_{wk}$, given in section I.3, provided the conditions (i)–(iii) remain satisfied. [We may also try to find the general solution of C_{wk}, T_{wk} and P_{wk} without condition (iii); in this case, there exists a trivial solution: $C_{wk} = 1$, $P_{wk} = C_{st} P_{st}$ and $T_{wk} = T_{st}$.]

35. In some cases, the total interaction Hamiltonian H cannot be expressed as a linear sum $(H_{st} + H_{\gamma} + H_{wk})$. For example, there may be derivative couplings in either H_{st} or H_{wk}, which would generate an induced electromagnetic current with an amplitude proportional to ef or eg where f and g are, respectively, the appropriate strong and weak interaction coupling constants. The mathematical definitions of H_{st}, H_{γ}, H_{wk} are, then, given by $H_{st} = \lim_{e=g=0} H$, $H_{\gamma} = \lim_{f=g=0} H$ and $H_{wk} = \lim_{e=f=0} H$.

36. To show this, let us consider first the functional relation between Q_K and I_z, where I_z is the z-component of the iso-spin operator. Since the eigen-values of both Q_K and I_z are additive quantum numbers, there can only be linear relations between these two operators; i.e., $Q_K = aI_z + b$. From $[Q_K, C_{st}] = [H_{st}, Q_K] = [H_{st}, C_{st}] = 0$, one can easily prove that all known particles have the eigen-value $Q_K = aI_z + b = 0$. Consequently, $a = 0$, and Q_K is independent of I_z. By extending this argument to include other components of the iso-spin operator $\underset{\sim}{I}$, we establish that $[Q_K, I] = 0$. Similarly, it can be shown that Q_K is a singlet under the SU_3 transformations.

In general, Q_K being an SU_3 singlet, or an iso-scalar, does not mean that K_μ should transform in the same way. [For example, Eq. (63) gives an octet current K_μ, but the Q_K is zero, and, therefore, a singlet.] However, if we assume that the electromagnetic current K_μ is given only by the minimal electromagnetic interaction of the spin $\frac{1}{2}$, or 0, particles, then K_μ must also be a singlet under the SU_3 transformations and the iso-spin transformations, and $K_\mu \neq 0$ implies that $Q_K \neq 0$.

It may be instructive to give an explicit example. Let us assume the existence of a unitary triplet of spin 0 or $\frac{1}{2}$ particles (a_1^+, a_2^o, a_3^o) where all the a's are of $Q_K = +\frac{1}{2}$, the a_1 and a_2 are of $I = \frac{1}{2}$, $N = S = 0$, and a_3 is of $I = 0$, $N = 0$ but $S = -1$. The total charge Q is given by $Q = I_z + \frac{1}{2}(N + S) + Q_K$. This triplet has, therefore, total charges (+, 0, 0). It becomes one with total charges (0, +, +) under C_{st} and a different set with total charges (−, 0, 0) under T_{st}. The C_{st} and T_{st} invariances, then, generate from the triplet (a_1^+, a_2^o, a_3^o) a total of four different sets of triplets. Nevertheless, the current K_μ, given by the minimal electromagnetic interaction of all these triplets, remains a unitary singlet.

37. For details, see J. Bernstein, G. Feinberg and T. D. Lee, loc. cit.; G. Feinberg, Phys. Rev. (to be published); T. D. Lee, Phys. Rev. 139, B1415 (1965); cf. also J. Prentki and M. Veltman, Phys. Letters 17, 77 (1965); S. L. Glashow and C. M. Sommerfield, Phys. Rev. Letters 15, 78 (1965).

38. This particular reaction was suggested to me by L. Lederman and M. Schwartz (private communication).

39. We note that reactions (77) satisfy the selection rule $\Delta N = 0$, $\Delta Q_K = +1$. The decay $a_F^+ \to \Lambda^o + \ell^+ + \nu_\ell$ satisfies the selection rule $\Delta(N + Q_K) = 0$, and the decay $a_F^+ \to \bar{\Lambda}^o + \ell^+ + \nu_\ell$ satisfies the selection rule $\Delta(N - Q_K) = 0$. Because

of the stability of the known baryons, these selection rules cannot co-exist. Consequently, if both $a_F^{\pm}$ and $a_B^{\pm}$ exist, then one of them cannot have such decay modes.

40. If $a^{\pm}$ is stable, then a lower limit $m_a > 5$ Bev can be set by using the recent experimental results of D. E. Dorfan, J. Eades, L. M. Lederman, W. Lee and C. C. Ting, Phys. Rev. Letters 14, 999 (1965).

PHYSICS TODAY VOL 19 NO 3

Space Inversion, Time Reversal and Particle-Antiparticle Conjugation

As we expand our observation, we extend our concepts. Thus the simple symmetries that once seemed self-evident are no longer taken for granted. Out of studies of different kinds of interactions we are learning that symmetry in nature is some complex mixture of changing plus into minus, running time backward and turning things inside out.

by T. D. Lee

THE MORE WE LEARN about symmetry operations—space inversion, time reversal and particle-antiparticle conjugation—the less we seem to understand them. At present, although still very little is known about the true nature of these discrete symmetries, we have, unfortunately, already reached the unhappy state of having lost most of our previous understanding. Let us, therefore, review the gradual evolution of our past concepts of these discrete symmetry operations.

P and T in classical physics

In classical mechanics, each particle is described by its space-time coördinates **r** and t, and every particle is assumed to be different from every other particle. The space-inversion (P) and time-reversal (T) invariances in classical mechanics simply mean that the dynamical laws remain unchanged under

$$\begin{aligned} \text{P:}\quad & \mathbf{r} \to -\mathbf{r}, \quad t \to t \\ \text{T:}\quad & \mathbf{r} \to \mathbf{r}, \quad t \to -t \end{aligned} \qquad (1)$$

Suppose we are given a record, say a movie record, of the motion of a system of particles. If P invariance holds, then by examining only the movie it is not possible to decide with certainty whether the movie represents the true sequence or whether it represents the mirror image of the true sequence. Similarly, the T invariance implies that if any movie of the motion of a system of particles is run backwards, then the time-reversed sequence also represents a possible solution of the dynamical equations.

For a macroscopic system with a large number of particles, although the time-reversed sequence is always a possible one if T invariance holds, it is, in general, an improbable one. Thus although we cannot know for sure whether such a movie is being shown in its time-reversed order or not, we may try to guess. If the number of particles is sufficiently large, our guess will *almost always* be right. It is only in this statistical sense that we can differentiate for a macroscopic system any time-ordered sequence of events from its time-reversed sequence, and, thereby, determine the direction of our macroscopic time.

If the system contains only a very small number of particles, it is not possible, even in a statistical sense, to differentiate a time-ordered sequence from its time-reversed sequence (provided that T invariance holds). As we shall see, this last statement has to be *modified* in quantum mechanics.

Nonconservation of parity

The symmetry operations of P and T in quantum mechanics were first studied and analysed by Eugene Wigner of Princeton.[1] Both operations were

In 1957 the author shared the Nobel Prize in physics with C. N. Yang for the discovery of nonconservation of parity in weak interactions. This article is an adaptation of a lecture delivered on 15 Nov. at the Fourth Annual Science Conference of the Belfer Graduate School of Science, Yeshiva University, New York. Professor Lee's lecture will appear in the proceedings of the conference later this year.

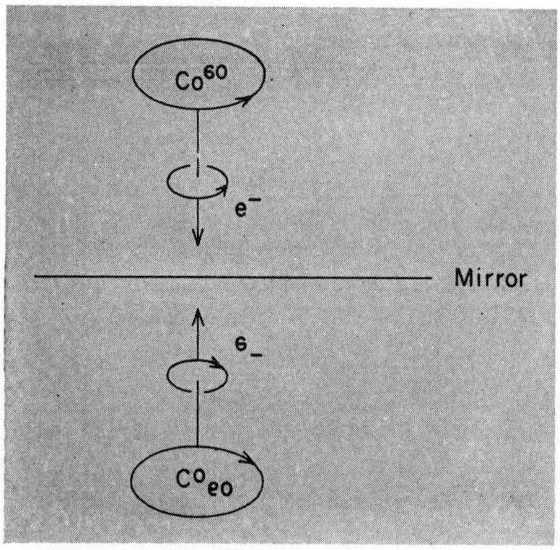

PARITY NONCONSERVATION. Cobalt-60 decays into nickel-60 plus an electron and an antineutrino. When the original nucleus is polarized, the emitted electron is found to have a left-handed spin, and its preferred direction of motion is opposite to the polarization direction of the original cobalt nucleus. The mirror image (without charge conjugation) is not realized in nature; therefore right-left symmetry (that is, what we call "space-inversion symmetry") is violated. —FIG. 1

successfully applied to the atomic system, which involves only the electromagnetic interaction; later, these symmetry operations were extended to include other phenomena in which not only the electromagnetic interaction but also the strong and the weak interactions participate. It is through these applications in particle physics that the validity of these discrete symmetries was questioned,[2,3] and the questioning led to the discovery of the nonconservation of parity.

The first experiment[4] on parity nonconservation (that is, space-inversion asymmetry) was made on beta decay by Professor C. S. Wu of Columbia, in collaboration with Ernest Ambler, Raymond W. Hayward, Dale D. Hoppes and R. P. Hudson of the National Bureau of Standards (see figure 1). From the same experiment it was deduced that charge-conjugation symmetry is also violated. Immediately afterward the same noninvariance properties were established for pi and mu decay.[5]

In quantum mechanics, the space-inversion operator P is a unitary operator and its eigenvalue is the parity.

If space-inversion symmetry holds, the parity must be conserved. Under P, the state of a particle with momentum **k** and helicity λ (defined to be its spin component along the direction of **k** and in units of $\hbar$) transforms to a state of the *same* particle but with momentum $-\mathbf{k}$ and helicity $-\lambda$. The operator P satisfies

$$P|\mathbf{k},\lambda\rangle = \eta_P|-\mathbf{k},-\lambda\rangle \tag{2}$$

Where η_P is a phase factor; P also satisfies similar equations for the multiparticle states. Here, the identity of a particle is defined through *all* of its interactions which include, in particular, its mass, charge and spin.

The suggestion that our known interactions are not strictly invariant under space-inversion symmetry was, at the beginning, based on the theta-tau puzzle[6]

$$K^+ = \begin{cases} \theta^+ \rightarrow \pi^+ + \pi^0 \\ \tau^+ \rightarrow \pi^+ + \pi^+ + \pi^- \end{cases} \tag{3}$$

The two particles θ^+ and τ^+ were found to be of the same mass and lifetime, suggesting that they are two decay modes of the same particle, the K^+ meson. However, the parity of the pion had been previously determined through both its strong interaction and its electromagnetic interaction to be -1.

$$P_{st}(\pi) = P_\gamma(\pi) = -1 \tag{4}$$

The subscripts st and γ indicate that the determinations are through H_{st}, the strong interaction and H_γ, the electromagnetic interaction, respectively. Based on the Dalitz analysis, the three pions in the tau decay mode are found to be in a zero-spin state. The same spin value can also be determined through the various production processes for θ^+ and τ^+. The parity that is determined by strong and electromagnetic interactions must, therefore, be different for the two final states, $+1$ for the two-pion state and -1 for the three-pion state; consequently we must have parity nonconservation if the θ^+ particle is identical with τ^+. There now exist numerous experiments that establish that both H_{st} and H_γ are invariant under the same space-inversion operation $P_{st}=P_\gamma$, but H_{wk}, the weak interaction, is not. Thus, it is *not possible* to construct a space inversion operator P that commutes with the *total* interaction Hamiltonian H.

At present, the best evidence for H_{st} and H_γ being invariant under $P_{st}=P_\gamma$ is from experiments in nuclear physics. These experiments[6] establish that for a nuclear level the magnitude of the parity-nonconserving amplitudes is smaller than that of the corresponding parity-conserving amplitudes by a factor of about 10^6.

Indistinguishable particles

If the identity of a particle could be taken for granted it would be possible to define P, the pure space inversion, and T, the pure time reversal, unambiguously. However, the distinguishability between different particles depends on all of their interactions, and degeneracies often arise if some of the interactions are absent. If the P and T symmetries are not valid for all interactions, then their definitions can only be given when some of the interactions are absent; consequently, such definitions are interaction dependent.

If strong and electromagnetic interactions could be switched off, it would not be possible to discover any parity nonconservation from the weak interaction alone. In fact under these hypothetical conditions many otherwise different particles would become degenerate and indistinguishable. As a result of such indistinguishability between particles, one can find a solution for the space-inversion operator under which the presently accepted form of the weak interaction is invariant.

For example, if $H_{st}=H_{\gamma}=0$, we might infer from reaction 3 that parity is conserved and the parity of the pion is +1. Indeed all known weak interactions are consistent with the assumption that H_{wk} is invariant under a different space-inversion symmetry operation called P_{wk}.[7] Inasmuch as we regard H_{wk} as violating conservation of parity where parity=$P_{st}=P_{\gamma}$, we could also regard H_{wk} as parity conserving where parity is P_{wk} and attribute the observed nonconservation of parity to violation of P_{wk} invariance by strong and electromagnetic interactions.

Time-reversal invariance

The question whether our known interactions are or are not invariant under time reversal was raised[3] when the possibility of parity nonconservation was being studied. After the discovery that parity is *not* conserved, several experiments were performed to test time-reversal invariance in both strong[8] and weak interactions, [9,10] and these experimental results were all consistent with time-reversal invariance. Recently, however, there has appeared an indirect evidence from the K_2^0 decay that time-reversal symmetry is, like space-inversion symmetry, only approximately valid for the known interactions.

Before discussing this indirect evidence, let us first review the meaning of time-reversal invariance. In quantum mechanics, the time-reversal operator T is an antiunitary operator.

If the theory is invariant under time reversal, then from a solution $\Psi(t)$ of the Schroedinger equation

$$H\Psi(t)=-\frac{\hbar}{i}\frac{d\Psi(t)}{dt} \tag{5}$$

we can generate a different solution $\Psi_T(t)$ of the same equation. The Ψ_T is related to Ψ by

$$\Psi_T(t)=T\Psi(-t)=U_T\Psi^*(-t) \tag{6}$$

U_T is a unitary operator in Hilbert space and the asterisk denotes complex conjugation.

An important consequence of time-reversal invariance is the *reciprocity relations* between the *transition probabilities:* the magnitude of the S-matrix element for any transition $a\rightarrow b$ is equal to that of $b_T\rightarrow a_T$

$$|\langle b|S|a\rangle|=|\langle b_T|S|a_T\rangle| \tag{7}$$

Here $|a_T\rangle=T|a\rangle$ and $|b_T\rangle=T|b\rangle$. Direct tests of such reciprocity relations have been made for several strong reactions,[8] and these tests give good evidence that the strong interaction is invariant under a (certain) time-reversal operation called T_{st}. The upper limit on the ratio of the magnitude of the time-reversal-noninvariant amplitude to that of the time-reversal-invariant amplitude is about 2% in the proton-triton-deuterium reactions $p+t\rightleftarrows d+d$.

As an illustration of how time-reversal invariance can be tested in the weak interaction, we may consider the example of Λ^0 decay

$$\Lambda^0\rightarrow N+\pi \tag{8}$$

The final $N+\pi$ system can be in either the $s_{1/2}$ or the $p_{1/2}$ spin-orbital state with relative amplitudes A_s and A_p, respectively; the total isospin of the final state is, predominantly, $I=1/2$. If the weak interaction satisfies the same T_{st} invariance, then the relative phase ϕ, defined by

$$\frac{A_s}{A_p}=\left|\frac{A_s}{A_p}\right|e^{i\phi} \tag{9}$$

is given by

$$\phi=(\delta_s-\delta_p)\ \text{or}\ (\delta_s-\delta_p)+\pi \tag{10}$$

where δ_s and δ_p are, respectively, the $s_{1/2}$ and $p_{1/2}$ phase shifts of the strongly interacting $N+\pi$ system in the $I=1/2$ state. The experimental results are $\delta_s-\delta_p\approx+7$ deg and[9] $\phi_{exp}=15\pm20$ deg which are consistent with H_{wk} being invariant under the same time-reversal invariance as the strong interaction. The same conclusion is also reached by a similar, but more accurate, experiment on beta decay.[10]

The relative phase θ between the Gamow-Teller coupling constant g_A and the Fermi constant g_V has been measured by M. T. Burgy and his collaborators.[10] If the weak interaction satisfies time-reversal symmetry, $\theta=0$ or 180 deg. The experimental value is 180 ± 8 deg. [In some theoretical models, however, the weak-interaction strangeness-conserving current J_μ of the nonleptons is assumed to satisfy

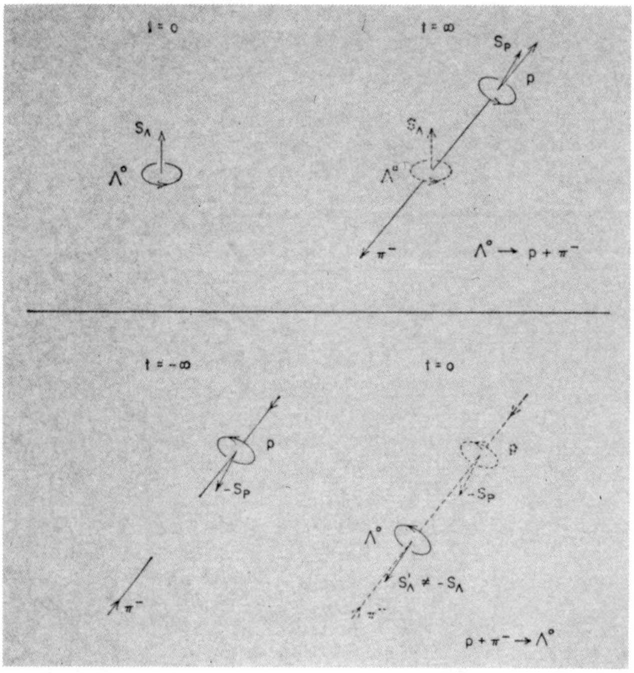

TIME REVERSAL. In the decay of a neutral lambda into a proton and a negative pion, if the initial lambda is assumed to be completely polarized along a unit vector s_Λ, then the final proton, at any given momentum k in the rest system of the original lambda, must be completely polarized, say, along s_p. In the reversed reaction when a proton combines with a pion to form a neutral lambda, the initial proton has reversed momentum —k and is completely polarized along the reversed direction $-s_p$; the final lambda is also completely polarized, but its polarization direction $s_{\Lambda'}$ is, in general, different from $-s_\Lambda$, even if time-reversal invariance holds. In these graphs all spin directions are drawn *in accordance* with time-reversal *invariance.* In quantum mechanics even a microscopic system is described by an infinite number of time-dependent variables; thus the time-reversed solution for any scattering problem is, in general, an improbable one (assuming that time reversal is an exact symmetry). —FIG. 2

charge symmetry: $J_\mu^* = -\exp(i\pi I_y)\, J_\mu \exp(-i\pi I_y)$ where I_y is the y component of the isospin operator. Under this assumption, independently of time-reversal invariance, g_A/g_V must be real.]

We note that in the Λ^0 decay, if the initial Λ^0 is completely polarized, say, along the unit vector $\mathbf{s}_\Lambda$ in its rest system, then at any given momentum $\mathbf{k}$ the final nucleon must also be completely polarized along a direction $\mathbf{s}_N$ which is uniquely determined by $\mathbf{s}_\Lambda$, $\mathbf{k}$ and the amplitudes A_s and A_p. Let us now consider the reversed reaction.

$$N + \pi \rightarrow \Lambda^0 \tag{11}$$

in which the initial nucleon is polarized along $-\mathbf{s}_N$ and its momentum is $-\mathbf{k}$. If the system obeyed *classical* mechanics, time-reversal invariance would imply that the final Λ^0 in the reversed reaction must be completely polarized along the reversed direction $-\mathbf{s}_\Lambda$. For the quantum-mechanical system, although the final Λ^0 in the reversed reaction does remain completely polarized, its direction is, in general, *different* from $-\mathbf{s}_\Lambda$ even if time-reversal invariance holds. This elementary property is demonstrated in figure 2.

To produce the final Λ^0 with its polarization along $-\mathbf{s}_\Lambda$, in case time-reversal invariance holds, we must not use just the $N+\pi$ state with the reversed spin and momentum, but we should start with an initially *coherent* mixture of the appropriate $s_{1/2}$ and $p_{1/2}$ incoming wave of $N+\pi$. Mathematically, such an initial state can be easily obtained by applying the antiunitary operator T onto the final state $\Psi(t=+\infty)$ in the decay $\Lambda^0 \rightarrow N+\pi$. Physically, however, it seems virtually impossible ever to construct the desired coherent time reversed state $T\Psi(t=+\infty)$ for a direct testing of the symmetry (or violation-of-symmetry) of the time-reversal operation. Here lies an important difference between the time-reversal operation in classical mechanics and that in quantum mechanics. In both cases time-reversal invariance means that the time-reversed solutions are always dynamically possible solutions. In classical mechanics such a time-reversed solution becomes an improbable one only for a macrosopic system. In quantum mechanics even a microscopic system is described by an infinite number of variables (that is, by a continuous function of space-time); thus the time-reversed solution for any scattering problem is, in general, an improbable one.

For all practical purposes the only *direct* and tangible test of time-reversal invariance seems to be the *reciprocity relations between various differential cross sections;* that is, equation 7 restricted to those states $|a\rangle$ and $|b\rangle$ that represent asymptotically the appropriate initial and final states in which every particle has a definite, but arbitrary, set of

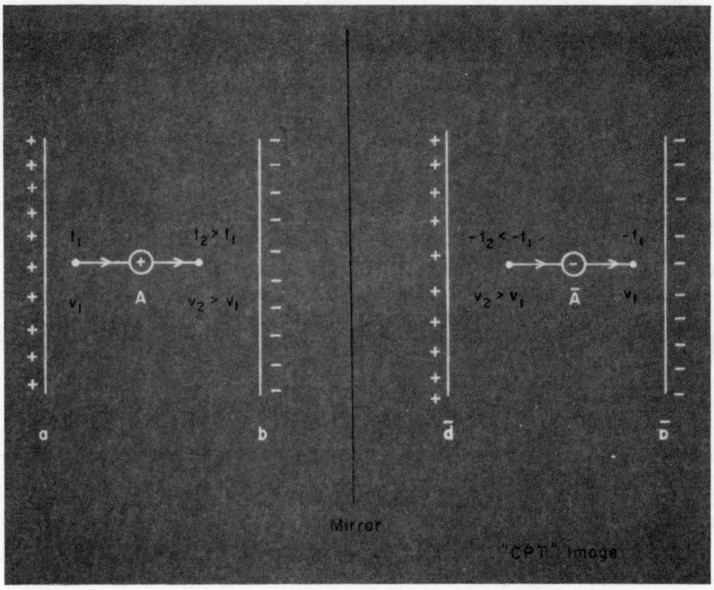

CPT SYMMETRY. The figure shows the motion of a positively charged particle A in an electric field and its "CPT" image. v_1 and v_2 are, respectively, the initial (at time t_1) and the final (at time t_2) velocities of A. CPT invariance requires for the existence of any particle an antiparticle of the same mass, opposite charge, opposite baryon number and opposite lepton number. It is important to note that the antiparticle state of A is *different* from the usual charge-conjugation state of A because charge conjugation is not an exact symmetry. —FIG. 3

spin and momentum. In this connection it is relevant to mention that all the presently known tests of time-reversal symmetry, such as equation 10, can be derived directly by using the reciprocity relations between the appropriate differential cross sections.[11] The violation of time-reversal invariance means, simply, that such reciprocity relations are not valid.

CPT invariance and particle-antiparticle relations

To understand the recent indirect evidence of time-reversal noninvariance, it is necessary to review the CPT theorem[12] and the experimental evidence for its validity.

In the framework of a local field theory, it can be shown that if a theory is invariant under the continuous group of Lorentz transformations without any discrete element (such as space inversion and time reversal), the theory is automatically invariant under a discrete symmetry operation, called "CPT." If CPT invariance holds for the total Hamiltonian H, the matrix element of H between any two states $|A\rangle$ and $|B\rangle$ is related to that between their CPT-conjugate states $|\bar{A}\rangle$ and $|\bar{B}\rangle$

$$\langle B|H|A\rangle = \langle \bar{B}|H|\bar{A}\rangle^* \qquad (12)$$

where

$$|\bar{A}\rangle = CPT|A\rangle$$
$$|\bar{B}\rangle = CPT|B\rangle \qquad (13)$$

From CPT invariance it follows that if A is a stable particle, then its CPT conjugate $\bar{A}$ is also a stable particle with exactly the same mass. This can be easily proved by setting A=B in equation 12.

The CPT operator is an antiunitary operator; furthermore it relates the state A with a momentum **k** and a helicity λ to that of $\bar{A}$ with the same momentum **k** but the opposite helicity $-\lambda$. It can be easily shown that the electromagnetic fields **E** and **H** are invariant under CPT. Thus by comparing the energy spectrum in an external electromagnetic field, one can prove that A and $\bar{A}$ have opposite charges but otherwise the same electromagnetic form factors (see figure 3). Within the same framework of local field theory it can also be proved that the baryon numbers, or the lepton numbers, of the states A and $\bar{A}$ must be equal in magnitude but opposite in sign. (We assume implicitly that charge, baryon number and lepton number are all strictly conserved.) In the following the state $\bar{A}$ is called the "*antiparticle* state" of A.

Among the mass equalities between particles and antiparticles, the most accurate one is that between K^0 and $\bar{K}^0$

$$\langle K^0|H|K^0\rangle = \langle \bar{K}^0|H|\bar{K}^0\rangle \qquad (14)$$

From the experimental mass difference[13] Δm, between K_1^0 and K_2^0, it is found that equation 14 holds to the accuracy $|\Delta m/m_K| \approx 10^{-14}$. Thus, we should regard H_{st}, H_γ and the strangeness conserving nonleptonic part of H_{wk} to be invariant under CPT.

For the other parts of H_{wk}, evidences for CPT invariance come from the lifetime equalities between the states A and $\bar{A}$ which decay through the weak interaction. Such lifetime equalities hold, at least, to the lowest order in H_{wk} and can be easily proved by substituting H_{wk} for H in equation 12.

The present upper limits on such lifetime differences $\Delta\tau$ are given for muons,[14] pions[15] and K mesons[16] by

$$\left|\frac{\Delta\tau}{\tau}\right| < \begin{cases} 0.001 \text{ for } \mu^{\pm} \\ 0.08 \text{ for } \pi^{\pm} \\ 0.15 \text{ for } K^{\pm} \end{cases} \tag{15}$$

Among these, the limit for the strangeness-nonconserving part of the weak interaction is rather inaccurate.

In the following we shall assume that CPT invariance holds; therefore for each particle state A there exists an antiparticle state $\overline{A}$ that has the same physical mass, the opposite charge, the opposite baryon number (or the opposite lepton number) and, if A is unstable, the same lifetime.

The C_{st} symmetry

As I remarked earlier, the strong interaction is found experimentally to be invariant under a space inversion P_{st} (to an accuracy of about 10^{-6}), a time reversal T_{st} (to an accuracy of a few percent) and the CPT operation (to an accuracy of about 10^{-14}). We can define an operator C_{st}

$$C_{st} \equiv (CPT) T_{st}^{-1} P_{st}^{-1} \tag{16}$$

The strong interaction is expected to satisfy the C_{st} invariance to the same accuracy as the T_{st} invariance. Under C_{st} we must have at least approximately, $p \rightarrow \bar{p}$ and $n \rightarrow \bar{n}$, but with their momenta and helicities unchanged; consequently we also have $\pi^+ \rightarrow \pi^-$, $\pi^0 \rightarrow \pi^0$, etc., since the transformation properties of these mesons are determined by those of baryons and antibaryons.

The C_{st} symmetry has also been directly tested[17] by studying the equality between the energy distributions of π^+ and π^- in the annihilation of protons by antiprotons

$$\bar{p} + p \rightarrow \pi^+ + \pi^- + \cdots \tag{17}$$

The result puts an upper limit on the C_{st}-noninvariant amplitude; it cannot be more than about 1% of the C_{st}-invariant amplitude. A similar upper limit of about 2% is obtained by studying the energy distributions of K^+ and K^- in the same (proton-antiproton) annihilation experiment.

However, the weak interaction is known to violate C_{st} invariance. This can be inferred by using the properties that H_{wk} has large violations of P_{st}, but it is (at least to a good approximation) invariant under $C_{st}P_{st}T_{st} = CPT$ and T_{st}. The same conclusion can also be directly reached by considering K^0 decays. Let K_1^0 be the neutral K meson with the short lifetime and K_2^0 that with the long lifetime. If C_{st} symmetry were conserved in their weak decays, K_1^0 and K_2^0 would be eigenstates of C_{st}, and their C_{st} eigenvalues must be of opposite signs. But in the decays $K_1^0 \rightarrow 2\pi$ and $K_2^0 \rightarrow 3\pi$ the final 2π and 3π are both in the $C_{st} = +1$ states. Thus C_{st} invariance must be violated, and the C_{st} symmetry is not exact. If $|p\rangle$ is the physical proton state, the state $CPT|p\rangle$ is the physical antiproton state, but the state $C_{st}|p\rangle$ is only *approximately* the antiproton state.

Two-pion decay of neutral K mesons

It was discovered last year by Christenson, Cronin, Fitch and Turlay[18] that the long-lived component K_2^0 of the neutral K meson has a two-pion decay mode

$$K_2^0 \rightarrow \pi^+ + \pi^- \tag{18}$$

Since both C_{st} and P_{st} (in the center-of-mass system) interchange π^+ and π^-, the final (two-pion) system in this decay must be of $C_{st}P_{st} = +1$. On the other hand, the same long-lived component K_2^0 also decays into three pions

$$K_2^0 \rightarrow \pi^+ + \pi^- + \pi^0 \tag{19}$$

All pions produced are observed to be predominantly in the *s* state. Thus, the final three-pion state is (or at least *predominantly* is) of $C_{st}P_{st} = -1$.

Now, K_2^0 is, by definition, a particle with a definite lifetime. That it can decay into different states with opposite $C_{st}P_{st}$ values shows conclusively that $C_{st}P_{st}$ is not conserved in the K_2^0 decays. The $C_{st}P_{st}$ violation is, therefore, established independently of the detailed theory of the K_1^0, K_2^0 system.

The observed $C_{st}P_{st}$ noninvariance implies that

$$[H, C_{st}P_{st}] \neq 0 \tag{20}$$

where H represents the total interaction. In this decay, both the initial and the final states are eigenstates of $H_{st} + H_\gamma$ and the transition is due to H_{wk}. This means that the $C_{st}P_{st}$ violation could be due to the strong interaction[19] or to the electromagnetic interaction[20] or to the weak interaction,[21] or a combination of these interactions, or the presence of some new interactions, such as the superweak interaction,[22] which, however, I will not discuss further here. From this single experiment, it is not possible to decide which interaction is responsible for this violation.

If we *assume* the validity of CPT invariance in the K_2^0 decay, then it follows from equations 16 and 20 that T_{st} invariance is also violated; that is

$$T_{st} H T_{st}^{-1} \neq H \tag{21}$$

The amount of the observed $C_{st}P_{st}$ violation in

the K_2^0 decay is a small one, characterized by a parameter

$$|\varepsilon| = \left| \frac{\text{Rate}\ (K_2^0 \rightarrow \pi^+ + \pi^-)}{\text{Rate}\ (K_1^0 \rightarrow \pi^+ + \pi^-)} \right|^{1/2} \approx 2 \times 10^{-3} \qquad (22)$$

K_1^0 is the short-lived component of the neutral K meson. The smallness of this parameter is, perhaps, one of the most puzzling features of this new discovery. To explain ε, we may invoke a small C_{st}-noninvariant term in H_{st}, or, if ε is a typical example of a $C_{st}P_{st}$-noninvariant amplitude, a small $C_{st}P_{st}$-noninvariant term in H_{wk}. There is no difficulty in doing either of these; one is only puzzled by the smallness of such violation and by the multitudes of different ways to construct such a violation.

An alternative, and perhaps more attractive, possibility is to assume that H_{st} and H_{wk} are invariant under $C_{st}P_{st}$, but H_γ has a large violation of C_{st} invariance. In this case all strong and weak processes can have a small C_{st}- and $C_{st}P_{st}$-noninvariant amplitude through virtual emission and absorption of photons. In particular K_2^0 can decay into two pions with a fractional amplitude $\varepsilon \approx \alpha/\pi$, (where α is the fine-structure constant) and this gives a natural explanation of the smallness of the observed $C_{st}P_{st}$-noninvariant amplitude.

This last possibility naturally raises many questions; among these the most urgent one is whether the hypothesis of H_γ being noninvariant under C_{st} is already in contradiction with some other known experiments. In an extensive study[20] that I have made in collaboration with Jeremy Bernstein and Gerald Feinberg, we find that there exists at present no experimental evidence that H_γ of the nonleptons is, or is not, invariant under C_{st}, nor is there any evidence that H_γ satisfies, or does not satisfy, the same time-reversal invariance T_{st} as the strong interaction. If the electromagnetic interaction is noninvariant under C_{st}, since H_γ is experimentally found to be invariant under both P_{st} and CPT, it follows that H_γ must also violate T_{st} invariance.

The possibility that H_γ may violate C_{st} and T_{st} invariances is, of course, only a theoretical possibility. Nevertheless, the present absence of evidence that H_γ is, or is not, invariant under C_{st} and T_{st} should provide sufficient incentive for further experimental efforts in this direction. Various tests have been proposed, and some of these are already in progress.

A possible mismatch pattern for C_i, P_i and T_i where i stands for wk, γ or st

H_{st}	C_{st}	T_{st}	P_{st}
	$\nparallel$	$\nparallel$	$\parallel$
H_γ	C_γ	T_γ	P_γ
	$\nparallel$	$\nparallel$	$\nparallel$
H_{wk}	C_{wk}	T_{wk}	P_{wk}
		$\parallel$	
		T_{st}	

$$C_{st}\,P_{st}\,T_{st} = C_\gamma\,P_\gamma\,T_\gamma = C_{wk}\,P_{wk}\,T_{wk} = \text{CPT}$$

What is charge conjugation?

The electromagnetic interaction of the leptons is well known to be invariant under *charge conjugation* C_γ. Furthermore we know that the minimal electromagnetic interaction of any system of spin-zero and spin-one-half particles is always invariant under charge conjugation; it is also invariant under a time reversal T_γ and a space inversion P_γ. The electromagnetic interaction of the leptons is well described by such a minimum interaction. It seems, therefore, aesthetically appealing to *assume* that there exists a charge conjugation operation C_γ under which all electromagnetic currents change sign and that *all electromagnetic interactions,* including that of the nonleptons, are invariant under this *charge conjugation symmetry* C_γ. From the experimental evidences of P_{st} and CPT invariances, we know that H_γ must also be invariant under P_γ and T_γ where

$$P_\gamma = P_{st}$$

$$C_\gamma P_\gamma T_\gamma = C_{st} P_{st} T_{st} = \text{CPT}$$

The question whether the electromagnetic interaction does, or does not, satisfy the C_{st} symmetry can be simply viewed as whether the charge-conjugation operator C_γ is, or is not, the same as the operator C_{st}. (The C_{st} may, for example, be regarded as the baryon-number conjugation, and can, in principle, be different from the charge conjugation.) If the electromagnetic interaction does violate the C_{st} symmetry, then $C_{st} \neq C_\gamma$ and, therefore $T_{st} \neq T_\gamma$.

As I have already remarked, the presently known form of weak interaction H_{wk} is invariant under its own space inversion P_{wk}; it can also be shown that H_{wk} is invariant under a time reversal T_{wk} and a C_{wk} conjugation.[7] Indeed, all our experimental results are consistent with the *assumption* that each of these interactions H_i is invariant under its own C_i, P_i and T_i where i=st, γ and wk, and

$$C_i P_i T_i = \text{CPT} \qquad (23)$$

The well known parity nonconservation is due to the mismatches

$$P_{wk} \neq P_{\gamma} = P_{st} \qquad C_{st} \neq C_{wk} \neq C_{\gamma}$$

It remains an open question whether the recent discovery of the $C_{st}P_{st}$ nonconservation in $K_2^0 \rightarrow \pi^+ + \pi^-$ can also be attributed to a similar mismatch—one between C_{st} and C_{γ}. A possible pattern of such mismatches is given in the table on page 30.

Our concept of "C" started with the charge conjugation C_{γ} determined by the electromagnetic interaction of the electron.[23] Later, the operator C_{γ} was extended to other interactions and was called "particle-antiparticle conjugation." After the discovery of parity nonconservation, it was already known that charge-conjugation invariance cannot be extended to the weak interaction and that the concept of particle and antiparticle rests, instead, on CPT invariance. Nevertheless, the hypothesis that charge-conjugation invariance C_{γ} is applicable to the strong interaction was assumed without question, and that the electromagnetic interaction and the strong interaction satisfy the same time-reversal invariance was taken for granted.

The progress of science has always been the result of a close interplay between our concepts of the universe and our observations of nature. The former can only evolve out of the latter, and yet the latter are also conditioned to a remarkable degree by the former. As we expand our fields of observation, naturally, we also extend our basic concepts. At times, these two factors, the concept and the observation, may become so interlocked that even some of the fundamental principles used in an entire domain of familiar phenomena may, to our chagrin, turn out to have no actual experimental basis. The history of these discrete symmetries has been a particularly rich one, full of such surprises.

References

1. E. P. Wigner, Z. Physik **43**, 624 (1927) ; Gött. Nach. Math. Naturw. Kl. 546 (1932) .
2. T. D. Lee, C. N. Yang, Phys. Rev. **104**, 254 (1956) .
3. T. D. Lee, R. Oehme, C. N. Yang, Phys. Rev. **106**, 340 (1957) .
4. C. S. Wu, E. Ambler, R. W. Hayward, D. D. Hoppes, R. P. Hudson, Phys. Rev. **105**, 1413 (1957) .
5. R. L. Garwin, L. M. Lederman, M. Weinrich, Phys. Rev. **105**, 1415 (1957) ; J. I. Friedman, V. L. Telegdi, Phys. Rev. **105**, 1681 (1957) .
6. F. Boehm, E. Kankeleit, Calt-63-13 (preprint); Yu. G. Abov, P. A. Krupchitsky, Yu. A. Oratovsky, *Comptes Rendus du Cong. Intern. Phys. Nucl.*, (Paris, 1964); L. Grodzins, F. Genovese, Phys. Rev. **121**, 228 (1961); R. E. Segel et al., Phys. Rev. **123**, 1382 (1961); D. E. Alburger et al., Phil. Mag. **6**, 171 (1961); R. Haas, L. B. Leipuner, R. K. Adair, Phys. Rev. **116**, 1221 (1959) ; F. Boehm, U. Hauser, Nuclear Physics **14**, 615 (1959); D. A. Bromley et al., Phys. Rev. **114**, 758 (1959).
7. T. D. Lee, Phys. Rev. **140**, B959 (1965) ; article in *Proc. Oxford International Conf. Elementary Particles*, The Rutherford High Energy Laboratory (1965).
8. L. Rosen, J. E. Brolley Jr., Phys. Rev. Letters **2**, 98 (1959); D. Bodansky et al., Phys. Rev. Letters **2**, 101 (1959) ; A. Abashian, E. M. Hafner, Phys. Rev. Letters **1**, 255 (1958) ; C. F. Hwang et al., Phys. Rev. **119**, 352 (1960) .
9. J. W. Cronin, O. E. Overseth, in *Proc. International Conf. High Energy Physics*, CERN, 453 (1962) .
10. M. T. Burgy, V. E. Krohn, T. B. Novey, G. R. Rings, V. L. Telegdi, Phys. Rev. Letters **1**, 324 (1958) .
11. T. D. Lee, pp. 5–16 of *Preludes in Theoretical Physics (in Honor of V. F. Weisskopf)* (A. De-Shalit, H. Feshbach, L. Van Hove, eds.), North Holland Publishing Co., Amsterdam (1966) .
12. W. Pauli, *Niels Bohr and the Development of Physics*, Pergamon Press, London (1955) ; J. Schwinger, Phys. Rev. **91**, 720, 723 (1953); **94**, 1366 (1953); G. Lüders, Kgl. Danske Videnskab. Selskab. Mat-fys. Medd. 28, No. 5 (1954) .
13. B. Aubert et al., Phys. Letters **10**, 215 (1964) ; U. Camerini et al., Phys. Rev. **128**, 362 (1962); J. H. Christenson et al., at Conf. Fundamental Aspects Weak Interactions, Brookhaven National Laboratory (1963) ; J. H. Christenson, Tech. Report 34, Princeton University (1964) ; V. L. Fitch et al., Phys. Rev. Letters **15**, 73 (1965) .
14. R. A. Lundby, Phys. Rev. **125**, 1686 (1962) ; M. Eckhause, Carnegie Inst. Tech. Report 9286 (1962) ; F. J. M. Farley et al., in *Proc. International Conf. High Energy Physics*, Geneva (1962) ; M. A. Meyer et al., Nevis Report 114, Columbia University (1963) .
15. J. Jakobson, A. Schulz, J. Steinberger, Phys. Rev. **81**, 894 (1951) ; R. P. Durbin, H. H. Loar, W. W. Havens Jr., Phys. Rev. **88**, 179 (1952) ; H. L. Anderson et al., Phys. Rev. **119**, 2050 (1960) ; J. Ashkin et al., Nuovo Cimento **16**, 490 (1960) ; A. Barkas, A. Rosenfeld, *Proc. International Conf. High Energy Physics*, Rochester (1960) .
16. P. Nordin Jr., Phys. Rev. **123**, 2168 (1961) ; B. Bhowmik et al., Nuovo Cimento **20**, 857 (1961) ; W. H. Barkas et al., Phys. Rev. **124**, 1209 (1961) ; A. M. Boyarski et al., Phys. Rev. **128**, 2398 (1962) ; F. Crawford, in *Proc. International Conf. High Energy Physics*, Rochester (1957).
17. C. Baltay, N. Barash, P. Franzini, N. Gelfand, L. Kirsch, G. Lütjens, J. C. Severiens, J. Steinberger, D. Tycko, D. Zanello, Phys. Rev. Letters **15**, 591 (1965) .
18. J. H. Christenson, J. W. Cronin, V. L. Fitch, R. Turlay, Phys. Rev. Letters **13**, 138 (1964) ; A. Abashian et al., Phys. Rev. Letters **13**, 243 (1964) .
19. T. D. Lee, L. Wolfenstein, Phys. Rev. **138**, B1490 (1965) ; J. Prentki, M. Veltman, Phys. Letters **15**, 88 (1965) ; L. B. Okun (preprint) .
20. J. Bernstein, G. Feinberg, T. D. Lee, Phys. Rev. **139**, B1650 (1965); see also S. Barshay, Phys. Letters **17**, 78 (1965).
21. (Of numerous papers on various possible ways to violate $C_{st}P_{st}$ invariance in the weak interactions, we list but a few.) R. G. Sachs, Phys. Rev. Letters **13**, 286 (1964); R. G. Sachs, S. B. Treiman, Phys. Rev. Letters **8**, 137 (1962) ; N. Cabibbo, Phys. Rev. Letters **12**, 137 (1964); S. Glashow, Phys. Rev. Letters **14**, 35 (1965); L. Wolfenstein, Phys. Rev. Letters **15**, 196 (1965). See also F. Salzman, Phys. Letters **15**, 91 (1965). (The authors postulated an electric dipole moment for the hypothetical intermediate boson $W^{\pm}$ of the weak interaction.)
22. L. Wolfenstein, Phys. Rev. Letters **13**, 562 (1964); T. D. Lee, L. Wolfenstein, loc cit in ref. 19.
23. H. Weyl, *The Theory of Groups and Quantum Mechanics* (translation of the 2d edition of *Gruppentheorie und Quantenmechanik*) E. P. Dutton and Co. (1931).

Reprinted from THE PHYSICAL REVIEW. Vol. 148. No. 4. 1385–1404. 26 August 1966
Printed in U. S. A.

Space Inversion, Time Reversal, and Other Discrete Symmetries in Local Field Theories*

T. D. LEE AND G. C. WICK
Department of Physics, Columbia University, New York, New York
(Received 28 February 1966)

The general algebraic relations between space inversion, time reversal, and the internal symmetry group are analyzed within the framework of a Lorentz-invariant local field theory. The problem of unitary representations of the full Poincaré group including the space and time reflection operators has been studied by Wigner, and the representations are classified into 4 cases. It is shown that, with the added assumption of the local field theory, Wigner's cases 2, 3, and 4 either do not not occur or can be reduced to his case 1. The concept of minimal group extension is introduced and the related mathematical analysis is given. The symmetry properties under space inversion, time reversal, and other discrete operators such as charge conjugation are analyzed separately for each of the three known interactions: strong, electromagnetic, and weak.

I. INTRODUCTION

OUR views of discrete symmetries such as space inversion P, time reversal T, and charge conjugation C have undergone great changes in recent years. Since 1957, it has been well established[1] that both P and C symmetries are only approximately valid. Rather naturally, this discovery led immediately to speculations as to the possible need of a profound revision of some of our concepts. Until the recent discovery[2] of the $\pi^+\pi^-$ decay mode of the long-lived K_2^0 meson, however, it was possible to retain intact the structure of the full Poincaré group and in particular to believe in the essential symmetry between left and right by using CP, instead of P. From a fundamental point of view, therefore, the strong evidence now existing of a CP violation in K_2^0 decay, and the concurrent indirect conclusion that T is also not an exact symmetry, are a more decisive blow to our notions on geometric symmetry principles than earlier results.

Strictly speaking, since the very existence of symmetry operators such as T and P, or CP, is deduced from the alleged equivalence of certain reference systems, the presence of phenomena violating that equivalence implies that the operators themselves cannot be exactly defined. The language employed most often in describing the situation is in fact somewhat inconsistent. One says: there is a P or CP, etc., operator and in specific theories one often proceeds to give exact definitions of these operators; subsequently one assumes a Hamiltonian containing terms which do not commute with P, CP or T, etc. This is the same as saying, however, that these operators do not satisfy exactly the multiplication laws of the coordinate transformations they are alleged to represent within the full Poincaré group. More explicitly, the transformation

$$\mathbf{r} \rightarrow -\mathbf{r}, \quad t \rightarrow +t,$$

corresponding to P or CP, should commute with a time translation, represented infinitesimally by H, and the transformation

$$\mathbf{r} \rightarrow +\mathbf{r}, \quad t \rightarrow -t,$$

corresponding to T, transforms the time translation $t \rightarrow t+\tau$ into $t \rightarrow t-\tau$ which, because of the anti-unitary nature[3] of T, again implies commutativity of the corresponding operators. Thus the "exact" definition of the symmetry operators is a purely formal convention, that in fact does not satisfy the basic geometrical requirements. One can similarly think of other possibilities of "exact" definitions. For example, one might start with the *physical* single-particle state $|\mathbf{k},\lambda\rangle$, and define P and T to be, respectively, the unitary and anti-unitary operators that satisfy

$$P|\mathbf{k},\lambda\rangle = \eta_P|-\mathbf{k}, -\lambda\rangle, \tag{1.1}$$

and

$$T|\mathbf{k},\lambda\rangle = \eta_T|-\mathbf{k}, +\lambda\rangle, \tag{1.2}$$

where $\mathbf{k}$ and λ are, respectively, the momentum and the helicity of the particle, and η_T and η_P are phase factors which may depend on λ. (Helicity is defined to be the spin component along the direction of $\mathbf{k}$.)

In equations such as (1.1) and (1.2), the states $|\mathbf{k},\lambda\rangle$ and $|-\mathbf{k}, \pm\lambda\rangle$ refer to the same physical particle. The equations, however, have an unambiguous meaning only for an exactly stable particle. The state vector of an unstable particle must sooner or later develop components corresponding to the decay products. No satisfactory generalization of (1.1) or (1.2) exists for such

* This research was supported in part by the U. S. Atomic Energy Commission.

[1] The first experiment that conclusively established the approximate nature of P and C symmetries was made on β decay by C. S. Wu, E. Ambler, R. W. Hayward, D. D. Hoppes, and R. P. Hudson, Phys. Rev. **105**, 1413 (1957). This was immediately followed by the observation of the same noninvariance properties in π, μ decays by R. L. Garwin, L. M. Lederman, and M. Weinrich, Phys. Rev. **105**, 1415 (1957) and by J. J. Friedman and V. L. Telegdi, Phys. Rev. **105**, 1681 (1957). The possibility that P, C, and T are only approximate symmetries was suggested theoretically by T. D. Lee and C. N. Yang, Phys. Rev. **104**, 254 (1956) and further discussed by T. D. Lee, R. Oehme, and C. N. Yang, Phys. Rev. **106**, 340 (1957).

[2] J. H. Christenson, J. W. Cronin, V. L. Fitch, and R. Turlay, Phys. Rev. Letters **13**, 138 (1964). See also A. Abashian *et al.*, Phys. Rev. Letters **13**, 243 (1964).

[3] E. P. Wigner, Gött. Nach. Math. Naturw. Kl., p. 546 (1932). For a general discussion of anti-unitary operators see, e.g., E. P. Wigner, *Group Theory* (Academic Press Inc., New York, 1959), Chap. 26.

a state, if the decay violates either space-inversion or time-reversal symmetry. Similar difficulties arise for multiparticle states, if the interaction involves symmetry violations.[4]

The notion that a certain "fuzziness" is attached to the definition of the symmetry operators P, PC, T is not new and is not likely to be questioned; but the situation is clearly an unsatisfactory one from a fundamental point of view. The following is an attempt to characterize the ambiguity in a mathematically more precise way; if this can be done, at least some order can be brought into the subject.

There is already a completely different source of ambiguity in the definition of symmetries which has been analyzed in some detail. This is the possible existence of superselection rules.[5] Such rules imply that the Hilbert space of a system (say a system of interacting fields) should split into "noncoherent subspaces." Thus, every symmetry operator can be multiplied (without detectable consequences) in each subspace by a phase factor, which is independent of the phase factors in the other subspaces. It is of course quite possible that there are no superselection rules, the rules now generally accepted being the result of an inadequate sensitivity of our experimental techniques. This latter point of view may in fact present some conceptual advantages, but it clearly leads to the same physical results as the superselection rules, so long as the violation is either unobserved or can be neglected.

It seems practical at the present stage to accept the view that there are superselection rules associated with N (baryon number), Q (electric charge) and perhaps also the leptonic numbers. The independent phase factors in each noncoherent subspace mentioned previously are adequately replaced by the gauge group generated by these quantities. It is then easy to summarize the situation by stating that each geometrical symmetry, if it exists, is represented not just by one operator, but by a coset of the gauge group (as will be discussed in more detail later). The *geometrical* group then appears as the quotient of the complete symmetry group by the above mentioned gauge-group. This formulation has been used, for example, by Michel for the full Poincaré group.[6] This discussion is still based, however, on the assumption that the discrete space-time symmetries are exact.

The approximate nature of the symmetries suggests, in our opinion, that the process of redefinition must be carried a step further. Let us consider a model Hamiltonian H which may be constructed from the real Hamiltonian H_{total} by setting some of its parameters equal to zero:

$$H=\lim_{g\to 0} H_{\text{total}}, \tag{1.3}$$

where g represents a set of parameters which can be certain coupling constants and maybe mass differences. By a suitable choice of the set of parameters g, both space inversion and time reversal can be regarded as exact symmetries for the solutions of the model Hamiltonian H.

As an example, we may take $H=H_{\text{free}}+H_{\text{st}}$ where

$$H_{\text{free}}+H_{\text{st}}=\lim_{e=G_{\text{wk}}=0} H_{\text{total}}. \tag{1.4}$$

The parameters e and G_{wk} are, respectively, the electron charge and the weak-interaction coupling constant. In this limit, the mass differences between particles within the same isospin multiplet are all assumed to be zero. The H_{free} is the free-particle Hamiltonian, and H_{st} the strong-interaction Hamiltonian. The existing experimental results strongly support the assumption that, so far as this model Hamiltonian H is concerned, both space inversion and time reversal are exact symmetries. Equations (1.1) and (1.2) may still be used to define the operators P and T provided $|\mathbf{k},\lambda\rangle$ is regarded as the single-particle eigenstate of the model Hamiltonian. Similar requirements for P and T can be easily extended to the multiparticle eigenstates of H. However, in the limit $e=G_{\text{wk}}=0$, it is not possible to distinguish between different members of the same isospin multiplet. Thus, if P is the unitary space-inversion operator, an equally good choice can be the product SP where S is any isospin transformation operator (or, any other internal symmetry operator). The same holds for the time-reversal operator T.

In general, for any given model Hamiltonian H, there exists an *internal symmetry group* $\mathcal{G}$:

$$\mathcal{G}=\{S\}, \tag{1.5}$$

where the group element S is called an internal symmetry operator. S will be defined later (in Sec. II) to be any unitary operator that satisfies

$$SHS^{-1}=H, \tag{1.6}$$

[4] For example, for the multiparticle states one might try to define P to be the unitary operator which transforms, say, any physical incoming-wave state $|\mathbf{k}_1,\lambda_1;\mathbf{k}_2,\lambda_2\cdots\rangle^{\text{in}}$ to $|-\mathbf{k}_1,-\lambda_1;-\mathbf{k}_2,-\lambda_2\cdots\rangle^{\text{in}}$, multiplied by an appropriate phase factor. Such an operator, which we may call P^{in}, has only matrix elements between states of the same energy; therefore, it commutes with the total Hamiltonian. The observed nonconservation of parity implies that a different operator, called P^{out}, would result if the incoming-wave states are replaced by the outgoing-wave states. At any finite time t, neither P^{in} nor P^{out} transforms the field operators at $\mathbf{r}$ to those at $-\mathbf{r}$. These two operators P^{in} and P^{out} are, respectively, connected only with the transformation properties of the initial and the final configurations of the system; such operators are, in general, rather useless (unless space inversion were an exact symmetry).

[5] G. C. Wick, A. S. Wightman, and E. P. Wigner, Phys. Rev. **88**, 101 (1952). See also G. Feinberg and S. Weinberg, Nuovo Cimento **14**, 571 (1959).

[6] L. Michel, in *Group Theoretical Concepts and Methods in Elementary Particles*, edited by F. Gürsey (Gordon and Breach Publishers, New York, 1964), pp. 135–200. A discussion of group extensions is also given in this paper. See also F. Kamber and N. Straumann, Helv. Phys. Acta **37**, 563 (1964), and L. C. Biedenharn, J. Nuyts, and H. Ruegg, Commun. Math. Phys. (to be published).

and is not connected with any coordinate transformations. We note that if H is invariant under the charge conjugation C, then, but only then, by definition, $\mathcal{G}$ contains C. The structure of $\mathcal{G}$ depends on the particular H under consideration. A full discussion of the properties of the space-inversion operator and the time-reversal operator cannot be given without studying their algebraic relations with the internal symmetry group $\mathcal{G}$.

In Sec. II, the general defintions of the space-inversion operator and the time-reversal operator are given for any Lorentz-invariant local field theory. It follows from these general definitions that if a unitary operator $\mathcal{P}$ is a space-inversion operator, then any member in the set $\{S\mathcal{P}\}$ can also be used as a space-inversion operator. Similarly, if an anti-unitary operator $\mathcal{T}$ satisfies this general definition of the time reversal operator, then any member in the set $\{S\mathcal{T}\}$ also satisfies the same definition. For clarity, we use the script letters $\mathcal{P}$ and $\mathcal{T}$ to denote an arbitrary choice of these operators. The general algebraic relations between these discrete elements $\mathcal{P}$, $\mathcal{T}$ and the internal-symmetry group $\mathcal{G}$ are then investigated.

As we shall see later, an immediate consequence of the usual "CPT" theorem[7] is that for any model Hamiltonian H, if $\mathcal{P}$ exists, then $\mathcal{T}$ exists and vice versa. Furthermore, for any $\mathcal{P}$ in the set $\{S\mathcal{P}\}$ there exists a $\mathcal{T}$ in the set $\{S\mathcal{T}\}$ such that

$$\mathcal{P}^2=\mathcal{T}^2, \tag{1.7}$$

$$\mathcal{P}\mathcal{T}=(-1)^{2J}\mathcal{T}\mathcal{P}, \tag{1.8}$$

and

$$\mathcal{P}^{-1}\mathcal{T}=\mathcal{J}, \tag{1.9}$$

where J is the spin of the local field operator and $\mathcal{J}$ is the usual CPT operator which satisfies

$$\mathcal{J}^2=(-1)^{2J}, \tag{1.10}$$

and

$$\mathcal{J}S=S\mathcal{J}, \tag{1.11}$$

for all internal-symmetry operators S. Thus, the algebraic relations between $\{S\mathcal{T}\}$ and the internal symmetry group $\mathcal{G}$ are completely determined if those between $\{S\mathcal{P}\}$ and $\mathcal{G}$ are known (and vice versa).

The set of operators $\{S\mathcal{P}\}$ and $\{S\}$ forms a new group $\mathcal{E}$ which we shall call a *minimal extension* of $\mathcal{G}$; $\mathcal{G}$ is an invariant subgroup of $\mathcal{E}$ and the factor group is the two-element group Z_2. The mathematical analysis of the minimal extension for any given group $\mathcal{G}$ is an elementary case of the general theory, and is given in Sec. III. All such minimal extensions can be classified into two general types A and B, depending on whether the correspondence $S\leftrightarrow\mathcal{P}S\mathcal{P}^{-1}$ is an inner or outer automorphism of $\mathcal{G}$. Some simple illustrative examples are given in Sec. IV.

The problem of unitary representations of the full Poincaré group including the space and time-reflection operators has been discussed by Wigner,[8] and the representations are classified into 4 cases, depending on the algebraic properties of the space-time reflection operators; these classifications are, however, made without explicitly considering the internal symmetry group. Equations (1.7)–(1.11) show that, with the added assumption of the local-field theory, Wigner's cases 2, 3 and 4 either do not occur or can be reduced to his case 1 by using different operators in the sets $\{S\mathcal{P}\}$ and $\{S\mathcal{T}\}$.

The notion of minimal extension is also useful in studying the question of possible discrete elements in the internal symmetry group. It has been suggested[9] recently that each of the three known interactions (the strong interaction H_{st}, the electromagnetic interaction H_γ, and the weak interaction H_{wk}) is invariant under its own C_i, P_i and T_i where $i=$ strong, γ, or weak. In all these cases, the internal symmetry group $\mathcal{G}_i$ of the particular H_i is itself a minimal extension of another group $\mathcal{G}_i{}^0$ which does not contain the discrete element C_i.

Applications to these interactions are given in Secs. V and VI. The general solutions of C_{wk}, P_{wk}, and T_{wk}

---

[7] W. Pauli, *Niels Bohr and the Development of Physics* (Pergamon Press, London, 1955); J. Schwinger, Phys. Rev. **91**, 720 (1953); **94**, 1366 (1953). G. Lüders, Kgl. Danske Videnskab. Selskab Mat. Fys. Medd. **28**, No. 5 (1954). See also R. Jost, Helv. Phys. Acta **30**, 409 (1957); R. F. Streater and A. S. Wightman, *PCT, Spin and Statistics, and All That* (W. A. Benjamin, Inc., New York, 1964).

It should be pointed out that in the framework of the present paper, the expression "CPT" may be misleading because either C does not exist as a symmetry, or if C exists then there is no clear distinction between our script letter $\mathcal{P}$ and $C\mathcal{P}$ (likewise for $\mathcal{T}$). Therefore, the CPT theorem could be called the $\mathcal{P}\mathcal{T}$ theorem.

[8] E. P. Wigner, in *Group Theoretical Concepts and Methods in Elementary Particles*, Ref. 6, pp. 37–80. See also R. M. F. Houtappel, H. Van Dam, and E. P. Wigner, Rev. Mod. Phys. **37**, 595 (1965). The identification of Wigner's four cases in terms of the language of field theory is not without certain possibilities of ambiguity. Here, we assume that each symmetry operator $\mathcal{P}$, or $\mathcal{T}$, or $\mathcal{J}$, is normalized so that the vacuum state is an eigenstate with eigenvalue $+1$. Let the signs ϵ_θ, ϵ_I be defined by $\mathcal{J}^2=\epsilon_I(-1)^{2J}$ and $\mathcal{P}\mathcal{T}=\epsilon_\theta\epsilon_I(-1)^{2J}\mathcal{T}\mathcal{P}$. [By using (1.9), this implies $\mathcal{T}^2=\epsilon_\theta\mathcal{P}^2$.] The four cases 1, 2, 3, 4 correspond to $(\epsilon_\theta,\epsilon_I)=(1,1)$, $(1,-1)$, $(-1,1)$ and $(-1,-1)$, respectively. Equations. (1.7)–(1.9) show that the choice $\epsilon_\theta=\epsilon_I=1$ is always possible.

The above normalization condition merely gives a convenient handle, to connect the transformation properties, under $\mathcal{P}$, $\mathcal{T}$, and $\mathcal{J}$, of a composite system to the product of the transformations of its constituents. The existence of a "composition law" expressing the connection between the space-time reflection properties of a multiparticle state and those of the single particle states is, of course, a natural consequence of local-field theory, but we must emphasize that *some* statement about this connection is an integral part of any usable formulation of the transformation laws; for example it is needed if one considers only the asymptotic states of any physical reaction.

The analysis of the possible irreducible representations or co-representations of the group of reflection operators, and the ensuing reduction, of the subspace of all single-particle states, into types of irreducible spaces, may be a logical step in a certain way of analyzing the problem. But the distinction between inequivalent representations is devoid of observable consequences, unless it is coupled with a specific prescription for the "composition law." We have adopted the rules of field theory, applied if one wishes only to the "in" and "out" states of S-matrix theory, as by far the most natural and simple way to formulate the necessary connections.

[9] T. D. Lee, Phys. Rev. **140**, B959 (1965).

are analyzed. It can be shown that C_{wk}, P_{wk}, and T_{wk} also commute with the SU_3-invariant part of the strong interaction. The various physical consequences of these discrete symmetries are discussed.

The operator C_i is an element of $\mathcal{G}_i$. An interesting, though perhaps academic, question is whether C_i (in particular C_{st}) can be continuously connected to the unit element of $\mathcal{G}_i$. In Sec. VII, it is found that, at least in some explicit models, this is possible. For example, if we consider all Feynman graphs which contain only the usual strong interaction vertices between the nucleons and any $G=+1$ mesons such as ρ, η^0, η'^0 etc., then the internal symmetry group automatically contains a subgroup SO_4 which, in turn, contains the usual isospin group SU_2. The C_{st} is one of the elements of this SO_4 group; consequently, it is connected to the unit element.

For practical applications, this SO_4 symmetry is violated by the strong interaction between the nucleons and the pions. Some possible uses of this broken SO_4 symmetry are discussed.

II. GENERAL DISCUSSION OF $\mathcal{P}$ AND $\mathcal{T}$

1. Spin-0 and Spin-½ Fields

We consider first a model Hamiltonian H which depends on n Hermitian[10] spin 0 fields $\phi_1(x), \cdots, \phi_n(x)$ and m Hermitian spin-½ fields $\psi_1(x), \cdots, \psi_m(x)$ where

$$\phi_a(x)=\phi_a^\dagger(x)\,,$$
$$[\psi_a(x)]_\mu=[\psi_a^\dagger(x)]_\mu\,, \tag{2.1}$$

$x=(\mathbf{r},t)$, and $a=1, 2, \cdots, n$ (or m). Each ψ_a is a 4-component spinor, and its spinor components are labeled by μ which varies from 1 to 4. The $\phi_a(x)$ and $\psi_a(x)$ satisfy the usual equal-time commutation and anticommutation relations:

$$[\phi_a(\mathbf{r},t),\phi_b(\mathbf{r}',t)]=[\phi_a(\mathbf{r},t),\psi_b(\mathbf{r}',t)]=0\,, \tag{2.2}$$

and, if there is no derivative coupling,

$$[\phi_a(\mathbf{r},t),(d/dt)\phi_b(\mathbf{r}',t)]=i\delta_{ab}\delta^3(\mathbf{r}-\mathbf{r}')\,,$$

and

$$[\psi_a(\mathbf{r},t)]_\mu[\psi_b(\mathbf{r}',t)]_\nu+[\psi_b(\mathbf{r}',t)]_\nu[\psi_a(\mathbf{r},t)]_\mu =\delta_{ab}\delta_{\mu\nu}\delta^3(\mathbf{r}-\mathbf{r}')\,; \tag{2.3}$$

otherwise, these relations become more complicated. Our general discussions, of course, do not depend on whether derivative couplings exist or not. For clarity, we assume here that the masses of the spin-½ fields are all different from zero. The zero-mass case will be discussed in Appendix A.

Definition 1. A unitary operator S in the Hilbert space is called an *internal-symmetry operator* if the Hamiltonian H is invariant under the transformation

$$S\phi(x)S^{-1}=\tilde{s}_0\phi(x)\,,$$

and

$$S\psi(x)S^{-1}=\tilde{s}_{1/2}\psi(x)\,, \tag{2.4}$$

where

$$\phi=\begin{bmatrix}\phi_1\\ \vdots\\ \phi_n\end{bmatrix},\quad \psi=\begin{bmatrix}\psi_1\\ \vdots\\ \psi_m\end{bmatrix},$$

and s_0, $s_{1/2}$ are, respectively, $(n\times n)$ and $(m\times m)$ matrices. From the Hermiticity condition and the commutation relations, both s_0 and $s_{1/2}$ must be real and orthogonal. Throughout the paper, the dagger and tilde denote, respectively, Hermitian conjugation and transposition. In (2.4), the transpose is introduced so that if the matrices s_J and s_J' are associated with the internal-symmetry operators S and S', then the matrix associated with (SS') is (s_Js_J'), where $J=0$ or $\frac{1}{2}$. For any given matrices s_0 and $s_{1/2}$, if S satisfies (2.4), so does $e^{i\theta}S$. Throughout our discussion we assume that there exists a *single* lowest eigenstate of H, called the vacuum state $|\text{vac}\rangle$. The arbitrary phase factor in $(e^{i\theta}S)$ will always be chosen so that

$$S|\text{vac}\rangle=|\text{vac}\rangle\,. \tag{2.5}$$

Definition 2. The group of *all* the internal symmetry operators is defined to be the *internal-symmetry group* $\mathcal{G}$:

$$\mathcal{G}=\{S\}\,. \tag{2.6}$$

Definition 3. The theory is said to be invariant under a *space inversion* if there exists a unitary operator $\mathcal{P}$ such that

$$\mathcal{P}\phi(\mathbf{r},t)\mathcal{P}^{-1}=\tilde{u}_0{}^P\phi(-\mathbf{r},t)\,,$$
$$\mathcal{P}\psi_\mu(\mathbf{r},t)\mathcal{P}^{-1}=\tilde{u}_{1/2}{}^P[i\gamma_4\psi(-\mathbf{r},t)]_\mu\,, \tag{2.7}$$

and

$$\mathcal{P}H\mathcal{P}^{-1}=H\,,$$

where $u_0{}^P$ and $u_{1/2}{}^P$ are, respectively, $(n\times n)$ and $(m\times m)$ matrices, $\psi_\mu(x)$ is a $(m\times 1)$ matrix whose components are $[\psi_1(x)]_\mu, \cdots, [\psi_m(x)]_\mu$. Throughout the paper, the Majorana representation of the Dirac matrices will be used; i.e.,

$$\gamma_1,\ \gamma_2,\ \gamma_3 \text{ are real}\,,$$

and

$$\gamma_4,\ \gamma_5 \text{ are imaginary}\,. \tag{2.8}$$

All the γ's are (4×4) Hermitian matrices. It follows from Eqs. (2.1)–(2.3) that $u_J{}^P$ must be real and orthogonal, where $J=0$ or $\frac{1}{2}$. For any given matrices $u_0{}^P$ and $u_{1/2}{}^P$, (2.7) only determines the unitary operator $\mathcal{P}$ up to a phase factor. We will choose this phase factor so that

$$\mathcal{P}|\text{vac}\rangle=|\text{vac}\rangle\,. \tag{2.9}$$

Definition 4. The theory is said to be invariant under a *time reversal* if there exists an anti-unitary operator

[10] Any non-Hermitian field can always be decomposed into a sum of two Hermitian fields. The use of Hermitian fields simplifies some of the general discussions, and there is no loss of generality.

$\mathcal{T}$ such that

$$\mathcal{T}\phi(\mathbf{r},t)\mathcal{T}^{-1}=\tilde{u}_0{}^T\phi(\mathbf{r},-t),$$
$$\mathcal{T}\psi_\mu(\mathbf{r},t)\mathcal{T}^{-1}=\tilde{u}_{1/2}{}^T[\gamma_1\gamma_2\gamma_3\psi(-\mathbf{r},t)]_\mu, \quad (2.10)$$

and

$$\mathcal{T}H\mathcal{T}^{-1}=H,$$

where in order to satisfy Eqs. (2.1)–(2.3), the matrices $u_0{}^T$ and $u_{1/2}{}^T$ must both be real and orthogonal.

Under $\mathcal{T}$, the state $|\text{vac}\rangle$ remains the same vacuum state, hence

$$\mathcal{T}|\text{vac}\rangle=e^{ia}|\text{vac}\rangle \quad (2.11)$$

where e^{ia} is a phase factor. Since $\mathcal{T}$ is an anti-unitary operator, independently of the phase factor e^{ia} we must have

$$\mathcal{T}^2|\text{vac}\rangle=|\text{vac}\rangle. \quad (2.12)$$

For any given matrices $u_0{}^T$ and $u_{1/2}{}^T$, if $\mathcal{T}$ satisfies (2.10), so does $e^{-ia}\mathcal{T}$. Thus we can always choose the phase factor so that

$$\mathcal{T}|\text{vac}\rangle=|\text{vac}\rangle. \quad (2.13)$$

[An alternative way to transform away the phase factor e^{ia} in Eq. (2.11) is to adopt $e^{\frac{1}{2}ia}|\text{vac}\rangle$ as the vacuum state.]

We note that if a unitary operator $\mathcal{P}$ satisfies the definition of the space-inversion operator, then any member in the set of unitary operators $\{S\mathcal{P}\}$ also satisfies the definition of space inversion, where S is any internal symmetry operator. Conversely, if $\mathcal{P}$ and $\mathcal{P}'$ both satisfy the definition of space inversion, then $\mathcal{P}'\mathcal{P}^{-1}$ must be a member of the internal symmetry group $\mathcal{G}$. Thus, the set $\{S\mathcal{P}\}$ contains all possible solutions of the space-inversion operators. Similarly, if an anti-unitary operator $\mathcal{T}$ satisfies the definition of the time-reversal operator, so does any member in the set $\{S\mathcal{T}\}$. The set $\{S\mathcal{T}\}$ also contains all possible time-reversal operators.

In the following, we assume that the local-field theory under consideration is invariant under the inhomogeneous proper Lorentz transformation. The *CPT* theorem[7] states that the theory must also be invariant under an anti-unitary operator $\mathcal{J}$ where

$$\mathcal{J}\phi(\mathbf{r},t)\mathcal{J}^{-1}=+\phi(-\mathbf{r},-t),$$
$$\mathcal{J}\psi_\mu(\mathbf{r},t)\mathcal{J}^{-1}=[i\gamma_5\psi(-\mathbf{r},-t)]_\mu, \quad (2.14)$$

and

$$\mathcal{J}H\mathcal{J}^{-1}=H. \quad (2.15)$$

The anti-unitary operator $\mathcal{J}$ is usually called the *CPT* operator. From (2.15), it follows that

$$\mathcal{J}|\text{vac}\rangle=e^{i\theta}|\text{vac}\rangle.$$

Consequently,

$$\mathcal{J}^2|\text{vac}\rangle=|\text{vac}\rangle. \quad (2.16)$$

The phase factor $e^{i\theta}$ can be transformed to unity by either transforming $\mathcal{J}\to e^{-i\theta}\mathcal{J}$ or by using $e^{\frac{1}{2}i\theta}|\text{vac}\rangle$ as the vacuum state. For convenience, we adopt the choice

$$\mathcal{J}|\text{vac}\rangle=|\text{vac}\rangle. \quad (2.17)$$

The operator $\mathcal{J}^2$ is unitary, and it satisfies

$$\mathcal{J}^2\phi(\mathbf{r},t)\mathcal{J}^{-2}=+\phi(\mathbf{r},t),$$
$$\mathcal{J}^2\psi(\mathbf{r},t)\mathcal{J}^{-2}=-\psi(\mathbf{r},t);$$

therefore, we may write

$$\mathcal{J}^2=(-1)^{2J}. \quad (2.18)$$

Any state of total angular momentum J is an eigenstate of $(-1)^{2J}$; the eigenvalue is $+1$ or -1 depending on whether J is an integer or a half-integer. The operator $\mathcal{J}^2$ is, by definition, an element of the internal-symmetry group $\mathcal{G}$.

By using the above definitions of $\mathcal{P}$, $\mathcal{T}$, $\mathcal{J}$ and S, one can easily verify that

$$\mathcal{J}S=S\mathcal{J}, \quad (2.19)$$
$$\mathcal{J}\mathcal{P}=(-1)^{2J}(\mathcal{P}\mathcal{J}), \quad (2.20)$$

and

$$\mathcal{J}\mathcal{T}=(-1)^{2J}(\mathcal{T}\mathcal{J}). \quad (2.21)$$

The operators $\mathcal{P}^2$, $\mathcal{T}^2$, $\mathcal{P}S\mathcal{P}^{-1}$, and $\mathcal{T}S\mathcal{T}^{-1}$ are all members of the internal-symmetry group $\mathcal{G}$, the operator $(\mathcal{P}\mathcal{J})$ belongs to the set $\{S\mathcal{T}\}$ and the operator $(\mathcal{T}\mathcal{J})$ belongs to the set $\{S\mathcal{P}\}$.

Theorem 1. If $\mathcal{P}$ exists, then $\mathcal{T}$ exists, and vice versa. Furthermore, for every given $\mathcal{P}$ there exists a $\mathcal{T}$ in the set $\{S\mathcal{T}\}$ such that

$$\mathcal{T}^2=\mathcal{P}^2, \quad (2.22)$$

and

$$\mathcal{T}\mathcal{P}=(-1)^{2J}\mathcal{P}\mathcal{T}. \quad (2.23)$$

Proof. For any given $\mathcal{P}$ in the set $\{S\mathcal{P}\}$, we can choose $\mathcal{T}=\mathcal{J}\mathcal{P}$. By using the definitions of $\mathcal{T}$, $\mathcal{P}$ and $\mathcal{J}$, we find

$$u_J{}^P=u_J{}^T, \quad (2.24)$$

where $J=0$ or $\frac{1}{2}$. Theorem 1, then, follows. So far, only the spin-0 and spin-$\frac{1}{2}$ fields are considered. As we shall see, Theorem 1 is valid for any Lorentz-invariant local-field theory.

2. Generalizations to Higher Spin Fields

All the above considerations can be readily applied to higher spin fields. Let $\chi_1, \chi_2, \cdots, \chi_n$ be n Hermitian field operators. Each χ_a describes a spin-J field. The general definitions of S, $\mathcal{P}$, $\mathcal{T}$ can be most easily obtained by noting that the formal transformation properties of the spin-J fields are the same as that of the appropriate derivatives (or the appropriate sum of derivatives) of the spin-0, or spin-$\frac{1}{2}$, fields. For example, the transformation properties of a set of n $J=1$ fields $[\chi_a(x)]_\mu$ are formally the same as those of $\partial\phi_a(x)/\partial x_\mu$ where $a=1, \cdots, n$ and ϕ_a is a spin-0 field. The definitions of S, $\mathcal{P}$, $\mathcal{T}$, and $\mathcal{J}$ are the same as before except that Eqs.

(2.4), (2.7), (2.10), and (2.14) are replaced by

$$S\chi_\mu(\mathbf{r},t)S^{-1}=\tilde{s}_1\chi_\mu(\mathbf{r},t)\,, \tag{2.25}$$

$$\mathcal{P}\chi_\mu(\mathbf{r},t)\mathcal{P}^{-1}=\tilde{u}_1{}^P[-\eta_{\mu\nu}\chi_\nu(-\mathbf{r},\,t)]\,, \tag{2.26}$$

$$\mathcal{T}\chi_\mu(\mathbf{r},t)\mathcal{T}^{-1}=\tilde{u}_1{}^T[\eta_{\mu\nu}\chi_\nu(\mathbf{r},\,-t)]\,, \tag{2.27}$$

$$\mathcal{J}\chi_\mu(\mathbf{r},t)\mathcal{J}^{-1}=-\chi_\mu(-\mathbf{r},\,-t)\,, \tag{2.28}$$

where s_1, $u_1{}^P$ and $u_1{}^T$ are all $(n\times n)$ real orthogonal matrices,

$$\chi_\mu=\begin{bmatrix}(\chi_1)_\mu\\ \vdots\\ (\chi_n)_\mu\end{bmatrix},$$

μ, ν vary from 0 to 3, $(\chi_a)_\mu=(\chi_a)_\mu{}^\dagger$, and $\eta_{\mu\nu}=1$ if $\mu=\nu\neq0$, -1 if $\mu=\nu=0$, and zero otherwise. [Throughout, we adopt $x_0=t$. Sometimes μ, ν may vary from 1 to 4. In such cases, $x_4=ix_0=it$.]

All previous considerations can be directly applied to the $J=1$ case, and in a similar way to any other spin-J case. The phases of these operators S, $\mathcal{P}$, $\mathcal{T}$, and $\mathcal{J}$ will always be chosen so that

$$S|\text{vac}\rangle=\mathcal{P}|\text{vac}\rangle=\mathcal{T}|\text{vac}\rangle=\mathcal{J}|\text{vac}\rangle=|\text{vac}\rangle. \tag{2.29}$$

One can readily verify that, for any spin-J case, Eqs. (2.18)-(2.21) remain valid; and for any given $\mathcal{P}$, the general validity of Theorem 1 can be established by choosing the element $\mathcal{T}=\mathcal{J}\mathcal{P}$ in the set $\{S\mathcal{T}\}$. The theory is, then, invariant under the group

$$\{\mathcal{G},\mathcal{G}\mathcal{P},\mathcal{G}\mathcal{T},\mathcal{G}\mathcal{J}\}. \tag{2.30}$$

The algebraic relations between $\{S\mathcal{T}\}$ and $\mathcal{G}$ are completely determined by those between $\{S\mathcal{P}\}$ and $\mathcal{G}$, and vice versa. Thus, the symmetry property of any Lorentz-invariant local-field theory can be studied in two steps by examining (i) the structure of the internal-symmetry group $\mathcal{G}$ and (ii) if $\mathcal{P}$ exists, the algebraic relations between $\{S\mathcal{P}\}$ and $\mathcal{G}$.

We note that while $\mathcal{P}$ is not a member of $\mathcal{G}$, both $\mathcal{P}^2$ and $\mathcal{P}S\mathcal{P}^{-1}$ are. The set of operators $\{S,S\mathcal{P}\}$, therefore, form a group $\mathcal{E}$ which is a minimal group extension of $\mathcal{G}$. The mathematical problem of classifying all possible minimal extensions of a given group $\mathcal{G}$ will be discussed in the next section.

III. MINIMAL EXTENSIONS

Consider a given group $\mathcal{G}$ of finite order n. The order of any group $\mathcal{E}$ which contains $\mathcal{G}$ as a subgroup must be a multiple of n. If the order is $2n$ we call $\mathcal{E}$ a "minimal extension."

Putting it differently, $\mathcal{G}$ possesses in $\mathcal{E}$ only one coset, which is therefore a right- and left-coset at the same time; $\mathcal{G}$ is therefore an invariant or normal subgroup. This definition does not require $\mathcal{G}$ to be of finite order and is the one preferred in the following.

An obvious minimal extension is the direct product of $\mathcal{G}$ by the cyclical group Z_2, consisting of the elements 1 and ρ, with $\rho^2=1$. There are, however, in general also other "minimal extensions" and our purpose is to analyze them.

Let us designate by Latin characters

$$a,\ b,\ c,\ \cdots \tag{3.1}$$

the elements of the subgroup $\mathcal{G}$. Let ρ be an arbitrarily picked element of the coset of $\mathcal{G}$. The elements of the coset are then

$$a\rho,\ b\rho,\ c\rho,\ \cdots. \tag{3.2}$$

We could also write them as ρa, ρb, $\cdots$ etc. but $\rho a=a'\rho$ where, in general, $a'\neq a$. In fact, $a'=\rho a\rho^{-1}$ defines an automorphism F of the group $\mathcal{G}$; we write

$$a'\equiv Fa=\rho a\rho^{-1}. \tag{3.3}$$

In that notation, F appears as an "operator" on the elements of $\mathcal{G}$. An automorphism, of course, preserves the group laws

$$Fa\cdot Fb=F(ab);\quad Fa^{-1}=(Fa)^{-1}. \tag{3.4}$$

In addition, the square of ρ must be an element of $\mathcal{G}$, which we designate by f:

$$\rho^2=f. \tag{3.5}$$

Thus, to a given minimal extension $\mathcal{E}$, we associate an automorphism F and an element f of the group $\mathcal{G}$. However, F and f are not uniquely determined by $\mathcal{E}$, since they depend on the choice of the element ρ. If we replace ρ by another element ρ' of the set (3.2), say $\rho'=g\rho$, the automorphism F is replaced by F'

$$F'a=\rho'a\rho'^{-1}=g(\rho a\rho^{-1})g^{-1}=g(Fa)g^{-1}.$$

The transformation

$$a\rightarrow gag^{-1} \tag{3.6}$$

is an "inner automorphism" of $\mathcal{G}$, which we designate by I_g and treat as an operator: $a\rightarrow I_g a$. We then see that changing ρ to $\rho'=g\rho$ induces the transformation[11]

$$F'=I_gF. \tag{3.7}$$

Likewise, $f=\rho^2$ becomes $f'=\rho'^2=g\rho g\rho=g(\rho g\rho^{-1})\rho^2$. One has, therefore,

$$f'=g(Fg)f. \tag{3.8}$$

Obviously, the pair F', f' is associated to the extension $\mathcal{E}$ in the same sense as the pair F, f.

We now ask: given an automorphism F and an element f of the group $\mathcal{G}$, is there a minimal extension $\mathcal{E}$, to which F and f are associated in the way described? We first notice that F and f are subject to certain conditions. The transformation F^2 must be an inner automorphism:

$$F^2a=\rho(\rho a\rho^{-1})\rho^{-1}=\rho^2a\rho^{-2}=faf^{-1}.$$

[11] One says that F and F' differ only by an inner automorphism or "belong to the same class." An inner automorphism thus belongs to the same class as the "identity automorphism:" $a\rightarrow a$.

In fact, F^2 is more precisely the automorphism generated by f

$$F^2 = I_f. \tag{3.9}$$

Moreover, the element f is obviously a fixed-point of F:

$$Ff = \rho f \rho^{-1} = f \tag{3.10}$$

since $f = \rho^2$. Thus F and f cannot be chosen arbitrarily. We can show, however, that an $\mathcal{E}$ exists, if (3.9) and (3.10) are satisfied.

To begin with, the structure of $\mathcal{E}$ is completely defined once the subgroup [i.e., the set (3.1) and the corresponding multiplication table] and the associated pair F, f are given. By definition, $\mathcal{E}$ consists of the subgroup $\mathcal{G}$ and its coset (3.2), where ρ satisfies Eq. (3.5). We may regard $a\rho$, $b\rho$, $\cdots$ etc. as "symbols" for the elements of the coset. The multiplication table for $\mathcal{E}$ follows from Eqs. (3.3) and (3.5), since F is given. In fact, we have

$$a \cdot b\rho = (ab)\rho, \tag{3.11a}$$

$$a\rho \cdot b = a(\rho b \rho^{-1})\rho = (a \cdot Fb)\rho. \tag{3.11b}$$

The right-hand sides are elements of the set (3.2) unambiguously defined by the multiplication table for $\mathcal{G}$. Finally, the product of two elements of the coset is

$$a\rho \cdot b\rho = a(\rho b \rho^{-1})\rho^2 = a \cdot Fb \cdot f, \tag{3.12}$$

which is the product of three elements of set (3.1) and thus belongs to $\mathcal{G}$.

To complete the proof that $\mathcal{E}$ exists, we must show that the multiplication law thus defined is associative and that each element of $\mathcal{E}$ has an inverse. Each element a of (3.1) has, of course, an inverse a^{-1} since $\mathcal{G}$ is a group; an element $a\rho$ of (3.2) has the inverse $\bar{a}\rho$ where $\bar{a} = f^{-1} \cdot Fa^{-1}$. One easily verifies, in fact, by means of (3.12), (3.4), and (3.10), that

$$a\rho \cdot \bar{a}\rho = a \cdot F\bar{a} \cdot f = a \cdot Ff^{-1} \cdot F^2 a^{-1} \cdot f$$
$$= a \cdot f^{-1} \cdot I_f a^{-1} \cdot f = 1, \tag{3.13a}$$

$$\bar{a}\rho \cdot a\rho = \bar{a} \cdot Fa \cdot f = f^{-1} \cdot Fa^{-1} \cdot Fa \cdot f = 1, \tag{3.13b}$$

where "1" is the unit element in the set (3.1). We verify the associative law for the product of three elements of (3.2) and leave the other cases to the reader. We have, because of (3.12), (3.4), and (3.9)

$$(a\rho \cdot b\rho) \cdot c\rho = (a \cdot Fb \cdot f) \cdot c\rho = (a \cdot Fb \cdot f \cdot c)\rho,$$

$$a\rho \cdot (b\rho \cdot c\rho) = a\rho \cdot (b \cdot Fc \cdot f) = aF(b \cdot Fc \cdot f)\rho$$
$$= (a \cdot Fb \cdot F^2 c \cdot Ff)\rho,$$
$$= (a \cdot Fb \cdot f \cdot c)\rho.$$

The result is the same. That the minimal extension thus defined is associated to the pair F, f is, of course, trivally implied by the construction.

We have thus seen that a pair F, f satisfying conditions (3.9) and (3.10) defines a minimal extension. Two pairs F, f and F', f' related by the transformation (3.7), (3.8) where g is some element of (3.1) may be said to define equivalent extensions. In fact, the groups $\mathcal{E}$ and $\mathcal{E}'$ constructed according to the above procedure only differ in the choice of the element ρ in the coset (3.2).

Our problem is to classify all possible *inequivalent* extensions. To this end we may use a transformation (3.7) and (3.8) to reduce F, f to the simplest possible form. Firse we make a major distinction between case A: F is an inner automorphism of $\mathcal{G}$, or case B: F is an outer automorphism.

Case A: One can choose g in Eq. (3.7) so that F is transformed into the identity automorphism ϵ. Let us therefore assume that $F = \epsilon$ to begin with. Condition (3.10) is now satisfied by any element of the group, but (3.9) tells us that $I_f = \epsilon$, i.e., that f commutes with all elements of $\mathcal{G}$ (belongs to the center). A possible value of f is, of course, $f = 1$; also if $f = f_1^2$ where f_1 is another element of the center, we can apply a transformation (3.7), (3.8) with $g = f_1^{-1}$, which leaves $F = \epsilon$ unchanged, and transforms f to $f = 1$. Thus, after all transformations we may have case A_1: $F = \epsilon$, $f = 1$. This means $\rho^2 = 1$ and ρ commutes with all elements of the group. This is the obvious solution: the direct product of $\mathcal{G}$ by the cyclic group of order 2, consisting of the elements 1 and ρ, with $\rho^2 = 1$.

There is also case A_2: $F = \epsilon$, $f = f_1^2$. In this case, there is an f_1 satisfying this equation outside the center, but not one in the center. Then by using a transformation (3.7), (3.8), with $g = f_1^{-1}$, we transform to the values

$$F = I_g, \quad f = 1.$$

Here I_g is an inner automorphism generated by an element g such that g^2 is an element of the center which is different from the square of any element of the center (in particular $g^2 \neq 1$). The product laws (3.11) and (3.12) can now be summarized by writing

$$a\alpha \cdot b\beta = (aF_\alpha b)\alpha\beta, \tag{3.14}$$

where α and β are two-valued symbols with the possible values 1 and ρ, $F_\alpha = \epsilon$, if $\alpha = 1$, and $= F$ if $\alpha = \rho$. This solution is known as a semidirect product. Another case of semidirect product will be encountered later.

Finally, we have case A_3: $F = \epsilon$, $f \neq f_1^2$. That is, f belongs to the center; there is no "square root of f" in the group $\mathcal{G}$. Clearly the existence of Cases A_2 and A_3 depends on whether the center of $\mathcal{G}$ has elements that have a square root only outside the center, or no square root at all.

Case B: Clearly F cannot be transformed to $F = \epsilon$, but only to other members of the same class[11] of outer automorphisms. According to (3.9), F^2 must be an inner automorphism. If this is true for an element of a certain class, it is also true for every other element of the same class. In fact, if $F = I\psi$ where I is an inner automorphism, $F^2 = I\psi I\psi$. But: $\psi I \psi^{-1} = I'$ which is another inner automorphism, since the inner automorphisms are an invariant subgroup of the group of all the automorphisms; therefore, $F^2 = II'\psi^2 = I''\psi^2$. Hence the assertion. (We have in fact proved that the squares of all elements of the same class belong to one class.) Therefore, there is not in general a solution for every class of outer automorphism.[12] We shall assume from now on that F belongs to an admissible class. In particular, there may be in the class an F such that $F^2=\epsilon$, the identity automorphism. Then, clearly, $f=1$ is a possible solution of Eqs. (3.9) and (3.10). We have then case B_1: F is an outer automorphism, such that $F^2=\epsilon$; $f=1$. This case leads to a semidirect product, with a multiplication law exactly similar to Eq. (3.14). Only the nature of F is different.

There is next case B_2: F is an outer automorphism and $F^2=\epsilon$; f satisfies (3.10), but it cannot be transformed to unity. The element f is an element of the center. Last, we have case B_3: F and f satisfy Eqs. (3.9) and (3.10). By using transformation (3.8) f cannot be transformed to any element of the center. Consequently, F^2 cannot be transformed to ϵ.

In the following sections, we will encounter several examples of cases A_1, A_2 and B_1, B_2, but not A_3 and B_3.

IV. SOME EXAMPLES

Before discussing the realistic cases, we first illustrate the results of the previous sections by considering some simple examples.

1. Example: $\mathcal{G}=U_1$

The elements of the U_1 group are $e^{i\theta}$ where θ varies from $-\pi$ to $+\pi$. The U_1 group is Abelian; therefore, it has no other inner automorphism except the identity automorphism. For problems of physical interest, we consider only the automorphism

$$e^{i\theta} \to e^{ig(\theta)}, \tag{4.1}$$

where $g(\theta)$ is a continuous function of θ. Since $e^{i\theta_1}\cdot e^{i\theta_2}=e^{i(\theta_1+\theta_2)}$, we must have

$$g(\theta_1)+g(\theta_2)=g(\theta_1+\theta_2). \tag{4.2}$$

Thus, $g(\theta)$ is a linear and homogeneous function of θ; i.e.,

$$g=K\theta. \tag{4.3}$$

In order to satisfy Eq. (3.9),

$$K^2=1. \tag{4.4}$$

The group U_1 has only one continuous outer automorphism $e^{i\theta}\to e^{-i\theta}$, and in order to satisfy Eq. (3.10) f must be $+1$ or -1.

The minimal extension $\mathcal{E}$ of U_1 falls into the following three cases:

$$A_1:\quad F=\epsilon,\quad f=1;$$

$$B_1:\quad F^2=\epsilon,\quad f=1;$$

$$B_2:\quad F^2=\epsilon,\quad f=-1.$$

To study the physical content of each of these different cases, let us consider a local-field theory whose internal-symmetry group consists *only* of U_1. To each element $e^{i\theta}$ there is an internal-symmetry operator $S(\theta)$. It is convenient to define the Hermitian operator Q by

$$Q=-i(dS/d\theta)\quad \text{at}\quad \theta=0. \tag{4.5}$$

It follows from the group property of U_1 that

$$S(\theta)=e^{iQ\theta}. \tag{4.6}$$

The operator Q commutes with the Hamiltonian H, and since $S(0)=S(2\pi)$ the eigenvalues of Q must be integers. The automorphism $Fe^{i\theta}=e^{ig}$ becomes $\mathcal{P}S(\theta)\mathcal{P}^{-1}=S(g)$. The three cases of minimal extensions can be written in the following alternative forms:

$$A_1:\quad \mathcal{P}Q-Q\mathcal{P}=0, \tag{4.7}$$

$$\mathcal{P}^2=1; \tag{4.8}$$

$$B_1:\quad \mathcal{P}Q+Q\mathcal{P}=0, \tag{4.9}$$

$$\mathcal{P}^2=1; \tag{4.10}$$

$$B_2:\quad \mathcal{P}Q+Q\mathcal{P}=0, \tag{4.11}$$

$$\mathcal{P}^2=S(\pi)=(-1)^Q. \tag{4.12}$$

Case A_1 corresponds to a theory which, in the usual language, is invariant under $\mathcal{P}=P$ and $\mathcal{T}=CT$. The theory is, by assumption, not invariant under the usual C where C, should it exist, must anticommute with Q.

The operator $\mathcal{I}^2$ must be a member of the center of the internal symmetry group $\mathcal{G}$ which is assumed to be U_1. Assuming that the theory contains some half-integer spin fields, then by using Eq. (2.18), all half-integer spin-particle states in this theory must belong to states of odd Q, and all integer spin-particle states belong to that of even Q. Otherwise, $\mathcal{G}$ cannot be U_1. Thus,

$$\mathcal{I}^2=S(\pi)=(-1)^Q=(-1)^{2J}. \tag{4.13}$$

Case B_2 corresponds to a usual theory which is invariant under $\mathcal{P}=CP$ and $\mathcal{T}=T$. For both cases A_1 and B_2, the *Hermitian* field operators associated with $Q\neq 0$ must exist in pairs, say ψ_1 and ψ_2. Under $S(\theta)$, the ψ_1 and ψ_2 transform as

$$S(\theta)\psi(x)S^{-1}(\theta)=\mathfrak{s}\psi(x), \tag{4.14}$$

$$\psi=\begin{pmatrix}\psi_1\\ \psi_2\end{pmatrix}, \tag{4.15}$$

where

$$s=\exp(iq\tau_2\theta), \tag{4.16}$$

$$\tau_2=\begin{pmatrix}0 & -i\\ i & 0\end{pmatrix}, \tag{4.17}$$

[12] It is known that the classes may be considered as elements of a factor group A/G where A is the group of automorphisms, G the subgroup of inner automorphisms. The problem reduces, therefore, to finding, in this factor group, the classes that are "square roots of unity" or the involuntary elements of the factor group.

and $q\neq 0$ is an eigenvalue of Q. The states of $Q=q$ are degenerate with those of $Q=-q$, and this degeneracy[13] is independent of $\mathcal{P}$, or $\mathcal{T}$, invariance.

Case B_1 is an unusual case.[14] An explicit model is given in Appendix B, in which the Hermitian field operators of half-integer spins must occur in double pairs. The theory now has an additional degeneracy because of $\mathcal{P}$, or $\mathcal{T}$, invariance.

2. Example: $\mathcal{G}=\{U_2, U_2\cdot C\}$

Next, we consider another example where the internal symmetry group contains an SU_2 group whose generators are called the "isospin" $\mathbf{I}$, a U_1 group whose generator is called the "hypercharge" Y, and a discrete element called the "charge-conjugation" operator C. The operators $\mathbf{I}$, Y, and C satisfy the same commutation relations as those of the corresponding operators in the realistic case:

$$[I_i,I_j]=i\epsilon_{ijk}I_k, \tag{4.18}$$

$$[I_i,Y]=0, \tag{4.19}$$

$$[G,I]=0, \tag{4.20}$$

$$GY+YG=0, \tag{4.21}$$

and

$$G=C\exp(i\pi I_2),\quad C^2=1, \tag{4.22}$$

where i, j, k vary from 1 to 3, $\epsilon_{ijk}=+1$, or -1, or 0 depending on whether (ijk) is an even permutation of (123), or an odd permutation, or otherwise.

The element of the U_2 group is represented by $\exp(i\mathbf{I}\cdot\boldsymbol{\theta}+iY\alpha)$ where $\boldsymbol{\theta}$ and α are real parameters. In order that it is a U_2 group and not the direct product $SU_2\times U_1$, we assume that[15] all field operators with half-integer I have odd Y; i.e.,

$$(-1)^Y=(-1)^{2I}. \tag{4.23}$$

The elements of the U_2 group and the coset $U_2C=U_2G$ form the internal symmetry group $\mathcal{G}=\{S\}=\{U_2,U_2G\}$.

The group $\mathcal{G}$ is a minimal extension (Case B_1) of the U_2 group.

In order to study the minimal extension of $\mathcal{G}$ due to $\mathcal{P}$ invariance, we first establish the following Theorem:

Theorem 2. If the theory is invariant under the space inversion, then among the set $\{S\mathcal{P}\}$ there is a space inversion operator $\mathcal{P}$ which satisfies

$$[\mathcal{P},\mathbf{I}]=[\mathcal{P},Y]=[\mathcal{P},C]=0.$$

A consequence of $[\mathcal{P},\mathbf{I}]=0$ is that different members of the same isospin multiplet must have the same transformation under $\mathcal{P}$.

Proof. Let us assume that at least in the neighborhood of $\boldsymbol{\theta}=0$ and $\alpha=0$, the transformed operator $\mathcal{P}\exp(i\mathbf{I}\cdot\boldsymbol{\theta}+iY\alpha)\mathcal{P}^{-1}$ is continuous and differentiable in the parameters θ_i and α. In this neighborhood, we have

$$\mathcal{P}\exp(i\mathbf{I}\cdot\boldsymbol{\theta}+iY\alpha)\mathcal{P}^{-1}=\exp(i\mathbf{I}\cdot\boldsymbol{\theta}'+iY\alpha'). \tag{4.24}$$

By differentiating with respect to θ_i and α, and by setting $\boldsymbol{\theta}=0$ and $\alpha=0$, (4.24) becomes

$$\mathcal{P}I_i\mathcal{P}^{-1}=a_{ij}I_j+b_iY, \tag{4.25}$$

and

$$\mathcal{P}Y\mathcal{P}^{-1}=c_iI_i+dY, \tag{4.26}$$

where

$$a_{ij}=(\partial\theta_j'/\partial\theta_i)_0,\quad b=(\partial\alpha'/\partial\theta_i)_0,$$
$$c_i=(\partial\theta_i'/\partial\alpha)_0,\quad d=(\partial\alpha'/\partial\alpha)_0$$

are all real numbers, and all repeated indices are to be summed over. By using (4.25) and the commutation relation (4.18), we find

$$a_{li}a_{mj}\epsilon_{ijk}I_k=\epsilon_{lmn}(a_{nk}I_k+b_nY).$$

Thus,

$$b_n=0,$$

and the (3×3) matrix (a_{ij}) must be real orthogonal and with determinant$=1$. Consequently, there exists an element $\exp(i\mathbf{I}\cdot\boldsymbol{\theta}_a)$ in $\mathcal{G}$ such that

$$\exp(i\mathbf{I}\cdot\boldsymbol{\theta}_a)I_i\exp(-i\mathbf{I}\cdot\boldsymbol{\theta}_a)=a_{ij}I_j.$$

By choosing $\exp(-i\mathbf{I}\cdot\boldsymbol{\theta}_a)\mathcal{P}$ as $\mathcal{P}$, we establish

$$[\mathcal{P},I_i]=0. \tag{4.27}$$

The commutation relation (4.19) requires that

$$c_i=0$$

in Eq. (4.26); i.e.,

$$\mathcal{P}Y\mathcal{P}^{-1}=dY. \tag{4.28}$$

Now, $\mathcal{P}^2$ is a member of $\mathcal{G}$. Since the square of any element in $\mathcal{G}$ commutes with Y, so must $(\mathcal{P}^2)^2$; therefore, $d^4=1$. The real constant d must be $+1$ or -1. If $d=+1$, $\mathcal{P}$ commutes with Y. If $d=-1$, we can select $G\mathcal{P}$ as the new $\mathcal{P}$. Thus,

$$[\mathcal{P},Y]=0. \tag{4.29}$$

The operator $\mathcal{P}G\mathcal{P}^{-1}$ anticommutes with Y, so it must belong to the set $\exp(i\mathbf{I}\cdot\boldsymbol{\theta}+iY\alpha)G$. It also commutes

---

[13] In the literature, the mass degeneracy between states of charge $Q=\pm q$ (e.g., e^+ and e^-) is often established by using invariance under $\mathcal{J}$ (i.e., the usual "CPT" operator). It is important to notice that in such a proof Q-conservation is implicitly assumed. In a Lorentz-invariant local-field theory, $\mathcal{J}$ invariance is automatically satisfied; the degeneracy between ψ_1 and ψ_2, or between states $Q=\pm q$, is insured by the invariance under the internal symmetry group U_1. Such degeneracy can be removed if one adds to the Hamiltonian an additional $\mathcal{J}$-invariant term H_1 which violates the U_1 invariance (i.e., Q conservation). As an example, we may mention the well-known case of K_1^0 and K_2^0 which are degenerate if $H_{wk}=0$, and, therefore, the strangeness S is conserved. H_{wk} does not alter the $\mathcal{J}$ invariance, but it violates S conservation. As a result, K_1^0 and K_2^0 do have different masses.

[14] If the theory contains only integer spin fields, then instead of (4.13) and (4.31) we have $\mathcal{J}^2=1$. In example 1, case B_1 corresponds to a usual theory which is invariant under $\mathcal{P}=$"CP" and $\mathcal{T}=$"T," while case B_2 corresponds to an unusual theory. Similarly, if there are only integer spin fields in the theory, then in example 2, case A_1 corresponds to a usual "C," "P," "T" separately invariant theory, and case A_2 corresponds to an unusual theory.

[15] Relations such as these have been emphasized by L. Michel, Ref. 6.

with **I**. By using Eq. (4.18), we find

$$\mathcal{P}G\mathcal{P}^{-1}=e^{iY\alpha}G. \tag{4.30}$$

We may choose $e^{-\frac{1}{2}iY\alpha}\mathcal{P}$ as the new $\mathcal{P}$, and this new $\mathcal{P}$ commutes with G. The theorem is, then, proved.

Since $\mathcal{P}$ commutes with all elements in $\mathcal{G}$, the automorphism $\mathcal{P}S\mathcal{P}^{-1}$ corresponds to $F=\epsilon$. The center of the group $\mathcal{G}$ consists of two elements: 1 and $(-1)^Y=(-1)^{2I}$. Thus, according to the discussion given in Sec. III, the minimal extension can only be one of the two cases:

case A_1

$$\mathcal{P}^2=1,$$

and

case A_2

$$\mathcal{P}^2=(-1)^Y=(-1)^{2I}.$$

The element $\mathcal{J}^2$ belongs to the center of the internal-symmetry group, where $\mathcal{J}$ is given by (2.14). Hence, it is either 1 or $(-1)^Y$. If the theory contains some half-integer spin fields, then we must have

$$\mathcal{J}^2=(-1)^{2J}=(-1)^Y; \tag{4.31}$$

i.e., all half-integer spin particles are of odd Y and all integer spin particles are of even Y. We may define

$$P=\mathcal{P}\exp(i\tfrac{1}{2}\pi Y); \tag{4.32}$$

P satisfies

$$[P,\mathbf{I}]=[P,Y]=0, \tag{4.33}$$

and

$$PC=(-1)^{2J}CP. \tag{4.34}$$

For case A_1,

$$P^2=(-1)^Y \tag{4.35}$$

and for case A_2

$$P^2=1. \tag{4.36}$$

Thus, case A_2 corresponds to a usual theory with separate C, P and T invariances. Case A_1 corresponds to an unusual theory[14] which has an additional degeneracy due to $\mathcal{P}$ or $\mathcal{T}$ invariance, somewhat like the case B_1 of the preceding example. An explicit model of such a case A_1 is given in Appendix B.

3. Remarks

1. Both case B_1 of example 1 and case A_1 of example 2 are theories in which $\mathcal{P}$ (or $\mathcal{T}$) invariance implies additional degeneracies. At present, there does not exist in nature any known example of such degeneracies. However, since $\mathcal{P}$ and $\mathcal{T}$ invariances are approximate symmetries, such degeneracy, if it exists, must also be only approximate. It is difficult to predict whether such approximate degeneracies might find some applications in the future.

2. In physical problems, one often deals with the U_n and SU_n groups. The above examples are concerned only with $n=1$ and 2. To study the general case of arbitrary n, it is useful to know the following mathematical theorem[16]:

> The group of all $(n\times n)$ unitary matrices u with determinant 1 is the SU_n group. If $u\rightarrow F(u)$ is an automorphism of SU_n and $n\geqq 3$, then there exists a v in SU_n such that for every u in SU_n
>
> $$F(u)=vuv^{-1},$$
>
> or, for every u
>
> $$F(u)=vu^*v^{-1}$$
>
> where u^* is the complex conjugate of u.

We will not discuss the detailed analysis of the minimal extensions of these groups. Such analysis is greatly simplified by using the above theorem. For example, consider a theory which satisfies $\mathcal{P}$ invariance and whose internal-symmetry group $\mathcal{G}$ is the SU_n group ($n\geqq 3$). Let the (n^2-1) generators of the SU_n group be $F_1, F_2, \cdots, F_{n^2-1}$. The above theorem shows that there must exist an element $\mathcal{P}$ in the coset $\mathcal{G}\mathcal{P}$ such that either

$$[\mathcal{P},F_i]=0, \tag{4.37}$$

for all $i=1, 2, \cdots, (n^2-1)$, or

$$\mathcal{P}F_i\mathcal{P}^{-1}=-F_i^*, \tag{4.38}$$

for all i.

Similar conclusions can be obtained if the internal symmetry group $\mathcal{G}$ is the U_n group. Let $F_1, F_2, \cdots, F_{n^2}$ be the n^2 generators of the U_n group where $F_1, F_2, \cdots, F_{n^2-1}$ satisfy the same algebraic relations as those of the SU_n group, and F_{n^2} commutes with all other F_i. If the theory satisfies $\mathcal{P}$ invariance, then there exists an element $\mathcal{P}$ in the coset $\mathcal{G}\mathcal{P}$ such that either (4.37) holds for all $i=1, 2, \cdots, (n^2-1)$, or (4.38) holds; in addition, $\mathcal{P}$ either commutes, or anticommutes, with F_{n^2}. Thus, $\mathcal{P}^2$ must belong to the center of the internal symmetry group. The detailed classification of the theory into the different cases of minimal extensions depends on the element $\mathcal{P}^2$ and the commutation relations between $\mathcal{P}$ and the various generators. Furthermore, depending on the relation between $\mathcal{P}^2$ and $\mathcal{J}^2=(-1)^{2J}$, the theory may, or may not, contain some unusual degeneracies similar to those in the above case B_1 of example 1 and case A_1 of example 2.

In a similar way and by following the arguments used in example 2, we can readily extend such analysis to the case in which the internal symmetry group $\mathcal{G}$ is itself a minimal extension of either the SU_n or the U_n group (due to the presence of other discrete internal-symmetry operators, such as the charge-conjugation operator).

---

[16] In proving this theorem, one does not have to assume that the automorphism preserves continuity. We wish to thank Professor E. Kolchin (private communication) for showing us the proof.

V. WEAK INTERACTIONS AND SU_3-INVARIANT STRONG INTERACTIONS

In this section, we consider the following Hamiltonian H:

$$H=H_{\text{free}}+H_{\text{st}}(SU_3)+H_{\text{wk}}\,, \tag{5.1}$$

where H_{wk} is the usual weak interaction, $H_{\text{st}}(SU_3)$ is the SU_3 invariant part[17] of the usual strong interaction and H_{free} is the free-particle Hamiltonian in which the masses of the different hadrons within the same SU_3 multiplet are considered to be the same, and the masses of all leptons are set to be zero; i.e.,

$$m_e=m_\mu=m_{\nu_e}=m_{\nu_\mu}=0\,. \tag{5.2}$$

The H_{wk} can be separated into three parts:

$$H_{\text{wk}}=H_{ll}+H_{lh}+H_{hh}\,, \tag{5.3}$$

where H_{ll}, H_{hh} and H_{lh} describe respectively the pure leptonic weak interaction, the pure hadronic weak interaction, and the weak interaction between leptons and hadrons. Among these, only the leptonic part is reasonably well understood:

$$H_{ll}=(G_\mu/\sqrt{2})\int j_\lambda{}^* j_\lambda d^3r\,, \tag{5.4}$$

and

$$H_{lh}=\int [J_\lambda{}^* j_\lambda+J_\lambda j_\lambda{}^*]d^3r\,, \tag{5.5}$$

where G_μ is real and denotes the usual μ-decay coupling constant,

$$j_\lambda=i\psi_e{}^\dagger\gamma_4\gamma_\lambda(1+\gamma_5)\psi_{\nu_e}+i\psi_\mu{}^\dagger\gamma_4\gamma_\lambda(1+\gamma_5)\psi_{\nu_\mu}\,, \tag{5.6}$$

$$j_\lambda{}^*=i\psi_{\nu_e}{}^\dagger\gamma_4\gamma_\lambda(1+\gamma_5)\psi_e+i\psi_{\nu_\mu}{}^\dagger\gamma_4\gamma_\lambda(1+\gamma_5)\psi_\mu\,, \tag{5.7}$$

ψ_e, ψ_μ, ψ_{ν_e}, and ψ_{ν_μ} are the field operators for e^-, μ^-, ν_e, and ν_μ, J_λ is the hadron current, $J_\lambda{}^*=J_\lambda{}^\dagger$ if $\lambda\neq 4$ and $J_4{}^*=-J_4{}^\dagger$. Apart from some general selection rules and certain matrix elements, the detailed forms of the hadron-dependent parts, J_λ and H_{hh}, are not known.

For clarity, a simplified model of the hadrons will be considered first. The symmetry properties derived for the simple model can be applied to the realistic case, and it leads to some consequences which may be tested experimentally.

1. A Simple Model

In this model, the hadrons are assumed to consist of only three *hypothetical* spin-$\frac{1}{2}$ fields ψ_1, ψ_2 and ψ_3 which correspond to p, n and Λ^0 in the Sakata model.[18] The ψ_1, ψ_2 and ψ_3 are of equal, but nonzero, masses; their strong interactions are invariant under the U_3 group of transformations between these three fields. For the weak interaction, we will consider only H_{ll} and H_{lh}; i.e.,

$$H_{\text{wk}}=H_{ll}+H_{lh}\,. \tag{5.8}$$

Furthermore, the hadron current in H_{lh} is assumed to be

$$J_\lambda=i2^{-1/2}G_\mu\psi_b{}^\dagger\gamma_4\gamma_\lambda(1+\gamma_5)\psi_a \tag{5.9}$$

where

$$\psi_a=\psi_1\,, \tag{5.10}$$

$$\psi_b=(\cos\theta)\psi_2+(\sin\theta)\psi_3\,, \tag{5.11}$$

and θ is the equivalent of the Cabibbo angle in this model.[19] It is useful to introduce

$$\psi_c=-(\sin\theta)\psi_2+(\cos\theta)\psi_3\,. \tag{5.12}$$

The fields ψ_a, ψ_b and ψ_c are all of unit baryon number, $N=1$; their charges are, respectively, $+e$, 0 and 0.

For the leptons, since only the zero-mass limit will be considered, we may require

$$\gamma_5\psi_\alpha=\psi_\alpha \tag{5.13}$$

where $\alpha=e$, μ, ν_e and ν_μ. Each ψ_α describes a two-component field. The Hamiltonian H is given by Eq. (5.1). In this model, H depends only on 4 zero-mass two-component fields ψ_e, ψ_μ, ψ_{ν_e}, ψ_{ν_μ} and 3 four-component hadron fields ψ_a, ψ_b and ψ_c. It is clear that H commutes with the charge Q, the baryon number N and the two usual lepton numbers L_e and L_μ where the eigenvalues of L_e and L_μ are given by

$$\begin{aligned} L_l&=-1 && \text{for } l^-, \nu_l\,,\\ &=+1 && \text{for } l^+, \bar\nu_l\,,\\ &=0 && \text{for all other particles}\,, \end{aligned} \tag{5.14}$$

and $l=e$ or μ; H also commutes with another Hermitian operator S' whose eigenvalues are given by

$$\begin{aligned} S'&=-1 && \text{for } c\,,\\ &=1 && \text{for } \bar c\,,\\ &=0 && \text{for all other particles}\,, \end{aligned} \tag{5.15}$$

and c, $\bar c$ refer to the particles described, respectively, by ψ_c and $\psi_c{}^\dagger$. For a general discussion of the symmetry properties, it is convenient to use, in place of Q, L_e, and L_μ, three alternative Hermitian operators L, M, and T_3:

$$L=L_e+L_\mu\,, \tag{5.16}$$

$$T_3=\tfrac{1}{2}(L_\mu-L_e)\,, \tag{5.17}$$

and

$$Q=\tfrac{1}{2}[L+M+N+S']\,. \tag{5.18}$$

The operators L, M, N, S', T_3 mutually commute.

Theorem 3. The internal symmetry group $\mathcal{G}_{\text{wk}}$ consists of a connected subgroup $\mathcal{G}_{\text{wk}}{}^0=U_2\times U_1\times U_1\times U_1$ and a coset $\mathcal{G}_{\text{wk}}{}^0C_{\text{wk}}$:

$$\mathcal{G}_{\text{wk}}=\{\mathcal{G}_{\text{wk}}{}^0,\mathcal{G}_{\text{wk}}{}^0C_{\text{wk}}\} \tag{5.19}$$

---

17 M. Gell-Mann, Phys. Rev. **125**, 1067 (1962); Y. Ne'eman, Nucl. Phys. **26**, 222 (1961).

18 Z. Maki M. Nakagawa, Y. Ohnuki, and S. Sakata, Progr. Theoret. Phys. (Kyoto) **23**, 1174 (1960).

19 N. Cabibbo, Phys. Rev. Letters **10**, 531 (1963).

where the element of $\mathcal{G}_{\rm wk}{}^0$ is given by

$$\exp[i(\mathbf{T}\cdot\boldsymbol{\theta}+L\xi+M\eta+N\zeta+S'\phi)],$$

$\boldsymbol{\theta}$, ξ, η, ζ, ϕ are all real parameters. The element $C_{\rm wk}$ and the generator $\mathbf{T}$ satisfy the following algebraic relations:

$$C_{\rm wk}M+MC_{\rm wk}=0, \tag{5.20}$$

$$[C_{\rm wk},\mathbf{T}]=[C_{\rm wk},L]=[C_{\rm wk},N]=[C_{\rm wk},S']=0, \tag{5.21}$$

$$[T_i,T_j]=i\epsilon_{ijk}T_k, \tag{5.22}$$

$$C_{\rm wk}{}^2=1, \tag{5.23}$$

and $\mathbf{T}$ commutes with L, M, N, and S'.

The group $\mathcal{G}_{\rm wk}$ is a minimal extension of $\mathcal{G}_{\rm wk}{}^0$, and it belongs to case B_1. Table I lists the eigenvalues $\mathbf{T}^2=T(T+1)$, T_3, L, M, N, S' for the various particles in the model.

TABLE I. Quantum numbers for the various particles in the simple model. The field operators for the particles a, b, c are given by Eqs. (5.10)–(5.12).

Particles	T	T_3	L	M	N	S'
e^-	$\frac{1}{2}$	$\frac{1}{2}$	-1	-1	0	0
μ^-	$\frac{1}{2}$	$-\frac{1}{2}$	-1	-1	0	0
ν_e	$\frac{1}{2}$	$\frac{1}{2}$	-1	1	0	0
ν_μ	$\frac{1}{2}$	$-\frac{1}{2}$	-1	1	0	0
a	0	0	0	1	1	0
b	0	0	0	-1	1	0
c	0	0	0	0	1	-1

Proof. Eq. (5.13) and the anticommutation relations between the lepton fields require that any internal symmetry operator S must satisfy

$$S\psi S^{-1}=u\psi, \tag{5.24}$$

where

$$\psi=\begin{bmatrix}\psi_e\\ \psi_\mu\\ \psi_{\nu_e}\\ \psi_{\nu_\mu}\end{bmatrix},$$

and u is a (4×4) *unitary* matrix. It is convenient to introduce the matrices $\boldsymbol{\sigma}$ and $\boldsymbol{\rho}$:

$$\rho_1=\begin{bmatrix}0&0&1&0\\0&0&0&1\\1&0&0&0\\0&1&0&0\end{bmatrix},\quad \rho_2=\begin{bmatrix}0&0&-i&0\\0&0&0&-i\\i&0&0&0\\0&i&0&0\end{bmatrix},$$

$$\sigma_1=\begin{bmatrix}0&1&0&0\\1&0&0&0\\0&0&0&1\\0&0&1&0\end{bmatrix},\quad \sigma_2=\begin{bmatrix}0&-i&0&0\\i&0&0&0\\0&0&0&-i\\0&0&i&0\end{bmatrix},$$

and

$$\rho_3=-i\rho_1\rho_2,\quad \sigma_3=-i\sigma_1\sigma_2. \tag{5.25}$$

In terms of ψ and these matrices, the lepton currents can be written as

$$j_\lambda=i\psi^\dagger\gamma_4\gamma_\lambda(\rho_1+i\rho_2)\psi, \tag{5.26}$$

and

$$j_\lambda{}^*=i\psi^\dagger\gamma_4\gamma_\lambda(\rho_1-i\rho_2)\psi. \tag{5.27}$$

Since H_{lh} is a linear function of j_λ and $j_\lambda{}^*$, the invariance property $SH_{lh}S^{-1}=H_{lh}$ implies that j_λ must become a linear combination of j_λ and $j_\lambda{}^*$ under S; i.e., under the transformation (5.24),

$$u^\dagger\rho_1 u=A_{11}\rho_1+A_{12}\rho_2, \tag{5.28}$$

and

$$u^\dagger\rho_2 u=A_{21}\rho_1+A_{22}\rho_2. \tag{5.29}$$

The coefficients A_{ij} form a (2×2) matrix A. The Hermiticity conditions of ρ_1 and ρ_2 require A to be real, and their commutation relations require A to be orthogonal. Thus, A must be of the form

$$\begin{pmatrix}\cos\eta & \sin\eta\\ \mp\sin\eta & \pm\cos\eta\end{pmatrix}, \tag{5.30}$$

where η is real. Let us define u':

$$u'=u\exp(-\tfrac{1}{2}i\eta\rho_3)$$

if $\det A=+1$, and

$$u'=u[\exp(-\tfrac{1}{2}i\eta\rho_3)]\rho_1$$

if $\det A=-1$; then,

$$u'^\dagger\rho_1 u'=\rho_1,$$

and

$$u'^\dagger\rho_2 u'=\rho_2. \tag{5.31}$$

All (4×4) unitary matrices can be written as a linear function of the sixteen (4×4) matrices: 1, ρ_i, σ_j, and $\rho_i\sigma_j$. In order to satisfy (5.31), u' must be a linear function of only 1 and σ_j; consequently, u is of the form

$$u=\exp[i(\tfrac{1}{2}\boldsymbol{\theta}\cdot\boldsymbol{\sigma}+\xi+\eta\rho_3)], \tag{5.32}$$

or

$$u=\rho_1\exp[i(\tfrac{1}{2}\boldsymbol{\theta}\cdot\boldsymbol{\sigma}+\xi+\eta\rho_3)]. \tag{5.33}$$

Under the transformation (5.32), $j_\lambda\to e^{-2i\eta}j_\lambda$ and $j_\lambda{}^*\to e^{2i\eta}j_\lambda{}^*$; under (5.33), $j_\lambda\to e^{2i\eta}j_\lambda{}^*$ and $j_\lambda{}^*\to e^{-2i\eta}j_\lambda$.

For the hadron fields, it can be shown that, for the present case, we need only consider the transformations

$$S\begin{pmatrix}\psi_a\\ \psi_b\end{pmatrix}S^{-1}=v\begin{pmatrix}\psi_a\\ \psi_b\end{pmatrix}$$

and

$$S\psi_c S^{-1}=e^{i\phi}\psi_c. \tag{5.34}$$

In order that the current J_λ is transformed into a linear combination of J_λ and $J_\lambda{}^*$, the (2×2) matrix v must be of the form

$$v=\exp[-i(\eta\tau_3+\zeta)] \tag{5.35}$$

or

$$v=\tau_1\exp[-i(\eta\tau_3+\zeta)], \tag{5.36}$$

where

$$\tau_1=\begin{pmatrix}0&1\\1&0\end{pmatrix}\quad\text{and}\quad \tau_3=\begin{pmatrix}1&0\\0&-1\end{pmatrix}. \tag{5.37}$$

Under the transformation (5.35), $J_\lambda \to e^{-2i\eta}J_\lambda$ and $J_\lambda^* \to e^{2i\eta}J_\lambda^*$; under (5.36), $J_\lambda \to e^{2i\eta}J_\lambda^*$ and $J_\lambda^* \to e^{-2i\eta}J_\lambda$. Theorem 3 is, then, proved.

The discrete element C_{wk} is identified with the transformation $u=\rho_1$, $v=\tau_1$ and $\phi=0$ in (5.24) and (5.34). Under C_{wk}

$$C_{\text{wk}} j_\lambda C_{\text{wk}}^{-1} = j_\lambda^*,$$
$$C_{\text{wk}} J_\lambda C_{\text{wk}}^{-1} = J_\lambda^*. \tag{5.38}$$

While the symmetry group $\mathcal{G}_{\text{wk}}$ has been discussed in the literature,[20,21] Theorem 3 establishes that, at least for this simple model, Eq. (5.19) gives all possible internal-symmetry operators.

Next, we investigate the question of $\mathcal{P}$ and $\mathcal{T}$ invariance for the model. Let us define

$$\mathcal{P}[\psi_n(\mathbf{r},t)]_\alpha \mathcal{P}^{-1} = (i\gamma_4)_{\alpha\beta}[\psi_n^\dagger(-\mathbf{r},t)]_\beta, \tag{5.39}$$

$$\mathcal{T}[\psi_n(\mathbf{r},t)]_\alpha \mathcal{T}^{-1} = (\gamma_1\gamma_2\gamma_3)_{\alpha\beta}[\psi_n(\mathbf{r},-t)]_\beta, \tag{5.40}$$

where n denotes the various particles e, μ, ν_e, ν_μ, a, b and c; the subscripts α and β denote the spinor indices which vary from 1 to 4. The γ_μ matrices are in the Majorana representation (2.8). Applying $\mathcal{P}$ and $\mathcal{T}$ to j_λ and J_λ, one finds

$$\mathcal{P} j_\lambda(\mathbf{r},t)\mathcal{P}^{-1} = j_\lambda^\dagger(-\mathbf{r},t),$$
$$\mathcal{P} J_\lambda(\mathbf{r},t)\mathcal{P}^{-1} = J_\lambda^\dagger(-\mathbf{r},t),$$
$$\mathcal{T} j_\lambda(\mathbf{r},t)\mathcal{T}^{-1} = -j_\lambda(\mathbf{r},-t),$$

and

$$\mathcal{T} J_\lambda(\mathbf{r},t)\mathcal{T}^{-1} = -J_\lambda(\mathbf{r},-t). \tag{5.41}$$

The Hamiltonian H is, therefore, invariant under $\mathcal{P}$ and $\mathcal{T}$. (In this model, H_{st} is assumed to be invariant under $\mathcal{P}$ and $\mathcal{T}$.)

In the above discussion, because of familiarity we use the field operators ψ_e, ψ_μ, $\cdots$ which are all non-Hermitian. Identical results can be derived if, instead, the appropriate Hermitian operators are used. Such a treatment would conform more to the general notations used in Sec. II; the details are given in Appendix A.

It is convenient to introduce

$$P_{\text{wk}} \equiv C_{\text{wk}}^{-1} e^{-i\pi T_2}\mathcal{P}, \tag{5.42}$$

and

$$T_{\text{wk}} \equiv \mathcal{T} e^{i\pi T_2}. \tag{5.43}$$

Thus,

$$[P_{\text{wk}},\mathbf{T}] = [P_{\text{wk}},M] = 0, \tag{5.44}$$

$$P_{\text{wk}}L + LP_{\text{wk}} = P_{\text{wk}}N + NP_{\text{wk}} = P_{\text{wk}}S' + S'P_{\text{wk}} = 0, \tag{5.45}$$

$$P_{\text{wk}}^2 = (-1)^{N+S'}, \tag{5.46}$$

and

$$C_{\text{wk}} P_{\text{wk}} T_{\text{wk}} = \mathcal{g} \tag{5.47}$$

where $\mathcal{g}$ is given by Eq. (2.14).

The minimal extension of the internal symmetry group $\mathcal{G}_{\text{wk}}$ is

$$\mathcal{E} = \{\mathcal{G}_{\text{wk}}, \mathcal{G}_{\text{wk}}\mathcal{P}\} = \{\mathcal{G}_{\text{wk}}, \mathcal{G}_{\text{wk}} P_{\text{wk}}\} \tag{5.48}$$

which belongs to case B_2.

2. Applications of C_{wk}, P_{wk} and T_{wk} Symmetries

The above C_{wk}, P_{wk} and T_{wk} symmetries can be readily applied to the Hamiltonian (5.1) for the general case, without the triplet model of the hadrons and without the special assumptions (5.8) and (5.9). In place of Eq. (5.9), the hadron current J_μ is assumed to transform like an octet member[19] under the SU_3 transformations:

$$J_\lambda = \cos\theta (O_\lambda)_1{}^2 + \sin\theta (O_\lambda)_1{}^3, \tag{5.49}$$

where $(O_\lambda)_1{}^2$ and $(O_\lambda)_1{}^3$ transform, respectively, like π^- and K^- under the usual SU_3 group of transformations. From Eq. (5.49), it follows that J_λ^* must also transform like an octet member under the SU_3 transformations. Without any further assumption, J_λ^* might belong to a *different* octet from J_λ, and their matrix elements would not be connected by any SU_3 transformation. The requirement of C_{wk} symmetry, however, links J_λ with J_λ^*.

To illustrate this, let us consider the matrix elements of J_λ between two single-baryon states. Such matrix elements can be reduced to the usual D_μ and F_μ by using the following relation:

$$\langle B_l{}^i(p')|(O_\lambda)_m{}^j|B_n{}^k(p)\rangle = d_{lmn}{}^{ijk}U'^\dagger D_\lambda U + f_{lmn}{}^{ijk}U'^\dagger F_\lambda U, \tag{5.50}$$

where all indices vary from 1 to 3, $|B_n{}^k(p)\rangle$ is the (k,n)th member of a baryon octet with 4-momentum p_μ, $\langle B_l{}^i(p')|$ is the Hermitian conjugate of $|B_i{}^l(p')\rangle$,

$$d_{lmn}{}^{ijk} = (4/9)\delta_l{}^i\delta_m{}^j\delta_n{}^k + \delta_n{}^i\delta_l{}^j\delta_m{}^k + \delta_m{}^i\delta_n{}^j\delta_l{}^k - \tfrac{2}{3}(\delta_l{}^i\delta_n{}^j\delta_m{}^k + \delta_n{}^i\delta_m{}^j\delta_l{}^k + \delta_m{}^i\delta_l{}^j\delta_n{}^k), \tag{5.51}$$

$$f_{lmn}{}^{ijk} = \delta_n{}^i\delta_l{}^j\delta_m{}^k - \delta_m{}^i\delta_n{}^j\delta_l{}^k, \tag{5.52}$$

$\delta_j{}^i$ is the Kronecker δ symbol, U and U' are, respectively, the appropriate free Dirac spinors with 4-momentum p and p'. From Lorentz invariance, we find

$$D_\lambda = i\gamma_4[\gamma_\lambda D_1{}^V + i(p'+p)_\lambda D_2{}^V + i(p'-p)_\lambda D_3{}^V] + i\gamma_4\gamma_5[\gamma_\lambda D_1{}^A + i(p'-p)_\lambda D_2{}^A + i(p'+p)_\lambda D_3{}^A], \tag{5.53}$$

and

$$F_\lambda = i\gamma_4[\gamma_\lambda F_1{}^V + i(p'+p)_\lambda F_2{}^V + i(p'-p)_\lambda F_3{}^V] + i\gamma_4\gamma_5[\gamma_\lambda F_1{}^A + i(p'-p)_\lambda F_2{}^A + i(p'+p)_\lambda F_3{}^A], \tag{5.54}$$

where the functions $D_i{}^V$, $D_i{}^A$, $F_i{}^V$, $F_i{}^A$ $(i=1,\cdots 3)$ depend only on $q^2=(p'-p)^2$. The octet current hypothesis (5.49) is satisfied for arbitrary twelve *complex* functions $D_i{}^V$, $\cdots$, $F_i{}^A$.

---

[20] G. Feinberg and F. Gürsey, Phys. Rev. **128**, 378 (1962).

[21] T. D. Lee, Nuovo Cimento **35**, 954 (1965). See also Ref. 9.

By using Eq. (5.38), the $C_{\rm wk}$ symmetry requires that

$$D_1^\alpha,\ D_2^\alpha,\ F_1^\alpha,\ F_2^\alpha \text{ are real},$$

and

$$D_3^\alpha,\ F_3^\alpha \text{ are imaginary}, \tag{5.55}$$

where $\alpha=V$, or A. Expression (5.55), at least in principle, can be tested experimentally.

In order that $C_{\rm wk}$ commute with $H_{\rm st}(SU_3)$, we may identify the transformation of hadrons under $C_{\rm wk}$ to be the same as that under the SU_3 group element

$$\begin{bmatrix} 0 & \cos\theta & \sin\theta \\ \cos\theta & -\sin^2\theta & \sin\theta\cos\theta \\ \sin\theta & \sin\theta\cos\theta & -\cos^2\theta \end{bmatrix}, \tag{5.56}$$

where θ is the Cabibbo angle. Thus, if $C_{\rm wk}$ symmetry holds, by applying (5.56) to (5.49), one finds

$$J_\lambda{}^* = \cos\theta(O_\lambda)_2{}^1+\sin\theta(O_\lambda)_3{}^1. \tag{5.57}$$

Conversely, if Eq. (5.57) holds, then $C_{\rm wk}$ symmetry follows and the condition (5.55) is valid.

We note that in Eq. (5.57) the strangeness-conserving part $(O_\mu)_2{}^1$ of the hadron current satisfies the charge symmetry condition; i.e., $(O_\mu)_2{}^1$ is related to $(O_\mu)_1{}^2$ by a 180° isospin rotation. Consequences of the charge-symmetry condition have been discussed in the literature.[22]

The $C_{\rm wk}$ symmetry can also be applied to the pure hadronic weak interaction H_{hh}. For example, if H_{hh} transforms like an octet $\mathcal{H}_j{}^i$ under SU_3, then $C_{\rm wk}$ symmetry requires H_{hh} to be of the form

$$(\cos^2\theta)\mathcal{H}_3{}^3+(\sin^2\theta)\mathcal{H}_2{}^2-\sin\theta\cos\theta(\mathcal{H}_3{}^2+\mathcal{H}_2{}^3), \tag{5.58}$$

which connects the strangeness-nonconserving part of H_{hh} to the strangeness-conserving part. [In (5.58), the trace $\mathcal{H}_i{}^i$ is taken to be zero.] Similar results can also be obtained if different SU_3 transformation properties of H_{hh} are assumed.

The $C_{\rm wk}$ symmetry is broken by the mass difference between l^- and ν_l, by the SU_3-violating strong interaction and by the electromagnetic interaction.

At present, it remains an open question whether the weak interaction does, or does not, satisfy the $\mathcal{T}$ invariance.[23] If we assume that it does satisfy $\mathcal{T}$ invariance, then Theorem 1 states that it also satisfies $\mathcal{P}$ invariance, and we may define $P_{\rm wk}$ and $T_{\rm wk}$ by using Eqs. (5.39)–(5.43).

A consequence of $\mathcal{T}$ invariance is that the 12 form factors $D_i{}^V, \cdots, F_i{}^A$ in Eqs. (5.53) and (5.54) must all be real. Thus, if both $\mathcal{T}$ invariance and $C_{\rm wk}$ symmetry hold, we must have

$$D_3{}^V=D_3{}^A=F_3{}^V=F_3{}^A=0. \tag{5.59}$$

[22] See T. D. Lee and C. S. Wu, Ann. Rev. Nucl. Sci. **15**, 408 (1965).

[23] See, e.g., the discussions given by T. D. Lee, in *Proceedings of the Oxford International Conference on Elementary Particles, 1965* (Rutherford High Energy Laboratory, Harwell, England, 1966).

VI. APPLICATIONS TO OTHER INTERACTIONS

1. Electromagnetic Interaction

Let us consider the model Hamiltonian

$$H=H_{\rm free}+H_\gamma, \tag{6.1}$$

where H_γ is the electromagnetic interaction and $H_{\rm free}$ is the free Hamiltonian of all charged particles and the electromagnetic field. For simplicity, we assume that H_γ is given by the minimal electromagnetic interaction[24] of spin-0 and spin-$\frac{1}{2}$ charged particles only; the masses of these charged particles are all assumed to be *different* and nonzero. Each charged particle is described by two Hermitian fields, say, $\psi_{\alpha,1}$ and $\psi_{\alpha,2}$ where $\alpha=1,\cdots,N$ labels the different particles. The explicit form of (6.1) is well known; its symmetry properties will be summarized in the following.

It can be readily verified that the internal symmetry group $\mathcal{G}_\gamma$ of (6.1) must contain a subgroup

$$\mathcal{G}_\gamma{}^0 = U_1\times U_1\times\cdots\times U_1. \tag{6.2}$$

For each charged particle α there is a group $U_1=\{S_\alpha(\theta)\}$:

$$S_\alpha(\theta)\begin{pmatrix}\psi_{\beta,1}\\ \psi_{\beta,2}\end{pmatrix}S_\alpha{}^{-1}(\theta)=\exp(i\delta_{\alpha\beta}\tau_2\theta)\begin{pmatrix}\psi_{\beta,1}\\ \psi_{\beta,2}\end{pmatrix}, \tag{6.3}$$

where τ_2 is given by Eq. (4.17) and $\delta_{\alpha\beta}$ is the Kronecker δ symbol. The group $\mathcal{G}_\gamma{}^0$ is the direct product of these U_1 groups. Let Q_α be the generator of the group $\{S_\alpha(\theta)\}$. The total charge of the system is (in units of e)

$$Q=\sum Q_\alpha.$$

The group U_1 generated by Q insures that states with $Q=+q$ and $-q$ are degenerate.[13]

The Hamiltonian (6.1) conserves not only the total charge, but also each Q_α separately; it is also invariant under the usual *charge-conjugation* operator C_γ where

$$C_\gamma Q_\alpha+Q_\alpha C_\gamma=0 \tag{6.4}$$

for all α, and

$$C_\gamma{}^2=1. \tag{6.5}$$

The internal symmetry group $\mathcal{G}_\gamma$ is a minimal extension of $\mathcal{G}_\gamma{}^0$:

$$\mathcal{G}_\gamma=\{\mathcal{G}_\gamma{}^0,\mathcal{G}_\gamma{}^0C_\gamma\} \tag{6.6}$$

and the extension belongs to case B_1.

The minimal electromagnetic interactions of spin-0 and spin-$\frac{1}{2}$ particles are well known to satisfy both $\mathcal{P}$ and $\mathcal{T}$ invariances. Furthermore, there exists an element $\mathcal{P}$ in the coset $\mathcal{G}_\gamma\mathcal{P}$ which satisfies (2.7) with

[24] The minimal electromagnetic interaction of a charged particle can be generated by replacing $(\partial/\partial x_\mu)\to(\partial/\partial x_\mu)-ieA_\mu$ in its free Lagrangian, where e is its charge. For the spin-0 or spin-$\frac{1}{2}$ particles, such a minimal electromagnetic interaction is always invariant under C_γ, P_γ, and T_γ. But, this is not true for the spin-1 particles. See T. D. Lee, Phys. Rev. **140**, B967 (1965).

$u_0{}^P = u_{1/2}{}^P = 1$. Consequently, $\mathcal{P}$ also satisfies

$$[\mathcal{P}, Q_i] = [\mathcal{P}, C_\gamma] = 0, \tag{6.7}$$

and

$$\mathcal{P}^2 = (-1)^{2J}. \tag{6.8}$$

Thus, the minimal extension

$$\mathcal{E} = \{\mathcal{G}_\gamma, \mathcal{G}_\gamma \mathcal{P}\} \tag{6.9}$$

belongs to case A_2.

We note that $\mathcal{P}$ commutes with C_γ; therefore, it does not correspond to the customary choice of the parity operator. Let us define the parity operator P_γ and the time-reversal operator T_γ by

$$P_\gamma = \mathcal{P} \prod_\beta S_\beta(\pi/2), \tag{6.10}$$

and

$$C_\gamma P_\gamma T_\gamma = \mathcal{I}, \tag{6.11}$$

where $\mathcal{I}$ is given by (1.9), and the product $\prod_\beta$ extends over all half-integer spin fields. The operator P_γ satisfies

$$[P_\gamma, Q_\alpha] = 0, \tag{6.12}$$

$$P_\gamma C_\gamma = (-1)^{2J} C_\gamma P_\gamma, \tag{6.13}$$

and

$$P_\gamma{}^2 = 1. \tag{6.14}$$

The electromagnetic interaction of the two charged leptons e and μ is well known to be invariant under C_γ, P_γ, and T_γ; it is also invariant under the group $\{S_e(\theta)\} \times \{S_\mu(\theta)\}$ where $S_e(\theta)$ and $S_\mu(\theta)$ are given by Eq. (6.3) with $\alpha = e$ and μ, respectively. The electromagnetic interaction of the hadrons is found to be invariant[25] under P_γ, $\mathcal{I}$ and the various U_1 groups generated by charge, hypercharge, and baryon number. It seems esthetically appealing to assume that it is also invariant under C_γ and T_γ. However, this is only a theoretical supposition and is, as yet, without experimental basis.[26]

[25] The best experiments establishing that H_{st} and H_γ are invariant under the same parity operator, $P_{st} = P_\gamma$, are from nuclear transitions: F. Boehm and E. Kankeleit, California Institute of Technology Report No. Calt-63-13 (unpublished); Yu. G. Abov, P. A. Krupchitsky, and Yu. A. Oratovsky, *Comptes Rendus du Congrès Internationale de Physique Nucleaire, Paris, 1964*; (Editions du Centre National de la Recherche Scientifique, Paris, 1965); L. Grodzins and F. Genovese, Phys. Rev. **121**, 228 (1961); R. E. Segel *et al.*, *ibid.* **123**, 1382 (1961); D. E. Alburger *et al.*, Phil. Mag. **6**, 171 (1961); R. Haas, L. B. Leipuner, and R. K. Adair, Phys. Rev. **116**, 1221 (1959); F. Boehm and U. Hauser, Nucl. Phys. **14**, 615 (1959); D. A. Bromley *et al.*, Phys. Rev. **114**, 758 (1959).

Evidence for C_{st} and T_{st} invariances of H_{st} comes from the $\bar{p}p$ annihilation measurements [C. Baltay *et al.*, Phys. Rev. Letters **15**, 591 (1965)], the pp scattering experiment [A. Abashian and E. M. Hafner, Phys. Rev. Letters **1**, 225 (1958); C. F. Hwang, T. R. Ophel, E. H. Thorndike, and R. Wilson, Phys. Rev. **119**, 352 (1960)], and the reciprocity relations [L. Rosen and J. E. Brolley, Jr., Phys. Rev. Letters **2**, 98 (1959); D. Bodansky *et al.*, *ibid.* **2**, 101 (1959)].

[26] If the electromagnetic interaction of the hadrons is not invariant under C_γ and T_γ, then through virtual emissions and absorptions of photons all strong-interaction processes would have small C_{st}- and T_{st}-violating amplitudes (proportional to α). However, the converse statement is not true. Such a small C_{st} and T_{st} noninvariant amplitude could appear if there is a mismatch between C_{st} and C_γ, even though the electromagnetic interaction is invariant under C_γ and T_γ. For a detailed discussion, see Ref. 9.

2. Strong Interactions

We now turn to the symmetry problem of the strong interaction. Let

$$H = H_{\text{free}} + H_{st}, \tag{6.15}$$

where H_{st} contains both the SU_3-invariant and the SU_3-breaking strong interactions, and H_{free} contains the free-particle Hamiltonian of all hadrons. The masses of different particles in the same isospin multiplet are assumed to be the same. It is well known that H commutes with the isospin $\mathbf{I}$, the charge Q and the baryon number N. The internal symmetry group, denoted by $\mathcal{G}_{st}$, must contain these generators $\mathbf{I}$, Q and N. [For simplicity, we do not include the leptons, nor the electromagnetic field, in H_{free}.] It is useful to introduce the hypercharge

$$Y = 2(Q - I_3). \tag{6.16}$$

All known hadrons satisfy[15]

$$(-1)^{2I} = (-1)^Y, \tag{6.17}$$

and

$$(-1)^{2J} = (-1)^{2N}. \tag{6.18}$$

Thus, Y and $\mathbf{I}$ are the generators of a U_2 group, N is the generator of a U_1 group; the element $\mathcal{I}^2 = (-1)^{2J}$ belongs to this U_1 group. The internal symmetry group $\mathcal{G}_{st}$ must contain a subgroup $\mathcal{G}_{st}{}^0$ which is the direct product of these two groups:

$$\mathcal{G}_{st}{}^0 = U_2 \times U_1. \tag{6.19}$$

In addition, H is invariant under a discrete internal symmetry operator[27] G which satisfies the following algebraic relations:

$$G^2 = (-1)^Y, \tag{6.20}$$

$$[G, \mathbf{I}] = 0, \tag{6.21}$$

and

$$GY + YG = GN + NG = 0. \tag{6.22}$$

Sometimes, it is convenient to use C_{st}, which is related to G by

$$C_{st} = G \exp(-i\pi I_2). \tag{6.23}$$

Although it was always accepted that C_{st} anticommutes with Q, this relation has been questioned recently.[9]

We note that if

$$C_{st} Q + Q C_{st} = 0, \tag{6.24}$$

and if H does not have any further internal symmetry properties, then $\mathcal{G}_{st}$ is a minimal extension of $\mathcal{G}_{st}{}^0$:

$$\mathcal{G}_{st} = \{\mathcal{G}_{st}{}^0, \mathcal{G}_{st}{}^0 C_{st}\}, \tag{6.25}$$

where the coset $\mathcal{G}_{st}{}^0 C_{st} = \mathcal{G}_{st}{}^0 G$. Since $C_{st}{}^2 = 1$, this minimal extension belongs to case B_1.

On the other hand, if

$$C_{st} Q + Q C_{st} \neq 0, \tag{6.26}$$

[27] T. D. Lee and C. N. Yang, Nuovo Cimento **3**, 749 (1956); L. Michel, *ibid.* **10**, 319 (1953).

then the group $\mathcal{G}_{st}$ must be a different one. Let us define

$$Q_J=\tfrac{1}{2}(Q-C_{st}QC_{st}^{-1}), \quad (6.27)$$

and

$$Q_K=\tfrac{1}{2}(Q+C_{st}QC_{st}^{-1}). \quad (6.28)$$

Thus,

$$C_{st}Q_J+Q_JC_{st}=0, \quad (6.29)$$

and

$$C_{st}Q_K-Q_KC_{st}=0. \quad (6.30)$$

Since H commutes with Q and C_{st}, it must also commute with Q_J and Q_K. The internal symmetry group $\mathcal{G}_{st}$ now contains, in addition to the generators $\mathbf{I}$, N, and $Q=Q_J+Q_K$, also (Q_J-Q_K). The exact structure of $\mathcal{G}_{st}$ depends on the specific form of the model Hamiltonian. In Appendix C, a *nonderivative-coupling* model of such a strong interaction is given. For the model,

$$\mathcal{G}_{st}=\{\mathcal{G}_{st}^{0\prime},\mathcal{G}_{st}^{0\prime}C_{st}\}, \quad (6.31)$$

and

$$\mathcal{G}_{st}^{0\prime}=U_2\times U_1\times U_1, \quad (6.32)$$

where the generators of the U_2 group are $\mathbf{I}$ and $2(Q_J-I_3)$, the generators of the two U_1 groups are, respectively, N and Q_K.

There exists strong evidence[25] that $(H_{\text{free}}+H_{st})$ satisfies $\mathcal{P}$ invariance, and consequently also $\mathcal{T}$ invariance; furthermore, in the set $\{S\mathcal{P}\}$ there exists an element, called P_{st}, which is the same operator as P_γ defined by $(H_{\text{free}}+H_\gamma)$. We have

$$P_{st}=P_\gamma, \quad (6.33)$$

and it satisfies

$$[P_{st},\mathbf{I}]=[P_{st},Q]=[P_{st},N]=0, \quad (6.34)$$

$$P_{st}C_{st}=(-1)^{2J}C_{st}P_{st}, \quad (6.35)$$

and

$$P_{st}^2=1. \quad (6.36)$$

Let us define

$$\mathcal{P}=P_{st}\exp(\tfrac{1}{2}i\pi N). \quad (6.37)$$

By using (6.18) and (6.34)–(6.36), we find $\mathcal{P}$ commutes with all internal-symmetry operators, and

$$\mathcal{P}^2=(-1)^N. \quad (6.38)$$

Thus, the minimal extension

$$\mathcal{E}=\{\mathcal{G}_{st},\mathcal{G}_{st}P_{st}\} \quad (6.39)$$

belongs to case A_2.

It is customary to define T_{st} by

$$C_{st}P_{st}T_{st}=\mathcal{J}. \quad (6.40)$$

The strong interaction is invariant under C_{st}, P_{st}, and T_{st}. The weak interaction H_{wk} violates both C_{st} and P_{st} symmetries, and the electromagnetic interaction H_γ is invariant under P_{st}. The important questions, whether H_γ is invariant under C_{st} and T_{st}, and whether H_{wk} is invariant under T_{st} and $(C_{st}P_{st})$, are, at present, still unanswered.

VII. REMARKS ON C_i AND THE EXAMPLE OF SO_4 SYMMETRY

As discussed in the two preceding sections, our present knowledge is consistent with the assumption that the model Hamiltonians (5.1), (6.1), and (6.15) are separately invariant under the group

$$\{\mathcal{G}_i,\mathcal{G}_iP_i,\mathcal{G}_iT_i,\mathcal{G}_i\mathcal{J}\}, \quad (7.1)$$

where $i=$ weak, γ, and strong, respectively. The group $\mathcal{G}_i$ is the internal symmetry group, and it consists of a connected invariant subgroup $\mathcal{G}_i^0$ and a coset $\mathcal{G}_i^0C_i$:

$$\mathcal{G}_i=\{\mathcal{G}_i^0,\mathcal{G}_i^0C_i\}. \quad (7.2)$$

From the definitions of $\mathcal{P}$, $\mathcal{T}$, and $\mathcal{J}$, it is clear that in (7.1) no element in the cosets $\mathcal{G}_iP_i$, $\mathcal{G}_iT_i$, and $\mathcal{G}_i\mathcal{J}$ can be continuously connected to the unit element. The same is also true for the elements in the coset $\mathcal{G}_i^0C_i$. However, it is less transparent why it should not be possible to connect the element C_i, or the elements in the coset $\mathcal{G}_i^0C_i$, to the unit element in a continuous way. As we shall see, this is perhaps due to the particular forms of (5.1), (6.1) and (6.15), rather than due to any general underlying principle.

To illustrate this, we shall discuss some explicit strong interaction models. These models are all invariant under the usual C_{st}, P_{st} and T_{st}, but unlike in (7.2), the element C_{st} can be continuously connected to the unit element.

Let us consider the virtual transitions

$$N\rightleftarrows N+\rho, \quad (7.3)$$

and

$$N\rightleftarrows N+\eta^0 \quad (\text{or } \eta'^0), \quad (7.4)$$

where N stands for n or p and ρ for $\rho^\pm$ or ρ^0. Such transitions can be represented, at least phenomenologically, by vertices of the form

$$ig_\rho[N^\dagger(x)\gamma_4\gamma_\mu\boldsymbol{\tau}N(x)]\boldsymbol{\varrho}_\mu(x), \quad (7.5)$$

and

$$ig_\eta[N^\dagger(x)\gamma_4\gamma_5N(x)]\eta(x), \quad (7.6)$$

where g_ρ and g_η are the appropriate strong-interaction coupling constants which, because of Hermiticity, must be real, $\boldsymbol{\tau}$ are the usual (2×2) Pauli matrices for isospin [which are the same as those given by (4.17) and (5.37)],

$$N(x)=\begin{pmatrix}\psi_p(x)\\ \psi_n(x)\end{pmatrix} \quad (7.7)$$

is the field operator for n and p, $\boldsymbol{\varrho}_\mu(x)$ and $\eta(x)$ are, respectively, the field operators for ρ and η (or η'). The fields $\boldsymbol{\varrho}_\mu(x)$ and $\eta(x)$ are Hermitian, but $\psi_n(x)$ and $\psi_p(x)$ are not. It is clear that both interactions (7.5) and (7.6) commute with the isospin $\mathbf{I}$ and the baryon number N; they also satisfy the usual C_{st}, P_{st} and T_{st} invariances. As we shall see, these two interactions are also invariant under a much bigger symmetry group.

To study the full extent of the symmetry group, we follow the general procedure outlined in Sec. II and express ψ_p and ψ_n in terms of Hermitian fields:

$$\psi_p = 2^{-1/2}(\psi_{p1} + i\psi_{p2}), \tag{7.8}$$

$$\psi_n = 2^{-1/2}(\psi_{n1} + i\psi_{n2}), \tag{7.9}$$

where

$$\psi_{aj} = \psi_{aj}^{\dagger}, \tag{7.10}$$

$a = n$ or p and $j = 1$ or 2. If S is an internal symmetry operator, then according to (2.4),

$$S\psi S^{-1} = \mathfrak{s}_{1/2}\psi, \tag{7.11}$$

where

$$\psi = \begin{pmatrix} \psi_{p1} \\ \psi_{p2} \\ \psi_{n1} \\ \psi_{n2} \end{pmatrix}, \tag{7.12}$$

and $\mathfrak{s}_{1/2}$ is a (4×4) real orthogonal matrix. Any such matrix with determinant $= +1$ can always be expressed in the form

$$\exp[i(\mathbf{I}\cdot\boldsymbol{\theta} + \mathbf{I}'\cdot\boldsymbol{\theta}')], \tag{7.13}$$

where $\boldsymbol{\theta}$ and $\boldsymbol{\theta}'$ are real parameters, the components of $\mathbf{I}$ and $\mathbf{I}'$ are related to those of $\boldsymbol{\rho}$ and $\boldsymbol{\sigma}$ matrices introduced in (5.25) by

$$2\mathbf{I} = (-\sigma_2\rho_1, \rho_2, -\sigma_2\rho_3), \tag{7.14}$$

and

$$2\mathbf{I}' = (\rho_2\sigma_1, \rho_2\sigma_3, -\sigma_2). \tag{7.15}$$

These operators are so chosen that $\mathbf{I}$ is the same as the isospin operator,

$$N = 2I_3', \tag{7.16}$$

$$G = \exp(-i\pi I_2'), \tag{7.17}$$

and

$$C_{st} = \exp[-i\pi(I_2 + I_2')]. \tag{7.18}$$

where N is the baryon number, and C_{st} and G are the same operators as those used in (6.20)–(6.23).

It can be easily verified that $\mathbf{I}$ commutes with $\mathbf{I}'$, and they are the generators of two independent SU_2 groups. Their direct product $SU_2 \times SU_2$ has an invariant subgroup Z_2 whose elements are 1 and $(-1)^{2I+2I'}$; its factor group is the SO_4 group which comprises all such (4×4) matrices given by (7.13). The irreducible representations of the SO_4 group can be labeled as (I,I') where $I(I+1)$ and $I'(I'+1)$ are, respectively, the eigenvalues of the operators $\mathbf{I}^2$ and $\mathbf{I}'^2$, and

$$(-1)^{2I+2I'} = 1. \tag{7.19}$$

The nucleon belongs to the representation $(\frac{1}{2},\frac{1}{2})$.

Theorem 4. This SO_4 symmetry is invariant under any virtual transition

$$N \rightleftarrows N + (G = +1 \text{ meson}), \tag{7.20}$$

but is violated by

$$N \rightleftarrows N + (G = -1 \text{ meson}). \tag{7.21}$$

Both (7.3) and (7.4) are examples of (7.20), and the usual pion-nucleon vertex is an example of (7.21).

Proof. Under the SO_4 group of transformations, any meson transforms like an appropriate nucleon-antinucleon system; therefore, it belongs to $I_3' = 0$ and $I' = 0$, or 1. From (7.17), it follows that the additional condition $G = +1$ implies $I' = 0$. Hence, reaction (7.20) conserves $\mathbf{I}'$. Since $\mathbf{I}$ conservation is implicitly assumed, reaction (7.20) is invariant under the entire SO_4 group of transformations. Any $G = -1$ meson transforms like the $I_3' = 0$ member of an $I' = 1$ multiplet. Thus, reaction (7.21) violates $\mathbf{I}'$ conservation. Theorem 4 is, then, proved.

It can be shown that this SO_4 symmetry is also violated by the known interactions between the strange particles and the nucleon. The usefulness of this SO_4 symmetry depends on whether we can separate out in any physical process the contributions due to the $G = +1$ mesons (ρ, η, etc.) from the symmetry-breaking terms such as those due to the $G = -1$ mesons (π, ω, etc.). [For example, in the theoretical study of the vector and axial-vector form factors of the nucleons, the contributions due to $G = +1$ and $G = -1$ mesons can be conveniently separated.] In the following, such separations are assumed to be useful, and we will consider those Feynman graphs which consist of *only* SO_4-invariant vertices.

For practical applications, it is convenient to use, instead of (7.12),

$$\begin{pmatrix} p \\ n \\ \bar{n} \\ -\bar{p} \end{pmatrix}, \tag{7.22}$$

as the base vectors of a (4×4) representation of the SO_4 group. In (7.22), the field operators for $\bar{n}$ and $\bar{p}$ are $\psi_n^{\dagger}$ and $\psi_p^{\dagger}$, respectively. In this representation, the generators $\mathbf{I}$ and $\mathbf{I}'$ become simply $\frac{1}{2}\boldsymbol{\sigma}$ and $\frac{1}{2}\boldsymbol{\rho}$, respectively [instead of (7.14) and (7.15)].

First, let us consider reactions

$$N + \bar{N} \to \text{mesons}, \tag{7.23}$$

and

$$N + \text{mesons} \to N + \text{mesons}, \tag{7.24}$$

where "mesons" represents an arbitrary distribution of $G = +1$ mesons. In (7.23), only the $I' = 0$ amplitudes contribute; in (7.24), on both sides only the nucleon carries a nonzero $I' = \frac{1}{2}$. Thus, all useful consequences of SO_4 invariance can be obtained by using only the usual $\mathbf{I}$ conservation and G symmetry.

However, for reactions such as

$$N + N \to N + N + \text{mesons} \tag{7.25}$$

and

$$N + \bar{N} \to N + \bar{N} + \text{mesons}, \tag{7.26}$$

the full SO_4 symmetry can yield additional relations between different amplitudes. The states NN and $N\bar{N}$

TABLE II. Eigenstates of I, I_3, I', and I_3' where I and I' are the generators of the SO_4 group discussed in Sec. VII. The order of the nucleons and antinucleons in the state, say, $\alpha\beta$ (where α and β can be p, n, $\bar{p}$ or $\bar{n}$) indicates that α has momentum $\mathbf{k}$ and helicity λ, and β has momentum $\mathbf{k}'$ and helicity λ'.

I	I_3	I'	I_3'	State
1	1	1	1	pp
1	1	1	0	$2^{-1/2}(p\bar{n}+\bar{n}p)$
1	1	1	-1	$\bar{n}\bar{n}$
1	0	1	1	$2^{-1/2}(pn+np)$
1	0	1	0	$\frac{1}{2}(n\bar{n}+\bar{n}n-p\bar{p}-\bar{p}p)$
1	0	1	-1	$-2^{-1/2}(\bar{p}\bar{n}+\bar{n}\bar{p})$
1	-1	1	1	nn
1	-1	1	0	$-2^{-1/2}(n\bar{p}+\bar{p}n)$
1	-1	1	-1	$\bar{p}\bar{p}$
0	0	1	1	$2^{-1/2}(pn-np)$
0	0	1	0	$\frac{1}{2}(-n\bar{n}+\bar{n}n-p\bar{p}+\bar{p}p)$
0	0	1	-1	$2^{-1/2}(\bar{p}\bar{n}-\bar{n}\bar{p})$
1	1	0	0	$2^{-1/2}(p\bar{n}-\bar{n}p)$
1	0	0	0	$\frac{1}{2}(n\bar{n}-\bar{n}n-p\bar{p}+\bar{p}p)$
1	-1	0	0	$2^{-1/2}(-n\bar{p}+\bar{p}n)$
0	0	0	0	$\frac{1}{2}(n\bar{n}+\bar{n}n+p\bar{p}+\bar{p}p)$

can be resolved into eigenstates of I, I_3, I' and I_3'; these eigenstates are listed in Table II. By using the explicit forms of these eigenstates, it is straightforward to obtain the various consequences of the SO_4 symmetry for reactions (7.25) and (7.26). For example,

$$A(p+p\to p+p+\eta^0)=\tfrac{1}{2}[A(p+\bar{n}\to p+\bar{n}+\eta^0)+A(p+\bar{n}\to\bar{n}+p+\eta^0)], \quad (7.27)$$

where A denotes the appropriate amplitude, and the state $(p+\bar{n}+\eta^0)$ is related to $(\bar{n}+p+\eta^0)$ by an exchange of the momenta and helicities of p and $\bar{n}$. Relations such as (7.27) cannot be derived by using only I and G.

Apart from such possible applications which is bound to be limited because of known violations of this SO_4 symmetry, the above discussions at least give a general class of strong interaction models in which the element C_{st} is connected, in a continuous way, to the unit element of the internal symmetry group.

ACKNOWLEDGMENTS

We wish to thank Professor G. Feinberg and Professor R. Serber for discussions.

APPENDIX A

In this Appendix, we will illustrate the use of Hermitian fields for a set of zero-mass spin-$\frac{1}{2}$ particles. For definiteness, let us take the simple model discussed in Sec. V. In this model, only the zero-mass limit of all known leptons is considered. Each lepton α is described by a non-Hermitian field ψ_α which satisfies Eq. (5.13); it can also be described by a Hermitian field ϕ_α defined by

$$(\phi_\alpha)_\lambda=(\psi_\alpha)_\lambda+(\psi_\alpha{}^\dagger)_\lambda, \quad (A1)$$

where $\alpha=e$, μ, ν_e or ν_μ, λ denotes the spinor index and it varies from 1 to 4. From (2.8), (5.13), and (A1), one finds

$$\psi_\alpha=\tfrac{1}{2}(1+\gamma_5)\phi_\alpha \quad (A2)$$

and that the components of $\psi_\alpha{}^\dagger$ are identical with those of $\frac{1}{2}(1-\gamma_5)\phi_\alpha$. The fields ψ_α and ϕ_α, therefore, represent two equivalent descriptions of the same two-component theory[28] for a spin-$\frac{1}{2}$ particle. By using (A1) and (A2), any transformation of ψ_α can be readily translated into a corresponding transformation of ϕ_α, and vice versa.

Consider, e.g., the gauge transformation

$$\psi_\alpha\to e^{i\xi}\psi_\alpha. \quad (A3)$$

The same transformation in terms of ϕ_α becomes

$$\phi_\alpha\to e^{i\gamma_5\xi}\phi_\alpha. \quad (A4)$$

As another example, we may consider the SU_2 transformation

$$\psi\to(\exp i\boldsymbol{\sigma}\cdot\boldsymbol{\theta})\psi \quad (A5)$$

where

$$\psi=\begin{pmatrix}\psi_e\\ \psi_\mu\\ \psi_{\nu_e}\\ \psi_{\nu_\mu}\end{pmatrix},$$

and $\boldsymbol{\sigma}$ is given by (5.25). In terms of ϕ_α, (A5) becomes

$$\phi\to e^{i(\gamma_5\sigma_1\theta_1+\sigma_2\theta_2+\gamma_5\sigma_3\theta_3)}\phi, \quad (A6)$$

where

$$\phi=\begin{pmatrix}\phi_e\\ \phi_\mu\\ \phi_{\nu_e}\\ \phi_{\nu_\mu}\end{pmatrix}.$$

Similarly, the transformation

$$\psi\to e^{i\rho_3\eta}\psi$$

becomes

$$\phi\to e^{i\gamma_5\rho_3\eta}\phi,$$

where ρ_3 is given by (5.25). Thus, Theorem 3 can be proved by following the same proof as that given in Sec. V, but, using the Hermitian fields ϕ_α.

APPENDIX B

Some explicit models will be given in this Appendix to illustrate the two unusual cases discussed in Sec. IV (case B_1 of example 1 and case A_1 of example 2).

[28] The use of a non-Hermitian field for the two-component spin-$\frac{1}{2}$ theory was first introduced by H. Weyl, Z. Physik **56**, 330 (1929); it was later applied to the leptons by T. D. Lee and C. N. Yang, Phys. Rev. **105**, 1671 (1957), L. Landau, Nucl. Phys. **3**, 127 (1957) and A. Salam, Nuovo Cimento **5**, 299 (1957). The use of a Hermitian field to describe a two-component spin-$\frac{1}{2}$ particle was made by E. Majorana, *ibid.* **14**, 171 (1937). See the discussion given by K. M. Case, Phys. Rev. **107**, 307 (1957). Both descriptions can be applied to the nonzero mass case as well as the zero mass case. That the Majorana theory is applicable to the nonzero mass case is obvious, since the Dirac equation is real in the Majorana representation (2.8). That the Weyl description is also applicable to the nonzero mass case follows from Eq. (A2). For a detailed discussion, see the review article by T. D. Lee and C. S. Wu (Ref. 22). If the mass is zero, then transformation (A3), or (A4), is an internal symmetry transformation; otherwise, it is not.

1. First, we consider the example $\mathcal{G}=U_1$. Let us introduce one spin-1 Hermitian field ϕ_μ and four spin-$\frac{1}{2}$ Hermitian fields ψ_1, ψ_2, ψ_3, and ψ_4. Under the internal symmetry operation (4.6), these fields transform as follows:

$$S(\theta)\phi_\mu S^{-1}(\theta)=\phi_\mu$$
$$S(\theta)\psi S^{-1}(\theta)=e^{i\sigma_2\theta}\psi \tag{B1}$$

where σ_2 is given by (5.25) and

$$\psi=\begin{bmatrix}\psi_1\\ \psi_2\\ \psi_3\\ \psi_4\end{bmatrix}. \tag{B2}$$

As an example of case B_1, we may assume that the total Hamiltonian is $H=H_{\text{free}}+H_{\text{int}}$ in which the interaction Hamiltonian is

$$H_{\text{int}}=\int j_\lambda\phi_\lambda d^3r. \tag{B3}$$

The current j_λ is given by

$$j_\lambda=il^\dagger\gamma_4(g_1+g_2\gamma_5)\gamma_\lambda l'+il'^\dagger\gamma_4(g_1^*+g_2^*\gamma_5)\gamma_\lambda l + ig_3(l^\dagger\gamma_4\gamma_\lambda l-l'^\dagger\gamma_4\gamma_\lambda l'), \tag{B4}$$

where $l(x)$, $l'(x)$ are non-Hermitian fields related to $\psi_1, \cdots, \psi_4$ by

$$l(x)=2^{-1/2}[\psi_1(x)+i\psi_2(x)], \tag{B5}$$

and

$$l'(x)=2^{-1/2}[\psi_3(x)+i\psi_4(x)]. \tag{B6}$$

The coupling constants g_1, g_2 are *not* relatively real, *nor* relatively imaginary, but the coupling constant g_3 is real. The interaction Hamiltonian H_{int} is invariant under the transformation (B1). In order that the total-Hamiltonian H is invariant under $S(\theta)$ we must have the bare mass (consequently, also the physical mass) of ψ_1 to be the same as that of ψ_2 and the mass of ψ_3 to be the same as that of ψ_4. Then, the internal symmetry group $\mathcal{G}$ is $U_1=\{S(\theta)\}$.

To satisfy $\mathcal{P}$ invariance, we assume all four fields $\psi_1, \cdots, \psi_4$ to be degenerate. It can be verified that H is invariant under any space-inversion operator in the coset $\mathcal{G}\mathcal{P}$ where

$$\mathcal{P}\phi_\mu(\mathbf{r},t)\mathcal{P}^{-1}=-\eta_{\mu\nu}\phi_\nu(-\mathbf{r}, t), \tag{B7}$$

$$\mathcal{P}\psi(\mathbf{r},t)\mathcal{P}^{-1}=\tilde{u}_{1/2}{}^P(i\gamma_4)\psi(-\mathbf{r}, t). \tag{B8}$$

The matrix $u_{1/2}{}^P$ is given by

$$u_{1/2}{}^P=\begin{bmatrix}0&0&0&1\\0&0&1&0\\0&-1&0&0\\-1&0&0&0\end{bmatrix},$$

and $\eta_{\mu\nu}=1$ if $\mu=\nu\neq 4$, $\eta_{\mu\nu}=-1$ if $\mu=\nu=4$, and zero otherwise. In terms of $l(x)$ and $l'(x)$, (B8) becomes

$$\mathcal{P}l(\mathbf{r},t)\mathcal{P}^{-1}=\gamma_4\bar{l}'(-\mathbf{r}, t),$$

and

$$\mathcal{P}l'(\mathbf{r},t)\mathcal{P}^{-1}=-\gamma_4\bar{l}(-\mathbf{r}, t),$$

where

$$\bar{l}(x)=2^{-1/2}[\psi_1(x)-i\psi_2(x)],$$

and

$$\bar{l}'(x)=2^{-1/2}[\psi_3(x)-i\psi_4(x)].$$

The operator $\mathcal{P}$ satisfies Eqs. (4.9) and (4.10), which are the conditions for case B_1. From Theorem 1, it follows that the theory is invariant under any time-reversal operator in the coset $\mathcal{G}\mathcal{T}$. In this case, the unusual degeneracy between $l(x)$ and $l'(x)$ is due to $\mathcal{P}$ or $\mathcal{T}$ invariance.

2. For the second example, $\mathcal{G}=\{U_2,U_2C\}$, it is case A_1 that has an unusual degeneracy.[14] As an illustration, let us consider a model which consists of one spin 0, $Y=0$, $I=1$ triplet $\boldsymbol{\phi}$ and two spin $\frac{1}{2}$, $Y=1$, $I=\frac{1}{2}$ doublets N and N' where

$$\boldsymbol{\phi}=(\phi_1,\phi_2,\phi_3), \tag{B9}$$

$$N=2^{-1/2}\begin{pmatrix}\psi_1+i\psi_2\\ \psi_3+i\psi_4\end{pmatrix}, \tag{B10}$$

and

$$N'=2^{-1/2}\begin{pmatrix}\psi_1'+i\psi_2'\\ \psi_3'+i\psi_4'\end{pmatrix}. \tag{B11}$$

The fields $\phi_1, \cdots, \psi_1, \cdots, \psi_1', \cdots, \psi_4'$ are all Hermitian fields. The different fields in the same isomultiplet are assumed to be degenerate. Let the interaction Hamiltonian be given by

$$H_{\text{int}}=\int \mathbf{J}\cdot\boldsymbol{\phi}d^3r, \tag{B12}$$

where

$$\mathbf{J}=N^\dagger\gamma_4(g_1+ig_2\gamma_5)\boldsymbol{\tau}N+N'^\dagger\gamma_4(g_1-ig_2\gamma_5)\boldsymbol{\tau}N' + ig_3(N^\dagger\gamma_4\gamma_5\boldsymbol{\tau}N'+N'^\dagger\gamma_4\gamma_5\boldsymbol{\tau}N),$$

the coupling constants g_1, g_2, g_3 are all real, and $\boldsymbol{\tau}=(\tau_1,\tau_2,\tau_3)$ are the usual three (2×2) Pauli matrices. It can be verified that the internal symmetry group is $\mathcal{G}=\{U_2,U_2C\}$ where $U_2=\{\exp(i\mathbf{I}\cdot\boldsymbol{\theta}+iY\alpha)\}$, and C satisfies

$$CN(x)C^{-1}=\bar{N}(x)=2^{-1/2}\begin{pmatrix}\psi_1-i\psi_2\\ \psi_3-i\psi_4\end{pmatrix},$$

$$CN'(x)C^{-1}=\bar{N}'(x)=2^{-1/2}\begin{pmatrix}\psi_1'-i\psi_2'\\ \psi_3'-i\psi_4'\end{pmatrix},$$

and

$$C\boldsymbol{\phi}(x)C^{-1}=\boldsymbol{\phi}(x).$$

To satisfy $\mathcal{P}$ invariance, we assume that the two isospin doublets $N(x)$ and $N'(x)$ are degenerate. The theory is invariant under any space-inversion operator in the coset $\mathcal{G}\mathcal{P}$ where

$$\mathcal{P}\phi(\mathbf{r},t)\mathcal{P}^{-1}=\phi(-\mathbf{r},t),$$
$$\mathcal{P}N(\mathbf{r},t)\mathcal{P}^{-1}=i\gamma_4N'(-\mathbf{r},t),$$

and

$$\mathcal{P}N'(\mathbf{r},t)\mathcal{P}^{-1}=-i\gamma_4N(-\mathbf{r},t).$$

The operator $\mathcal{P}$ commutes with **I**, Y, and C, and it satisfies $\mathcal{P}^2=1$, which is the condition for case A_1. By using Theorem 1, one finds that the theory is also invariant under any time-reversal operator in the coset $\mathcal{G}\mathcal{T}$.

Both of these examples are quite artificial; they are constructed to illustrate the physical contents of these unusual cases. At present, no such degeneracy which is the result of $\mathcal{P}$, or $\mathcal{T}$, invariance has been observed.

APPENDIX C

In this Appendix, we will give an explicit example[29] of a strong interaction with no derivative coupling, and the internal symmetry group is (6.31). Let us assume, in addition to the known hadrons, the existence of three hypothetical charged particles, called $a_1^{\pm}$, $a_2^{\pm}$, and $a_3^{\pm}$. The masses of these three particles are all *different.* There exist strong interactions between the $a_i^{\pm}$ particles and an $I=0$ meson, say the known η_0' (which is also an SU_3 singlet); these interactions are described by the Hamiltonian density $H_{\text{st}}{}^a$:

$$H_{\text{st}}{}^a=i\phi(g_1\psi_2{}^\dagger\gamma_4\gamma_5\psi_3+g_2\psi_3{}^\dagger\gamma_4\gamma_5\psi_1+g_3\psi_1{}^\dagger\gamma_4\gamma_5\psi_2)+\text{H.c.},\quad \text{(C1)}$$

where ϕ and ψ_i are, respectively, the fields of η_0' and a_i $(i=1, 2, 3)$. The operator ϕ is Hermitian, but ψ_i is not. The three coupling constants g_1, g_2, g_3 can be arbitrary, provided that their product is *not* real; i.e.,

$$(g_1g_2g_3)^*\neq g_1g_2g_3. \quad \text{(C2)}$$

Let $H_{\text{st}}{}^0$ be the usual strong interactions of all known hadrons; $H_{\text{st}}{}^0$ is independent of ψ_i, and it contains, among other terms, the interaction between the nucleons and η_0'. We may write

$$H_{\text{st}}{}^0=if\psi_N{}^\dagger\gamma_4\gamma_5\psi_N\phi+\cdots, \quad \text{(C3)}$$

where the nucleon field operator is

$$\psi_N=\begin{pmatrix}\psi_p\\ \psi_n\end{pmatrix},$$

and f is the appropriate coupling constant.

[29] For an explicit example with derivative couplings, see Ref. 9.

All known hadrons have $Q_K=0$; they also have their usual quantum numbers I, I_3, N and $Q=Q_J$. For the $a_i^{\pm}$, we assign $I=N=Q_J=0$ and $Q_K=\pm1$.

It can be verified that (C1) and (C3) conserve[30] **I**, N, Q_J, and Q_K. The $H_{\text{st}}{}^0$ is known to satisfy the C_{st} symmetry where C_{st} is given by Eq. (6.23). Under C_{st}, all known hadrons transform in the usual way; e.g.,

$$C_{\text{st}}\psi_N(x)C_{\text{st}}{}^{-1}=\psi_N{}^\dagger(x),$$
$$C_{\text{st}}\phi(x)C_{\text{st}}{}^{-1}=\phi(x),$$

etc. The interaction $H_{\text{st}}{}^a$ is also invariant under C_{st}, provided $\psi_i(x)$ is unchanged under C_{st}; i.e.,

$$C_{\text{st}}\psi_i(x)C_{\text{st}}{}^{-1}=\psi_i(x), \quad \text{(C4)}$$

where $i=1$, 2 or 3. Thus, the internal symmetry group $\mathcal{G}$ is (6.31). Both $H_{\text{st}}{}^0$ and $H_{\text{st}}{}^a$ satisfy $\mathcal{P}$ invariance, and therefore also $\mathcal{T}$ invariance. Furthermore, there is an operator P_{st} in the coset $\mathcal{G}\mathcal{P}$, which satisfies

$$P_{\text{st}}\psi_i(\mathbf{r},t)P_{\text{st}}{}^{-1}=\gamma_4\psi_i(-\mathbf{r},t),$$
$$P_{\text{st}}\phi(\mathbf{r},t)P_{\text{st}}{}^{-1}=-\phi(-\mathbf{r},t),$$

and

$$P_{\text{st}}\psi_N(\mathbf{r},t)P_{\text{st}}{}^{-1}=\gamma_4\psi_N(-\mathbf{r},t). \quad \text{(C5)}$$

The strong interaction $(H_{\text{st}}{}^0+H_{\text{st}}{}^a)$ is invariant under C_{st}, P_{st} and T_{st} where T_{st} is given by Eq. (6.40).

The electromagnetic current of the $a_i^{\pm}$ is

$$J_\lambda{}^a=ie\sum_{i=1}^{3}\psi_i{}^\dagger\gamma_4\gamma_\lambda\psi_i, \quad \text{(C6)}$$

and the electromagnetic current $J_\lambda{}^0$ of all known hadrons is given by the usual expression

$$J_\lambda{}^0=ie\psi_p{}^\dagger\gamma_4\gamma_\lambda\psi_p+\cdots. \quad \text{(C7)}$$

It can be shown that the electromagnetic interaction $(J_\lambda{}^0+J_\lambda{}^a)A_\lambda$ is invariant[31] under C_γ, P_γ, and T_γ, where

$$C_\gamma\psi_N(x)C_\gamma{}^{-1}=\psi_N{}^\dagger(x),$$
$$C_\gamma\psi_i(x)C_\gamma{}^{-1}=\psi_i{}^\dagger(x), \quad \text{(C8)}$$

$(i=1, 2, 3)$, $P_\gamma=P_{\text{st}}$, and

$$C_\gamma P_\gamma T_\gamma=C_{\text{st}}P_{\text{st}}T_{\text{st}}=\mathcal{J}.$$

The charge-conjugation operator C_γ anticommutes with both Q_J and Q_K, but C_{st} anticommutes only with Q_J. Because of condition (C2), the strong interaction is not invariant under C_γ and T_γ. By using (C4), one finds that the electromagnetic interaction is not invariant under C_{st} and T_{st}.

[30] We assume that the remaining part $\cdots$ in (C3) is invariant under P_{st}, T_{st} and the group of transformations (6.31); we also assume that the remaining part $\cdots$ in (C7) is invariant under C_γ, P_γ, and T_γ.

Session 5

OTHER ASPECTS OF WEAK INTERACTIONS: HIGH ENERGY NEUTRINO PHYSICS AND QUESTIONS OF C, P, T NONINVARIANCE

T. D. Lee, Rapporteur

I. High Energy Neutrino Physics

There has been no new discovery made during the last two years by using the high energy neutrinos from the accelerators. For the elastic cross section, all results are still consistent with the crude approximation that the axial-vector form factor is approximately the same as the vector form factor. The lower limit of the mass of the intermediate boson, which has still not been observed, is $m_w > 2$ BeV. The μ neutrino remains different from the e neutrino.

There are two contributions given at the Conference on the cosmic ray neutrino reactions. The first is the Kolar gold field experiment[1] reported by Dr. Menon. There are 11 possible cosmic-ray neutrino events of which 4 are sure. The rate is $\approx 8 \times 10^{-13}$ events per cm^2 per second per steradian. The second is the Case-Witwatersrand experiment[2] reported by Dr. Reines. There are 17 sure events and the rate is $\lesssim 5 \times 10^{-13}$ events per cm^2 per second per steradian.

Both results are consistent with the hypothesis that these neutrinos are generated from the cosmic rays in the atmosphere. There is no evidence of any anomalous neutrino source or any anomalous high energy neutrino cross section, nor any evidence of the intermediate boson with $m_w \lesssim 2$ to 3 BeV.

II. Phenomenological Analysis of CP-Nonconserving Interactions

Next, we discuss the present theoretical status of CP nonconservation. The experimental situation was beautifully summarized by Professor Fitch yesterday; after two years, there remains only the $K_2^0 \to \pi^+\pi^-$ decay[3] that shows CP nonconservation. In other reactions, such as η^0 decay, the experimental situation on CP violation is unclear and confused. I am happy to say that it is quite the opposite on the theoretical side. Two years ago there was almost no theory predicting CP nonconservation; now, there are far too many different kinds of CP-violating theories.

We may write for the total Hamiltonian

$$H = H_+ + H_- ,$$

where $$(CP)H_\pm(CP)^{-1} = \pm H_\pm .$$

From $K_2^0 \to \pi^+\pi^-$, we know only that the CP = -1 interaction H_- exists, but of its other properties, very little is known. The different types of CP-nonconserving interaction H_- can be classified[4] according to its selection rule with respect to the strangeness S. (See the table.)

	Coupling constant of H_-	Selection rule of H_-
Superweak (Ref. 5)	$\approx 10^{-9}\ G_{wk}$	$\Delta S = \pm 2$ allowed
Weak (Ref. 6)	$10^{-1} - 10^{-3}\ G_{wk}$	$\Delta S = \pm 1$ allowed $\Delta S \neq \pm 2$
Nonweak (Ref. 7, 8)	e (electromagnetic) $10^3\ G_{wk}$	$\Delta S = 0$ only P conserved C nonconserved

Now, the K_1^0 and K_2^0 states are related to the K^0 and $\bar{K}^0$ states (provided CPT theorem holds) by

$$K_1^0 = [(1+\epsilon)K^0 + (1-\epsilon)\bar{K}^0] \times 1/\sqrt{2(1+|\epsilon|^2)}$$

and

$$K_2^0 = [(1+\epsilon)K^0 - (1-\epsilon)\bar{K}^0] \times 1/\sqrt{2(1+|\epsilon|^2)} ,$$

where $\bar{K}^0 = (CPT)K^0$.

If H_- is Wolfenstein's superweak interaction, then neglecting effects $0(F/G_{wk}) \sim 0(10^{-9})$, only in the decays of K_1^0 and K_2^0 can one detect any CP violations; furthermore, all such CP violations are characterized by the single parameter ϵ, which, in this case, has already been measured:

$$|\epsilon| = |\eta_{+-}| = (1.83 \pm .12) \times 10^{-3}$$

and $$\arg \epsilon = \tan^{-1} 2(m_2 - m_1)/\gamma_1 \approx 42° .$$

It should, however, be emphasized that in any case, from the various known decay amplitudes of K_1^0 and K_2^0, one can establish[9] that $\arg \epsilon = 42° \pm (\lesssim 15°)$, independent of the underlying mechanism of H_-, whether it is weak or nonweak. In general, the parameter η_{+-} is given by $\eta_{+-} = \epsilon + \epsilon'$, where ϵ' is due to the CP-violating amplitude in the I = 2 two-pion state. Although $\eta_{+-} = \epsilon$ in the superweak case, a great variety of other possibilities would also lead to $\eta_{+-} \approx \epsilon$; for example, if one assumes that only $\Delta I = 1/2$ transitions violate CP. Thus, it is difficult to use measurements on, say, $\arg \eta_{+-}$ to establish the validity of the superweak case, unless such a phase is measured to an accuracy of within $\approx 1°$.

From a phenomenological point of view, the superweak interaction gives definite predictions about all CP-violating amplitudes, namely, zero for all processes except the decays of the neutral K mesons. If this is the case, then for all practical purposes, we have already determined all that could be measured concerning CP-violating parameters, and it will take a very long time for us to learn anything more about the CP-violating interaction. The superweak interaction has a certain appeal, because, then, we do not have to think any more.

If the CP-violating interaction belongs to the

"weak" case, then there should be other weak processes besides K_1^0, K_2^0 decays that exhibit some small CP violations; if the CP-violating interaction belongs to the "nonweak" case, then in addition, there should be some strong processes that exhibit small C or T violations; if the CP-nonconserving interaction is the electromagnetic case, then, in addition, there should be some electromagnetic processes that have sizable C, T noninvariant amplitudes. To be sure, none of these effects can be detected easily, and this accounts for the present remarkable lack of new experimental information.

From the theoretical side, the difficulty is the other way around. It is too easy to add a small C- or CP-violating term to the present strong or weak interaction. The difficulty is to develop some principles which might enable one to make selections among this multitude of such possibilities. In case the CP-violating interaction is the electromagnetic interaction, some relatively exact statements can be made, because of charge conservation and the possibility of a minimal interaction principle.

III. Some Properties of Q_K and η^0 Decay

Let us discuss the case in which the strong interaction is assumed to be invariant under a particle-antiparticle conjugation operator $C \equiv C_{st}$,

$$[C_{st}, H_{st}] = 0,$$

but

$$[C_{st}, H_\gamma] \neq 0.$$

Let

$$\mathcal{J}_\mu = \text{hadronic electromagnetic current},$$

$$\mathcal{J}_\mu = J_\mu + K_\mu,$$

$$C_{st} = -1 : \ J_\mu = \tfrac{1}{2}(\mathcal{J}_\mu - C_{st}\mathcal{J}_\mu C_{st}^{-1}),$$

$$+1 : \ K_\mu = \tfrac{1}{2}(\mathcal{J}_\mu + C_{st}\mathcal{J}_\mu C_{st}^{-1})$$

Since $\mathcal{J}_\mu$ is absolutely conserved, and C_{st} is conserved under H_{st}, it follows that, under H_{st}, J_μ and K_μ are separately conserved. Thus, the charges

$$Q_J = -i \int J_4 d^3 r$$

and

$$Q_K = -i \int K_4 d^3 r$$

must be separately conserved quantities under H_{st};

$$[Q_J, C_{st}] = 0$$

and

$$\{Q_K, C_{st}\} = 0.$$

The total charge is $Q = Q_J + Q_K$.

It is easy to show[10] that $Q_K \mid \text{known particle}\rangle = 0$. This follows simply from the fact that all known particles change the sign of their charges under C_{st}. Thus, particles with $Q_K \neq 0$ must be some unknown particles called $a^\pm$:

$$Q_K \mid a^\pm\rangle \neq 0.$$

Under fairly general conditions, it can also be shown that as operators[11] Q_K = isoscalar, $Q_J = I_z + Y/2$, even though the isospin of the $a^\pm$ particles is not zero; i.e., the commutation relations between Q_K and the isospin vector $\underset{\sim}{I}$ are just like that between the usual hypercharge Y and $\underset{\sim}{I}$.

If we assume that K_μ and J_μ transform under $\underset{\sim}{I}$ like Q_K and Q_J, then we find K_μ = isoscalar, and J_μ = isoscalar + isovector.

We list the various consequences if K_μ is isoscalar:

1. To the lowest order in α, because of isospin conservation, $\eta^0 \not\to \pi^0 + e^+ + e^-$.

2. To the lowest order in α, there is no time-reversal violation effect in $\Sigma^0 \to \Lambda^0 + e^+ + e^-$.

3. In $\eta^0 \to \pi^+ + \pi^- + \pi^0$, the final $C_{st} = -1$ three-pion state must be of I = 0, the $\pi^\pm$ asymmetry parameter is given by $A \approx (kR)^6 \sin(\delta_1 - \delta_0)$, where δ_1, δ_0 are the strong-interaction 3π eigen phase shifts in the I = 1 and I = 0 states, respectively. Furthermore, this asymmetry would change sign between the neighboring sextants in the Dalitz plot.

The interaction radius R is, of course, not known. If we take $R \sim 1/m_\eta$, then we find $|A| < 10^{-3}$. Because of the factor $(kR)^6$, this upper limit is very sensitive to the value of R asssumed.

4. In $\eta^0 \to \pi^+ + \pi^- + \gamma$, the $\pi^\pm$ asymmetry has an upper limit $|A| < [\sim 1.5\,\%]$ if $R \sim 1/m_\eta$.

Thus, none of these is particularly sensitive to the isoscalar current K_μ. More sensitive tests can be made by studying the inelastic collision of e^- on polarized targets,[12] the reciprocity relation[13] on $\gamma N \rightleftarrows \pi N$, and[14] $\gamma d \rightleftarrows np$. If there exists a strangeness-conserving nonleptonic weak interaction, then one may try to measure the magnitude of the electric dipole moment of the neutron, and use it as a test[15] of T invariance of the electromagnetic interaction.

If, on the other hand, one finds experimentally that the final 3π in the $\eta^0 \to \pi^+\pi^-\pi^0$ can be in an I = 2 (which has $C_{st} = -1$) state, the $\pi^\pm$ asymmetry is given approximately by $A \sim (kR)^2 \sin(\delta_1 - \delta_2)$, which can be much larger ($\lesssim$ 5 %) than the previous $I_{3\pi} = 0$ case, even if $R \approx 1/m_\eta$. This then would lead to the interesting conclusion not only that C violation is not weak, but if it is due to the electromagnetic interaction, then the isospin transformation property of K_μ is not proportional to its charge Q_K. In the same connection, it has been suggested by Dr. Okun[16] and discussed by Dr. Kabir in this Conference that one may also ask whether the isospin property of J_μ is, or is not, proportional to Q_J; i.e., whether the well-known selection rule $\Delta I = 0, 1$ of the electromagnetic interaction is really a valid one.

Independent of the result of 3π asymmetry in η^0 decay, any proposal in which $K_\mu \neq 0$ has to encounter the question: Why is the particle-antiparticle conjugation determined by the strong interaction different from the usual charge-conjugation operation? This leads us to discuss a simple model of all interactions.

IV. A Simple Model

At present, we know that

$$H_{wk} = G_{wk}\Sigma(V_\lambda^{wk} + A_\lambda^{wk})^\dagger \cdot (V_\lambda^{wk} + A_\lambda^{wk}),$$

$$H_\lambda = e\, V_\lambda^\gamma \cdot A_\lambda,$$

where A_λ denotes the electromagnetic field, and V_λ^γ the electromagnetic current. In general, there are various particles with different spins; V_λ^γ can be written as

$$V_\lambda^\gamma = V_\lambda^\gamma(\text{spin } 0) + V_\lambda^\gamma(\text{spin } 1/2) + V_\lambda^\gamma(\text{spin } 1) + \cdots .$$

We observe that, at least for the free particle and at the zero mass limit for spin 1/2 particles, these currents obey the following two properties:

(a) they are all bilinear functions of field operators with the minimum number of derivatives,

(b) $\partial V_\lambda^i/\partial X_\lambda = \partial A_\lambda^i/\partial X_\lambda = 0$. For V_λ^γ, (a) is identical with the minimal principle of electromagnetic interaction; (b) holds for the leptonic part of $(V_\lambda + A_\lambda)^{wk}$ only at the zero mass limit; the same applies to the hadronic part of A_λ^{wk}.

In this model, let us propose that

$$H_{st} = G_{st}(A_\lambda^{st})_j^i \cdot (A_\lambda^{st})_i^j,$$

where i and j vary from 1 to 3, and all repeated indices are summed over. The V_λ^γ is related to V_λ^{wk} and A_λ^{st} is related to A_λ^{wk}.

What we will show is that if A_λ^{st} consists of some charged spin-1 field,

$$A_\lambda^{st} = A_\lambda^{st}(\text{spin } 1/2) + A_\lambda^{st}(\text{spin } 1),$$

then $C_{st} \neq C_\gamma$. To show this, let us recall that the currents which satisfy the above conditions (a) and (b) for different spin fields are

$$\text{spin } 0, \quad V_\lambda = i\left(\frac{\partial \phi^\dagger}{\partial X_\lambda}\phi - \phi^\dagger \frac{\partial \phi}{\partial X_\lambda}\right),$$

$$\text{spin } 1/2, \quad V_\lambda = i\psi^\dagger \gamma_4 \gamma_\lambda \psi,$$

$$A_\lambda = i\psi^\dagger \gamma_4 \gamma_\lambda \gamma_5 \psi,$$

$$\text{spin } 1, \quad V_\lambda = i\left(\frac{\partial \phi_\mu^\dagger}{\partial X_\lambda}\phi_\mu - \phi_\mu^\dagger \frac{\partial \phi_\mu}{\partial X_\lambda}\right) + \cdots,$$

$$A_\lambda = \epsilon_{\lambda\rho\mu\nu}\frac{\partial}{\partial X_\rho}(\phi_\mu^\dagger \phi_\nu).$$

If one defines the charge-conjugation operator as

$$C_\gamma \phi C_\gamma^{-1} = \phi^\dagger,$$

$$C_\gamma \psi C_\gamma^{-1} = \psi^\dagger \quad \text{(Majorana representation)},$$

$$C_\gamma \phi_\mu C_\gamma^{-1} = \phi_\mu^\dagger,$$

then one finds

	spin 0	spin 1/2	spin 1
$C_\gamma V_\lambda C_\gamma^{-1}$	$-V_\lambda$	$-V_\lambda$	$-V_\lambda$
$C_\gamma A_\lambda C_\gamma^{-1}$	---	$+A_\lambda$	$-A_\lambda$

Thus, if there exists a sum of such spin 1/2 and spin 1 currents in the strong-interaction current A_λ, then the strong interaction cannot be invariant under the charge-conjugation operator (provided that the spin 1 field is charged). To be specific, let us assume the existence of a spin 1/2 unitary triplet field ψ_i and a charged unitary singlet spin 1 field ϕ_μ:

$$(A_\lambda^{st})_j^i = i\psi_i^\dagger \gamma_4 \gamma_\lambda \gamma_5 \psi_j + \delta_j^i \epsilon_{\lambda\rho\mu\nu}\frac{\partial}{\partial X_\rho}(\phi_\mu^\dagger \phi_\nu)$$

$$H_{st} = G_{st}(A_\lambda^{st})_j^i (A_\lambda^{st})_i^j.$$

Define

$$C_{st}\psi_i C_{st}^{-1} = \psi_i^\dagger, \qquad C_{st}\phi_\mu C_{st}^{-1} = \phi_\mu.$$

Then, we have

$$C_{st}(A_\lambda^{st})_j^i C_{st}^{-1} = (A_\lambda^{st})_i^j$$

and

$$C_{st} H_{st} C_{st}^{-1} = H_{st}.$$

In this model, the usual octet part of the $(A_\lambda^{st})_j^i$ current is related to the hadronic part of A_λ^{wk}, and it is easy to verify that H_i is invariant under C_i, P_i, T_i, where i = st, γ or wk, and $C_i P_i T_i$ = same. The C_i operator is defined by

$$C_{st}(A_\lambda^{st})_j^i C_{st}^{-1} = (A_\lambda^{st})_i^j,$$

$$C_\gamma (V_\lambda^\gamma) C_\gamma^{-1} = -V_\lambda^\gamma,$$

$$C^{wk}(V_\lambda + A_\lambda)^{wk} C_{wk}^{-1} = (V_\lambda + A_\lambda)^{wk^\dagger}.$$

Because of the difference between these currents we find

$$C_{st} \neq C_\gamma \neq C_{wk} \neq C_{st},$$

and because $C_i P_i T_i$ = same, we find the various nonconservations of P_i and T_i. This model is only a simple illustration of how C_{st} can be different from C_γ.

In this model, the C_{st}-violating current is simply the minimal vector current of the spin 1 field, and it is an isoscalar.

Let $|a^\pm\rangle$ denote the particle state associated with such a spin 1 field ϕ_μ. It is clear that $a^\pm$ does not correspond to any known particle. If one is in the spirit of Gell-Mann's "chimeron," one may say that $a^\pm$ is also the intermediate boson of the weak interaction, although I will not assume it to be also a magnetic monopole.

V. Some General Remarks

In conclusion, I would like to make some general remarks about the question of C, P, T nonconservation. Our views of such discrete symmetries have undergone great changes in recent years. Although P and C symmetries have been well established since 1957 as only approximately valid, until the recent discovery of the 2π decay of the long-lived K_2^0 meson it was still possible to believe the essential symmetry of right and left by using CP, instead of P. From a fundamental point of view, therefore, the discovery of CP violation in K_2^0 decay and the indirect conclusion of time-reversal asymmetry are a more decisive blow to our notions on geometric symmetry principles than earlier results.

Indeed, the fact that P, CP, and T symmetries are violated means that these operators are not defined. To see this, let us recall that the space-inversion operator P (or, CP) in the Hilbert space should represent the coordinate transformation P (or CP): $\vec{r} \to -\vec{r}$, $t \to +t$. Now, the time translation operator is $e^{-iH\tau}$: $\vec{r} \to +\vec{r}$, $t \to t + \tau$. From a geometrical point of view, it is obvious that the space inversion $\vec{r} \to -\vec{r}$ must commute with the time translation $t \to t + \tau$. On the other hand, $[H, P] \neq 0$ (or $[H, CP] \neq 0$) implies that the alleged space-inversion operator P (or CP) fails to satisfy the multiplication law of the coordinate transformations that it is supposed to represent; thus, the fact that the space-inversion symmetry is being violated shows that the operator P (or CP) cannot be exactly defined. The same arguments can be applied equally well to the time-reversal operator.

We may, however, give an approximate definition of P, C, and T by replacing the total interaction H by a certain part H_i, such that within this approximation H_i is invariant under C_i, P_i, and T_i. It is clear that operators C_i, P_i, T_i thus defined depend on the approximation $H \approx H_i$. We know that if $H \approx H_{st}$, then there exists a C_{st}, P_{st}, and T_{st} operator. The (C_{st} P_{st}) and T_{st} symmetries thus defined must be violated by some other interactions. But at present it is not clear by what part of the interaction, whether it is electromagnetic or whether it is by some other interactions.

The evolution of our views on these discrete symmetry operators has been one of the most remarkable ones in physics. Apart from its physical information and philosophical implications, it certainly has demonstrated to us in a most forceful way the necessity of keeping an open-minded approach in physics. The search for the origin of CP violation is certainly a difficult one; yet, this difficulty is precisely its challenge. Let me hope that in the next Conference, we may have not one but many solid pieces of experimental information on CP violation to discuss, instead of merely theoretical speculations, as I have given you today.

Footnotes and References

1. C. V. Achar, M. G. K. Menon, V. S. Narasimham, P. V. Ramana Murthy, B. V. Sreekantan, K. Hinotani, S. Miyake, D. R. Creed, J. L. Osborne, J. M. B. Pattison, and A. W. Wolfendale, Detection of Muons Produced by Cosmic Ray Neutrinos Deep Underground, Paper 5b.1.

2. F. Reines, M. F. Crouch, T. L. Jenkins, W. R. Kropp, H. S. Gurr, and G. R. Smith, Phys. Rev. Letters 15, 429 (1965).

3. J. H. Christenson, J. W. Cronin, V. L. Fitch, and R. Turlay, Phys. Rev. Letters 13, 138 (1964).

4. T. D. Lee and L. Wolfenstein, Phys. Rev. 138, B1490 (1965).

5. The possibility of a superweak interaction was first pointed out by L. Wolfenstein, Phys. Rev. Letters 13, 562 (1964). The general implication of the superweak interaction was discussed by Lee and Wolfenstein (Ref. 4).

6. The possibility that the CP violation is due to the weak interaction is an obvious one after the experimental discovery of $K_2^0 \to \pi^+\pi^-$. The precise mechanism of such a weak CP violation has been the subject of many theoretical papers. We give only an incomplete list of these papers: R. G. Sachs, Phys. Rev. Letters 13, 286 (1964); T. N. Truong, Phys. Rev. Letters 13, 358a (1964); 17, 153 (1966); S. L. Glashow, Phys. Rev. Letters 14, 35 (1965); E. P. Shabalin, J.E.T.P. Letters 2, 276 (1965); N. Cabibbo, Phys. Letters 12, 137 (1964); L. Wolfenstein, Phys. Letters 15, 196 (1965); Okun and Zeldovic, Phys. Letters 16, 319 (1965). See also F. Salzmann and G. Salzmann, Phys. Letters 15, 91 (1965), and B. A. Arbuzov and A. T. Filippov, Phys. Letters 20, 537 (1966).

7. J. Bernstein, G. Feinberg, and T. D. Lee, Phys. Rev. 138, B1650 (1965); see also S. Barshay, Phys. Letters 17, 78 (1965).

8. T. D. Lee and L. Wolfenstein (Ref. 4); L. B. Okun, Soviet J. Nucl. Phys. 1, 670 (1965); J. Prentki and M. Veltman, Phys. Letters 15, 88 (1965).

9. See, especially, L. Wolfenstein, Nuovo Cimento 62A, 17 (1966).

10. T. D. Lee, Phys. Rev. 140, B959 (1965).

11. For a brief discussion, see footnote 20 of Ref. 10. A general proof is given elsewhere.

12. N. Christ and T. D. Lee, Phys. Rev. 143, 1310 (1966); see also C. A. Piketty, Nuovo Cimento 44A, 39 (1966); B. Desplanques and G. Karpman, Nuovo Cimento 38, 627 (1965).

13. N. Christ and T. D. Lee, Phys. Rev. 148, 1520 (1966).

14. S. Barshay, Phys. Rev. Letters 17, 49 (1966).

15. G. Feinberg, Phys. Rev. 140, B1402 (1965).

16. A. D. Dolgov, A. B. Kaidalov, I. Yu. Kobzarev, L. B. Okun, and V. I. Zakharov, On Possible Violation of the $\Delta T = 0$, 1 Rule for Electromagnetic Current, Paper 5a.5.

Discussion

Barshay (Rutgers): Professor Lee, could you comment on the possible decay modes of this new particle, should it exist? Could it have also strong or electromagnetic decays?

Lee: I do not know if it exists. If the hypothetical spin 1 particle is also the weak intermediate vector boson, it will decay just like the usual one. It could not decay by a strong or electromagnetic decay. K_μ is conserved in the strong interactions and also in the electromagnetic interactions in the model.

Truong (Brown): I would like to ask what average P-wave $\pi\pi$ phase shift do you use to estimate the asymmetry in $\eta^0 \to \pi^+\pi^-\gamma$ decay?

Lee: What I use is an upper limit, not an estimation, which is 30°. This is as large as I think you can have. Therefore, if it's much smaller you can scale accordingly.

Truong: I think that the value of 30° is too large if one extrapolates the $\pi\pi$ P-wave phase shift from the ρ resonance. The asymmetry should be much smaller, as you estimated. There is another process where the $\pi\pi$ interaction is large and has much larger asymmetry. Barrett and I have calculated the asymmetry in $\eta' \to \pi^+\pi^-\gamma$ decay and found that it is large.

Lee: In the η' there is one difficulty. There is probably another neutral particle, the X', which has the same mass as the η' but a different isospin value.

Truong: There is another way of testing C violation in electromagnetic interactions, namely trying to detect an effect in weak interactions other than $K_L \to 2\pi$. One should try to find asymmetry in spectra and partial rate of τ^+ vs τ^-, τ'^+ vs τ'^- and also $\pi^+\pi^-$ asymmetry in $K_L^0 \to \pi^+\pi^-\pi^0$.

Feinberg (Columbia): A measurement of the electric dipole moment of the neutron to an accuracy of 10^{-24} e·cm would be very useful in distinguishing between the different alternatives for explaining the CP violation. In the hypothesis of superweak interactions, one expects an electric dipole moment of less than 10^{-28} e·cm. In the hypothesis of 10^{-2} violation in weak interactions, one expects a dipole moment of about 10^{-23} e·cm. In the hypothesis of a large electromagnetic T violation, one expects a dipole moment of about 10^{-20} e·cm, so these alternatives can be easily distinguished by the prospective measurements to be done at Brookhaven and Oak Ridge.

Lee: Yes, so the future looks very bright. [It is important to note that these estimations depend sensitively on the additional assumption that there exists a $\Delta S = 0$, nonleptonic weak interaction which has a coupling constant comparable in magnitude to that of β decay. If such an interaction does not exist, then the expected value of the neutron electric dipole moment would be $\approx 10^{-26}$ e·cm if the T violation is due to the electromagnetic interaction, and $\approx 10^{-28}$ to 10^{-29} e·cm if the T violation is due to the weak interaction.]

Nauenberg (Santa Cruz): Have you looked into the effect of the $\pi^{\pm}\pi^{0}$ mass difference on the $\pi^{+}\pi^{-}$ asymmetry in η decay under the assumption that the C-violating transition is $\Delta I = 0$?

Lee: I believe there was one discussion during the Conference by Dr. Shaw. The idea is that the pions have a mass difference, so that even if the C violation is $\Delta I = 0$ you can mix in some $\Delta I = 2$ through this mass difference. To begin with, let us ask what will be the asymmetry even if you have pure $\Delta I = 1$ for K_μ and $\Delta I = 0, 1$ for J_μ. In this case, the final 3π can be in an $I = 2$, $C = -1$ state, and the $\pi^{\pm}$ asymmetry parameter can be estimated to be $A \approx (kR)^2 \sin(\delta_1 - \delta_2)$, where δ_1 and δ_2 are the 3π eigen phase shifts for the $I = 1$ and $I = 2$ final states. If K_μ is pure $\Delta I = 0$, then $A \approx (kR)^6 \sin(\delta_1 - \delta_0)$. What Dr. Nauenberg wants to point out, which has been discussed by Dr. Shaw, is that through the pion mass difference there can be an $I = 2$ three-pion final state even if K_μ satisfies the $\Delta I = 0$ rule; this will give a $\pi^{\pm}$ asymmetry $A \approx (\Delta m/m)(kR)^2 \sin(\delta_1 - \delta_2)$, which is the order of 10^{-3} if we take the interaction radius to be of the order of m_η^{-1}. Now, one can push this a little, but it seems extremely hard to get something of the order of a few percent.

Frazer (San Diego): I would like to comment on the sensitivity of time-reversal tests in photo- or electroproduction of pions in the region of the $N^*(1238)$. Here we are in the unusual situation of having a quite successful dynamical theory—the Chew-Goldberger-Low-Nambu theory. If one examines this theory, he finds that T invariance is not a necessary assumption. The weaker assumption of Hermitian analyticity (which is about as well-founded as crossing) suffices. Therefore, any T violation must fit into the discrepancy between the CGLN theory and experiment. Perhaps the situation would be more favorable at the higher-energy resonances.

Lee: I think this remark is a relevant one. On the other hand, one can also look apart from the theoretical point of view and examine the experimental situation—the question of measuring, say, the relative phase in the M1(3/2) and the E1(1/2). Now, we may ask: Independent of theory, can we measure the relative phase? The answer is yes. If you measure, to the order of 10% accuracy at the right energy and the right angle, the reciprocity, then you would measure this relative phase angle to the order of 10° accuracy. Dr. Frazer stressed that by using the Chew-Goldberger-Low-Nambu theory, one can get a relatively small deviation for this phase angle at the first resonance even if there is a basic large T violation.

Barshay: I comment on Professor Frazer's remark. I think reciprocity should be tested experimentally, not theoretically—in $\gamma + d \rightleftarrows \eta + p$ and $\gamma + p \rightleftarrows \pi^+ + \eta$, wherever it is possible, at energies about 150 MeV.

Todorov (Dubna): I would like to hear Professor Lee's opinion about the possibility of explaining CP violation by an interaction with strength $e\cdot G_{wk}$. I guess you are classifying this possibility in the group of weak-type interactions. I am referring to the paper by Arbuzov and Filippov (Paper 5a.1, submitted to Physics Letters), in which such a theory is proposed on a geometrical ground.

Lee: I think this is one of the many possible theories. I should mention in this connection that Salzmann and Salzmann have proposed a similar theory by allowing the weak vector boson to have an electric dipole moment. One cannot rule out such theories on the basis of the present experimental data. The situation reminds me of a cosmology; we have only one experiment but many theories. I regard this situation as unhealthy.

Omnes (Strasbourg): Since the intermediate boson is not observed, the rather remote possibility for its being a Regge pole is worth mentioning. It could have spin 1 for zero mass, zero residue, and would respect the locality of weak interactions. It would also explain the seemingly constant neutrino total cross section at high energy. It would never be produced as an observable particle. Analogous considerations can be made in conventional field theory with vanishing form factors at zero momentum transfers. You will never find it.

Michel (Bures-sur-Yvette): I would like to defend the usefulness of P and T even if they are not conserved. For free particles, or in a given reaction, P can be defined kinematically. P and T are elements of the Poincaré group and we know their action on space, time, energy, momentum, and polarization. (T exchanges, of course, initial and final particles in a reaction.)

The confusion you spoke about when considering invariance of Lagrangians can be avoided if one remembers that the discrete operations are defined up to a phase. As is well known to theorists who have worked on it, to prove that a Lagrangian is CPT-invariant and CP-violating, you have to prove that you can find an involutive transformation which has the action-defined CPT and, whatever choice of phase no CP involution can be performed. A likely possibility considered by you and many is that the full Lagrangian is CP-violating and any subpart of it is not, so you cannot blame CP violation for particular interactions, but the coexistence of all of them.

Lee: Kinematically you can, of course, have the coordinate transformations: space inversion and time inversion, but when you have a Lagrangian which is not invariant under any definition of C, P, T then in the Hilbert space, the unitary and the antiunitary operators C, P, T are also not defined (except their product CPT, which is well defined). [For the free particles, P (or CP) and T can indeed be defined. This is because the usual free-particle Lagrangian is invariant under P (or CP) and T. This cannot be (exactly) true for any physical reaction. Of course, one may select, among the infinite variety of possibilities, a specific ad hoc operator "P" in the Hilbert space (applicable to all states) and insist that it correspond to the geometrical transformation $\vec{r} \to -\vec{r}$, $t \to +t$. The fact that this operator "P" does not commute with the time translation operator $e^{-iH\tau}$ shows that the identification of "P" with the space-inversion transformation is an artificial one, except as an approximation.]

Telegdi (Chicago): I have two questions. I am somewhat embarrassed about the first one. I understand so little, so that's why I'm asking you. There was a recent letter by Rolnick which roughly makes the following statement: If one combines your ideas of C violation and some form or other of SU(6) (I have to make that statement carefully since I don't know what's meant by it), if one embraces these ideas one reaches the conclusion that particles of fractional charge are unavoidable — like quarks. Would you like to comment on that?

Lee: No comment.

Gottfried (Cornell): The content of Rolnick's letter is as follows. Let us adopt Professor Lee's suggestion that a nonzero Q_K exists, and amend the Gell-Mann–Nishijima formula accordingly. It is then rather natural to look for a strong-interaction symmetry group that has one generator, Q_K, even under C, as well as the two commuting SU(3) generators T_3 and Y, which are odd under C. However, because C must be an automorphism of this group, one can show that none of the rank-three Lie groups has the properties just delineated. In view of this, there does not seem to be any reason to worry about the question of particles with noninteger charge mentioned by Telegdi.

Telegdi: Second question. After Kabir presented some of his own work and Okun's remarks about this part of the current with isospin 2, there was a small discussion among several people who attended this talk. The quintessence of this talk was that known information on the selection rules in nuclear physics already rules out such currents. Perhaps Professor Radicati could comment on this.

Radicati (Pisa): What do you want me to comment about? I agree with what Telegdi said. There is evidence in nuclear physics coming from the validity of the E1 isotopic spin selection rule that there is no $\Delta I = 2$ component in the electromagnetic current.

Lee: That, I think, is very important. It shows, perhaps, that the idea that the isospin transformation properties of the current and the charge are the same is a useful one.

Kabir (Rutherford): Whatever support the assumed $\Delta I = 0$, 1 selection rule for electromagnetic interaction may receive from selection rules for nuclear γ transitions, this may not be very relevant to the question of possible terms in the electromagnetic current transforming as $|\Delta I| > 1$ objects, because (a) individual nucleons can only have $|\Delta I| \leq 1$ interactions, so that to the extent that the electromagnetic interaction of nuclei is well described by the sum of the interactions of individual nucleons, nuclear interactions are not sensitive tests, and (b) if the $|\Delta I| > 1$ parts of the electromagnetic interaction correspond to vanishing charge, their effects are more likely to appear at momentum transfers greater than those commonly occurring in nucleon transitions.

Low (MIT): In connection with $|\Delta I| = 2$ electromagnetic currents, it is of interest to ask whether $|\Delta I| = 2$ β-decay currents exist. One can examine this by comparing transitions from the two ends of an I = 2 multiplet to an I = 1 multiplet underneath. The data at present appear to show a 1-standard-deviation discrepancy. Also here we must remember that single nucleons can't produce a $\Delta I = 2$ transition, so that many nucleon effects are needed so that the amplitude can be quite suppressed.

A. Goldhaber (Berkeley): What is the basis for the choice of radius in the η decay? This is crucial to the upper limit on C-violating effects.

Lee: **There is no real basis. It is only an estimate.** If it is much larger than m_η^{-1}, say m_π^{-1}, then kR can be of order unity and the magnitude of the effect can be quite large. However, there is a way to distinguish experimentally $\Delta I = 0$ from $\Delta I = 2$ from the asymmetry of the spectrum. If it is $\Delta I = 0$, the asymmetry must change sign from sextant to sextant, since the C = -1 three-pion state must be a completely antisymmetric one. If it is $\Delta I = 2$, then it will not. By looking at the experimental spectrum, one can then work backwards and see which it is. I believe at present the experimental situation is quite unclear.

Ne'eman (Tel Aviv): Since you haven't sketched your proof of the nonexistence of the isovector possibility, does your proof, in fact, also exclude the possibility of a complete octet multiplet, and: does it also mean that it has to be a unitary singlet in the limit of SU(3)?

Lee: Right, the operator Q_K is also a unitary singlet.

Källen (Lund): I should like to comment about the definition of C, P, and T. It appears to me that the point made by Michel is that there is no difficulty in defining these symmetry operations at a given time. Your statement is that if you perform these operations at different times you get different results. This is certainly correct, but it appears to me that the statement that you cannot define C, P, or T as a consequence of this is a bit extreme. If you only state at which time you perform the symmetry operations, these concepts are perfectly well defined.

Lee: When you have, at every different time, a different definition, that means you have an infinite number of definitions. When anything has an infinite number of definitions, that's as good as being undefined.

Källen: Well, then you define the P operation at a given time; then you know what you're doing.

Lee: You may.

Amaldi (Istituto di Sanità, Rome): Would you comment about the use of the $\phi \to \omega + \gamma$ decay as a measure of CP nonconservation in the hypothesis that K_μ is an isoscalar?

Lee: The $\phi \to \omega + \gamma$ decay is consistent with the $\Delta I = 0$ rule; therefore it is possible in principle. However, the question is the rate. It is not, at this moment, clear what the rate might be. [The $\phi \to \omega + \gamma$ is, however, forbidden by SU(3) symmetry if K_μ transforms like a unitary singlet.]

Weisskopf (MIT): I have a simple question. You said that your new particle, which you called ϕ, might be the W meson, or decay like the W; but if it does so, where does the new kind of charge Q_K go then? It should be conserved, you said.

Lee: Yes, it is conserved by the strong and electromagnetic interactions, but it is not conserved by the weak interaction. You see, in a strict sense, there is only one conservation law for the electromagnetic current; that's the conservation of the total charge. Only because the strong interactions are invariant under C_{st} which is not equal to C_γ do we have two conservation laws. But these separate conservation laws are only approximate.

Nauenberg: It has been repeatedly stated at this Conference that the superweak theory of C violation and most other theories predict the phase of ϵ to be ≈42°. Isn't it true, however, that the sign of ϵ is <u>not</u> predicted?

Lee: To begin with, there is an ambiguity in the sign of ϵ, but it can be determined, since $\eta_{+-} = \epsilon$.

Yang (Stony Brook): In answer to Nauenberg, the sign is not theoretically determined.

Wolfenstein (Carnegie Tech): There is a sign ambiguity. If we write $\epsilon = \epsilon_0 e^{i\delta}$ with δ between +90° and -90°, then the superweak interaction predicts δ but ϵ_0 may be positive or negative and theory cannot tell the sign. This sign determines whether the interference is constructive or destructive, and has now been determined by experiment.

Lee: From the superweak interaction, one determines only the tangent of the phase of ϵ, and there are always two solutions which differ by π. In the same theory, $\epsilon = \eta_{+-}$, and the phase of η_{+-} has now been determined experimentally in the interference experiments.

M. Goldhaber (Brookhaven): What can $2\pi^0$ relative to $\pi^+\pi^-$ decays of K_2^0 teach us to decide among the various models?

Lee: Thank you very much. I meant to mention this point. For the K_2 decay, there are two parameters. One is η_{00}, the amplitude for $K_2 \rightarrow 2\pi^0$ divided by the amplitude for $K_1 \rightarrow 2\pi^0$. If the superweak theory is right, then $\eta_{00} = \eta_{+-} = \epsilon$. However, in many other models, this can also be true, or approximately true. I think only in one specific model is this absolutely not true; that is the model in which the violation is $\Delta I = 3/2$ only. Also, many models do not give you an exact prediction, but only a range of values. The $\Delta I = 3/2$ model is, however, definite, and it gives a factor-of-2 difference; i.e., $\eta_{00} = -2\,\eta_{+-}$.

Sachs (Argonne): I just want to repeat something I said in the discussion. And that is that this equation $\eta_{00} = \eta_{+-} = \epsilon$ is valid for a very wide class of models, not just the superweak model. The only cases in which it's not valid are the cases in which there is a violation in $\Delta I = 3/2$.

Reprinted from
Proc. XIIth Int. Conf. High Energy Physics Berkeley, 1966
Univ. California Press, pp. 75–81

C, P, T Symmetries*

T. D. LEE
Columbia University, New York, N.Y.

1. General Discussion

1) Symmetry Principles and Their Violations

Symmetry principles in physics are directly connected with "non-observables." For example, the impossibility of measuring the absolute space-time position leads to the four-dimensional translational symmetry, which, in turn, implies the conservation of energy and momentum. The impossibility of measuring the absolute spatial direction leads to the three-dimensional rotational symmetry and, therefore, the conservation law of angular momentum.

With our growing experimental techniques, more physical quantities become measurable. What were thought to be non-observables might turn out to be actually measurable, and hence the breakdown of symmetry principles.

The discrete symmetry operators C, P and T are related, respectively, to the supposed non-measurability of particle-antiparticle, right-left, and the absolute sign of t. As examples of the violations of these symmetry principles, we may recall:[1]

(i) $\theta-\tau$ Puzzle

$$K^+ \to \begin{cases} \pi^+\pi^0 & \theta\text{-mode} \\ \pi^+\pi^+\pi^- \text{ (or } \pi^+\pi^0\pi^0) & \tau\text{-mode} . \end{cases}$$

From $\pi^0 \to 2\gamma$ or from the p-state capture $\pi + N \rightleftarrows N$ one infers that

$$P(\pi^0) = -1 . \qquad (1)$$

Since K^+ has spin 0 and since the three pions in the τ-mode are found to be (predominantly) in the S-states, one concludes that the parity of θ and τ-modes must be different; i.e.,

$$P(\tau) = P(\theta) \cdot P(\pi^0) = -P(\theta) \qquad (2)$$

which shows the violation of parity conservation.

(ii) μ-decay

In μ^+-decay

* This research was supported in part by the U.S. Atomic Energy Commission.

$$\mu^+ \to e^+ + \bar{\nu}_\mu + \nu_e \,, \tag{3}$$

all e^+ are observed to have their spins parallel to their momenta, thus showing the separate violation of C and P symmetries.

(iii) In a similar way, by comparing the final sates in

$$K_2{}^0 \to \pi^+\pi^- \quad \text{and} \quad 2\pi^0$$

and

$$K_2{}^0 \to \pi^+\pi^-\pi^0 \quad \text{and} \quad 3\pi^0 \,, \tag{4}$$

one concludes the violation of CP symmetry. From the difference in the decay rates of

$$K_2{}^0 \to \pi^+ + l^- + \bar{\nu}_l$$

and

$$K_2{}^0 \to \pi^- + l^+ + \nu_l \tag{5}$$

one observes the violation of C and CP symmetries.

The time reversal asymmetry is, at present, only inferred indirectly through the use of CPT symmetry. In this respect, it is useful to recall that the experimental validity of CPT symmetry for the weak interaction is only through the measurement of lifetime equalities between $\pi^+\pi^-$, K^+K^-; its experimental accuracy is not much beyond the level of CP violation observed in K decay.

2) Time Reversal Symmetry

Time reversal symmetry implies that for every dynamical solution there exists another corresponding time-reversed solution of the same dynamical equations. For a macroscopic system the time-reversed solution is in general a highly improbable solution; it is this improbability that gives rise to our determination of the sign of the macroscopic time.

As we shall see, assuming time reversal symmetry, the time-reversed solution in quantum mechanics, even for a microscopic system (except for free particles) is in general highly improbable. Let us take the example of μ^+ decay.

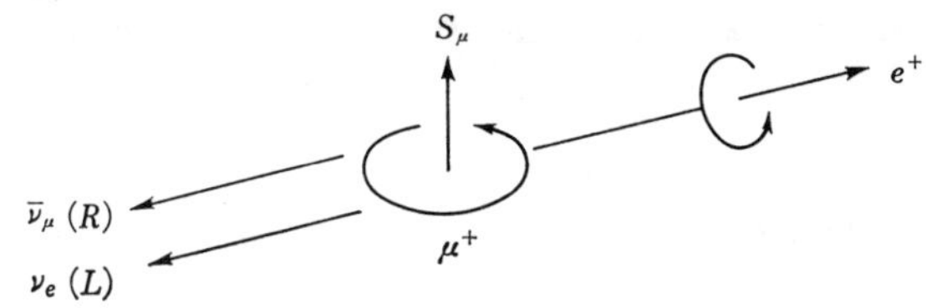

Fig. 1.

For simplicity, let us take the special case in which the momentum of $\bar{\nu}_\mu$ is parallel to that of ν_e. Now let us consider the inverse reaction

$$e^+ + \nu_e + \bar{\nu}_\mu \to \mu^+ \tag{6}$$

by reversing the momenta and the spins of e^+, ν_e and $\bar{\nu}_\mu$. It is easy to see that the spin of μ^+ formed is parallel to the direction of e^+ momentum which is, in general, *different* from the reverse of the initial μ-spin in the decay $\mu^+ \to e^+ + \nu_e + \bar{\nu}_\mu$.

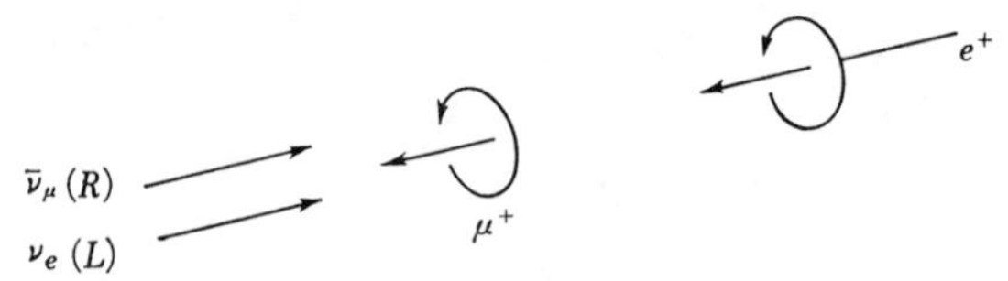

Fig. 2.

The above considerations can be applied to any reaction

$$a + b \cdots \to a' + b' + \cdots . \tag{7}$$

Even for a microscopic system, the time-reversed solution $T\psi(t)$ requires the construction of a coherent incoming spherical state which is extremely difficult to construct. For all practical purposes, time reversal symmetly implies "reciprocity"; i.e.,

$$\begin{aligned} &|<a'(k_a', \sigma_a'), b'(k_b', \sigma_b') \cdots |S| a(k_a, \sigma_a), b(k_b, \sigma_b) \cdots >| \\ &\quad = |<a(-k_a, -\sigma_a), b(-k_b, -\sigma_b) \cdots |S| a'(-k_a', -\sigma_a'), \\ &\qquad\qquad b'(-k_b', -\sigma_b') \cdots >| \end{aligned} \tag{8}$$

where S is the S-matrix and (k_i, σ_i) denotes the momentum and spin of the ith particle. Violation of time reversal symmetry means simply the violation of such reciprocity.

3) Implication of Non-conservation

Lut us consider, e.g., the case of parity non-conservation. In the literature, one often proceeds by first defining a space inversion operator P, and then producing a Hamiltonian H which does not commute with P,

$$[H, P] \neq 0 . \tag{9}$$

It is important to note that there is an element of self-inconsistency in this procedure. To see this, let us recall that the space inversion operator P is a unitary operator in the Hilbert space, and by definition represents the coordinate transformation-space inversion.

	Coordinate Transformation	Hilbert Space Representation
Space Inversion:	$t \to t,\ \boldsymbol{r} \to -\boldsymbol{r}$	P
Time Translation:	$t \to t+\tau,\ \boldsymbol{r} \to +\boldsymbol{r}$	$e^{-iH\tau}$.

For the coordinate transformation, it is obvious that space inversion should commute with time translation, the failure, shown by (9), points out clearly that P does not satisfy the multiplication laws of the coordination transformation that it is supposed to represent. The violation of space inversion symmetry shows that the operator P *cannot be exactly defined.*

Rather, one may make a dynamical approximation of replacing H by, say, the strong interaction Hamiltonian H_{st}. A resulting P_{st} can then be defined. The definition of $P = P_i$ depends, then, on the approximation

$$H \cong H_i$$

and

$$[H_i, P_i] = 0 \ .$$

To illustrate the interaction-dependent nature, let us consider again the θ-τ puzzle. We may define a $P_{wk}(\pi^0) = +1$; therefore $P_{wk}(\theta) = P_{wk}(\tau)$. The K^+ decay then conserves P_{wk}, but $\pi^0 \to 2\gamma$ and $\pi + N \to N$ violates P_{wk}. In (1), $P(\pi^0) = -1$ where $P = P_{st} = P_\gamma$ and it is violated by H_{wk}.

Identical considerations can be extended to CP and T. Furthermore, it can be shown explicitly that all present evidence is consistent with the view that each of the three known interactions H_i ($i = st$, wk and γ) is invariant under its own C_i, P_i and T_i.

4) Group Extension

The property that the definitions of these discrete symmetry operators depend on the particular dynamical approximation raises some interesting new questions. Consider the approximation $H_\gamma = H_{wk} = 0$ and

$$H_{\text{total}} = H_{st} \ . \tag{10}$$

We know that P_{st} exists where

$$[P_{st}, H_{st}] = 0 \ . \tag{11}$$

However, in this approximation, there is no difference between a proton and a neutron. Should the operator P_{st} commute with the isospin generator? If so, why? If not, what is the most general definition of P_{st}?

This general problem is solved in a recent paper[2)] by Wick and myself.

Let $\mathscr{G}$ be the internal symmetry group of H_{st}. For example, $\mathscr{G}=SU_2$ group of iso-spin rotation, or $\mathscr{G}=SU_3$ if we are interested only in the SU_3 conserving part of the strong interaction.

$$\mathscr{G}=\{u\} \tag{12}$$

where u is the member of $\mathscr{G}$. From any particular solution $P_{st}=P$ we can generate a set

$$\{uP\}\ . \tag{13}$$

It is clear that the transformation

$$u \rightarrow PuP^{-1} \tag{14}$$

is a mapping of the group $\mathscr{G}$ into itself; i.e., it is an automorphism. Since for every group $\mathscr{G}$, the different possibilities of automorphism can be classified; the algebra between P and $\mathscr{G}$ is therefore also determined.

As an example, we may consider the SU_2 group.

Theorem 1. If

$$u \rightarrow F(u) \tag{15}$$

is an automorphism of the group

$$\mathscr{G}=\{u\}=SU_2 \tag{16}$$

then there exists a fixed v in $\mathscr{G}$ such that

$$F(u)=vuv^{-1} \tag{17}$$

for every u in $\mathscr{G}$.

Such automorphism is called inner-automorphism. From this theorem, one immediately sees that for the iso-spin SU_2 group, if we happen to choose a P which does not commute with the iso-spin, since for every such P there exists a fixed iso-spin rotation v such that

$$PuP^{-1}=vuv^{-1}$$

for all iso-spin rotations u, one can redefine a

$$P_{st}=v^{-1}P$$

then

$$P_{st}uP_{st}^{-1}=u$$

or

$$[P_{st}, \boldsymbol{I}]=0\ . \tag{18}$$

There exists a more general theorem applicable to other SU_N groups.

Theorem 2. If $u \rightarrow F(u)$ is an automorphism of the group

$$\mathscr{G}=\{u\}=SU_N \tag{19}$$

where $N \geqslant 3$ and u are the $N \times N$ matrices with determinant 1, then there exists a fixed v in $\mathscr{G}$ such that for every u in $\mathscr{G}$

$$F(u)=vuv^{-1} \tag{20}$$

or, for every u

$$F(u)=vu^*v^{-1} . \tag{21}$$

The new possibility (21) is an outer automorphism; for the parity case, it corresponds to the choice CP. [For the SU_2 group, $u^*=\sigma_2 u \sigma_2^{-1}$. Thus the mapping $u \rightarrow vu^*v^{-1}$ is also an inner-automorphism. Thus, one can construct $G=C\exp(iI_2\pi)$ which commutes with the iso-spin generators $\boldsymbol{I}$.]

2. Possible Existence of C_{st} and T_{st} Violating Electromagnetic Interaction

1) General Discussion

Throughout this lecture we shall assume that the electromagnetic interaction H_γ is invariant under its own C_γ, T_γ and P_γ. Similarly, the strong interaction H_{st} is also invariant under its own C_{st}, T_{st} and P_{st}. The former is a theoretical assumption, while the latter is an experimental fact. Furthermore, from the mass difference between K_1^0 and K_2^0

$$|\Delta m| \sim 10^{-14}\, m_K , \tag{22}$$

one infers that

$$C_{st}P_{st}T_{st} = C_\gamma P_\gamma T_\gamma \equiv CPT ; \tag{23}$$

from the absence of any parity-mixing amplitude in the known nuclear wave functions, one concludes that

$$P_\gamma = P_{st} . \tag{24}$$

At present, it remains an open question whether

$$C_{st} \stackrel{?}{=} C_\gamma ,$$

or

$$T_{st} \stackrel{?}{=} T_\gamma . \tag{25}$$

In this lecture we will assume

$$C_{st}H_{\gamma}C_{st}^{-1} \neq H_{\gamma}$$

and

$$T_{st}H_{\gamma}T_{st}^{-1} \neq H_{\gamma} . \tag{26}$$

Therefore,

$$C_{st} \neq C_{\gamma}$$

and

$$T_{st} \neq T_{\gamma} . \tag{27}$$

2) **Some Necessary Properties of K_μ**

We may resolve the hadronic electromagnetic current $\mathscr{J}_\mu$ into two parts

$$\mathscr{J}_\mu = J_\mu + K_\mu \tag{28}$$

where

$$J_\mu = \frac{1}{2}(\mathscr{J}_\mu - C_{st}\mathscr{J}_\mu C_{st}^{-1})$$

and

$$K_\mu = \frac{1}{2}(\mathscr{J}_\mu + C_{st}\mathscr{J}_\mu C_{st}^{-1}) . \tag{29}$$

The assumption that $C_{st}H_{\gamma}C_{st}^{-1} \neq H_{\gamma}$ implies

$$K_\mu \neq 0 . \tag{30}$$

Since

$$[H_{st}, C_{st}] = 0 \tag{31}$$

and $\mathscr{J}_\mu$ is conserved under both H_{st} and H_γ, it follows that under H_{st}

$$\frac{\partial J_\mu}{\partial x_\mu} = \frac{\partial K_\mu}{\partial x_\mu} = 0 . \tag{32}$$

Thus, the charge associated with each of these two currents must respectively be conserved; i.e.,

$$[Q_J, H_{st}] = [Q_K, H_{st}] = 0 \tag{33}$$

where

$$Q_J = -i\int J_4 d^3r \tag{34}$$

and

$$Q_K=-i\int K_4 d^3r\ . \tag{35}$$

The total hadron charge Q is

$$Q=Q_J+Q_K\ . \tag{36}$$

From (29), one has

$$\{Q_J,\ C_{st}\}=0 \tag{37}$$

and

$$[Q_K,\ C_{st}]=0 \tag{38}$$

where { } denotes the anti-commutator and [ ] the commutator.

Theorem. $\qquad Q_K\mid\text{all known particles}>=0\ .$

Proof. Consider the subspace spanned by

$$\begin{pmatrix} p \\ n \\ \bar{n} \\ \bar{p} \end{pmatrix} \tag{39}$$

where p, n, $\bar{n}$, $\bar{p}$ denote the usual one-nucleon and one-antinucleon state. From $[Q_K, H_{st}]=0$, it follows that the state, say, $Q_K|p>$ must be degenerate with $|p>$, apart from possible electromagnetic corrections. Thus $Q_K|p>$ must remain in the sub-space (39). In the same sub-space,

$$Q=Q_J+Q_K=\begin{pmatrix} 1 & & & \\ & 0 & & \\ & & 0 & \\ & & & -1 \end{pmatrix} \tag{40}$$

and C_{st} is

$$C_{st}=\begin{pmatrix} & & & 1 \\ & & 1 & \\ & 1 & & \\ 1 & & & \end{pmatrix}; \tag{41}$$

therefore, $\{Q, C_{st}\}=0$. From (37), one finds $\{Q_K, C_{st}\}=0$ which, together with (38), implies $Q_K=0$ in the sub-space (39).

Next, one considers the sub-space spanned by

$$\begin{pmatrix} \Lambda^0 \\ \bar{\Lambda}^0 \end{pmatrix}, \tag{42}$$

in which $Q=Q_J+Q_K=0$ and

$$C_{st}=\begin{pmatrix} 0 & 1 \\ 1 & 0 \end{pmatrix}.$$

The same reasoning leads to the conclusion that $Q_K=0$ also in the subspace (42). The Theorem can then be proved by considering all possible compound states of nucleons, antinucleons, Λ^0 and $\bar{\Lambda}^0$.

From the Theorem, it follows that if $|A>$ and $|B>$ are any two states composed only of *known* particles, then

$$<B|K_\mu|A> = \frac{\partial}{\partial x_\nu} <B|M_{\mu\nu}|A> \tag{43}$$

where $M_{\mu\nu} = -M_{\nu\mu}$. (43) resembles in form the electric current generated by magnetic moments.

3) **Classification**

All such C_{st}, T_{st} violating electromagnetic interactions can be separated into two classes:

A. There exist states, say $|a>$, such that

$$Q_K|a> \neq 0\,. \tag{44}$$

B. $Q_K=0$ always. (45)

Example of B:

$$K_\mu = i\lambda \frac{\partial}{\partial x_\nu}(\phi_\nu\omega_\mu - \phi_\mu\omega_\nu) \tag{46}$$

where ϕ_μ, ω_μ are the field operators of the known ϕ^0 and ω^0 mesons [both of $C_{st}=-1$, but $C_\gamma(\phi) = -C_\gamma(\omega)$].

Example of A:

Let $a^\dagger$ be an $I=0$, unitary singlet, $Q=+1$, spin$=1$ particle, and a_μ its associated field operator.

$$K_\lambda = ie\left[\frac{\partial a_\nu^*}{\partial x_\lambda}a_\nu - a_\nu^*\frac{\partial a_\nu}{\partial x_\lambda} - \left(\frac{\partial a_\nu^*}{\partial x_\nu}\right)a_\lambda + a_\lambda^*\frac{\partial a_\nu}{\partial x_\nu}\right] + i\mu\frac{\partial}{\partial x_\nu}[a_\lambda^* a_\nu - a_\lambda a_\nu^*] \tag{47}$$

where

$$a_\nu^* = \begin{cases} a_\nu^\dagger & \text{for } \nu\neq 4 \\ -a_4^\dagger & \text{for } \nu=4\,, \end{cases}$$

e is the electric charge and μ its magnetic moment. K_λ is the usual minimal electromagnetic current.[3] The strong interaction of $a^\dagger$ with, say, the nucleon is of the form

$$\mathscr{L}_{st} = g(N^\dagger\gamma_4\gamma_\lambda\gamma_5 N)\cdot\varepsilon_{\lambda\mu\nu\sigma}\frac{\partial}{\partial x_\mu}(a_\nu^* a_\sigma)\,. \tag{48}$$

In order that $\mathscr{L}_{st}$ is invariant under

$$C_{st}NC_{st}^{-1}=N^{\dagger}\ , \tag{49}$$

one must have

$$C_{st}a_{\mu}C_{st}^{-1}=a_{\mu}\ . \tag{50}$$

In (49), we have chosen, for convenience, the Majorana representation of the γ_{μ} matrices.

It is easy to see that the electromagnetic interaction is invariant under C_{γ} where

$$C_{\gamma}a_{\mu}C_{\gamma}^{-1}=a_{\mu}^{*}\ . \tag{51}$$

Thus, one has

$$[H_{\gamma}, C_{\gamma}]=0\ ,$$

$$[H_{st}, C_{st}]=0\ ,$$

$$[H_{\gamma}, C_{st}]\neq 0\ ,$$

$$[H_{st}, C_{\gamma}]\neq 0\ ,$$

$$C_{st}K_{\mu}C_{st}^{-1}=+K_{\mu}\ ,$$

but

$$C_{\gamma}K_{\mu}C_{\gamma}^{-1}=-K_{\mu}\ .$$

Independent of the example, one can establish some further necessary consequences of Q_K for the class A type C_{st} violating interactions.

4) Some Further Properties of Q_K

Assume that there exist "*a*-particles" with

$$Q_K(a)\neq 0 \tag{52}$$

and the iso-spin of a is not necessarily zero

$$\boldsymbol{I}(a)\neq 0\ .$$

Nevertheless, as we shall see,

$$Q_K \text{ remains an iso-scalar.} \tag{53}$$

Proof. Consider the set of "*a*-particles" with different $I_z(a)$. Since Q_K and I_z are both additive conserved quantum numbers, one must have

$$Q_K(a)=\lambda I_z(a)+\lambda'$$

where λ is a constant and λ' can be either a constant or another additive quantum number. Under *CPT*,

$$a \to \bar{a} \equiv CPT \text{ conjugate of } a \ ,$$

$$Q_K(a) \to -Q_K(a) = Q_K(\bar{a}) \ ,$$

$$I_z(a) \to -I_z(a) = I_z(\bar{a}) \ .$$

Hence,

$$Q_K(\bar{a}) = \lambda I_z(\bar{a}) - \lambda'$$

and

$$Q_K(a\bar{a}) = \lambda I_z(a\bar{a}) \ . \tag{54}$$

The a-particle cannot be produced singly from known particles through strong interactions, but

$$\text{known particles} \rightleftarrows a\bar{a} + \text{known particles}$$

must occur; or

$$\text{known particles} \rightleftarrows a\bar{a} \ (\text{or, } 2a + 2\bar{a}, \cdots) \tag{55}$$

must occur. Thus

$$Q_K(a\bar{a}) = Q_K \, (\text{known particles}) = 0 \ ,$$

or

$$\lambda I_z(a\bar{a}) = \lambda I_z \, [\text{known particle states in (55)}] = 0 \ . \tag{56}$$

Consider any reaction of the form (55) with $|\boldsymbol{I}_{\text{total}}| \neq 0$. We can choose $(I_z)_{\text{total}} \neq 0$. From (56), we conclude

$$\lambda = 0 \ ,$$

and Q_K is an iso-scalar. Now, if all such reactions (55) can only occur with $I_{\text{total}} = 0$, then the theory is invariant under the $SU_2 \times SU_2$ group generated independently by $\boldsymbol{I}(a)$ and the total isospin of the known particles, $\boldsymbol{I}$ (known particles). Thus, $Q_K(a)$ is an iso-scalar with respect to the isospin SU_2 group generated by $\boldsymbol{I}$ (known particles), which also leads to (53).

In an identical way, one can also prove that Q_K is an SU_3 unitary singlet.

If one makes the *additional* assumption that the algebra between $\boldsymbol{I}$ and K_μ is the same as that between $\boldsymbol{I}$ and Q_K, then

$$K_\mu \text{ is an iso-scalar} \tag{57}$$

and it is also an SU_3 unitary singlet.

5) Experimental Consequences of K_μ

(i) The decay

$$K_2{}^0 \to 2\pi \tag{58}$$

becomes possible through the third order effect $H_{wk}K_\mu J_\nu$. Since J_ν has $|\Delta I|=0$ and 1 amplitude, H_{wk} has $|\Delta I|=1/2$ amplitude, both $|\Delta I|=1/2$ and 3/2 amplitudes in the transition (58) can occur, and carry comparable magnitudes, giving

$$|\varepsilon| \sim |\varepsilon'| \sim 0(\alpha)\ . \tag{59}$$

(ii) The electromagnetic form factor of the nucleon

$$\langle N|K_\mu(x)|N\rangle = 0$$

by Hermiticity and $(\partial K_\mu/\partial x_\mu)=0$.

(iii) If K_μ exists, then π^0 can decay into 3γ.

$$\left[\frac{\text{Rate }(\pi^0 \to 3\gamma)}{\text{Rate }(\pi^0 \to 2\gamma)}\right]_{\text{th}} \sim \alpha(k_\gamma R)^8 \frac{(\text{phase space})_{3\gamma}}{(\text{phase space})_{2\gamma}} < 10^{-8}\ . \tag{60}$$

An example of such an effective Lagrangian is

$$\mathscr{L}_{\text{eff}} \propto e^3 \frac{\partial \phi_\pi}{\partial x_\alpha} F_{\mu\nu}\varepsilon_{\mu\nu\rho\sigma} \frac{\partial^2 F_{\rho\sigma}}{\partial x_\beta \partial x_\gamma} \frac{\partial F_{\alpha\beta}}{\partial x_\gamma}\ . \tag{61}$$

The present experimental value is

$$\left[\frac{\text{Rate }(\pi^0 \to 3\gamma)}{\text{Rate }(\pi^0 \to 2\gamma)}\right]_{\text{exp}} < 10^{-5}\ . \tag{62}$$

(iv) $\eta^0 \to \pi^+\pi^-\gamma$

Through $(J_\mu)_{I=1}$, $I_{2\pi}$ can be 1 (i.e., p-state), and through $(K_\mu)_{I=0}$, $I_{2\pi}$ can be 0 (i.e., d-state). Let E_+ and E_- be the energy of π^+ and π^- in the rest system of η^0. The asymmetry parameter A is defined to be

$$A = \frac{N(E_+ > E_-) - N(E_+ < E_-)}{N(E_+ > E_-) + N(E_+ < E_-)}\ . \tag{63}$$

$$|A_{\text{th}}| \sim |\tfrac{1}{3}(kR)^2 \sin(\delta_p - \delta_d)| \sim |\sin(\delta_p - \delta_d)| \cdot 3\times 10^{-2}\ , \tag{64}$$

$$|A_{\text{exp}}| = (+1.5 \pm 2.5)\% \tag{65}$$

from R. A. Bowen et al., Phys. Letters **24B**, 206 (1967).

(v) Neutron Electric Dipole Moment

If there exists a $\Delta S=0$ non-leptonic $P=-1$, $CP=+1$ weak interaction, then there exists a neutron electric dipole moment, through such an interaction and K_μ. At present it is not clear what is the magnitude of such an interaction. We note that in the approximation.

$$(H_{wk})_{\Delta S\neq 0,\ \text{non-leptonic},\ P=-1}=K_1{}^0(x) \tag{66}$$

then, through the usual octet hypothesis

$$(H_{wk})_{\Delta S\neq 0,\ \text{non-leptonic},\ P=-1}=0\ . \tag{67}$$

[Since $\pi(x)$, $\eta^0(x)$ have $CP=-1$.] Thus, in this approximation, the neutron electric dipole moment $\cong 0$, even though $K_\mu(x)$ may exist.

[Recently, V. M. Lobashov et al., Phys. Letters **25B**, 104 (1967) reported a direct observation of parity non-conservation in the gamma transition of ^{181}Ta.]

(vi) If $K_\mu(x)$ is an iso-scalar, then to the lowest order in α,

$$\eta_0 \not\rightarrow \pi^0 e^+ e^-\ , \tag{68}$$

there are no T_{st} violating effects in

$$\Sigma^0 \rightarrow \Lambda^0 e^+ e^- \tag{69}$$

and the asymmetry parameter A in

$$\eta^0 \rightarrow \pi^+\pi^-\pi^0$$

is

$$A \sim |(kR)^6 \sin(\delta_{I=1}-\delta_{I=0})| \sim 10^{-3}\cdot|\sin(\delta_{I=1}-\delta_{I=0})| \tag{70}$$

if $(kR)\sim 1/3$. The present experimental value [A. M. Cnops et al., Phys. Letters **22**, 546 (1966)] is

$$A_{\exp} = (0.3\pm 1.1)\%\ . \tag{71}$$

6) Application of Field Current Identity

If one applies the idea of field-current identity and assumes the algebra between K_μ and $\boldsymbol{I}$ to be the same as that between Q_K and $\boldsymbol{I}$, then the existence of K_μ current implies that it is also an $I=0$ meson field. It would be natural to expect the existence of a corresponding strongly interacting meson, which may be called h-meson (h for hawaion), with the following properties;

$$\left.\begin{matrix} J=1 & P=-1\ , \\ I=0 & C=+1\ . \end{matrix}\right\} \tag{72}$$

Consequently, $G=+1$. The field operator of h is $K_\mu(x)$. The h-meson has strong decays

$$h \to 2\rho, \quad \bar{K}K\pi, \quad \pi\pi\eta, \quad 4\pi, \quad \text{etc.} \tag{73}$$

provided its mass is sufficiently large. It cannot decay into $2\eta^0$ because of Bose statistics; it cannot decay into 2π and $\bar{K}K$ except in violation of C_{st} conservation; it also does not decay into 3π except in violation of G conservation.

References

1) For references to these experimental works, see the article by Professor V. L. Fitch in the "Proceedings to the Second Hawaii Topical Conference in Particle Physics", University of Hawaii Press, Honolulu, 1968, and the review article by T. D. Lee and C. S. Wu, Annual Review of Nuclear Science **15**, 381 (1965), **16**, 471 (1966).

2) For further details, see T. D. Lee and G. C. Wick, Phys. Rev. **148**, 1385 (1966).

3) T. D. Lee, Phys. Rev. **140**, B967 (1965).

WEAK INTERACTIONS AND CPT ASYMMETRIES

T. D. LEE
Columbia University,
New York, N. Y.,
United States of America

Abstract

WEAK INTERACTIONS AND CPT ASYMMETRIES. 1. Introduction; 2. General symmetry properties of weak interactions; 3. CP asymmetry; 4. Phenomenological Lagrangian and its selection rules; 5. Limitation of Fermi's theory and higher-order weak interactions.

1. INTRODUCTION

When Abdus Salam asked me to review in this Symposium the present status of weak interactions and CPT asymmetries, I thought this would be an easy task. It is well known that the status of this year does not differ too much from that of last year, which in turn is about the same as that of the year before last, etc. Therefore, I could simply refer you to the annual report of the 1967 International Conference on High Energy Physics, which is also quite similar to the 1966 annual report. However, such a description would not do justice to the enormous efforts during the past few years. Indeed, a number of new experimental results have been reported. The apparent slowness of progress lies in the curious, but perhaps profound, difficulty that in spite of the small probable errors quoted in some of these publications, very similar experiments performed by different groups, and sometimes even by the same group but at different times, may give contradictory results.

It should be emphasized that such a phenomenon of mutual contradiction has been an almost integral part of weak interaction research for quite some time. A classic example is the Michel ρ-parameter [1] in μ-decay

$$\mu^- \longrightarrow e^- + \bar{\nu}_e + \nu_\mu$$

$$\mu^+ \longrightarrow e^+ + \nu_e + \bar{\nu}_\mu$$

The normalized momentum spectrum of $e^\pm$ is given by

$$N(x) = 6x^2\left[2(1-x) + \frac{4}{9}(-3+4x)\rho\right]$$

where

$$\int_0^1 N(x)\,dx = 1$$

and x is related to the magnitude of the electron momentum p_e and the muon mass m_μ ($\hbar = c = 1$) by

$$x = \frac{p_e}{\frac{1}{2}m_\mu}$$

As one can see from Fig.1, different ρ-values give very different shapes of the electron spectrum. The theoretical value based on the two-component neutrino theory [2] gives

$$\rho_{th} = \frac{3}{4}$$

which, as shown in Fig. 2, is in excellent agreement with the present experimental value [3]

$$\rho_{exp} = 0.747 \pm 0.005$$

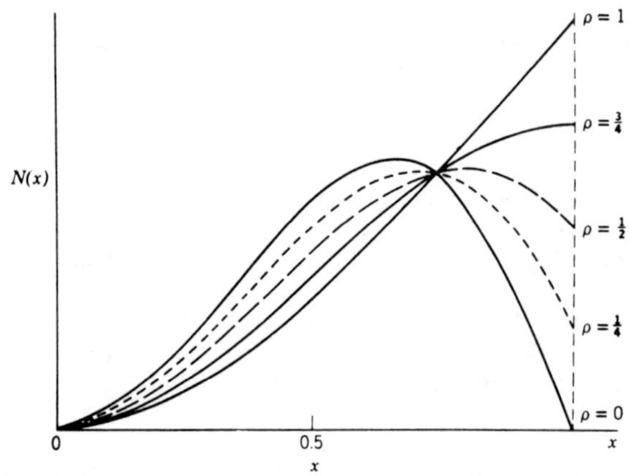

FIG. 1. Shapes of the electron spectrum for various values of ρ.

The theoretical value was calculated near the end of 1956. Figure 3 shows that before 1956 some rather different experimental values have been reported. (The first correct ρ-value — within one standard deviation — was determined by Sargent et al. [4].) Similar fluctuations and changes have also been observed in many other experiments, for example the measurements on the π_{e2} decay rate, the experiments on angular correlations between the β-particle and the recoiled ^{6}He nucleons, the more recent results on the $2\pi^0$ decay of the long-lived K-meson, etc.

Such evolutionary changes certainly make the task of a reviewer difficult. In this summary paper, an effort will be made to distinguish the different degrees of reliability of various experiments and theories. I will, now and then, follow the admirable example of the Guide Michelin, and use the rating symbols of *, ? and X. These ratings are quite subjective; their yearly changes might give a more truthful representation of our understanding of the subject.

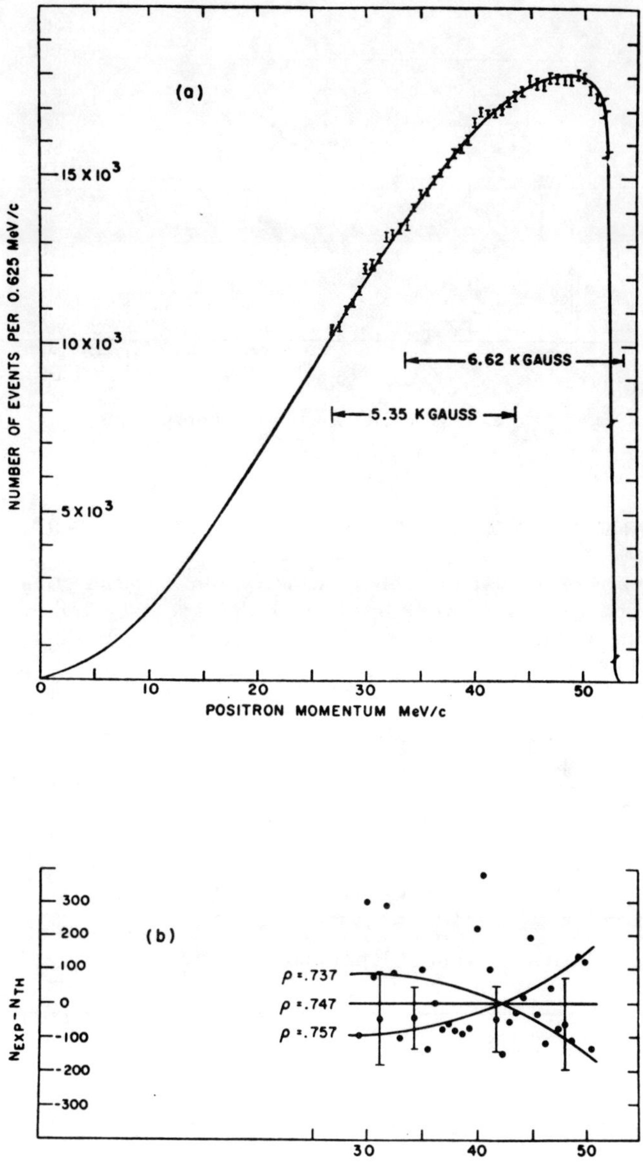

FIG. 2. a) Number of events per 0.625 MeV/c ; b) experimental value of ρ.

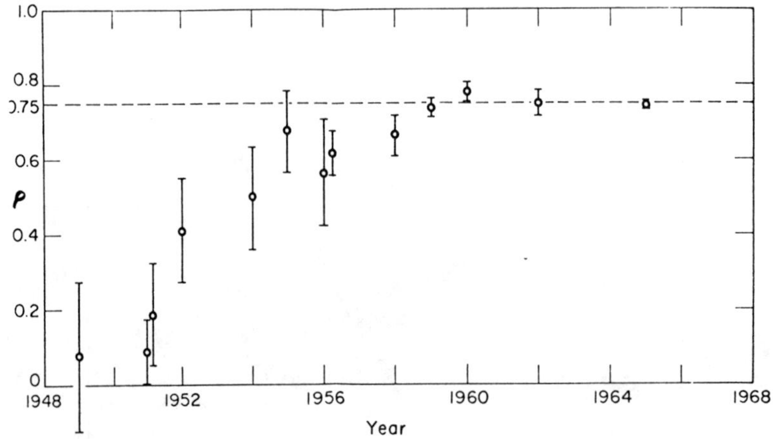

FIG. 3. Experimental values of ρ reported since 1948.

2. GENERAL SYMMETRY PROPERTIES OF WEAK INTERACTIONS

To illustrate the use of such a rating system, we may first discuss those general properties of the weak interaction that are expected to hold to <u>all</u> orders of the weak interaction (for details see, for example, the review article by Lee and Wu [5]).

Rating	Property
***	Lorentz invariance Charge conservation Baryon number conservation
**	CPT symmetry Two-component theory of the neutrino
*	Conservation of lepton numbers, L_e and L_μ μ-e symmetry
?	$U_2 \times U_2$ symmetry among leptons

All these symmetry properties are consistent with our present knowledge of the weak interaction. The best evidence of CPT symmetry comes from the mass equality [1]

$$\langle K^0 | H | K^0 \rangle = \langle \overline{K}^0 | H | \overline{K}^0 \rangle$$

[1] The K_1^0 and K_2^0 mass difference has been measured by many groups, see Ref. [6] .

which is accurate to 1 part in 10^{14}, where H is the total Hamiltonian; in addition, there are the lifetime measurements [7]

$$\left|\frac{\Delta\tau}{\tau}\right| < \begin{cases} 10^{-3} & \text{for } K^{\pm} \\ 10^{-3} & \text{for } \mu^{\pm} \\ 6 \times 10^{-3} & \text{for } \pi^{\pm} \end{cases}$$

where $\Delta\tau$ is the difference between the lifetime τ of the particle and that of its antiparticle. One notes that the CPT operation used in quantum field theory is by no means an obvious one. The requirement of CPT invariance states that when one changes particles to antiparticles, but keeping their momenta fixed and helicities reversed, all matrix elements of the interaction should become their respective complex conjugates. None of these experiments has any bearing on the particular anti-unitary nature of CPT symmetry; this accounts for its ** rating in the above list, instead of ***.

Before 1956, the space inversion P and the particle-antiparticle conjugation C would probably have belonged to the *** rank. Now both are removed from the list. Before 1964, CP would probably have been rated as, perhaps, a * symmetry; it is now also removed from the list.

Throughout the lecture, C, P and the time reversal operator T denote, respectively, the usual C_{st}, P_{st} and T_{st}, determined by the strong interaction.

3. CP ASYMMETRY

The first experiment observing CP asymmetry was reported by Christenson et al. [8], who found that the long-lived component K_2^0 of the neutral K-meson has a 2π decay mode

$$K_2^0 \rightarrow \pi^+ + \pi^-$$

Since both the particle-antiparticle conjugation operator C and the space inversion P (in the centre-of-mass system) interchange π^+ and π^-, the final (2π) system in this decay must be of CP = +1. On the other hand, the same long-lived component K_2^0 also decays into 3π,

$$K_2^0 \rightarrow \pi^+ + \pi^- + \pi^0$$

in which the π^+ and π^- are observed to be predominantly in the S-state. Thus, the final 3π-state is (or, at least, predominantly is) of CP = -1.

Now, K_2^0 is, by definition, one with a definite lifetime. The fact that it can decay into different states with opposite CP values shows conclusively that CP is not conserved in the K_2^0 decays. The CP violation is, therefore, established independently of the detailed theory of the K_1^0, K_2^0 system.

More recently, CP violation has also been observed in

$$K_2^0 \longrightarrow 2\pi^0$$

$$K_2^0 \longrightarrow \begin{cases} \pi^- + e^+ + \nu_e \\ \pi^+ + e^- + \bar{\nu}_e \end{cases}$$

and

$$K_2^0 \longrightarrow \begin{cases} \pi^- + \mu^+ + \nu_\mu \\ \pi^+ + \mu^- + \bar{\nu}_\mu \end{cases}$$

So far, there is no direct observation of time reversal asymmetry. If CPT invariance is assumed, then CP asymmetry implies T asymmetry.

The amplitudes of CP violation are uniformly small. For the 2π decay modes of K_2^0, it is useful to define

$$\eta_{+-} = \frac{\text{Amplitude } (K_2^0 \rightarrow \pi^+\pi^-)}{\text{Amplitude } (K_1^0 \rightarrow \pi^+\pi^-)}$$

and

$$\eta_{00} = \frac{\text{Amplitude } (K_2^0 \rightarrow 2\pi^0)}{\text{Amplitude } (K_1^0 \rightarrow 2\pi^0)}$$

The values $|\eta_{+-}|^2$ and $|\eta_{00}|^2$ are directly obtainable from the observed decay rates. The present experimental values are

$$*** \quad |\eta_{+-}|^2 = (3.6 \pm 0.2) \times 10^{-6}$$

and

$$|\eta_{00}|^2 = (24 \pm 5) \times 10^{-6} \quad \text{(Ref. [9])}$$

$$(19 {}^{+9.5}_{-6.9}) \times 10^{-6} \quad \text{(Ref. [10])}$$

$$(-2 \pm 7) \times 10^{-6} \quad \text{(Ref. [11])}$$

The last value is obtained through a subtraction of 102 expected background events from a total of 100 possible candidates. One may mention that the earlier experiment on $|\eta_{00}|^2$ has just lost its ** rating; the authors are encouraged to improve the experiment and to return to its former high standing.

In addition, the phase of η_{+-} has been measured. Its present value [12] is reported to be

$$\text{phase } (\eta_{+-}) = (51.2 \pm 11)^\circ$$

For the leptonic three-body decay modes, the CP violation is usually expressed in terms of the parameter

$$4 \operatorname{Re} \epsilon = \frac{\text{Rate } (K_2^0 \to \pi^- \ell^+ \nu_\ell)}{\text{Rate } (K_2^0 \to \pi^+ \ell^- \bar{\nu}_\ell} - 1$$

The present experimental value [13, 14] of the right-hand side is

$$(8.1 \pm 2.7) \times 10^{-3} \qquad \text{for } \ell = \mu$$

and

$$(4.48 \pm 0.72) \times 10^{-3} \qquad \text{for } \ell = e$$

At present, there are certain doubts as to the precise nature of the CP violating interaction H_-, defined by

$$H_- = \tfrac{1}{2} [H - (CP) H (CP)^{-1}]$$

where H is the total Hamiltonian. Table I summarizes the present theoretical status.

TABLE I. SUMMARY OF PRESENT STATUS OF CP VIOLATION

		Coupling constant of H_-	Selection rule
?	Super-weak [15]	$\sim 10^{-9}\, G_{wk}$	$\Delta S = \pm 2$ allowed
?	Milli-weak [16] [a]	$\sim 10^{-3}\, G_{wk}$	$\Delta S = \pm 1$ allowed $\Delta S \neq 2$
?	Electromagnetic [17]	e	$\Delta S = 0$ only P conserved C not conserved
?	Milli-strong [18]	$\sim 10^{3}\, G_{wk}$	

[a] The possibility that the CP violation is due to the weak interaction is an obvious one after the experimental discovery of $K_2^0 \to \pi^+\pi^-$. The precise mechanism of such a weak CP violation has been the subject of many theoretical papers. Ref. [16] gives only an incomplete list of these papers.

If H_- is the super-weak interaction, then neglecting effects $O(10^{-9})$, only in the decay of K_1^0 and K_2^0 can one detect any CP violations; furthermore, all such CP violations are determined by the state-vectors of K_1^0 and K_2^0. If one assumes CPT invariance, the K_1^0 and K_2^0 states are related to the states K^0 and $\bar{K}^0$ by

$$K_1^0 = [2(1 + |\epsilon|^2)]^{-\frac{1}{2}} [(1+\epsilon) K^0 + (1-\epsilon)\bar{K}^0]$$

and

$$K_2^0 = [2(1 + |\epsilon|^2)]^{-\frac{1}{2}} [(1+\epsilon) K^0 - (1-\epsilon)\bar{K}^0]$$

where

$$\overline{K}^0 = (CPT)K^0$$

In the super-weak case

$$\epsilon = \eta_{+-} = \eta_{00}$$

in addition [19]

$$\text{phase } \epsilon = \tan^{-1}\frac{2(m_2 - m_1)}{\gamma_2 - \gamma_1} \cong 44°, \text{ or } 224°$$

where m_i and γ_i are, respectively, the mass and the width of K_i^0 (i=1, 2). These values are quite consistent with the present experimental situation. In this case, for all practical purposes, we have already determined all that could be measured concerning CP violating parameters. The super-weak interaction has a certain appeal, because, then, we do not have to think any more.

If the CP violations are due to a milli-weak interaction, then there should be other weak processes besides K_1^0 and K_2^0 decays that exhibit some milli-CP-violating amplitudes. If the CP non-conservation interaction belongs to the milli-strong case, then, in addition, there should be some strong processes that exhibit some small C or T violations. If the CP non-conserving interaction is the electromagnetic interaction, then, in addition, there should be some electromagnetic processes that have sizable C, T violating amplitudes.

For example, the decay

$$\eta^0 \rightarrow \pi^+\pi^-\pi^0$$

is forbidden by the known strong interaction, but allowed by the known electromagnetic interaction. If the CP violating interaction belongs either to the electromagnetic case or to the milli-strong case, then in this η^0 decay there could be a small $\pi^\pm$ asymmetry $\sim$1% to 0.1%, depending on the isospin selection rule of the CP violating interaction. The asymmetry parameter A is defined to be

$$A = \frac{N(E_{\pi^+} > E_{\pi^-}) - N(E_{\pi^+} < E_{\pi^-})}{N(E_{\pi^+} > E_{\pi^-}) + N(E_{\pi^+} < E_{\pi^-})}$$

The recent experiment [20] of 36 500 η^0 events gives

$$A = (1.5 \pm 0.5) \times 10^{-2}$$

as compared to the previous values [21]

??? $A = (7.2 \pm 2.8) \times 10^{-2}$ (1351 events)

and [22]

$$A = (0.3 \pm 1.0) \times 10^{-2} \qquad \text{(10 665 events)}$$

Through interactions which violate P and T, the neutron can acquire an electric dipole moment. The present experimental value is

$$\mu(n) = (-2 \pm 3) \times 10^{-22} \text{e cm}$$

reported by Miller et al. [23], and

$$\mu(n) = (+2.4 \pm 3.9) \times 10^{-22} \text{ e cm}$$

reported by Shull and Nathans [24]. While these results are still consistent with all four general cases of CP violating interactions, they have ruled out many specific theoretical models of CP violation.

Needless to say, all of these experiments are exceedingly difficult; in my opinion, many certainly deserve to be in the * class.

4. PHENOMENOLOGICAL LAGRANGIAN AND ITS SELECTION RULES

We now turn to other aspects of weak interactions. All known weak processes can be classified phenomenologically into three classes: purely leptonic, semi-leptonic and non-leptonic, described respectively by the following three terms of the effective Lagrangian

$$\mathscr{L}_{wk} = \mathscr{L}_{\ell\ell} + \mathscr{L}_{\ell h} + \mathscr{L}_{hh}$$

where the subscript ℓ denotes leptons and h hadrons. Among these, the μ-decay part of $\mathscr{L}_{\ell\ell}$ [5] is known explicitly:

$$** \quad (\mathscr{L}_{\ell\ell})_{\mu\text{-decay}} = \frac{G_{wk}}{\sqrt{2}} (j_\lambda)^*_\mu (j_\lambda)_e + \text{h.c.}$$

where G_{wk} is the Fermi constant $\cong 10^{-5} m_N^{-2}$,

$$(j_\lambda)_\ell = i\psi_\ell^\dagger \gamma_4 \gamma_\lambda (1 + \gamma_5) \psi_{\nu_\ell}$$

$(j_\lambda)^*$ is the Hermitian conjugate of (j_λ) times +1, or -1, depending on whether $\lambda \neq 4$ or $\lambda = 4$, $\ell = e$ or μ, ψ_ℓ describes the lepton field and $\gamma_1, \gamma_2 \, .. \, \gamma_5$ are the Dirac matrices. The semi-leptonic reaction is well described by the V-A interaction form [25, 26]

$$* \quad \mathscr{L}_{\ell h} = \frac{G_{wk}}{\sqrt{2}} (J_\lambda + S_\lambda)^* j_\lambda + \text{h.c.}$$

where

$$j_\lambda = (j_\lambda)_\mu + (j_\lambda)_e$$

J_λ and S_λ are, respectively, the strangeness conserving and the strangeness non-conserving part of the hadron current, which can be further decomposed into a vector and an axial-vector part

$$J_\lambda = V_\lambda + A_\lambda$$

and

$$S_\lambda = V_\lambda^S + A_\lambda^S$$

The explicit form of $\mathscr{L}_{hh}$ is still not known. An attractive idea is to express the entire weak interaction Lagrangian [26] as

$$? \quad \mathscr{L}_{wk} = \frac{G_{wk}}{\sqrt{2}} \mathscr{J}_\lambda^* \mathscr{J}_\lambda$$

where

$$\mathscr{J}_\lambda = J_\lambda + S_\lambda + j_\lambda$$

Table II presents the various selection rules satisfied by the $\mathscr{L}_{wk}$. (The references for the various selection rules are quite numerous. We give only a partial list of the theoretical papers [27] .)

TABLE II. SELECTION RULES SATISFIED BY THE $\mathscr{L}_{wk}$

* $\Delta S \neq 2$ rule for $\mathscr{L}_{hh}$ Isotriplet: V_λ, $(J_\lambda^{el})_{I=1}$, V_λ^*
$\Delta S \neq 2$ rule for $\mathscr{L}_{\ell h}$ $SU_2 \times SU_2$ algebra for $Q_V = \int V_4 \, d^3r$, $Q_A = \int A_4 \, d^3r$, etc. $\lvert \Delta \vec{I} \rvert = \frac{1}{2}$ rule for S_λ (i.e. for $\mathscr{L}_{\ell h}$) { $\lvert \Delta \vec{I} \rvert = \frac{1}{2}$ rule for $\mathscr{L}_{hh}$, or { ? $\mathscr{L}_{hh} = \frac{G_{wk}}{\sqrt{2}} (J_\lambda + S_\lambda)^* (J_\lambda + S_\lambda)$ in violation of $\lvert \Delta \vec{I} \rvert = \frac{1}{2}$ $\mathscr{L}_{\ell h} \sim SU_3$ octet
? $\Delta Q = \Delta S$ rule $\mathscr{L}_{hh} \sim SU_3$ octet $SU_3 \times SU_3$ algebra for $\int S_4 \, d^3r$, etc. algebra of the local currents

5. LIMITATION OF FERMI'S THEORY AND HIGHER-ORDER WEAK INTERACTIONS

The use of the phenomenological Lagrangian was originally introduced by Fermi [28] to describe β-decay, in a spirit quite similar to that of his pseudo-potential in nuclear physics. The lowest-order matrix elements of the phenomenological Lagrangian $\mathscr{L}_{wk}$ give directly the observed reaction amplitudes; the higher-order elements are, in general, divergent and should be discarded. To search for the underlying equations of the phenomenological Lagrangian is the basic problem of weak interactions.

We note that even as a phenomenological Lagrangian, the usual (current $\times$ current) form has to be modified; the rule of using only the lowest-order matrix elements of $\mathscr{L}_{wk}$ must be changed at high energy. For example, although the low-energy μ-decay process is well described by the phenomenological Lagrangian

$$(\mathscr{L}_{\ell\ell})_{\mu\text{-decay}} = \frac{G_{wk}}{\sqrt{2}} (j_\lambda)^*_\mu \ (j_\lambda)_e$$

the same phenomenological Lagrangian implies

$$\nu_\mu + e^- \rightarrow \nu_e + \mu^-$$

occurs only in S-wave, and its cross-section [29] is

$$\sigma = \frac{4}{\pi}(G_{wk} p_\nu)^2$$

where p_ν is the momentum of ν_μ in the centre-of-mass system. Since from unitarity

$$\sigma < \frac{\pi}{2}\left(\frac{1}{p_\nu}\right)^2$$

the above expression <u>cannot</u> be valid if

$$p_\nu > 300 \text{ GeV}$$

Thus, modification must come in at a momentum Λ, beyond which the above phenomenological Lagrangian should be modified. The unitarity limit gives

$$\Lambda < 300 \text{ GeV}$$

It has been speculated that perhaps weak interactions are transmitted by a spin-one charged boson $W^\pm$, and the above phenomenological Lagrangian is really the second-order effect through emissions and absorptions of $W^\pm$. In such a case, one may identify

$$? \quad \Lambda = m_W$$

The present experimental results from high-energy neutrino reactions require

$$m_W > 2.5 \text{ GeV}$$

The charged vector boson theory is unrenormalizable. The higher-order weak interaction effects remain divergent. Yet, the mass difference Δm_K between K_1^0 and K_2^0 appears in every way to be a higher-order weak interaction effect $\propto G_{wk}^2$. From what is essentially a dimensional analysis, one may estimate

$$\frac{|\Delta m_K|}{m_K} \sim \left(\frac{G_{wk}}{4\pi}\right)^2 \theta^2 (\Lambda')^4$$

where θ is the Cabibbo angle, and Λ' is some cut-off energy parameter. By comparing with the experimental value of $\Delta m_K / m_K$, one finds

$$\Lambda' \sim O(m_N)$$

A direct perturbation calculation using either $W^{\pm}$-theory or Fermi theory gives the ridiculous answer $\Lambda' = \infty$. The usual assumption of current-algebra makes it unlikely that $\Lambda' \sim O(m_N)$ can be due to the strong interaction. It seems quite likely that this Λ' parameter is connected with the previous limit Λ of the phenomenological Lagrangian. At present, there are many theoretical attempts trying to understand the higher-order weak interaction effect; as yet, none is a * idea.

REFERENCES

[1] MICHEL, L., Proc. phys. Soc. (London) A63 (1950) 514; Phys. Rev. 86 (1952) 814.

[2] LEE, T.D., YANG, C.N., Phys. Rev. 105 (1957) 1671.
SALAM, A., Nuovo Cim. 5 (1957) 299.
LANDAU, L., Nucl. Phys. 3 (1957) 127.

[3] BARDON, M., NORTON, P., PEOPLES, J., SACHS, A., LEE-FRANZINI, J., Phys. Rev. Lett. 14 (1965) 449.

[4] SARGENT, C.P., RINEHART, M., LEDERMAN, L.M., ROGERS, K.C., Phys. Rev. 99 (1955) 885.

[5] LEE, T.D., WU, C.S., A. Rev. nucl. Sci. 15 (1965) 381.

[6] ALFF-STEINBERGER et al., Physics Lett. 20 (1966) 207.
BOTT-BODENHAUSEN et al., Physics Lett. 20 (1966) 212.
AUBERT et al., Physics Lett. 10 (1964) 215; 11 (1964) 267; 17 (1965) 59.
BALDO-CEOLIN et al., Nuovo Cim. 38 (1965) 684.
FRANZINI et al., Phys. Rev. 140 (1965) B127.
GOOD et al., Phys. Rev. 124 (1961) 1223.
FUJII et al., Phys. Rev. Lett. 13 (1964) 253.
CAMERINI et al., Phys. Rev. 128 (1962) 352.
FITCH et al., Nuovo Cim. 22 (1961) 1160.
PICCIONI, O., Proc. Int. Conf. Weak Interactions, Argonne National Lab. (1965) 230.
CHRISTENSON, J.H., Tech. Rep. 34, Princeton Univ. (1964).

[7] $K^{\pm}$ lifetime
LOBKOWICZ et al., Phys. Rev. Lett. 17 (1966) 548.
FORD et al., Phys. Rev. Lett. 18 (1967) 1214.

NORDIN, P., Jr., Phys. Rev. 123 (1961) 2168.
BHOWMIK et al., Nuovo Cim. 20 (1961) 857.
BARKAS et al., Phys. Rev. 124 (1961) 1209.
BOYARSKI et al., Phys. Rev. 128 (1962) 2398.
CRAWFORD, F., Proc. Int. Conf. High Energy Physics, Rochester (1957).

$\mu^{\pm}$ lifetime

LUNDBY, R.A., Phys. Rev. 125 (1962) 1686.
ECKHAUSE, M., Carnegie Inst. tech. Rep. 9286 (1962).
FARLEY et al., Proc. Int. Conf. High Energy Physics, Geneva (1962).
MEYER et al., Nevis Rep. 114, Columbia Univ. (1963).

$\pi^{\pm}$ lifetime

LOBKOWICZ et al., Phys. Rev. Lett. 17 (1966) 548.
AYRES et al., Physics Lett. 24B (1967) 483.
BARDON et al., Phys. Rev. Lett. 16 (1966) 775.
JAKOBSON, SCHULZ, STEINBERGER, Phys. Rev. 81 (1951) 894.
DURBIN, LOAR, HAVENS, Phys. Rev. 88 (1952) 179.
ANDERSON et al., Phys. Rev. 119 (1960) 2050.
ASHKIN et al., Nuovo Cim. 16 (1960) 490.
BARKAS, A., ROSENFELD, A., Proc. Int. Conf. High Energy Physics, Rochester (1960).

[8] CHRISTENSON, J.H., CRONIN, J.W., FITCH, V.L., TURLAY, R., Phys. Rev. Lett. 13 (1964) 138.
ABASHIAN, A. et al., Phys. Rev. Lett. 13 (1964) 243.

[9] CRONIN, J., KUNZ, P., RISK, W., WHEELER, P., Phys. Rev. Lett. 18 (1967) 25.

[10] GAILLARD, J.M., KRIENEN, F., GALBRAITH, W., HUSSRI, A., JANE, M., LIPPMAN, N., MANNING, G., RATCLIFFE, T., DAY, P., PASHAM, A., SHERWOOD, A., PAYNE, B., FAISSNER, H., REITHER, H., Phys. Rev. Lett. 18 (1967) 20.

[11] FITCH, V. et al., presented at meeting of Am. phys. Soc., Chicago (January 1968) (private communication).

[12] BENNETT, S., NYGREN, D., SAAL, H., SUNDERLAND, J., STEINBERGER, J. (to be published).
FITCH, V.L., et al., Phys. Rev. Lett. 15 (1965) 73.
ALFF-STEINBERGER, C. et al., Physics Lett. 21 (1966) 595.
BOTT-BODENHAUSEN, M. et al., Physics Lett. 20 (1966) 212.
FIRESTONE, A. et al., Phys. Rev. Lett. 16 (1966) 556.
RUBBIA, C., STEINBERGER, J., Physics Lett. 24 (1967) 531.

[13] DORFAN, D., ENSTROM, J., RAYMOND, D., SCHWARTZ, M., WOJCICKI, S., MILLER, D.H., PACIOTTI, M., Phys. Rev. Lett. 19 (1967) 987.

[14] BENNETT, S., NYGREN, D., SAAL, H., STEINBERGER, J., SUNDERLAND, J., Phys. Rev. Lett. 19 (1967) 997.

[15] WOLFENSTEIN, L., Phys. Rev. Lett. 13 (1964) 562.

[16] SACHS, R.G., Phys. Rev. Lett. 13 (1964) 286.
TRUONG, T.N., Phys. Rev. Lett. 13 (1964) 358a; 17 (1966) 153.
GLASHOW, S.L., Phys. Rev. Lett. 14 (1965) 35.
SHABALIN, E.P., Soviet Phys. JETP Lett. 2 (1965) 276.
CABIBBO, N., Physics Lett. 12 (1964) 137.
WOLFENSTEIN, L., Physics Lett. 15 (1965) 196.
OKUN, ZELDOVIC, Physics Lett. 16 (1965) 319.
SALZMANN, F., SALZMANN, G., Physics Lett. 15 (1965) 91.
ARBUZOV, B.A., FILIPPOV, A.T., Physics Lett. 20 (1966) 537.

[17] BERNSTEIN, J., FEINBERG, G., LEE, T.D., Phys. Rev. 138 (1965) B1650.
BARSHAY, S., Physics Lett. 17 (1965) 78.

[18] LEE, T.D., WOLFENSTEIN, L., Phys. Rev. 138 (1965) B1490.
OKUN, L.B., Soviet J. nucl. Phys. 1 (1965) 670.
PRENTKI, J., VELTMAN, M., Physics Lett. 15 (1965) 88.

[19] LEE, T.D., WOLFENSTEIN, L., Phys. Rev. 138 (1965) B1490.

[20] GORMLEY, M., HYMAN, E., LEE, W., NASH, T., PEOPLES, J., SCHULTZ, C., STEIN, S., Phys. Rev. Lett. (to be published).

[21] BALTAY, C. et al., Phys. Rev. Lett. 16 (1966) 1224.

[22] CNOPS, A. et al., Physics Lett. 22 (1966) 546.

[23] MILLER, P.D. et al., Phys. Rev. Lett. 19 (1967) 381.

[24] SHULL, C.G., NATHANS, R., Phys. Rev. Lett. 19 (1967) 384.

[25] MARSHAK, R., SUDARSHAN, G., Padua-Venice Int. Conf. (1957); Phys. Rev. 109 (1958) 1860.
SAKURAI, J.J., Nuovo Cim. 7 (1958) 649.

[26] FEYNMAN, R., GELL-MANN, M., Phys. Rev. 109 (1958) 193.

[27] $\Delta S \neq 2$ rule
OKUN, L.B., PONTECORVO, B.M., Zh. éksp. teor. Fiz. 32 (1957) 1587 (transl. Soviet Phys. JETP 5 (1957) 1297).
$\Delta Q = \Delta S$ rule and the iso-triplet hypothesis of the vector current
FEYNMAN, R., GELL-MANN, M., Phys. Rev. 109 (1958) 193.
$|\Delta \vec{I}| = \frac{1}{2}$ rule
GELL-MANN, M., PAIS, A., Proc. Int. Conf. High Energy Physics, Pergamon Press, London (1955).
MARSHAK, R., OKUBO, S., SUDARSHAN, E.C.G., TEUTSCH, W.B., WEINBERG, S., Phys. Rev. 112 (1958) 665.
SU_3 transformation properties of the weak interaction
CABIBBO, N., Phys. Rev. Lett. 10 (1963) 513.
Current algebra
GELL-MANN, M., Phys. Rev. 125 (1962) 1067; Physics 1 (1964) 63.
ADLER, S.L., Phys. Rev. Lett. 14 (1965) 1051.
WEISSBERGER, W.I., Phys. Rev. Lett. 14 (1965) 1047.

[28] FERMI, E., Z. Phys. 88 (1934) 161.

[29] LEE, T.D., Bull. Am. phys. Soc. 5 (1960) 64.

Symmetry Principles in Physics.

T. D. Lee

Columbia University - New York, N. Y.

1. – Introduction.

Symmetry considerations have played an important role since the beginning of physics. A great deal of our understanding of nature can be formulated through symmetry principles. However, especially over the past decade, we have learned that many of these symmetry considerations turn out to be not strictly correct. Why do natural laws have a connection with symmetries; but often with a slight amount of asymmetry rather than with perfect symmetry? For example, physical laws are almost symmetrical with respect to left and right. However, in the weak interactions where

$$\frac{\text{weak forces}}{\text{strong forces}} \sim 10^{-6},$$

the right-left symmetry is broken. Similar symmetry breaking effects have been observed for many other trasformations; these are summarized in Table I.

TABLE I. – *Symmetry violations in strong interaction processes* (*e.g.*, np → np).

Symmetry	Violation amplitude (relative magnitude)
P (space inversion)	10^{-6}
C (particle anti-particle conjugation)	10^{-3} - 10^{-6}
T (time reversal)	10^{-3} - 10^{-15}
CP	10^{-3} - 10^{-15}
SU_2 (iso-spin)	10^{-2}
SU_3	10^{-1}

This situation may appear to some people as aesthetically disturbing. Why should nature be slightly asymmetrical?

Before starting the technical discussion of these symmetries and asymmetries in physics, let us recall that the word « symmetry » has two different meanings even in our daily language. According to the 1949 Webster's Dictionary (*), these two definitions are:

> **sym'me·try** (sĭm'ĕ·trĭ), *n.* [F. or L.; F. *symmétrie* (now *symétrie*), fr. L., fr. Gr. *symmetria*, fr. *syn-* + *metron* a measure.] **1.** *Now Rare.* Due or balanced proportions; beauty of form arising from such harmony. **2.** Correspondence in size, shape, and relative position, of parts that are on opposite sides of a dividing line or median plane.

It is of interest to note the italic *Now Rare* for the first definition. Can it be anticipating the broken symmetry principles that physicists have now discovered about nature? Indeed, as we shall see, perhaps beauty should be associated with a slight asymmetry, rather than with total symmetry.

The concept of beauty is, of course, quite subjective. Which is a more beautiful object, one with total symmetry, or one with a slight asymmetry? The answer is clearly open to debate. However, we may take a look at some of the well-known art pieces. For example, both the Greek statue and the mosaic from S. Apollinare in Classe near Ravenna, shown in Figs. 1 and 2, emphasize bilateral symmetry. The painting of Poplars by Monet in Fig. 3 suggests a discrete space-translational symmetry. While the beauty of symmetry is clearly demonstrated in each case, this beauty is greatly enhanced by the presence of slight asymmetries.

In physics, our concern is with the laws of nature. The concept « beauty » comes in only because of our belief that nature is beautiful. The concepts « symmetry » and « asymmetry » are applied to the various transformations which are connected with both space-time co-ordinates and the interactions between elementary particles. Both symmetry and asymmetry should be formulated in terms of precise mathematical language, so that their implications can lead to predictions which can be tested experimentally.

There are four main groups of symmetries, or broken symmetries, that are found to be of importance in physics.

1) Permutation symmetry: Bose-Einstein and Fermi-Dirac statistics,
2) continuous space-time transformations: translations, rotations, accelerations, etc.,

(*) *Webster's New Collegiate Dictionary*, 1949 edition (G. and C. Merriam Co., Springfield).

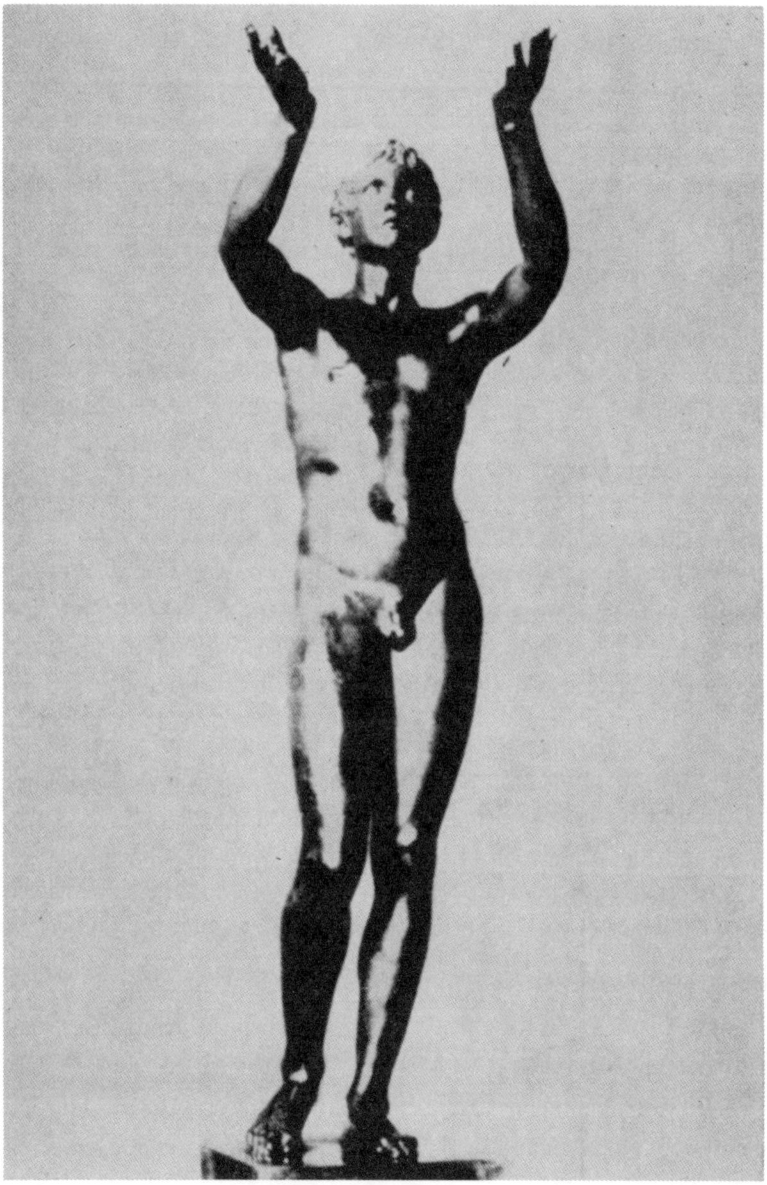

Fig. 1. – Praying Boy, Greek sculpture. Reprinted from H. WEYL: *Symmetry* (Princeton University Press, Princeton, New Jersey, 1952).

Fig. 2. – Mosaic from S. Apollinare in Classe.

Fig. 3. – Poplars by Claude Monet (on display at the Metropolitan Museum of Art).

3) discrete transformations: space inversion P, time reversal T, particle-antiparticle conjugation C, G-parity, etc.,

4) unitary transformations:
U_1-symmetries: conservation laws of charge Q, baryon number N, and lepton numbers L_e and L_μ, SU_2 (isospin) symmetry, and SU_3 symmetry.

Among these, the symmetries connected with the first two groups of transformations are, at present, believed to be exact. In the third group only the product CPT is perhaps exact, but each individual discrete symmetry operation is not. In the fourth group only the U_1-symmetries are exact.

2. – Symmetries, nonmeasurables and conservation laws.

The root af all symmetry principles in physics lies in the assumption that it is *impossible* to measure certain basic quantities; these will be called nonmeasurables in the following discussion. For example, we may consider the interaction energy V between two particles at positions $\boldsymbol{r}_1$ and $\boldsymbol{r}_2$. The physical assumption that it is not possible to measure an absolute position leads to the mathematical conclusion that the interaction energy V should be unchanged under a space translation

$$\boldsymbol{r}_1 \to \boldsymbol{r}_1 + \boldsymbol{\Delta}$$

and

$$\boldsymbol{r}_2 \to \boldsymbol{r}_2 + \boldsymbol{\Delta}\,.$$

Therefore, the interaction energy V is a function only of the relative distance $(\boldsymbol{r}_1 - \boldsymbol{r}_2)$, *i.e.*,

$$V = V(\boldsymbol{r}_1 - \boldsymbol{r}_2)\,. \tag{1}$$

From this, we deduce that the total momentum of this system of two particles must be conserved, since its rate of change is equal to

$$-(\boldsymbol{\nabla}_1 + \boldsymbol{\nabla}_2)V$$

which, due to (1), is zero.

This simple example illustrates the close connection between three aspects of a symmetry principle: the assumption of a nonmeasurable, the implied invariance under the connected mathematical transformation and the physical consequence of a conservation law. In an entirely similar way, we assume absolute time to be a nonmeasurable; the physical laws must then be invariant under a time translation

$$t \to t + \tau$$

which results in the conservation law of energy. By assuming absolute direction to be a nonmeasurable, we derive rotation invariance and obtain the conservation law of angular momentum.

Similar reasoning extends to all other symmetry considerations. The special theory of relativity assumes that absolute velocity is a nonmeasurable, and the general theory of relativity assumes that the difference between an acceleration and a gravitational field is not a measurable quantity. The permutation symmetry rests on the impossibility of observing any difference between identical particles.

In order to derive the conservation law of electric charge, we assume that it is not possible to measure the relative phase between states of different charges; therefore, one must have invariance under an arbitrary multiplication of phase factor $e^{i\theta}$ between states of different charges. Since $e^{i\theta}$ is a simple 1×1 unitary matrix, this invariance is called « U_1-symmetry ». It implies that the transition matrix elements between states of different charges must be zero, for otherwise there could be interference between two states of different charges, and their relative phase would be measurable. This U_1-symmetry then leads to the well-known conservation law of electric charge. Similarly, the impossibility of measuring relative phases between states of different baryon numbers implies the conservation of baryon number; the assumption that relative phases between states of different lepton numbers are nonmeasurables results in the conservation of lepton number.

3. – Symmetry violations.

Violations of symmetries arise when what were thought to be nonmeasurables turn out to be actually measurable. Let us take as a first example the question of right-left symmetry.

The concept that nature (*i.e.*, physical law) is symmetrical with respect to right and left dates back to the early history of physics. Of course, in our daily life left and right are quite distinct from each other. Our hearts, for example, are usually on our left sides. The word « right » means also correct, while the word « sinister » in its Latin root means left. In English, one says right-left, but in Chinese, 左 (left) always precedes 右 (right). However, such asymmetry in daily life is attributed to either the accidental asymmetry of our environment or the initial condition in organic life.

The principle of the symmetry between right and left has been found to be true in classical physics, in atomic physics, and in many areas of nuclear physics. Yet, in 1957 it was discovered that the laws of nature are in fact not symmetrical with respect to right and left. The apparent symmetry previously found in macroscopic physics, atomic physics, and nuclear physics is only an approximate one. For example, the neutrino emitted in a π^+ decay has its spin always antiparallel to its momentum. Although the initial π^+, being of zero spin, is in a totally spherically symmetrical state, nevertheless we may use its final decay product to give an absolute definition of left versus right,

The same applies to other violations of symmetries. Previously, in electromagnetic theory, the sign of electric charge could only be defined in terms of that of a test charge. Now, because of C and CP asymmetries, one may give an absolute definition of the sign of electric charge. For example, in the decay of the long-lived neutral K meson, K_L^0, the decay rates [1, 2] to e^+ and μ^+ are different from those to e^- and μ^-:

$$\frac{\text{rate}(K_L^0 \to e^+\pi^-\nu_e)}{\text{rate}(K_L^0 \to e^-\pi^+\bar{\nu}_e)} = 1.00315 \pm 0.0003$$

and

$$\frac{\text{rate}(K_L^0 \to \mu^+\pi^-\nu_\mu)}{\text{rate}(K_L^0 \to \mu^-\pi^+\bar{\nu}_\mu)} = 1.00405 \pm 0.00135\,.$$

Such slight differences in decay rates enable one to give an absolute definition of the sign of electric charge, without the use of a test charge.

As discussed in the previous Section, the validity of all symmetry principles rests on the theoretical hypotheses of nonmeasurables. Some of these hypotheses may indeed be correct in a fundamental sense, some may simply be due to the limitations in our present abilities to measure things. As we improve our experimental techniques, our domain of observations naturally becomes enlarged. It should not be too surprising that we may even succeed in observing some of those supposedly nonmeasurables, and therein lies the root of symmetry breaking.

In this sense, we should be prepared for the eventual possibility that we might be able to measure absolute space-time positions, absolute direction, and absolute velocity, and even relative phases between states of different charges, different baryon numbers, and different lepton numbers. Even if these were possible, it should be expected that such discoveries could lead only to small symmetry breakings in all of the presently known physical phenomena, because, otherwise, these supposedly nonmeasurables would have been measured long ago.

Just as in most of our artistic creations, the harmony and beauty of symmetry is always enhanced by the presence of a small degree of asymmetry. From an aesthetic point of view, it is rather satisfying to find nature also has a similar preference in small symmetry violations.

REFERENCES

[1] J. Steinberger: *Proc. CERN Topical Conf. Weak Interactions* (1969), p. 291.
[2] D. Dorfan *et al.*: *Phys. Rev. Lett.*, **19**, 987 (1967).

Reprinted from:

PHYSICAL REVIEW D VOLUME 4, NUMBER 1 1 JULY 1971

CP Nonconservation and Inequalities Between $\mu^+\mu^-$ and 2γ Decay Rates of K_S^0 and K_L^0 †

N. Christ* and T. D. Lee

Columbia University, New York, New York 10027

(Received 27 April 1971)

Assuming that the absorptive part of $K_L^0 \to 2\mu$ is due only to the on-mass-shell 2γ intermediate states, several inequalities between 2μ and 2γ decay states of K_S^0 and K_L^0 are derived without the assumption of *CP* conservation. These inequalities, together with the present experimental upper bound on $K_L^0 \to 2\mu$, imply that the branching ratio for $K_S^0 \to 2\mu$ should be greater than 5×10^{-7}.

I. INTRODUCTION

It has been pointed out by several authors[1,2] that if the absorptive part of $K_L^0 \to \mu^+\mu^-$ is assumed to be due only to the on-mass-shell 2γ intermediate state and if, in addition, *CP* conservation holds, then the usual quantum electrodynamics leads to the inequality

$$\frac{\text{rate}(K_L^0 \to 2\mu)}{\text{rate}(K_L^0 \to 2\gamma)} > \lambda^2, \tag{1}$$

where[1,3]

$$\lambda^2 = \frac{1}{2}\frac{\alpha^2}{v_\mu}\left(\frac{m_\mu}{m_K}\right)^2\left(\ln\frac{1+v_\mu}{1-v_\mu}\right)^2 \cong 1.2\times10^{-5} \tag{2}$$

and

$$v_\mu \cong 0.9 \tag{3}$$

is the velocity of $\mu^\pm$ in the rest system of the kaon. The present experimental upper bound on the branching ratio[4] is

$$\frac{\text{rate}(K_L^0 \to \mu^+\mu^-)}{\text{rate}(K_L^0 \to \text{all})} < 1.8\times10^{-9}, \tag{4}$$

while, according to (1), the theoretical lower bound for the same branching ratio should be $\cong 6\times10^{-9}$.

At first sight, this discrepancy may not seem to be too disturbing since in $K_L^0 \to \mu^+\mu^-$, besides 2γ there are also $2\pi\gamma$, 3π, and other on-mass-shell intermediate states, and furthermore, *CP* conservation is known to be violated. However, difficulty does arise on a dynamical level. At present, attempts to include $2\pi\gamma$, 3π, and other intermediate states in the absorptive part lead only to a small correction to the above theoretical lower bound,[5] and it seems quite difficult to explain the large difference between the theoretical and experimental bounds on $K_L^0 \to \mu^+\mu^-$ by using any simple theoretical model.

The purpose of this note is to examine the alternative possibility, i.e., the effect of *CP* nonconservation. As we shall see, there are definite tests which can be used to trace whether the present discrepancy is due to *CP* nonconservation or due to other reasons. In order to separate out the implications of different theoretical hypotheses, we shall assume, throughout our subsequent discussions, (i) that the absorptive part of the $K_L^0 \to 2\gamma$ amplitude is zero, (ii) that the absorptive part of the $K_L^0 \to \mu^+\mu^-$ amplitude is due only to the on-mass-shell 2γ intermediate state, and (iii) that both *CPT* invariance and quantum electrodynamics are valid, but *CP* conservation is not. As we shall see, under these assumptions, the lower bound given above by (1) no longer holds, and it is replaced by

$$[\text{rate}(K_S^0 \to \mu^+\mu^-)]^{1/2} \geq (\text{Re}\,\epsilon)^{-1}\{\lambda v_\mu[\text{rate}(K_L^0 \to 2\gamma)]^{1/2} - [\text{rate}(K_L^0 \to \mu^+\mu^-)]^{1/2}\} \tag{5}$$

and

$$[\text{rate}(K_S^0 \to \mu^+\mu^-)]^{1/2} \leq (\text{Re}\,\epsilon)^{-1}\{\lambda[\text{rate}(K_L^0 \to 2\gamma)]^{1/2} + [\text{rate}(K_L^0 \to \mu^+\mu^-)]^{1/2}\}, \tag{6}$$

where[6]

$$\text{Re}\,\epsilon \equiv \tfrac{1}{2}\langle K_L^0 | K_S^0\rangle \cong 1.4\times10^{-3},$$

and λ and v_μ are given by Eqs. (2) and (3), respectively. For clarity, the derivation of the above formulas will be given in Sec. II.

We observe that by using (4) and the experimental value[7]

$$\frac{\text{rate}(K_L^0 \to 2\gamma)}{\text{rate}(K_L^0 \to \text{all})} = (4.7 \pm 0.6)\times 10^{-4}\,,$$

the branching ratio for $K_S^0 \to \mu^+\mu^-$ should lie within

$$1\times 10^{-5} \geqslant \frac{\text{rate}(K_S^0 \to \mu^+\mu^-)}{\text{rate}(K_S^0 \to \text{all})} \geqslant 5\times 10^{-7}\,, \tag{7}$$

which is consistent with its present experimental upper bound[8] of 7.3×10^{-6}.

The validity of the above inequality (7) can be determined by an order-of-magnitude improvement of the present experimental limit on the branching ratio of $K_S^0 \to \mu^+\mu^-$. A violation of (7) can be used as a conclusive proof that the discrepancy between (1) and (4) cannot be attributed to CP nonconservation; other reasons must be sought to explain this difference. On the other hand, the validation of (7) implies that $[\text{rate}(K_S^0 \to \mu^+\mu^-)]^{1/2}$ is larger than $[\text{rate}(K_L^0 \to \mu^+\mu^-)]^{1/2}$ by, at least, a factor $\sim O(10^3)$; this strongly suggests that these decay rates do depend sensitively on the small CP-violating parameter $\text{Re}\epsilon$, and therefore CP nonconservation does play an important role in both $\mu^+\mu^-$ and 2γ decays of K_S^0 and K_L^0. Of course, even if this does turn our to be correct, one must still seek the detailed CP-violating mechanism that does produce such a large difference between the 2μ decay rates of K_S^0 and K_L^0.

II. INEQUALITIES

Let us define four complex numbers $a_\pm$ and $b_\pm$, related to the four decay amplitudes of K^0 and $\bar{K}^0$ into $\mu^+\mu^-$ in $CP=+1$ and -1 states by

$$\text{amp}(K^0 \to \mu^+\mu^-, CP=\pm 1) \equiv b_\pm + i a_\pm \tag{8}$$

and

$$\text{amp}(\bar{K}^0 \to \mu^+\mu^-, CP=\pm 1) \equiv \pm(b_\pm^* + i a_\pm^*)\,. \tag{9}$$

[The $\pm$ sign in (9) is introduced so that if CP were conserved and if we had defined $|\bar{K}^0\rangle = CP|K^0\rangle$, then $a_\pm$ and $b_\pm$ would all be real.] Throughout this section, we shall include in all amplitudes the appropriate phase-space factors, so that the squares of their magnitudes are simply the transition rates. It is convenient to choose the phases of the initial and final states in these decays such that

$$CPT|K^0\rangle = |\bar{K}^0\rangle \tag{10}$$

and for *free* $\mu^+\mu^-$ states

$$CPT|\mu^+\mu^-, CP=\pm 1\rangle = \pm|\mu^+\mu^-, CP=\pm 1\rangle\,; \tag{11}$$

CPT invariance, then, implies that in (8) and (9) $a_\pm$, $a_\pm^*$ are the absorptive parts of the decay amplitudes and $b_\pm$, $b_\pm^*$ the corresponding nonabsorptive parts.

The decay amplitudes

$$L_\pm \equiv \text{amp}(K_L^0 \to \mu^+\mu^-, CP=\pm 1)$$

and

$$S_\pm \equiv \text{amp}(K_S^0 \to \mu^+\mu^-, CP=\pm 1)$$

can now be readily expressed in terms of the constants $a_\pm$, $b_\pm$ and their complex conjugates through the usual relations

$$|K_S^0\rangle = [2(1+|\epsilon|^2)]^{-1/2}[(1+\epsilon)|K^0\rangle + (1-\epsilon)|\bar{K}^0\rangle]$$

and

$$|K_L^0\rangle = [2(1+|\epsilon|^2)]^{-1/2}[(1+\epsilon)|K^0\rangle - (1-\epsilon)|\bar{K}^0\rangle]\,.$$

Let $(L_\pm)_{\text{abs}}$ denote the absorptive part of $L_\pm$, i.e., the $a_\pm$-dependent part of $L_\pm$. One has then

$$(L_+)_{\text{abs}} = \sqrt{2}\,(-\text{Im}a_+ + i\epsilon\,\text{Re}a_+) \tag{12}$$

and

$$(L_-)_{\text{abs}} = \sqrt{2}\,(i\,\text{Re}a_- - \epsilon\,\text{Im}a_-)\,, \tag{13}$$

in which, for clarity of presentation, we have *neglected* all terms that are proportional to ϵ^2. (The same approximation applies to all subsequent discussions as well.) By using the explicit expressions for $L_\pm$ and $S_\pm$, one can readily verify that these absorptive parts also satisfy

$$(L_+)_{\text{abs}} - \text{Re}L_+ = -\text{Re}(\epsilon S_+) + i\epsilon\,\text{Im}S_+ \tag{14}$$

and

$$(L_-)_{\text{abs}} - i\,\text{Im}L_- = -i\,\text{Im}(\epsilon S_-) + \epsilon\,\text{Re}S_-\,. \tag{15}$$

It is convenient to regard the right-hand sides of (14) and (15) as the two components of a 2-dimensional complex vector, whose magnitude is given by

$$[\,|-\text{Re}(\epsilon S_+) + i\epsilon\,\text{Im}S_+|^2 + |-i\,\text{Im}(\epsilon S_-) + \epsilon\,\text{Re}S_-|^2]^{1/2} = (\text{Re}\epsilon)[\text{rate}(K_S^0 \to 2\mu)]^{1/2}\,. \tag{16}$$

Similarly, one may regard the two terms on the left-hand sides of (14) and (15) as the appropriate components of two other complex vectors: The magnitudes of these two complex vectors are given, respectively, by

$$L_{\text{abs}} \equiv [\,|(L_+)_{\text{abs}}|^2 + |(L_-)_{\text{abs}}|^2]^{1/2} \tag{17}$$

and

$$[(\text{Re}L_+)^2 + (\text{Im}L_-)^2]^{1/2}\,. \tag{18}$$

Since (18) is less than or equal to

$$[\text{rate}(K_L^0 \to 2\mu)]^{1/2}\,,$$

the usual triangular inequality, implied by Eqs.

(14) and (15), leads to

$$[\text{rate}(K_L^0 \to 2\mu)]^{1/2} \geq |L_{\text{abs}} - (\text{Re}\epsilon)[\text{rate}(K_S^0 \to 2\mu)]^{1/2}|. \tag{19}$$

In an identical way, through the interchange of K_L^0 and K_S^0, one can establish

$$[\text{rate}(K_S^0 \to 2\mu)]^{1/2} \geq |S_{\text{abs}} - (\text{Re}\epsilon)[\text{rate}(K_L^0 \to 2\mu)]^{1/2}|, \tag{20}$$

where, in a manner similar to (17), S_{abs} denotes the total magnitude of the absorptive amplitudes in $K_S^0 \to 2\mu$ decay.

From the assumptions (i) and (ii), stated in Sec. I, it follows that the absorptive parts $(L_+)_{\text{abs}}$ and $(L_-)_{\text{abs}}$ of the $K_L^0 \to 2\mu$ amplitudes are proportional to the appropriate 2γ decay amplitudes of K_L^0:

$$\text{amp}(K_L^0 \to 2\gamma, CP=+1) = (\lambda v_\mu)^{-1}(L_+)_{\text{abs}} \tag{21}$$

and

$$\text{amp}(K_L^0 \to 2\gamma, CP=-1) = \lambda^{-1}(L_-)_{\text{abs}}, \tag{22}$$

where the factors λv_μ and λ denote, respectively, the $2\gamma \to \mu^+\mu^-$ amplitudes in $CP=+1$ and -1 channels; to the lowest order in α, they are given by (2) and (3). The quantity L_{abs}, related to the absorptive amplitudes in $K_L^0 \to 2\mu$ decay by (17), therefore satisfies the inequality

$$\lambda v_\mu \leq [\text{rate}(K_L^0 \to 2\gamma)]^{-1/2} L_{\text{abs}} \leq \lambda. \tag{23}$$

From (19) and (23), the inequalities (5) and (6) follow readily. (If Reϵ were zero, then (19) and (23) would lead to

$$[\text{rate}(K_L^0 \to 2\mu)]^{1/2} \geq \lambda v_\mu [\text{rate}(K_L^0 \to 2\gamma)]^{1/2},$$

which would reduce to (1) if, in addition, CP conservation were valid.)

The inequalities (4) and (5) imply that

$$[\text{rate}(K_L^0 \to 2\mu)]^{1/2} < (\text{Re}\epsilon)[\text{rate}(K_S^0 \to 2\mu)]^{1/2}.$$

Thus, upon neglecting terms $\sim O(\epsilon^2 |S_\pm|)$, the inequality (20) becomes simply

$$[\text{rate}(K_S^0 \to 2\mu)]^{1/2} \geq S_{\text{abs}}. \tag{24}$$

Hitherto, our assumptions (i) and (ii) have concerned only K_L^0 decay amplitudes. Because of the dominant $K_S^0 \to 2\pi$ decay mode, the extension of similar assumptions to K_S^0 decays is expected to be a worse approximation. For completeness, we note that if, in addition to (i) and (iii) stated in Sec. I, one makes the assumption that (iv) the $K_S^0 \to 2\gamma$ decay amplitude has also a zero absorptive part, then it is possible to derive two more inequalities:

$$[\text{rate}(K_S^0 \to 2\gamma)]^{1/2} \leq (\text{Re}\epsilon)^{-1}[\text{rate}(K_L^0 \to 2\gamma)]^{1/2} \tag{25}$$

and

$$[\text{rate}(K_S^0 \to 2\gamma)]^{1/2} \geq (\text{Re}\epsilon)[\text{rate}(K_L^0 \to 2\gamma)]^{1/2}. \tag{26}$$

Furthermore, if one makes the additional assumption that (v) the absorptive part of the $K_S^0 \to 2\mu$ amplitude is due only to the on-mass-shell 2γ intermediate state, then from (24), one obtains,[9] independently of CP conservation,

$$[\text{rate}(K_S^0 \to 2\mu)]^{1/2} \geq \lambda v_\mu [\text{rate}(K_S^0 \to 2\gamma)]^{1/2}. \tag{27}$$

We emphasize that these additional inequalities (25)–(27) are based on assumptions which are less certain than those required for (5) and (6).

ACKNOWLEDGMENTS

We wish to thank C. Baltay for helpful discussions, and G. Farrar and S. Treiman for communicating their results to us before publication.

---

†Research supported in part by the U. S. Atomic Energy Commission.

*Alfred P. Sloan Foundation Fellow.

[1]C. Quigg and J. D. Jackson, L. R. L. Report No. UCRL-18487 (unpublished); L. M. Sehgal, Nuovo Cimento 45, 785 (1966), and Phys. Rev. 183, 1511 (1969).

[2]This decay has also been discussed by M. A. Baqi Bég, Phys. Rev. 132, 426 (1963); A. Pais and S. B. Treiman, *ibid.* 176, 1974 (1968).

[3]A similar expression has also been derived previously in connection with η decay by D. A. Geffen and B. L. Young, Phys. Rev. Letters 15, 316 (1965), and C. G. Callan and S. B. Treiman, *ibid.* 18, 1083 (1967). See also S. D. Drell, Nuovo Cimento 11, 693 (1959) and S. M. Berman and D. A. Geffen, Nuovo Cimento 18, 1192 (1960) for similar expressions derived for π^0 decay.

[4]A. R. Clark, T. Elioff, R. C. Field, H. J. Frisch, R. P. Johnson, L. T. Kerth, and W. A. Wenzel, Phys. Rev. Letters 26, 1667 (1971).

[5]B. R. Martin, E. de Rafael, and J. Smith, Phys. Rev. D 2, 179 (1970). In this paper, CP conservation is assumed, and the fractional uncertainty of the lower bound (1) due to contributions of all other intermediate states is estimated to be less than 20%. For the $2\pi\gamma$ intermediate state, a more detailed calculation has been recently made by G. R. Farrar and S. B. Treiman, *ibid.* 4, 257 (1971); they find the resulting fractional change of the lower bound (1) to be ~10%, if the decay form factor in $K \to 2\pi\gamma$ is either a constant or proportional to the electromagnetic pion form factor. For the 3π intermediate state, at present there is no model-independent theoretical statement, but we have considered a simple model

$$K_L^0 \to 3\pi \text{ (on-mass-shell)} \to \pi^0 \text{ (off-mass-shell)} \to 2\gamma.$$

This yields an absorptive amplitude whose magnitude is found to be only ~10^{-3} times the magnitude of the observed amplitude for $K_L^0 \to 2\gamma$; therefore, the resulting fractional change in the lower bound (1) is expected to be of a similarly small magnitude.

[6]See the review talk by J. Steinberger in *Topical Conference on Weak Interactions, CERN, Geneva, Switzerland, 1969* (CERN, Geneva, 1969), p. 291.

[7]M. Banner, J. W. Cronin, J. K. Liu, and J. E. Pilcher, Phys. Rev. 188, 2033 (1969).

[8]B. D. Hyams *et al.*, Phys. Letters 29B, 521 (1969).

[9]Apart from a misprint that gives the factor $\frac{1}{2}$ as 2, the inequality (27) is the same as the inequality (4.1) in the paper by Martin *et al.* (Ref. 5). Martin *et al.* also discussed modifications to this inequality due to other on-mass-shell intermediate states.

Reprinted from:

PHYSICAL REVIEW D VOLUME 8, NUMBER 4 15 AUGUST 1973

A Theory of Spontaneous T Violation*

T. D. Lee

Department of Physics, Columbia University, New York, New York 10027

(Received 11 April 1973)

A theory of spontaneous T violation is presented. The total Lagrangian is assumed to be invariant under the time reversal T and a gauge transformation (e.g., the hypercharge gauge), but the physical solutions are not. In addition to the spin-1 gauge field and the known matter fields, in its simplest form the theory consists of two complex spin-0 fields. Through the spontaneous symmetry-breaking mechanism of Goldstone and Higgs, the vacuum expectation values of these two spin-0 fields can be characterized by the shape of a triangle and their quantum fluctuations by its vibrational modes, just like a triangular molecule. T violations can be produced among the known particles through virtual excitations of the vibrational modes of the triangle which has a built-in T-violating phase angle. Examples of both Abelian and non-Abelian gauge groups are discussed. For renormalizable theories, all spontaneously T-violating effects are finite. It is found that at low energy, below the threshold of producing these vibrational quanta, T violation is always quite small.

I. INTRODUCTION

In this paper we discuss a theory of spontaneous T violation. To illustrate the theory, we shall first discuss a simple model in which the weak-interaction Lagrangian, as well as the strong- and electromagnetic-interaction Lagrangians, is assumed to be invariant under (1) the time reversal T and (2) a gauge transformation, e.g., that of the hypercharge Y. Yet the physical solutions are required to exhibit both T violation and Y nonconservation. In its construction, the model is similar to those gauge-group spontaneous symmetry-violating theories[1-4] that have been extensively discussed in the literature. The only difference is that one now has, in addition, the spontaneous violation of a discrete symmetry.[5] As we shall see, there exists a general class of such spontaneously T-violating theories. (*Note added in proof.* There actually exist two general classes of such spontaneously T-violating theories, depending on whether the spin-0 fields belong to a complex representation or a real representation of the gauge group. In this paper, except in Appendix D, only the class of complex representation is discussed. For real representation, such as that in the Georgi-Glashow model, one may follow the method developed in Appendix D. The details will be given in a separate paper.) The simple model serves only as a prototype which nevertheless embodies most of the essential features.

In addition to the known matter fields, the model consists of two independent spin-0 $Y=1$ complex fields ϕ_1, ϕ_2 and a neutral spin-1 gauge field B_μ. Under the hypercharge gauge transformation $e^{iY\Lambda}$, we have

$$\phi_k \to e^{i\Lambda}\phi_k$$

and (1)

$$B_\mu \to B_\mu + f^{-1}\frac{\partial\Lambda}{\partial x_\mu},$$

where f is the hypercharge coupling constant and the subscript $k=1$ and 2. As usual, T is assumed to commute with Y,

$$TYT^{-1}=Y. \tag{2}$$

This gives then a well-defined difference between T and either CT or CPT. Since T is an antiunitary operator, we can always choose the phase of ϕ_k such that

$$T\phi_k T^{-1}=\phi_k. \tag{3}$$

To avoid irrelevant complications, we assume the theory *not* to be symmetric under any linear transformation which mixes ϕ_1 and ϕ_2, so that the right-hand side of (3) must remain ϕ_k.

As will be discussed in the next section, the spontaneous T-violation mechanism can be introduced by assuming a T-invariant potential energy $V(\phi)$ between ϕ_1 and ϕ_2 which has a minimum at the c-number point

$$(\phi_1,\phi_2)=2^{-1/2}(\rho_1 e^{i\theta},\rho_2), \tag{4}$$

where $\rho_1>0$, $\rho_2>0$, and $\theta\neq 0$ or π. This minimum point therefore defines a triangle where ρ_1 and ρ_2 form two sides and θ the angle in between. Because of quantum effects there must be fluctuations of ϕ_1 and ϕ_2 around their average values. These fluctuations can be shown to correspond to the vibrations of the triangle. The entire ϕ_1, ϕ_2 system can then be visualized as a triangular molecule which is defined by both its shape and its three vibrational modes of oscillation in the plane of the triangle. For convenience of nomenclature, we shall refer to this ϕ_1, ϕ_2 complex simply as the *triangle*.

In the absence of the gauge field B_μ, there would be a zero-mass boson, in accordance with the Goldstone theorem.[1] In the present case, this Goldstone boson corresponds simply to the rotational degree of freedom of the triangle. Because of the Higgs mechanism,[2] the presence of the gauge field B_μ eliminates the zero-mass boson. As a result, B_μ acquires a mass, and the would-be Goldstone boson becomes, as usual, the longitudinal mode of B_μ. The detailed description of the triangle and its interaction with the gauge field is given in Sec. II.

While the Lagrangian is assumed to be T-invariant, its solution, as characterized by the triangle, carries a phase angle $\theta \neq 0$ or π. Therefore, it has a built-in T violation, somewhat analogous to the two-component neutrino theory which carries a built-in screw direction. We recall that just on the basis of the two-component neutrino theory alone, but without any appropriate interaction, one cannot distinguish[6] between P and CP, and consequently there is no observable parity-violation effect. Here, one has a similar situation. Both the gauge field and the vibrational levels of this triangular molecule are of zero average hypercharge, $\langle Y\rangle = 0$. Thus, although these vibrational levels are not invariant under T, there is *no* violation of the reciprocity relations, since for states with $\langle Y\rangle = 0$ reciprocity relations can be derived by using CT invariance alone. In order that violations of the reciprocity relations be observed, there must be states with $\langle Y\rangle \neq 0$, such as K^0, $\bar{K}^0$, etc.

Once this triangle is allowed to interact with known particles with $\langle Y\rangle \neq 0$, T violation becomes a natural consequence. However, the existence of the triangle does not determine the exact form of its interactions (just as the interaction of a neutrino is not specified by the two-component theory). As a pure illustration, we consider in Sec. III a particularly simple form in which the usual T-invariant $\Delta Y = \pm 1$ weak-interaction Lagrangian $L_\pm$ is replaced by

$$\mathcal{L}_{\text{int}} = (g_1\phi_1 + g_2\phi_2)L_- + \text{H.c.} \tag{5}$$

Because of the transformation property (1), this new Lagrangian clearly conserves Y. It is also T-invariant, provided g_1 and g_2 are relatively real. Through the virtual emission and absorption of the triangle, violations of reciprocity relations can occur among the known particles. As we shall see, this can give rise to $K_L^0 \to 2\pi$, and (if we assume the threshold energy for producing these triangles is $\gtrsim$ a few GeV) the resulting CP violation in K_L^0-K_S^0 decays is of the superweak form.

As will also be discussed in the subsequent sections, in addition to the direct exchange of the triangle between the matter fields there is still another important mechanism which can violate the reciprocity relations via the coupling between the matter field and the gauge field. In this mechanism, the triangle propagates only in a loop diagram, and as a result one may have violations of the Furry theorem, i.e., the loop diagram connecting an odd number of the gauge-field quanta may now be nonzero. Such a loop diagram can in turn produce T violations among the matter fields.

In Sec. IV, we examine some generalizations of the model to other gauge groups, either Abelian or non-Abelian, but we restrict our discussion only to renormalizable theories. In all these cases, the general mechanism of T violation remains the same, and the basic structure of the triangle remains intact, though its interaction with the known matter fields can be quite different. Because in these cases the spontaneous T violation is tied to the spontaneous gauge-symmetry violations[7] of the weak and electromagnetic interactions, at low energy the magnitude of T violation among known particles always turns out to be very small, either milliweak or superweak. Furthermore, since such theories are renormalizable, all spontaneously T-violating effects are finite and computable, at least in principle. In any case, one feels that whatever the eventual gauge theory may be for the weak and electromagnetic interactions, it should contain T violation as an integral part. The mechanism of spontaneous T violation discussed in this paper may provide just such a needed possibility.

II. THE TRIANGLE AND THE GAUGE FIELD

In this section we consider the simple system of spontaneous T violation mentioned in the Introduction. The system consists of two complex spin-0 fields ϕ_1, ϕ_2 and a gauge field B_μ. The most general form of a gauge-invariant, T-invariant, and renormalizable Lagrangian density is

$$\mathcal{L}(B,\phi) = -\sum_{k=1,2}\left[\left(\frac{\partial}{\partial x_\mu} + ifB_\mu\right)\phi_k^\dagger\right] \times\left[\left(\frac{\partial}{\partial x_\mu} - ifB_\mu\right)\phi_k\right] - \frac{1}{4}\left(\frac{\partial}{\partial x_\mu}B_\nu - \frac{\partial}{\partial x_\nu}B_\mu\right)^2 - V(\phi), \tag{6}$$

where the dagger denotes the Hermitian conjugate, and the potential energy $V(\phi)$ is given by

$$V(\phi) = -\lambda_1 \phi_1^\dagger \phi_1 - \lambda_2 \phi_2^\dagger \phi_2 + A(\phi_1^\dagger \phi_1)^2 + B(\phi_2^\dagger \phi_2)^2 + C(\phi_1^\dagger \phi_1)(\phi_2^\dagger \phi_2) + \tfrac{1}{2}[(\phi_1^\dagger \phi_2)(D\phi_1^\dagger \phi_2 + E\phi_1^\dagger \phi_1 + F\phi_2^\dagger \phi_2) + \text{H.c.}], \quad (7)$$

where the eight constants $\lambda_1, \lambda_2, A, \ldots, F$ are all real so that T invariance holds.

In the spirit of renormalization, the renormalized values of these constants can be arbitrarily assigned. Following the standard treatment of spontaneous symmetry-breaking mechanism for the gauge group,[1] we assume

$$\lambda_1 \text{ and/or } \lambda_2 > 0. \quad (8)$$

As we shall see, the spontaneous T violation can be induced by imposing

$$D > 0. \quad (9)$$

In addition, in order for $V(\phi)$ to have a lower bound, we require

$$A - \frac{E^2}{8D} > 0,$$

$$B - \frac{F^2}{8D} > 0,$$

and

$$\left(A - \frac{E^2}{8D}\right)\left(B - \frac{F^2}{8D}\right) > \frac{1}{4}\left(C - D - \frac{EF}{4D}\right)^2. \quad (10)$$

As usual, all the above conditions refer to the renormalized constants.

Let us first locate the minimum of the function $V(\phi)$ in its c-number form. In the tree approximation, this minimum determines the vacuum expectation values of ϕ_1 and ϕ_2:

$$\langle \phi_1 \rangle_{\text{vac}} = 2^{-1/2} \rho_1 e^{i\theta}$$

and

$$\langle \phi_2 \rangle_{\text{vac}} = 2^{-1/2} \rho_2 .$$

Because of (8) the minimum is not at the origin, and because of the gauge invariance of the Lagrangian we can always transform one of the vacuum expectation values, say $\langle \phi_2 \rangle_{\text{vac}}$, to be real and not negative. It is straightforward to obtain the necessary and sufficient condition for both $\rho_1 > 0$ and $\rho_2 > 0$. [See Appendix A for further details.] Similarly, one can readily verify that because of (9)

$$\cos\theta = -(4D\rho_1\rho_2)^{-1}(E\rho_1^2 + F\rho_2^2), \quad (11)$$

in which the constants are chosen to satisfy $-1 < \cos\theta < 1$. Equation (11) has two solutions: θ and $-\theta$. By using (3), one sees that either solution is not invariant under T, and therefore one has a spontaneous T violation. The T invariance of the Lagrangian ensures that both solutions exist, and that they transform into each other under T.

The normal modes of this system can be derived by expanding the operators ϕ_1 and ϕ_2 around their vacuum expectation values. We write

$$\phi_1 = 2^{-1/2}(\rho_1 + R_1 + iI_1)e^{i\theta}$$

and (12)

$$\phi_2 = 2^{-1/2}(\rho_2 + R_2 + iI_2),$$

where ρ_1, ρ_2, and θ are, as before, c-numbers, but R_1, R_2, I_1, and I_2 are Hermitian fields. If the coupling constant f between the gauge field B_μ and ϕ_1, ϕ_2 were zero, then the Goldstone theorem would apply and there should be one normal mode, called the Goldstone boson G, that has a zero mass. It can be easily verified that in the tree approximation G is given by

$$G = (\rho_1^2 + \rho_2^2)^{-1/2}(\rho_1 I_1 + \rho_2 I_2). \quad (13)$$

(This can also be established by using the geometrical considerations given below.) The remaining three normal modes, which will be referred to as t_1, t_2, and t_3, are linear combinations of the fields R_1, R_2 and

$$I \equiv (\rho_1^2 + \rho_2^2)^{-1/2}(\rho_2 I_1 - \rho_1 I_2). \quad (14)$$

This linear relation may be written as

$$t = \begin{pmatrix} t_1 \\ t_2 \\ t_3 \end{pmatrix} = U \begin{pmatrix} R_1 \\ R_2 \\ I \end{pmatrix}, \quad (15)$$

where U is a (3×3) real orthogonal matrix.

As already mentioned in the Introduction, the description of the system can be characterized by a triangular molecule. For example, Fig. 1 gives a schematic picture of such a triangular molecule where the two sides are ρ_1 and ρ_2, respectively, and the angle in between is θ. In the plane of the triangle, a triangular molecule also has three normal modes of vibration, each of which is a linear combination of the displacements $\delta\rho_1 = R_1$, $\delta\rho_2 = R_2$, and $\delta\theta = \rho_1^{-1} I_1 - \rho_2^{-1} I_2$, i.e.,

$$\delta\theta = \frac{(\rho_1^2 + \rho_2^2)^{1/2}}{\rho_1\rho_2} I$$

(as illustrated in Fig. 1). Under the gauge transformation $e^{iY\alpha}$, the entire triangle rotates an angle α. Thus, the Goldstone boson G corresponds simply to the rotational degree of freedom of the triangle; this then leads to Eq. (13).

The configuration and vibration of a triangular

molecule depend on nine real parameters: three for the shape of the triangle, three for the Eulerian angles that specify the real orthogonal matrix U, and three for the frequencies (or masses) of the normal modes. In the present case, the function $V(\phi)$ depends only on eight constants $\lambda_1, \lambda_2, A, \ldots, F$. This imposes a constraint

$$\sum_{a=1}^{3}(\rho_1 U_{a_1}+\rho_2 U_{a_2})U_{a_3} m_a^2=0, \qquad (16)$$

where m_a is the mass of the normal mode t_a. Since the coupling constant $f\neq 0$, the zero-mass Goldstone boson is removed through the Higgs mechanism.[2] G now joins the two transverse components of B_μ to form a single massive neutral spin-1 boson B. In the tree approximation, the mass of B is

$$m_B^2=f^2(\rho_1^2+\rho_2^2). \qquad (17)$$

The Lagrangian (6) is constructed to be invariant under the gauge transformation (1). Therefore one has the current conservation

$$\frac{\partial j_\mu}{\partial x_\mu}=0,$$

where

$$j_\mu=i\sum_{k=1,2}\left(\frac{\partial\phi_k^\dagger}{\partial x_\mu}\phi_k-\phi_k^\dagger\frac{\partial\phi_k}{\partial x_\mu}\right),$$

and the spatial integral of its time component is Y. The Lagrangian (6) is T-invariant; in addition, it is symmetric under the particle-antiparticle conjugation C and the space inversion P. The parity of B_μ is -1; the parity of ϕ_1 must be the same as that of ϕ_2, but it can be either $+1$ or -1, since the Lagrangian is an even function of ϕ_k. Under C, one has

$$C\phi_k C^\dagger=\phi_k^\dagger,$$

$$CB_\mu C^\dagger=-B_\mu,$$

and consequently

$$CYC^\dagger=-Y.$$

[If one wishes, one may also set $C\phi_k C^{-1}=-\phi^\dagger$ for both $k=1$ and 2.]

The normal modes t_1, t_2, and t_3 are not eigenstates of C nor of T. As an example of C violation or T violation, we may consider diagrams for

$$nB_\mu\to mB_\mu, \qquad (18)$$

where $n+m$ is an *odd* number. Because of loop diagrams in which the propagators are those of the triangles, the amplitudes for these C-violating processes can be nonzero. (However, see Appendix A for a list of special circumstances under which some of these amplitudes may happen to be zero.)

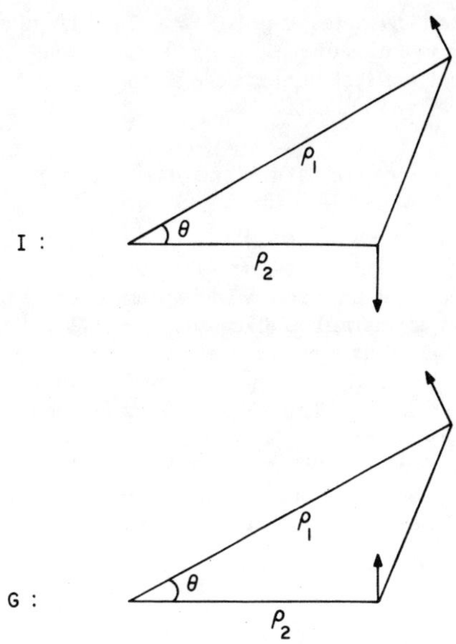

FIG. 1. A schematical drawing of the triangle; I represents one of its vibrational degrees of freedom defined by (14), and G represents its rotational degree of freedom defined by (13).

By using (1) and (12)–(15), one can readily verify that both the gauge field and the normal modes of the triangle are of zero average hypercharge; i.e., $\langle Y\rangle=0$. Thus, for the system of the triangle and the gauge field alone, one can always introduce a new "time-reversal" operator T_n and a new "particle-antiparticle conjugation" operator C_n such that

$$T_n t_a T_n^{-1}=t_a,$$

$$C_n t_a C_n^{-1}=t_a,$$

$$T_n G T_n^{-1}=G,$$

$$C_n G C_n^{-1}=G,$$

$$T_n B_\mu T_n^{-1}=B_\mu,$$

and

$$C_n B_\mu C_n^{-1}=B_\mu.$$

Since the Lagrangian $\mathcal{L}(B,\phi)$ can be written as a real function of these Hermitian fields, it must be invariant under the new C_n and T_n. Reaction (18) does not violate either T_n invariance or C_n invariance. The reciprocity relations are then maintained. However, under T_n one now has $\phi_1\to\phi_1^\dagger$ and $\phi_2\to\phi_2^\dagger$, and under C_n $\phi_1\to\phi_1$ and $\phi_2\to\phi_2$. Thus, the hypercharge Y neither commutes with T_n nor anticommutes with C_n. Nevertheless, this is totally acceptable, since in this simple system all the eigenstates are of $\langle Y \rangle = 0$. In order to observe violation of reciprocity relations, one should enlarge the system to include some known particles with $\langle Y \rangle$ nonzero.

III. VIOLATIONS OF RECIPROCITY RELATIONS

To illustrate how violations of reciprocity relations may occur, we discuss the example of a particular weak-interaction Lagrangian given by (5). [Other forms will be discussed in Sec. IV.] For clarity, let us consider first only the $P=-1$ part of the usual $\Delta Y = \pm 1$ nonleptonic weak interaction Lagrangian $L_\pm$. The operator $L_\pm$ is in general rather complicated, *not* a single canonical field; but so far as its transformation properties are concerned, $L_\pm$ is the same as the appropriate K^0 or $\bar{K}^0$ meson field. Thus we may write

$$L_- \sim K^0$$

and

$$L_+ \sim K^{0\dagger},$$

where ~ indicates that both sides have the same transformation properties. According to (5), with the inclusion of the triangle the corresponding weak interaction becomes

$$\mathcal{L}_{\text{int}} \sim (g_1\phi_1 + g_2\phi_2)K^0 + \text{H.c.}, \tag{19}$$

where g_1 and g_2 are both real so that T invariance holds. It is clear that (19) is also invariant under the hypercharge gauge transformation. By using (12), one may rewrite (19) in the form

$$\mathcal{L}_{\text{int}} \sim \Gamma K_1^0 + g(K_1^0 \chi_1 + K_2^0 \chi_2), \tag{20}$$

where Γ and g are both real and positive;

$$\Gamma^2 = g_1^2\rho_1^2 + g_2^2\rho_2^2 + 2g_1g_2\rho_1\rho_2\cos\theta \tag{21}$$

and $g = (g_1^2 + g_2^2)^{1/2}$. The K_1^0 and K_2^0 meson fields are defined by

$$K_1^0 = 2^{-1/2}(K^0 e^{i\alpha} + K^{0\dagger}e^{-i\alpha}),$$

$$K_2^0 = i2^{-1/2}(K^0 e^{i\alpha} - K^{0\dagger}e^{-i\alpha})$$

and α is given by

$$\Gamma e^{i\alpha} = g_1\rho_1 e^{i\theta} + g_2\rho_2. \tag{22}$$

The χ_1 and χ_2 fields are related to R_1, R_2 and I_1, I_2 by

$$\chi_1 = g^{-1}[g_1\cos(\theta-\alpha)R_1 - g_1\sin(\theta-\alpha)I_1 + g_2\cos\alpha R_2 + g_2\sin\alpha I_2]$$

and (23)

$$\chi_2 = g^{-1}[g_1\sin(\theta-\alpha)R_1 + g_1\cos(\theta-\alpha)I_1 - g_2\sin\alpha R_2 + g_2\cos\alpha I_2].$$

Under a hypercharge gauge transformation, the relative phase between K^0 and $\bar{K}^0$ meson states changes; therefore, we can always choose their relative phase so that K_1^0 represents the usual $CP=+1$ component, and K_2^0 the usual $CP=-1$ component. The first term ΓK_1^0 in (20) gives rise to the CP-conserving transition $K_1^0 \to 2\pi$. The second term in (20) leads to the CP-violating transition

$$K_1^0 \rightleftharpoons K_2^0;$$

its amplitude is determined by the Fourier transform of the contraction[8]

$$g^2 \chi_1^{\cdot}(x)\chi_2^{\cdot}(0). \tag{24}$$

With this CP violation, there is automatically also a violation of the reciprocity relation. Similarly, one may phenomenologically include the $P=+1$ part of the usual $\Delta Y = \pm 1$ nonleptonic weak interaction and, if one wishes, also the usual $\Delta Y = \pm 1$ semileptonic weak interaction in the Lagrangian (5). The former gives, among other transitions, the CP-conserving reaction $K_2^0 \to 3\pi$, and the latter gives all the usual CP-conserving semileptonic $\Delta Y \neq 0$ transitions. It is easy to show that in both cases there is, in addition, a CP-violating transition amplitude which also depends linearly on (24).

The magnitude of the T-violating amplitude (24) depends on the detailed characteristics of the triangle: both its shape and its vibrational modes t_a. It is of interest to search for the maximum of (24). As will be shown in Appendix B, if the coupling constants f, Γ, g_1, g_2 and the masses m_B, m_1, m_2, m_3 are fixed, then by varying the shape and the vibrational modes of the triangle, under the constraint (16), the maximum value of the Fourier transform of (24) at zero 4-momentum transfer, and for $m_1 > m_2 > m_3$, is found to be

$$\tfrac{1}{2}g[g^2 - (f\Gamma/m_B)^2]^{1/2}(m_3^{-2} - m_1^{-2}). \tag{25}$$

The corresponding vibrational modes of the triangle are given by (B10) and (B11) in Appendix B, and the shape is determined by (17), (21), and (B15). As an illustration, we may mention the special case in which $g_1 = g_2$ and $m_2^2 = \frac{1}{2}(m_1^2 + m_3^2)$; then the maximum T-violating amplitude (25) can be realized if the triangle is isosceles. If in addition we assume a right-angle isosceles triangle, then according to Eq. (B20) in Appendix B the maximum T-violation amplitude is

$$\left(\frac{1}{2\sqrt{2}}\right) g^2(m_3^{-2} - m_1^{-2}). \tag{26}$$

In any case, (25) or (26) is proportional to g^2 and is therefore of second order in the weak interaction. For example, by using (17) and (26) one finds the magnitude of the CP-violating amplitude

$K_1^0 \rightleftarrows K_2^0$ to be of the order of

$$\sim\Gamma^2 f^2 m_B^{-2}(m_3^{-2} - m_1^{-2}).$$

Since Γ denotes the first-order weak interaction constant, one expects this CP-violating amplitude to be of the superweak strength.[9]

As discussed earlier, there is another mechanism through which T violations can be produced, and that is via the coupling between the matter fields and the gauge field. Such T-violation effects are at least proportional to f^6; furthermore, such a mechanism conserves Y. Thus, if f^2 is arbitrarily set to be $\sim\alpha$, the fine-structure constant, one expects it to generate a $\Delta Y = 0$ but C, T-violating weak (or milliweak) interaction among known particles. For K_L^0, K_S^0 decays, it may add to the above $\Delta Y = \pm 2$ $K_1^0 \rightleftarrows K_2^0$ transition an amplitude $\sim\Gamma^2\alpha^3$. Hence, the superweak character of CP violation in K decay remains the same.

This simple interaction Lagrangian (5) is not intended to be a realistic theory of weak interactions because it contains many defects. It leaves out all $\Delta Y = 0$ weak reactions, and, since (without introducing additional gauge fields) the usual weak interaction Lagrangian $L_\pm$ is not renormalizable, it is also nonrenormalizable. Nevertheless, this simple example does illustrate how, through virtual emissions and absorptions of the triangle, T violations, and consequently also reciprocity violations, can be observed among known particles.

IV. OTHER APPLICATIONS

The above theory of spontaneous T violation can be applied to a large class of interactions, which can be quite different from the simple model discussed in the previous section. To illustrate these possibilities, we consider the following two examples of renormalizable theories.

A. An Abelian Example

We may identify the transformation (1) not with the hypercharge, but with a different gauge, say $e^{iN\Lambda}$ where for the known particles N is the number of left-handed charged leptons. So far as the descriptions of the triangle and the gauge field are concerned, the discussion given in Sec. II remains intact, except that ϕ_1 and ϕ_2 are now considered to be of $N=1$ (instead of $Y=1$). Of course, the discussion given in Sec. III has to be modified.

To study the interaction with matter fields in this new case, let us introduce a left-handed charged lepton field $l_L(x)$ and a right-handed charged lepton field $l_R(x)$ which satisfy

$$\gamma_5 l_L(x) = l_L(x)$$

and (27)

$$\gamma_5 l_R(x) = -l_R(x).$$

Throughout the paper, all Dirac matrices $\gamma_1, \gamma_2, \ldots, \gamma_5$ are Hermitian. By definition, $l_L(x)$ is of $N=1$ and $l_R(x)$ of $N=0$. The total Lagrangian density of the system can be written as

$$\mathcal{L}(l, B) + \mathcal{L}(B, \phi) + \mathcal{L}(\phi, l), \tag{28}$$

where $\mathcal{L}(B,\phi)$ is given by (6),

$$\mathcal{L}(l, B) = -l_L^\dagger \gamma_4\gamma_\mu\left(\frac{\partial}{\partial x_\mu} - ifB_\mu\right) l_L - l_R^\dagger\gamma_4\gamma_\mu \frac{\partial}{\partial x_\mu} l_R,$$

and

$$\mathcal{L}(\phi, l) = -(g_1\phi_1 + g_2\phi_2) l_L^\dagger \gamma_4 l_R + \text{H.c.}, \tag{29}$$

where g_1 and g_2 are both real so that T invariance holds. The total Lagrangian (28) is also invariant under the gauge transformation $e^{iN\Lambda}$. In addition, it is invariant under a second gauge transformation,

$$l_L \to e^{i\xi} l_L$$

and

$$l_R \to e^{i\xi} l_R.$$

By using (12), we may rewrite (29) in the form

$$\mathcal{L}(\phi, l) = -m_l \psi_l^\dagger \gamma_4 \psi_l - (g/\sqrt{2})(\chi_1 \psi_l^\dagger \gamma_4\psi_l - i\chi_2 \psi_l^\dagger\gamma_4\gamma_5\psi_l), \tag{30}$$

where χ_1 and χ_2 are exactly of the same expressions given by (23) in the previous section, $g = (g_1^2 + g_2^2)^{1/2}$ as before,

$$\psi_l(x) = l_L(x) + e^{i\alpha} l_R(x), \tag{31}$$

$$m_l e^{i\alpha} = 2^{-1/2}(g_1\rho_1 e^{i\theta} + g_2\rho_2), \tag{32}$$

and therefore

$$m_l^2 = \tfrac{1}{2}(g_1^2\rho_1^2 + g_2^2\rho_2^2 + 2g_1g_2\rho_1\rho_2\cos\theta). \tag{33}$$

Since $\psi_l^\dagger\gamma_4\psi_l$ is of $P=1$, $C=1$, and $T=1$ while $i\psi_l^\dagger\gamma_4\gamma_5\psi_l$ is of $P=-1$, $C=1$, and $T=-1$, through the direct emission and absorption of the triangle, there is a P-, T-violating effect in the $(l+l)$ scattering amplitude. To lowest order, the amplitude is proportional to (24), exactly as before. According to (25), and after replacing Γ by $\sqrt{2}\,m_l$, one finds the maximum value of the Fourier transform of (24), at the zero 4-momentum transfer, to be

$$\frac{g[g^2 - 2(f m_l/m_B)^2]^{1/2} m_T\Delta_T}{(m_T^2 - \frac{1}{4}\Delta_T^2)^2}, \tag{34}$$

where m_T is the mean mass of the vibrational modes of the triangle

$$m_T = \tfrac{1}{2}(m_1 + m_3)$$

and $\Delta_T = (m_1 - m_3)$ is the corresponding difference. The other mass m_2 lies between $m_T + \frac{1}{2}\Delta_T$ and $m_T - \frac{1}{2}\Delta_T$. If in addition we assume the triangle to be a simple right-angle isosceles triangle and $g_1 = g_2 = 2^{-1/2}g$, then (26) holds; the maximum T-violation amplitude (34) becomes

$$\frac{g^2 m_T \Delta_T}{\sqrt{2}\,(m_T^{\,2} - \frac{1}{4}\Delta_T^{\,2})^2}\,. \tag{35}$$

Moreover, there can also be T-violating effects due to the direct coupling f between B_μ and l, just as before. We emphasize that although in this example both the gauge group and the interaction are quite different from those in Sec. III, the basic mechanism of T violation is identical.

B. A Non-Abelian Example

Let us first consider the Weinberg model of the leptons.[3] The group is $SU_2 \times U_1$. There are four gauge fields $\vec{A}_\mu$ and B_μ. The usual l-neutrino field ν_l and the right- and left-handed charged lepton fields l_R and l_L form an SU_2 doublet and an SU_2 singlet:

$$L = \begin{pmatrix} \nu_l \\ l_L \end{pmatrix}$$

and (36)

$$R = l_R\,.$$

In order to have spontaneous T violation, we assume that there are two SU_2-doublet spin-0 fields

$$\phi_1 = \begin{pmatrix} \phi_1^+ \\ \phi_1^0 \end{pmatrix}$$

and (37)

$$\phi_2 = \begin{pmatrix} \phi_2^+ \\ \phi_2^0 \end{pmatrix},$$

where the superscript denotes the electric charge. Both ϕ_1 and ϕ_2 are assumed to transform like the product $R^\dagger L$ under the $SU_2 \times U_1$ group; therefore, their coupling to the gauge fields is completely determined by the requirements of gauge invariance. The most general form of a renormalizable, gauge-invariant, and T-invariant potential energy $V(\phi)$ is now given by, instead of (7),

$$\begin{aligned} V(\phi) = &-\lambda_1\phi_1^\dagger\phi_1 - \lambda_2\phi_2^\dagger\phi_2 + A(\phi_1^\dagger\phi_1)^2 + B(\phi_2^\dagger\phi_2)^2 \\ &+ C(\phi_1^\dagger\phi_1)(\phi_2^\dagger\phi_2) + \overline{C}(\phi_1^\dagger\phi_2)(\phi_2^\dagger\phi_1) \\ &+ \tfrac{1}{2}[(\phi_1^\dagger\phi_2)(D\phi_1^\dagger\phi_2 + E\phi_1^\dagger\phi_1 + F\phi_2^\dagger\phi_2) + \text{H.c.}], \end{aligned} \tag{38}$$

which contains nine constants; all these constants are assumed to be real so that T invariance holds. The only formal difference between (38) and (7) is the $\overline{C}$ term. We assume that both (8) and (9) are valid, and in addition

$$D > \overline{C}\,. \tag{39}$$

In the tree approximation, the minimum of the c-number function $V(\phi)$ determines the vacuum expectation values of ϕ_1 and ϕ_2. As will be shown in Appendix C, the additional condition (39) ensures that the minimum of $V(\phi)$ occurs at

$$\langle\phi_1^+\rangle_{\text{vac}} = \langle\phi_2^+\rangle_{\text{vac}} = 0\,,$$

$$\langle\phi_1^0\rangle_{\text{vac}} = 2^{-1/2}\rho_1 e^{i\theta}\,,$$

and (40)

$$\langle\phi_2^0\rangle_{\text{vac}} = 2^{-1/2}\rho_2\,,$$

which again defines a triangle. Both ρ_1 and ρ_2 are assumed to be >0, and $\theta \neq 0$ or π. So far as the neutral (but complex) fields ϕ_1^0 and ϕ_2^0 are concerned, the discussion is exactly the same as previously given in Sec. II, except that the constant C in Sec. II is now replaced by $C + \overline{C}$.

We may expand, as before in (12),

$$\phi_1^0 = 2^{-1/2}e^{i\theta}(\rho_1 + R_1' + iI_1)$$

and (41)

$$\phi_2^0 = 2^{-1/2}(\rho_2 + R_2 + iI_2)\,.$$

The three vibrational modes of the triangle t_1, t_2, and t_3 are given by (15), and they correspond respectively to three neutral bosons of masses m_1, m_2, and m_3. There are now three Goldstone bosons; besides the neutral one

$$G^0 = (\rho_1^{\,2} + \rho_2^{\,2})^{-1/2}(\rho_1 I_1 + \rho_2 I_2)\,, \tag{42}$$

which is identical to the G given by (13), there are two charged ones:

$$G^\pm = (\rho_1^{\,2} + \rho_2^{\,2})^{-1/2}(\rho_1\phi_1^\pm e^{\mp i\theta} + \rho_2\phi_2^\pm)\,, \tag{43}$$

where $\phi_k^- = (\phi_k^+)^\dagger$ and $k = 1$ or 2. Through the Higgs mechanism, these three Goldstone bosons (just as in the usual Weinberg model) join the gauge fields to form a set of three massive spin-1 intermediate bosons $W^\pm$ and W^0. In addition, there are also two massive charged spin-0 bosons

$$H^\pm = (\rho_1^{\,2} + \rho_2^{\,2})^{-1/2}(\rho_2\phi_1^\pm e^{\mp i\theta} - \rho_1\phi_2^\pm)\,, \tag{44}$$

and their masses are

$$m_H^{\,2} = \tfrac{1}{2}(D - \overline{C})(\rho_1^{\,2} + \rho_2^{\,2})\,.$$

For simplicity, we shall assume the mass m_H to be much greater than the masses m_1, m_2, and m_3 of the neutral spin-0 bosons. Therefore, all $H^\pm$-exchange processes will be ignored in the following discussions. So far as the mechanism of T violation is concerned, one has then exactly the same basic structure as before. The triangle is again characterized by its shape and its vibrational modes, and with the same constraint (16).

The interaction between the lepton fields and the gauge fields is determined by the requirement of gauge invariance; it is exactly the same as in the usual Weinberg model. The interaction between the spin-0 fields ϕ_1, ϕ_2 and the lepton fields is now given by

$$\mathcal{L}(l,\phi) = -(g_1\phi_1 + g_2\phi_2)L^\dagger\gamma_4 R + \text{H.c.}, \qquad (45)$$

where g_1 and g_2 are again assumed to be real so that T invariance holds. This Lagrangian is clearly also invariant under the $SU_2 \times U_1$ gauge transformation; it describes an interaction between the charged lepton $l^\pm$ and ϕ_k^0 which is exactly the same as (29) and which can again be rewritten as (30). Therefore there is a P-, T-violating amplitude in $l+l$ scattering that is proportional to (24).

The same P-, T-violating amplitude also leads to an electric dipole moment for the charged lepton. For definiteness, let us assume (35) holds; one then has a simple right-angle isosceles triangle. In this case, one finds

$$\rho_1^2 = \rho_2^2 = (2\sqrt{2}\,G_F)^{-1},$$

$$g_1^2 = g_2^2 = 2\sqrt{2}\,G_F m_l^2,$$

where $G_F \cong 10^{-5} m_N^{-2}$ is the Fermi constant. The P-, T-violating amplitude (35) becomes

$$\frac{4G_F m_l^2 m_T \Delta_T}{(m_T^2 - \frac{1}{4}\Delta_T^2)^2}. \qquad (46)$$

The electric dipole moment $eD(l)$ of $l^\pm$ can then be readily evaluated. By using (30), (24), and (46), we find

$$D(l) = \frac{G_F m_l^3}{8\pi^2}[m_3^{-2}J(\epsilon_3) - m_1^{-2}J(\epsilon_1)], \qquad (47)$$

where $m_1 = m_T + \frac{1}{2}\Delta_T$ and $m_3 = m_T - \frac{1}{2}\Delta_T$ denote respectively, as before, the largest and the smallest mass of the vibrational modes of the triangle, $\epsilon_1 = (m_l/m_1)^2$, $\epsilon_3 = (m_l/m_3)^2$, and

$$J(\epsilon) = \epsilon^{-1} + (2\epsilon^2)^{-1}\left\{\ln\epsilon + \frac{1-2\epsilon}{(1-4\epsilon)^{1/2}} \times \ln\left[\frac{1+(1-4\epsilon)^{1/2}}{1-(1-4\epsilon)^{1/2}}\right]\right\}.$$

For $\epsilon \ll 1$,

$$J(\epsilon) \cong \ln(1/\epsilon) - \tfrac{3}{2} + O(\epsilon\ln\epsilon).$$

If m_1 and m_3 are arbitrarily set to be ~15 GeV and ~10 GeV, respectively, then $D(\mu)$ is $\sim 1.3\times10^{-25}$ cm and $D(e)$ is $\sim 3.6\times10^{-32}$ cm. At present, both are too small to be detected.

The extension of the Weinberg model to hadrons is not without arbitrariness. The direct coupling between hadrons and the spin-0 fields ϕ_1^0 and ϕ_2^0 has the same form as that in (45), except that L and R now refer to the appropriate hadron fields. Such a coupling is usually assumed to conserve the isospin. Thus, similar to (46), in the hadron-hadron scattering there is a P-, T-violating, but $|\Delta\vec{I}| = 0$, amplitude which is given by

$$\frac{4G_F m_h^2 m_T \Delta_T}{(m_T^2 - \frac{1}{4}\Delta_T^2)^2} \qquad (48)$$

and which can lead to an electric dipole $eD(h)$ of the order of

$$D(h) \sim \frac{G_F m_h^3 \Delta_T}{4\pi^2 m_T^3}, \qquad (49)$$

where m_h denotes the corresponding hadronic mass.

The present experimental limit on the electric dipole moment of the neutron[10] is $D(n) < 10^{-23}$ cm. If we arbitrarily set $m_h \sim m_N$ (m_N is the nucleon mass), then (49) gives

$$\frac{m_N^2 \Delta_T}{m_T^3} \lesssim 2\times10^{-3}, \qquad (50)$$

which implies that the P-, T-violating amplitude (48) in a $|\Delta\vec{I}| = 0$ hadronic scattering process is

$$\lesssim 10^{-2} G_F. \qquad (51)$$

For the $\Delta Y = \pm1$ processes, the corresponding T-violating amplitude should be at least smaller by an additional factor $G_F m_W^2 \sim \alpha$, i.e.,

$$\lesssim 10^{-4} G_F. \qquad (52)$$

For the $\Delta Y = \pm2$ processes, some special constructions must be introduced to make the usual T-conserving amplitude in the K_L^0, K_S^0 mass-difference calculation $\sim G_F m_N^2$ (not $\sim G_F m_W^2$) times smaller than the corresponding T-conserving $\Delta Y = \pm1$ amplitude. It seems reasonable to expect that relative to (52), a similar factor $\sim G_F m_N^2$ also applies for the corresponding T-violating $\Delta Y = \pm2$ amplitude, and that would lead to a T-violating amplitude

$$\lesssim 10^{-9} G_F \qquad (53)$$

in the mass matrix of the K^0-$\bar{K}^0$ complex. Since (52) seems to be smaller than the milliweak strength, one may expect the CP-violating phenomena in the K decay to be dominated by (53); the result would be of the observed superweak character.[9]

For the $|\Delta\vec{I}| = 0$ processes, the T-violating amplitude can be of the milliweak strength, and this may have important experimental consequences. As discussed earlier (and also in Appendix A), there exist other T-violating diagrams in which the T violation is generated via the direct coupling between the matter fields and the spin-1 intermediate bosons. In addition, T violation can also occur due to exchanges of the charged spin-0 bosons $H^{\pm}$. However, a full investigation lies outside the scope of this paper.

ACKNOWLEDGMENTS

I am indebted to G. C. Wick for his valuable suggestions, and I also wish to thank N. Christ and A. Mueller for helpful discussions.

APPENDIX A

1. We first discuss the vacuum expectation values of ϕ_1 and ϕ_2:

$$\langle \phi_1 \rangle_{\text{vac}} \equiv 2^{-1/2}\rho_1 e^{i\theta}$$

and (A1)

$$\langle \phi_2 \rangle_{\text{vac}} \equiv 2^{-1/2}\rho_2 .$$

In the tree approximation, ρ_1, ρ_2, and θ can be determined by setting the minimum of the c-number function $V(\phi)$ at $(\phi_1, \phi_2) = 2^{-1/2}(\rho_1 e^{i\theta}, \rho_2)$. According to (7),

$$V = -\tfrac{1}{2}(\lambda_1\rho_1^2 + \lambda_2\rho_2^2) + \tfrac{1}{4}(p\rho_1^4 + q\rho_2^4 + 2r\rho_1^2\rho_2^2) + \tfrac{1}{2}D\rho_1^2\rho_2^2(\cos\theta - \delta)^2 , \tag{A2}$$

where

$$p = A - (8D)^{-1}E^2 ,$$

$$q = B - (8D)^{-1}F^2 ,$$

$$r = \tfrac{1}{2}\left(C - D - \frac{EF}{4D}\right),$$

and

$$\delta = -(4D\rho_1\rho_2)^{-1}(E\rho_1^2 + F\rho_2^2) .$$

The function V must have a lower bound, and therefore (10) holds; i.e.,

$$p > 0 ,$$

$$q > 0 ,$$

and (A3)

$$pq > r^2 .$$

Since $D > 0$, the minimum of V is at

$$\cos\theta = \delta , \tag{A4}$$

and since (λ_1 and/or λ_2) > 0, this minimum is not at the origin but at

$$\rho_1^2 = (pq - r^2)^{-1}(q\lambda_1 - r\lambda_2)$$

and (A5)

$$\rho_2^2 = (pq - r^2)^{-1}(p\lambda_2 - r\lambda_1) .$$

In order that $\theta \neq 0$ or π, and both ρ_1 and ρ_2 are >0, we require, in addition to (8) and (9),

$$|\delta| < 1 ,$$

$$q\lambda_1 > r\lambda_2 ,$$

and (A6)

$$p\lambda_2 > r\lambda_1 .$$

We note that if $r > 0$, then both λ_1 and λ_2 must be >0; but if $r < 0$, then at least one of them (either λ_1 or λ_2) must be >0, but the other one could be <0, provided (A6) is satisfied.

2. The expansion of ϕ_1 and ϕ_2 around their vacuum expectation values

$$\phi_1 = 2^{-1/2}(\rho_1 + R_1 + iI_1)e^{i\theta}$$

and (A7)

$$\phi_2 = 2^{-1/2}(\rho_2 + R_2 + iI_2)$$

leads to

$$V(\phi) = V_{c\text{-No.}} + V_{\text{quad}} + V_{\text{cub}} + V_{\text{quart}} , \tag{A8}$$

where the subscripts refer to, respectively, a c-number expression, a quadratic function of R_k, I_k, and corresponding cubic and quartic functions. (The linear function is absent because of the minimum condition.) To obtain the normal modes t_1, t_2, and t_3, we need only to diagonalize V_{quad}:

$$\begin{aligned} V_{\text{quad}} = {} & [A\rho_1^2 + \tfrac{1}{2}\cos\theta(D\rho_2^2\cos\theta + E\rho_1\rho_2)]R_1^2 + [B\rho_2^2 + \tfrac{1}{2}\cos\theta(D\rho_1^2\cos\theta + F\rho_1\rho_2)]R_2^2 \\ & + [C - D(1 + \cos^2\theta)]\rho_1\rho_2 R_1 R_2 + \tfrac{1}{2}D(\rho_1^2 + \rho_2^2)\sin^2\theta\, I^2 \\ & - (4\rho_1\rho_2)^{-1}\sin\theta(E\rho_1^2 - F\rho_2^2)(\rho_1^2 + \rho_2^2)^{1/2}(\rho_2 R_1 - \rho_1 R_2)I , \end{aligned} \tag{A9}$$

where ρ_1, ρ_2, and θ are given by (A4) and (A5). It is convenient to introduce I, G, R, and R', where G and I are defined by (13) and (14), respectively, and

$$R \equiv (\rho_1^2 + \rho_2^2)^{-1/2}(\rho_2 R_1 - \rho_1 R_2)$$

and (A10)

$$R' \equiv (\rho_1^2 + \rho_2^2)^{-1/2}(\rho_1 R_1 + \rho_2 R_2) .$$

The constraint (16) is derived by noting that the product $R'I$ is absent in V_{quad}, and the Goldstone boson (13) is determined by observing that G is absent in $V_{\rm quad}$.

3. We shall now derive a set of conditions under which certain C-, T-violating diagrams must be zero if the system contains only ϕ_k and B_μ. (As explained before, even if such diagrams are not zero there is no violation of reciprocity relations without other fields.) We define

$$\phi_1' \equiv \rho^{-1}(\rho_2\phi_1 e^{-i\theta} - \rho_1\phi_2),$$
$$\phi_2' \equiv \rho^{-1}(\rho_1\phi_1 e^{-i\theta} + \rho_2\phi_2) \qquad \text{(A11)}$$

where

$$\rho = (\rho_1{}^2 + \rho_2{}^2)^{1/2}.$$

From (A1) we find

$$\langle\phi_1'\rangle_{\rm vac} = 0$$

and (A12)

$$\langle\phi_2'\rangle_{\rm vac} = 2^{-1/2}\rho,$$

which are both real. In terms of ϕ_1' and ϕ_2', (A7) becomes

$$\phi_1' = 2^{-1/2}(R + iI)$$

and (A13)

$$\phi_2' = 2^{-1/2}(\rho + R' + iG).$$

The function $V(\phi)$, defined by (7), can now be written as

$$V = -\lambda_1'\phi_1'^\dagger\phi_1' - \lambda_2'\phi_2'^\dagger\phi_2' - (\lambda_3'\phi_1'^\dagger\phi_2' + \text{H.c.})$$
$$+ A'(\phi_1'^\dagger\phi_1')^2 + B'(\phi_2'^\dagger\phi_2)^2 + C'(\phi_1'^\dagger\phi_1')(\phi_2'^\dagger\phi_2')$$
$$+ \tfrac{1}{2}[(\phi_1'^\dagger\phi_2')(D\phi_1'^\dagger\phi_2' + E'\phi_1'^\dagger\phi_1' + F'\phi_2'^\dagger\phi_2') + \text{H.c.}], \qquad \text{(A14)}$$

where the new constants $\lambda_1', \lambda_2', \ldots, F'$ can be readily expressed in terms of the original eight real parameters $\lambda_1, \lambda_2, \ldots, F$; e.g.,

$$\lambda_1' = \rho^{-2}(\lambda_1\rho_2{}^2 + \lambda_2\rho_1{}^2),$$
$$\lambda_2' = \rho^{-2}(\lambda_1\rho_1^2 + \lambda_2\rho_2{}^2),$$
$$\lambda_3' = \rho^{-2}(\lambda_1 - \lambda_2)\rho_1\rho_2,$$

etc.

Because of Hermiticity only λ_3', D', E', and F' may have imaginary parts. By using (A4), (A5), and (A11), we find

$$\text{Im}\lambda_3' = 0,$$
$$\text{Im}F' = 0,$$
$$\text{Im}\, D' = -(2\rho_1\rho_2)^{-1}(E\rho_1{}^2 - F\rho_2{}^2)\sin\theta, \qquad \text{(A15)}$$

and

$$\text{Im}E' = 4\rho^{-2}\rho_1\rho_2 D\sin 2\theta.$$

In addition, there are three equalities:

$$\rho_1\rho_2(\lambda_1' - \lambda_2') + (\rho_1{}^2 - \rho_2{}^2)\lambda_3' = 0,$$
$$\rho^2 B' = \lambda_2',$$

and (A16)

$$\rho^2 F' = 4\lambda_3'.$$

These three equalities imply that among the eleven new real parameters λ_1', λ_2', λ_3', A', B', C', $\text{Re}D'$, $\text{Im}D'$, $\text{Re}E'$, $\text{Im}E'$, and F' only eight are independent. In terms of these new fields, (A9) becomes

$$V_{\rm quad} = \tfrac{1}{2}[-\lambda_1' + \tfrac{1}{4}\rho^2(C' + \text{Re}\,D')]R^2 + \tfrac{1}{4}\lambda_2' R'^2$$
$$+ \tfrac{1}{2}[-\lambda_1' + \tfrac{1}{4}\rho^2(C' - \text{Re}\,D')]I^2 + 2\lambda_3' RR'$$
$$+ \tfrac{1}{2}\rho^2(\text{Im}\,D')IR. \qquad \text{(A17)}$$

The coupling between ϕ_k' and the gauge field B_μ has the same covariant form as that between ϕ_k and B_μ; e.g., the current operator j_μ remains given by

$$j_\mu = i\sum_{k=1,2}\left(\frac{\partial\phi_k'^\dagger}{\partial x_\mu}\phi_k' - \phi_k'^\dagger\frac{\partial\phi_k'}{\partial x_\mu}\right). \qquad \text{(A18)}$$

The corresponding interaction Lagrangian is

$$-fB_\mu\left(G\frac{\partial R'}{\partial x_\mu} - R'\frac{\partial G}{\partial x_\mu} + I\frac{\partial R}{\partial x_\mu} - R\frac{\partial I}{\partial x_\mu}\right)$$
$$- f m_B B_\mu{}^2 R' - \tfrac{1}{2}f^2 B_\mu{}^2(I^2 + R^2 + R'^2 + G^2). \qquad \text{(A19)}$$

In addition to this interaction, we also have $V_{\rm cub}$ and $V_{\rm quart}$ in (A8). In a power-series expansion in f, we regard all masses to be of the zeroth order, and therefore

$$V_{\rm cub} \sim O(f)$$

and (A20)

$$V_{\rm quart} \sim O(f^2).$$

We note that because of (A12) and (A15), in order to have C, T violations [e.g., $nB_\mu \rightleftarrows mB_\mu$ when $n + m$ is an odd number] one must have $\text{Im}\,D' \neq 0$ and/or $\text{Im}E' \neq 0$. The following theorems can then be easily established:

Theorem 1. If $\theta = \frac{1}{2}\pi$ and if I is a normal mode, then (without other fields) there is no C, T violation.

Proof. If I is a normal mode, then the coefficient of IR in (A17) must be zero. Hence $\text{Im}\,D' = 0$. If $\theta = \frac{1}{2}\pi$, then according to (A15) $\text{Im}E' = 0$. The theorem is then established. Thus, for example, the amplitude for $nB_\mu \rightleftarrows mB_\mu$ must be zero if $n + m$ is an odd number.

The same theorem can also be proved by noting that in this case, by using (A4) and (A9), one has $E = F = 0$. Hence, a new time-reversal operation may be defined, under which $\phi_1 \to -\phi_1$ and $\phi_2 \to \phi_2$, instead of (3). The vacuum expectation values $\langle\phi_1\rangle_{vac}=i2^{-1/2}\rho_1$ and $\langle\phi_2\rangle_{vac}=2^{-1/2}\rho_2$ are compatible with this new time-reversal operation, and therefore without other matter fields there is no T violation.

Theorem 2. If the normal modes t_1, t_2, and t_3 are all degenerate (i.e., $m_1=m_2=m_3$), then there is also no C, T violation.

Proof. Because of the degeneracy, we may choose the normal modes to be $t_1=R$, $t_2=I$, and $t_3=R'$. From (A17), one sees that the absence of RR' and IR coupling gives $\lambda_3'=0$ and $\mathrm{Im}\,D'=0$. The degeneracy $m_1=m_2$ gives $\mathrm{Re}\,D'=0$. These together with (A16) imply

$$\lambda_3'=F'=D'=0.$$

There is, therefore, only one term in (A14) that depends on the relative phase between ϕ_1' and ϕ_2', and it is proportional to E'. We may rotate $\phi_1'\to e^{i\beta}\phi_1'$, $\phi_2'\to\phi_2'$; this does not alter their vacuum expectation values, nor the coupling between ϕ_k' and B_μ, but it can transform E' to real. Once E' becomes real, one may define the time reversal to be the antiunitary operator under which $\phi_1'\to\phi_1'$ and $\phi_2'\to\phi_2'$. Since (A14) contains only real parameters and since $\langle\phi_1'\rangle_{vac}$ and $\langle\phi_2'\rangle_{vac}$ are also both real, the theorem is proved.

Remark. The condition of Theorem 2 can be weakened: We need only $t_1=R$, $t_2=I$, $t_3=R'$, and $m_1=m_2$, but m_3 can be different.

Theorem 3. If R' is a normal mode, then to order f^3 the amplitude of $B_\mu(k)\rightleftarrows B_\mu(p)+B_\mu(q)$ vanishes for arbitrary virtual momenta k, p, and q.

Proof. From (A17), it follows that if R' is a normal mode then $\lambda_3'=0$, which implies $F'=0$, on account of (A16). In (A14) there are only two terms, one proportional to D' and the other to E', that depend on the relative phase between ϕ_1' and ϕ_2'. Just as in the previous proof, we may rotate $\phi_1'\to e^{i\beta}\phi_1'$ and $\phi_2'\to\phi_2'$, but this time to make D' real. All C-, T-violating effects must then be proportional to $\mathrm{Im}E'$. It is easy to verify that in (A14) the $\mathrm{Im}E'$ term is of the form

$$\tfrac{1}{4}(\mathrm{Im}E')(R^2+I^2)(I\rho+IR'-GR).$$

Because of (A20), V_{quart} does not contribute to the lowest-order f^3 diagrams for $B_\mu(k)\rightleftarrows B_\mu(p)+B_\mu(q)$, but V_{cub} may. Since R' is assumed to be a normal mode, V_{cub} can contribute to such diagrams only if it contains at least one factor of R' [in which case the V_{cub} vertex can link with the $-fm_BB_\mu{}^2R'$ vertex in (A19) through the R' propagator]. However, the cubic-function part of the above $\mathrm{Im}E'$ term does not contain any R' factor; hence, the theorem.

For a general triangle, the actual calculation of the transition $B_\mu(k)\rightleftarrows B_\mu(p)+B_\mu(q)$ is rather complicated, but we hope to give some of the details in a separate publication.

APPENDIX B

To establish (25), we first express χ_1 and χ_2, defined by (23), as linear functions of the normal modes t_1, t_2, t_3 and the Goldstone mode G:

$$\chi_1=a_1t_1+a_2t_2+a_3t_3 \tag{B1}$$

and

$$\chi_2=b_1t_1+b_2t_2+b_3t_3+\gamma G \tag{B2}$$

where a_i, b_j, and γ are constants. From (22), it follows that

$$g_1\rho_1\sin(\theta-\alpha)-g_2\rho_2\sin\alpha=0$$

and

$$g_1\rho_1\cos(\theta-\alpha)+g_2\rho_2\cos\alpha=\Gamma;$$

these together with (23) require χ_1 to be independent of G and the constant γ in (B2) given by

$$\gamma=(g\rho)^{-1}\Gamma, \tag{B3}$$

where $g=(g_1{}^2+g_2{}^2)^{1/2}$ and $\rho=(\rho_1{}^2+\rho_2{}^2)^{1/2}$. Let us define

$$a=\begin{pmatrix}a_1\\a_2\\a_3\end{pmatrix}$$

and (B4)

$$b=\begin{pmatrix}b_1\\b_2\\b_3\end{pmatrix}.$$

By using (23), one derives

$$\tilde a a=1,$$

$$\tilde a b=0,$$

and (B5)

$$\tilde b b=1-\gamma^2=1-(g\rho)^{-2}\Gamma^2,$$

where the tilde denotes the transpose.

The Fourier transform of (24) at the zero momentum is

$$g^2\tilde a M^{-2}b \tag{B6}$$

where

$$M^2=\begin{pmatrix}m_1{}^2&0&0\\0&m_2{}^2&0\\0&0&m_3{}^2\end{pmatrix}. \tag{B7}$$

We may search for the maximum of (B6) by varying a and b, but keeping M^2 and the three orthonormal relations given in (B5) fixed. It is straightforward to show that for $m_1>m_2>m_3$,

$$|g^2\tilde{a}M^{-2}b| \leq \tfrac{1}{2}g^2[1-(g\rho)^{-2}\Gamma^2]^{1/2}(m_3^{-2}-m_1^{-2}), \tag{B8}$$

and the equal sign holds when

$$a_1 = a_3 = 2^{-1/2},$$
$$b_1 = -b_3 = -2^{-1/2}[1-(g\rho)^{-2}\Gamma^2]^{1/2}, \tag{B9}$$

and

$$a_2 = b_2 = 0.$$

Clearly, the maximum value for $|g^2\tilde{a}M^{-2}b|$ remains the same, if one changes $t_1 \to -t_1$, or $t_3 \to -t_3$, or $a \to -a$, or $b \to -b$.

For any given shape of the triangle ρ_1, ρ_2, and θ, one finds that the maximum in (B8) can be realized if the normal modes are given by

$$\left.\begin{matrix} t_1 \\ t_3 \end{matrix}\right\} = 2^{-1/2}\{-[\beta \pm (\alpha^2+\beta^2)^{-1/2}\alpha]I + [\alpha \mp (\alpha^2+\beta^2)^{-1/2}\beta]R + \gamma R'\} \tag{B10}$$

and

$$t_2 = (\alpha^2+\beta^2)^{-1/2}[\gamma\beta I - \gamma\alpha R + (\alpha^2+\beta^2)R'], \tag{B11}$$

where t_1 assumes the upper sign in (B10) and t_3 the lower sign, γ is given by (B3), I, R, and R' are defined by (14) and (A10),

$$\alpha = (\rho g\Gamma)^{-1}[(g_1^2-g_2^2)\rho_1\rho_2 - g_1 g_2(\rho_1^2-\rho_2^2)\cos\theta], \tag{B12}$$

$$\beta = (g\Gamma)^{-1}\rho g_1 g_2 \sin\theta, \tag{B13}$$

and therefore

$$\alpha^2 + \beta^2 + \gamma^2 = 1. \tag{B14}$$

The only condition is that the constraint (16) should hold. Because of (B10) and (B11), this constraint can also be written as

$$\tfrac{1}{2}(\alpha^2+\beta^2)^{-1/2}\alpha(m_1^2-m_3^2) = \beta[m_2^2 - \tfrac{1}{2}(m_1^2+m_3^2)]. \tag{B15}$$

One can readily verify that the solutions (B10) and (B11) together with (B9) indeed lead to χ_1 and χ_2 defined by their original expression (23).

As an explicit example, we may consider the special case $g_1 = g_2 = 2^{-1/2}g$ and $\rho_1 = \rho_2 = 2^{-1/2}\rho$; i.e., an isosceles triangle. From (B12), one sees that $\alpha = 0$; therefore, (B15) implies

$$m_2^2 = \tfrac{1}{2}(m_1^2 + m_3^2). \tag{B16}$$

The angle θ and the side $2^{-1/2}\rho$ of the isosceles triangle are determined by (17) and (21), which can now be written as

$$\Gamma^2 = \tfrac{1}{2}g^2\rho^2(1+\cos\theta) \tag{B17}$$

and

$$m_B^2 = f^2\rho^2. \tag{B18}$$

If in addition we assume $\theta = \frac{1}{2}\pi$, i.e., a right-angle isosceles triangle, then $\Gamma = 2^{-1/2}g\rho$; hence, $\beta = \gamma = 2^{-1/2}$, the normal modes become

$$t_1 = \tfrac{1}{2}(-I - \sqrt{2}R + R'),$$
$$t_2 = \frac{1}{\sqrt{2}}(I + R'), \tag{B19}$$
$$t_3 = \tfrac{1}{2}(-I + \sqrt{2}R + R'),$$

and the corresponding maximum T-violation amplitude in (B8) is

$$2^{-3/2}g^2(m_3^{-2} - m_1^{-2}). \tag{B20}$$

As another example we may take the limiting case $\theta = 0$, or π. From (B13) it follows that $\beta = 0$; hence (B15) implies $m_1^2 = m_3^2$, and therefore according to (B8) the maximum T-violation amplitude is zero, as it should be.

In general, if the coupling constants f, g_1, g_2, and Γ and the masses m_1, m_2, m_3, and m_B are given, then in order to realize the maximum T-violation amplitude in (B8) the shape parameters ρ_1, ρ_2, and θ are determined by (17), (21), and (B15), and the vibrational modes t_1, t_2, and t_3 by (B10) and (B11).

APPENDIX C

In this appendix we discuss the minimum of the c-number potential energy $V(\phi)$ defined by (38); this minimum is assumed to be at

$$\phi_1 = 2^{-1/2}\begin{pmatrix} 0 \\ \rho_1 e^{i\theta} \end{pmatrix} \tag{C1}$$

and

$$\phi_2 = 2^{-1/2}\begin{pmatrix} \sigma \\ \rho_2 \end{pmatrix}, \tag{C2}$$

where θ is real, and σ, ρ_1, ρ_2 are all real and ≥ 0. Since V is invariant under the $SU_2 \times U_1$ gauge transformation, we can always transform the upper component of ϕ_1 to zero, and both components of ϕ_2 to real and non-negative. Equation (38) can then be written as

$$V = -\tfrac{1}{2}\lambda_1\rho_1^2 - \tfrac{1}{2}\lambda_2(\sigma^2+\rho_2^2) + \tfrac{1}{2}D\rho_1^2\rho_2^2[(\cos\theta-\Delta)^2 - \Delta^2] + \tfrac{1}{4}[A\rho_1^4 + B(\sigma^2+\rho_2^2)^2 + C\rho_1^2(\sigma^2+\rho_2^2) + (\bar{C}-D)\rho_1^2\rho_2^2], \tag{C3}$$

where

$$\Delta = -(4D\rho_1\rho_2)^{-1}[E\rho_1^2 + F(\sigma^2+\rho_2^2)]. \tag{C4}$$

For $D > 0$, the minimum of V is at

$$\cos\theta = \Delta. \tag{C5}$$

Keeping (C5) satisfied, we find

$$\frac{\partial V}{\partial \sigma^2} = \frac{\partial V}{\partial \rho_2^2} + \tfrac{1}{4}(D - \bar{C})\rho_1^2 . \qquad \text{(C6)}$$

Since $D > \bar{C}$, at $\partial V/\partial \rho_2^2 = 0$ the derivative $\partial V/\partial \sigma^2$ is always positive; hence to obtain the minimum of V we require

$$\sigma^2 = 0 . \qquad \text{(C7)}$$

The function V then reduces to (A2) discussed in Appendix A provided the constant C in (A2) is replaced by $C + \bar{C}$.

APPENDIX D

It is possible to have a spontaneous T violation, but without a spontaneous gauge-symmetry violation. Let us consider a simple example which consists of a spin-$\frac{1}{2}$ field ψ and a single Hermitian spin-0 field ϕ. The Lagrangian density $\mathcal{L}$ is assumed to be renormalizable, and it is invariant under T, C, and P:

$$\mathcal{L} = -\frac{1}{2}\left(\frac{\partial \phi}{\partial x_\mu}\right)^2 - V(\phi) - \psi^\dagger \gamma_4 \left(\gamma_\mu \frac{\partial}{\partial x_\mu} + m\right)\psi - ig\psi^\dagger \gamma_4 \gamma_5 \psi \phi , \qquad \text{(D1)}$$

where the potential $V(\phi)$ is given by

$$V(\phi) = -\tfrac{1}{2}\lambda \phi^2 + \tfrac{1}{4} A \phi^4 . \qquad \text{(D2)}$$

From Hermiticity, the parameters m, g, λ, and A must be real. It can be readily verified that the Lagrangian $\mathcal{L}$ is invariant under T, C, and P under which

$$T\phi(\vec{r}, t)T^{-1} = -\phi(\vec{r}, -t) , \qquad \text{(D3)}$$

$$C\phi(\vec{r}, t)C^{-1} = \phi(\vec{r}, t) , \qquad \text{(D4)}$$

and

$$P\phi(\vec{r}, t)P^{-1} = -\phi(-\vec{r}, t) ; \qquad \text{(D5)}$$

the corresponding transformations of ψ are standard. In addition, $\mathcal{L}$ is invariant under the simple gauge transformation

$$\psi \to e^{i\beta}\psi \text{ and } \phi \to \phi . \qquad \text{(D6)}$$

As we shall see, the solution may violate T invariance, but it remains gauge-invariant.

To generate a spontaneous T violation (but *without* a spontaneous gauge-symmetry violation), we assume the (renormalized) coupling constants to satisfy

$$\lambda > 0 \text{ and } A > 0 . \qquad \text{(D7)}$$

Thus, the vacuum expectation value of ϕ is not zero;

$$\langle \phi \rangle_{\text{vac}} = \rho \neq 0 . \qquad \text{(D8)}$$

Since according to (D3)–(D5) ϕ is of $P = -1$, $CP = -1$, and $T = -1$, such a nonzero vacuum expectation value implies spontaneous violations of P, CP, and T.

In the tree approximation, ρ is determined by the minimum of the c-number function V. Therefore,

$$\rho^2 = A^{-1}\lambda . \qquad \text{(D9)}$$

We may write

$$\phi = \rho + \chi . \qquad \text{(D10)}$$

The potential V becomes

$$V(\chi) = V_0 + \tfrac{1}{2}\mu^2 \chi^2 + A\rho \chi^3 + \tfrac{1}{4} A \chi^4 , \qquad \text{(D11)}$$

where $V_0 = -\frac{1}{4}A^{-1}\lambda^2$ and $\mu^2 = 2\lambda$. In order to render the quadratic expression $-\psi^\dagger \gamma_4 (m + ig\rho\gamma_5)\psi$ in a more familiar form, we perform a unitary transformation U:

$$U\psi U^\dagger = [\exp(-\tfrac{1}{2} i \gamma_5 \alpha)]\psi \qquad \text{(D12)}$$

and

$$U\phi U^\dagger = \phi ,$$

where

$$\sin\alpha = M^{-1} g\rho , \quad \cos\alpha = M^{-1} m \qquad \text{(D13)}$$

and

$$M = (m^2 + g^2 \rho^2)^{1/2} . \qquad \text{(D14)}$$

The Lagrangian $\mathcal{L}$ becomes

$$U\mathcal{L}U^\dagger = -\frac{1}{2}\left(\frac{\partial \chi}{\partial x_\mu}\right)^2 - V(\chi) - \psi^\dagger \gamma_4 \left(\gamma_\mu \frac{\partial}{\partial x_\mu} + M\right)\psi - g\psi^\dagger \gamma_4 (\sin\alpha + i\gamma_5 \cos\alpha)\psi \chi . \qquad \text{(D15)}$$

Since $\psi^\dagger \gamma_4 \psi$ is of $P = 1$, $C = 1$, and $T = 1$ while $i\psi^\dagger \gamma_4 \gamma_5 \psi$ is of $P = -1$, $C = 1$, and $T = -1$, the Lagrangian (D15) satisfies spontaneous T, P, and CP violations, but the gauge invariance (D6) remains preserved.

From (D9) one sees that there are two solutions of $\langle \phi \rangle_{\text{vac}}$: ρ and $-\rho$. Either solution is not invariant under T, P, and CP. But since the Lagrangian is invariant under T, P, and CP both solutions must exist, and they should transform into each other under either T, or P, or CP.

*Research supported in part by the U. S. Atomic Energy Commission.

[1]J. Goldstone, Nuovo Cimento 19, 154 (1961); J. Goldstone, A. Salam, and S. Weinberg, Phys. Rev. 127, 965 (1962).

[2]P. W. Higgs, Phys. Lett. 12, 132 (1964); Phys. Rev. Lett. 13, 508 (1964); Phys. Rev. 145, 1156 (1966); F. Englert and R. Brout, Phys. Rev. Lett. 13, 321 (1964); G. S. Guralnik, C. R. Hagen, and T. W. B. Kibble, Phys. Rev. Lett. 13, 585 (1964); T. W. B. Kibble, Phys. Rev. 155, 1554 (1967).

[3]S. Weinberg, Phys. Rev. Lett. 19, 1264 (1967); 27, 1688 (1971); Phys. Rev. D 5, 1412 (1972); 5, 1962 (1972); A. Salam and J. C. Ward, Phys. Lett. 13, 168 (1964); A. Salam, in *Elementary Particle Theory*, edited by N. Svartholm (Almqvist, Stockholm, 1968).

[4]G. 't Hooft, Nucl. Phys. B33, 173 (1971); B35, 167 (1971); Phys. Lett. 37B, 195 (1971); B. W. Lee, Phys. Rev. D 5, 823 (1972); B. W. Lee and J. Zinn-Justin, *ibid.* 5, 3121 (1972); 5, 3127 (1972); 5, 3155 (1972); H. Georgi and S. L. Glashow, Phys. Rev. Lett. 28, 1494 (1972); J. Prentki and B. Zumino, Nucl. Phys. B47, 99 (1972); J. D. Bjorken and C. H. Llewellyn Smith, Phys. Rev. D 7, 887 (1973).

[5]There also exist in the literature discussions of other theories of spontaneous CP violation, different from those considered in this paper: T. D. Lee, Phys. Rev. 137, B1621 (1965); Roger Dashen, Phys. Rev. D 3, 1879 (1971); J. Nuyts, Phys. Rev. Lett. 26, 1604 (1971); 27, 361(E) (1971); M. A. B. Bég, Phys. Rev. D 4, 3810 (1971). See also Appendix D.

[6]For the free neutrino, the Weyl theory and the Majorana theory are clearly equivalent. For further discussions, see K. M. Case, Phys. Rev. 107, 307 (1957).

[7]The spontaneous T violation is logically an independent proposition; it does not have to be tied to any spontaneous gauge-symmetry violations. See Appendix D for a simple example.

[8]See the definition given by G. C. Wick, Phys. Rev. 80, 268 (1950).

[9]We recall that in the superweak theory [L. Wolfenstein, Phys. Rev. Lett. 13, 502 (1964)], the CP-violating element in the mass matrix between K_1^0 and K_2^0 is only $\sim 10^{-8}$ eV, which is $\sim 10^{-7} G_F^2 m_N^5$, where G_F is the Fermi constant of β decay and m_N the nucleon mass.

[10]W. B. Dress, P. D. Miller, and N. F. Ramsey, Phys. Rev. D 7, 3147 (1973).

CO - 2271 - 18

SPONTANEOUS T VIOLATION

T. D. Lee
Columbia University, New York, N.Y. 10027

A talk presented at the Topical Seminar on Weak Interactions, Trieste, Italy, June 29, 1973.

This research was supported in part by the U. S. Atomic Energy Commission.

1. Introduction

Ever since the discovery of CP violation by Christenson, Cronin, Fitch and Turlay[1] about nine years ago, the difficulty has always been that there are far too few facts on the experimental side, but far too many speculations on the theoretical side. This talk will add one more to the long list of theoretical speculations.

So far, CP violations have only been observed in the neutral K-decay. All such CP violations can be characterized by one complex number ϵ , defined by

$$\epsilon \equiv \langle K_2^o \mid K_S^o \rangle = \langle K_1^o \mid K_L^o \rangle$$

where K_1^o and K_2^o refer respectively to the $CP = +1$ and -1 states, while K_S^o and K_L^o are the observed short- and long-lived neutral K-meson states. Experimentally, one finds ϵ to be extremely small:

$$|\epsilon| \cong 2 \times 10^{-3} \ .$$

This naturally leads to two fundamental questions:

1. Why should CP symmetry be violated?
2. Why should the violation be so small?

In today's discussion, I would like to examine a general class of theories in which the total Lagrangian

$$\mathcal{L} = \mathcal{L}_{strong} + \mathcal{L}_{\gamma} + \mathcal{L}_{weak}$$

is assumed to satisfy

(a) invariance under CP and T,

(b) renormalizability

and (c) invariance under certain weak and electromagnetic gauge transformations.

We note that if (a) holds, then by identifying CP with space inversion and T with time reversal, the full Poincaré group of space-time transformations, including these discrete ones, may be regarded as exact symmetries of all interactions. The dynamical equation is completely symmetrical with respect to T and CP, though its particular solution can be non-invariant. Such a mechanism will be referred to as the spontaneous T violation. Esthetically, to have a CP-invariant and a T-invariant interaction may appear to be a rather pleasing possibility; it also gives an answer to our first question. Furthermore, the requirement (b) of renormalizability makes all such CP and T non-invariant effects of the solution finite and therefore, at least in principle, computable. In addition, as we shall see, because of the condition (c) the magnitude of CP violation becomes naturally small at low energy; thus, it provides a possible answer to our second question.

2. A Simple Example

To illustrate the phenomenon of spontaneous T violation, let us first examine a simple example[2] which satisfies our conditions (a) and (b), but without the condition (c). In this way, we may also demonstrate the logical independence of the spontaneous T violation from the well-known mechanism of spontaneous gauge-symmetry violation[3-5].

This simple example consists of a spin $\frac{1}{2}$ Dirac field ψ and a spin 0 Hermitian field ϕ. The Lagrangian density is assumed to be invariant under T, C and P

$$\mathcal{L} = -\tfrac{1}{2}\left(\frac{\partial\phi}{\partial x_\mu}\right)^2 - V(\phi) - \psi^\dagger\gamma_4\left(\gamma_\mu\frac{\partial}{\partial x_\mu} + m\right)\psi$$

$$- i g \psi^\dagger \gamma_4 \gamma_5 \psi \phi \tag{1}$$

where on account of renormalizability the most general functional form of the potential $V(\phi)$ is given by

$$V(\phi) = -\tfrac{1}{2}\lambda\phi^2 + \tfrac{1}{4}\kappa\phi^4 .$$

From Hermiticity, the parameters m, g, λ and κ must be real. It can be readily verified that the Lagrangian $\mathcal{L}$ is invariant under T, C and P under which ψ transforms in the standard way, and ϕ transforms according to

$$T\phi(\vec{r}, t)T^{-1} = -\phi(\vec{r}, -t) ,$$

$$C\phi(\vec{r}, t)C^{-1} = \phi(\vec{r}, t) , \tag{2}$$

and

$$P\phi(\vec{r}, t)P^{-1} = -\phi(-\vec{r}, t) .$$

To generate a spontaneous T violation, we assume that the (renormalized) coupling constants λ and κ satisfy

$$\lambda > 0 \qquad \text{and} \qquad \kappa > 0 \ .$$

Thus, the vacuum expectation value of ϕ is not zero;

$$\langle \phi \rangle_{vac} = \rho \neq 0 \ .$$

Since according to (2), ϕ is of $T = -1$, $CP = -1$ and $P = -1$, a non-zero vacuum expectation value of ϕ implies a spontaneous violation of T (and also of CP and P). The T symmetry of the Lagrangian requires that if $\langle \phi \rangle_{vac} = \rho$ is a solution, then $\langle \phi \rangle_{vac} = -\rho$ must also be a solution. These two solutions transform into each other under T, but by itself either solution is not invariant under T. (The same applies to CP and P.)

In the tree approximation, ρ is determined by the minimum of the c. number function V. Therefore

$$\rho^2 = (\lambda/\kappa) \ . \tag{3}$$

Because of quantum effects, the field ϕ fluctuates around its vacuum expectation value. We may write

$$\phi = \rho + \delta\phi \ . \tag{4}$$

In terms of $\delta\phi$, the potential V becomes

$$V = V_0 + \tfrac{1}{2}\mu^2 (\delta\phi)^2 + \kappa\rho(\delta\phi)^3 + \tfrac{1}{4}\kappa(\delta\phi)^4$$

where $V_0 = -\tfrac{1}{4}\kappa^{-1}\lambda^2$ and $\mu^2 = 2\lambda$. In this case, since there is no spontaneous gauge-symmetry violation, one does not encounter any Goldstone boson[3]. The mass μ of the field fluctuation is not zero.

In order to exhibit more clearly the T-violating character of the solution, we may perform a unitary transformation under which ϕ is unchanged but

$$\psi \to \left[\exp\left(-\tfrac{1}{2} i \gamma_5 \alpha\right)\right] \psi \,. \tag{5}$$

Therefore, the quadratic expression

$$\psi^\dagger \gamma_4 (m + i g \rho \gamma_5)\psi \to \psi^\dagger \gamma_4 M \psi$$

where $M = (m^2 + g^2\rho^2)^{\frac{1}{2}}$, provided that the angle α is given by $\tan\alpha = m^{-1} g\rho$. Correspondingly, the Lagrangian density $\mathcal{L}$ becomes

$$-\tfrac{1}{2}\left[\frac{\partial}{\partial x_\mu}(\delta\phi)\right]^2 - V - \psi^\dagger\gamma_4\left(\gamma_\mu \frac{\partial}{\partial x_\mu} + M\right)\psi$$

$$- g\,\psi^\dagger\gamma_4(\sin\alpha + i\gamma_5\cos\alpha)\psi \cdot (\delta\phi) \,. \tag{6}$$

Since the operator $\psi^\dagger \gamma_4 \psi$ is of $P = 1$, $C = 1$ and $T = 1$ while the operator $i\,\psi^\dagger \gamma_4 \gamma_5 \psi$ is of $P = -1$, $C = 1$ and $T = -1$, any exchange of the $\delta\phi$-quantum would give an interference term between s and p, which violates T, P and CP; e.g., the T-violating amplitude A_- in the diagram

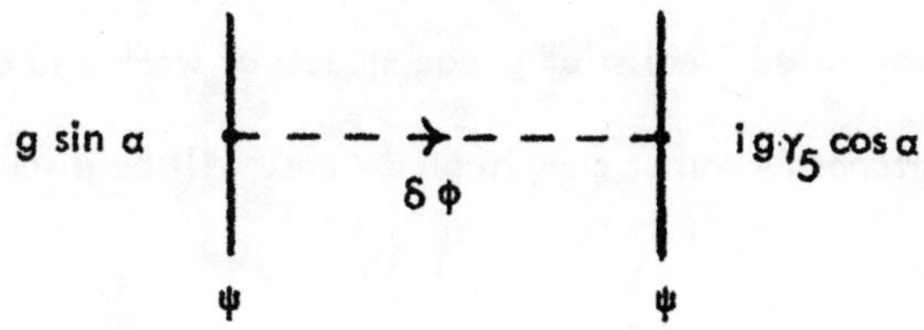

is given by

$$A_- = g^2 \sin\alpha \,\cos\alpha \cdot \frac{1}{k^2 + \mu^2} \tag{7}$$

where k denotes the 4-momentum transfer.

In this simple example, the magnitude of A_- is, of course, entirely arbitrary. As we shall see, this arbitrariness may be limited if the spontaneous T violation is coupled with the spontaneous gauge-symmetry violation of the weak and electromagnetic interactions.

3. A Second Example

In order to illustrate how a discrete symmetry, such as T or CP, may be combined with a continuous gauge-symmetry so that both symmetries can be broken spontaneously and simultaneously, let us consider another simple example. While this second example is also <u>not</u> realistic, it nevertheless does contain most of the essential features. [All the so-called "realistic" gauge models of weak and electromagnetic interactions are unfortunately rather complicated; these will be discussed in the next section.]

We assume that the Lagrangian density $\mathcal{L}$ of all interactions satisfies the three conditions (a), (b) and (c), stated in the introduction. In this simple model, the gauge-symmetry group is assumed to be the SU_2 group, like that in the Georgi-Glashow model[6]. The generators of the group are

$$Q^{wk}, \quad Q^{\gamma} \quad \text{and} \quad (Q^{wk})^{\dagger}$$

which are related to the weak interaction current J_{μ}^{wk} and the electromagnetic current J_{μ}^{γ} by

$$Q^{wk} = \int J_0^{wk}\, d^3r \ , \qquad Q^{\gamma} = \int J_0^{\gamma}\, d^3r \ ,$$

and † denotes the Hermitian conjugate. Corresponding to these three generators, there is a triplet of spin 1 gauge fields $\vec{W}_{\mu}$:

$$W_{\mu}^{+}, \ W_{\mu}^{o} = A_{\mu}, \ W_{\mu}^{-} \ .$$

In addition, there is a triplet of spin 0 Hermitian fields $\vec{\phi}$:

$$\phi^+ \, , \quad \phi^o \, , \quad \phi^-$$

and a triplet of spin $\frac{1}{2}$ hadron fields

$$\psi = \begin{pmatrix} p^+ \\ n^o \\ q^- \end{pmatrix} .$$

[The realistic Georgi-Glashow model[7] would require at least two such triplets plus two more singlets, and in addition some assorted lepton fields.]

The Lagrangian density is

$$\mathcal{L} = \mathcal{L}_{\phi W} + \mathcal{L}_{W\psi} + \mathcal{L}_{\psi\phi} \tag{8}$$

where

$$\mathcal{L}_{\phi W} = -\tfrac{1}{4} \vec{W}_{\mu\nu}^2 - \tfrac{1}{2} (\partial_\mu \vec{\phi})^2 - V(\phi) \, ,$$

$$\mathcal{L}_{W\psi} = -\psi^\dagger \gamma_4 (\gamma_\mu \partial_\mu + m) \psi \tag{9}$$

and

$$\mathcal{L}_{\psi\phi} = -i g \psi^\dagger \gamma_4 \gamma_5 \vec{I} \psi \cdot \vec{\phi}$$

in which the components of $\vec{I}$ form the (3 X 3) Hermitian matrix representation of the generators of the SU_2 group,

$$\vec{W}_{\mu\nu} = \frac{\partial}{\partial x_\mu} \vec{W}_\nu - \frac{\partial}{\partial x_\nu} \vec{W}_\mu + e(\vec{W}_\mu \times \vec{W}_\nu) \; ,$$

$$\partial_\mu \vec{\phi} = \left(\frac{\partial}{\partial x_\mu} + e \vec{W}_\mu \times\right) \vec{\phi} \; , \tag{10}$$

$$\partial_\mu \psi = \left(\frac{\partial}{\partial x_\mu} - i e \vec{I} \cdot \vec{W}_\mu\right) \psi$$

and

$$V(\phi) = -\tfrac{1}{2} \lambda \vec{\phi}^2 + \tfrac{1}{4} \kappa (\vec{\phi}^2)^2 \; .$$

In this Lagrangian, all constants e, g, λ and κ are real because of Hermiticity. One can easily verify that the above Lagrangian is invariant under T, C, P and the SU_2 transformation, and is also renormalizable[8].

Just as in the previous example, we require the renormalized constants λ and κ to be both positive definite, so that $\langle \vec{\phi} \rangle_{vac}$ is not zero. Defining the direction of $\langle \vec{\phi} \rangle_{vac}$ to be the neutral direction, we may write

$$\langle \phi^o \rangle_{vac} = \rho > 0$$

and

$$\langle \phi^\pm \rangle_{vac} = 0 \; . \tag{11}$$

Since ϕ^o is the third component of a vector, any solution with $\langle \phi^o \rangle_{vac} \neq 0$ clearly is not invariant under a general SU_2 transformation. In addition, because of the pseudoscalar interaction in $\mathcal{L}_{\psi\phi}$, ϕ^o is of $T=-1$, $P=-1$ and $C=+1$; such a solution is also not invariant under T, P and CP.

So far as the Lagrangian $\mathcal{L}_{\phi W}$ is concerned, the structure is identical to that of the usual Higgs mechanism. Since $\mathcal{L}_{\phi W}$ is an even function of $\vec{\phi}$, the Goldstone-Higgs treatment makes no distinction between ϕ^{o} of $T = -1$ or $+1$. Thus, just as in the usual discussion, the charged spin 0 field $\phi^{\pm}$ combines with the charged gauge field $W_{\mu}^{\pm}$ (which has only the transverse components) to form a single massive spin 1 charged intermediate boson field of mass

$$m_W = e\rho . \tag{12}$$

The neutral gauge field A_{μ} remains massless; it is the photon field. Let G be the weak interaction constant in this hypothetical example, then one has

$$\frac{e^2}{4 m_W^2} = \frac{G}{\sqrt{2}} . \tag{13}$$

[The realistic Georgi-Glashow model would have an additional factor $\sin^2 \beta$ on the left-hand side.]

In order to exhibit more clearly the T-violating character of the solution, let us concentrate on the ϕ^{o}-interaction part of $\mathcal{L}_{\psi\phi}$. By using (9), one may write

$$\mathcal{L}_{\psi\phi} = -ig(p^{\dagger}\gamma_4\gamma_5 p - q^{\dagger}\gamma_4\gamma_5 q)\phi^{o} + \cdots .$$

Just as in the first example, similarly to (4) we may expand

$$\phi^{o} = \rho + \delta\phi^{o} . \tag{14}$$

The transformation (5) is slightly modified because of the SU_2 character

$$\psi \to \left[\exp\left(-\tfrac{1}{2} i \gamma_5 \alpha I_3\right)\right] \psi \ ; \tag{15}$$

i.e., $p \to \left[\exp\left(-\tfrac{1}{2} i \gamma_5 \alpha\right)\right] p$, $n \to n$ and $q \to \left[\exp\left(\tfrac{1}{2} i \gamma_5 \alpha\right)\right] q$ where $\tan \alpha = m^{-1} g \rho$. The masses of p, n and q become

$$m_p^2 = m_q^2 = m_n^2 + (g\rho)^2 \tag{16}$$

and $m_n = m$. Correspondingly, after the transformation, the $\delta\phi^o$-part of the interaction is given by

$$\begin{aligned} \mathcal{L}_{\psi\phi} &= -g\left[p^\dagger \gamma_4 (\sin\alpha + i\gamma_5 \cos\alpha) p\right] \cdot \delta\phi^o \\ &\quad - g\left[q^\dagger \gamma_4 (\sin\alpha - i\gamma_5 \cos\alpha) q\right] \cdot \delta\phi^o \\ &\quad + \ldots \ . \end{aligned} \tag{17}$$

Thus, just as in the previous example, the exchange of the $\delta\phi^o$ quantum leads explicitly to T, P and CP violations. In the single $\delta\phi^o$-exchange diagram the T-violating amplitude A_- remains as given by (7). However, because the spontaneous T violation is coupled to the spontaneous gauge-symmetry violation of the weak interaction, one finds that by using (12), (13) and (16), at $k = 0$

$$A_- \cong 2^{\frac{3}{2}} G \cdot \frac{\Delta^3}{\mu^2 m_p} \tag{18}$$

where μ is the mass of the neutral spin 0 boson, $\Delta^2 = m_p^2 - m_n^2$. By definition, the ratio Δ/m_p is < 1. Independently of the question of T invariance, in order that the usual weak interaction be of the V-A character, it must be dominated by the exchanges of the intermediate spin 1 boson, not the spin 0 boson; consequently, μ must be fairly large. Thus, it seems reasonable to assume $(\Delta^3/\mu^2 m_p) \ll 1$, and therefore the T-violating amplitude A_- is much smaller than the Fermi constant G. Furthermore, the selection rule for the T-violating amplitude is the same as that for the mass matrix; i.e., it satisfies the $\Delta Y = 0$ rule, where Y is the hypercharge. If we assume

$$\frac{\Delta^3}{\mu^2 m_p} \sim 10^{-3} ,$$

then the theory gives a milliweak $\Delta Y = 0$ T-violating interaction.

In order to make a "realistic" model, with the inclusion of the Cabibbo angle, the usual V-A current, etc., one must increase the number of hadron fields to eight[7], and the number of lepton fields also to eight. In addition, one must at least double the neutral spin 0 Hermitian fields; otherwise, there will be difficulties. For example, among the lepton fields, there is a triplet representation

$$\psi_\ell = \begin{pmatrix} X^+ \\ X^o \\ e^- \end{pmatrix} .$$

If there is only a single spin 0 triplet $\vec{\phi}$, the interaction between ψ_ℓ and $\vec{\phi}$ would have the same form as $\mathcal{L}_{\phi\phi}$ given by (9). By going through the same argument that leads to (16), one would have obtained the unacceptable conclusion that the heavy electron X^+ should have the same mass as that of the electron e^-. The avoid this difficulty, we must have at least two neutral spin 0 fields ϕ_1^o and ϕ_2^o; one of them, say ϕ_1^o, has a pseudoscalar interaction, and the other, ϕ_2^o, a scalar interaction. Therefore, ϕ_1^o is of $T = -1$ and ϕ_2^o is of $T = +1$. We assume

$$< \phi_1^o >_{Av} = \rho_1 \neq 0$$

and

$$< \phi_2^o >_{Av} = \rho_2 \neq 0 \quad . \tag{19}$$

Under T, $< \phi_1^o >_{Av}$ becomes $-\rho_1$, but $< \phi_2^o >_{Av}$ remains the same. The T invariance of the interaction requires both solutions $+\rho_1$ and $-\rho_1$ to exist; but, as before, either solution is not T-invariant, and it leads to a spontaneous T violation.

Under the SU_2 transformation, ϕ_1^o and ϕ_2^o must belong to two different representation spaces, which may or may not be of the same dimension. For simplicity, we may consider the special case where ϕ_1^o belongs to a triplet $\vec{\phi}_1$, and ϕ_2^o to another triplet $\vec{\phi}_2$, different from $\vec{\phi}_1$. The field $\vec{\phi}_1$ may interact with the Fermions through a pseudoscalar interaction like that in (9), and $\vec{\phi}_2$ through a scalar interaction like that in the original Georgi-Glashow model[6-7]. The discussion of such a "realistic" Georgi-Glashow model with sponteneous T violation is straightforward, but somewhat complicated because of the relatively large number of fields involved. We note that the order of magnitude of the T-violating amplitude remains as given by (18), provided

$<\phi_1^o>_{vac}$ is not too small; e.g., $<\phi_1^o>_{vac}$ may be either comparable in magnitude to $<\phi_2^o>_{vac}$, or larger in magnitude than $<\phi_2^o>_{vac}$.

4. General Discussion

Because of the large variety of different gauge models of weak and electromagnetic interactions available, we will not discuss any one in detail. Rather, we shall try to study the general question: how can each of these models be enlarged to include a spontaneous T violation, and what would be the general pattern of such a T violation? We assume the total Lagrangian to satisfy the three conditions (a), (b) and (c), stated in the Introduction. Let $\mathcal{G}$ be the relevant gauge-symmetry group. Because of the condition (c), the neutral spin 0 field ϕ^o must belong to a representation of $\mathcal{G}$.

We distinguish two different cases:

(i) ϕ^o belongs to a real representation,

and (ii) ϕ^o belongs to a complex representation.

In case (i), as illustrated in our second example, a single Hermitian ϕ^o with a non-zero vacuum expectation value is sufficient to generate a spontaneous T violation. [Although in the "realistic" Georgi-Glashow model, because of other requirements there are actually two such Hermitian ϕ^o fields.]

In case (ii), the situation is quite different. The complex representation means that ϕ^o must be a complex field. The gauge invariance of the interaction implies that the time reversal operator T is not defined up to a gauge transformation factor. Therefore, if there is only a single ϕ^o field, one has in general

$$T \phi^o T^{-1} = e^{ia} \phi^o \tag{20}$$

where a is arbitrary. Let the vacuum expectation value of ϕ^o be

$$\langle \phi^o \rangle_{vac} = \rho e^{ib}$$

where ρ is real. By choosing $a = -2b$, one can always arrange to have

$$T \langle \phi^o \rangle_{vac} T^{-1} = \langle \phi^o \rangle^*_{vac} = \langle T \phi^o T^{-1} \rangle_{vac} . \quad (21)$$

Therefore, with only a single complex ϕ^o field, it is not possible to generate a spontaneous T violation. The simplest way to produce a spontaneous T violation is to double the number of neutral spin 0 fields, so that there are two such complex fields, say ϕ_1^o and ϕ_2^o. These two neutral fields must belong to two separate representation spaces, which may or may not be equivalent. In general, the relative phase between ϕ_1^o and ϕ_2^o is no longer arbitrary. For definiteness let us assume

$$T \phi_1^o T^{-1} = e^{ia} \phi_1^o$$

and

$$T \phi_2^o T^{-1} = e^{ia} \phi_2^o$$

with the <u>same</u> phase angle a. The vacuum expectation values of ϕ_1^o and ϕ_2^o may be written as

$$\langle \phi_1^o \rangle_{vac} = \rho_1 e^{i(\theta + b)}$$

and $$< \phi_2^o >_{vac} = \rho_2 e^{ib}$$

where ρ_1 and ρ_2 are both real and > 0 . Just as in (20), the angle a is arbitrary since the time reversal operator T is not defined up to a gauge transformation factor. Furthermore, the absolute phase of ϕ_1^o and ϕ_2^o is not determined; therefore, the angle b is also arbitrary. Thus, we can always set

$$a = b = 0 . \tag{22}$$

The vacuum expectation values of ϕ_1^o and ϕ_2^o may be represented geometrically by a triangle[9]:

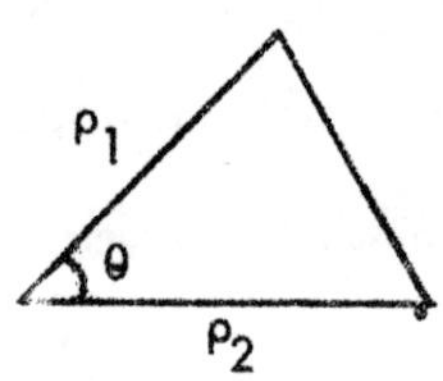

where $< \phi_1^o >_{vac} = \rho_1 e^{i\theta}$ and $< \phi_2^o > = \rho_2$. Under T , $\theta \to -\theta$. If $\theta \neq 0$ or π, then the T-invariance of the interaction implies that both solutions θ and $-\theta$ must exist, but either solution is not T-invariant.

The quantum fluctuations of ϕ_1^o and ϕ_2^o about their vacuum expectation values correspond precisely to the vibrations of this triangle. Since there are three degrees of freedom: $\delta\rho_1$, $\delta\rho_2$ and $\delta\theta$, there must be three normal modes, just like the vibrations of a triangular molecule. The virtual exchange of these vibrational quanta can lead to explicit T and CP violations, similar to the exchange of the $\delta\phi^o$-quantum discussed before in the simple example.

As mentioned before, in the Georgi-Glashow model, although the spin 0 field belongs to a real representation [i.e., case (i)], because of the other dynamical reasons, in order to construct a "realistic" theory, we also need to double the neutral spin 0 field; their vacuum values are given by (19), which can also be represented by a triangle, but with

$$\theta = 90^o \tag{23}$$

and the corresponding fluctuation $\delta\theta = 0$.

In this connection, we may also regard our simple example of only one Hermitian neutral field ϕ^o (discussed in the previous section) as a further degeneration of the triangle, one with $\rho_1 \neq 0$, but $\rho_2 = 0$ and the corresponding fluctuation $\delta\rho_2 = 0$ also. In this sense, the triangular representation offers the general characterization of the spontaneous T violatic , applicable to all ch cases.

In the general case, if ρ_1 and ρ_2 are of the same order of magnitude, one has

$$e\rho_1 \sim e\rho_2 \sim m_W ,$$

$$(e/m_W) \sim G \sim 10^{-5} m_N^{-2} ,$$

and therefore, just as in (18), at low energy the T- and CP-violating amplitude is of the order of

$$A_- \sim G \cdot \frac{\Delta^3}{\mu^2 m} \tag{24}$$

where Δ^2 is some appropriate $(\text{mass})^2$-difference between the hadrons (or leptons), m is the appropriate hadron (or lepton) mass, and μ is the mass of the relevant spin 0 quantum. Since in any reasonable model (Δ/m) is expected to be < 1 and μ has a lower bound, the T-violating amplitude A_- has an upper bound. At low energy, A_- must be quite small. For example, if $(\Delta^3/\mu^2 m)$ is assumed to be $\sim 10^{-3}$, then the T-violating amplitude A_- is of the milliweak strength.

To lowest order, T violations can be produced via either the exchange of the neutral massive spin 0 boson (i.e., the vibrational quantum of the triangle), or the charged (non-Goldstone) massive spin 0 boson[10] (if it exists in the theory). In the former case, there is usually the selection rule $\Delta Y = 0$, and in the latter case, it typically satisfies the $\Delta Y = 0$ or ± 1 rule. In both cases, the T-violating amplitude is given by (24).

For $\Delta Y = \pm 2$ processes, even the usual T-conserving amplitude requires some special engineering, and so does the T-violating amplitude. If one wishes, one can arrange to have the T-violating amplitude in the neutral K-decays be of the superweak character[11]. While the T-violating effect remains small at low energy ($\ll G\, m_N^2 \sim 10^{-5}$), at high energy, depending on the process, it could become $\sim 0(\alpha)$ (since the dimensionless coupling constant in any such model is the fine structure constant α). In this sense, it appears possible to regard the usual phenomenological classifications of CP violations as due to millistrong, electromagnetic, milliweak and superweak interactions not as separate alternatives, but as different manifestations of a single basic mechanism. At different energy regions and for different physical processes, the same T-violating mechanism may produce effects that seem to be of different apparent strengths.

5. Remarks

The general geometrical representation of the T-non-invariant solution by a triangle carries a built-in T-violating phase angle $\theta \neq 0$ or π, somewhat analogous to the two-component neutrino theory which has a built-in screw direction, therefore violating P. We recall that the two-component theory by itself does not determine the details of its interactions; the present theory also allows a large variety of possible interaction forms. Nevertheless, one feels that whatever the eventual gauge theory may be for the weak and electromagnetic interactions, it should contain T violation as an integral part. The mechanism of spontaneous T violation discussed in this talk may provide just such a needed possibility.

References

1. J. H. Christenson, J. W. Cronin, V. L. Fitch and R. Turlay, Phys. Rev. Letters 13, 138 (1964).

2. See Appendix D of T. D. Lee, Phys. Rev. D (in press).

3. J. Goldstone, Nuovo Cimento 19, 154 (1961); J. Goldstone, A. Salam and S. Weinberg, Phys. Rev. 127, 965 (1962).

4. P. W. Higgs, Phys. Letters 12, 132 (1964), Phys. Rev. Letters 13, 508 (1964) and Phys. Rev. 145, 1156 (1966); F. Englert and R. Brout, Phys. Rev. Letters 13, 321 (1964); G. S. Guralnik, C. R. Hagen and T. W. B. Kibble, Phys. Rev. Letters 13, 585 (1964); T. W. B. Kibble, Phys. Rev. 155, 1554 (1967).

5. S. Weinberg, Phys. Rev. Letters 19, 1264 (1967) and 27, 1688 (1971), Phys. Rev. D5, 1412 (1972) and D5, 1962 (1972); A. Salam and J. C. Ward, Phys. Letters 13, 168 (1964); A. Salam in Elementary Particle Theory, ed. N. Svartholm (Almquist and Forlag A. B. Stockholm, 1968).

6. H. Georgi and S. L. Glashow, Phys. Rev. Letters 28, 1494 (1972).

7. B. W. Lee, J. R. Primack and S. B. Treiman, Phys. Rev. D7, 510 (1973).

8. G. 't Hooft, Nucl. Phys. B33, 173 (1971), B35, 167 (1971) and Phys. Letters 37B, 195 (1971); G. 't Hooft and M. Veltman, Nucl. Phys. B44, 189 (1972), B50, 318 (1972); B. W. Lee, Phys. Rev. D5, 823 (1972); B. W. Lee and J. Zinn-Justin, Phys. Rev. D5, 3121, 3127, 3155 (1972). H. Georgi and S. L. Glashow, Phys. Rev. Letters 28, 1494 (1972); J. Prentki and B. Zumino, Nucl. Phys. B47, 99 (1972); J. D. Bjorken and C. H. Llewellyn Smith, Phys. Rev. D7, 887 (1973).

9. T. D. Lee, Phys. Rev. D, loc. cit.

10. This occurs because any transformation of the form (5), or (15), will make an originally (i.e., before the transformation) T-conserving scalar or pseudo-scalar interaction become T-violating (after the transformation).

11. In order to achieve the superweak character, we assume that CP violations in the neutral K-decays are dominated by the mass matrix, not by the $\Delta Y = \pm 1$ decay transition matrix elements. This can be most easily done by assuming the mass of the charged massive (non-Goldstone) spin 0 boson to be larger than that of the neutral spin 0 boson. So that to the lowest order only the $\Delta Y = 0$ T-violating process is of importance, and that could be of the milliweak strength. Thus, all T-violating $\Delta Y = \pm 1$ transitions have amplitudes weaker than the milliweak strength, and therefore can be neglected in the neutral K-decays.

PHYSICS REPORTS (Section C of Physics Letters) 9, no. 2 (1974) 143–177. NORTH-HOLLAND PUBLISHING COMPANY

CP NONCONSERVATION AND SPONTANEOUS SYMMETRY BREAKING*

T.D. LEE

Columbia University, New York, N.Y. 10027, USA

Received 4 October 1973

Contents:

1. *CP* nonconservation	145
1.1. Present status	145
1.2. Two fundamental questions	147
2. Spontaneous *T* violation	148
2.1. A simple example	148
2.2. A second example	151
2.3. General classification	155
3. A generalized Georgi–Glashow model	159
3.1. Lagrangian	159
3.2. Vertices	163
3.3. Lowest order diagrams	166
3.4. ϵ-parameter	170
3.5. Electric dipole moment	173
Acknowledgement	174
Appendix	174
References	176

Abstract:

The observed *CP* violation is assumed to be due to the spontaneous symmetry-breaking mechanism; the Lagrangian is *CP* invariant but its particular solution is not. The general classification of such theories when coupled with different unified gauge models of the weak and electromagnetic interactions is given. All such theories lead naturally to a basically milliweak *CP* noninvariant solution. The possibility that for most weak transitions the result may resemble a superweak theory is analysed, and possible experiments to distinguish these two different types of theories are discussed. Detailed calculations for various *CP* violating amplitudes are carried out for a generalized Georgi–Glashow model.

* This research was supported in part by the U.S. Atomic Energy Commission.

1. *CP* nonconservation

1.1. *Present status*

Although the theoretical possibility of *CP* nonconservation and time reversal asymmetry was raised [1] as early as 1956, at about the same time parity violation was suggested, the actual experimental discovery came much later. In 1964 Christenson, Cronin, Fitch and Turlay [2] observed that the long-lived neutral kaon K_L^o can also decay into 2π:

$$K_L^o \to \pi^+\pi^- . \qquad (1.1)$$

This is of great significance since K_L^o is known to decay into 3π with the final pions predominantly in the s states. The *CP* value of the $\pi^+\pi^-$ final state is clearly +1, but that of the $\pi^+\pi^-\pi^o$ is -1 (or, at least predominantly -1). The fact that a single non-degenerate state K_L^o which has a definite lifetime can decay into states with different *CP* values establishes the violation of *CP* conservation.

At present, *CP* violations have been observed *only* in the various decay modes of the $K^o - \bar{K}^o$ system; besides the $\pi^+\pi^-$ decay, there are [3–7] the decay $K_L^o \to 2\pi^o$ and the charge asymmetry measurements [8, 9] in the $K_{\pi l 3}$ decay. Furthermore, from the detailed analysis [10] of various K decay amplitudes and especially the recent upper bound [11]

$$\mathrm{rate}(K_S^o \to 3\pi^o)/\mathrm{rate}(K_L^o \to 3\pi^o) < 1.2, \qquad (1.2)$$

one can conclude that the time reversal symmetry *T* is also violated, while the product *CPT* symmetry remains valid.

From these experimental results, it follows that the state vectors of K_L^o and K_S^o are *not* eigenstates of *CP*. If one assumes *CPT* invariance, then these state vectors can depend only on one complex parameter [1]:

$$K_S^o = [2(1+|\epsilon|^2)]^{-1/2}\,[(1+\epsilon)K^o + (1-\epsilon)\bar{K}^o], \qquad (1.3)$$

and

$$K_L^o = [2(1+|\epsilon|^2)]^{-1/2}\,[(1+\epsilon)K^o - (1-\epsilon)\bar{K}^o], \qquad (1.4)$$

where $\bar{K}^o$ is defined to be the *CPT* conjugate of K^o. If *CP* invariance were true, then K_L^o should be orthogonal to K_S^o, and therefore ϵ would be zero. Thus, the parameter ϵ characterizes the extent of *CP* nonconservation in the description of the K_L^o and K_S^o state vectors.

The phase angle $\arg((1-\epsilon)/(1+\epsilon))$ depends on the relative phase between K^o and $\bar{K}^o$ which can be arbitrarily chosen. If we adopt the phase convention that the transition amplitude $K^o \to 2\pi$ in the isospin $I = 0$ final state is real, then the phase angle ϵ becomes approximately determined [12–14]

$$\arg\epsilon \approx \tan^{-1}\{2(m_L - m_S)/(\gamma_S - \gamma_L)\}, \qquad (1.5)$$

where m_L, γ_L and m_S, γ_S refer to the mass and the width of K_L^o and K_S^o respectively. The magnitude of ϵ can then be measured from the charge asymmetry results in the $K_{\pi l3}$ decay. By using [1, 8–10]

$$\frac{\text{rate}\,(K_S^o \to \pi^- l^+ \nu_l)}{\text{rate}\,(K_S^o \to \pi^+ l^- \bar{\nu}_l)} \approx \frac{\text{rate}\,(K_L^o \to \pi^- l^+ \nu_l)}{\text{rate}\,(K_L^o \to \pi^+ l^- \bar{\nu}_l)} \approx 1 + 4\,\text{Re}\,\epsilon, \tag{1.6}$$

one finds

$$|\epsilon| \approx 2 \times 10^{-3}. \tag{1.7}$$

The present experimental situation can best be summarized in terms of the simple phenomenological description that all observed T and CP violations are completely determined by the single small parameter $\epsilon \neq 0$; i.e., to a good approximation, one may regard the transition matrix elements between K^o, $\bar{K}^o$ and 2π, 3π, $\pi l \nu$, etc., as T and CP invariant. For example, in this approximation one has

$$\eta_{+-} \approx \eta_{oo} \approx \epsilon, \tag{1.8}$$

where

$$\eta_{ab} \equiv \text{amp}(K_L^o \to \pi^a \pi^b)/\text{amp}\,(K_S^o \to \pi^a \pi^b). \tag{1.9}$$

Equation (1.8) is in good agreement with the present measurements of η_{+-} and η_{oo} [15, 16].

Since the K^o and $\bar{K}^o$ system is determined not only by the weak interaction but also by the strong and the electromagnetic interactions, there are several alternative theoretical possibilities concerning the origin of the CP violation [17]:

(i) superweak theory [18]: the CP violating interaction is assumed to violate the hypercharge conservation by two units and its strength is $\sim 10^{-9}\, G_F$ where G_F is the usual Fermi coupling constant for the weak interaction. Because of the extreme smallness of the superweak coupling constant, one may safely neglect any higher order effect. Equations (1.5) and (1.8) are then exact [17];

(ii) milliweak possibility: the CP violating interaction is assumed to violate the hypercharge conservation by one unit, and its strength is $\sim 10^{-3}\, G_F$. We note that the deviation from (1.8) is given by [19]

$$|\eta_{+-} - \epsilon| = |2^{-1/2}\, \text{Im}\, A_2/A_0| \equiv |\epsilon'|, \tag{1.10}$$

where A_0 and A_2 refer respectively to the transition amplitudes of $K^o \to 2\pi$ in the $I = 0$ and 2 states. At present, the origin of the usual $|\Delta I| = \frac{1}{2}$ rule is quite unclear. If one assumes that the same rule applies to the CP violating amplitude as well, then any such milliweak theory would imply the magnitude of $|\text{Im}\, A_2/A_2| \sim 10^{-3}$. Consequently, one has $|\text{Im}\, A_2/A_0| \sim 5 \times 10^{-5}$, which is $\ll |\epsilon|$ and is, therefore, consistent with (1.8). On the other hand, if the milliweak CP violating interaction does not satisfy the $|\Delta I| = \frac{1}{2}$ rule, then an additional explanation is needed for (1.8);

(iii) millistrong possibility [17, 20]: the *CP* violating interaction strength is assumed to conserve the hypercharge and its strength is $\sim 10^{-3}$ times the strong interaction strength. It is also possible that such a millistrong force may be electromagnetic in origin [21]. In any case, if such a millistrong force conserves isospin also, then one expects (1.8) to hold at least to the level where the $|\Delta I| = \frac{1}{2}$ rule holds for the usual (*T* conserving) weak interaction.

1.2. *Two fundamental questions*

At present, the full implications of *CP* nonconservation and *T* violation are far from being understood. There are at least two fundamental questions:

1. Why should *T* symmetry and *CP* symmetry be violated?
2. Why should the violation be so small?

In order to provide some possible answers to these questions, we shall examine in this paper a general class of theories [22] in which the total Lagrangian (which includes all interactions, strong, electromagnetic and weak) is assumed to satisfy:

(a) invariance under *CP* and *T*,
(b) renormalizability, and
(c) invariance under certain weak and electromagnetic gauge transformations.

We note that if (a) holds, then by identifying time reversal with *T* and space inversion with *CP* (not *P*), the full Poincaré group of space-time transformations, including these discrete ones, may be regarded as *exact* symmetries of all interactions. The dynamical equation is completely symmetrical with respect to *T* and *CP*, though its particular solution can be non-invariant. Such a mechanism will be referred to as the *spontaneous T violation*, or spontaneous *CP* violation. In physics, it is a familiar practice to assume that the dynamical equation is symmetric under a relatively large group of transformations, while its particular solution does not have to be an invariant under the same transformations. For example, rotational invariance of the physical law by no means implies that all physical solutions should be spherically symmetrical. From an esthetical point of view, it appears pleasing to assume a *CP*-invariant and *T*-invariant interaction, but to attribute all observed *CP* violations merely to the asymmetrical nature of the particular solution. Such a possibility gives an answer to our first question.

As we shall see, the requirement (b) of renormalizability makes all such *CP* and *T* noninvariant effects of the solution finite and therefore, at least in principle, computable. In addition, as we shall also show, because of the condition (c) the magnitude of *CP* violation must be quite small at low energy. This then provides a possible answer to our second question.

We emphasize that the above three conditions (a), (b), and (c) are logically independent. For example, as discussed by Pais [23], *CP* violation can be introduced into a gauge theory [24, 25] without assuming the spontaneous *CP* violation mechanism. However, since the entire approach of the unified gauge theory of the electromagnetic and weak interactions [26–30] is based on the spontaneous symmetry-breaking mechanism, it seems natural, and also esthetically more appealing, to include *CP* violation as an integral part of the same mechanism.

2. Spontaneous *T* violation

2.1. *A simple example*

To illustrate the phenomenon of spontaneous T violation, let us first examine a simple example which satisfies our conditions (a) and (b), but without the condition (c). In this way, we may also demonstrate the logical independence of the spontaneous T violation from the well-known mechanism of spontaneous gauge-symmetry violation [24–26].

This simple example consists of a spin $\frac{1}{2}$ Dirac field ψ and a spin 0 Hermitian field ϕ. The Lagrangian density is assumed to be invariant under T, C and P

$$L = -\tfrac{1}{2}\left(\frac{\partial\phi}{\partial x_\mu}\right)^2 - V(\phi) - \psi^\dagger\gamma_4\left(\gamma_\mu \frac{\partial}{\partial x_\mu} + m\right)\psi - \mathrm{i}g\,\psi^\dagger\gamma_4\gamma_5\psi\phi, \tag{2.1}$$

where on account of renormalizability the most general form of the potential $V(\phi)$ is a quartic function, given by

$$V(\phi) = -\tfrac{1}{2}\lambda\phi^2 + \tfrac{1}{4}\kappa\phi^4 .$$

From Hermiticity, the parameters m, g, λ and κ must be real. It can be readily verified that under T, C and P, ϕ transforms according to

$$T\phi(\boldsymbol{r},t)T^{-1} = -\phi(\boldsymbol{r},-t), \tag{2.2a}$$

$$C\phi(\boldsymbol{r},t)C^{-1} = \phi(\boldsymbol{r},t), \tag{2.2b}$$

and

$$P\phi(\boldsymbol{r},t)P^{-1} = -\phi(-\boldsymbol{r},t). \tag{2.2c}$$

To generate a spontaneous T violation, we assume that the (renormalized) coupling constants λ and κ satisfy

$$\lambda > 0, \qquad \text{and} \qquad \kappa > 0.$$

Thus, as shown in fig. 1, the vacuum expectation value of ϕ is not zero;

$$\langle\phi\rangle_{\text{vac}} = \rho \neq 0.$$

Since according to (2.2), ϕ is of $T = -1$, $CP = -1$ and $P = -1$, a non-zero vacuum expectation value of ϕ implies a spontaneous violation of T (and also of CP and P). The T symmetry of the Lagrangian requires that if $\langle\phi\rangle_{\text{vac}} = \rho$ is a solution, then $\langle\phi\rangle_{\text{vac}} = -\rho$ must also be a solution. These two solutions transform into each other under T, but by itself either solution is not invariant under T. (It is also not invariant under CP and P, though it is invariant under C and CPT.)

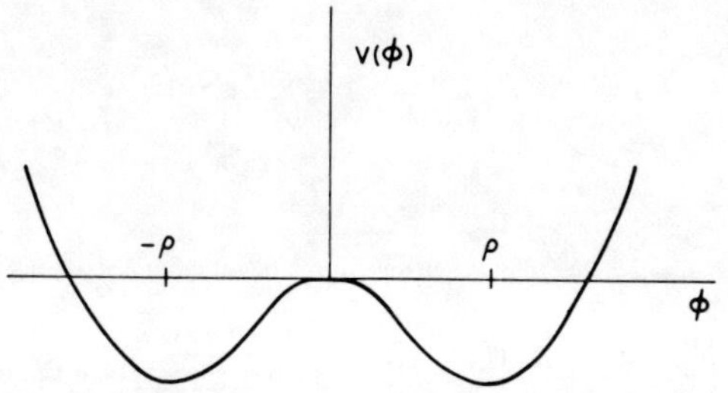

Fig. 1. The potential $V(\phi) = -\frac{1}{2}\lambda\phi^2 + \frac{1}{4}\kappa\phi^2$ for $\lambda > 0$ and $\kappa > 0$.

In the tree approximation, ρ is determined by the minimum of the c. number function V. Therefore

$$\rho^2 = (\lambda/\kappa). \tag{2.3}$$

Because of quantum effects, the field ϕ fluctuates around its vacuum expectation value. We may write

$$\phi = \rho + \delta\phi. \tag{2.4}$$

In terms of $\delta\phi$, the potential V becomes

$$V = V_0 + \tfrac{1}{2}\mu^2(\delta\phi)^2 + \kappa\rho(\delta\phi)^3 + \tfrac{1}{4}\kappa(\delta\phi)^4,$$

where $V_0 = -\frac{1}{4}\kappa^{-1}\lambda^2$ and $\mu^2 = 2\lambda$. In this case, since T is a discrete symmetry and there is no spontaneous gauge-symmetry violation, one does not encounter any Goldstone boson [24]. The mass μ of the field fluctuation $\delta\phi$ is not zero.

In order to exhibit more clearly the T-violating character of the solution, we may perform a unitary transformation under which ϕ is unchanged but

$$\psi \to [\exp(-\tfrac{1}{2}\mathrm{i}\gamma_5\alpha)]\,\psi. \tag{2.5}$$

Therefore, the quadratic expression

$$\psi^\dagger\gamma_4(m + \mathrm{i}g\rho\gamma_5)\psi \to \psi^\dagger\gamma_4 M\psi, \tag{2.6}$$

where

$$M = (m^2 + g^2\rho^2)^{1/2},$$

provided that the angle α is given by $\tan\alpha = m^{-1}g\rho$. Correspondingly, the Lagrangian density L becomes

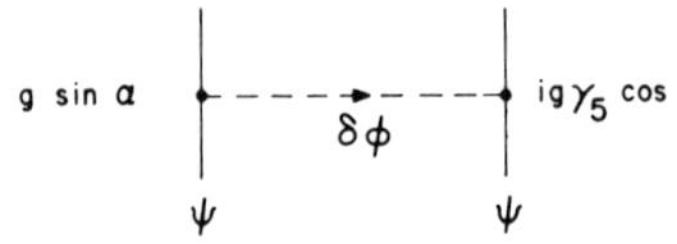

Fig. 2. A T violating scattering diagram due to the exchange of the quantum of $\delta\phi$, where $\delta\phi \equiv \phi - \langle\phi\rangle_{\text{vac}}$.

$$-\tfrac{1}{2}\left[\frac{\partial}{\partial x_\mu}(\delta\phi)\right]^2 - V - \psi^\dagger\gamma_4\left(\gamma_\mu\frac{\partial}{\partial x_\mu} + M\right)\psi - g\psi^\dagger\gamma_4(\sin\alpha + i\gamma_5\cos\alpha)\psi\cdot(\delta\phi). \tag{2.7}$$

Since the operator $\psi^\dagger\gamma_4\psi$ is of $P = 1$, $C = 1$ and $T = 1$ while the operator $i\psi^\dagger\gamma_4\gamma_5\psi$ is of $P = -1$, $C = 1$ and $T = -1$, any exchange of the $\delta\phi$-quantum would give an interference term between these two operators that violates T, P and CP; but the product CPT remains invariant. For example, the T and CP violating amplitude A_- of the diagram given in fig. 2 is

$$A_- = g^2 \sin\alpha\cos\alpha\,(k^2 + \mu^2)^{-1}, \tag{2.8}$$

where k denotes the 4-momentum transfer.

This simple example illustrates the basic mechanism of a spontaneous T violation. One assumes that the ground state of the system (called the vacuum state) has a nonzero expectation value $\langle\phi\rangle_{\text{vac}}$, where ϕ is a $T = -1$ spin 0 field; thus, the vacuum state is noninvariant under T, even though the interaction satisfies T invariance. The fact that $\langle\phi\rangle_{\text{vac}} \neq 0$ also makes the vacuum state resemble more closely a "medium", and this is why the fermion field ψ in moving through this "medium" acquires a (mass)^2-shift Δ^2, given by

$$\Delta^2 \equiv M^2 - m^2 = g^2\langle\phi\rangle^2_{\text{vac}}. \tag{2.9}$$

For small Δ, and at zero momentum transfer ($k = 0$), the T violating amplitude (2.8) becomes simply

$$A_- \approx g^2\,\Delta/\mu^2 M. \tag{2.10}$$

The T invariance of the interaction implies that the vacuum state must have a double degeneracy. It is interesting to examine the barrier penetration between these two degenerate ground state solutions: $\langle\phi\rangle_{\text{vac}} = \rho$ and $-\rho$. By assuming the entire system to be in a finite volume Ω and by following the standard procedure of quantization, one can easily show that as $\Omega \to \infty$, this barrier penetration probability becomes zero exponentially. On the other hand, in any realistic cosmological model the volume Ω should be finite. In particular, if one assumes an oscillating cosmological model in which Ω changes periodically, then when Ω is of a microscopic dimension, the system should undergo transitions between these two solutions $\langle\phi\rangle_{\text{vac}} = \rho$ and $-\rho$. At different cycles of such Ω oscillations, the system may acquire different solutions, sometimes with $\langle\phi\rangle_{\text{vac}} = \rho$ and sometimes $-\rho$. An examination of such cosmological consequences is exremely interesting, but lies outside the scope of the present paper.

In this simple example, the magnitude of the T violating amplitude A_- is, of course, entirely arbitrary. As we shall see, this arbitrariness may be limited if the spontaneous T violation is coupled with the spontaneous gauge-symmetry violation of the weak and electromagnetic interactions.

2.2. *A second example*

In order to illustrate how a discrete symmetry, such as T or CP, may be combined with a continuous gauge-symmetry so that both symmetries can be broken spontaneously and simultaneously, let us consider another simple example. While this second example is also *not* realistic, it nevertheless does contain most of the essential features. (All the so-called "realistic" gauge models [26–30] of weak and electromagnetic interactions are unfortunately rather complicated; these will be discussed in the next section.)

We assume that the Lagrangian density L of all interactions satisfies the three conditions (a), (b) and (c), stated in section 1.2. In this simple model, the gauge-symmetry group is assumed to be the SO_3 group, like that in the Georgi-Glashow model [27]. The generators of the group are

$$Q^{\mathrm{wk}}, \quad Q^{\gamma} \quad \text{and} \quad (Q^{\mathrm{wk}})^{\dagger}, \tag{2.11}$$

which are related to the weak interaction current J_μ^{wk} and the electromagnetic current J_μ^γ by

$$Q^{\mathrm{wk}} = \int J_0^{\mathrm{wk}}\, \mathrm{d}^3 r, \qquad Q^{\gamma} = \int J_0^{\gamma}\, \mathrm{d}^3 r, \tag{2.12}$$

and $\dagger$ denotes the Hermitian conjugate. Corresponding to these three generators there is a triplet of spin 1 gauge fields $\boldsymbol{W}_\mu$:

$$W_\mu^+, \qquad W_\mu^{\mathrm{o}}, \qquad W_\mu^-, \tag{2.13}$$

where W_μ^{o} is the usual electromagnetic field A_μ. In addition, there is a triplet of spin 0 Hermitian fields $\boldsymbol{\phi}$:

$$\phi^+, \qquad \phi^{\mathrm{o}}, \qquad \phi^-,$$

and a triplet of spin $\frac{1}{2}$ hadron fields

$$\psi = \begin{pmatrix} p^+ \\ n^{\mathrm{o}} \\ q^- \end{pmatrix}.$$

(The realistic Georgi-Glashow model [27] would require many additional fields.)

The Lagrangian density is

$$L = L_{\phi W} + L_{W\psi} + L_{\psi\phi}, \tag{2.14}$$

where

$$L_{\phi W} = -\tfrac{1}{4} W_{\mu\nu}^2 - \tfrac{1}{2}(\partial_\mu \boldsymbol{\phi})^2 - V(\phi), \tag{2.15}$$

$$L_{W\psi} = -\psi^\dagger \gamma_4 (\gamma_\mu \partial_\mu + m)\psi, \tag{2.16}$$

and

$$L_{\psi\phi} = -\mathrm{i}g\,\psi^\dagger \gamma_4 \gamma_5 \boldsymbol{I}\psi \cdot \boldsymbol{\phi}, \tag{2.17}$$

in which the components of $\boldsymbol{I}$ form the (3×3) Hermitian matrix representation of the generators of the SO_3 group,

$$\boldsymbol{W}_{\mu\nu} = \frac{\partial}{\partial x_\mu} \boldsymbol{W}_\nu - \frac{\partial}{\partial x_\nu} \boldsymbol{W}_\mu + e(\boldsymbol{W}_\mu \times \boldsymbol{W}_\nu), \tag{2.18}$$

$$\partial_\mu \boldsymbol{\phi} = \left(\frac{\partial}{\partial x_\mu} + e\boldsymbol{W}_\mu \times\right)\boldsymbol{\phi}, \tag{2.19}$$

$$\partial_\mu \psi = \left(\frac{\partial}{\partial x_\mu} - \mathrm{i}e\boldsymbol{I}\cdot \boldsymbol{W}_\mu\right)\psi, \tag{2.20}$$

and

$$V(\phi) = -\tfrac{1}{2}\lambda \boldsymbol{\phi}^2 + \tfrac{1}{4}\kappa(\boldsymbol{\phi}^2)^2 . \tag{2.21}$$

In this Lagrangian, all constants e, g, λ and κ are real because of Hermiticity. One can easily verify that the above Lagrangian is invariant under T, C, P and the SO_3 transformation, and is also renormalizable [31, 32].

Just as in the previous example, we require the renormalized constants λ and κ to be both positive definite, so that $\langle \boldsymbol{\phi} \rangle_{\mathrm{vac}}$ is not zero. Defining the direction of $\langle \boldsymbol{\phi} \rangle_{\mathrm{vac}}$ to be the neutral direction, we may write

$$\langle \phi^{\mathrm{o}} \rangle_{\mathrm{vac}} \neq 0,$$

and

$$\langle \phi^{\pm} \rangle_{\mathrm{vac}} = 0. \tag{2.22}$$

Since ϕ^{o} is the third component of a vector, any solution with $\langle \phi^{\mathrm{o}} \rangle_{\mathrm{vac}} \neq 0$ clearly is not invariant under a general SO_3 transformation. In addition, because of the pseudoscalar interaction in $L_{\psi\phi}$, ϕ^{o} is of $T = -1$, $P = -1$ and $C = +1$; such a solution is also not invariant under T, P and CP.

So far as the Lagrangian $L_{\phi W}$ is concerned, the structure is identical to that of the usual Higgs mechanism [25]. Since $L_{\phi W}$ is an even function of $\boldsymbol{\phi}$, the usual Goldstone-Higgs treatment makes no distinction between ϕ^{o} of $T = -1$ or $+1$. To understand the Higgs mechanism, let us consider the specific term $-\tfrac{1}{2}(\partial_\mu \boldsymbol{\phi})^2$ in the Lagrangian (2.15). By using (2.19) and the fact that $\langle \boldsymbol{\phi} \rangle_{\mathrm{vac}} \neq 0$, we may write

$$-\tfrac{1}{2}(\partial_\mu \boldsymbol{\phi})^2 = -\tfrac{1}{2} e^2(\boldsymbol{W}_\mu \times \langle \boldsymbol{\phi} \rangle_{\rm vac})^2 - e\langle \boldsymbol{\phi} \rangle_{\rm vac} \cdot (\boldsymbol{W}_\mu \times \partial \boldsymbol{\phi}/\partial x_\mu) + \dots . \tag{2.23}$$

Let us examine the first term in the above expression: because of the vector product, only the components of $\boldsymbol{W}_\mu$ that are perpendicular to the direction of $\langle \boldsymbol{\phi} \rangle_{\rm vac}$ are involved. One sees that these components acquire a mass $m_{\rm W}$ given by

$$m_{\rm W}^2 = e^2 \langle \boldsymbol{\phi} \rangle_{\rm vac}^2, \tag{2.24}$$

while the component of $\boldsymbol{W}_\mu$ that is parallel to $\langle \boldsymbol{\phi} \rangle_{\rm vac}$ remains massless and is identified to be the photon field A_μ; therefore, the direction of $\langle \boldsymbol{\phi} \rangle_{\rm vac}$ is indeed the neutral direction, as given by (2.22). The second term $\langle \boldsymbol{\phi} \rangle_{\rm vac} \cdot (\boldsymbol{W}_\mu \times \partial \boldsymbol{\phi}/\partial x_\mu)$ couples only the charged components $W_\mu^\pm$ with $\partial \phi^\pm/\partial x_\mu$. It is straightforward to verify that $\partial \phi^\pm/\partial x_\mu$ now combines with $W_\mu^\pm$ (which has only transverse components) to form a single massive spin 1 charged intermediate boson field; $\partial \phi^\pm/\partial x_\mu$ now becomes the longitudinal mode of the spin 1 field. Let $G_{\rm F}$ be the Fermi weak interaction constant in this simple example; one has

$$e^2/4m_{\rm W}^2 = G_{\rm F}/\sqrt{2}. \tag{2.25}$$

(The realistic Georgi-Glashow model [27] would have an additional factor $\sin^2 \beta$ on the left-hand side.)

In order to exhibit more clearly the T violating character of the solution, let us concentrate on the $\phi^{\rm o}$-interaction part of $L_{\psi\phi}$. By using (2.17), one may write

$$L_{\psi\phi} = -{\rm i}g(p^\dagger \gamma_4 \gamma_5 p - q^\dagger \gamma_4 \gamma_5 q)\phi^{\rm o} + \dots .$$

Just as in the first example, similarly to (2.4) we may expand

$$\phi^{\rm o} = \langle \phi^{\rm o} \rangle_{\rm vac} + \delta\phi^{\rm o}. \tag{2.26}$$

The transformation (2.5) is slightly modified because of the SO_3 property

$$\psi \to [\exp(-\tfrac{1}{2}{\rm i}\gamma_5 \alpha {\rm I}_3)]\, \psi; \tag{2.27}$$

i.e., $p \to [\exp(-\frac{1}{2}{\rm i}\gamma_5 \alpha)]p$, $n \to n$ and $q \to [\exp(\frac{1}{2}{\rm i}\gamma_5 \alpha)]q$ where $\tan \alpha = m^{-1} g\rho$. The masses of p, n and q become

$$m_{\rm p}^2 = m_{\rm q}^2 = m_{\rm n}^2 + (g\rho)^2, \tag{2.28}$$

and $m_{\rm n} = m$. Correspondingly, after the transformation, the $\delta\phi^{\rm o}$-part of the interaction is given by

$$L_{\psi\phi} = -g[p^\dagger\gamma_4(\sin\alpha + \mathrm{i}\gamma_5\cos\alpha)p]\cdot\delta\phi^{\mathrm{o}}$$
$$-g[q^\dagger\gamma_4(\sin\alpha - \mathrm{i}\gamma_5\cos\alpha)q]\cdot\delta\phi^{\mathrm{o}} + \dots . \tag{2.29}$$

Thus, just as in the previous example, the exchange of the $\delta\phi^{\mathrm{o}}$ quantum leads explicitly to T, P and CP violations. In the single $\delta\phi^{\mathrm{o}}$-exchange diagram the T-violating amplitude A_- remains as given by (2.8). However, because the spontaneous T violation is coupled to the spontaneous gauge-symmetry violation of the weak interaction, one finds that by using (2.24), (2.25) and (2.28), at $k = 0$

$$A_- \approx 2^{3/2}G_{\mathrm{F}}\cdot\Delta^3/\mu^2 m_{\mathrm{p}}, \tag{2.30}$$

where μ is the mass of the neutral spin 0 boson and $\Delta^2 = m_{\mathrm{p}}^2 - m_{\mathrm{n}}^2$. By definition, the ratio Δ/m_{p} is < 1.

Although the magnitude of A_- still depends on the unknown mass ratio $(\Delta^3/\mu^2 m_{\mathrm{p}})$, certain limits can already be estimated. The mass shift Δ must reflect in some way certain observed hadron mass differences; therefore Δ cannot be too large. In addition, independently of the question of T invariance, in order that the usual weak interaction be of the V–A character, it must be dominated by the exchanges of the intermediate spin 1 boson, *not* the spin 0 boson; consequently, μ must be fairly large. Thus, it seems reasonable to assume $(\Delta^3/\mu^2 m_{\mathrm{p}}) \ll 1$, and therefore the T violating amplitude A_- is much smaller than the Fermi constant G_{F}.

This simple example illustrates the main feature of combining the spontaneous T violation with the spontaneous gauge symmetry violation of the weak and electromagnetic interactions. Because $\langle\phi\rangle_{\mathrm{vac}} \neq 0$, the vacuum acts as a "medium"; it gives the hadron field a mass shift $\Delta = g\langle\phi^{\mathrm{o}}\rangle_{\mathrm{vac}}$ and the intermediate boson field a mass $m_{\mathrm{W}} = e\langle\phi^{\mathrm{o}}\rangle_{\mathrm{vac}}$, which then leads the T violating amplitude A_- from eq. (2.10) of our first example to the above expression (2.30). Furthermore, since $g\langle\phi^{\mathrm{o}}\rangle_{\mathrm{vac}}$ contributes to the mass matrix of the hadrons, any exchange of $\delta\phi^{\mathrm{o}}$ quantum satisfies the same selection rule as that particular piece of the mass matrix. As we shall see in subsequent sections, in a more realistic model this implies the selection rule $\Delta Y = 0$ where Y is the hypercharge. If one assumes

$$\Delta^3/\mu^2 m_{\mathrm{p}} \sim 10^{-3},$$

then it gives a milliweak $\Delta Y = 0$ T-violating interaction.

In order to make a "realistic" model, with the inclusion of the Cabibbo angle, the usual V–A current, etc., one must at least increase the number of hadron fields to eight [27, 28], and the number of lepton fields also to eight. In addition, one must at least double the neutral spin 0 Hermitian fields; otherwise, there will be difficulties. For example, among the lepton fields, there is a triplet representation

$$\psi_l = \begin{pmatrix} X^+ \\ X^{\mathrm{o}} \\ e^- \end{pmatrix}.$$

If there is only a single spin 0 triplet $\boldsymbol{\phi}$, the interaction between ψ_l and $\boldsymbol{\phi}$ would have the same form as $L_{\psi\phi}$ given by (2.17). By going through the same argument that leads to (2.28), one would have obtained the unacceptable conclusion that the heavy electron X^+ should have the same mass as that of the electron e^-. To avoid this difficulty, we must have at least *two* neutral spin 0 Hermitian fields ϕ_I^o and ϕ_R^o; one of them, say ϕ_I^o, has a pseudoscalar interaction, and the other, ϕ_R^o, a scalar interaction. Therefore, ϕ_I^o is of $T = -1$ and ϕ_R^o is of $T = +1$. We assume

$$\langle \phi_I^o \rangle_{vac} = \rho_I \neq 0$$

and

$$\langle \phi_R^o \rangle_{vac} = \rho_R \neq 0. \tag{2.31}$$

Under T, $\langle \phi_I^o \rangle_{vac}$ becomes $-\rho_I$, but $\langle \phi_R^o \rangle_{vac}$ remains the same. The T invariance of the interaction requires both solutions $+\rho_I$ and $-\rho_I$ to exist; but, as before, either solution is not T invariant, and it leads to a spontaneous T violation.

Under the SO_3 transformation, ϕ_I^o and ϕ_R^o must belong to two different representation spaces, which may or may not be of the same dimension. For simplicity we may consider the special case where ϕ_I^o belongs to a triplet $\boldsymbol{\phi}_I$, and ϕ_R^o to another triplet $\boldsymbol{\phi}_R$, different from $\boldsymbol{\phi}_I$. The field $\boldsymbol{\phi}_I$ may interact with the fermions through a pseudoscalar interaction like that in (2.17), and $\boldsymbol{\phi}_R$ through a scalar interaction like that in the original Georgi-Glashow model [27, 28]. The discussion of such a "realistic" Georgi-Glashow model with spontaneous T violation is straightforward, but somewhat complicated because of the relatively large number of fields involved. The details will be given in section 3.

2.3. *General classification*

Because of the large variety of different gauge models of weak and electromagnetic interactions available, in this section we shall try to study only the general question: how can each of these models be enlarged to include a spontaneous T violation, and what would be the general pattern of such a T violation? We assume the total Lagrangian to satisfy the three conditions (a), (b) and (c), stated in section 1.2. Let G be the relevant gauge-symmetry group. The neutral spin 0 field ϕ^o must belong to a representation of G.

We distinguish two different cases:

(i) ϕ^o belongs to a real representation, and

(ii) ϕ^o belongs to a complex representation.

In case (i), as illustrated in our second example, a single Hermitian ϕ^o with a non-zero vacuum expectation value is sufficient to generate a spontaneous T violation (although in the "realistic" Georgi-Glashow model, as mentioned before, because of other requirements there are actually two such Hermitian ϕ^o fields).

In case (ii), the situation is quite different. The complex representation means that ϕ^o must be a complex field. The gauge invariance of the interaction implies that the time reversal operator T is not defined up to a gauge transformation factor. Therefore, if there is only a single ϕ^o field, one has in general

$$T\phi^o T^{-1} = e^{i\alpha}\phi^o, \tag{2.32}$$

where a is arbitrary. Let the vacuum expectation value of ϕ^o be

$$\langle\phi^o\rangle_{vac} = \rho e^{ib},$$

where ρ is real. By choosing $a = -2b$, one can always arrange to have

$$T\langle\phi^o\rangle_{vac}T^{-1} = \langle\phi^o\rangle^*_{vac} = \langle T\phi^o T^{-1}\rangle_{vac}. \tag{2.33}$$

Therefore, with only a single complex ϕ^o field, it is not possible to generate a spontaneous T violation. The simplest way to produce a spontaneous T violation is to double the number of neutral spin 0 fields, so that there are two such complex fields, say ϕ_1^o and ϕ_2^o. These two neutral fields must belong to two separate representation spaces, which may or may not be equivalent. In general, the relative phase between ϕ_1^o and ϕ_2^o is no longer arbitrary. For definiteness let us assume

$$T\phi_1^o T^{-1} = e^{ia}\phi_1^o,$$

and

$$T\phi_2^o T^{-1} = e^{ia}\phi_2^o$$

with the *same* phase angle a. The vacuum expectation values of ϕ_1^o and ϕ_2^o may be written as

$$\langle\phi_1^o\rangle_{vac} = \rho_1 \exp\{i(\theta + b)\},$$

and

$$\langle\phi_2^o\rangle_{vac} = \rho_2 e^{ib},$$

where ρ_1 and ρ_2 are both real and > 0. Just as in (2.32), the angle a is arbitrary since the time reversal operator T is not defined up to a gauge transformation factor. Furthermore, the absolute phase of ϕ_1^o and ϕ_2^o is not determined; therefore, the angle b is also arbitrary. Thus, we can always set

$$a = b = 0,$$

and consequently

$$\langle\phi_1^o\rangle_{vac} = \rho_1 e^{i\theta}, \qquad \text{and} \qquad \langle\phi_2^o\rangle_{vac} = \rho_2. \tag{2.34}$$

These vacuum expectation values may be represented geometrically by the triangle given in fig. 3. Under T, $\theta \to -\theta$. If $\theta \neq 0$ or π, then the T invariance of the interaction implies that both solutions θ and $-\theta$ must exist, but either solution is not T invariant.

The quantum fluctuations of ϕ_1^o and ϕ_2^o about their vacuum expectation values correspond pre-

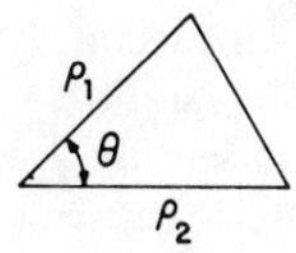

Fig. 3. Representation of $\langle\phi_2^\mathrm{o}\rangle_\mathrm{vac} = \rho_2$ and $\langle\phi_1^\mathrm{o}\rangle_\mathrm{vac} = \rho_1 e^{i\theta}$ by a triangle.

cisely to the vibrations of this triangle. Since there are three degrees of freedom: $\delta\rho_1$, $\delta\rho_2$ and $\delta\theta$, there must be three normal modes, just like the vibrations of a triangular molecule. The virtual exchange of these vibrational quanta can lead to explicit *T* and *CP* violations, similar to the exchange of the $\delta\phi^\mathrm{o}$-quantum discussed before in the simple example.

As mentioned before, in the Georgi-Glashow model, although the spin 0 field belongs to a real representation (i.e., case (i)), because of other dynamical reasons, in order to construct a "realistic" theory, we also need to double the neutral spin 0 field; their vacuum values are given by (2.31), which can again be represented by a triangle, but with

$$\theta = 90^\circ,$$

and the corresponding fluctuation $\delta\theta = 0$.

If one wishes, one may also regard our first simple example of only one Hermitian neutral field ϕ^o (discussed in section 2.1) as a further degeneration of the triangle, with the length of one of its sides approaching zero. In this sense, the triangular representation offers the general characterization of the spontaneous *T* violation, applicable to all such cases. The detailed description of such a triangle is given in ref. [22].

In general, if the two sides of the triangle ρ_1 and ρ_2 are of comparable magnitude, one has

$$e\rho_1 \sim e\rho_2 \sim m_\mathrm{W},$$

$$(e/m_\mathrm{W})^2 \sim G_\mathrm{F},$$

and therefore, just as in (2.30), at low energy the *T* and *CP* violating amplitude is of the order of

$$A_- \sim G_\mathrm{F} \cdot \Delta^3/\mu^2 m, \tag{2.35}$$

where Δ^2 is some appropriate (mass)^2-difference between the hadrons (or leptons), m is the appropriate hadron (or lepton) mass, and μ is the mass of the relevant spin 0 quantum. As noted before, in any reasonable model, Δ cannot be too large because of the observed hadron spectrum, and μ cannot be too small since the observed weak interaction is dominated by the exchange of an intermediate boson of spin 1, not spin 0. In addition, the ratio (Δ/m) is by definition < 1. Thus, A_- must be quite small. If we assume $\Delta \lesssim 1$ GeV and $\mu \gtrsim 10$ GeV, then it seems reasonable to expect $(\Delta^3/\mu^2 m) \sim 10^{-3}$ and therefore that the *T* violating amplitude A_- is milliweak.

A typical pattern of such a theory is that any exchange of spin 0 bosons leads to *T* and *P* violation while any exchange of spin 1 bosons between the usual V–A currents gives only *C* and *P* violation. This is because a transformation of the form $\psi \to \exp(-\frac{1}{2} i\gamma_5 \alpha)\,\psi$ leaves the vector

$i\psi^\dagger\gamma_4\gamma_\mu\psi$ and the axial-vector $i\psi^\dagger\gamma_4\gamma_\mu\gamma_5\psi$ invariant, but not the scalar $\psi^\dagger\gamma_4\psi$ and the pseudoscalar $i\psi^\dagger\gamma_4\gamma_5\psi$. To lowest order, T violation can be produced via the exchange of the neutral massive spin 0 boson; as already mentioned in section 2.2, such a T violation usually satisfies the selection rule $\Delta Y = 0$. In all the realistic models that we have examined, there is always the existence of some genuine (i.e., non-Goldstone) charged massive spin 0 bosons. Through the exchange of these charged bosons, T violation can also be generated. As we shall see in the next section, such exchange typically satisfies the $\Delta Y = 0$ and ± 1 rule. In both cases, the order of magnitude of the T violating amplitude is given by (2.35). If we assume the ratio $(\Delta^3/\mu^2 m)$ to be $\sim 10^{-3}$, then one has a *milliweak P, T* violating theory which satisfies the rule:

$$\Delta Y = 0, \qquad \text{and} \qquad \pm 1. \tag{2.36}$$

In principle, since all T violating effects in the theory are finite, one can calculate, e.g., the ϵ parameters in K_L^o and K_S^o states, and the electric dipole moment D_E of the neutron. However, in practice, there are many serious difficulties: there is no confidence that any of the presently available gauge theories is the correct one (especially when it is applied to hadrons). In any of these theories, one does not know how to properly include the strong interaction; reliable calculations can only be made for free quark-like hadron fields. The connection between a calculation for these free fields and that for the physical hadrons is extremely uncertain. Furthermore, the theoretical expressions for ϵ and D_E are in terms of some appropriate mass difference Δ of these quark-like hadrons and the mass μ of the relevant spin 0 boson, which are all unknown. Nevertheless, it seems possible that the order of magnitude of, say, $(\Delta^3/\mu^2 m)$ in (2.35) may be obtained by using the observed ϵ parameter in the $K_L^o - K_S^o$ system, and from that one may estimate the order of magnitude of D_E for the neutron. As remarked before, without any detailed calculation, one may well expect the resulting theory to be of the milliweak type. Just as in any milliweak theory, the order of magnitude of D_E may be estimated to be

$$D_E \sim |(4\pi)^{-1} G_F e m_N \epsilon| \sim e \times 10^{-23}\ \text{cm}, \tag{2.37}$$

which is consistent with the present experimental value [33]. To see how such a crude estimation may survive a more detailed calculation, we shall in the next section consider a specific model: the "realistic" Georgi-Glashow model, but generalized to include the spontaneous T violation mechanism. Both ϵ and D_E will be calculated explicitly.

Before discussing the details of such calculations, we note that in any "realistic" model, it is necessary to postulate the existence of some "charmed" hadrons [34]. If one wishes, one may arrange to have T violation *only* when there are either real or virtual "charmed" hadrons. (See, e.g., eq. (3.51) below.) In all observed lowest order weak transitions between known particles, it is a good approximation to neglect the "charmed" hadrons. Thus, even though the theory is basically of a milliweak character, so far as all known weak decays are concerned, the predictions can be almost identical to those of a superweak theory; i.e., in all lowest order $\Delta Y = 0$ and ± 1 weak reactions the T violating transition amplitudes can be much less than 10^{-3} times the corresponding T conserving amplitudes. However, as we shall see in section 3.5, the order of magnitude of the electric dipole moment of the neutron, or proton, should retain its milliweak character, and this may well offer the best experimental possibility for differentiating such a theory from a genuine superweak theory.

3. A generalized Georgi-Glashow model

In order to give a quantitative illustration of the general remarks made at the end of the previous section, we shall consider in some detail a specific model, which is chosen because of some of its relatively simple features. (Of course, none of these "realistic" gauge models [26–30] is truly simple, especially when applied to hadrons.)

3.1. *Lagrangian*

This model is a generalization of the Georgi-Glashow model [27, 28]. In this model, the basic gauge group of the weak and electromagnetic interaction is SO_3, with its three generators given by the usual weak interaction charge Q^{wk}, the electric charge Q^{γ} and the Hermitian conjugate $(Q^{wk})^{\dagger}$. Correspondingly, there is a triplet of spin 1 gauge field $\boldsymbol{W}_{\mu}$, just as in (2.11) and (2.13). There are at least eight quark-like hadron fields grouped into two triplets ψ, ψ' and two singlets χ, χ'. These fields are related to the physical (but still quark-like) states p^+, n^o, λ^o, q^- and p'^+, n'^o, λ'^o, q'^- as follows:

$$\psi = \begin{pmatrix} p \\ n_c \sin\beta + n' \cos\beta \\ q \end{pmatrix}_L + \begin{pmatrix} [\exp(i\alpha_p)]p \\ n' \\ [\exp(i\alpha_q)]q \end{pmatrix}_R ,$$

$$\psi' = \begin{pmatrix} p' \\ \lambda_c \sin\beta + \lambda' \cos\beta \\ q' \end{pmatrix}_L + \begin{pmatrix} [\exp(i\alpha_{p'})]p' \\ \lambda' \\ [\exp(i\alpha_{q'})]q' \end{pmatrix}_R , \tag{3.1}$$

$$\chi = (n_c \cos\beta - n' \sin\beta)_L + n_R ,$$

and

$$\chi' = (\lambda_c \cos\beta - \lambda' \sin\beta)_L + \lambda_R ,$$

where

$$n_c = n \cos\theta_c + \lambda \sin\theta_c ,$$
$$\lambda_c = -n \sin\theta_c + \lambda \cos\theta_c , \tag{3.2}$$

θ_c is the Cabibbo angle and β is the Georgi-Glashow angle, the subscripts L and R refer respectively to the usual $\frac{1}{2}(1+\gamma_5)$ and $\frac{1}{2}(1-\gamma_5)$ projections of the Dirac spinors, and the angles $\alpha_p, \alpha_{p'}, \alpha_q, \alpha_{q'}$, are due to the spontaneous T violation which will be discussed later. (See (3.19) below.)

There are two SO_3-triplets of spin 0 Hermitian field: $\boldsymbol{\phi}_R$ and $\boldsymbol{\phi}_I$. Under T,

$$T\boldsymbol{\phi}_R T^{-1} = +\boldsymbol{\phi}_R, \qquad \text{but} \qquad T\boldsymbol{\phi}_I T^{-1} = -\boldsymbol{\phi}_I. \tag{3.3}$$

The total Lagrangian density can be written as

$$L = L(W, \mathrm{h}) + L(\mathrm{h}, \phi_R) + L(\mathrm{h}, \phi_I) + L(\phi, W), \tag{3.4}$$

in which

$$L(W, \mathrm{h}) = -\psi^\dagger\gamma_4(\gamma_\mu\partial_\mu + m_0)\psi - \psi'^\dagger\gamma_4(\gamma_\mu\partial_\mu + m_0')\psi'$$
$$-\chi^\dagger\gamma_4(\gamma_\mu\partial/\partial x_\mu + m_1)\chi - \chi'^\dagger\gamma_4(\gamma_\mu\partial/\partial x_\mu + m_1')\chi', \tag{3.5}$$

where

$$\partial_\mu = \partial/\partial x_\mu - \mathrm{i}e\boldsymbol{I}\cdot\boldsymbol{W}_\mu, \tag{3.6}$$

and $\boldsymbol{I}$ is the 3×3 matrix representation of the SO_3 generators. The interaction Lagrangian $L(\mathrm{h}, \phi_R)$ between the hadron fields and $\boldsymbol{\phi}_R$ is identical to the usual one given in ref. [28]. We may write

$$L(\mathrm{h}, \phi_R) = -(g_R\psi^\dagger\gamma_4\boldsymbol{I}\psi + g_R'\psi'^\dagger\gamma_4\boldsymbol{I}\psi')\cdot\boldsymbol{\phi}_R + \ldots \tag{3.7}$$

where ... contains all other *CP* invariant and SO_3 invariant couplings between $\boldsymbol{\phi}_R$ and the hadron fields, and in addition it also contains a simple quadratic coupling term between χ and χ', but without $\boldsymbol{\phi}_R$. The interaction between $\boldsymbol{\phi}_I$ and the hadron fields is assumed to be given by

$$L(\mathrm{h}, \phi_I) = -\mathrm{i}(g_I\psi^\dagger\gamma_4\gamma_5\boldsymbol{I}\psi + g_I'\psi'^\dagger\gamma_4\gamma_5\boldsymbol{I}\psi')\cdot\boldsymbol{\phi}_I, \tag{3.8}$$

where g_I and g_I' are real by Hermiticity. For simplicity, we assume $\boldsymbol{\phi}_I$ *not* directly interacting with the singlet fields χ and χ'. The Lagrangian density $L(\phi, W)$ is given by

$$L(\phi, W) = -\tfrac{1}{4}W_{\mu\nu}^2 - \tfrac{1}{2}(\partial_\mu\boldsymbol{\phi}_R)^2 - \tfrac{1}{2}(\partial_\mu\boldsymbol{\phi}_I)^2 - V(\phi), \tag{3.9}$$

where

$$\boldsymbol{W}_{\mu\nu} = \frac{\partial}{\partial x_\mu}\boldsymbol{W}_\nu - \frac{\partial}{\partial x_\nu}\boldsymbol{W}_\mu + e(\boldsymbol{W}_\mu\times\boldsymbol{W}_\nu),$$
$$\partial_\mu\boldsymbol{\phi}_a = \left(\frac{\partial}{\partial x_\mu} + e\boldsymbol{W}_\mu\times\right)\boldsymbol{\phi}_a, \tag{3.10}$$

the subscript a denotes R or I, and $V(\phi)$ is the potential function between $\boldsymbol{\phi}_I$ and $\boldsymbol{\phi}_R$. The entire Lagrangian is assumed to be renormalizable and invariant under *T, CP* and SO_3 transformations; i.e., it satisfies the three conditions (a), (b) and (c) given in section 1.2. The most general form of $V(\phi)$ is

$$V(\phi) = -\tfrac{1}{2}\lambda_R \boldsymbol{\phi}_R^2 - \tfrac{1}{2}\lambda_I \boldsymbol{\phi}_I^2 + \tfrac{1}{4}A_R(\boldsymbol{\phi}_R^2)^2 + \tfrac{1}{4}A_I(\boldsymbol{\phi}_I^2)^2 + \tfrac{1}{4}B(\boldsymbol{\phi}_R^2)(\boldsymbol{\phi}_I^2) - \tfrac{1}{4}C(\boldsymbol{\phi}_R \cdot \boldsymbol{\phi}_I)^2, \tag{3.11}$$

in which the six constants λ_R, λ_I, ..., C are real by Hermiticity.

To insure that $V(\phi)$ has a minimum, we assume

$$A_R > 0, \qquad A_I > 0,$$

and

$$A_R A_I > \tfrac{1}{4}(B-C)^2. \tag{3.12}$$

In order that the minimum of $V(\phi)$ occurs at the values of $\boldsymbol{\phi}_R$ and $\boldsymbol{\phi}_I$ given by

$$\langle \boldsymbol{\phi}_R \rangle_{vac} = \boldsymbol{\rho}_R \neq 0,$$

and

$$\langle \boldsymbol{\phi}_I \rangle_{vac} = \boldsymbol{\rho}_I \neq 0, \tag{3.13}$$

we assume

$$A_I \lambda_R > \tfrac{1}{2}(B-C)\lambda_I,$$

and

$$A_R \lambda_I > \tfrac{1}{2}(B-C)\lambda_R. \tag{3.14}$$

Consequently, at least one of the two constants λ_R and λ_I should be greater than zero. In addition, we impose

$$C > 0, \tag{3.15}$$

so that

$$\boldsymbol{\rho}_R \parallel \boldsymbol{\rho}_I.$$

Due to the nature of the vector product $\boldsymbol{W}_\mu \times \boldsymbol{\phi}_a$ in (3.10), among the three components of $\boldsymbol{W}_\mu$, only the two that are $\perp$ to $\langle \boldsymbol{\phi}_a \rangle_{vac}$ acquire a mass m_W, given by

$$m_W^2 = e^2(\rho_R^2 + \rho_I^2), \tag{3.16}$$

where ρ_a (a = R and I) denotes the magnitude of $\langle \boldsymbol{\phi}_a \rangle_{vac}$. The $\parallel$ component of $\boldsymbol{W}_\mu$ remains of zero mass; therefore, it is the photon field A_μ. The direction of $\langle \boldsymbol{\phi}_a \rangle_{vac}$ then defines the neutral direction. As usual, all the above conditions (3.12), (3.14) and (3.15) refer to the renormalized constants. Because ϕ_I^o is odd under T and CP, (3.13) implies that our solution is not invariant under T and CP; it is also not invariant under a general SO_3 transformation, since $\boldsymbol{\phi}_I$ and $\boldsymbol{\phi}_R$ are both vectors.

By setting $\boldsymbol{\phi}_R$ and $\boldsymbol{\phi}_I$ equal to their vacuum expectation value and by using (3.5), (3.7) and (3.8), one finds the masses of p^+, q^-, p'^+ and q'^- to be given by, respectively

$$
\begin{aligned}
m_p^2 &= (m_0 + g_R \rho_R)^2 + (g_I \rho_I)^2, \\
m_q^2 &= (m_0 - g_R \rho_R)^2 + (g_I \rho_I)^2, \\
m_{p'}^2 &= (m'_0 + g'_R \rho_R)^2 + (g'_I \rho_I)^2, \\
m_{q'}^2 &= (m'_0 - g'_R \rho_R)^2 + (g'_I \rho_I)^2,
\end{aligned}
\tag{3.17}
$$

where, because of (3.1), m_0 and m'_0 are also related to the masses of n' and λ' by

$$
m_0 = m_{n'} \cos\beta, \qquad \text{and} \qquad m'_0 = m_{\lambda'} \cos\beta. \tag{3.18}
$$

The angles $\alpha_p, ..., \alpha_{q'}$ in (3.1) are given by

$$
\begin{aligned}
\alpha_p &= \sin^{-1}(g_I \rho_I / m_p), & \alpha_q &= -\sin^{-1}(g_I \rho_I / m_q), \\
\alpha_{p'} &= \sin^{-1}(g'_I \rho_I / m_{p'}), \quad \text{and} & \alpha_{q'} &= -\sin^{-1}(g'_I \rho_I / m_{q'}).
\end{aligned}
\tag{3.19}
$$

To find the normal modes of the spin 0 fields, we introduce new field variables

$$
R^o \equiv \phi_R^o - \langle \phi_R^o \rangle_{vac}, \qquad I^o \equiv \phi_I^o - \langle \phi_I^o \rangle_{vac},
$$

$$
H^\pm \equiv (\rho_R^2 + \rho_I^2)^{-1/2} [\rho_I \phi_R^\pm - \rho_R \phi_I^\pm],
$$

and

$$
G^\pm \equiv \mp i(\rho_R^2 + \rho_I^2)^{-1/2} [\rho_R \phi_R^\pm + \rho_I \phi_I^\pm]. \tag{3.20}
$$

These fields represent the various quantum fluctuations of the two vector fields $\boldsymbol{\phi}_R$ and $\boldsymbol{\phi}_I$ around their respective vacuum expectation values $\boldsymbol{\rho}_R$ and $\boldsymbol{\rho}_I$, which are parallel. By definition, R^o and I^o denote the oscillations along the direction $\boldsymbol{\rho}_a$ (a = R or I). By decomposing $G^\pm$ to two Hermitian fields G_1 and G_2:

$$
G^\pm = 2^{-1/2}(G_1 \mp i G_2),
$$

one sees that G_1 and G_2 are the two nonzero components of the vector

$$
\boldsymbol{G} = (\rho_R^2 + \rho_I^2)^{-1/2} (\boldsymbol{\rho}_I \times \boldsymbol{\phi}_I + \boldsymbol{\rho}_R \times \boldsymbol{\phi}_R),
$$

which is ⊥ to $\boldsymbol{\rho}_a$. It is straightforward to verify that geometrically G_1 and G_2 represent the overall SO_3 rotations of $\boldsymbol{\phi}_R$ and $\boldsymbol{\phi}_I$ as an ensemble; they are the two Goldstone [24] degrees of freedom in this model. The fields H^+ and H^- represent the quantum fluctuations due to the relative rotation between $\boldsymbol{\phi}_R$ and $\boldsymbol{\phi}_I$.

Recalling that $R^o = I^o = H^\pm = 0$ denotes the minimum of $V(\phi)$, one finds

$$V(\phi) = \text{constant} + \tfrac{1}{2} C(\rho_R^2 + \rho_I^2) H^+ H^- + A_R \rho_R^2 (R^o)^2 + A_I \rho_I^2 (I^o)^2 + (B-C)\rho_R \rho_I R^o I^o + \ldots \tag{3.21}$$

where ... consists of only cubic and quartic expressions of these new field variables. Thus, the normal modes consist of two (genuine) neutral spin 0 bosons, called ϕ_1^o and ϕ_2^o, whose (mass)^2 are the eigenvalues of the matrix

$$m_\phi^2 = \begin{pmatrix} 2A_R \rho_R^2 & (B-C)\rho_R \rho_I \\ (B-C)\rho_R \rho_I & 2A_I \rho_I^2 \end{pmatrix}. \tag{3.22}$$

Both ϕ_1^o and ϕ_2^o are coherent mixtures of R^o and I^o determined by the eigenstates of m_ϕ^2. There is also a (genuine) charged spin 0 boson $H^\pm$ whose mass is

$$m_H^2 = \tfrac{1}{2} C(\rho_R^2 + \rho_I^2). \tag{3.23}$$

As expected, the quadratic expression of $V(\phi)$ is independent of $G^\pm$. In accordance with the Higgs mechanism [25], $G^\pm$ becomes the longitudinal mode of the spin 1 intermediate boson.

The inclusion of leptons is straightforward, just as in the usual Georgi-Glashow model [27]. Since our main interest lies in the $K_L^o - K_S^o$ decay, the details of the lepton part will not be given here.

3.2. *Vertices*

It is convenient to group the eight quark-like hadron states $p^+, n^o, \ldots, \lambda'^o, q'^-$ into four doublets

$$P = \begin{pmatrix} p^+ \\ p'^+ \end{pmatrix}, \qquad Q = \begin{pmatrix} q^- \\ q'^- \end{pmatrix},$$
$$N = \begin{pmatrix} n^o \\ \lambda^o \end{pmatrix}, \quad \text{and} \quad N' = \begin{pmatrix} n'^o \\ \lambda'^o \end{pmatrix}. \tag{3.24}$$

The interaction between these hadron states and the other integer-spin normal modes $W_\mu^\pm$, $H^\pm$, ϕ_1^o and ϕ_2^o can be readily derived by using (3.4). In the following we list a few of these interaction vertices:

1. The interaction vertex between $W_\mu^\pm$ and $\overline{\text{P}}$N, or $\overline{\text{Q}}$N, is given by the Lagrangian density

$$L_{\bar{\mathrm{P}}\mathrm{NW}} = -\mathrm{i}(\tfrac{1}{2} e \sin\beta)\, W^{+}_{\mu} P^{\dagger} \gamma_4 \gamma_\mu (1+\gamma_5) c N + \text{h.c.}$$

and

$$L_{\bar{\mathrm{Q}}\mathrm{NW}} = +\mathrm{i}(\tfrac{1}{2} e \sin\beta)\, W^{-}_{\mu} Q^{\dagger} \gamma_4 \gamma_\mu (1+\gamma_5) c N + \text{h.c.} \tag{3.25}$$

where W^{+}_{μ} and W^{-}_{μ} annihilate, respectively, the positively and the negatively charged intermediate boson, and c is the matrix for the Cabibbo rotation [35, 36]

$$c = \begin{pmatrix} \cos\theta_{\mathrm{c}} & \sin\theta_{\mathrm{c}} \\ -\sin\theta_{\mathrm{c}} & \cos\theta_{\mathrm{c}} \end{pmatrix}. \tag{3.26}$$

One sees that, as expected, these interaction vertices conserve CP and T.

2. The interaction vertices between $G^{\pm}$ and these hadron states are determined by

$$L_{\bar{\mathrm{P}}\mathrm{NG}} = +\mathrm{i}\left(\frac{e\sin\beta}{2m_{\mathrm{W}}}\right) G^{+} P^{\dagger} \gamma_4 \,[(1+\gamma_5) m_{\mathrm{P}} c - (1-\gamma_5) c m_{\mathrm{N}}] N + \text{h.c.},$$

and

$$L_{\bar{\mathrm{Q}}\mathrm{NG}} = -\mathrm{i}\left(\frac{e\sin\beta}{2m_{\mathrm{W}}}\right) G^{-} Q^{\dagger} \gamma_4 \,[(1+\gamma_5) m_{\mathrm{Q}} c - (1-\gamma_5) c m_{\mathrm{N}}] N + \text{h.c.},$$

where m_{P}, m_{Q} and m_{N} are the mass matrices,

$$m_{\mathrm{P}} = \begin{pmatrix} m_{\mathrm{p}} & 0 \\ 0 & m_{\mathrm{p}'} \end{pmatrix}, \qquad m_{\mathrm{Q}} = \begin{pmatrix} m_{\mathrm{q}} & 0 \\ 0 & m_{\mathrm{q}'} \end{pmatrix}, \quad \text{and} \quad m_{\mathrm{N}} = \begin{pmatrix} m_{\mathrm{n}} & 0 \\ 0 & m_{\lambda} \end{pmatrix}. \tag{3.28}$$

As it should, the $G^{\pm}$-transition amplitude between N and P, or N and Q, conserves CP and T, like the corresponding $W^{\pm}$-transition amplitude.

3. The interaction vertices between $H^{\pm}$ and these hadron states are given by

$$L_{\bar{\mathrm{P}}\mathrm{NH}} = \tfrac{1}{2}\left(\frac{e\rho_{\mathrm{I}}}{m_{\mathrm{W}}\rho_{\mathrm{R}}}\right)\sin\beta \;\; H^{+} P^{\dagger} \gamma_4 \,[(1+\gamma_5) M_{\mathrm{P}} c - (1-\gamma_5) c m_{\mathrm{N}}] N + \text{h.c.}$$

and

$$L_{\bar{\mathrm{Q}}\mathrm{NH}} = \tfrac{1}{2}\left(\frac{e\rho_{\mathrm{I}}}{m_{\mathrm{W}}\rho_{\mathrm{R}}}\right)\sin\beta \;\; H^{-} Q^{\dagger} \gamma_4 \,[(1+\gamma_5) M_{\mathrm{Q}} c - (1-\gamma_5) c m_{\mathrm{N}}] N + \text{h.c.} \tag{3.29}$$

where

$$M_{\mathrm{P}} = m_{\mathrm{P}} - \mathrm{i}\rho_{\mathrm{I}}^{-1}(\rho_{\mathrm{I}}^2 + \rho_{\mathrm{R}}^2) \begin{pmatrix} g_{\mathrm{I}} \exp(-\mathrm{i}\alpha_{\mathrm{p}}) & 0 \\ 0 & g_{\mathrm{I}}' \exp(-\mathrm{i}\alpha_{\mathrm{p}'}) \end{pmatrix}$$

and

$$M_{\mathrm{Q}} = m_{\mathrm{Q}} + \mathrm{i}\rho_{\mathrm{I}}^{-1}(\rho_{\mathrm{I}}^2 + \rho_{\mathrm{R}}^2)\begin{pmatrix} g_{\mathrm{I}}\exp(-\mathrm{i}\alpha_{\mathrm{q}}) & 0 \\ 0 & g'_{\mathrm{I}}\exp(-\mathrm{i}\alpha_{\mathrm{q}'}) \end{pmatrix}. \tag{3.30}$$

Thus, the $H^{\pm}$-transition amplitude violates T and CP conservation, reflecting the T noninvariant character of our solution.

4. The interaction between the neutral spin 0 fields ϕ_1^{o}, ϕ_2^{o} and the charged hadrons also violates T and CP. For example,

$$\begin{aligned} L_{\overline{\mathrm{P}}\mathrm{P}\phi^{\mathrm{o}}} &= -p^{\dagger}\gamma_4[\exp(-\mathrm{i}\alpha_{\mathrm{p}}\gamma_5)](g_{\mathrm{R}}R^{\mathrm{o}} + \mathrm{i}g_{\mathrm{I}}\gamma_5 I^{\mathrm{o}})p \\ &\quad -p'^{\dagger}\gamma_4[\exp(-\mathrm{i}\alpha_{\mathrm{p}'}\gamma_5)](g'_{\mathrm{R}}R^{\mathrm{o}} + \mathrm{i}g'_{\mathrm{I}}\gamma_5 I^{\mathrm{o}})p', \\ L_{\overline{\mathrm{Q}}\mathrm{Q}\phi^{\mathrm{o}}} &= +q^{\dagger}\gamma_4[\exp(-\mathrm{i}\alpha_{\mathrm{q}}\gamma_5)](g_{\mathrm{R}}R^{\mathrm{o}} + \mathrm{i}g_{\mathrm{I}}\gamma_5 I^{\mathrm{o}})q \\ &\quad +q'^{\dagger}\gamma_4[\exp(-\mathrm{i}\alpha_{\mathrm{q}'}\gamma_5)](g'_{\mathrm{R}}R^{\mathrm{o}} + \mathrm{i}g'_{\mathrm{I}}\gamma_5 I^{\mathrm{o}})q', \end{aligned} \tag{3.31}$$

where R^{o} and I^{o} are linear combinations of the normal modes ϕ_1^{o} and ϕ_2^{o}, depending on the matrix (3.22). In contrast, the interaction between ϕ_1^{o}, ϕ_2^{o} and the neutral hadrons n and λ conserves T and CP:

$$L_{\overline{\mathrm{N}}\mathrm{N}\phi} = -(\rho_{\mathrm{R}}^{-1}\sin^2\beta)R^{\mathrm{o}}(m_{\mathrm{n}}n^{\dagger}\gamma_4 n + m_{\lambda}\lambda^{\dagger}\gamma_4\lambda). \tag{3.32}$$

This is because in (3.8), for the sake of simplicity, we assume Φ_{I} coupled only to ψ and ψ'.

If one defines the hypercharge Y for λ to be -1 and the others zero, then the T and CP violating matrix elements (3.29) via $H^{\pm}$-transitions satisfy the $\Delta Y = 0$ and ± 1 rule, while those via ϕ_1^{o} and ϕ_2^{o} transitions satisfy the $\Delta Y = 0$ rule, in agreement with the general conclusion given in section 2.3.

5. The interaction between the neutral charmed hadrons n', λ' and $W_{\mu}^{\pm}$ violates T and CP conservation. We find

$$L_{\overline{\mathrm{P}}\mathrm{N}'\mathrm{W}} = -\tfrac{1}{2}\mathrm{i}eW_{\mu}^{+}P^{\dagger}\gamma_4\gamma_{\mu}[(1+\gamma_5)\cos\beta + (1-\gamma_5)\exp(-\mathrm{i}\alpha_{\mathrm{P}})]N' + \mathrm{h.c.},$$

and

$$L_{\overline{\mathrm{Q}}\mathrm{N}'\mathrm{W}} = +\tfrac{1}{2}\mathrm{i}eW_{\mu}^{-}Q^{\dagger}\gamma_4\gamma_{\mu}[(1+\gamma_5)\cos\beta + (1-\gamma_5)\exp(-\mathrm{i}\alpha_{\mathrm{Q}})]N' + \mathrm{h.c.}, \tag{3.33}$$

where

$$\alpha_{\mathrm{P}} = \begin{pmatrix} \alpha_{\mathrm{p}} & 0 \\ 0 & \alpha_{\mathrm{p}'} \end{pmatrix}, \quad \text{and} \quad \alpha_{\mathrm{Q}} = \begin{pmatrix} \alpha_{\mathrm{q}} & 0 \\ 0 & \alpha_{\mathrm{q}'} \end{pmatrix}. \tag{3.34}$$

In the above expression, the intermediate boson $W_\mu^\pm$ interacts with a rather unfamiliar looking current: consists of a (V + A), as well as a (V − A), current.

3.3. *Lowest order diagrams*

In order to evaluate the weak interaction transition amplitudes for the *observed* hadrons, one faces a serious defect of the model: the absence of the strong interaction. Strictly speaking, our calculations are limited to the scattering matrix of free quark-like hadrons which have only electromagnetic and weak interactions. All extensions to the observed hadrons are therefore extremely tentative; at best, they are only order of magnitude estimates.

In the following, we regard the usual SU_3 octet mesons as the appropriate bound states of $p\bar{n}$, $\lambda\bar{p}$, etc., consisting only of p, n, λ and their conjugate states. For example, we have the familiar expressions

$$K^0 = (n\bar{\lambda}), \qquad \bar{K}^0 = (\lambda\bar{n}), \qquad \pi^+ = (p\bar{n}), \text{ etc.}$$

Because these quark-like hadrons are of integer-charges, the known SU_3-octet baryons are assumed to be bound states of these octet mesons and another neutral quark-like hadron, called Z^0, where Z^0 is a singlet under SU_3 and it has a unit baryon number, a unit hypercharge and a half-integer spin. Equivalently, the physical baryons may also be regarded as three-body bound states of these quark-like hadrons; for example,

$$\begin{aligned} &\text{physical neutron} = (Z^0\bar{\lambda}n), \\ &\text{physical proton} \;\; = (Z^0\bar{\lambda}p), \text{ etc.} \end{aligned} \tag{3.35}$$

For simplicity, we assume Z^0 to be the ninth quark-like hadron, different from n′ and λ′; it has *no* weak and electromagnetic interaction. (The alternative assumption that Z^0 is either n′ or λ′ does not change any calculations of the lowest order transition amplitudes for these known hadrons. However, for higher order amplitudes, since n′ and λ′ have some rather unusual interactions such as those given by (3.33), one would encounter unnecessary complications which might be best avoided.)

For practical computations it is convenient to adopt a gauge in which $W_\mu^\pm$ and $G^\pm$ are treated as independent degrees of freedom. The propagator of $W_\mu^\pm$ has the same form as that used in the ξ-limiting process [37]

$$S_{\mu\nu} = -\mathrm{i}(k^2 + m_W^2)^{-1}(\delta_{\mu\nu} + m_W^{-2}k_\mu k_\nu) + \mathrm{i}(k^2 + \xi^{-1}m_W^2)^{-1}(m_W^{-2}k_\mu k_\nu). \tag{3.36}$$

The propagator of $G^\pm$ is given by

$$D = -\mathrm{i}(k^2 + \xi^{-1}m_W^2)^{-1}. \tag{3.37}$$

There is no mixing between $W_\mu^\pm$ and $G^\pm$ in the propagator. The parameter ξ is an *arbitrary* positive constant. The *S*-matrix is independent of ξ, and this independence serves as a useful test for the correctness of the matrix elements. Formally, this gauge can be achieved by adding to the original Lagrangian, the integral of (3.4), an additional term

$$-\int \frac{\xi}{2}\left[\frac{\partial W_\mu}{\partial x_\mu} + \xi^{-1} e(\boldsymbol{\rho}_R \times \boldsymbol{\phi}_R + \boldsymbol{\rho}_I \times \boldsymbol{\phi}_I)\right]^2 d^4x - \mathrm{i}\ \mathrm{trace}\ \ln(I + e\Delta^{-1}N), \tag{3.38}$$

where I is the unit matrix given by

$$\langle i, x | I | j, y\rangle = \delta_{ij}\delta^4(x-y), \tag{3.39}$$

in which i and j vary from 1 to 3 denoting the coordinates of the SO_3-transformation space, x and y are the usual 4-dimensional space-time coordinates, and δ_{ij} is the usual Kronecker symbol. The matrices Δ and N are given by

$$\langle i, x | \Delta | j, y\rangle = \left[-\frac{\partial^2}{\partial x_\mu^2}\delta_{ij} + e^2\xi^{-1}(\rho_R^2 + \rho_I^2)(\delta_{ij} - \hat{n}_i\hat{n}_j)\right]\delta^4(x-y), \tag{3.40}$$

where $\hat{n}$ is the unit vector parallel to $\boldsymbol{\rho}_R$ and $\boldsymbol{\rho}_I$, and

$$\langle i, x | N | j, y\rangle = \epsilon_{ijk}\frac{\partial}{\partial x_\mu}\{[W_\mu(x)]_k\, \delta^4(x-y)\} \tag{3.41}$$

$$+ e\xi^{-1}[\delta_{ij}(\boldsymbol{\rho}_R\cdot\boldsymbol{\phi}_R + \boldsymbol{\rho}_I\cdot\boldsymbol{\phi}_I) - (\boldsymbol{\rho}_R)_j(\boldsymbol{\phi}_R)_i - (\boldsymbol{\rho}_I)_j(\boldsymbol{\phi}_I)_i - (\delta_{ij} - \hat{n}_i\hat{n}_j)(\rho_R^2 + \rho_I^2)\ \delta^4(x-y),$$

where ϵ_{ijk} is +1 or −1 if (ijk) is an even or odd permutation of (123), and 0 otherwise. The last term in (3.38) is related to the so-called Fadeev-Popov ghost [38, 39]. For the matrix elements considered in this paper, this term makes no contribution.

In the following we give the various lowest order weak interaction transition elements that are relevant to the known hadrons. Only diagrams whose external lines are p, n and λ need be considered. These diagrams are listed in fig. 4.

The usual weak interaction of the known hadrons is given by the $W^\pm$-exchange diagram (i) and the $G^\pm$-exchange diagram (ii) in fig. 4. For on-mass-shell external lines, the sum of these two diagrams is independent of ξ. We find

$$\text{(i)} + \text{(ii)} = \mathrm{i}\,2^{-1/2}G_F(k^2 + m_W^2)^{-1}[m_W^2 j_\mu j_\mu^* + s_+ s_+^\dagger], \tag{3.42}$$

where k is the 4-momentum transfer,

$$\begin{aligned}
G_F &= (2\sqrt{2}m_W^2)^{-1}e^2\sin^2\beta,\\
j_\mu &= \mathrm{i}p^\dagger\gamma_4\gamma_\mu(1+\gamma_5)(n\cos\theta_c + \lambda\sin\theta_c),\\
j_\mu^* &= \mathrm{i}(n\cos\theta_c + \lambda\sin\theta_c)^\dagger\gamma_4\gamma_\mu(1+\gamma_5)p,\\
s_+ &= \cos\theta_c p^\dagger\gamma_4[m_p(1+\gamma_5) - m_n(1-\gamma_5)]n + \sin\theta_c p^\dagger\gamma_4[m_p(1+\gamma_5) - m_\lambda(1-\gamma_5)]\lambda,
\end{aligned} \tag{3.43}$$

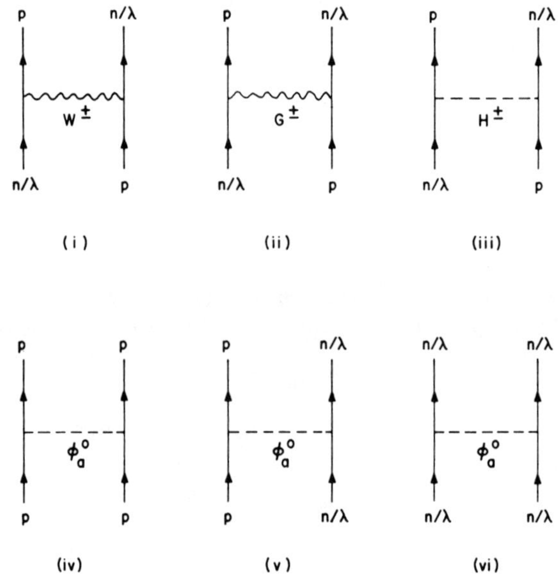

Fig. 4. Some second order weak interaction diagrams in the generalized Georgi-Glashow model (see (3.42), (3.44) and (3.46) for their values).

and $s_+^\dagger$ is its Hermitian conjugate. These two diagrams violate P and C invariance, but conserve CP and T; they also satisfy the usual $\Delta Y = 0$ and ± 1 rule.

The $H^\pm$ exchange diagram (iii) in fig. 4 generates an additional weak interaction among the known hadrons. By using (3.29), we find

$$\text{(iii)} = \mathrm{i}\, 2^{-1/2} G_F (k^2 + m_H^2)^{-1} (\rho_R^{-1} \rho_I s_+ - \mathrm{i} s_-)(\rho_R^{-1} \rho_I s_+^\dagger + \mathrm{i} s_-^\dagger), \tag{3.44}$$

where

$$s_- = g_I \rho_R^{-1} (\rho_I^2 + \rho_R^2) \exp(-\mathrm{i}\alpha_p)\, p^\dagger \gamma_4 (1 + \gamma_5)(\cos\theta_c n + \sin\theta_c \lambda). \tag{3.45}$$

Thus, (iii) satisfies the $\Delta Y = 0$ and ± 1 rule; it violates P, C and T invariance, but remains invariant under CPT.

The exchanges of ϕ_1° and ϕ_2° bosons lead to diagrams (iv), (v) and (vi) in fig. 4. By using (3.31) and (3.32), we find

$$\begin{aligned} \text{(iv)} &= \mathrm{i} x^\dagger (k^2 + m_\phi^2)^{-1} x, \\ \text{(v)} &= \mathrm{i}[x^\dagger (k^2 + m_\phi^2)^{-1} y + \text{h.c.}], \\ \text{(vi)} &= \mathrm{i} y^\dagger (k^2 + m_\phi^2)^{-1} y, \end{aligned} \tag{3.46}$$

where m_ϕ is the (2×2) matrix given by (3.22)

$$x = \begin{pmatrix} g_R p^\dagger \gamma_4 (\cos \alpha_p - i \sin \alpha_p \gamma_5) p \\ g_I p^\dagger \gamma_4 (\sin \alpha_p + i \cos \alpha_p \gamma_5) p \end{pmatrix}, \tag{3.47}$$

and

$$y = \begin{pmatrix} \rho_R^{-1} \sin^2 \beta (m_n n^\dagger \gamma_4 n + m_\lambda \lambda^\dagger \gamma_4 \lambda) \\ 0 \end{pmatrix}. \tag{3.48}$$

Except for (vi), which is invariant under *C, P* and *T*, the other two diagrams violate *P, T* and *CP* invariance, but remain invariant under *C* and *CPT*; they all satisfy the $\Delta Y = 0$ rule.

Remarks

1. In the following, for order of magnitude estimations we shall assume

$$m_H, m_\phi \sim O(m_W),$$

and

$$m_W \gg \text{“quark” mass} \gg g_R \rho_R \quad \text{or} \quad g_I \rho_I, \tag{3.49}$$

since $g_R \rho_R$ and $g_I \rho_I$ represent some appropriate "quark" mass shifts. If we assume each of these inequalities carries a factor ~ 10, then to lowest order, the ratio of the *T* violating amplitude to that of the usual *T* conserving amplitude is typically

$$\sim (g_I \rho_I) m_p / m_H^2 \qquad \text{or} \qquad (g_I \rho_I)(g_R \rho_R)/m_\phi^2, \tag{3.50}$$

which is $\sim 10^{-3}$ or 10^{-4}.

2. An interesting possibility is

$$g_I = 0 \qquad \text{but} \qquad g_I' \neq 0. \tag{3.51}$$

This condition implies that *all the above lowest order weak interaction diagrams are T conserving. T* violation occurs only in higher order diagrams, in which there is a nonzero probability of having a virtual "charmed" quark component in the known hadron state. Because of (3.17) and (3.18), (3.51) implies a "quark" mass relation

$$m_p + m_q - 2 m_{n'} \cos \beta = 0. \tag{3.52}$$

3.4. ϵ-parameter

In the usual analysis [1] of the $K^o - \overline{K}^o$ system one defines a mass matrix M and a decay matrix Γ, both Hermitians. The physical K^o_L and K^o_S states are eigenvectors of the equation

$$(\Gamma + iM)|K^o_j\rangle = (\tfrac{1}{2}\gamma_j + m_j)|K^o_j\rangle\,, \tag{3.53}$$

where j = L and S. Assuming CPT invariance, the solutions K^o_S and K^o_L are given by (1.3) and (1.4), where to a good approximation, as discussed in section 1.1, the phase of the ϵ-parameter is given by (1.5), and its magnitude is

$$|\epsilon| \approx 2^{-3/2}\left|\frac{\mathrm{Im}\langle\overline{K}^o|M|K^o\rangle}{\mathrm{Re}\langle\overline{K}^o|M|K^o\rangle}\right|\,. \tag{3.54}$$

If T invariance were true, then M would be real and ϵ should be zero.

In the present model, because of the lack of strong interaction, we can only compute the scattering matrix S for

$$n\overline{\lambda} \rightleftharpoons \lambda\overline{n}. \tag{3.55}$$

To simulate the K-meson, we define $|K^o\rangle$ and $|\overline{K}^o\rangle$ to be, respectively, the properly normalized free states $|n\overline{\lambda}\rangle$ and $|\lambda\overline{n}\rangle$, which have zero kinetic energy in their rest frame and a spin-parity $= 0-$ (in the convention that the relative parity between n and λ is +1). The rest energy of these states K^o and $\overline{K}^o$ is therefore $(m_n + m_\lambda)$.

The S-matrix can be decomposed into a sum of a $T = +1$ part S_+ and a $T = -1$ part S_- where

$$S_+ \equiv \tfrac{1}{2}[S + TS^\dagger T^{-1}]\,, \qquad S_- \equiv \tfrac{1}{2}[S - TS^\dagger T^{-1}]\,, \tag{3.56}$$

and T is the usual antiunitary time reversal operator. The ϵ parameter in our model is then given by

$$|\epsilon| \approx 2^{-3/2}\left|\frac{\langle n\overline{\lambda}|S_-|\lambda\overline{n}\rangle}{\langle n\overline{\lambda}|S_+|\lambda\overline{n}\rangle}\right|\,, \tag{3.57}$$

at zero kinetic energy.

The lowest order T conserving diagrams for the $\Delta Y = \pm 2$ process (3.55) are fourth order ones via exchanges of two $W^\pm$:

$$n\overline{\lambda} \rightleftharpoons W^+W^- \rightleftharpoons \lambda\overline{n}, \tag{3.58}$$

where, for convenience of notation, $W^\pm$ represents either $W^\pm_\mu$ or $G^\pm$. These diagrams have been calculated by Lee, Primack and Treiman [28], and the resulting S_+ is given by

$$S_+ \approx -i(24\pi^2)^{-1} G_F^2 \sin^2\theta_c \cos^2\theta_c\,[3(\Delta m_p + \Delta m_q)^2 - 5(\Delta m_p - \Delta m_q)^2]$$
$$\times\,[\lambda^\dagger\gamma_4\gamma_\mu(1+\gamma_5)n]\,[\lambda^\dagger\gamma_4\gamma_\mu(1+\gamma_5)n]\,, \tag{3.59}$$

where

$$\Delta m_{\mathrm{p}} = m_{\mathrm{p'}} - m_{\mathrm{p}} \qquad \text{and} \qquad \Delta m_{\mathrm{q}} = m_{\mathrm{q'}} - m_{\mathrm{q}}. \tag{3.60}$$

The lowest order T violating diagrams for the same process are also fourth order ones, but via exchanges of $\mathrm{W}^{\pm}$ and $\mathrm{H}^{\mp}$:

$$\mathrm{n}\bar{\lambda} \rightleftharpoons \mathrm{W}^{\pm}\mathrm{H}^{\mp} \rightleftharpoons \lambda\bar{\mathrm{n}}. \tag{3.61}$$

These diagrams are given in fig. 5. It is convenient to assume (3.49), and to consider

$$\Delta m_{\mathrm{p}}, \qquad \Delta m_{\mathrm{q}}, \qquad (\Delta g_{\mathrm{R}})\rho_{\mathrm{R}} \equiv (g'_{\mathrm{R}} - g_{\mathrm{R}})\rho_{\mathrm{R}}, \qquad \text{and}$$

$$(\Delta g_{\mathrm{I}})\rho_{\mathrm{I}} \equiv (g'_{\mathrm{I}} - g_{\mathrm{I}})\rho_{\mathrm{I}} \quad \text{to be of comparable magnitude.} \tag{3.62}$$

Furthermore, we shall regard ρ_{I} to ρ_{R} to be of the same order of magnitude, and also Δg_{I}, Δg_{R}, $(g'_{\mathrm{I}} + g_{\mathrm{I}})$ and $(g'_{\mathrm{R}} + g_{\mathrm{R}})$ to be of the same order of magnitude (although g'_{I} may be much larger than g_{I}, leaving the possibility (3.51) open). Such an estimate is consistent with the view that all

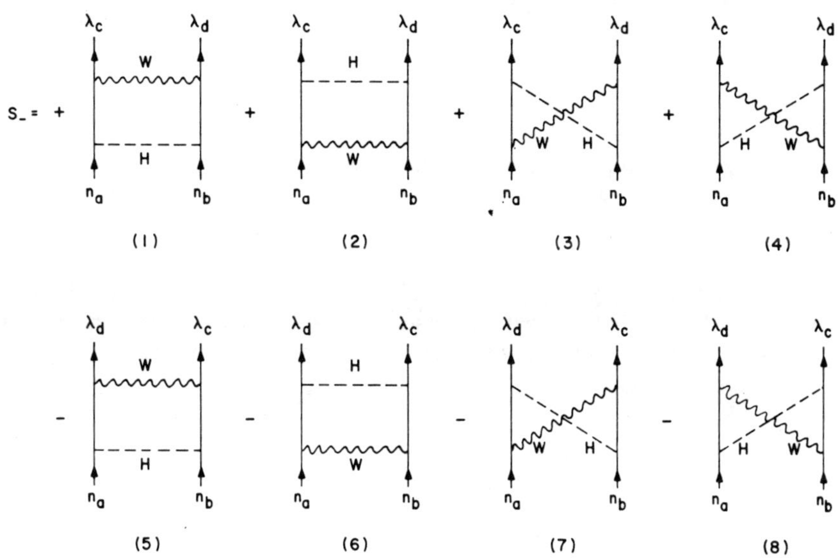

Fig. 5. Diagrams for the T violating scattering S_- of the $\Delta Y = \pm 2$ reaction $\mathrm{n_a n_b} \rightleftharpoons \lambda_c \lambda_d$. Each wavy line can be either $\mathrm{W}^{\pm}$ or $\mathrm{G}^{\pm}$, and each solid internal line can be either p^{+}, or p'^{+}, or q^{-}, or q'^{-}. Thus, each diagram refers to 16 different assignments, making a total of 128 possibilities.

these quark-like hadron masses are ~ a few GeV, all their mass differences and mass shifts (due to the Goldstone-Higgs mechanism) are of an order of magnitude smaller, while all integer-spin boson masses (except the photon's) are an order of magnitude bigger. With this in mind, it is only necessary to evaluate S_- for the reaction (3.61) to order $(\Delta g_{\rm I})(\Delta m_i)$, neglecting $(g'_{\rm I}+g_{\rm I})(\Delta m_i)(\Delta m_j)$ where i and j can be p or q.

The evaluation of the T violating scattering matrix element is straightforward though somewhat tedious. The resulting S_- is given in the Appendix. It leads to (at small values of external 4-momenta)

$$|\epsilon| = 2^{-3/2}\left|\frac{(\Delta g_{\rm I})\rho_{\rm I}(\Delta m_{\rm p}-\Delta m_{\rm q})(\rho_{\rm I}^2+\rho_{\rm R}^2)(m_{\rm n}^2-m_\lambda^2)}{[3(\Delta m_{\rm p}+\Delta m_{\rm q})^2-5(\Delta m_{\rm p}-\Delta m_{\rm q})^2]\rho_{\rm R}^2 m_{\rm H}^2}\right| . \tag{3.63}$$

To have an understanding of this expression, we may consider the special case $m'_0 = m_0$; this implies that the mass differences $\Delta m_{\rm p}$ and $\Delta m_{\rm q}$ are entirely due to the Goldstone-Higgs mechanism. By using (3.17) and neglecting higher order corrections, we find

$$\Delta m_{\rm p} = -\Delta m_{\rm q} = (\Delta g_{\rm R})\rho_{\rm R} . \tag{3.64}$$

Consequently, (3.63) reduces to

$$|\epsilon| = \left| \rho_{\rm R}^{-3}\rho_{\rm I}(\rho_{\rm I}^2+\rho_{\rm R}^2)\left(\frac{\Delta g_{\rm I}}{\Delta g_{\rm R}}\right)\frac{m_{\rm n}^2-m_\lambda^2}{20\sqrt{2}m_{\rm H}^2}\right| . \tag{3.65}$$

If we set $\rho_{\rm I}\sim\rho_{\rm R}$, $\Delta g_{\rm I}\sim\Delta g_{\rm R}$, then from the observed value $|\epsilon|\approx 2\times 10^{-3}$ it follows that

$$m_{\rm H}^2 \sim 25\sqrt{2}\,|m_{\rm n}^2-m_\lambda^2| . \tag{3.66}$$

Thus, if $m_{\rm n}$ and m_λ are in the range of ~ 5 GeV, their difference of the order of 1 GeV, then $m_{\rm H}$ is ~ 20 GeV, quite consistent with our general order of magnitude considerations.

We emphasize that since ϵ depends on the difference $\Delta g_{\rm I}$, one has the attractive possibility of setting $g_{\rm I} = 0$, and therefore $g'_{\rm I} = \Delta g_{\rm I}$ while leaving ϵ unchanged. In this case, as already discussed in sections 2.3 and 3.3, the milliweak character of T violation resides only in the "charmed" hadrons. In all known $\Delta Y = 0$ and ± 1 weak processes, one can certainly neglect the contributions of "charmed" hadrons. If we consider only the lowest order diagrams for these weak processes, then all amplitudes are T conserving, and that resembles a superweak theory. In particular, in the 2π decay of K_L^o and K_S^o, we have

$$\epsilon' = 0, \tag{3.67}$$

where ϵ' is defined in (1.10). As mentioned before at the end of section 2.3, to find the difference between such a theory and a *bona fide* superweak theory, one must consider other higher order diagrams, such as those for the electric dipole moment (see eq. (3.73) below).

3.5. *Electric dipole moment*

In this model, it is quite simple to calculate the electric dipole moment of the quark-like hadrons n, λ, etc. However, without properly taking into account the strong interaction, one cannot determine with any certainty the electric dipole moment of the physical baryon. For definiteness, let us adopt (3.35); we consider the physical neutron N to be a composite ($Z^o n\bar{\lambda}$). As an approximation, we may expect the electric dipole moment of N to be linearly related to those of its constituents

$$D_E(N) \approx f_n D_E(n) + f_{\bar{\lambda}} D_E(\bar{\lambda}), \tag{3.68}$$

where f_n and $f_{\bar{\lambda}}$ are constants, assumed to satisfy $|f_n| + |f_{\bar{\lambda}}| \leqq 1$. $D_E(n)$ and $D_E(\bar{\lambda})$ are respectively the electric dipole moment of n and $\bar{\lambda}$. (The electric dipole moment of Z^o is by assumption zero.) If in this composite state the spin of N is completely correlated with that of Z^o (e.g., $n\bar{\lambda}$ is in a singlet state), then we have $f_n = f_\lambda = 0$, and consequently

$$D_E(N) \approx 0. \tag{3.69}$$

If on the other hand the spin of N is completely correlated with that of n, or $\bar{\lambda}$, then we expect $D_E(N) \approx D_E(n)$, or $D_E(\bar{\lambda})$.

The diagrams for the electric dipole moment of n are given in fig. 6. We find

$$\begin{aligned} D_E(n) = &-(4\sqrt{2}\pi^2)^{-1} eG_F(m_H \rho_R)^{-2} \rho_I(\rho_I^2 + \rho_R^2) m_n \\ &\times [m_p g_I \cos^2\theta_c \cos\alpha_p F(n,p) + m_{p'} g'_I \sin^2\theta_c \cos\alpha_{p'} F(n,p') \\ &+ m_q g_I \cos^2\theta_c \cos\alpha_q F(n,q) + m_{q'} g'_I \sin^2\theta_c \cos\alpha_{q'} F(n,q')], \end{aligned} \tag{3.70}$$

where

$$F(n,p) = -\frac{m_H^2}{2m_n^2} \ln\frac{m_H^2}{m_p^2} + \frac{m_H^2}{2m_n^2} R^{-1}(m_H^2 - m_p^2 + m_n^2) \ln\frac{m_H^2 + m_p^2 - m_n^2 + R}{m_H^2 + m_p^2 - m_n^2 - R}, \tag{3.71}$$

and

$$R = [(m_H + m_p)^2 - m_n^2]^{1/2} [(m_H - m_p)^2 - m_n^2]^{1/2}.$$

For $m_H \gg m_n$ and m_p, the function F becomes

$$F(n,p) = \ln(m_H^2/m_p^2) - 1 + O[(\ln(m_H^2/m_p^2))\, m_H^{-2}(m_p^2 \text{ or } m_n^2)]. \tag{3.72}$$

Similarly, one finds the electric moment of λ to be given by the same expression (3.70), provided n is replaced by λ and θ_c by $\frac{1}{2}\pi - \theta_c$.

It is of interest to examine the special case $g_I = 0$. As noted before, in this case all lowest order

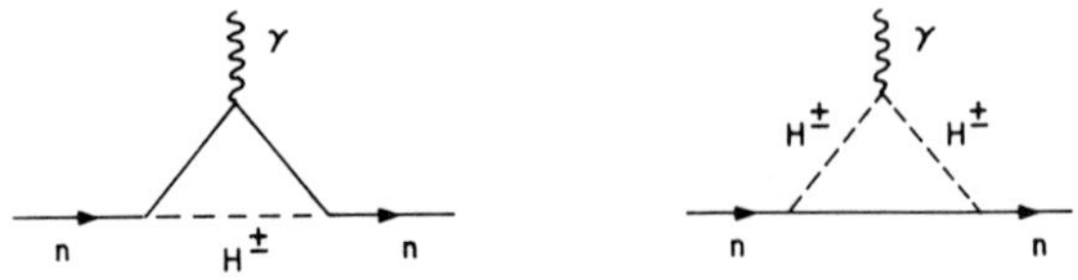

Fig. 6. Diagrams for the electric dipole moment of the quark-like hadron n°. The solid internal line can be either p^+, or p'^+, or q^-, or q'^-. Thus, each diagram represents four possibilities.

weak processes among known hadrons are T conserving. Nevertheless, the electric dipole moment may be quite different from that of *bona fide* superweak theory. For order of magnitude estimations, we may set, in addition to $g_I = 0$, also $m_{p'} \approx m_{q'}$, $\rho_I \approx \rho_R$ and the angles $\alpha_{p'}$, $\alpha_{q'}$ much less than 1. Equation(3.70) becomes

$$D_E(n) \approx -(\sqrt{2}\pi^2)^{-1} eG_F m_n m_p m_H^{-2} (g_I' \rho_I) \sin^2\theta_c \; F(n, p). \tag{3.73}$$

If we assume m_H to be an order of magnitude larger than m_n or m_p, and the mass shift $g_I'\rho_I$ to be ~, or less than, 1 GeV, then the product $m_n m_p m_H^{-2}(g_I'\rho_I)$ is $\lesssim 10^{-2}$ GeV; therefore, the magnitude of $D_E(n)$ becomes $\lesssim e \times 10^{-23}$ cm. It is reasonable to expect the electric dipole moment $D_E(N)$ of the physical neutron to be also to be of a similar order of magnitude, retaining the milliweak character of the theory. Of course, as shown in (3.69), $D_E(N)$ may also be much smaller, depending on the spin correlation between N and its constituents.

Acknowledgement

Part of this work was done at the Institut für Theoretische Physik der Universität Wien and at CERN during the summer of 1973. I especially wish to thank Professors W. Thirring and B. Zumino for their kind hospitality. I also wish to thank Professor G.C. Wick for many helpful discussions.

Appendix

In this appendix we calculate the T violating amplitude for the scattering

$$n_a + n_b \to \lambda_c + \lambda_d, \tag{A.1}$$

where a, b, c, d denote the appropriate 4-momenta of the initial and final particles. The diagrams are given in fig. 5. Let us introduce

$$\sigma_\mu = (a+b)_\mu, \qquad s = -\sigma_\mu^2,$$
$$\tau_\mu = (a-d)_\mu, \quad \text{and} \quad t = -\tau_\mu^2. \tag{A.2}$$

In evaluating these diagrams we shall assume Δm_p, Δm_q small, and keep only the first order terms in Δm_p and Δm_q, but neglect all quadratic and higher order terms in these mass differences. The coupling constants g_I and g'_I are assumed to be small, and therefore all higher order terms in g_I and g'_I will also be neglected. In addition, we assume m_W and m_H to be much bigger than the masses of the quark-like hadrons.

Let L_α be the effective Lagrangian density for the T violating part of the αth diagram in fig. 5. For the first two diagrams, we find

$$L_1 + L_2 = \mathrm{i}K \cdot f(s)\,(\Delta m_p - \Delta m_q)$$
$$\times \{m_n[\lambda_c^\dagger \gamma_4\gamma_\nu(1+\gamma_5)n_a \cdot \lambda_d^\dagger\gamma_4\gamma_\nu\phi(1-\gamma_5)n_b + \lambda_c^\dagger\gamma_4\gamma_\nu\phi(1-\gamma_5)n_a \cdot \lambda_d^\dagger\gamma_4\gamma_\nu(1+\gamma_5)n_b]$$
$$- m_\lambda[\lambda_c^\dagger\gamma_4\gamma_\nu(1+\gamma_5)n_a \cdot \lambda_d^\dagger\gamma_4\phi\gamma_\nu(1+\gamma_5)n_b + \lambda_c^\dagger\gamma_4\phi\gamma_\nu(1+\gamma_5)n_a \cdot \lambda_d^\dagger\gamma_4\gamma_\nu(1+\gamma_5)n_b]\}, \quad (A.3)$$

where

$$K = (24\pi^2)^{-1}(m_H\rho_R)^{-2}\,\rho_I(\rho_I^2 + \rho_R^2)\,(G_F \sin\theta_c \cos\theta_c)^2\,(g'_I - g_I), \quad (A.4)$$

$$f(s) = s^{-2}\,3m_p m_q \left\{ (m_p^2 - m_q^2)\ln(m_p^2/m_q^2) - 2s + [s(m_p^2 + m_q^2) - (m_p^2 - m_q^2)^2] \right.$$
$$\left. \times [R(s)]^{-1} \ln \frac{m_p^2 + m_q^2 - s + R(s)}{m_p^2 + m_q^2 - s - R(s)} \right\}, \quad (A.5)$$

$$R(s) = [s - (m_p + m_q)^2]^{1/2}\,[s - (m_p - m_q)^2]^{1/2}, \quad (A.6)$$

the initial and final Dirac spinors are denoted by n_a, n_b, λ_c and λ_d respectively, and $\phi = -\mathrm{i}\gamma_\mu\sigma_\mu$. At $s = 0$ and $m_p = m_q$, the function $f(s)$ is 1.

The sum of the effective Lagrangian densities for the third and fourth diagrams is given by

$$L_3 + L_4 = \mathrm{i}K \cdot [f_p(t)\Delta m_p - f_q(t)\Delta m_q]$$
$$\times \{m_n[-\lambda_c^\dagger\gamma_4\gamma_\nu t(1-\gamma_5)n_a \cdot \lambda_d^\dagger\gamma_4\gamma_\nu(1+\gamma_5)n_b + \lambda_c^\dagger\gamma_4\gamma_\nu(1+\gamma_5)n_a \cdot \lambda_d^\dagger\gamma_4\gamma_\nu t(1-\gamma_5)n_b]$$
$$+ m_\lambda[-\lambda_c^\dagger\gamma_4\gamma_\nu(1+\gamma_5)n_a \cdot \lambda_d^\dagger\gamma_4 t\gamma_\nu(1+\gamma_5)n_b + \lambda_c^\dagger\gamma_4 t\gamma_\nu(1+\gamma_5)n_a \cdot \lambda_d^\dagger\gamma_4\gamma_\nu(1+\gamma_5)n_b]\}, \quad (A.7)$$

where

$$f_p(t) = t^{-2}\,6m_p^2 \left\{ -1 + 2m_p^2[R_p(t)]^{-1} \ln \frac{R_p(t) - t}{R_p(t) + t} \right\}, \quad (A.8)$$

$$R_p(t) = [t(t - 4m_p^2)]^{-1}, \quad (A.9)$$

$\not{\tau} = -i\gamma_\mu \tau_\mu$, and $f_q(t)$ is given by the same expression (A.8) but with the subscript p replaced by q. At $t = 0$, the function $f_p(t)$ is 1. The other four diagrams in fig. 5 are the same as the first four provided one interchanges the final states λ_c with λ_d.

For the $K^o \rightleftharpoons \bar{K}^o$ transition, we have to consider the crossed reaction

$$n + \bar{\lambda} \rightleftharpoons \lambda + \bar{n}, \tag{A.10}$$

at zero kinetic energy in their rest frame. Therefore, $\sigma_\mu = 0$. In fig. 5, only diagrams 3, 4, 7 and 8 contribute. For simplicity, we set $f_p(t)$ and $f_q(t)$ equal to their value at $t = 0$; the result leads to eq. (3.63).

References

[1] T.D. Lee, R. Oehme and C.N. Yang, Phys. Rev. 106 (1957) 340.
[2] J.H. Christenson, J.W. Cronin, V.L. Fitch and R. Turlay, Phys. Rev. Letters 13 (1964) 138.
[3] D.F. Bartlett et al., Phys. Rev. Letters 21 (1968) 558.
[4] J.M. Gaillard et al., paper 139 contributed to the 14th Intern. Conf. on High Energy Physics, Vienna, 1968 (CERN, Geneva, 1968).
[5] M. Banner et al., Phys. Rev. Letters 21 (1968) 1107.
[6] R.J. Cence et al., paper 476 contributed to the 14th Intern. Conf. on High Energy Physics, Vienna, 1968 (CERN, Geneva, 1968).
[7] J. Chollet et al., Topical Conference on Weak Interactions (CERN, 1969) p. 309.
[8] D. Dorfan et al., Phys. Rev. Letters 19 (1967) 987.
[9] S. Bennett et al., Phys. Rev. Letters 19 (1967) 993.
[10] J. Steinberger, Topical Conference on Weak Interactions (CERN, 1969) p. 291.
[11] V.V. Barmin et al., paper 828b contributed to the XVI Intern. Conf. on High Energy Physics, National Accelerator Laboratory, Batavia, 1972.
[12] L. Wolfenstein, Nuovo Cimento 62A (1966) 17.
[13] J.S. Bell and J. Steinberger, Proc. Oxford Intern. Conf. on Elementary Particles (Rutherford Laboratory, 1965) p. 195.
[14] T.D. Lee and C.S. Wu, Annual Review of Nuclear Sciences 16 (1966) 511.
[15] G. Geweniger et al., Proc. Aix-en-Provence Conf. on Elementary Particles, to be published.
[16] T.A. Lasinski et al., Review of Particle Properties, Revs. Mod. Phys. 45 (1973).
[17] T.D. Lee and L. Wolfenstein, Phys. Rev. 138 (1965) B1490.
[18] L. Wolfenstein, Phys. Rev. Letters 13 (1964) 180.
[19] T.T. Wu and C.N. Yang, Phys. Rev. Letters 13 (1964) 180.
[20] L.B. Okun, Soviet J. Nucl. Phys. 1 (1965) 670;
J. Prentki and M. Veltman, Phys. Letters 15 (1965) 88.
[21] J. Bernstein, G. Feinberg and T.D. Lee, Phys. Rev. 138 (1965) B1650.
[22] T.D. Lee, Phys. Rev. D8 (1973) 1226; Topical Conference on Weak Interactions, Trieste, 1973 (Columbia University preprint CO-2271-18, unpublished).
[23] A. Pais, Phys. Rev. Letters 29 (1972) 1712; 30 (1973) 114(E).
[24] J. Goldstone, Nuovo Cimento 19 (1961) 154;
J. Goldstone, A. Salam and S. Weinberg, Phys. Rev. 127 (1962) 965.
[25] P.W. Higgs. Phys. Letters 12 (1964) 132; Phys. Rev. Letters 13 (1964) 508; and Phys. Rev. 145 (1966) 1156;
F. Englert and R. Brout, Phys. Rev. Letters 13 (1964) 321;
G.S. Guralnik, C.R. Hagen and T.W.B. Kibble, Phys. Rev. Letters 13 (1964) 585;
T.W.B. Kibble, Phys. Rev. 155 (1967) 1554.
[26] S. Weinberg, Phys. Rev. Letters 19 (1967) 1264; 27 (1971) 1688; and Phys. Rev. D5 (1972) 1412; D5 (1972) 1962;
A. Salam and J.C. Ward, Phys. Letters 13 (1964) 168;
A. Salam, in Elementary Particle Theory, ed. N. Svartholm (Almquist and Forlog A.B., Stockholm, 1968).
[27] H. Georgi and S.L. Glashow, Phys. Rev. Letters 28 (1972) 1494.
[28] B.W. Lee, J.R. Primack and S.B. Treiman, Phys. Rev. D7 (1973) 510.

[29] J. Prentki and B. Zumino, Nucl. Phys. B47 (1972) 99.
[30] J.D. Bjorken and C.H. Llewellyn Smith, Phys. Rev. D7 (1973) 887.
[31] G. 't Hooft, Nucl. Phys. B33 (1971) 173; B35 (1971) 167; and Phys. Letters 37B (1971) 195; G. 't Hooft and M. Veltman, Nucl. Phys. B44 (1972) 189; B50 (1972) 318.
[32] B.W. Lee, Phys. Rev. B5 (1972) 823; B.W. Lee and J. Zinn-Justin, Phys. Rev. D5 (1972) 3121, 3127, 3155.
[33] W.B. Dress, P.D. Miller and N.F. Ramsey, Phys. Rev. D7 (1973) 3147.
[34] S. Glashow, J. Iliopoulos and L. Maiani, Phys. Rev. D2 (1970) 1285.
[35] N. Cabibbo, Phys. Letters 10 (1963) 513.
[36] M. Gell-Mann and M. Levy, Nuovo Cimento 16 (1960) 705.
[37] T.D. Lee and C.N. Yang, Phys. Rev. 128 (1962) 885.
[38] L.D. Faddeev and V.N. Popov, Phys. Letters 25B (1967) 29; Kiev Report No. ITP-67-36, unpublished.
[39] K. Fugiwara, B.W. Lee and A.I. Sanda, Phys. Rev. D6 (1972) 2923.

Permissions

Birkhäuser Boston would like to thank the original publishers of the papers of T. D. Lee for granting permission to reprint specific papers in this collection:

V. Quantum Field Theory

[12] Reprinted from *Phys. Rev.* **94,** © 1954 by the American Physical Society.
[13] Reprinted from *Phys. Rev.* **95,** © 1954 by the American Physical Society.
[65] Reprinted from *Proc. Nat. Acad. Sci.* **49,** © 1963 by the National Academy of Sciences.
[69] Reprinted from *Phys. Rev.* **133,** © 1964 by the American Physical Society.
[100] Reprinted from *Phys. Rev.* **157,** © 1967 by the American Physical Society.
[101] Reprinted from *Phys. Rev. Letters* **18,** © 1967 by the American Physical Society.
[103] Reprinted from *Phys. Rev.* **163,** © 1967 by the American Physical Society.
[117] Reprinted from *Nucl. Phys.* B**9,** © 1969 by North-Holland Physics Publishing.
[118] Reprinted from *Nucl. Phys.* B**10,** © 1969 by North-Holland Physics Publishing.
[121] Reprinted from *Phys. Rev.* D**2,** © 1970 by the American Physical Society.
[129] Reprinted from *Phys. Rev.* D**3,** © 1971 by the American Physical Society.
[150] Reprinted from *Phys. Rev.* D**9,** © 1974 by the American Physical Society.
[173] Reprinted from *Phys. Rev.* D**19,** © 1979 by the American Physical Society.
[174] Reprinted from *Phys. Rev.* D**22,** © 1980 by the American Physical Society.
[175] Reprinted from *A Festschrift for Maurice Goldhaber,* edited by G. Feinberg, A. W. Sunyar and J. Weneser, *Transactions of the New York Academy of Sciences* **40,** © 1980 by the New York Academy of Sciences.

VI. Symmetry Principles

[14] Reprinted from *Phys. Rev.* **98,** © 1955 by the American Physical Society.
[16] Reprinted from *Phys. Rev.* **100,** © 1955 by the American Physical Society.
[20] Reprinted from *Nuovo Cimento* **3,** © 1956 by Editrice Compositori.
[21] Reprinted from *Phys. Rev.* **104,** © 1956 by the American Physical Society.
[21a] Reprinted from *Phys. Rev.* **106,** © 1957 by the American Physical Society.
[24] Reprinted from *Phys. Rev.* **105,** © 1957 by the American Physical Society.
[25] Reprinted from *Phys. Rev.* **106,** © 1957 by the American Physical Society.
[27] Reprinted from *Rochester Conference 1957,* © 1957 by Wiley-Interscience.
[28] Reprinted from *Phys. Rev.* **106,** © 1957 by the American Physical Society.
[32] Reprinted from BNL Report 443 (T-91), © 1957 by Brookhaven National Laboratory.
[33] Reprinted from *Proceedings of the Rehovoth Conference on Nuclear Structure,* © 1958 by North-Holland Publishing Company.
[35] Reprinted from *Phys. Rev.* **109,** © 1958 by the American Physical Society.
[38] Reprinted from *Le Prix Nobel 1957,* © 1958 by the Nobel Foundation.
[41] Reprinted from *Phys. Rev.* **112,** © 1958 by the American Physical Society.
[53] Reprinted from *Physics Today* **13,** © 1960 by the American Institute of Physics.
[54] Reprinted from *Phys. Rev.* **122,** © 1961 by the American Physical Society.

[73] Reprinted from *Phys. Letters* **12,** © 1964 by North-Holland Publishing Company.
[75] Reprinted from *Nuovo Cimento* **35,** © 1965 by Editrice Compositori.
[77] Reprinted from *Phys. Rev.* **137,** © 1965 by the American Physical Society.
[78] Reprinted from *Phys. Rev.* **138,** © 1965 by the American Physical Society.
[80] Reprinted from *Phys. Rev.* **139,** © 1965 by the American Physical Society.
[85] Reprinted from *Preludes in Theoretical Physics,* edited by A. DeShalit, H. Feshbach and L. Van Hove, © 1966 by North-Holland Publishing Company.
[89] Reprinted from *Proceedings of the Oxford International Conference on Elementary Particles,* © 1965 by Rutherford Appleton Laboratory.
[91] Reprinted from *Physics Today* **19,** © 1966 by the American Institute of Physics.
[93] Reprinted from *Phys. Rev.* **148,** © 1966 by the American Physical Society.
[97] Reprinted from *Proceedings of the XIII International Conference on High Energy Physics,* © 1966 by the University of California Press.
[110] Reprinted from *Fundamental Particle Physics,* edited by G. Takeda and Y. Hara, © 1968 by the Benjamin/Cummings Publishing Company.
[114] Reprinted from *Contemporary Physics, Vol. II,* © 1969 by the International Atomic Energy Agency.
[122] Reprinted from *Evolution of Particle Physics,* edited by M. Conversi, © 1970 by Academic Press.
[137] Reprinted from *Phys. Rev.* D**4,** © 1971 by the American Physical Society.
[146] Reprinted from *Phys. Rev.* D**8,** © 1973 by the American Physical Society.
[147] Reprinted from *Topical Seminar on Weak Interactions,* © 1973 by the International Centre for Theoretical Physics, Trieste.
[149] Reprinted from *Phys. Reports* **9**C, © 1974 by North-Holland Physics Publishing.